Annual Conference of the Australasian Corrosion Association 2017

Corrosion and Prevention 2017

Sydney, Australia
12 - 15 November 2017

Volume 1 of 2

ISBN: 978-1-5108-5699-8

Printed from e-media with permission by:

Curran Associates, Inc.
57 Morehouse Lane
Red Hook, NY 12571

CURRAN ASSOCIATES INC. proceedings .com

Some format issues inherent in the e-media version may also appear in this print version.

Printed by Curran Associates, Inc. (2018)

Australasian Corrosion Association
P.O. Box 112 Kerrimuir
Victoria 3129
Australia

Phone: 61 3 9890 4833
Fax: 61 3 9890 7866

aca@corrosion.com.au

Additional copies of this publication are available from:

Curran Associates, Inc.
57 Morehouse Lane
Red Hook, NY 12571 USA
Phone: 845-758-0400
Fax: 845-758-2633
Email: curran@proceedings.com
Web: www.proceedings.com

TABLE OF CONTENTS

VOLUME 1

VOLUME 2

CORROSION & PREVENTION 2017

CORROSION & PREVENTION 2017

In 2017, the ACA Newcastle Branch will host the annual ACA conference in Sydney at the newly opened International Convention Centre.

The annual ACA conference is a 3 day gathering of world experts on corrosion mitigation. This will be a premium networking event as well as a source for the latest information concerning corrosion mitigation. Entitled Corrosion & Prevention 2017, the conference will comprise a program of keynote speakers and presentations under a range of industry 'streams', integrated with an exhibition which will showcase the latest products and services of the corrosion mitigation industry.

Over 500 delegates and visitors are expected to attend from industries such as; protective coatings, water, defence, building and construction, mining, oil & gas, cathodic protection, power and more.

CORROSION & PREVENTION 2017

THE AUSTRALASIAN CORROSION ASOCIATION INC.

The Australasian Corrosion Association Inc. (ACA) was established in 1955 to service the needs of Australian and New Zealand companies, organisations and individuals involved in the fight against corrosion. It is dedicated to protecting the environment, ensuring public safety and reducing the economic impact of corrosion.

A membership based, not-for-profit, industry association, the ACA promotes the co-operation of academic, industrial, commercial and governmental organisations in relation to corrosion and its mitigation and for disseminating information upon all aspects of corrosion and its prevention by promoting lectures, symposia, publications and other activities.

The strength of the ACA lies in the great breadth of industry experience and academic training of its members; the common interest of which is the desire to increase effectiveness in controlling corrosion. Consequently, education and training, technical meetings and conferences and the distribution of the quarterly technical publication *Corrosion & Materials* are regular and necessary activities to disseminate information and technological advance. The ACA encourages its members to become involved in other industry based organisations and utilises its membership base to achieve appropriate representation to bodies such as Australasian Standards organisations and State Electrolysis Committees.

The ACA holds an annual conference in November, providing a forum for industry professionals to present and discuss the latest in corrosion prevention methodology, products and cutting edge research. Approximately 90 peer reviewed papers are presented on a range of issues within technical streams such as infrastructure, concrete and concrete structures, coatings, cathodic protection, science and more.

On joining the ACA, members may select membership of up to five of ACA's eight Technical Groups. These Groups provide a structure for technical activities within a segment of industry or technical expertise, and were created from and formed for the benefit of members and Industry.

The eight Technical Groups are:

- Applicators
- Concrete Structures & Buildings
- Cathodic Protection
- Coatings
- Mining Industry
- Oil & Gas
- Research
- Water & Water Treatment

Membership of the ACA is available to anyone with an interest in corrosion and its control. There are four categories of membership: Corporate, Individual, Student and Retired.

For further information on the ACA, please refer to www.corrosion.com.au or contact the ACA direct on +61 3 9890 4833.
For conference information go to www.acaconference.com.au

CORROSION & PREVENTION 2017

AUTHOR EMAILS

Vahid Afshari	Vahid.Afshari@ghd.com
Asif Baloch	asif.baloch@my.jcu.edu.au
Anthony Betts	anthony.betts@dit.ie
Les Boulton	lboulton@wave.co.nz
Stephen Bowditch	bowdits@chesterton.com
Richard Brodribb	rbrodribb@brodribb.com.au
Andy Caddy	andy@absafe.com.au
Ricardo Cardoso	rcardoso@andersal.com.au
Graham Carlisle	grahamc@ias-group.com.au
Rainier Catubig	rainier.catubig@deakin.edu.au
Chee Chan	chee.chan@smec.com
Igor Chaves	igor.chaves@newcastle.edu.au
Boris Chernov	Chernov@msun.ru
James Cull	jim.cull@elorane.com.au
Timothy Davison	timdavison@ae-sys.com
Nick Doblo	nick.doblo@apa.com.au
Patrick Dodds	p.c.dodds@swansea.ac.uk
James Edy	554907@swansea.ac.uk
Raed El Sarraf	raed.elsarraf@opus.co.nz
Harvey Flitt	hflitt@bigpond.com
Robert Francis	robfrancis766@gmail.com
Sarah Furman	sarah.furman@aecom.com
Jim Galanos	jimgalanos@cceng.com.au
Oumaïma Gharbi	oumaima.gharbi@monash.edu
Ian Godson	ian.godson@infracorr.com
Warren Green	wgreen@vinsi.com.au
Graeme Gummow	graeme.gummow@g2sol.com.au
Khaled Habib	khaledhabib@usa.net
Andrew Hackett	andrew.hackett@kbr.com
Robert Jeffrey	robert@pacifictesting.com.au
David Johnstone	david.johnstone@akzonobel.com
Michael Karlaftis	kimw@marineandcivil.com.au
George Keller	geo.keller@verizon.net
Inamullah Khan	i.khan@bcrc.com.au
Steven Knight	steven.knight2@baesystems.com
Simon Krismer	simon.krismer@au.bureauveritas.com

CORROSION & PREVENTION 2017

AUTHOR EMAILS

Gordon Lacey	Gordon.Lacey@dstdryer.com.au
Majid Laleh	mlaleh@deakin.edu.au
Meifeng Li	meifeng.li@student.unsw.edu.au
Chun-Qing Li	chunqing.li@rmit.edu.au
Mengxia Liang	mengxia.liang@uon.edu.au
Jane Lim	jane.lim@shawcor.com
Steven Loftus	sloftus@rosen-group.com
Willie Mandeno	willie.mandeno@opus.co.nz
Brian Martin	Brian@MartinCorrosion.com.au
Jonathan McCray	jonathan.mccray@arup.com
Jack Mclean	jmclean@freyssinet.com.au
Will McLean	will@gaa.com.au
Robert Melchers	rob.melchers@newcastle.edu.au
Seyed Mostafavi	seyed.mostafavi@stork.com
Will Nash	will.nash@monash.edu
Thuan Nguyen	thuan.nguyen@unsw.edu.au
Alex Nielsen	523533@swansea.ac.uk
Derek Northwood	dnorthwo@uwindsor.ca
Matthew O'Keeffe	matthew.okeeffe@akzonobel.com
Torrill Pape	Torill.Pape@aecom.com
Chris Partington	chris.partington@akzonobel.com
Sanjid Ahmed Patwary	sanjid.patwary@monash.edu
Andrew Peek	APeek@ghd.com
Robert Petersen	robert.petersen@newcastle.edu.au
Yustina Pusparizkita	ymppraxedis@gmail.com
Marianne Repacholi	mrepacholi@fbrice.com.au
Justin Rigby	justin.rigby@docopro.com
Rihan Rihan	rihanr@gmail.com
Mike Rutherford	conspectus@bigpond.com
Bertil Sandberg	bertil.sandberg@swerea.se
Guilherme Sander Malab	guilherme.sandermalab@monash.edu
Nestor Sequera	nestor@snintegrity.com.au
Ambrish Singh	vishisingh4uall@gmail.com
Zohreh Soltani Asadi	Zohreh.SoltaniAsadi@uon.edu.au
Amy Spark	amy.spark@aecom.com
Nick Subotsch	nick@epigen.com.au

CORROSION & PREVENTION 2017

AUTHOR EMAILS

Aaron Sudholz	aaron.sudholz@dst.defence.gov.au
David Sunjaya	davidsunjaya@cceng.com.au
Mike Yongjun Tan	mike.tan@deakin.edu.au
Sebastian Thomas	sebastian.thomas@monash.edu
Wayne Thomson	wayne@teml.com.au
Narendra Tripathi	narendra.tripathi@rms.nsw.gov.au
Adrian Vinnell	adrian.vinnell@aurecongroup.com
Scott Wade	swade@swin.edu.au
Ke Wang	bxc@deakin.edu.au
Weigang Wang	s3468547@student.rmit.edu.au
William Ward	w.ward@pcte.com.au
Aifang Wei	awei@deakin.edu.au
Tony Wells	tony.wells@newcastle.edu.au
Jen Wood	Jen.Wiltshire@latrobe.edu.au
Yun Xie	yun.xie1@student.unsw.edu.au
Yunze Xu	xuyunz@deakin.edu.au

CORROSION & PREVENTION 2017

REVIEWERS

Each manuscript published in these proceedings has been peer reviewed by industry experts selected from ACA's review panel. The ACA acknowledges the valuable contribution reviewers make to technical events such as these.

Bruce Ackland
Mohammad Ali
Fikry Barouky
Nick Birbilis
Candice Blackney
Frederic Blin
Les Boulton
Richard Brodribb
Alan Bryson
Neil Campbell
Francis Carroll
Rainier Catubig
Xiaobo Chen
Nick Critchley
Matthew Dafter
Mike Dinon
Brad Dockrill
Peter Dove
John Duncan
Oscar Duyvestyn
Daniel Fabijanic
Harvey Flitt
Robert Francis
Sarah Furman
Estela Garcez
Will Gates
Bill Gerritsen
John Grapiglia
Warren Green
Graeme Gummow
Bruce Hinton
Robert Jeffrey
Brian Kinsella
Steven Knight
David Lake
Katerina Lepkova
Laura Macchuca
Ian MacLeod
Alastair Macleod
Brian Martin
Grant McAdam
Rob Melchers
Greg Moore
Wayne Neil
David Nicholas
Derek Northwood
Andrew Peek
Raman Singh
Anthony Somers
Graham Sussex
Mike Yongjun Tan
Sebastian Thomas
John Triebe
Facundo Varela
Scott Wade
Chris Weale
Geoff White
Geoffrey Will
Kerryn Wilson
David Young
Jianqiang Zhang

SPONSORS

MAJOR SPONSOR

JOTUN AUSTRALIA
www.jotun.com

Jotun is one of the world's leading manufacturers of paints, coatings and powder coatings with over 9600 employees with 68 companies and 33 production facilities on all continents. Jotun has agents, branch offices and distributors in more than 90 countries. Jotun Australia Pty Ltd supplies: Protective, Marine, Fire Protection Coatings and Powder Coatings nationally, providing a high level of local service, manufacturing locally while supported by Jotun's global strength to provide best practice solutions. Industries supported include: Offshore Oil & Gas Production, Hydrocarbon Processing Industries (HPI), Energy, Mining, Infrastructure, Pipelines and maintenance. Marine Dry Dockings, Marine Vessel New Builds, Tank Coatings & Linings.

Ted Riding | Technical Manager
Phone: + 61 3 9314 0722

CLICK ON AD FOR A CLOSER LOOK

CORROSION & PREVENTION 2017

SPONSORS

SUPPORTING SPONSOR & WIRELESS INTERNET SPONSOR

3C CORROSION CONTROL COMPANY
www.3ccc.se

3C Corrosion Control Company

With a strong international focus and an extensive know-how in cathodic protection and sea water applications, 3C is a highly specialized company with strategic partners on a global level.

Focusing on Power Generation, Oil & Gas and Desalination industries, our mission is to offer practical solutions that secure highest operational reliability. We build long-term partnerships with international industrial contractors to whom we offer our expertise and reliability.

3C develop a range of products for cathodic protection, such as Switch Mode rectifiers, remote controls and monitoring systems, customized MMO anodes and probes for corrosion detection.

Strong customer focus, highest quality and extensive know-how. That's 3C.

Bjorn Lindell
Phone Number: +46 418 411 900
Email Address: bjorn.lindell@3ccc.se

CLICK ON AD FOR A CLOSER LOOK

SPONSORS

SUPPORTING SPONSOR

FREYSSINET

www.freyssinet.com

Freyssinet
Sustainable Technology

Offering integrated technical solutions in construction and structural repair, Freyssinet possesses an unparalleled array of civil engineering expertise in construction methods, post-tensioning, cable-stayed structures, structural accessories, repair, strengthening, cathodic protection, earthquake protection, and structural maintenance – on assets including bridges, tunnels, stadiums, power plants, LNG tanks, buildings, wharves, water tanks, and more.

Freyssinet offers extensive in-house design and technical capabilities with professional engineers and corrosion specialists including NACE and ACA certified Corrosion Technologists and Technicians – the most widely recognised qualifications for personnel demonstrating experience in designing, installing, commissioning, monitoring and maintaining CP systems.

CLICK ON AD FOR A CLOSER LOOK

CORROSION & PREVENTION 2017

SPONSORS

SUPPORTING SPONSOR

MARINE & CIVIL MAINTENANCE

www.marineandcivil.com.au

MCM Marine & Civil Maintenance

Marine & Civil Maintenance Pty Ltd (MCM) is an engineering rehabilitation specialist. We rejuvenate critical ageing infrastructure such as wharves, bridges, buildings and industrial facilities by re-engineering and extending the asset's life through optimised Whole of Life solutions.

Over the last 16 years, MCM has built an enviable record of delivering high-quality engineered solutions in concrete repairs, cathodic protection, protective coatings, structural strengthening and other specialist activities. Embedded within our workforce, we bring this essential expertise and experience to ensure the successful completion of complex rehabilitation projects, whatever the location or scope

Blane McGuiness | Strategy & Development Manager
Phone Number: +61 2 9809 0374
Email Address: blanem@marineandcivil.com.au

CLICK ON AD FOR A CLOSER LOOK

SPONSORS

SUPPORTING SPONSOR

RUSSELL FRASER SALES | RFS

www.rfsales.com.au

Russell Fraser Sales
inspection & testing equipment

Russell Fraser Sales (RFS) has served Australia's inspection and testing community since 1993 and along with its team of 10 staff, RFS has sustained its position as Australia's most trusted NDT equipment supplier by staying ahead of the changing needs of industry. For the corrosion mitigation industry RFS supplies 3D Laser Scanners for pipeline integrity assessment, Ultrasonic Thickness Gauges, Ultrasonic Phased Array, Holiday Detectors, Thermal Imagers, Videoscopes and more; many of which will be showcased on stand #14. RFS has also launched an innovative approach to NDT product video reviews on YouTube with the series "unboxing NDT" offering a detailed look at precisely what's in the box.

Phone Number: +61 2 9545 4433
Email Address: rfs@rfsales.com.au

CLICK ON AD FOR A CLOSER LOOK

CORROSION & PREVENTION 2017

MEDIA PARTNERS

cea hea wwm CORROSIONPEDIA

INFRASTRUCTURE UTILITY NATIONAL RESOURCES REVIEW

CORROSION & PREVENTION 2017

SPONSORS

ACA ANNUAL DINNER & AWARDS SPONSOR

DENSO AUSTRALIA

www.densoaustralia.com.au

Denso Australia is an industry leader in corrosion prevention and sealing technology and has been manufacturing and supplying corrosion prevention solutions to the Australian market since 1961.

The Denso range of global brands include Protal, SeaShield, Denso, PCS, Archo Rigidon and Sylglas.

Denso's extensive product range is suitable for use in a wide range of highly corrosive environments with operating temperatures ranging from sub-zero to 250°C, across the pipeline, marine, tanking and structural markets. Denso will be showcasing a range of corrosion prevention products at its stand.

CORROSION & PREVENTION 2017

SPONSORS

WELCOME RECEPTION SPONSOR

DULUX PROTECTIVE COATINGS

www.duluxprotectivecoatings.com.au

Dulux Protective Coatings provides protection against corrosion, chemical attack, abrasion and impact damage in diverse, harsh and corrosive environments with our broad range of products. Heavy duty inorganic zinc silicate and zinc-rich epoxy primers, high build epoxies, chlorinated rubbers, epoxy-acrylic and polyurethane topcoats, micaceous iron oxide coatings, polyurea and heat resisting product lines are available from over 230 outlets across Australasia.

Dulux Protective Coatings' experienced technical consultants provide tailor made solutions for whatever your project demands, even for the most demanding and difficult conditions.

Call your local Dulux Protective Coatings Representative
or Dulux Customer Service on 13 23 77. Contact us today
at www.duluxprotectivecoatings.com.au

Phone Number: 13 23 77

CORROSION & PREVENTION 2017

SPONSORS

FIRST TIME DELEGATE & YCG FUNCTION SPONSOR

GALVANIZERS ASSOCIATION OF AUSTRALIA & GALVANIZING ASSOCIATION OF NEW ZEALAND
www.gaa.com.au and www.galvanizing.org.nz

galvanizers ASSOCIATION OF AUSTRALIA

GA NZ GALVANIZING ASSOCIATION

The Galvanizing Association of New Zealand (GANZ) and Galvanizers Association of Australia (GAA) represent all major hot dip galvanizers in New Zealand and Australia. The associations work together to promote existing and new uses of hot dip galvanizing to end-users and specifiers, to encourage constant improvement in the environmental practices of its members, and to circulate knowledge of the process to fabricators and engineers.

The Associations provide free technical publications and information on all aspects of hot dip galvanizing; including the process, application, durability, design and painting of hot dip galvanized steel.

John Notley | Chairman | GANZ
Phone Number: +64 215 728 272
Email Address: enquiry@galvanizing.org.nz

CORROSION & PREVENTION 2017

SPONSORS

COFFEE SPONSOR

ZINGA
www.zinga.com.au

ZINGA

Zinga is a unique liquid zinc galvanizing film for the recoating of weathered galvanized surfaces, abrasive blasted new steel or suitably prepared, previously painted steel.

Zinga can be rejuvenated in-field with minimal cleaning and surface preparation before recoating. The process can be repeated indefinitely. Service life is predictable on a similar basis to hot dipped galvanizing. It will depend on the coating thickness and the local zinc usage rate.

Zinga will not flake, blister or peel, plus the cathodic protection qualities prevent under-creep if the coating is damaged.

Grant Weatherburn | General Manager
Phone Number: +61 418 120 550
Email Address: gweatherburn@zinga.com.au

SPONSORS

FAREWELL FUNCTION SPONSOR

AUSTRALASIAN INDUSTRIAL WRAPPINGS & COATINGS
www.aiwc.com.au

AIWC is a manufacturer and distributor of petrolatum, bitumen, butyl, elastomeric and epoxy coatings.

After 19 years of manufacturing for the pipeline and corrosion industry, AIWC have further diversified and for over the last decade the company have been providing one stop shop solutions to the pipeline and marine industry.

Amongst some of the agencies AIWC represent are Viscotaq with their world renowned elastomeric tape coatings and Tinker & Rasor with their holiday detection equipment and corrosion monitoring products.

These products complement AIWC's pipeline and marine coatings which ensure AIWC are one of the market leaders with our worldwide distribution network.

CORROSION & PREVENTION 2017

SPONSORS

WATER BOTTLE SPONSOR

BLASTONE
www.BlastOne.com

Originally established over 40 years ago to provide technical consulting, BlastOne has grown to become a single source supplier of blasting equipment, abrasives and know-how to customers all over the world. BlastOne operates internationally from several offices across Australia, New Zealand, North America and UK.

We stand behind our brand claim of Superior Performance. It's something we define as 'Performance3' – the result of combining superior know-how with superior abrasives and superior equipment. In short, it's something that delivers greater cost-efficiencies for our customers.

Jim Gooden | Global Technical Director – Corrosion Control
Phone Number: +61 8 8292 2000
Email Address: jim.gooden@blastone.com

SPONSORS

WIRELESS INTERNET SPONSOR

3C CORROSION CONTROL COMPANY
www.3ccc.se

With a strong international focus and an extensive know-how in cathodic protection and sea water applications, 3C is a highly specialized company with strategic partners on a global level.

Focusing on Power Generation, Oil & Gas and Desalination industries, our mission is to offer practical solutions that secure highest operational reliability. We build long-term partnerships with international industrial contractors to whom we offer our expertise and reliability.

3C develop a range of products for cathodic protection, such as Switch Mode rectifiers, remote controls and monitoring systems, customized MMO anodes and probes for corrosion detection.

Strong customer focus, highest quality and extensive know-how. That's 3C.

Bjorn Lindell
Phone Number: +46 418 411 900
Email Address: bjorn.lindell@3ccc.se

CORROSION & PREVENTION 2017

SPONSORS

APP SPONSOR

PACIFIC CORROSION CONSULTANTS (A DIVISION OF LTH LTD)
www.linetech.co.nz/services/asset-inspection/

Pacific Corrosion Consultants (PCC) is a division of LTH Ltd, the parent company for a group of specialist consultancy teams with its foundations in the power engineering field circa 1994. We have New Zealand's largest NACE certified independent coatings inspection and condition assessment team. We can also develop procedures, specifications, training materials, and maintenance strategies.

Where there are protective coating issues, we can provide coating failure investigations. Our transmission tower climbing capability together with rope access (IRATA), confined space, and high-resolution UAV (drone) capabilities allow us to get access to challenging structures throughout NZ, Australia, and the Pacific.

Mike Boardman | Director | Principal Consultant
Phone Number:+64 421 906 684
Email Address: mike.boardman@pacific-corrosion.com

EXHIBITORS

EXHIBITORS

3C CORROSION CONTROL COMPANY

Booth Numbers: 12 & 15
www.3ccc.se

With a strong international focus and an extensive know-how in cathodic protection and sea water applications, 3C is a highly specialized company with strategic partners on a global level.

Focusing on Power Generation, Oil & Gas and Desalination industries, our mission is to offer practical solutions that secure highest operational reliability. We build long-term partnerships with international industrial contractors to whom we offer our expertise and reliability.

3C develop a range of products for cathodic protection, such as Switch Mode rectifiers, remote controls and monitoring systems, customized MMO anodes and probes for corrosion detection.

Strong customer focus, highest quality and extensive know-how. That's 3C.

Bjorn Lindell
Phone Number: +46 418 411 900
Email Address: bjorn.lindell@3ccc.se

EXHIBITORS

ABSAFE

Booth Number: 26
www.absafe.com.au

Absafe is a specialist contractor that provides diagnostics, repair and maintenance to all forms of the built environment. Our forte is difficult access; whether it be at height, confined space, or otherwise extreme conditions. Highly experienced, triple accredited and award-winning.

Engineering
- Diagnostic and remedial
- Design
- Fabrication

Coatings
- Industrial Steel
- Commercial Facades
- Abrasive Blasting

Concrete Repair
- Patch Repairs
- Shotcreting
- Crack Injections

Access
- Rope Access
- Rigging
- Environmental Protection

We also operate a retail division: Absafe Specialist Equipment, which provides world-class equipment solutions to meet even the most rigorous demands in the field.

Andy Caddy | Director
Phone Number: +61 3 9486 3777
Email Address: andy@absafe.com.au

EXHIBITORS

ABSS

Booth Numbers: 67 & 68
www.abss.net.au

ABSS are industry leaders in the surface finishing sector throughout Australia. Our offices are located in Melbourne and Sydney with distributors located all over Australia, Canada & USA.

We are an Australian Manufacturer of quality Shot Peening Equipment, Shot Blast Machines, Turbine Wheel Blasting Machines, Abrasive Sandblasting Cabinets, Sandblast
Pots and Sand Blast Rooms. We carry a large range of Sandblasting Abrasive Media including Steel Shot, Steel Grit, Glass Bead, Garnet, Ceramic, Plastic, Aluminium Oxide and Organic Blast Media.

We also hold a large range of accessories including Sandblast Hoses, Sandblasting Nozzles, Operator Sandblast Helmets and Abrasive Media Feed Valves.

EXHIBITORS

AEGIS

Booth Number: 27
www.aegis.net.au

For more than 80 years Aegis has designed and manufactured test equipment for utility companies worldwide. Aegis strives to maximise the life of a utility network, seeking continuous reliable operation and performance. From contractors to network owners, Aegis provides test and measurement solutions for daily use.

Aegis is transforming the corrosion protection industry by providing 24/7 monitoring of your protection systems. You no longer need to wonder whether your network is being protected, let the DataCell product range do it for you. View your network on any smart device, react to alarms only as required, and proactively maintain your asset efficiently.

Matthew Mitchell, Steve Watt
Phone Number: 1300 723 447
Email Addresses: m.mitchell@aegis.net.au, s.watt@aegis.net.au

EXHIBITORS

AERIS ENVIRONMENTAL

Booth Number: 22
www.aeris.com.au

Aeris Environmental brings 'clean, green technology' to drive measurable improvements and changes in the performance of key assets in climate controlled environments.

Australian made and patented technology with a long pedigree in health care and hospitals, Aeris solves real world problems more effectively than conventional toxic technologies. It is uniquely based on validated clean-tech solutions.

AerisGuard™ Corrosion Protection Plus demonstrates superior technical benefits:

- Salt spray resistance 20,000+ hours more than three times that of our nearest competitor
- Extend asset life
- Repairable Coating – Blue or Clear
- Applied to new or existing assets
- 5 year warranty when applied by trained applicators
- 10 year warranty when applied by OEM

EXHIBITORS

AIRBLAST AUSTRALIA

Booth Number: 32
www.airblastaustralia.com.au | www.blastrac.com.au

For more than 40 years, Airblast has been the world leader in providing blasting and painting solutions to the anticorrosion industry. With an unparalleled network of offices around the world Airblast works closely with our customers and distribution partners providing field proven equipment as well as developing customised solutions for your specific application. With world class products come class leading solutions, with an unparalleled arsenal of surface treatment products including, Airblast Abrasives, Airblast blast room facilities, Airblast high production wheel blasting machines, Dust management, abrasive recycling, Blastrac grinding/scarifying and Graco.

With branches and Agents located across the country ready to help with all your surface preparation and coating application solutions, we have the best products to achieve the best results.

David Pocock or Ian Beets | Blastrac Surface Prep Equipment
Phone Number: +61 8 9271 2265 or +61 407 945 694
Email Address: david@airblastwa.com.au
or ian.beets@blastrac.com.au

EXHIBITORS

AZKONOBEL – INTERNATIONAL

Booth Numbers: 2 & 3
www.international-pc.com

AkzoNobel creates everyday essentials to make people's lives more liveable and inspiring. As a leading global paints and coatings company and a major producer of specialty chemicals, we supply essential ingredients, essential protection and essential colour to industries and consumers worldwide. Backed by a pioneering heritage, our innovative products and sustainable technologies are designed to meet the growing demands of our fast-changing planet, while making life easier. Headquartered in Amsterdam, the Netherlands, we have approximately 45,000 people in around 80 countries, while our portfolio includes well-known brands such as Sikkens, International, Interpon and Eka. Consistently ranked as a leader in sustainability, we are dedicated to energizing cities and communities while creating a protected, colorful world where life is improved by what we do.

Karen Palin | Marketing Support Specialist
Phone Number: +61 7 3727 5148
Email Address: karen.palin@akzonobel.com

CORROSION & PREVENTION 2017

EXHIBITORS

ANODE ENGINEERING

Booth Number: 5
www.anodeengineering.com

Anode Engineering Pty Ltd is a specialist product and service provider to the wider Australian, New Zealand and Pacific Rim asset integrity industry. Our long standing association with the industry has seen us develop relationships and corresponding expertise in providing cutting edge products, services and technology in corrosion management and prevention. With offices in Brisbane, Melbourne, Auckland and Wellington, our team are able to lead the industry in design, installation, commissioning and management of corrosion management and sytems including:

Anodes – CP systems, Instrumentation & Materials – Coating Assessments, Equipment & Services – Survey Services & Equipment – Remote Monitoring Systems – Corrosion Monitoring – Earthing/Surge Equipment – CP Backfills – Exothermic Welding – Holiday Detectors – MIJ's & Flange Isolation – Pile Protection & Anti-fouling – Nut & Bolt Protection – Staff Training – Vapour Phase Corrosion Inhibitors – Pipe Supports.

Wayne Burns | Managing Director
Phone Number: +61 7 38015521
Email Address: sales@anodeengineering.com

EXHIBITORS

AMAC

Booth Number: 45
www.amacgroup.com.au

Since 1974, we have been manufacturing cathodic protection systems in Melbourne. We specialise in producing the highest quality Aluminium, Magnesium and Zinc sacrificial anodes as well as impressed current anodes and related cathodic protection materials and equipment. Our continued accreditation to ISO9001 (since 1997) is testament to our product quality.

We continue design and manufacture products to suit every application and industry, while providing customers with a friendly and sincere service. Nothing is impossible for our diverse manufacturing team, and we welcome your enquiries.

Please come and have a chat at our stand in the exhibition.

Mark Rigg | General Manager
Phone Number: +61 3 9729 8888
Email Address: markrigg@amacgroup.com.au

CORROSION & PREVENTION 2017

EXHIBITORS

ANTI CORROSION TECHNOLOGY

Booth Number: 43
www.anticorrosiontechnology.com

Anti Corrosion Technology (ACT) is an Australian-based engineering company that specializes in providing sustainable solutions in materials engineering and corrosion control by senior specialists for the oil & gas, water, mining, marine and power generation industries. ACT is also the authorised and sole distributor of STOPAQ systems in New Zealand and Australia offering unique and cost effective solutions for external pipeline corrosion protection including field joint coating, factory applied mainline coating, flange and valve protection, HDD, steel pipe casings under road & rail crossings as well as corrosion protection of tank bottoms and offshore/submerged structures such as platform risers, wharf piles and jacket legs.

Simon Ghobrial | Managing Director
Phone Number: +61 7 3344 4434
Email Address: info@anticorrosiontechnology.com

CORROSION & PREVENTION 2017

EXHIBITORS

AUSTRALASIAN INDUSTRIAL WRAPPINGS & COATINGS PTY LTD

Booth Number: 44
www.aiwc.com.au

AIWC is a manufacturer and distributor of Petrolatum, bitumen, butyl, elastomeric and epoxy coatings.

After 19 years of manufacturing for the pipeline and corrosion industry, AIWC have further diversified and for over the last decade the company have been providing one stop shop solutions to the pipeline and marine industry.

Amongst some of the agencies AIWC represent are Viscotaq with their world renowned elastomeric tape coatings and Tinker & Rasor with their holiday detection equipment and corrosion monitoring products.

These products complement AIWC's pipeline and marine coatings which ensure AIWC are one of the market leaders with our worldwide distribution network.

CORROSION & PREVENTION 2017

EXHIBITORS

AWS SERVICES

Booth Number: 72
www.awsservices.com.au

AWS Services is a multifaceted company based in Boronia, Melbourne with over 20 years of experience. Our expertise covers civil, commercial and residential sectors, and we offer a long range of services such as remedial works, waterproofing, caulking, plastering, painting, general construction etc.

Carlos Abreu
Phone Number: +61 425 720 268
Email Address: office@awsaus.com.au

EXHIBITORS

BLASTONE

Booth Numbers: 36, 37 & 38
www.BlastOne.com

Originally established over 40 years ago to provide technical consulting, BlastOne has grown to become a single source supplier of blasting equipment, abrasives and know-how to customers all over the world. BlastOne operates internationally from several offices across Australia, New Zealand, North America and UK.

We stand behind our brand claim of Superior Performance. It's something we define as 'Performance3' – the result of combining superior know-how with superior abrasives and superior equipment.

In short, it's something that delivers greater cost-efficiencies for our customers.

Jim Gooden | Global Technical Director – Corrosion Control
Phone Number: +61 8 8292 2000
Email Address: jim.gooden@blastone.com

CORROSION & PREVENTION 2017

EXHIBITORS

BLYGOLD OCEANIA

Booth Number: 56
www.blygold.com.au

Blygold prevents and stops corrosion in air conditioning & refrigeration systems with our unique anti-corrosion coating products. Blygold Oceania is part of Blygold International, being established for over 40 years with a reputation for excellence and quality of products and services. Blygold services many clients such as retail, airports, hospitals, hotels, oil & gas, mining and many other industries.

Blygold Treatment Benefits:
- save up to 20% on energy
- triple the lifespan
- increase reliability
- provide cleaner air

Specifications Blygold treatment:
- salt and acid resistant over 11,000 hours
- coating 25 to 30 microns
- full coverage, fins, copper tubes plus the connection between the fins and tubes
- metal pigmentation, conductivity stays high
- can be applied and maintained on site

Mark Weir | Managing Director
Phone Number International: +61 7 3807 0660
Phone Number Australia: 1300 271 115
Email Address: admin@blygold.com.au

EXHIBITORS

BRISTLE BLASTER AUSTRALIA PTY LTD

Booth Numbers: 52, 53 & 54
www.bristleblaster.com.au

Bristle Blaster® Australia Pty Ltd is the proud Monti Partner for Australia, New Zealand, Papua New Guinea and New Caledonia. Monti is the manufacturer of Bristle Blaster® Surface Preparation Technologies.

Bristle Blaster® Australia Pty Ltd supplies and supports the Bristle Blaster® throughout a variety of industries, including pipeline, oil & gas, marine / shipyards, bridge maintenance, mining and construction, to mention a few. Based in Melbourne, we work closely with select local partners throughout the region to provide on-site service, training and support.

Contact the team at Bristle Blaster® Australia today to find out more about products and services.

Phone Number: + 61 3 9737 2800
Email Address: info@bristleblaster.com.au

CORROSION & PREVENTION 2017

EXHIBITORS

CARBOLINE

Booth Number: 30
www.carbolineaustralia.com.au

Carboline is dedicated to supplying high performance coatings, linings and fireproofing products around the world through continuous technological improvements and superior service.

Our commitment to provide excellence in service and in product means our customers have access to the best possible coatings and solutions for superior corrosion and asset protection.

Altex Coatings is part of the Carboline global network. Since 1954 Altex has become one of the largest privately owned manufacturers of high performance coatings in Australasia. Supplying a wide range of coating types and systems for almost every major industry sector, with a strong focus in technology, training, staff and customer education. Altex Coatings Limited is a Resene Group company.

Ricky Collins
Phone Number: +64 075 411 221
Email Address: ricky@carboline.co.nz

EXHIBITORS

CATHODIC ANODES AUSTRALASIA

Booth Number: 71
www.cathodicanodes.com.au

Since commencing manufacturing in 1984, we have established ourselves as Australia's largest galvanic anode manufacturer. Our foundry is designed based on Lean Manufacturing principals, where work flow efficiency gains and minimal down time has led to significantly shorter lead times and reduced energy use.

Our anodes protect billions of dollars' worth of infrastructure throughout Australia, the South Pacific and South East Asia including industrial anodes for offshore oil rigs, floating FPSOs, mining, construction and pipelines as well as for the protection of infrastructure such as ports, wharves, sheet pile walls and desalination plants. Our anodes also protect ships, yachts, marine motors, tanks and hot water systems.

Brent Linde | Business Development Manager
Phone Number: +61 7 5476 9788
Email Address: brent@cathodicanodes.com.au

CORROSION & PREVENTION 2017

EXHIBITORS

CATHODIC PROTECTION CO LIMITED

Booth Number: 50
www.cathodic.co.uk

Cathodic Protection Co Limited, founded in 1950, was one of the first companies in the United Kingdom set up specifically to provide equipment and engineering services for the cathodic protection of pipelines, storage tanks, well casings, and marine structures for the oil, gas, petrochemical, water and power industries.

Following several successful years of growth, the company continues to expand its operations in Australasia, with its full range of cathodic protection and marine anti-fouling services and materials.

EXHIBITORS

CERAMISPHERE

Booth Number: 59
www.ceramisphere.com

Ceramisphere is a micro-encapsulation company based in Sydney, Australia. We manufacture a series of new smart anticorrosive pigments: Inhibispheres™. This innovative and patented technology brings a range of unique benefits: the potential for metal free and environmentally friendly corrosion protection, the simplification of the paint system by reducing the number of coatings required for optimum protection, the ability to formulate inhibitors otherwise incompatible with the paint and the extension of the corrosion protection in time. In addition, Inhibispheres™ are very easy to incorporate in a wide range of paint formulations with no impact on the final (mechanical, optical and chemical) paint properties.

Ron Cozijnsen | Business Development Manager
Phone Number: +61 2 8011 1364
Mobile Number: +61 414 029 382
Email : roncozijnsen@ceramisphere.com

CORROSION & PREVENTION 2017

EXHIBITORS

CHEMCO SPECIALITY COATINGS AUSTRALIA

Booth Number: 64
www.chemcosc.com.au

Chemco Specialty Coatings Australia is a world leader in innovative coating solutions. Our wide range and product adaptability covers many different industries, with our products used extensively in heavy industrial applications and challenging conditions. We focus on providing long term corrosion solutions, coating inspection and refurbishment of corroded equipment.

Our capabilities in industrial coatings specification, design review, project management and protective coatings application have been refined from over 30 years of experience.

Our expertise encompasses workshop based repairs or rehabilitation of industrial equipment and mobilisation to site, to reduce costs due to the onset of corrosion and improve operation efficiencies in the long-term.

CORROSION & PREVENTION 2017

EXHIBITORS

CORROSION CONTROL ENGINEERING

Booth Number: 39
www.cceng.com.au / www.cceng.co.nz

Corrosion Control Engineering (CCE) is Australasia's leader in Cathodic Protection engineering, field services and materials supply. CCE has more NACE certified CP Specialists than any other company in Australia and operates 7 regional offices across Australia and New Zealand which are managed by Principal Engineers, each of whom have in excess of 30-40 years' experience in the corrosion control industry.

The CCE team consists of qualified professional engineers and technicians who have specialised in the field of corrosion prevention and Cathodic Protection for many years, and who are certified with the ACA and/or NACE International. CCE also supplies a wide range of CP and pipeline related products and materials to the industry.

CCE specialises in the design, supply, installation and monitoring of cathodic protection systems for:

- Pipelines, plants & facilities
- Ports, wharves and jetties
- Storage tanks & reservoirs
- Steel reinforced concrete
- Marine vessels
- Offshore structures

Jason Paterson
Phone Number: +61 499 499 301
Email Address: jasonpaterson@cceng.com.au

CORROSION & PREVENTION 2017

EXHIBITORS

DECORA GROUP PTY LTD

Booth Number: 35
www.decora.co.nz

The Decora Group has been manufacturing and distributing anti corrosive coatings for over 30 years. Located in Auckland, New Zealand the manufacturing facility produces an extensive range of decorative and industrial coatings. It is ISO 9001 and APAS certified, assuring quality coatings are produced to the most stringent quality control standards.

The Decora Group distributes ZRC, Camrex Camovin and Procoat ranges of anti-corrosive coatings to Australia, New Zealand and the Pacific Islands. For over 30 years our industry leading anti-corrosive coatings have been specified by architects and engineers for thousands of corrosion protection projects.

The Decora Group offers superior service, experienced technical back up, combined with industry leading product performance for extended longevity of corrosion protection.

Derek McClelland
Phone Number: +64 9813 5900
Mobile Number: +64 274 940 721
Email Address: derek@decora.co.nz

EXHIBITORS

DENSO AUSTRALIA

Booth Numbers: 40 & 47
www.densoaustralia.com.au

Denso Australia is an industry leader in corrosion prevention and sealing technology and has been manufacturing and supplying corrosion prevention solutions to the Australian market since 1961.

The Denso range of global brands include Protal, SeaShield, Denso, PCS, Archo Rigidon and Sylglas.

Denso's extensive product range is suitable for use in a wide range of highly corrosive environments with operating temperatures ranging from sub-zero to 250 degrees celsius, across the pipeline, marine, tanking and structural markets. Denso will be showcasing
a range of corrosion prevention products at its stand.

CORROSION & PREVENTION 2017

EXHIBITORS

DETECTION SOLUTIONS

Booth Number: 19
www.detectionsolutions.com.au

Detection Solutions is a supplier of essential locating and specialist equipment to various industries such as plumbing, gas, power, construction, and also to local councils and government departments.

The Company has been in operation for over 25 years, hence enjoys an excellent reputation, with much repeat business from satisfied clients.

Our equipment works! Call us now to discuss your specific needs.

Phone Number: 1300 885 383
Email Address: info@detectionsolutions.com.au

CORROSION & PREVENTION 2017

EXHIBITORS

DIRECT CONNECTIONS

Booth Number: 28
www.directconnections.com.au

Direct Connections is an innovative R&D company, designing and manufacturing equipment to support the corrosion mitigation industry.

We have engineered and produced numerous devices, although our main focus has been on Dataloggers for monitoring CP Systems.

Our new range of BLE Loggers allows unsurpassed flexibility with wireless control, using Smartphones, Tablets, Laptops, or even remote monitoring from anywhere in the world.

Minimising maintenance costs and downtime, the dataloggers are totally field-serviceable and field-calibratable. Their IP67 rated enclosure permits effortless access to the standard AA batteries, which power the datalogger's continual use for up to 2 years.

Nick Papas
Phone Number: +61 419 711 701
Email Address: nick.p@directconnections.com.au

CORROSION & PREVENTION 2017

EXHIBITORS

DOCOPRO

Booth Number: 33
www.docopro.com

Docopro, an online document generation service, is now offering quality assurance and specification templates for industrial use. The initial template sets, which include Inspection and Test Plans (ITP) and Product Inspection Forms (PIF), are designed for the protective coatings industry and comply with coating systems standards in Australia, the U.S. and internationally.

The templates are downloadable on the company's website, docopro.com, in a format for electronic entries or as printouts for manual entries. The templates can be personalized with a company's name and logo, project name and description, and contact information.

Email Address: info@docopro.com

EXHIBITORS

DST DRYER

Booth Number: 61
www.dstdryer.com.au

DST is one of the world's leading suppliers of desiccant dehumidifiers, represented in more than 40 countries. With strong ties to the Corrosion, Grit Blasting, Military, Museums and Power Stations, DST strives to maintain high standards of quality and level of humidity. DST has been using dry air as a method of preventing corrosion during surface preparation and coating, during the grit blasting process and to ensure readiness in military and power stations. Using dry air as a preventative method of reducing corrosion and mould is a cost-effective way of ensuring high standards of product and stable humidity conditions.

CORROSION & PREVENTION 2017

EXHIBITORS

DEHUMIDIFICATION TECHNOLOGIES

Booth Number: 62
www.RentDH.com

DeHumidification Technologies, LP provides humidity and temperature control solutions to multiple industries in the US, Canada, Australia and Thailand. In addition to the highest quality equipment for every job, DH Tech is proud to have the most experienced and highly trained technical staff in the industry. Owners, Ken Armstrong and Brian Battle work closely with employees to deliver unparalleled customer service.

Ken Armstrong
Phone Number: 1-800-Rent-DH (736-834)
Email Address: karmstrong@rentdh.com

EXHIBITORS

DULUX PROTECTIVE COATINGS

Booth Number: 55
www.duluxprotectivecoatings.com.au

Dulux Protective Coatings provides protection against corrosion, chemical attack, abrasion and impact damage in diverse, harsh and corrosive environments with our broad range of products. Heavy duty inorganic zinc silicate and zinc-rich epoxy primers, high build epoxies, chlorinated rubbers, epoxy-acrylic and polyurethane topcoats, micaceous iron oxide coatings, polyurea and heat resisting product lines are available from over 230 outlets across Australasia.

Dulux Protective Coatings' experienced technical consultants provide tailor made solutions for whatever your project demands, even for the most demanding and difficult conditions.

Call your local Dulux Protective Coatings Representative or Dulux Customer Service on 13 23 77. Contact us today at www.duluxprotectivecoatings.com.au.

Phone Number: 13 23 77

CORROSION & PREVENTION 2017

EXHIBITORS

DUOGUARD AUSTRALIA

Booth Number: 17
www.duoguard.com.au

DuoGuard Australia are the sole distributors for Concrete Preservation Technologies range of DuoGuard and PatchGuard range of Hybrid and Galvanic anodes for complete leading edge corrosion solutions.

DuoGuard Australia also rely upon the Technical support provided by CPT research and development of new technologies for the treatment, assessment and monitoring of steel reinforced concrete structures and assets.

With over 40 years experience in corrosion management, we offer an innovative and flexible range of cutting edge products designed to effectively treat steel reinforced structures such as Wharves, Bridges, Bridge supports, culverts, high rise apartments and many more. We have installed our products into over 250 projects in 15 countries around the globe.

David Hadley | General Manager
Phone Number: +61 419 632 241
Email Address: david.hadley@duoguard.com.au

EXHIBITORS

DURATEC AUSTRALIA

Booth Number: 63
www.duratecaustralia.com.au/

Duratec Australia is a solutions-based contractor specialising in the protection and remediation of steel and concrete structures.

With ten branches across Australia, we service the defence, mining, oil and gas and marine industries, along with government assets and commercial infrastructure.

Duratec's team of specialist structural, civil and corrosion engineers are committed to providing clients with best practice methodologies, material selection and innovative solutions to maximise the life of our clients' assets.

Our services:

- Concrete repair
- Flooring systems
- Cathodic protection
- Waterproofing
- Blasting & painting
- High performance coatings
- Structural strengthening
- Precision grouting
- Joint sealing
- Acid protection
- Specialist access systems
- Technical surveys

CORROSION & PREVENTION 2017

EXHIBITORS

FAVCOTE | REZITECH SERVICES

Booth Number: 24
www.favcote.com | www.rezitech.com.au

Rezitech Services in alliance with Favcote are please to present the Belzona range of wear and chemical resistant metal, steel, rubber concrete coatings. Our companies supply and apply Belzona protective coatings which have many uses in all process industries including cold curing, wear resistant coatings providing similar wear rate to basalt tiles. Belzona protective coatings are used for internal lining of pipes, tanks, separators, production vessels, pylons and pumps for protection against erosion, corrosion and abrasive attack. Favcote are an experienced coatings applicator and abrasive blasting contractor experienced in the use of Sponge-Jet low dust abrasive blasting equipment. Sponge-Jet technology reduces the level of airborne particles typically experienced with more traditional dry blasting methods by up to 98%.

Bruno Favretti | Graeme Norris
Mobile numbers: +61 418 252 345 | +61 405 183 546
Email Addresses: Bruno@favcote.com.au |
graemenorris@rezitech.com.au

EXHIBITORS

FOSROC

Booth Number: 48
www.parchem.com.au

Fosroc is an international leader in delivering Constructive Solutions for projects across a broad range of market segments. Our commitment to customer service and technical support is second to none.

We work closely with architects, structural engineers, contractors and owners to best understand their requirements and develop a bespoke solution for a construction project, adding value and becoming more than just a materials supplier.

Parchem Construction Supplies is a leading manufacturer and supplier of products and equipment to the Australian and New Zealand concrete and construction markets and is the sole licensee of the Fosroc brand in Australia and New Zealand.

Phone Number: 1300 787 737
Email Address: sales@parchem.com.au

CORROSION & PREVENTION 2017

EXHIBITORS

FREYSSINET

Booth Numbers: 8 & 11
www.freyssinet.com

Offering integrated technical solutions in construction and structural repair, Freyssinet possesses an unparalleled array of civil engineering expertise in construction methods, post-tensioning, cable-stayed structures, structural accessories, repair, strengthening, cathodic protection, earthquake protection, and structural maintenance – on assets including bridges, tunnels, stadiums, power plants, LNG tanks, buildings, wharves, water tanks, and more.

Freyssinet offers extensive in-house design and technical capabilities with professional engineers and corrosion specialists including NACE and ACA certified Corrosion Technologists and Technicians – the most widely recognised qualifications for personnel demonstrating experience in designing, installing, commissioning, monitoring and maintaining CP systems.

EXHIBITORS

HUMISCOPE

Booth Number: 25
www.humiscope.com.au

At Humiscope, we provide customised air treatment services to a diverse range of industries. Our key product lines are commercial and industrial dehumidifiers/humidifiers.

Humiscope strive to set standards of excellence in our industry in all aspects of selection, design and customer service. We accomplish this through quality work and clear communication between all team members including clients, mechanical contractors and consultants. Our wealth of technical expertise and over 20 years' experience help us to deliver the best possible solution to our customers and their budgets.

Melissa Le Roux
Phone Number: +61 451 557 622
Email Address: melissa@humiscope.com.au

CORROSION & PREVENTION 2017

EXHIBITORS

IMATECH

Booth Number: 51
www.imatech.com.au

Imatech is one of the most innovative solution providers in the area of industrial wear protection solutions.

With over 30 years of combined practical experience with the best available products in the world, we have successfully expanded our presence in international markets and gained a reputation for continually delivering ground-breaking technological solutions in Asia-Pacific, South Africa, the Americas and Canada.

Our main area of operation is the mining & minerals processing Industry, where we provide a diverse range of solutions including engineered reinforced polymer composites, rubber-ceramic wear liners, rubber repair solutions, engineered abrasion resistant pipelines and pump efficiency solutions.

Phone Number: +61 2 8853 3000
Email Addresss: info@imatech.com.au

EXHIBITORS

INTERPROVINCIAL CORROSION CONTROL

Booth Number: 46
www.rustrol.com

ICCC

Interprovincial Corrosion Control features the most certified DC-Decoupling Device with the most extensive range of protection. The fail-safe, maintenance-free Rustrol® Model: DCD & Model: SSP are certified under the ATEX Directive, IECEx Scheme and KCs Marking for intended use in Gas & Dust explosive atmospheres, and bear the CE marking. The Rustrol® Model: SSP & Model: DCD effectively block the DC needed for cathodic protection, while simultaneously providing an effective AC continuity grounding path for AC fault current, lightning protection, mitigation of induced AC voltages, and also reduce and/or prevent the potential shock hazard to personnel and equipment, and AC Corrosion.

Please visit www.Rustrol.com to review the extensive range of Corrosion Control Products offered by ICCC.

CORROSION & PREVENTION 2017

EXHIBITORS

JOTUN

Booth Numbers: 9 & 10
www.jotun.com

JOTUN

Jotun is one of the world's leading manufacturers of paints, coatings and powder coatings with over 9,600 employees with 68 companies and 33 production facilities on all continents. Jotun has agents, branch offices and distributors in more than 90 countries. Jotun Australia Pty Ltd supplies: Protective, Marine, Fire Protection Coatings and Powder Coatings nationally, providing a high level of local service, manufacturing locally while supported by Jotun's global strength to provide best practice solutions. Industries supported include Offshore Oil & Gas Production, Hydrocarbon Processing Industries (HPI), Energy, Mining, Infrastructure, Pipelines and maintenance. Marine Dry Dockings, Marine Vessel New Builds, Tank Coatings & Linings.

Ted Riding | Technical Manager
Phone Number: + 61 3 9314 0722

EXHIBITORS

KK&S INSTRUMENTS

Booth Number: 60
www.kks.com.au

The Best in Quality Service and Products for Non destructive testing. As an Australian owned company, KK&S Instruments established itself as a leader in materials analysis and has been providing Australian Industry with state of the art Non-Destructive Testing Technologies since 1977.

KK&S specialize in Coating Thickness Measurement & Analysis along with; Ultrasonic Thickness Gauging, Eddy Current Testing, Material Analysis, Ultrasonic Testing, Ferrite Measurement, Hardness Testing, Borescopes, Magnetic Inspection, X-ray, as well as Energy Dispersive x-ray fluorescence on a wide variety of products from base metals & Plastics through to Gold Assay.

Products provided through KK&S Instruments are on an exclusive basis from worldwide leaders in their field. Companies such as Karl Deutsch, Institute for Dr Foerster, Helmut Fischer, Henke Sass Wolf, Parker Research Corporation, Technisonic, Gilardoni, to name a few.

Our qualified staff has for over 35 years conducted Demonstration, Training and Service support on our wide range of NDT products to expand our business throughout Australia and regionally to incorporate New Zealand, Papua New Guinea, New Caledonia and Fiji.

Continuing with our long history in Non-Destructive Testing KK&S has completely updated our website, in order to provide the complete overview of our product range and the areas in which they are applicable. KK&S hopes the information contained on our website will assist you in contacting us and be a useful evaluation tool for now & in the future.

Phone Number: +61 2 8850 3755
Email Address: contact@kks.com.au

CORROSION & PREVENTION 2017

EXHIBITORS

LET'S CLEAN PTY LTD

Booth Number: 65
www.letsclean.com.au

Supplier of Innovative Equipment, Products and Techniques to the Restorative, Conservation and Heritage Cleaning Industry as well as other Specialised Cleaning requirements. We are now manufacturing, here in Australia, the New "CAPTIVE HEAD WASHING SYSTEM" for Reducing Salt Loads in porous materials like Brick – Masonry – Concrete – Timber – Sandstone to a depth of up to 150 mm, and it's proven it works.

The principle is that the hand held wash head contains a low pressure (70psi) spray nozzle which is connected to a water supply. A 'flexible' skirt encloses the head and seals the unit against the surface so that the attached vacuum system creates a negative pressure within the wash head which is greater than the capillary effect of the substrate. Waste water is collected into the vac tank which enables the operator to analyse for soluble salts by electrical conductivity.

With over 25 years in business we have become experts in paint stripping, removing sealers & varnish, removing mould, brick colouring, old fashioned corrosion control, efflorescence removal, façade cleaning, fire damage restoration cleaning, anti-graffiti coatings and designing special purpose cleaning tools techniques.

Products & Equipment includes "Blue Vac System", "JOS" , "Heritage No1" natural paint removal poultice, "Soy Gel", "Wonder Strip", "Super Strip", Mastic remover, "Rustmasters" and accessory Products.

To help with your projects we can send our experienced staff to carry out on site testing, producing as many test patches as required, using equipment & products which would be used to carry out the actual job. This service has proved invaluable in drawing up accurate specifications and costing at the pre-contract stage of a project.

Phone Number: +61 2 9451 8422

EXHIBITORS

MARINE & CIVIL MAINTENANCE

Booth Number: 13
www.marineandcivil.com.au

MCM Marine & Civil Maintenance

Marine & Civil Maintenance Pty Ltd (MCM) is an engineering rehabilitation specialist. We rejuvenate critical ageing infrastructure such as wharves, bridges, buildings and industrial facilities by re-engineering and extending the asset's life through optimised Whole of Life solutions.

Over the last 16 years, MCM has built an enviable record of delivering high-quality engineered solutions in concrete repairs, cathodic protection, protective coatings, structural strengthening and other specialist activities. Embedded within our workforce, we bring this essential expertise and experience to ensure the successful completion of complex rehabilitation projects, whatever the location or scope.

Blane McGuiness | Strategy & Development Manager
Phone Number: +61 2 9809 0374
Email Address: blanem@marineandcivil.com.au

CORROSION & PREVENTION 2017

EXHIBITORS

MRJ INDUSTRIAL SERVICES

Booth Number: 28
www.mrjindustrial.com.au

MRJ Industrial Services Pty Ltd

MRJ Industrial Services provide services and products in the construction industry and are a company built on over 25 year's industry experience.

MRJ have partnered with leaders in the industry bringing the latest technologies to the Australian market from many corners of the globe.

MRJ offer supply, supply and install, training (on or offsite) and full specification services to architects, engineers, asset owners and end users with service based integrity, leaning on the vast experience of our international product partners.

MRJ's technologies are spread over a wide variety of market sectors including but not limited to mining, construction, marine, civil, commercial and residential.

Phone Number: +61 3 6268 6027 or +61 419 348 846
Email Address: admin@mrjindustrial.com.au

EXHIBITORS

MUNTERS

Booth Number: 16
www.munters.com.au

Munters is a global leader in energy efficient air treatment solutions based on expertise in humidity and climate control technologies. We can deliver state of the art dehumidification equipment for short-term or long-term climate control to any location throughout Australia and New Zealand.

For over 60 years, Munters has engineered and manufactured equipment to solve humidity and climate control problems. Our technicians will install, set and monitor equipment throughout the duration of your project. Our technical expertise enables us to advise you on your optimal requirements, giving you reassurance that the job will be completed efficiently and cost-effectively.

We provide:

- Corrosion Control
- Surface Preparation & Protective Coating
- Utilities (Oil & Gas, Power Stations)

Khalid Shaikh | National Project Manager
Phone Number: +61 2 8843 1588
Email address: Khalid.Shaikh@munters.com.au

CORROSION & PREVENTION 2017

EXHIBITORS

MURPHYS SPRAY & BLAST EQUIPMENT

Booth Number: 66
www.sprayandblast.com.au

Murphys Spray and Blast Equipment was incorporated in 1984 and currently employs 17 staff. Now part of the Wagner Group of companies, the head office is situated in Wangara Western Australia, with representation in Victoria and New South Wales.

MSBE is the Authorised Australian distributor for Blast Vision products, Titan Speeflo spray equipment and Torbo Engineering Wet Abrasive Blasting equipment. Additionally MSBE is a Diamond distributor for Carlisle Finishing Brands which includes brands such as DeVilbiss, Binks, Samson, Arnold & Ransburg. Through their extensive distribution network, customers can access many leading brands covering all your spray and blast needs.

Brendan Murphy | General Manager
Phone Number: +61 8 9309 1599
Email Address: sales@sprayandblast.com.au

CORROSION & PREVENTION 2017

EXHIBITORS

NACE INTERNATIONAL

Booth Number: 49
www.nace.org

NACE International, The Worldwide Corrosion Authority, serves more than 36, 000 members in 130 countries and is recognized as the premier authority for corrosion control solutions. NACE International's global membership includes: engineers, inspectors, technicians, scientists, business owners, CEOs, researchers, educators, students, and other corrosion professionals. Located in Houston, Texas, with offices in the U.S., China, Malaysia, Saudi Arabia, and Brazil, the organization serves all industries impacted by corrosion.

NACE International is the key resource for corrosion prevention and mitigation, technical knowledge and information, conferences and exhibitions, industry standards, reports, publications, and the most specified technical training and certification programs worldwide.

Phone Number: +1 281 228 6200
Email Address: firstservice@nace.org

CORROSION & PREVENTION 2017

EXHIBITORS

NATIONAL FACILITY FOR PIPELINE COATING ASSESSMENT

Booth number: 42
www.deakin.edu.au/ifm/our-facilities/national-facility-for-pipeline-coating-assessment

The National Facility for Pipeline Coating Assessment (NFPCA) is an independent and NATA accredited testing facility jointly developed by the Energy Pipelines Cooperative Research Centre and Deakin University. It provides the technical capability and capacity required to perform a wide range of standard and custom-designed materials performance tests such as:

- Assessing the performance and failure of pipeline coatings;
- Evaluating the environmental degradation of materials such as plastic pipes;
- Testing corrosion susceptibility and behaviour of metals and alloys;
- Research on protective coatings, cathodic protection, corrosion monitoring, corrosion inhibitors etc.

EXHIBITORS

NDT EQUIPMENT SALES

Booth Number: 57
www.ndt.com.au

NDT Equipment Sales has been trading since 1992 and is a leading supplier of equipment for the inspection of valuable plant assets, including the detecting of Corrosion Under Insulation (CUI). Our extensive range of products include: gamma and x-ray equipment, radioactive sources, ultrasonic equipment, hardness testers, eddy current systems, magnetic particle equipment and consumables. We are proud to be distributors for a range of manufacturers, including: QSA Sentinel (OpenVisionTM), Lixi Inc. (The ProfilerTM), NOVO DR (Digital Radiography), Flyability (Elios Drone) and Elektrophysik, amongst others.

Rod Martin | Manager
Phone Number: +61 2 9524 0558
Email Address: ndt@ndt.com.au

CORROSION & PREVENTION 2017

EXHIBITORS

NMT ELECTRODES

Booth Number: 41
www.nmtelectrodes.com

NMT® Electrodes is a well-established name in providing high quality, cost effective solutions and products in the field of Cathodic Protection. NMT® Electrodes' Mixed Metal Oxide (MMO) and Platinised titanium Anodes are the products of choice and are used and specified worldwide for use in concrete, underground storage tanks, pipelines, offshore and marine installations among numerous other applications. NMT® Electrodes' full range of tubular, wire, ribbon, rod, mesh, mesh ribbon and plate anodes are available from its Australian and South African facilities.

Kim van Loggerenberg
Phone Number: +61 8 9256 4499 | +61 418 810 396
Email Address: kim@nmtelectrodes.com

EXHIBITORS

OLYMPUS

Booth Number: 70
www.olympus-ims.com

Olympus is a world-leading manufacturer and distributor of high-quality optical, electronic and precision engineering products, for scientific, medical and industrial applications. The Olympus industrial leading edge testing technologies include remote visual inspection, microscopy, ultrasound, phased array, eddy current, eddy current array, X-ray fluorescence and diffraction.

The Olympus products include ultrasonic flaw detectors and thickness gages, videoscopes, borescopes, microscopes, high speed video cameras, in-line and advanced nondestructive testing systems, XRF and XRD analyzers, interferometers, and a large selection of industrial scanners, probes, software programs, and instrument accessories.

Phone Number: 1300 132 992
Email Address: info@olympus.com.au

CORROSION & PREVENTION 2017

EXHIBITORS

OMNIFLEX

Booth Number: 4
www.omniflex.com.au

Omniflex has been designing and manufacturing electronic products and systems for the automation and control industry since 1965. We specialise in providing solutions to industry in the fields of Remote Monitoring and Cathodic Protection (CP) Systems for both galvanic and impressed current applications.

Omniflex's PowerView CP system integrates world leading technology in distributed CP systems with full web-based remote monitoring, testing and control to provide significant cost saving in both installation and ongoing monitoring.

The PowerView CP system harnesses our depth of experience in harsh environments to bring reliable enterprise wide solutions to the challenges associated with installing and managing cathodic protection systems in difficult locations spread over remote locations.

EXHIBITORS

PAPWORTHS CONSTRUCTION TESTING EQUIPMENT

Booth Number: 18
www.pcte.com.au

PCTE

PCTE specialises in the sale and hire of testing equipment for construction materials. That includes field testing equipment (non-destructive testing equipment), laboratory test equipment as well as geotechnical and structural monitoring sensors.

For corrosion investigations on concrete structures, they carry a broad range of equipment vital to determining the extent of corrosion such; as cover meters, potential mapping equipment and resistivity meters. PCTE also supply a range of corrosion activity monitoring probes and data logging equipment.

Finally, PCTE supply coating inspection equipment for use over ferrous or nonferrous metals, or with a concrete substrate.

William Ward
Phone Number: +61 418 381 709
Email Address: w.ward@pcte.com.au

CORROSION & PREVENTION 2017

EXHIBITORS

PHILLRO INDUSTRIES

Booth Number: 20
www.phillro.com.au

Phillro
INDUSTRIES P/L

Founded in 1971, Phillro Industries have grown to be a respected specialised supplier of high quality products and equipment including:

- NDT (Ultrasonic Non Destructive Testing)
- Enterprise training Coatings fluid transfer pumps
- Spraying and Finishing Guns, Hoses and Systems
- Spray Booth Filtration
- Associated Safety products
- Solvent Recovery
- Food and Pharmaceutical Fluid Transfer
- Coatings Inspection
- Concrete Inspection
- Metal Detection

They are also trained and authorised service and repairers of Major Brands such as Graco & Ciemme. As the Principal Australasian/Oceania Importer Distributor for Elcometer Ltd, Phillro are the only Elcometer authorised Service centre in the region.

Phillro's activities include focussed Product Training, Sales, Service and Hire of Equipment manufactured by the world's most respected Brands.

Paul Jenkins
Phone Number: 1300 503 610
paulj@phillro.com.au

EXHIBITORS

PPG

Booth Number: 6
www.ppg.com

PPG Protective & Marine Coatings (PPG) has products that protect customers' assets in the world's most demanding conditions and environments. Our exceptional heritage has resulted in a range of brands that are tried and trusted to deliver consistent performance in their given markets.

All brands have PPG Protective & Marine Coatings as an endorsement, benefiting from unparalleled levels of experience and expertise in coatings technology. Working closely with our customers, we blend technical and business skills creating solutions to meet the constantly changing demands in all sectors.

As part of PPG Industries, we are uniquely placed to offer a complete service, so you will enjoy the resources, stability and capability of a worldwide organization. Our business is built on firm foundations, enabling us to provide not only market-leading products and services, but also a level of support unmatched in the market.

Graeme Gunston
Phone Number: +61 417 717 937
Email Address: gunston@ppg.com

EXHIBITORS

PRODIGITEK

Booth Number: 29
www.prodigitek.com

As the Australian representative for BioLogic Science Instruments, ProDigitek, based in Seven Hills NSW, offers the most cost effective Potentiostats to the scientific community involved in Corrosion and Coatings research, aimed toward developing products to improve corrosion prevention. Whether in the Laboratory or the field, solutions provided by ProDigitek include high end research equipment to Universities, Government and Industry.

When you buy from ProDigitek you don't just buy the equipment but also over 30 years of expertise in the use of such equipment which is also supported by a global support team with several years of field experience.

EXHIBITORS

RUSSELL FRASER SALES | RFS

Booth Number: 14
www.rfsales.com.au

Russell Fraser Sales (RFS) has served Australia's inspection and testing community since 1993 and along with its team of 10 staff, RFS has sustained its position as Australia's most trusted NDT equipment supplier by staying ahead of the changing needs of industry. For the corrosion mitigation industry RFS supplies 3D Laser Scanners for pipeline integrity assessment, Ultrasonic Thickness Gauges, Ultrasonic Phased Array, Holiday Detectors, Thermal Imagers, Videoscopes and more; many of which will be showcased on stand #14. RFS has also launched an innovative approach to NDT product video reviews on YouTube with the series "unboxing NDT" offering a detailed look at precisely what's in the box.

Phone Number: +61 2 9545 4433
Email Address: rfs@rfsales.com.au

EXHIBITORS

SAVCOR PRODUCTS AUSTRALIA

Booth Number: 58
www.savcorproducts.com.au

Savcor Products Australia (SPA) is one of the leading distributors of brand name corrosion products in Australia.

The company provides expert technical support for a wide
range of corrosion products including materials for cathodic protection systems.

SPA also focuses on fast responses to enquiries, and can often deliver various products directly from stockpiles, which substantially reduces the waiting time for customers.

EXHIBITORS

SIKA TECHNICAL EXCELLENCE

Booth Number: 7
www.sika.com

BUILDING TRUST

With 84 ICRI awards in 16 years, Sika is the global company with the most concrete repair projects awarded by the International Concrete Repair Institute (ICRI).

Sika is a global leader in speciality chemical product supply for the building and construction industry, Sika Australia prides itself on being technically driven and project-motivated. Our product line features high-quality concrete repair mortars, Latest technology corrosion control solutions including corrosion inhibitors and sacrificial and hybrid anodes, concrete admixtures, specialty grouts, sealants and adhesives, structural strengthening systems, industrial flooring as well as roofing and waterproofing systems.

CORROSION & PREVENTION 2017

EXHIBITORS

SN INTEGRITY PTY LTD

Booth Number: 69
www.sensornetworkscorp.com

SN Integrity

SN Integrity Pty Ltd has been established as an engineering company in Australia to provide smart solutions for asset integrity. We are Australian and New Zealand business partners for USA Company Sensor Networks Inc. We offer the very best minds in the highly specialized fields of ultrasonic and remote visual technologies. We deliver smarter solutions with a refreshingly personalized approach to the world's critical asset management applications. We specialize in the design and fabrication of industrial ultrasonic transducers and tooling for demanding in-situ test and inspection applications.

Telephone number: +61 8 7220 6670
Mobile Number: +61 414 470 985
Email Address: nestor@snintegrity.com.au

EXHIBITORS

SVENIC

Booth Number: 21
www.svenic.com.au

SVENIC
AUSTRALIA
The Complete Solution Since 1984

Svenic will be exhibiting its Coating and Foam (CAF) Spray System. This portable and lightweight system sprays paints, coatings, sealants, adhesives, foams, polyurethanes, and polyureas using 2 part cartridges and static mixers. It is excellent for use in confined spaces, repair jobs, and smaller applications. Material wastage is significantly reduced. CAF requires minimal set up time, training, clean up and machinery maintenance. More expensive 2 part pumps are not required.

Svenic also provides filling, labelling, and assembly of two part cartridges, syringes, and other containers filled with sealants, adhesives, coatings and foams.

We also supply:

- Crack Injection Components and Accessories
- Manual, Air and Battery Operated Dispensers
- Advise, testing and package suitability
- 2 part cartridges and static mixers

Contact Person: Peter Coundouris
Phone Number: +61 7 5539 2255
Email Address: peter@svenic.com.au

CORROSION & PREVENTION 2017

EXHIBITORS

TECHNOSEAL

Booth Number: 34
www.tecnoseal.it

TECNOSEAL

Tecnoseal is one of the world's leading suppliers of sacrificial anodes for cathodic protection in the leisure boat market, and for shipping, offshore and industrial applications. With over 30 years of experience, we produce and supply from our foundry in Grosseto (Tuscany) zinc, aluminum and magnesium anodes with our own peculiar passion for seeking top quality perfection of the product and the customer service.

Phone Number: +61 7 5529 7050
Email Address: sales@tecnoseal-usa.com

EXHIBITORS

UNIVERSAL CORROSION CONTROL

Booth Number: 1
www.unicc.com.au

CANUSA-CPS
UCC

Leading solution providers for corrosion and degradation problems across Oil and Gas, Ports and Harbour, Water, Energy, Mining, Refining and Chemical Industries.

Phone Number: +61 417 163 161
Email Address: d.anderson@unicc.com.au

CORROSION & PREVENTION 2017

EXHIBITORS

VALSPAR PAINT | WATTYL PROTECTIVE COATINGS

Booth Numbers: 31 & 23
www.wattylpc.com

wattyl
Protective Coatings

Wattyl Protective Coatings is a leading paint and coatings manufacturer in Australasian, specializing in high performance protective and marine coating solutions across a vast range of industries. Markets we operate in are: oil and gas, mining. water and waste, power, marine, OEM, as well as infrastructure and fireproofing. Wattyl's engineered solutions are designed to optimize the life of our customers' assets, provide maximum cost effectiveness, and deliver brilliant aesthetics. Wattyl's Epinamel® epoxies, Poly U™ polyurethanes, Galvit® zinc primers and, Seapro® anti-fouling range are well known brands that have been protecting major assets for over 40 years, a testament to the enduring performance of our product capabilities. As part of the Sherwin Williams group, Wattyl is committed to bringing the latest technologies, expertise and experience from the world's largest coatings company – straight to you.

Jasmine Moey | Marketing Manager
Phone Number: 13 21 01
Email Address: wattylprotectivecoatings@valspar.com

ACA CORPORATE MEMBERS – PLATINUM

as at October 11 2017

3C Corrosion Control Company, AECOM, AIWC, ALS Industrial, Anode Engineering, apa, ATTAR Advanced Technology Testing and Research, Blygold, Bureau Veritas, carboline, CAKE Commercial Services, CCE Corrosion Control Engineering, Curtin University, DHTech, Denso Corrosion Prevention, Dulux Protective Coatings, Duratec Australia, EPTEC, Freyssinet Sustainable Technology, HEMPEL, IMATECH Innovative Industrial Solutions, Industrial Galvanizers (A valmont Company), International, Jemena, JOTUN, MCM Marine & Civil Maintenance, MEP instruments (The right chemistry), McElligotts, PPG (We protect and beautify the world), PQR Quality Solutions, RUST-OLEUM Industrial Brands, Santos, SPA Savcor Products Australia Pty Ltd, SA Water, select solutions, Sika Building Trust, SMEC, SRG, Canusa-CPS UCC, vinsi partners, water Care services limited, wattyl Protective Coatings, XYPEX, ZINGA

CORROSION & PREVENTION 2017

ACA CORPORATE MEMBERS – GOLD

as at October 11 2017

A&E Systems, Advisian, Altitude Resurfacing, ATOTECH, Australian PPPS, baam consulting, Corrocoat, DEHN, epic energy, GHD, Giovenco Industries, Australian GMA Garnet, GPC Goldfields Protective Coatings, Independent Maintenance Consultants, infracorr, Intertek FSI Division, ITW Polymers & Fluids, KAEFER NOVACOAT, Metserve, MURPHYS Spray & Blast Equipment, neptune, OCEANEERING, OPUS, ORONTIDE, OSD Pipelines, PanBlast, parchem, peerless, Qenos, QUEST Integrity Group, Remedia Consulting Pty Ltd, RFS NDT, stanwell, STEULER KCH, TBS Corporation, United Industrial Solutions

CORROSION & PREVENTION 2017

ACA CORPORATE MEMBERS – SILVER

as at October 11 2017

- A S Harrison & Co Pty Limited
- Abrasiflex Pty Ltd
- ALS Industrial Pty Ltd
- Applus
- Artens
- Arup Pty Ltd
- Aurecon Australasia Pty Ltd
- Australian Maritime Systems Ltd
- Connect Group
- Cox Steel Coating Pty Ltd
- Crest Coatings Australia Pty Ltd
- CSIRO CMSE
- CSP Coatings Ltd
- Extrin Consultants
- Favcote Pty Ltd
- Galvanizers Association of Australia
- GPR Electrical & Corrosion Protection Service
- Hertel Industrial Services
- Inspection & Consultancy Services Ltd
- Intertek Technical Services Pty Ltd
- M Brodribb Pty Ltd
- Melbourne Water
- Oliver Technologies Pty Ltd
- Olympus Australia
- Omniflex
- Port of Newcastle
- Pumpline - Airblast
- Rhino Linings Australasia Pty Ltd
- RKF Engineering Services
- Rotafab Pty Ltd
- South Coast Surface Protection
- SVT
- Tranzblast Coating Service Aust
- Veolia
- VicRoads Technical Services
- WA Stroud Ltd
- Water Corporation of Western Australia
- WOMA Australia Pty Ltd

CORROSION & PREVENTION 2017

ACA CORPORATE MEMBERS – BRONZE

as at October 11 2017

- A.HAK Industrial Services
- ABSAFE Pty Ltd
- Absolute Blast Ltd
- Acacia Corrosion Control
- Action Alliance Group
- Advanced Aqua Blasting
- AFL Services
- Airservices Australia
- Allunga Exposure Laboratory
- AMAC Corrosion Protection Pty Ltd
- Amog Pty Ltd
- Andersal Pty Ltd
- Anti Corrosion Technology
- ARC West Group Pty Ltd
- ArmorGalv (Aust) Pty Limited
- ASC Pty Ltd
- AsClear Pty Ltd
- Asset Alignment Pty Ltd
- Asset Integrity Australasia Pty Ltd
- ATCO Gas Australia
- Australian Inhibitor
- Australian Maritime Safety Authority
- Ausvic Pty Ltd
- AWS Services P/L
- BAE Systems Australia
- BASF Australia Ltd
- Bastion Ltd
- Bells Property Services
- BP Refinery (Kwinana) Pty Ltd
- BRANZ Limited
- Brisbane Abrasive Blasting
- Bristle Blaster Australia Pty Ltd
- Broadspectrum Services
- Bundaberg Sandblasting Pty Ltd
- C. P Plating Pty Ltd
- Caltex
- Caltex Refineries (QLD) Ltd
- Cameleon Paints
- Cathodic Anodes Australasia
- CBI Constructors
- Central Coast Metal Protectives Pty Ltd
- Central Highlands Water
- Centreport Limited
- City West Water
- Clarkes Painting Services
- Clavon Pte Ltd
- Coating Industries Australia Pty Ltd
- Complete Steel Projects
- Contract Resources Pty Ltd
- CORE Water Management Solutions Pty Ltd
- Corrosion Electronics Pty Ltd
- Corrotek Consultants
- Costin Roe Consulting
- Couplertec Electronic Rustproofing
- Covestro Pty Ltd
- CTI Consultants Pty Ltd
- Dalrymple Bay Coal Terminal Pty Ltd
- Dapcor Building Services Pty Ltd
- Deakin University
- Decora Group Ltd
- Department of State Development (DSD)
- Department of Transport and Main Roads
- Dept of Infrastructure, Energy & Resources
- Dept of Transport
- DH Corrosion-DPC Coatings
- Diagnostech Pty Ltd
- Doito Pty Ltd
- Doogood Specialised Coatings Pty Ltd
- Downer Edi Engineering Power Pty Ltd
- DST Group
- Dukes Painting Services Pty Ltd
- Dulux NZ Ltd
- Eddysonics Pty Ltd
- Energy Infrastructure Management Pty Ltd
- Energy Safe Victoria
- Energyworks Ltd
- Esso Australia Ltd
- First Gas Limited
- Fitzroy Engineering
- Fremantle Sailing Club
- Gippsland Water
- Gisborne Abrasive Blast & Coatings 2000 Ltd
- Gladstone Ports Corporation Ltd
- GORODOK Pty Ltd
- GPA Engineering
- Gummow Pty Ltd (T/A G Squared Solutions)
- Highline Rope Access

CORROSION & PREVENTION 2017

ACA CORPORATE MEMBERS – BRONZE

as at October 11 2017

- Horiso Pty Ltd
- Hort Enterprises Pty Ltd
- Hunter H20
- Hychem International Pty Ltd
- Hydro Tasmania
- Independent Maintenance Services
- Inductabend Pty Ltd
- Innovative Asset Solutions Group (IAS Group)
- Innovative Corrosion Management Pty Ltd
- Integral Engineering Pty Ltd
- Intercept Australia Pty Ltd
- Ionode Pty Ltd
- Iontech Engineering Solutions Pty Ltd
- IPCO
- Keppel Prince Engineering Pty Ltd
- KGB Protective Coatings
- Korvest Ltd - Galvanising Division
- Lasserat Australia
- Linetech Consulting
- Linkup Paints BOP Ltd
- Liquid Gas Services
- Liquigas Ltd
- Loy Yang Power Ltd
- Mac Coatings Pty Ltd
- Magnatex
- Marden Corrosion Services Pty Ltd
- Master Blasters
- Materials & Testing Laboratories
- Mattioli Bros Pty Ltd
- McBerns Pty Ltd
- McCoy Engineering Pty Ltd
- MEnD Consulting
- Metal Spray Suppliers (NZ) Ltd
- Metal Spray Supplies Australia
- Methanex New Zealand Ltd
- Metrocorp Technologies
- Metz Specialty Materials Pty Ltd
- Michael Reardon
- Mills Sign & Painting Service
- Monadelphous Group Limited
- Multicoat Pty Ltd
- Nalco Australia Pty Ltd
- Napier Sandblasting Co Limited
- NDT Equipment Sales Pty Ltd
- New Zealand Steel Ltd
- NMT Electrodes (Australia) Pty Ltd
- NOF Metal Coatings Asia Pacific Co., Ltd
- Norblast Industrial Solutions Pty Ltd
- North Australian Centre For Oil & Gas
- Northport Ltd
- Nor-West Protective Coatings
- NZ Refining Co Ltd
- Origin Energy
- Orrcon Steel Pty Ltd
- Osborne Cogeneration
- Pacific Quality Corrosion Control
- Pacific Remedial Solutions
- Paint N Colour
- Paint Supplies Pty Ltd
- Paynter Dixon Qld Pty Ltd
- PD Remedial
- PF Systems
- Phillro Industries Pty Ltd
- Phoenix Australasia Pty Ltd
- Pipe Lining & Coating Pty Ltd
- PIVOTROM Pty Ltd
- Polynter Group Ltd
- Port of Portland
- Powered Aus
- Powerlink QLD
- Prendos New Zealand Ltd
- Preservation Technologies
- ProDigitek - BioLogic
- Professional Diving Services
- Prompt Engineering
- Pumpspec Ltd
- QLD Painters & Maintenance Services Pty Ltd
- Quality Maritime Surveyors Pty Ltd
- Queensland Bulk Water Supply Authority t/a Seqwater
- Queensland Sugar Limited
- Quick Strip Dustless Blasting Services
- R.Stahl Australia Pty Ltd
- Reinforced Earth Pty Ltd
- Rema Tip Top
- Remedy Asset Protection
- Reno Blast
- Renroc Group Australia

CORROSION & PREVENTION 2017

ACA CORPORATE MEMBERS – BRONZE

as at October 11 2017

- Resene Paints Ltd
- Resource Coatings Pty Ltd
- Rheem Australia Pty Ltd
- Rightway Industrial Pty Ltd
- RM Watson Pty Ltd
- Roads and Maritime Services
- Rochele Painting
- Rust-oleum Industrial Coatings
- S & L Steel Fab Pty Ltd
- Samaras Structural Engineers
- Schmidts Pty Ltd
- SICC SERVICES Pty Ltd
- Siemens Rail Automation
- Sika (NZ) Ltd
- SLH Contracting Ltd
- Solutions 4 Steel Pty Ltd
- South East Water Limited
- Southern Ports Authority
- Specialised Coating Systems
- State Steel Treatment (AMW MIH Pty Ltd)
- Steel Mains Pty Ltd
- Steelpipe Limited
- Stott Industrial Supplies
- Sulco Limited
- Summit Fertilizers
- SunWater Limited
- Surface Removal Specialists
- Sydney Trains
- Syntech Distributors Ltd
- Tasmanian Ports Corporation Pty Ltd
- TasNetworks
- TasWater
- Tcorr Inspection
- Thermoshield Batty
- Tincone Pty Ltd
- TOD Consulting
- Total Paint Protection (Walz)
- Traditional Coatings
- Transpower New Zealand Ltd
- Trelleborg Marine Systems Pty Ltd
- Tristar Australia Pty Ltd
- UGL Kaefer
- Undersea Construction Ltd
- Valicote Pty Ltd
- Vappro Corrosion Control Australia Pty Ltd
- VDM Metals Australia Pty Ltd
- Vincent Painting
- Viva Energy
- VT Industrial Coatings
- WAG Pipeline Proprietary Ltd
- Wannon Region Water Corporation
- Wellington Water Ltd
- WGASA Pty Ltd (Wallbridge Gilbert Aztec)
- Wood Group
- Woodside Energy Ltd
- WSP
- Yarra Valley Water
- Zinfra Network Services

CORROSION PREVENTION COATINGS: FROM NOAH'S ARK TO NANOTECHNOLOGY

Brian S. Skerry

The Sherwin-Williams Company

SUMMARY: Throughout history, protective coatings have been used for two purposes: improved aesthetics and asset protection. The focus of this review is on coatings that provide corrosion protection to metallic substrates, and which are used in many market segments, such as general infrastructure, transportation, marine, and energy production. The global market size for these coatings is estimated at over $20 billion in 2017. Protective coatings can provide effective corrosion prevention by a combination of film barrier properties, galvanic electrochemical activity, or by the use of inhibitive pigments. The evolution of new coating technologies has also been accompanied by the need to develop more reliable test methods for determining whether a coating will be effective for its intended use. Recently developed test methods are described here, including advances in quantitative measurement techniques, such as electrochemical impedance and noise techniques. In the future, emerging technologies, such as self-stratifying polymers, nanotechnology, and microencapsulated coating ingredients that function only when needed, hold great promise for the development of the next generation of protective coatings. Finally, increasing global environmental concerns and regulatory compliance needs will continue to drive the development of ever higher performing and more sustainable protective coatings. The prospects for future protective 'smart' coatings look bright.

Keywords: Protective coatings, metal corrosion, overview, electrochemical testing

1 HISTORICAL PERSPECTIVE

As evidenced by the Biblical description of Noah's use of pitch to protect his ark [1], protective coatings have been in use for millennia. Natural polymeric products, such as pitch blend, egg white, gum arabic, and beeswax, were known to the ancient Egyptians and Greeks and were used for various applications. In AD 77, Pliny suggested the use of white/red vegetable pitch to protect iron from rusting. During the 14th century, vegetable drying oils were used, and in 1676, John Smith published an early patent using a rosin/linseed oil mixture for the protection of military armor. In 1830, Thenard de la Rive was the first to recognize that the corrosion of iron was an electrochemical reaction, and in 1835, Davy first suggesting using zinc for the protection of marine structures.

In general, until the beginning of the 20th century, relatively little had been done to improve upon the natural materials available for the production of protective coatings. However, starting with the work of Baekland with phenolic resins in the early 1900s, the development of synthetic polymers for coatings and paints has proceeded rapidly. During the 20th century, rapidly accelerated developments in the area of protective coatings produced the acrylics, vinyls, polyesters, epoxies, urethanes, silicones, fluoropolymers, and others in use today. An introduction to the development of these polymers and resins systems has been published by Prane [2].

Today's urgent need for effective corrosion protecting paints is apparent from the fact that the total global sales of protective and marine coatings was estimated at approximately $20.9 billion USD in 2015 [3], and recent market studies project that the global market size for anti-corrosion coatings will approach $30 billion USD by 2021 [4]. An overview of all types of protective coatings technologies and pre-treatment methods available today is beyond the scope of this paper. Rather, in this work, focus is given to coatings for metal substrates in market segments including infrastructure, power and energy generation, as well as marine and offshore.

In this paper, a brief review of corrosion science basics is followed by an overview of protective coatings technologies in widespread use today and the mechanisms by which coatings function to provide corrosion protection. A review of coatings performance testing methods is given, and finally a brief summary of an anticipated technology roadmap to the future is presented.

2 CORROSION OF METALS

In order to understand how paints and coatings can protect metallic substrates from corrosion, it is first helpful to review the principles underlying the phenomenon of corrosion in normal service conditions. Metallic corrosion is generally considered to be inevitable and can be described as the spontaneous reversion of a metal to its preferred natural valence state. With the exceptions of the 'noble' metals such as gold or platinum, most metals occur in nature in chemically combined states in the form of oxides, sulfides, carbonates and others. Considerable amounts of energy are required to produce metals from their naturally occurring ore structures. Thus, it should not be surprising that freshly extracted metal, as well as finished metallic products that are in comparatively high energy states tend to deteriorate when exposed to environments which contain chemical species that allow the metal to return to their preferred lower energy state forms, Figure 1.

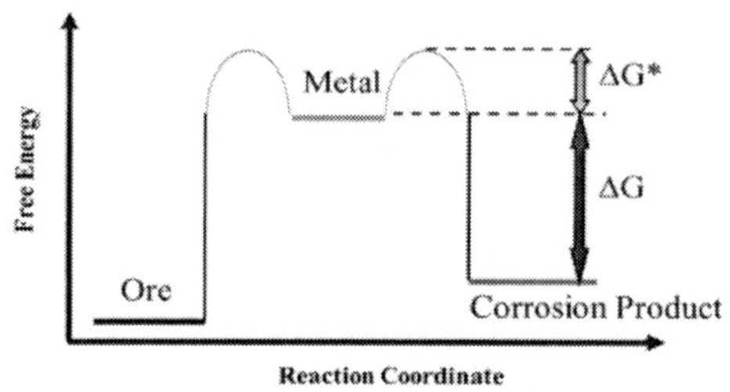

Where ΔG and ΔG^* represent changes in the free energy of products and reactants in standard and non-standard states.

Figure 1. Schematic of Relative Energy States for Metals

A simple corrosion cell in the case of iron (or steel) when exposed to aqueous conditions, is shown in Figure 2.

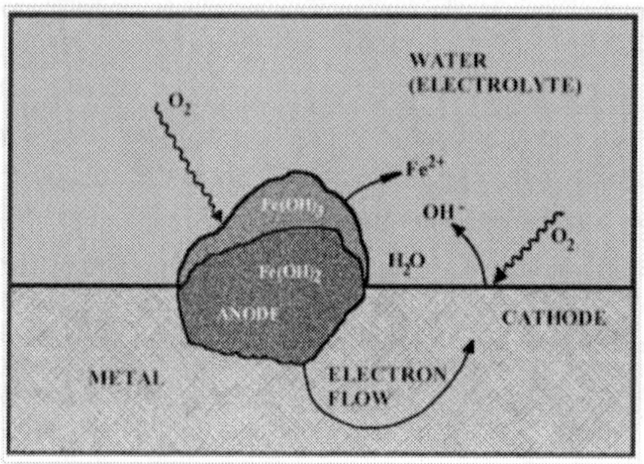

Figure 2. Corrosion cell for iron in aqueous conditions

In this case the anodic reaction is oxidation of the metal, and the cathodic reaction is the reduction of available oxygen, as follows;

$$Fe \rightarrow Fe^{2+} + 2e^-$$

$$O_2 + 2H_2O + 4e^- \rightarrow 4OH^-$$

The overall process is represented by the addition of the two half reactions;

$$2Fe + 2H_2O + O_2 \rightarrow 2Fe^{2+} + 4\ OH^- \rightarrow 2Fe(OH)_2$$

Ferrous hydroxide precipitates from solution but is unstable in oxygenated solutions, and is further oxidised to ferric hydroxide:

$$2Fe(OH)_2 = H_2O + ½\ O_2 \rightarrow 2Fe(OH)_3$$

The final product is the familiar rust formed by the partial dehydration of the hydroxide.

This example of corrosion not only illustrates the electrochemical nature of corrosion mechanisms, but also implies that certain prerequisite conditions must exist for corrosion to occur, summarised as follows:

1. Electronic conductivity of the metal
2. Electrolytic conductivity in the environment
3. Availability of an oxidizing agent (a cathodic reactant)
4. Ability to transfer charge at the metal interface.

By preventing any one of these necessary conditions, corrosion effectively may be controlled. It is this principle on which the design and development of protective coatings is based. Coatings essentially control or limit one or more of these prerequisite conditions for corrosion to occur.

Corrosion can be described as a function of three factors; the chemical environment, the thermodynamics of the corrosion process, and the kinetic rates of the corrosion reactions, all of which play a role in the outcome of corrosion and corrosion prevention.

The chemical environment experienced by a substrate determines what chemical corrosion products can be formed, and can be characterized in terms of pH, as summarised in Table 1.

Table 1. Corrosion Products as a Function of Chemical Environment

Chemical Environment		**Corrosion Products**
Neutral solutions of pH 5-9 ***e.g., rainwater, river and sea water, moist soils***	CORROSION →	Sparingly soluble products *e.g., iron hydroxides, basic zinc, lead or copper carbonates, aluminum hydroxide*
Acid solutions pH<5 ***e.g., cleaning agents, wood acids***	CORROSION →	**Rapid destruction of metals**
Alkaline solutions pH>9	CORROSION →	**Amphoteric metals attacked *(aluminum, copper, zinc);* Iron corrosion slow at high pH *precipitation of carbonate films***

Whether or not a corrosion reaction is possible depends on the principles governing the thermodynamics of the process under consideration. For a given electrode reaction such as

$$Ox + ne^- \rightarrow Red$$

The electrode potential, E, can be shown to depend on the concentrations of reduced and oxidised species (the Nernst equation):

$$E = E^0 - \frac{RT}{nF} \ln \frac{[Ox]}{[Red]} \qquad (1)$$

Where E^o is the standard electrode potential (*i.e.*, for a pure metal with metal ions at unit activity in Solution),
R is the gas constant,
T the absolute temperature, n the change of ionic charge in the electrode process,
F represents Faraday's constant.

Using the Nernst equation and measureable solubility data for various metal compounds, diagrams of potential vs. pH have been drawn up for many metals [5]. These potential/pH diagrams have proven very useful in gaining a rapid indication of which reactions are thermodynamically possible. Wranglen provides a helpful guide in understanding thermodynamic expectations in the case of an iron specimen immersed in an oxygen free solution at approximately pH 6, Figure 3 [6].

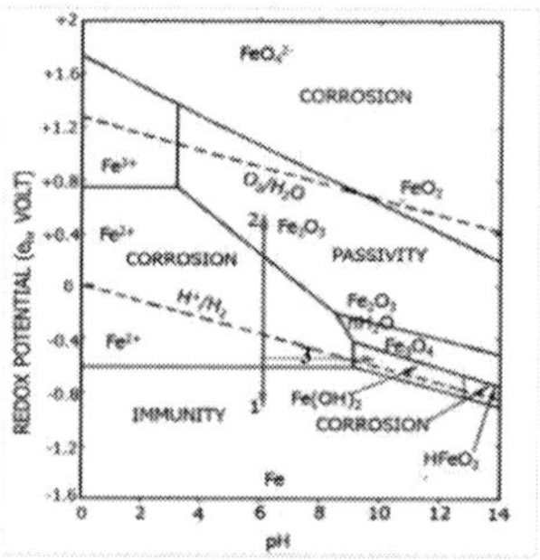

Figure 3. Potential – pH Diagram for Iron at 10^{-6} gram-atoms per litre @ 25°C

In this case, the iron corrosion potential would occur between the horizontal line (representing the equilibrium between the metal ions and its ions in solution) and the sloping, dashed line (corresponding to the hydrogen electrode potential). Three possible actions could be taken to remove the specimen from the corrosive region (denoted by arrows 1, 2, and 3).

1. The electrode potential may be lowered by making the iron the cathode in a galvanic cell so the specimen attains corrosion immunity.
2. The electrode potential may be raised by adding oxidizing agents, which makes the specimen passive
3. The pH may be raised to provide a degree of specimen passivity.

These potential/pH diagrams are extremely useful in predicting possible corrosion reactions, but there are limitations. Influence of other ions that may be present (*e.g.*, chlorides) is not included. They do not take into consideration any overriding influence of highly localized effects (*e.g.*, pH), and no indication is provided regarding the rates of reactions possible.

For rates of reaction considerations, these are determined by the kinetics of the corrosion process in question. Providing that the reaction rate is not controlled by diffusion or resistance effects, the rate of the corrosion process is controlled by the activation overpotential;

$$K = A^{exp}\left[\frac{\Delta G^*}{RT}\right] \qquad (2)$$

where ΔG* is the activation energy

$$\text{And } \Delta G^* = nF\,\Delta E^* \qquad (3)$$

where ΔE* is the activation overpotential, F is the Faraday constant, and n is the change of ionic charge in the electrode process.

In a corroding system, the corrosion potential, E_{corr}, results from the mixed combination of the two half cell reactions (anodic and cathodic) as described above, and the relative rates of possible reactions can be assessed by measuring so called polarization curves in a laboratory electrochemical arrangement. From polarization measurements, Evans developed graphical representations of corrosion process, known as Evans diagrams, which can be used to assess electrochemical effects of chemical species to inhibit overall corrosion rates [7]. Effects on cathodic and/or anodic processes can be obtained by these procedures, and this information can be helpful in selecting additives to formulate corrosion protective coating systems.

In summary, for corrosion to occur, any activity depends on the chemical and physical nature of the environment, and whether the reaction is thermodynamically favourable. The extent of corrosion depends on the kinetic rates of the reaction processes. In combination, these factors determine the likelihood and extent of any corrosion that could occur in a given system. Protective coatings provide a means for influencing these factors to effect the control and mitigation of metallic corrosion.

3 COATINGS TECHNOLOGIES USED FOR CORROSION PROTECTION

Coatings technologies for most industrial and marine type applications include the following: alkyds, epoxies, polyurethanes (2-component, or moisture cured systems), baking phenolics, polyesters and vinyl esters. More recent developments have included polyureas, polyaspartics, silicones, and polysiloxanes. Effective corrosion protection for industrial steel structures can often be obtained using zinc-rich coatings (either organic or inorganic based). Several key reference documents provide insight into which types of protective coatings work best for selected industrial applications. [8, 9]. A particularly useful guide to selection criteria for these different coatings technologies can also be found in a more recent guide published by the Society for Protective Coatings (SSPC) in the USA [10]. All such paints and coatings have several factors in common, including that they are formulated using well established principles of resin, pigment, and additive combinations.

The mechanisms by which paints and coatings are thought to reduce or prevent corrosion of metallic substrates are complex. The basic components of a coating – polymeric binder, pigments, additives – provide a protective barrier for the substrate. Corrosion inhibiting additions to the coating affect the kinetics of anodic or cathodic processes at the substrate surface, and galvanic coatings introduce sacrificial metal pigments that corrode more readily than the substrate.

For simplicity, protective coatings are often classified into three basic categories: barrier coatings (*e.g.,* thick films, usually pigmented), active inhibitive coatings (containing corrosion inhibitive pigments), or sacrificial galvanic coatings (*e.g.*, zinc-rich primers). These coatings categories are described in more detail below, but it is important to note that the corrosion inhibitive functionality of protective coatings is often some combination of the effects of various mechanisms, rather than clearly separated mechanistic functions.

3.1 Barrier Coatings

The primary component of any paint film is the polymeric binder, which essentially establishes most of the physico-chemical properties of the final coating formulation. Specifically for corrosion protection, the need to limit water, oxygen, and corrosive ion ingress into the coating is of paramount significance. The resistance to permeation of those substances through the coating is affected greatly by the polymeric film forming properties of the coating. Typical solventborne protective coatings (*e.g.* epoxies, polyurethanes) form highly crosslinked polymer networks, whereas waterborne (*e.g.*, acrylic latex) coatings form films via the coalescence of distinct polymer particles. These differing film formation mechanisms are illustrated in Figure 4.

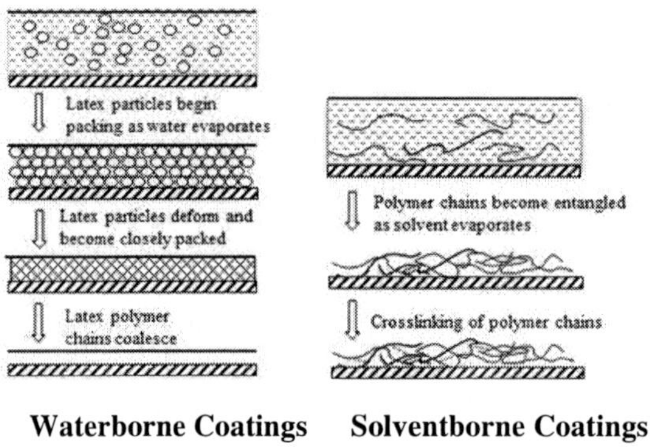

Figure 4. Film Formation in Waterborne and Solventborne Coatings

Although water based coatings are thought to be less effective for high performance corrosion control applications, recent work by resin suppliers and research consortia has made progress towards developing successful water based coatings options for severely corrosive environments [11, 12].

Simple dry film thickness can be a key factor for long term corrosion protection. For example, long term successful corrosion protection by old technology coal tar enamels and bituminous films have been observed historically when using paint film thicknesses at 2.5 mm or greater [8, p.348]. On the other hand, even coatings of 100 – 200 um dry film thickness, in layers of 30-40 um, remain permeable to H_2O and O_2. For corrosion reactions at the metal surface, there is often sufficient water and usually enough oxygen present for active corrosion [13]. The strength of the physico-chemical attachment to the metallic substrate (*i.e.*, dry and/or wet adhesion) also affects how the coating functions in preventing the spread of incipient corrosion.

In addition to the choice of binder system, pigment type and concentration can also influence the permeation of water, oxygen, and corrosive ions through the coating. Figure 5 illustrates how lamellar shaped pigments, such as micaceous iron oxides, can increase the tortuosity of the water pathways through the paint when compared with the ease of water transport in the presence of more granular shaped pigments.

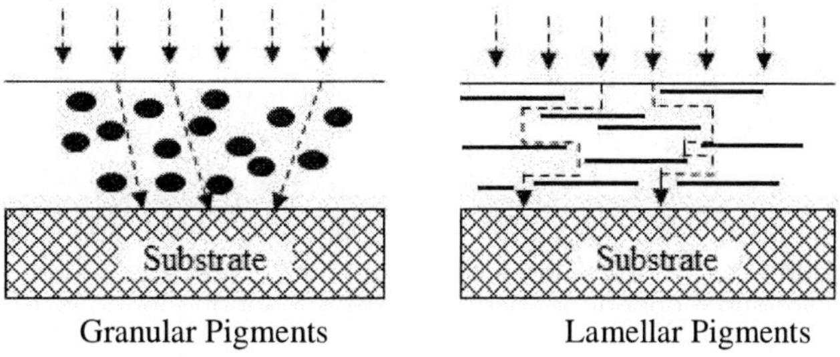

Figure 5. Effect of Pigment Shape on Water Transport Pathways

In the case of primers intended for corrosion protection, it is well established that the combination of coating ingredients (pigments and resin) produce different performance properties depending on the level of pigment volume concentration (PVC) to resin content. Property variations in corrosion resistance as well as other characteristics (porosity, blistering tendency, gloss) change drastically at the critical pigment/resin ratio (CPVC) where just not enough resin is present to completely coat all pigment surfaces. Corrosion resistance and gloss decline significantly as CPVC is approached. Conversely, porosity and susceptibility to blistering increase significantly above CPVC.

3.2 Active Inhibitive Coatings

In addition to creating a physical barrier, protective coatings also act to control anodic and/or cathodic processes that would otherwise occur. Lead and chromate pigments were once widely used as corrosion inhibitors in primers for metal corrosion prevention, although environmental concerns have rendered these pigments unacceptable today. However, the mechanisms by which these pigments functioned were in line with the principles of corrosion processes described above. Red lead in the form of Pb_3O_4 mixed with PbO was somewhat soluble in water, and formed soaps with the fatty acids generated by hydrolysis or oxidative degradation of the drying oils (*e.g.,* linseed oil) used as the paint vehicle (binder). Lead compounds so formed acted as a passivating agent by oxidising ferrous ions to ferric ions, followed by co-precipitation of mixed ferric, and lead salts. These highly insoluble and stable corrosion products effectively limited further corrosion. Chromate ions, when used in protective coatings formulations, also effectively acted as an oxidising agent, at least in relatively neutral pH conditions, to elevate the redox potential of corroding iron or steel into a thermodynamic range of passivity, as illustrated in the example in Figure 3 (arrow 2), above. Oxidizing chromate anions are similarly effective in protecting alloys of aluminium, so much so, that they are still used in aerospace and military applications today. However, with further regulatory restrictions anticipated, all chromate based inhibitors will likely soon be replaced by selected alternative candidate inhibitors, such as mixed rare earth organophosphates inhibitors based on technology developed in Australia [14].

The effectiveness of these pigments, acting primarily to control the anodic corrosion reactions, continues to stimulate research to develop new paint pigments that would work in similar ways. Many new pigments have been commercialized in recent decades, including other oxidising anions, such as molybdates, tungstates and vanadates. In contrast to the oxidising anions, sparingly soluble inorganic materials that are considered as non-oxidising pigments can also be effective in controlling metallic corrosion, but these effectively stifle the kinetics of the anodic reaction process (*e.g.,* phosphates, borates, silicates). Other sparingly soluble inhibitors used in paints tend to migrate to cathodic reaction sites and effectively slow the kinetics of the overall corrosion process by forming insoluble deposits that are more insoluble than the corrosion products that otherwise would form in their absence (*e.g.,* zinc hydroxide, calcium phosphate).

Many other inhibitors are used in paints that exhibit combination effects to slow corrosion, including ion exchanging and ion scavenging compounds. Effective corrosion inhibitors in paints are not restricted to inorganic pigments and can include many forms of organic compounds (*e.g.,* benzoates, amines, sulphonates, thiols, *etc).* Such compounds tend to exhibit hydrophobic film forming properties, where complexes are formed at anodic and cathodic corroding areas, effectively producing a resistive function to slow corrosion processes.

3.3 Sacrificial Galvanic Coatings

One class of coatings used for corrosion protection of iron and steel substrates that is described as functioning by galvanic activity is that of zinc-rich primers. In this coating material, zinc metal is used as a pigment to function in an analogous way to zinc galvanizing. Zinc-rich primers are typically required to contain high levels of powdered zinc (commonly > 85 wt% zinc) in the finished formula. This is because the zinc metal particles must be in physical contact (greater pigment concentration than the CPVC) in order to provide for electrical connectivity between the zinc particles and the substrate surface. The porosity of these coatings above the CPVC allows for relatively rapid, but controlled ingress of water, with associated dissolved salts, to complete the electrical circuit. Since the zinc is anodic to steel, zinc corrodes preferentially, thus reducing the redox potential of steel to the region of immunity, as described above in Figure 3 (arrow 1). Zinc particles can be formulated in an inorganic based binder system, usually a silicate, or alternatively can be used in an organic binder, often epoxy based resins. A detailed overview of these technologies can be found in the literature [9].

Although inorganic and organic zinc-rich primers function by complex corrosion protection mechanisms involving metallic galvanic activity combined with some barrier protective paint film properties, it is important to note that the practical lifetimes of such coatings are not limited simply by the reservoir of zinc available in the coating. After initial electrochemical consumption of zinc occurs, partially hydrated zinc oxide formed fills pores in the paint film, and reinforces the barrier effect of the coating to keep further water, oxygen and corrosive ions from contacting the underlying steel substrate surface.

4 COATINGS TESTING AND EVALUATION METHODS

While coating design is guided by our understanding of corrosion science and film properties, there is, of course, still a need to demonstrate a coating's effectiveness before relying on it to protect valuable assets. Test methods to evaluate the effectiveness of coatings for corrosion prevention have developed extensively over several decades, and further progress continues to be made today. Accelerated test methods have been developed to approximate the expected service conditions, recreate the paint degradation mechanisms that would occur under those conditions, and accelerate that degradation to mimic the effect of long-term exposure in a reasonably short period of time. Examples of test methods and procedures widely used to generate quantitative performance data for coatings in atmospheric service are noted below.

4.1 Test Methods for Atmospheric Service Conditions

Valid test methods for assessing or predicting long-term performance properties of protective coatings systems for metallic substrates have long been sought by coatings' end users and manufacturers alike. For corrosion in natural atmospheres, new paint formulations have historically been evaluated by long term (multi-year) outdoor exposures in natural environments. However, due to the inherent variability occurring in all natural exposure testing programs (changing weather, temperature, humidity, sunlight conditions, *etc*), and the lengthy time frames needed to observe any meaningful performance differentiation amongst competing formulations (often many years), the global coatings industry, including customers and manufacturers, have long since moved away from relying on exterior exposure testing programs in favor of decisions made on the basis of accelerated laboratory-based testing methods.

Demonstrably improved correlations between accelerated testing results and long term natural exterior exposures have increased the industry's confidence in accelerated testing methods. (An exception would be testing by use of neutral salt-fog exposures, such as ASTM B-117 method [15], which has been largely discredited over the years. Results obtained using this method never correlated well with practical experience [16, 17], resulting in that test being described as "notoriously unreliable" [18].)

Recently developed laboratory-based accelerated test methods for performance assessments have been shown to provide acceptable correlation with results obtained from long term exterior exposures [19 – 22]. For example, Figure 6(a) below, published by the Q-Lab Company in Cleveland, OH, USA, illustrates the closeness of the light emitting UVA bulbs to that of natural sunlight, and Figure 6(b) shows results obtained for a 2-component epoxy-polyamide primer/top-coat paint system, compared with results obtained after natural exposure testing for 27-months at an aggressive outdoor coastal site in the USA.

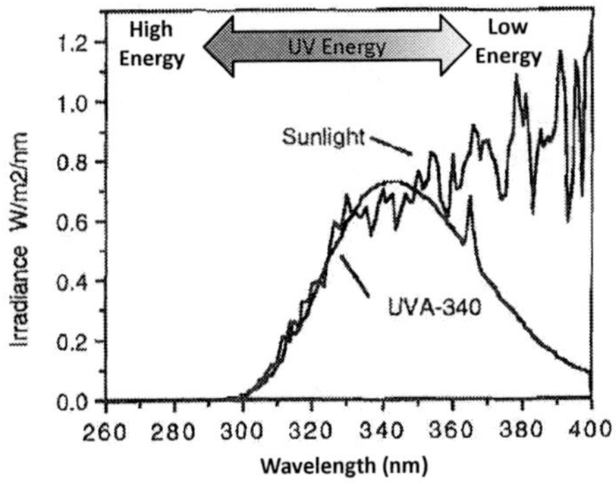

Figure 6(a). Comparison of UV Irradiance intensity of Natural (Sunlight) and Accelerated (UVA-340) Exposure

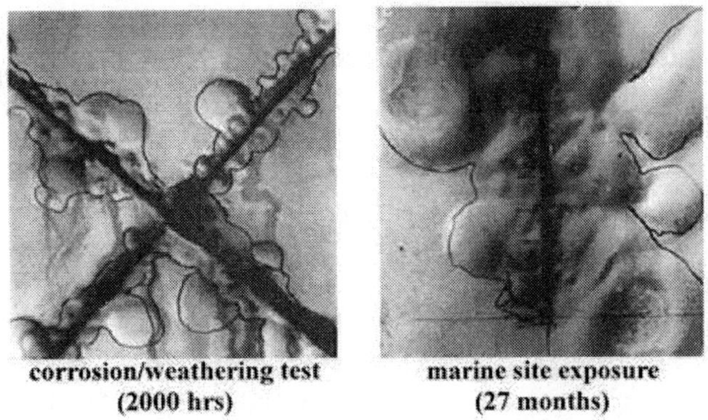

Figure 6(b). Comparison of ASTM D5894-16 Testing (2,000 hrs) *vs.* Marine Site Exposure (27-Months)

The blistering and scribe line corrosion patterns occurring in the accelerated test panels and the exterior test panels are the same. Further analysis in this example showed that the corrosion products formed in the accelerated test exhibited the same structure, morphology and chemical composition characteristics as the exterior exposed panel. Changes in color and gloss of the exterior exposed panels were also closely matched by the accelerated test panels. Correlation studies conducted by multiple organizations over the years have further corroborated the close relationship existing between these accelerated test results and long term exterior data. In effect, a 3-month accelerated lab test program will closely mimic 2-years of exterior exposures in an aggressive coastal environment [19,20].

Based on the above results, a widely adopted test method in current use is a cyclic corrosion/weathering procedure in which coated metal samples, along with control painted samples are placed in test chambers that use accelerated wet/dry corrosion cycles combined with UV-condensation weathering cycles in one complete test regimen. This method is described in ASTM D5894-16 (ISO 11997-2:2013) *Standard Practice for Cyclic Salt Fog/UV Exposures of Painted Metal (Alternating Exposures in a Fog/Dry Cabinet and a UV/Condensation Cabinet)*[23]. The UV exposure cycle has been carefully designed to mimic natural sunlight, and the test can be conducted in a laboratory using standard corrosion and weathering testing equipment (see Figure 7).

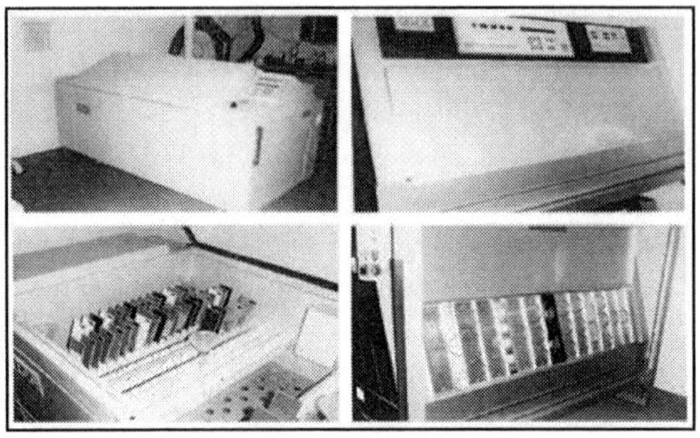

Figure 7. Laboratory Equipment Required to Run Accelerated Corrosion/Weathering Testing of Paints

This method has been in widespread industry use for evaluating coatings performance properties since the 1990's. Global oil and gas production companies rely on this cyclic corrosion method, or very closely related procedures, such as the ISO 20340 cyclic corrosion test [24], to project coatings performance and to specify and select coatings for their high performance applications. In the United States, most state Departments of Transportation now use this test practice as the basis for paints and coatings performance qualifications and approvals for all new construction and refinish coatings for bridge and highway structures.

4.2 Electrochemical Test Methods

Since corrosion processes are electrochemical in nature, and since paints function to inhibit corrosion processes by the principles of electrochemistry described above, it is clear that the application of quantitative electrochemical measurements to measure the effectiveness of paints in preventing corrosion is highly applicable. Opinions vary as to which specific electrochemical test methods are the best choices for assessing corrosion-protective performance properties of protective coatings [25], but two of the most effective such methods with a long history of success for this work are electrochemical impedance spectroscopy (EIS) and electrochemical noise measurements (ENM). Both of these methods have significant benefits and advantages. Used in combination with accelerated testing, these non-destructive electrochemical methods can provide valuable data on the condition and protective functionality of coatings over time.

4.2.1 Electrochemical Impedance Spectroscopy (EIS)

Experimental procedures for EIS have long been established for paint testing [26, 27]. In this laboratory based measurement method, the painted metal surface is treated as an electrochemical interface and quantitative measurements are made of coating and corrosion parameters such as coating and charge transfer resistances and capacitance values. A test measurement parameter often cited as significant for corrosion prevention by paint is the EIS impedance obtained at 0.1 Hz, and it is assumed that if this value exceeds approximately 1×10^6 ohms.cm^2, then the coating is providing adequate corrosion protection to the underlying metal substrate. Figure 8 provides a simple representation of the electrochemical interface, and Figure 9 illustrates for an epoxy based coating how measured impedance signals change progressively with paint film degradation over time.

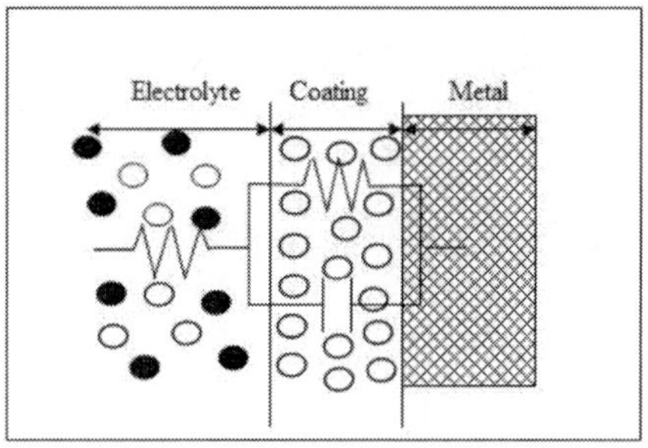

Figure 8. Schematic of the Metal/Paint Electrochemical Interface

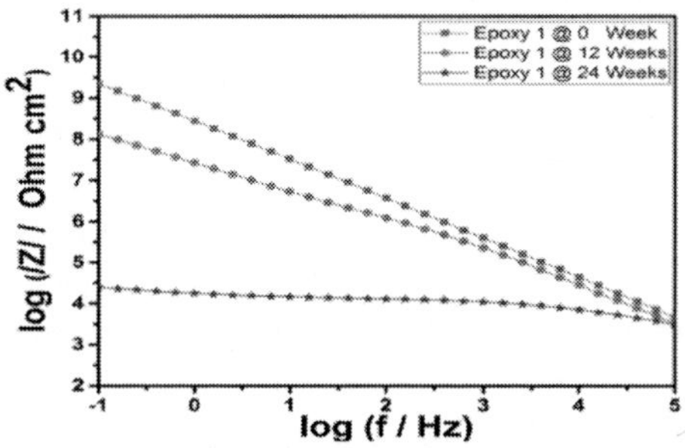

Figure 9. EIS Changes with Time as a Function of Paint Degradation

To assess performance attributes over time of coatings intended for atmospheric service conditions, these electrochemical measurements can be taken at intervals during exposures according to the preferred test method, for example ASTM D-5894.

4.2.2 Electrochemical Noise Measurements (ENM)

Electrochemical noise measurements have also been applied to protective coating systems for many years, and the standardized laboratory procedures are well established for this method as well [28-30]. During measurements of corrosion occurring under intact protective coatings, characteristic patterns of voltage noise and current noise emerge as the protective coating degrades over time of test exposure by allowing corrosion to proceed under the paint film. High frequency transient noise is apparent when a coating is providing effective corrosion protection; lower frequency voltage and noise patterns develop as a coating provides progressively less corrosion protection over the time of the test period. Figure 10 shows a schematic of the standard electrochemical noise experimental arrangement in the laboratory.

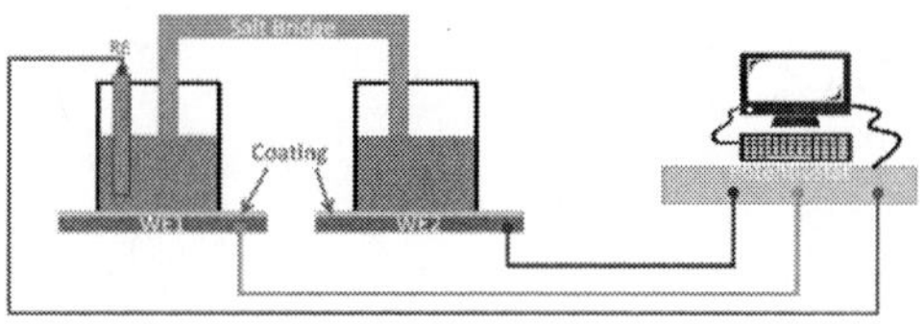

Figure 10. ENM Standard Laboratory Test Arrangement

Figure 11 illustrates typical voltage and current noise patterns for a high impedance paint providing good corrosion protection (*e.g.*, epoxy), compared with a lower impedance paint that is not expected to provide the same level of corrosion protection (*e.g.*, acrylic-latex).

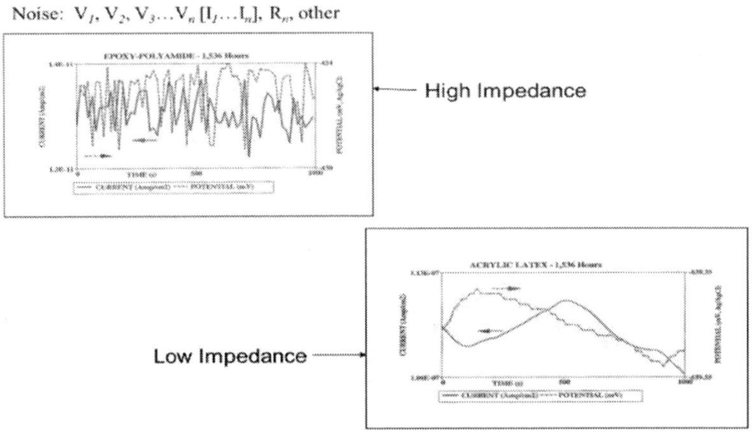

Figure 11. Representative Voltage and Current Noise Transient Data for High and Low Impedance Paints

When the noise transient data are analysed by simple statistics, the ratio of the standard deviation of the current noise divided by the standard deviation of the voltage noise over a selected time window provides a measure of the standard deviation of resistance noise, which can be simply described as the resistance noise value, Rn. It has been shown that the values of Rn from electrochemical noise measurements correlate well with the low frequency limit of impedance (realistically at the 0.1 Hz frequency) as measured by the EIS technique [25].

As with EIS, when using ENM measurement methods, it should be noted that these measurements should be taken as a result of exposures to the preferred test method, whether accelerated atmospheric testing (*e.g.*, ASTM D-5894), or in testing designed to mimic in–service operating conditions. In fact, these electrochemical measurement techniques (EIS, ENM) are particularly well suited to instrumented painted metal test panels in order to generate quantitative performance testing data on coatings when exposed either in standardized test methods or in customized laboratory analogue experiments for specific in-service test conditions. Figure 12 illustrates some customized electrochemical test apparatus arrangements for assessing underwater marine and immersed tank lining evaluations, as used in the author's laboratory.

Figure 12. Customized Test Chamber Arrangements for Electrochemical Assessments of Painted Electrodes

A further benefit recently developed using the ENM method is that valid ENM data can be obtained without the need to make a physical electrical connection to the metallic substrate under the paint film . This newly developed methodology is described as the 'no connection to substrate' (NCS) method as described in the recent work of Mabbutt, Mills, and Woodcock [31]. In this measurement procedure, the electrode connection arrangement is illustrated here in Figure 13. This arrangement now allows for possible ongoing monitoring corrosion measurements to be made on painted surfaces without damaging the paint to generate the information needed. This, in turn, allows for the possibility of ENM measurements to be made in-situ on functioning engineered structures (*e.g.,* highway bridges, offshore drilling platforms, *etc.*).

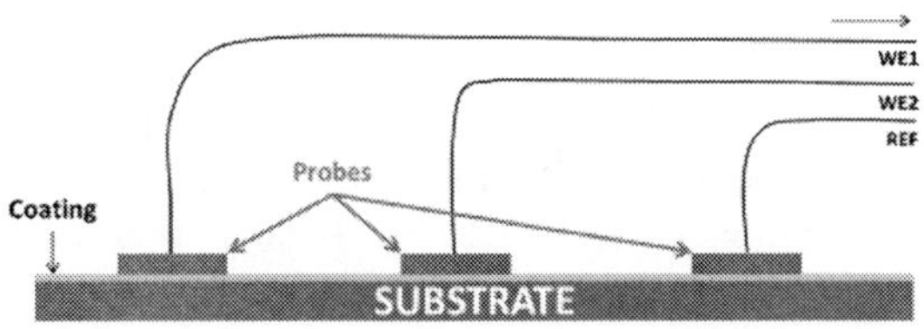

Figure 13. Electrode Arrangement for ENM Using No Connection to the Substrate (NCS) Method

5 ROADMAP TO THE FUTURE – TECHNOLOGY INNOVATION

The global paint industry has indeed exhibited a rich history of technology development and innovation over the last century. However, industry trends and customer preferences demand further improvements in paint system performance in ever harsher environments, longer service lifetimes with lower total costs, and greater ease of application in ever broader ranges of temperature and humidity weather conditions, Ever more stringent global environmental regulations also force innovation in coatings design, aiming for lower or zero VOCs, 100% solids formulations, and solvent free systems.

To meet these ever more stringent requirements for paint performance properties, the industry has focused on new, innovative technologies to produce higher performing products. Examples of emerging technologies applicable to protective coatings include polymers that can self-stratify into multiple layers, potentially obviating the need for separate primers and top-coats; paint additive materials that can encapsulate active ingredients such as corrosion inhibitors, but which are only used as and when needed; and nanomaterials that can enhance thermal, electrical and physical properties of a coating. These are broad categories with significant overlap of concepts and materials, but overall, paints formulated using any or all of these new technology approaches all function according to the principles for corrosion control described earlier in this paper. Although by no means a comprehensive overview, some brief examples are noted below.

5.1 New Polymer Designs and Layer Stratification

In recent years, considerable progress has been made in designing high performance polymer structures. One example is water borne acrylic-latex polymers that can self-stratify during the curing process, to produce more durable and longer lasting paint films. Figure 14 gives an example of such a hybrid acrylic-latex polymer network in cross-section, where there is an indication of two types of polymer forming one paint film but structured in individual layers.

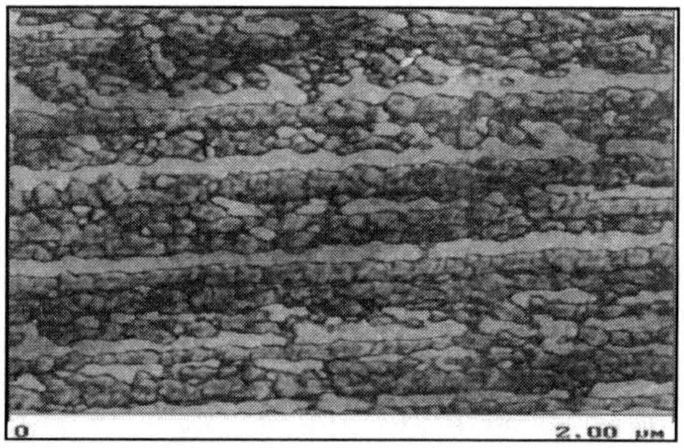

Figure 14. Self-stratifying Polymer Composite

In this case, fluoro and silicon/silicone modified polymer layers can be separated due to differences in surface energy characteristics between the two combined latex designs. In turn, the stratified matrix can be designed to have specific coating performance properties, such as enhanced UV durability in some layers, or enhanced hydrophobicity characteristics in secondary layers to reduce moisture permeation, water vapour, and corrosive species transmission rates through the coating.

5.2 Micro-Encapsulation and 'Smart Coatings'

As coatings design moves forward, recent emphasis has been given to the development of so called 'smart coatings' that can potentially heal themselves if they become damaged in service. Much work has been published on the concept of using micro-encapsulation techniques to create embedded capsules that can break when needed and release a payload intended for resin/binder self-healing or corrosion protection. Encapsulation is a particularly attractive method for on-demand delivery of corrosion inhibitors. Several capsule trigger mechanisms have been investigated, including resin release after physical rupture of the capsules when a coating is damaged, as well as the use of polymer microcapsules and inorganic microparticles that release corrosion inhibitors when corrosion reactions generate high pH conditions at predominantly cathodic reaction sites [32]. Another published trigger release mechanism is simple diffusion of water through the coating, which allows dissolution of selected corrosion inhibitors to be released from porous microstructured pigment particles. In this work, a combined emulsion and sol-gel chemistry approach produces spheres of selected sizes where the inhibitive active material is homogeneously distributed in a pore network inside the particles. Diffusion of water in the paint releases the active inhibitor at selected release rates [33]. Some representative microcapsule characteristics are illustrated in Figure 15 [32,33].

Polymer Microcapsules [32] Inorganic microparticles [32] Ceramic microparticles [33]

Figure 15. Examples of microencapsulation particles for use in coatings

The above are some examples of new technology approaches to developing self-repairing, corrosion controlling 'smart coatings'. There are many others published in the open literature. It seems likely that continuing research and development work globally will result in effective new paints and coatings for corrosion protection applications that would have been unheard of in the past

5.3 Nanotechnology

Although a broad term, there are many examples of small nano-scale additives or pigment like materials that have been added to paint formulations to improve product performance for corrosion protection applications. Polymer nanocomposite coatings have been described in the literature based on Al_2O_3, Fe_3O_4, $Ce(NO_3)_3$, and others [34], as have epoxy based paints with dispersed nanoparticles of polyanilines [35]. The advent of nanotechnology has also been a strong impetus for work conducted using carbon nanotubes, and other functionalized flake-like nanoparticles which are intended to influence the electrochemical corrosion process at the same time as providing plate-like improved barrier properties of the coating.

Recent investigations have explored the possible use of plate-like graphene particles for corrosion protection [36]. The graphene properties that could contribute to corrosion-resistance include high impermeability to water, oxygen, and corrosive species of the plate like graphene nanoparticles, high surface areas of the particles themselves, (providing that they align successfully in a finished paint dispersion), and high levels of electronic conductivity induced by the graphene platelets themselves. The electronic conductivity factors are not yet well understood, but can be described as enhanced electron delocalization effects, possibly allowing the protective coating to provide an alternative pathway for electronic transport rather than remaining in the metal substrate only. Whether this electron delocalization effect is likely to reduce metallic corrosion rates, or possibly increase corrosion is yet to be clearly determined in the laboratory by practical coating corrosion measurements. An illustration of the shape and form of graphene particles is shown in Figure 16 [37].

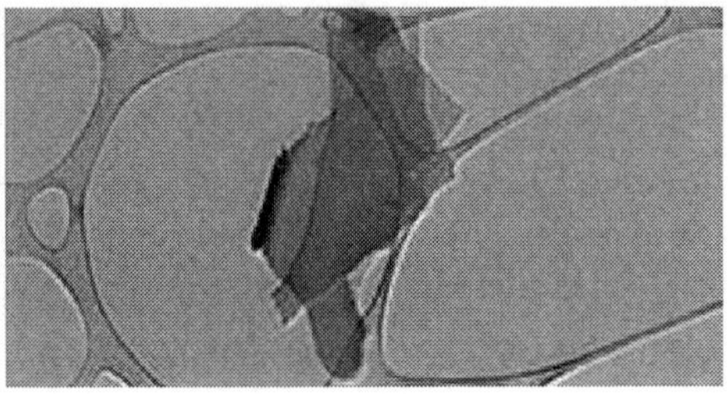

Figure 16. Example of Graphene Nanoparticles [37]

Similar suggestions for electronic delocalization effects have been applied to claimed corrosion resistance improvements produced by use of inherently conductive polymers, such as doped or undoped forms of polyaniline. [38, 39]. Some measure of 'self-healing' after coating scribe damage by inherently conductive polymers when tested by a rotating disc electrode experiment in aerated 5% NaCl was also reported by Kendig [40].

Whether or not laboratory based performance claims for these innovative technologies translate successfully to actual paint formulations for in-service operating conditions for protective coatings remains under investigation in many industrial and academic organizations today. However, with the advent of more effective coating test methods, such as cyclic corrosion/weathering testing for atmospheric service coatings, or improved mimicking of in-service operating conditions during laboratory testing of immersion grade coatings, particularly when combined with quantitative electrochemical measurements using EIS or ENM methods, the effectiveness of new polymer designs, micro-encapsulation, or nanotechnology solution options should be relatively easy to assess in the future.

6 CONCLUSIONS

1. Metallic corrosion is a natural process by which engineered metallic structures will degrade to a thermodynamically preferred energy state. This process cannot be eliminated, but it can be ameliorated by the application of protective paints and coatings designed with principles of electrochemistry in mind.
2. Progressively more cost effective paint and coatings products have been developed during many decades of research and development, evolving from naturally occurring paint raw materials and ingredients to synthetic materials developed on sound scientific principles.
3. Recent decades of research have produced laboratory test methods that can mimic in-service operating conditions for intended paint applications such that new materials and paint products can be assessed quickly and with ever greater levels of confidence that these products will function as intended for their service operating conditions.
4. Electrochemical science also has provided test methods for laboratory and future in-service corrosion measurements that provide quantitative assessments of the effectiveness of protective coatings in slowing or inhibiting metallic corrosion processes.
5. New technology developments in areas of polymer design, micro-encapsulation based 'smart coatings', and nanotechnology offer new paint performance possibilities, and indicate that the future prospects for corrosion protective coatings are bright indeed.

7 REFERENCES

1. Genesis 6:14, Revised Standard Version
2. Prane, J. A. Federation Series on Coatings Technology : Introduction to Polymers and Resins. Philadelphia: Federation of Societies for Coatings Technology; 1986
3. Anonymous. Facts in Five: Anti-Corrosive Coatings. European Coatings Journal. 2015; 11, p.26
4. Mordor Intelligence. Anti-Corrosion Coatings Market - Global Analysis, Current Trends, Forecast to 2021. February 2017. Retrieved on July 19, 2017, from https://www.mordorintelligence.com/industry-reports/anti-corrosion-coatings-market,.
5. Pourbaix, M. Atlas of Electrochemical Equilibria in Aqueous Solutions. Oxford: Pergamon Press;1966
6. Wranglen, G. An Introduction to the Corrosion and Protection of Metals. Stockholm: Inst. Metallskydd; 1972, p.66
7. Evans, U. R. Introduction to Metallic Corrosion. London: E.Arnold;1950
8. Hare, C. H. Protective Coatings, Fundamentals of Chemistry and Composition. SSPC 94-17. Pittsburgh: Technology Publishing Company; 1994
9. Munger, C.G.; Vincent, L.D. Corrosion Prevention by Protective Coatings, 2nd Edn. Houston: NACE International; 1999
10. Drisko, R. W., Ed. Selecting Coatings for Industrial and Marine Structures. SSPC 01-08. Pittsburgh: SSPC; 2008
11. Dehan, V., et al. High Performance Water-Based Barrier Coatings for the Corrosion Protection of Structural Steel. Submitted to Steel Construction Design & Research, June 2017
12. Skerry, B. S., Hilton, S. L., Sibbald, M. S., Martin-Fabiani, I., Fortini, A., Sear, R. P., et al. Advanced Water Borne Barrier Coatings for the Corrosion Protection of Structural Steel. European Technical Coatings Conference, May 25 – 26, 2016 (ETCC 2016).
13. Mayne, J. E. O, Off. Dig. 1952; 24, p.127
14. Markley, T. A., Mardel, J. I., Hughes, A. E., Hinton, B. R. W., Glenn, A. M., Forsyth, M. Materials and Corrosion. 2011; 62, (9), P.836
15. ASTM B117-16 Standard Practice for Operating Salt Spray (Fog) Apparatus. Annual Book of ASTM Standards, Vol. 06.01, West Conshohocken: ASTM International; 2016
16. W. Funke, J. Oil and Colour Chemists' Assoc. 1979; 62, p.17
17. Appleman, B. R., Campbell P. G. J. Coatings Technology. 1982; 54, p.17
18. Townsend, H. and McCune, D., "Round-Robin Evaluation of a New Standard Laboratory Test for Cosmetic Corrosion," SAE Technical Paper 970734, 1997, doi:10.4271/970734,
19. Simpson, C.H., Ray, C.J, and Skerry, B. S. J.Prot. Coatings & Linings, 1991; 8 , 28
20. Skerry, B. S., Alavi, A., and Lindgren, K. I. J. Coatings Technology, 1998; 60, p.97
21. Boocock, S.K. J. Prot. Coatings & Linings. 1994;11, p.51
22. Cleveland Soc. Coatings Technology. J. Coatings Technology. 1994; 65, p.49
23. Annual Book of ASTM Standards, vol. 06.01 (2017), ASTM D5894-16. Standard Practice for Cyclic Salt Fog/UV Exposures of Painted Metal (Alternating Exposures in a Fog/Dry Cabinet and a UV/Condensation Cabinet),
24. ISO 20340:2009, "Paints and varnishes – Performance requirements for protective paint systems for offshore and related structures." Geneva, Switzerland; 2009
25. Mills, D. J., Jamali, S. S. Prog. Org. Coatings. 2017; 102, p.8
26. Mansfeld, F. Kendig, M. W. Electrochemical Impedance Tests for Protective Coatings, ASTM STP 866. West Conshohocken: ASTM International;1985; p.122
27. Chen C-T., Skerry, B. S. Corrosion. 1991; 47 p.379
28. Chen C-T., Skerry, B. S. Corrosion. 1991; 47, p.598
29. Skerry, B. S., Eden D.A. Prog. Org. Coatings. 1991; 19, p.379
30. Moon, M., Skerry, B. S. J. Coatings Tech. 1995; 67, p.35
31. Mabbutt, S.J., Mills, D.J. , Woodcock, C.P. Prog. Org. Coatings. 2007; 59, p.192
32. Pearman, B. P. , Li, W., Buhrow , J. , Zhang, X., Surma, J., Fitzpatrick, L., et al. Int. Workshop on Environment and Alternative Energy, Kennedy Space Center, FL, USA, 2014. Retrieved on July 19, 2017, from https://ntrs.nasa.gov/archive/nasa/casi.ntrs.nasa.gov/20140017326.pdf

33. Reproduced courtesy of Ceramisphere Pty Ltd. image retrieved on July 19, 2017, from http://www.ceramisphere.com/industrial/what-are-they/
34. Wang, Y. , Lim, S., Luo, J.L., Xu, Z. H. Wear. 2006; 260, p. 976
35. Riaz, U., Nwaoha, C., Ashraf, S.M. Progress in Organic Coatings. 2014; 77, p.743-756
36. Aneja, K.S., Bohm, S., Khanna, A. S., Mailika Bohm, H. L. Nanoscale. 2015; 7, p. 17879
37. Reproduced courtesy of Applied Graphene Materials Ltd, UK, 2017; retrieved from http://www.appliedgraphenematerials.com/products/graphene-nanoplatelets/
38. Talo, A. et al. Synth Met. 1999; 102, p.1394
39. Holness, R. et al. J. Electrochem. Soc. 2005; 152 (2), p. 73
40. Kendig, M. Prog. Org. Coatings. 2003; 47, p. 183

The Advantage of Heat Resistant Composites

M.Buckhurst , Dr. S.J. Kaspersen
Jotun AS, Norway

SUMMARY: It is well known that materials can be chosen to perform specific tasks. It is also commonly understood that we can optimise the performance of said material by combining it with another specialist material, thereby creating a duplex system or a composite. When it comes to corrosion protection at high temperature, by nature we almost always create a duplex system of Carbon Steel with a polymer or an alternative metal on top. However, it is possible to take it further by creating a composite material that is applied onto the carbon steel, that adopts the benefits of the polymer, the added material and the backbone of the carbon steel structure. This paper looks at the different ways one can create composites for the high temperature environment in order to achieve the different needs and benefits desired to keep equipment safe and operational. We consider different resin bases and alternative components, ratios and orientation to reinforce and strengthen the material.

Keywords: Protective Coating, Composites: Ceramics; Glass spheres; Glass flake

1. INTRODUCTION

Corrosion protection of carbon steel in general is done by using a paint or coatings. Traditional paint thinking is that the resin itself is the dictating factor for functional performance. Although pigments, solvents and additives certainly play a part in the system and the methodology that is used – the full extent to which these additional components effect performance after application is often underestimated, and only the obvious benefit is considered. It is also apparent that the fact that it is applied on a substrate (in this case carbon steel) the synergy between these two materials as a duplex system is providing a positive benefit to the system, however this will not be discussed or presented in this paper.

1.1 The challenge

1.1.1 Limited resin bases available for protection at higher temperatures

Most resin systems used have in one form or another been around for many years – with minor or major adjustments to make small improvements in the desired performance. The challenge often observed is that the modifications have a positive effect where needed and a negative performance on another for example heat performance versus adhesion.

1.1.2 Understanding of Paint technology

Although in materials science the use of composite technology and thinking of the combination of two materials combined to create a third more beneficial material is common. This idea is often over looked in terms paints and coatings. Thus, the use of pigments in a more creative manner than the traditional, barrier, Pacifying or galvanic uses is over looked.

2. DISCUSSION

2.1 Thinking Differently

If we consider how paint technology is traditionally thinking, pigments are referred to as 'Finely ground, natural or synthetic, inorganic or organic, insoluble particles that, when dispersed in a liquid vehicle to make paint, may provide color and other properties, including opacity, hardness, durability, and corrosion resistance. The term is used to include extenders as well as white or colored pigments.' [1]

There is of course room in this description for other functional properties, but the strength of this description and thinking takes us to the three standard functions of pigments:

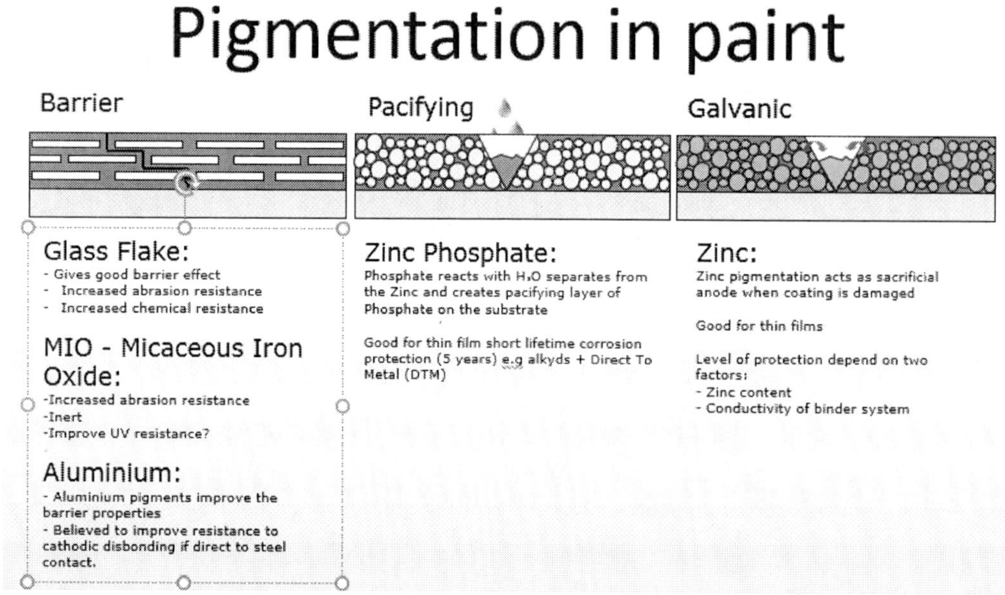

Figure 1: Traditional Functional Pigments

If we consider Materials as an industry the use of composites is common. Regardless of industry we see plastics used to perform extreme conditions, whether it be for a hobby or heavy industry in a chemical plant

Figure 2: Use of composite plastics to resist internal or external pressures and the chemical attack from seawater or hydrocarbon mixtures.

In industry, we see composites in other materials too, for example steel rebars in concrete providing structural strength and stability, where the concrete alone could not be utilised alone.

Figure3: Carbon steel rebars in concrete

So why not think of paints and coatings as materials and utilizing certain pigments as structural integrity of the coating?

Figure 4: Glass flake in an epoxy coating

2.2 Choice of Resin

Although there are new resins coming into the market every year, often due to commercial reasons completely new technology may not be the viable option for a paint company, although from time to time they come. They may also have technical reasons such as application difficulties or porosity problems that make them less than ideal.

Thus, we often come back to base technologies that we well proven in real life – but may also be modified, improved and adjusted to meet some of our requirements. In paints and coatings the most tried and tested industrial coating technology is in fact Epoxy Resin technology.

Currently there are three main technology groups within epoxy main technology,

- Pure Epoxy (Bisphenol A)
- Phenolic Epoxy (Bisphenol F)
- Novolac (a high crosslinked Phenolic)

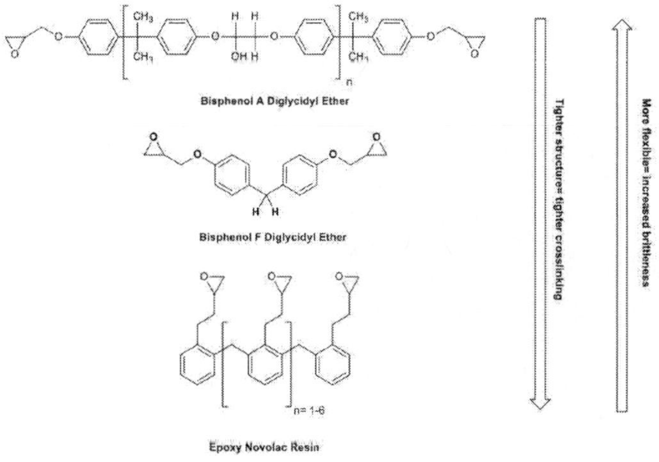

Figure 5: Epoxy resins with increasing crosslinking

With higher levels of crosslinking comes greater chemical resistance and to some extent heat resistance. But in both cases it is often with a penalty. For example if you look at chemical absorption it is reduced with high crosslinking but at the same time, the ability to release absorbed material is reduced for the same reason. We found with constant heat testing that we could improve heat resistance by blending the three resins. This gave us some improvement up to about 230 degC but was at risk of cracking if over applied.

2.3 Choosing the correct reinforcement pigment

The terminology for paint compared to materials science is a little weak in this area a composite is a 'combination of two or more materials differing in form on a macro scale. Each still retaining individual identities and exhibit an interface between them.' [1].

If we accept that macro is still the scale at which we talk about pigments in coatings then we can talk about reinforcement pigments in coatings creating a composite material. Choosing the correct reinforcement pigment in terms of composite is a matter of choosing a material with the correct physical properties and the correct geometry to achieve the desired benefit. In the case of the resin base we have chosen, we had a problem of micro cracking should the temperature increase and/or the final dry film thickness exceed the required amount to provide corrosion protection. For us to manage this we needed a flexible pigment that was long and planar in structure. The materials could be various. Ceramic pigment exhibit fantastic properties but were general of a spherical shape or similar. Micaceous Iron Oxide is ideal as it is low cost and reasonably inert. But the length of the particle is relatively short and under testing reduced cracking but did not sufficiently support the resin base. Glass flakes are excellent in this situation as we could choose the type size and length of the particle. Thus, we optimized the combination of material and geometry to get our desired result

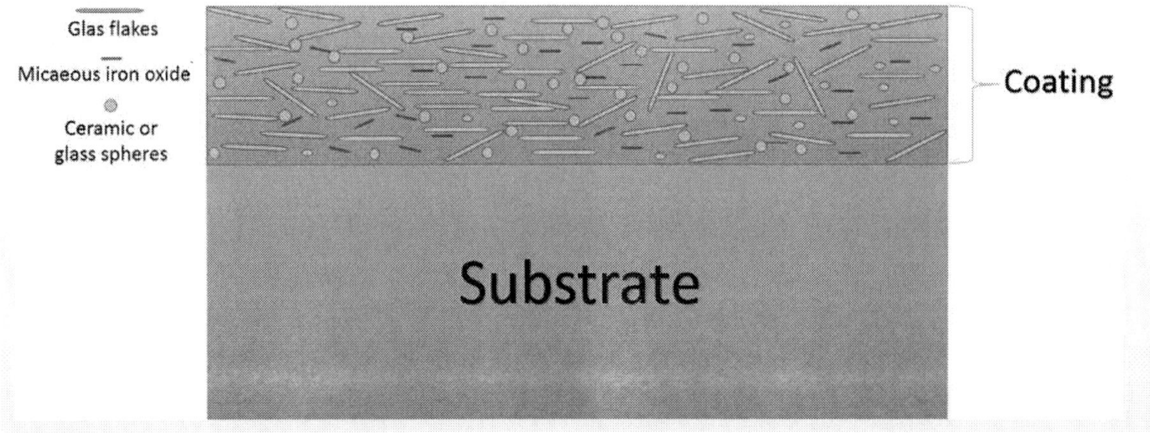

Figure 6: Geometry of various pigments

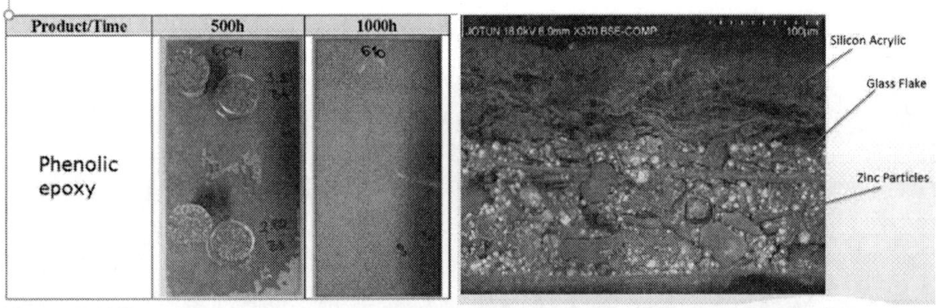

Figure 7: Effect of no reinforcement particles and the geometry of different particles in an electron microscope image of a coating

Finally there is a final question, how much is the correct amount? So far the only way to find this out is to do practicle testing. Starting with the correct pigment and resin base that has been chosen and varying the amount of pigment added.

In this case we found the correct amount was in fact 25% by weight. More gave no benefit and we suspect, but not proven, could give disadvantage technically, so we held at this level. We tested by running the two highest temperatures of 230 degC and 250 degC continuously and adjusting the percentage of glass flake

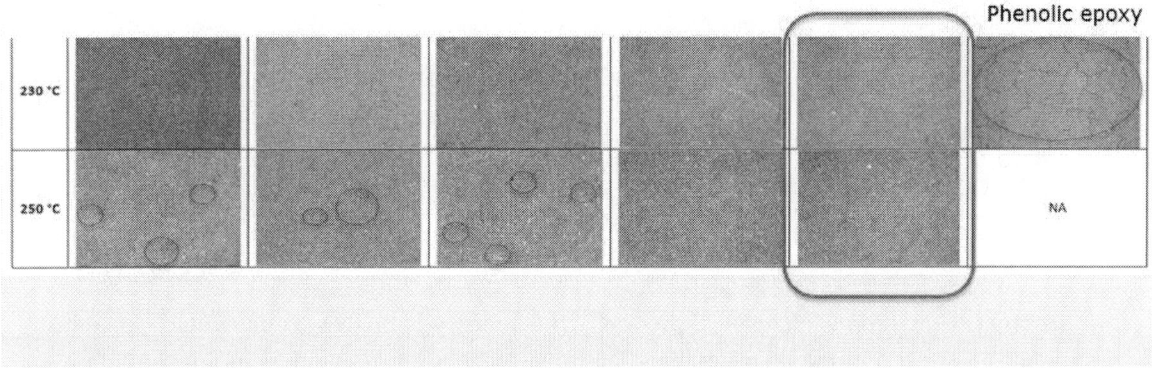

Figure 8: Glass flake varied, circles are micro cracks in the paint film. The boxed picture shows no cracks.

3. CONCLUSION

With a small mindset change considering similar industries, we can create new versatile functional coating products that are more like materials as composites to protect high temperature substrates and give long life in varying conditions. Adjusting according to the required scenario and without loosing other beneficial properties.

4. ACKNOWLEDGMENTS

Jotun Paint school for use of the diagrams, Stine Floberg

Jotun Paints Speciality Laboratory for Scanning Electron Microscope pictures, Dr. Kaspersen

5. REFERENCES

[1] SSPC 00-07 Protective Coatings Glossary, Terms from Coating of Industrial Steel and Concrete Structures, Failure Analysis, and Regulations

6. AUTHOR DETAILS

M Buckhurst is the Global Concept Director for Jotun AS. He has been in Jotun since 1990 and since 2010 in his current position. He is responsible for systems and solutions supplied for the Hydrocarbon Processing Industry (Oil, Gas and chemicals onshore)

Controlling Corrosion with Chemistry

Maria Forsyth

Institute for Frontier Materials and ARC Centre of Excellence for Electromaterials Science Deakin University, Geelong, VIC 3216, Australia.

Abstract: Corrosion is a chemical process and chemistry is the best counterattack to control it and to extend lifetime of materials – from materials used in engineering infrastructure to biomedical implants. Chemistry has for a long time been a very important tool for the engineer to achieve longer design life. The specific form of chemistry can range from designing metallic coatings (such as in ZincAlum), creating oxides on a metal substrate (as in anodizing), using new chemistry to create protective polymeric coatings or chemical additives to a given environment to produce inhibited surfaces. For example, when chromates are added to epoxy pigments which then protect the underlying metal substrate such as aluminium alloys on aircraft; phosphates, quaternary ammonium salts, etc., are added in water treatment formulations; or indeed anti-oxidant additions in polymer coatings to protect against UV degradation – these all assist with controlling corrosion. Even cathodic protection is essentially a chemical process where the hydroxyl ion build up, and subsequent increase in pH at the metal surface under CP can lead to passivation of the metal substrate. In microbial attack, the bacteria modify the chemistry of the environment at a metal surface so that they flourish, for example through formation of a biofilm, while the metal substrate rapidly corrodes. In mitigation of MIC, chemistry can again be used; anti-microbials can kill the bacteria and avoid formation of a biofilm. Recently, new dual active inhibitor systems that can act as inhibitors for both abiotic corrosion as well as being anti-microbials, are being designed to more effectively combat such corrosion.

1. Preamble

Percival Faraday Thompson – a chemist, a metallurgist, a corrosion scientist and an educator. As others before me, when I was asked to give the PF Thompson memorial lecture at this year's ACA meeting I wanted to find out more about this man. Arguably a chemist first and foremost and apparently, a self-taught metallurgist and corrosionist later in his career, PF Thompson wrote many papers across a vastly diverse range of topics. He first became a lecturer in Chemistry in 1906 at what was then the Melbourne Men's Working College (now RMIT) and moved to the University of Melbourne in 1918. He was appointed as President of the Society of Chemical Industry of Victoria in 1925 and although his early work focused on analytical chemistry and chemical analysis, he started to investigate and publish in areas we now consider corrosion science in the early 1930's [1,2].

PF Thompson recognised the importance of chemistry in understanding and combating corrosion. In his 1932 lecture to the Australasian Institute of Mining and Metallurgy he remarked that corrosion 'is a matter which has primarily concerned the engineer, but the paucity of his training in chemistry and electrochemistry has been a stumbling block to a better understanding of its mechanism'. Sadly, I think this is still true today and it is up to the educators to change that – those teaching chemistry to use corrosion as a very important case study with which to teach fundamental chemical concepts, and those teaching engineering to ensure that the electrochemical nature of corrosion is well understood.

In reading about PF Thompson's life, I was pleased to see some parallels with the path that I have taken as an educator. In the spirit of PFT, trying to develop an understanding of the underpinning science to solve important problems, to educate the younger generation of science and engineering leaders and also encouraging them to think about and communicate their science in different forums – from school children to University graduates and from their peers to the general community. Demonstrations are by far the best way to achieve effective communication. Through visualizing processes and the use of experiments we can simplify difficult concepts. At this point I want to divert and acknowledge another great Australian corrosion scientist and educator as well as the first PF Thompson memorial lecturer – Professor Ed Potter [2] Ed's magnesium clock, which has been used to start the ACA annual meetings for decades now, is no more than the harnessing of a corrosion process to provide energy; a primary magnesium battery. Energy generation and storage are key challenges today as we transition away from fossil fuels. This opens up remarkable new opportunities and challenges for our corrosion community – whether it be in understanding and controlling corrosion of metals used for molten salts in thermal storage applications, to controlling interfacial electrochemical processes on Al and Mg alloys for next generation battery storage, or to designing more robust pipeline infrastructure to enable storage and transportation of solar fuels such as hydrogen and ammonia. All of this requires knowledge of materials performance, interfacial science and, above all, chemistry. This brings me back to PF Thompson – a chemist at heart– with chemistry the underpinning science in our field of corrosion.

2. Corrosion – a chemical reaction involving electrons and ions.

Corrosion, defined as the degradation of a material in a given environment, is a chemical process. We generally think of metallic corrosion being an electrochemical process which requires both an oxidation reaction (loss of electrons, eg. the dissolution of Fe) at an anode and a reduction reaction (gain of electrons, eg. the reduction of O_2 or H^+). To support these two electrochemical reactions, an ionic conduction pathway is necessary via the electrolyte. These definitions are clarified in Figure 1.

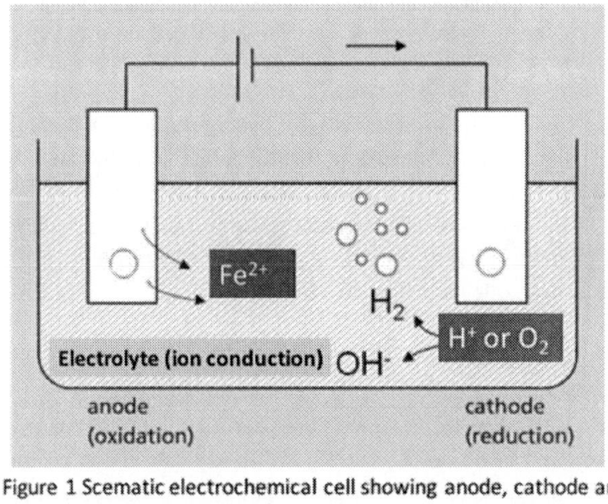

Figure 1 Scematic electrochemical cell showing anode, cathode and electrolyte which all required to support a corrosion process.

If either of the anodic or cathodic reactions is stifled or if the electrolytic pathway is absent (or highly resistive) then corrosion will be mitigated. This then gives us many chemical handles with which to control corrosion, remembering that engineering alloys are thermodynamically unstable and all we can do is slow down the inevitable process of a metal returning to its stable oxidised (or corroded) state. We thus can control corrosion by controlling the kinetics of the electrochemical reactions and we can do this by putting barriers in the way of electron and ion transport, meanwhile taking care not to create localised corrosion environments. Barriers can be created by either physically placing a coating on the surface of a metal (this can be organic or metallic), by creating a conversion coating which is highly resistive to electrons and/or ions, or by adding one or more chemical species to the environment which can adsorb onto the metal surface thereby inhibiting the electrochemical process. We will discuss each of these chemical barriers and their effect on the corrosion processes in what follows.

3. Physical Barriers - Coatings

One of the primary functions of a coating is to act as a physical barrier restricting the ingress of aggressive species such as chloride ions, acid, oxygen and water. Coatings have evolved from the coal tar enamels introduced in the early 1940's and used extensively till the 1960's, to polyethylenes, liquid epoxy, fusion bonded epoxies and layered coatings more recently introduced [3,4]. The chemical nature of the coating depends on the application and its function. For example, in some cases the coating not only has to protect against the aggressive environment (i.e. provide an efficient barrier) but also needs to be able to withstand high electrical currents without cathodic disbondment under cathodic protection conditions as well as avoid shielding of cathodic protection currents should a defect arise. Here chemistry is needed to control the adhesion of the coating to the underlying metal substrate and avoid blistering; and to control the crosslink density and hence the mechanical properties and permeability to water, amongst other key criteria. In this paper, however, we will not focus on polymer chemistry. Whilst coatings are often the first form of defense against a corrosive environment, they can degrade or be damaged during use and hence are unable to act as a sufficient barrier (see Figure 2).

Thus, in many applications, coatings are multi layered and multi-functional. Figure 3 illustrates a typical three pronged process to provide protection to reactive metals. The underlying metal surface is first treated to form a conversion coating, an inhibited primer is then applied and finally a topcoat protective barrier coating completes the protective system.

If the layers are undamaged, then the key function of the coating system is to keep out the aggressive environment. However, when a defect arises, a dangerous situation arises whereby the exposed metal can preferentially undergo attack. It is under these circumstances that the inhibited primer and the conversion coating need to act. Historically, chromate conversion coatings (CCC) have been coupled with epoxy primer that contains a $SrCr_2O_7$ inhibitor pigment that has sparing solubility and leaches out as needed to reinforce the damaged surface film and re-passivate the metal substrate.

Figure 2. Example of a coating failure on steel showing extensive failure of the coating and subsequent rusting of the metal (from https://www.corrosionpedia.com)

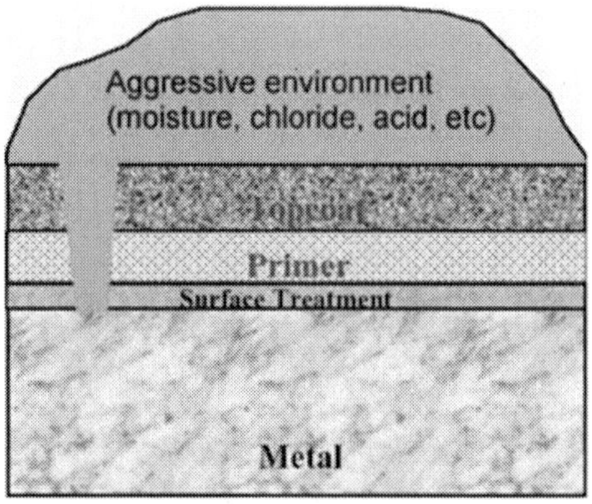

Figure 3. Typical three layered protective coating systems for metals such as Al alloys, which include a conversion layer, inhibited primer and barrier topcoat.

4. Conversion Coatings

Despite being used for many decades, the precise mechanism of action of chromate is still not completely understood. Hagans and Haas [5] give a detailed description of the process of a chromate conversion coating (CCC) on aluminium alloys and comment that the reactivity of the metal itself assists in the very rapid formation of a complex surface film consisting of an amorphous mixture of hydrated aluminium and chromium oxides. They also discuss the role played by the addition of fluoride ions in developing thick protective films arising from the ability to solubilise the aluminium oxide allowing the redox and deposition of new, more protective oxide. Furthermore, the F^- ion continues to act as a solubiliser for some portion of the developing oxide, thus allowing greater penetration of the solution to the metal surface and allowing transport of ions from the surface into the growing film. In this way, the final protective film will be a heterogeneous multi-valent (ie. both Cr^{3+} and Cr^{6+} co-exist) mixed oxide consisting of both chromium and aluminium oxides. Interestingly, as we discuss below with other treatments that have been investigated by us based on ionic liquid pretreatments or protective films formed from rare earth inhibitor compounds, this complex heterogeneous film formation also appears. Furthermore, controlling the reactivity (or activation) of the underlying metal to be protected using heat, pickling or acid additions has also been shown to play a significant role in the formation of the protective film [6-10].

Kending and Bucheit [11] suggest a possible mechanism of action of the CCC involving a polymerization process to form a ($Cr[OH]_3$) polymer consisting of edge and corner shared Cr^{3+} octahedral units. This condenses onto the metal substrate and, through further reaction of this polymer, an additional Cr^{6+} reservoir builds up within the coating leading to Cr^{3+}-O-Cr^{6+} bonds in the coating. Thus, this complex hydrated chromium hydroxide layer contains both Cr^{6+} and Cr^{3+} (Figure 4) which results in its excellent protective properties.

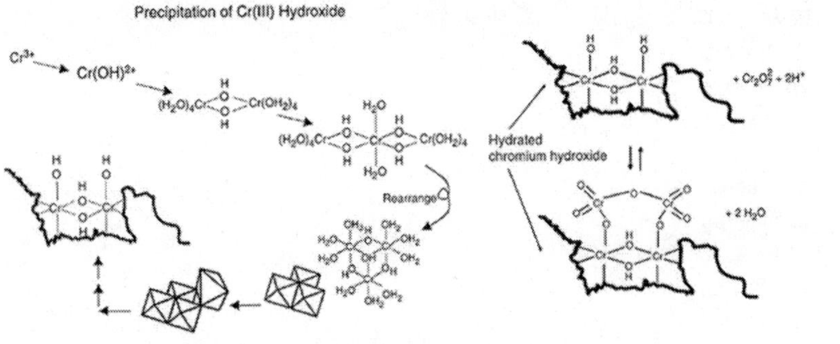

Figure 4. Schematic representation of the hydrolysis-polymerization-precipitation mechanism for $Cr(OH)_3$ backbone formation. (b) Condensation of Cr(IV) on the Cr(III) backbone (from reference 11)

It has proven very difficult to match the performance of chromate conversion coatings although Hughes et al [12] have developed a cerium oxide conversion coating for aluminium alloys that has shown promise as an environmentally friendlier alternative to chromate. In the presence of peroxide, the cerium conversion coating, consisting predominantly of Ce(IV) hydroxides can deposit on the metal surface, and further local increases in the pH near the surface due to the reduction of O_2 lead to the formation of cerium oxide (equations 1 and 2) although this simplistic chemical picture is in fact not the precise route to the formation of the CeO_2 in the presence of H_2O_2; the latter is elegantly explained by Scholes *et al* in reference [13].

$$2Ce^{3+}(aq) + H_2O_2(aq) + 2OH^-(aq) \rightarrow 2Ce(OH)_2^{2+}(aq) \qquad \{\text{equation 1}\}$$
$$Ce(OH)_2^{2+}(aq) + 2OH^-(aq) \rightarrow Ce(OH)_4(s) \rightarrow CeO_2(s) + 2H_2O \qquad \{\text{equation 2}\}$$

CCC has also been used for magnesium alloys however, as with Al alloys, this has been phased out and alternative methods have been investigated as reviewed by Pommiers *et al* [14]. These include, permanganate, vanadium, rare earth and phosphate conversion coatings.

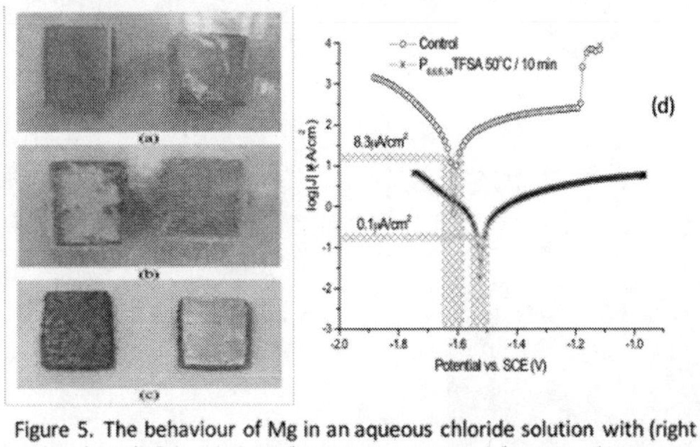

Figure 5. The behaviour of Mg in an aqueous chloride solution with (right hand photos and without (left hand photos) IL pretreatment as a function of immersion time (a-c) and a comparison of potentiodynamic polarisation data for AZ31 with and without IL pretreatment (d). The IL here is trihexyl, tetradecyl phosphonium bistrifluoromethane sulfonylamide adapted in part from [15]

We have also focused on the use ionic liquid conversion coatings for Mg alloys. Early work showed that immersion of Mg metal or AZ31 alloys in a trihexyl tetradecyl phosphonium bistrifluoromethane sulfonylamide ($P_{66614}NTf_2$) produced a complex film on the Mg alloy surface, likely consisting of MgF_2 and other magnesium salts from the breakdown of the ionic liquid moieties [15,16]. This surface coating resulted in an 80-fold decrease in corrosion rate. Figure 5 illustrates the effect of immersing Mg in a chloride solution before and after immersion with the $P_{66614}NTf_2$ ionic liquid. The bubbling on the untreated bare surface and the subsequent corrosion of the metal is evident in the left-hand images whereas minimal corrosion is observed post treatment right-hand images. The polarisation data in Figure 5d also attest to the effectiveness of this ionic liquid surface treatment with the ennoblement of the AZ31 surface accompanying the 80-fold decrease in corrosion current density. Subsequent research showed that replacing the NTf_2 anions with organophosphate anions also resulted in protective coatings which were more stable and biocompatible, the latter being important for protection of implants such as Mg based stents.

Figure 6 shows the structure of some of these ionic liquids which have been demonstrated to form protective surface films [6-8]. The shorter chain phosphonium cation (P_{1444}) is biologically more benign than the longer chained P_{66614} as demonstrated by Zhang *et al* [8] The P_{1444} cation coupled with the diphenylphosphate anion [P_{1444}dpp] was shown to form thin surface films which could control the morphology of corrosion in simulated body fluid, such that it was less localised and more uniform, an important parameter for bioabsorbable Mg implants. Figure 7 shows the performance of AZ31 in a simulated body fluid before and after treatment with [P_{1444}dpp] as well as the presence of the film on the alloy surface that was detected using FIB-SEM (Figure 7c). In this case Pt is deposited on top of the film to protect it during sample preparation and also to make the film more evident. A schematic demonstrating the interaction of the IL anion with the Mg alloy surface is also shown in Figure 7d. Spectroscopic characterization of the surface using FTIR spectroscopy as well as SEM-EDXS and ToFSIMS analysis of the treated Mg surfaces suggest the formation of a magnesium organo-phosphate surfaces with the ionic liquid cation somehow entrained in the film. Details of this can be found in the original publications [6-10].

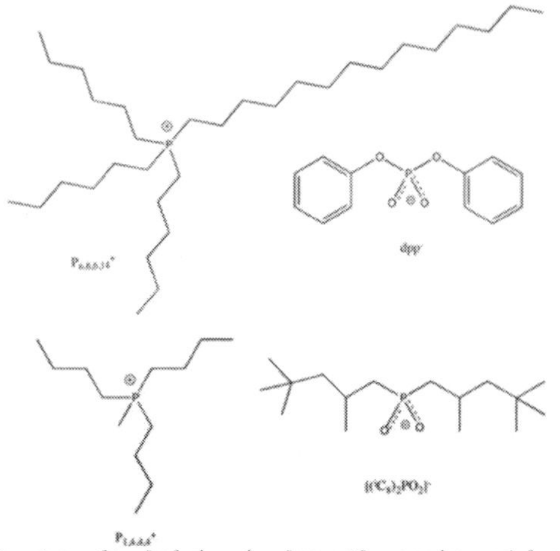

Figure 6. Chemical structure of typical phosphonium cations and organiphosphate anions that have been used in ionic liquid pretreatments of Mg alloys [Y. Zhang etal Surf. And Coat. Tech. 2016]

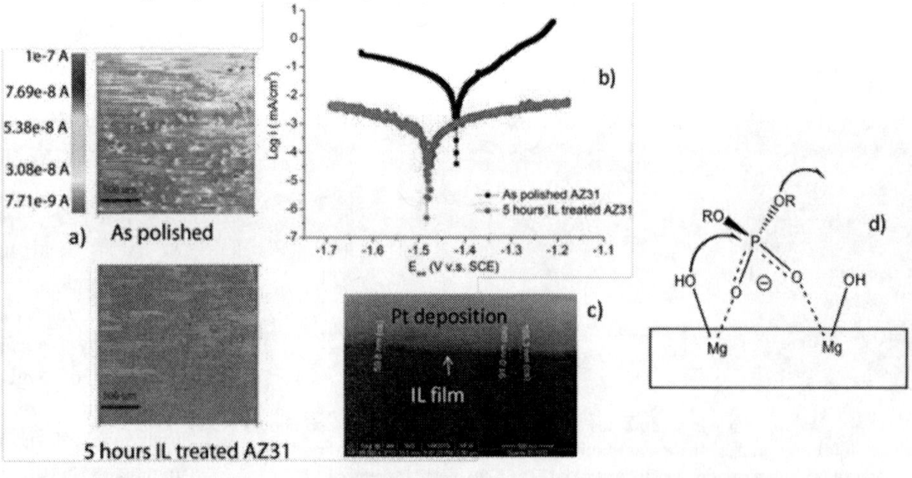

Figure 7. SECM after 24 hour immersion of AZ31 alloy in SBF with and without IL pretreatment (a) polarisation data in SBF before and after IL treatment (b) FIB-SEM cross section image of the thin IL surface film after IL pretreatment (c) and schematic of the possible interaction of the organophosphate anion with the Mg surface.

5. Chemical inhibitors

The addition of inhibitor compounds to solutions and paint formulations can also reduce corrosion by impeding either the anodic, cathodic or both. Molybdates, phosphates and nitrites, are examples of inhibitor species that act by reducing the rate of anodic dissolution [17,18]. They react to form a passive barrier layer on the surface of the metal, or stabilise the existing passive film. However, if the concentration of an anodic inhibitor is insufficient the inhibitor can only partially protect the corroding surface and thus can result in higher, localised corrosion rates; i.e. exposed local anodic regions and a large passive area that can act as the cathode leads to an acceleration of pitting corrosion.

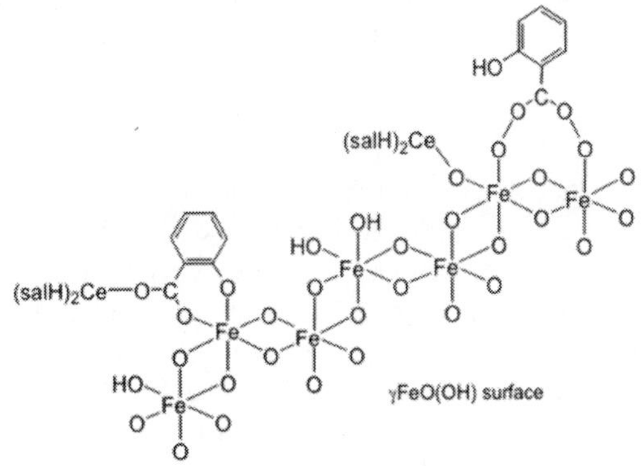

Figure 8. Schematic representation of the chemisorption and interaction between cerium salicylate inhibitor and the steel surface. This model assumes the adsorption occurs on the underlying iron oxide surface.

On the other hand, cathodic inhibitors deposit onto cathodic sites on the metal surface and impede cathodic corrosion reactions, in particular the reduction of oxygen to hydroxide ions. Examples of cathodic inhibitors include rare earth cations and zinc cations [19-21]. Such compounds do not promote pitting corrosion if they are at lower concentrations, but they do become less effective. Compounds that can impede both the anodic and cathodic reactions are known as mixed inhibitors, of which chromates are one example. In recent years, we have shown that a combination of a rare earth metal coupled with carboxylate or organophosphate ligands can also result in mixed inhibition on steel and aluminium alloys respectively [22-25]. These multifunctional inhibitors combine the inhibition properties of the rare earth metal and those of the organic component to produce compounds which display synergistic behaviour i.e the inhibition efficiency is better than either of the two individual components alone. Spectroscopic characterisation of the inhibitor solutions using electrospray mass spectroscopy has provided a correlation between the extent of inhibition and the nature of the chemical complexes present in the solution although more work is required to fully understand the effect of speciation [26].

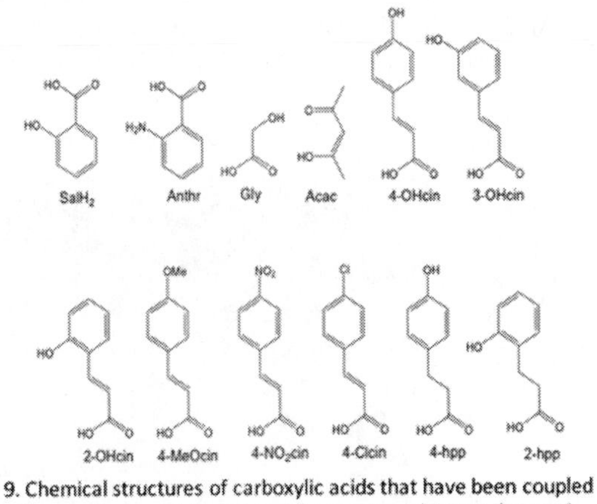

Figure 9. Chemical structures of carboxylic acids that have been coupled with rare earth metal cations in the search for improved corrosion inhibitors for steel.

Early work combining Ce^{3+} or Ln^{3+} with carboxylates such as salicylate and anthranilate [22,23] demonstrated that these REM organocarboxylate compounds provided high inhibition efficiency for steel in chloride contaminated aqueous systems, displaying synergistic behaviour when coupled together. Together with spectroscopic evidence of the surface chemistry and chemical modelling of possible reaction products between the inhibitor and the steel surface, Behrsing and Wilson [27] showed that the inhibition process involved chemisorption and the surface reactions produced a bimetallic compound involving both the rare earth metal and Fe^{3+} ions as demonstrated schematically in Figure 8. Similar behavior was found for Ce^{3+} combined with organophosphate anions such as diphenyl phosphate or dibutyl phosphate by Ho [28] and Markley [29].

Blin [30], Lee [25] and co-workers investigated a series of substituted aromatic carboxylates based on the cinnamic acid and propionic acid together with different REM cations. In those works, it was found that the nature of both the organic ligand and the lanthanide cation species influenced the efficacy of inhibition in the REM carboxylate compound. Figure 9 gives the chemical structures of some of the carboxylates investigated to date. Subsequently, Seter *et al* confirmed that there is a correlation between the species present in the solution (and presumably therefore at the metal surface) and the inhibition efficiency, and that these species depended on the pH and inhibitor concentration [26]. There is still much to understand about the role of chemistry and speciation in achieving high levels of inhibition. Whilst there is no one unique compound to combat all corrosion for all metals (as it seemed we had in the case of chromate inhibitors) it is evident that clever design of new compounds will result in series of inhibitors that will match the high performance of chromates.

Indeed, the nature of rare earth organic inhibitors lends themselves to chemical modification which can modify and improve their properties. We have discussed only a few examples in this paper, however there are numerous additional examples in the literature that show how simply changing the position of a functional group on a benzene ring can markedly change the inhibitor efficiency of a compound. The nature of the REM cation itself also modifies the inhibition properties. Therefore, the vast array of chemical modifications that can be made to the organic components of these multifunctional inhibitors together with the choice of REM cation itself, means that a number of new REM inhibitors are possible.

6. Microbiologically influence corrosion and how to stop it?

MIC is often a more aggressive form of corrosion that can lead to catastrophic failure of structures within a matter of months even if the structural design life is expected to be decades [31]. It is estimated that MIC accounts for at least 20% of all corrosion failures [32]. In this type of corrosion, microorganisms colonise metal surfaces forming a biofilm that consists of bacterial cells embedded in a highly hydrated, extracellular polymeric substance (EPS)[31,33]. Due to the localised nature of this bacteria colony, MIC can rapidly lead to deep pitting, compromising structural integrity. Typically, colonies consist of a number of species, both aerobic and anaerobic, acting cooperatively. For example *Pseudomonas aeruginosa* is considered to be the pioneer coloniser for biofilm formation, under which form the oxygen-free areas required for sulfate-reducing bacteria (SRB), widely believed to be responsible for the bulk of damage caused by MIC [31].

Various organic compounds with antimicrobial properties have been considered to combat MIC. For example, surfaces that are functionalised with quaternary ammonium or phosphonium salts have high positive charge densities and thus exert a strong electrostatic interaction with negatively charged bacteria. The cation can diffuse through the cell wall and disrupt the cytoplasmic membrane causing cell death [34]. Combining an active cation such as this with an active anion provides the opportunity to create a "dual active" compound, thus designing synergistic effects (Figure 10).

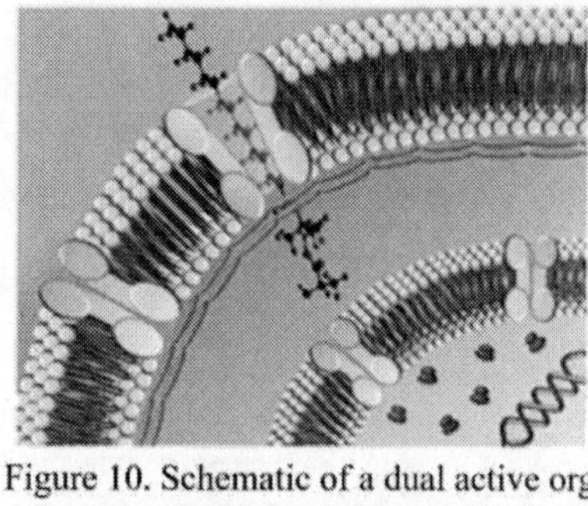

Figure 10. Schematic of a dual active organic salt permeating through bacterial cell wall, disrupting the cell and ultimately causing death. From Seter et al Chem Comm 2013

In particular, quaternary ammonium cations, when combined with a pharmaceutically active or biocompatible anion, have been shown to exhibit synergistic dual (antimicrobial and anticorrosion) activity. We have investigated a range of such compounds and demonstrated that cetrimonium salts based on carboxylate anions such as nalidixate and lactate perform well both with respect to decreasing bacterial survivability and mitigating corrosion [35,36]. Figure 11a illustrates the effect of the cetrimonium lactate compound on the cell survivability for the bacteria *Enterobacter cloacae*. Similar behavior was observed for other bacteria and other organic salts as presented in reference [35]. Figure 11b shows the effect of incorporating cetrimonium nalidixate salt (CN) into a varnish coated onto mild steel and subsequently exposed to solutions containing various bacteria. The control sample shows extensive evidence of biofilm formation and corrosion in contrast to the relatively clean surfaces found for in the presence of the CN inhibitor.

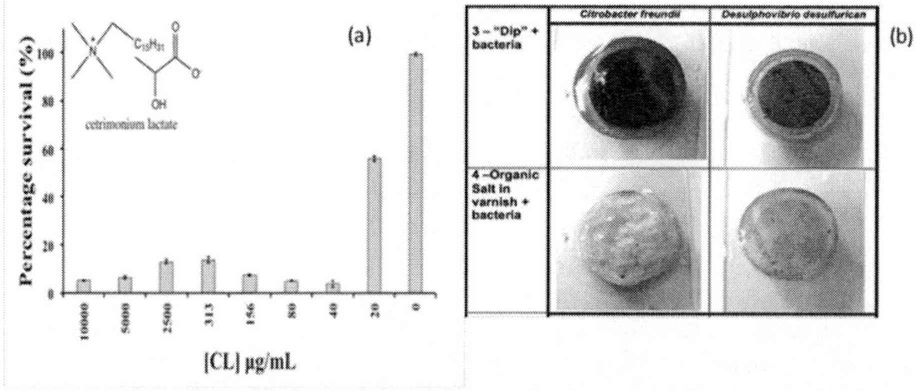

Figure 11. (a) Percentage survival of *Enterobacter cloacae* in the presence of cetrimonium lactate (CL) and (b) the effect of organic inhibitor on biofilm formation and corrosion when exposed to two types of bacteria, citrobacter freundii (left panel) and desulphovibrio (right panel). Adapted in part from Seter et al [34]

7. Conclusions

This paper is by no means meant to provide an exhaustive summary of coatings or inhibitors but merely seeks to demonstrate how important chemistry can be in combating corrosion. Having a good grasp of chemical and electrochemical concepts can assist the corrosion engineer in finding suitable mitigation strategies for protecting metals in aggressive environments. Corrosion is increasingly becoming a multidisciplinary field with the need not only for metallurgists, coating engineers and electrochemists but also surface chemists, synthetic chemists and microbiologists. This diverse and exciting field continues to be important both from the perspective of educating the next generation of corrosion scientists and engineers and also with respect to providing new arsenal in the continual 'battle' against corrosion which is such a cost burden on industry and government.

8. Acknowledgements

First and foremost I want to acknowledge the amazing team of students, postdocs and colleagues I have worked with in the corrosion inhibitor field over the past years (many of whose papers are cited here). In particular my current collaborators in this field including Professors Bruce Hinton, Tony Hughes, Doug Macfarlane, Glen Deacon, Peter Junk, Mike Tan, and Drs. Patrick Howlett, Anthony Somers, Rainier Catubig, Wayne Neil as well as the numerous students. I'd also like to thank Warren Green and Timothy Khoo for discussions and reading of the manuscript and to acknowledge the Australian Research Council, DSTG, CSIRO and Deakin University for funding the various stages of research discussed herein.

9. References

1. Green WK and Hinton BRW. P F Thompson - Australia's Pioneer in Corrosion Research and Education. Corrosion & Materials. The Australasian Corrosion Association Inc. 40(4) (August 2015) 24-31.
2. Potter EC, The corrosion scene in Australia. Australasian Corrosion Engineering, 16(3) (1972) 21-9.
3. Bud Senkowski E. Meeting Demands of Gas Exploration: The Evolution of Pipeline Coatings. JPLC. Painsquare. (November 2015)
4. Lyons CJ., A Two-Pack Waterborne Polyurethane Topcoat for Military Aircraft. CoatingsTech. American Coatings Association. 14(6) (June 2017).
5. Hagans PL. Chromate conversion coatings. ASM Handbook Volume 5: Surface Engineering: ASM International; 1994. p. 405-11.

6. Huang PP, Latham JA, MacFarlane DR, Howlett PC and Forsyth M, A review of ionic liquid surface film formation on Mg and its alloys for improved corrosion performance. Electrochim Acta, 110 (2013) 501-10.
7. Sun J, Howlett PC, MacFarlane DR, Lina J and Forsyth M, Synthesis and physical property characterisation of phosphonium ionic liquids based on P(O)(2)(OR)(2)(-) and P(O)(2)(R)(2)(-) anions with potential application for corrosion mitigation of magnesium alloys. Electrochim Acta, 54(2) (2008) 254-60.
8. Zhang Y, Liu X, Jamal SS, Hinton BRW, Moulton SE, Wallace GG, et al., The effect of treatment time on the ionic liquid surface film formation: Promising surface coating for Mg alloy AZ31. Surf Coat Technol, 296 (2016) 192-202.
9. Howlett PC, Gramet S, Lin J, Efthimiadis J, Chen XB, Birbilis N, et al., Conversion coatings of Mg-alloy AZ91D using trihexyl(tetradecyl) phosphonium bis(trifluoromethanesulfonyl)amide ionic liquid. Science China-Chemistry, 55(8) (2012) 1598-607.
10. Efthimiadis J, Neil WC, Bunter A, Howlett PC, Hinton BRW, MacFarlane DR, et al., Potentiostatic Control of Ionic Liquid Surface Film Formation on ZE41 Magnesium Alloy. Acs Applied Materials & Interfaces, 2(5) (2010) 1317-23.
11. Kendig MW and Buchheit RG, Corrosion inhibition of aluminum and aluminum alloys by soluble chromates, chromate coatings, and chromate-free coatings. Corrosion, 59(5) (2003) 379-400.
12. Hughes AE, Gorman JD, Miller PR, Sexton BA, Paterson PJK and Taylor RJ, Development of cerium-based conversion coatings on 2024-T3 Al alloy after rare-earth desmutting. Surf Interface Anal, 36(4) (2004) 290-303.
13. Scholes FH, Soste C, Hughes AE, Hardin SG and Curtis PR, The role of hydrogen peroxide in the deposition of cerium-based conversion coatings. Applied Surface Science, 253(4) (2006) 1770-80.
14. Pommiers S, Frayret J, Castetbon A and Potin-Gautier M, Alternative conversion coatings to chromate for the protection of magnesium alloys. Corrosion Sci, 84 (2014) 135-46.
15. Birbilis N, Howlett PC, MacFarlane DR and Forsyth M, Exploring corrosion protection of Mg via ionic liquid pretreatment. Surf Coat Technol, 201(8) (2007) 4496-504.
16. Forsyth M, Howlett PC, Tan SK, MacFarlane DR and Birbilis N, An ionic liquid surface treatment for corrosion protection of magnesium alloy AZ31. Electrochemical and Solid-State Letters, 9(11) (2006) B52-B5.
17. El-Ghaffar MAA, Youssef EAM and Ahmed NM, Preparation and characterisation of high performance yellow anticorrosive phosphomolybdate pigments. Pigment & Resin Technology, 35(1) (2006) 22-9.
18. Kalendova A, Comparison of the efficiencies of anticorrosive pigments based on chemically modified phosphates. Anti-Corros Methods Mater, 49(5) (2002) 364-72.
19. Hinton BRW, Corrosion inhibition with rare-earth-metal salts. J Alloy Compd, 180 (1992) 15-25.
20. Hinton BRW and Wilson L, The corrosion inhibition of zinc with cerious chloride. Corrosion Sci, 29(8) (1989) 967-85.
21. Bethencourt M, Botana FJ, Calvino JJ, Marcos M and Rodriguez-Chacon MA, Lanthanide compounds as environmentally-friendly corrosion inhibitors of aluminium alloys: A review. Corrosion Sci, 40(11) (1998) 1803-1819.
22. Forsyth M, Seter M, Hinton B, Deacon GB and Junk P, New 'Green' Corrosion Inhibitors Based on Rare Earth Compounds. Australian Journal of Chemistry, 64(6) (2011) 812-819.
23. Forsyth M, Wilson K, Behrsing T, Forsyth C, Deacon GB and Phanasgoankar A, Effectiveness of rare-earth metal compounds as corrosion inhibitors for steel. Corrosion, 58(11) (2002) 953-960.
24. Seter M, Girard GMA, Lee WW, Deacon G, Junk P, Hinton B, et al., The influence of organic structure and rare earth metal cation on the corrosion efficiency observed on AS1020 steel compared with La(4OHCin)3. 2(1) (2015) 1-15.
25. Lee WYW. Rare earth hydroxycinnamates and hydroxycinnamate-derivative complexes as corrosion inhibitors, Thesis (Ph.D.) Victoria, Australia: Monash University; 2010.
26. Seter M, Hinton B and Forsyth M, Understanding speciation of lanthanum 4-hydroxy cinnamate and its impact on the corrosion mechanism for AS1020 steel. J Electrochem Soc, 159(4) (2012) C181-C189.
27. Deacon GB, Forsyth CM, Behrsing T, Konstas K and Forsyth M, Heterometallic CeIII-FeIII-salicylate networks: models for corrosion mitigation of steel surfaces by the [prime or minute]Green[prime or minute] inhibitor, Ce(salicylate)3. Chemical Communications(23) (2002) 2820-2821

28. Forsyth M, Forsyth CM, Wilson K, Behrsing T and Deacon GB, ATR characterisation of synergistic corrosion inhibition of mild steel surfaces by cerium salicylate. Corrosion Sci, 44(11) (2002) 2651-2656.
29. Ho D, Brack N, Scully J, Markley T, Forsyth M and Hinton B, Cerium dibutylphosphate as a corrosion inhibitor for AA2024-T3 aluminum alloys. J Electrochem Soc, 153(9) (2006) B392-B401.
30. Markley TA, Forsyth M and Hughes AE, Corrosion protection of AA2024-T3 using rare earth diphenyl phosphates. Electrochim Acta, 52(12) (2007) 4024-4031.
31. Blin F, Leary SG, Deacon GB, Junk PC and Forsyth M, The nature of the surface film on steel treated with cerium and lanthanum cinnamate based corrosion inhibitors. Corrosion Sci, 48(2) (2006) 404-419.
32. Seter M, Thomson MJ, Chong A, MacFarlane DR and Forsyth M, Cetrimonium Nalidixate as a Multifunctional Inhibitor to Combat Biofilm Formation and Microbiologically Influenced Corrosion. Australian Journal of Chemistry, 66(8) (2013) 921-929.
33. Videla HA and Herrera LK, Microbiologically influenced corrosion: looking to the future. International Microbiology, 8(3) (2005) 169-180.
34. Enning D and Garrelfs J, Corrosion of Iron by Sulfate-Reducing Bacteria: New Views of an Old Problem. Applied and Environmental Microbiology, 80(4) (2014) 1226-1236.
35. Xue Y, Xiao HN and Zhang Y, Antimicrobial Polymeric Materials with Quaternary Ammonium and Phosphonium Salts. International Journal of Molecular Sciences, 16(2) (2015) 3626-3655.
36. Seter M, Thomson MJ, Stoimenovski J, MacFarlane DR and Forsyth M, Dual active ionic liquids and organic salts for inhibition of microbially influenced corrosion. Chemical Communications, 48(48) (2012) 5983-5985.

Professor Maria Forsyth is an Australian Academy of Science fellow, Australian Laureate Fellow, an Alfred Deakin Professorial Fellow at Deakin University in Australia as well as the Associate Director of the ARC Centre of Excellence for Electromaterials Science (ACES), Deputy Director of Institute for Frontier Materials (IFM) and Chair of Electromaterials and Corrosion Science and at Deakin University where she leads the research effort in metal-air batteries and the development and understanding of new electrolyte systems. Her research area informs the broad field of electromaterials science with application to both corrosion and energy related technologies. One key interest of her research is in the area of selective transport in materials for energy storage applications. Specifically, her work has focused on understanding the phenomenon of charge transport at metal/electrolyte interfaces and within novel electrolyte materials. Such materials have included a range of novel ionic liquids, polymer electrolytes and plastic crystals. She leads collaborative projects in lithium and sodium battery technologies funded through recent Australian Research Council grants. Professor Forsyth is a co-author of over 400 journal publications. She has delivered more than 20 invited and plenary talks in the past 5 years and currently has more than 14000 citations with an h index of 62 (Web of Science August 2015).

Professor Forsyth and her group have had significant and ongoing interaction with industry, both domestic and international, for over 20 years. This is based on their strong reputation for providing a technical and expertise resource in the fields of materials engineering, especially as related to electrochemical properties of materials relevant to corrosion and battery technologies. In the area of corrosion Professor Forsyth's work in corrosion monitoring and mitigation has had impact in a number of fields from aerospace structures, to water treatment, water pipelines, desalination, oil and gas pipes and other civil and marine infrastructure. Professor Forsyth and her group have also made significant contributions to technological advances in the energy sector and have been involved with projects relating to the development of lithium ion and lithium metal secondary batteries with Pacific Lithium Ltd, Evonik GmbH and CSIRO, Cytec, Toyota and LG.

EFFECTIVE CORROSION MANAGEMENT OF REINFORCED CONCRETE ASSETS

P.C. Robery[1], J.P.M. Segers[2]
[1]University of Birmingham & RFEL, Solihull, UK; [2]SegCorr Ltd, Shropshire, UK

SUMMARY: The history of reinforced concrete usage in Europe over the last 60 years offers a chronology of ambition and achievement, but with a sobering sprinkling of construction projects that did not go as well as was intended in terms of durability. Exposure to chloride ion continues to present industry with the greatest challenges. To the present day in Europe, industry has continued to amend its durability performance criteria for reinforced concrete to account for chloride ion, as the latest edition of BS8500 1:2015 sets out – in the UK, for CEM I (OPC) in seawater, we now recommend 90mm cover and a cube strength Class of 55N/mm^2, compared with CP110:1972 that required 50N/mm^2 cube strength and only 50mm cover. The challenge for effective corrosion management is to provide robust design solutions that allow reinforced concrete to perform reliably in hostile exposure conditions that are outside the provisions of standards on the basis of either the climate or the long-life expectancy. This paper explores the practical aspects of the durability of reinforced concrete, how the performance can be modelled and enhanced for extreme environments and how performance can be assured by effective corrosion management strategies that include impressed current cathodic protection.

Keywords: Reinforced Concrete, Durability, Seawater, De-icing salt, Standards, Cathodic Protection.

1. INTRODUCTION

Over the past 60 years in Europe, reinforced concrete infrastructure has had a chequered history in terms of successful durability performance in service [1]. Early reinforced concrete structures were carefully designed and built, with trials and mock-ups to prove concepts, resulting in buildings like Weaver's flour mill [2] (Figure 1) and civil engineering structures such as the Mulberry Harbour units (Figure 2), lasting for many decades without any appreciable signs of deterioration. However, as cement technology was not that advanced, the cement contents needed were possibly higher than those used today, to gain strength quickly.

Figure 1: Weaver's flour mill built 1898, survived to 1979 [3]

Figure 2: Mulberry Harbour Units after Hartcup [4]

Increasing popularity of a material that behaves like a "liquid rock", contributed to reinforced concrete deterioration [6]. Much UK road infrastructure between the 1950s and 1980s was built with designers and contractors neither having a full understanding of durability design for reinforced concrete nor the importance of factors including cement content and type, compaction, cover and curing; it was mainly the structural strength was considered in the design process.

Concrete was used for increasingly ambitious and impressive structures, producing slender and delicate shapes, with patterns and colours, including white cement. These structures increasingly benefitted from the high early strength gain of more finely ground ordinary Portland cement ("OPC"), which Europe now terms CEM I, to reduce cement contents and costs. But what a cost saving that proved not to be, as industry realised it had built with a concrete prone to carbonation and chloride ingress – we were even adding calcium chloride to reinforced concrete to accelerate setting and using seawater as mixing water on some structures in the late 1970s, without realising chloride-induced corrosion risk increased the higher the free chloride/hydroxide ion ratio becomes [9], there being no absolute chloride ion "threshold" for corrosion to occur.

The lessons that had to be re-learned from the early research of the 1920s, were that cement content, cement type, water/cement ratio and cover had to be considered from the perspective of reinforced concrete durability, as well as structural strength, and that workmanship was vitally important [35]. Limits were introduced in CP110 in the early 1970s [5] to help combat the effects of atmospheric carbonation of reinforced concrete and limits then followed on chloride ion content. As explained in this paper, these limits have been tightened progressively as the severity of our prevailing exposure environments have been realised, and to address industry's demands for longer-lasting assets with "*indefinite*" service life.

2. BUILIDNG BUILDABILITY

Today, few construction engineers will question the ability of well-built concrete buildings to perform over a normal service life in Europe when carbonation is the deterioration mechanism, provided construction is to a high standard of quality. But even today, failures continue to occur because of poor quality construction, meaning that the design service life will not be achieved. Practice "issues" that still occur include: poor formwork alignment and sealing; low cover to the reinforcement; lower insitu strength than specified; unstable mix design; and poor mixing, placing and compaction. The "issues" lead to bleed runs and honeycombing (Figure 3), mix segregation, restricted flow through reinforcement, grout loss from formwork (Figure 4), plastic shrinkage cracking, poor surface curing and restraint to thermal or drying shrinkage movement. How, therefore can industry be confident the reinforcement will be protected by concrete alone over the design service life?

Figure 3: Bleed runs and poor compaction with honeycombing at the base of a column

Figure 4: Grout leakage at form joint, leading to honeycombing and corrosion of rebar

In Europe, these "issues" have their root cause in industry design and practice [6]:

- Structural designs and details that produce an impossibly difficult fit for the reinforcement within the formwork and the chosen workability of the concrete - the structural designer fails to consider the flow properties of concrete, but these are matters well known to concrete technologists;
- Move to increasing procurement by design and construct, where "value engineering" and cost saving compromises can impact on the durability of the design;
- Lack of adequate training of the site operatives – the attitude seems to be "it is only concrete", so how difficult can it be to pour it?
- Lack of experienced independent supervision, in particular the loss of the traditional "Clerk of Works" and no client's Engineer in a supervisory role (particularly in design and construct projects).

As an industry, we still do not pay adequate care and attention to the basics: the design, formulation and use of concrete in the relatively benign exposure environments of UK buildings and warehousing. Ultimately, the end user of the concrete is the most important person on the project if high quality concrete infrastructure is to be consistently produced.

3. THE CHLORIDE ION

Superimpose severe chloride environments on top of these buildability "issues", and the reduction in durability performance can have (and has had) devastating consequences. In Europe, heavily-loaded bridges are regularly exposed to de-icing salts and, as innumerable papers testify, the post-war roads programme led to considerable numbers of bridge assets being built between 1950 and 1980, including Spaghetti Junction (Figure 5). Unfortunately, these designs suffered from inadequate durability provisions against de-icing salts based on the codes and standards of the time. Combined with buildability issues, many bridges have failed to achieve the intended 120-year designs service; in some cases, only 20-years being achieved before major maintenance is required to address pitting corrosion of reinforcement (Figure 6). That is not to say that all bridges failed, as some well-built assets have performed well under chloride ion exposure, but the warning signs were there.

Figure 5: Spaghetti Junction, Birmingham, UK [7]

Figure 6: Corrosion in Bridge Half-Joints, M1, UK

Marine structures have performed similarly, with some lasting well, but others failing prematurely. A good example of durable concrete is the Tongue Sands Fort and Mulberry Harbour units (Figure 2), built during 1939-1945, made with OPC reinforced concrete. The units have largely survived exposure in seawater for over 70-years, with little evidence of chloride penetration and corrosion [8]. More recent construction has deteriorated quickly in seawater and as illustrated in Figure 7, the benefits of rapid precast pile construction can only provide a lasting structure if the durability assessment is correct.

Figure 7 also illustrates a key characteristic of reinforcement corrosion, namely that corrosion effectively stops at around mid-tidal range. At this point, the concrete remains permanently saturated with seawater (and it is no surprise this is where barnacles and mussels thrive), and oxygen access through the chloride-laden concrete pores is so low that appreciable rates of reinforcement corrosion do not occur.

Figure 7: Kurnell Wharf, Sydney, c 1985

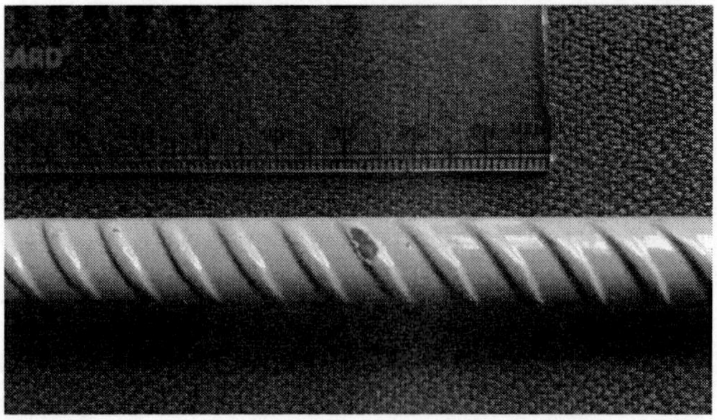

Figure 8: Typical handling damage on epoxy-powder coated rebar

These same European design approaches and construction techniques were also being exported around the world, resulting in reinforced concrete structures being built in the Arabian Gulf with less suitable cements, poor quality aggregates and brackish water, and exposed to high temperatures, saline ground conditions and seawater spray. These extreme environments were far worse than those found in European countries, with high temperatures and high rates of evaporation, yet standard UK or ASTM codes were used for design, leading to short design service life (<30 years) due to corrosion of reinforcement.

The problems caused by one of the commonest ions on the planet, have cost the global economy countless billions of dollars for repair, unplanned maintenance and re-building of infrastructure. Yet the facts of the matter are that when reinforced concrete was in its infancy, researchers knew about the risks: Loov cites two quotations from the 1920s that illustrate what we forgot in the 1950s to 1980s [35]: researchers warned "*Reinforced concrete is not a material that should be recklessly used*" and "*Do not bring reinforced concrete in contact with seawater*".

Macro-environmental factors exacerbated the chloride ion problem, such as the "hollow leg" effect, where evaporation of water from the occurs on the inside of unwaterproofed tubular structures. Early examples of hollow structures include North Sea offshore platforms such as MCP01 and mass rapid transit systems such as the Hong Kong MTR, which were found to be vulnerable to salt build-up at the level of the internal reinforcement. Even mildly brackish groundwater can be pulled through the reinforced concrete section and the chloride ion concentrated at the inside cover zone due to evaporation, requiring consideration of waterproof coatings or other measures.

In Europe, the answer to chloride has been the progressive tightening of codes and standards, particularly for concrete in seawater. With advances in our understanding of reinforced concrete corrosion [9], the European cement standard EN 197-1 [10] now includes a wide range of options for blended cement compositions, and durability provisions require consideration of cement content and type, cover, water/cement ratio and strength for given exposure conditions. Table 1 is reproduced from TR 70 [2], illustrating the changes in UK provisions for seawater exposure in the "Class XS3" upper tidal, splash and spray zone; the UK now requires 90mm nominal cover for Portland cement concrete, but recognises the benefits from using alternative cementing materials ("ACM") such as ground granulated blastfurnace slag ("ggbs"), at levels between 36% and 95% by mass of CEM I, as providing equivalent durability at lower strength and cover.

Table 1: Summary of reinforced concrete durability requirements for exposure class XS3 (seawater upper tidal, splash and spray zones) based on TR 70 [2]

Code of Practice / British Standard	Portland cement / CEM1				Blast-furnace cement / CEM III/A*			
	mcc kg/m^3	mwc	$f_{ck,cube}$ MPa	c_{nom} mm	mcc kg/m^3	mwc	$f_{ck,cube}$ MPa	c_{nom} mm
CP 110-1:1972	330	0.45	50	50	330	0.45	40	50
BS 6349-1:1984	400	0.42	40	50	400	0.42	40	50
BS 8110-1:1985	400	0.45	50	50	400	0.45	50	50
BS 8110-1:1997	400	0.45	50	50	400	0.45	50	50
BS 6349-1:2000	400	0.40	50	60	360	0.50	37	50
BS 8500-1:2002	360	0.40	50	60	360	0.40	45	50
BS 8500-1:2006	380	0.40	50	60	380	0.40	45	50
BS 6349-1-4:2013	360	0.35	50	90	380	0.35	50	55
BS 8500-1:2015	380	0.35	55	90	380	0.35	50	55

Assumes maximum aggregate size 20mm
XS3 is equivalent to exposure classes previously described as severe, very severe or extreme
* Assumes CEM III/A has GGBS content of at least 40%, representative of UK Portland blast-furnace cements 1925-1990
mcc = Minimum cement and ACM by mass equivalent to the definitions of EN 197-1 [10]
mwc = Maximum free water/cement ratio
$f_{ck,cube}$ = Characteristic cube compressive strength
c_{nom} = Nominal cover, assuming required minimum cover plus an allowance of 10mm.

Table 1 shows the latest versions of BS 6349-1-4 [11] and BS 8500-1 [12] have significantly tightened the requirements for concrete made with just CEM I under conditions of severe chloride ion exposure Class XS3, which UK technical committees consider is problematic; 90mm nominal cover is now recommended for XS3 exposure Class, while only 55mm is needed for CEM IIIA, where the GGBS content is at least 46%. If CEM IIIB is used (66-85% GGBS), then for the same 55mm nominal cover the characteristic cube compressive strength can be reduced to 45MPa.

BS 6349-1-4 [11] and BS 8500-1 [12] have also tightened the requirements for concrete that is permanently immersed in seawater (Class XS2): BS8500-1 2006 [13] previously permitted 30mm minimum cover for any composition of concrete; but the latest codes require, in this "oxygen-starved" situation that exists below the waterline, higher strengths and increased minimum nominal cover, again depending on the cement type (Table 2):

Table 2: Comparison between BS8500-1 from 2006 [13] and 2015 [12] and BS 6349 [11] for XS2

<table>
<tr><th>Cement/combination types</th><th>BS 8500-1: 2006</th><th>BS 8500-1: 2015</th><th>BS 6349-1-4: 2013</th></tr>
<tr><td>CEM I</td><td>f_{ck} C40/50, c_{nom} = 40mm</td><td>f_{ck} C45/55, c_{nom} = 70mm</td><td>f_{ck} C40/50, c_{nom} = 70mm</td></tr>
<tr><td>CEM III/A (<46%)</td><td rowspan="2">f_{ck} C35/45, c_{nom} = 40mm</td><td>f_{ck} C35/45, c_{nom} = 50mm</td><td>f_{ck} C35/45, c_{nom} = 50mm</td></tr>
<tr><td>CEM III/A (≥46%)</td><td>f_{ck} C40/50, c_{nom} = 45mm</td><td rowspan="2">f_{ck} C40/50, c_{nom} = 45mm</td></tr>
<tr><td>CEM III/B</td><td>f_{ck} C32/40, c_{nom} = 40mm</td><td>f_{ck} C35/45, c_{nom} = 45mm</td></tr>
</table>

Given the oxygen-starved conditions it is not immediately obvious why cover and strength needs to be increased, although as recent studies have shown, if bare bar does exist below water, such as at voidage or honeycombing in the concrete, then the resulting concentration cell can produce severe corrosion. Providing extra cover for sea walls and caissons is therefore a good idea to counter workmanship issues from placing concrete using methods such as tremie or below bentonite.

4. NON-CEMENTITIOUS ENHANCEMENTS

Various durability design enhancements have been proposed, including powder-epoxy resin coating of reinforcement, waterproofers, corrosion inhibitors and surface coatings, each having its own particular issues.

- ***Powder-epoxy coatings*** have proven performance under laboratory conditions, but with less care and attention from standard handling (Figure 8), the protective coating can easily become chipped and vulnerable. Also, when bent at site, the coated bars can split over ribs, as the Florida Keys failures demonstrated [14];
- ***Hydrophobic admixtures*** in the concrete have proven to be successful and are commonly specified for basements below the water table. Part of the success has, in the author's opinion, been from independent supervision of the works by the manufacturer, as part of the warranty, which can only be good. However, treating large volumes of concrete, as used in large bases and caissons, when it is only the cover zone that needs the protection, makes the process expensive for large pours. Also, resistance to water under hydrostatic pressure is limited, making application under water more of a problem.
- ***Corrosion inhibitors*** have been available in various admixed forms, with calcium nitrite being used in the US since before 1970 [15], at which time the combination set accelerator and corrosion inhibitor was very popular and recommended as a '*proven corrosion-inhibiting admixture*' to combat chloride-induced corrosion. More recent research has indicated that calcium nitrite may not be the panacea that it was once thought, as under-dosing of calcium nitrite can increase the severity of pitting corrosion to a condition worse than if no inhibitor had been used at all [16], although corrosion inhibitors can be effective as part of a comprehensive corrosion management solution [17];
- ***Surface coatings*** have remained the protection of last resort, as unlike steelwork, concrete should not need to be given additional surface protection provided the durability strategy is appropriate and the concrete construction is to a high standard, which it rarely is for the reasons explained above. With the decision to apply a coating, then a maintenance commitment arises, with the need to maintain and re-apply the coating at intervals of between 10 and 20 years, depending on the exposure environment [18, 19].

No single solution provides the panacea for durable reinforced concrete, so a strategy is needed for hostile exposure, both at new build, and following deterioration in service.

5. STRATEGY FOR HOSTILE EXPOSURE

By the mid-1990s, on new construction projects such as the Copenhagen Metro [20] and Singapore metro extension [17], the author was using deterioration modelling to predict corrosion risk for service life well over 100 years. Alternative "multi-barrier protection strategies" began to emerge that incorporated within the design provision for additional corrosion monitoring measures, supported by provision for cathodic protection should it be needed later in service, such as by checking electrical continuity of reinforcement. The structures currently being built across the Pearl River Estuary to create the 29.6km long bridge and tunnel HKZM Crossing, are the latest testament to how much more confident the industry has become about asset durability.

Recognition also is seen in codes such as EN 206-1 [21] that in hostile exposure conditions, concrete durability requires more than just a prescriptive approach to mix proportioning and cover. Durable construction needs performance-related methods that consider the locally available materials and environment and encourages, where appropriate, deriving the equivalent durability of the proposed concrete against control mixes. Specifically, EN 206-1 [21] requires consideration of: pre-construction durability assessments, using the locally available concreting materials; modelling the design service life; and then putting into place a thorough quality assurance plan to ensure the reinforcement has the correct cover, and the concrete is mixed, placed, compacted and cured as intended by the durability design. Several software packages are now available for modelling durability, including Life 365 [22, 23] and DuraCrete [24], driven by electrically-accelerated chloride migration testing such as NT Build 492 [25]. A recent review of the different modelling approaches is given by Pillai [26]. This is basically the *equivalent durability procedure* (EDP) to CEN/TR 16563 [27].

Also, the past 20 years have taught us not to build and forget our infrastructure, particularly where it is in extreme exposure environments. We now know and accept we must actively monitor and protect our infrastructure over the service life and that reinforced and prestressed concrete is not so durable that it would last indefinitely; it is not just the conception, design and construction phases that are important, but continued satisfactory performance in service must also be assured. To do this effectively, we need a plan to manage the assets, monitor changes and take proactive steps to address deficiencies, using tools and techniques that give early warning of problems and thereby help owners make informed finance-led decisions as to the best time to intervene.

Most landmark structures such as the HKMZ Crossing, already incorporate sensors into the concrete to provide computer-aided structural health monitoring, which has been used increasingly over the last 20 years. There is now an almost endless array of systems to assist, inform and assure asset operators about the continued 'fitness for purpose' of their assets, as explained by Brownjohn [28]. Sensors can detect: load behaviour, such as defections, dynamic response, strain and inclination; environmental exposure, including wind speed, seismic effects, precipitation and temperature; and deterioration, such as the effect of sea salt, causing resistivity, half-cell and linear polarisation changes. Sensors built into the structure from the start offer much more certainty in terms of life prediction than retrofitted sensors when unexpected problems develop.

Greater assurance can also be gained from using active corrosion control measures, including sacrificial and impressed current corrosion control systems from the start. Cathodic protection system design is possible through the latest European standard BS EN 12,696 [29]. The system can be designed to manage corrosion in new structures (commonly called *Cathodic Prevention*), or as explained below to halt the damage being caused by de-icing salt penetration into existing structures such as roads and marine structures to provide substantial life extensions of up to 50 years.

6. EXPERIENCE WITH CATHODIC PROTECTION AT RETROFIT

One of the most difficult forms of chloride ion corrosion to address is where penetration occurs into inaccessible components, one of the most common being the bridge half joint. Some half-joints (e.g. Figure 6) have collapsed unexpectedly, particularly in North America, as undetected corrosion has taken place within the concrete under saturated conditions and consequently, there is little obvious signs of corrosion taking place. With ongoing loss of bar cross-section due to pitting corrosion, eventually the structural element can be structurally weakened, leading to collapse.

The authority responsible for highways in England (Highways England) uses impressed current CP systems on many of its most severely deteriorated bridges and viaducts, including Spaghetti Junction (Figure 5) and the M1, the first inter-urban motorway to be built in the UK (opened between 1959 & 1968). Several of the viaduct sections were built using a half-joint system, where column-supported sections were constructed and a closure section lowered onto bearings, such as in the Trent Valley Floodplain Viaduct (TVFV) near Nottingham, England (Figure 9). With penetration of de-icing salt into the joint from above, corrosion of the supporting bearing is commonly reported, with rust staining and delamination of the concrete cover.

Figure 9: Typical half joint at the TVFV, M1 motorway, exhibiting sound drop in slab, but deteriorated bearing

The cross section is shown in Figure 10, illustrating the drop-in slab (right hand side) bearing on a heavily-reinforced bearing section [30]. The gap between the two reinforced concrete sections is typically 20-30mm wide, making it difficult to inspect visually by borescope. Reinforcement density is so congested, techniques such as impulse radar are ineffective at mapping the construction and standard cores of 100mm diameter cannot be used without risk of cutting bar.

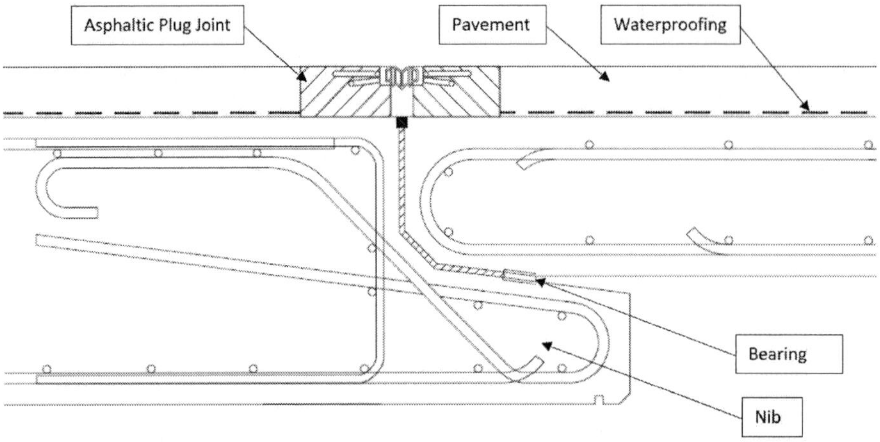

Figure 10: Section through Trent Valley Floodplain Viaduct, constructed in 1966 [30]

In such circumstances, non-destructive testing has to be used to identify areas at risk from corrosion. The most common technique is still the half-cell potential method [31], sometimes modified and used down a core hole, purpose cut to receive the half-cell attached to a rod. Chloride ion content determination from incremental dust drillings are taken at high negative potential areas and then analysed to EN 14,629 [32] to determine corrosion risk, which in the case of the half joint in Figure 9, was "high".

The objectives of the repair approach to the TVFV [30] included whether the repair would:

- Control and mitigate against chloride-induced pitting of the embedded reinforcement
- Limit the requirement for lane closure of the motorway for repairs and rehabilitation
- Limit the extent of repairs to spalled and damaged concrete only, not chloride contaminated sound concrete
- Provide longevity for the installed repairs and corrosion mitigation system
- Permit future plug joint replacement to maintain water tightness of the joints and associated road user delay cost
- Function under ongoing joint leakage conditions
- Be accessible within the expansion joint's mating surfaces and allow repairs and future investigation activities

BS EN 1504 Part 9 [33] was used in conjunction with the above repair objectives and the site access constraints to identify suitable repair options, principles and methods. The analysis identified cathodic protection would be the most appropriate method to control corrosion in a situation with heavy chloride ion contamination, and yet keep the motorway fully operational by working only from beneath, as had been done for many years previously on the Midland Links.

Given the high steel density, the high initial current demand needed to passivate steel in highly contaminated concrete, the reduced areas for installation, the long design life requirements and the limited track record, galvanic anode options were discounted for this installation. With the possibility of future joint leakage and the difficulty accessing and inspecting the joints, impressed current cathodic protection (ICCP) was selected, which offered several specific advantages when compared with the other repair approaches. Also, street lighting to the motorway provided a power supply that for some installations is not the case.

ICCP of steel in concrete is a very effective method of stopping and controlling corrosion in chloride contaminated concrete and includes the following principles:

- The negative charge of the steel surface which occurs as a direct result of the applied cathodic protection repels the negatively charged chloride ions away from the steel surface toward the positively charged anode.
- The excess of energy at the steel surface leads to a cathodic reaction occurring on all surfaces of the steel including the base of the pits, this reaction reduces oxygen and water to form hydroxide ions at the steel surfaces, so lowering the free chloride/hydroxide ion ratio and thereby the corrosion risk should the CP system be switched off.

- The production of hydroxide within pits at anode sites on the steel surface neutralises the acid generated by galvanic corrosion and as a result the pH increases, promoting restoration of a passive film and stabilisation within the base of the pits.

Figure 11 shows the CP installation arrangement at a typical half joint on the TVFV.

The discrete anode systems consisted of 19mm diameter mixed metal oxide coated titanium tubes grouted into a 28mm diameter drilled holes targeted from engineering drawings to mix the reinforcement as far as possible. The anodes were spaced at 300mm intervals between the reinforcement along the line of the half joint. The discrete anode tail and cabling was set within a chase cut into the cover region of the concrete to prevent vandalism and a negative connection welded to the reinforcement. To prevent shorting at areas where the reinforcement was close to or possibly struck during drilling of the cut holes, locally the anode was electrically insulated.

Embedded reference electrodes were also installed at strategic locations to evaluate the performance of the anode systems and enable system adjustment to obtain optimum protection levels at all of the reinforcing bar surfaces, as shown in Figure 11.

Given the location and difficulty of access, all ICCP system cables were run to non-metallic junction boxes fixed to the soffits of the viaduct using multicore cables and a non-metallic cable trays, linking to the power and control units. The design identified that access to the control equipment was a critical element for future operation and control. To improve access without the need for mobile elevated work platforms or underbridge access where the viaduct was over water, the power and control equipment was installed within a vandal-resistant enclosure on a purpose-built concrete plinth located on the motorway hard shoulder. Given the risk and precautions required to access the equipment on the hard shoulder of a live motorway, the design included fully remote operated and control functionality within the power and control equipment, using mobile phone technology [30].

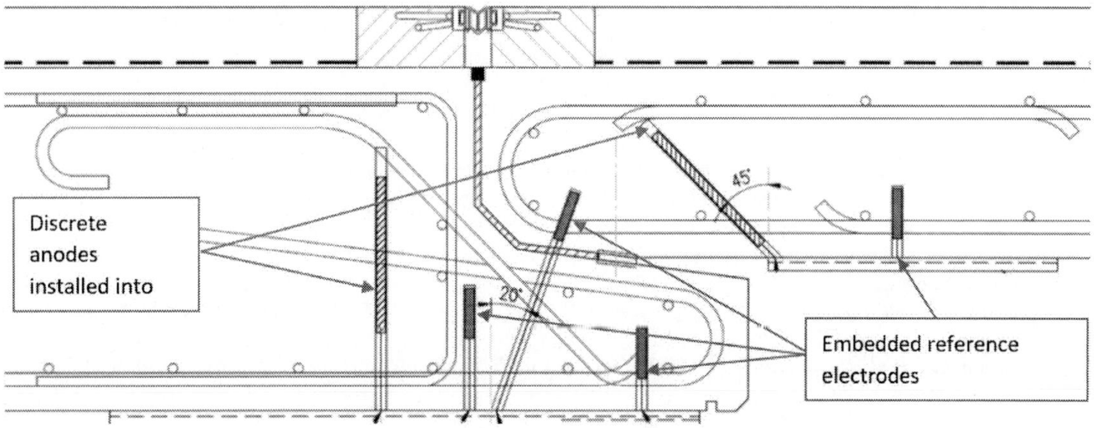

Figure 10: Installation of ICCP on the TVFV Half-Joints [30]

The functionality of the communication system included:

- automated monitoring (polarisation decay tests) could take place at scheduled intervals;
- remote adjustment of the system, based on the monitoring data;
- automatic e-mail the performance data in report format;
- sending alarm monitoring of the complete system and AC supply, powered by UPS backup, to identify outages and whether operatives would need to attend site for maintenance.

The design was undertaken to ensure a minimum 50-year design life of the embedded anodes and cabling, with junction boxes, control and operation systems, power supplies and external cabling designed for replacement when necessary.

The ICCP designs were undertaken using BS EN ISO 12,696 [29] and Highways England's document BA 83/02 [34]. The complete ICCP and concrete repair designs were subject to third party design review and approval by Highways England as a "*Departure from Standard*", which is a standard process used in the UK to denote the change from the original design.

7. CONCLUSION

This paper has set out some of the recent history of reinforced concrete usage in Europe. It explored how the codes have changed over the last 60 years, and how prior to 1980, the industry complacently allowed concrete structures to be designed for strength alone, assuming the concrete cover would protect the steel forever. However, this proved not to be the case and since the early 1980s, researchers, designers and standards bodies have been playing catch-up to offer an effective design for durability approach.

We can now confidently design for durable concrete buildings in normal environments, with adjustments to minimum cement contents, minimum cover, water/cementitious ratio, as well as meeting the required strength.

Preventing chloride-induced corrosion in hostile exposure conditions still poses a challenge, especially as most landmark structures have exposure conditions and life expectancies that are outside the normal provisions in the codes.

As a result, European codes have been regularly updated to increase durability provisions, providing robust design solutions that allow reinforced concrete to perform reliably in standard environments. For more severe exposure, measures include durability modelling, based on pre-construction durability assessments with tools such as DuraCrete, to provide assurance the service life will be achieved. As necessary, increased assurance can be provided by electrochemical measures such as cathodic prevention. Finally, the construction must follow a quality assurance plan, to ensure the reinforcement has the correct cover, and the concrete is mixed, placed, compacted and cured as intended.

Engineers also have to manage the legacy structures, which were built to lesser durability standards and are actively corroding as a consequence. Measures such as cathodic protection offer the best long-term approach for heavily-loaded structures such as bridges and ports. As the case study illustrates, cathodic protection offers a solution for the most difficult construction configurations such as half joints. These solutions can be implemented without closing essential road space.

8. ACKNOWLEDGEMENTS

The authors would like to thank the concrete and corrosion prevention community whose enthusiastic discussions in recent years have helped formulate some of the concepts in this paper.

This paper is also dedicated to the memory of the man who first generated the lead author's interest in concrete technology while at the University of Leeds, the late Professor Adam Neville, CBE FREng.

9. REFERENCES

1. Robery, P.C., '*Preserving the life of infrastructure through effective monitoring and intervention*', 2nd R.N. Raikar International Conference and Banthia-Basheer International Symposium on Advances in Science and Technology of Concrete, 18-19 December 2015, Mumbai, India
2. The Concrete Society, '*Historical approaches to the design of concrete buildings and structures*, Technical Report 70, July 2009
3. http://www.explore-gower.co.uk/explore/swansea/144-weaver-s-flour-mill, Copyright Brian Whittle, licenced for re-use under http://creativecommons.org/licenses/by-sa/2.0/
4. Hartcup, G., '*Code Name Mulberry – The planning, building and operation of the Normandy Harbours*', Publ. David and Charles, UK, 1977
5. BS CP 110: '*Code of practice for the structural use of concrete*', BSI, 1972
6. Robery, P.C., "*The Ability of Concrete to Resist Hostile Environments*", The Institute Of Concrete Technology Annual Technical Symposium, 6th April 2017
7. http://i.dailymail.co.uk/i/pix/2012/05/24/article-2149198-13436BEF000005DC-581_964x637.jpg
8. CIRIA UEG TN.5/1, '*Concrete in the Oceans: Marine durability survey of the tongue sands tower*', Report 5652, CIRIA UEG, London, 1979
9. Broomfield, J. P., '*Corrosion of steel in concrete – understanding, investigation and repair*', 2nd Edition, Taylor & Francis, 2007, 271p
10. BS EN 197-1, '*Cement - Part 1: Composition, specifications and conformity criteria for common cements*', BSI, 2001
11. BS 6349-1-4, '*Maritime works – Part 1-4: General – Code of Practice for Materials*', BSI, 2013
12. BS 8500-1, '*Concrete - Complementary British Standard to BS EN 206, Part 1: Method of specifying and guidance for the specifier*', BSI, 2015
13. BS 8500-1, '*Concrete - Complementary British Standard to BS EN 206, Part 1: Method of specifying and guidance for the specifier*', BSI, 2006 (superseded by [11])
14. Saguees, A.A., Powers, R.G. and Kessler, R., '*Corrosion Processes and Field Performance of Epoxy-Coated Reinforcing Steel in Marine Substructures*', United States: NACE International, Houston, TX (United States), 1994
15. Berke, N. S. & Rosenberg, A., 'Technical Review of Calcium Nitrite Corrosion Inhibitor in Concrete', Transport Research Board, n 1211, 1989, pp 18-27, https://trid.trb.org/view.aspx?id=308410 viewed 23/3/17
16. Anstice, D. J., '*Corrosion inhibitors for the rehabilitation of reinforced concrete*', PhD thesis, 2000, available through http://ethos.bl.uk/OrderDetails.do?uin=uk.bl.ethos.322144 (viewed 23/3/17)

17. Nmai, C., et al., '*Application of High-Performance Concrete in Singapore's North-East and Changi International Airport MRT Lines*', 5th CANMET / ACI International Conference on Recent Advances in Concrete Technology, Singapore, 2001
18. Robery, P.C., '*Requirement of coatings*', Journal of the Oil and Colour Chemists' Association, v 71, n 12, Dec 1988, pp 403-406
19. Almusallam, A. A., et al, '*Effectiveness of surface coatings in improving concrete durability*', Cement and Concrete Composites, v 25, n 4-5, May-July 2003, Pages 473-481
20. Geiker, M., '*Durability Design of Concrete Structures Minimising Total Life Cycle Costs -Considerations and Examples*', Proc Int Conf on Concrete Durability and Repair Technology, University of Dundee, UK, 8-10 September 1999
21. BS EN 206-1: '*Concrete - Part 1: Specification, performance, production and conformity*', BSI, 2013
22. ACI 365.1R-00, '*Service-Life Prediction – state-of-the-art report*', ACI Committee 365, 2000, 44p
23. Life-365 Consortium: http://life-365.org/consortium.html, accessed 24/3/17
24. DuraCrete, '*Modelling of Degradation*', EU-Project (Brite EuRam III) No. BE95-1347, Probabilistic Performance based Durability Design of Concrete Structures, Report, vol. 4–5, 1998
25. NT BUILD 492, '*Concrete, Mortar and Cement-based Repair Materials: Chloride Migration Coefficient from Non-steady-state Migration Experiments',* Nordtest, Esbo, Finland, 1999
26. Pillai, R.G and Annapareddy, A., '*Service Life Prediction Models for Chloride-Laden Concrete Structures: A Review and Nomographs*', International Journal of 3R's, Vol. 4, No. 2, 2013, pp563-580
27. PD CEN/TR 16563, '*Principles of the equivalent durability procedure*', BSI, 2013
28. Brownjohn, J.M.W., '*Structural health monitoring of civil infrastructure*', Phil. Trans. R. Soc. A, 15 February 2007 vol. 365 no. 1851, pp 589-622
29. BS EN ISO 12,696:2012. Cathodic Protection of Steel in Concrete, British Standards Institution.
30. Segers, J.P.M., Robery, P.C. et al, "*Sustainable life extension of deteriorating M1 motorway bridge half joints using impressed current cathodic protection*", Proc. 16th European Bridge Conference, Edinburgh UK, 23-25 June 2015.
31. ASTM C876-09, *Standard test method for corrosion potentials of uncoated reinforcing steel in concrete*, American Society for Testing and Materials, reaffirmed 2009.
32. BS EN 14,629:2007, Products and systems for the protection and repair of concrete structures – Test methods - Determination of chloride content in hardened concrete, British Standards Institution.
33. British Standards Institution, BS EN 1504. Products and systems for the repair and protection of concrete structures-Definitions, requirements, quality control and evaluation conformity Part 1-10.
34. BA 83/02, February 2002. Cathodic Protection for use in reinforced concrete highway structures, Highways England, UK.
35. Loov, R.E., Reinforced Concrete at the Turn of the Century, Concrete International, Dec 1991, pp 67-73

10. AUTHOR DETAILS

Peter Robery is RAEng Visiting Professor in Forensic Engineering, University of Birmingham and director of Robery Forensic Engineering Ltd.

He is a chartered civil engineer with over 38 years of experience in the field of durability, inspection, testing, maintenance and repair. Former employment includes Technology Fellow (concrete materials) with CH2M HILL and Director with AECOM.

Diverse project experience from around the world, including repairs to Spaghetti Junction, and durability specification advice for parts of the new HKMZ crossing.

He is author of over 80 papers, presentations and articles in the field of concrete performance and remains actively involved with the development of EN1504 on materials for concrete protection and repair.

Paul Segers is an independent corrosion and cathodic protection consultant and director of SegCorr Ltd.

A Chartered Engineer and Fellow of the Institute of Corrosion with 20 years of industrial corrosion control and cathodic protection experience, certificated to BS EN 15257 Level 3 Senior CP Engineer for all application sectors, to NACE Corrosion and Cathodic Protection Specialist and holds a Master's degree in Corrosion Control Engineering.

He has worked in the UK, Africa, Middle East and Far East, providing specialist technical expertise into deterioration mechanisms, failure investigation, corrosion prevention, corrosion control and mitigation, infrastructure life extension design and whole life planning.

Corrosion of emerging materials

N. Birbilis, Y. Qiu, R.L. Liu, Y. Yan, S. Thomas, O. Gharbi

Department of Materials Science and Engineering, Monash University, Victoria, Australia

**E-mail address:* nick.birbilis@monash.edu

Abstract

As technologies rapidly advance, materials are being utilised in previously unconsidered environments and unique loading scenarios. Similarly, the emergence of new materials, and the ability to produce them, has dramatically increased in the past decade – owing to a widespread emergence of computational materials design and additive manufacturing. Such aspects will be discussed quite generally, along with specific examples of the durability of the following emerging materials: (i) additively manufactured (i.e. 3D printed) metals and alloys, (ii) High entropy alloys, and (iii) ultra lightweight alloys.

1. INTRODUCTION

Materials, in particular, metals and alloys, have been the biggest factor in the development and modernisation of societies if one considers major milestones from the Bronze Age to Iron Age. Indeed, scientists and engineers have successfully unravelled the periodic table since those 'ages', in addition to developing a number of Corrosion Resistant Alloys (CRAs) in the past century. Notably, we have just crossed 100 years since Harry Brearly was credited with the invention of stainless steel [1], which by all accounts was in part a fortuitous discovery [2] (although stainless steel would undoubtedly exist today based on the empirical work of numerous industrial researchers in the UK, France and Germany in the early 20th century). Using the benefit of hindsight, it is possible to argue (perhaps even if somewhat controversially) that relatively minor advances have occurred in the field of metallurgy in the past century. This is not a statement intended to belittle what have been exceptional advances in key foundational principles and the engineered development of alloys presently in use in everything from smartphones to aircraft – but a statement intended to rationalise the pace of materials design relative to say, computer science. Turning our attention to the present day, an era of ubiquitous computing is here [3] (and here to stay). In fact, the present trend in manufacturing (known as Industry 4.0 [4]) is unfolding to reveal that once 'physical' and 'engineering' technologies have been wholly disrupted (and now driven) by cyber-physical systems, cloud computing, the internet of things and cognitive computing. Whilst Industry 4.0 is an 'industrial' revolution, the pervasive nature of computing in everything from simulations, design, and deployment of machine learning [5-6], is also relevant to the manner in which materials design is approached. For example, alloy production trials can be largely minimised by calculation of phase diagrams - which can now be based on first principles calculations [7], meaning compositions that significantly deviate from previously (empirically) explored alloys, may be computationally explored. In the context of corrosion, first principles calculation of Pourbaix Diagrams is possible for an enhanced understanding of how 'surfaces' respond in a given environment [8]; in addition to calculation of Pourbaix Diagrams for complex alloys [8-9] (not previously possible). Neural networks and machine learning can nowadays inform alloy design [10], and indeed, the methods by which alloys may be produced have seen a revolution with the advent of Additive Manufacturing (AM).

Additive manufacturing of metals and alloys using laser fusion allows design complexity and the production of complex net shape components. However, from a metallurgical perspective the local melting temperatures in the two main methods of metal AM (which include selective laser melting and direct laser deposition) permit very rapid local (super) heating and concomitant rapid cooling. These concepts won't be elaborated herein, however as one example, the conventional cast production of aluminium alloys that contain transition metal alloying additions is not possible owing to lack of liquid or solid solubility. Furthermore, the incompatible melting points for constituent metals may differ by as much as 2000 ^{o}C, making casting implausible. In the case of AM, metals with wide melting differences and no mutual solubility can be melted and alloyed.

The coming decades are likely to be characterised by two factors that are relevant to corrosion of metals and alloys. These include (i) the utilisation of engineering alloys in previously unconsidered environments (for example, routine space flight, new renewable energy systems, personalised medicine and diverse consumer products), and (ii) the use of previously un-used alloy systems that are likely to include a number of lightweight alloys (e.g. magnesium or aluminium based), additively manufactured alloys (which may include previously unknown compositions) and so-called compositionally complex alloys (further described below).

2. COMPOSITIONALLY COMPLEX ALLOYS AND HIGH ENTROPY ALLOYS

The past decade has seen the emergence of what has been named a new class of metallic alloys, termed "High Entropy Alloys" (HEAs) [11]. Such alloys were named HEAs on the basis of a high entropy of mixing, owing to the criteria that a HEA is a HEA if it consists of 5 or more principal alloying elements in near equi-atomic concentrations [12]. By virtue of their complex compositions and the difficulty in production of such complex compositions by traditional casting, HEAs were not previously explored (as the mutual solubility of most metal-metal combinations does not permit conventional casting, with HEAs nominally prepared by arc melting). The interest in HEAs was further accelerated by the discovery that many HEAs (consisting of 5 metallic elements) were capable of forming simple solid solutions of high hardness and thermal stability. As a consequence, intense research has now followed in the field of HEAs, exploring wide combinations of multi-metal alloying. It has also now been determined – from the proliferation of HEA production, that not all 5 element metal blends are capable of forming simple solid solutions; with many HEAs having multiphase structures, numerous intermetallic compounds, and a matrix structure that may be cubic or hexagonal [12]. Examples of the microstructure developed in the AlCoCrFeNi and AlTiZrMoNbTa (Compositionally Complex Alloys – CCAs) are given in Figure 1.

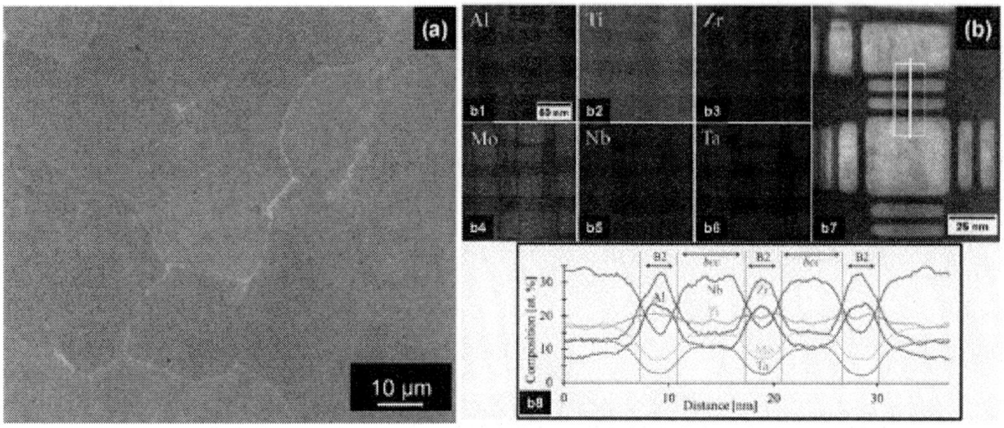

Figure 1. *(a) SEM micrograph of an as-cast Al0.3CoCrFeNi showing a single-phase microstructure (reprinted with permission from [13]), (b1–b6) XEDS elemental maps of Al, Ti, Zr, Mo, Nb and Ta, respectively, recorded in STEM using a Super-XTM detector, (b7) A STEM-HAADF image with a white line identifying the location of the EDXS linescan shown in (b8 (reprinted with permission from [14]).*

As a result of this, a more generic name has been given to such alloys, namely CCAs. Of particular relevance to the context of the present paper, the corrosion behaviour of HEAs / CCAs is unique, on the basis that they exhibit extreme passivity in a wide range of environments. A compilation of the corrosion potentials and pitting potentials of numerous CCAs is provided in Figure 2, whereby stainless steels are also provided as a reference point.

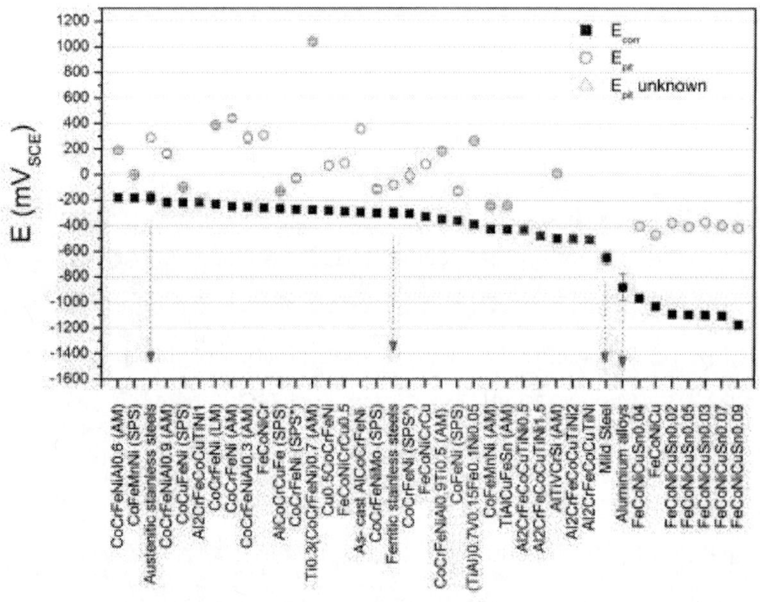

Figure 2. *A summary of the E_{corr} and E_{pit} values of different CCAs in quiescent 0.6 M NaCl at 25 °C (from [15]. Herein, LM denotes CCAs manufactured by laser melting, AM denotes arc melting and SPS denotes spark plasma sintering. The alloy CoCrFeNi (SPS*) was produced by spark plasma*

sintering and held at 1000 °C for 5 min, while CoCrFeNi (SPS^) was firstly held at 500 °C for 5 min then sintered at 1000 °C for another 5 min

This corrosion resistance of CCAs is anathema to conventional wisdom on the basis that: 1) HEAs / CCAs are comprised of numerous metals, many of which may have inherently poor (or no) passivity, and (2) microstructural heterogeneity, which can be extensive in CCAs, and nominally is thought to result in localised corrosion. As a result, from a corrosion perspective, the corrosion performance of compositionally complex multi-metal blends is poorly understood (as noted in a recent review by Qiu [15]), and merits closer attention.

3. ADDITIVELY MANUFACTURED ALLOYS

At present, additive manufacturing (a.k.a. 3D printing) of numerous alloys is routine, including numerous Ti-alloys, a range of Ni-alloys (including Hastelloy® variants), aluminium alloys (including age hardenable Al-alloys [16]), and stainless steels [17-18]. The present paper does not seek to cover the variety and breadth of this topic, however it merits comment that from a corrosion perspective, there are a number of (very) unique aspects to the performance of AM prepared alloys. For example, the following considerations are relevant in the corrosion of AM alloys, (1) the solidification is rapid, such that appreciable grain refinement is common, (2) thermal exposure upon solidification is insufficient to permit 'conventional' phase transformations in many cases, (3) the extremely high local temperature of the melt allows for annihilation of intermetallic phases, but also impurities, (4) residual stress and unique dislocation structures may exist in AM alloys, and, (5) there exists the possibility that porosity in the vicinity of 0.1 to 1% may be present.

The ability for AM prepared 316L to display a wider passive window and higher pitting potential relative to wrought 316L is demonstrated in Figure 3, and rationalised by point (3) above, whereby MnS inclusions have been demonstrated to be annihilated. The annihilation of MnS results in the resolutionising of S, whilst Mn is associated with oxygen in the form of inert oxide particles (Figure 4).

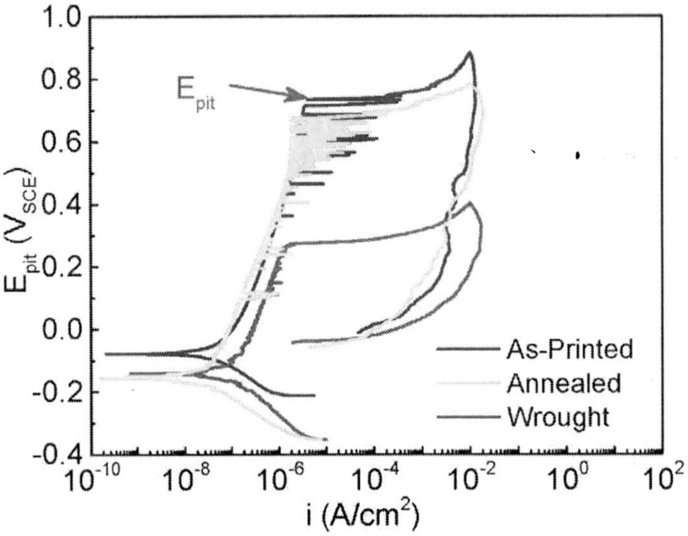

Figure 3. *Representative cyclic potentiodynamic polarisation curves for the as-printed, selective laser melting annealed and wrought 316L stainless steels in 0.6 M NaCl [18]*

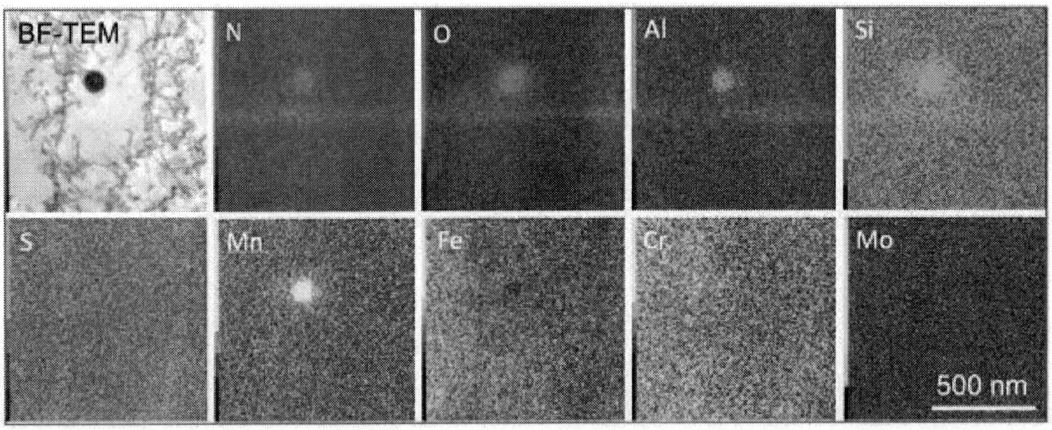

Figure 4 *TEM-EDS mapping of the cellular sub-structure and inclusion in the as-printed 316L stainless steel [18].*

4. LIGHTWEIGHT ALLOYS

Lightweight and ultra-lightweight alloys are increasingly in demand for the purposes of energy reduction in transportation (air, auto, rail), but also in the context of consumer products which are carried (i.e. laptops, smartphones, etc.). The most 'ultra' lightweight of alloys are those which incorporate Li which has the atomic number of 3 and a very low density (0.534 g/cm^3). When combined with the lightest of the engineering metals, namely Al or Mg, then alloys of engineering relevance and low density are possible. The now third generation Al-Li alloys (which are Al-Cu-Li based) have rapidly penetrated the aerospace sector and also show appreciable corrosion resistance relative to the incumbent aerospace alloys [19-20]. However, it merits attention that recently reported corrosion resistant Mg-Li alloys are possible [21]. Whilst Mg is a highly reactive metal (in aqueous environments) and Li even more reactive, the combination of the metals in proportions that include Mg with >10.5 wt. % of Li, permits the evolution of a protective surface film. The surface film in such cases in a dynamically developed lithium carbonate (Li_2CO_3), that is subsequently insoluble in aqueous conditions and offers significant corrosion protection. Presently, further research is underway in order to rationalise and explore the phenomenon of dynamically grown surface films for corrosion protection, as shown by Hou et. al. [22]. In this context, what is meant by a dynamically forming film, is a film that is not predicted to form by thermodynamic analysis (i.e. Pourbaix diagrams), but which forms as a result of kinetic (dissolution) developments at the alloy surface. Such developments release metal ions (often incongruently) and generate local surface pH that can result in the formation of hitherto un-anticipated species (such as Li_2CO_3 in the case of Mg-Li alloys).

5. SUMMARY

The present era has seen an emergence of 'new materials' being utilised or considered for engineering applications. It is a unique period in the context of insertion of new metallic materials – and a period where the durability of such emerging materials is to an appreciable extent, unknown. There does not exist a long service record for additively manufactured alloys (in corrosive environments), nor for HEAs or CCAs, and similarly, not for Li-rich Mg alloys. As a consequence, research into understanding the degradation of emerging materials is critical to their success in service. The coming decade promises to provide even more new materials, as processes such as additive manufacturing permit previously unexplored alloys, and the development of so-called multi-materials.

6. ACKNOWLEDGEMENTS

We gratefully acknowledge the Monash Centre for Electron Microscopy and the Monash Centre for Additive Manufacturing. The authors acknowledge support from the Australian Research Council, and NB is supported by Woodside Energy. The collaborators and collaborating organisations with which the authors work are also acknowledged, in particular PPG, the US Army Research Laboratory and Baosteel.

7. REFERENCES

1. "A non-rusting steel". *New York Times*. 31 January 1915.
2. J. Waldman, Rust: The longest war. Simon and Schuster, USA. 2016.
3. Greenfield, A. Everyware: the dawning age of ubiquitous computing. New Riders. USA. 2006. pp. 11–12.
4. https://en.wikipedia.org/wiki/Industry_4.0 (accessed 3 October 2017)
5. Alpaydin, E., Introduction to machine learning. MIT press, USA. 2014.
6. Hermann, M., Pentek, T. Otto, B., Design principles for industrie 4.0 scenarios. In *System Sciences (HICSS), 2016 49th Hawaii International Conference on* (pp. 3928-3937). IEEE.
7. Liu, Z-K. "First-principles calculations and CALPHAD modeling of thermodynamics." *Journal of phase equilibria and diffusion* 30, no. 5 (2009): 517.
8. Williams K.S., Labukas J.P., Rodriguez-Santiago V., Andzelm J.W., First principles modeling of water dissociation on Mg (0001) and development of a Mg surface Pourbaix diagram. Corrosion. 2014 Jun 25;71(2):209-23.
9. Yuwono, J.A., Birbilis, N., Williams, K.S., and Medhekar. N.V., "Electrochemical Stability of Magnesium Surfaces in an Aqueous Environment." *The Journal of Physical Chemistry C* 120, no. 47 (2016): 26922-26933.
10. Birbilis, N, Cavanaugh, M.K., Sudholz, A.D., Zhu, S.M., Easton, M.A., and Gibson, M.A., "A combined neural network and mechanistic approach for the prediction of corrosion rate and yield strength of magnesium-rare earth alloys." *Corrosion Science* 53, no. 1 (2011): 168-176.
11. Murty, B.S., Yeh, J-W. and Ranganathan. S., *High-entropy alloys*. Butterworth-Heinemann, 2014.
12. Zhang, Y., Zuo, T. T., Tang, Z., Gao, M. C., Dahmen, K. A., Liaw, P. K., & Lu, Z. P. (2014). Microstructures and properties of high-entropy alloys. *Progress in Materials Science*, *61*, 1-93.
13. Shun, T. T., Du, Y. C. Microstructure and tensile behaviors of FCC Al0.3CoCrFeNi high entropy alloy. J. Alloy Compd. 479, 157–160 (2009).
14. Jensen, J. K., Welk, B.A., Williams, R.E.A., Sosa, J.M., Huber, D.E., Senkov, O.N. Viswanathan, G.B. and Fraser, H.L., "Characterization of the microstructure of the compositionally complex alloy Al 1 Mo 0.5 Nb 1 Ta 0.5 Ti 1 Zr 1." *Scripta Materialia*121 (2016): 1-4.
15. Qiu, Y., Thomas, S., Gibson, M. A., Fraser, H.L., and Birbilis, N. (2017). Corrosion of high entropy alloys. *npj Materials Degradation*, *1*(1), 15.
16. Martin, J.H., Yahata, B.D., Hundley, J.M., Mayer, J.A., Schaedler, T.A. and Pollock, T.M., 2017. 3D printing of high-strength aluminium alloys. *Nature*, *549*(7672), p.365.
17. Sander, G., Thomas, S., Cruz, V., Jurg, M., Birbilis, N., Gao, X., Brameld, M. and Hutchinson, C.R., 2017. On The Corrosion and Metastable Pitting Characteristics of 316L Stainless Steel Produced by Selective Laser Melting. *Journal of The Electrochemical Society*, *164*(6), pp.C250-C257.
18. Chao, Q., Cruz, V., Thomas, S., Birbilis, N., Collins, P., Taylor, A., Hodgson, P.D. and Fabijanic, D., 2017. On the enhanced corrosion resistance of a selective laser melted austenitic stainless steel. *Scripta Materialia*, *141*, pp.94-98.
19. Rioja, R.J., Liu, J., The Evolution of Al-Li Base Products for Aerospace and Space Applications. Metall. Mater. Trans. A, 43 (2012), pp. 3325–3337
20. Gharbi, O., Birbilis, N. and Ogle, K., 2017. Li reactivity during the surface pretreatment of Al-Li alloy AA2050-T3. *Electrochimica Acta*, *243*, pp.207-219.
21. Xu, W., Birbilis, N., Sha, G., Wang, Y., Daniels, J.E., Xiao, Y. and Ferry, M., 2015. A high-specific-strength and corrosion-resistant magnesium alloy. *Nature materials*, *14*(12), pp.1229-1235.
22. Hou, L., Raveggi, M., Chen, X.B., Xu, W., Laws, K.J., Wei, Y., Ferry, M. and Birbilis, N., 2016. Investigating the Passivity and Dissolution of a Corrosion Resistant Mg-33at.% Li Alloy in Aqueous Chloride Using Online ICP-MS. *Journal of The Electrochemical Society*, *163*(6), pp.C324-C329.

THE SYNERGISM OF ELECTROCHEMICAL AND MECHANICAL FACTORS IN MATERIALS DEGREADATION

B. Lu[1, 2], J.L. Luo[1]

[1]Department of Chemical and Materials Engineering, University of Alberta, Edmonton, Alberta, T6G 1H9, Canada, [2]Materials Engineering Department, Southwest Research Institute, San Antonio, TX, USA, 78238

SUMMARY: In heavy oil wells, carbon steel liner pipe often suffers from severe corrosion due to the corrosive environment with solid particles and bitumen. This paper provides an overview of our work on erosion corrosion behaviors of metals. A physical model was developed based on non-equilibrium thermodynamics, electrochemistry and dislocation kinetics of erosion-corrosion of carbon steels in flowing slurry. The effects of hydrodynamics of flowing slurry on anodic dissolution rates of metals were theoretically and experimentally evaluated. The experimental results show that the material loss rate due to corrosion-enhanced erosion increases linearly with the logarithm of anodic current density, as predicted by the theoretical model. The in-situ nanoindentation shows that the presence of anodic current on surface results in the surface softening caused by an increase in the dislocation mobility in the surface layer. Both theoretical analysis and experimental evaluation show that corrosion-induced surface softening is a linear function of logarithm of current density, and the correlation of relative hardness degradation vs. anodic current density is almost independent of the polarization behavior of the materials.

Keywords: Erosion-corrosion, Steel, Mechanistic studies, Physical models.

1. INTRODUCTION

Erosive fluids can destroy the surface films on metals, and remove small pieces of material, leading to a significant increase in corrosion rate. For instance, carbon steel pipe carrying water is usually protected by a rust scale and its corrosion rates are typically < 1mm/y. Fluid-induced forces[1] or flowing-enhanced dissolution[2] can damage or destroy rust scale, and increase corrosion rates to an order of 10 mm/y in addition to the any erosion of underlying metal [3]. Meanwhile, corrosion can degrade in mechanical properties of surface layer and promote mechanical erosion in fluids [4]. This conjoint action of erosion and corrosion is known as erosion-corrosion [5]. Erosion-corrosion encompasses a wide range of flow-induced corrosion [6]. It is also regarded as a subject within the broader area of tribo-corrosion which covers all aspects of tribologically induced interactions with electrochemical processes [7]. Erosion-corrosion was rated in the top 5 most prevalent forms of corrosion damage in the oil and gas production [8] and cause an immense economic loss [9,10].

In industrial practice, selection of structure materials with good performance/cost efficiency is of critical importance to reduce loss due to erosion-corrosion. A material performance map is regarded as an effective tool to do so, which demonstrates correlation between erosion-corrosion performance (or mechanisms) and material (or process) parameters [11,12]. Since service conditions leading to erosion-corrosion are very different in various cases, the erosion–corrosion mechanism can change and the performance of structural materials can vary significantly with changes in operating parameters (temperature, flowing velocity, impingement angle etc.) and properties of target materials (hardness, strength, microstructure, composition etc.) and fluids (viscosity, density, fluid chemistry, concentration and size of erodent etc.). It is technically difficult to demonstrate the dependence of the transitions of erosion-corrosion mechanism or material performance on various variables. A plausible progress on this aspect was reported by a group led by Stack [11,12], who made an effort combine different variables in dimensionless groups and constructed erosion-corrosion maps using certain combinations of those variable groups. The material performance map is fully based on experimental measurements and hence its predictive function depends on the experimental and analytical methods employed. Therefore, a comprehensive understanding of mechanisms of synergistic effect is critical to improve the erosion resistance in corrosive media. As pointed out by Poulson [10], complexity in predicting erosion-corrosion is mainly due to existence of synergistic effects between the mechanical and electrochemical variables. Several good articles were present to review this topic from different view angles [2, 6,7,9]. This paper will summarize the progresses achieved by the authors' group in this aspect.

2. EROSION, CORROSION AND THEIR SYNERGISM

2.1 Assessment Method

Two different material loss mechanisms are involved in erosion-corrosion of metals: mechanical erosion and electrochemical corrosion. The mechanical erosion relates to plastic deformation and rupture in surface layer. Debris of metal are removed from the surface by various mechanical forces before being ionized. The electrochemical corrosion relates to metal being dissolved into the slurry after it is ionized. Therefore, the total material loss rate $\dot{w}$ is the sum of material loss rates caused by erosion, $\dot{e}$, and corrosion, $\dot{c}$,

$$\dot{w} = \dot{e} + \dot{c} \tag{1}$$

The total material loss of material in corrosive fluids is normally larger than the sum of those caused by pure mechanical erosion and pure electrochemical corrosion. According to standard of ASTM G119, the pure mechanical erosion, $\dot{e}_0$ is defined as the erosion in an inert environment and the pure electrochemical corrosion, $\dot{c}_0$, is the corrosion under erosion-free condition. The additional wastages of erosion and corrosion components caused by the synergistic effects are regarded as the corrosion-enhanced erosion, $\dot{e}_c$, and the erosion-enhanced corrosion, $\dot{c}_e$ [13]. Therefore, erosion and corrosion can be expressed using Eqs. (2) and (3), respectively.

$$\dot{e} = \dot{e}_0 + \dot{e}_c \tag{2}$$

$$\dot{c} = \dot{c}_0 + \dot{c}_e \tag{3}$$

The erosion-corrosion mechanism is affected by all the factors which control corrosion and all the factors which affect erosion. In combination, the damage is synergistic and can be extremely aggressive. The synergism of erosion and corrosion, $\dot{s}$, is expressed as the sum of $\dot{e}_c$ and $\dot{c}_e$ [13]:

$$\dot{s} = \dot{e}_c + \dot{c}_e \tag{4}$$

The synergism of erosion and corrosion often contributes to such a large part of the total material loss that it cannot be ignored in service lifetime assessment in engineering. The corrosion in erosive liquid can be determined using the standard procedures that used in erosion-free condition, such as the one to measure the linear polarization resistance (ASTM G59) [14] and the one to generate the potentiodynamic curves (ASTM G5) [15]. The pure mechanical erosion rate in corrosive slurries should be conducted under the same hydrodynamic conditions under cathodic protection. ASTM G119 recommended polarizing the specimen to one volt cathodic with respect the open circuit potential to guarantee a fully protected condition. However, caution must be taken because hydrogen embrittlement may occur in some materials under the cathodic protection. Besides, the gas bubbles produced by the hydrogen evolution may affect the hydrodynamic conditions. A recent study indicated that the erosion rates under cathodic protection in the slurries prepared by dilute acidic solutions are much higher than those in neutral and alkaline slurries [31,32]. In line with ASTM G119 [13], the following dimensionless factors can be defined to describe the degree of synergism:

$$\text{Total synergism factor} = \frac{\dot{w}}{\dot{w} - \dot{s}} = 1 + \frac{\dot{s}}{\dot{e}_0 + \dot{c}_0} \tag{5}$$

$$\text{Corrosion augmentation} = \frac{\dot{c}}{\dot{c}_0} = 1 + \frac{\dot{c}_e}{\dot{c}_0} \tag{6}$$

$$\text{Erosion augmentation} = \frac{\dot{e}}{\dot{e}_0} = 1 + \frac{\dot{e}_c}{\dot{e}_0} \tag{7}$$

Although efforts have been made, it is still difficult to build an integral model of erosion-corrosion [16,17,18] because a large amount of factors are involved in the erosion-corrosion processes including the metallurgical features of material [19,20,21], the hydrodynamics of fluid [22,23] and flow field [24], the characteristics of erodent [25,26,27,28], the temperature [29,30] and corrosivity of media [31,32]. During impingement, the sand degradation may result from the broken of sand particles and/or the bluntness of particle corner or edge, leading to a reduced erosion rate. If the effect of sand degradation is excluded, the erosion rate under a given hydrodynamic condition is independent of time [17, 33]. The total material loss rate resulting from a cavitating liquid or impingement of liquid droplets is a function of time. There is an incubation time within which the rate of material loss is negligible. After the incubation, the material loss rate increases rapidly, reaches a peak value and then reduces to a steady value gradually [34,35].

2.2 Theoretical Considerations

During erosion-corrosion, at least two irreversible processes take place simultaneously, namely, electrochemical reactions resulting in corrosion and surface plastic deformation produced by erodent impingements [4,25]. The flux of corrosion can be characterized by anodic current density, i_A, and the dynamically plastic deformation can be evaluated by the plastic strain rate, $\dot{\gamma}_P$. Materials loss may be caused by either fluid-induced mechanical forces (erosion) or electrochemical reactions (corrosion). In accordance with non-equilibrium thermodynamics, the enhanced fluxes of these two irreversible processes due to the synergisms can be formulated with Onsager's relations [4,25,36].

$$\dot{\gamma}_P = L_P F_P + L_{CP} F_C = \dot{\gamma}_{P0} + L_{CP} F \tag{8}$$

$$i_A = L_{PC} F_P + L_C F_C = L_{PC} F_P + i_{A0} \tag{9}$$

$\dot{\gamma}_{P0}$ is the plastic strain rate when corrosion is absent, i_{A0} is the anodic dissolution current density in flowing electrolyte free of solid particle, L_P, L_C, L_{CP} and L_{PC} are Onsager's coefficients, F_P and F_C are the generalized driving forces for the plastic deformation and anodic dissolution, respectively. In light of the theories of dislocation dynamics [37], $\dot{\gamma}_P$ under a given applied shear stress is proportional to the probability to activate (Frank-Read) dislocation sources, which is proportional to $\nu \cdot \exp[(E_D - \tau_{P-N} V_D)/R_D T]$. The affinity of plastic deformation $A_P = \tau_{P-N} V_D$, ν is the vibrational frequency of dislocation source, E_D is the potential barrier for activating the dislocation source, τ_{P-N} is the Peierls-Nabarro stress, i.e., the flowing shear stress component for overcoming short range obstacles in microstructure of metals, V_D is the activation volume of dislocation source, $R_D = kN_{\max}$ where k is the Boltzmann constant and $N_{\max}$ is the maximum dislocation density in a material; T is the absolute temperature. Accordingly, the driving force of the irreversible slip of dislocations, F_P, is formulated by [4,36]

$$F_P = R_D \left(\exp \frac{A_P}{R_D T} - 1 \right) \tag{10}$$

It is known from electrochemistry that the driving force for the electrochemical dissolution, F_C, is given by [4,36]

$$F_C = R \left(\exp \frac{\alpha \tilde{A}}{RT} - 1 \right) \tag{11}$$

where $\tilde{A} = \sum_j \nu_j \tilde{\mu}_j$ is the affinity of electrochemical dissolution, respectively, ν_j are the stoichiometric coefficients in corrosion reactions, $\tilde{\mu}_j$ are the electrochemical potentials. R is the gas constant and α is the charge transfer coefficient for the anodic reaction.

The second term in Eq. (8) and the first term in Eq. (9) represent the synergistic effects between the two irreversible processes above mentioned. The positive Onsager's coefficients indicate that the synergistic effects will lead to the enhanced plastic deformation and accelerated anodic dissolution. Above equations provide a theoretical foundation, in the following sections for quantitative assessment of synergism of erosion and corrosion.

3. CORROSION IN EROSIVE LIQUIDS

3.1 Electrochemical Corrosion of Non-Passivated Target

When corrosion is controlled by the mass transfer of dissolved oxygen or in the boundary layer of the liquid at the electrolyte/electrode or diffusion of some other soluble species away from the surface [10], the corrosion rate is formulated as follows [2,38,39,40,41],

$$Sh = \alpha Re^{\beta} Sc^{\gamma} \tag{12}$$

The non-dimensional parameters in Eq. (12) are Sherwood number $Sh = Kd/D$, Reynolds number $Re = Ud/\nu$ and Schmidt number $Sc = \nu/D$; where α, β and γ are constants depending upon the flow conditions and the geometry of the test devices; K is the specific mass transfer coefficient, d is the specific size depending on the geometry of test device, D is the diffusion coefficient of the species of which diffusion in the boundary layer controls the corrosion process, U is the flow velocity and ν is the kinematic viscosity of the fluid.

Eq. (12) was originally established in the rotating disk electrode (RDE) system based on kinetics of electrochemical reaction [42]. When the electrochemical reaction over the RDE surface is mass transfer control, $\alpha = 0.791$, $\beta = 0.7$ and $\gamma = 0.356$. Eq. (12) was extended to various systems. In a straight pipe, d in Eq. (12) could be the pipe diameter. Corrosion rate, $\dot{c} = K \Delta C$, as the corrosion is dominated by the mass transfer process in the boundary layer at the electrode/electrolyte interface. ΔC is the concentration driving force or concentration drop of species within the boundary layer, of which the diffusion controls the corrosion process. Thus

$$\dot{c} = K\Delta C \propto K = \alpha Re^{\beta} Sc^{\gamma} D\Delta C/d \propto Re^{\beta} \tag{13}$$

Eq.(13) was validated experimentally, such as the test data shown in Fig. 1. Dissolved oxygen is often believed to be the species in the flowing electrolyte controlling the corrosion process. In real pipe system, the change of surface roughness due to erosion corrosion can also affect the β-value [2,39]. For a mass transfer-controlled corrosion reaction, the value for β may range from 0.5 to 1 [6]. When the corrosion reaction at the target surface is solely controlled by the diffusion of dissolved oxygen within the boundary layer, the corrosion rate is proportional to the limited current density of dissolved oxygen [2,39,40,41] and the corrosion rate $\dot{c}$ in the is given by

$$\dot{c} = \eta K C_0 \propto K = \alpha Re^{\beta} Sc^{\gamma} D/d \tag{14}$$

Where C_0 is the dissolved oxygen concentration in bulk liquid phase and K is the mass transport coefficient. Theoretically, $\eta = 1$ but experimental measurements showed $\eta \leq 1$. Postlethwaite et al. [39] believed that only 2/3 of dissolved oxygen reaching the wall is used in oxidizing the iron into ferrous ions and that the rest is used in the oxidization of the ferrous ions to ferric ions close to the wall, so that $\eta = 2/3$ [43].

As shown in Fig. 1, $\beta \approx 0.75$ for corrosion of carbon steel exposed to flowing tailing water free of solid particle, a dilute solution with chemical composition shown in the table inserted in figure, and 0.55 in following tailing slurries [44]. If we assume electrochemical corrosion in flowing tailing water is dominated by mass transfer of dissolved oxygen in liquid phase, $\beta \approx 0.75$ for a system of carbon steel - flowing water. The presence of solid particles in liquid phase is unlikely to change mass transport of dissolved oxygen significantly, suggesting that the corrosion rate in the flowing slurry is not proportional to but a power function of the mass transport coefficient [2,10]

$$\dot{c} \propto K^{b} \propto Re^{b\beta} \tag{15}$$

where $b(> 0)$ is an experimental constant. $b \neq 1$ implies that some factors other than the mass transport may affect corrosion process. According to Poulson [2] $b < 1$ is due to corrosion controlled by the partial activation of target surface. The local activation may result from the removal of corrosion scale in fluid and/or dynamic plastic strain in the surface layer caused by impingement of solid particle. For carbon steel exposed to the tailing water, corrosion scale is loose and less protective, dynamic plastic strain of surface layer is likely the dominant mechanism. According to Eq. (9), anodic dissolution rate of plastic straining electrode will increases linearly with plastic strain rate. The plastic strain caused by solid particle impingement is non-uniform and time-dependent. It is technically difficult to formulate its value but reasonable to assume that the average plastic strain rate is a power function of erosion rate [4,25],

$$\bar{\dot{\gamma}}_p \propto \dot{e}^{n} \tag{16}$$

Thus the corrosion augmentation defined by Eq. (6) will be correlated to erosion rate as follows

$$\text{Corrosion augmentation} = 1 + \epsilon \frac{\dot{e}^{n}}{\dot{c}_0} \tag{17}$$

ϵ is an experimental constant. Above relation is validated by the experimental data summarized in Fig. 2.

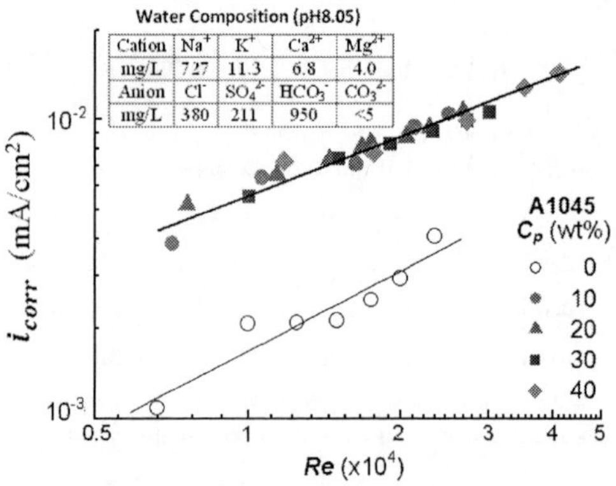

Fig. 1. The dependence of corrosion rate on the Reynolds number. (Target material: carbon steel A1045; slurry: tailing water of oil sand production + silica sand, RCE system) 0.55/slurry, 0.75 solution [44].

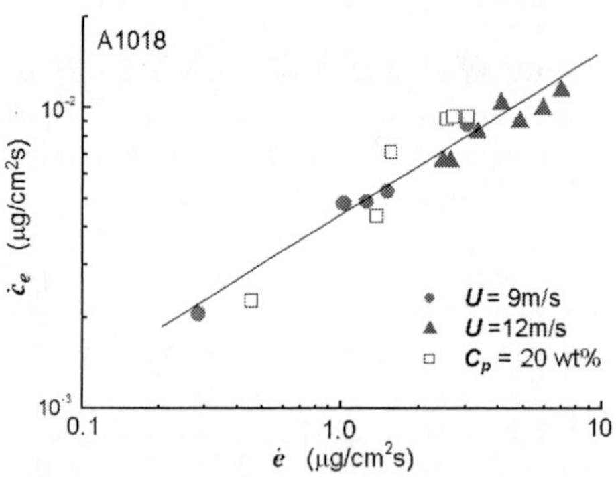

Fig. 2. Dependence of corrosion augmentation in the tailing slurry on erosion rate [4].

3.2 Electrochemical Corrosion of Passivated Target in Flowing Slurry

When a corrosion scale exists on the surface, the apparent mass transport coefficient K is formulated as follows [45]

$$\frac{1}{K}=\frac{1}{K_B}+\frac{1}{K_F} \tag{18}$$

where K_B and K_F are the mass transport coefficients in the boundary layer and the corrosion scale, respectively. If the metal is under the passive condition, the mass transfer in the passive film will be much slower than that the liquid phase, $K_B>>K_F$, and $K \approx K_F$. If the fluid does not induce the breakdown of passive film, the corrosion of iron-based alloys is controlled by the diffusion of oxygen vacancy within the passive film [46,47] and hence the corrosion rate is controlled by the density and diffusion coefficient of oxygen vacancy density within the passive film [48]. Frequently, the passive film is partially destroyed by erosive fluid. In this situation, anodic dissolution rate will be a function of hydrodynamics of slurry and repassivation of target.

When the kinetic energy of solid particles in flowing slurry exceeds a threshold value, the particle impingement will remove a small piece of passive film and produce a crater. It will lead to a sharp rise of local corrosion current over the crate surface to a peak value, i_{peak}. Then the local current will decay with time ($\hat{t}$) because of repassivation [49]. As a result, the corrosion current density over target surface that is impacted by slurry is no longer uniform and the average corrosion current density will depends on the rate of passive film removal and repassivation kinetics. A predictive model has been developed for corrosion rate at the steady state for a target subjected to slurry impingement on the basis of following hypotheses [29,50].

(1) The target material is uniform;

(2) All solid particles in slurry are identical in size, mass, geometry, velocity and moving direction, so that the craters generated by solid particle impingement are identical in size and geometry and uniformly distribute over the electrode surface that is given as $\overline{A}_{crater}$;

(3) Passive film on a new crater surface is fully destroyed by the solid particle impingement;

(4) Under a given test condition, the local current density response over the crater surface is uniquely formulated as $i = i(\hat{t})$. $\hat{t}$ is time elapsed after passive film is ruptured by particle impingement;

(5) The repassivation–kinetics is independent of hydrodynamics of slurry and is not affected by repeated impingements of solid particles.

Consequently, the average anodic current density will be given by [29,50,51]

$$\bar{i}=\int_0^1 i(\hat{t})\,d\hat{A} \tag{19}$$

In accord with the hypotheses (4) and (5), the current density at any location on the electrode surface at any moment is determined by the repassivating period of this point. In Eq. (19), $\hat{A}$ is the surface area fraction where the local current density is $i(\hat{t})$. As the repassivating time $\hat{t}$ increases, $\hat{A}(\hat{t})$ will decrease due to successive impingements of solid particles. $d\hat{A}/d\hat{t}$ is proportional to correlated to $\hat{A}(\hat{t})$ because the probability of target surface to be impacted by a solid particle is proportional to the surface area. Defining $\dot{A}_e$ as the generation rate fresh (crater) surface of a target with unit surface area, we will have

$$\frac{d\hat{A}}{d\hat{t}}=-\dot{A}_e\hat{A} \tag{20}$$

Considering $d\hat{A}/d\hat{t}\big|_{\hat{t}=0}=\dot{A}_e$

$$\hat{A}=\exp\left(-\dot{A}_e\hat{t}\right) \tag{21}$$

Accordingly

$$d\hat{A}=-\dot{A}_e\exp\left(-\dot{A}_e\hat{t}\right)d\hat{t} \tag{22}$$

Substituting Eq. (22) into Eq. (19), we have

$$\bar{i} = \dot{A}_e \int_0^{\infty} i(t) \exp\left(-\dot{A}_e \hat{t}\right) d\hat{t} \tag{23}$$

In accord with hypotheses 1 through 3

$$\dot{A}_e = \frac{C_p V_p \sin\theta \, \bar{A}_{crater}}{\bar{m}_p} \tag{24}$$

where i is a function of the repassivation kinetics, C_p and $\bar{m}_p$ are the concentration (kg/m^3) and average mass (kg) of solid particle, respectively, V_p and θ are the impingement velocity and the impingement angle of slurry. The time-dependent anodic current density over the fresh surface $i(\hat{t})$ is the characteristic of the repassivation kinetics. The kinetic mode of repassivation may vary with the material-environment system and hydrodynamics of corrosion media, which is commonly determined with the single particle impingement [51,52], scratch electrode [29,53] and strained electrode methods [54]. Since $\bar{A}_{crater}$ can be directly measured from the eroded surface, $\bar{i}$ will be predictable after the repassivation kinetics, $i(\hat{t})$, is known.

The single particle impingement measurement shown in Fig.3 indicated that repassivation of 304 stainless steel obeyed the bi-exponential law [29,51],

$$i = i_S + i_1 \exp\left(-\frac{\hat{t}}{\tau_1}\right) + i_2 \exp\left(-\frac{\hat{t}}{\tau_2}\right) \tag{25}$$

where the second term in Eq. (25) (i_1, τ_1) relates to certain quickly decaying processes such as the formation of a passive film with monolayer thickness on a bared crater surface, and the third term (i_2, τ_2) relates to a slowly decaying process for growth of a passive film [55], $i_1 + i_2 = i_{peak}$, i_{peak} is the peak response of local current density over the crate surface to the particle impingement; i_S is the stable current density in the flowing water free of sand.

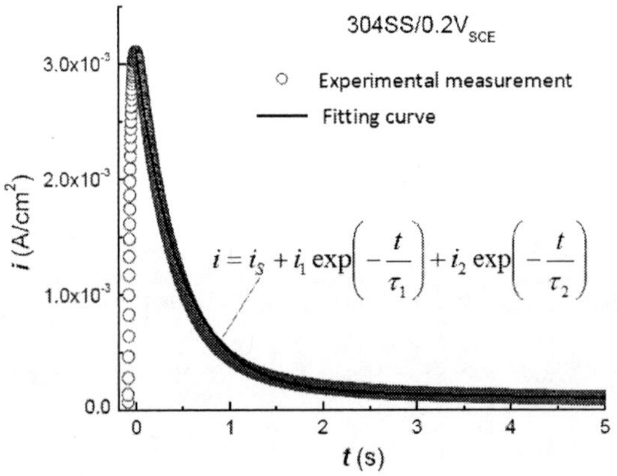

Fig. 3. A typical current response of 304 stainless steel target to single particle impingement [51].

The experimental measurements in Fig. 4 indicated that the repassivation kinetics of stainless steel was not affected by hydrodynamics of slurry.

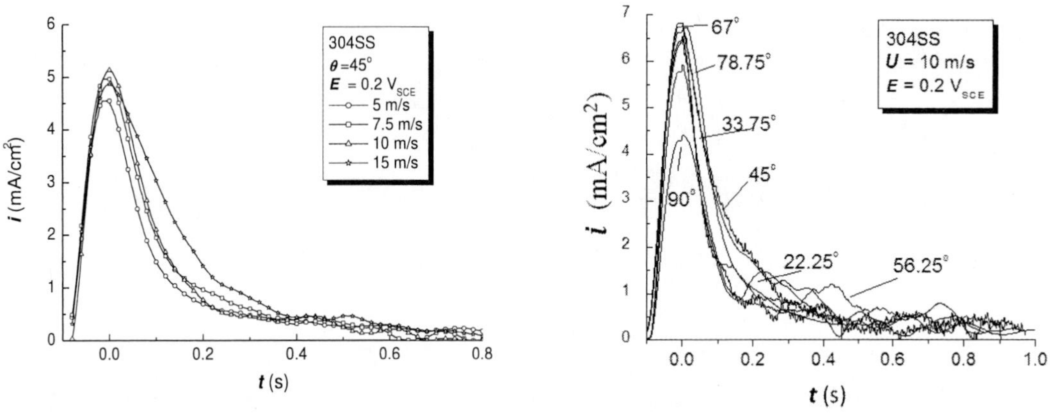

(a) Effect of impingement velocity. (b) Effect of impingement angle.

Fig. 4. Current density response to solid particle impingement with different velocities and angles [51].

In this case, the corrosion current density in flowing slurry is formulated as follows by inserting Eq. (25) into Eq.(23) and integrating [29,51]

$$\bar{i} = i_S + i_1\frac{\lambda_1}{1+\lambda_1} + i_2\frac{\lambda_2}{1+\lambda_2} \tag{26}$$

The non-dimensional parameters are $\lambda_1 = \tau_1\dot{A}_e$ and $\lambda_2 = \tau_2\dot{A}_e$ that represent the combined effects of the hydrodynamic conditions and repassivation kinetics. An example in Figs. 5, 6, and 8 indicates that Eq. (26) gives a good prediction to slurry hydrodynamics on corrosion current density of 304SS in the flowing slurries. i_S can be regarded as the corrosion rate under the erosion-free condition, so that the corrosion augmentation defined by ASTM G119 is given by

$$\text{Corrosion augmentation} = \frac{\bar{i}}{i_S} = \frac{\dot{c}}{\dot{c}_0} = 1 + \frac{i_1}{i_S}\frac{\lambda_1}{1+\lambda_1} + \frac{i_2}{i_S}\frac{\lambda_2}{1+\lambda_2} \tag{27}$$

The measurements using scratching electrode [29,53] and strained electrode methods [54] indicated that the repassivation of carbon steel the borate buffer solution followed the power law,

$$i = i_S + i_{Peak}\left(\frac{\hat{t}}{\tau_0}\right)^{-m} \qquad \hat{t} > \tau_0,\ 0 < m < 1 \tag{28}$$

where i_S is the current density at the steady state; i_{Peak} is the current density immediately after passive film rupture; m and τ_0 are experimental constants. Substituting Eq. (28) into Eq. (23), we will have [29,53]

$$\bar{i} = i_s + i_{peak}(1 - e^{-\lambda_0}) + i_{peak}\,\lambda_0^m\gamma(m,\lambda_0) \tag{29}$$

where $\lambda_0 = \tau_0\dot{A}_e$ is a dimensionless parameter. $\gamma(m,\lambda_0)$ is the gamma function

$$\gamma(m,\lambda_0) = \int_{\lambda_0}^{\infty} u^{-m}e^{u}du \tag{30}$$

According to electrochemical measurements: $\tau_0 \le 0.01$ s [29,53,54]. In practical application, $V_p \le 10$ m/s, $C_P \le 500$ kg/m^3, the sand particle diameter is less than 0.5 mm, the density of sand particle is 2,560 kg/m^3, and $\bar{A}_{crater} \le 1\times10^{-10}$m^2. Substituting these value into Eq. (24), we have $\lambda_0 < 0.07 << 1$. Thus, $1 - e^{-\lambda_0} \approx 1 - \lambda_0$ and

$\gamma(m,\lambda_0) \approx \Gamma(m) = \int_0^{\infty} u^{-m}e^{u}du \approx \frac{1}{1-m}$ [29,53]. Considering $0 < m < 1$, Eq.(29) can be approximately simplified into

$$\bar{i} \approx i_S + \frac{i_{peak}}{1-m}\lambda_0^m \tag{31}$$

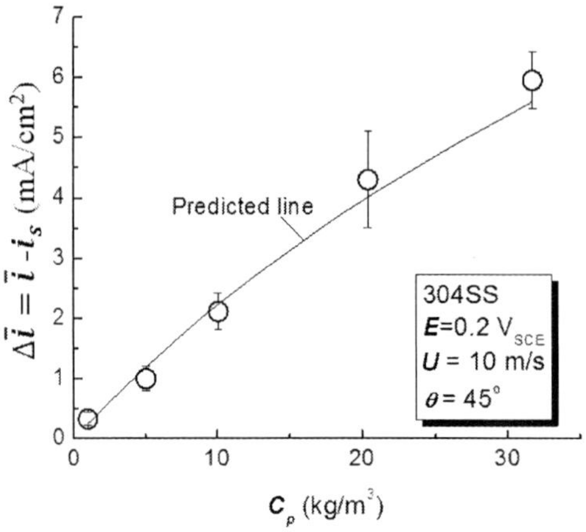

Fig. 5. Comparison between theoretical prediction and experimental results of corrosion current density of 304SS under slurry impingement (Slurry: Tap water + sand, Impingement velocity: 5m/s, Impingement angle: 45°) [51].

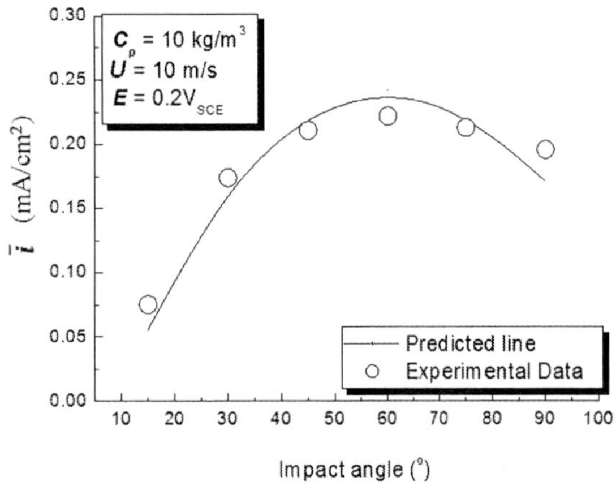

Fig. 6. Comparison between theoretical prediction and experimental results of corrosion current density of 304SS under slurry impingement with different impact angles (Slurry: Tap water + sand, Impingement velocity: 5m/s) [51].

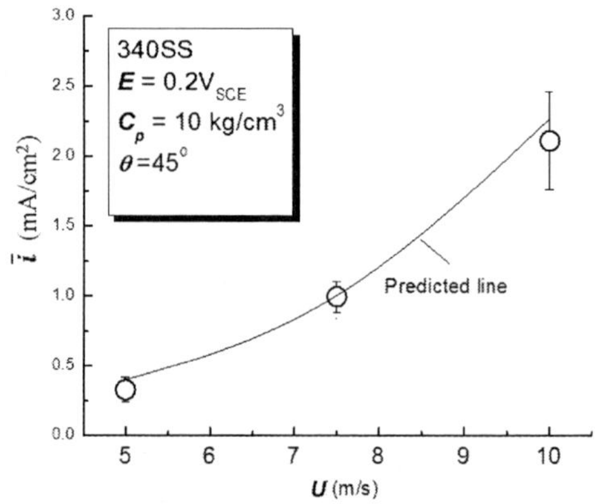

Fig. 7. Effect of slurry impingement velocity on corrosion of 304SS [51].

Eq. (31) can be used to estimate corrosion rate in flowing slurry when $\lambda_0 = \tau_0 \dot{A}_e$ is known [29,51]. When erosion-corrosion experiments were carried out with slurry pipe loop or RCE systems, only a portion of solid particles in slurry have a chance to impact the target. Moreover, the distribution of solid particles in flowing slurry depends heavily on the flowing velocity. It will be non-uniform under action of gravity or centrifugal force because sand particles have a higher density than water does. The flowing velocity over the target surface may also differ from the nominal flowing velocity of slurry. As such, it will be technically difficult to estimate $\dot{A}_e$ with Eq. (24).

In the case that the particle impingement is the sole source of erosion, the total material loss rate caused by the erosion $\dot{e}$ should be proportional to the product of solid particle impingement frequency and the average volume of a material removed by a single impingement, $\bar{V}_{crater}$. To correlate corrosion rate in flowing slurry to erosion wastage, we defined the surface/volume ratio of crater as $y = \bar{A}_{crater} / \bar{V}_{crater} = \rho \dot{A}_e / \dot{e}$ where ρ is the density of target material. Thus, λ_0 is given by

$$\lambda_0 = \frac{\tau_0 y \dot{e}}{\rho} \tag{32}$$

where i_{peak} and $\tau_0$can be determined from measurements of scratch electrode or single particle impingement experiment; the surface/volume ratio of crater y can be estimated from crater geometry measurements over the eroded target surface. Substituting Eq.(32) into Eq.(31), the average corrosion current density can be formulated by

$$\bar{i} \approx i_S + \frac{i_{peak}}{1-m}\left(\frac{\tau_0 y\dot{e}}{\rho}\right)^m \tag{33}$$

As typically demonstrated by Fig. 8, Eq. (33) gives reasonably good predictions to the dependence of anodic current density measured with the RCE systems on the erosion wastage, as well as the effect of chloride concentration and applied potential. More examples for validating the model described above can be found in Refs.[29,48,50,51,52,53]. It was demonstrated that Eq.(33) was also applicable for estimate the effects and erodent concentration and impingement velocity.

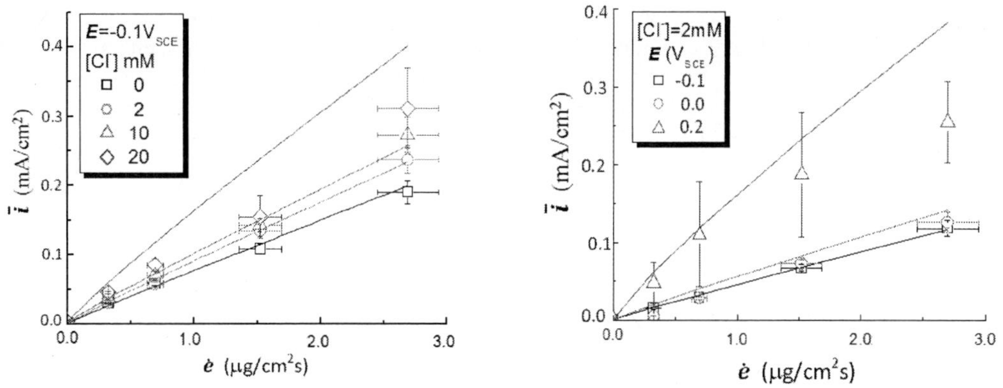

(a) Effect of chloride concentration. (b) Effect of applied potential.

Fig. 8. Comparison between theoretical prediction and experimental results of corrosion current density of 1045 carbon steel (Test system: RCE; slurry: buffered borate solution (pH8.4) + 30 wt% silica sand, U = 12 m/s) [53].

4. EROSION IN CORRODING LIQUIDS

4.1 Erosion Resistance and Mechanical Properties of Target Materials

Many mechanical erosion models have been established to correlate the erosion resistance of target materials to their mechanical properties and hydrodynamic parameters [56,57]. A detailed literature review on this aspect is out of scope of this article. Meng and Ludema [57] provide an exhaustive overview up to 1995, found 182 equations and selected 28 for special study. Lyczkowski and Bouillard [58] gave one up to 2002. As pointed out by Tsai et al. [59], over a fairly wide range of variables, at least in flowing slurries, the overall dependence of the particle and target hardness (H_P and H respectively),on erosion is approximately given by $\dot{e}_0 \propto H_P^{1/2}/H$. Generally, the erosion resistance of target materials increases with their hardness if no substantial change takes place in the erosion mechanisms and the power law erosion rate of materials can give a reasonably good fit to the correlation between the mechanical erosion rate and surface hardness [4,7,60,61].

$$\dot{e}_0 = \kappa_H H^{-n_H} \tag{34}$$

where κ_H and $n_H(>0)$ are experimental constants depending heavily on the erosion mechanisms.

4.2 Corrosion-Induced Degradation of Surface Mechanical Properties Caused by Anodic Dissolution

Generally, in Eq. (8), $L_{CP} \geq 0$. It suggests the existence of chemo-mechanical effect, i.e., anodic dissolution present on the surface will induce an extra dislocation flux in surface layer. Accelerated creep due to anodic dissolution was observed in austenitic steel [62,63], copper [64], magnesium alloy [65], cadmium [36] and brass [66]. The researchers found that the creep rate of samples at room temperature increased as anodic current was applied [36,62,64,66] and would recover its original value after the current-off [36,64]. Gutman [36], Revie and Uhlig [64] believed the anodic generation of a supersaturation of subsurface vacancies was responsible for this phenomenon. Jones et al. [62,67] reported the experimental evidence that the anodic dissolution promoted formation and diffusion of vacancies in the surface layer. The generation of vacancies could attenuate the strain-hardening subsurface and enhance the localized surface plasticity [62].

The injection of vacancies from surface due to the anodic dissolution is likely to bring about a chemical potential gradient for the dislocations in the surface layer [66]. The chemical potential gradient may result in an additional dislocation flux in this layer [36]. This has been confirmed by the experimental observations [36,68,69]. Due to the fact that the mean free path of dislocation possesses a limited value, only the dislocation situated in the surface layer with a depth less than or equal to the mean free path can reach the surface, so that the chemo-mechanical effect can affect mainly the mechanical properties in this surface layer. This point has been confirmed by the fact that the anodic dissolution–enhanced creep is more readily observed in stressed thin wires or foils [36,63,64]. Substituting Eqs. (10) and (11) into Eq.(8)

$$\dot{\gamma}_p = L_p R_D \left[\exp\left(\frac{\tau V_D}{R_D T}\right) - 1\right] + L_{CP} R \left[\exp\left(\frac{\alpha \tilde{A}}{RT}\right) - 1\right] \tag{35}$$

Imagining two tensile experiments under the constant strain rate control: test 1 is conducted in an inert environment and test 2 is performed in a corroding medium. In line with Eq. (9), the anodic current density during the testing is $i_A \left\{= L_C R \left[\exp\left(\frac{\alpha \tilde{A}}{RT}\right) - 1\right]\right\}$. The flowing stress of surface layer produced by the strain hardening in the inert environment is denoted as τ_1 and that measured in the corroding medium is τ_2. Owing to the chemo-mechanical effect, $\tau_2 < \tau_1$. Let $\Delta\tau_A = \tau_2 - \tau_1$. Adjust the anodic current density to eliminate strain hardening effect so that the plastic strain rate induced by the anodic dissolution in the corroding environment is equal to the one produced by the mechanical force in test 1. In this case, $\tau_2 = 0$ and $\Delta\tau_A = \tau_1$ and Eq. (35) will become

$$L_P R_D \left[\exp\left(\frac{\Delta\tau_A V_D}{R_D T}\right) - 1\right] = L_{CP} R \left[\exp\left(\frac{\alpha A}{RT}\right) - 1\right] = \frac{L_{CP}}{L_C} i_A \tag{36}$$

According to Eq. (36), the change of flow stress in surface layer induced by corrosion $\Delta\tau_A$ is given by [4,36]:

$$\Delta\tau_A = -\frac{R_D T}{V_D} \ln\left(\frac{i_A}{i_{th}}\right) \tag{37}$$

$i_{th} \left(= \frac{i_A^0}{R_D} \frac{L_C L_P}{L_{CP}}\right)$ is the threshold current density to cause the surface strength degradation, where i_A^0 is the anodic current density at the equilibrium state. Eq. (37) indicates that the anodic dissolution present at surface will reduce the surface strength. Because the effect of corrosion-induced surface strength will decay with increasing depth to surface [36], it is difficult to measure $\Delta\tau_A$ directly. If the yield behaviour of target material is assumed to obey von-Mises flowing rule and that Tabor's factor is equal to 3 [70], we will have

$$\Delta Hv = Hv_c - Hv = -B \ln\left[\frac{i_A}{i_{th}}\right] \tag{38}$$

where Hv and Hv_c is the surface hardness value measured in air and corroding environment, respectively; $\mathrm{B} = \mathrm{R_D T}/\sqrt{3}\mathrm{V_D}$ and i_{th} the threshold current to cause the surface strength degradation. Eq. (38) has been validated by the in-situ microhardness measurement, as is typically shown in Fig. 9, where microhardness decreases about 15% when anodic current density of 1 mA/cm^2 is applied.

Experimental evidence indicated that the typical value of crater depth generated by slurry erosion was around 30~40 nm, which is much smaller than the indent depth generated by microhardness measurement (~ µm). Therefore, nanoindentation is the most suitable technique at the this moment to evaluate the effects of various factors on surface hardness although the indent depth generated by nanoindentation may be up to more than a few hundred nm, still larger than the crater depth produced by slurry erosion, as typically shown in Fig. 10. The surface hardness measured with nanoindentation method a function indent depth. Because a thinner layer of material was involved in the plastic deformation process, the relative surface hardness drop was be up to 38% ~45% when applied anodic current density is same (Fig.10) [4, 25,31, 71,72].

Generally, the surface hardness is reduced as the indentation depth, contact radius or plastic zone size is increased. This is attributed to an indentation size effect, i.e., the resistance of material against the plastic deformation increases with decreasing volume of material. It is known that plastic deformation of crystals results from slip of dislocations and the quantity of dislocations in crystal increases with volume of material and, further, during micro-nanoindentation, sharp deformation gradients are developing in the neighborhood of the indenter's tip. One way to evaluate the indentation size effect is to resort to mechanism based dislocation arguments by utilising, for example, Ashby's concept of geometrically necessary dislocations, the density of which is directly related to plastic strain gradients [73]. Following this approach, Nix and Gao [70] proposed a dislocation model that could offer a semi-quantitative interpretation for the indentation size effect observed in indentation measurement. Another mechanics based approach is to utilise a gradient plasticity formulation, for example, the one advanced by Aifantis and co-workers [74], which, in conjunction with Johnson's cavity model, seems to be more flexible in interpreting the indentation size effect. For very small indentation depths, plastic flow does not occur until the equivalent strain, in the gradient elasticity region surrounding the hydrostatically pressurized core beneath the indenter's tip, reaches a critical yield value. In this latter case, the observed indentation size effect is interpreted through gradient elasticity considerations [75].

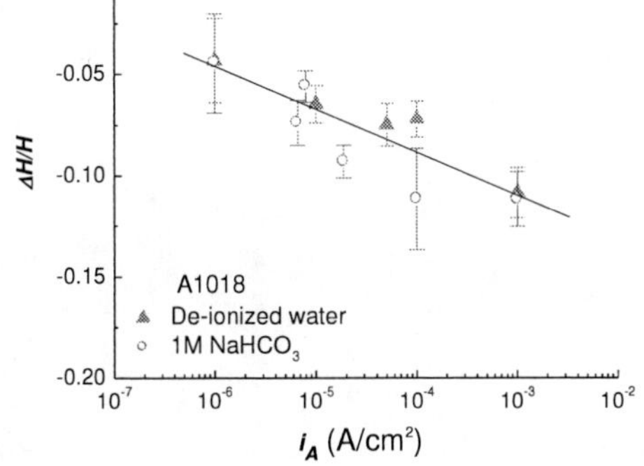

Fig. 9. Typical relationship between the normalized hardness drop and anodic current density [4].

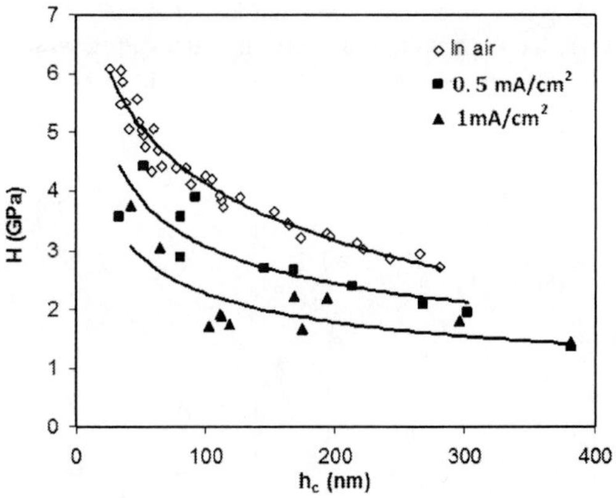

Fig. 10. Effect anodic current density on the *in-situ* surface hardness measured with nano-indentation technique[72].

4.3 Corrosion- Enhanced Erosion due to Anodic Dissolution

If the increasing erosion rate caused by the anodic dissolution-induced hardness degradation is the only mechanism of corrosion-enhanced erosion, the increment of erosion rate $\Delta\dot{e}$ due to the presence of anodic dissolution can be defined as the corrosion-enhanced erosion, namely,

$$\dot{e}_c = \Delta\dot{e}_0 = \dot{e} - \dot{e}_0 \tag{39}$$

Substituting Eq.(34) into Eq.(39) gives

$$\dot{e}_c = \Delta\dot{e}_0 = \kappa_H (Hv_c^{-n_H} - Hv^{-n_H}) = \dot{e}_0\left[\left(\frac{Hv_c}{Hv}\right)^{-n_H} - 1\right] \approx -n\dot{e}_0\left(\frac{\Delta Hv}{Hv}\right) \tag{40}$$

where $\Delta Hv = Hv_c - Hv \leq 0$. In the approximation at right end of Eq. (40), the first Taylor polynomial was adopted. The wastage due to corrosion-enhanced erosion can be correlated to anodic current density by inserting Eq. (38) into Eq. (40) [4,25,31],

$$\frac{\dot{e}_c}{\dot{e}_0} \approx Z\log\left(\frac{i_A}{i_{th}}\right) \tag{41}$$

Z is an experimental constant. It has been shown the prediction of Eq. (41) agrees well with the experimental results obtained of carbon steels, as shown in Fig.11 [4, 25].

The practical engineering, the slurry pipe is normally operated under open circuit potential (OCP). The experimental results indicated that the corrosion-enhanced erosion at the OCP was predictable using the curve obtained under galvanostatic control as the corrosion current density at the OCP is known. According to Eq. (41), the erosion will be enhanced by the chemo-mechanical effect when corrosion occurs simultaneously and the erosion augmentation defined by ASTM G119 $(=1+\dot{e}_c/\dot{e}_0)$ will be approximately a linear function of the logarithm of anodic current density.

$$\text{Erosion augmentation} = 1+\frac{\dot{e}_c}{\dot{e}_0} = 1+Z\log\left(\frac{i_A}{i_{th}}\right) \tag{42}$$

As shown in Fig.11, the corrosion-enhanced erosion is also affected by the concentration of solid particles in slurry, as flowing velocity as well, when the anodic current density is held unchanged. A possible mechanism may be attributed to the non-uniform distribution of anodic dissolution rate in micro-scale. The transient current peaks resulted from the impingement events of solid particle and the local anodic current density i_A^L over the fresh surface of wear scares during scratching by the impingement of sand particles was much higher than the average current density on the whole electrode surface i_A [55,76,77]. The corrosion-enhanced erosion occurs while sand particles are scratching the electrode surface, i.e., it is be governed by the local current densities. The difference between i_A^L and i_A decreases with increasing impingement frequency of solid particles. Another important mechanism may be related to the enhanced plastic deformation in surface layer due to the increases in frequency impingement and momentum of solid particles, which will lead to an increase of surface hardness due to the strain hardening effect. The both effects can be assessed with the erosion rate. As is indicated by Fig.12, the impact the solid concentration and flowing velocity on the corrosion-enhanced erosion can be predicted when the normalized wastage ratio $\dot{c}/\dot{e}_0$ is employed to replace anodic current density.

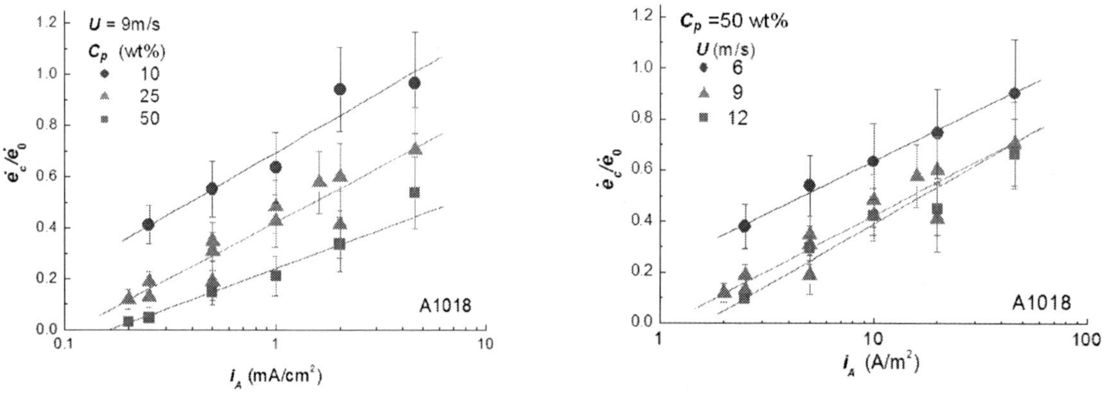

(a) Effect of solid particle concentration (b) Effect of flowing velocity

Fig. 11. Effect of anodic current density on normalized corrosion-enhanced erosion rate [4].

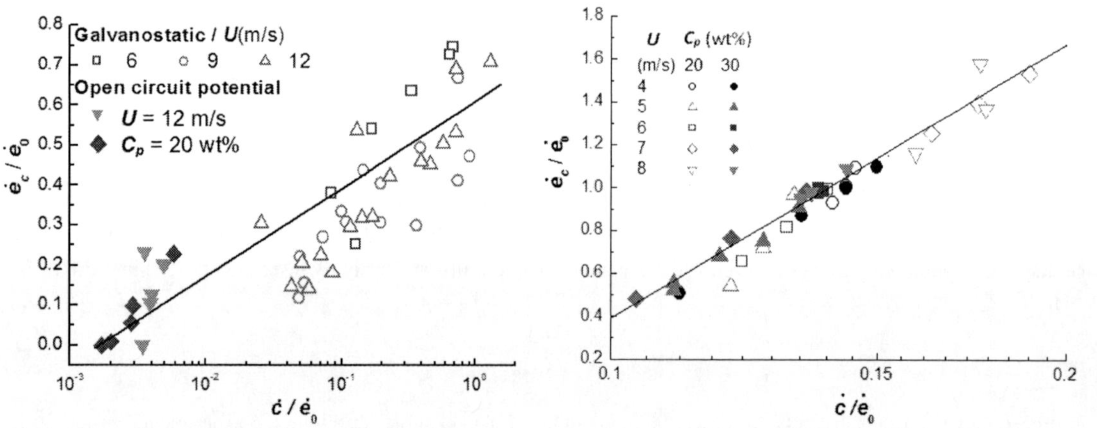

(a) A1018 CS [4] (b) A1045CS [29]

Fig. 12. Correlation between normalized mechanical erosion rate and wastage ratio $\dot{c}/\dot{e}_0$ under galvanostatic control [4,31].

5. SUMMARY

Erosion-corrosion is big concern in oil and gas production. The physical model developed in authors' group can be used to quantitatively evaluate the synergism of electrochemical and mechanical factors in erosion-corrosion. The effects of hydrodynamics of flowing slurry on anodic dissolution rates of metals were theoretically and experimentally evaluated. The economic loss due to erosion-corrosion can be minimized by utilizing proper design, material selection and chemical control.

6. ACKNOWLEDGMENTS

The authors would like to appreciate financial support from the Natural Sciences and Engineering Research Council of Canada (NSERC), the research contributions from Dr. Junfang Lu, Dr. Haixia Guo, Mr. Ke Wang, and Ms. Xuemei Wan.

7. REFERENCES

1 G. T. Burstein and K. Sasaki, Electrochim. Acta, 46 (2001) 3675-3683.

2 B. Poulson, Corrosion Science, 23 (1983) 391-430.

3 J. Postlethwaite, M. H. Dobbin and K. Bergevin, Corrosion, 42 (1986) 514-524.

4 B. T. Lu and J. L. Luo, J. Phys. Chem. B, 110 (2006) 4217-4231.

5 ASTM G 15-05, Standard Terminology Relating to Corrosion and Corrosion Testing, Annual Book of ASTM Standards, Vol.03.02, Wear and Erosion, Metal Corrosion, West Conshohocken, PA, 2005, pp.65-68.

6 J. Postlethwaite and S. Nesic, Erosion-Corrosion in Single and Multiphase Flow, Uhlig's Handbook, 2nd Edition by W. Revie, 2000, John Wiley & Sons, Inc., 249-272.

7 R. J. K Wood, Wear, 261 (2006) 1012-1023.

8 P. McLintyre, Marine Corrosion Club Meeting, Aberdeen, April, 1999.

9 R. J. K. Wood, T. F. Jones, N. J. Miles and J. Ganeshalingam, Wear, 250 (2001) 770-778.

10 B. Poulson, Wear, 233-235 (1999) 497-405.

11 M. M. Stark, Intern. Mater. Rev., 50 (2005) 1-17.

12 M. M. Stark, N. Corlett and S. Turgoose, Wear, 233-235 (1999) 535-541.

13 ASTM G119-04, Standard Guide for Determining Synergism between Wear and Corrosion. West Conshohocken, PA, 2006, 519-523.

14 ASTM G59-97, Conducting Potentiodynamic Polarization resistance Measurement, West Conshohocken, PA, 2005, pp.519-523.

15 ASTM G5-94, Making Potentiostatic and Potentiodynamic Anodic Polarization Measurements, West Conshohocken, PA, 1994.

16 J. Postlethwaite, Corrosion, 37 (1981) 1-5.

17 A. Neville and X. Hu, Wear, 250-251 (2001) 1284-1294.

18 Y. Li, G.T. Burstein, and I.M. Hutchings, Wear, 186-187 (1995) 515-522

19 P. Christodoulou, A. Drotlew and W. Gutowski, Wear, 211 (1997) 129.

20 V. A. Pugsley, and C. Allen, Wear, 233-235 (1999) 93.

21 M. C. Wang, S. Z. Ren, X. B. Wang and S. Z. Li, Wear, 160 (1993) 259.

22 H. M. Clark, Wear, 152 (1992) 223-240.

23 W. Blatt, W. Kohloey, U. Lotz and E. Heitz, Corrosion, 45 (1989) 793.

24 I. M. Hutchings, The Erosion of Materials By Liquid Flow, MTI Publication No. 25, Materials Technology Institute of the Chemical Process Industries, Inc., 1986.

25 B. T. Lu, J. F. Lu and J. L. Luo, Corros. Sci. 53 (2011) 1000-1008.

26 M. Bjordal, E. Bardal, T. Rogne and T. G. Eggen, Surface and Coatings Technology, 70 (1005) 215.

27 S. Das, D. P. Mondal, O. P. Modi and R. Dasgupta, Wear, 231 (1999) 195.

28 B. K. Prasad, Wear, 238 (2000) 151.

29 B. T. Lu and J. L. Luo, Electrochimica Acta, 53 (2008) 7022-7031.

30 A. Neville and T. Hodgkiess, British Corrosion Journal, 32 (1997) 197.

31 B. T. Lu, K. Wang, X. M. Wan, J. L. Luo, Tribology International, 83 (2015), 146-155.

32 K. Wang, B. T. Lu, J. L. Luo, Canadian Metallurgical Quarterly, 50 (2011) 181-185.

33 ASTM G76-05, Standard Test Method for Conducting Erosion Tests by Solid Particle Impingement Using Gas Jet, West Conshohocken, PA, 2006, pp.310-315.

34 ASTM G134-95, Standard Test Method for Erosion of Solid Materials by a Cavitating Liquid Jet, West Conshohocken, PA, 2006, pp.559-571.

35 ASTM G72-04, Standard Practice for Liquid Impingement Erosion Testing, West Conshohocken, PA, 2006, pp.273-290.

36 E. M. Gutman, Mechanochemistry of Materials, Cambridge International Science Publishing, Great Abington, Cambridge, UK, 1998.

37 J. Friedel, Dislocations, Pergamon Press, New York, USA, 1964.

38 D.C. Silverman, Corrosion, 60 (2004) 1003-1024.

39 J. Postlethwaite, M. H. Dobbin and K. Bergevin, Corrosion, 42 (1986) 514-521.

40 T.V. Chen, A. A. Moccari and D. D. Medonald, Corrosion, 48 (1992) 239-255.

41 E. Heitz, Corrosion, 47 (1991) 135-145.

42 J.S. Newman and K. E. Thomas-Alyea, Electrochemical Systems, Wiley-IEEE, 2004, 378-400.

43 C. Wang, A. Neville, S. Ramachanfran and V. Jovancicevic, Wear, 258 (2005) 649-658.

44 C.W. Chen, T. Yu, B. T. Lu, J. L. Luo, Corrosion in Waster Slurry Transportation, unpublished data, University of Alberta, 2004, Edmonton, Alberta, Canada.

45 E. Heitz, Corrosion, 47(1991) 135-145.

46 D.D. Mcdanold, J. Electrochem. Soc., 196 (1992) 3434-3449.

47 P. Andrews, I. F. Illson and S. J. Matthews, Wear, 233-235 (1999) 568-574.

48 H. X. Guo, B. T. Lu, J. L. Luo, Electrochimica Acta, 52 (2006) 1108-1116.

49 K. Sasaki and G. T. Burstein, Philosophical Magazine Letters, 80 (2000) 489-493.

50 B. T. Lu, J. L. Luo and H. Y. Ma, J. Electrochem. Soc., 154 (2007) C159-C168

51 B. T. Lu, L. C. Mao, J. L. Luo, Electrochimica Acta, 56 (2010) 85-92.

52 F. Mohammadi, J. L. Luo, B. T. Lu, A. Afacan, Corrosion Science, 52 (2010) 2331-2340.

53 B. T. Lu, J. L. Luo, H. X. Guo, L. C. Mao, Corrosion Science, 53 (2011) 432-404.

54 R. W. Staehle, Corrosion Science, 49 (2007) 7-19.

55. R. Oltra, B. Chapey, and L. Renaud, Wear, 186-187 (1995) 533.

56 A. Levy, Gas-Solid Particle Erosion and Erosion-corrosion of Metals, Uhlig's Handbook, 2nd Edition by W. Revie, 2000, John Wiley & Sons, Inc., 273-293.

57 H.C. Meng and K. C. Ludema, Wear, 181-183 (1995) 443-457.

58 R.W. Lyczkowski and J. Bouillard, Progress in Energy and Combustion Science, 28 (2002) 543-602.

59 W. Tsai, J. A. C. Humphrey, I. Cornet and A. Levy, Wear, 68 (1981) 289-303.

60 G. Sundararajan, Wear, 186-187 (1995) 129-144.

61 B.T. Lu, J. F. Lu and J. L. Luo, Corros. Sci._ 53 (2011) 1000-1008

62 D. A. Jones, Corrosion, 52 (1996) 356.

63 Y. L. Huang, J. Mater. Sci. Tech., 17 (2001) S77-S80.

64 R.W. Revie, and H.H. Uhlig, Acta Metall., 22 (1974) 69.

65 E. M. Gutman, A. Eliezer, Ya Unigovski, and E. Abramov, Mater. Sci. Eng., A302 (**2001**) 63.

66 B. Gu, W.Y. Chu, L.J. Qiao, and C.M. Hsiao, Corrosion Science, 36 (1994) 1437.

67 D.A. Jones, and A.F. Jankowski, Scripta Metall. Mater., 22 (1993), 701.

68 K. Lian, and E.I. Meletis, Corrosion, 52 (1996) 347.

69 D.J. Duquette, Corrosion, 46 (1990) 434.

70 W. D. Nix and H. Gao, J. Mech. Phys. Solid, 46 (1998) 411-425.

71 H. X. Guo, B. T. Lu and J. L. Luo, Electrochimica Acta, 51 (2005) 315-323

72 H. X. Guo, B. T. Lu, J. L. Luo, Electrochemistry Communications, 8 (2006) 1092-1098.

73 M. F. Ashby, Phil. Mag. 21 (1970) 399-424.

74 K. E. Aifantis, W. A. Soer, T. Th. M. De Hossen, J. R. Willis, Acta Mater., 54 (2006) 5077-508

75 G. P. Mokios, E. C. Aifantis, J. Mech. Behav. Mater., 20 (2011) 35-40.

76 R. J. K. Wood, J. A. Wharton, A. J. Speyer, and K. S. Tan, Tribology International, 35 (2002), 631.

77 C. W. Chen, B. T. Lu and J. L. Luo, In Environmental Degradation of Materials and Corrosion Control in Metals, Proc. COM'2003. Luo, J., Elboujdaini, M., Shoesmith, D. and Patnaik, P. C., Eds., Vancouver, Canada, 2003, p.43.

8. AUTHOR DETAILS

Baotong Lu, Got his Ph. D Degree in 1991. Conducted research on fatigue and fracture of metallic, ceramic and composite materials for 17 years when he worked in universities in China. Conducted research, since 2000, on corrosion theory, environmentally assisted cracking, erosion corrosion in university of Alberta, Canada, and Southwest Research Institute, USA. His current research area is modeling of corrosion fatigue crack growth and stress corrosion cracking (including sulfide stress cracking) in oil and gas production and transportation.

Dr. Jing-Li Luo is a professor in the Department of Chemical and Materials Engineering at the University of Alberta, Canada. Dr. Luo received her B. Eng. Degree from the Department of Physics and Chemistry of Metals, University of Science and Technology Beijing in 1982 and her Ph.D in Materials Science and Engineering at McMaster University, Canada in 1992. Dr. Luo has conducted extensive research on corrosion control and the development of green energy systems and has published over 280 papers in refereed journals and holds five US patents. She has served as a member of International Corrosion Council since 2005.

MEASUREMENT OF ALUMINUM OXIDE FILM IN SITU DURING ANODIZATION OF ALUMINUM BY FABRY-PÉROT INTERFEROMETRY

Khaled Habib[1], Waleed Mohammad[2], Farzia Karim[3], Joydeep Dutta[4]

[1]Materials Science &Photo-Electronics Lab, RE, EBR, KIS R, P.O.Box24885 Safat 13109 Kuwait
[2]Center of Research in Optoelectronics, Communication and Control System, School of Engineering, Bangkok University, Pathum Thani 12120, Thailand
[3]Center of Excellent in Nanotechnology, School of Engineering and Technology, Asian Institute of Technology, Pathum Thani12120, Thailand
[4]Chair of Functional Materials division at the KTH Royal Institute of Technology, Isafjordsgatan 22, Stockholm, Sweden

SUMMARY: A novel experimental scheme for real time measurement of aluminum oxide film during anodization was developed for the first time. The scheme was established based on a combination of a fiber optic sensor of Fabry-Pérot interferometry and direct current (DC) electrochemical methods. The scheme was assembled in a way to simultaneously anodize the aluminum samples and to measure the thickness of the aluminum oxide film. The anodization process of aluminum sample was carried out in 2% sulfuric acid (H_2SO_4) solution by the DC methods at room temperature. The estimated thickness of the aluminum oxide film by the novel scheme was verified by scanning electron microscopy (SEM) and holographic interferometry. This study shows that real time measurement of the thickness of aluminum oxide film by Fabry-Pérot interferometry (2.136μm) is feasible as it closely matched the thickness determined by SEM (2.1μm). The obtained values of the resistance by Fabry-Pérot interferometer (2.4 $\times 10^9$ Ohms) and by holographic interferometry (2.54x10^9 Ohms) were found twice the obtained value of the resistance by EIS (1.25x10^9 Ohms).

Keywords. Aluminum, Thickness of Oxide Film, Fabry-Pérot interferometry, Holographic Interferometry, Electrochemical Impedance Spectroscopy (EIS), and Sulfuric Acid.

1. INTRODUCTION

There are DC electrochemical methods of not only monitoring the growth of the aluminum oxide films [1-5], but also of measuring the corresponding resistance of the developed oxide films during the anodization processes of the aluminum in solutions. In addition, there are a few number of non-contact methods [6-8] of monitoring the growth of the aluminum oxide films. The noncontact methods are solely dependent on light as a quantum tool to traverse the information of the anodized surface of aluminum samples in aqueous solutions to an optical sensor for monitoring the growth of the aluminum oxide films. There are two types of optical sensors. The first type is the phase shift-optical sensor, i.e., holographic interferometry based on time or spatial domain. This type of optical sensor is dependent on a laser (coherent) light [6-8]. The other type is the spectral-optical sensor, i.e., Fabry–Pe´rot interferometry [9] based on frequency domain. This type of optical sensor is independent of a laser (coherent) light. In other words, the spectral optical sensor can work with a white (non-coherent) light. In fact, the phase shift-optical sensor such as holographic interferometry was not only used for monitoring the growth of the aluminum oxide films [6-8], but also it was used in the electrochemical field [10]. In contrast, the spectral optical sensor such as Fabry-Pérot interferometry has been only used to characterize optical thin films on glass/metallic substrates in three separate investigations [11-13] without attempting to measure the corresponding resistance to the formed films.

The present study has addressed for the first time the measurement of the resistance of aluminum oxide film during the anodization processes of aluminum sample in situ in aqueous solution, by Fabry–Pe´rot interferometry, using a noncoherent light, i.e., white light. The technique of Fabry-Pérot interferometric [9] is an efficient approach that can measure the resistance of thin films in situ in aqueous solutions. In addition, the technique, like other optical interferometry techniques [6-8, 10], is an environmentally friendly method in which no pollution, no contamination waste would be generated.

2. THEORETICAL ANALYSIS

The relationship between the electrical resistance of an oxide film and the thickness of the oxide film on a metal sample can be described from literature as follows [6-7]:

$$R= \rho\, L/A \tag{1}$$

where:

R is the electrical resistance of the oxide film,
ρ is the electrical resistivity of the Al_2O_3 film ($\rho = 2.5\text{x}10^{13}$ Ohm • cm) in 2%H_2SO_4 [10].
A is area of the sample ($A = 22.5\ cm^2$),
L is the thickness of the oxide film which can be obtained by Fabry-Pérot interferometry [11,13].

From Eq. (1) one can determine the electrical resistance from the thickness of the oxide film by Fabry-Pérot interferometry. This can be accomplished by substituting the obtained value of the thickness (L) of the oxide film in Eq.1. Detail procedures on obtaining the value of L is described elsewhere [11,13].

3. EXPERIMENTAL DETAILS

A fiber optic sensor based on a combination of the real-time Fabry–Pe´rot interferometry and DC electrochemical methods has been established to monitor the growth of aluminum oxide film during the anodization of the aluminum samples. Subsequently, the resistance of aluminum oxide film can be obtained during the formation of the film. The proposed system is schematically described in Figure 1a, b,c.

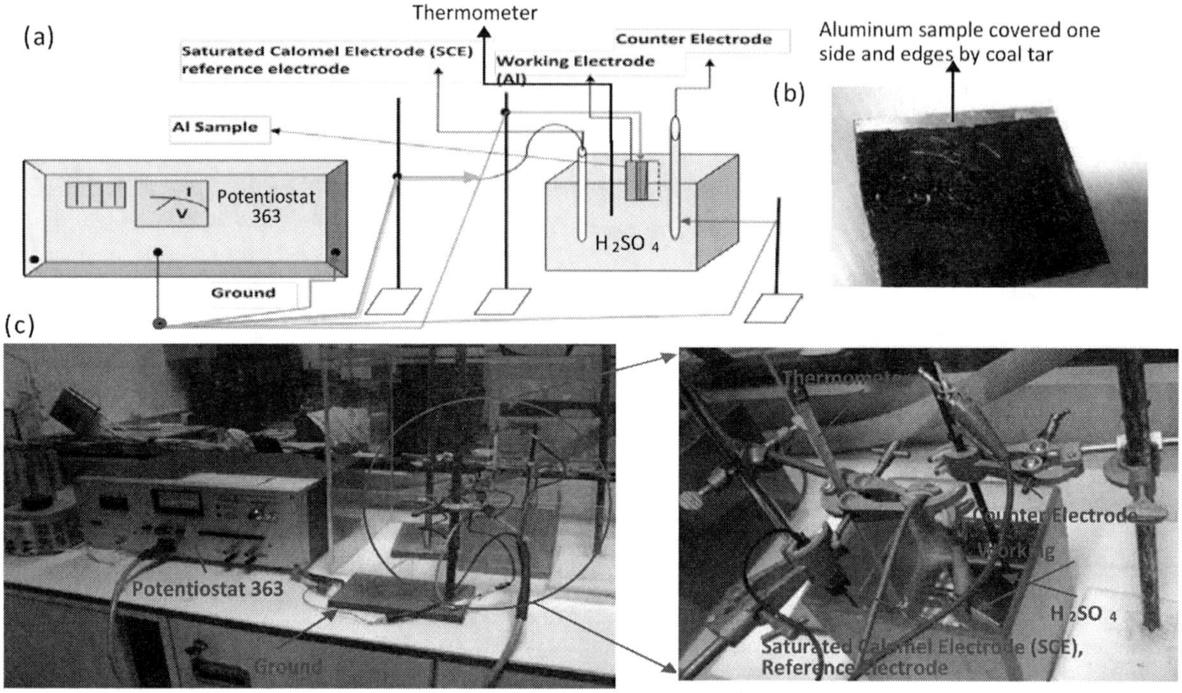

Figure.1 (a) The schematic diagram of experimental set up for anodization (b) Aluminum sample covered one side and edges by coal tar (c) whole experimental setup for anodization of aluminum.

The key steps of the spectral-optical sensing of the real-time Fabry–Pe´rot interferometry are briefly illustrated as the following:

1- During the DC anodization process, light reaches the aluminum sample through a fiber and collimator (to reduce the distance of incident light). The fiber and the collimator lens are inserted in a glass tube, as shown in Figure 2a.
2- The fiber and the collimator are inserted into a glass tube to protect them from the acid solution, while the aluminum sample in the solution, as shown in Fig.2a,b.
3- When light reaches the sample, multiple reflections occur inside the aluminum oxide film causing Fabry–Pe´rot effect [9].
4- A three-port beam splitter was used to assist the transmission of the incident light onto the sample and collect the reflected wave after the multiple reflections occur in the sample, as shown in Fig.2b.
5- For this purpose a chromium coated glass slide was used as a beam splitter, as shown in Fig.2b.

a b

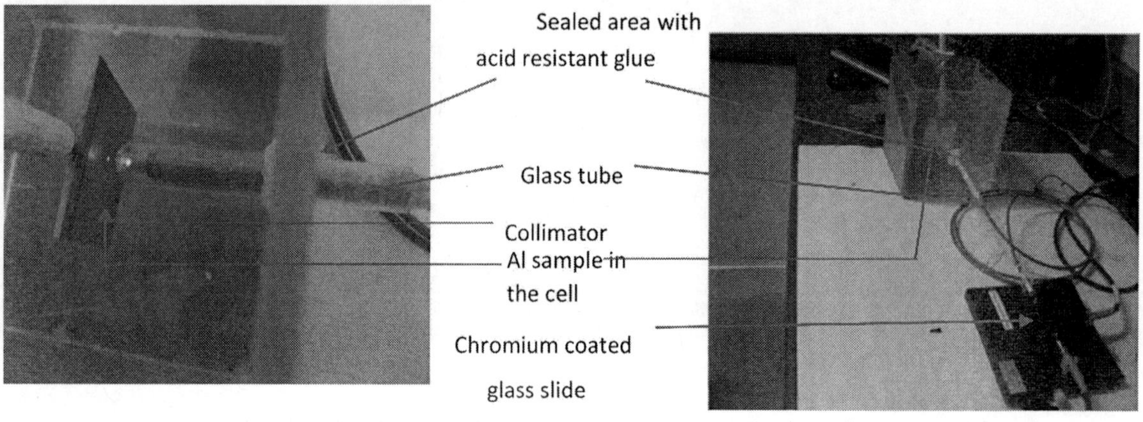

Figure.2a a top view of electrochemical cell along with aluminum sample in the cell, sample holder, glass tube illuminating the sample, and a sealed area with acid resistance glue at the intersection where the glass tube assembled into the cell. Fig.2b a side view of the electrochemical cell illustrating the connection between the cell and a three-port beam splitter; the chromium coated glass slide.

In this experiment, metallic samples of pure aluminum (99.7%) were used. The aluminum samples with dimension 5 cm x 5 cm x 0.15 cm were then cut from pure aluminum sheet. A surface treatment of the aluminum samples was done by a procedure called chemical mechanical polishing (CMP). The procedure of CMP is basically annealing and electropolishing of the aluminum samples to remove uneven surfaces. First, all the samples were polished with silicon carbide particle papers (2500 grade). Subsequently, the samples were mechanically polished first by 300 nm and then by 50 nm alumina nanoparticles to smoothen the surface. The nanoparticles were suspended in a slurry solution, in which the solution could be controlled to induce a pressure, i.e., mechanically polishing effect, on the surface of the samples [14]. The schematic diagram of experimental set up of the electropolishing of the aluminum samples is given in Figure 3a,b.
In order to reduce surface defects and mechanical stresses as well as re-crystallization of aluminum [15], the samples were annealed in the high temperature furnace at 400 ^oC for 1 h under atmospheric condition. Electro-polishing was done in a mixture of perchloric acid ($HClO_4$) and ethanol (C_2H_5OH) (1:4 by volume) [16] at controlled temperature (0–15 ^0C). The applied voltage was varied according to the requirement of current density on the specific area of the samples to obtain a mirror finished surface of aluminum samples. The current-time transient graph was recorded by GDM 396 multimeter communicating with a computer. Thereafter, a coal tar (black) epoxy (polyamide cured) was applied on one side and all edges of the samples(Fig.1b) in order to protect these surfaces from anodization. Anodization was then carried out in the presence of 2% H_2SO_4 solution at room temperature. A graphite electrode was used as a counter electrode and a potentiostat (EG&G Princeton Applied Research model 363) was used as a DC power source, see Fig.1a. Prior to connecting the electrochemical cell, while the sample immersed in the electrolyte, the corrosion potential (open circuit potential) was monitored by a voltmeter with respect to a Saturated Calomel Electrode (SCE), as a reference electrode, see Fig.1a. With respect to the recorded open circuit potentials (OCP), 9 V potential had been applied and anodic current was recorded during the anodization process. At the same time, a reflection spectrum was recorded after a specific time interval (every one or two minutes) using Fabry–Pe´rot interferometry according to Fig.2a,b.

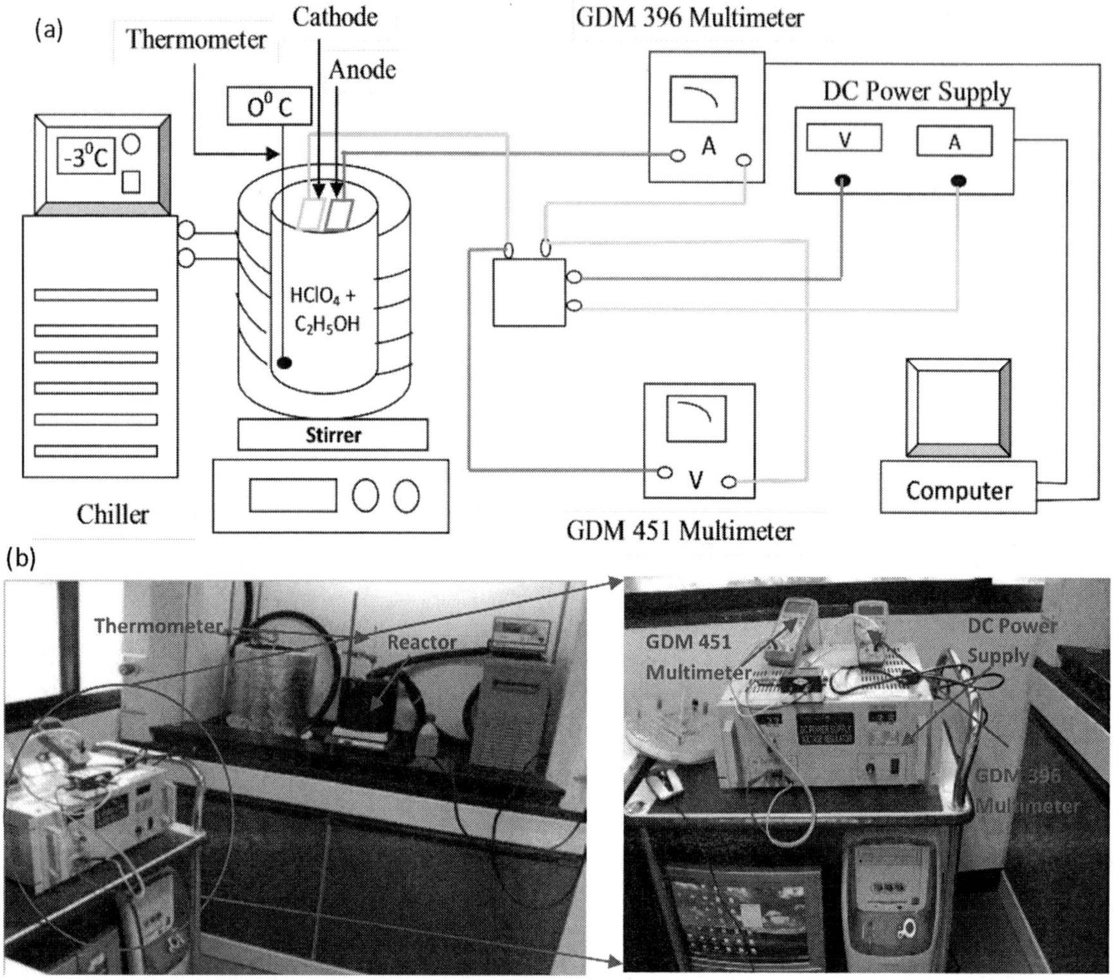

Figure 3 (a) The diagram of experimental set up (b) whole experimental setup for electropolishing of aluminum.

4. RESULTS

Figure 4 illustrates a plot of the elapsed anodization time versus the thickness of the aluminum oxide film in 2% H_2SO_4 solution. The thickness of the aluminum oxide film was formed during 90 mins of anodization, of 9 V of applied potential. The final thickness of the aluminum oxide film (2.136μm) was verified by scanning electron microscopy (SEM), 2.1μm, see the imposed SEM image in Fig.4. Figure 5 illustrates a plot of the corresponding resistance values of Al_2O_3 film thickness during anodization of aluminum in 2% H_2SO_4 solution by Fabry-Pérot interferometer. This study shows that real time measurement of the thickness of aluminum oxide film is feasible as it closely matched the thickness determined by SEM. In addition, the corresponding resistance (2.4 x10^9 Ohms) to the final thickness of the aluminum oxide film was found in an agreement with the resistance value that was measured by holographic interferometry, 2.54x10^9 Ohms [17]. On the contrary, the corresponding resistance (2.4 x10^9 Ohms) to the final thickness of the aluminum oxide film was found twice the value of the resistance that was measured by the electrochemical impedance spectroscopy (EIS), 1.25x10^9 Ohms [17]. The obtained values of the resistance by the holographic interferometry and by Fabry-Pérot interferometer were twice the obtained value of the resistance by EIS due to the fact that those techniques work as zero-resistance Ammeters, i.e., function via electromagnetic principles, Compared to the EIS, which works as a resistance Ammeter, i.e., function via electronics principles [17].

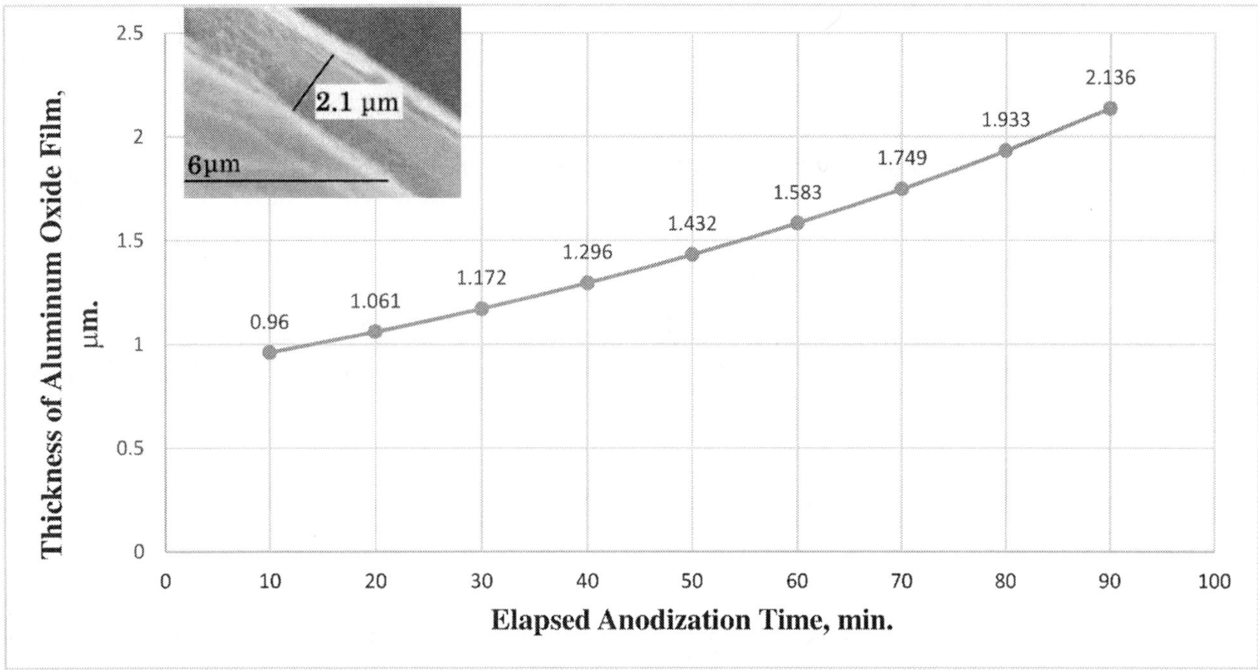

Fig.4 Measurements of Al_2O_3 film thickness during anodization of aluminum in 2% H_2SO_4 solution by Fabry-Pérot interferometer and SEM. The SEM illustrates the final thickness of aluminum sample in 2% H_2SO_4 solution.

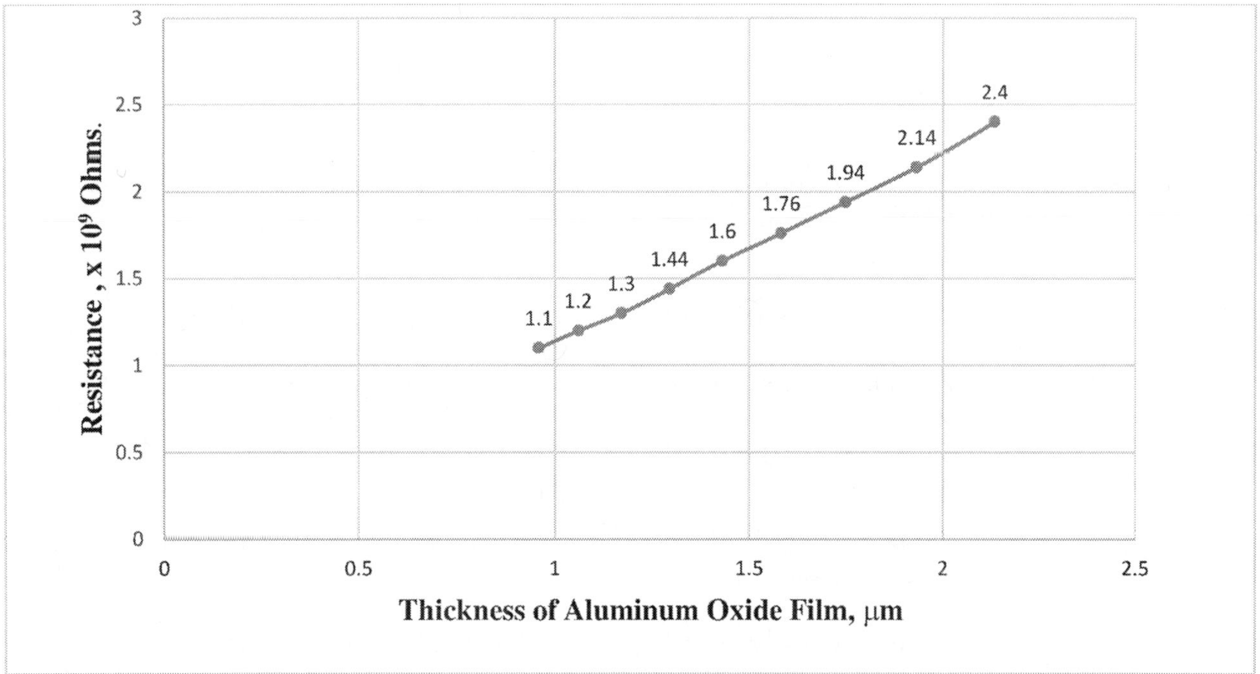

Fig.5 The corresponding resistance values of Al_2O_3 film thickness during anodization of aluminum in 2% H_2SO_4 solution by Fabry-Pérot interferometer.

5. CONCLUSIONS

The following conclusions are drawn from the present investigation:

1- A combination of Fabry-Pérot interferometry and the DC electrochemical method have been used in situ to measure the thickness of aluminum oxide films in 2% sulfuric acid solution.
2- The corresponding resistance (2.4 x10^9 Ohms) to the final thickness of the aluminum oxide film was found in an agreement with the resistance value that was measured by holographic interferometry.
3- The obtained values of the resistance by Fabry-Pérot interferometer (2.4 x10^9 Ohms) and by holographic interferometry (2.54x10^9 Ohms) were found twice the obtained value of the resistance by EIS(1.25x10^9 Ohms).
4- This is explained due to the fact that the optical techniques work as zero-resistance ammeters, i.e., function via electromagnetic principles, compared to the EIS, which works as a resistance ammeter, i.e., function via electronics principles.

6. REFERENCES

1-Liu G, Muller DB(2012), Addressing sustainability in the aluminum industry: a critical review of life cycle assessments, J. Cleaner Prod. 35,108–117.

2-Jirang CUI, Roven H J (2010), Recycling of automotive aluminum, Trans. Nonferrous Met. Soc. China 20, 2057–2063.

3-Takahashi Sakairi M, Kikuchi T (2009), Modern aspects of electrochemistry, in: Su-II Pyun, J.-W. Lee (Eds.), Progress in Corrosion Science and Engineering I, vol. 46, Springer, New York, p. 59 (Chapter 2).

4-Huang W, Zhou C, Liu B, Wang M, Wang H, Ma N(2012), Improvement in the corrosion resistance of TiB2/A356 composite by molten-salt electrodeposition and anodization, Surface Coat. Technol. 206, 4988–4991.

5-Tao X, Jianmin C, Jiazheng Z, Hongxin D(1996), The pore-enlargement and self-lubrication treatment of anodic oxide film of aluminum, Wear 196, 214–218.

6-Habib K(2001), In situ measurement of oxide film growth on aluminum samples by holographic interferometry, Corros. Sci. 43 (3) 449–455.

7-Habib K(2004), Measurement of the electrical resistance of aluminum samples in sulphuric acid solutions by optical interferometer techniques, Int. J. Light Electron Opt. (OPTIK) 115 (4) 145–150.

8-Habib K(2010), AC impedance–emission spectroscopy for determining the electrochemical behaviour of anodised aluminium in aqueous solutions, Nondestr. Test. Eval. 25 (3) 181–188.

9-Hernandez G(1986) , Fabry–Pe´ rot Interferometers, Cambridge University Press, Cambridge, UK.

10-Habib K(2012),"Measurement of surface resistivity and surface conductivity of anodised aluminum by optical interferometry techniques", Optics and Laser Technology 44,318-321

11- Karim F, Bora T, Chaudhari M, Habib K, Mohammed W, and Dutta J(2013), Optical-Fiber based sensor for in situ monitoring of cadmium sulfide thin-film growth, Opt. Lett. 38 (24)5385–5388.

12-Jiang M, Li Q, Wang J, Jin Z, Sui Q, Ma Y , Shi J, Zhang F, Jia L, Yao W, Dong W(2013), TiO2 nanoparticle thin film-coated optical fiber Fabry–Perot sensor, Opt. Express 21, 3083.

13- Karim F, Bora T, Chaudhari M, Habib K, Mohammed W, and Dutta J(2016)," Measurement of aluminum oxide film by Fabry–Pe´rot interferometry and scanning electron Microscopy" Journal of Saudi Chemical Society ,in press, http://dx.doi.org/10.1016/j.jscs.2016.02.001.

14-Kuo H K and Tsai W T(2000) , "Electrochemical behavior of aluminum during chemical mechanical polishing in phosphoric acid base slurry", Journal of The Electrochemical Society147 (1) 149-154.

15- Jessensky O, Muller F, Go U(1998)," Self-organized formation of hexagonal pore arrays in anodic alumina", Appl. Phys. Lett. 72 (10) 1173.

16-Gonzaa J A ^ Lez, Loa^ Pez V, Bautista A, Otero E, Noa^ Voa X R(2012), Characterization of porous aluminum oxide films from A. C. impedance measurements, Appl. Electrochem. 29, 229, 229-23.

17-Habib K (2016), "Opto-electrochemical spectroscopy of metals in aqueous solutions", Review of Scientific Instruments 87, 034102; doi: 10.1063/1.4944478.

8. AUTHOR DETAILS

Professor Khaled Habib is a senior research scientist with the Materials Science & Photo-Electronics Lab, of Renewable Energy Program, at Energy and Building Research center, of Kuwait Institute for Scientific Research, in Kuwait. Mr. Habib is responsible about integrating the optical and the electrochemical systems in the present work.

Dr. Waleed Mohammad is a research scholar with Center of Research in Optoelectronics, Communication and Control System, School of Engineering, Bangkok University, in Pathum Thani, Thailand. Mr. Mohammad was responsible about the setup of the Fabry-Perot interferometry of the early work of the present findings.

Engineer. Farzia Karim is a post-graduate student with the Center of Excellent in Nanotechnology, School of Engineering and Technology, Asian Institute of Technology, in Pathum Thani, Thailand.
Mrs. Farzia was responsible about conducting the early work of the present findings, as a part of her MS thesis's work.

Professor Joydeep Dutta is currently a Chair of Functional Materials division at the KTH Royal Institute of Technology, in Stockholm, Sweden. Mr.Dutta was the chairman of the Center of Excellent in Nanotechnology, School of Engineering and Technology, Asian Institute of Technology, in Pathum Thani, Thailand, while Mrs. Farzia was conducting her MS thesis's work. Mr. Dutta was an academic advisor of the early work of the present findings.

COMPREHENDING COATING ADHESION: PART 1: TO STICK OR NOT TO STICK?

Rob Francis,
R A Francis Consulting Services, Ashburton, Victoria, Australia

SUMMARY: The inability of an epoxy coating to stick to a surface such as polyethylene, for example in a pipeline field joint coating, is puzzling to coating specialists used to their excellent adhesion to most other surfaces. This paper looks at surface energy properties of coatings and how this can explain such behaviour. It also explains why abrasive blasting such surfaces can make adhesion worse and how surface oxidation can be used on such low energy surfaces can improve properties. Pressure sensitive adhesives, such as tapes and shrink sleeves, can be used on low energy surfaces because their viscosity and mechanical properties enable them to provide good bonding. Finally, there is a brief review of methods used to measure adhesion of coatings, pointing out their limitations especially regarding the inability of test results to relate to actual interfacial adhesion.

Keywords: Coatings, adhesion, surface energy, polyethylene, epoxy, adhesion testing.

1. INTRODUCTION

Liquid epoxies are the workhorse for the protective coating industry, providing economic protection to steel and other substrates in a wide range of atmospheric, submerged and buried environments. One of their major advantages over other generic types is their excellent adhesion to steel substrates and previous coatings. However, if applied to polyethylene, as for example, field joint coatings on three-layer polyethylene pipeline coatings as shown schematically in Figure 1, they do not adhere. Yet there is no problem applying epoxies to steel or fusion bonded epoxy pipeline coatings. Such behaviour is puzzling to specifiers and applicators used to dealing with epoxies on steel.

Figure 1: Liquid epoxy field joint coatings can have problems adhering to polyethylene mainline pipeline coatings.

This paper looks at the factors that determine whether or not a given coating will adhere to a substrate. It discusses why epoxy usually shows good adhesion to most substrates and why there are problems with polyethylene and related polymers. It discusses the reasons abrasive blasting the surface does not assist adhesion in this case and in fact can make it worse, and looks at treatments for such surfaces to try to improve adhesion. There are coatings that can successfully adhere to polyethylene and similar surfaces and their mechanism of bonding is reviewed. Finally, there is brief review of methods that are used for measuring coating adhesion and why results bear little relationship to actual interface bonding forces. First some background on adhesion fundamentals.

2. BASICS OF COATING ADHESION

Consider a situation where a paint is applied to a surface. To adhere to a substrate it must first wet, or stick to, the substrate. Wetting can be defined as adhesion of a liquid to a solid. The coating will wet the substrate only when the relative surface energies of the coating and the substrate are appropriate. All surfaces have surface energy (also known as surface free energy or surface tension for liquids) as a result of the atoms or molecules tending to orient themselves to minimise the surface area. In a liquid, droplets will tend to form a spherical shape which has the smallest surface area for a given volume. Solids also have a surface energy and try to minimise their surface area by causing an applied liquid to spread across this surface. However, the surface tension of the liquid will oppose this property by attempting to minimise its own surface area. As a result, there is a competition between these two tendencies. To determine if a given liquid will wet a given surface, we need to look at the relative surface energies of each. The two components cannot be looked at separately.

Surface energy is defined as the amount of work or energy required to create a unit area of surface and has units such as Joules per square metre (J/m^2). Surface tension is the force per unit length acting upon an imaginary line drawn on the surface. The SI units for surface tension are Newton per metre (N/m), which is numerically equivalent and used interchangeably with Joules per square metre. Because N/m is rather large, surface tension is commonly expressed as mN/m which is equivalent to the old unit of dyne/cm used in literature in the past and still widely quoted. Similarly, surface energy is usually quoted in mJ/m^2. For liquids, the term surface tension rather than surface energy is more commonly used.

A polar molecule such as water has a relatively high surface tension because of the hydrogen bonding forces exerted on the surface molecules by the internal molecules beneath it. A non-polar liquid, such as hexane, will not have these forces and have a lower surface tension. The surface tension of liquids can be measured relatively easily using, for example, the capillary rise method. The end of a capillary is immersed into the solution and the height the solution rises is related to the surface tension. Surface energy of solids is more complex and usually measured indirectly using the contact angle method discussed below.

A drop of liquid placed on a surface may spread out or ball up, depending on how it wets the surface. Raindrops bead up on a freshly waxed car because the surface energy of the water is higher than that of the wax. In contrast, if the surface energy of the liquid is less than that of the substrate, such as soapy water on an oil-free surface, it will spread out and wet the substrate thus making the intimate contact necessary for good bonding. If it wets the surface, the contact angle (θ) between the liquid and the surface is very small, as shown in Figure 2(a). A liquid which completely wets the surface will have a contact angle of 0°. A liquid which shows poor wetting properties will have an angle greater than 90°, as shown in Figure 2(b).

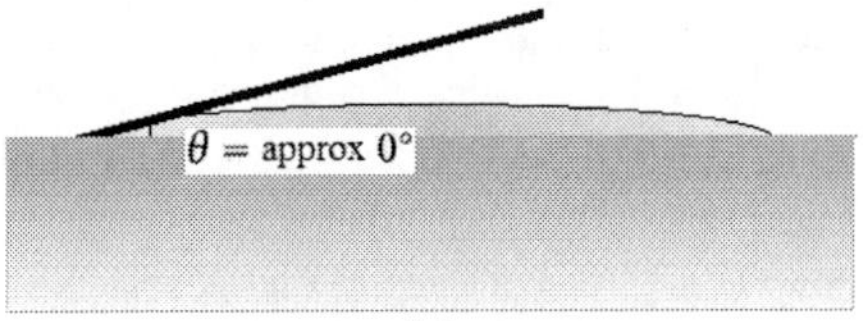

(a)

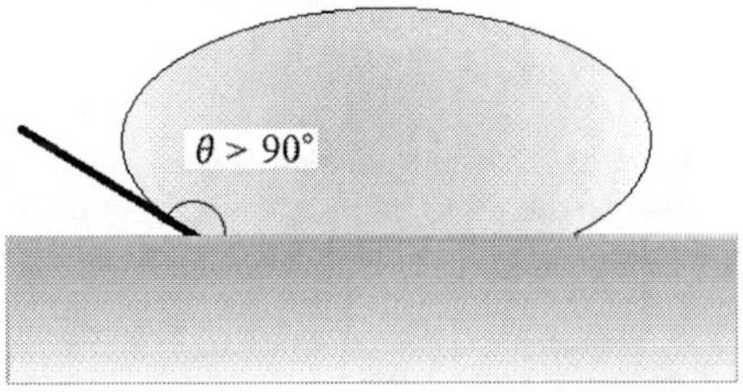

(b)

Figure 2: Contact angle θ for a liquid on a solid with (a) good wettability and (b) poor wettability.

For a drop of liquid on a surface shown in Figure 2, the substrate and the liquid in air will each have a different surface free energy value. For wetting, the surface energy (tension) of the liquid being applied must be lower than the surface energy of the solid. Practically, the surface tension of the liquid would need to be at least 10 mJ/m^2 lower than the surface energy of the substrate. The surface energy of a substrate depends on the type of material as well as its cleanliness and roughness. Table 1 gives surface energy figures for some surfaces, coatings and related materials. On typical well-prepared metals, the surface energy is of the order of several hundred millijoules per square metre. So, even if not perfectly clean, they have much greater surface energy than typical paints (20 to 70 mJ/m^2) and wetting of such surfaces is not normally a problem. However, plastics such as polyethylene and PTFE are low energy surfaces and the same coating would not wet the surface. Adding solvent to a coating will lower surface tension and viscosity, enabling better flow over a surface, but still not allow wetting of a low energy surface. Modern coatings with low or no solvent have higher surface energy and viscosity and are more likely to have difficulty wetting out a surface. It is worth noting that actual surface energy figures of metal surfaces are an order of magnitude less than ideal values.

Table 1: Surface energy of a substrates, coatings and related materials.

High surface energy		Medium surface energy		Low surface energy	
Surface	**mJ m-^{2}**	**Surface**	**mJ m-^{2}**	**Surface**	**mJ m-^{2}**
Metals (ideal)	500 - 5000	Epoxy paints	~50	Polyethylene	31
Metal oxides	200 - 500	Alkyd enamel	45	Polypropylene	30
Metals (realistic in air)	50 - 500	Polyurethane paint	43	Mineral spirits	24
Water	73	Polyester	43	Methyl isobutyl ketone (MIBK)	24
		Polyvinyl chloride (PVC)	40	Ethanol	22
		Acrylic emulsion	32	Hydrocarbons	~20
				PTFE	18

From a practical point-of-view, relative surface energy studies can be carried out placing drops of coating on a substrate and watching the spreading. If the droplet remains as a small ball (high contact angle) wetting is poor and adhesion problems must be expected. If the drop spreads across the surface (low contact angle) the coating will wet the surface and should have good adhesion. However, sometimes solvents or additives can create an atmosphere above the drop that may cause a drop that initially wets the surface to ball up and dewet the surface.

3. ROUGHENING OF LOW ENERGY SURFACES

Table 1 shows that surfaces such as polyethylene, polypropylene and PTFE are inherently low energy (18 to 31 mJ/m^2) and cannot be wetted by typical coatings such as epoxies. Is there anything that can overcome this inherent incompatibility between epoxies and low energy surfaces? Abrasive blasting is a quick and economical means of cleaning and preparing metals and other surfaces to improve coating adhesion, so it would the obvious first method to look at to see if this can allow wetting of a low energy surface.

Surface roughness will have an effect on wetting, contact angles and adhesion. This effect can be through a number of different mechanisms. Roughness is generally agreed to improve wettability and increase dry adhesion of paints on metals by increasing surface area and providing resistance to shear stresses in coating curing and during service, but it can have a negative effect on wetting of low energy surfaces. This unusual behaviour can again be explained in terms of wetting and contact angles.

For most organic liquids with a contact angle less than 90°, such as epoxy on steel as shown in Figure 3(a), the effect of roughening is to make the contact angle lower so the effect of roughening is to improve wettability. However, if the surface is contacted by a liquid with a contact angle greater than 90°, such as epoxy on polyethylene, then roughening has the opposite effect of increasing contact angle and reducing wettability, as shown in Figure 3(b). Therefore, the wet adhesion of an epoxy on a roughened surface of polyethylene is less than that on a smooth surface. Trapping of air pockets and difficulty filling surface features, especially with viscous materials, with a poorly wetting liquid can also reduce adhesion on low energy surfaces. In addition, gap filling may be competing with curing of the coating. One reason fast curing epoxies, for example, may not have the same adhesion as their slow curing equivalents are that the fast curing epoxy does not have time to fill the crevices on the substrate. Overall, roughening a low energy surface by blast cleaning will not improve wetting of the surface by paint coatings and have a detrimental effect on adhesion.

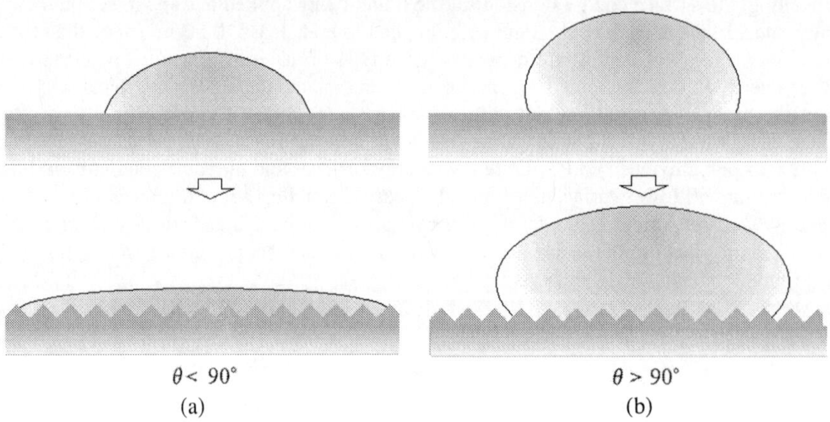

Figure 3: Effect of surface roughness on contact angle.

Abrasive blasting cannot be used to prepare low energy surfaces. The next possibility is to look at changing the chemical rather than physical properties at the surface.

4. TREATMENT OF POLYETHYLENE SURFACES

The introduction of more polar groups, such as hydroxyl or carbonyl groups, on the polyethylene or polypropylene surface will raise the surface energy of a substrate. This requires surface treatments usually involving surface oxidation, such as by propane torches, corona discharge techniques or acid etching. Figure 4, from the work of Rauhut (1), shows that flame treatment and chromic acid etching can raise the surface energy of polyethylene to about 40 mJ/m^2. At the same time, a significant improvement in adhesion is achieved. In addition, such process will remove contaminants and weak boundary layers. However, it should be noted that such surface modification may be complex, will typically add to time spent and cost and may have environmental or health and safety issues. Furthermore, excessive treatment can destroy adhesion.

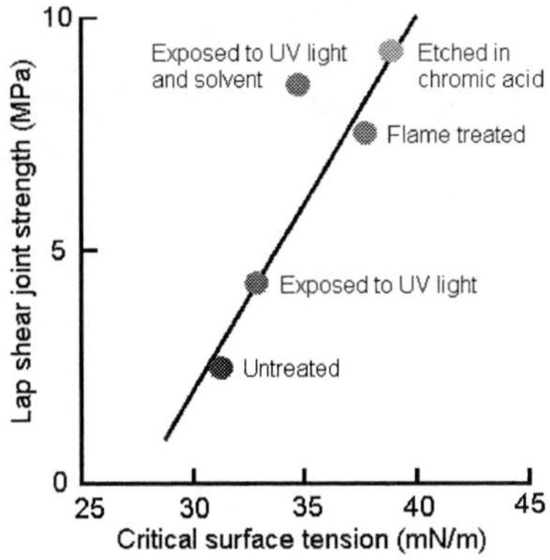

Figure 4: Lap shear joint strength versus critical surface energy for LDPE pre-treated by various methods (1).

The main surface treatments are:

- Corona discharge or plasma treatment is achieved by applying a high voltage of 5 to 50 kV which oxidises the surface. An 'active' gas such as oxygen or inert gas such as nitrogen may flood the gap between the electrode and plastic surface. UV radiation along with ions, radicals and electrons generated in the corona attack the surface oxidizing it to a depth of approximately 5 nanometres within a few seconds increasing to ten times this depth with further treatment. After treatment, the surface contains numerous reactive polar species which raises the surface energy, but ablation, oxidation and cross-linking of the surface layer will also take place. UV radiation alone, as used in the work described in Figure 4, will cause similar surface modification.
- Flame treatment produces similar polar species to those created during corona discharge treatment, modifying the surface to a slightly greater depth. The process can overheat surfaces so is not used on thin film.
- Chromic acid preferentially etches regions of low and zero crystallinity of the plastic simultaneously oxidizing and etching the surface. It is much slower than corona discharge or flame treatments, but creates much better mechanical keying as root-like cavities may form in the surface.

Is the significant improvement in adhesion due to the approximately 10 mJ/m^2 increase in surface energy? A quick back-of-the-envelope calculation shows that this is nonsense. Surface energy of 10 mJ/m^2 is equivalent to 0.01 N/m surface tension, or a gravity load of about 1 gram (~0.01 Newton) removing a one metre wide strip of the material. This is far less than the force required to remove a sticky note! Yet the actual joint strength has increased many times with the flame or chemical treatment. So this improvement in joint strength is not due to the small increase in surface energy from surface modification from the chemical or other treatments; but rather from some other factor. In fact, what is happening is that the treatment opens up the crystalline surface of the polyethylene making it more amenable to intermingling and entanglement of the polymer chains of the adhesive. The more the chains are intermingled, the harder it is to pull them apart. However, if opened up too much, the rough surface can reduce adhesion. Also, time can cause crystallinity to heal itself and the adhesion of a treated surface can drop away over time. The increase in surface energy does assist initial wettability, but it does not cause the increase in adhesion. Improvements in wettability and bonding arise from complex topography rather than polar groups introduced by surface oxidation (2).

5. COATINGS FOR LOW ENERGY SURFACES

Because of the difficulty in wetting, alternative coatings are required if good adhesion to polyethylene or other low energy surfaces is required, such as a field joint coating for a polyolefin mainline coating. AS 4822 (3) Appendix K recommends cold applied polymeric tapes and heat shrink sleeves field joint coatings for polyethylene mainline coatings. Polymeric tapes and heat shrink sleeves are examples of pressure sensitive adhesives, and the adhesion mechanism is more complex than simple wetting of the surface.

Pressure sensitive adhesives (PSAs) have a backing film with a special adhesive that can bond to a variety of surfaces. The adhesive strength can range from easily removable *Post-It* notes to high-strength, high adhesion, corrosion protection tapes and sleeves. They can adhere with relatively light pressure, the adhesives remain permanently tacky at room temperature and the materials are elastic enough to be handled by hand. There is a fine balance between the adhesive flow, which is just sufficient to flow into surface crevices, and cohesion of the adhesive which resists flow. Another advantage of pressure sensitive tapes is that the bond is immediate so there is no clamps required or cure time. PSAs will usually have operating temperature limitations and, by remaining soft, may have limited resistance to shear forces. Unlike many PSAs, corrosion protection tapes and sleeves are required to be permanent.

PSAs usually bond to surfaces through dispersion (van der Waal's bonds). Pressure-sensitive adhesives are unique among adhesives in that they do not cure or undergo a chemical or physical change during the bonding process. The adhesive material when the tape is applied is identical in properties to that there when the tape is doing its job or when it is being peeled off. An epoxy, on the other hand, is liquid when applied but solidifies to resist deformation in service. PSAs are examples of viscoelastic materials that show liquid-like behaviour for bond formation along with solid-like behaviour to resist debonding. When the adhesive is applied to the substrate and pressure is applied, it makes immediate contact for initial adhesion as it acts as a viscous liquid but because of the elasticity, resists separation when stressed like a solid.

In theory, PSAs must meet the same requirements regarding contact angle and surface energy so they can wet the substrate and they will stick better to higher energy substrates such as metals or PVC than low energy surfaces. But there is a lot more than just getting the surface energy right. Pressure sensitive adhesives make use of their unique viscoelastic nature of the adhesive to wet the surface. They are soft enough to deform to achieve intimate contact with the surface so it is their low modulus rather than viscosity which promotes wetting. Figure 5 shows how adhesion and cohesion properties change with peel speed (strain rate). At very low peel speeds, the adhesive acts like a liquid and very low peel strength is measured. However, as the peel rate increases, the adhesive becomes more viscous and the forces are now used up in pulling the adhesive apart. The adhesive fails in a cohesive manner and the required peel force increases. At very high strain rates, the adhesive acts largely as a glassy solid and easily fractures, resulting in much easier removal.

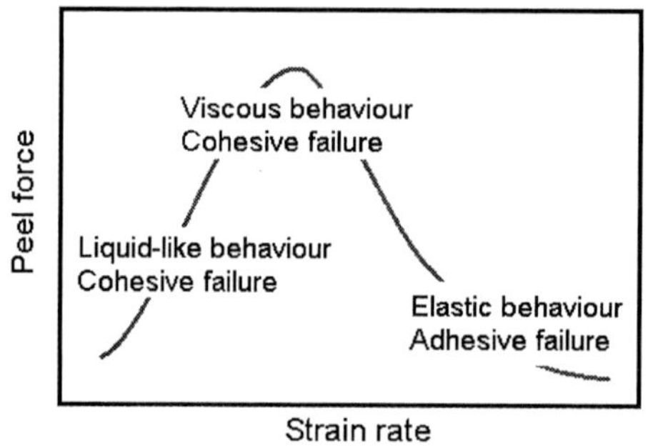

Figure 5: Relationship between peel force and visco-elastic properties.

The mechanism of PSA adhesion is different from other adhesion combinations. In many PSAs, it is interaction between the molecule chains some small distance from the surface that provides the strength, not the adhesion of the chains to the surface. The important fact is that they remain liquid at the interface. Even if the surface energy of the adhesive is much greater than the substrate, the backing tape will cause the adhesive to wet out the substrate by eliminating the air at the air/adhesive interface (4).

Roughening the substrate can reduce adhesion with PSAs. The high viscosity of the adhesive does not get a chance to fill the surface features on the substrate of a roughened surface. As a result, there can be built-in defects which can initiate failure. The surface must clean and free from dust, oil and moisture, but roughness can have a negative effect.

6. MEASURING ADHESION OF COATINGS

There are many methods for measuring the "adhesion" of a coating with for example, a recent draft standard ISO/TR 19402 (5) listing 44 different methods. These are often adapted from test methods for adhesives, and can be divided into three main techniques, as shown schematically in Figure 6, with tensile, shear or peel loading operating. The different approaches will give widely varying results for many reasons. Firstly, each method applies a load on the coating in different directions and the loading direction will usually vary over the area of the testing sample. It is difficult if not impossible to generate stresses only in the desired direction, and actual forces are usually some combination of tensile, compression and shear. Further, the rapid loading rates inherent in the equipment and methods used will amplify the measured failure strength compared to that achieved under normal viscoelastic processes in polymers. Also, equipment or methods that apply an uneven manual load will produce lower readings than consistent, automatic equipment.

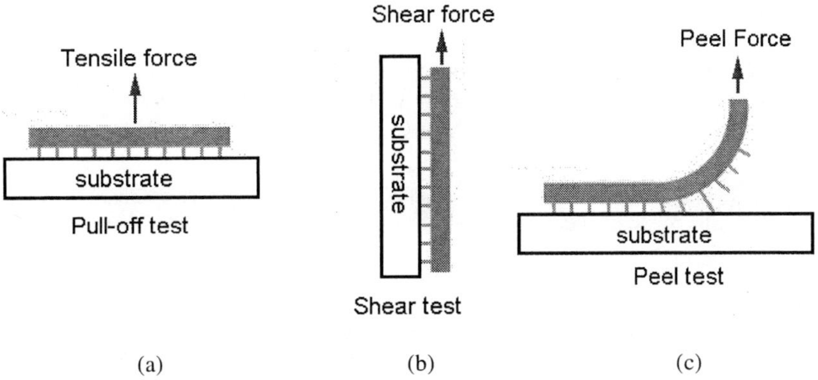

Figure 6: Three common test types for measuring adhesion: pull-off, shear and peel.

The basic features of the three test types are described below:

- Pull-off or tensile stress is created when a load is applied perpendicular to the surface. Although this is the major component of most test methods used for measuring adhesion of coatings, such forces are rare on a coating in service. A major problem is that if the load is applied slightly off-axis, then stress direction changes to peel, and failure mechanism and force changes. This seemingly simple test can yield widely diverse results depending on the apparatus and methodology used.
- Shear stresses operate when a force is applied parallel to the coating surface. Such stresses would be applied to a coating when, for example, the substrate expands and contracts at a different rate to the coating such as from temperature changes. This property is difficult to test on most coatings, although methods are available to assess such stresses in tapes and sleeves (for example AS 4822 Appendix H) as shear is an important stress imposed by movement of pipelines underground.
- Peel loads concentrate all the stress at the interface where the coating pulls from the surface. But stress distribution is very complex and there is in fact a compression zone just ahead of the peel front causing local changes to the adhesive. Excessive extension of the backing can contribute to failure at the adhesive/ substrate interface. So factors such as the modulus and strength of the backing and thickness and strength of adhesive are more important than interfacial adhesion. Furthermore, values are in terms of force per unit length, which cannot be compared with results from tensile or shear values which are determined in terms of force per unit area. Although it does not represent conditions that a coating may experience in-service, the peel test has been found to provide a reasonable indication of how well a coating has been applied by determining how well it is bonded to the substrate.

These basic principles underlie numerous test methods ranging from simple qualitative picking at coating with a sharp knife, through to measuring the pull-off strength using sophisticated tensile testing equipment. Part 2 of this paper looks closely at pull-off testing along with an examination of the method, results obtained and how they may be interpreted.

Practical adhesion values measured for coatings are almost always orders of magnitude greater than thermodynamic work of adhesion as determined by wetting experiments. The measured adhesion strength will be influenced by many factors independent of forces at the coating/ substrate interface including elastic, viscoelastic and plastic deformation which occurs during testing and energy involved in molecular bond breaking. There will be relaxation of the internal stress which will always decrease the inherent adhesion strength. These stresses are usually many times the adhesion from molecular interactions between the film and the substrate. So measured adhesion strength depends on the chemical and physical properties of the coating and substrate, the loading rate, temperature, joint geometry and many other factors. The measuring technique can also introduce factors which results in values obtained by various methods being not directly comparable.

Adhesion testing methods do not, in fact, measure the adhesion of a coating to a substrate or another coating, and the results from the various methods cannot be compared with one another. There are other problems relating such testing to coating performance:

- The coating is subject to very high stress and strain levels during testing which it would rarely see in practice. In service, coatings have been observed to delaminate under much lower stress levels than measured values.
- As discussed, rapid loading during testing results in a different failure mechanism and value from that which would be observed with slow testing. In service, loading rates on coatings are likely to be very slow.
- The coating under test will be a very small sample compared to coating applied in practice and variations in properties such as thickness, defects and contamination will cause local stress concentrations which have a significant effect on results.
- The small sample, uneven loading, strain rate variations, visco-elastic behaviour and many other factors cause wide scatter in the data produced.
- The substrate thickness and flexibility can affect results, especially with field testing.
- The peel test only applicable to tough, flexible coatings and initiating a strip for peel testing can be difficult in coatings with strong adhesion

- In pull-off testing, if load is even slightly off centre, results will be completely different from results with normal loading. In addition, the glue can influence results.

Results from adhesion tests should not be used to explain any fundamental property of a coating system, but methods can be used to monitor changes in adhesion due to weathering, surface treatments, application variables, etc.

7. CONCLUSIONS

a) Adhesion of coatings to substrates can generally be explained in terms of surface energy differences. Epoxies for example have lower surface energy than metallic substrates so easily wet and adhere to the surface. However, polyethylene has a low surface energy and epoxies will not wet such surfaces.

b) Roughening the surface improves wetting of high energy surfaces but has a detrimental effect on low energy surfaces.

c) Low energy surfaces can be treated to improve adhesion of coatings, but the improvement in adhesion is due to opening up the surface allowing molecular intermingling rather than the small improvement in surface energy.

d) Tapes and shrink sleeves adhere to low energy surfaces due to their high viscosity, by remaining liquid and their visco-elastic properties.

e) There are many methods for measuring coating "adhesion", but they do not in fact measure interfacial adhesion. Such tests apply complex stresses and cannot be used to explain fundamental adhesion properties of coatings.

8. REFERENCES

1 Rauhut, H.W. Adhesives Age 13(1), p. 34 (1970).

2 Abbott, S "Adhesion Science: Principles and Practice", DEStech Publications, Inc., (2015)

3 AS 4833-2008, "External field joint coatings for steel pipelines", Standards Australia, Sydney, 2010.

4 Paul, C W, "How Thermodynamics Drives Wet-out in Adhesive Bonding: Correcting Common Misconceptions", Jnl Adhesion Science and Technology 22 (2008) 31–45.

5 ISO/TR 19402, Paints and varnishes -- Adhesion of coatings (standard under development at time of writing), International Organization for Standardization, Geneva, 2016.

9. AUTHOR DETAILS

Rob Francis has over 40 years' years' academic, industrial and consulting experience in corrosion and protective coatings. Dr Francis has a B.Sc. in metallurgy and a Ph.D. in corrosion science. He has authored or co-authored over forty technical papers or presentations on corrosion and coatings. He edited the publication "Inorganic Zinc Coatings", for the ACA in 2013. He has been awarded the JPCL editor's award twice and was made a JPCL Top Thinker in 2012. He was awarded the Victor Nightingall award for outstanding contributions to the protective coating industry by the ACA in 2014 and made an ACA Life Member in 2016.

COMPREHENDING COATING ADHESION: PART 2: PULL-OFF ADHESION TESTING

Rob Francis,
R A Francis Consulting Services, Ashburton, Victoria, Australia.

SUMMARY: Pull-off adhesion testing is often specified and used to quantify adhesion of coatings to substrates or within a coating system. There are a number of commercial devices available for this purpose. However, investigation of the methods used shows that the devices can give significantly different results and are subject to substantial levels of error. Furthermore, failure is often cohesive within a coating and the result gives no indication of coating adhesion. This paper looks at pull-off testing devices, results obtained and provide guidelines to evaluate their meaning and significance.

Keywords: Coatings, adhesion, pull-off testing, cohesion.

1. INTRODUCTION

Pull-off 'adhesion' testing of coatings appears to be more regularly discussed and specified in recent years. Manufacturers have produced new testing models, more papers are being published on the subject and pull-off requirements are increasingly being required in coating specifications. Is this increased interest based on its ability to provide useful information on performance of coating systems, or is this a case of a test method being more attractive because it can produce a number, even if that number has little meaning? This paper looks closely at pull-off testing, what is measured and what the results mean.

2. PRINCIPLES OF PULL-OFF TESTING

Common perception is that the "adhesion strength" as determined using a pull-off tester is a measure of how tightly the coating layers adhere to the substrate or between one another. In fact, the majority of figures obtained from such testing and the actual failure path show that adhesion strength is not being measured but rather some other property independent of the level of bonding between various interfaces. To properly interpret the results, it is necessary to have an understanding of how pull-off testing is actually carried out.

For such testing, a portable adhesion tester, commonly called a pull-off tester is used, usually to standards such as ASTM D4541 (1), AS 3894.9 (2) or ISO 4624 (3). ISO 16276-1 (4) also covers such testing, but refers to the measurement of "fracture strength", which includes both adhesion and cohesion failure. Unlike ISO 4624, this standard covers testing on site, with instructions for interpretation of results and acceptance/ rejection criteria. In all cases, the test method consists of securing a loading fixture (commonly called a dolly) to the coating surface with an adhesive (glue) as shown in Figure 1. The testing apparatus is then affixed to the dolly and aligned to pull the loading fixture perpendicular to the coating surface. The force is gradually increased until the loading fixture detaches, which will be that at the weakest interface. The user ideally then reports not only the pull-off strength, but also the location of the break in the coating system, i.e., adhesion failure between the primer and substrate or between coats, cohesive within a certain coating layer, glue failure, etc.

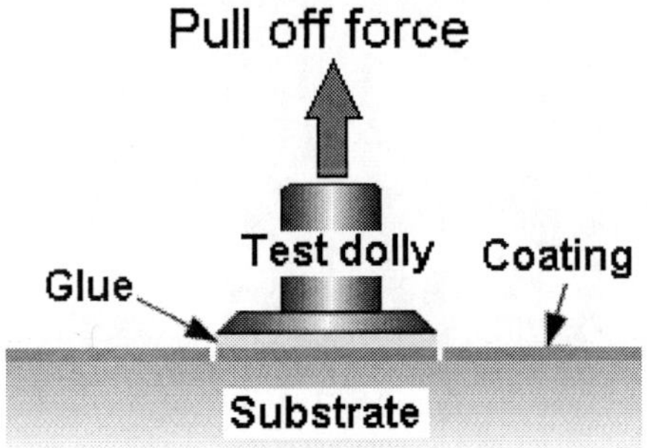

Figure 1: Pull-off Adhesion Test.

3. PRACTICAL PULL-OFF TESTING

There are a number of devices on the market which use this principle for carrying out such tests, but there are differences between them which can significantly affect results. In some models, the load is applied by compressing a spring (mechanical), while in others the load is applied pneumatically or hydraulically. In earlier models, the grip for engaging the dolly is fixed, while in later models it can move, ensuring alignment is more perpendicular to the load. Because of these differences, the pull-off gauges available do not give comparable results. ASTM D4541 recognises this and divides them into different types, each with its own test method, as shown in Table 1. Type I gauges are generally used for concrete and were dropped from the standard in 2004, although results from this device can be found in the literature.

Table 1: Pull-off Adhesion Testing Devices according to ASTM D4541

ASTM D4541 Type	Test method	Example	Load mechanism	Dolly alignment	Repeatability Limit (%)	Reproducibility Limit (%)
Type I	A	Dyna Z5	Mechanical	Fixed		
Type II	B	Elcometer 106	Mechanical	Fixed	64.7	76.0
Type III	C	HATE, Elcometer 108	Hydraulic	Self-aligning	33.8	65.9
Type IV	D	PATTI	Pneumatic	Self-aligning	14.8	28.4
Type V	E	Positest AT	Hydraulic	Self-aligning	27.8	34.1
Type VI	F	PAT Handy, PAT GM03	Hydraulic	Self-aligning	17.5	23.0

In the 2009 version of ASTM D4541, the pull-off strength of four painted panels was determined using five devices in a round-robin survey. Average readings with each type are given in Figure 2. The repeatability and reproducibility limits in Table 1 give the maximum acceptable percentage difference between results before they should be considered significant; repeatability for a single operator with a single device and reproducibility between different operators. These are also plotted as error bars on Figure 2, with the repeatability limits the lesser values. These high figures indicate that test results are subject to considerable uncertainty, an issue also discussed by Schilling (5). For example, if one operator gets a result of 7MPa with a Type II gauge under certain conditions, then a second result under different conditions needs to be less than (7-(7 x .65)=) 2.5 MPa or more than (7+(7 x .65)=) 11.6 MPa for the change to be considered as significant. Even the most repeatable gauge, the Type IV, would need results outside the range of 6 to 8 MPa to be considered significant. When comparing results to those obtained by other operators, the expected spread of results will be even greater. Clearly, small differences in pull-off values within or between investigations, especially higher values, should not be considered as significant.

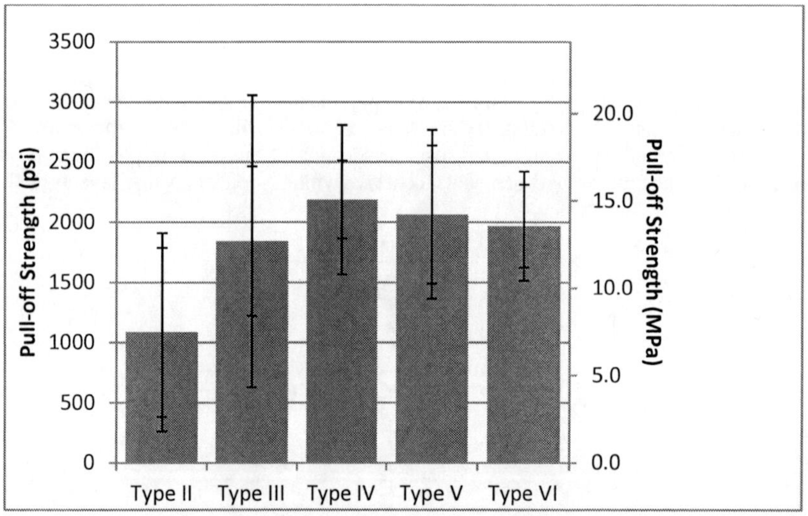

Figure 2: Pull-off results for test panels and repeatability and reproducibility limits for five pull-off testing methods from Tables 1 to 5 in ASTM D4541-2009.

The Type II fixed alignment gauge appears to give greater variability than the self-aligning gauges so the scatter is probably due to higher probability of shear forces operating with the fixed alignment gauge. The variability obtained with the Type III gauge, although self-aligning, may be due to the different loading mechanism through the centre of the testing dolly which again may introduce shear stresses. Cunningham and Steele (6) also found that fixed alignment gauges tend to create variable results with more 'scatter' around the mean result than self-aligning gauges.

Some of the reasons for the varying results include misalignment of the apparatus or loading fixture so that it was not perpendicular to the surface, variation of the area stressed due to improper application of the glue, holidays in the glue caused by voids or inclusions, improperly prepared surfaces, and sliding or twisting of the test fixture during the initial glue curing. It should be noted that it is very difficult to apply a perfectly axially-centric load, even with self-aligning grips. Other factors are glue runs underneath the loading fixture on vertical surfaces and knocking the loading fixture off prior to or during the attachment of the test apparatus. Almost anything that can go wrong in performing the test will give low results, not higher results. Experienced field users find that five to ten percent of pulls end up being invalid. Therefore, applying at least three test fixtures per location in the field is a must when performing this test.

As well as differences in repeatability and reproducibility, different gauges give different values of pull-off strength for the same test coating and panel. The round robin survey quoted in the 2009 version of D4541 and plotted in Figure 2 shows that Type III, Type IV, Type V and Type VI instruments give comparable results, definitely within the uncertainty discussed above. However, the popular Type II instruments give pull-off strengths of about half that of the other instruments. It is clearly critical to identify the instrument that is used in any investigation.

An investigation of a number of pull-off test methods (7) confirmed that the Type II Elcometer Model 106 gave considerably lower figures than pneumatic or hydraulic methods. Cunningham and Steele (6) also noted instruments based on hydraulics appear to achieve higher results than mechanical (spring) devices, probably because the application of force was smoother in the case of hydraulic instruments. They described an experiment where a mechanical tester was modified by removing the spring and replacing it with hydraulics operated by an electric pump. The average results were twice those previously obtained. The significant factor appears to be the smoothness of force application rather than the absence of self-alignment. These finding suggest that values obtained from a Type II unit need to be doubled when compared to results from one of the other types of gauges.

4. WHAT DOES THE PULL-OFF TEST ACTUALLY MEASURE?

The pull-off test does not measure adhesion. It can be viewed as a simplified tensile test which measures fracture strength of the weakest bond in the substrate/ coating/ glue/ dolly system. If the fracture is between layers, the strength determined will be a measure of adhesion, but if fracture is within a coating, then the result will be a measure of the coating's tensile strength. However, the tensile strength as determined from the pull-off test will not be the same as that determined using tensile testing instrumentation for a number of reasons. For example, for some polymers the rapid loading (strain) rate of manual pull-off testing will result in a increased strength measurement. (Ductility decreases and modulus increases also but these cannot be determined by pull-off testing, only the ultimate tensile strength). The sensitivity of polymers to strain rate depends on the type of polymer: for brittle polymers the effect is relatively small, whereas for rigid, ductile polymers and elastomers, the effects can be quite substantial if the strain rate covers several decades. For other polymers, rapid loading can decrease strength, especially of cracks and other defects are present and failure changes from ductile to brittle. However, it is the nature and design of the test sample that will result in the greatest variation. Standard tensile testing uses a dogbone-shaped sample where the stress is restricted to run parallel to the applied force along the central part of the sample and there is no significant stress concentration at the ends of the sample altering stress distribution. The stress distribution across the plug of coating in pull-off testing is not uniform unlike the dogbone-shaped specimen, with stress concentration at the edge giving higher localised stresses resulting in considerable variation and scatter in results, as well as lower strength figures.

Much published work quotes the results of pull-off tests, claiming that they are values of 'coating adhesion' without noting the location of the failure (8) or combining and comparing results at any failure location, whether adhesion, cohesion, within the glue or some combination (9) without distinguishing between the meaning of the results. The pull-off value for a cohesion failure will not give any indication of the adhesion properties of the coating; it will only give some measure of the tensile strength of the coating where the failure occurs. Yet a number of papers claim to investigate the effects of various parameters on coating adhesion by quoting these pull-off cohesion failure strengths. Even the results of work that looks at comparisons, say before and after weathering, indicate nothing about adhesion strength if the failure is not an adhesion failure.

A further issue in analysing results from pull-off testing is that it is unlikely to correlate to the long term corrosion protection offered by a coating. There is no doubt that coatings must adhere well in order to resist the advance of water and other aggressive species across the coating surface, but there is no indication that high pull-off values will provide improved corrosion protection (10). A slow-curing coating may show poor initial adhesion or cohesion, but these properties may rapidly improve. Alternatively, a coating may show good initial adhesion but moisture penetration may cause corrosion at the interface leading to rapid reduction in adhesion. Pull-of strength measured soon after application will rarely provide an indication of longer term performance.

5. PULL-OFF TEST RESULTS

Despite these concerns, there are a large number of reports and papers presenting the results of pull-off testing and useful information can be obtained from such work detailed below. From the above discussion, useful conclusions can really only be obtained from work that clearly identifies testing unit used and the location of fracture. Furthermore, triplicate or better testing should have been carried out for useful results. Results without such information are of little value.

5.1 Effect of the substrate

The substrate can be an important consideration. Testing an epoxy, for example, applied to concrete will usually fail in the concrete, but the same epoxy over steel will fail in the epoxy, assuming there is no weaker interface. O'Dea et al (11) investigated various pull-off testing methods and standards for concrete. For such work, the dolly diameter is usually much greater, 50 mm or more. There are different standards for testing, and often different testing units are used so results cannot be compared with the other work discussed in this paper which covers steel substrates. Of interest were the results from different testers using different methods, but all gave low values of 1.4 to 3.4 MPa as the concrete substrate is typically the region where failure arises. The fixed alignment unit gave lower figures, but not always.

5.2 Cutting around dolly

Before looking at typical results, it is worth looking at the influence of cutting around the dolly, a factor often considered important. Some have said this has no effect (12), others have shown it only has an effect with thicker coatings (6) while Baek et al noted an effect only with thin coatings (8), although the effect was minor. ISO 4624 claims it can have "a big influence in the adhesion of the paint system" (clause 5.4 in 2002 version). ASTM D4541 states that "Scoring around the fixture violates the fundamental in-situ test criterion that an unaltered coating be tested" with the additional comment in the 2009 version that "Scoring is only recommended for thicker films, that is thicknesses greater than 500 microns, reinforced coatings and elastomeric coatings". Fletcher and Barnes (13) investigated the effect of a number of variables on pull-off results. With 15 investigations using a Type V tester and results varying from 8.5 to 10.6 MPa (failure in the coating or glue or both), little effect of cutting could be determined. Unfortunately the type of coating and its thickness were not given with this work. It is most likely that the effect of scoring is often absorbed by the uncertainty noted above, but may be important in thicker coatings.

5.3 Glue failure

Clearly the glue should have a greater cohesive or adhesive strength than the coating being investigated, but fractures within the glue have been reported. Different glues will have different strengths which could influence results. Standard two-part epoxies are considered to have highest strength, with cyanoacrylates somewhat lower. A number of investigations have shown that both give high strengths greater than 10 MPa (with Type III or Type V devices). Fast drying epoxies have been found to have the lowest strength (around 5 MPa) with similar test units. Zhang and Myers (9) used a standard epoxy glue and the maximum pull-off strength (coating cohesion failure) reported was 18.6 MPa, indicating glue strength must be greater than this figure. Nilsen and Scheie (7) tested a number of cyanoacrylate glues using a PATTI (Type IV) device, reporting pull-off strengths above 25 MPa. Fletcher and Barnes (13), using a Type V device compared two glues provided with test kits (assumed to be epoxies) and both gave pull-off strengths of about 10 MPa. The authors noted "the crucial factor is that [they] have sufficient strength to carry out a successful adhesion test" which should not be a problem, unless perhaps fast drying epoxies are used. Another issue with the glue is that it is important that it does not modify coating properties, although this has not been reported.

5.4 Coating (Cohesion) strength before and after weathering

Biddle (14) found that the Type II pull off strength of uncured or partially cured inorganic zinc primer under an epoxy gave cohesion failure values of 1.5 to 2 MPa with failure in the primer, increasing to 3.2 MPa or higher for cured IZS primer, with the break occurring in the epoxy film.

Schwab and Drisko (12) applied a number of coating systems to blast cleaned panels and measured the pull-off strength using a modified tensile testing machine which, due to the slow and steady load application, would give results similar to or higher than a self-aligning hydraulic or pneumatic unit. They exposed the panels to 8000 plus hours in a salt fog cabinet and measured the pull-off strength again. They also assessed the degree of coating breakdown as a blistering rating from 1 to 10 (10: best, 0: worst) after exposure to salt spray. Figure 3 shows the initial coating fracture strength for epoxy, vinyl, alkyd and acrylic latex ranges from 6 to 18 MPa, but a much lower figure was obtained for the inorganic zinc silicate (IZS)/ vinyl system, which adhesively failed at the primer/ top coat interface. The alkyd and acrylic latex systems failed early during accelerated testing and were removed from the test. After salt fog exposure, pull-off strength dropped considerably for three of the remaining systems, but increased slightly for the inorganic zinc/ vinyl system. There was no relationship between pull-off strength (either before or after weathering) and degree of blistering. This work gives an indication of the fracture strength of a number of coating types and shows that, apart from IZS, weathering causes a significant drop in strength.

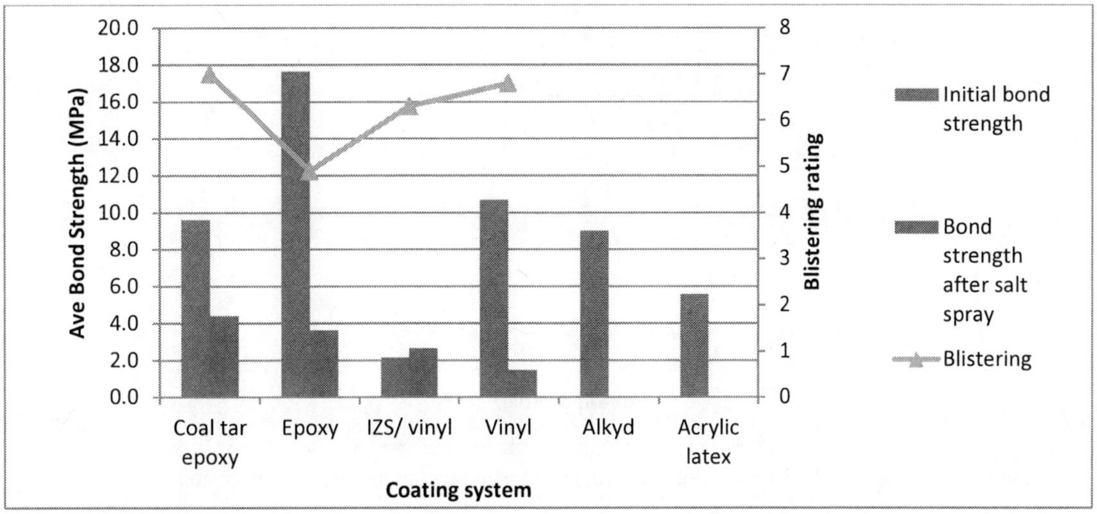

Figure 3: Pull-off strength and blistering as a function of coating system before and after salt fog exposure.

Similar results were reported by Zheng and Myers (15), who carried out a large number of pull-off tests (ASTM D4541 Type V testing unit) on a range of coating systems before and after accelerated testing (freeze-thaw and salt spray). Thick (0.5 to 1mm) polyurea systems showed good initial strength of around 13.8 MPa, failing by cohesion within the thick topcoat. On salt spray exposure, pull-off strength dropped to 2.8 to 4.1 MPa, with adhesion failure between substrate and primer. Despite the thick topcoat, corrosive spray had penetrated to the primer/ substrate interface and caused the primer to lose adhesion.

Baek et al (8) carried out pull-off 'adhesion' testing but reported no adhesion failures, so was in effect using the method to determine fracture strength of coatings. They used a Type IV pneumatic instrument (PATTI) with 13mm dollies on a polyamine epoxy. They found that tensile strength increased slightly from around 12.5 to 14.5 MPa with curing from 7 to 56 days at 23°C. The samples were cured in air for four weeks then soaked in water for 14 days at 50°C. Pull-off strength dropped to around 8.5MPa immediately after soaking, but recovered to around 13MPa after about 800 hours of drying.

Bajat and Dedic (16) determined the adhesion of epoxy primers using a Type II tester. They did not record failure locus, but it appears to be cohesion failure. The zinc rich epoxy gave a pull-off strength of 7.5 MPa, a zinc chromate epoxy 5 MPa and a non-inhibitive epoxy primer 4.5 MPa. They measured adhesion after water soaking and recorded a reduction in coating strength with water saturation of the coating, dropping to 1 to 2 MPa for all three coatings. They did not record a change in failure mode, so it is assumed to still fail cohesively. Figures twice these would be expected with other testing units.

Curran (17) investigated pull-off strength when looking at polysiloxane alternatives to polyurethane systems used at NASA Kennedy Space Center. Testing was carried out to ASTM D4541, but the test unit type was not given although it appears to be hydraulic. As part of the investigation, adhesion of the IZS primer was investigated, with pull-off values of 2.8 to 6.2 MPa for a range of commercial solvent-borne inorganic zinc primers, failing cohesively. Interestingly, adhesion was tested after heating to 400°C for 24 hours, for coatings used in regions of rocket exhaust. Post-heat adhesion increased significantly to 11.6 to 17.9 MPa as a result of continued curing of this type of coating. Adhesion testing of the candidate topcoats showed a range of pull-off strengths and failure modes, but where failure occurred cohesively in the IZS primer, values of 12.4 to 14.5 MPa were observed, indicating the primer was thoroughly cured before topcoating.

These results suggest that cohesion strength of epoxy coatings, using a hydraulic or pneumatic tests should be at least 10 MPa, although other workers have quoted figures two or three times this. Other coatings will probably have slightly lower cohesive strength. Improperly cured inorganic zinc will show low cohesive strength (usually less than 5 MPa) but will increase on curing and weathering. Other generic types will show a marked reduction in cohesive strength on weathering or water soaking.

5.5 Adhesion strength

Very few pull-off investigations actually quote true adhesion strength. Most failures, such as detailed above, are cohesive. One useful investigation was that of Islam et al (18) who looked at the effect of different methods of surface preparation on the pull-off strength of chopped strand mat impregnated epoxy resin applied to steel. Pull-off tests were carried out to ASTM D4541 and, although the type of tester is not mentioned, it appears to be a Positest (Type V) model. The authors found a relationship between the percentage adhesion failure and the pull-off strength (Figure 4). For a purely cohesive (resin) failure, pull-off strength was around 14 MPa, and for a purely adhesive failure, the strength dropped to just over 1 MPa. The pull-off strength increases as the failure becomes more cohesive. It would appear that, if the failure is more than a few per cent cohesive, then the pull-off strength will noticeably increase.

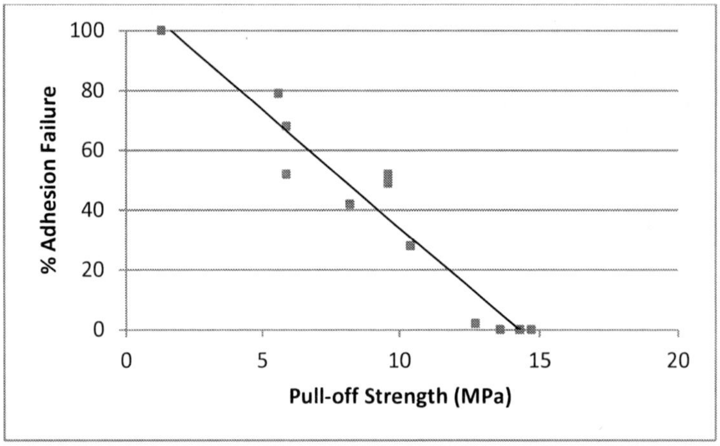

Figure 4: Relationship between pull-off strength and % adhesion failure for one investigation (18).

Bilotta et al (19) carried out adhesion tests on intumescent coatings using a Type IV tester and found low adhesion between the primer and intumescent coat. The thickness of the intumescent coat had some influence, with pull-off values of 4 to 8 MPa for coatings of around 500 microns, but down to 1 MPa when the intumescent coat was applied at 1500 microns.

Higher values for adhesion are uncommon, but have been reported. For example, Curran (17) reported two results which showed adhesion failure between the mid and topcoat, with relatively low values of about 6.2 MPa, but two similar adhesion failures gave pull-off strengths of 11.7 MPa and 13.8 MPa, significantly higher pull-off adhesion values than appear to have been recorded elsewhere. Yang (20) investigated adhesion of coatings to galvanized surfaces using, it would appear, a Type V testing unit. There were a number of largely adhesion failures between the epoxy intermediate and galvanized surfaces resulting in pull-off strengths of 4.1 to 13.8 MPa for epoxy/ polyurethane systems applied to 'mechanically-prepared' galvanizing, which appears to have been prepared with a fine disc sander. For whip blasted galvanizing, adhesion failures between the primer and the zinc were not observed, with all failures cohesion or adhesion between one of the other layers, and pull-off figures of 10.3 to 17.2 MPa.

It would appear that 'poor' adhesion gives pull-off strengths of approximately 1 to 5 MPa, but there may be rare cases where there is an 'adhesion' failure at values expected for cohesion or glue failures, which will signify 'good' adhesion.

5.6 Summary

Figure 5 combines many of the above results (along with some others) to show the range of failure modes and pull-off strengths that may be observed. These can be divided into a number of alternatives. Substrate and glue failures have not been considered, and the strength values are those obtained with hydraulic or pneumatic testing units. Pull-off values obtained with a mechanical (Type II) unit would be about half those noted.

1. True adhesion failures may be between the first coat and substrate or between coats. If adhesion is poor, pull-off values of less than approximately 3.5 MPa will typically be observed, usually much less than this. 'Good' adhesion will show values of double this, usually more.
2. Whether an adhesion value of 3.5MPa or less is a problem depends on the situation. If the coating is exposed to the atmosphere, then it is unlikely to be subject to significant tensile stresses and there are unlikely to be problems. However, if the coating is subject to tensile stresses, then poor adhesion could be a problem. For example, if the coating is to be overcoated with a product with high shrinkage stresses, such as some epoxies, then it may lift. Low adhesion values should be considered when selecting maintenance paint systems.
3. The values obtained for cohesion failures are an indication of the tensile strength of the coating, and provide no information on adhesion strength, other than observing it must be greater than the value obtained. Sound epoxies will give pull-off values of at least 10 MPa, but values of two or three times this may be observed. Other generic types (other than inorganic zinc, see below) may be slightly weaker.
4. Weathered or water-saturated coatings will fail cohesively or adhesively at lower pull-off values, usually between 1 and 10 MPa depending on how degraded the coating has become.
5. Insufficiently cured inorganic zinc silicate will fail cohesively at low pull-off values of approximately 1 to 5 MPa, depending on the extent of curing. Properly cured or weathered IZS will fracture at increasing greater pull-off values as the coating cures and hardens.

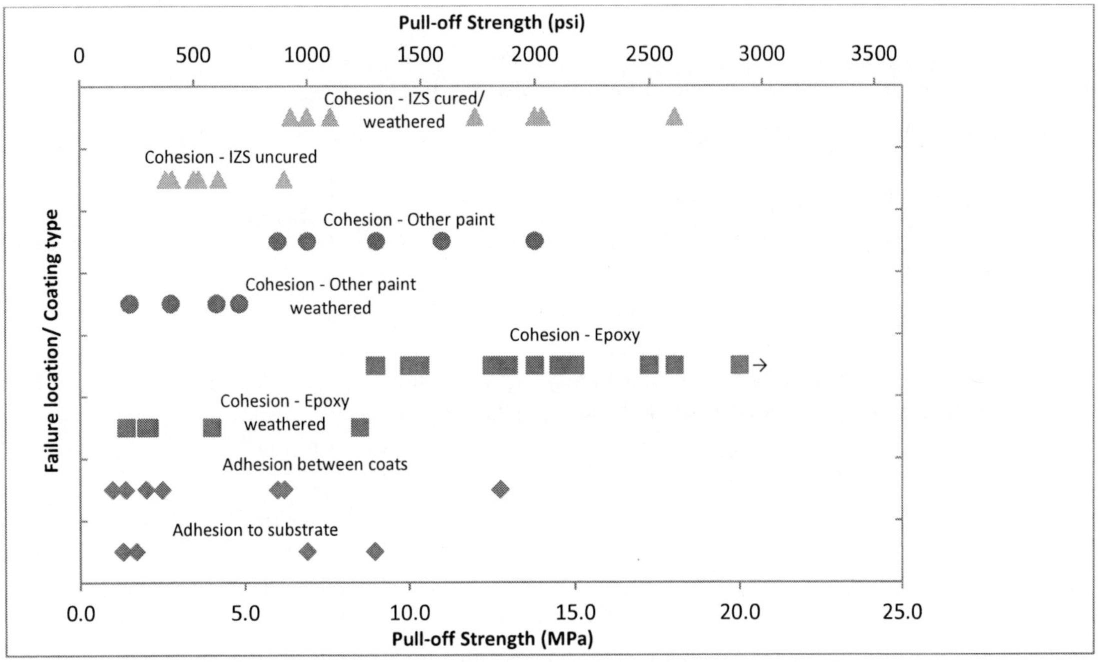

Figure 5: Summary of pull-off strength results for different failure locations and coating types.

6. CONCLUSIONS

Pull-off testing is commonly specified or used to quantify the adhesion of coatings to substrates, or adhesion between coats. However, close investigation of the methods used, and published results, show that it does have a number of weaknesses which limit its value. These findings can be considered as the 'Terms and Conditions' for those carrying out pull-off testing and include:

a) There are a number of commercial devices available, but results can differ significantly between them. For example, the mechanical, spring-operated device will give pull-off values about half that of pneumatic or hydraulic types.
b) The relatively poor repeatability and reproducibility of the test methods makes obtaining and interpreting meaningful results difficult.
c) There is no convincing evidence that such results provide any indication of long-term coating performance.
d) Most failures will be cohesion within the coating, which is a measure of coating tensile strength, but does not relate at all to coating adhesion.
e) An adhesion failure at a relatively low pull-off strength (<3.5 MPa) is an indicator of poor coating adhesion. However, low pull-off strength is unlikely to be an issue, except when selecting coatings for overcoating and repair.

7. REFERENCES

1 ASTM D4541 "Standard Test Method for Pull-Off Strength of Coatings Using Portable Adhesion Testers"

2 AS 3894.9, "Site testing of protective coatings - Determination of adhesion"

3 ISO 4624, "Paints and varnishes -- Pull-off test for adhesion"

4 ISO 16276-1 "Corrosion protection of steel structures by protective paint systems — Assessment of, and acceptance criteria for, the adhesion/cohesion (fracture strength) of a coating — Part 1: Pull-off testing"

5 M Schilling, "Coating Adhesion Testing in Accordance with ASTM D4541—Sticky Business", JPCL, April 2004, p18.

6 T Cunningham and J Steele, "Measuring adhesion of coatings to concrete and steel", NACE Conf 2000, paper 00614. Also, Materials Performance, January 2001, pages 36 – 40.

7 R Nilsen and S Scheie, "Testing the Accuracy of Adhesion/Cohesion Equipment", PCE, November 1996, p24.

8 Y H Baek "Reliability on Coating Pull-off Adhesion Strength Test", NACE Conference 2009, paper 09007.

9 W Zheng and J J Myers, "Adhesion Tests And Failure Modes Study On Structural Steel Coatings", SSPC Conference, San Antonio, 2013

10 S G Croll, C A Vetter and B D Keil, "Variability of Pipe Coating Pull-off Adhesion Measurements on Cylindrical Steel Pipelines", Pipelines 2012: Innovations in Design, Construction, Operations, and Maintenance, ASCE 2012.

11 V O'Dea et al, "Pull-Off Adhesion Strength Testing of Lining Systems on Concrete", JPCL, Vol 33, p64-74, Jan 2016

12 L K Shwab and R W Drisko, "Relation of Steel Surface Profile to Coating Performance", Corrosion Control by Organic Coatings, H L Leidheiser, ed, pages 222-226, NACE, 1981.

13 J F Fletcher and D J Barnes: "Pull-Off Adhesion Testing of Coatings – Improve Your Technique" SSPC Conference 2015. Also, Corrosion & Prevention 2015, Adelaide, paper 095.

14 G J Biddle, "Paint Coating Selection for Atmospheric Exposure", ACA Conference 31, Paper C2, Sydney, Australia, 1991.

15 W Zheng and J J Myers, "Adhesion Tests And Failure Modes Study On Structural Steel Coatings", Conference: 2013 The Society for Protective Coatings (SSPC) Annual Meeting, San Antonio, Texas.

16 J B Bajat and O Dedic, Adhesion and corrosion resistance of epoxy primers used in the automotive industry", J. Adhesion Sci. Technol., Vol. 21, No. 9, pp. 819–831 (2007)

17 J Curran, "Validation of Alternatives to Aliphatic Isocyanate Polyurethanes", NASA Technical Report KSC-2007-225, 2007.

18 M S Islam, L Tong and P J Falzon, "Influence of metal surface preparation on its surface profile, contact angle, surface energy and adhesion with glass fibre prepreg", International Journal of Adhesion & Adhesives 51 (2014) 32–41.

19 A Bilotta, "Tests On Intumescent Paints For Fire Protection Of Existing Steel Structures", IFireSS – International Fire Safety Symposium, Coimbra, Portugal, 20th-23rd April 2015.

20 S C Yang et al, "Measurement of Adhesion Properties between Topcoat Paint and Metallized/Galvanized Steel with Surface Energy Measurement Equipment", New England Transport Consortium Report NETCR93, Sept 2013.

8. AUTHOR DETAILS

Rob Francis has over 40 years' years' academic, industrial and consulting experience in corrosion and protective coatings. Dr Francis has a B.Sc. in metallurgy and a Ph.D. in corrosion science. He has authored or co-authored over forty technical papers or presentations on corrosion and coatings. He edited the publication "Inorganic Zinc Coatings", for the ACA in 2013. He has been awarded the JPCL editor's award twice and was made a JPCL Top Thinker in 2012. He was awarded the Victor Nightingall award for outstanding contributions to the protective coating industry by the ACA in 2014 and made an ACA Life Member in 2016.

LIFECYCLE COST STUDY FOR MAINTENANCE PAINTING AT COAL EXPORT TERMINAL

M. Rutherford
Conspectus (Qld) Pty Ltd, Queensland

SUMMARY: Asset owners spend hundreds of millions of dollars every year to remediate corrosion and deterioration of steel structures. The corrosion and deterioration of steel structures in marine environments has emerged as a most severe and demanding challenge for the owners of these assets.

What is required is the capability to assess the corrosion degradation of structures and implement an effective remediation and maintenance solution committed to the philosophy that all structural remediation methods must be founded upon thorough investigation of the cause and extent of the deterioration and distress. This can be done by means of a through asset condition assessment and corrosion audit. Once this is completed we can then proceed to produce a complete maintenance painting program for the lifecycle of the structures.

It is a crucial part of asset management to determine the remaining lifecycle of an asset and the capability of the asset to meet the designed performance and level of service requirements.

Being unaware of the current condition of an asset may lead to the premature failure of the asset leaving limited options to the facility owner with replacement being the most expensive option. Unforeseen failure of an asset provides major consequences that constitute a risk to business operations or potential loss to the organisation.

Keywords: Corrosion, Steel, Marine, Asset Maintenance, Protective Coating.

1. INTRODUCTION

This Lifecycle Cost Study case history relates to the provision of corrosion and protective coating consultancy services to provide a long range protective coating maintenance plan for the steel structures within a Coal Export Terminal. The aim was to prepare a document that outlines the current condition and anticipated durability of the asset, maintenance priority parameters and provide sufficient information regarding surface area of each individual asset to compile a cost modelling document for maintenance works over a thirty (30) year period to assist the client prepare long term budget estimates. The completed study has become an on-going working document.

Asset condition assessments combine the processes of periodic inspection and testing and the assessment and interpretation of the resultant data to provide an indication of the current condition of a specific asset as to the determination of the requirement for remedial action.

Asset condition assessments determine the physical state of an asset that may affect the performance of the asset and the ability of the asset to provide the required level of service. Gathered information assists the determination of the remaining useful life of an asset and the scheduling of remediation requirements that are required to reinstate the level of service that is provided by the asset to the desired standard.

The benefits of knowing the current condition of an asset are:

- The ability to plan and manage the delivery of the required level of service of the asset.
- Avoiding premature asset failure providing the option of cost effective remediation.
- Providing an accurate estimate of future expenditure that is required.
- Determination and refinement of maintenance and rehabilitation strategies.

Figure 1 – Assessment of Stacker Reclaimer Mast

Figure 2 – Assessment of Shiploader Conveyor

Figure 3 – Under Berth Inspection Unit

The Corrosion and Protective Coating Consultancy Services provided consisted of the following procedures and tasks:

- Review of the client's Drawing Register, previous coating and condition reports, historical coating application costing data, completed maintenance coating works records, protective coating application history.
- Preliminary discussions with current protective coating contractors, protective coating manufacturers currently suppling the clients projects, quantity surveyors and the facility maintenance staff.
- Determination of surface areas of all structural assets by a professional Quantity Surveying organisation with surface areas being determined from details taken from drawings sourced from the clients Drawing Register, physical measurements taken on site and equipment manufacturer information supplied by the client.

- Assessment of the current condition of assets to determine protective coating system durability, measurements of structural elements provided by the Quantity Surveyor for each asset, past budget estimates and actual costs for the treatment of structural assets.

Formulation of parameters for the calculation of future budget requirements including preparation of an accounting model in which figures are presented, assessment of the cost model by an independent professional accountant, presentation of draft Long Range Protective Coating Maintenance Cost model to the client, discussion regarding the document construction intended method of use and implementation of the recommendations and amendments and adjustments to the document to accommodate any client modifications required.

The Maintenance Painting Forecasts were provided for the following on-shore and off-shore structures and assets:

- Yard, Trestle and Berth Conveyors, Rail Receival Dump Stations, Yard and Berth Transfer Towers, Surge Bins, Stackers and/Stacker Reclaimers, Approach Trestles, Berths, Mooring Dolphins, Bridges, Shiploaders and Sample Plants.

Figure 4 – Approach Trestle and Trestle Conveyors

Figure 5 – Stacker Reclaimer Boom and Boom Conveyor

Figure 6 – Shiploader and Shiploader Conveyors

Figure 7 – Berth and Berth Truss Bridge and Conveyors

Figure 8 – Berth Truss Bridge Coating Failure

Figure 9 - Conveyor Column Severe Corrosion

2. LIMITATIONS

The Long Range Protective Coating Maintenance Cost Modelling for the structural assets at the Coal Export Terminal were based on known and available Australian and International reference standards, coating and corrosion surveys conducted in the past and during this project, historical coating works records provided by the client and the use of then currently available coating materials and application costs

Any likely variations to the figures provided in the cost modelling due to future environmental conditions, developments in technology, legislation, acceptable practices and materials that may influence coating life and budget estimates were unable to be predicted.

The cost estimates prepared by the consultancy and the quantity surveyor involved with this project were based on the assets outlined above and the Drawings provided by the client. In many instances, the provided Drawings reference other Drawings that were not provided. Access to these Drawings was essential in undertaking the Scope of Works and most of the referenced Drawings were made available to verify the works involved and determine if the additional works could be covered by the allowances made in the proposal. Additional site and take off time that exceeded that allowed in cost estimate to cover the work required for the short supplied Drawings was negotiated as a variation to the proposal.

3. SCHEDULE

The schedule to undertake the works was developed through discussion and negotiation with the client and the quantity surveyor.

The agreed schedule was to gather all available information from the client including current and missing drawings, historical protective coating records and other information that would assist with the development of the Long Range Cost Modelling. Review. Drawings were essential in determining surface area take-offs for all structural components of the relevant structures and protective coating records assisted assessment of past protective coating life cycles for on-shore and off-shore structures.

The format for the draft Long Range Protective Coating Maintenance Cost Model was discussed with the client's staff to enable integration with current and proposed maintenance programs utilised. This included sourcing the required details for all structures and assets for discussion with the independent quantity surveyor to provide support during the completion of the quantity surveying works. Site visits undertaken independently by the consultant and/or with the quantity surveyor or others were scheduled as required to ensure accuracy of surface areas of structural sections.

All structural asset measurements supplied by the quantity surveyor were collated together with the durability assessment report and budget estimates were completed in-house. All information was combined into one document for further development and forecasting with cost modelling and maintenance priority forecasting prepared in line with each asset. A Draft Report was then prepared.

Project Progress Reports were prepared monthly outlining the status of the works completed to the end of each calendar month including internal revision of the draft report and preparation for release approval.

The Draft Report was presented to the client with discussion regarding the development and the intended methodology for the use of the document. The detail provided was evaluated and amended or modified rto suit client requirements.

4. DISCUSSION

The capabilities of Lifecycle Cost Study are based on results determined by Asset Condition and Corrosion Assessment Surveys that are undertaken on an annual basis.

Asset Condition and Corrosion Assessments combine the processes of periodic inspection and testing, the assessment and interpretation of the resultant data to provide an indication of the current condition of a specific asset as to the determination of the requirement for remedial action.

Asset maintenance to be undertaken over the balance of structure service life is a major challenge to provide reliable and sustainable facility operation. Operating ageing facilities efficiently and safely requires an asset maintenance cycle to include inspection, diagnosis, evaluation and implementation of remediation processes.

It is a crucial part of asset management to determine the remaining lifecycle of an asset and the capability of the asset to meet the designed performance and level of service requirements. Asset Condition and Corrosion Assessments determine the physical state of an asset that may affect the performance of the asset and the ability of the asset to provide the required level of service.

In today's environment, the preventative maintenance of ageing structures is far more preferable to the construction of new structures once original design lives have been exceeded. Asset Condition and Corrosion Assessment gathered information assists the determination of the remaining service life of an asset, the scheduling of remediation requirements that are required to reinstate the level of service that is provided by the asset to the desired standard.

Being unaware of the current condition of an asset may lead to the premature failure of the asset leaving limited options to the facility owner with replacement being the most expensive option. Unforeseen failure of an asset provides major consequences that constitute a risk to business operations or potential loss to the organisation. The benefits of knowing the current condition of an asset are the ability to plan and manage the delivery of the required level of service of the asset, avoiding premature asset failure providing the option of cost effective remediation, providing an accurate estimate of future expenditure that is required and determination and refinement of maintenance and rehabilitation strategies.

Assessment of damaged or deteriorated structures should only be made by qualified and experienced people and the process of should always include the aspects of the condition of the structure including all visible, non-visible and potential damage and defects, a review of the past, current and future exposure and service functions/requirements.

With most damaged or deteriorated structures, the facility owner has a number of options which will effectively decide the appropriate remediation and protection strategy that will meet the future service and exposure requirements of the structure. These options will include to do nothing, downgrade the capacity or function of the structure, prevent or reduce further damage without repair, improve, strengthen or refurbish the structure, reconstruct all or part of the structure or demolish the structure.

Proper remediation methodology begins with inspection and testing to identify the type and extent of defects and degradation mechanisms and the overall condition and quality of the structure. Remediation projects are prone to increasing in volume and costs once work has commenced – investing in comprehensive and accurate Asset Condition and Corrosion Assessments has proven cost effective in the long term.

Often there is limited information on "as built" with drawings and construction records being partial at best and more than often incorrect.

Understanding of structure is critical in being able to provide comprehensive report on all aspects of the construction envelope. Prior to diagnosing the causes of defects or failure within a structure it is important to understand that defects result from several factors: design, construction practices, materials, the environment, loading applied to structure.

The Asset Condition and Corrosion Assessment was intended to form the foundation for short-term maintenance strategies in which structural elements of the facility are prioritised aligned with the visual degree of deterioration and degradation.

All structural sections were documented in the format as shown in Table 1 below:

Table 1 – Document Format for Structural Sections

Overall Structure	Client Drawing Reference Number, Asset ID, Structure, Total Surface Area, Operating Environment, Past Coating Details, Assessment Methods and Overall Condition Assessment
Section Item Each documented line item has a reference number allocated to assist cross referencing with site drawings and prioritising maintenance	**Investigation Details:** Item Number, Section Description, Last Coating Applied (date and method), Drawing Reference and Estimated Surface Area for Repair. **Visual and Physical Site Testing Conducted:** Defect Photo Numbers, Coating and Corrosion Defects (blistering, flaking, delamination, area percentage of corrosion), m2 per section, Adhesion Testing, Section Condition Description, Summary Recommendations (scaffold requirements, maintenance surface preparation requirements and maintenance priority.

Interpretations for Coating Condition were based on NACE Paper 477 as shown in Table 2 below:

Table 2 – Interpretations for Coating Condition

Coating Condition Rating	Coating % Failure
G = Good Condition	<2% Failure
F = Fair Condition	3%-5% Failure
P = Poor Condition	5%-10% Failure
VP = Very Poor Condition	>10% Failure

Interpretations for Maintenance Priority were based on relevant sections of NACE Paper 477 (overall condition, coating age and system type), AS 1508 Method 408.2/4/5, Method 481.1.19, Method 401.5 and SSPC Vis 2. These interpretations were condensed into six main Ratings as shown in Table 3 below:

Table 3 – Interpretations for Maintenance Priority

Maintenance Priority Rating	Coating % Damage
A	<3% Coating Damage – No Corrosion
B	3%-5% Coating Damage – No Corrosion
C	5%-10% Coating and Corrosion Damage
D	10%-15% Coating and Corrosion Damage
E	15%-<20% Coating and Corrosion Damage
F	>20% Coating and Corrosion Damage

Interpretations for Overall Condition were based on Coating Condition, Corrosion Severity, Coating Defects and Combined Information with ratings as shown in Table 4 below:

Table 4 – Interpretations for Overall Condition

Condition Rating	Overall Condition
G	Good Condition
F	Fair Condition
P	Poor Condition
UA	Urgent Action/Engineering Decision Required

The Site Information Gathering Document utilised for the project works is as shown in Table 5 below:

Table 5 – Site Coating Condition and Priority Maintenance Information Gathering Document

Coal Load Out Site Coating Condition and Priority Maintenance Information Gathering Document

Date	28th July 2008	Layout Drawing Ref No	24378	Asset ID	7P	Structure	Conveyor 7P	Total Surface Area M²	3864m²
Substrate	STEEL								
Client									

Opperating Environment / Past Coating Details / Assessment Methods & Overall Condition Assessment

Operating Environment – Classification NACE Paper No 477		Coating Details – Known Coating System Applied to Structure or Section					Coating & Corrosion Defects – Assess Coating & Corrosion Defects as outlined below							Overall Structure Condition
Off Shore Immersed		Last Coating Applied & Date	Year	HPS	Spot / Whip / Full	Full	Blisters as per AS1580 Method 481.1.10 Rating and Density				Corrosion as per SSPC Vis 2			Based on overall condition
Off Shore Atmospheric		System Products Applied	Measure DFT as per AS 3894.3 Method B				Blisters	Cracking	Delamination %	Flaking	Corrosion	Pitting Depth		Good/Fair/Poor/ VeryPoor
Sea Coast Marine		No records provided	1st coat	DFT Specified Microns			Rating & Density	%	15%	%	20%	MM		Poor
Sea Coast Marine Heavy		HPX	2nd Coat	DFT Specified Microns			Assess Adhesion and Hard Dry site testing in accordance with standards outlined below							
Industrial	X		3rd Coat	DFT Specified Microns			Pull off Test as per AS 1580 Method 408.5				Cross Cut Test Standard AS 1580 Method 408.4			
Severe Mining / Heavy Ind	X		4th Coat	DFT Specified Microns			Adhesion Knife Test AS 1580 Method 408.2				Hard Dry Test AS 1580 Method 401.5			
Other (Nominate)			Total Average Existing Coating DFT		585um		Maintenance Priority Predictions to be based on NACE Paper No 477.							

Site Section Investigation Details / Visual and Physical Site Testing Conducted / Summary Recommendations

Section Item No & Description: Item No	Description	Last Coating Applied Method Spot/Whip/Full: Date	Method	Drawing Ref.	Estimated Repair Surface	Defect Photos: Numbers	Blisters Rating	Coating and Corrosion Defects (Total Overall Coating & Corrosion Damage as % of Section): Flaking (%)	Delamination (%)	Corrosion (%)	M² Per Section	Adhesion / Hardness (Good / Fair / Poor / UA): Cross Cut	Pull Off	EIC Y/N	Section Condition Description: Good / Fair / Poor / UA	Scaffold Req: Suspended EWP Other	Maintenance: Spot / Whip / Full Blast Other	Priority: A,B,C,D,E,F
130, 131, 176	Top bottom Beam/Trestle Beam/Columns	N/A	N/A	1	15	2552, 2553,2609			10	20	370			Y	Poor	H	Spot	F
														Y				
130, 131, 176	Top bottom Beam/Trestle Beam/Columns	N/A	N/A	2	15	2552, 2553,2609			10	20				Y	Poor	H	Spot	F
														Y				
130, 131, 176	Top bottom Beam/Trestle Beam/Columns	N/A	N/A	3	15	2552, 2553,2609			10	20				Y	Poor	H	Spot	F
														Y				
130, 131, 176	Top Beam/Trestle Beam/Columns	N/A	N/A	4	15	2552, 2553,2609			10	20				Y	Poor	H	Spot	F
														Y				
130, 131, 176	Top Beam/Trestle Beam/Columns	N/A	N/A	5	15	2552, 2553,2609			10	20				Y	Poor	H	Spot	F
														Y				
130, 131, 176	Top Beam/Trestle Beam/Columns	N/A	N/A	6	15	2552, 2553,2609			10	20				Y	Poor	H	Spot	F
														Y				
130, 131, 176	Top Beam/Trestle Beam/Columns	N/A	N/A	7	15	2552, 2553,2609			10	20				Y	Poor	H	Spot	F
														Y				
130, 131, 176	Top Beam/Trestle Beam/Columns	N/A	N/A	8	15	2552, 2553,2609			10	20				Y	Poor	H	Spot	F
														Y				
130, 131, 176	Top Beam/Trestle Beam/Columns	N/A	N/A	9	15	2552, 2553,2609				5				Y	Fair	H	Spot	F
														Y				
132, 134, 177	Top Beam/Trestle Beam/Columns/Bottom	N/A	N/A	10	5	2554, 2555, 2610				5				Y	Fair	H	Spot	F
														Y				
132, 134, 175, 168, 169	Trestle Beam/X Beam/X Brace/Top Bottom Beam-Ocean/Road	N/A	N/A	11	35	2554, 2556, 2557,2608,2602,2601				20				Y	Poor	H	Full Blast	E

The Long-Range Maintenance Cost Model was presented in the format shown in Table 6 below:

Table 6 – Long-Range Maintenance Cost Model Format for Presentation

Key:

FB Full blast steelwork to class AUS 2.4

SW Spot blast up to 25% and whip blast remaining sound coating

SO Spot blast only

*Figures are indicative ONLY and include coating material ar

Item No.	Section Description	Dwg Ref / Section No.	Surface area M²	Scaff B/H	Encap Y/N	Section Cond.	2009 (Indexation Factor 3.00%) Blast Type $/M²	2010 (Indexation Factor 3.50%) Blast Type $/M²	2011 (Indexation Factor 3.25%) Blast Type $/M²	Bl
130, 131, 176	Top,bottom Beam/Trestle Beam/Columns-Ocean/Road	1	400	H	Y	F	FB $250 $163,564	0	0	
130, 131, 176	Top,bottom Beam/Trestle Beam/Columns-Ocean/Road	2	400	H	Y	F	FB $250 $163,564	0	0	
130, 131, 176	Top,bottom Beam/Trestle Beam/Columns-Ocean/Road	3	400	H	Y	F	FB $250 $163,564	0	0	
130, 131, 176	Top Beam/Trestle Beam/Columns-Ocean/Road	4	380	H	Y	F	FB $250 $155,386	0	0	
130, 131, 176	Top Beam/Trestle Beam/Columns-Ocean/Road	5	380	H	Y	F	FB $250 $155,386	0	0	
130, 131, 176	Top Beam/Trestle Beam/Columns-Ocean/Road	6	380	H	Y	F	FB $250 $155,386	0	0	
130, 131, 176	Top Beam/Trestle Beam/Columns-Ocean/Road	7	380	H	Y	F	FB $250 $155,386	0	0	
130, 131, 176	Top Beam/Trestle Beam/Columns-Ocean/Road	8	380	H	Y	F	FB $250 $155,386	0	0	
130, 131, 176	Top Beam/Trestle Beam/Columns-Ocean/Road	9	380	H	Y	F	FB $250 $155,386	0	0	
132, 134, 177	Top Beam/Trestle Beam/Columns/Bottom Beam-	10	380	H	Y	F	FB $250 $155,386	0	0	
132, 134, 175, 168,	Trestle Beam/X Beam/X Brace/Top,Bottom Beam-	11	380	H	Y	E	0	0	FB $250 $155,763	
132, 134, 175, 168,	Trestle Beam/X Beam/X Brace/Top,Bottom Beam-	12	380	H	Y	E	0	0	FB $250 $155,763	
132, 134, 175,	Trestle Beam/X Beam/X Brace/Top,Bottom Beam-	13	380	H	Y	E	0	0	FB $250 $155,763	

5. CONCLUSIONS

A Lifecycle Cost Study is an essential tool to develop fact and data related decisions in corrosion related areas and has a critical and essential role to play in asset management decision making processes.

The results determined in this case study present an example of how the Lifecycle Cost Study plays an essential role in asset management decision making, maintenance prioritising and its impact on operation of the facility. The developed Lifecycle Cost Study model provided a decision support system and determined that Lifecycle Cost Studies have an enormous potential for decision making support for asset management.

The Lifecycle Cost Study was completed and presented to the client. The Lifecycle Cost Study document has been adjusted following Annual Condition and Corrosion Assessments in the years following completion and is still currently being used as an asset maintenance planning tool by the client.

6. REFERENCES AND ACKNOWLEDGMENTS

The Lifecycle Cost Study concept, planning, development, initiation and completion of this case study was undertaken by Corprev Corrosion Engineers and Associates Pty Ltd and Conspectus (Qld) Pty Ltd.

7. AUTHOR DETAILS

Mike has three decades experience in providing a wide range of specialist asset maintenance engineering services for facility owners, engineers, consultancies, manufacturers and remedial contractors. Extensive experience has been attained locally and internationally in specialist asset maintenance services for all forms of concrete and steel remediation and protection within the water and wastewater treatment, marine and port, power generation, mining, refining, construction and remedial and restoration industries.

Mike was the 2013 recipient of the ACA Foundation's International Travel Scholarship and he is a member of the ACA Applicator Technical Group.

PROTECTING UNDERGROUND STRUCTURES AGAINST CORROSION - THE ROLE OF STAINLESS STEELS

L. H. Boulton
Consultant to the Nickel Institute, Auckland, New Zealand

SUMMARY: In recent years there have been cases of underground structures fabricated from conventional engineering materials suffering failures due to corrosion. Examples of underground corrosion problems include electric power transmission towers, buried LPG storage tanks, and metal collars on waste water pipelines. The cause of failure of underground structures is sometimes due to the material utilised and in other cases projects being driven by cost-saving during the design stage.

An established approach for the construction of structures designed for a long life employs the principles of life cycle costing. Life cycle costing (LCC) is a tool for dealing with the cost impact of selecting engineering materials for a construction project. LCC analysis carried out on an underground structure often favours the selection of stainless steel as the most durable material requiring low maintenance. In recent times stainless steels have been employed in a wide range of underground service environments such as road tunnels and power cable tunnels. When properly specified the appropriate grade of stainless steel will likely exceed the expected design life of the underground structure.

The paper presents cases where corrosion has not affected the performance of underground structures because the correct grade of stainless steel was specified at the design stage resulting in a long service life. An outline is given of the selection process for the appropriate grade of stainless steel to be utilised for underground structures including tunnels.

Keywords: Underground structures, corrosion of buried structures, tunnels, stainless steels

1. INTRODUCTION

Corrosion of construction materials used for an underground structure can lead to premature failure of plant or equipment. Such failure necessitates the replacement of structural components, at high cost in operational downtime, labour costs and replacement material costs. It is imperative that the material selection process for construction of buried structures and underground infrastructure is robust, in order to avoid unexpected deterioration of structural elements during service.

The first stage in evaluating candidate materials for underground infrastructure is to assess the service environment at the construction site. A corrosivity assessment will help to determine if the environment will be corrosive and the degree of corrosivity that is anticipated. The environmental corrosivity assessment may involve testing and evaluation of the properties of ground soil, ground water, and the atmospheric conditions if the structure is an underground tunnel.

The corrosion rate of materials in an underground structure will vary with the material and the environmental conditions that prevail at the construction site. Factors to be considered in a site corrosivity assessment include: soil chemistry, soil physics, moisture levels, humidity, temperature, and exposure to emissions if the structure is a tunnel. The environmental assessment should take into account whether the structure is in contact with backfill, subject to flood water, exposed to contaminated air, or possibly a number of these conditions prevailing simultaneously.

Material selection for the construction of underground infrastructure is a more difficult task than the material selection process for an above ground structure. This is due to the corrosive and/or hazardous exposure conditions encountered during underground construction. The construction materials selected must be *fit for purpose* when the underground structure is put into service. It is necessary to carry out proper due diligence during the material selection stage for an underground project. The material selection process is vital to the success of the project once the structure goes into service and it is exposed to the prevailing corrosive environment.

In recent years there has been an increasing awareness that life cycle costs (or *whole of life* costs) and not just the initial costs should be considered when selecting construction materials for underground structures [1]. Life cycle costing (LCC) takes into account:

- Initial costs.
- Operating costs including maintenance.
- Residual value.

Stainless steels are generally considered to be an expensive option for engineering applications. However, LCC has shown that using a corrosion-resistant alloy, such as a stainless steel, for critical engineering applications can help to minimise future maintenance, downtime and replacement costs. The choice of stainless steel can save costs which far outweigh the initial higher material cost.

The initial cost of structural stainless steel products is higher than the cost of an equivalent mild steel product, depending upon the grade of stainless steel that is selected. However, costs can be offset by the omission of surface coatings on mild steel that are needed at regular intervals over the service life. In addition, the higher corrosion resistance of stainless steel can offer additional benefits, including:

- Reduced inspection frequency and associated costs.
- Lower maintenance costs.
- Provision of a long service life.

Stainless steel (SS) also has a high salvage value, that is, value at the end of structure's life. However, this factor is rarely a deciding factor for a structure with a long projected life of possibly 100 years. LCC uses the standard accountancy principle of *discounted cash flow* to reduce all the costs to present day values. The discount rate encompasses inflation, interest rates, and sometimes a risk factor. This permits a realistic comparison to be made of the material options available, and potential long term benefits of using SS, to be accurately assessed against other engineering materials. Frequently, LCC analysis of candidate materials for a large project favours the selection of an appropriate grade of SS.

There is documentation available that assesses corrosion performance of engineering materials such as mild steel, cast iron, galvanised steel, lead, and aluminium alloys, used in the construction of buried infrastructure [2]. There is also information available on SS applications in underground tunnels around the world [3,4,5]. However, there is a dearth of information and scant data on the corrosion performance of various grades of SS buried in different soil types.

A study carried out in New Zealand (1992-95) investigated the behaviour of five grades of SS buried in seven soil types for applications of metal components utilised in the water industries. The SS samples were interred in the various soil types along with carbon steel (CS) reference samples and four common grades of copper alloys.

The samples of 3mm SS plates with TIG welding to join two halves of the sample plates (*Figure 1*) were buried at one metre depth at the various test sites. A full set of samples (CS/SS/Cu alloys) was exhumed annually for metallurgical laboratory examination during the 3-year study. The grades of SS that were buried for subsequent evaluation and testing were:

1) Austenitic SS grades 304L and 316L.
2) Duplex (austenitic-ferritic) SS grades 2304, 2205 and super-duplex 2507.

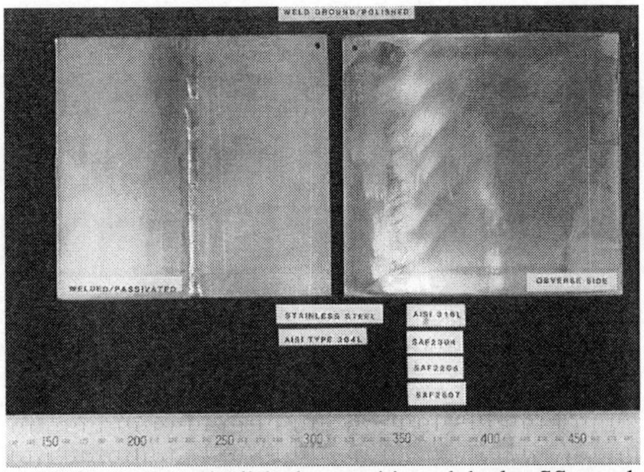

Figure 1. Five grades of welded, passivated/polished austenitic and duplex SS sample plates before burial [7]

The seven soil types and a common soil property, the soil resistivity, are given in *Table 1.*

Table 1. Soil Corrosivity and Resistivity at Test Sites

Site No.	Soil type	Resistivity (Ω-cm)*	Soil corrosivity
1	Clay (yellow)	900	Very corrosive
2	Sandy soil	500000	Non-corrosive
3	Reclaimed soil	3400	Corrosive
4	Peaty soil	14400	Mildly corrosive
5	Volcanic soil	11800	Mildly corrosive
6	Pumice	286500	Non-corrosive
7	Geothermal soil	20100	Mildly corrosive

*Wenner four-pin method for measurement of soil resistivity

Results of the study revealed that the SS sample plates with SS welds were all in original condition after burial for three years. There were two exceptions: the lean grades 304LSS and 2304SS, showed slight localised corrosion at mechanically engraved sample numbers on the SS plates, at two sites. The minor corrosion observed on the SS plate samples developed as a result of the crevice corrosion mechanism [6,7].

After exhumation at the seven soil test sites the CS reference samples gave general corrosion rates of between 0.02 mm/yr (site 2: sandy soil) to 0.07 mm/yr (site 7: geothermal soil). The CS corrosion rate results obtained were consistent with known data on the corrosion rates of CS buried in soils [8]. The copper alloys (60/40 brass; DR brass; LG2/AB2 bronze) showed corrosion rates from zero in pumice soil to 0.015 mm/yr in geothermal soil.

The sample exposure time in the various soils was limited, however, the results obtained for the buried SS samples showed that properly selected grades of stainless steel give excellent corrosion performance compared to conventional construction materials if an engineering structure made from the correct grade of SS is buried in the ground.

2. CASE STUDY

2.1 Durability of buried concrete jacking pipe collar materials

Pre-cast concrete jacking pipe segments are installed by underground thrusting of pipe segments into place in order to construct new pipelines for waste water collection projects. The pipe joints are made using annular collars from different materials depending upon the soil corrosivity and the service life expectancy. An assessment has been carried out to evaluate the durability of the concrete jacking pipe collar materials when buried in a range of different soil types.

The purpose of the assessment was to ensure that a chosen collar material was fit for purpose when a new pipeline is installed underground with pipe segments connected using collars made from the most appropriate material. A typical pre-cast concrete jacking pipe segment with a collar is shown in *Figure 2.*

Figure 3 shows two concrete pipe segments with metal collars in place during underground pipe-jacking to construct a new waste water pipeline.

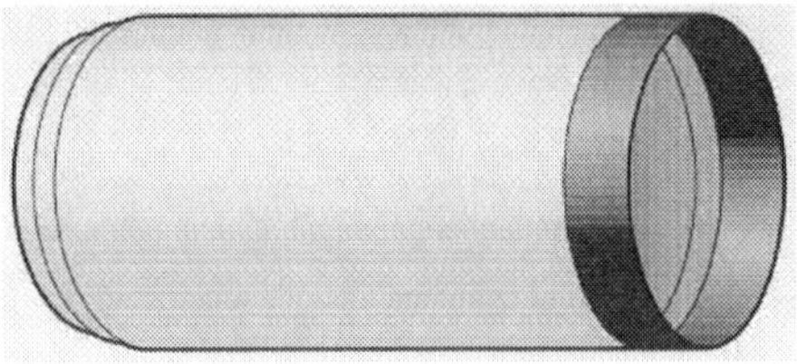

Figure 2. A pre-cast concrete jacking-pipe with its steel collar at the right end.

Figure 3. Concrete jacking-pipe segments being thrust underground.

2.1.1 Jacking-pipe service conditions

The corrosion rate for CS buried in typical loamy soils in New Zealand is approximately 0.02 mm/year [8]. A durability allowance can be factored into the design of the collar thickness for buried CS pipe collars to ensure that the expected design life of the pipeline is achieved.

The corrosion rates of welded *austenitic* stainless steel (SS) collar materials, such as grade 304 SS and grade 316 SS, can safely be assumed to be *zero* when buried in soils, based on the research results [6,7]. The welds on the SS pipe collars must be pickled and passivated after pipe collar fabrication which involves SS welding.

Fibreglass reinforced polymer (FRP) has a history of good durability when buried in soil for applications such as underground fuel storage tanks. However, cut edges on FRP collars for pipes need to be resin-coated to stop *wicking* of moisture that can lead to premature deterioration of the FRP collar material.

2.1.2 Soil Types

The soil types encountered in most parts of New Zealand are not generally very corrosive towards buried metals. However there are some exceptions to this such as burial in geothermal soils. Clay, peat, reclaimed soil and backfill can be more aggressive than are the common loam soils and sandy soils. For the purposes of the assessment the common NZ soils were classified into six types. The approximate soil corrosivity increases for the soil types, as shown below. The loamy soil types were less corrosive, whereas clay, peat and reclaimed soil were more corrosive.

- Loamy-sand
- Loamy-gravel
- Loam
- Clay
- Peat
- Reclaimed fill

↓ *Increasing moisture content and corrosivity*

The outcomes of the durability appraisal of the collar materials are given in *Table 2.* The life expectancy, given in years, is the *estimated* service life for the various collar materials for concrete jacking pipes. The estimated life expectancy of the pipe collar materials only allowed for external corrosion on the collars in the soil types. A newly installed pipeline was assumed to be buried about 5-6 metres below ground at the *water table* level.

The estimates of the expected life of the collars did not imply that failure of the collar material would occur after the service life period expired. The collar service life expectancy was the *best estimate* made using research results, soil properties and soil corrosivity for buried metals [6,7].

The assessment of the pipe collar durability demonstrated that grade 304 SS, grade 316 SS and FRP collars were the most corrosion-resistant materials available to avoid the anticipated corrosion damage mechanisms when a pipeline is buried in soil.

The life expectancy of pipe collars fabricated from grade 304 SS was 80-100 years; grade 316 SS was 100 years; FRP was 100 years. There was also a potential for these collar materials to provide satisfactory service on a buried waste water pipeline for more than 100 years.

Table 2. Durability Table: Pipe Collar Bands

Durability Table: Pipe Collar Bands

Collar Band Material #	Loam with sand	Loam with gravel	Loam	Clay	Peat	Reclaimed Soil or Fill
Mild steel standard thickness	>60 yr	>60 yr	>50 yr	60 yr	50 yr	50 yr
Mild steel + 1mm corrosion allowance*	100 yr	100 yr	100 yr	>80 yr	>80 yr	>80 yr
Mild steel + 2mm corrosion allowance*	>100 yr	>100 yr	>100 yr	100 yr	100 yr	100 yr
Mild steel + epoxy coating**	>80 yr	>80 yr	>80 yr	80 yr	80 yr	80 yr
Hot-dip galvanised steel **	>80 yr	>80 yr	>80 yr	80 yr	60 yr	60 yr
Hot-dip galvanised steel + epoxy coating**	>80 yr	>80 yr	>80 yr	80 yr	80 yr	80 yr
Fibreglass reinforced plastic (FRP)	>100 yr	>100 yr	>100 yr	>100 yr	>100 yr	>100 yr
Grade 304 stainless steel	100 yr	100 yr	100 yr	80 yr	80 yr	80 yr
Grade 316 stainless steel	>100 yr	>100 yr	>100 yr	100 yr	100 yr	100 yr

Notes on pipe collar durability:
The estimated life assumes that soil is moist and only general corrosion was occurring during service life.
* **Corrosion** allowance is the estimated increase in bare steel collar thickness required to give longer service life.
** Coatings applied on external metal collar surfaces are likely to be damaged during installation of the pipework.
Galvanised coatings on steel collars should comply with AS/NZS4680: *After-Fabrication Hot Dip Galvanising*.
The estimated life for steel collars with a surface coating assumes that localised corrosion does not occur.
If localised corrosion occurs the risk of failure increases and the safety factor for steel collar service life is diminished.

3. DURABILITY OF MATERIALS IN UNDERGROUND TUNNELS

The construction of underground tunnels provides an efficient way of improving the flow of road vehicles and rail traffic in large cities around the world. Tunnels can also provide a conduit for safely conducting high voltage power supply cables and other critical underground services into the CBD of large cities. The conditions that tunnels pass through can vary widely from drilling through rock, passing under a river and passing under a harbour. Each tunnel is unique when it comes to selection of the materials that are used in the tunnel construction and the ancillary equipment. The materials chosen for tunnels must be able to function satisfactorily for long periods of time with minimal maintenance under service conditions that are likely to be corrosive [5].

The most common type of transport tunnel is the road tunnel. The atmosphere that prevails inside road tunnels is almost always corrosive. The atmosphere can vary in temperature from cool to hot (14°C to 28°C) with varying relative humidity (40-95% RH). The tunnel environment likely contains corrosive gases such as sulphur oxides (SOx), nitrogen oxides (NOx), hydrogen sulphide (H_2S), dust, soot, carbonaceous debris, and water soluble chlorides.

The effect of these gases and vapours can cause corrosion on the tunnel infrastructure. Specific tunnel structures that can be affected by corrosion include power cable trays, cable support brackets and wall anchor points for tunnel linings. To counteract the corrosive nature of a road tunnel atmosphere and to minimise the devastating effect of a fire, tunnel engineers deploy SS in various applications such as fire exit doors, sprinkler systems, tunnel linings, anchor fixings, ceiling suspension hangers, lighting structures, signage hangers, cable ladders and cable support brackets.

The grades of SS that are employed in tunnels include austenitic grades 304 SS, 316 SS, 310 SS, and high alloys containing 6% Mo such as the proprietary super-austenitic grade 254SMO. Recently the duplex grade 2205 SS has been utilised for applications in road tunnels. The conditions and applications that exist inside an individual tunnel, and the degree of maintenance that is carried out, can determine which SS grades are specified for specific road tunnel applications.

3.1 Stainless steel utilised in power cable tunnels

Underground power cable tunnels are being constructed as part of energy supplier upgrading and future-proofing of electricity infrastructure in cities such as Auckland, New Zealand, and Sydney, Australia. The concrete tunnels (3-4 metre width) carry high voltage power transmission cabling into the CBD of the two coastal cities to ensure the safe delivery of electricity to commercial consumers. The atmosphere that prevails inside tunnels is hot and humid, the service environment is corrosive. Selection of the most appropriate materials to ensure long life in power cable tunnels is an important part of the engineering design for tunnel construction to ensure that the internal structures in the tunnels will provide a service life of more than 100 years (*Figure 4*).

Power cable support bracket arms fabricated from SS are fixed to the circular tunnel walls with SS concrete anchor bolts. The bracket support arms carry the power cables along the full length of the 9km long Vector tunnel in Auckland. Grade 304L SS was employed for the welded cable bracket arms shown in *Figure 4*. In addition, the Vector tunnel has a 304 SS fire sprinkler system in the ceiling and 26 metre high ventilation shafts above ground that are made from 900mm grade 316L SS pipe. The ventilation shafts allow cooling air to be drawn down into the underground tunnel. In the 2km long power cable tunnel that carries the power into the Sydney CBD, grade 316L SS is utilised for the cable bracket arms, shown in *Figure 5*.

Figure 4. 304L SS power cable support brackets installed during construction of Vector tunnel, Auckland.

Figure 5. 316L SS power cable support brackets installed in Sydney tunnel.

3.2 Stainless steel utilised in road tunnels

The twin 2.5km long Waterview Tunnels in Auckland utilise grade 304L SS support brackets on the walls in the tunnel invert culverts located below the main road tunnels, carrying power cable trays that are attached to the tunnel invert concrete walls (*Figures 6 & 7*). The environment in the twin tunnel invert culverts is hot and humid, and the atmosphere is corrosive.

Figure 6. Twin 2.5 km long concrete-lined Waterview tunnels, Auckland

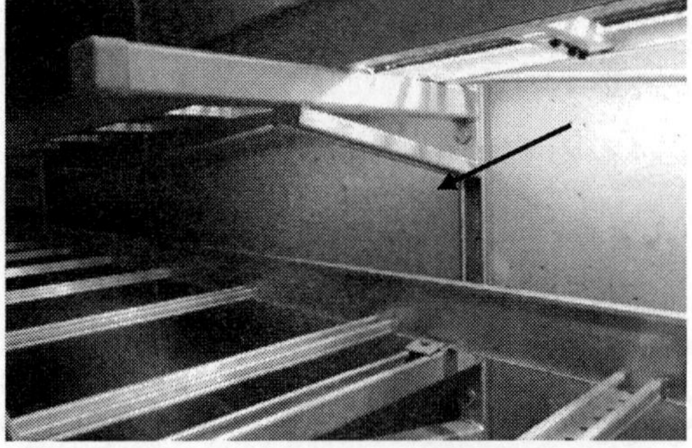

Figure 7. Waterview Tunnels utilise 304 SS support brackets for cable trays in tunnel inverts.

The North-South bypass Tunnels in Brisbane are 4.8 km long road tunnels built under the Brisbane River to reduce traffic congestion into the CBD. The tunnels have a ventilation system for fumes and heat extraction comprising 100 jet-fans that rapidly extract fumes through a longitudinal duct in the tunnel ceilings. A suspension fixing system made from grade 2205 duplex SS supports large concrete slabs which form the ventilation duct (*Figure 8*). The duplex 2205 SS hangers meet the mechanical property and corrosion resistance requirements to provide long term performance with no maintenance in the corrosive atmosphere that could prevail in the ducts. Within the Brisbane road tunnels there are also 30,000 light gauge duplex 2205 SS posts supporting the internal lining installed on the concrete tunnel walls.

Figure 9 shows an application of grade 316 SS sheetmetal utilised for the construction of the ventilation shaft ducting in the *Chunnel* between Britain and France. Contaminated air from the tunnel is extracted via the ventilation ducting. The atmosphere in the tunnel is corrosive and grade 316 SS provides the required corrosion resistance to ensure that the ventilation shaft ducts do not suffer from unseen corrosion on the internal ducting surfaces.

Figure 8. Duplex 2205 SS hangers support concrete slabs comprising a ceiling duct in North-South bypass tunnels, Brisbane.

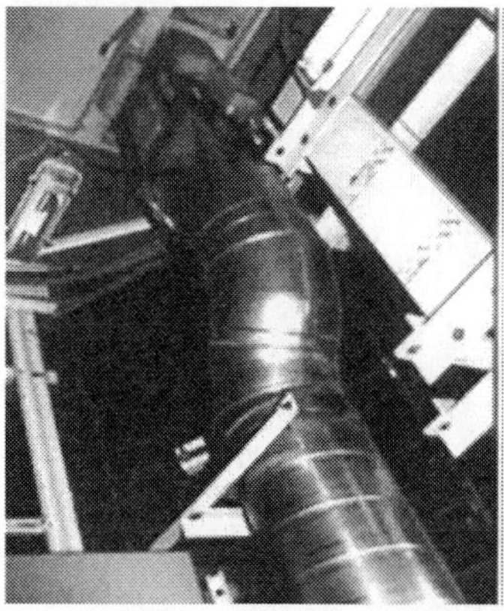

Figure 9. Grade 316 SS air ventilation shaft ducting utilised in the *Chunnel* (UK-France).

3.3 Fixings and fasteners utilised in tunnels

Fasteners are widely employed for fixing of specialised components in tunnels, such as power cable support brackets, tunnel linings, signage and lighting hangers. The fixings are exposed daily to the corrosive environment that prevails in underground tunnels. From European experience [9] it has been found that SS is the best material for fixings in road tunnels. However, there are only a few grades of SS that have adequate mechanical strength, corrosion resistance, and fire resistance, to provide the durability required in the corrosive atmospheres in tunnels.

Tunnel designers take into account an acceptable level of risk due to the corrosive impact of the conditions in a tunnel when selecting fasteners for structural applications. There is a possibility that sudden failure of a low alloy SS fastener (e.g. grade 304 SS) due to *stress corrosion cracking* could result in the failure of a complete structure. This prospect then presents the risk of a fatal accident occurring in a road tunnel with consequential health and safety issues, operational issues, and financial losses that would accompany such an undesirable event.

In some European road tunnels only super-austenitic SS with 6% molybdenum (6Mo alloys[1]) are mandated for use in critical tunnel applications [9]. Applications where super-austenitic 6Mo SS fasteners are specified include suspended ceilings under tunnel roofs, suspension of ventilation jet-fans, overhead road signage, overhead signal systems, tunnel linings for fire protection, and noise control wall cladding panels. More recently the application of the high-strength duplex 2205 SS fasteners has found applications for fixings of tunnel wall claddings in new road tunnels [5].

Under some circumstances fixings can be made from lower grade SS. For example, grade 316 SS is utilised for fixing of tunnel wall linings in the refurbished Mont Blanc tunnel [10] shown in *Figure 10.* The fixing material selection issue is often encountered in tunnel design, mainly due to the influence that the increased cost of a high grade SS has on the overall cost of an installation. Accordingly, tunnel engineers use the *life cycle costing* approach to reach a prudent decision regarding selection of a suitable alloy which meets the fire and corrosion resistance requirements of an application.

Alloy selection for fixings has to be carefully assessed by materials engineers with the mechanical properties, corrosion resistance and fire resistance clearly in focus. If the tunnel designer decides on an alternative to super-austenitic SS for critical fasteners, then the operating concept for the tunnel has to make allowance for routine monitoring and inspection of the structure which may include periodic replacement of the fixings.

---

[1] 6Mo alloys include EN1.4529, alloy 926, alloy 254SMO and alloy 6XN.

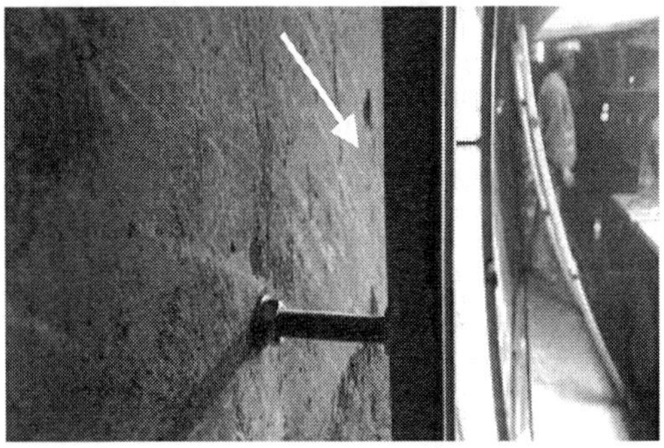

Figure 10. 316 SS anchor bolts used to fix fire-resistant cladding to concrete wall in Mont Blanc road tunnel

4. SELECTION OF STAINLESS STEELS FOR UNDERGROUND STRUCTURES

In the majority of structural applications utilising SS for underground structures it is the corrosion resistance of SS that is being exploited. However other properties will need to be taken into considerations also such as durability, mechanical properties, aesthetics, fire resistance and low maintenance. Good design should ensure that the grade of SS selected gives trouble-free performance during service.

Designers are aware that some grades of SS can be subject to various types of localised corrosion under adverse conditions in tunnel environments. The basic principles of material selection are kept in mind when specifying SS for structural applications. These principles are well documented and take into account: corrosivity of the service environment; the fabrication route; the surface finish required; maintenance of the structure by the asset owner [11].

The first step in selecting a grade of SS is to carefully characterise the service environment including anticipated or unexpected deviations from the service conditions. This is important because it is prudent to always select a grade of SS that will have sufficient inherent corrosion resistance to accept an upset in normal service conditions without suffering some form of unexpected corrosion damage. For example, seasonal changes of the humidity and temperature inside road tunnels combined with corrosive vehicle emissions are factored into the selection of the construction materials. It is important for tunnel designers to have an understanding of the corrosion damage mechanisms that can affect certain SS grades utilised in an underground operating environment. For example, the likelihood and consequences of *stress corrosion cracking* occurring on austenitic SS fasteners installed in road tunnels are documented [9].

Under operating conditions where a particular grade of SS needs to be used with caution in road tunnels, it is advisable to seek specialist advice. In some cases a lower grade of SS can be utilised with care, but the lower grade SS choice will involve more routine inspection and appropriate maintenance.

Assessing the suitability of the grade of SS for fabrication of an underground structure is best approached by referring to past experience of using SS in similar applications and/or environments. Publications are available that provide industry guidance on the selection of the correct grade of SS for specific corrosive environments [1,12,13]. As is normal practice consideration would be given to the SS grade mechanical properties, availability of product forms, ease of fabrication, surface finish, and the cost.

5. DISCUSSION

The utilisation of SS alloys for underground applications is an extension of the calculated outcomes from applying life cycle costing to material selection for an above ground structure. Material selection is a process whereby the designers have to decide between two options: either utilise lower grade materials with more frequent maintenance outages, or specify higher grade materials that require low maintenance and have a long service life. In recent years the trend towards ensuring that *sustainability* of infrastructure is impinging on engineering projects. Sustainability favours the option of choosing higher quality construction materials.

The modern approach to materials selection is also driven by the unacceptability of any failure occurring due to corrosion. The selection of an appropriate grade of SS ensures that failure of underground structures is much less likely, maintenance costs are reduced, and the life expectancy of a structure is extended for many years. To ensure that costly remediation is not required on structures during their service life in tunnels designers are specifying SS for any critical tunnel application.

The characteristics of SS that meet the rigorous requirements for critical structural components in tunnel construction include:

- SS does not emit toxic fumes at elevated temperatures that prevail in tunnel environments.
- SS retains a high proportion of its strength at temperatures in excess of 900°C, if fire occurs.
- SS requires no additional surface protection to resist the corrosive environments encountered in tunnels.
- SS is available in a range of grades and various forms that provide long life with minimum maintenance costs.
- SS provides long term operation of tunnel equipment that may affect user safety within a tunnel.

When the above properties and benefits of employing SS are taken into account designers can be confident that an underground structure being constructed will perform to the asset owner's expectations.

6. CONCLUSIONS

- Over the years there have been cases of underground structures fabricated from conventional engineering materials that have undergone failure due to corrosion.

- Research shows that correctly selected grades of SS are not vulnerable to corrosion in most types of soil and backfill when a structure made of SS, such as a pipeline, is buried.

- The utilisation of SS instead of conventional engineering materials has resulted in considerable improvement in the life expectancy of critical plant and equipment in underground structures.

- With the expansion of road, rail and power networks in large cities there is a requirement for more underground tunnels which may have to be constructed under rivers and harbours. Transport tunnels and power cable tunnels are corrosive environments where the properties of SS are eminently suitable and in some cases mandatory.

- Essential safety features employed in modern transport tunnels include fire fighting equipment, escape routes, ventilation, lighting and signage, all of which operate in a corrosive environment and must be totally reliable during service.

- SS has high corrosion resistance, performs well at high temperatures, emits no toxic fumes in a fire, has high durability, good reliability, and low maintenance.

- Examples of SS being utilised for fixing applications in underground tunnels include power cable racking, ventilation ducting, water deluge systems, fire resistant wall linings, lining fasteners, lighting support systems, signage hangers, and tunnel ceiling suspension systems.

7. REFERENCES

1. *Design Manual for Structural Stainless Steel*, 3rd Ed, Euro Inox, 2006.

2. *Corrosion: Materials,* ASM Handbook Vol.13B, ASM International, Ohio, USA, 2005.

3. *Stainless Steel in Tunnel Construction and Applications,* International Stainless Steel Federation (ISSF), Technical Brochure, Belgium, 2012.

4. *Stainless Steel in Infrastructure,* International Stainless Steel Federation (ISSF), Technical Brochure, Belgium, 2012.

5. *Stainless Steel in Tunnel Construction and Applications,* D.J.Cochrane, B. Heritier, & A. Kosmac, Technical Article, Euro Inox, 2012.

6. *Underground Corrosion of Engineering Alloys in the Water Industry,* L. H. Boulton, Water NZ Conference Proceedings, Auckland, New Zealand, 1995, pp 126-131.

7. *Aspects of Underground Corrosion of some Engineering Alloys*, L. H. Boulton, Corrosion & Prevention Conference Proceedings, Australasian Corrosion Association, Paper 18, Adelaide, Australia, 1994.

8. *Corrosion of Mild Steel Plates in some New Zealand Soils after 20 years*, H.R.Penhale, New Zealand Journal of Science, Vol. 27, pp 57-68, 1984.

9. *Corrosion of fasteners and connecting parts in road tunnel atmospheres*, Technical Document, Hilti Corporation, Liechtenstein, p.21, 2016.

10. *Nickel* magazine, Vol.18, No.2, p.12, February, 2003.

11. *Code of Practice for the Fabrication of Stainless Steel Plant and Equipment,* 3rd Ed, NZ Stainless Steel Development Association, Auckland, NZ, 2015.

12. *Stainless Steels,* J. R. Davis Ed., Specialty Handbook, ASM International, Ohio, USA, 1996.

13. *Corrosion Handbook for Stainless Steels,* 11th Ed, Outokumpu, Finland, 2015.

8. AUTHOR DETAILS

Les Boulton has forty five years experience in materials engineering and corrosion consultancy. He is Principal Consultant of Les Boulton & Associates Ltd, Materials and Corrosion Consultants, based in Auckland, NZ. Les has been a Consultant to the international Nickel Institute, Brussels, Belgium, for 19 years. He is the author of papers on stainless steel performance and corrosion of engineering materials. Les has been a Member of ACA for 45 years and he is an ACA Life Member. Les is a past ACA Australasian President, past-President of ACA NZ Branch and a current member of the ACA Awards Committee. He presently serves on the ACA Auckland Division Committee and the ACA NZ Branch Committee.

RESTORATION CONSIDERATIONS FOR METAL ROOF STRUCTURES

R. Cardoso, H. van den Heuvel
Andersal Pty Ltd

SUMMARY: Andersal P/L was engaged to assist develop a technical specification further to a consultant Engineer's initial recommendations and then to execute a trial restoration on two large metal roof structures showing some advanced stages of corrosion on a 44 storey high rise building within 500m of the ocean. The principle modes of corrosion were observed to be General Corrosion and Galvanic Corrosion. Andersal identified the following areas as requiring improvement:

- Access and maintenance
- Protective coatings and materials
- Drainage and waterproofing

Consultation with experts, analysis and testing within these fields resulted in a range of recommendations put forward that addressed issues requiring immediate action in a manner that would significantly prolong the serviceable life of the structures. The recommendations described in the case study highlight the importance of:

- Engaging a NACE qualified professional to assess site specific environmental conditions, specification of appropriate protective coating systems and inclusion of performance monitoring as part of a periodic maintenance schedule.
- Investing in higher grade materials such as stainless steel and Aluminium to reduce maintenance requirements, improve structural performance and prolong serviceable life of structures.
- Separating dissimilar metals with non-conductive polymer packers to eliminate galvanic corrosion
- Installing long term waterproofing solutions such as high density polymer sheet membranes that are not reliant on sealants
- Having an established methodology (mechanical where possible) to remove and replace any structural component on a roof structure.

Keywords: Stainless Steel, Aluminium, Protective Coating, Surface Preparation, Waterproofing

1. INTRODUCTION

The remedial works were undertaken on a 44 Storey high-rise building situated in Sydney CBD within 500m of the ocean. Construction of the building was completed in 2000 and by 2010 corrosion and failure of protective coatings was reported by building management on the two large architectural roof structures shown in Figure 1, called the Western Sail and Eastern Louvre Roof.

The location of the roof is considered to be in a temperate zone (Sydney), with a high level of pollution, a high level of humidity (contributed from cooling towers) and with substantial airborne chlorides from the nearby ocean. Therefore categorized with a corrosivity category of C4 (High).

Figure 1. Western Sail and Eastern Louvre Roof

1.1 Purpose

This paper discusses some of the technical challenges Andersal investigated and the solutions developed to achieve the requirements of the original Project Deed. Key areas of research and development included:

- Access and maintenance (Section 2)
- Protective coatings and materials (Section 3)
- Drainage and waterproofing (Section 4)

1.2 Background and Scope of Work

1.2.1 Western Sail

The Western Sail is an architectural feature approximately 32m in height and 75m in length comprising aluminium and glass cladding panels fixed to a steel frame. The frame structure, shown in Figure 2 and in detail in Figure 4, is comprised of large steel vertical columns and horizontal circular hollow sections (CHS). External cladding panels are connected to the frame via steel connectors called dog bone cleats (DBC), shown in Figure 3 and in detail in Figure 5.

The scope of work on the Western Sail included replacement of the corroded DBCs and restoration of deteriorated protective coatings on the CHS members.

Figure 2. Western Sail

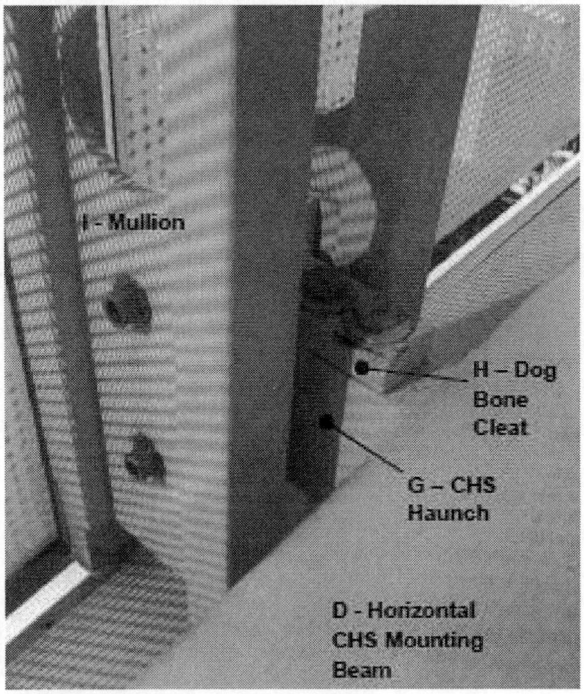

Figure 3. Cladding Panel Connection

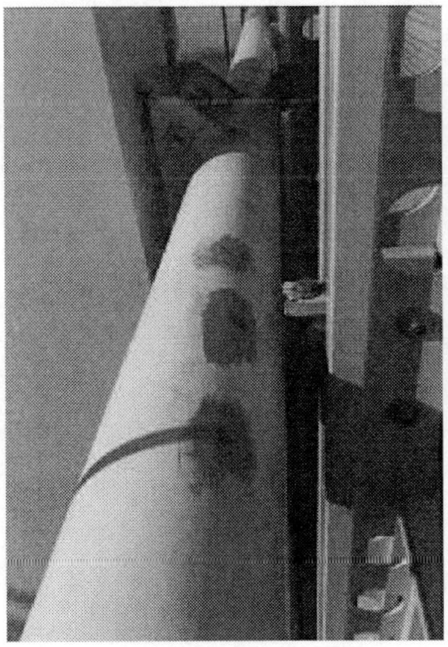

Figure 4. CHS Coating Failure

Figure 5. Corroded Dog Bone Cleat (DBC)

1.2.2 Eastern Louvre Roof

The Eastern Louvre Roof is an architectural feature that provides shade for the top floor tenancies via fixed and motorised aluminium louvre fins shown in Figure 6. These fins are connected to steel louvre rafters to form louvre panels. The louvre panels are supported by stub columns which are welded to large steel primary rafters in turn are supported by internal steel columns shown in Figure 7 and detailed in Figure 8.

Figure 6. External Side of Eastern Louvre Roof

Figure 7. Internal Side of Eastern Louvre Roof

The scope of work on the Eastern Louvre Roof was principally to replace corroding louvre rafters, restore protective coatings on stub columns and prevent on-going water ingress into the tenancies through the base of stub columns as detailed in Figure 8.

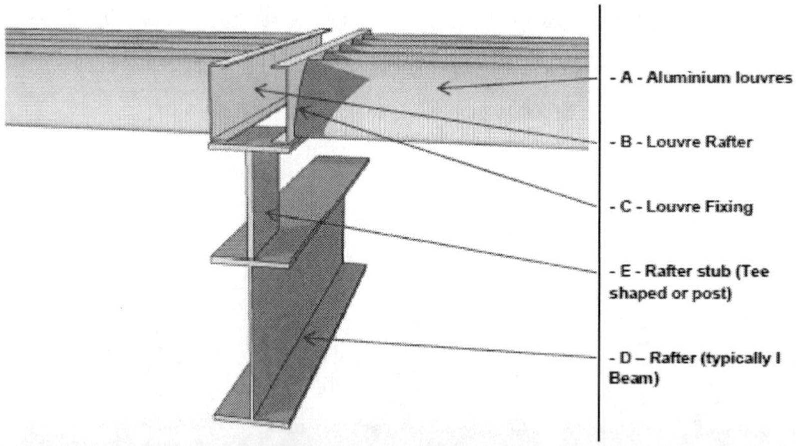

Figure 8. Eastern Louvre Roof Section

2. ACCESS AND MAINTENANCE

This section outlines the process and trials undertaken by Andersal to finalise the methodology that would enable access to all areas of the roof structures, in order to install replacement elements and implement a practical maintenance program.

2.1 Challenges

Access for inspection and refurbishment of components of the architectural roof features on top of the 44 storey building was always going to be a challenge. Some of the key access considerations were:

- Protective coating manufacturers recommend that coating surfaces are as a minimum washed and degreased regularly. Furthermore, it is recommended coatings are monitored by visual inspection and testing at regular intervals so that areas requiring localised repairs i.e. mechanical scrapes or delamination of coatings occurring during construction or operation phases or from non-conforming surface preparation and coating application are identified and rectified. Due to the difficult access conditions, this minimum maintenance requirement is not feasible at many locations on the roof structure.
- Given the geometry of the DBC components, the only way to restore their protective coating was to remove them. However at the time of construction, cladding panels were connected to the supporting sail structure in a sequence that 'sandwiched' the DBCs preventing removal.
- The louvre fins on the Eastern Louvre roof impede access to supporting rafters and stub columns preventing installation of an adequate waterproofing system.
- Some elements were not accessible for bristle-blasting to the specified surface preparation.

2.2 Experiments and Results

Table 1 summarises key experiments undertaken by Andersal and their outcomes.

Table 1. Access and Maintenance Trial

Key Access and Maintenance Issues	Outcome
Abseiling was trialled as the primary means of access as shown in Figure 9	- Abseiling allowed workers to access every work front on the two structures. - Abseiling appears to be a more flexible and cost effective option when compared to other options such as scaffolding - Some items of work such as dismantling louvres on the Eastern Roof manually were not feasible by hand
A bracing system was designed and installed around DBC locations effectively replacing the DBC support	The bracing system was an adequate temporary support replacement allowing the destructive removal of the DBC on the Western Sail.
New designs of DBCs were trialled using 3D modelling and printing. Refer Figure 11	3D printing allowed prototypes to be checked without having to fabricate actual DBCs
Belt linishing tools were used to clean corrosion in areas not accessible to the bristle-blaster as shown in Figure 12	A NACE inspector and Protective Coatings manufacturer were engaged to test the adequacy of the surface profile. It was agreed that the belt linishing tool was capable of cleaning to bare metal whilst maintaining a workable surface profile
A winch was attached to the building maintenance unit (BMU) jib effectively converting it into a crane to lift louvre panels and shown in Figure 10	The BMU conversion to a crane enabled 800 kg louvre panels to be lifted in bulk to a safe working area at ground level. This in turn exposed the stub columns and glazing underneath which required painting and waterproofing respectively.

2.3 Conclusions

Upon completion of trials the following conclusions and recommendations were made to the client:

- The combination of abseiling and converting the BMU to a crane enabled access to all scope of work items.
- By optimising the DCB design using 3D modelling and printing, a new DBC was fabricated with dimensions that enabled it to be installed and removed non-destructively.
- A linishing belt was able to clean corroded areas not accessible to the bristle blaster back to bare metal however the surface profile was found to be too smooth and not in accordance with AS 1627.4 Sa2½ between 40 and 70μm

Rz. As a result, the paint specification was revised for areas inaccessible to the bristle-blaster to ensure the 15 year protective coating warranty was still able to be achieved.

- A site specific maintenance specification for protective coatings was prepared.
- Looking to the future, given the difficulty and high risk nature of changing out and maintaining structural members on a building top, it is recommended, where possible, replacement items comprise higher grade materials or offer improved corrosion resistance to prolong the next major maintenance as long as possible.

Figure 9. Abseiling Western Sail

Figure 10. BMU conversion to Crane

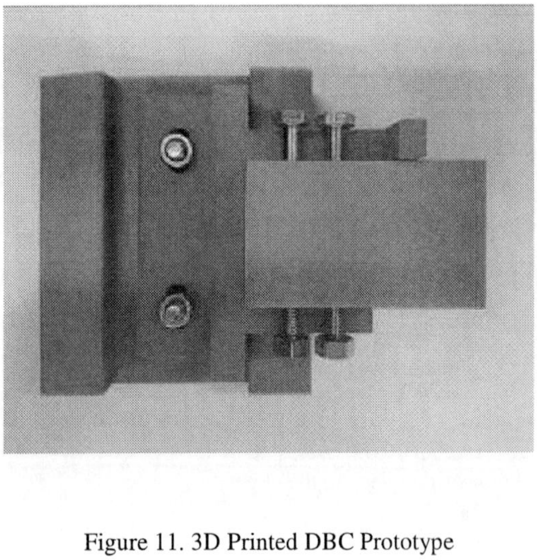

Figure 11. 3D Printed DBC Prototype

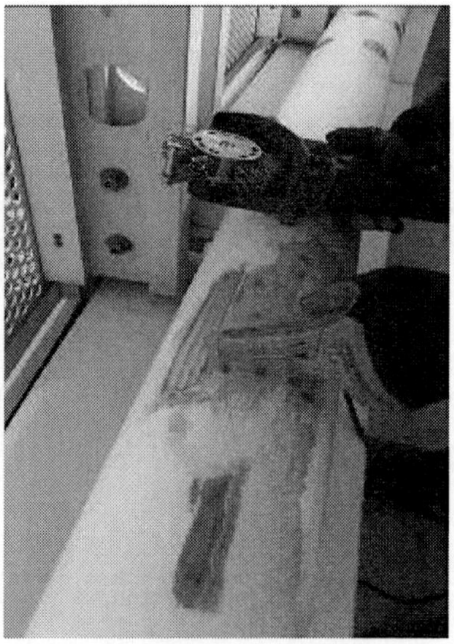

Figure 12. Bristle Blasting a CHS

3. PROTECTIVE COATINGS AND MATERIALS

This section describes the challenges and research undertaken in order to meet the required serviceability for new materials and protective coatings of 10 years and 15 years respectively.

3.1 Challenges

Protective coating failure was the principle reason for this restoration project and so the following considerations were critical to understand and address.

- Variation in severity of coating failure which potentially indicates inconsistency in protective coating application during the construction phase and/or varying environmental conditions across the structures. Figure 14 shows a protective coating failure where a top coat has been applied directly to an unprepared surface.
- Galvanic corrosion due to dissimilar metals (Figure 13).
- Widespread failure of the coating system associated with inadequate surface preparation of member edges prior to painting i.e., radius < 2mm during the construction phase.
- To replace deteriorated steel components with a more corrosion resistant material that can achieve nominal structural performance requirements and be fabricated within project budget.

Figure 13. Galvanic Corrosion on CHS

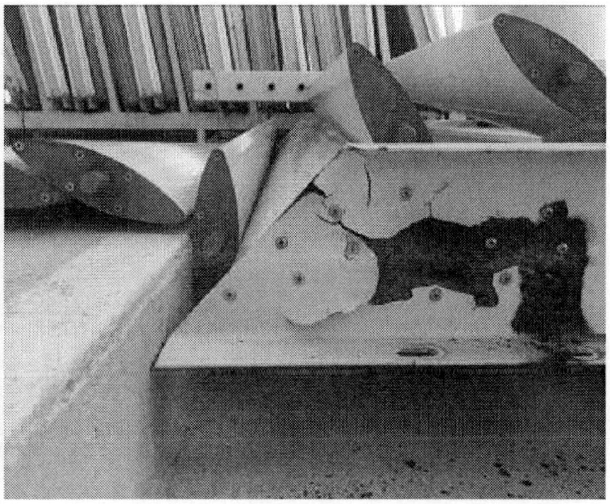

Figure 14. Failed Protective Coating on Louvre Rafters

3.2 Experiments and Results

Table 2. summarises key experiments undertaken by Andersal and their outcomes.

Table 2. Protective Coating and Material Selection Trials

Key Corrosion Prevention Issues	Outcomes
The specified protective coating system was trialled on the West Sail CHS.	Blistering due to possible contamination from mist spray from adjacent cooling towers between the different protective coats was observed
Structural analysis of the DBC was carried out to determine strength requirements for replacement DBCs.	A limit state force requirement of 57KN was calculated for the DBCs
A prototype welded stainless steel 316 DBC was fabricated	Welding 5 different plates together proved difficult in maintaining low dimensional tolerances required by the DBC to fit into a restricted space.
Casting DBC in stainless steel CF8M (equivalent to 316) was trialled. (Figure 17)	- The required dimensional accuracy of the DBC was achieved through casting. - Material costs increased.

An independent NATA approved laboratory was engaged to carry out destructive testing An independent NATA approved laboratory was engaged to carry out radiograph testing to check for porosity or discontinuities in the cast DBC.	The A4-800 high tensile stainless steel bolting assembly failed at a pull force of 171 kN (Figure 15) before the DBC failed. This cast stainless steel DBC had an improved factor of strength of 3 when compared to original design requirements. Tests results showed that cast DBC had no recordable discontinuities as shown in Figure 16 and were in accordance with Class 1 acceptance criteria detailed in AS 3507-1987.
Various packer materials were considered for separation of dissimilar metals causing galvanic corrosion.	Polypropylene satisfied selection criteria.
Custom extruded aluminium sections were trialled to replace old steel louvre rafters on the Eastern Roof. In order to achieve similar strength to the old louvre rafters aluminium grade 6061-T6 was used	Aluminium sections were lighter resulting in easier manoeuvrability however the material costs increased. Extruding allowed new sections to be made with edges over 2 mm in radius to assist longevity of applied protective coatings.
New aluminium rafters were powder coated (Figure 18).	Powder coating new rafters out of the factory was a more efficient way of coating the sections in comparison to manually painting them. Cost wise powder coating is considered to be on par with painting.
Round sharp edges (i.e., < 2mm radius).	Rounding sharp edges on site was not seen as feasible given the number of areas (some inaccessible) and the inability to guarantee a compliant 2 mm radius using hand tools.

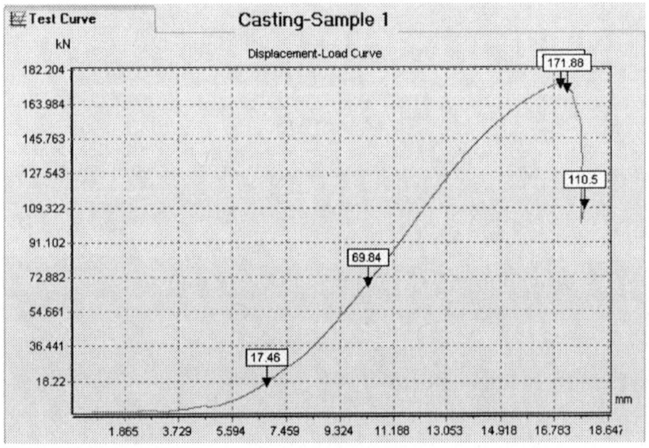

Figure 15. DBC Destructive Test Result

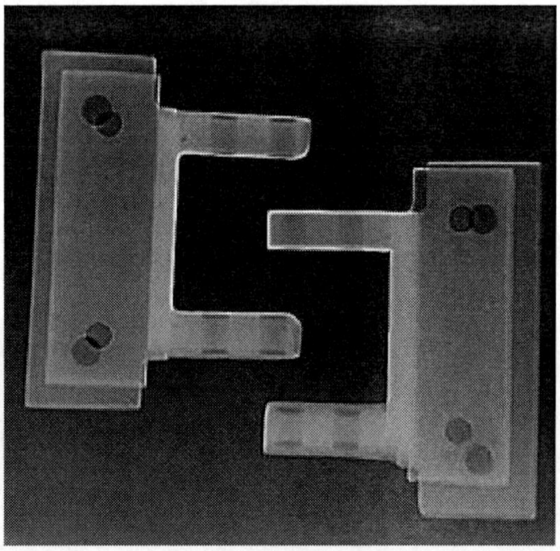

Figure 16: Radiographic Image of DBC

Figure 17. Cast Stainless Steel DBC

Figure 18. Powder Coated Aluminium Rafters

3.3 Conclusions

Upon completion of trials the following conclusions and recommendations were made to the client:

- The recommended protective coating comprised 1st primer coat of 75 μm DFT single component moisture-cure urethane (with zinc and micaceous iron oxide pigments), 2nd coat of 75 μm DFT single component moisture cure urethane (based on micaceous iron oxide), 3rd coat of 75μm DFT single component moisture cure urethane (based on micaceous iron oxide) and 4th coat of 75μm DFT single pack acrylic polysiloxane.
- Blistering observed in the first two coats was noted by NACE CIP inspectors as being typical of outgassing. This either occurs due to moisture on the surface (either before or immediately after application) or due to excessive film thickness application. Formation of carbon dioxide is a normal by product of the curing reaction of moisture cured urethanes. An excessively thick coating film will prevent the CO_2 from being released from the film and cause bubbles within (and on the surface) of the coating film. As a result, repairs undertaken adjacent to cooling towers are recommended to be encapsulated for their full curing period to prevent blistering.
- It was noted by the NACE CIP inspector that the atmospheric effect of the cooling towers created a micro-climate resulting in the Environmental Classification of the roof area being increased to C4-High. Consequently, the coating system was upgraded to a 4-coat system.
- A cast stainless steel CFM8 (316 equivalent) DBC was adopted as it satisfied the required dimensional accuracy, strength and durability requirements. The DBCs were also electro-polished as a further means of corrosion protection and to reduce surface tea staining.
- Polypropylene washers were recommended to separate dissimilar metals as they satisfied strength, UV and chemical resistance requirements.
- The most challenging and costly work item of the project was to replace louvre rafters on the Eastern Roof using the BMU. It is in the client's benefit to delay repeating this maintenance activity for as long as possible and so it was recommended that Aluminium 6061-T6 be used for new rafters including powder coating with a 20 year warranty.
- It was determined that rounding edges by hand tools insitu could not guarantee the required 2mm radius edge requirement. Given the quantity of non-compliant edges, it was deemed unfeasible to undertake this item of work. Rather, it was preferred by the client to address areas of concern through ongoing maintenance and ensuring that all newly fabricated elements satisfied the 2mm edge radius requirement.

4. DRAINAGE AND WATERPROOFING

Rain water ingress via the Eastern Louvre roof was reported at several locations in the top floor tenancy. Andersal was requested to determine where the leaks were coming from and detail an adequate waterproofing solution.

The Western Sail and Eastern Louvre Roof Structures were also observed to have areas prone to ponding water which were visibly corroding at higher rates than adjacent areas.

4.1 Challenges

The following items were seen to be key in addressing to improve the structures' ability to manage rain water

- Given the slope of the Eastern louvre roof and over 10m internal ceiling height, leaks could appear several metres away from the ingress location making it difficult to pin point repair locations.
- Installation of a waterproofing system around stub columns that would not be visible against the surrounding roof.
- Selection of a waterproofing system that meets a service life of 15 years.

4.2 Experiments and Results

Table 3. summarises key experiments undertaken by Andersal and their outcomes.

Table 3. Drainage and Waterproofing Trials

Key Drainage and Waterproofing Issues	Outcomes
Water testing of eastern glazed roof	- Water ingress was determined to be at locations where stub columns penetrated the glazed roof, however there was some difficulty establishing accurate ingress locations. - The polyurethane sealants applied around the stub columns were observed to have failed in many locations.
Ponding and poor drainage acerbating deterioration of steelwork	Ponding water was visibly accelerating corrosion of structural connections such as welds.
Trial of high density polymer sheet membrane over one run of stub columns to investigate efficacy to waterproof the detail.	Water testing of the membrane showed it was effective in preventing water ingress via stub columns.

4.3 Conclusions

Upon completion of trials the following conclusions and recommendations were made to the client:

- A high density polymer sheet membrane system (Figures 19 and 20) was recommended to waterproof all stub columns across the roof structure. Selection of the final membrane type required consideration of its achievable service life so as to prevent unnecessary work lifting louvre panels prematurely to conduct maintenance. A 15 year service life to equal that of the refurbished coating system was required.
- An acrylic paint was applied to the membrane in a matching colour to the adjacent roof so that the membrane is not visible against the roof
- Acrylic / Bituminous sealing tapes and UV stable Polyurethane have been recommended to be used in areas prone to ponding water and crevice corrosion.

Figure 19. Installing High Polymer Membrane

Figure 20. Waterproofed Stub Column

5. CONCLUSIONS

This project presented Andersal with some unique challenges resulting in key considerations that can be applied to metal roof restoration with similarly difficult access:

- Maintenance of the life of the asset must be considered to have the same level of importance as the material selection and assembly of the structure. Simple tasks such as washing down protective coatings, monitoring and carrying out spot repairs will prolong or even avoid the need for costly major maintenance operations.

- Perimeter access and craneage during the construction period of a building are not possible to replicate after the building handover. Therefore a plan should be in place to allow practical access to structural elements as well as provisions to remove and replace them in non-destructive ways.

- Metals such as stainless steel 316 and aluminium 6061-T6 cost more than steel however the upfront cost is almost negligible compared to the cost of major maintenance projects. These materials should be considered and used as construction materials during the design phase as well as being considered as replacement materials for restoration works.

- Qualified NACE CIP professionals should be consulted at all stages of a project to develop site specific protective coating systems, pre-empt and diagnose modes of coating failure.

- A minimum serviceable life should be set for an asset and include all structures, elements and components (including waterproofing).

- During the design and construction phase, simple measures such as separating dissimilar metals and sealing crevices will help the structure achieve its design service life.

6. ACKNOWLEDGEMENTS

Andersal would especially like to acknowledge the following persons whose input and expertise contributed to a successful project delivery.

- Tully Allsop, Practical Access Solutions
- Jon Kutek, Incospec Corrosion Engineers
- Oscar Duyvestyn, Aecom Australia
- Colin Lim, Aecom Australia
- Peter Whalan, PPG Protective Coatings
- Wayne Robinson, ARL Laboratory Services

7. AUTHOR DETAILS

R. Cardoso Project Manager at Andersal P/L since 2012. He is responsible for procurement, design, planning, and site management of restoration projects.

IMPACT OF PATH DEPENDENCE ON CORROSION OF COMMERCIAL AND MILITARY SYSTEMS

D. Dunmire[1], G. Keller[2]
[1]U.S. Department of Defense Corrosion Policy and Oversight Office, Pentagon, DC, USA; [2]LMI, McLean, VA, USA

SUMMARY: Path dependence is a phenomenon displayed by complex systems during their emergence and performance. Typhoons and willy-willys are examples of natural systems that are influenced by path dependence, where their emergence and initial direction depend on original climatic conditions; and changes in their speed, power, and direction depend on random changes in climatic conditions as they proceed on their destructive paths. During their development and operation, most technological systems are also influenced by path dependence. System designs, production plans and operational applications usually consist of a pattern of sequential steps that form paths, where the performance of each step affects one or more succeeding steps in that path. This is particularly true of systems vulnerable to material degradation.

During a technological system's evolution, the final outcome depends on the specific development paths followed – it is path dependent. Initial conditions define the starting point and direction of a system's design path. During subsequent design, production and operation, positive feedback at any step influences the systems engineer, producer or operator to continue on that path. Negative feedback is likely to influence a change to an alternative path. On some occasions, the design, production path, operating sequence or support system can be "locked-in" due to unforeseen conditions that eliminate other practical alternatives. Path dependence can produce positive effects and accelerate achievement of the best system to combat corrosion. But continuing along a path of positive feedbacks could cause better solutions to be overlooked. And lack of feedback is almost certain to eliminate any change in a design, production or operational path. This paper describes underlying mechanisms that create, dictate and maintain path dependence; discusses the effects of path dependence on corrosion of commercial and military systems during design, production and operation; and suggests actions to address its effects.

Keywords: path dependence, corrosion, initial conditions, feedback, lock-in, systems design.

1. INTRODUCTION

Why is understanding path dependence so important? The design, production, operation and maintenance of every commercial and military system is affected by this phenomenon. From the first step in a system's planning process, to the last step taken to scrap or decommission a system, every decision that was made – or should have been made – at each step in the system's path influences the performance quality, cost, sustainability and survivability of that system. This influence of path dependence begins when the design path for each element of a system's development, production, assembly, operation or maintenance is determined; the decision points along that path are designated; and the decision criteria are defined. The most critical impact of path dependence occurs if decision points are ignored; decision criteria are ineffective or not properly applied; or erroneous decisions are made. The impact of path dependence, and the strategies to ensure positive outcomes and avoid negative effects, are applicable to all commercial and military systems. And, it is particularly applicable to systems that are vulnerable to material degradation, otherwise known as corrosion.

For example, material degradation of operational systems costs the United States Department of Defense over $20 billion each year. This corrosion affects mission performance, readiness, logistics support, and safety of warfighting systems and supporting infrastructure throughout the Army, Navy, Air Force, Marine Corps and Coast Guard. This annual $20 billion price tag does not include the cost of corrosion prevention efforts during system or facility acquisition – this price tag is the downstream cost to inspect and repair or replace components that are deteriorating or damaged due to operational and environmental conditions; or to periodically restore some components' resistance to material degradation. While some of these downstream corrosion costs are unavoidable, more effective management and engineering design efforts during system and infrastructure acquisition could prevent downstream corrosion and offset a significant portion of the annual $20 billion operational system corrosion cost [1]. Understanding how path dependence influences corrosion of these military systems, and applying strategies to avoid its negative effects, could significantly reduce that cost, increase system availability, reduce the size of the logistics footprint and improve warfighting system safety.

Material degradation also significantly affects performance, availability, maintenance requirements, safety and life-cycle costs of commercial systems around the world. The 2016 NACE IMPACT Study estimates the annual global cost of corrosion to be US$2.5 trillion [2]. In Australia alone, corrosion cost is likely well above $30 billion each year [3]. The NACE IMPACT Study suggests that up to 35% corrosion cost savings are possible [2]. Several broad underlying factors contribute to such reduction in performance, availability and safety while increasing maintenance efforts and overall cost. These factors include:

- Lack of sufficient decision-maker education, experience and comprehension regarding the causes and effects of corrosion
- Incomplete or misinformed executive direction to program managers, systems engineers, production managers and logisticians
- Effects of path dependence on system design, development, production, operation and maintenance.

The first two factors are often influenced by path dependence dynamics. Therefore, this paper focuses on path dependence and its impact on corrosion management, corrosion engineering and downstream corrosion maintenance and repair.

2. PATH DEPENDENCE FUNDAMENTALS

Path dependence is a phenomenon displayed by complex systems during their emergence and performance. Typhoons and willy-willys are examples of natural systems that are influenced by path dependence, where their emergence and initial direction depend on original climatic conditions; and changes in their speed, power, and direction depend on random changes in climatic conditions as they proceed on their destructive paths.

Much like the progression of a typhoon, the initiation and growth of corrosion displays path dependence. Initial conditions require an anode, cathode, electrolyte and an electric path between the electrodes. These conditions, as well as the initial direction of corrosion growth, can occur randomly. Once corrosion is initiated, its path and growth rate are dictated by physical conditions along its path. Somewhere along the path, random conditions could cause a change in direction or a phase transition, where the growth rate rapidly increases and material failure results. Unless the corrosion is detected and treated early enough in its progression, costly repairs or material replacement will be required.

The above examples describe path dependence where the sequence of events is continuous. But in most cases, the sequence of events is discrete. For example, consider how path dependence influences automobile travel. The driver selects the route to reach a given destination. The route represents a series of discrete steps, where each step is dictated by the pattern of successive intersections and selected turns. If the driver encounters no traffic problems, the expected route is followed. But heavy traffic or an accident will cause the driver to evaluate alternatives and perhaps take an alternate route at one of the intersections. This decision might allow the driver to avoid traffic problems, or it could cause even greater travel problems. Thus, the final route and the time required is path dependent.

During their development and operation, most technological systems are also influenced by path dependence. System designs, production plans and operational applications usually consist of a pattern of sequential steps, where the performance of each step affects one or more succeeding steps. This is particularly true of commercial systems and infrastructure during design and manufacturing; and military weapon systems and facilities during the acquisition process.

3. PATH DEPENDENCE DURING ACQUISITION

During acquisition of commercial systems and infrastructure, as well as military weapon systems and facilities, path dependence is a sequential system development characteristic where initial conditions, random events, and positive feedback all conspire to dictate the path taken to complete the development process [4]. The literature [4-6], the experience of many systems developers and engineers, and one's intuition indicate that system design follows a distinct path – a sequence of discrete steps, each of which depends on the prior step. During a system or facility design evolution, the specific design outcome and subsequent performance is determined by the set of design search paths followed – it is path dependent [7].

Consider what happens when a new system or product is specified. The initial conditions that influence the design concept and the design path frequently depend on an existing system or product that resembles or will be replaced by the newly specified one. The systems engineers are quite likely to replicate the current system's design path and incorporate many of the same design concepts, technologies and production techniques. When this occurs, the initial conditions are path dependent, since the starting point and direction of the new acquisition process are based on prior experience and methodology. As a simple example, consider the design of vacuum cleaners. For years, upright vacuum cleaners had bags that collected dust and debris and had to be replaced when full. Most brand names featured the same vacuum cleaner configuration, vacuuming method and bag disposal process. The only differences were size, weight, suction strength, external appearance and "bells and whistles." Each new model of a vacuum cleaner was based on the prior model. Then, someone devised the bagless "wind tunnel" technology concept, which meant new initial conditions and a different design path. This significant change in initial conditions and direction produced better vacuuming performance and a much better debris disposal method.

Some organizations pursue disruptive technologies that can avoid dependence on entrenched technologies and design processes – in other words, they jump completely off an existing design and development path. The lesson here is that dependence on a prior design concept and development path often suppresses the opportunity for businesses or consumers to acquire a significantly better system, product or service at a reasonable price.

Once an acquiring organization specifies the initial conditions and concept of a system acquisition, the systems engineer chooses, either randomly or based on prior experience, that starting configuration and design path that appears to offer the best chance of establishing and improving design fitness. After the initial step, the chosen path is divided and subdivided into a series of parallel paths, dictated by the configuration of the system under development. The path dependence behavior described in the text can be applied to each parallel path until it merges with another path and continues until all paths converge to produce the completed system. Path dependence dynamics, shown below in Figure 1, impact the acquisition process at each step on the design and production path. System acquisition continues on that initial path as long as positive feedback indicates the correct design choices were made. If negative feedback indicates that a specific design feature will not be effective, the systems engineer seeks and chooses a different approach (path). A design obstacle also might dictate a change in direction or a retreat to a previous design configuration, so the systems engineer takes another path. If a chance occurrence threatens the design, the systems engineer may work around that threat or change direction. But, the further the system acquisition proceeds down the design path, the fewer the available alternative approaches. Hopefully, the acquiring organization has developed an effective system design and is able to produce a system that performs as specified. However, when cost limitations prohibit resolution of previously undetected technical problems, or unforeseen external conditions create a situation where there are no remaining alternatives, the acquiring organization is "locked-in" to the design configuration at that point [7].

Path dependence can produce positive or negative effects. Positive feedback could accelerate achievement of the best design. On the other hand, continuing along a path of positive feedbacks could cause the systems engineer to miss better solutions. Negative effects are often attributable to false positive or false negative inspection and test results. False positives will influence the systems engineer or manufacturer to continue along the current path when an alternate path might be better. False negatives will influence a change to an alternate path when the current path should be maintained. The impact of such errors may not be felt until far downstream in design or production. This impact can be particularly acute during system or facility acquisition, where underlying mechanisms that create, dictate and maintain path dependence can negatively affect design, performance and affordability, particularly in systems or facilities that are composed of components vulnerable to material degradation [7].

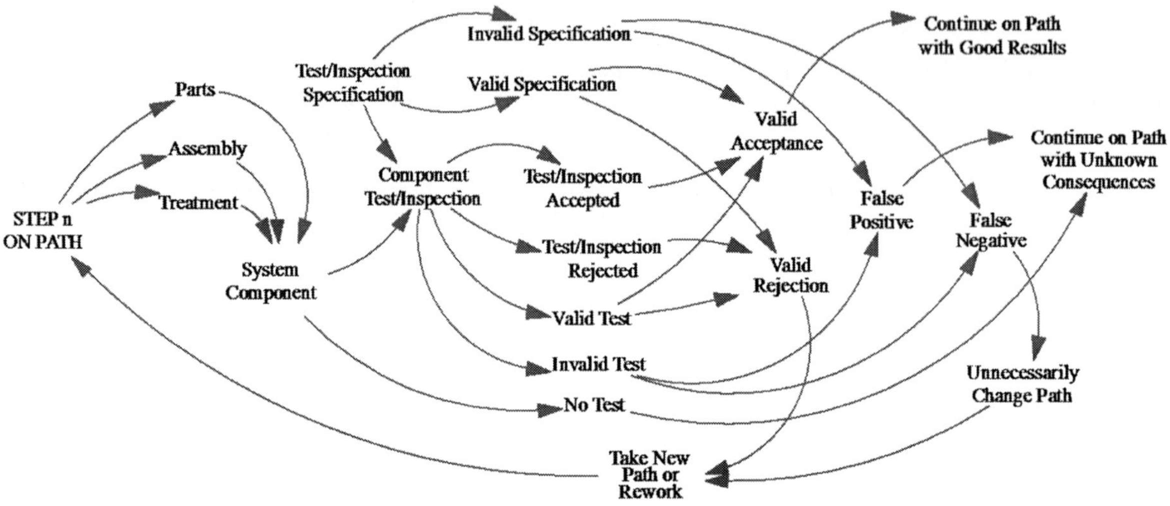

Figure 1. Impact of Path Dependence on a Single Step in a System's Production or Construction Process

Figure 1 illustrates the impact of path dependence during a single step in an acquisition path. In this example, the nth component is assembled and, if needed, treated to prevent or resist material degradation. If a valid test or inspection specification has been established, and the test results are valid, the component will be properly accepted or rejected. If it is accepted, the process continues to the next step on the predetermined path with expected good results. If the test or inspection results are not accepted, the decision must be made to revise the path at that point or rework the component so it passes the test or inspection. However, if either the specification is invalid or the test or inspection is not properly performed, the result might be a false positive (accepted but should have been rejected) or false negative (rejected but should have been accepted). A false positive means proceeding on the predetermined path without knowing the consequences of doing so. The result could be a failure far downstream from this test or inspection. A false negative means a likely change in the predetermined path, where the new path might not be as productive. Not specifying or performing a needed test or inspection yields the same result as accepting a false positive test or inspection result.

4. IMPACT OF PATH DEPENDENCE ON MATERIAL DEGRADATION

Vulnerability to corrosion is directly associated with material selection, system or facility design, production methods, and testing requirements, as well as the environment. Decisions made at the outset of an acquisition cycle are critical, for they dictate the initial design and material selection path. The choice between retaining old system concepts and materials and developing new system concepts and more effective materials is influenced by the system engineers' knowledge and experience. Systems engineers sometimes use a variant of an old design concept. Even new design concepts are assimilated from combinations of prior experience and existing technologies. So, resulting choices in design and material selection are almost always path dependent. The same path dependent decision process is applied to development and application of production and test specifications [7].

The process of acquiring a commercial, industrial or military system or facility usually consists of several phases: conceptualization, initial design, detailed design and preliminary production. Not only does path dependence affect each of these phases, but these effects are frequently carried forward to succeeding phases and finally to operation of the end product. Thus, the outcome of every step in each phase can be influenced by the factors and dynamics described in Section 3 and illustrated in Figure 1.

During conceptualization, a system concept that includes a suboptimal alternative for corrosion prevention could be chosen. A better alternative might be overlooked if positive feedback, based on inappropriate requirements or poor measures of effectiveness, indicates the selected concept provides the best design. Similarly, a systems engineer could conclude that the entire corrosion prevention concept is or is not feasible based on the same type of feedback errors, such as erroneous performance data that indicate a similar operational system would or would not be an appropriate exemplar for the system being designed [7].

The initial design phase is particularly important because of the impact on system corrosion. Establishing corrosion prevention parameters and quantifying performance measurements are the most critical steps associated with corrosion prevention and control. Measuring the wrong parameters will surely put the design on the wrong path. And, if measures of effectiveness or measures of performance do not accurately reflect performance goals, trade-off decisions regarding corrosion prevention methods and alternatives may be faulted; substandard material performance accepted; or superior material performance rejected [7].

The fact that corrosion prevention and control is often overlooked during conceptualization and initial design sets the stage, during remaining phases, for lock-in to materials, detailed designs, and test methods that frequently cause downstream corrosion initiation, acceleration and subsequent damage. In the detailed design phase, the development of each component features an initial condition and path direction. Systems engineers make choices based on the specification, but often rely on prior similar designs, previous experience, or shared knowledge to establish initial conditions and path direction. During the detailed design and preliminary production phases, each in-process and completed item test presents an opportunity to evaluate the fitness of a component. If metrics and test methods are effective, the design or resulting product is likely to be accurately assessed. If metrics and/or test methods are ineffective or non-existent, the design or resulting product may be erroneously accepted or rejected. This error will roll up with integration at each level, and continued positive feedback will compound the error until it is detected, mitigated or never observed. Mitigation might be accidental and never perceived, but if the error is detected far up the technological development path, it may be difficult and costly to trace the error to its source and even more costly and time consuming to change the design path or production process.

Corrosion-related design errors will not be detected unless proper testing is applied and effectively analyzed. The further the design and integration progresses, the more vulnerable the design path or resulting product becomes to lock-in. **Unfortunately, when corrosion-related design errors remain undetected, the system containing the errors are introduced to the operating community with a built-in high probability of corrosion that reduces performance, availability and safety; and significantly increases the need for increased logistic resources with accompanying costs** [7].

In addition to the effects of path dependence during system and facility acquisition as cited in previous paragraphs, path dependence affects the performance of maintenance and repairs on commercial and military systems and facilities during their operational use. The same type of errors that prevail during systems and facility acquisition can also occur during maintenance and repair. Corrosion test or inspection methods, as well as maintenance and repair processes, can be poorly designed or improperly applied. Inspection results could be erroneously read or interpreted, which could cause ineffective repairs or replacements of parts or materials. Frequently, such conditions are not found until far downstream, sometimes with potentially dangerous consequences.

5. EXAMPLES OF COMMERCIAL SYSTEMS IMPACTED BY PATH DEPENDENCE

a. Bridge Design.

In 2010, corrosion design problems plagued the new east span of the San Francisco-Oakland Bay Bridge. The bridge, already billions of dollars over budget and years behind schedule, was found to have corrosion in thousands of steel tendons that are integral to the suspension system; corrosion caused by exposure to salty mist, rainfall and debris introduced into unsealed ducting around the tendons. After inspecting the corroded tendons, leading experts in tendon corrosion expressed concern with the durability of the bridge. The experts said the span's construction violated universally accepted techniques meant to prevent corrosion. One expert opined that the California Department of Transportation use of erroneous testing triggered faulty conclusions, and caused a dilemma when corrosion problems were finally discovered, implying path dependence lock-in to the existing condition [8]. While this example reflects problems with civilian infrastructure, it mirrors potential and actual military facility corrosion issues [9].

The lesson learned is that poor design, inadequate testing and erroneous or no feedback along the bridge design and construction path led to severe downstream corrosion with potential safety issues in a highly corrosive environment.

b. Copper Mine Design

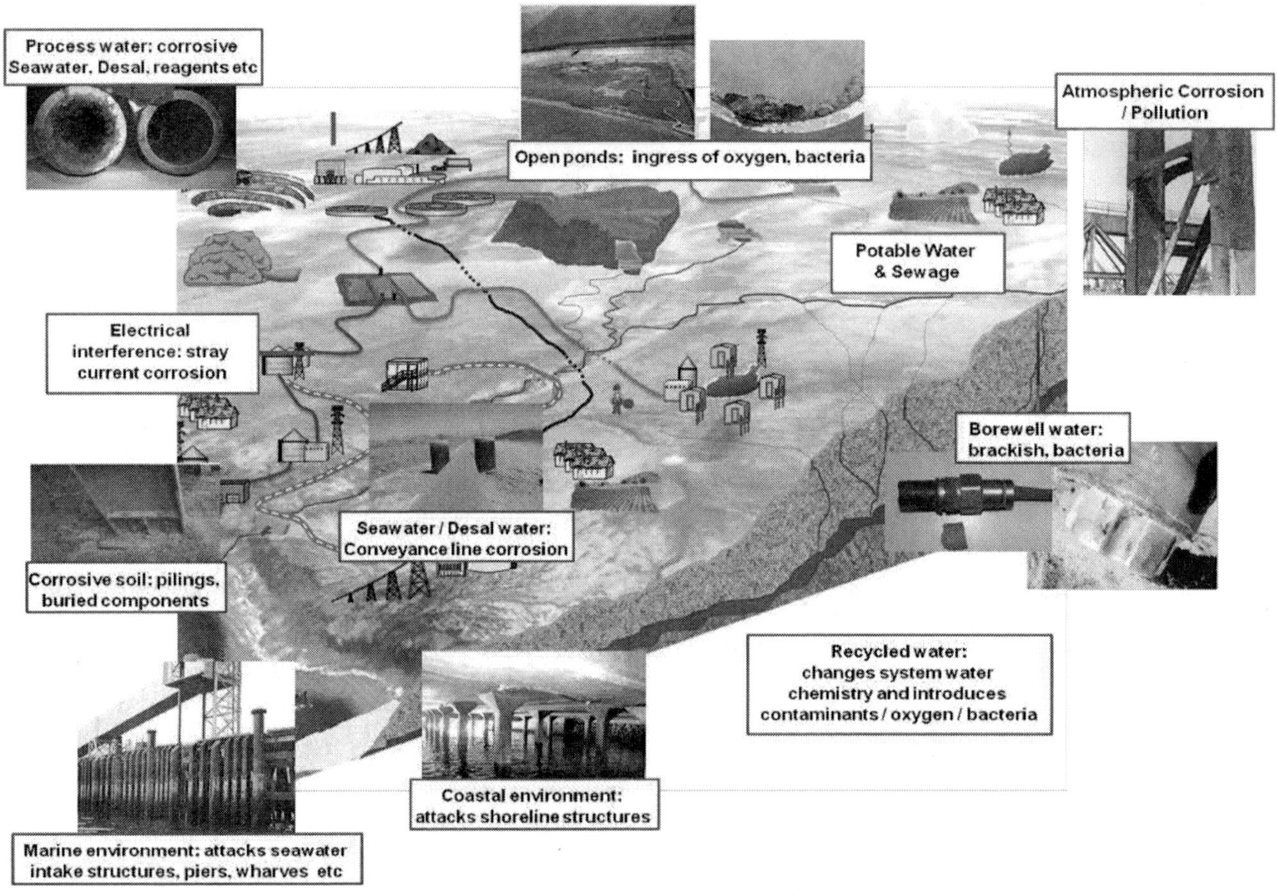

Figure 2. Path Dependent Corrosion Challenges for a Typical Mine [10]

Inadequate engineering design of a seawater conveyance line for a copper mine in South America resulted in internal corrosion of that conveyance line. The system design called for an unlined pipeline. Although an epoxy liner was included on some pump components, the liner later disbonded, probably due to issues with pump instability and accumulation of debris. These deposits subsequently caused failure of pipeline filters. In addition, the seawater was neither filtered nor treated with corrosion inhibitors or biocides at its source or at the mine. After the pipeline was hydrotested with seawater during commissioning, the seawater was left in the pipeline for three months and became stagnant, thus accelerating pipeline corrosion. Corrosion rate measurements revealed that the pipeline corrosion rate was about seven time the assumed design rate. This led to significant concerns regarding the pipeline integrity as well as the effect of exposure to the seawater by other system components such as tanks and process piping. These and other path dependent corrosion challenges are depicted in Figure 2.

The lesson learned is that, if the potential corrosion challenges and associated risks and were considered during the seawater pipeline design, systems engineers could have developed a corrosion management plan that considered the downstream path dependence impact; and included actions and tests to avoid these technical issues, along with the adverse schedule and budget consequences [10].

6. EXAMPLES OF MILITARY SYSTEMS IMPACTED BY PATH DEPENDENCE

In 2010, the U.S. Department of Defense (DoD) filed a report regarding the grounding of the F22 fighter fleet and reduction in capability of the F35 fighter fleet due to the serious impact of corrosion. This condition was induced by the path dependence sequence and negative effects described above [11, 12]. Both aircraft designs included no specified corrosion requirements, and suffered from lack of independent testing of designs and reduced climatic testing. During design trade-offs, corrosion protection was sacrificed in favor of manufacturing and performance goals [12].

The F-22 acquisition process included unrealistic testing methods, and used similarity to prior aircraft designs to qualify some F-22 systems' designs, rather than directly testing the F-22 designs. F-22 coatings were selected using inadequate testing methods. Likewise, system-level verification of corrosion requirements was inadequate [12].

Figure 3a. F-22 Raptor

Figure 3b. F-35 Lightning II

The F-35 acquisition was based on immature systems engineering processes that adversely affected the corrosion test specifications. Although the F-35 design and production reflected an improved test approach based on lessons learned from the F-22 acquisition, there was no overall system verification process [12].

These two examples might be the "tip of the iceberg." In their review of the DoD F-22/F-35 report, the Government Accountability Office (GAO) observed that effective corrosion prevention and control of many weapon systems depend on key corrosion practices cited in the report, and that generic factors related to current acquisition practices could contribute also to corrosion in other weapon systems [13].

The lessons learned from the F-22 and F-35 acquisition are that inadequate design and material selection, insufficient corrosion test requirements, and unsatisfactory or no testing along the design, development, production and testing acquisition path resulted in significant material degradation of several aircraft systems and grounding or restricted operation of first-line fighter aircraft. The broader lesson is that the impact of erroneous or no feedback during each step in the system acquisition path could cause downstream corrosion of many current and future weapon systems and increase operational and maintenance costs, reduce warfighting system availability and induce potential safety issues.

7. STRATEGIES TO REDUCE IMPACT OF PATH DEPENDENCE ON MATERIAL DEGRADATION

Executives, managers, systems engineers and systems engineers associated with the acquisition of commercial and military systems and facilities, need to apply strategies that take advantage of path dependence and avoid its potential pitfalls. While these strategies should be applied to all elements of the acquisition and operation processes, they are particularly critical in the selection, treatment, assembly, and testing of materials that are susceptible to material degradation in any future operational environment.

Reducing the negative effects of path dependence should begin with the sequence of steps in the design path. Avoid the tendency to start at the beginning of the process to determine the design paths. Rather, start at the final step – the outcome of the design effort – and work your way backwards, step by step, where each step is defined by the outcome demanded of its succeeding step. Of course, more than one preceding step is often required to produce a succeeding step. Thus, parallel paths will emerge and subsequently converge as preceding steps begin to satisfy more than one succeeding step. This process ensures that needed steps are not overlooked, while false paths are avoided.

Once the complete design path itself is established, systems engineers and program managers should analyze each design phase to identify critical variables in the design path, and to evaluate the impact of positive and negative feedbacks associated with those variables. They should:

- Establish clear, measurable requirements and high fidelity metrics for evaluating system quality along the design path.
- Perform inspection and tests with reliable test equipment and in accordance with established test requirements.
- Select initial conditions (including technologies and products) that are likely to have a positive influence in the design path and yield desired system quality.
- Plan for frequent feedback from inspections and tests during prototype, preliminary and full production.
- Perform planned analysis throughout the acquisition phases to assess system quality along the design path.
- Quantitatively assess alternative paths when feedback indicates declining quality, or when obstacles force a path change. One technique, termed a "long jump," is to look far afield to identify and evaluate alternate designs or technologies that might offer a high-quality solution and overall economies.
- Evaluate the effect of random external influences that could occur along the design path and develop plans and techniques to reduce or eliminate adverse impacts if changing the design path could reduce system fitness. For example, identify an alternate source or design should a single source supplier of a critical product component be lost.
- Occasionally execute a "long jump" to investigate new, exemplar or dominant technologies for applicability to the product or service under development.

To implement the above strategies, executives and managers should take the following actions:

1. Produce and deliver a corrosion prevention and control plan that conforms to established corrosion prevention and control planning guidance [14].
2. Document and deliver affordability trade-off studies that analyze performance, availability, safety and cost impact of corrosion-caused material degradation during system or facility operation, compared to the operational effects of implementing established corrosion prevention and control processes and procedures during acquisition.
3. Articulate corrosion-related measures of effectiveness, measures of performance, test methods and procedures in a Test and Evaluation Master Plan, and see that they are properly applied during system acquisition.
4. Document and deliver well-planned, tailored operational-phase corrosion control and maintenance requirements and procedures.

8. CONCLUSIONS

Path dependence is a persistent characteristic of sequential activities. Its consequences are dictated by initial conditions, and by the existence and nature of feedback mechanisms along the chosen path. In the commercial and military system acquisition sequence, these initial conditions include design, material and production decisions made very early in the acquisition cycle.

Acquisition cycle feedback mechanisms primarily consist of inspection and test results throughout the acquisition process. If inspection and testing is not performed, no feedback occurs and the path continues as if all conditions are positive. If ineffective or faulted inspection and test requirements are prescribed, erroneous feedback can trigger erroneous decisions regarding either continued development on the current path or changing direction. Likewise, if inspection or test procedures are inaccurate, or equipment is faulty, erroneous feedback can occur with the same potential for and effects from erroneous decisions. As erroneous decisions accumulate and multiply, recovery from negative effects becomes increasingly difficult, and undesirable lock-in to an inferior design with poor performance and high operating cost might result.

The impact of path dependence on corrosion of infrastructure and military systems is particularly acute. Historically, corrosion prevention and control activities often are not seriously considered, if at all; or they are reduced or eliminated by trade-off decisions to fund design changes for increased performance. Part of the enormous annual cost of corrosion during facility and system operation is due to acquisition cycle path dependence-driven decisions that directly contribute to serious post-acquisition corrosion and resulting performance, availability, safety, and cost issues.

Fortunately, the recent standard published jointly by NACE and SSPC, "Corrosion Prevention and Control Planning," describes corrosion planning requirements that can reduce the impact of path dependence during commercial, industrial and military system acquisition [15].

9. REFERENCES

1. Hummel, R., *Alternate Futures for Corrosion and Degradation Research*. 2014, Arlington, Virginia: Potomac Institute Press.
2. *International Measures of Prevention, Application, and Economics of Corrosion Technologies Study*, G. Jacobson, Editor. 2016, NACE: Houston.
3. Ratcliff, S., *Research shows corrosion costs the local economy*. 2009, Media Centre: Curtin University.
4. Sterman, J., *Business dynamics: systems thinking and modeling for a complex world*. 2000, Boston: Irwin/McGraw-Hill.
5. Arthur, W.B., *Increasing Returns and Path Dependence in the Economy*. 1994, Ann Arbor: The University of Michigan Press.
6. Frenken, K., *Innovation, evolution and complexity theory*. 2006, Northampton, MA: Edward Elgar.
7 Keller, G., *Modeling and Measuring Affordability as Fitness*, in *Industrial and Systems Engineering Thesis*. 2012, Virginia Polytechnic Institute and State University: Blacksburg, Virginia.
8. Piller, C., *Corrosion plagues new Bay Bridge span*, in *The Sacramento Bee*. 18 May 2013
9. Report: *Facility and Infrastructure Corrosion Evaluation Study*. 2013, Department of Defense Office of Corrosion Policy and Oversight Washington DC.
10. Coull, Z., *Corrosion Management Planning: Lessons Learned for Seawater Conveyance in Mining*, in Proc. *NACE Corrosion Conference*. 2017: New Orleans.
11. Report: *Corrosion Evaluation of the F-22 Raptor and F-35 Lightning II Joint Strike Fighter*. 2010, Department of Defense Office of Corrosion Policy and Oversight: Washington DC.
12. Peeler, D.D., *Corrosion Evaluation of F-35 Lightning II and F-22 Raptor*. 2010, SAE: see http//www.sae.org/events/dod/ presentations/2010/B3Peeler.pdf
13. *Defense Management: DoD Needs to Monitor and Assess Corrective Actions Resulting from Its Corrosion Study of the F-35 Joint Strike Fighter*, United States Government Accountability Office, Editor. 2010: Washington, DC.
14. *Corrosion Prevention and Control Planning Guidebook for Military Systems and Equipment*, D.o. Defense, Editor. 2014, Office of Corrosion Policy and Oversight: Washington DC.
15. NACE SP21412-2016/SSPC-CPC 1 Standard, *Corrosion Prevention and Control Planning*. 2016: Houston/Pittsburgh.

10. AUTHOR DETAILS

Daniel Dunmire, DoD Director of Corrosion Policy and Oversight, has served the Office of the Secretary of Defense since 1984. His civilian Federal service started as a Presidential Management Fellow in the Office of the Under Secretary of Defense for Acquisition, Technology and Logistics where he has thrice received the OSD Medal for Exceptional Civilian Service. He served four years in the Army and is a retired Corps of Engineers Reservist. He received his Masters Degree in Public Administration from the University of Alabama, Birmingham. Mr. Dunmire is a doctoral candidate seeking a Ph.D. in Public Administration from Virginia Tech.

George Keller is a Senior Consultant at LMI, a government consulting firm, where he has been employed since 2006. He has provided broad consulting support to the DoD Office of Corrosion Policy and Oversight since 2003 in the areas of strategic planning, requirements development, project selection, communications and outreach, and training and education. He is a retired Air Force officer with 21 years of service. He received his Master's Degree in Industrial Engineering from Arizona State and his Doctor of Philosophy Degree in Industrial and Systems Engineering from Virginia Tech. He has been a NACE member since 2004.

INFRASTRUCTURAL HEALTH MONITORING USING CORROSION SENSORS: PAST, PRESENT AND FUTURE

Mike Yongjun Tan[1,2]
[1]Institute for Frontier Materials, [2]School of Engineering
Deakin University, Waurn Ponds, Geelong, VIC3220, Australia

SUMMARY: Infrastructural health monitoring (IHM) using corrosion sensors (or probes) is a practical approach to enhancing the safety, durability and economic operation of industrial and civil infrastructures such as underground and submerged oil and gas pipelines. A well designed IHM system is capable of acquiring in-situ and site-specific corrosion data, facilitating early detection and warning of corrosion damage, forecasting and prioritising site inspection and maintenance requirements. The need for IHM is particularly apparent when developing life extension strategies for older infrastructures. Unfortunately the application of corrosion sensors is not always successful, in particular when corrosion sensors are used for monitoring corrosion of infrastructures exposed to complex environmental conditions. This paper aims to provide a brief historical view of corrosion monitoring sensors, and discuss issues and limitations of major forms of corrosion sensors. It is shown that currently the prime challenge in corrosion monitoring is difficulties in simulating and evaluating localised corrosion mechanisms, especially when corrosion mechanisms change with time. Discussion has been extended to future development needs for enhancing the reliability and trustworthiness of corrosion monitoring sensors. Cases have been presented to illustrate potential future corrosion sensor applications with improved design and installation methods as well as with new big data based IHM information technology platforms.

Keywords: Structure health monitoring, pipeline corrosion, corrosion monitoring, corrosion probe

1. INTRODUCTION

Infrastructural health monitoring (IHM) is the application of structural health monitoring technologies to industrial and civil infrastructures such as pipelines and bridges that are vital for the provision of the nation's essential services and the maintenance of its economic activities. Compared to all-purpose structural health monitoring that covers a wide range of structures such as aircrafts, IHM focuses on the application of sensors to perform in-situ, continuous or regular measurement and analysis of key structural and environmental parameters under infrastructural operating conditions for the purpose of warning impending abnormal states or accidents at an early stage and giving maintenance and rehabilitation advice [1]. A typical IHM system includes three major components: a sensor system for data acquisition; a data transmission, storage, mining system for feature extraction and extraction; and a health evaluation system for decision-making based on identified features in combination with identified models [2]. The first step of cultivating an IHM system for a particular infrastructure is usually the identification of major structural integrity, durability and reliability concerns, followed by the selection of sensors or probes that are able to collect data/parameters related to these concerns. Over the past decades, many specialised sensors or probes, such as fibre optic sensors, piezoelectric actuators, vibration sensors, leakage detecting sensors, pressure, flow, and seismic sensors and corrosion detection sensors/probes have been developed and utilised in many infrastructural systems for monitoring various structural and environmental parameters such as stress, displacement, acceleration, wind speed, temperature, corrosion etc. IHM was originated mainly from major bridges, probably due to high awareness of the economic and social effects of aging, deterioration and extreme events on civil infrastructures [3,4]. In the case of long-span bridges, for instance, major concerns are often wind load, temperature, traffic load, geometric configuration, strain, and global dynamic characteristics, therefore sensors selected to be installed on bridges are often anemometers, temperature, strain and accelerometer sensors [4,5]. In the case of underground and submerged oil and gas pipelines, corrosion is usually among the major structural integrity concerns because of the corrosive environmental condition and the relatively poor corrosion resistance of steel pipelines. Corrosion sensors or probes are therefore needed for providing visibility of corrosion and material degradation processes occurring on these 'invisible' and 'hidden' infrastructural assets.

Corrosion sensors are usually referred to as devices such as a fibre optic corrosion sensor that detect certain external stimuli and respond in a distinctive manner, while corrosion probes such as electrochemical corrosion probes could be described as devices that generate and detect internal changes in the probe exposed to a corrosive media. The difference between sensors and probes is often not clear, and therefore 'sensor' is used as a general term in this paper for describing data acquisition devices in IHM systems. An IHM system with suitable corrosion sensors, if used properly, should be able to provide in-situ and site-specific corrosion information for the early warning of structural failure and for infrastructural life prediction. However currently the fact is that successful application of corrosion monitoring sensors in complex infrastructural systems has been actually limited despite of the fact that major corrosion induced accidents are frequently reported around the world [6-9] and the fact that smart sensors have been identified as a critical national need for infrastructure transformational research in the United States [10]. This paper presents a brief review of past and current corrosion sensor techniques used for infrastructure health monitoring in order to identify weaknesses in current IHM systems and forecast future development needs. Particular focus is on difficulties, challenges and limitations in current corrosion sensor application in oil and gas pipeline systems. The main purpose is to find ways that IHM could contribute much more significantly to the management of infrastructure assets especially 'invisible' underground and offshore infrastructures.

2. A HISTRICAL VIEW OF CORROSION MONITORING SENSORS

The lack of visibility of corrosion and material degradation processes, especially localised forms of corrosion, occurring on 'invisible' infrastructures is believed to be a major contributor to corrosion induced gas pipeline explosion and oil spill accidents [6,7,10,11]. Traditionally historical inspection data are used as the main source of information for forecasting the degradation and service lives of buried pipelines. Industries such as oil and gas pipeline industry collect corrosion and materials degradation data through survey and inspection using methods such as field excavation, DCVG and in-line inspection tools (intelligent *pigs*, i.e. pipeline inspection gauges). These methods are expensive (e.g. pigging of a pipeline can cost $1million or more). These historical survey and inspection data are often used with asset management tools that are usually computer software developed based on probabilistic models. The hypothesis behind these models is that statistical analysis of historical survey and inspection data could allow for structural lifetime assessment and prediction. For instance Li et al. [12] used a Monte Carlo simulation technique to calculate the remaining life of a structure. Lee et al. [13] presented an intelligent failure prediction system for oil and gas structure using an Euclidean-Support Vector Machines classification approach. Senouci et al. [14] developed a fuzzy-based model to predict the failure type of oil pipelines using historical data of pipeline accidents. Peng et al. [15] developed a fuzzy artificial neural network model, which is based on a failure tree and fuzzy number computing model, for predicting the failure rates of long-distance oil/gas pipelines and for identifying distressed pipeline segments. These asset management tools are useful in providing an overall assessment of the aging of a structure; however the success of these tools is heavily dependent upon the availability and reliability of structural condition data. Unfortunately high quality and timely data are often not available because time based routine inspections using various condition assessment tools are performed only on a periodic basis (e.g. pigging are usually done only once every 5-15 years). Another issue is that these asset management tools are often not suitable for infrastructural systems that are under localised corrosion attack and that are under the effect of dynamically changing environmental conditions.

A practical approach to acquiring in-situ and site-specific corrosion data from 'invisible' infrastructures is using various corrosion monitoring sensors. Over the past decades, many corrosion monitoring techniques have been developed. They can be generally divided into two distinct groups: off-line 'retrospective' techniques that detect cumulative corrosion damage; and on-line 'instantaneous' techniques that continuously measure the prevailing corrosion rates. The most widely applied retrospective corrosion monitoring/testing techniques include weight loss coupon, ultrasonic thickness measurement, radiography, electrical resistance probe, field signature method, and non-destructive inspection '*pigs*'. A summary of these techniques can be found in references [16,17].

Weight loss coupon testing is the simplest and most widely used corrosion testing and monitoring method. A metal coupon of known metallurgy, size, shape and weight is exposed to a corrosive environment and is inspected for corrosion after a period of time (e.g. every 90 days). Corroded coupons are subjected to visual and optical or microscopic examination, and measurement of weight loss due to corrosion. Corrosion products may be analysed by methods such as SEM, EDAX, XRD and infrared or Raman spectroscopy. Corrosion coupons are an excellent source of corrosion information if monitoring is continuously maintained and results are well accumulated and documented. However corrosion coupon tests have well-known limitations: they are considered to be offline, labour intensive, and not easily configured for automation and control systems. They require long exposure period to generate field test results and during this period corrosion may have already occurred to an industrial structure. They require periodic removal of test specimen from the corrosive environment which is cumbersome and may alter the progress of localised corrosion. They only detect the cumulative corrosion damage at the end of the exposure period and provide little information on specific events that may have triggered this damage. Although corrosion coupon test appears to be an easy task, there are in fact problems that often lead to unsuccessful and misleading results. For instance in the oil and gas pipeline, corrosion coupon tests were found fail to produce corrosion rate values comparable to actual pipe wall loss due to their misplacement [18].

A major challenge in corrosion coupon testing is to overcome difficulties in simulating true localised corrosion conditions and mechanisms. For instance, if a test interval of 30 days is selected for corrosion coupon tests in an oil and gas pipeline, it may lead to reporting of falsely high corrosion rates since this short exposure test may overlook changing corrosion mechanism with the formation of a protective deposit, which is vital to the ongoing corrosion behaviour of the pipeline. This short term test may also fail to detect any localised corrosion that requires an initiation period. Extension of the coupon test to 90 days may overcome some of these problems; however a 90 day test may still be far too short for microbiological build-up that often leads to localised biological corrosion in actual piping systems. Thus, a detailed understanding of the corrosion mechanism is essential for selecting appropriate coupon test parameters and durations. Another challenge in corrosion coupon test is difficulties in detecting galvanic corrosion and crevice corrosion. In an oil pipeline, for instance, corrosion coupons are typically isolated from the pipe surface; they are unable to detect galvanic corrosion effects due to the formation of a large cathodic layer of iron carbide or iron oxide on the pipe surface. As a result, a major localised corrosion mechanism responsible for a significant amount of metal loss in most piping systems is not measured. On the other hand, the corrosion coupon test is also unable to simulate or detect crevice corrosion such as underdeposit corrosion which occurs in some oil flowlines. These cases have been discussed previously in reference [17].

Electrical resistance (ER) probe is often referred to as an 'intelligent' weight-loss coupon test. The ER probe monitoring corrosion by measuring the electrical resistance of a thin metal test wire (sensor element) since the resistance of the wire increases as the wire becomes thinner due to corrosion dissolution. ER corrosion monitoring has been applied in industry to provide an indicator of environmental corrosivity for more than four decades. An advantage of ER probe is that it provides cumulative metal loss values without the need to remove the samples from the service environment. Another advantage of ER probes is that the technique is applicable to both conductive and non-conductive corrosion environments. A major disadvantage of the ER technique is that it is unable to detect localised corrosion since localised corrosion may neither lead to significant metal dissolution, nor noticeable change in electric resistance. It also has similar limitations as corrosion coupon tests in simulating the localised corrosion mechanisms. ER measurements generally do not respond rapidly to a change in corrosive conditions, and it was reported to take four days to respond to 1 mpy corrosion rate in the internal corrosion monitoring of subsea oil and gas production equipment [19]. Field signature method (FSM) is also considered to be a cumulative corrosion monitoring technique developed from the commonly used principle of electrical resistance determination. FSM impress a current through several electrodes (sensing pins) on the outer surface of a pipeline and the potential drops between these electrodes are measured. The sensing pins are typically separated by a distance of 2-3 times wall thickness. Changes in the geometry in the form of cracks, general corrosion, erosion corrosion, or pitting will disturb the potential field in the metal. The change in the measured potential field pattern is related to a change in pipe wall thickness, and thus the development of corrosion or cracks can be detected. The FSM technique is non-intrusive and is used to measure corrosion damage over a relatively large section of a structure. Possible limitations of the technique include its relatively slow response to corrosion. It was reported to take one year to respond to 1 mpy corrosion rate [19].
Physics based corrosion assessment methods are also retrospective in nature. Ultrasonic testing utilises high frequency sonic waves that transverse the thickness of a specimen and return to the probe to determine the thickness of the specimen. Ultrasonic probes are widely used for field inspection of cracks and corrosion damage. If sufficient numbers of test points are taken over a wide enough range, it can provide reliable corrosion inspection data. Automatic ultrasonic scanning and recording techniques combined with computer techniques are also able to produce three-dimensional maps of the corroded surface. The technique responds relatively slowly to corrosion, however, and it was reported to take four years to respond to 1 mpy corrosion rate [19]. Radiography makes use of the penetrating quality of short wave electromagnetic beams to image corrosion and to determine pit depths and the degree of thinning due to corrosion. Radiography is limited to detect narrow cracks that are oriented in directions other than parallel to the electromagnetic beams. Techniques such as eddy current are used to visualise initial cracks caused by stress corrosion or corrosion fatigue. Eddy current technique detects surface cracks, pits or other defects by measuring disturbed eddy currents on a material surface. Based on these techniques and their combination, equipment known as a '*pig*' has been developed for internal examination of pipes and tubes. The '*pig*' follows the flowing medium in the pipeline and records corrosion related data for analysis after it is removed from the pipe. A disadvantage of pipeline inspection using the '*pigs*' is that they are very expensive (e.g. pigging of a pipeline can cost $1million or more) and disruptive.

Although these cumulative corrosion monitoring techniques are obviously very useful in corrosion inspection; they are able to detect corrosion only when sufficient corrosion has occurred to cause an accumulated change in the bulk material properties. For this reason the response time for these techniques is usually undesirably long. They are used only to provide off-line average corrosion data over a rather long interval of weeks, months or sometimes years. Field corrosion inspections occur relatively infrequently, usually coinciding with routine shutdown and maintenance. Peak corrosion rates are usually not detected and, most importantly, the specific time periods of peak corrosion rates are not identified. Another obvious limitation of these techniques is practical difficulties of knowing the location of corrosion damage prior to the assessment. These locations could be estimated through engineering analysis, however with the increasing complexity of industrial structures such as multi-phase oil and gas flowlines and buried pipelines, the potential damage locations are not known or are too numerous to be economically inspected using these damage detection techniques.

Corrosion rates always change with time and corrosion can accelerate rapidly and grow exponentially once initiated, and thus it is important to identify the specific time periods and environment conditions of peak corrosion rates. For this reason on-line 'instantaneous' techniques are important to continuously measure the prevailing corrosion rates and provide quantitative data for use as a process variable for integrated corrosion management systems. Instantaneous corrosion testing and monitoring techniques are usually electrochemical in nature. They include corrosion potential measurement, potentiodynamic polarisation, linear polarisation resistance (LPR), electrochemical impedance spectroscopy (EIS), electrochemical noise analysis (ENA) and many others. Electrochemical techniques rely on electrochemical corrosion theory and the measurement of electrochemical potentials and currents that are fundamentally related to the thermodynamics and kinetics of corrosion reactions. Electrochemical corrosion testing and monitoring can be fast, sensitive and versatile. Modern electrochemical instruments can measure very low corrosion rates, if properly used accurate corrosion measurements can be made in a matter of minutes in most industrial processes [20].

Corrosion potential can be easily measured by recording the potential difference between a corroding electrode and a stable reference electrode. It can be used in conjunction with Pourbaix diagrams as a basic indicator of corrosion thermodynamic status, for instance, to predict active or passive behaviour. Corrosion potential monitoring is often useful in understanding cathodic protection, anodic protection, general corrosion and stray current corrosion problems. Unfortunately the corrosion potential value on its own does not provide information on the rate of corrosion. The LPR and EIS techniques are employed to measure corrosion rates using the Stern and Geary equation and have been discussed in numerous publications. Although LPR is useful in determining corrosion kinetics, care should be taken since this technique is accurate only under several fundamental assumptions. In principle, it only applies to a uniform corrosion system with a stable corrosion potential. The corrosion process should involve only one anodic and one cathodic reaction, with both under activation control. The Tafel constants should be known; and there should be only negligible solution resistance. For this reason, application of LPR to practical corrosion systems often provides only semi-quantitative corrosion data with variable accuracy. EIS can be used as an alternative technique for determining polarisation resistance. It has some advantages in measuring corrosion under organic coatings and in studying corrosion inhibition mechanisms, for instance, it has been used to study the formation and destruction of corrosion inhibitor films and corrosion scales under simulated oil and gas production environment [21-23]. ENA can also be used to determine polarisation resistance [24-28], although the prime attraction of ENA is its possibility of early detection and warning of localised corrosion by detecting 'noise signatures' [29,30]. For instance, it has been used to monitor the formation and destruction of corrosion inhibitor films [28]. The noise resistance measurement has the advantage that there is no need to apply a perturbation to the test system by an externally imposed polarisation, and the instrumental system is simple. The noise signatures method was proposed to detect localised corrosion by recognising characteristic noise patterns (often referred to as noise signatures) in the time domain [29] or in the frequency domain [30]. Although some controversial issues still exist in the interpretation of electrochemical noise data, electrochemical noise has been recognised to be a rich source of information on the corrosion process and localised corrosion. For instance, the noise signatures are well recognised to be valuable indicators of localised breakdown of passive film, the incubation, initiation, propagation and repassivation processes of localised corrosion [31]. However, it should be noted that quantitative analysis of localised corrosion using ENA parameters such as pitting factor remain controversial and requires further development. Although electrochemical techniques such as LPR, EIS and ENA have been widely used in estimating the rates of general or uniform corrosion, they have major limitation in measuring localised corrosion. In fact, conventional electrochemical corrosion testing techniques have major limitations in measuring localised forms of corrosion, which is responsible for some 70-90% of all corrosion failures.

It should be pointed out that limitations and problems associated with corrosion testing and monitoring have also been frequently reported in the literature [32-38]. For instance, Srinivasan and Kane [32,33] reported poor correlations between actual corrosion behaviour of gas wells in the Gulf Coast region and the North Sea and corrosion monitoring data obtained from various techniques including iron counts, caliper surveys, down-hole mounted coupons, radioactive sleeves, electrical corrosion probes and inspection of pulled tubing. They found that actual pitting rates were between 2 and 15 times the general corrosion rates monitored for various cases, and in some cases the differences in corrosion rates obtained using different techniques varied by an order of magnitude in value. Papavinasam *et al* [34,35] also reported significant concerns on some testing methodologies for evaluating corrosion inhibitors for oil and gas pipeline applications. On the other hand, Tan *et al.* [36-38] analysed fundamental limitations in some electrochemical methods, especially in the measurement of localised corrosion. These reports confirm difficulties and complexities in accurate corrosion measurement and monitoring using corrosion sensors.

Variously designed corrosion sensors have been developed and applied in various industry applications including those reviewed by Varela et al.[39] on corrosion monitoring techniques designed for the pipeline industry. A common weakness in corrosion sensor application is that principles and issues associated with the practical application of these sensors are often not sufficiently examined and corrosion sensors are not always used successfully in the industry. In some cases corrosion sensors are used with an expectation of monitoring corrosion in a similar manner as using a thermometer to measure temperature. Inevitably these practices could lead to confusing and misleading results, and consequently successful application of corrosion monitoring in complex systems such as buried oil and gas pipelines has been limited.

3. RECENT PROGRESSES IN CORROSION MONITORING SENSORS

Extensive research has been carried out to enable the capability of corrosion monitoring sensors in monitoring localised corrosion attacks in complex environmental conditions such as in highly resistive and inhomogeneous soil media and under the effect of dynamically changing environmental conditions. This has been proven challenging because localised corrosion rapidly attacks a small portion of a metallic structure as the result of fast local electrochemical dissolution. Conventional corrosion sensors such as resistance probes and LPR probes have difficulties in measuring local electrochemical dissolution. Localised corrosion remains one of the most common causes of pre-mature infrastructure failure and a major unsolved issue in corrosion science, and it is believed to be a fundamental factor contributing to the lack of predictability of infrastructure failure. A rational approach to preventing unexpected pre-mature failure of critical infrastructures is to monitor the initiation and propagation of localised corrosion using sensors.

A successful IHM system using corrosion sensors should be able to detect data related to a targeted critical corrosion mechanism in order to determine the effects of corrosion mechanisms (e.g. corrosion under sand deposits, pitting, and crevice corrosion) on corrosion kinetics and patterns. A basic principle that could underpin the use of corrosion sensors is that corrosion and materials failure are not accidental occurrences, they occur as the result of fundamental thermodynamic instability of a metal or a material in a specific environment. Therefore corrosion and materials failure occurring on a structure such as a pipeline would also occur on a sensor made of the same material and exposed to the same condition. A properly designed sensor and measurement method should be able to detect such thermodynamic instability and reaction kinetics from the sensor, and therefore facilitate the monitoring and prediction of corrosion. However this applies to localised corrosion only if the sensor surface effectively simulates a local corrosion mechanism occurring on the corroding infrastructure surface. It is important to correctly simulate a true corrosion mechanism because the corrosion severity and pattern is often determined by the corrosion mechanism, especially localised form of corrosion mechanism. For this reason, sufficient attention should be given to the understanding of localised corrosion mechanism, the possibility of corrosion mechanism change, and possible transitions from general to localised corrosion. This is a major challenge since designing or selecting sensors that are able to simulate a complicated corrosion mechanism in a particular environment-material combination is often difficult. This challenge is more acute when corrosion is affected by many inter-related variables such as non-uniform temperature and pressure, heterogeneous metallurgy, inhomogeneous soil or solution chemistry and thermo-mechanical conditions, local mechanical stress, soil and water composition, oxygen level, humidity, salinity, pH, temperature, stray currents, and biological organisms, as well as stray currents, coating defects, coating disbondment and cathodic shielding.

Over the past decade there are various research and development effort for developing corrosion sensors to monitor localised corrosion due to the interaction of local environment and materials. As examples, several electrochemical sensor have been developed based on an electrochemically integrated multi-electrode array known as the wire beam electrode (WBE) [16,17,36-38]. The WBE was initially developed to detect and monitor local defects in organic coating films [39]. Its application has later been extended into the study of localised corrosion including crevice corrosion [40], water-line corrosion [36] and localised carbon dioxide corrosion in an oil flowline [41,42]. Two important characteristic of the WBE method that are particularly valuable for the investigation of pipeline corrosion are, (i) WBE is applicable to high resistance underground or multi-phase pipeline environment, which has been demonstrated in a previous study [42,43]; (ii) WBE can map corrosion processes on an instantaneous and continuous basis [44]. In order to achieve the highest possible sensitivity of detecting early signs of localised corrosion, the WBE can be used in conjunction with electrochemical noise analysis since ENA has the advantage of providing early warning of localised corrosion initiation. The combined WBE-ENA technique is capable not only of measuring the corrosivity of the environment and the rates of corrosion, but also detecting the occurrence of localised corrosion well before localised corrosion becomes visible [37]. The WBE has also been used in various corrosion research [45-49]. Here we describe several localised corrosion sensor designs to illustrate current status of corrosion monitoring sensor development. The capability of the electrode array based corrosion probes in detecting the initiation and propagation of localised corrosion and coating failure is illustrated in these cases.

Case study: Designing a sensor for monitoring underdeposit corrosion (UDC)
The challenge of simulating and measuring complex localised corrosion mechanism is achievable. This can be illustrated by a practical case of developing a corrosion sensor for monitoring evaluating the performance of underdeposit carbon dioxide corrosion inhibitors under simulated oil flowline conditions [50]. Corrosion inhibitors are used to prevent internal oil pipeline failure due to pitting and mesa corrosion under solid deposits such as sand and biofilms. A problem is that the efficiency of corrosion inhibitors is often unquantified in the field because it is considered to be nearly impossible to assess by normal corrosion testing techniques [51]. Underdeposit carbon dioxide corrosion is believed to be controlled by factors including galvanic effects between a large cathode (structure surface) and a small anode (surface under deposits), failure of inhibitors to penetrate the deposits and the retention of aggressive species in the deposits. A corrosion sensor should effectively simulate these controlling factors and measure their effects on corrosion rates and patterns. Tan *et al* [50] designed a corrosion monitoring sensor based on the WBE, where the WBE working surface was partially covered with a rubber 'O' shaped ring filled with sand to simulate a localised underdeposit corrosion environment. This partially covered WBE becomes a new sensor that effectively measured and monitored very different corrosion behaviour in a CO_2 saturated brine environment with and without the presence of corrosion inhibitor imidazoline. This case clearly illustrates the successful detection of complex underdeposit corrosion and the effects of inhibitors.

Case study: Designing a sensor for monitoring dynamic corrosion under anodic stray currents
An electrochemically integrated multi-electrode array based sensor has also been designed to facilitate the in-situ monitoring and visualisation of electrochemical processes occurring on buried steel surfaces under cathodic protection (CP) and anodic transient conditions [52]. The WBE method has been applied for the first time as a new sensor for detecting

localised corrosion initiation under various dynamic anodic transient influences. More details on the sensor design and data analysis methods can be found elsewhere [52]. This corrosion sensor worked under CP has enabled significant progress to systematically categorise and quantify the level and nature of damage of pipeline as a result of CP excursions. A common phenomenon that was observed from these sensors is that shortly after an anodic transient was applied to a CP protected steel surface, anodic current and corrosion activity dropped dramatically from an initial anodic current peak value. This has been explained by the passivity of steel under CP induced high pH condition. Another phenomenon observed by inspecting the occurrence of local anodic currents in WBE maps was that localised corrosion initiation occurred after a critical duration. This critical duration could be explained by the breakdown of passivity under the effects of anodic transient induced pH and surface chemistry changes. This work suggests that the WBE sensor can be used as an effective tool for studying localised corrosion initiation under the effect of complex factors, as well as for the in-situ monitoring of stray current corrosion of buried steel structures [52].

Case study: Designing a sensor for monitoring coating cathodic disbondment under overprotection
Cathodic disbondment is a major form of electrochemically induced coating failure that frequently takes place at the metal/coating interface on cathodically protected steel infrastructure such as pipelines. Extensive research over the past decades has developed good understanding of the phenomenon, however currently there is no technique that can be used to perform in-situ monitoring of its occurrence in the field. Traditional methods of evaluating cathodic disbondment of pipeline coatings are based on ex-situ visual inspection of excavated pipes. The electrode array has been designed as a probe for monitoring and evaluating the cathodic disbondment of defective coatings [53]. It is illustrated that local electrochemical impedance measurements using an electrode array sensor have significantly improved sensitivity for monitoring the propagation of cathodic disbondment of defective coatings compared with the conventional overall electrochemical impedance and local current measurements approaches. This new approach also provides the opportunity of eliminating the effects of the low impedance coating defect regions on the visibility of higher impedance regions deep in the disbond coating, facilitating the probing of electrode processes and mechanisms in selected regions of heterogeneous electrode surfaces.

Case study: Designing a sensor for monitoring of localised corrosion processes under disbonded coatings
A new sensor [54-56] has also been designed to monitor the distribution of corrosion under disbonded coatings by measuring electrochemical currents over an electrode array surface partially covered by a crevice that simulates a disbonded coating. The sensor has been evaluated using immersion tests at open circuit potential (OCP) and under CP conditions. Under both OCP and CP conditions, anodic as well as cathodic current densities were detected within the crevice. Corrosion patterns were estimated based on the current density distributions from two different methods [56]. The acceptable level of correlation with the corrosion damage observed at the array surface at the end of the tests suggests that the sensor surface has the potential to monitor localised corrosion under disbonded coatings. Using sensors to simulate and detect early stages of corrosion or to measure corrosion susceptibility under disbonded coatings could provide a valuable and inexpensive means of obtaining in situ monitoring information on the health of a structure.

4. FUTURE OUTLOOK TO CORROSION MONITORING SENSORS

In order to make corrosion sensors more useful for IHM systems, several major weakness have to be overcome. The improper design and use of corrosion sensors that fail to simulate various corrosion mechanisms and possible changes in corrosion mechanisms with the extension of corrosion processes is a major reason that false corrosion rates and patterns are reported from corrosion monitoring. The incorrect installation and application of corrosion sensors/probes is another source of corrosion monitoring error. Another challenge is to develop or select suitable measurement and data analysis techniques that are able to detect and analyse data from corrosion sensors that contain 'predictor features' signifying the occurrence of corrosion and materials failure.

Future development to enhance the reliability of corrosion sensors
It is a highly challenging task to design reliable corrosion sensors, especially localised corrosion sensors, which are able to effectively simulate corrosion behaviour in actual service environments and reliably evaluate the effects of various factors on corrosion processes, rates and mechanisms. It is well appreciated that corrosion sensors needs to simulate the actual service exposure environment; however relatively less considerations have been given to the effects of environmental parameters on corrosion patterns and mechanisms. It is not uncommon to receive misleading test results due to inappropriate selection of testing parameters and measuring techniques. This challenge is more acute when corrosion is affected by many inter-related variables such as non-uniform temperature and pressure, heterogeneous metallurgy, inhomogeneous soil or solution chemistry and thermo-mechanical conditions, local mechanical stress, coating defects, and cathodic cathodic potential and excursions. These effects may have not yet fully quantified and need to be better understood and used in corrosion sensor design and application. In order to fully realise the advantage of corrosion sensors for providing site-specific and in-situ warning of unexpected structure failures, corrosion sensors may need to be placed at strategic and 'worst-case scenario' high risk locations of a structure. For a buried pipeline, for instance, typical high risk structure sites could be those with high stray current activities, low soil resistivity, high underground water level, high concentration of corrosive species, and those highly corrosion rate areas identified by pigging, field survey and historical excavations. Sensors embedded at these strategic sites can be used to collect real-time and site specific data that would contain critical 'predictor features' and parameters needed for modelling and predicting localised corrosion, coating disbondment and degradation.

Future development to enhance the data analysis and IF platform for IHM systems
Another development trend is a big data based IT platform for corrosion data management and analysis. Corrosion sensors should be a part of the data sources to help overcome weaknesses in asset management models. Although sensors can provide useful in-situ data from selected locations of an asset, there is a need to integrate data from limited monitoring sites into the whole database by suitable models in order to provide fuller coverage of a huge structure (e.g. a 1000km underground pipeline). An information platform would enable the integration of various data inputs and allow industry to cost effectively gather information on the in-service integrity of assets/infrastructure, gain high levels of confidence in the condition of the asset, timely maintenance, safety and continuous availability/operation of the asset. A web-based information platform that can linkup multiple industries and multi-disciplinary areas of research would be extremely useful for infrastructure health monitoring, failure prediction and life extension. However such a web based information platform is complex to develop because it requires expertise from several disciplines.

Other future development needs to enhance the IHM system
For wide application of corrosion sensors based IHM system, there are needs of future developments in terms of instrumentation, data acquisition, communication systems and data mining and presentation procedures for diagnosis of infrastructural 'health'. These have been discussed in reference [1-5] including,

1. The GPS has provided new possibilities for direct measurement of infrastructure and should be used in IHM.
2. An exhaustive cost–benefit and reliability analysis will be required to show assist infrastructure operators/owners that benefits outweigh costs and that the IHM system is reliable.
3. A better understanding of the design, implementation and early operation of the instrumentation system; as well as the quality of the instruments and their survival rate, the expected performance and critical locations in the infrastructure are needed to plan for fewer but more reliable sensors.
4. More research on optimal placement of sensors, and careful planning using scenarios of detectable degradation or damage, together with examination of similar operational SHM systems, will lead to more efficient sensor deployment.
5. Data collection, storage and communications: Wireless or land links are a necessity for remote unmanned installations; dialup modems or permanently connected leased lines may be used. Robust low-cost wireless systems play an increasing communications role but due to present limited data capacity, data compression or pre-processing will be necessary unless data are slowly sampled static signals. Successful SHM procedures should incorporate the means to compensate for or filter out the environmental and noise effects or at least establish confidence levels for anomaly detection against noise [5].
6. Data mining and information presentation: One of the most significant issues with SHM is converting data to information, an issue addressed in detail elsewhere in this set of papers. Not to be overlooked is the charting or presentation of information to operators who are very unlikely to be familiar with the sophisticated underlying numerical procedures [5].

5. CONCLUDING REMARKS

A major problem in today's management of infrastructure assets especially 'invisible' underground and offshore infrastructures is the lack of information regarding corrosion and materials degradation and their effects on the safety, reliability and durability of these assets. The lack of information regarding the interaction of local environment and materials hinders our ability to provide sufficient warning and maintenance of 'hidden' assets. A practical approach to overcoming weaknesses in conventional corrosion and materials degradation data and asset management models is structural health monitoring using corrosion probes/sensors. Corrosion sensors, if designed, installed and used properly, are practical tools to cost effectively acquiring in-situ and site-specific corrosion data that are critical for predicting the occurrence of localised corrosion and materials failure occurring at specific high risk structure locations of infrastructures. Although there are weaknesses in existing corrosion sensors and their application practices, corrosion monitoring is a cost effective, practical and useful method for enhancing the safety, reliability and durability of civil and industry infrastructures. The principles, issues, advantages and disadvantages of corrosion sensor technologies for corrosion testing and monitoring has been briefly reviewed, with particular focus on their application in the oil and gas pipeline environmental conditions. It is found that a major challenge in corrosion monitoring is difficulties in simulating localised corrosion mechanisms. Cases have been presented to illustrate that conventional electrochemical methods have limitation in measuring localised corrosion, and that the electrochemically integrated multi-electrode arrays are considered to be able to monitor localised corrosion. Several cases have been described to illustrate the importance and difficulties of simulating and monitoring localised corrosion.

6. REFERENCES

1. G.W. Housner, L.A. Bergman, T.K. Caughey, A.G. Chassiakos, R.O. Claus, S.F. Masri, et al. Structural control: past, present, and future. Journal of Engineering Mechanics 1997; 123(9):897–971.
2. Hong-Nan Li, Dong-Sheng Li, Gang-Bing Song, Recent applications of fiber optic sensors to health monitoring in civil engineering, Engineering Structures 26 (2004) 1647–1657.
3. P.C. Chang, A. Flatau and S.C. Liu - Structural health monitoring, 2003 - journals.sagepub.com.
4. J.M. Ko and Y.Q. Ni, Engineering structures, Volume 27, Issue 12, October 2005, Pages 1715-1725.
5. J. M. W. Brownjohn, Structural health monitoring of civil infrastructure, Phil. Trans. R. Soc. A (2007) 365, 589–622.
6. Li, X. et al., Materials science: Share corrosion data, Nature, 527, 2015, 441–442).

7. Gosch, E, *WA gas supply cut 30pc by blast*, *The Australian,* 7 June 2008 / The Western Australian, Report by the National Offshore Petroleum Safety Authority, page 13, 11 October 2008.
8. see *http://en.wikipedia.org/wiki/I-35W_Mississippi_River_bridge* (last visited 30/06/2017).
9. Koch, G. et al. Corrosion Cost and Preventive Strategies in the United States (NACE International, 2002).
10. *National Institute of Standards and Technology report, http://www.nist.gov/tip/prev_competitions/upload/ci_wp_031909.pdf,* (last visited 30/06/2017).
11. Alan Bailey, Petroleum News, Vol. 11, No. 20 Week of May 14, 2006
12. S. Li, S. Yu, H. Zeng and J. Li, R. Liang, *Journal of Petroleum Science and Engineering*, 65 (2009), 162–166
13. L. H. Lee, R. Rajkumar, L. H. Lo, C. H. Wan and D. Isa, *Expert Systems with Applications*, 40 (2013), 1925–1934
14. A.Senouci, M. S. El-Abbasy and T. Zayed, M.ASCE, *Journal of Infrastructure Systems*, 20 (2014), December 2014
15. X. Peng, P. Zhang, L. Chen, *Proceeding CSIE '09 Proceedings of the 2009 WRI World Congress on Computer Science and Information Engineering* - Volume 05, Pages 651-655, IEEE Computer Society Washington, DC, USA ©2009
16. M.Y. Tan , F. Varela and M. Forsyth, Structural health monitoring: A practical approach to achieving in-situ and situ-specific warning of pipeline corrosion and coating failure, *Proceedings Corrosion & Prevention 2015*, Editor(s): Australasian Corrosion Association, Adelaide, 15 - 18 November 2015, paper 32, 10pp.
17. Y.J. Tan, K Lepkova, S Bailey and R Gubner, A critical review on various corrosion testing and monitoring techniques, *Proceedings of the 2009 Annual Conference of the Australasian Corrosion Association*, Editor(s): Australasian Corrosion Association, Adelaide, paper no.44, Coffs Harbour, 15-18 November 2009
18. Corrview, The limitation of corrosion coupons, *http://www.corrview.com/corr_04.htm* (visited June 2009)
19. M.W. Joosten, K.P. Fischer, R. Strommen, K.C. Lunden, 'Internal corrosion monitoring of subsea oil and gas production equipment", *Materials Performance*, April 1995, pp.44-48
20. R. Kane, 'A new approach to corrosion monitoring', Chemical Engineering, pp34-41 (June 2007) (*www.che.com*)
21. B. Kinsella, Y.J. Tan, and S. Bailey, 'Studies of CO2 corrosion product scales using electrochemical impedance spectroscopy and surface characterisation techniques, Corrosion – NACE (USA), no.10, vol.54, p835-842 (1998)
12. S. Bailey, Y. J. Tan, and B. Kinsella, 'Use of electrochemical impedance spectroscopy for the study of CO2 corrosion protection by batch treatment inhibitors', British Corrosion Journal (UK), vol.32, no.1, p49-55 (1997)
23. Y. J. Tan, S. Bailey and B. Kinsella, 'Investigations on the formation and destruction processes of corrosion inhibitor films using electrochemical impedance spectroscopy', Corrosion Science (UK), vol.38, p1545 – 1561(1996).
24. D. A. Eden, K. Hladky, D. G. John, and J. L. Dawson, *Corrosion 86*, Paper 274, NACE, Houston, TX (1986).
25. U. Bertocci, C. Gabrielli, F. Huet, and M. Keddam, *J. Electrochem. Soc.*, 144, 31 (1997).
26. F. Mansfeld and H. Xiao, *J. Electrochem. Soc.*, 140, 2205 (1993).
27. H. Xiao and F. Mansfeld, *J. Electrochem. Soc.*, 141, 2332 (1994).
28. Y. J. Tan, S. Bailey, and B. Kinsella, *Br. Corros. J.*, 32, 212 (1997).
29. K. Hladky and J. L. Dawson, *Corros. Sci.*, 21 (1981) 317
30. K. Hladky and J. L. Dawson, *Corros. Sci.,* 22 (1982), 231
31. R. A. Cottis, *Corrosion,* 57 (2001) 265-285
32. R. D. Kane and S. Srinivasan, IntelCorr/96 Technical Sessions, *http://www.corrosionsource.com/events/intercorr/techsess/papers/session6* (visited June 2009)
33. S. Srinivasan and R. D. Kane 'Prediction of Corrosivity of CO_2/H_2S Production Environments', Corrosion/96, Paper No. 11, Denver, CO, 1996
34. S. Papavinasam, R. W. Revie, M. Attard, A. Demoz and K. Michaelian, 'Comparison of laboratory methodologies to evaluate corrosion inhibitors for oil and gas pipelines', Corrosion-NACE, vol. 59, no. 10, pp897-912
35. S. Papavinasam, R. W. Revie, M. Attard, A. Demoz and K. Michaelian, 'Comparison of techniques for monitoring corrosion inhibitors in oil and gas pipelines', Corrosion-NACE, vol. 59, no. 12, pp1096-1111
36. Y. J. Tan, 'Monitoring localised corrosion processes and estimating localised corrosion rates by means of a wire beam electrode', Corrosion-NACE (USA), no.5, vol.54, p403-413 (1998)
37. Y. J. Tan, S. Bailey, B. Kinsella, and A. Lowe, 'Mapping corrosion kinetics using the wire beam electrode in conjunction with electrochemical noise resistance measurements', *J. Electrochem. Soc.*, 147, p530-540 (2000)
38. F. Varela, M.Y. Tan, M. Forsyth, *Corrosion engineering science and technology*, **50** (2015), 226-235
39. Y. J. Tan, Prog. Org. Coat., 19, p89 (1991)
40. Y. J. Tan, Corrosion (NACE), 50, p266 (1994)
41. Y. J. Tan, Corrosion Science: A Retrospective and Current Status, G. S. Frankel, H. S. Isaaca, J. R. Scully and J. D. Sinclair (eds.), The Electrochemical Society 201st Meeting, PV2002-13, Philadelphia (2002)
42. Y. J. Tan et al. 'Novel Corrosion Experiments Using the WBE: (I)-(IV)', Corros. Sci., 48, p23-78 (2006)
43. Y. J. Tan, 'Experimental methods designed for measuring corrosion in highly resistive and inhomogeneous media', *Corrosion Science*, 53 (2011), 1145
44. Y. J. Tan, Sensing electrode inhomogeneity and electrochemical heterogeneity using an electrochemically integrated multielectrode array, *Journal of the Electrochemical Society,* 156 (2009), C195
45. N.D. Budiansky, F. Bocher, H. Cong, M. F. Hurley, and J. R. Scully, Use of coupled multi-electrode arrays to advance the understanding of selected corrosion phenomena, *Corrosion,* 63 (2007), 537
46. A. Legat, Monitoring of steel corrosion in concrete by electrode arrays and electrical resistance probes, *Electrochimica Acta*, 52 (2007), 7590
47. W. Zhang, B. Hurley and R. G. Buchheit, Characterization of chromate conversion coating formation and breakdown using electrode arrays, *Journal of the Electrochemical Society,* 149 (2002), B357

48. Q. D. Zhong and Z. Zhang, Study of anti-contamination performance of temporarily rustproofing oil using wire beam electrode, *Corrosion Science*, 44 (2002), 2777
49. L. T. Yang, N. Sridhar, C. S. Brossia and D. S. Duna, Evaluation of the coupled multielectrode array, *Corrosion Science*, 47(2005), 1794
50. Y.J. Tan, Y. Fwu and K. Bhardwaj, *Corrosion Science*, 53, 1254 (2011)
51. NACE Task Group TG 380, Under Deposit Corrosion (UDC) testing and mitigation methods, draft report distributed to members for Review and Comment (March 2009)
52. Y. Huo, M.Y. Tan and M. Forsyth, Visualising dynamic passivation and localised corrosion processes occurring on buried steel surfaces under the effect of anodic transients, Electrochemistry communications 66:21-24 (2016)
53. F. Mahdavi, M.Y. Tan and M. Forsyth, Communication - an approach to measuring local electrochemical impedance for monitoring cathodic disbondment of coatings, Journal of the electrochemical society 163(5):C228-C231 (2016)
54. F. Varela , MYJ Tan and M. Forsyth, An electrochemical method for measuring localized corrosion under cathodic protection, ECS electrochemistry letters 4(1):C1-C4 (4 pages) 2015
55. F. Varela , M. Tan and M. Forsyth, Electrochemical method for studying localized corrosion beneath disbonded coatings under cathodic protection, Journal of the electrochemical society 162(10):C515-C527 (13 pages) 01 Jan 2015
56. F. Varela , M. Tan and M. Forsyth, Understanding the effectiveness of cathodic protection under disbonded coatings, Electrochimica acta 186:377-390 (14 pages) 20 Dec 2015

7. AUTHOR DETAILS

Dr Mike Yongjun Tan is a Professor in Applied Electrochemistry and Corrosion Technologies at Deakin University in Australia. Dr Tan's principal teaching and research interests are in corrosion science and engineering and their applications for enhancing the reliability and durability of civil and industrial infrastructures. His work over the past two decades spanned the range from fundamental study on issues such as electrochemical heterogeneity and passivity, and applied research on issues such as corrosion testing and monitoring, localised corrosion prediction, corrosion inhibition and anti-corrosion coatings, to industry engagement and contracted research on issues such as pitting corrosion of external and internal oil pipeline, and corrosion of desalination pipeline weldments. He is also a Research Program Leader of the Energy Pipelines Cooperative Research Centre Australia. The goal of this Coatings and Corrosion research program is to cost effectively extend the life of pipeline infrastructure by mitigating corrosion and environmentally assisted degradation of pipelines. He is the author of some 170 publications and approximately 100 industry reports todate, and a research book entitled 'Heterogeneous Electrode Processes and Localized Corrosion' (2013 John Wiley & Sons). He is also the Asia Pacific Editor for the Corrosion Engineering, Science and Technology (Formerly known as British Corrosion Journal, since 1965).

MARINE ASSET PROTECTION SYSTEMS - CATHODIC PROTECTION AND MARINE PILE WRAPS WORKING TOGETHER

W J Thomson
Thomson Engineering Management, Queensland

SUMMARY: The use of both impressed and galvanic cathodic protection (CP) systems to protect submerged steel marine structures from corrosion is well understood. However it is only a viable method of corrosion prevention when the structure is fully immersed in the electrolyte (seawater). The portions of the steel structures subject to tidal, splash and spray effects require a different approach to corrosion prevention.

Marine pile wraps have been successfully used to prevent corrosion for over 40 years. The opportunities now however are to consider them in conjunction with CP systems.

This paper presents marine pile corrosion prevention solutions by considering a combination of impressed or galvanic current CP and marine pile wraps.

From coupon and electrical resistance corrosion monitoring systems, cable security, CP structural support and dielectric shielding properties, the combination presents the corrosion engineer a multitude of benefits, not only economically, but practically.

This paper highlights the benefits of considering the two corrosion prevention systems together.

Keywords: Cathodic protection, marine barrier wraps, corrosion prevention, marine infrastructure.

1. INTRODUCTION

Steel structures exposed to seawater are prone to complex corrosion actions. These include: the interaction of dissolved oxygen with the various materials from which the structure is composed of; passivating of the materials; breakdown of this passivation; microfouling; and macrofouling (1). Cathodic protection (CP) of steel infrastructure and assets in marine environments is universally used method for arresting and preventing corrosion (2). Figure 1 illustrates the different corrosion zones of a typical marine pile structure highlighting the diverse environments the steel is exposed to. In order for the corrosion engineer to prevent the formation of the corrosion cell "principle" (3), they must consider a variety of corrosion prevention / management systems. Other than material selection, corrosion protection systems available include: impressed current cathodic protection (ICCP); galvanic (or sacrificial); or even a combination of both. They can be limited in that they only operate effectively in the driven, submerged and lower tidal zones. The electrolyte, the seawater, whilst being the source of the corrosion is also required as part of the corrosion protection system (2).

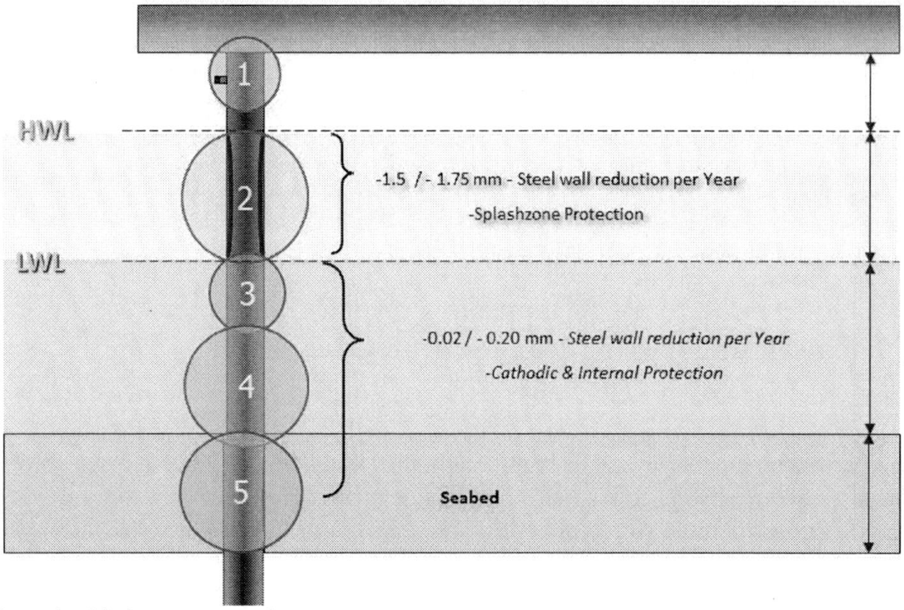

Figure 1 – Marine asset corrosion zones.

In order to ensure that the whole asset remains protected from below the seabed to the wharf soffit or headstock, another system needs to be considered, often in conjunction with CP. Barrier systems in the form of coatings or wraps are commonly used in zones 1, 2 and 3 where CP has little or no effect. Furthermore, exposed steel under the wharf superstructure is prone to being in a high corrosion zone due to the moist, unwashed and deposit rich micro-environment it finds itself in (4). Barrier systems that rely on a painted or epoxy coating are limited in their applicability and longevity in these zones. This is due to surface preparation, curing, and physical application limitations. Marine barrier wraps, a multi-part or monolithic combination of a corrosion inhibitor gel infused layer and multi-layer mechanical outer jacket are a viable option to coat a cleaned steel surface. A typical structure of a marine barrier wrap structure is illustrated in Figure 2.

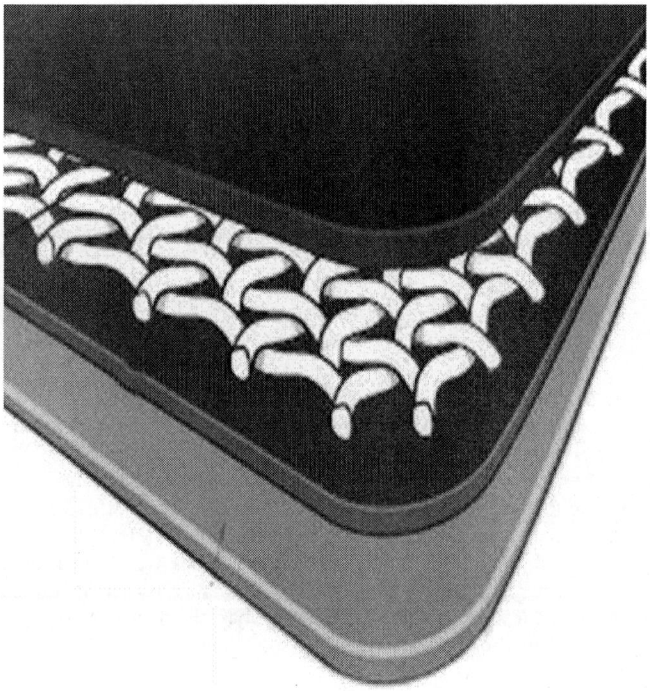

Figure 2 – Typical monolithic marine barrier wrap profile.

Too often CP and barrier systems are considered in isolation. This paper intends to describe the practical benefits of considering these systems together. Importantly, the electro-chemical, mechanical, structural, installation and protection factors of these systems are potentially symbiotic.
This paper commences with a brief description of CP and marine wraps and then addresses the complementary benefits related to total asset corrosion prevention.

2. CATHODIC PROTECTION OF MARINE STRUCTURES.

The intention of this section is to describe the role CP plays in protection of the corrosion zones as illustrated in Figure 1. Firstly, in the fully submerged zones current demand to maintain a stable steel structure varies with dissolved oxygen levels (5). Piles may have been coated prior to installation, and as they age so the current demand increases as the exposed surface area increases. On smaller or new marine infrastructure a sacrificial system (usually aluminium anodes) can be installed. The aluminium sacrifices itself as the more galvanic of the metals to the steel in the pile (mostly composed of lesser galvanic iron). ICCP though, is more commonly used in both new and retrofit installations. An ICCP system's components are more complex than a galvanic system, yet current is derived from a constant controllable source and easier managed than a submerged sacrificial anode. Components of ICCP systems include cables and connections of both positive and negative circuits, mixed metal oxide (MMO) anodes, transformer rectifier units (TRU), test points, connections, clamps, suspension brackets and other associated equipment. Some of these are illustrated in Figures 3-6.

Figure 3 – TRU control panel

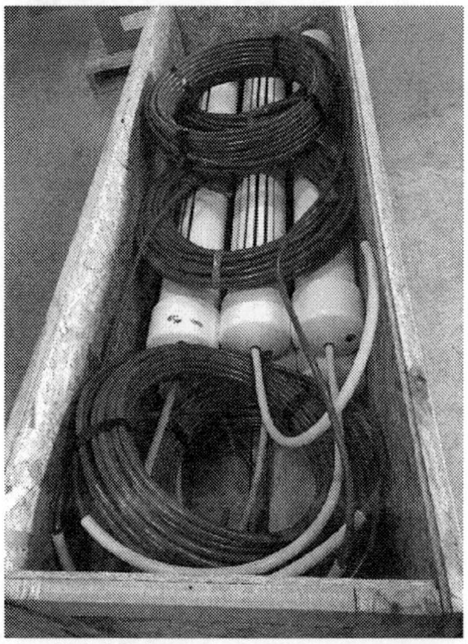

Figure 4 – Shielded marine MMO anodes packed for shipping

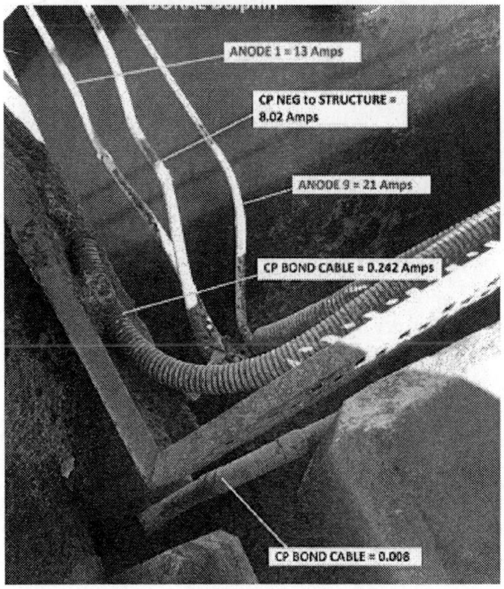

Figure 5 – Wharf anode cables

Figure 6 – Wharf CP test point

The CP designer is required to consider manufacture, installation, commissioning and maintenance of the CP system. Often the latter point is overlooked with cables hanging in exposed areas, anodes placed in areas exposed to boat wash for example, introducing operational and maintenance risk to the performance of the corrosion prevention system.

3. MARINE BARRIER WRAPS

The barrier wrap is a robust solution to retrospectively establishing a corrosion prevention mechanism on marine (and other) infrastructure. It relies on a hoop tension induced during installation to force a corrosion inhibiting gel onto the surface of the pile and keep it there for the duration of its lifespan. An example is illustrated in Figure 7. It is important that the outer mechanical layer maintains an even stretch around its circumference for the duration of its expected lifespan. The modulus of elasticity is maintained through a combination of urethane and scrim fabric layers (as illustrated in Figure 2) such that the wrap can be removed and re-installed for inspection purposes and of course for retrofitting systems such as CP and corrosion monitoring systems.

Figure 7 – Typical example of a marine wrap in situ

Most importantly marine barrier wraps enable the asset owner to protect the asset in the corrosion zones where the CP has limited, or no, corrosion prevention action. The barrier wrap due to its ease of installation, robustness, flexibility and simplicity represents a suitable complementor to the CP system itself. The complementary functions include how: it works with the CP corrosion prevention system to ensure complete asset corrosion protection; how wraps can improve the efficiency of CP systems; how wraps incorporate structural requirements of the CP system; and how it incorporates corrosion monitoring systems and ease of asset protection inspections.

4. CP AND WRAPS: COMPLEMENTARY CORROSION PREVENTION SYSTEMS.

Figure 1 illustrates the various corrosion zones of a typical marine asset. Zones 4 and 5 are able to be protected by CP as these areas are fully submerged at all times. Depending on the nature of the environment such as marine growth, wave and wash action, Zone 3 may be fully or partially protected. Zone 3 can vary from being inconsequential in protected harbours or can be protected intermittently on offshore structures. These considerations must be taken into account when designing the depth below MLW / LWL the marine wrap extends to. In protected harbours, this may be as low as 300mm or in offshore locations, could be 500mm or so below the lower level of the wave trough. The wrap can then be extended to the high corrosion Zones, 1 and 2. In situations where regular access is difficult and there is no natural freshwater washing due to rainfall of the asset, the marine wrap then protects the marine asset for a significant length of time without the need for regular maintenance.

4.1 The CP / Wrap tradeoff

Where current demand (approximately 120mA / m^2 of exposed steel) of the marine structure may require excessive power systems beyond the supply capabilities at the location, the wrap may provide the CP engineer with a power management option. By extending the marine wrap beyond Zone 3 to cover or partially cover Zone 4, the current demand can be reduced. In remote places, for example islands without power supply, a low maintenance option is to extend the marine wrap to the seabead and install a sacrifical CP system to protect the lower current demand (20mA/m^2) in this environment. The added benefit of this is the required size (and cost) of the transformer rectifier unit required to deliver current in an ICCP system.

4.2 Wraps as dielectric shields

Where MMO anodes are required to be placed strategically to offer maximum CP protection of pile structures, they are often designed to be mounted on or near a pile. The wrap itself offers several advantages in this regard. Firstly, it reduces the current demand, secondly it provides a support structure on which the anode can be mounted and offers the flexibity to position the anode to achieve maximum corrosion prevention as well as ease of maintenance. The wrap also offers the system dielectric shielding properties. The structure on which the anode is mounted offers the CP current a path of least resistsance. In order to mitigate this effect, the dielectric shield prevents this from occuring as wraps can provide a dielectric strength of up to 275 kV/mm. An example of this arrangement is illustrated in Figure 8. An added benefit is that wraps with robust flanges offer the CP engineer an opportunity to install anode brackets mounted on this feature.

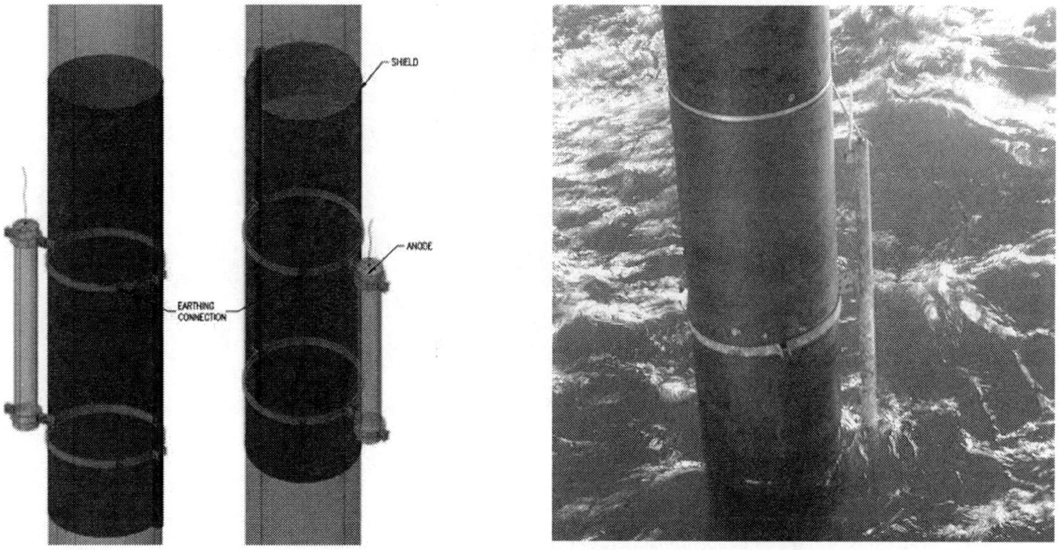

Figure 8 – Anode and marine wrap mounting to illustrate its use as a dielectric shield

4.3 Wraps provide CP system integrity

Often, the CP installation is undertaken in isolation from other wharf systems. Marine barrier wraps are no exception and this is an opportunity lost by the CP designer. Figure 9 illustrates typical intallations where the anode is installed where risk of entanglement and damage is heightened.

Figure 9 – Typical anode and cabling site installation

By utilising the strucure of the wrap itself, the loose cables can be secured inside the wrap either during installation or later an as a retrofit. The choice of marine wrap is important particularly in relation to retrofit applications. A wrap that maintains its stretch for the duration of its life is more appropriate for retrofit applications. Those wraps that plasticise post installation are not suitable for retrofit applications. A general rule is to select a wrap that exhibits an extension range of 5-7%, a maximum extension to rupture of 45% and an extension without creep of 12% (ATSM D-751). This allows the wrap to accommodate the cabling associated with CP installations. Figure 10 illustrates this feature.

Figure 10 – Marine wrap accommodating CP cabling

4.4 Anode wraps to accommodate corrosion management systems.

When installing a CP system and marine wrap combination, the asset manager often wants to have information on the performance of the system. Often, a diving company is contracted to undertake visual, ultrasonic thickness (UT), and marine reference electrode inspections. Whilst this data is useful, it can be costly to gather. An option to the asset manager is to install corrosion monitoring systems such as corrosion coupons and electrical resistance (ER) probes. The two systems are common but offer different mechanisms of obtaining corrosion data from which to determine the rates of corrosion. In the case of coupons (Figure 11) the data is obtained from the weight lost (or not) after a period of time under the marine wrap. Electrically connecting a coupon to the bare steel can provide the asset owner with a "yardstick" with which to undertand the effectiveness of CP protection. This can be juxtaposed with a coupon that is isolated from the CP system and thence the effectivenss of the CP system can be determined.

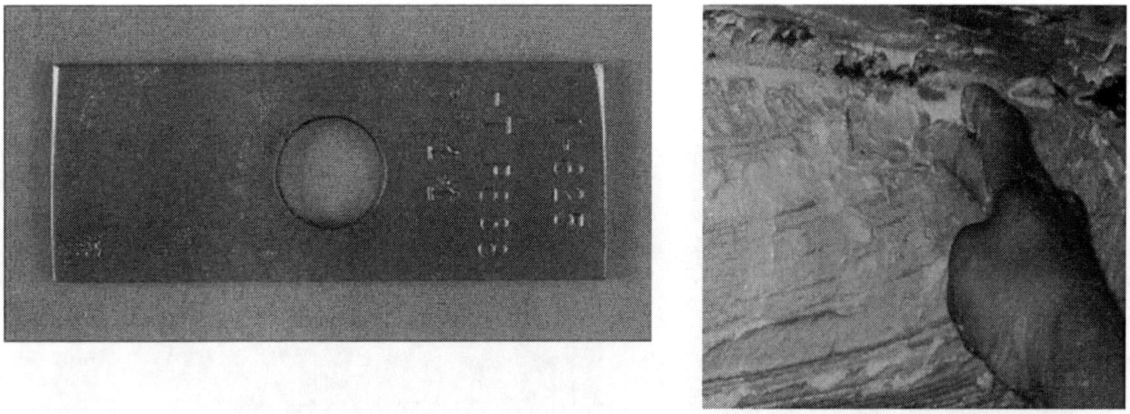

Figure 11 – A typical weight loss corrosion coupon and its placement under a marine wrap

ER probes can be connected to dataloggers to gather more accurate CP system perfomance information. Data can be relayed to remote monitoring systems. Once again, their practical use for CP monitoring is greatly enhanced by incorporating them in the physical structure of the marine wrap. Figure 12 illustrates a typcial ER probe (note the green wire used to electrically connect to the CP system).

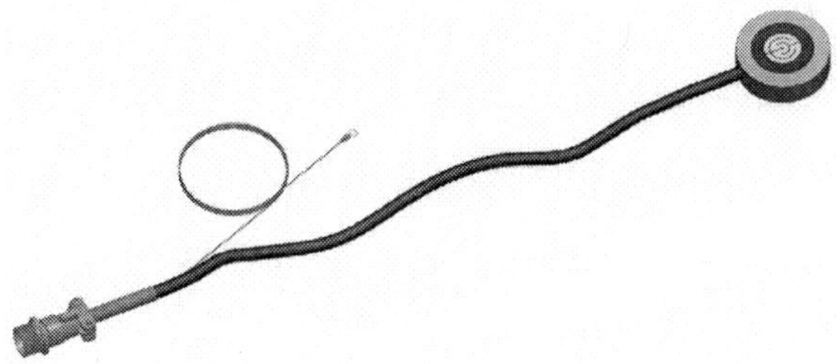

Figure 12 – Electrical resistance probe

5. DISCUSSION

The marine asset is exposed to a high corrosive environment and in order to maintain the structure's integrity, the corrosion engineer is faced with various corrosion zones to manage. Effective corrosion management is dependent on the correct methodologies chosen for a particular environment, including the particular corrosion zones (1). In order to ensure that no "bad design" is implemented (3), each zone potentially requires diverse corrosion prevention solutions, or at least nuances of these. Traditionally, a CP engineer approaches the solution from a sacrificial, ICCP point of view or a combination of these to prevent corrosion where this system would be effective. A coatings engineer would do the same for corrosion prevention barriers. This paper was prepared in order to highlight the potential benefits of considering the two systems together in part (or whole) to manage the corrosion prevention of the asset.
Several functions of the marine barrier wrap when working in tandem with the CP system are highlighted above. It is important though to select the most appropriate marine barrier wrap for the function it is required to perform. For instance, the marine barrier wrap must be structurally sound enough to support anode brackets, yet flexible to accommodate cables and anodes where there may be strong wash or tidal flows.

6. CONCLUSION

The flexibility of the marine wrap system offers the corrosion engineer with an opportunity to effectively manage corrosion zones 1-5 on the asset whilst also providing a structural means to safely attach CP system components. The marine wrap system also offers the corrosion engineer the opportunity to minimise current demand in remote areas where power supply may be impractical and a combination of wraps and a sacrificial CP system would suffice. In considering the two systems together, the corrosion engineer is able to offer the asset owner with the most effective corrosion prevention system possible.

7. REFERENCES

(1) Shifler DA. Understanding material interactions in marine environments to promote extended structural life. Corrosion science 2005;47(10):2335.
(2) Ashworth V. 4.18 - Principles of Cathodic Protection. In: Cottis B, Graham M, Lindsay R, Lyon S, Richardson T, Scantlebury D, et al, editors. Shreir's Corrosion Oxford: Elsevier; 2010. p. 2747-2762.
(3) Sarandily A. INFLUENCE OF CORROSION ON DESIGN OF MARINE STRUCTURES. Corrosion Management 2012(106):18.
(4) Heidersbach R. Marine environments and corrosion. Materials Performance 2004;43(10):6.
(5) Phull B. 2.18 - Marine Corrosion. In: Cottis B, Graham M, Lindsay R, Lyon S, Richardson T, Scantlebury D, et al, editors. Shreir's Corrosion Oxford: Elsevier; 2010. p. 1107-1148.

8. AUTHOR DETAILS

Wayne Thomson is an asset integrity management specialist at Thomson Engineering Management. Wayne is secretary of the ACA Queensland Branch and has served on the New Zealand Branch Committee and ACA Council. Wayne started out his working life in the nuclear industry as a Mechanical Engineer. He has worked in the energy, infrastructure, chemical process and packaging industries in South Africa, Italy, New Zealand and Australia. He has an MBA and PhD in business, the focus of his thesis is sustained growth of engineering firms in the energy industry.

GAMIFICATION OF TRAINING FOR SPECIFYING INSPECTION AND REMEDIATION OF CONCRETE STRUCTURES

W. Ward[1], B. McGuiness[2], R. Barnes[1], D. Veloso[2]
[1]Papworths Construction Testing Equipment, Belrose NSW, Australia
[2]Marine and Civil Maintenance, Gladesville NSW, Australia

SUMMARY: This paper examines a sample event design run by the Young Corrosion Group committee of the NSW Branch of the Australasian Corrosion Association. The event is a technical training event but delivered as a game. The proposed event places attendees into teams responsible for selecting inspection methods for a concrete structure after an initial visual inspection report. Based upon the results of this inspection they receive teams will complete a specification for remediation of the structure.

The event will be presented in detail, with each design aspect discussed regarding proposed outcome, content mastery to be acquired, game theory behind its construction and the specific choices made to suit this approach to a corrosion professional audience.

Keywords: Training, Concrete, Inspection, Remediation

1. INTRODUCTION

Formal Training Courses and Technical events are a vital tool in the ongoing professional development of Corrosion Professionals. This paper examines a sample event design to be run by the Young Corrosion Group Committee of the NSW Branch of the Australasian Corrosion Association, this event is a technical training event delivered as a game. This paper will discuss the design intent and background in game theory of this game to make a more successful training event and to properly convey the topic to its attendees.

2. DEFININITONS

2.1 Gamification

The paper Deterding (2011) is written as attempt to supply a formal definition of the term "gamification", its conclusion is that "Gamification is the use of game design elements in non-game contexts", which will be used as the definition of gamification in this paper. The proposed event this paper discusses is an attempt to use game design elements to improve a training event

Further to this the paper Nicholson (2015) separates "reward-based gamification" [e.g. the arbitrary point scores and ranking tables used by fitness activity trackers, such as awarding 6 fuel points for a run, and ranking people by collected fuel points] from "meaningful gamification" which aims to engage the intrinsic motivations of people taking part in the activity. In the context of the event this paper discusses a reward-based system that might be used would be to provide a score such as one receives after a test is marked and rank participating teams by score, the section 5.3 in discussion examines how tweaks to this feedback mechanism were designed to potentially improve training outcomes.

2.2 Continuing Professional Development

Continuing professional development (CPD) is an important responsibility of people working within a professional career. Engineers Australia (2017) specifically defines CPD as "anything that helps you expand your knowledge, maintain up-to-date technical skills and progress your engineering career". Formal and obvious sources of CPD include conferences, seminars, training courses, presentation and site visits, less formal avenues may be industry committee or volunteer work, presenting papers and writing articles.

One of the aims of this paper is to compare the performance of differing forms of CPD, limited mainly to seminar or training course type content. Specifically excluded from this analysis will be conference seminars and papers, these focus instead on the presentation and dissemination of the results of new research, this paper looks at techniques to improve education on already well understood topics.

3. TRAINING EVENT CASE STUDIES

This section consists of case studies of common event formats used for training or CPD activity, along with a discussion of benefits and disadvantages from the viewpoint of this papers Authors who have attended or presented during these events.

3.1 Formal training course without practical work

This format is when a training course only discusses the theory of its main topic, perhaps with the presenter using relevant equipment or case studies to illustrate points, but not with any hands-on time for attendees. The case study discussed here will be a training course in NDT equipment presented by a manufacture on a yearly basis. The key distinction here and in the next example is that a training course is typically more thorough than a presentation, with attendees expected to be competent to complete the task the training course discusses at completion.

This multi day training course presents information on the use of non-destructive evaluation techniques for Structures, Foundations/Pavements and Geophysics over three days. This course offers 7.5 hours of CPD time a day, evenly split between theoretical topics and simple demonstrations of testing techniques. The topics for each day were presented in blocks of a few linked techniques, followed by a short demonstration and then a break. This course did not provide any assessment procedure.

One of the benefits of this approach are the volume of information that can be covered, a topic can be introduced with a presentation covering its purpose and theory quickly, then an attendee can review the content later at their own leisure to reinforce their learning. Adding a quick demo also allows for basic operation insight, such as correct set up of equipment, but not interpretation of collected data or troubleshooting.

The key disadvantage is that an intensive training course requires a motivated audience, people who choose to engage fully with the topics, work hard with their revision and practice afterwards will gain the full benefit, but without any built in component of hands on activity can miss important information. For this reason, another disadvantage with this form of CPD is that if it is desirable to confirm competence it is critical to undertake an assessment process such as an end of course quiz or project, otherwise there is no way for the instructor to see that attendees can perform the tasks taught.

3.2 Formal training course with practical work

A formal training event can be combined with practical experience in the area taught, for examples a course in visual inspection may include inspection work of samples with known defects, or a coating application course may involve preparing a surface and applying the materials in question. This case study examines a half day training course on Non-Destructive Testing (NDT) tools for Corrosion Inspection.

The 3.2 case study training course was intended to place the NDT techniques discussed within context. Presentations cover first theory and use of individual techniques, along with showing them as part of a larger investigation. During a second half of the event after presentations are complete attendees collect data using the taught NDT techniques and provided these results to assessors along with answers to written questions.

An advantage of this method is that with enough instructors each attendee can be assessed on their real-world performance. While the case study provided assessment questions for a large course number, a smaller course may choose to assess only at their hands-on work and interpretation of the data.

Disadvantages of this approach during short, seminar style events such as this case study 3.2 are that unless the course is very long it can be difficult to cover a physical task in all detail, instead only presenting a rough outline. If attendees do not then go on to review the full theory and practice of a physical operation from their course notes and other training materials they may work only from the hands on, missing important steps or nuance of a technique or task.

3.3 Technical Overview Presentation

Technical overview presentations explain the theory of subject, they may use small scale case studies to illustrate details of the theory, but do not present a holistic view of a project. The case study here is a single presentation on non-destructive tests at the ACA event "Corrosion in the Water & Wastewater" in Victoria.

Presentation 3.2 presents a number of evaluation techniques suited to issues of corrosion of steel in concrete, or the degradation of the concrete material itself. Each topic consists of an overview of the technique in around 1-2 slides, 2-3 slides of sample data and a conclusion on its suitability to water and wastewater applications and possible uses. For example, for the task of determining the remaining thickness of a cement lining in a steel pipe using Ground Penetrating Radar (GPR) a very quick overview of GPR was presented, enough to understand it is a visual technique that responds to metals best, sample data of its use for un-reinforced concrete over steel plates, and then a quick discussion of setup for accurate measurements.

An advantage of this method is the ability to cover topics quickly, by presenting either a task or technique in its most basic terms this form of CPD quickly increases general knowledge of attendees, especially where multiple presentations are offered by different perspectives in the industry.

As with most presentation type event it is difficult or impossible to assess how much learning is retained be attendees. It also requires them to conduct their own additional research if they wish to utilise topics introduced as typically slides or course notes are not available.

3.4 Case Study Presentation

A case study presentation is where a complete task or projected is presented as an example. The case study for this form is a presentation on projects completed during concrete remediation work at a past YCG event which discusses a number of remedial projects, repairing structures exposed to a marine environment.

Each topic was presented as a story, describing the repair work to be completed, showing progress images and discussing interesting technical notes that occurred during the project. As the audience of the YCG event was people new or looking for work within the Corrosion Industry no theory was presented during this event.

Advantages of a case study presentation is that a natural, narrative structure can easily maintain attention of attendees and the insight from real life projects is often invaluable to people who are not exposed to the reality of site work, or provide examples of issues they may face in future projects. Case studies are also one the of the quickest form of knowledge sharing between people doing the same kind of work, or between disciplines.

The limits of the case study approach are that there is no formal training component and assessment of learning is difficult.

4. EVENT DETAILS

The event "NSW YCG: In the Shoes of a Corrosion Engineering Team - Specifying the Inspection and Repair of Concrete Structures" (YCG-Event) to be presented later in 2017 is planned to place attendees into teams responsible for selecting inspection methods for a concrete structure after an initial visual inspection, then based upon the results of this inspection completing a specification for remediation of the structure.

The following section will offer an overview of the planned event, with examples of printed documentation, presentation content and other tools used to create the event. Its structure will be explained but the intended outcomes are presented during section 5. Discussion.

4.1 Target Audience

The proposed event is for a Young Corrosion Group audience within the Australasian Corrosion Association. The Young Corrosion Group is a sub group within the ACA for new members of the corrosion industry and those under 35 years of age. Further to this in NSW the YCG events are often advertised and supported by local universities, so the topics of many of the events must assume limited or nil prior knowledge for students who attend the event.

Previous events presented for this audience include a Case Study Presentation focused on a job profile and example projects by people who already work in the industry in various roles such as site technicians, project managers and corrosion consultant and a Technical Overview Presentation on cathodic Protection.

4.2 Event Schedule

The proposed schedule for the event is shown in the table below;

Activity	Description
Welcome and distribution of Visual Inspection Report & Reference Document	Attendees are greeted and organised into teams pairing experienced professional with new members or students. Each team receives a one page visual inspection document for each element under consideration Every team member receives a reference document
Presentation on Inspection Techniques	Technical Overview Presentation on Inspection Techniques for corrosion in concrete
Selection of Inspection Options	Teams complete and submit an Inspection Options Form
Receive Inspection Reports and Networking Break	Teams receive their Inspection Report documents and have a short break to consider the results
Presentation on Remediation Techniques	Technical Overview of Remediation Techniques for corrosion in Concrete
Selection of Remediation Options	Teams complete and submit a Remediation Option Form
Presentation of Results and Team Scores	Teams are given their scores; a few examples are given of correct and incorrect choices by teams and 2 samples of a "Perfect" game are presented

4.3 Presentations

Presentation by two speakers will be offered to attendees. One will be Technical Overview presentation on inspection / testing methods relevant to corrosion in concrete, the second on remediation methods relevant to corrosion in concrete.

A short list of topics to cover during the testing method presentation would be; Compressive Strength [Core Sample Tests], Compressive Strength [Schmidt hammer], Breakout for visual inspection and measurement, Drummy Survey, sonic testing such as Ultrasonic Pulse Velocity, Impact Echo or Ultrasonic Pulse Echo, Carbonation Depth measurement using phenolphthalein, Chloride profiles by lab analysis, Sulphate Content by lab analysis, Half Cell Potential mapping, Resistivity, Corrosion Rate, Cover Meter Survey, Ground penetrating radar, Xray and Thermal Camera. This list was mainly selected from NDT methods discussed in the Australian standards for cathodic protection and the new section of the Australian Bridge Code AS 5100.8, along with other tests commonly seen in inspection reports. This is a large scope but is intended to highlight the cross over between similar techniques and that both simple techniques and more advanced techniques have a place during inspection.

A short list of topics to cover during the remediation method presentation are: Patch repair of damaged areas, Impressed Current Cathodic Protection [CP], Sacrificial/Galvanic Anode CP, Hybrid CP, Sacrificial anodes in patch repair, Corrosion inhibitors and Surface coatings. These methods were selected to be those commonly used to prevent the start of corrosion or rebar in a concrete structure using a barrier method or those that will slow or halt its progress once corrosion has begun. It specifically does not include any strengthening techniques or go into detail on the type of patch repairs required as a means of reducing the scope to corrosion in concrete. It also does not make any technical judgment between different techniques, for the training event each technique is judged only on potential and estimated costs, any cathodic protection technique is considered adequate to halt corrosion, any barrier method to stop chloride or carbonation ingress etc.. This is to simplify scoring and to remain agnostic to technology.

4.4 Game Documents

4.4.1 Reference Document

As attendees will be viewing two presentations with many different testing and remediation techniques they will be given a summary document repeating information from the presentations and adding additional details where useful. This document is their main reference on remediation methods as they will not have any recommendations for remediation on either their visual inspection or testing reports.

4.4.2 Visual Inspection Report

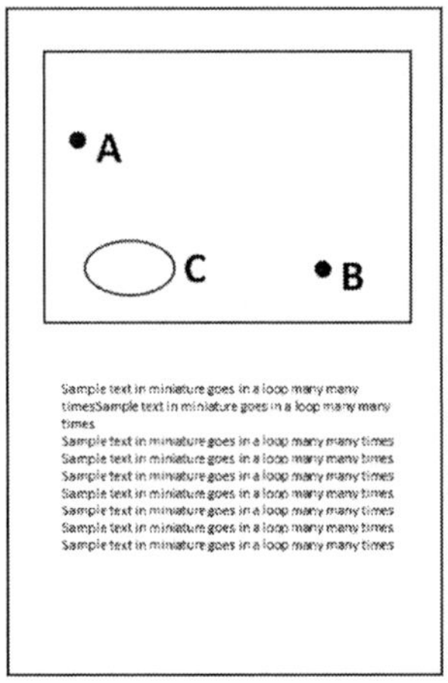

Figure 1 – Visual Inspection of Structure

Before the first presentation attendees are issued a set of Visual Inspection reports for the structures they will be working with. This inspection report provides a description of the structure, its environment and a diagram of its geometry. Further to this marked on the diagram are points of interest representing typical areas of the element, along with defects.

During the latter sections of the game attendees will use these areas of interest to select the scope of point investigations such a compression testing samples or chloride profiles. The structure diagram will be reused to present all mapping results such as Half Cell Potential Surveys in context.

4.4.3 Inspection Option Form

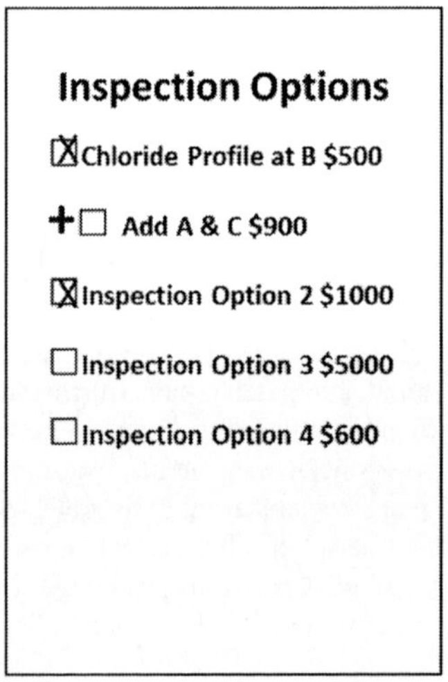

Figure 2 – Sample Inspection Options Form

The inspection options form is the first decision point for attendees, presented after the first presentation on inspection methods it allows them to choose between all the options discussed in the presentation. For "point" tests which are normally completed only in selected areas from a sample they will have the choice of the previously defined points of interest to sample, mapping tests will simply be a yes or no choice.

4.4.4 Inspection Reports

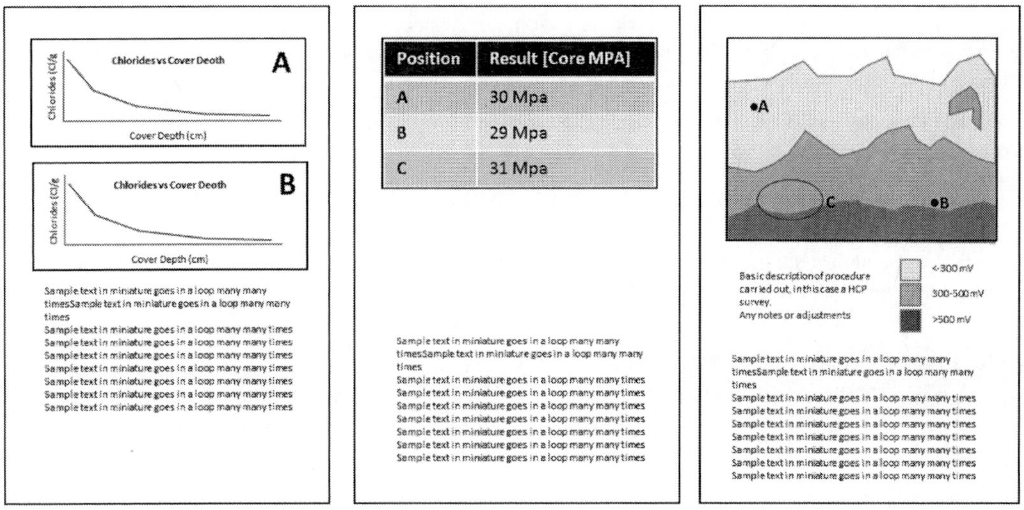

Figure 3 – Inspection report samples

Inspection reports are pre-prepared for all tests methods, again as a single page document to allow quick review and interpretation. Where results are presented for points of interest they will receive results for all points they chose to have tested, along with a description of data interpretation taken from the earlier presentation.

4.4.5 Remediation Option form

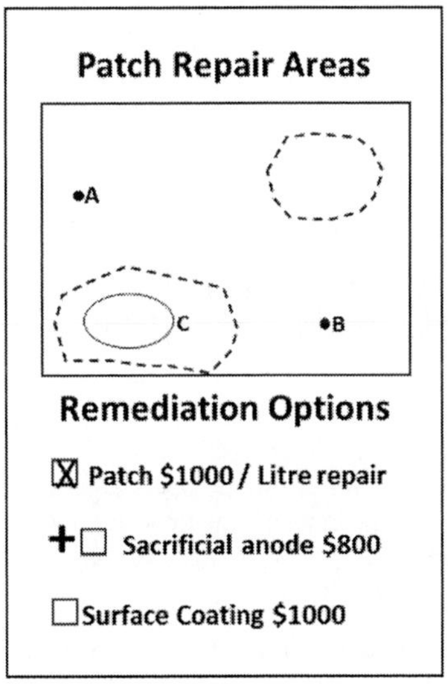

Figure 4 – Selection of remediation options and areas for patch repair

The remediation Option form will present attendees with two tasks, first they must define regions for concrete repair, using a method based upon the Australian Concrete Repair Association's [ACRA] "ACRA's Standard Method of Measurement of Concrete Repairs".

Secondly they will have the opportunity to specific as additional measures all remediation methods discussed during the presentation.

4.5 Scoring

Arbitrary scoring methods are not desirable during Gamification, to enhance the roleplaying aspect of working as a corrosion engineer attendees will instead be charged a monetary value for each inspection technique selected and for the completion of their desired remediation works.

This cost, plus a defined penalty value for additional life cycle costs will form a "Golf" score of total lifecycle cost, where the lower the total value for inspection, remediation, and lifecycle cost the better.

5. DISCUSSION

This discussion section will demonstrate the thought process behind choices made in YCG-Event's design and some comparison between other event and the final intent of the vents design.

5.1 Training with Demo Equipment vs. Gamification

The most similar event to YCG-Event was the Formal Training Course with Practical Work. They share many of the same challenges and benefits, both introduce a direct engagement with the topic of the presentations and both introduce an element of assessment.

The practical component of the formal training combines the theory of operation and operation techniques with hands on work, allowing an attendee to see the nuance of using real equipment to complete a task. YCG-Event's use of interpretation of data shifts the focus away from specific equipment and site techniques, to a higher-level interpretation of data and the choice of techniques and remediation after the task is completed.

Both approaches encourage the attendee to better engage with the topic of the presentations, YCG-Event with the specification and data interpretation stage, a training event with practical work with the remediation or site investigations. There is no reason elements of the interpretation and specification game could not be incorporated in future formal training sessions, perhaps as a component of a more formal assessment of learning where a course is intended not as an introduction but as a training to complete a task. Combined with hand on experience with equipment this should further improve retention and the ability to complete the task for which the training course teaches.

5.2 Starting at inspection rather than remediation

The game structure used could be applied equally to an event which talks only about remediation, attendees being provided all preliminary test reports.

If this game structure was used for an event that only talks about inspection, it is no longer suited, the scoring methods used require a life cycle cost to be estimated, which is not possible without including remediation choices.

The decision was made to include both topics during an entry level training event as to avoid overspecialisation and to ensure that inspection techniques are presented as part of the overall process of remediation, this is a similar approach to recent standards that present inspection as well as remediation as part of their scope.

5.3 Use of lifecycle cost rather than a more arbitrary point system

Scheider (2015) describes feedback loops which effect user behaviour, placing their interaction with a points system in the second part of the loop during which the behaviours that are "scored" are linked back to the intended behaviour the training event is hoping to reinforce and encourage the user's actions to follow this path. The choice to use a point system based around inspection, repair and lifecycle cost was intended to encourage certain behaviour during the event.

During the selection of inspection methods the following key behaviours were desirable; Use of well understood destructive and NDT inspection methods, the selection of relevant methods, and a lack of duplication between similar methods.

During the selection of remediation methods the key behaviours were; Use of suitable remediation methods for the structures location and expected design life, lack of duplication of between similar methods and application of remediation to all required areas.

For example, a cover meter survey [Medium cost] will determine cover depths to reinforcing, a vital point of information to compare to chloride profile with depth and carbonation depth. If the cover depth is shallower than either of these the structure is likely to already be experiencing corrosion even if damage is not evident from a visual inspection. An alternative method, Ground Penetrating Radar [Medium-High Cost] also provides cover depth information, along with a more detailed review of the internal geometry of the concrete, but it is not as precise as a cover meter survey.

In this example if a user chooses to do a cover meter survey they have picked arguably the most suitable method, it is more precise and of slight lower cost than a GPR survey. If they choose a GPR survey they pay a little more, reducing their "golf" score of Lifecyle cost slightly, but will have the information to select a suitable remediation, avoiding a very high "emergency remediation" cost penalty to their lifecycle result for missing areas during the remediation stage. Selecting both is still cheaper than the emergency remediation loading, but is a much bigger addition to their lifecycle costs, a simple penalty for duplication of effort.

By couching the scores in monetary terms, and using this to define their level of success attendees are incentivised to balance their selections between suitability and cost, rather than only on performance. This better captures the estimation and quote or tender request procedures seen during real project, and the very real effect cost has on decision making.

5.4 Refinements

An important part of preparing a Request for Tender document is to provide potential firms with enough information for useful, comparable estimates. ACRA's Standard Method of Measurement of Concrete Repairs document is a means of outlining the expected scope of repairs at the time of FRT, rather than relying upon each tendering company to prepare their own scope. Using this standard definition, it becomes easier to compare tenders received as there is less risk of over or under estimating of scope and the ability to look at a cost per volume of repair in additional to overall tender cost.

It was decided to ask attendees to draw up a volume of repairs on their final remediation form to highlight the importance of defining such details in a tender, and to highlight another professional organisation with close links to the corrosion industry.

Two "Perfect" games are presented during the wrap up rather than one to highlight that there are typically several options in a complex project and good outcomes can be achieved using different approaches, it was also important in continuing to remain agnostic between different inspection or remediation techniques that are similar in result on intent.

5.5 Expected Benefits of an event with Gamification

The purpose of applying gamification techniques to an entry level training event is both to improve its appeal to the proposed audience and to improve retention and present teachable moments for attendees that go beyond the theory of the topic to practical decisions made in the work place.

The event promotion does not market it as a game, but an interactive training event. It describes the information attendees will receive and the tasks asked of them. This avoids any stigma to the use of "game" techniques and allows a potential attendee to see the topics discussed as if it were a normal event, but with something a little different.

The main reason for the event structure is to reinforce in memory the topics discussed during the technical presentations. By presenting a topic, asking an attendee to use this knowledge, then reinforcing its correct use at the end there are now three opportunities for an attendee to remember a piece of information vs one for an event which only contains technical presentations. As a bonus, the fact that a choice can be wrong, but not simply be marked incorrect on an assessment means that feedback comes in practical terms as a teachable moment, rather than just what the correct answer is without context. The quality of this feedback will be dependent upon the explanations given by the presenters during the wrap up.

Finally, as previously discussed the scoring structure adds an overlay of real decision making, attendees now must think commercially as well as technically. For student attendees, this is an opportunity to better understand decisions made in business vs research work and for junior professionals or technical staff to gain context on why choices are made by project managers or senior engineers.

6. REFERENCES

Deterding, S. (2011). Gamification: Toward a Definition. In CHI 2011 (pp. 12–15).

Cite as Nicholson, S. (2015). A RECIPE for Meaningful Gamification. To be published in Wood, L & Reiners, T., eds. Gamification in Education and Business, New York: Springer. Available online at http://scottnicholson.com/pubs/recipepreprint.pdf

Scheider, S., Kiefer, P., Weiser, P., & Raubal, M. (2015). Score design for meaningful gamification. In CHI 2015 (pp. 1–4).

Engineers Australia. (n.d.). Continuing Professional Development (CPD). Retrieved May 8, 2016, from https://www.engineersaustralia.org.au/Training-And-Development/Continuing-Professional-Development

7. AUTHOR DETAILS

Experienced with Non-destructive testing equipment and instrumentation, W Ward is the NSW manager at Papworths Construction Testing Equipment and is also responsible for training and sales support. He has a Bachelor of Engineering in Mechanical/Mechatronic engineering.

B McGuiness is an experienced engineer specialising in the estimating, project management, design and installation of electrochemical corrosion protection systems of concrete and steel structures, condition assessments of concrete structures and ICCP system monitoring.

Throughout his career, he has worked on numerous large scale structural remediation projects both nationally and internationally including Hong Kong, New Zealand, India, China and Papua New Guinea.

R Barnes has been working with non-destructive evaluation and testing equipment for more than 10 years as both consultant and the managing director of Papworths Construction Testing Equipment.

D Veloso has been working with Marine and Civil Maintenance as a Project Engineer since the completion of his Material Science and Engineering degree at the University of New South Wales.

During this time, he has worked as a project planner, project auditor and assisted with estimating and tendering.

WHAT DOES "CP" MEAN NOW FOR STEEL IN CONCRETE?

W Green[1], J Katen[1] and B Dockrill[2]
Vinsi Partners[1], Sydney, Australia, Vinsi Partners[2], Newcastle, Australia

SUMMARY: "Hybrid corrosion protection (HCP)" is a recently coined description for an electrochemical treatment method for steel reinforced concrete structures. For galvanic anode systems descriptions like "galvanic protection" are being used to describe performance. Those using these descriptions imply cathodic protection (CP) is being afforded, so what does CP mean now for steel in concrete? CP has always meant cathodic protection and discussion provided in the paper shows that "corrosion protection" or "galvanic protection" does not mean CP. A CP system, for steel in concrete, works by converting the reinforcing steel (or prestressing steel) surface into a cathode, hence the name, and moving the anodic reaction onto an alternative material. Cathodic protection is provided to steel in concrete when the acceptance criteria of AS 2832.5 or ISO 12696 are met. These acceptance criteria are proven over many decades and have repeatedly demonstrated reliability. It is also commented that the informative (non-normative) notes to the acceptance criteria of the ISO 12696 concrete CP standard are not an additional performance or protection criterion, but rather a reference to a possible assessment of corrosion risk (rate) where galvanic anode systems do not meet acceptance/protection criteria.

Keywords: Reinforced concrete, Cathodic protection, Criteria, Hybrid treatment, Galvanic anodes.

1. INTRODUCTION

Cathodic protection (CP) of reinforced concrete can be achieved by impressed current and galvanic means. CP has been applied to reinforced concrete structures and buried prestressed concrete pipelines since the 1950s (Unz, 1955; 1956). Richard Stratfull first applied impressed current CP to atmospherically exposed steel in concrete in a successful trial in 1959 (Stratfull, 1959). Stratfull continued to investigate and experiment with impressed current CP, and in 1973 undertook the first full scale installation on a concrete bridge deck in California (Stratfull, 1974). Direct experience with impressed current CP of concrete structures in Australia dates back to the 1980s (Grapiglia & Green, 1995). In the UK the use of impressed current CP on large scale infrastructure projects commenced in 1989 principally with the implementation of contracts on the Midland Links Motorway Viaducts in the West Midlands, with the first major project commissioned in 1991 (Preston et al, 2017). Around the same time large scale concrete repair and refurbishment projects were being undertaken in the Middle East, for example at Mina Zayed in the United Arab Emirates described by Wyatt el al (1991). These projects also included the use of ICCP to prevent future deterioration of the reinforced concrete.

CP itself of course dates back to 1824 when Humphry Davy postulated that it would be possible to prevent corrosion of copper sheathing on ships by connecting it to zinc, tin or iron (Davy, 1824). Davy's work defined the two ways in which we apply CP to this very day, by using galvanic anodes or where current is forced (impressed) from an anode through the electrolyte onto a structure (cathode) using a DC power supply (Ackland, 2005). The anode is consumed (including impressed current anodes, albeit very slowly) while the cathode is protected if there is sufficient current to provide the requisite polarisation (Ackland, 2005). A CP system, for steel in concrete, works by converting the reinforcing steel (or prestressing steel) surface into a cathode, hence the name, and moving the anodic reaction onto an alternative material (Grapiglia & Green, 1995). Cathodes don't corrode and reduction reactions only occur on their surfaces. Anodes corrode, oxidation reactions occur at the metal surface and metal loss results.

CP has for nearly 200 years therefore meant cathodic protection. Recently CP is being termed "corrosion protection" for steel in concrete (Christodoulou et al, 2016; Glass et al, 2016). It is considered necessary therefore to explore the literature where CP has been termed "corrosion protection" and thereby examine whether cathodic protection has or has not been provided to reinforced concrete (RC) in such instances. Discussion of the criteria for CP of concrete is also considered relevant prior to such an examination and some other theoretical aspects including corrosion rate estimation of cathodically polarised steel also need to be explored.

Galvanic anode systems for concrete, the monitoring thereof, corrosion risk assessment associated with and CP assessment are also considered to warrant discussion under the circumstances of what does CP now mean for steel in concrete.

Application of cathodic current to steel in concrete is of course beneficial, unless it is excessive thereby leading to the risk of hydrogen embrittlement of prestressing steel or impacting of the bond for conventional reinforcement. However, the extent of cathodic current application needs to be sufficient to effect CP or else merely a reduction in corrosion rate (risk) is achieved. For a RC structure in a marine environment that is genuinely suffering from chloride induced pitting, as a structure owner, one would want to be confident that "a reduction in corrosion rate (risk)", provision of "corrosion protection" or provision of "galvanic protection" is adequate. As a structure owner it may be more appropriate to have certainty and confidence that corrosion is being stopped by CP (to long-accepted protection criteria) so as to sustain (and extend) service life (cost-effectively).

2. CATHODIC PROTECTION CRITERIA

2.1 Background

A vital part of the design of a system of CP for reinforced concrete is a method of determining whether the scheme is in fact applying an appropriate level of protection and that the whole of the reinforcing steel is being converted to a cathode and thereby that corrosion is stopped. A number of criteria have therefore been developed that enable a corrosion engineer to decide whether the structure is being protected. The criteria to be discussed include:

a) The potential criterion.
b) The 300 mV shift criterion.
c) The 100 mV potential decay.

The criteria proposed in the "concrete CP standard", AS 2832.5 (2008), as well as international standards for "concrete CP", will conclude this section.

2.2 The Potential Criterion

This criterion has long been used for the protection of coated steel structures which are buried underground. For steel exposed to soil conditions it was found many years ago that polarisation to a potential more negative than -0.85 V copper/saturated copper sulphate (CSE) renders the metal immune from corrosion and this has long been used as a CP criterion in the pipeline industry (Cherry & Green, 2017).

In an attempt to examine the protection of reinforcing bar in an environment similar to that which it would experience when immersed in concrete, Hausmann (1969) examined the corrosion of steel bars exposed to a saturated calcium hydroxide environment containing a range of chloride ion concentrations. His results are shown in Figure 1. He found that a potential more negative than -710 mV to CSE was sufficient to prevent corrosion of the bars. Hausmann (1969) also concluded that "corrosion of steel can be arrested in chloride contaminated concrete if sufficient current is applied to shift the polarisation potential to a minimum value of -710 mV to a CSE electrode". Cherry & Green (2017) propose that under these circumstances it is probable that the cathodic current would be concentrated on those regions of the steel surface where the passive film had broken down and would generate a sufficient concentration of hydroxyl ions to cause the passive film to be reformed.

Hausmann (1969) further suggested that it would be dangerous to adopt the -0.85 V CSE criterion used for buried pipelines since the over-protection so obtained would lead to the evolution of hydrogen and the disruption of the bond between the concrete and the steel. He was also concerned that with the high strength steel used for prestressed structures hydrogen evolution could lead to embrittlement.

Vrable (1976) examined the electrochemical behaviour of rebar steel in high pH solutions and found that hydrogen evolution does not occur at potentials more positive than about -1170 mV (CSE) and that a potential, measured at the surface of the metal to be protected of -770 mV CSE was sufficient to prevent corrosion in all circumstances.

Over-protection is however more serious in the case of prestressed structures reinforced by high tensile steel. The susceptibility of steel to hydrogen embrittlement increases as the yield strength of the steel increases. In general high tensile steel with a strength greater than 1000 MPa is considered susceptible to hydrogen embrittlement in severe environments (Cherry & Green, 2017). For such structures, AS 2832.5 (2008) for concrete cathodic protection, limits the potential of prestressing steel to no more negative than –900 mV (vs Ag/AgCl/0.5M KCl) so as to avoid hydrogen embrittlement.

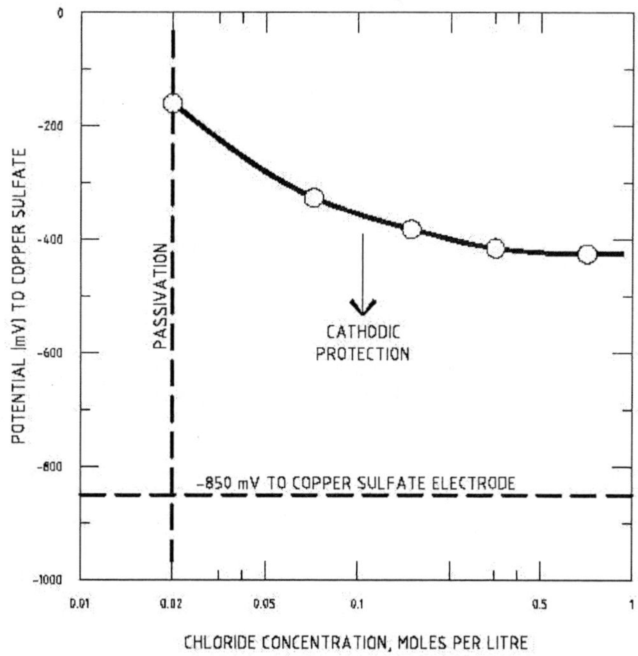

Figure 1 Corrosion and its relation to potential in saturated $Ca(OH)_2$ solution (Hausmann, 1969)

2.3 The 300 mV Shift Criterion

This criterion was also adopted from the pipeline industry. It suggests that a shift in the potential of the reinforcing bar of at least 300 mV (in the negative direction) from that which it demonstrated immediately prior to the application of the cathodic protection current indicates that the current density that has been applied is adequate to prevent further corrosion. This criterion takes no account of the IR drop in the concrete (or resistance potential where such is defined in AS 2832.5 (2008) as that part of a measured electrode potential attributable to the passage of current through the resistance between the cathode surface and a reference electrode) (Cherry & Green, 2017).

During free corrosion, current is flowing off the anodic areas of the steel surface and so the actual interface potential will be negative to that measured at the concrete surface, whereas when cathodic protection is applied the current will be flowing on to the metal surface and so the interface potential may be considerably positive to that measured at the concrete surface. The actual swing at the interface may therefore be less than that measured at the surface by an unknown amount (Cherry & Green, 2017).

The negative 300 mV shift criterion of the surface potential is consistent with the results of Hausmann (1969), where polarisation to -700 mV CSE (the actual swing at the metal surface would as described above be less) could place the metal in the range described by Hausmann (1969) as adequate to stop the corrosion.

2.4 The 100 mV Decay Criterion

A major shortcoming of the 300mV shift criterion is that it fails to recognise that the potential measured at the concrete surface is a function of the relative areas of cathodic and anodic activity beneath the reference electrode. As cathodic protection proceeds, the areas of anodic activity decrease so that the rest potential of the structure becomes more positive (less negative) and the application of 300 mV to the originally more negative rest potential may represent gross overprotection (Cherry & Green, 2017).

The 100mV decay criterion attempts to overcome the disadvantage of the 300mV criterion by referring the shift always to the instantaneous value of the rest potential. It suggests that a minimum negative polarisation of 100mV is necessary to ensure that the current density applied is adequate to prevent (stop) further corrosion. In this case the polarisation is measured as the immediate rise in potential when the current is switched off. Since "this shift must be unaffected by any IR drops in the concrete due to the cathodic protection current", this implies that it must be measured by an instantaneous off technique which excludes the IR drop in the concrete. This criterion effectively allows for the change in pH which has arisen around the cathode (reinforcement) as a result of the application of the cathodic protection (Cherry & Green, 2017).

The 100 mV decay criterion is consistent with the results of Cherry and Kashmirian (1983) who suggested that a 100 mV shift represents the application of sufficient current density to a reinforcing bar to move the bar into the perfectly passive region of the potential-pH (Pourbaix) diagram. The criterion also allows for an initial high current period whilst the pH in the region of the anodic areas is building up to protective values and then a lower current period when the sole function of the cathodic protection is to maintain the whole of the surface in the perfectly passive range appropriate to the level of chloride which is still present adjacent to the interface.

Glass (1999) indicates that the 100 mV potential decay criterion can be used to monitor development of steel passivity induced by a cathodic current and that the achievement of 100 mV of negative potential shift by the low current densities normally applied when cathodically protecting steel in concrete indicates that, in theory, the steel corrosion rate is negligible.

Broomfield (1992) reports that in a large survey of impressed CP systems in North America, no failures appear to have been uncovered when the 100 mV decay criterion was met.

Bennett and Broomfield (1997) propose that for cathodically protected, atmospherically exposed reinforced concrete, a polarisation decay or development of 150 mV or more in anodic areas is needed for "worst case conditions". However, they note that given that most structures are not worst case, that complete depolarisation is not usually observed, and that depolarisation values usually exceed the minimum required, a polarisation (potential) decay of at least 100 mV is a reasonable, practical objective.

2.5 AS 2832.5 Criteria

Australian Standard AS 2832.5 (2008) "Cathodic protection of metals – Part 5: Steel in concrete structures", Section 2, details criteria for establishing whether a CP system is providing adequate protection to a concrete structure. The standard does not rank the applicable criteria and each is equally relevant.

There is an overriding safety criterion in the standard namely that no 'instantaneous off' steel/concrete potential (measured between 0.1 s and 1.0 s after switching the DC circuit open) shall be more negative than -1100mV for plain reinforcing steel or more negative than -900mV for prestressing steel with respect to Ag/AgCl/0.5M KCl. This is to ensure no compromising of bond in the case of plain reinforcing steel and no hydrogen embrittlement risk in the case of prestressing steel.

The standard then requires that initial and continuous adjustment of the CP system shall then be based on meeting one of the following criteria:

a) **Potential decay criterion**. A potential decay over a maximum of 24 h of at least 100 mV from the instantaneous off potential.
b) **Extended potential decay criterion**. A potential decay over a maximum of 72 h of at least 100 mV from the instantaneous off potential subject to a continuing decay and the use of reference electrodes (not potential decay sensors) for the measurement extended beyond 24 h.
c) **Absolute potential criterion**. An instantaneous off potential (measured between 0.1 s and 1.0 s after switching the DC circuit open) more negative than –720 mV with respect to Ag/AgCl/0.5M KCl.
d) **Absolute passive criterion**. A fully depolarised potential, or a potential which is continuing to depolarise over a maximum of 72 h after the cathodic protection system has been switched off, which is consistently less negative than –150 mV with respect to Ag/AgCl/0.5M KCl.

The standard then also requires compliance with at least one of the above criteria shall be achieved within 6 months or alternatively, within a longer period as agreed with structure's owner.

Furthermore, the standard requires that compliance with at least one of the above criteria is necessary and that the compliance shall be maintained on a continuous basis.

The standard notes that the criteria given in a), b), c) and d) have been found to indicate adequate polarisation which will lead to the maintenance or re-establishment of protective conditions for the steel within the concrete (such a note being normative like the criteria and the above other requirements).

Within part of the informative Appendix A section of AS 2832.5 (2008) is noted the following:

- If environmental conditions, which favour corrosion of reinforcement, are likely to occur during the lifetime of the structure, or during service, cathodic protection is one method of preventing corrosion of steel in concrete.
- Sufficient corrosion protection will be achieved if specific criteria of protection are met at representative points on the structure.
- The criteria of protection in this Standard are based on electrochemical considerations regarding corrosion processes and on practical experience.

2.6 ISO 12696 Criteria

The international standard for CP of steel in concrete is ISO 12696 (BSI, 2016). ISO 12696 states 'Prestressing steel may be sensitive to hydrogen embrittlement and, due to the high tensile loading, failure can be catastrophic. It is essential that caution is exercised in any application of cathodic protection to prestressed elements.' It further states 'No instantaneous off steel/concrete potential more negative than −1100 mV (with respect to Ag/AgCl/0.5M KCl) shall be permitted for plain reinforcing steel or -900 mV (with respect to Ag/AgCl/0.5M KCl) for prestressing steel'.

In terms of criteria of protection, ISO 12696 states for any structure, any representative steel in concrete location shall meet any one of the criteria given in items a) to c) namely:

a) an instantaneous off potential more negative than -720 mV with respect to Ag/AgCl/0.5M KCl; or
b) a potential decay over a maximum of 24 h of at least 100 mV from instantaneous off; or
c) a potential decay over an extended period (typically 24 h or longer) of at least 150 mV from instantaneous off subject to continuing decay and the use of reference electrodes (not potential decay probes) for the measurement extended beyond 24 h.

2.7 Other Standards

Broomfield (2017) indicates that the first major international standard for cathodic protection of steel in concrete was originally published by NACE International in 1990 (NACE, 1990). This was followed by the European CEN standard BS EN 12696 in 2000. He advises that although the NACE standard has been reapproved and republished several times it has not changed significantly. However, the heavily revised and expanded BS EN ISO 12696 (BSI, 2016) now covers the cathodic protection of steel embedded in concrete in atmospherically exposed, buried and submerged service, as well as both galvanic and impressed current cathodic protection. Broomfield (2017) also notes that NACE has published standards for concrete pipes (NACE, 2014a) and buried and submerged RC structures generally (NACE, 2014b). There are also several state-of-the-art reports on evaluating RC cathodic protection systems (NACE, 2016a), CP of prestressing steel in concrete (NACE, 2002) and CP of masonry buildings incorporating structural steel frames (NACE, 2010).

2.8 CP Criteria are Proven

Concrete CP criteria have been in existence for many decades and have been proved time and time again to show that corrosion to steel reinforcement, whether it be conventional reinforcement or prestressing steel, as well as whether the cause of corrosion is chlorides, carbonation or leaching, is basically stopped by cathodic protection.

It is not possible, given the numerous variables within steel in concrete, exposure environments, element type, treatment period, etc that an applied potential or required current could be determined to predict theoretical criteria for CP achievement (Solomon, 2005). Due to this limitation, protection criteria being applied for RC CP systems are largely based on accumulated practical experience and are criteria which have repeatedly demonstrated reliability (Solomon, 2005).

3. "HYBRID CORROSION PROTECTION" SYSTEM

3.1 Background

It is noted in the informative (non-normative) notes to the cathodic protection criteria of ISO 12696 (BSI, 2016) that "if a corrosion risk is identified with a galvanic system" then one of the actions that is possible to take is "applying a temporary electrochemical treatment (using the galvanic anodes, if possible) and a temporary power supply". It is assumed therefore that this action is the "hybrid treatment system".

Dodds et al (2017) describe the "hybrid corrosion protection (HCP)" system as a discrete zinc anode system that is drilled into pre-drilled cavities within reinforced concrete, refer Figure 2. They describe it as combining both an impressed and galvanic system whereby the hybrid anodes are manufactured using 18 mm zinc cylinders, ranging from 42 mm to 220 mm long, with an integrated titanium connector wire. A "coating" is applied to the surface of the zinc anode in order to keep the anode "active" throughout its design life. Initially the system is connected to a temporary, constant 9 V DC power supply, typically for a period of at least 1 week, depending on the type of reinforcement (e.g. mild steel or prestressed), to deliver a charge to the reinforcing steel.

Christodoulou and Kilgour, (2013) describe an initial impressed current phase of "typically 8 to 12 weeks" rather than 1 week. Christodoulou et al (2016) advise of "an initial temporary energisation phase" of 7 to 28 days. Thompson et al (2012) advise between 7 and 14 days from an external power supply of between 6 and 12 V DC.

Figure 2 Installation of the hybrid anode system to a typical reinforced concrete pier (Dodds et al, 2017)

A minimum of 50 kC/m^2 of steel surface area is passed during the initial impressed current phase according to Christodoulou & Kilgour (2013) and Christodoulou et al (2016). The hybrid anode is then disconnected from the temporary DC power supply and connected directly to the steel reinforcement. This latter treatment phase continues for the remainder of the anodes' working life and provides a "relatively low current to the steel reinforcement" (Dodds et al, 2017).

Dodds et al (2017) advise that the hybrid anode system is installed at a spacing dependent on steel density (per m^2 of concrete), and the individual anodes are connected in series with "insulated" titanium wire. Different sizes of anode are available and are named based on the expected charge output during the anode's life (e.g. D350 and D550 anodes relate to an output of 350 and 500 kC, respectively). The size of the anode is chosen based on the design life required, average current density and steel density within the concrete structure (Dodds et al, 2017).

According to Christodoulou et al (2016), hybrid anodes are usually designed to provide a residual service life of 30-50 years "based on an expected current demand from the surrounding reinforcing steel in a structure".

In terms of terminology for this treatment method, Holloway et al (2012) indicate that the the UK patent holders for the technology published a paper in 2008 describing the system and its principles for providing protection (Glass et al, 2008). In the Glass et al (2008) paper the system is referred to as "hybrid electrochemical treatment", "hybrid treatment" or "hybrid corrosion treatment". More recently the terminology is "hybrid corrosion protection (HCP)" (Christodoulou et al, 2016; Glass et al, 2016).

On the other hand, Christodoulou and Kilgour (2013) call it "hybrid cathodic protection" and do so even though there is no data and assessment in the paper where the "hybrid cathodic protection" system to prestressed concrete substructure elements of two bridges complies with cathodic protection (CP) criteria and CP has been achieved, refer section 3.3 below for further discussion.

To add further confusion, Thompson et al (2012) term it "hybrid electrochemical protection (HEP)".

3.2 Corrosion Risk Assessment

Preston et al (2017) point out that in the informative (non-normative) notes to the criteria of ISO 12696 (BSI, 2016) an assessment of corrosion risk to reinforcement is permitted when, specifically, with galvanic anodes, cathodic protection performance criteria are not met. Preston et al (2017) note that it is clear within the Standard that this assessment of corrosion risk is not an additional performance criterion, it is a reference to a possible additional assessment of corrosion rate where cathodic protection (CP) systems do not meet the protection/performance criteria.

Now it is not an error above by saying galvanic anodes when this section of the paper is discussing hybrid "corrosion protection" because, another of the informative (non-normative) notes to the criteria of ISO 12696 (BSI, 2016) provides actions to take if a corrosion risk is identified with a galvanic anode system for concrete, namely:

i. turning the galvanic system into an impressed current system (if this is possible);

ii. installing more galvanic anodes; or,
iii. applying a temporary electrochemical treatment (using the galvanic anodes, if possible) and a temporary power supply.

As mentioned previously, item iii. above, presumably describes the "hybrid corrosion protection (HCP)" system.

Preston et al (2017) advise that in ISO 12696 (BSI, 2016) that the recommended method of determining corrosion risk presented in the Standard where a galvanic anode system is used is corrosion rate assessment using the Butler Volmer (BV) equation:

$$i_{appl} = i_{corr}(\exp\left(\frac{2.3\Delta E}{\beta_c}\right) - \exp\left(-\frac{2.3\Delta E}{\beta_a}\right)) \qquad (1)$$

Where:

i_{appl} is the applied current density
[Note: Preston et al (2017) advise it is necessary to measure current and calculate the current density using the steel surface area under consideration, which in practice requires a considerable assumption];
i_{corr} is the corrosion rate;
ΔE is the electrode potential shift; and
β_a and β_c are constants (anodic and cathodic Tafel slopes respectively).

Use of the BV equation seems to also be the only method of performance evaluation of a "HCP" system (e.g. Thompson et al (2012); Christodoulou & Kilgour, 2013; Christodoulou et al, 2016; Glass et al, 2016; Dodds et al, 2017). Assessment to cathodic protection (CP) criteria, as per section 2 above, has not been undertaken.

The UK patent holders (Glass et al, 2007) in one of their first published papers (Glass et al, 2008) about the system also describe performance assessment to the BV equation rather than the accepted and proven cathodic protection (CP) criteria of section 2 above.

Thompson et al (2012) provide some details on how the corrosion rate of a "hybrid electrochemical protection (HEP)" system is determined namely:

- "The test first measures the current applied to a certain steel area, the user must calculate the steel area it is applied to."
- "The instant off potential must then be recorded (reference potential measure between 0.1 and 0.5 s of the system being switched off)."
- "The difference between the instant off potential and the depolarised potential (between 4 and 24 hrs) is then used in the (Butler Volmer) equation as ΔE."

It is beyond the scope of this paper to explore the various concerns that apply to the corrosion risk (rate) assessment method of the hybrid treatment method, the use of the BV equation, etc suffice to say, for example, that Angst & Büchler (2015) describe various uncertainties in the results of corrosion rate measurements in galvanic corrosion cells, in relation to concrete, including errors within presently used standard formulae. Flitt et al (2015) comment that the Wagner and Traud equation has been erroneously called the BV equation and that the use of the BV equation to estimate reinforcement corrosion rate as a means of corrosion risk assessment in the case of cathodic protection systems that don't meet protection criteria, is not correct as it eliminates important electrokinetic and corrosion kinetic information. Preston et al (2017) comment that practical field experience has shown that using the BV approach is not straightforward, refer section 4.3 below. Elsener (2002) also, for example, advises that an error to a factor of 10 can arise for corrosion rate estimations of chloride induced (pitting).

3.3 Cathodic Protection Assessment

Glass et al (2008) in one of the first published papers about the hybrid treatment system describe laboratory trials applied to chloride contaminated reinforced concrete blocks and the achievement of in excess of 100 mV potential decay. However, for a hybrid treated bridge in the same paper, emphasis is placed on corrosion rate estimation using the BV equation and no data is provided showing whether cathodic protection (CP) was achieved to accepted criteria.

Thompson et al (2012) advise that "the monitoring of a HEP system is similar to an ICCP system in that the current is recorded and a depolarisation test is conducted but the evaluation of the results is carried out in a different manner. ICCP systems are evaluated using the 100mV depolarisation criteria set out in the AS 2832.5. (Other criteria are used also). HEP however uses a much smaller current in the galvanic phase and therefore produces a much smaller potential shift and can't be evaluated in the same way."

Christodolou and Kilgour (2013) describe "hybrid cathodic protection" treatment of the prestressed concrete substructure elements of the Kyle of Tongue Bridge in Sutherland, Northern Scotland and the Tiwai Point Bridge in Ivercargill, Southland, New Zealand. For Kyle of Tongue, the authors state that "hybrid cathodic protection was used to arrest existing corrosion activity to the prestressed concrete beams of the superstructure and extend their service life for a 30 year period". For the Tiwai Point Bridge, the authors state that "a trial hybrid cathodic protection system was developed to provide corrosion prevention to the prestressed concrete piles within the tidal zone with a targeted service life of 50 years". Performance assessment by Christodoulou and Kilgour (2013) of both systems has been made solely on the basis of corrosion rate estimation and utilisation of the BV equation. No data is provided to enable performance assessment to cathodic protection (CP) criteria, even for the impressed current phase (which was of 8 to 12 weeks duration).

Christodoulou et al (2016) now describe the hybrid treatment system as "hybrid corrosion protection (HCP)" for two existing reinforced concrete bridges as part of the State Highway 16 Motorway in Auckland. Systems performance assessment is again on the basis of corrosion current density calculations based on applied currents and applied potential shifts and insertion into the BV equation. In this paper some potential depolarisation (decay) results at a limited number of locations for one time during the impressed current phase only are provided. Some depolarisations (decays) are more than the CP criteria requirement of 100 mV but the data is limited in quantum and location, the time during which the data was taken during the impressed current stage is not provided and no data is provided for the galvanic phase of the treatment.

Glass et al (2016) also describe the hybrid treatment process as "HCP" and discuss "long term" performance (i.e. dating back to 2006) on three reinforced concrete bridges. Emphasis is again placed on steel corrosion rate data, presumably by the same methods as previous, i.e. on the basis of corrosion current density calculations based on applied currents and applied potential shifts and insertion into the BV equation. There is yet again no data to enable systems assessment to CP criteria. However in this paper, there is also some steel potential data (presumably instant off potentials) that show a trend to less negative (more positive) potential with time of up to 9-10 years but it is not known from the paper where the reference electrodes that are providing such potential data are positioned, i.e. at corroding locations, in repair areas, etc. Glass et al (2008) have proposed previously that such a trend is indicative of steel passivity.

More recently Dodds et al (2017) discuss the performance of the "hybrid corrosion protection" system installed to reinforced concrete substructure elements of six bridges, 1-8 years after treatment. Performance assessment in all instances is again by corrosion rate estimation and there is no data provided to enable assessment to CP criteria. Some steel on-potential data with time up to 8 years is provided and shows a trend towards less negative (more positive) values but again it is not known from the paper where the reference electrodes that are providing such potential data are positioned, i.e. at corroding locations, in repair areas, etc.

4. GALVANIC ANODE SYSTEMS

4.1 Background

There are various galvanic anode types for the provision of cathodic current to reinforcing steel (or prestressing steel) in concrete, refer Table 1 (Green, 2014).

Table 1. Concrete galvanic anodes by type (Green, 2014)

Surface Applied	Encapsulated	Immersed/Buried
• Arc sprayed zinc • Arc sprayed Al-Zn • Arc sprayed Al-Zn-In • Adhesive zinc-sheet	• Zinc mesh within jackets • Zinc mesh in cementitious overlay • Discrete zinc anode in patch repairs • Discrete zinc cylinders • Discrete rolled zinc sheet • Discrete zinc strips	• Cast zinc alloys • Cast aluminium alloys • Cast magnesium alloys

4.2 Monitoring of Galvanic Anode Systems

As mentioned previously in section 3.2, Preston et al (2017) point out that it is recognised that reinforced concrete cathodic protection systems utilising galvanic anodes embedded within concrete frequently do not have the required ability to collect the data necessary to assess if the system meets the performance criteria in ISO 12696 (BSI, 2016). They then advise that as such these systems cannot be considered compliant with ISO 12696; although some manufacturers of these systems they say claim otherwise.

Preston et al (2017) go on further to advise that, as above, in the informative (non-normative) notes to the criteria of ISO 12696 (BSI, 2016) an assessment of corrosion risk to reinforcement is permitted when, specifically, with galvanic anodes, CP performance/protection criteria are not met. They note that, as above, it is clear within the Standard that this assessment of corrosion risk is not an additional performance/protection criterion, it is a reference to a possible additional assessment of corrosion rate where CP systems do not meet the performance/protection criteria. Then, if the CP system does not meet the criteria and the structure is found to be at risk of corrosion there is then a requirement to supplement the galvanic anode system.

Broomfield (2017) advises that if a galvanic system is designed and installed to the requirements of BS EN ISO 12696 then monitoring is a requirement. He states this means that a monitoring system should be built-in at least providing the ability to monitor current and steel potentials at selected locations. Broomfield (2017) then further comments that it is easier to ignore this requirement on a galvanic system then on an ICCP system but it is important to ensure that structures are being protected and that anodes and connections are still functioning as designed.

4.3 Corrosion Risk Assessment

As mentioned previously at section 3.2 above, Preston et al (2017) advise that in ISO 12696 (BSI, 2016) that the recommended method of determining corrosion risk presented in the Standard where a galvanic anode system is used is corrosion rate assessment using the BV equation. As discussed at section 3.2, another of the informative (non-normative) notes to the criteria of ISO 12696 (BSI, 2016) provides actions to take if a corrosion risk is identified with a galvanic system.

Preston et al (2017) further comment that practical field experience has shown that using the BV approach is not straightforward. Many galvanic anode systems are not designed with any monitoring provisions and there is no ability to isolate the anodes from the reinforcement, no method to measure the anode currents or no reference electrodes embedded in order to measure potential shift.

Furthermore they advise that when systems have been designed with the necessary 'anode segments' to permit such measurements the sources of error in the measurement and analysis are many. When using hand-held instrumentation to measure currents from galvanic anodes in concrete there can be considerable variances in the measurements obtained, in particular as current can rapidly vary when cables are disconnected and a meter then connected in series. The selected meter range itself can also give rise to a significant error as the shunt resistance increases with decreasing range.

Aside from the practical difficulty in taking such measurements, Preston et al (2017) go on further to say that a particular source of error is then how the steel area under test is determined when the distribution of polarisation from the anodes may vary. Records of the steel area associated with each monitoring segment are also required to be documented and stated in the project operation and maintenance manual, something that in practice unfortunately does not happen they say. They comment further that even if records are available though, this would also be an approximation only.

4.4 Cathodic Protection Assessment

Preston et al (2017) comment that the lack of performance data for galvanic anode systems is now something that needs to be considered by all parties, clients, designers, contractors and manufacturers. Clarity as to the claims of whether such systems can deliver 'full' cathodic protection to the criteria in ISO 12696 (BSI, 2016) and the claims that manufacturers make to possible customers in the concrete repair contracting industry or the facility owners, who are often not CP experts, also need to be clarified. Clear and updated guidance is required as to what can, or cannot be achieved using galvanic anodes embedded in concrete.

4.5 Galvanic Anode Materials

Broomfield (2017) advises that there are no test methods for galvanic anodes which means that designers must rely on their own experience, the reported experience of others, or carry out trials before applying novel anodes or applying them in novel situations. He states that BS EN ISO 12696 (2016) gives guidance on applying galvanic anodes to create a full distributed and monitored system and that there is also a NACE Standard (2016b) and a NACE state-of-the-art report (NACE, 2005).

Preston et al (2017) advise that there is no standard or test method that defines what constitutes a galvanic anode for use in concrete. They consider that current best practice for the specification of products is presented in the model specification in UK Technical Report 73 (Concrete Society, 2011). This requires an anode to meet several key performance characteristics:-

i) It shall have proven technical merit;
ii) The galvanic anode system, including the galvanic metal element, the activating agent and the backfill where needed, shall not present a corrosion risk to the steel at the end of the service life of the system; and
iii) The consumption of a galvanic anode in an anode system embedded in cavities in the concrete shall not give rise to expansive forces that disrupt the concrete cover.

Preston et al (2017) go on to say that it is necessary for a client or specifier to determine if a product meets these characteristics, for which there are no defined criteria or test methods, along with a need for them to assess if the products can deliver what is needed by the structure. If it is to be a fully effective CP system, meeting the protection criteria in ISO 1296 (BSI, 2016), then this is unlikely to be fulfilled. If it is a system able to provide some lesser degree of corrosion protection for more than a decade or two this is also unlikely to be fulfilled.

Furthermore, Preston et al (2017) comment that increasingly, new products will be brought to market and prior to use it will be necessary to assess their merit. They quite rightly advise that all zinc anodes in concrete require some form of activation agent to ensure that they do not passivate, which will prevent their operation. They also note that it is also necessary to ensure anodes do not passivate in storage or prior to installation and propose that present procedures do not specify the importance of ensuring that this does not occur.

5. CONCLUSIONS

Cathodic current application to steel in concrete is beneficial, unless it is excessive thereby leading to the risk of hydrogen embrittlement of prestressing steel or impacting of the bond for conventional reinforcement. However, the extent of cathodic current application needs to be sufficient to effect cathodic protection (CP) or else merely a reduction in corrosion rate (risk) is achieved.

CP has always and will always mean cathodic protection and making all of the reinforcement a cathode, hence the name.

It is proposed that an electrochemical treatment for reinforced concrete providing "corrosion protection", "hybrid corrosion protection (HCP)" or "galvanic protection" is an indicator that cathodic protection (CP) is not being provided by such treatments, consequently corrosion of steel reinforcement is not being stopped, merely a reduction in corrosion rate (risk) is being effected.

For a reinforced concrete structure in a marine environment that is genuinely suffering from chloride induced pitting, as a structure owner, one would want to be confident that "a reduction in corrosion rate (risk)", provision of "corrosion protection" or provision of "galvanic protection" is adequate. It may be more appropriate to have certainty and confidence that corrosion is being stopped by CP (to long-accepted protection criteria) so as to sustain (and extend) structure service life (cost effectively).

6. ACKNOWLEDGMENTS

The authors would like to thank colleagues for their assistance with the paper.

7. REFERENCES

Ackland, B (2005), Cathodic Protection – Black Box Technology?, Proc. Corrosion & Prevention 2005 Conf., Australasian Corrosion Association Inc., Gold Coast, November, Paper 004.

Angst, U and Buechler, M (2015), On the Applicability of the Stern–Geary Relationship to Determine Instantaneous Corrosion Rates in Macro-cell Corrosion, Materials and Corrosion, 66, 10, 1017-1028.

Bennett, J and Broomfield, J P (1997), Analysis of Studies on Cathodic Protection Criteria for Steel in Concrete, Materials Performance, 36, 12, 16-21.

Broomfield, J P (1992), Field Survey of Cathodic Protection on North American Bridges, Materials Performance, 31, 9, 28-33.

Broomfield, J P (2017), Up-to-Date Overview of Repair & Protection Aspects for Reinforced Concrete, In Reinforced Concrete Corrosion, Protection, Repair and Durability, Eds W K Green, F G Collins and M A Forsyth, Australasian Corrosion Association Inc, Melbourne.

BSI (2016), BS EN ISO 12696, Cathodic Protection of Steel in Concrete, British Standards Institute, London.

Cherry, B and Green, W (2017), The Corrosion and Protection of Concrete Structures and Buildings, Australasian Corrosion Association and Australasian Concrete Repair Association, Training Course Notes, Version 1.0, Melbourne.

Cherry, B W and Kashmirian, A S (1983), Cathodic Protection of Steel Embedded in Porous Concrete, British Corrosion Journal, 18, p 194.

Christodoulou, C and Kilgour, R (2013), The World's First Hybrid Corrosion Protection Systems for Prestressed Concrete Bridges, Proc. Corrosion & Prevention 2013 Conf., Australasian Corrosion Association Inc., Brisbane, November, Paper 076.

Christodoulou, C, Corbett, P and Coxhill, N (2016), Service Life Extension of State Highway Bridges – New Zealand's First Hybrid Corrosion Protection Application, Proc. Corrosion & Prevention 2016 Conf., Australasian Corrosion Association Inc., Auckland, November, Paper 65.

Concrete Society (2011), Technical Report No. 73, Cathodic Protection of Steel in Concrete, Surrey.

Davy, H (1824), On the Corrosion of Copper Sheeting by Sea Water, and on the Methods of Preventing this Effect; and on their Application to Ships of War and Other Ships, Philosophical Transactions, 114, 151-158.

Dodds, W, Christodoulou, C and Goodier, C (2017), Hybrid Corrosion Protection – An Independent Appraisal, Proc. of the Institution of Civil Engineers-Construction Materials, Forthcoming.

Elsener, B (2002), Macrocell Corrosion of Steel in Concrete – Implications for Corrosion Monitoring, Cement & Concrete Composites, 24, 65-72.

Flitt, H, Will, G and Green, W (2015), Electrokinetic Review and Appraisal of Steel Reinforcement Corrosion in Concrete, Proc. Corrosion & Prevention 2015 Conf., Australasian Corrosion Association Inc., Adelaide, November, Paper 16.

Glass, G K (1999), Technical Note: The 100-mV Potential Decay Cathodic Protection Criterion, Corrosion, March, 286-290.

Glass, G K, Roberts, A C and Davison, N (2007), Treatment Process for Concrete, UK Patent GB2426008B, November.

Glass, G K. Robert A C and Davidson N (2008), Hybrid Corrosion Protection of Chloride-Contaminated Concrete, Proc. of the Institute of Civil Engineers, Construction Materials, 161 (CM4), pp 163-172.

Glass, G K, Davidson, N, Roberts, A C (2016), Hybrid Corrosion Protection of Reinforced Concrete Structures, Proc. Corrosion & Prevention 2016 Conf., Australasian Corrosion Association Inc., Auckland, November, Paper 164.

Hausmann, D A (1969), Criteria for Cathodic Protection of Steel in Concrete Structures, Materials Protection, October, 23.

Holloway, L, Green, W, Karajayli, P and Birbilis, N (2012), A Review of Sacrificial and Hybrid Cathodic Protection Systems for the Mitigation of Concrete Reinforcement Corrosion, Proc. Corrosion & Prevention 2012 Conf., Australasian Corrosion Association Inc., Melbourne, November, Paper 062.

NACE (1990), Impressed Current Cathodic Protection of Reinforcing Steel in Atmospherically Exposed Concrete Structures, Recommended Practice 0290-2000, NACE International, Houston.

NACE (2002), State-of-the-Art Report: Criteria for Cathodic Protection of Prestressed Concrete Structures, Publication SP01102, NACE International, Houston.

NACE (2005), Sacrificial Cathodic Protection of Reinforced Concrete Elements – A State of the Art Report, Publication 01105, Task Group 047, Houston.

NACE (2010), Cathodic Protection of Masonry Buildings Incorporating Structural Steel Frames, Publication 01210, NACE International, Houston.

NACE (2014a), Cathodic Protection to Control External Corrosion of Concrete Pressure Pipelines and Mortar-Coated Steel Pipelines for Water or Waste Water Service, Standard Practice 0100 (formerly RP0100), NACE International, Houston.

NACE (2014b), Cathodic Protection of Reinforcing Steel in Buried or Submerged Concrete Structures, Standard Practice 0408, NACE International, Houston.

NACE (2016a), State-of-the-Art Report on Evaluating Cathodic Protection Systems on Existing Reinforced Concrete Structures, Publication 01116, NACE International, Houston.

NACE (2016b), Sacrificial Cathodic Protection of Reinforcing Steel in Atmospherically Exposed Concrete Structures, Standard Practice 0216, NACE International, Houston.

Preston, J, Wyatt, B S and Spring, I T (2017), Cathodic Protection of Reinforced Concrete: Is There Anything Still to Learn?, In Reinforced Concrete Corrosion, Protection, Repair and Durability, Eds W K Green, F G Collins and M A Forsyth, Australasian Corrosion Association Inc, Melbourne.

Solomon, I (2005), Corrosion and Electrochemical Protection of Reinforced Concrete Structures, Australasian Corrosion Association, Training Course Notes, Version 1.1, Melbourne.

Stratfull, R F (1959), Progress Report on Inhibiting the Corrosion of Steel in a Reinforced Concrete Bridge, Corrosion, 15, 6, June, 65-68.

Stratfull, R F (1974), Transportation Research Record 500, 1-15, TRB, Washington, DC.

Standards Australia (2008), AS 2832.5-2008 Cathodic Protection of Metals - Part 5: Steel in Concrete Structures, Sydney.

Thompson, L, Godson, I and Heath, J (2012), Introduction to Design Techniques Used for Hybrid Corrosion Protection Systems, Proc. Corrosion & Prevention 2012 Conf., Melbourne, November, Paper 148.

Unz, M (1955), Proposed Methods for Cathodic Protection of Composite Structures, Corrosion, 11, 2, February, 40-43.

Unz, M (1956), Intrinsic Protection of Water Mains, Corrosion, 12, 10, October, 66.

Vrable, J B (1976), Chloride Corrosion of Steel in Concrete, ASTM STP 629, Eds D E Tonini and S W Dean, American Society for Testing and Materials, p124.

Wyatt, B S, Schols, A J H, Leppard, N W and John, D G (1991), Repair and Cathodic Protection of Reinforced Concrete Substructures of Four Wharves at Mina Zayed, Abu Dhabi – Part II – Concrete Repairs and Application of Cathodic Protection System, Proc. UK Corrosion 91 Conf., 22-24 October, Manchester.

8. AUTHOR DETAILS

Warren Green is a Director and Corrosion Engineer at Vinsi Partners, based in Sydney. He is a Fellow and Chartered Professional Engineer (CPEng) with Engineers Australia. Warren has over 30 years of experience in corrosion engineering and materials technology covering marine, infrastructure, industrial, civil and building structures. He is also an Adjunct Associate Professor in the Institute for Frontier Materials at Deakin University in Melbourne, an Industry Lead with the Australian Centre for Infrastructure Durability (ACID) in Melbourne and an Adjunct Teaching Fellow in the School of Chemical and Petroleum Engineering at Curtin University in Perth.

Jack Katen is a Graduate Materials Engineer at Vinsi Partners, based in Sydney. He is developing competences and experience in durability assessment and planning, condition assessment of structures, determination of remedial, protection and maintenance options for structures and cathodic protection design and monitoring.

Brad Dockrill has over 30 years of experience as a professional engineer after gaining a Bachelor of Engineering, Civil Engineering (Hons 1) from the University of Newcastle. He is a Partner and Structural Engineer at Vinsi Partners, based in Newcastle. Brad's experience includes the design of commercial, residential, large industrial and mining structures. He has also had experience with the condition assessment and development of remedial, maintenance and strengthening strategies of these structure types as well as marine and civil infrastructure.

POLARISATION RESISTANCE AND CORROSION OF STEEL IN SOIL MEDIA

James P. Cull
Elorane Geophysical Surveys

SUMMARY: Polarisation resistance data provide critical constraints for estimates of structural integrity based on corrosion rates in soil. However several pitfalls have been identified in the application of standard electrochemical methods. Packing fractions, moisture, and galvanic resistance must be assessed prior to subjective Tafel analysis. Better results can be obtained using a transient galvanic pulse with numerical modelling of the observed response. Multiple determinations for a bentonite synthetic soil indicate drift in Rp values consistent with water content. Weight loss methods confirm the associated corrosion rates assuming general oxidation but a ten-fold increase is required where pitting is significant.

Keywords : steel; LPR, polarization; weight loss; pitting corrosion

1. INTRODUCTION

Long-term corrosion of iron-based structures embedded in soil may involve many different chemical mechanisms depending on the local mineralogy and electrolytic complexity at any specific site. Consequently there are no simple methods for routine estimates of structural integrity and risk analysis which are required to ensure the provision of essential services in the water industry (e.g. [1], [2]). Instead pipeline managers are often forced to rely on statistical modelling based on failure history or extrapolation of isolated inspection data. Better constraints are now required to refine these models and avoid the potential catastrophic failure

The effects of soil corrosion can be determined directly using the linear polarisation (LPR) technique (e.g. [3], [4]). Soil samples can be readily extracted for laboratory tests using electrochemical systems of great precision (e.g. [5], [6]) with negligible impact on operating systems. Consequently previous methods for risk analysis, commonly based on a cumulative score for qualitative environmental factors (e.g. the DIPRA 10 point scale, [7]), can be greatly improved. However strict quality controls are essential to obtain accurate LPR data prior to more detailed statistical analysis. Relevant ASTM standards have been previously framed (ASTM G59) but there have been few examples of their application [8].

This report provides an overview of the LPR method and the accuracy of the data obtained. In particular laboratory methods are verified using both analogue components and synthetic soils to investigate data quality subject to sampling errors, environmental drift, and surface alteration. LPR data obtained using transient pulse techniques are compared with weight-loss observations, surface corrosion and deep pitting over an extended period.

2. LPR METHODS

Electrochemical methods used to obtain LPR data have been reviewed by Mansfeld [5]. Characteristic (Tafel) slopes can be extracted from observations of current-voltage relationships based on minor excursions from the rest potential (E_0); these define quasi-linear segments according to the expression

$$\mathbf{Rp = (\Delta E/\Delta i) = B/i_{corr}} \qquad (1)$$

where Rp is the polarisation resistance (ohm-cm2), i_{corr} is the current density (amp/cm2), and B is a constant derived from the positive (anode) and negative (cathode) Tafel segments according to the expression

$$\mathbf{B = (\beta a^*\beta c) / (2.3^*\{\beta a+\beta c\})} \qquad (2)$$

The determination of ba and bc often involves a subjective element to identify a suitable linear segment and precise measurements are complicated by several factors including internal resistivity (iR drop), passivation, excessive scan rates, and complex surface character. However typical values can be assumed for most applications with B=26 (mV). In this case it is a simple matter to calculate the corrosion current (i_{corr}) following observations of Rp (equation 1).

At this stage corrosion rates are readily calculated from observations of i_{corr} (equ 1) along with charge transfer rates determined according to Faraday's law including a term for the metal density of an active species (e.g. [3], [9]; ASTM G59).
In this case (CR (mm/yr) = K * Icor (EW/ρ) which simplifies to the expression

$$\mathbf{CR\ (mm/yr) = 0.0116 * i_{corr}} \qquad (3)$$

where K is a constant = (0.00327 mm g uA^{-1} cm^{-1} y^{-1}); EW is equivalent weight (dimensionless = 27.92 for Fe), and ρ is density of working electrode (= 7.86 g/cm3).

In order to avoid the previous subjective elements in measurements of Rp, Deo et al [6] have suggested further improvements based on curve fitting for a transient pulse response. These methods have the advantage that individual samples can be tested for both polarisation resistance and galvanic resistance in a single short-term measurement with minimum disturbance to the geochemical equilibrium. Consequently data can be obtained over a prolonged period for investigations of repetition, stability and temporal drift.

Considerable care is required during sample preparation. The working electrode must be fully conditioned by polishing and cleaning to ensure a consistent electrical contact with the soils under test. Suitable polishing grits and cleaning agents are well documented for this purpose. However preparation and stabilisation methods for the soil electrolytes are much more complicated. For example the data in Fig 1 indicate immediate drift in the corrosion potential (E0) following the initial loading of a test cell and the onset of oxidation at the working electrode.

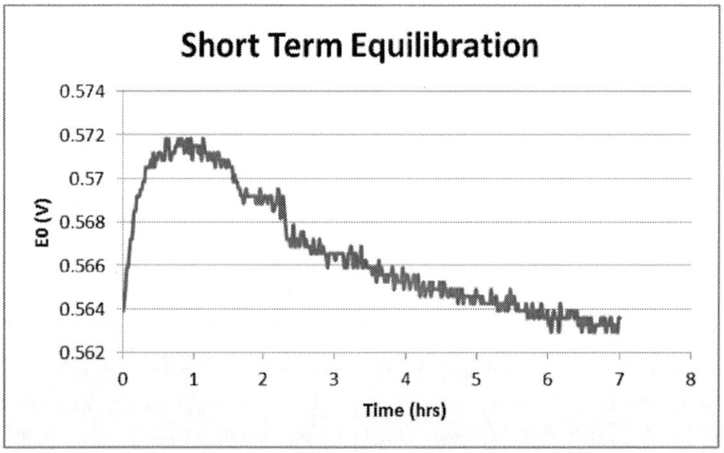

Figure 1. Dipole double-layer stabilisation rates and drift in E_0. Data obtained relative to Ag/AGCl probe (3M).

Moisture levels are a critical factor in corrosion estimates since ion mobility is essential for oxidation. Consequently the nature of the soil electrolyte must be fully recognised. In particular packing fractions and load pressure determine the moisture content at the metal/soil interface. Loose packing results in potential air-gaps and a reduction in the effective area of the working electrode. Similarly excessive packing loads result in unrealistic matric pressures and the expulsion of fluids from the LPR test cell. Consequently a wilt-point protocol is often adopted to ensure a consistent ranking between soil samples. Actual wilt points can be manipulated using dew-point meters but an equivalent 1.5 MPa matric load also ensure uniform grain contacts within each cell (Fig 2).

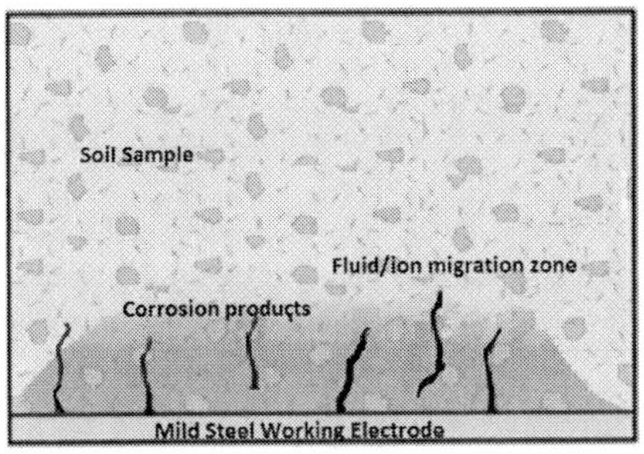

Figure 2. (a) Schematic of packing density and fluid saturation from constant matric pressure applied to soil at working electrode; (b) Residue of test cell showing migration of corrosion products.

3. DATA PRECISION

Simple analogue circuits can be used to provide an indication of error levels in the galvanic pulse method for Rp determinations. Data were obtained for this purpose using a Randle's RC equivalent circuit (Fig 3) connected to an Ecochemie µ autolab; circuit parameters were subsequently determined using NI Labview inversion modules to emulate the full transient response. Values of 376 ohm, 414 µF, and 159.7 ohm were obtained for **Rg, Cp, and Rp** respectively representing errors of **6.3%, 3.7%, and 2.8%** respectively.

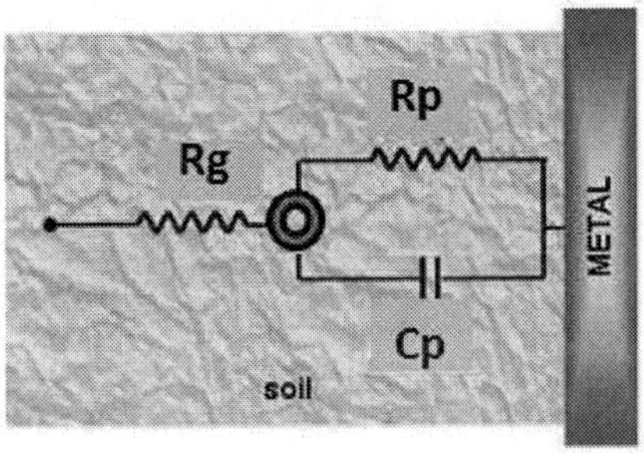

Figure 3. Randle RC equivalent circuit for test of LPR method. Test circuit values adopted for Rg, Cp, Rp are 150.2 ohm, 399 µF, and 387.1 ohm respectively.

Although analogue circuits can be used to provide an indication of system error in determinations of Rp, actual soil samples extracted from the field are subject to more complex conditions. In particular field samples may be poorly sorted with significant variations in mineralogy, salinity and moisture content. In addition the corrosion process results in a continual drift in the physical properties under test. Consequently different results may be obtained for successive readings regardless of the testing regime.

3.1 Synthetic Soils - Bentonite

Synthetic soils provide an opportunity for more realistic appraisals of error and precision. In addition a suitable standard is required to provide for verification and validation of results from different operators and laboratories. Several clays are available for this purpose from commercial outlets but modifications are required to provide more realistic synthetic soils. In particular the raw materials must be diluted using unreactive fractions to ensure more suitable geochemical stability.

Synthetic soils selected for the present study consist of Bentonite/sand mixtures. Suitable reactivity can be obtained using 10% Bentonite by weight (40g) mixed with washed silica sand (360g) as a buffer. Dry fractions of each must be well mixed to ensure a uniform distribution of clay prior to adding water (typically 50 ml for ~11% moisture content). Uniform conditions are ensured by rolling and kneading the final mixtures (however the final moisture content for each sample is determined by desiccation after any LPR measurements).

Bulk synthetic samples are stored for a minimum of 4 days to allow for moisture distribution and equilibration prior to any application. Results for a mixture of this type are indicated in Fig 4. Significant variations are obtained in values of Rp immediately after test cells are loaded; these trends indicate a dynamic response at the soil/steel boundary. A gradual drift can be expected with finite rates for electrochemical reactions and the evolution of a corresponding dipole double-layer during corrosion and ion migration.

4. LABORATORY DATA

After the initial equilibration of the bulk synthetic mixture LPR data were obtained for six separate sub-samples to test for data precision, temporal stability, and accuracy in comparison with weight-loss methods. The results are summarised in Fig 4-6. Only one sample was employed in day 1-3 to reduce the impact of sample disturbance but additional samples were employed on subsequent days. In addition samples were progressively extracted and disassembled for weight-loss observations. Consequently the data density varies over the period of observation.

Consistent trends were obtained in observations of Polarisation Resistance (Rp, Fig4a) and the associated corrosion rate (CR, Fig 4b). Average values of **998 ohm-cm2** were obtained for Rp over the full period with a standard deviation of 130 ohm-cm2. However linear projections to the origin (t0) suggest an Rp value of **927 ohm-cm2**.

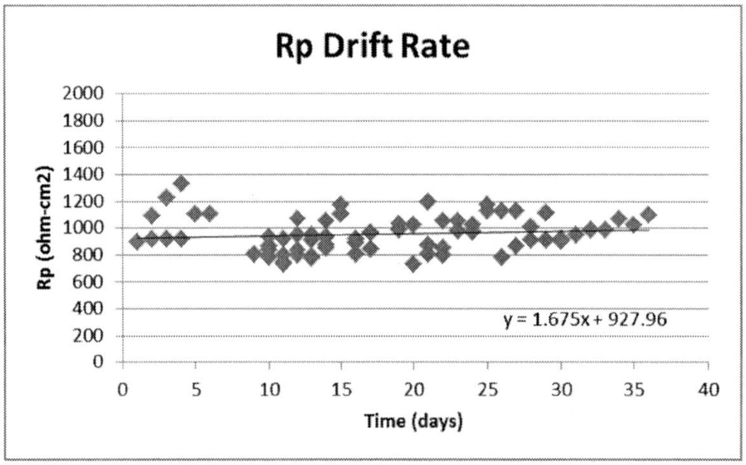

Figure 4a. Drift in values of Rp obtained for Bentonite/Sand synthetic soil mix.

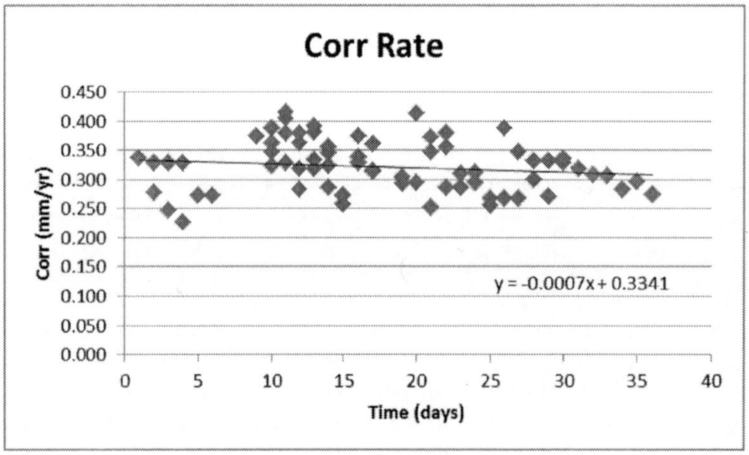

Fig 4b. Calculated corrosion rates (CR) conforming to values of Rp in Fig 4a.

4.1 Weight Loss Data

The current observations provide an opportunity to verify predictions of corrosion rate based on a comparison between Rp values and weight-loss determinations. At the completion of each measurement cycle the individual samples were disassembled and the electrodes were inspected. Typical electrodes are indicated in Fig 5 demonstrating the extent of corrosion and the onset of local pitting. Pit depths were determined by progressive polishing and observations of weight loss to eliminate surface defects. The final data are summarised in Table 1 (a,b) and Fig 6.

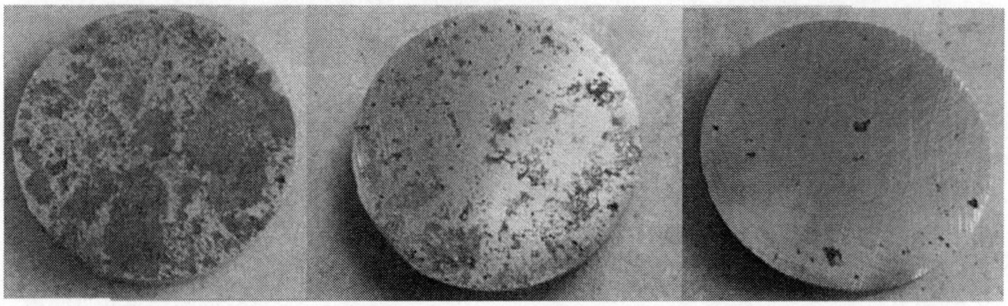

Figure 5. Surface corrosion of assembly (d) after 22 days and progressive removal of pits by polishing.

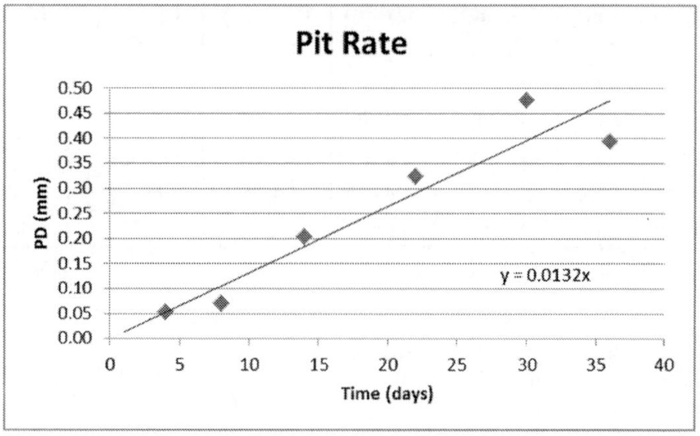

Figure 6. Pit depths measured by weight loss for successive assemblies (Table 1b).

The data in Table 1(a) provide reasonable support for previous estimates of general corrosion rates (surface oxidation) using representative Tafel constants along with observations of Rp according to equation (3). However significant modifications may be required to accommodate the extent of pitting and penetration rates summarised in Table 1(b). Similar results have been obtained previously for coupons buried in natural soils over an extended period [4] [10]. Consequently an additional factor (F>10) may be required to reflect higher penetration rates according to the expression

$$\textbf{PR (mm/yr)= F * 0.0116 * } i_{corr} \qquad (3)$$

To some extent the factor F reflects the pitting density (Fig 5) resulting in an increase in the effective current density compared to uniform surface oxidation. Andrade et al [4] recommend similar values of F (=10) for pitting of re-bar in concrete.

Table 1. (a) comparison of corrosion rates from weight loss and LPR methods.								
Assembly		Start		End		Corrosion	LPR	R1
cell	t (days)	w1(mg)	z1 (mm)	w2(mg)	z2 (mm)	CR (mm/yr)	CR (mm/yr)	Ratio
a	4	36689	4.055	36647	4.050	0.424	0.290	1.46
b	8	40911	4.522	40845	4.514	0.333	0.300	1.11
c	14	41729	4.612	41617	4.600	0.323	0.340	0.95
d	22	41613	4.599	41467	4.583	0.268	0.331	0.81
e	30	44293	4.895	44126	4.877	0.225	0.306	0.73
f	36	45565	5.036	45422	5.020	0.160	0.287	0.56

Table 1. (b) calculation of pitting rates from weight loss methods.								
Assembly		Start		Pit removal		Pitting	Rate	F1
cell	t (days)	w1(mg)	z1 (mm)	w3(mg)	z3 (mm)	PD(mm)	PR(mm/yr)	Multiple
a	4	36689	4.055	36214	4.003	0.052	4.791	11.3
b	8	40911	4.522	40277	4.452	0.070	3.197	9.6
c	14	41729	4.612	39890	4.409	0.203	5.299	16.4
d	22	41613	4.599	38691	4.276	0.323	5.358	20.0
e	30	44293	4.895	39993	4.420	0.475	5.782	25.7
f	36	45565	5.036	42016	4.644	0.392	3.977	24.8

4.2 Water Balance

In chemical terms iron is oxidised according to expressions of the type:

iron + oxygen + water >>> hydrated iron(III) oxide

$$4Fe + 3O_2 + nH_2O \;>>>\; 2Fe_2O_3.nH_2O \quad (4)$$

where n is a variable amount of water reflecting the extent of hydration.

The final corrosion or oxidation products may vary with the concentration of oxygen and water resulting in Fe(ii) or Fe(iii) compounds. These range from FeO, Fe_3O_4, and several Fe-OH species (orange-brown solid hydrated iron(III) oxides; rust). Consequently reductions in water content should be expected as corrosion proceeds in-situ.

In the current study the moisture content for each of the synthetic sand/bentonite samples was determined by desiccation at the completion of each measurement period. A consistent decrease in water content was observed for successive periods consistent with the reactions noted in equation 4. Some additional losses can be attributed to evaporation but all samples were housed in a high humidity enclosure to limit any atmospheric effects. Consequently the consumption of water is expected to reflect the process of ion exchange, oxidation and hydration of iron.

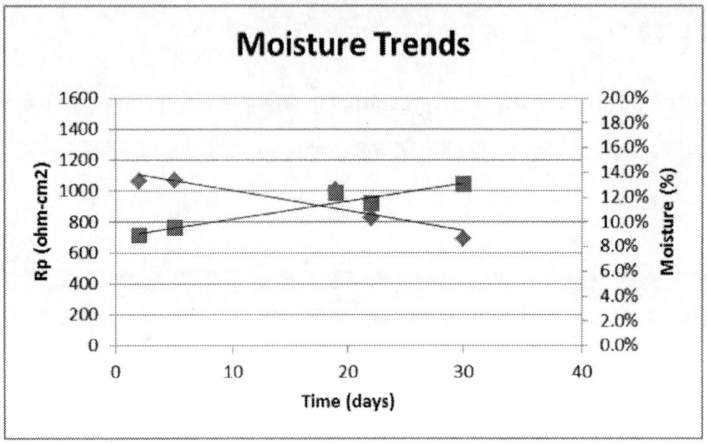

Figure 7. Increase in Rp values reflecting loss of moisture used in corrosion process.

The relationship between water consumption and Fe corrosion can be further demonstrated using molality expressions based on equation 4. Molality, or molal concentration, is a measure of the concentration of a solute in terms of the amount of substance in a specified mass of solvent.

As an example for sample (e) in Table 1(a) it is observed that Fe is oxidised at a rate of 0.005567g/day. Since Fe has an atomic weight of 55.85 g/mol, the corrosion rate can be expressed as (0.005567/55.85)*A = 6.0E+19 atoms/day (where A is Avogadro's constant 6.02E+23 atoms/mol). Similarly the data in Fig 7 indicate water consumption of 0.0660 g/day. Evaporation from a free water surface in the same enclosure was measured at 0.026 g/day indicating a net loss from the sample of 0.040 g/day. However since the atomic weight of water is 18 g/mol, the equivalent consumption is 1.34E+21 molecules/day.

Variations in polarisation resistance as a function of water content were further investigated using a second bulk sample of synthetic bentonite soil (10% wt). After saturation and mixing several sub-samples were extracted and partially dried over a range of periods prior to measurement. Actual moisture contents were determined by desiccation following on from the electrochemical determinations of polarisation resistance. The results are indicated in Fig 8 demonstrating a systematic variation with moisture content consistent with reductions in ion mobility.

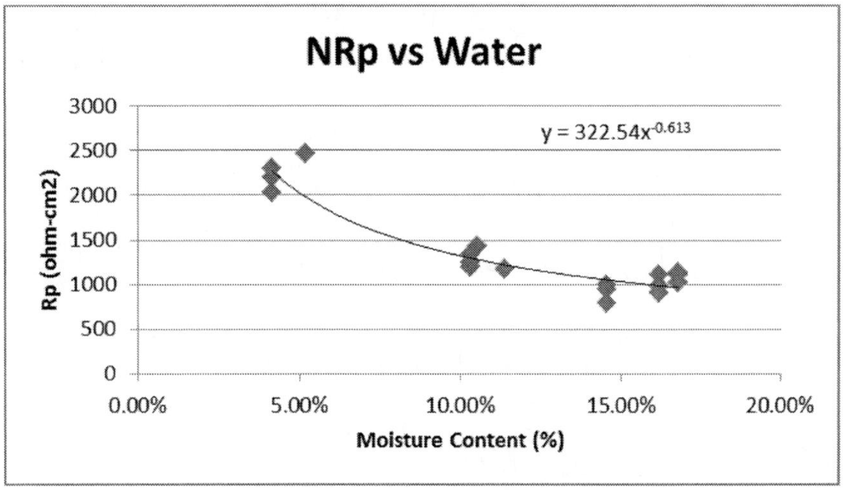

Figure 8. Variations in Rp values as a function of water content for Bentonite (10%) synthetic soils.

5. CONCLUSIONS

LPR methods can provide valuable constraints for numerical methods of risk analysis related to critical water mains and other buried infrastructure. However there are no stationary parameters associated with an active corrosion cell. Significant time is required to establish a dipole double-layer at the interface of a metal/soil boundary. Ion mobility is greatly modified by the nature of the electrolyte and the locale of the corrosion products which can in some cases lead to surface passivation. Consequently a continual drift is observed in corrosion potential (E0) and the associated values of polarisation resistance (Rp).

In spite of the short term drift in LPR parameters several authors have reported reasonable correlations with long-term corrosion rates [10]. Apart from generalised oxidation, pitting rates can be predicted with reasonable confidence from Rp data (Fig 4). More complex bi-modal models can be adopted to accommodate a range of factors which impact on ion mobility and corrosion rates (e.g. [11] [12]). However all associated electrochemical parameters are assessed to some extent in a single LPR response reflecting the structure of a dipole double-layer at the metal/soil interface.

Precise laboratory methods are required in order to obtain accurate LPR data. Apart from adequate preparation of the working electrode considerable care is required to ensure uniform packing fractions in the system test cell. In view of the sensitivity to moisture content all samples must be maintained in a controlled environment during equilibration of the cell and working electrode interface prior to LPR determinations. Assuming adequate precautions are maintained in sample preparation, LPR data can be obtained with a precision better than 10-15%. However any associated estimates of the local corrosion rate may require significant revisions (a 10 fold acceleration) if pitting mechanisms are prevalent.

6. REFERENCES

[1] B. Rajani, Y. Kleiner, Comprehensive review of structural deterioration of water mains: physically based models, Urban Water 3 (2001) 177-190.

[2] R.B. Petersen, R.E. Melchers, R.E., Long term corrosion of cast iron cement lined pipes. Conference, Corrosion and Prevention 2012.

[3] K. Kasahara, F. Kajiyama, F., Determination of underground corrosion rates from polarisation measurements, Corrosion 39 (1983) 475-480.

[4] C. Andrade, C. Alonso, Test methods for on-site corrosion rate measurement of steel reinforcement in concrete by means of the polarisation resistance method, Materials and Structures 37 (2004) 623-643.

[5] F. Mansfeld, Fundamental aspects of the polarization resistance technique - the early days, J. Solid State Electrochem. 13 (2009) 515-520.

[6] R. Deo, N. Birbilis, J.P. Cull, Measurement of corrosion in soil using the galvanostatic pulse technique, Corrosion Science, 80 (2014) 339-349.

[7] P. Davis, I. Allan, S. Burn, R. Graaff, Indentifying trends in cast iron pipe failure with GIS maps of soil environments, PIPES 2003 – Back to Basics: Design and Innovation, 21-23 October 2003.

[8] L. Wong, K. J. King, S. Martin, R.B. Rebak, Methods of Calculation of Resistance to Polarization (Corrosion Rate) Using ASTM G 59, Lawrence Livermore National Laboratory, CA 94550, Report (2004).

[9] R. Baboian, R. (ed), Corrosion tests and standards : Application and Interpretation. 2nd edition. ASTM International, (2005).

[10] E. McCafferty, Introduction to Corrosion Science, ISBN 978-1-4419-0455-3, Springer International 2010.

[11] R.E. Melchers, Modeling and prediction of long-term corrosion of steel in marine environments, Int. J. Offshore and Polar Engineering, 22 (2012) 257-263.

[12] G.S. Frankel, Pitting corrosion of metals – A review of the critical factors, J. Electrochemical Soc. 6 (1998) 2186-2198.

7. AUTHOR DETAILS

Jim Cull is a Director of Elorane Geophysical Surveys (EGV). Previously as Professor of Geophysics he has been responsible for several major research programs in geotechnical, regional and exploration geophysics at Monash University. More recently he has developed a range of NDT and geophysical methods for the water industry including LPR, ultrasonic, GPR, and electromagnetic tools for sub-surface imaging,corrosivity, and evaluations of pipeline integrity.

CONCRETE DURABILITY ASSESSMENT OF A CONCRETE GRAVITY STRUCTURE ON AUSTRALIA'S NORTH WEST SHELF

A Peek[1], L Holloway[2], S Long[1], C O'Brien[3] and R Paull[1]
[1]GHD, Perth, Australia, [2]MEnD Consulting, Perth, Australia, [3]Vermilion Oil & Gas Australia, Perth Australia

SUMMARY: There are three Concrete Gravity Structure (CGS) oil and gas platforms operating in Australian waters. One of these is the Wandoo B facility, owned and operated by Vermilion Oil & Gas Australia. Wandoo B is approaching its original 20 year design operational life since it was installed in March 1997. The platform is located in tropical waters on the North West Shelf oilfield, approximately 75 km northwest of Karratha, Western Australia. As part of the current life extension program, a detailed condition and remaining life assessment of the CGS was commissioned, including extensive NDT, concrete quality assessments and deterioration mechanism modelling. The inspection scope included securing approximately 39 core samples at various locations around the structure in the atmospheric and splash exposure zones. Inspection of the structure below water was completed by remotely operated vehicle (ROV) at areas of possible structural load related concrete cracking, where stresses are concentrated by design actions. The data from these investigations was then used to develop an understanding of the current condition of the reinforced concrete structure, assess the future durability risk, and develop an asset management strategy to support extension of the operating licence for a further 20 years. This paper provides a unique insight into the performance of reinforced concrete in a tropical offshore environment.

Keywords: Reinforced concrete, offshore, condition assessment, remaining life, chlorides, carbonation.

1. INTRODUCTION

Following from the success of concrete offshore platform jackets in the North Sea oil and gas fields, three concrete gravity structures (CGS's) were constructed in Australia during the mid-1990's: Wandoo B, West Tuna and Bream B. West Tuna and Bream B were constructed at Port Kembla, New South Wales, and operate in Bass Strait.

Wandoo B was designed by Arup's for the Wandoo Alliance and constructed at Bunbury, Western Australia in 1996. After launching, it was towed to its current position on the North-West Shelf oilfields, approximately 75 km northwest of the town of Karratha, and ballasted down onto the sea floor. The original operational life of the platform was 20 years from commencement of production in 1997. Vermilion Oil & Gas Australia (VOGA, a subsidiary of the Canadian company Vermilion Energy Inc.) acquired the platform in 2005. It was subsequently identified that an extension of the operating life by a further 20 years was required, and that a condition and remaining life assessment was needed to demonstrate continued safe operation and help support renewal of the operating license by the Australian regulator, NOPSEMA. Figure 1 shows the layout of the field.

All elevation levels in this paper are based on the Platform Datum of 0.00 mPD, which corresponds to the design Lowest Astronomical Tide (LAT) level at the platform location. The Wandoo B CGS comprises:

- A submerged rectangular caisson comprising 60 oil storage cells; segregated into two separate storage compartments and one buffer compartment and 16 open-topped cells containing crushed iron ore ballast. Roof, floor and wall slabs have extensive post-tensioned prestress. The oil storage cells free-flood with seawater during production and loadout operations, maintaining minimal pressure differentials across the walls. The soffit of the base slab rests nominally at seabed level (-53.7 mPD) with a short precast concrete skirt penetrating below the seabed, the caisson is 114.5 m long, 69.3 m wide and 16.77 m high.

- Four circular cross-section shafts extend from the base slab through the top slab to a total height of 68.7 m (top of cap slab at +15.0 mPD). The wall thickness progressively reduces from 575 mm at the caisson top slab to 400 mm at -21.0 mPD, then asymmetrically increases again. Above 0.5 mPD the shafts become ovoid in plan to create a corbel structure that supports the attachments for the deck. The shafts were slip-formed and have vertical post-tensioned prestress.
- Shafts 1 and 3 are permanently flooded to approximately mean sea level (MSL, +2.2 mPD) with limited ventilation of the headspace. Shaft 2 acts as a diesel fuel storage tank for onboard services. The interior of Shaft 4 is cyclically exposed to crude oil and seawater during crude loadout. There is no access available for inspection of the interior surfaces of the caisson cells or Shaft 2, and limited access to Shafts 1, 3 and 4.
- The concrete used in construction of the CGS was S50 grade, the binder being nominally 65% blast furnace slag.
- Various steel caissons and export risers are attached to the CGS, which are provided with a sacrificial anode cathodic protection (CP) system using aluminium alloy anodes. The reinforcement and other steel items embedded in the submerged areas of the CGS are bonded to the CP system, potentially receiving a level of protection.
- A scour protection blanket, consisting of an embankment of rock placed against the caisson wall to a design height of 4 m, is provided around the perimeter of the caisson.

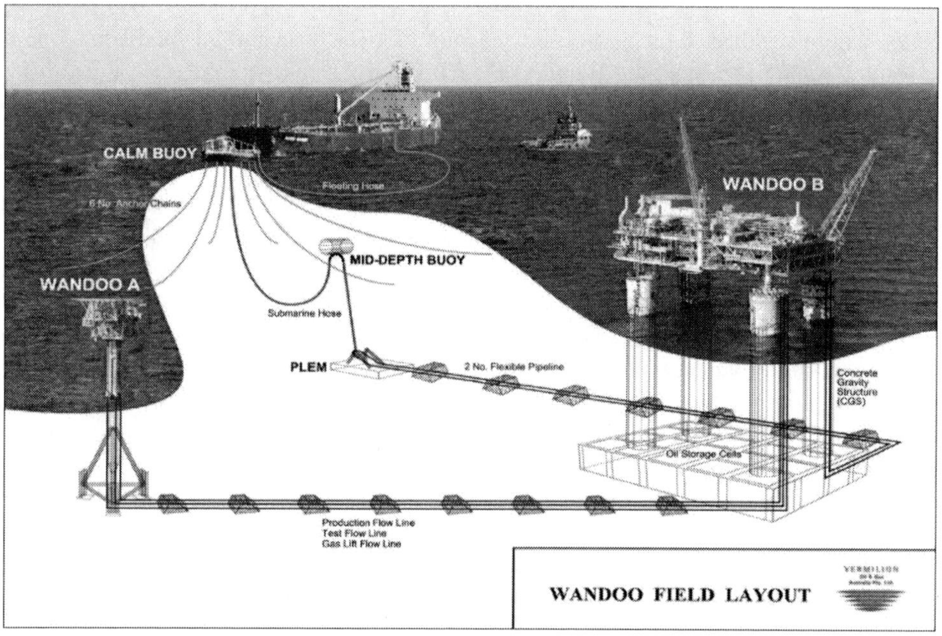

Figure 1: Schematic of the Wandoo Field layout (courtesy VOGA).

2. OVERVIEW OF ASSESSMENT PROGRAMME

The Authors have presented a detailed description of the development and implementation of the assessment programme elsewhere [1], a summary is provided below:

- Stage 1 Review of Standards and Guidelines: Of 23 documents reviewed, published by 11 standards organisations and technical bodies, only three documents specifically written with relevance to inspection of offshore concrete structures were located: ISO 19903 [2], DNV-OS-C502 [3] and NORSOK N-006 [4]. ISO 19902 [5] for offshore steel structures also provided useful information that could be adapted for the assessment of the CGS and development of a management system, and elements were drawn from 21 additional publications to inform the data collection and assessment process.
- Stage 2 Structural Review and Inspection Scope: For the conduct of in-situ sampling, inspection and testing it was expected that operational, weather, time and safe access constraints would affect the amount of information that could be collected. A system for prioritising locations for investigation was therefore necessary. A high-level structural review of criticality and loading was conducted to identify locations for examination, deduce the potential damage mechanisms, and define how evidence of any distress would be manifested. The original construction report, relevant basis of design reports, as-constructed drawings, and photo- and videographic records were recovered from archive. The high-level structural review was performed using relevant design standards, some of which are now obsolete since use in the design, together with finite element structural modelling methods, and a preliminary site inspection made. The basis for inspection priorities at the proposed locations were sample areas for visual inspection for possible structural load related concrete cracking, where stresses are concentrated by design actions. For elements above water, consideration was also given to relative vulnerability to reinforcement corrosion based on height above LAT and

factors such as documented construction issues and previous repairs. Separate detailed Scope of Work (SoW) instruction documents were prepared for components above and below water level.

- Stage 3a Site Inspections, Shafts Above Water: Investigation procedures were standard for in-service assessment of a reinforced concrete structure: Visual, crack width measurement, sounding for delaminations, half-cell (electrode) potential, covermeter, and sampling for carbonation and chloride concentration profile. All work was conducted by roped-access from the cap slab level. The SoW provided a prioritised list of locations as "drops" by shaft number in terms of distance below the cap slab and rotation in plan around the shaft, and detailing testing requirements at each. Some minor modifications were made to locations to address safe access issues. Where core samples were taken, three cores were extracted slightly removed from each other, with one being taken over a reinforcement bar to verify cover and corrosion state.
- Stage 3b Site Inspections, Shafts Below Water and Caisson: Use of a lightweight inspection ROV offered several significant advantages over divers due to the working depth and strength of currents. The SoW provided a prioritised list of locations for inspection, each defined in terms of element, depth and compass orientation. Due to the likely small size of defects, removal of marine growth and coralline deposits using a pressure water jet mounted on the inspection ROV was necessary, the SoW defining the length, width and orientation (vertical or horizontal) of strips to be cleaned at each location. A high quality live video feed during cleaning operations allowed adjustments to the work area according to features of interest, additional high definition (1080P) video scans were recorded after completion of cleaning.
- Stage 4 Remaining Life Assessment: A remaining life assessment was performed using the data collected from site, detailed in Section 3 below, and the processes described in Section 4. The final assessment report was independently reviewed and verified by Lloyd's Register.
- Stage 5 Inspection, Maintenance and Repair Plan (IMRP): ISO 19903 [2] does not provide guidance on maintenance management of offshore concrete structures, beyond recommendations for inspection subjects, nor describe levels of inspection activity in terms of frequency or complexity of investigation. However, ISO 19902 [5] provides a higher level of guidance, the structure of the IMRP was therefore based on the philosophy espoused in ISO 19902 [5] with levels of inspection adapted from the recommendations for offshore steel structures contained in Clause 23.6.2 and Appendix B, the process being designed to be wholly compatible with VOGA's existing risk assessment and asset management procedures and systems as currently used on the platform.

3. SITE INSPECTION RESULTS

3.1 Shafts Above Water Level

3.1.1 Inspection Methodologies

The inspections of the shafts above water level were primarily conducted by rope access. The rope access team included an engineer experienced in concrete inspection and testing. The testing regime included four "drops" at selected locations around each shaft. The final positions of the "drops" were selected to provide a representative coverage of the varying exposure conditions, and capture areas that visibly presented signs of deterioration or subject to potentially more aggressive exposures, such as those near water outfall caissons.

Each drop was completed by a two-person team working in parallel. This allowed a close visual inspection of the concrete surfaces across a window of approximately 3 m wide from the top of the shaft to the top of the black marine growth just above the high-water level. The final height at the bottom of the drop depended on the tide level at the time of the inspection and the sea state.

A detailed inspection methodology was prepared to meet the objectives of the work scope efficiently and without compromising data integrity. This methodology took into consideration the site conditions and difficulties in performing such works from ropes. The scope of work included; a close visual inspection, crack width measurements, delamination testing, localised cover meter testing, half-cell potential mapping, as well as the securing of core samples for chloride content analysis and carbonation depth testing.

To capture the survey data on site accurately, certain information was recorded on the surface of the concrete with permanent markers and then high quality photographs of the survey area were taken. The photographic records were captured systematically to allow easy review and post processing of the data. As the inspection team descended height marks were recorded on the surface of the concrete to provide scale and orientation to the images. The defects such as delaminations were then marked up on the concrete. At the location of covermeter testing these results were also directly recorded on the concrete. An overview photo of the area was then taken. The field of view was aligned to capture a window of approximately 0.5 m x 1 m. The images to the left and right of the drop were recorded, as well as detailed images of any defects, such as cracks or surface etching. This process for recording the key information avoided the need for making detailed sketches and limited the amount of note taking required. Overall this sped up the process and provided a more accurate account of the condition. After the survey, the photos were used to recreate elevation drawings of each shaft showing the location of the drop with marked up defects and feature maps.

The detailed non-destructive testing and core sampling was conducted on selected drops and primarily at two heights below the top of the shaft, between -3 m to -5 m (+10 mPD to +12 mPD) and -7 m to -10 m (+5 mPD to +8 mPD). The final height and location of the testing and sampling was decided on site based on the condition of the concrete and accessibility.

The covermeter surveys were conducted using two different instruments, a Hilti PS 38 and a Proceq Profoscope. At the indicated depths of cover, the PS 38 gave better detection and discrimination at indicated depths exceeding approximately 80 mm and up to approximately 120 mm. However, the wheeled search head of the PS 38 was more difficult to use in scanning areas of rougher surface texture, while the operation of the Profoscope was less affected by surface roughness. Cover depth measurements were calibrated against direct measurements of the cover at the location of breakouts. Covermeter measurements of the reinforcement were typically conducted over areas of approximately 0.5 m x 1 m.

Half-cell (electrode) potential measurements were recorded using a high impedance digital multimeter and proprietary Cu/satd $CuSO_4$ reference electrode. The potential measurements were recorded at 200 mm centres over an area of 1 m x 0.8 m.

Multiple core samples were secured at each detailed inspection point. The core samples were secured by wet diamond core drilling with a 50 mm or 75 mm diameter bit. The depth of the core samples was nominally 100 mm, with one sample at each location taken to expose the reinforcement. The condition and depth of the reinforcement was recorded photographically.

3.1.2 Results

The close visual inspection and delamination survey of the shaft revealed several characteristic defects. The extent and nature of the defects appeared to be dependent on the aspect of the shaft, and the height above the water level.

Figure 2: Example cold joint like defect at the concrete surface, likely related to uneven progress of the slipform used to construct the shaft.

Figure 3: Example of delaminated and spalled skim coat repairs from the time of construction.

At the top of the shaft, around the edge of the cap slab, the concrete was generally in good condition with only some evidence of minor cracking, poor compaction and spalling at the location of embedments. The most prevalent defects on the walls of the shaft were:

- Light to medium scaling of the concrete surface that covered large areas.
- Horizontal cracking and heavy scaling that had a similar appearance to cold joints (Figure 2) at localised positions.
- Past surface repairs that had subsequently degraded (Figure 3).

There were also other minor defects including but not limited to; delaminations, spalls at redundant embedments, and cracking.The defects of most interest were the apparent cold joints with associated cracking and heavy scaling as shown in Figure 2 and the superficial repairs from the time of construction that were showing signs of deterioration (Figure 3).

The apparent cold joint defects were predominately at the level of the shafts just below the bulge; where the shaft geometry bellows out and become ovoid in shape. Close inspection of these defects suggested that they were likely to be related to the slip form construction process and the transition from the cylindrical to ovoid section. It appeared that the plastic concrete had been torn as the forms were moved, thus creating an apparent crack in the concrete and a weaker matrix that was subsequently subject to greater erosion. Several core samples were taken through these typical defects to interrogate the nature of the cracking at depth. Figure 4 shows a core sample from the location of an apparent cold joint. There was evidence of cracking, porosity and poor consolidation within the first 40 mm of the concrete that is parallel to the surface. It is noted that out of 39 samples, this result had the maximum affected depth. Based on the number of cores taken, it was generally observed that these areas showed similar defects in the first 5 to 10 mm of the core.

Figure 4: Example of a core sample taken at a cold joint location showing discontinuity, porosity and evidence of poor consolidation of the concrete.

There was evidence of extensive superficial repairs to the concrete shaft walls from the time of construction. There were various types of skim coats and render-like finishes that had been applied to the concrete. The repair material in Figure 3 was one of the thicker types of finishes that were applied. It appeared that the repairs had been applied to rectify some of the as-constructed defects including the apparent cold joints or provided added protection to the concrete. The original construction report [7] recorded criteria and materials to be used in such repairs, however only the approximate extent of the thinnest polymer modified cementitious coating repair to marginally low cover was documented on drawings. The condition of these repairs varied and there was widespread evidence of cracking, delamination and spalling, suggesting that they offer limited ongoing supplementary cover protection.

Reinforcing steel was intersected at 21 of the core sample locations. There was no visible evidence of corrosion on the exposed surfaces of the rebar. This was typical of all reinforcing bars where the steel surface was exposed in a similar manner

The results from the covermeter testing are summarised in Section 4.1.1. The results were derived from a total of 69 measurements.

Half-cell potential mapping was completed at 8 separate locations. Most of the potential results were more positive than -200 mV relative to a $Cu/CuSO_4$ reference cell, with some localised areas between -200 mV and -350 mV. The main exception to this was for the potential readings recorded at an elevation of approximately -9.5 m on Shaft 1 Drop 4. It is important to note that this test location was below the level of the black marine growth and adjacent to an area where the concrete surface was regularly wet by a water outfall pipe situated adjacent to the drop. Therefore, it was considered that the more negative values were an indication of a lack of oxygen availability as opposed to corrosion activity.

3.2 Shafts and Storage Caisson Below Water level

3.2.1 Inspection Methodologies

Removal of marine growth and coralline material was performed using a water jet mounted on a lightweight inspection ROV, an Ocean Modules Sweden AB model V8 Sii, operating at a nominal nozzle pressure of 27.6 MPa (4,000 psi). The water jet was sufficient to remove all but the most adherent calcareous material without visible damage to the concrete.

The following areas were inspected by general (and close) visual inspection (GVI & CVI):

- External caisson and shaft walls.
- Intersection between shafts and caisson top slab.
- Vertical interfaces between storage cells.
- Exposed steel to concrete surface interfaces (i.e. embedded steel attachments).

Prioritisation of locations for inspection were identified during the high-level structural review conducted in Stage 2 of the work, as described in Section 2 above. Laser-light "dots" of known diameter projected onto the target surface allowed scaling of defect sizes from photographs (refer Figure 5).

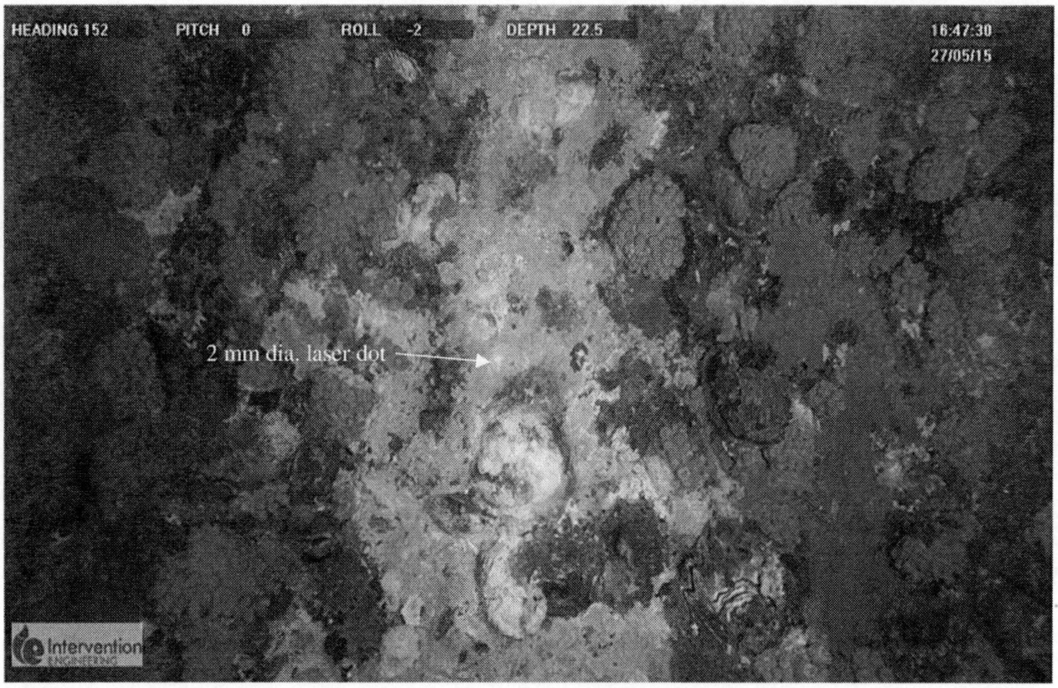

Figure 5: High resolution photograph of typical concrete surface on shaft walls at elevations above 23 m from the caisson top slab with laser light dot for scaling purposes.

In addition to the areas cleaned for close inspection, a general "fly-over" scan for issues such as dropped-object damage and evidence of loss of containment, and a "fly around" scan of the walls, scour protection blanket and adjacent seabed (for indications of scour and subsidence) were also performed as recommended in ISO 19903 [2].

3.2.2 Results

The underwater inspection by ROV identified that the condition of concrete of the caisson and shaft structures below water were generally sound at inspected areas. Typical features identified during the inspection were:

- Horizontal and vertical construction joints of the caisson structure.
- Metal embedment plates on caisson walls, similar to those observed on shaft walls above water.
- Small areas of previous repair undertaken at caisson walls and at shaft to caisson wall interface. Repairs appear sound and intact.

No significant defects were identified, and features such as restrained shrinkage cracks at the caisson top slab edge estimated to be in the order of 0.2 mm wide could be resolved in the high definition video feed (refer Figure 6a and 6b).

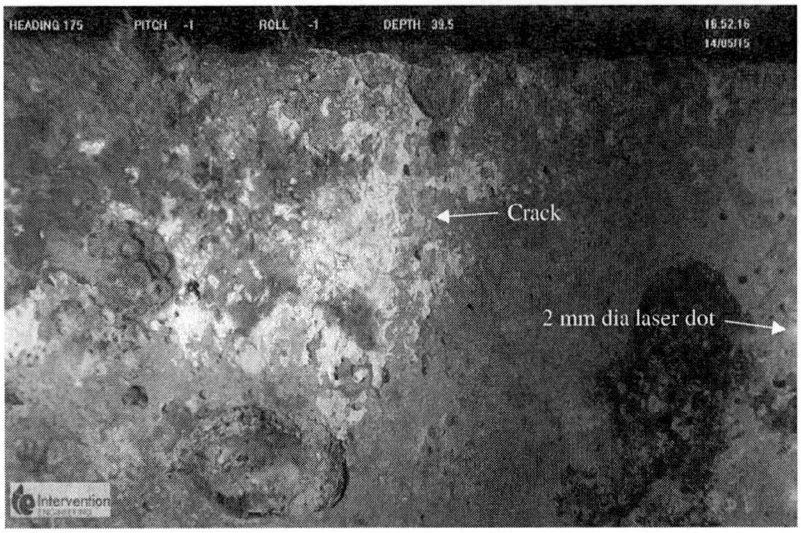

Figure 6a: ROV photo of reinforced concrete caisson top slab edge with a "fine" crack from construction, likely due to retrained shrinkage of the top slab.

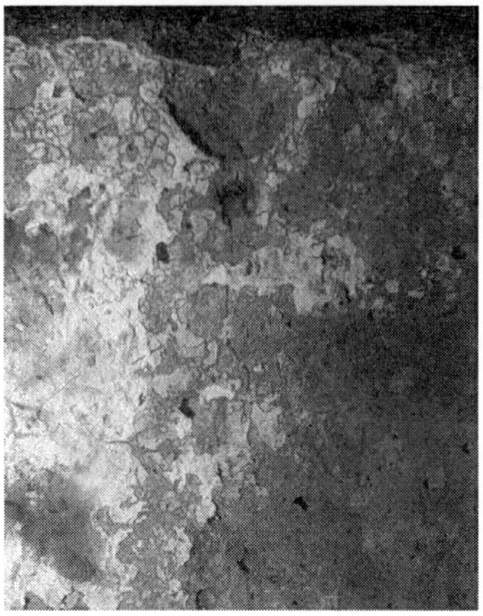

Figure 6b: Close up of crack shown in Figure 6a.

Visual inspection of the scour blanket observed localised areas of lower scour blanket heights, however generally all measured heights were greater than the minimum design scour blanket requirements. Scour blanket heights were consistent with previous survey data, and no evidence of scour was observed.

The seabed surrounding the scour blanket, including at localised areas without scour blanket, appeared even with no evidence of sediment build up, subsidence, or scour. No dropped objects were identified.

4. REMAINING LIFE ASSESSMENT

4.1 Shafts Above Water Level

4.1.1 Cover to Reinforcement

Design cover for the shafts above -3.6 mPD was 80 mm to the outer (horizontal) reinforcement. From 69 covermeter and taped cover values, a mean cover of 90 mm and standard deviation (σ) 16 mm was calculated. The lowest measured cover was 48 mm, and the 95% characteristic (mean – 1.65σ) low cover was 64 mm. Comparison of covermeter readings and taped cover measurements made at core holes showed acceptable correlation, with no difference between values. No systematic variation with height or between shafts was identified.

4.1.2 Concrete Quality

To obtain a semi-quantitative indication of general concrete quality, the excess voidage of the core samples was estimated by comparing the number and size of voids exposed on the surface of the cores with those displayed in BS EN 12504-1:2009 Figure NA.1 (a) to (e) [6].

The concrete core side surface porosity values ranged from 0% to 0.5%, with five cores having some localised positions up to 1.5%. Dispersion of aggregate was uniform, with no evidence of segregation.

4.1.3 Carbonation

Depths of carbonation measured by phenolphthalein tests on 36 core samples varied from 0 mm to 20 mm. No systematic variation with height or between shafts was identified. The mean calculated depth of carbonation was 3.4 mm with a standard deviation of 4.8 mm, the 95% characteristic (mean + 1.65σ) high carbonation depth was 8.7 mm. For the age at sampling of 18 years from 1997, applying a simple square-root time function yielded carbonation coefficients (k values) of 0.8 mm/yr$^{1/2}$ for the mean and 2.7 mm/yr$^{1/2}$ for the characteristic high carbonation depths respectively.

For modelling purposes, a generally pessimistic lower bound case was created by combining the characteristic minimum cover with the characteristic maximum carbonation rate. The carbonation modelling implications for the risk of future carbonation induced reinforcement corrosion in uncracked concrete, also applicable for the vast majority of the shafts above water level as cracks were typically less than 0.2 mm surface width, and excluding localised positions of slipform-related defects, were that no significant corrosion is expected until more than 60 years age.

4.1.4 Chloride Penetration

An independently verified GHD proprietary probabilistic model, based on Fick's Second Law of Diffusion, was used to model potential future damage due to reinforcement corrosion. This model uses a Reliability Index approach analogous to that used in structural design, a Reliability Index of 1.65 was used to estimate times to 5% risk of corrosion initiation and related cracking.

Chloride concentration profiles were determined in 35 core samples, and "diffusion" (more correctly penetration) coefficients derived using least-squares curve fitting of the data to Fick's Second Law for non-steady-state diffusion, at an age of 18 years from 1997. Initially the data was divided into two zones by height, nominally +5.8 mPD (predominantly wet, more exposed to wave action) and nominally +9.6 mPD (wet/dry exposure). As shown in Table 1, the shorter estimated times for the higher levels on the shaft are consistent with higher chloride penetration and corrosion rates expected under wet/dry cycling. However, a comparison of means test also showed that the data could be treated as a single pool, due to the relatively large standard deviations resulting from significant differences in concentration profiles between cores - including where taken in close proximity.

The predictive modelling used 95% characteristic high (mean + 1.65σ) diffusion coefficients with small attendant coefficients of variation to produce pessimistic lower-bound cases. The notional surface chloride concentrations (C_s values) used in the predictive modelling were mean values with their corresponding standard deviations, as the characteristic high values were unrealistic in the Author's experience, and significantly biased the resulting predictions. Cover value was the mean of 90 mm reported in Section 4.1.1 with a "minimum" -2σ value that is the default in the model used. For the first 30 years after construction, an ageing factor of 0.4 was applied for the high-slag binder.

Table 1 summarises the results of the modelling process in terms of the age of the CGS from placement on the North-west Shelf. The pessimistic lower-bound predicted times to events at abnormally low cover are consistent with the recorded appearance of a small number of minor corrosion-related defects on the shafts.

Table 1. Summary of chloride damage modelling results for shafts uncracked concrete (min cover = mean - 2σ)

Elevation	Diffusion Coefficient (measured, t = 18 years) ($D_t' = D_t + 1.65\sigma$)	Predicted time to initiation (yrs in service)		Time to 5% probability of initiation (yrs in service)	Time to 5% probability of cracking (yrs in service)
		Min cover 57 mm	Mean cover 90 mm		
+9.6 mPD	Mean D_t: 0.44×10^{-12} m²/sec C_s: 0.86%	42	>60	43	>60
	Characteristic D_t': 0.94×10^{-12} m²/sec C_s: 0.86%	15	47	22	48
+5.8 mPD	Mean D_t: 0.24×10^{-12} m²/sec C_s: 0.80%	>60	>60	>60	>60
	Characteristic D_t': 0.42×10^{-12} m²/sec C_s: 0.80%	43	>60	56	>60
All Samples	Mean D_t: 0.35×10^{-12} m²/sec C_s: 0.83%	53	>60	55	>60
	Characteristic D_t': 0.77×10^{-12} m²/sec C_s: 0.83%	20	58	30	58

4.2 Shafts and Caisson Below Water Level

With no evidence of structural damage or distress being identified during the inspection work, no further assessment was necessary. Of note was the observation from half-cell potential measurements made below water level that the sacrificial anode CP system on the flow lines and export risers was providing some level of protection to the embedded reinforcement, these components being designed to be electrically continuous. At any defect where the resistive path through the concrete is reduced, e.g. by a seawater filled crack or void, a higher level of local CP protection to steel reinforcement would be present.

5. DISCUSSION

The average measured depth of cover to reinforcement exceeded the design value by 10 mm, however the cover was found to be highly variable, with a 95th percentile lower bound of 64 mm and lowest measured value of 48 mm. This is consistent with the implementation, at construction, of various low-cover repairs [7], and the appearance of a small number of minor reinforcement corrosion related defects in service as reported previously [8] and herein. The most common repair for low cover identified at construction (specified for areas where cover was only marginally less than the specified thickness) was application of a polymer modified cementitious coating that, although consistent with the design standard used [9], had extensively degraded within the 18 years' service to date. The reasons for this relatively rapid degradation have not been studied in this work, however (while acknowledging that such repairs do not necessarily have to last the whole service life of the primary structure to achieve the desired effect on overall durability) the level of deterioration raises questions regarding the long-term efficacy of such repairs for low cover where a long "maintenance free" service period is specified in severe environments.

Much of the variation in cover, and a majority of the visible construction-related defects discussed in Section 3.1.2, are attributed to slow and inconsistent progress of the slipform during changes in pour geometry to construct the ovoid flare in the upper section of the shafts. Where progress of the form is stop/start or too slow in relation to the setting time of the concrete, a degree of binding and formation of lifting cracks in the concrete surface can occur, resulting in horizontal fissures in vertical elements as observed on the flared sections of the shafts.

Analysis of the chloride penetration data collected shows that application of the above protective coating under the stated criteria was superfluous, and that the early degradation of the coating has not prejudiced current service life requirements.

Analysis of the carbonation data collected shows that carbonation-induced corrosion of reinforcement does not present a risk to the integrity of the CGS within the expected operating life.

It is predicted that additional non-critical corrosion related defects will appear during the life extension period at areas of low cover and associated with surface-exposed embedded steel items. Local delaminations resulting from chloride-induced corrosion of reinforcement at areas of abnormally low cover identified during this work programme totalled approximately 1.7% of the area tested. Assuming that this is representative of the estimated 1,600 m^2 surface area of the shafts above MSL, there is expected to be approximately 27 m^2 of delaminations currently present. Management of these and future defects has been addressed in the IMRP.

6. CONCLUSIONS

The physical inspection interpretation associated with the completed scopes of works for the WNB CGS atmospheric and below water concrete shows that there is no significant structural integrity impairment that would limit the remaining life of the WNB CGS to less than the current intended operating life extension period of 20 years.

7. ACKNOWLEDGMENTS

The Authors wish to acknowledge their colleagues at GHD Pty Ltd and Vermilion Oil & Gas Australia who conducted the structural review, contributed to the preparation and review of reports, and facilitated the site work. The Authors also wish to acknowledge the Neptune Marine team who performed the roped-access work, and Jarrod Collings and the Intervention Engineering team who performed the ROV inspections.

8. REFERENCES

1 Peek A, Paull R, Long S and O'Brien C: Condition and remaining life assessment of the Wandoo B concrete gravity structure, (in) Proc. Concrete 2017/3rd International Congress on Durability of Concrete, Oct. 22-25, Adelaide, Australia, Paper 285

2 ISO 19903 2006-12-01 Petroleum and natural gas industries – Fixed concrete and offshore structures

3 DNV-OS-C502 September 2012: Offshore Concrete Structures

4 NORSOK N-006 March 2009 Assessment of structural integrity for existing offshore load-bearing structures

5 ISO 19902:2007 Petroleum and natural gas industries – Fixed steel offshore structures

6 BS EN 12504-1:2009 Testing concrete in structures. Cored specimens. Taking, examining and testing in compression

7 Wandoo Alliance WNB-1238-RZ-0001 Construction Project Report (1997)

8 Arup WPA-1233-RC-0001 Arup Wandoo B Concrete Inspection – Inspection Report and Recommendations for Remedial Work (2003)

9 BS 8110 Structural use of concrete. Code of practice for design and construction (various parts and dates, now obsolete)

9. AUTHOR DETAILS

Andrew Peek has over 28 years of experience in durability, deterioration, protection and testing of construction materials. Particular experience relates to the performance of reinforced concrete structures in aggressive environments. His experience includes durability planning, specification for new installations, remedial works, failure investigation, and the design and implementation of materials performance test programs. Andrew has authored over 40 technical papers and presentations to professional societies on this subject and is a co-author of the Concrete Institute of Australia Z7/07 Performance Tests to Assess Concrete Durability and contributed to Z7/05 Durability – Modelling of Reinforcement Corrosion in Concrete Structures.

Liam Holloway is Managing Director and Principal Engineer at MEnD Consulting. Liam has over 15 years' experience working as an engineer in the field of concrete durability, inspection and repair. He completed is PhD by research on the application of organic inhibitors for the mitigation of reinforcement corrosion in marine structures in 2001 at Monash University. Since that time Liam has worked as a consulting engineer and spent 8 years working as National Engineering Manager for a large remedial contractor. In 2014 Liam established MEnD Consulting, a small practice specialising in the investigation and remedial engineering for existing infrastructure.

Sharyn Long has over 12 years of experience in construction materials technology. Prior to GHD, Sharyn completed her PhD by research on solid state electrochemistry at Monash University in 2002. Since joining GHD, Sharyn has accumulated extensive experience in investigation and repair projects, durability planning, and developing specifications for new installations as well as remedial works. Sharyn has been involved in the delivery of remediation project works, including project management, tendering, contract administration and provision of technical support during remedial works.

Based in Perth, Australia, Colin O'Brien has over 25 years of experience in fixed and floating offshore structural design and assessment of greenfield and brownfield oil & gas facilities. Expertise in the management of asset technical integrity and the application of advanced structural and mechanical design assessment methods. Experienced in safety critical system extended life and design requalification studies, ultimate strength, reliability and probabilistic method assessments together with fitness for service and failure investigation studies

Rodney Paull has 36+ years' experience in durability design of new and investigation / upgrade of existing structures. He was resident in South East Asia and Hong Kong for 19 years managing Taywood Engineering offices and in Australia from 2003. Rodney is Chair of CIA Durability National Committee and had author input to new Recommended Practices Z7/01 Durability Planning; Z7/04 Durability Good Practice through Design, Concrete Supply and Construction and Z7/07 Performance Tests to Assess Concrete Durability. He is member of BD02-002 Committee reviewing Sections 3 & 4 of AS 3600 and CIA nominee on ACI 201 Durability of Concrete

EFFECTIVELY CAPTURE IN-FIELD DATA – A NEW WAY FORWARD FOR PROTECTIVE COATINGS PROJECTS

Justin Rigby
Docopro Pty Ltd, Melbourne, Australia.

SUMMARY: Asset management requires a process of inspection and documenting construction, maintenance and refurbishment works.
Most organisations have good systems for in-field inspection however there is difficulty in getting documentation when construction or maintenance activities take place.
Asset Owners who have a high level of awareness in the coatings field spend tremendous effort to develop coating specifications, which are specific to their projects.
However a disconnect occurs when quality responsibilities transfer to field staff.
Our aim is to;
1. Describe how inspection and test plans (ITP's) are;
a) Being used as the in-field quality document.
b) Treated as the Contractors document.
c) Not being prepared properly.
2. We will then describe how the in-field data is being lost because;
a) In lieu of a reasonable alternative, Specifiers use AS3894 part 10 through 14 for a suitable report form.
b) AS3894 series of report forms;
- Are not aligned with project specifications.
- Once complete, are usually incoherent and are difficult to reconstruct project chronologies.
3. Provide the audience with a new way forward, by;
a) Avoiding responsibility for ITP's whilst ensuring Contractors have improved documentation.
b) How to build a better suite of report forms that are aligned to each Specification.

The data obtained through use of aligned ITP's and report forms provides Asset Owners with documented histories and enables the organization to capture Asset knowledge which can be used to assist the asset integrity cycle.

Introduction

The life cycle of many industrial assets are determined by a combination of the environment they are operating within and the success of installed corrosion prevention systems.

Both concrete and steel make up the bulk of construction materials with the majority of concrete degradation being caused by corrosion of the steel reinforcing.
In protecting structures built of these materials, the primary corrosion prevention method is through the use of protective coatings acting as a barrier to any corrosive environment.

Given the structural properties and relative cost of steel and concrete in comparison to corrosion resistant materials often the added cost of protective coatings far outscores the increased cost of alternative materials.

So suddenly 'paint' plays the leading role in any life cycle assessments. To obtain and maintain asset durability relies on initial coating followed by a series of condition assessment, maintenance coating and eventual re-coating operations to maintain the protective coating integrity.

For effective Asset Management an Owners management team will maintain a register including; as built drawings, geographic information systems (GIS) and technical specifications.

Easting	Northing	Elevation (AHD)	Description
344727.44	5783687.53	15.96	New Coating
344727.33	5783686.77	15.96	Girth Weld Joint
344727.16	5783684.96	15.92	Repair Patch
344726.76	5783682.88	15.94	Stand

Figure 1 - Example of GIS information for a pipe line

However initial and maintenance projects all utilize labor intensive disciplines such as; scaffolding, environmental containment, surface preparation, and application of multiple coats of paint.

Many of the Owners we have worked with have a wealth of documentation regarding engineering, design and specification but what is lacking is the in-field data documenting 'what actually occurred onsite'.

It's usually here, where the 'fog' of construction activity obscures the in-field collection of data that becomes so valuable as the Asset moves through later coating cycles.

In this discussion I will;

1. Remind the industry's we serve of the importance of in-field data in focusing corrosion management systems to deliver optimum benefit to the Owner, and;
2. Show industry a new way forward in effectively collecting in-field data.

A description of Asset Management.

Asset management is a systematic process of deploying, operating, maintaining, upgrading, and disposing of assets cost-effectively.
Included within this framework, should include a Corrosion Management System (CMS) as described by the NACE Impact Study where a process of risk assessment, condition assessment and maintenance activities are required to maintain durability.

Many organizations are indeed struggling with rolling out corrosion management principles to contractors, suppliers, and vendors.[1]

[1] NACE International IMPACT Study

What is in-field data?

In-field data describes a process of data collection from the work face.

A construction or maintenance project should be very well assessed and documented through engineering, design and specification. These may include a site inspection but the bulk of the documentation is developed at the desk top.

Asking a Site Manager to deliver these concepts exactly as written is like constructing your flat pack furniture from Ikea™. You have all the instructions but site conditions, available resources and persons interpretations will show obvious distractions from the intended outcome.

In field data is the inspection, testing and reporting of the actual construction or maintenance processes that occurred onsite, in the field. It will verify which processes met with the specification but more importantly it will describe which deviations occurred from the process and the peculiarities' of the site conditions.

How can in-field data help you?

In-field data will help the organization reduce costs, reduce future maintenance, reduce integrity risks, improve work practices, improve asset utilization and longevity.

Subtle differences in construction and maintenance processes can result in different outcomes when assessing performance of the corrosion protection system in later years.

Often Owners of network assets have areas of protective coating that only last a few years whilst other areas last 4 or 5 times the duration. Effective asset management techniques will seek to understand where increased durability has been achieved in contrast to areas of early coating failures.

Reviewing in-field data will provide the organization with an explanation of the different application procedures used and with this to identify;

Figure 2 - Coating failure in a filter cell

1. Which methods and processes provide better durability on your network.
2. Identify practices leading to coating failure on your network.
3. Improvements to existing design and specification for reason of;
   a) Coating materials not being suitable given available site conditions and work practices.
   b) Coating materials not performing in the micro-environment conditions.

The global cost of corrosion is estimated to be US$2.5 trillion, which is equivalent to 3.4% of the global Gross Domestic Product (GDP) (2013). By using available corrosion control practices, it is estimated that savings of between 15 and 35% of the cost of corrosion could be realized.[2]

Allocate resources to capture in-field data and you will be able to save money, time and resources.

How in-field documentation is used currently

[2] NACE International IMPACT Study

In field documentation requires a re-invention of current use of Inspection and Test Plans (ITP) and Daily Report Forms.
Why? Because despite all the best intentions of Owners, Designers and Specifiers all of the quality information provided in project plans and specifications are left on the shelf at the job site and the primary quality document becomes the ITP.
Additionally daily report forms are a basic table that are part of an Applicators document suite and not related to your specification.

Inspection and Test Plan (ITP)

So lets focus on the ITP first.
This is a document that is handled like a political football, no-one wants to take responsibility for it as doing so may assign responsibility for project methodologies so most Specifications place the onus for development of ITP's with the Applicator.
Here one or a combination of events takes place;

Item	Ref to ASNZ	Activity	Test reference	Acceptance criteria	Verifying document	Frequency	Contract		
							Task	Initial	
6		Surface Preparation							
6.1	9.8.5	Soluble salts - Only if requested	AS 3894.6 Method A using swab test or brestle patch	Recommend: Average conductivity =<50 mg/ m2 and max =<70 mg/ m2 water soluble salt(s).	Docopro PIF ASNZ2312-EHB 7-DF001 or equivalent'	Suggest: 1 per 10m2 or 3 per 200m2 or 5 per 500m2. Also after Salts has been removed.			
6.2	4.2.2	Solvent wash to remove contaminants	AS/NZ 1627.1	No dirt /contaminants visible	Docopro PIF AS NZ2312-EHB7-DF001 or equiv'	Check 100% of surface area			
6.3	3.4	Substrate surface visual inspection	AS/NZ 2312.1	No defects/sharp edges/ gouges/ rough welds	Docopro PIF AS NZ2312-EHB7-DF001 or equiv'	Check 100% of surface area			
6.4		Dust level	AS/NZ 3894.6 Method C Tape test	< Rating 2	Dust tape sample & Docopro PIF CSADF001 or equiv'	Spot checks, but not less than 1 per 50m².			

Figure 3 - Inspection Test Plan loaded with Specification specific information.

1. The Applicator is not a Technical expert or even not great at developing documents, so the ITP that is supplied does not align with the specification.
2. The ITP is submitted to the Client for review. The Client has assigned a project manager who is typically a 'generalist' rather than having specific knowledge of coatings. Their task is to manage many trades and intermediaries' so they review the ITP, and do one of the following;
   a) The Project Manager reviews the ITP and see's it includes; blast, paint, weather, check (hopefully not in that order). So he approves the ITP.
   b) Alternatively the Project Manager may be coatings savvy and says, 'I have seen this all before, and it didn't work out so well', so he sends it back to the Applicator for revision.
3. The Applicator is frustrated by the process and eventually builds a satisfactory ITP but it may take many hours for both parties and numerous revisions.

Daily Report Forms

Next the supporting documents most Applicators use are a Daily Report From. These are the form site staff is asked to complete each day detailing weather conditions, batch numbers of materials and measurement readings for; surface profile and coating thickness.
These forms are successful at collecting basic data but usually don't have enough detail to re-construct project activities during project review. Most of the ones I have seen are not tailored to the daily activities and therefore are not considered relevant by those whose task is to complete them.

A new way forward

A renewed focus should be applied to the use of existing in-field documents including Inspection and Test Plans and Daily Report Forms.
The focus should be to;

1. Provide ITP's;

a) That Asset Owners have assurance are aligned with their specification.
b) That saves frustration of Applicators and Project Managers in trying to develop these documents from scratch.
c) Consider using a third party independent to develop these documents in collaboration with Project Manager and Applicator.

2. Change Daily Report Forms to Product Inspection Forms (PIF). A Product Inspection Form still collates all the information from site however presents it in a manner that makes the data align with project stages whilst also being easy to read and use by the persons on site.

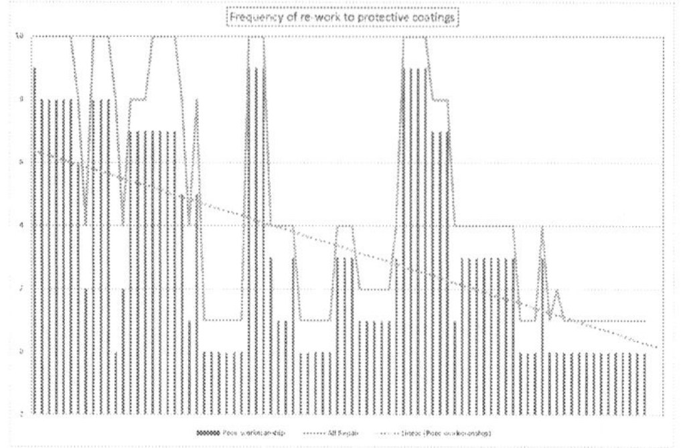

Figure 4 - Frequency of re-work to protective coatings

Daily reporting needs remodeling to produce forms that are relevant to the project tasks and reference the objectives of the Specification. This is the best chance you have to obtain coherent in-field data. The more useful these forms are to the site personnel, the more likely they are to fill the form out accurately. This can be done by;

1. Change the 'Daily' format into what we call a 'Product Inspection Form (PIF). The Daily Form includes information for all project stages on 1 page and each day the field personnel only fill out the areas that are relevant for that days task. Therefore most of the form is irrelevant for projects occurring over many days or weeks.
   We flip the script from being a series of one day forms to being a 'Product Inspection Form' (PIF) where the sequence follows the project. The field staff still completes items each day but a 5 page form can provide enough space for a 10 day project.
2. The Product Inspection Form is broken into sections including; project diary, surface condition and preparation, coating application, weather recording, a repairs page including surface preparation and repair application, a testing page to record measurements.
3. Each section needs to be expandable to allow users to 'print' extra sheets and insert into the PIF for long term projects.
4. Each section needs to include Specification requirements for the field staff such as; what is the tolerance, how is it measured, which Standard is being used and any noteworthy items; such as; when salt testing is required, what DFT tolerances are allowed. Including the relevant information from the Specification makes your form a handy reference for the site personnel and increases the likelihood of them using the form.

A recent study[3] reviewed 2 years of in-field data using the above principles found a 43% decrease in coating repairs.

[3] Document Templates - A Study for Quality Control in Protective Coatings by DocoPro Pty Ltd 2017

To put that into context, rarely does a coatings project achieve compliance within a single application. Often there are minor areas of re-work required to meet Specification and poor workmanship will require significant re-work.
This Study reviewed the frequency of re-work being identified on a project by project basis for a single asset using a single specification.

1. Applicators use of reporting increased 90%.
2. Applicators implemented secondary systems to reduce reportable repairs.
3. Re-work overall decreased by 43% and re-work due to poor workmanship decreased 60%.

Hard copy or electronic?

There are many 'apps' out there that take on a subscription fee but allow you to record in-field data on a tablet.
The alternative is a 'hard copy', printed form that is messy.
I believe in-field data needs to have both options;
Often site based tasks occur in a range of environmental conditions from;

a) Hot, dusty and humid to frigid, windy and wet.
b) With site constraints including; enclosed and dark, open and noisy, at heights with difficult access. Sometimes electronic items are not allowed into the site.

Here paper based recording is easier for operators to juggle the various environmental conditions whilst recording important data.
You can drop paper, you can get paper wet and dry it out, you can fold it up whilst using both hands to perform tasks or climb a ladder.

For those that are less tech savvy paper is an easy choice and note taking on paper helps provide a wealth of additional information such as; different hand writing means different persons filled out the form, you can see the process of note taking and corrections, you can see the sweat and dirt on the page. A hand written page tells a story behind the data.

Paper based forms easily become electronic. Do you want to see me turn a paper form into an electronic version? Easy take a photo of the completed page, or scan it and save to your computer.

However an electronic form has obvious benefits. It can be completed on a tablet and photos can be uploaded to the form. There are proprietary cases and harnesses to carry them whilst performing other tasks. The entered data is automatically saved to the cloud and when completed the form can be shared across offices and geographies.

The best system is an online form that can be printed as a hard copy for the in-field data to be recorded and the captured data is then entered into the online form from the comfort of a site office or wherever you set up your laptop.

Summary

Capturing In-field data helps an organization develop a history of their Asset but more importantly enables the Organization to review systems and materials that provide better corrosion control and durability outcomes and then use those learnings on future capital and maintenance projects.

To effectively capture in-field data you need to place the emphasis on the site personnel, those at the workforce who are living, sweating, and breathing your project.
They need documents that they feel are relevant, that help them achieve their outcome. Without relevance they don't volunteer the information rather they see it as another obligation.
So beefing up the ITP and the PIF's provides a place for the Applicator to reference Specification requirements whilst completing the raw data the Asset Owner often finds lacking.
Hard-copy or electronic?
I leave the decision up to you.

AUTHOR DETAILS

Justin Rigby is the Principal at DocoPro Ltd. Docopro, an online document generation service, offers quality assurance and specification templates for industrial use. These include Inspection and Test Plans (ITP) and Product Inspection Forms (PIF), which are designed for the protective coatings industry and comply with coating systems standards in Australia, the U.S. and internationally.

Justin has extensive experience in the Protective Coatings industry developed over 20 years and is a NACE CIP lecturer and serves as Chairperson of two of the ACA's Technical Groups.

Justin has authored a number of papers for publication in prestigious journals and has presented at international conferences in Australia with regard to coatings technologies.

He has a passion for project delivery and his focus is on introducing efficiencies and building team based strategies with enhanced performance and cost benefits for Clients and their suppliers.

GIPPSLAND WATER FACTORY – A CASE STUDY OVER 3 YEARS

Justin Rigby
Remedy Asset Protection, Melbourne, Australia.

SUMMARY: The Gippsland Water Factory (GWF) is a $230 million domestic facility that consists of two wastewater treatment plants sharing common infrastructure. It treats wastewater from nine towns in central Gippsland as well as industrial waste water from the Australian Paper mill at Maryvale. The plant serves the needs of more than 48,000 customers and 300 local businesses.

Apart from treating waste water the plant can produce up to eight million litres of recycled water per day for use by Australian Paper, thus safeguarding domestic supply of drinking water.

Built in 2010, a major part of the plant is the concrete structure that houses the Membrane Bio-Reactor (MBR) cells. These cells have been lined with an epoxy, however the lining has failed resulting in fragments of coating delaminating from the surface.

These fragments can cause damage to the delicate filter technology that this facility has won awards for, thus requiring partial shutdowns whilst maintenance works occur to the filters.

Since 2014 we have been involved in the refurbishment of four of these cells using two different coating systems. My presentation will focus on the practical elements of delivering these projects and provide assessment of each coating type in this environment.

Figure 1- GWF state of the art facility

Keywords: Concrete Repair, Protective Coatings, Coating Inspection, Wastewater.

1. INTRODUCTION

This is a unique project that is situated in a difficult environment. I will begin by describing the work chronologically so you can follow the learnings we gained along the way.

The first cell refurbishment occurred between March and May 2014 where we were engaged as the project subject matter expert (SME) to assist in the relining project for MBR cell #2 using a Vinyl Ester Resin and CSM fibre laminate. The duration of the relining project was 41 days.

In March - May 2015 we were re-engaged as the project SME for the relining project for MBR cell #10 using a polyurethane and geo-fabric reinforced system. The relining project was 18 days in duration.

Figure 2- MBR filtration cells

In 2016, cell #4 was relined using the same project delivery model.

The team again used the same system as the year before.
The duration of works reduced again to 12 days.

These MBR cells are within the sewerage treatment plant and the cells are in a bank of 11 live cells. Thus during the works, the individual MBR cell is isolated and a temporary containment hood is installed. Whilst the containment hood provides clean dry and conditioned air for the duration of works, it also provides an obstruction when terminating the coating at the top edge of the walls. Additionally there are permanent fixtures within the cell such as; bolts, fittings, and a semi imbedded mild steel pipe with numerous feed pipes.

All these factors add to provide a great depth of challenges when considering a seamless liner.
This is a project requiring thorough collaboration and relationships that can provide a fluid work process in order to deal quickly with new requests for changes to the specification.

Figure 3- MBR cell with equipment stripped out.

2. THE MBR CELL ARRANGEMENT

Each cell has a detailed arrangement of racks, filters and plumbing that are easily disassembled.
Once a MBR cells components are stripped out, the remaining surfaces essentially consist of a concrete cube (except the lid). As in life, things are never as simple as first thought, so unique features of our concrete cube include:

- *Wall 1* - Fixed studs in an arrangement of four, in three locations to provide a fixing point for mechanical equipment.
- *Wall 2* - A head board extending almost across the width of the wall with 36 fixings, and an outfall pipe located approximately in the center of the wall.
- Wall 3 - Fixed studs in an arrangement of four, in three locations to provide a fixing point for mechanical equipment.
- Wall 4 - A semi-imbedded 400 mm header pipe with two bolted flanges.
  a) The flanges are offset from the end walls by less than 100 mm.
  b) From the header pipe there are 15 feeder pipes.
  c) There is approx. 100 mm clearance around each of the feeder pipes.
  d) The embedded and exposed header pipe arrangement is recessed into the wall cavity except the flanges protrude beyond the wall elevation. Therefore, the recessed section of wall cavity has a 300 mm step back.
  e) This wall is lower in height compared to the other walls providing a weir for excess water to flow back into the raw water trough.

Figure 4- MBR cell and header pipe after removal of existing coatings.

- *Floor* - The floor has a recessed gutter along Wall 2 that provides a 70 mm step.
- *Overall* - The surface area is approximately 80 m^2.

3. THE WORK ENVIRONMENT

The Gippsland Water Factory has a staff culture where there is a great mix of a team-based approach to getting tasks done whilst having multiple layers of safety embedded into every work process.
The facility itself is well planned out and the MBR cells are easy to access from three access points via two walkways and a ladder. The cells are positioned in the base of the facility itself and there is overhead gantry cranes available to provide hoisting and positioning of equipment.
These filter cells are in two banks of six therefore to work on any one cell requires works to be adjacent to live wastewater. In keeping with GWF safety culture, a containment hood to be placed over the MBR cell when refurbishment works are taking place.
This provides:

a) A great work platform to position equipment and support staff.
b) A great method of fall restraint into active MBR cells or the work zone.
c) A method of protecting the work staff from live wastewater.
d) A method of containing abrasive media used during abrasive blasting from entering surrounding cells.
e) A method of sealing the cell and controlling the ambient air conditions.

4. THE TEAM

GWF management initiated the project after the existing epoxy coatings were deemed to be failing and an engineering firm provided assessment and was assigned the project lead. The firm also provided material selection, coating specification and project oversight. The engineering firm engaged our Company, Remedy Asset Protection as subject matter experts (SME).

Our services were used to provide the inspection and testing required under the specification as well as a consulting role in foreseeing issues that the specification did not deal with.

After a competitive tender process the contract was awarded to the successful tenderer.

5. THE PROJECT

The project was assessed as requiring removal of the existing coatings followed by application of a new coating system to line all concrete surfaces and embedded steel work. These requirements were described in a detailed project specification developed by the engineer, the contractor developed a detailed work method statement that provided detail on every surface and interface encountered.

6. THE MATERIALS

After a careful selection process, engineering selected a vinyl ester resin and chopped strand matting (CSM) fibre laminate system. Essentially this system is highly chemical resistant fibreglass, consisting of:

a) Vinyl ester primer (roll or brush applied) 25-75 microns thick.
b) Fused silica filled resin base coat, trowel applied to approx. 1.5 mm thick.
c) Two layers 450gsm CSM.
d) One layer of C glass tissue/veil.
e) Waxed top coat to assist proper resin cure (roller applied) approx. 250 microns thick.

Each layer of glass reinforced resin was to be rolled to remove entrapped air.
Total laminate thickness equalled 3-4 mm.

The features of such a coating are:

a) The base coat is filling the surface imperfections across the concrete surface.

b) The vinyl ester binder is providing a hard, chemical resistant surface.
c) The CSM is providing reinforcement of the resin binder to increase its thickness and cohesive strength.

Initially this product was used in the first cell successfully but latter cells were specified using a different product.

7. THE SURFACE PREPARATION

Surface preparation commenced with two weeks of arduous abrasive blasting to remove the existing coatings.

Figure 5- Failure of existing coatings to MBR cell.

Observing the abrasive blasting revealed:

a) The existing coating had weak cohesive strength as it would splinter off.
b) The existing coating had poor adhesion in localised areas, however this was most likely a symptom of the poor cohesive strength rather than poor surface preparation.
c) There had been multiple touch-ups performed over the years that had no adhesion.
d) The concrete substrate was not substantially contaminated prior to the existing coatings being applied.
e) The concrete substrate was sound with no evidence of surface movement.

Once surface preparation was completed it came time to plan the coatings application process.

8. THE COATING APPLICATION

In order to protect a concrete substrate from the products it is containing requires a seamless lining. If we have an edge where the liner terminates within the vessel there is potential for the contents to slowly or quickly seep below that edge and begin delaminating the coating.

Figure 6- Filling of flange to the header pipe followed by application of vinyl ester system.

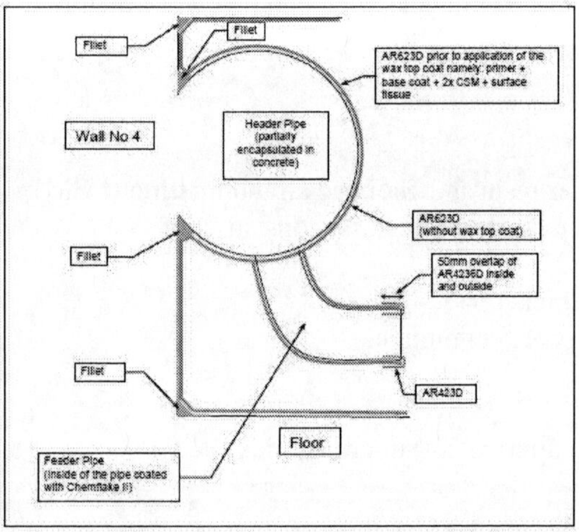

Figure 7- Schematic of coating interfaces around header pipe

Liquid coatings act as a seamless liner when applied correctly, however as described earlier, within this cell there are multiple interfaces that require unique solutions.

As a team, we were able to identify areas of concern ahead of the project schedule, thus:

1. Identifying the issues ahead of schedule to avoid delay to works.
2. Ensuring works did not progress to a stage of poor application over the interfaces.

So you can see how difficult an 80m² concrete cube can be to re-line given just a few mechanical installations.

Figure 8- MBR cell completed in the vinyl ester fiber laminate system.

9. USE OF DIFFERENT COATINGS

These projects began with the vinyl ester system being installed in cell #2 where there was no alteration to the structure except for filling in the cavity between the header pipe flanges and the adjacent walls.

In the following years, the team switched to a solventless, elastomeric aromatic polyurethane and geofab laminate system.

The polyurethane system differs from the vinyl ester in that the vinyl ester resin wets out each of the fabric layers to form a single shell of hard resinous material just like fiberglass.

Whereas the polyurethane system consists of:

1. A single application of coating to the substrate, called a glue coat at 500 μm DFT.
2. A heat set geo-fabric is embedded into the still wet glue coat.
3. An additional application of the coating is applied over the fabric at 2000 μm DFT

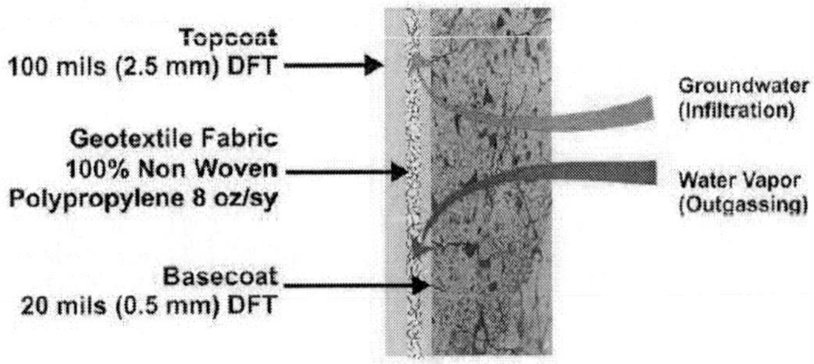

Figure 10- Cross section of polyurethane and geofab laminate system.

The geo-fabric layer is a heavy polypropylene, thus the two coating layers only wet through the fabric occasionally, leaving most of the fabric to provide a separate layer within the coating system.
This fabric layer:

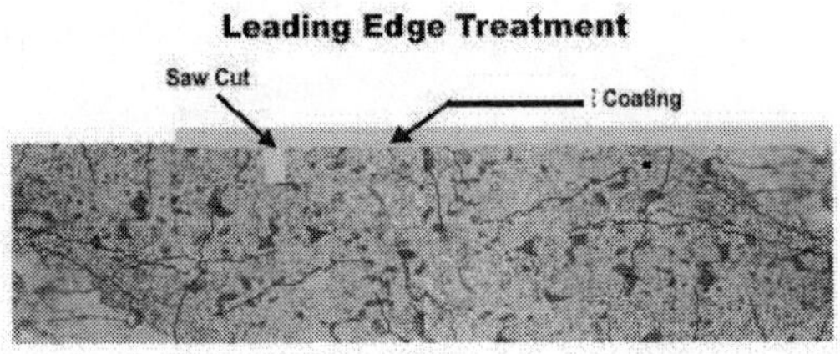

Figure 9- Cross section of leading edge treatment for both coating types.

a) Provides an alternative substrate for the coatings to lay up on. Therefore, concrete characteristics such as bug holes and cracks do not require filling.

b) Acts as an air trap during coating application.

During the coating of concrete, many liquid coatings have difficulty forming a film over the porous surface as air is “outgassing” from the porous concrete during the curing cycle. This results in

pinholes or discontinuities where the coating could not form back over where the air escaped from. Traditional practice is to apply the coating and then perform multiple cycles of inspection and spot repair until all pin holes have been rectified leaving a continuous protective coating.
The geo-fabric in this system helps prevent outgassing avoiding many of the pinholes.

c) In the event of wastewater egress through the concrete and under the coating, the fabric layer will act as a moisture trap, whereas traditional coatings would disbond from the surface.

Figure 12 - Completed MBR Cell in polyurethane and geofab laminate system.

d) In the event of concrete movement, the geo-fabric layer acts as a flexible reinforcement allowing the coating system to deform over the affected area.

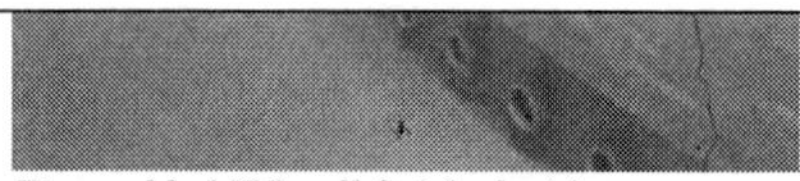

Figure 11- MBR cell finished with polyurethane and geo-fabric laminate system.

Table 1 - Description of coating differences

COATING DIFFERENCES		
Element	**Vinyl Ester**	**PU**
Coating properties	Hard	Elastomeric
Reinforcement properties	Wetted out to be incorporated within the laminate	Provides a separate layer within the laminate
Surface preparation	Remove all existing coatings	Suitable over sound coatings
Substrate rectification	Trowel on mortar to fill surface irregularities	Minimal
Pot life at 25°C	20 minutes	3 minutes
Hard dry at 25°C	1 day	1 day
Thickness	3 mm +	2.5 mm plus fabric
Safety	Use Sytrene (toxic)	Contains isocyanate (allergic)
Result	Hard, thick lining that has great mechanical adhesion	Flexible, thick lining that can bridge over defective substrates
Holidays	Did not produce any pinholes due to multiple layer application procedure	Still produced pin holes where spray application could only be applied in one direction
Labour required	Required a lot of additional effort to apply five layer coating system	Ease of application being effectively a two coat system
Failure modes	Concrete movement beyond the limits of the coating allowing cracks and disbondment	Impact damage to coating, allowing moisture through coating to the fabric to bare concrete (bug holes)

	Positive result
	Unique quality
	Disadvantage
	Neutral

11. INSPECTION CONCERNS

As described, the client had already been affected by defective coatings and there was no way they were willing to accept poor workmanship, so we put together an inspection regime that focused on the critical stages of:

1. Surface preparation.
2. Coating application.

3. Inspection and testing.

This means we need to focus on:

1. The concrete being prepared properly.
2. Ensuring the work area is prepared all the way to the limit of the temporary hood.
3. Verifying termination saw cuts are installed in the right spots.
4. Confirming the reinforcing fabric does not tent over the termination saw cuts.
5. Verifying the fabric sits snug around interfaces and internal corners.
6. Performing a porosity test to all areas, but that won't pick up on uncoated fibers protruding through the coating, so a full visual inspection is required.
7. When the containment hood is removed we ensure the coating interface is:
   - Sufficiently prepared onto the existing coatings and also onto the new coating.
   - Observe and test the application of coatings to the top of walls.
8. Repairs must be re-inspected and if you question this requirement, ask yourself why. Is it because it seems mundane and probably only relates to <3% of the surface area? Re-inspect repairs - it's amazing how many times they don't pass when everyone has confidence that it's just an easy fix.

Figure 13 - Inspection concerns including terminations.

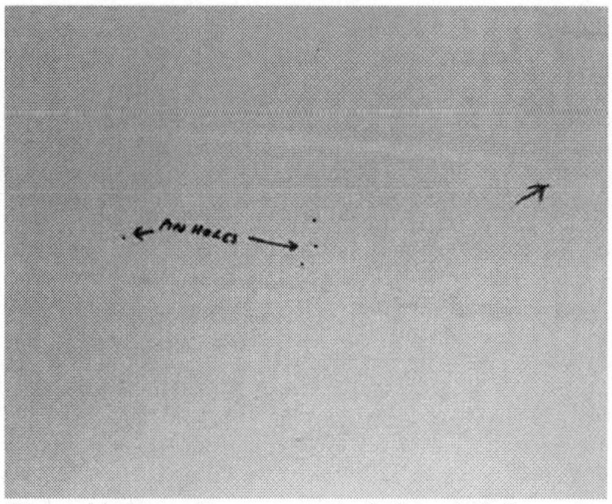

Figure 14- Inspection concerns including pin holes.

12. SUMMARY

We are proud to have been involved in this project over the course of four years and look forward to having the opportunity to observe two different coatings in the same environment.
Remember these works are remedial. The first time around, the owner relied on contractor warranties and didn't have the detailed specifications, tight working procedure documents, focused contractors and third-party coating inspection.

This resulted in defective coatings and left the owner without recourse to enforce project warranties due to the lack of project detail.

Detailed project appraisal and documentation provided at project commencement will save your project time and money. This will also protect your reputation for successful project management.
We hope this presentation has helped you focus on the importance of having the right procedures in place for your projects.

13. ACKNOWLEDGMENTS
Sajeev Dep, Maintenance & Optimisation Engineer, Gippsland Water Factory.

Gavin Pearson, GWF Works Engineer, Gippsland Water Factory.
Adrian Dal Pozzo, Managing Director, Valicote.

14. AUTHOR DETAILS

Justin Rigby is the Principal at Remedy Asset Protection (RAP).
Remedy Asset Protection is a consultancy for Clients that require specialists in Protective Coatings to provide expert assistance within their own team structures, including; coatings project advisory & coating inspection services.
Justin has extensive experience in the Protective Coatings industry developed over 20 years and is a NACE CIP lecturer and serves as Chairperson of two of the ACA's Technical Groups.
Justin has authored a number of papers for publication in prestigious journals and has presented at international conferences in Australia with regard to coatings technologies.
He has a passion for project delivery and his focus is on introducing efficiencies and building team based strategies with enhanced performance and cost benefits for Clients and their suppliers.

CATHODIC PROTECTION CRITERIA FOR STRAY CURRENT AFFECTED PIPELINES

Brian Martin
Brian Martin & Associates Pty Ltd

SUMMARY: This paper reviews the four cathodic protection criteria that may be used for stray current affected pipelines in accordance with the Australian Standard, two additional criteria being evaluated internationally, and the use of the Australian Standard Appendix B2 criterion. This criterion allows the use of Pipeline Off-Potential monitoring but is inappropriate for stray current affected pipelines. Field testing results are provided to compare the criteria.

Keywords: Cathodic protection, steel pipelines, stray current, telluric current, traction current, protection criteria.

1. INTRODUCTION

The clause "Potential Criteria" in the Australian Pipeline Cathodic Protection Standard AS2832.1:2015 [1] requires that pipelines shall be considered to be stray current affected when as a result of stray current affects the difference in potential between the most positive peak and the most negative peak in potential at a test point in a 24 hour period exceeds 200mV. Under these conditions the entire pipeline protection section shall be subject to a data logger or chart recording survey of potentials. Other sections in the Standard provide alternative criteria.

AS2832.1:2015 provides the following cathodic protection criteria that may be applied to stray current affected pipelines whether traction [2] or telluric [3] affected:

a) That the pipeline potential remain negative of -850mV $Cu/CuSO_4$, with an allowance for the voltage gradient error in the measurement, for at least 95% of a 24 hour period. Due to the complexities in determining the voltage gradient error in a stray current environment the Standard allows that the error may be estimated. The voltage gradient error at -850mV $Cu/CuSO_4$ is typically of the order of 50mV. To be conservative this is commonly interpreted as meaning that the pipeline should be negative of -1000mV $Cu/CuSO_4$ On-Potential for 95% of the 24 hour period.

b) The Off-Potential determined on coupons or resistance probes is -850mV $Cu/CuSO_4$ or more negative.

c) The depolarised potential determined on a coupon or resistance probes must be at least 100mV negative of the Off-Potential.

d) The corrosion determined on resistance probes does not exceed 5μm in one year.

Three additional protection criteria considered are:

e) The mean On-Potential over a 24 hour period is -1200mV $Cu/CuSO_4$ or more negative. This criterion is presently being evaluated in Europe. [4, 5, 6, 7, 8, 9 ,10]

f) Some laboratory and field trials have been carried out in Europe using Coulometric Oxidation as a protection criterion [11, 12]. Corrosion does not proceed when the steel surface is passive. If the surface is passive the passive film becomes the electrode surface and not the steel beneath it. Therefore if a substantial positive voltage is applied to the steel, such that the potential is in the oxygen evolution range, the reaction will be water oxidation to oxygen on the passive film surface. The instant Off-Potential after such polarisation would be that of water oxidation which is positive to $Cu/CuSO_4$. If there was no passive film the anodic reaction would be corrosion of the steel and the instant Off-Potential would be that of corroding steel; typically negative to $Cu/CuSO_4$. The procedure was originally developed for assessment of AC corrosion [6]. It has been presented as a protection criterion at a course in Adelaide where delegates carried out the Coulometric Oxidation evaluation themselves [13], and in work carried out at Deakin University [10, 14].

g) AS2832.1:2015 Appendix B2 refers to the use of Pipeline Off-Potential monitoring where the Off-Potential is determined from data logs of potential at test points when the cathodic protection systems are briefly interrupted for a period of 5 seconds. The data logging typically samples at between 5 and 10 second intervals.

In the field work the difference in the off-potential determined after 5 seconds of depolarisation and 0.1 seconds of depolarisation were compared. ISO 15589-1: 2003 [15] allows the reading to be taken within 1 second providing there is no rapid depolarisation. However 0.1 seconds is the preferred time limit [16].

The testing was carried out at four test points on a pipeline in Queensland that was well coated and was subject to telluric affects. The 24 hour data logging was carried out at a 3 second sampling rate. All other loggings were carried out at a 0.1 second sampling rate. The cathodic protection systems were synchronously interrupted for 5 seconds every 10 minutes. Test points 1 and 4 had resistance probes installed. Test points 2 and 3 had temporary coupons installed 40 hours before the testing.

2. FIELD TEST RESULTS

Results for Test Point 1

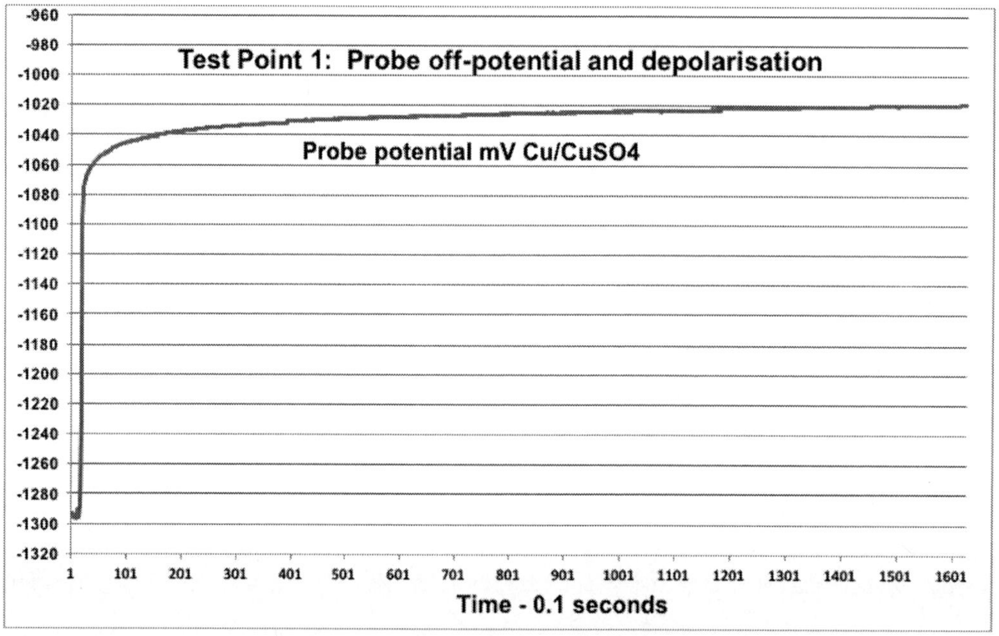

Figure 1: Test Point 1 Probe Off-Potential and Depolarisation

Figure 1 indicates that the Probe Off-Potential is approximately -1070mV $Cu/CuSO_4$, satisfying the protection criterion of -850mV $Cu/CuSO_4$. It also indicates that the probe depolarized from -1070mV $Cu/CuSO_4$ to -1020mV $Cu/CuSO_4$ in 2.7 minutes. Due to site issues it was not possible to wait for further depolarization. It is probable that it would depolarize the further 50mV in due course satisfying the protection criterion of 100mV.

The resistance probe was installed for a period of 5 months 17 days. The calculated annualized corrosion rate is 0.9μm/y; satisfying the protection criterion of less than 5μm/y.

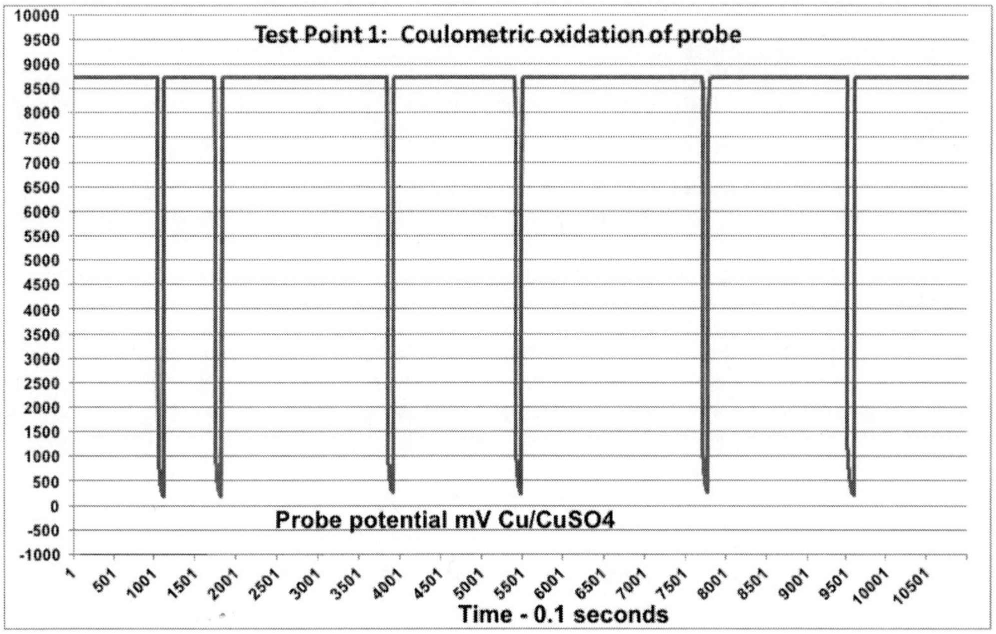

Figure 2: Test Point1 Coulometric Oxidation of Probe

Figure 2 indicates that the Off-Potential, after applying +9V to the probe, is positive to $Cu/CuSO_4$ indicating that a passive film is protecting the surface from corrosion and that the anodic reaction is water oxidation.

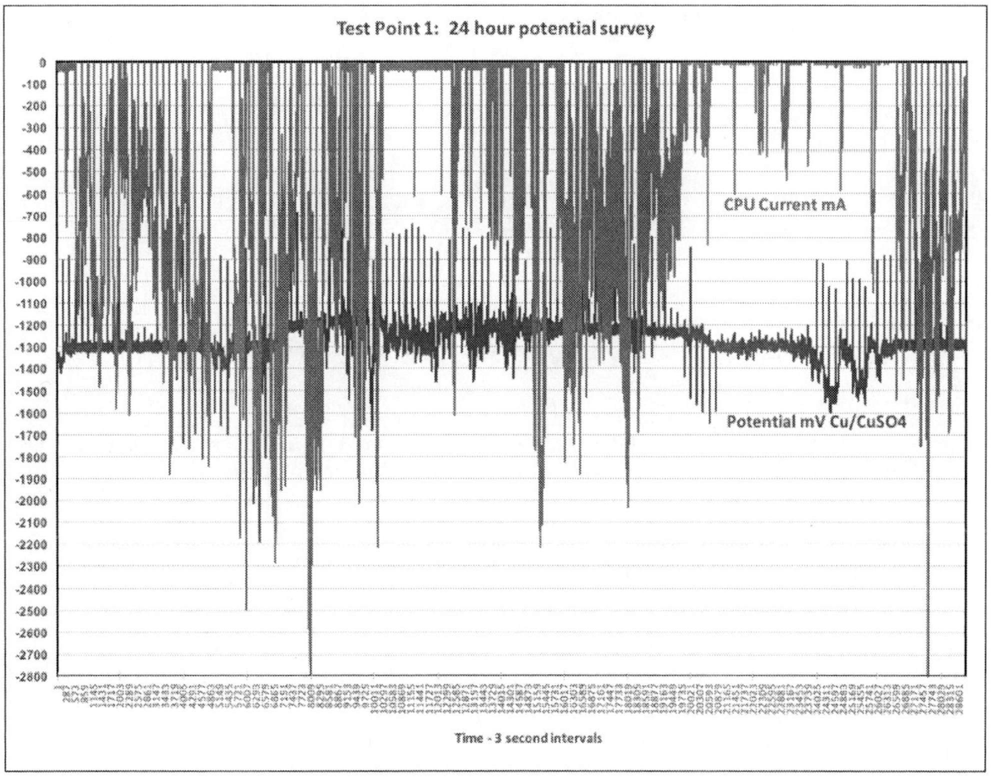

Figure 3 Test Point 1 24 hour Potential Survey

Figure 3 presents the results of the 24 hour data logging. The red graph presents the current output from the adjacent CP unit. The fluctuations in the current graph indicate that the telluric stray current effect was significant even though the On-

Potentials only varied between -1088mV $Cu/CuSO_4$ and -1567mV $Cu/CuSO_4$. The spikes in the blue potential graph are a result of the cathodic protection unit interruption. The positive spikes are the pipeline Off-Potentials that varied between -508mV $Cu/CuSO_4$ and -1088mV $Cu/CuSO_4$. The negative spikes varied between -1567mV $Cu/CuSO_4$ and -2496mV $Cu/CuSO_4$. The spikes only existed for 0.8% of the time and thus were ignored in determining compliance with the 95% protection criterion.

The results indicate that the pipeline was negative of -1177mV $Cu/CuSO_4$ for 95% of the time; satisfying the protection criterion of -1000mV $Cu/CuSO_4$.

Figure 3 also indicates that the mean On-Potential is -1270mV $Cu/CuSO_4$ satisfying the protection criterion of -1200mV $Cu/CuSO_4$.

Some of the spikes were only in the positive direction from the telluric effect, some only in the negative direction and some in both directions. The spikes in the negative direction are due to the cathodic protection unit operating in the constant potential mode over compensating when it is re-energised as shown in the graph below. Figure 4 shows the resistance probe current and pipeline potential during a 5 second CP unit interruption.

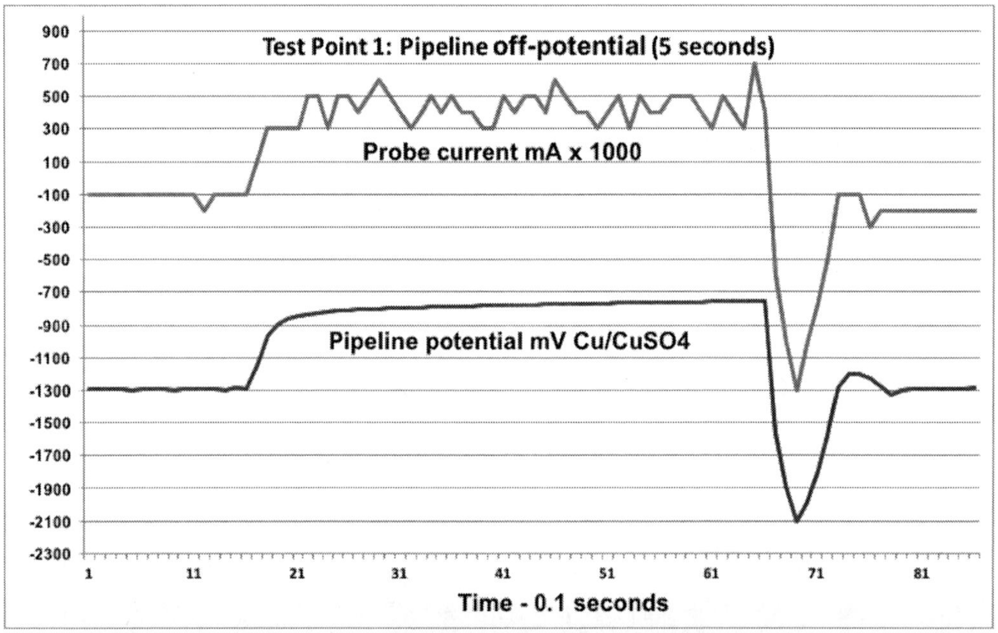

Figure 4 Test Point 1 Pipeline Off-Potential (5 seconds)

Initially the probe took 100μA to protect it to -1300mV $Cu/CuSO_4$ On-Potential. When the cathodic protection units were turned off the probe discharged 500μA of current to supply cathodic protection current to the pipeline. When the cathodic protection unit re-energised in the constant potential mode it initially over compensated and passed 1300μA to the probe driving its on-potential to -2100mV $Cu/CuSO_4$ before eventually settling down.

Spikes in the positive direction are due to the IR drop from the anodic transient not being mitigated by the cathodic protection unit. When there are no spikes it means there is a cathodic transient and so the cathodic protection unit is not outputting any current.

Figure 4 shows that a data logger sampling at 5 second intervals may give an off-potential of -760mV $Cu/CuSO_4$, whereas a data logger sampling at 0.1 second intervals would give a reading of -890mV $Cu/CuSO_4$.

Results for Other Test Points

All four test points gave similar results that, to reduce the length of this paper, are presented in tabular form in the Summary below. However the data logging graphs of pipe potential results are presented for information below.

Test Point 2
Figure 5 indicates that:
On-potentials varied between -1183mV Cu/$CuSO_4$ and -1717mV Cu/$CuSO_4$.
Off-potentials varied between -694mV Cu/$CuSO_4$ and -1183mV Cu/$CuSO_4$.

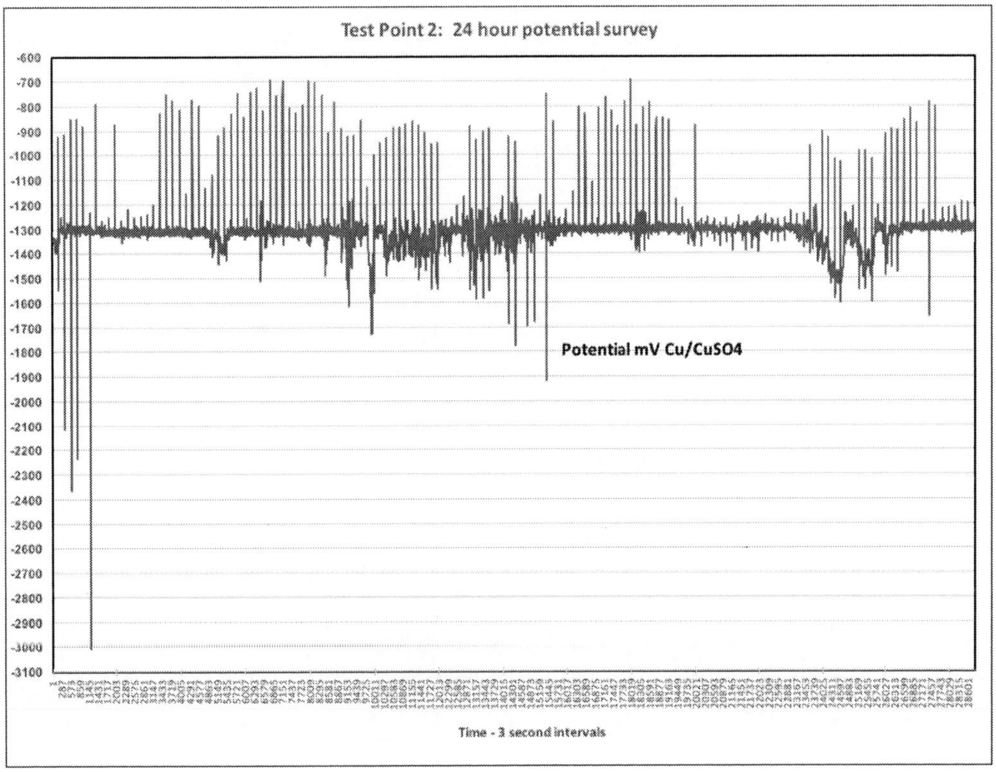

Figure 5 Test Point 2 24 hour Potential Survey

Test Point 3
Figure 6 indicates that:
On-potentials varied between -1266mV Cu/$CuSO_4$ and -1699mV Cu/$CuSO_4$.
Off-potentials varied between -814mV Cu/$CuSO_4$ and -1266mV Cu/$CuSO_4$.

Test Point 4
Figure 7 indicates that:
On-potentials varied between -1049mV Cu/$CuSO_4$ and -1430mV Cu/$CuSO_4$.
Off-potentials varied between -660mV Cu/$CuSO_4$ and -1049mV Cu/$CuSO_4$.

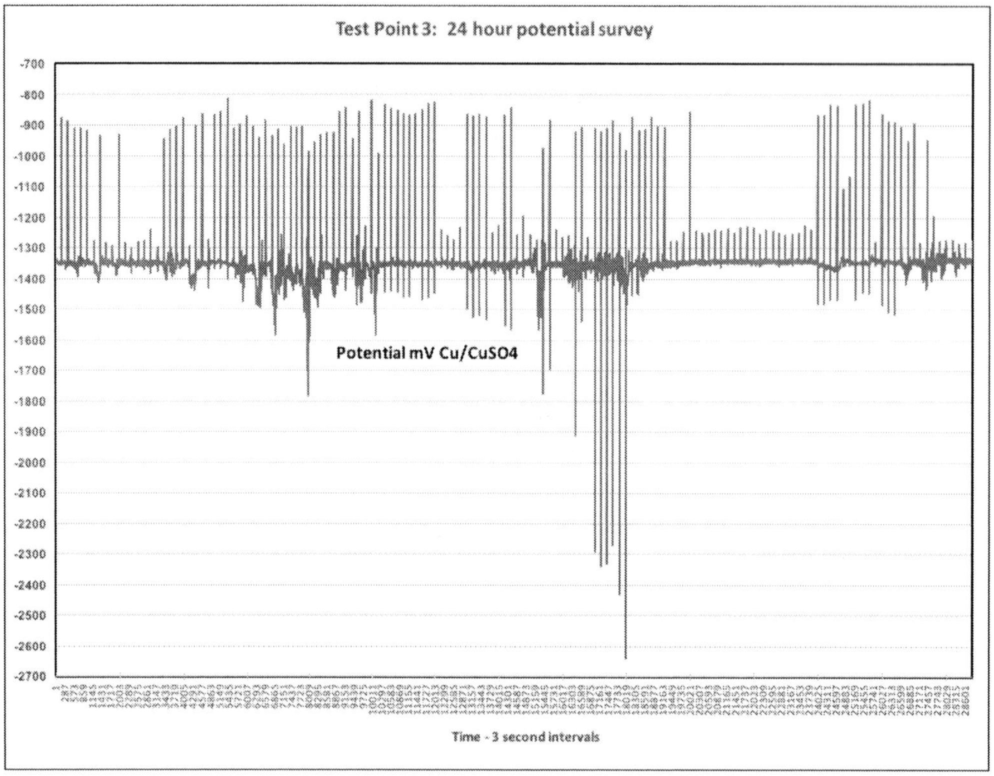

Figure 6 Test Point 3 24 hour Potential Survey

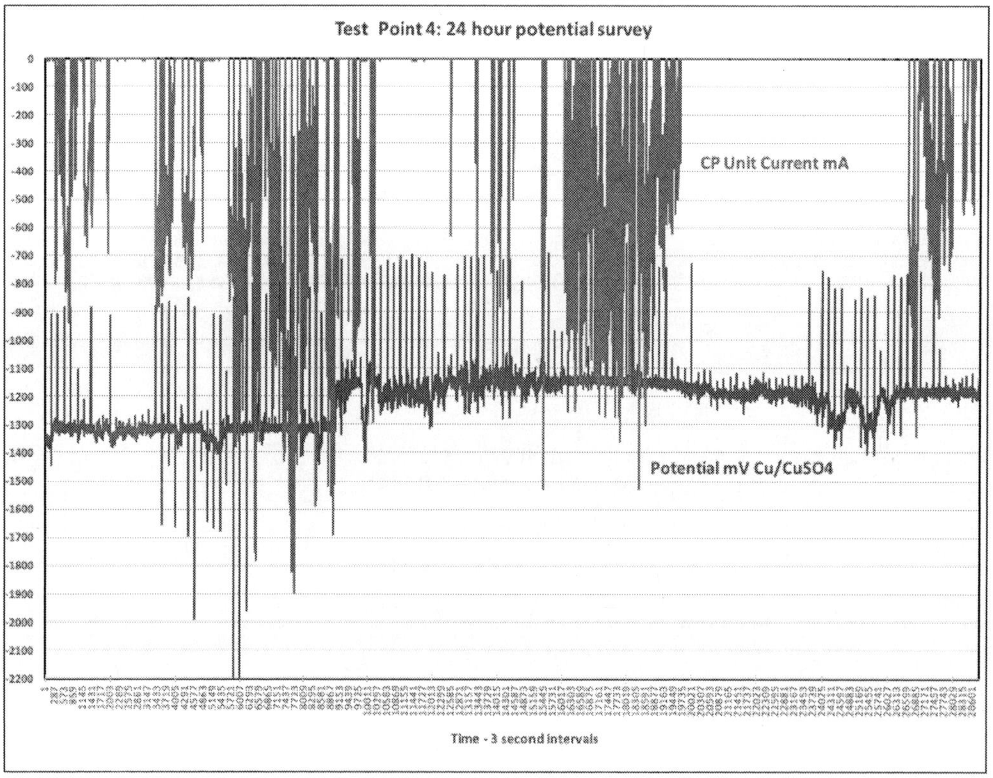

Figure 7 Test Point 3 24 hour Potential Survey

Summary of Results

Test	Criterion	Test Point 1	Test Point 2	Test Point 3	Test Point 4
Coupon off-potential	≤-850mV	-1070mV	-960mV	-1070mV	-1280mV
Coupon depolarization	≥100mV	50mV in 2.7min	100mV in 7.5min	100mV in 10min	115mV in 7.7min
Probe corrosion	≤5μm/y	0.9μm/y	N/A	N/A	2.4μm/y
Coulometric Oxidation	+ve to $Cu/CuSO_4$	+1500mV	+700mV	+500mV	+400mV
AS2832.1:2015 95% criterion	≤-1000mV	-1177mV	-1285mV	-1070mV	-1133mV
Mean on-potential	≤-1200mV	-1270mV	-1319mV	-1356mV	-1221mV
Pipeline on-potl Pipeline off-potl Spikes		-1088 to -1567 -508 to -1088 -508 to -2496	-1183 to -1717 -694 to -1183 -694 to -3010	-1266 to -1699 -814 to -1266 -814 to -2641	-1049 to -1430 -660 to -1049 -660 to -2210
Pipeline off-potl 5s and 0.1s	≤-850mV	-760mV -890mV	-800mV -925mV	-1130mV -1180mV	-900mV -1100mV

(Note: All potentials to $Cu/CuSO_4$.)

Table 1 Result Summary

3. DISCUSSION

General

There are three types of protection criteria; On-Potential, Off-Potential and direct measurements. Prior to the late 1960s only On-Potential criteria were used. Mears & Brown postulated that full cathodic protection was achieved when the local cell cathodes were polarized to the open circuit potential of the anodes [17]. Kuhn determined by practical tests that this occurred at -850mV $Cu/CuSO_4$ On-Potential [18]. This has subsequently been confirmed by others both in the field and in the laboratory [19]. However in the late 1960s there were a number of failures in Germany on a pipeline protected between -850mV and -1200 mV On-Potential. As the pipeline was in a rocky high resistivity environment the cause of the failure was attributed to the inclusion of the soil voltage gradient in the measurement circuit [20, 21]. This lead to a wide range of measurement techniques to overcome this "error" and the introduction of the Off-Potential [22, 23, 24, 25]].

Recent work has shown that if a pipeline is properly backfilled so that the alkaline environment around a defect is allowed to develop, the On-Potential criterion is satisfactory. Ohm's Law indicates that net current will flow onto defects, increase the pH and form a passive film formed on the steel surface preventing corrosion. An evaluation of the pipeline failures in Germany show that the backfill was rocky such that the soft water, in combination with bacterial activity, prevented the formation of an alkaline environment. When the backfill allows soft water to flow the pH will not provide passivation and so to provide protection the potential has to be moved into the immunity range and determined by potentials free of the voltage gradient. However this is negative of the water reduction potential so it requires very high current densities because water is not stable at these potentials. [4, 6 ,9]. Due to the calcium content of seawater it can form calcareous deposits at defects and allow a pH increase to provide protection to the steel, however in flowing soft waters providing cathodic protection requires extreme current densities and may not be achievable. A similar situation occurs on pipelines subject to bacterial corrosion; it may not be possible to increase the pH to provide passivity. Therefore temporary cathodic protection should be applied to a pipeline as soon as it is installed to increase the pH at defects and prevent the initiation of any bacterial corrosion [26].

The direct measurement criteria include electrical resistance probes and Coulometric Oxidation, neither of which relies on the determination of a protection potential.

Coupon Off-Potential

All test results complied with this criterion. However, cathodic protection polarises defects to different potentials. Larger defects and defects in well aerated higher resistivity soils can have substantially less negative potentials than smaller defects and defects in lower resistivity soils. Therefore all this criterion can indicate is that a defect of a given size in soil of a given resistivity and compaction is protected at the time of the measurement.

Coupon depolarisation

All test results complied with this criterion. However it has all of the limitations of the Coupon Off-Potential and it may take a significantly longer period of time to measure. There are also questions on its validity when applied to pipelines [27]

Resistance probe corrosion

All test results complied with this criterion. However it has all of the limitations of the Coupon Off-Potential plus the probes have a machined surface whereas the pipeline has a blast cleaned profile surface that can affect initial corrosion rates. Also work has been carried out in Europe that shows that as pipes have a thicker steel section than the probes, as corrosion proceeds the surface area of the exposed steel is increased. This decreases current density and decreases the corrosion penetration rate [28].

Coulometric oxidation

All test results complied with this criterion. However it needs more work to determine its limitations and whether it is a viable field technique.

95% Criterion

All test results complied with this criterion. This is the primary criterion that has been used successfully in Australia for decades. It is only valid if the pipeline has been properly backfilled, but this applies to all of the criteria. Its premise is that if the potential is negative of the protection criterion for most of the time, current will be flowing onto the pipeline for most of the time and thus the pH will increase and the surface made passive.

Mean On-Potential

All test results complied with this criterion. The -1200mV $Cu/CuSO_4$ criterion in Europe is also the maximum potential to avoid AC corrosion under most conditions, however -1100mV $Cu/CuSO_4$ and -1000mV $Cu/CuSO_4$ are also being considered [4]. The work at Deakin indicates that substantial periods of anodic polarision during a 24 hour period may not cause a loss of passivity and that a mean On-Potential criterion in the European range is a sufficiently conservative protection criterion [14].

Pipeline Off-Potentials

It is difficult to know whether the test results complied with this criterion because the Off-Potentials varied dramatically over short periods of time; from -508mV to -1088mV at one location. This technique is invalid for determination of protection status because:

a) If the pipeline is being subject to a positive telluric transient at the time of the measurement then turning the cathodic protection system off will include the voltage gradient error associated with the telluric transient giving a pessimistic view of the protection status. This can lead to unnecessary cathodic protection installation works.

b) The instant the cathodic protection system is turned off current flows in the soil from the more protected defects to the less protected defects. The voltage gradient error in the measuring circuit due to these equalisation currents can be of a similar magnitude to the error in the On-Potentials. This can give an optimistic view of the protection status. This can present a risk to pipeline integrity.

c) Even if the reference electrode is located directly over a defect, particularly in a high resistivity soil, a remote defect in a low resistivity soil will dominate the potential being measured due to the lower measuring circuit resistance of the defect in the lower resistivity soil. This gives an optimistic view of the protection status. This can present a risk to pipeline integrity.

d) The cathodic protection units are typically interrupted for 5 seconds every 10 minutes. The actual Off-Potential should be measured the instant the cathodic protection unit is interrupted; however the data loggers only sample at typically between 5 and 10 second intervals. The difference between taking the off reading at 0.1 seconds and 5 seconds after interruption, as shown above, can be the difference between -890mV and -760mV giving a pessimistic view of the protection status. This can lead to unnecessary cathodic protection installation works.

4. CONCLUSIONS

For cathodically protected pipelines subject to stray current effects:

- Coupons and electrical resistance probes may not provide an accurate evaluation of protection status.
- Pipeline Off-Potentials do not provide data suitable for determining protection status.
- On-Potentials recorded over an appropriate period of time do provide data satisfactory for determining protection status.
- None of the protection criteria are valid if the pipeline is not satisfactorily backfilled.

5. REFERENCES

1. Australian Standard AS2832.1: 2015 "Cathodic Protection of Metals – Pipes and Cables".
2. Martin B, "Stray current corrosion", Proceedings Australasian Corrosion Association Annual Conference 2001 Hobart.
3. Martin B, "Telluric effects on a buried pipeline", Corrosion, Vol 49, 4 pp343 1993
4. Buchler M & Martin B et al, "Cathodic protection of soil buried pipelines – a critical discussion of protection and threshold values", Materials and Corrosion, November 2016
5. Buchler M, "Cathodic Protection Strategy", Schweizerische Gesellschaft für Korrosionsschutz, 2017
6. Buchler M, "Cathodic Protection: A general discussion of the involved processes and their consequences for threshold values", Proceedings CEOCOR Conference 2010 Paper 2-13.
7. European Standard EN15280: 2013 "Evaluation of a.c. corrosion likehood of buried pipelines applicable to cathodically protected pipelines"
8. Buchler M, "Corrosion Protection Technical Report" 3R International, January 2014
9. Buchler M, "A new perspective on cathodic protection criteria promotes discussion", Materials Performance, January 2015
10. Varela F & Tan M et al "On the effect of short On-Potential transients", Australian Energy Pipeline CRC Report 2016
11. Büchler M, & Schöneich, H, "Investigation of Alternating Current Corrosion of Cathodically Protected Pipelines: Development of a Detection Method, Mitigation Measures, and a Model for the Mechanism", Corrosion 65, 578 (2009).
12. ISO 18086 Annex C "Corrosion of metals and alloys — Determination of AC corrosion — Protection criteria – Coulometric oxidation" 2015
13. Buchler M, "Advanced cathodic protection concepts course" Course proceedings Adelaide November 2015

14. Huo Y, "A study of the effects of potential excursions and the environment on the effectiveness of cathodic protection", 2016 Deakin University Doctor of Philosophy Thesis

15. International Standard ISO 15589-1: 2003 "Cathodic protection of pipeline transportation systems _ on-land pipelines"

16. Magnifico I & Soldevilla J, , "Remote monitoring instant off measurements – differences between using coupons and switching rectifiers in presence of stray currents", Proceedings CEOCOR Conference 2015 Paper 01.

17. Mears R & Brown R, Transactions Electrochemical Society: Vol 74, pp 519 1938.

18. Kuhn R, "Galvanic corrosion of cast iron pipes", Bureau of Standards Soil Corrosion Conference paper 73. Washington USA 1928.

19. Schwerdfeger W & McDorman O, "Potential and current requirements for the cathodic protection of steel in soils" Journal Res Nat Bureau Standards Vol 47, pp104 1951

20. von Baeckmann W & Baltes A, Europe & Oil 1971

21. von Baeckmann W, 12th International Gas Union Conference paper C37, 1973

22. Martin B "The ohmic component of potential measurement – laboratory determinations with a polarisation probe in aqueous environments: Materials Performance Vol 20, 1 pp52 1981

23. Martin B, "Potential measurement on buried pipelines" Proceedings Cathodic protection theory and practice – the present status: Coventry UK 1982

24. Martin B, "A method of determining pipeline polarized potentials in stray current areas using linear regression analysis" Industrial Corrosion Vol 3, 3 pp10 1985

25. Martin B, "Pipeline cathodic protection – a risk based technology" Proceedings CEOCOR Conference 2015 Stockholm, Sweden

26. Angst U et al., "Cathodic protection of soil buried steel pipelines – a critical discussion of protection criteria and threshold values", Materials and Corrosion 11, 9 (2016).

27. Joos H & Buchler M, "An objective discussion on cathodic protection criteria based on literature data" Proceedings CEOCOR Conference 2017 Luxembourg

28. Büchler M & Joos, D "AC-corrosion on cathodically protected pipelines: A discussion of the involved processes and their consequences on mitigation measures", in The European corrosion congress EUROCORR, Dechema e.V., (2016).

6. AUTHOR DETAILS

Brian Martin has been a Pipeline Corrosion Engineer for 48 years and for the last 21 years has consulted through his company Brian Martin & Associates Pty Ltd.

CORROSION MECHANISM FOR STRAY CURRENT AFFECTED CATHODICALLY PROTECTED PIPELINES

Brian Martin
Brian Martin & Associates Pty Ltd

SUMMARY: Cathodically protected pipelines subject to time varient stray current affects corrode at a significantly lower rate than indicated by a Faradaic evaluation of the current discharge. This paper reviews work by others on the reasons for this; the mechanism and the effects of the reversible ferrous/ferric reactions, and passive film formation, on corrosion rates. An evaluation of the mechanism may affect the criteria used for evaluating corrosion status of pipelines subject to variable stray current effects.

Keywords: Cathodic protection, steel pipelines, stray current, corrosion rate, passivity, protection criteria.

1. INTRODUCTION

Cathodically protected pipelines subject to variable stray current affects corrode at a significantly lower rate than indicated by a Faradaic evaluation of the current discharge. There is significant evidence for this:

a) The Australian Standard for cathodic protection of pipelines [1] provides a criterion that in a stray current area the pipeline can be positive of the protection criterion for 5% of the time without causing corrosion. This criterion has been successfully used in Australia for decades.

b) In Europe a mean On-Potential criterion over a 24 hour period of -1200mV Cu/$CuSO_4$ is presently being evaluated [2, 3, 4, 5, 6, 7, 8]. However mean On-Potential criteria of -1100mV Cu/$CuSO_4$ and -1000mV Cu/$CuSO_4$ are also being considered [2]. Therefore the pipeline can be positive of the On-Potential criterion for 50% of the time without causing corrosion.

c) A substantial amount of work has been carried out at Deakin University that shows a significant anodic transient can be applied to a cathodically protected pipeline without causing corrosion. Figure 1 is for a coupon in saturated sand that was protected to -950mV Cu/$CuSO_4$ for 24 hours and then had an 800mV anodic potential change applied. The graph of current shows that the passive film on the coupon maintained protection of the steel for 62 minutes before passivity broke down and the corrosion current increased [9]. The work also showed that when the anodic potential change applied was 500mV then 23 hours elapsed before the corrosion current increased.

Figure 1 Graph of corrosion current against time on a -950mV Cu/$CuSO_4$ cathodically protected coupon with a +800mV potential change applied.

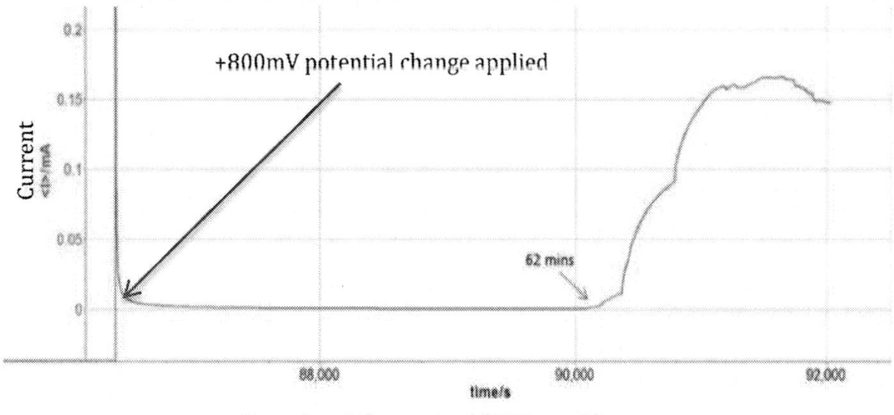

The effect of a pipeline losing its cathodic protection over a period of time, when subject to an anodic transient, is termed depolarization. The depolarization mechanism can be applied to potential transients from traction [10] and telluric [11] effects. It can also be applied to AC interference [12, 13] albeit at a higher frequency than DC stray current. The mechanism is presented in terms of Pourbaix Diagrams. While these present thermodynamic rather than kinetic data they are in more common use in the cathodic protection industry, primarily because cathodic protection tends to be a long term steady state condition reflecting thermodynamic conditions.

2. MECHANISM

Cathodic protection of steel is typically achieved by passivation due to an alkaline environment. It is rarely achieved by immunity due to the very high current density required to polarize the steel negative of the hydrogen evolution line where water is unstable [2, 4, 7]. In a stray current environment the potential may change rapidly but the change in pH at a defect on a pipeline in compacted backfill is a relatively slow diffusion process [9].

Figure 2 Pourbaix Diagram for iron in the passive range
(The potentials are electrode potentials and do not include the voltage gradient errors that may occur in field measurements.)

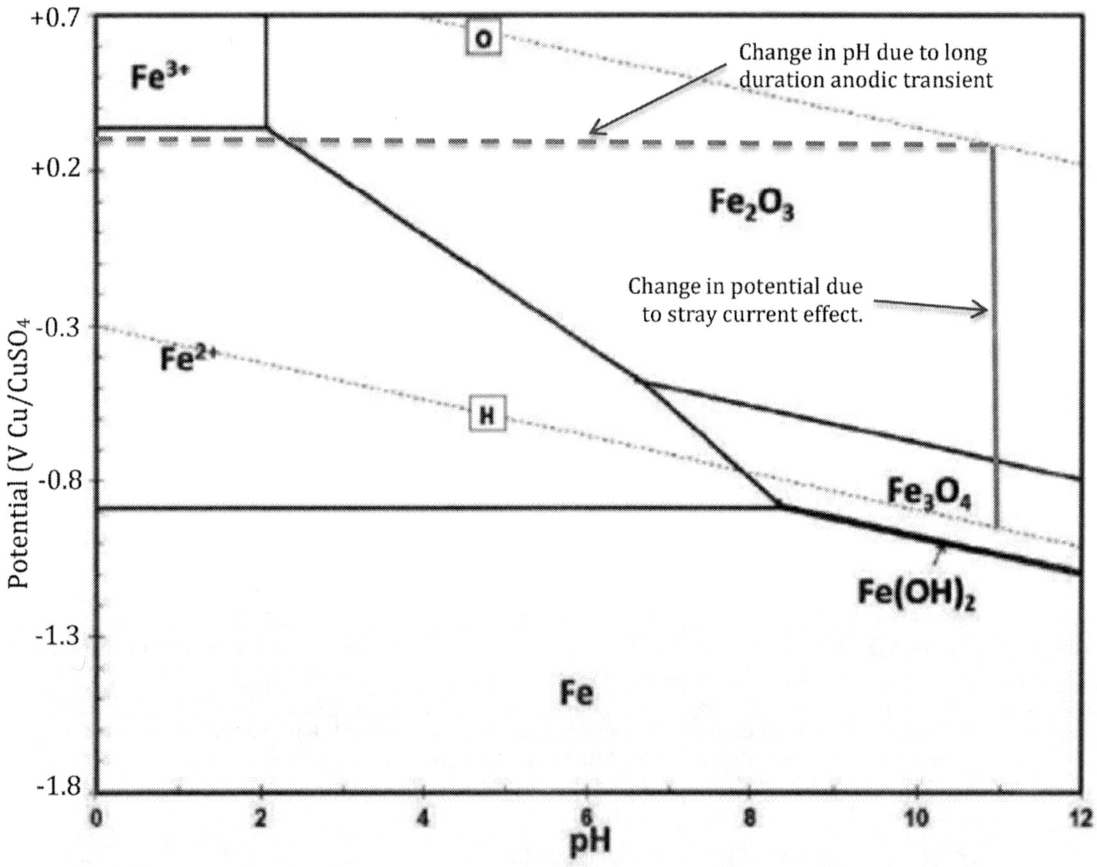

Figure 2 is a Pourbaix Diagram showing the effect of stray current on a cathodically protected pipeline. The cathodic protection provided an environment with a pH of 11. When a cathodic transient occurred the potential went as negative as the hydrogen line. It rarely goes significantly negative of the hydrogen line due to the very high current density required to polarize the steel negative of the hydrogen evolution line where water is unstable. During this time the reaction on the steel surface is the cathodic protection reaction that generates hydroxyl ions to maintain the high pH. The reactions are:

Equation 1: $2H_2O + O_2 + 4e^- > 4OH^-$ for near neutral conditions, or
Equation 2: $2H_2O + 2e^- > 2OH^- + H_2$ for acidic conditions.

When an anodic transient occurs the potential of the pipeline can go as positive as the oxygen line. It rarely goes positive of the oxygen line due to the very high current density required to polarize the steel positive of the oxygen evolution line where water is unstable. During this time the reaction on the steel generates hydrogen ions that can reduce the pH:

Equation 3: $2H_2O > 4H^+ + O_2 + 4e^-$

During this time there is no corrosion occurring because the steel remains passive due to the insoluble passive films in the Fe_3O_4 and Fe_2O_3 regions of the diagram [12, 13]. If however the positive transient remains for an extended period of time the pH will decrease and follow the dashed line into the soluble Fe^{+2} corrosion region of the diagram. This can take an extended period of time before it takes effect as indicated in Figure 3 where at Deakin University a 10mm diameter coupon was polarized to -1000mV Cu/$CuSO_4$ in 41Ωm saturated sand for 5 hours [9]. A 1000mV positive potential change was then applied. Corrosion current only increased after a period of 3 hours.

This is the basis of the European work on using the mean On-Potential as the protection criterion. A sufficiently negative mean on-potential means a mean cathodic current and hence net increase of pH. Due to the slow mass transport conditions this mean pH increase generates the conditions for passivity that are maintained during temporary anodic excursions. If the steel surface is passivated by the alkaline environment, and the mean gain and loss of hydroxyl ions due to stray current over a 24 hour period is zero, then passivity remains.

Figure 3 Depolarisation test

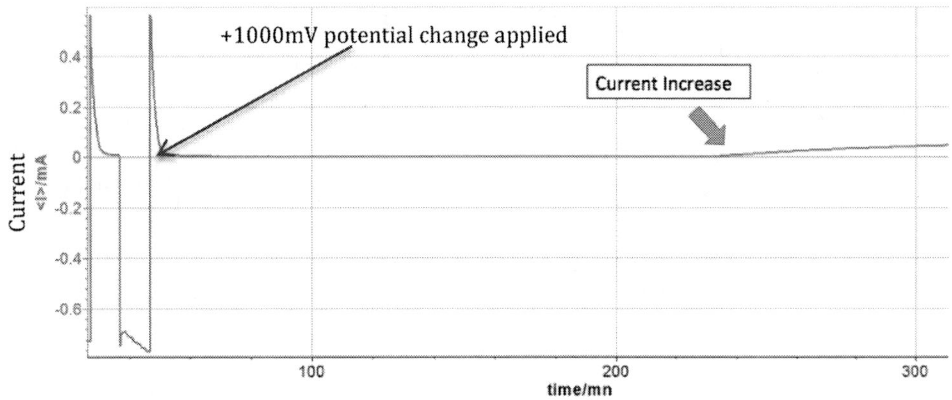

Figure 4 Coulometric oxidation

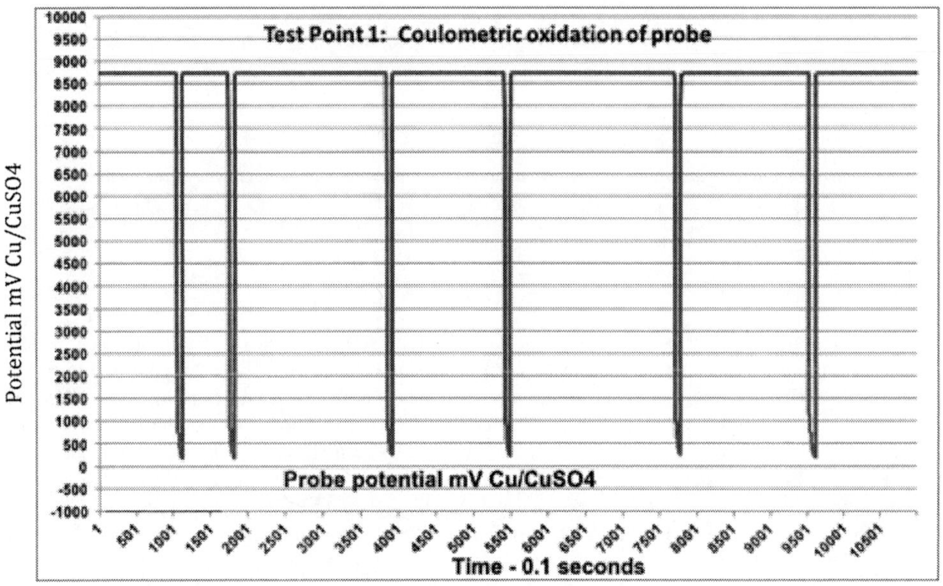

Corrosion does not proceed when the steel surface is passive. However passivity can occur at any pH above 2 and at any potential positive of -850mV Cu/$CuSO_4$ under certain circumstances as is indicated in the Pourbaix Diagrams. If the surface is passive the passive film become the electrode surface and not the steel beneath it [9, 12, 13, 14]. Therefore if a substantial positive voltage is applied to the steel, such that the potential is in the oxygen evolution range, the reaction will be water oxidation to oxygen. In this case an instant Off-Potential would be that of water oxidation; positive of the oxygen line and thus positive to $Cu/CuSO_4$. If there was no passive film the anodic reaction would be corrosion of the steel and the instant Off-Potential would be that of corroding steel; typically negative to $Cu/CuSO_4$ [9, 12, 14]. Figure 4 shows this technique applied in the field to a resistance probe on a cathodically protected pipeline in a stray current environment [15]. Voltages of +9V were applied to the probe. The instant Off-Potentials recorded were approximately +250mV $Cu/CuSO_4$ indicating that there was a passive film. The Off-Potential measured on the probe reflects the actual physical changes on the bare steel surface.

Figure 5 Pourbaix Diagram for iron in the active/passive range
(The potentials are electrode potentials and do not include the voltage gradient errors that may occur in field measurements.)

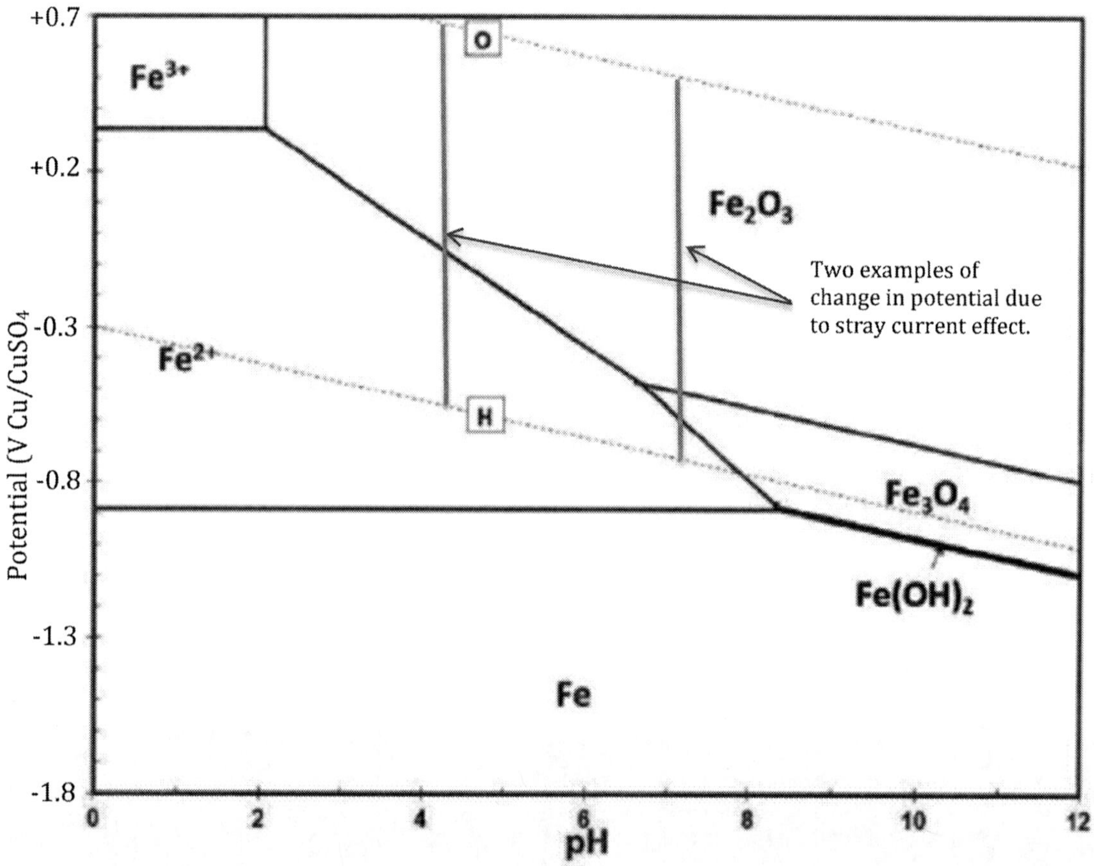

Figure 5 is a Pourbaix Diagram showing two examples of the effect of stray current on a pipeline that would not be cathodically protected to present criteria, but would not be corroding when the steel remains passive due to the insoluble passive films in the Fe_3O_4 and Fe_2O_3 regions of the diagram. However when a cathodic transient occurs the steel may move down into the soluble Fe^{+2} corrosion region of the diagram. Corrosion does not necessarily occur under these conditions. When the cathodic transient occurs the passive Fe^{III} film reacts to form Fe^{II}. When there is a anodic transient the Fe^{II} oxidises to Fe^{III} to add to the passive film. Under these conditions there is no corrosion, instead there is just oxidation and reduction occurring between the Fe^{II} and Fe^{III} valency states [4, 12, 13] allowing anodic transients to pass without causing corrosion. This can explain the lack of corrosion observed when the potentials are not in the cathodically protected range [16, 17] and the pH is not alkaline [4, 18]. However the capacity to maintain protection is significantly reduced at the neutral to acidic conditions where there is little or no buffer to support the passive film.

Figure 6 Pourbaix Diagram for iron in the passive/immunity range
(The potentials are electrode potentials and do not include the voltage gradient errors that may occur in field measurements.)

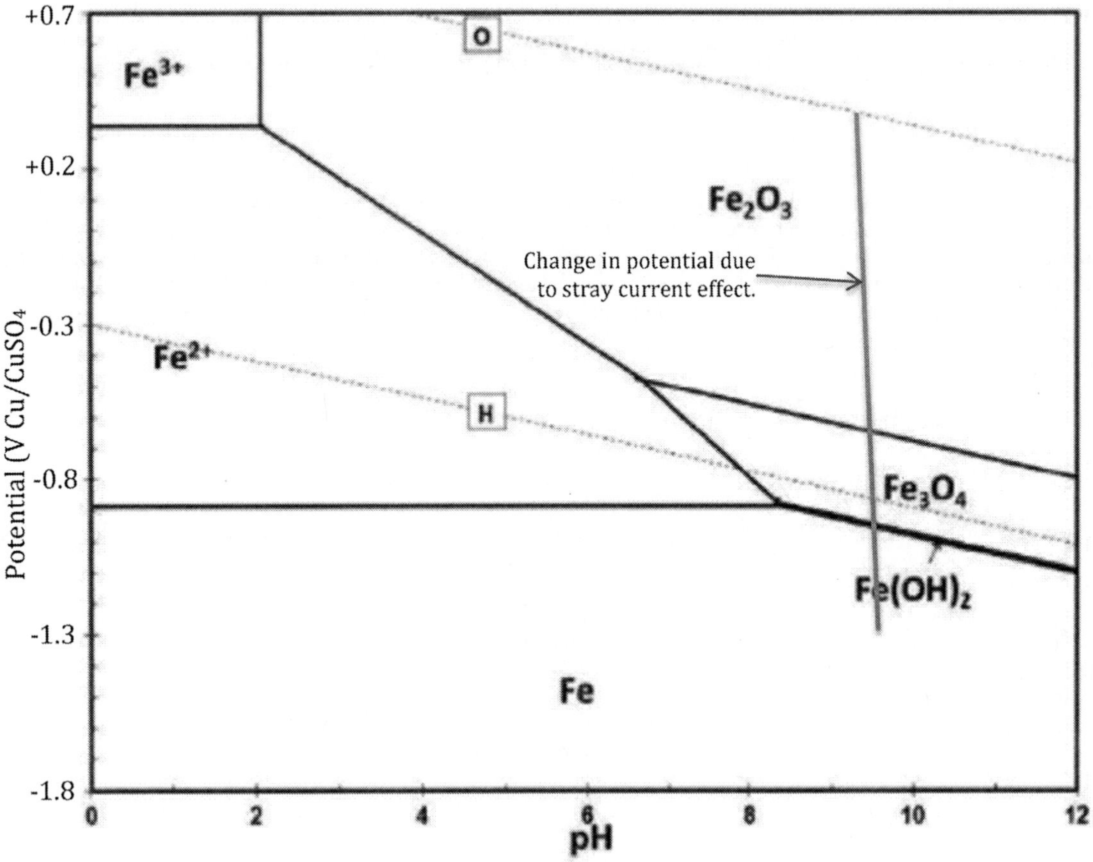

Figure 6 is a Pourbaix Diagram showing the effect of stray current on a pipeline when the potential varies between passivation and immunity. In this example the conditions comply with cathodic protection criteria but corrosion will occur. When the steel is in the Fe_3O_4 and Fe_2O_3 regions of the diagram the insoluble passive films are formed. When, under high current density conditions the potential moves negative of the hydrogen line into the $Fe(OH)_2$ and immune Fe regions, the passive film is reduced to $Fe(OH)_2$ which forms a porous rust layer [19]. When the potential moves back into the passive region there is corrosion to reform the passive film. The corrosion may only be one iron atom deep across the surface, which is not substantial and may be regarded as inconsequential for DC interference, however for AC interference where the frequency of change from immunity to passivity is high the corrosion rate can be of the order of 100mm per year [19].

3. DISCUSSION

The corrosion behavior of cathodically protected pipelines subject to variable stray current effects is complex. From an electrochemical view there is no simple pH or electrode potential criterion that will define the areas where corrosion does not occur. The amplitude and duration of the transients, both anodic and cathodic, can determine whether corrosion will occur. There are three domains where corrosion will not occur:

- If the electrode potential remains in the immunity area. However this requires a very high cathodic protection current density and no anodic transients that would take the potential into the corrosion or passive range.
- If the electrode potential remains between the hydrogen and oxygen lines, and the pH is above 9, corrosion will not occur. However if the duration of the anodic transient is substantial, or the cathodic transient goes into the immunity range, then corrosion can occur.

- If the electrode potential remains between the hydrogen and oxygen lines, and the pH is below 9 (and above approximately 2) corrosion will not occur however the capacity to maintain protection is significantly reduced in the neutral to acidic conditions where there is little or no buffer to support the passive film.

If we ignore AC interference and thus discount corrosion that occurs when the electrode potential moves between immunity and corrosion or passivation, and we ignore maintaining the potential in the immunity region due to the high current densities required, the matter is simplified: Corrosion is prevented if the steel surface spends sufficient time in the anodic passive range to provide sufficient Fe^{III} to allow for its reduction to Fe^{II} in the cathodic range before corrosion occurs.

Apart from testing for passivity using coulometric oxidation, there is no test that can indicate protection under these conditions, and coulometric oxidation is not a viable field technique [15]. In practical terms a solution would be to ensure that the pH of the steel surface remained sufficiently alkaline to maintain the passive film. If the On-Potential remained negative of the protection criterion then by Ohms Law current would charge onto the steel surface, which would generate alkaline conditions. Anodic transients would only take the steel surface further into the passive range which would not cause corrosion providing the transients were not so long as to reduce all of the Fe^{III} to Fe^{II}, However work has shown that under typical stray current conditions the On-Potential only has to be negative of the protection criterion for 50% of the time [15].

4. CONCLUSIONS

Pourbaix Diagrams can be used to consider the physical conditions on a cathodically protected steel surface because cathodic protection tends to be a long term steady state condition reflecting thermodynamic conditions. Stray current anodic transients are kinetic and if of short duration and do not affect the state of the passivating layer. Cathodic transients into the corrosion immunity region can cause corrosion over short periods of time to reform the passive layer, however the corrosion rate is low for DC interference but can be substantial for AC interference due to its higher frequency.

There is no practical test that can determine whether a pipeline is protected against corrosion under all conditions in a stray DC environment. This is because the Off-Potential of a passive surface is irrelevant for the assessment of its corrosion behavior. Also anodic current does not indicate a corrosion risk for a passive surface since the electrical charge is consumed by oxidation of Fe^{II} to Fe^{III} and not Fe to Fe^{II}. The most effective criterion is that the mean On-Potential of the well backfilled pipeline is negative of an On-Potential criterion for at least 50% of the time providing that none of the anodic transients are of substantial duration.

5. REFERENCES

1. Australian Standard AS2832.1: 2015 "Cathodic Protection of Metals – Pipes and Cables".
2. Buchler M & Martin B et al, "Cathodic protection of soil buried pipelines – a critical discussion of protection and threshold values", Materials and Corrosion, November 2016
3. Buchler M, "Cathodic Protection Strategy", Schweizerische Gesellschaft für Korrosionsschutz, 2017
4. Buchler M, "Cathodic Protection: A general discussion of the involved processes and their consequences for threshold values", Proceedings CEOCOR Conference 2010 Paper 2-13.
5. ISO 18086:2015 "Determination of AC corrosion -- Protection criteria"
6. Buchler M, "Corrosion Protection Technical Report" 3R International, January 2014
7. Buchler M, "A new perspective on cathodic protection criteria promotes discussion", Materials Performance, January 2015
8. Varela F & Tan M et al "On the effect of short On-Potential transients", Australian Energy Pipeline CRC Report 2016
9. Huo Y, "A study of the effects of potential excursions and the environment on the effectiveness of cathodic protection", 2016 Deakin University Doctor of Philosophy Thesis
10. Martin B, "Stray current corrosion", Proceedings Australasian Corrosion Association Annual Conference 2001 Hobart.

11. Martin B, "Telluric effects on a buried pipeline", Corrosion, Vol 49, 4 pp343 1993

12. Buchlker M & Schoneich H, "Investigation of AC corrosion of cathodically protected pipelines: the development of a detection method, of mitigation measures, and a model for the mechanism" Proceedings CEOCOR Conference 2010

13. Buchler M & Voute C et al, "Effect of variation of cathodic protection level over time on the AC corrosion process" Proceedings CEOCOR Conference 2012 Lucerne Paper 19.

14. ISO 18086 Annex C "Corrosion of metals and alloys — Determination of AC corrosion — Protection criteria – Coulometric oxidation" 2015

15. Martin B, "Cathodic protection criteria for stray current affected pipelines", Proceedings Australasian Corrosion Association Annual Conference 2017 Sydney.

16. Thomson N & Barlo T. "Fundamental process of cathodically protecting steel pipelines" Gas Research Conference USA 1983

17. Barlo T, "Field testing the criteria for cathodic protection of buried pipelines" PRCI Report PR-208-163 1994

18. Buchler M & Stalder F et al "A new electrochemical method for the detection of AC corrosion", Proceedings CEOCOR Conference 2004

19. Büchler M, & Schöneich, H, "Investigation of Alternating Current Corrosion of Cathodically Protected Pipelines: Development of a Detection Method, Mitigation Measures, and a Model for the Mechanism", Corrosion 65, 578 (2009).

6. AUTHOR DETALS

Brian Martin has been a Pipeline Corrosion Engineer for 48 years and for the last 21 years has consulted through his company Brian Martin & Associates Pty Ltd.

FINDINGS OF A 7 YEAR FIELD STUDY OF CONCRETE SEWER PIPE CORROSION

T. Wells, R. Melchers
Centre for Infrastructure Performance and Reliability,
The University of Newcastle, Callaghan, NSW 2308, Australia

SUMMARY: The long term corrosion of concrete sewer infrastructure costs the Australian water industry 100's of millions of dollars per year. However until recently little information was known about how the corrosion process progresses over time and how different environmental factors impact on the rate of deterioration. This information is vital in predicting the time to failure of existing reinforced concrete sewer systems, and also assessing strategies for effectively monitoring sewer corrosion potential and to minimise corrosion rates in the sewer. A recently completed 7 year field trial examined the corrosion behaviour of new and aged sewer pipe in Australian gravity sewers. The work enabled a detailed picture of the corrosion life of a pipe to be developed and led to a validated predictive model which is applicable over a wide range of sewer conditions. The findings of the study and the subsequent model are discussed in this paper.

Keywords: Concrete corrosion, sewers, microbial, model.

1. INTRODUCTION

The cost of managing concrete sewer pipe corrosion globally is estimated to be billions of dollars per year (Kaempfer and Berndt, 1999). In the USA alone the annual cost of sewage infrastructure corrosion is estimated to be approximately US$13.8 billion (Appendix K, Koch et al., 2001) while in Germany the cost of repairing sewer pipe damaged as a result of sulphuric acid attack is estimated to be in excess of $40billion (Hewayde et al., 2006).

Microbiologically induced corrosion (MIC) is primarily responsible for the internal corrosion experienced by concrete sewer pipe. MIC of reinforced concrete sewer pipe was first observed in the early 20th century by Olmstead and Hamlin, (1900) however it was not until the mid-1940s that the role of bacteria in the corrosion process was recognised (Parker, 1945a; Parker, 1945b). Later research revealed that many species of bacteria and fungi are involved (for example see Gu et al., 1998 and Nica et al., 2000). The reduction in biologically toxic metals entering the wastewater that occurred as a result of the introduction of clean water legislation in many countries in the 1970's combined with an increased use of high sulphur detergents and increased residence times in lengthening sewers has seen an increase in MIC activity over recent decades (Mansfeld et al., 1991; Milde et al., 1983; Sand et al., 1992).

The challenge posed by the increasing influence of MIC on concrete sewer pipes has prompted an increased international research effort to better understand and quantify the corrosion process (for example see Fujimoto et al., 2013; Herisson et al., 2013; Nielsen et al., 2014). Until recently however, the lack of a sound understanding of how the corrosion process evolves over the life of the sewer pipe and the link between sewer conditions and corrosion rates has hindered our ability to predict corrosion rates and to develop operational procedures that can minimize corrosion rates and thereby prolong the service life of existing concrete sewer networks (Wells and Melchers, 2011; Wells and Melchers, 2014).

A 7 years field study of concrete sewer pipe corrosion has recently been completed which has afforded a better understanding of the corrosion process inside working sewers. A first pass model enabling corrosion prediction has also been developed from knowledge of the biological and physical processes taking place. The findings, model and strategies for corrosion minimization are discussed in this paper.

2. METHODOLOGY

In situ corrosion of concrete sewer pipe was studied over a 7 year period (2010-2016) in a number of working gravity sewers located in Perth, Melbourne, and Sydney. The Perth sewers are part of a sealed system operating at very high H_2S concentrations and in which the concrete pipes are predominantly internally lined for protection. The sewers in Sydney and Melbourne, consisting of unlined concrete pipes, are naturally ventilated and operate t much lower H_2S levels. In each of these bare concrete test coupons were exposed, as described below. The test locations were carefully selected to enable concrete corrosion to be observed under a wide range of conditions, ranging from relatively benign to very aggressive. Table 1 lists the daily average sewer gas humidity, temperature and H_2S concentrations over the study period. It should be noted, however, that conditions varied over diurnal and seasonal time scales. This was particularly the case for H_2S (for example see Figure 1).

Table 1. Daily averaged conditions at the corrosion field sites (note: humidity averages for the two Melbourne sites is based on limited data)

	Average Daily conditions		
Site	H_2S level (ppm)	Humidity (%)	Temperature (°C)
Sydney North	3	94	22
Sydney South	2	94	21
Melbourne South Eastern Trunk Sewer (SETS)	2	100	20
Melbourne Western Trunk Sewer (WTS)	6	100	21
Perth Main Sewer	116	97	25
Perth Bibra Lake	423	91	27

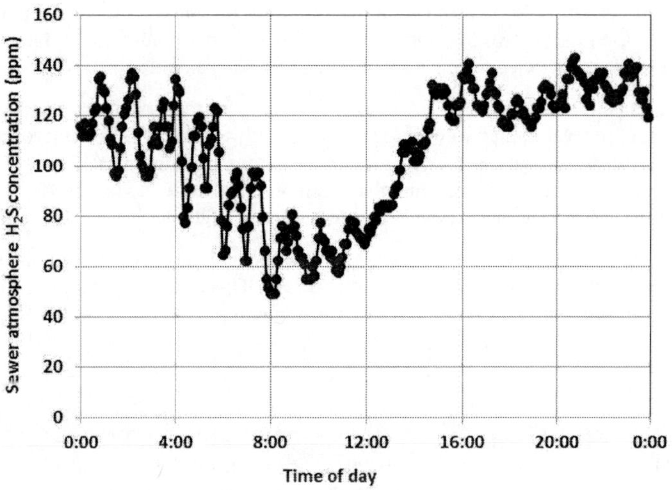

Figure 1 Diurnal variation in Perth Main Sewer H_2S concentration (20 March 2012)

Two sets of concrete coupons were exposed at each site. One set, ('new' coupons), were cut from newly manufactured 1.2m ID spun cast standard reinforced concrete sewer pipe with a 45-50% aggregate content. A second set, ('old' coupons), were cut from previously corroded reinforced concrete slabs that had served for 70 years as covers on a concrete sewer pit that was part of a typical operating sewer system carrying domestic, industrial and trade waste. New and old coupons exhibited a similar mineralogical composition, however, the old coupon alkalinity (as determined by the method of Snell and Ettre (1972)) was lower (0.15g $CaCO_3$/g concrete compared to 0.22g $CaCO_3$/g concrete) as some of the old coupon portlandite had previously been converted to carbonated forms (calcite and vaterite). The porosity of the old coupon concrete however was higher (11.4% versus 6% for new coupon concrete). Study of the new coupons enabled the corrosion process to be examined at the beginning of the pipe's service life while observation of the old coupon deterioration allowed more advanced corrosion behaviour to be investigated. Combined, the study of the two coupons allowed a picture of sewer pipe corrosion to be developed over the lifespan of the pipe.

Pairs of coupons comprising one new and one old sample were embedded in epoxy resin in stainless steel sample containers allowing side-by-side comparison of the corrosion behaviour of the two concrete types to be made. Samples at the Perth sites were mounted in an inverted position in stainless steel racks fitted to manholes at each site (Figure 2). At each of the Sydney and Melbourne sites the stainless steel sample containers were fixed in an inverted position along the crown of the sewer with the exposed surface facing down (Figure 3).

Figure 2 Samples mounted at Perth Bibra Lake.

Figure 3 Samples mounted at the crown of the Melbourne South Eastern Trunk Sewer.

At each site extensive environmental monitoring was undertaken to establish the average conditions as well as the range of diurnal and seasonal variations in sewer gas temperature, humidity and H_2S concentration in the sewer gas stream. H_2S concentrations and sewer gas temperatures were logged every 5 minutes over a 2 week period every 3 months throughout the first 4 years of the study period. After 4 years conditions were monitored in the 2 weeks prior to coupon removal. Sewer humidities were recorded at the two Perth sites at the same frequency as the other variables. However, at the Melbourne and Sydney sites repeated sensor failure meant that humidity measurements were relatively sparse. For this reason, humidity averages reported for the two Melbourne sites are based on a small pool of data (2-4 days) while humidity observations at the Sydney sites are restricted to four records each of 2 weeks duration.

Individually numbered coupon pairs were recovered every 6 to 8 months for the first 3 years and annually thereafter. Immediately after samples were recovered they were inspected under a low magnification microscope for changes in surface morphology and the average exposed surface pH determined from a minimum of 4 spot measurements. Finally, the depth of the corrosion product layer that had formed and the depth of sound concrete removed during the time of exposure were determined using photogrammetric analysis (see Wells and Melchers (2014).

Field observations were used in conjunction with a knowledge of the behaviour of bacteria implicated in causing corrosion to develop a better quantitative understanding of the corrosion process. This allowed a phenomenological model relating concrete corrosion loss in sewer pipes to local average sewer gas temperature, humidity and H_2S concentration to be developed.

3. RESULTS

Coupons were installed during the period September 2009 to June 2010. Melbourne WTS and SETS coupons were recovered after 7, 13, 18, 24, 32, 39, 63, and 72 months exposure with a further recovery at 84 months at Melbourne SETS only. Perth Main Sewer and Bibra Lake coupons were recovered after approximately 6, 14, 19, 26, 31, 37 and 48 months with a further recovery after 58 months at the Perth Bibra Lake site. Sydney coupons were recovered after 9, 12, 24, and 36 months exposure. All results reported here are the average values for each set of 2-3 coupons retrieved at each recovery. The trial at the Sydney sites was unfortunately stopped after 36 months due to an unintended contamination of the coupons.

3.1 Evolution of concrete surface pH

The corrosion of concrete sewer pipe is primarily brought about by the secretions of acidophilic (acid loving) bacteria (Islander et al., 1991). The species composition, population density and activity of the bacterial colonies existing on the exposed surface of the pipe are strongly influenced by the pH of the local environment, that is, the exposed surface of the pipe or coupon. At the time of installation, the surface of a concrete sewer pipes surface is highly alkaline (pH >10) and consequently there is a minimal presence of acidophilic bacteria. However over time the dissolution of CO_2 and H_2S into condensate lining the upper reaches of the sewer pipe surface produces acids that neutralize alkaline minerals such as Portlandite in the concrete matrix and reduce the overall surface pH of the pipe. As the surface pH falls, so does the nature, population and activity of the bacterial and fungal communities colonizing the pipe surface and consequently the dynamics of the corrosion process also change. Surface pH therefore serves as a useful indicator of the progress the corrosion process.

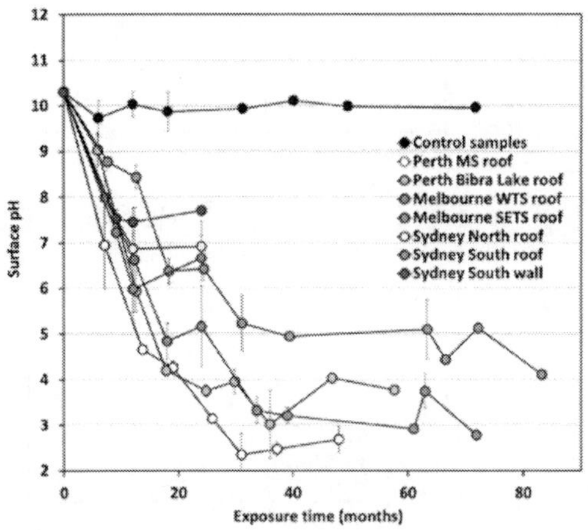

Figure 4 Evolution of new coupon surface pH

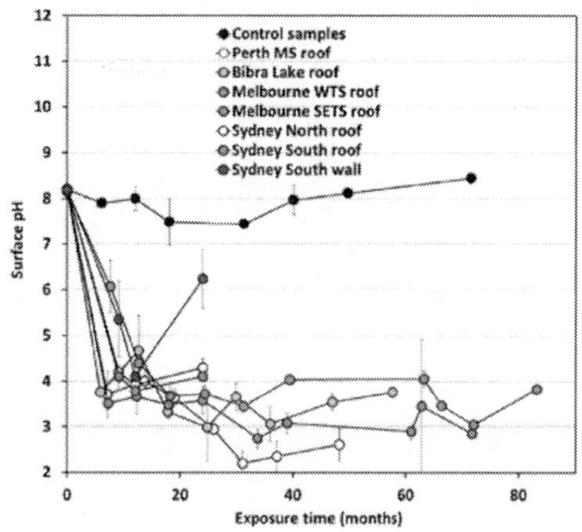

Figure 5 Evolution of old coupon surface pH

Figures 4 and 5 illustrate how the surface pH of the new and old coupons changed over time at each site. The surface pH of new coupons was pH=~10 at the commencement of the study. Over the first 18 months of exposure new coupons experienced a moderately rapid decline in surface pH values (0.3 to 0.35 pH units per month) at all sites except Melbourne SETS where the decline was somewhat slower (~0.2 pH units per month). After 18 months the rate of decline slowed at all sites so that after ~36 months the surface pH had stabilized at between pH=2.5 (Perth Main Sewer) to 5.0 (Melbourne SETS). Over the same period no significant variation was observed in control samples (average pH=10.0 +/- 0.3) held in a dry, warm environment at the Newcastle University laboratories.

The starting surface pH for old coupons was significantly lower (pH = ~ 8), a value consistent with the presence of various carbonated forms of the original portlandite mineral content. The decline of old coupon surface pH was more rapid than that of the new coupons such that after 6-12 months of exposure the surface pH at most sites had fallen to final, stable values of between pH=2.2 (Perth Main Sewer) and pH=4 (Melbourne SETS). Over the same period old concrete control samples showed little variation in surface pH (pH=8 +/- 0.5).

3.2 Corrosion Losses

Corrosion losses as a function of time for new and old coupons are shown in Figures 6-9. No data is shown for Sydney as the trial was stopped before any significant corrosion could take place.

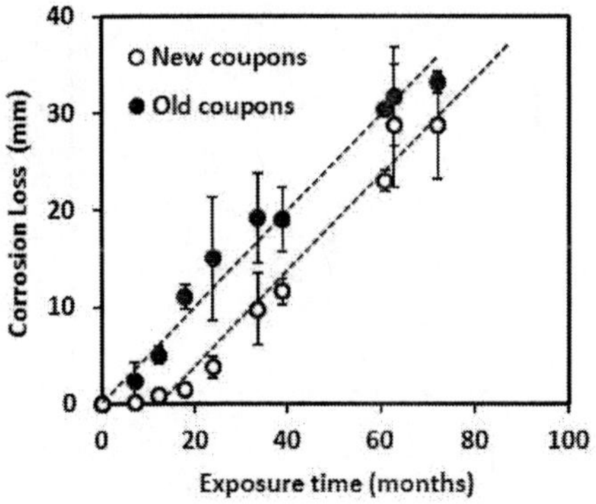

Figure 6 Melbourne WTS corrosion losses as a function of exposure period.

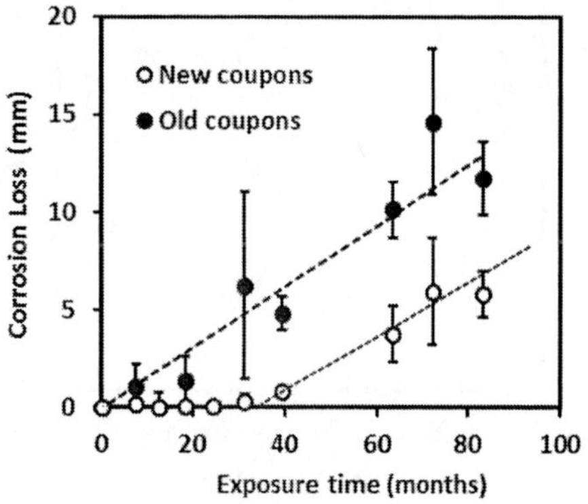

Figure 7 Melbourne SETS corrosion losses as a function of exposure period.

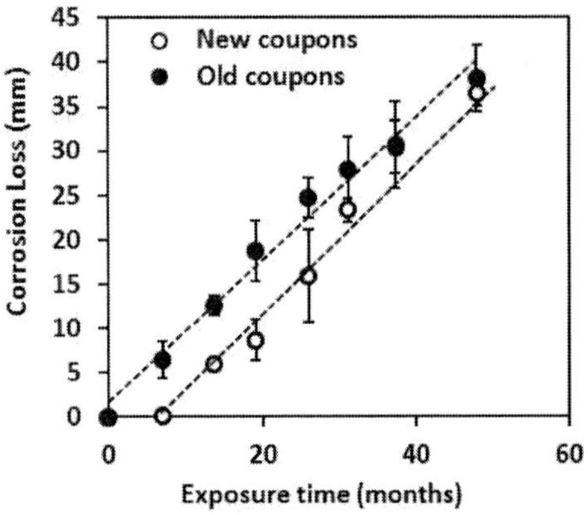

Figure 8 Perth Main Sewer corrosion losses as a function of exposure period.

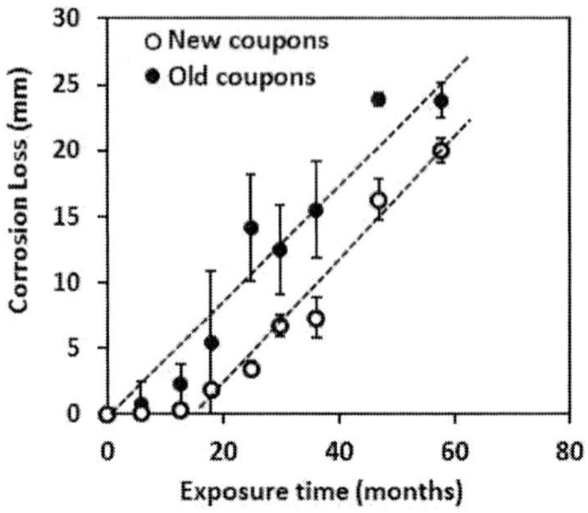

Figure 9 Perth Bibra Lake corrosion losses as a function of exposure period.

3.2.1 New coupons

The general pattern of mass loss experienced by new coupons, (open circle data in Figures 6-9), was the same at all sites Following installation of samples in the sewer, new coupons passed through an initial "induction" period during which no losses. This period lasted between 9 months (Perth Main Sewer) and 25 months (Melbourne SETS). New coupons at the two Sydney sites had experienced negligible corrosion by the time the trials there were terminated (data not shown). Given this it is reasonable to assume that the induction period was > 36 months at those sites. While no loss of material was observed during the induction period it is clear from Figure 4 that surface pH was being reduced from initial high values to more neutral levels during this time.

Concrete losses for the new concrete coupons at the end of the study period ranged from 36.4 mm at Perth Main Sewer (48 months), 20.0 mm (Perth Bibra Lake, 58 months), 28.8 mm (Melbourne WTS, 72 months), to 5.8 mm at the Melbourne SETS site (83 months). Irrespective of the local conditions once losses began to occur they accumulated in a linear manner over time at all sites, i.e. new coupons at all sites experienced a constant rate of corrosion once corrosion commenced. The point at which concrete losses began at all sites was found to coincide with the surface pH falling to pH=6 at all sites (for example see Figures 10 and 11).

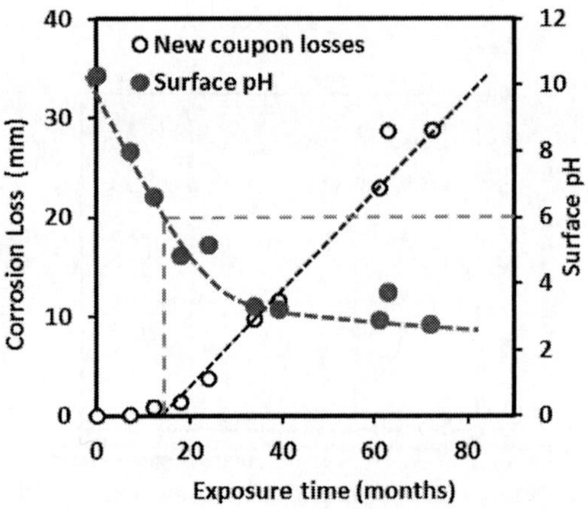

Figure 10 New coupon losses versus surface pH (Melbourne WTS)

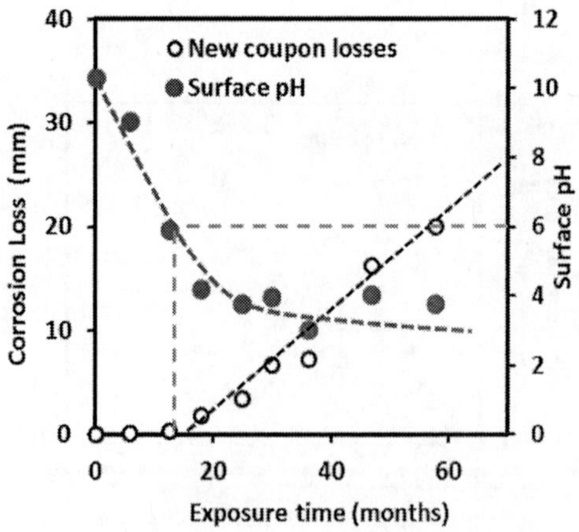

Figure 11 New coupon losses versus surface pH (Perth Bibra Lake)

3.2.2 Old Coupons

The general pattern of corrosion behaviour was the same for old coupons at all sites irrespective of the local conditions (see Figures 6-9). No induction period was observed for old coupons. Instead mass loss commenced almost immediately upon exposure to the sewer environment. As already noted for new coupons, rates of corrosion were found to constant at each site over the duration of the study period.

Corrosion losses for the old concrete coupons ranged, at the end of the study, from 38.1 mm at Perth Main Sewer (48 months), 23.8 mm (Perth Bibra Lake, 58 months), 33.2 mm (Melbourne WTS, 72 months), to as low as 11.8 mm, 2.7 and 2.0 mm at Melbourne SETS (83 months), Sydney Southern and Western Suburbs Ocean Outfall Sewer (SWSOOS) South and North (36 months) respectively.

The calculated corrosion rates for new and old coupons are listed in Table 2. Corrosion rates ranged from ~1.5mm/yr at the most benign site (Melbourne SETS) up to >10mm/yr observed at the Perth Main Sewer site. There was no significant difference between the rates experienced by the new and old coupons at each site.

Table 2. Corrosion rates observed for New and Old coupons over the course of the study

	New coupon		**Old coupon**	
	Rate of corrosion		**Rate of corrosion**	
Site	**(mm/yr)**	**+/- 90% ci**	**(mm/yr)**	**+/- 90% ci**
Perth Main Sewer	11.6	1.7	10.2	0.5
Perth Bibra lake	5.8	1.2	5.3	0.6
Melbourne SETS	1.4	0.6	1.9	0.4
Melbourne WTS	6.5	1.1	6.0	0.4
Sydney SWSOOS North			0.6	0.5
Sydney SWSOOS South			0.7	0.7

3.3 Corrosion layer development

The buildup of corrosion products, (primarily gypsum, $CaSO_4.2H_2O$), on the new and old coupon surfaces at the Perth and Melbourne sites is plotted in Figures 12-15. During the incubation phase only a very thin film, (<0.5mm), of corrosion product was observed, which XRD/XRF analysis confirmed was primarily composed of gypsum. Once new coupons began to incur mass loss the depth of corrosion product increased in a non-linear manner over time. The corrosion product had a white pasty appearance. XRD/XRF analysis of the corrosion product revealed that it was primarily composed of gypsum with entrained quartz (retained sand particles from the original cement matrix). Optical microscopy of the corrosion product revealed the presence of numerous cracks and voids throughout the layer particularly around the aggregate. By the end of the study new coupon corrosion layer depths ranged from ~9 mm on Perth Main Sewer coupons to ~2 mm for the Melbourne SETS coupons.

The old concrete coupons had a pre-existing ~2 mm hard crystalline layer of corrosion product covering the exposed surface. Upon exposure to the sewer environment this surface material was rapidly transformed to a soft, pasty consistency similar in appearance, chemistry and morphology to that present on the surface of new coupons. At all sites the surface appearance of new and old coupons gradually became indistinguishable with time. At all sites old coupons corrosion layer depth increased in a non-linear fashion with time in a manner similar to that observed for new coupons at the same site. By the end of the study period, average corrosion layer depths had reached ~15 mm at Perth Main Sewer (the highest in the study), 10 mm at Perth Bibra Lake, 9 mm at Melbourne WTS and 4 mm at Melbourne SETS.

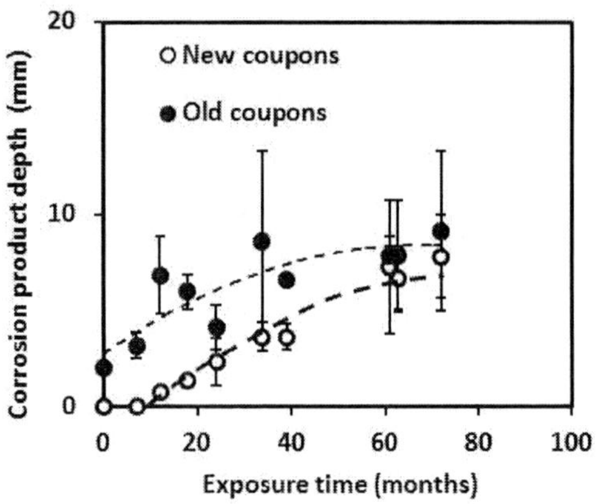

Figure 12 The build-up of corrosion product on new and old coupons (Melbourne WTS) as a function of exposure period.

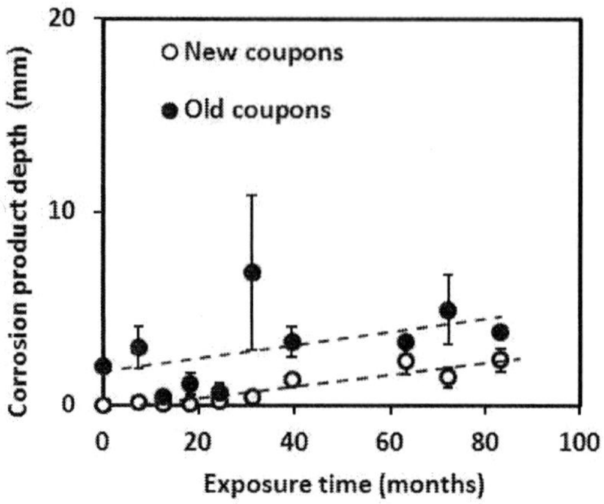

Figure 13 The build-up of corrosion product on new and old coupons (Melbourne SETS) as a function of exposure period.

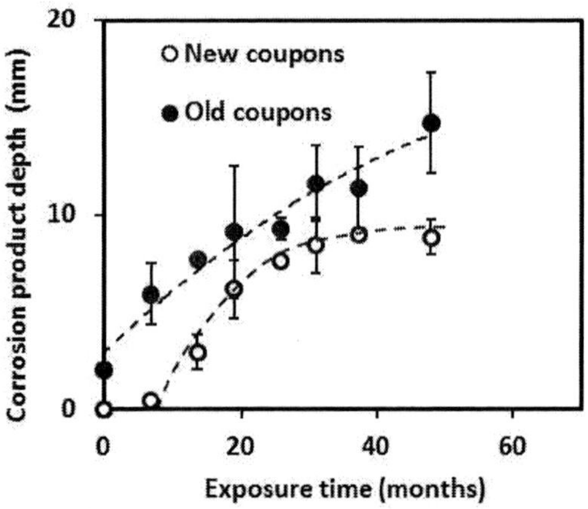

Figure 14 The build-up of corrosion product on new and old coupons (Perth Main Sewer) as a function of exposure period.

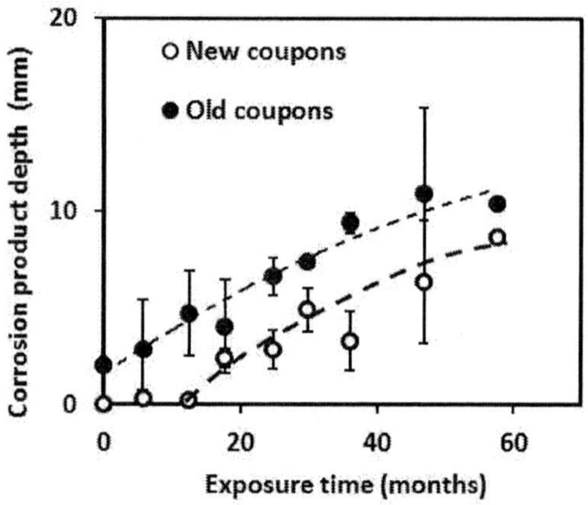

Figure 15 The build-up of corrosion product on new and old coupons (Perth Bibra Lake) as a function of exposure period.

4. DISCUSSION

4.1 Corrosion behaviour over time

Inspection of Figures 5-8 reveals that the corrosion process falls into two distinct stages. Immediately after installation the sewer pipe goes through an initial stage lasting between 3 and 36 months where it suffers little if any loss of material. This initial "incubation" period occurred at all sites irrespective of their relative aggressiveness– only the duration of the stage varies (the incubation period lasted longer at more benign sites).

When a concrete pipe is first installed the surface is very alkaline (pH>10). Under these conditions it is difficult for acidophilic bacteria and fungi (those responsible for producing the sulphuric acid that corrodes the pipe) to colonise the surface so no MIC can take place. Instead during the incubation phase the alkaline pipe surface reacts with acidic gases (CO_2 and H_2S) that have dissolved into the condensate present within the upper pipe pore voids to form mild inorganic acids. These acids neutralise alkaline minerals present in the cement binder of the pipe concrete, reducing the pH of the pipe surface. The more CO_2 and H_2S present in the sewer gas the faster this process takes place and the more rapidly the surface pH is reduced. Inspection of Figure 4 reveals that this initial neutralisation of the pipe surface happens over a period of a few months for aggressive sites (H_2S>10ppm) up to several years for less aggressive sites (H_2S<2ppm).

Once the surface pH falls to more neutral levels microbial colonies and fungal communities begin to colonise the pipe surface. When the surface pH falls to pH~6 acid producing bacteria dominate the bacterial population and the corrosion process enters the second more destructive phase. The transition between the two stages is clear at each site and is observed at the same surface pH regardless of the aggressiveness of the site (irrespective of how long it takes for surface conditions to reach this value).

By way of contrast the old coupons entered the more active second corrosion phase almost immediately upon installation in the sewer (Figures 6-9). This is readily explained by the fact that the starting surface pH levels for the pre-exposed coupons were already close to pH=6 before they were installed and as a result it only took a matter of weeks of exposure in the sewer before the surface was acidic enough to allow acidophilic bacterial colonisation and hence, mass loss to commence.

Once corrosion was under way losses experienced by new and old coupons increased in a linear manner over time at all sites. The slope of the loss curve (i.e. the corrosion rate) varied from site to site but in all cases was observed to be constant for a given site. It is also notable that the corrosion rates at each site are very similar for new coupons, (representing the pipe in its early stages of service), and for old coupons, (representing the sewer pipe at a more advanced stage) (Table 2). The similarity in corrosion rates suggests that the rate of corrosion for concrete sewer pipe at a given site is likely to remain constant over the entire service life of the pipe (provided the long term average sewer conditions do not change significantly). In this context it is of interest that long term corrosion behaviour has been observed to be a linear function of time during the whole of an 18 year field study conducted under moderately aggressive conditions (H_2S concentrations ranging from 5 to 40ppm) on actual concrete sewers near Melbourne, Australia (CSR Humes, 2001).

4.2 Formulation of a first pass sewer pipe corrosion model

In general the sewer pipe corrosion process can be represented as a bi-linear function with time (Figure 16) and can therefore be characterized using only an incubation time, t_i, and the long-term steady state corrosion rate, *C*, both of which are functions of the local sewer environmental conditions. The incubation time, t_i, is likely to be short compared to the lifespan of the pipe so only the corrosion rate is considered in the modelling process.

In practice it is likely that a number of factors will influence the rate of concrete corrosion. However, many are likely to be difficult to quantify particularly over the long term. To formulate a model to predict corrosion rates a small number of relatively easy-to-measure parameters were identified that could be used to develop a first pass corrosion model. They were: sewer gas temperature, H_2S concentration and sewer gas humidity. In formulating the corrosion model the impact of these three factors on the chemical and biological processes specific to the sewer corrosion process were considered.

The influence of sewer gas temperature on the chemical and biological processes involved in sewer pipe corrosion can be described using the common Arrhenius relationship:

$$k_r \propto e^{-E_a/RT} \tag{1}$$

where k_r, (mol s^{-1}), is the rate of the process in question at temperature − T (K), E_a is the activation energy for the process (J mol^{-1}) and R is the universal gas constant (= 8.314 J mol^{-1} K^{-1}).

Activation energies ranging from E_a =35 to 80 kJ mol^{-1} have been reported for abiotic and biotic oxidation and for uptake of H_2S under aerobic conditions in wastewater environments and in seawater (Millero et al., 1987). Activation energies of between 47 kJ mol^{-1} to 55 kJ mol^{-1} have been reported for the oxidation of elemental sulphur by *At. Thiooxidans* (Franzmann et al., 2005; Parker, 1945a). Taken together, these results show a considerable range in activation energies. An analysis of available corrosion data from this study and other historical data indicated that a midrange value of E_a=45 kJ mol^{-1} was a good estimate. The temperature dependence for the corrosion rate was therefore expressed as follows:

$$C_T \propto e^{\left(-45000/RT\right)} \tag{2}$$

Gas phase H_2S levels often are used as the sole indicator of the corrosivity of the environment inside a sewer pipe. Consideration of the data obtained in this study clearly shows that H_2S alone is not a good indicator (and therefore predictor) of concrete corrosion rates. For example, the rate of corrosion observed at the Perth Bibra Lake site (*C*~5.5 mm yr^{-1}) was lower than that at Melbourne WTS (*C*~6.2 mm yr^{-1}), despite Perth Bibra Lake having H_2S levels approximately 70 times higher. This demonstrates that other factors, such as temperature (see above) and relative humidity (see below) also influence corrosion activity.

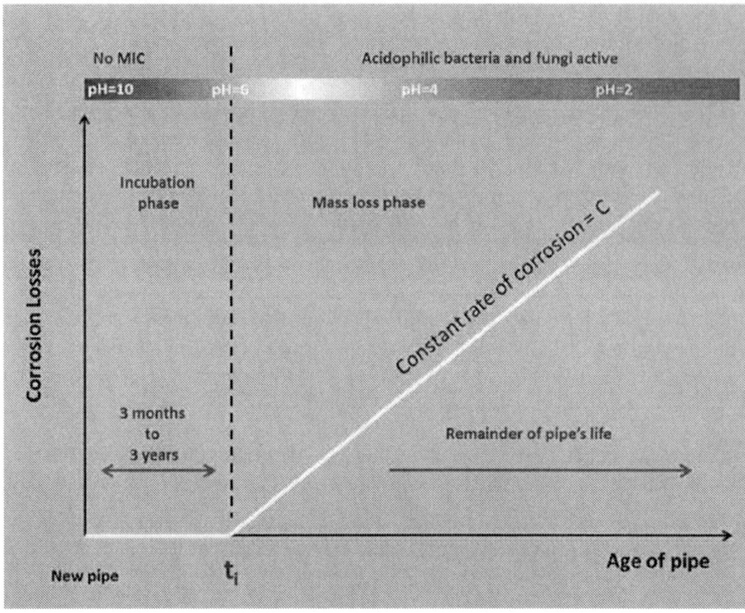

Figure 16 The evolution with time of concrete sewer pipe corrosion

Studies examining the oxidation of H_2S on corroding concrete surfaces suggest that the impact of H_2S on corrosion rates can be described using simple *n*th order kinetics with a process order of n=0.4 to 0.75 (Nielsen et al., 2014). Analysis of the corrosion rates observed at the Melbourne SETS, WTS and Perth Main Sewer sites, (where humidity levels were close to saturation) indicates a value of n=0.5 best describes the influence of H_2S on corrosion rates observed in the present study. A simple square root relationship was therefore proposed:

$$C_{H2S} = [H_2S]^{0.5} \tag{3}$$

where C_{H2S} represents the impact of H_2S factor in the corrosion model and $[H_2S]$ is the concentration of H_2S in the sewer gas phase (ppm).

Throughout the study humidity was found to have a large bearing on the rate of corrosion experienced by the concrete coupons. This impact is best seen in the significant suppression of corrosion activity at the Perth Bibra Lake site where, despite the very high H_2S levels, rates of corrosion were only moderate. The microbes responsible for corrosion occurring require a moist environment in which to live. The availability of moisture within the concrete pore space also influences the progress of ionic reactions and related processes involved in abiotic corrosion of the concrete. The relationship between humidity in the sewer gas space and the moisture content of the concrete sewer walls is therefore important.

In the present study the field samples were located at the crown of the sewers where waste water levels rarely reached. Hence concrete moisture content was dictated by capillary condensation of the surrounding water vapour into the pore structure and consequently moisture levels were sensitive to changes in sewer gas humidity. This is important as it is known that when relative humidity is less than ~85%, (a "dry pipe"), the moisture level within the concrete matrix is sufficiently low to halt acid production by bacteria. At higher humidity levels, water vapour initially condenses within smaller diameter pores and then progressively in larger diameter pores as humidity increases until eventually, at 100% humidity, the pore space is completely filled and a condensate film forms on the concrete surface of the pipe.

Two practical implications follow from the above. First, if humidity within the sewer is lowered sufficiently corrosion activity can be much reduced or even halted even at very high H_2S levels. This can be achieved for example, by increased ventilation, forced or natural. Second, although many sewers are regularly monitored for H_2S, monitoring of humidity is much less common. As a result, the sewer atmosphere humidity levels and their time dependent characteristics tend not to be well quantified. The Perth Bibra Lake example shows that such data are important as without adequate quantification of the humidity conditions, (or the resultant levels of concrete moisture content), making reasonable estimates of corrosion activity is likely to remain problematic.

When the pore filling process was considered (Wells and Melchers, 2015) the following relationship between humidity and corrosion rate was derived:

$$C_{RH} = \left(\frac{0.0016H - 0.1355}{1 - 0.00977H} \right) \tag{4}$$

where C_{RH} represents the humidity factor of the corrosion model and H is the relative humidity (%).

From the above considerations, a first pass model that allows prediction of the rate of corrosion for a concrete pipe is proposed by gathering together Eqns. (2), (3) and (4), as follows:

$$C = A \times [H_2S]^{0.5} \times \frac{(0.0016H - 0.1355)}{(1 - 0.00977H)} \times e^{\left(-45000/RT\right)} \tag{5}$$

where A is a scaling constant (= 207,000,000 mm $yr^{-1}ppm^{-1/2}$). The value of A was determined by calibrating Eqn. (5) to field data as follows Average humidity, temperature and H_2S conditions associated with corrosion measurements obtained in this study and elsewhere (Morton et al., 1989; Parker et al., 1965) were first inputted into Eqn. (5) along with nominal values of the constant A. The resultant rates of corrosion predicted by Eqn. (5) were then compared with the observed rates. The value of A was then varied to obtain the minimum least squared error between predicted and observed values of the concrete corrosion rates.

The rates of corrosion predicted by Eqn. (5) and those extracted from the field observations and some historical data sources are compared in Figure 17. It shows excellent agreement between predicted and observed corrosion rates over a wide range of conditions. Thus Eqn. (5) is shown to provide a good first-pass estimate of the rate of concrete sewer pipe corrosion over a wide range of conditions, provided the relevant long term average gas phase properties (temperature, humidity and H_2S concentration) are known or can be estimated with a sufficiently high degree of accuracy.

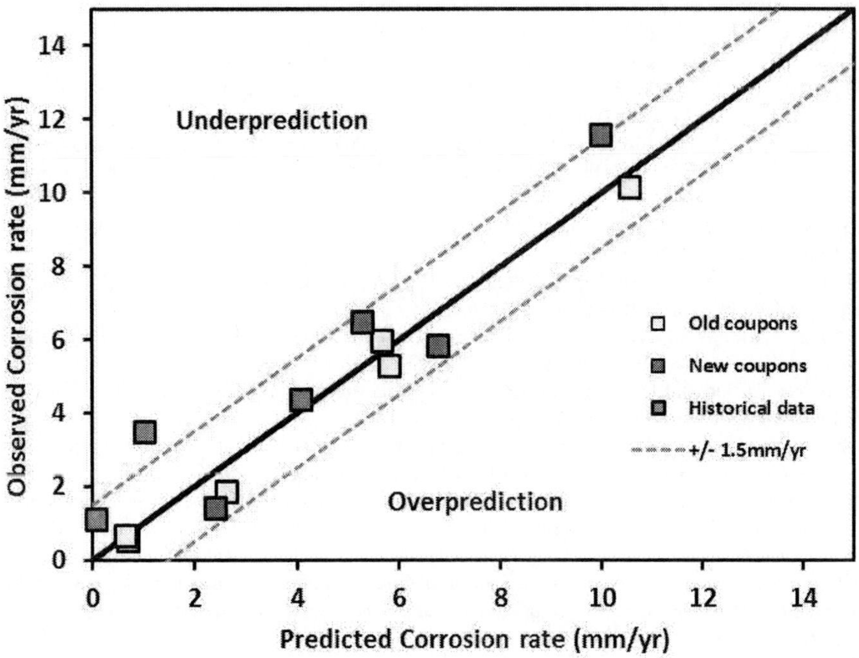

Figure 17 A comparison of observed and predicted rates of concrete sewer pipe corrosion

5. CONCLUSIONS

A 7 year field study of sewer pipe corrosion has shown that concrete sewer pipe corrosion is a bi-linear process beginning with an initial incubation period involving no mass loss. Once the pipe surface pH falls to pH to around 6 corrosion losses commence and continue at a constant rate with time for the remainder of the life of the pipe. A model predicting the rate of corrosion and the service life of concrete sewer pipes using local environmental conditions has been developed and calibrated over a wide range of conditions using the field data. Predicted concrete sewer pipe rates of corrosion showed an excellent with observed values from this and previous studies with discrepancies falling within a +/- 1.5mm/yr band.

6. ACKNOWLEDGMENTS

The authors acknowledge the financial and other support of the Australian Research Council as well as Industry Partners Barwon Regional Water Corporation, CH2M Hill, Gold Coast Water, Hunter Water Corporation, Melbourne Water Corporation, South Australia Water, South East Water Limited, Sydney Water Corporation, United Water International, Water Quality Research Australia, Western Australia Water Corporation for funding this Project. They also acknowledge the support of the other Research Partners; The University of New South Wales, The University of Queensland, The University of Sydney and Curtin University of Technology.

7. REFERENCES

CSR Humes, 2001. Pipeline materials in H_2S 1977 Werribee Pipeline. CSR Humes Information Series NDP0101. August '01.

Franzmann, P.D., Haddad, C.M., Hawkes, R.B., Robertson, W.J., Plumb, J.J., 2005. Effects of temperature on the rates of iron and sulfur oxidation by selected bioleaching Bacteria and Archaea: Application of the Ratkowsky equation. Minerals Engineering 18(13–14), 1304-1314.

Fujimoto, H., Mishina, F., Onishi, K., Miyaguchi, K., 2013. Evaluation of Sulfuric Acid Resistant Mortar as a Corrosion Protection Method for Sewerage Facilities, Third International Conference on Sustainable Construction Materials and Technologies., Kyoto, Japan.

Gu, J.D., Ford, T.E., Berke, N.S., Mitchell, R., 1998. Biodeterioration of concrete by the fungus Fusarium. International Biodeterioration & Biodegradation 41(2), 101-109.

Herisson, J., van Hullebusch, E.D., Moletta-Denat, M., Taquet, P., Chaussadent, T., 2013. Toward an accelerated biodeterioration test to understand the behavior of Portland and calcium aluminate cementitious materials in sewer networks. International Biodeterioration & Biodegradation 84(0), 236-243.

Hewayde, E., Nehdi, M., Allouche, E., Nakhla, G., 2006. Effect of geopolymer cement on microstructure, compressive strength and sulphuric acid resistance of concrete. Magazine of Concrete Research 58(5), 321-331.

Islander, R.L., Devinny, J.S., Mansfeld, F., Postyn, A., Hong, S., 1991. Microbial Ecology of Crown Corrosion in Sewers. Journal of Environmental Engineering-Asce 117(6), 751-770.

Kaempfer, W., Berndt, M., 1999. Estimation of service life of concrete pipes in sewer networks. In: M.A. Lacasse, D.J. Vanier (Eds.), 8th Conference of Durability of Building Materials and Components, Vol 1., pp. 36-45.

Koch, G., Brongers, M., Thompson, N., Virmani, Y., Payer, J., 2001. Report # FHWA-RD-01-156, Corrosion cost and preventive strategies in the United States., Office of Infrastructure Research and Development, Federal Highway Administration, McLean, VA.

Mansfeld, F., Shih, H., Postyn, A., Devinny, J., Islander, R., Chen, C.L., 1991. Corrosion Monitoring and Control in Concrete Sewer Pipes. Corrosion 47(5), 369-376.

Milde, K., Sand, W., Wolff, W., Bock, E., 1983. Thiobacilli of the Corroded Concrete Walls of the Hamburg Sewer System. Journal of General Microbiology 129(5), 1327-1333.

Millero, F.J., Hubinger, S., Fernandez, M., Garnett, S., 1987. Oxidation of H_2S in seawater as a function of temperature, pH, and ionic strength. Environ. Sci. Technol. 21(5), 439-443.

Morton, R., Caballero, R., Chen, C.-L., Redner, J., 1989. Study of sulfide generation and concrete corrosion of sanitary sewers, In-house report; Sanitation Districts of Los Angeles County Carson, California.

Nica, D., Davis, J.L., Kirby, L., Zuo, G., Roberts, D.J., 2000. Isolation and characterization of microorganisms involved in the biodeterioration of concrete in sewers. International Biodeterioration & Biodegradation 46(1), 61-68.

Nielsen, A., Hvitved-Jacobsen, T., Jensen, H., Vollertsen, J., 2014. Experimental Evaluation of the Stoichiometry of Sulfide-Related Concrete Sewer Corrosion. Journal of Environmental Engineering 140(2), 04013009.

Olmstead, W.M., Hamlin, H., 1900. Converting portions of the Los Angles outfall sewer into a septic tank. Engineering news 44(19), 317-318.

Parker, C.D., 1945a. The corrosion of concrete 1. The isolation of a species of bacterium associated with the corrosion of concrete exposed to atmospheres containing hydrogen sulphides. The Australian journal of experimental biology and medical science 23, 81-90.

Parker, C.D., 1945b. The corrosion of concrete 2. The function of *Thiobacillus concretivorus* (nov. spec.) in the corrosion of concrete exposed to atmospheres containing hydrogen sulphides. The Australian journal of experimental biology and medical science 23, 91-98.

Parker, C.D., Beer, A.D., Mitchell, I.G., Bryant, G.R., Ryan, J.M., 1965. Hydrogen sulphide corrosion of concrete sewers, Part 8, Attack on Concrete. Section 3: Influence of concrete composition on rate of deterioration. Melbourne and Metropolitan Board of Works technical paper No. A8. .

Sand, W., Dumas, T., Marcdargent, S., 1992. Tests for biogenic sulfuric acid corrosion in a simulation chamber confirm the on-site performance of calcium aluminate-based concretes in sewage applications. In: K.D. Basham (Ed.), Infrastructure: New materials and methods of repair - Proceedings of the Third Materials Engineering Conference. ASCE, New York, NY, pp. 35-55.

Snell, F.D., Ettre, L.S., 1972. Encyclopedia of industrial chemical analysis. Interscience Publishers Division, vol 15, New York.

Wells, P.A., Melchers, R.E., 2011. Microbial corrosion of sewer pipe in australia - initial field results, 18th International Corrosion Conference, Perth, W.A., Australia, pp. Paper no. 183.

Wells, T., Melchers, R.E., 2014. An observation-based model for corrosion of concrete sewers under aggressive conditions. Cement and Concrete Research 61–62, 1-10.

Wells, T., Melchers, R.E., 2015. Modelling concrete deterioration in sewers using theory and field observations. Cement and Concrete Research 77, 82-96.

8. AUTHOR DETAILS

Dr Tony Wells is a research academic in the Civil, Surveying and Environmental Engineering Department at the University of Newcastle, NSW, Australia a position he has held for the last 15 years. He has a BE (Chem. Eng.) from Newcastle University and a PhD from the University of New South Wales, Australia in 1991. He has managed and participated in research projects in areas as diverse as climate change, geomorphology, high pressure physics, and the corrosion of advanced polymeric materials, concrete sewer and cast iron water pipes.

Robert E Melchers is Professor of Civil Engineering at The University of Newcastle, Australia. He holds a BE and MEngSc from Monash University and a PhD from the University of Cambridge, UK. He is a Fellow of the Australian Academy of Technological Sciences and Engineering.

He was awarded the Marshall Fordham prize (ACA) in 1999, 2002 and 2007, the 2004 TP Hoar Prize (Institute of Corrosion, UK) (with Robert Jeffrey), the 2007 Guy Bengough Award (Institute of Materials, Minerals and Mining, UK), the 2009 ACA Corrosion Medal and the 2012 Jin S Chung Award (International Society of Offshore and Polar Engineers). His research interests include structural reliability and marine corrosion.

MODELLING THE ELECTROKINETICS OF CONCRETE STEEL REINFORCEMENT CORROSION

H. J. Flitt[1], G.W. Will[1] and W. Green[2]
[1] Queensland University of Technology, Queensland. [2] Vinsi Partners, Sydney

SUMMARY: The electrokinetics of the corrosion of steel reinforcement in concrete has been explained in the literature by the use of polarization curve schematics. Four states of corrosion can be delineated by these curves namely passive, pitting, general corrosion and active low potential corrosion. By using a modified previous developed electrokinetic model for iron corrosion the schematics can be replaced with more accurate representations of the corrosion states. Electrokinetic parameters can be generated by the deconvolution of measured polarization curves in synthetic pore solution. With the parameters various scenarios can be generated to show expected polarization behaviour at varied values of pH.

Keywords: Concrete, electrokinetics, Steel, Corrosion

1. INTRODUCTION

This paper describes the initial development of a computer application to model the electrokinetics of steel reinforcement corrosion embedded in concrete. Polarisation curves and linear polarisation resistance are experimental techniques routinely used to evaluated corrosion rate and status of reinforcement in concrete. The interpretation of these curves and the extraction of parameters such as Tafel slopes are often difficult particularly when the experimental can play such an important role.[1] We have adapted our iron SYMADEC program[2] to deal specifically with this corrosion system. The electrokinetic model uses parameters to dynamically generate polarization curves across the pH range. Active, passive and transpassive corrosion conditions can all be considered. The operator can also vary parameter such as oxygen concentration and pH and determine the effects. Of greater utility is the use of these ideal models to determine when non ideal behaviour is occurring and therefore identifying the presence of a new mechanism.

2. RESULTS AND DISCUSSION

The setting of parameters to use with SYMADEC concrete are obtained by matching a measured polarization curve by Page and Treadaway [2] in saturated calcium hydroxide with no chloride addition. Figure 1a left gives the experimental curve. Figure 1b gives the match [1] with the synthesized polarization using the resolved electrochemical and thermodynamic parameters. To test the SYMADEC concrete application the generation of the four states of corrosion detailed by Arup [3] are considered below.

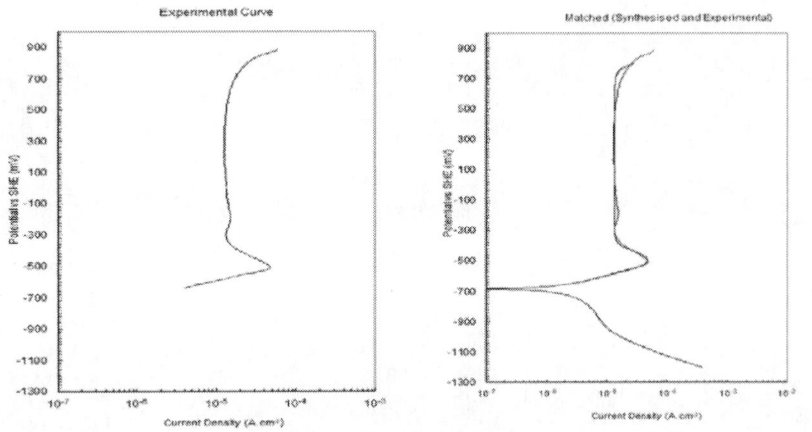

Figure 1a and 1b Experimental and matched polarization curve of steel in saturated calcium hydroxide

The simulated and the deconvoluted polarization curves are generated for each of the corrosion states.

2.1 Low Potential Corrosion

Low potential active corrosion occurs in a chloride free environment when steel access to oxygen is restricted such that a chemical passive film is not generated. The corrosion rate is lower than in the passive state. The passive film can be reinstated if the oxygen supply is sufficiently increased to produce an available cathodic current density greater than the anodic peak. The simulated curve for this corrosion state is given by Figure 1b.

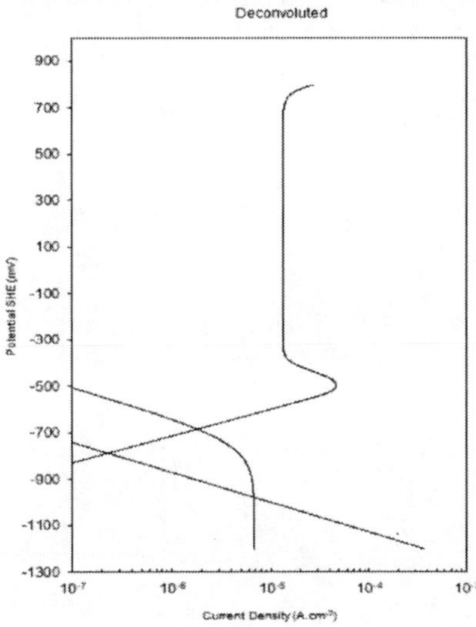

Figure 2 Deconvoluted curve for Figure 1b.

The corrosion is controlled by oxygen reduction and is not influenced by hydrogen evolution. The level of oxygen is sufficient to prevent limiting control.

2.2 Passivity

Passivity is the protected state of steel in concrete in an environment of high pH and the availability of oxygen. The usual open circuit potential range is 450 mV to 50 mV SHE. This corrosion state is maintained as long as the chloride level does not exceed a critical level.

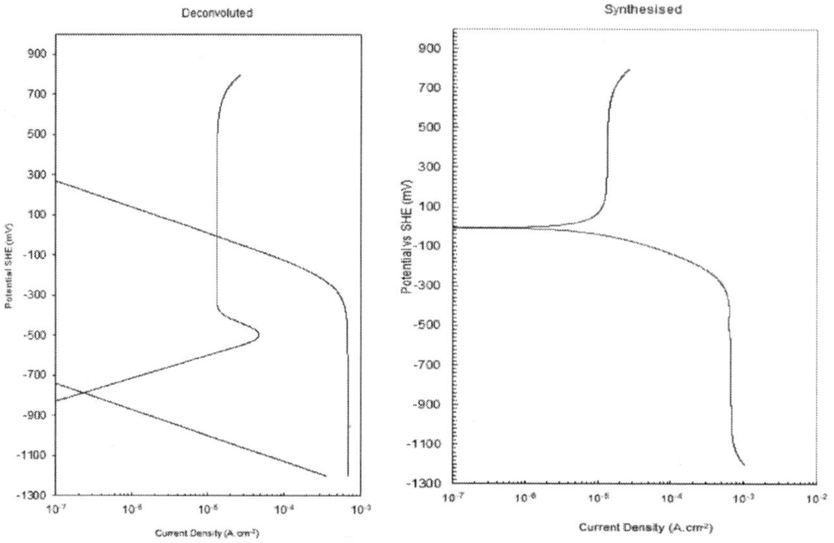

Figures 3a and 3b deconvoluted and synthesised curves for the passive state

Figures 3a and 3b above give the deconvoluted and synthesised curves for the passive state. The oxygen Tafel line intersects the passive current giving a corrosion potential in the range suggest by Arup [3].

2.3 Pitting

Pitting occurs when the corrosion potential exceeds the breakdown potential. Passive film breakdown requires the chloride level to reach a critical concentration. Figures 4a and 4b below give the deconvoluted and synthesised curves for the onset of pitting. The oxygen reductions Tafel line is slightly above the breakdown potential.

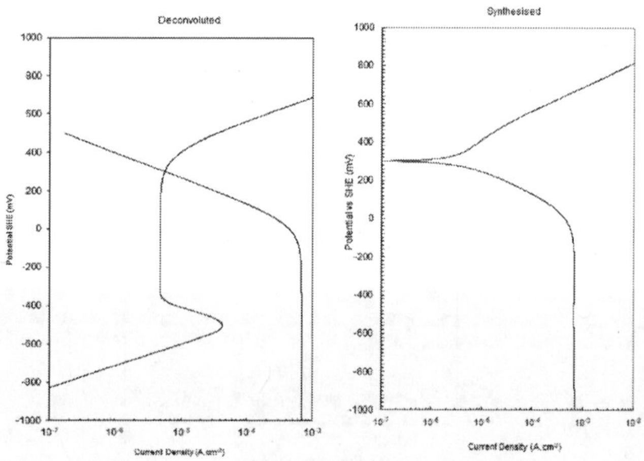

Figures 4a and 4b Pitting deconvoluted and synthesised curves

2.4 General Corrosion

General corrosion is the result of a complete loss of steel passivity. This can be the result of carbonation or large chloride ingress. The solution pH shifts to a near neutral value. The open circuit potential is typically in the range of -200 mV to -350 mV SHE. Figure 5a and 5b below give the deconvoluted and synthesised curves for active corrosion.

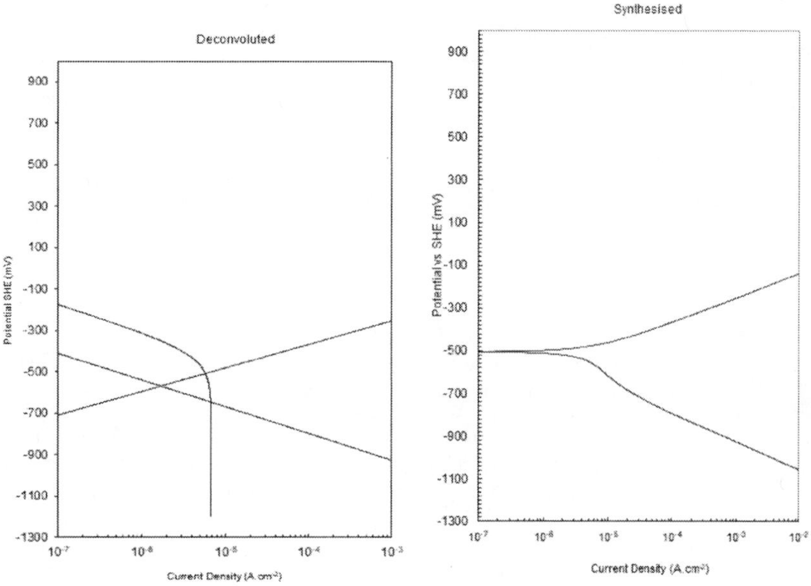

Figures 5a and 5b Pitting deconvoluted and synthesised curves

3. CONCLUSIONS

The four states of steel corrosion in concrete can be generated by SYMADEC concrete. The set of parameters obtained from matching a polarization curve in calcium hydroxide proved be to adequate and will contribute to a better and more accurate understanding of the system in the future.

4. REFERENCES

1. Zhen-TianChang, BrianCherry, MartonMarosszeky Corrosion Science 50, 2, 2008, pg. 357-363
2. H. J. Flitt, D. P. Schweinsberg, Corrosion Science 47, 2005, 2125.
3. C. L, Page, K. W. J. Treadaway, Nature 297, 1982, 109.
4. H. Arup, "The mechanisms of the protection of steel in concrete", Chapter 10. In Corrosion of Reinforcement in concrete construction, Ed. A Crane, Ellis Horwood Ltd, 19

5. AUTHOR DETAILS

Harvey Flitt is a corrosion consultant with over 40 years of experience

Assoc. Prof. Geoffrey will is a corrosion scientist at the Queensland University of Technology

Warren Green BAppSc (App Chem), Post Grad Dip Chem, MSc (Corr Sc & Eng), FIEAust CPEng NER, RPEQ

Warren has over 25 years experience in corrosion engineering and materials technology, in both private and public sector industries covering marine, infrastructure, industrial, civil and building structures.

WATER-BORNE INORGANIC ZINC SILICATE: TWO NZ BRIDGES

W L Mandeno
Opus International Consultants Ltd, Wellington, New Zealand

SUMMARY: In 2010, two new steel highway bridges were built for the New Zealand Transport Agency where a high-build water-borne inorganic zinc silicate (WBIZS) was used as the protective coating. On the Matahorua Gorge Bridge on SH2 the long life coating has performed as expected, but the same coating on the SH1 bridge over the Waikato River near Taupo, suffered a premature failure and has since been completely repainted in situ.

This paper reviews the specification selection and application methods used for the two bridges and the various investigations into the cause of the failure, with references to similar successes and failures of this coating system that have previously been reported. It is also a reminder of the importance of providing suitable environmental conditions to ensure early release of solvent and thus allow the initial cure to occur, so that the potential very long life of a high build WBIZS protective coating system can be realised.

Keywords: Steel Bridge, Water-borne Inorganic Zinc Silicate, Coating Failure

1. INTRODUCTION

In 2010, two new steel highway bridges [1] were built for the New Zealand Transport Agency (NZTA) where a high-build water-borne inorganic zinc silicate (WBIZS) was used as the protective coating to meet its requirements of an expected life of 40 years to first major maintenance painting of the primary structural elements. On the Matahorua Gorge Bridge on SH2 the long life coating has to date performed as expected, but the same coating on the SH1 bridge over the Waikato River near Taupo, suffered a premature failure and has since been completely repainted in situ.

Figure 1 Matahorua Gorge Bridge (NZTA photo)

Figure 2 Waikato River Bridge (NZTA photo)

1.1 Matahorua Gorge Bridge

This is a 137m long 3 span ladder deck plate girder bridge that crosses the Matahorua River. This formed part of a 3.1km roading realignment project 52km North of Napier, where the 420 tonne of steel work was fabricated and coated. The coating system specified was what is now designated as "IZS3" in AS/NZS 2312.1 [2], i.e. 125 microns of water-borne inorganic zinc silicate applied over an abrasive blast cleaned surface to Sa 2½ cleanliness standard.

1.2 Waikato River Bridge (WRB)

This is a 149m long network arch bridge with a central 100m span across the Waikato River approximately 1km upstream from the Wairakei Geothermal Power Station. This, together with an adjacent 444m long steel ladder bridge across the geothermal steam pipes, formed part of the East Taupo Arterial roading project that allowed SH1 to bypass the Taupo CBD. As part of a design/build contract, in 2008 both bridges were tendered to be coated with arc sprayed aluminium spray, the system designated as "TSA225S" in AS/NZS 2312 [3]. While this was used on the bridge over the steam pipes (that release H_2S containing bleed water), the protective coating system for the WRB steelwork was amended in January 2009 during the design process to the lower cost solvent-borne high build ethyl zinc silicate system designated as IZS4 (and in 2013 used on the plate girders of the 142m Horotiu Bridge and 150m Kaituna River Bridge, also both part of new NZ highway projects). However, following discussion with the coating supplier, this was later changed to the water-borne IZS3 system that was applied beginning in October 2009, with a red vinyl topcoat applied onto the bowstring arch chords.

Figure 3 TSA coated ladder bridge over steam pipes

Figure 4 Ladder bridge with power station in distance

2. SITE ATMOSPHERIC CORROSIVITY

2.1 Matahorua Gorge Bridge

This bridge is located 8.5km from Hawke Bay on New Zealand's east coast which is not subject to strong onshore winds from the Pacific Ocean so has been assessed as having an atmospheric corrosivity category (ACC) of its macroclimate of C2 as defined in Table 2.1 of AS/NZS 2312.1 (and formerly as "B: Low").

2.2 Waikato River Bridge

The bridge site is near the centre of the North Island of NZ and within an active geothermal zone. During the failure investigation a short term site specific assessment of the ACC was carried out, which was reported as finding that the first year corrosion rate of steel at the bridge site was 4.5 microns/year. This is less than that reported by BRANZ [4] for the Wairakei Power Station village at 20.4 microns/year, but is still within the same C2 ACC range.

3. APPLICATION

3.1 Matahorua Gorge Bridge

The Napier based applicator advised that application was carried out during the autumn and winter of 2010, and that;

"the process we used for applying for the bulk of the time was in two coats. However the first one was aimed at 40-50μm DFT (but in reality we were achieving 60-70μm) and the second being a further 75-85μm (and in reality we were getting 80-100μm). This approach was to have the first coat cure more fully before applying the second.

I do remember really struggling when the humidity was high and temperatures cold (as it was mostly done in winter) to get a cure on the first coat and we had big fans trying to get air movement over them, even then it was taking up to 3 days to cure before we could apply the next coat.

Also had one beam that was delivered to us at 4pm and had to be on a truck at 6am the next day, which was very cold and wet. We had to try a single coat application which had an average of about 200μm but was touch dry when it went out. The following day it was raining and by the time they had driven to site the zinc had been completely washed off the beam".

Following a recent inspection the WBIZS coating was reported to still be in excellent condition.

3.2 Waikato River Bridge

The New Plymouth based applicator was using the same brand of the WBIZS (and reportedly some from the same batch) as was later applied to the Matahorua bridge steelwork. However a different application procedure was used where the author was advised that a heavy first coat of 75 - 100µm DFT was applied and, after initial cure was achieved, this was followed by a lower build second coat with 20% water added. Coating was applied indoors without added ventilation during spring and summer in 2009, then delivered to site where it was stored outdoors waiting for erection. Within a year of erection, blisters and spot rusting was observed on the below deck steelwork, which progressively became more severe and more widespread.

Figure 5 WRB main beams on site (December 2009)

Figure 6 WRB secondary steelwork stacked on site

4. INORGANIC ZINC SILICATE (IZS) COATINGS

The excellent long term durability of single coat inorganic zinc systems has been well documented, e.g. by Munger [5] Szokolik [6], and Devanney [7]. Their chemistry is discussed in detail in several papers included in the ACA publication [8] edited by Dr Rob Francis. Potential problems with the water-borne version were described by Hemmings et al [9] in a paper to the 2001 ACA Conference which has been included in the second edition (2013) of the ACA publication.

The following is an edited extract from an internal Opus report on the premature failure prepared by the author in 2012 which hypothesised failure was due to lack of cure before exposure to moisture, and discussed WBIZS and the other types of inorganic zinc silicate (IZS) coatings.

There are six types of IZS listed in AS/NZS 3750.15 which include the original heat-cured (Type 1) a later chemical-cured (Type 2) and weld-thru pre-primer (Type 5). The most common type applied in NZ is the solvent-borne (Type 4) which is used both as a primer for high performance multi-coat systems in atmospheric exposure or as a 'stand-alone' coating (IZS1 in AS/NZS 2312) for structural steelwork in moderate environments or lining fuel storage tanks.

The remaining two are the conventional water borne (Type 3) and a "high-ratio" version (Type 6) developed by NASA for their coating rocket launch facilities at Cape Canaveral in Florida. The Type 6 builds more thickly without mud cracking (a minimum of 200 microns in AS/NZS 3750.15) and cures more quickly than the Type 3 version, but is more expensive to make and more difficult to apply. It was the basis of the IZS3 system in AS/NZS 2312 which also allows a multi-coat system of Type 3 (Referred to as Paint Reference Nos. C01b and C01c in this Standard).

One of the main problems with Type 3 IZS is that it needs to release the water used as its solvent before the silicate film will begin to cure and develop its strong adhesive, abrasion and corrosion resistant properties. This happens very slowly in cold, or humid and/or calm conditions, which is why the solvent borne type is usually preferred. Historically the water-borne Type 3 has rarely been used in NZ outside Central Otago, where the average RH is too low to cure the solvent-borne Type 4 IZS.

CG Munger was a paint chemist involved with Ameron's development of self-curing zinc silicates, and was regarded as 'the authority' on zinc silicates. His book *'Corrosion Prevention by Protective Coatings'* [10], in Table 6.7 gives the arbitrary but safe minimum evaporation rate of 0.02lb/sqft/hr to achieve satisfactory film formation of WBIZS. This means that in calm conditions and if the steel temperature is 90degF/32degC, the RH must be <76%. At this RH the surface temperature can be lowered to 60degF/15.6degC if there is air movement above 5 miles/hr. However in cooler calm conditions, e.g. 60 degF/15degC, the maximum allowable RH drops to 36%. Munger also says (pg. 157) about WBIZS that *"these materials are not effective under cold, highly humid conditions"*.

However if conditions are right, according to Szokolik, WBIZS give slightly better performance under accelerated testing than SBIZS, probably due to their higher zinc content, and Type 6 IZS have been successfully used by Esso as stand-alone and self-recoatable maintenance coatings on offshore gas platforms in the Bass Strait.

A formerly well-known problem with IZS is their risk of premature failure if put into service before they have fully cured. This was noted on the product data sheet for the Type 3 WBISZ coating used on the Waikato River Bridge, which states under Curing Schedule:

"Higher film thickness, insufficient ventilation, high humidity, or cooler temperatures will require longer cure times and could result in premature failure. Excessive humidity or condensation on the surface during curing can interfere with the cure."

The data sheet also gave Special Instructions Concerning Cure which stated;

"Like all water-based inorganic zinc primers trace amounts of alkalinity may remain within its film after cure. This alkaline residue can be detrimental to coating integrity when water is allowed to puddle on its surface. .... Trace amounts of alkaline residue may concentrate in a drying puddle and result in high pH values that dissolve the coating film."

In the ACA's publication of collected papers on IZS [8] there are many references to the unreliability of attaining water insolubility and problems with early exposure to rain which leach the unpolymerised alkali silicates out of the film. Munger also states (pg143);

"Although initial drying is important, rewetting by dew condensation or rain before sufficient insolubilization of the binder can also be detrimental to the coating. Spot water condensation on the coating can result in spot discoloration. General condensation on the coating surface with rundown can result in binder leaching and subsequent powdery coating. Rain can also result in washing and removal of the coating. The amount of damage to the coating is dependent on the degree of insolubilization attained when water contact occurs, the amount of water, and the contact time."

This suggests that application in multi-coats, where the second coat was thinned with an addition of 20% extra water (as recommended by the manufacturer), has a high risk potential for undercuring if the first coat had not fully cured. Szokolik, Hemmings and others have noted that WBIZS can develop an apparent surface cure which may allow the coating to pass the water-rub cure test specified in AS/NZS 3750.15. Szokolik recommends that prior to carrying out the test, the film should be fully saturated by the application of a wet cloth or sponge for 5–10 minutes.

It would appear that this knowledge acquired in last century, regarding the risk of a false positive to the cure test and formation of strongly alkaline residues in retained moisture, was not known or had been forgotten during application of WBIZS to the WRB steel work. The consequences of this are illustrated in the following photos.

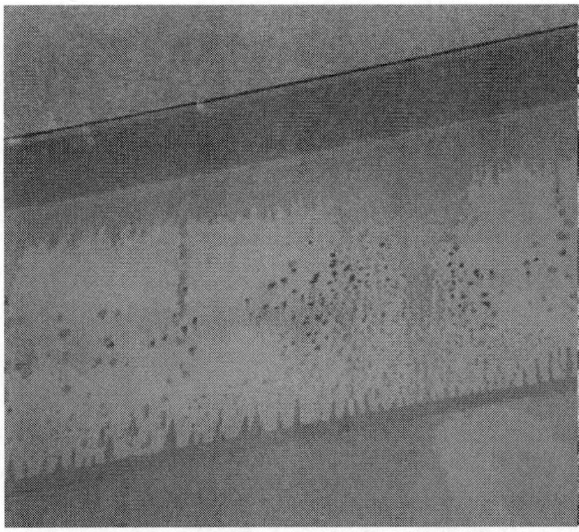

Figure 7 Freckle rust and blisters on web (Aug. 2011)

Figure 8 'Tide' marks on web & top flange (Jan. 2012)

Figure 9 Rust on top flange only (Jan. 2012)

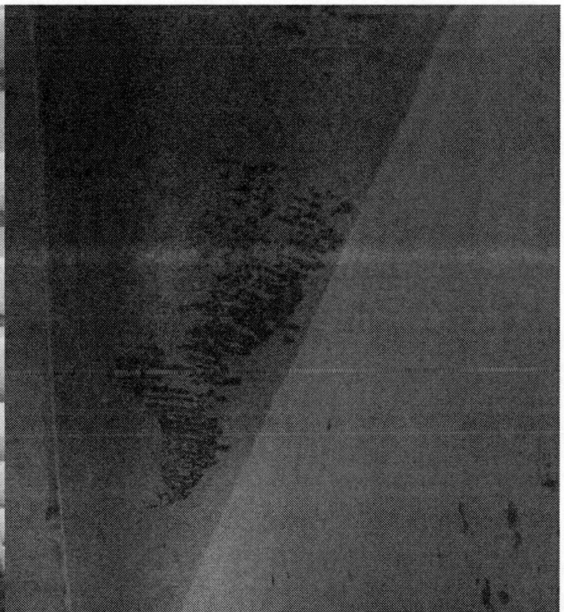

Figure 10 Breakdown on both surfaces (Jan. 2012)

Figure 11 Bottom flange of I-beam (Dec. 2013)

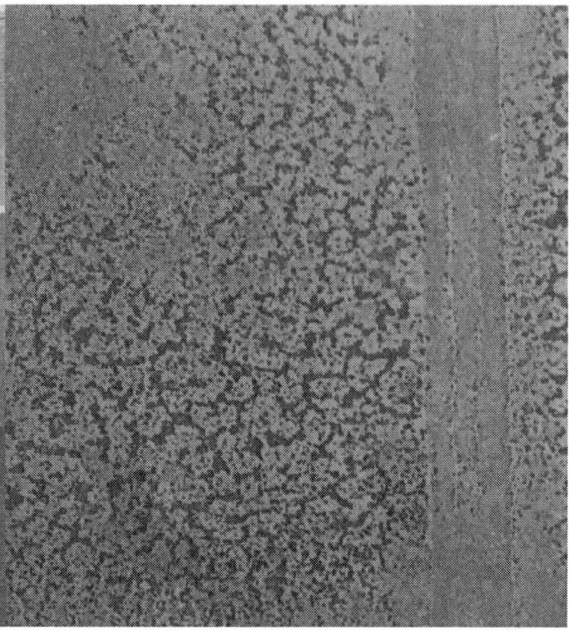

Figure 12 Close up of Figure 11

Figure 13 General view under deck (Dec. 2013)

Figure 14 General view under deck (Feb. 2015)

5. FAILURE INVESTIGATIONS

Following the first signs of premature failure appearing on the WRB, investigations were carried out by several other independent coating consultants over the following two years. Most noted that;

- the dry film thicknesses complied with the specification
- the failure locations were apparently randomly distributed, occurring on different faces of the steel members
- the vinyl coated WBIZS on the above deck steelwork was in excellent condition, apart from chalking of the red surfaces
- failures did not correlate well with variations in coating thickness
- much of the coating was still in good condition having a cohesive strength of >6 MPa
- failure was generally assumed to be due H_2S gas from the geothermal environment causing low pH condensate on steelwork during river fog conditions. (This despite observing that uncoated galvanizing was still in pristine condition and that corrosion was originating from under intact coating).
- IZS was therefore generally claimed to be an unsuitable system for this environment.

One consultant differed from the consensus and agreed with the author's hypothesis. He had investigated a similar failure pattern of WBIZS on steelwork under an Australian wharf in 2002. He concluded that failure was most likely due to localised alkaline attack resulting from early exposure to moisture that had allowed re-dissolution, then concentration, of alkali-metal hydroxides at the steel/coating interface.

Figure 15 WRB during river fog conditions (July 2012)

Figure 16 Unaffected galvanized bolts, pipe and grating on WRB beside failed WBIZS on PFC section (Feb. 2015)

6. REMEDIATION

In early 2016, the application sub-contractor for the WRB removed all the rust and weakly adherent coating by dry abrasive blasting within containment, after pressure washing at 3,500 psi and then sweep blasting all below deck steel surfaces. Both bare steel areas (~20% of the surface) and remaining WBIZS were primed with 75 microns of an epoxy zinc. All primed steel work was stripe coated, then fully coated with 125 microns of a high build epoxy mastic, and finished with 125 microns a high build polyurethane to give a system of equivalent durability to that designated as "PUR5" in AS/NZS 2312.1.

7. CONCLUSION

The Matahorua Gorge Bridge shows that WBIZS can be successfully applied in NZ to provide long durability. However the coating on the Waikato River Bridge failed with significant rust spotting, showing the importance of following correct curing conditions for this type of coating. As a result of its premature failure on the WRB, a warning on the need to provide optimum drying conditions when applying WBIZS, together with a reference to the Hemmings paper [9], were included in the NZ Transport Agency Bridge Painting Guide [11] which was first published in 2014.

8. ACKNOWLEDGMENTS

The author acknowledge the support of Opus International Consultants Ltd and the NZ Transport Agency to write and present this paper, and thanks the two applicators for their assistance in sharing with the wider protective coatings industry, the lessons they have relearned.

9. REFERENCES

1. Presland R., Chan M., Jackson R., & El Sarraf R. (2009) 'Composite Steel Bridge Solutions for the East Taupo Arterial', 7th Austroads Bridge Conference proceedings, Auckland.
2. Standards Australia and Standards New Zealand jointly, AS/NZS 2312.1:2014. 'Guide to the protection of structural steel against atmospheric corrosion by the use of protective coatings, Part 1: Paint coatings'.
3. Standards Australia and Standards New Zealand jointly, AS/NZS 2312:2002. 'Guide to the protection of structural steel against atmospheric corrosion by the use of protective coatings'.
4. Duncan JR & Cordner RJ, (1991) 'Atmospheric Corrosion Rates over Two Years Exposure at 156 Sites in New Zealand', IPENZ Transactions, Vol.18 No.1/GEN, Wellington. Available as BRANZ Reprint No. 112 (1992).
5. Munger CG (1984) 'Inorganic Zinc Coatings; a review'. Materials Technology, pp187-193.
6. Szokolik A (1995) 'Field testing inorganic zinc silicates'. Journal of Protective Coatings and Linings, pp55-67.
7. Devanney J (2006) 'The Short Happy Life of Waterborne Zinc'. Centre for Tankship Excellence, USA http://www.c4tx.org/ctx/pub/znsi.pdf
8. ACA (Edited by Francis RA) (1998) 'Sixty Years of Inorganic Zinc Coatings; History, Chemistry, Properties, Applications and Alternatives', Melbourne.

9. Hemmings E & Demirdjian N (2001) 'Water based inorganic zincs – performance vs practicability', ACA Corrosion & Prevention 2001 Conference proceedings, Newcastle.
10. Munger CG (Revised by Vincent LD) (1999) 'Corrosion Prevention by Protective Coatings' 2nd edition, NACE.
11. Mandeno WL & El Sarraf R, (2014) 'Protective coatings for steel bridges: A guide for bridge and maintenance engineers', NZ Transport Agency, Wellington.

10. AUTHOR DETAILS

Willie Mandeno is the former Technical Principal: Corrosion Engineering and Protective Coatings of Opus International Consultants Ltd, based in Wellington, New Zealand. He is a Chartered Professional Engineer, an IPENZ Fellow, an ACA Corrosion Medallist and ACA Life Member, and a NACE Corrosion Specialist and CIP Instructor.

CORROSION OF STEELS UNDER LOW D.O. CONCENTRATION AND ELEVATED TEMPERATURE

R.E. Melchers & R. Jeffrey
Centre for Infrastructure Performance and Reliability,
The University of Newcastle, Australia

SUMMARY: The immersion corrosion of mild steels, under low dissolved oxygen conditions in stagnant or low flow fresh waters at elevated temperatures, is of interest to a number of industries. Very few results are available and understanding of the variables involved is limited. Herein results are reported for an exposure trial conducted over 12 months under DO concentrations of 2mg/L at 70°C for coupons composed of steels with a range of carbon contents. Corrosion loss was determined from mass loss for recoveries at approximately 2 weeks intervals. Throughout the experiment there was no discernible difference in corrosion loss between the three different carbon contents. The corrosion losses for the coupons exposed to similar DO concentrations exhibited a bi-modal characteristic as a function of time, consistent with other results under low DO conditions. Unexpectedly, the corrosion products formed on the steel coupons were almost entirely black magnetite. Also, the inside of the exposure container and other fittings were coated with a mixture of brown-red hematite and goethite. Further, there were unusually large areas of localized corrosion around the fixing bolt and also around identification holes. These observations are considered to be the direct or indirect result of using commercial sodium hydrogen sulphite ($NaHSO_3$) as the deoxider. The implications are discussed.

Keywords: Corrosion, Steel, Low Dissolved Oxygen, Temperature.

1. INTRODUCTION

In several applications the immersion corrosion of steels under stagnant fresh water conditions is of interest, including in the oil and gas industries, for fire sprinkler systems and water storage systems. However, for the longer term exposures of interest to such industries and to infrastructure more generally data are very limited. The long-term exposures in the Panama Canal Zone (Southwell et al. 1960) considered fresh water but it is unlikely to have been close to stagnant. Standard handbooks such as Schumacher (1979), Kreysa & Eckermann (1992) and Uhlig & Revie (1985) provide only limited data, mainly corrosion rates measured over different periods of time (usually 12 months or less). Mercer & Lumbard (1995) reported results from controlled laboratory tests for corrosion of mild steel in fresh drinking water (hard Teddington water at ambient temperature) and also in triply distilled water (TDW) at various temperatures up to 70°C. The most comprehensive data, and interpreted trends for longer-term corrosion, were given by Melchers & Chernov (2010) for 2.5 years exposure in a water heating storage tank operated at 50-60°C and for a range of dissolved oxygen (DO) concentrations. The water in this case was drinking water and thus could be considered relatively pure.

The present paper is concerned with an attempt to replicate the above results so as to (1) independently check these earlier observations, (2) determine more accurately the corrosion loss trend with time for stagnant fresh water conditions and (3) study the development of pitting corrosion with time. Herein only the first two points are addressed. The next section describes the test programme, the results obtained and the observations made. As will be described, some of the observations were unexpected, for reasons considered in the discussion section. The implications of the observations, and the overall trend for corrosion loss are then discussed. It will become apparent that the results have interesting practical as well as theoretical implications.

2. TESTING PROGRAMME

2.1 Test Environment

To try to replicate the earlier test results as closely as possible it was decided to use ultra-pure water rather than TDW or local (soft) fresh drinking water. The ultra pure water was sourced from a local commercial supplier. Its composition is shown in Table 1 together with that of the local drinking water. The water quality analyses were performed by an independent certified laboratory (ALS Water, Newcastle).

Table 1. Composition of ultra-pure water and local tap water

	pH	Dissolved Oxygen (mg/L)	Chloride (mg/L)	Nitrate (mgN/L)	Nitrite (mgN/L)	Total Oxidised N (mgN/L)	Sulphate (mg/L)	Sulphide (mg/L)
Ultra-Pure water	5.8	7.83	<1.0	< 0.05	< 0.03	< 0.05	< 2	< 0.05
Local tap-water	7.18	96.3	2	0.013	0.002	0.015	27	< 0.05

The water for the experiment was held in an exposure drum composed of two 200L drums welded together end-to-end and externally heavily insulated as shown in Figure 1. Internally the drum was given 3 coats of heavy-duty epoxy protective coating. For maintaining the water temperature the drum was fitted with two 1000W band heaters connected in series, resulting in an effective heating power 500W. The heaters were located 380mm and 550mm inside and above the base of the drum. A temperature sensor was used to regulate the power supplied to the heaters. This used a Type-T thermocouple attached to the exterior metal surface of the drum, and mounted 1000mm from the base. The temperature was controlled using a Eurotherm P116 digital PID temperature controller. This equipment was able to maintain the temperature within 0.5°C of the 70°C set point at the heaters.

Figure 1. External insulation on the two 200L drums forming the exposure vessel.

Figure 2. One of 5 white plastic tubular coupon holders, with steel coupons held by nylon bolts and soft plastic spacers behind.

2.2 Test Coupons

For the test coupons, three different steels were used, as summarized in Table 2. The low carbon content steel was available only in the size show. The other two were distinguished from one another by a slightly different coupon thickness. Despite these differences, all coupons had similar total surface areas and similar total edge lengths. For conversion from mass loss to corrosion loss the total exposed surface area was used in each case.

Table 2. Size of coupons (100mm long) and their composition (%wt)*

	Cross-section (mm)	C	Mn	P	S	Si	Ni	Cr	Mo	Cu
'Pure' iron	38 x 6.4	0.01	0.05	0.007	0.004	0.00	0.02	0.02	0.00	0.01
Low carbon steel	50 x 4	0.04	0.19	0.005	0.013	0.00	0.04	0.11	0.02	0.04
'Mild' steel	50 x 2	0.06	0.20	0.009	0.009	0.00	0.01	0.01	0.00	0.02

*Analysed by Bureau Veritas, Newcastle, using an energy dispersive spectrography (EDS).

2.3 Experimental Procedures

Coupons were exposed in and recovered from the exposure tank in sets of three, one from each set of carbon content. This was facilitated by attaching the coupons to the inside surface of five 10L plastic cylinders. Each of the five cylinders had 6 sets of coupons (18 in all) attached using plastic fasteners and spacers. The cylinders were approximately 300 mm in depth with the coupons attached in two rows, 100 mm and 200 mm from the top (or bottom) of each cylinder (Fig. 2). In each cylinder the coupons were attached sequentially; 0.01 %C, 0.04 %C and 0.06% C, so that any recovery of three consecutive coupons included one from each carbon content sub-group. Hence there were three sets at 100 mm from the top of each cylinder and 3 sets 200 mm from the top of each cylinder. The cylinders themselves were immersed to various depths, commencing with one near the base of the drum and the others at water mid-depth and also just below the surface of the water. One set of (3) coupons was recovered nominally every two weeks for the whole period of the experiment (367 days).

Prior to the coupons being placed in the exposure tank they were cleaned of mill-scale using inhibited 50% HCl solution and, after thorough washing drying, weighed to the nearest 0.1 mg. After exposure, in most cases the corrosion product was almost entirely black magnetite that could be removed readily by washing with clean tap water and scrubbing the surface with a soft plastic brush. Later in the experimental programme there were some instances for which the coupon had to be cleaned of the majority of corrosion product with a flash dip (less than 10 seconds) in inhibited 50 % HCl solution. The corrosion products were analysed by The University of Newcastle Electron Microscope and X-Ray Unit employing a Philips X'Pert MPD XRD with associated standard equipment.

Apart from the reduction in DO caused by the elevated temperature, sodium hydrogen sulphite $NaHSO_3$) was added to the water prior to the commencement of the experiment to reduce the DO content further. The intent was to achieve a low DO content of around 1 mg/L to allow comparison to other test results. The dissolved oxygen (DO) content of the water that was actually achieved in the tank was measuring using a Horiba OM51 DO meter with a (9551-20D/100 D) DO electrode to obtain a DO profile from the base of the tank to the water surface at 100 mm vertical increments. The same Horiba instrument was used to measure the water temperature vertical profile along the centreline of the tank. The water quality in the test tank was sampled an analysed at the end of the experiment.

3. EXPERIMENTAL RESULTS

The corrosion losses estimated from mass loss and the size of the coupons and steel density are summarized in Fig. 3. Comparing the 3 different steels it is seen that there is negligible difference between the corrosion losses below about 120 μm and that for higher corrosion losses the 0.01%wt C steel tends to have corroded slightly more than the 0.04%wt C steel that in turn corroded more than the 0.06%wt C (mild) steel. Only the results for the 0.06%wt C (mild) steel are considered below.

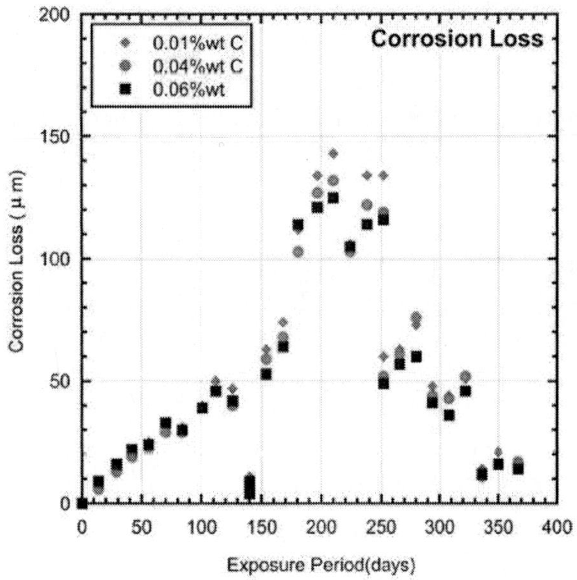

Figure 3. Corrosion loss as a function of exposure period for all coupons, irrespective of DO concentration.

The results of the analysis of the water n the test tank are shown in Table 3. The chloride, nitrate, nitrite, and total oxidized nitrogen are very low as expected, but compared to the water that was added to the tank (Table 1), it is clear that the pH and DO concentration are much lower and the concentrations of sodium, sulphate and in particular iron are much higher. The reasons for this are considered further below.

Table 3. Tank water quality analysis at conclusion of experiment

Analysis	Units	Value
pH		3.9
Chloride	mg/L	< 1
Iron	µg/L	14,500
Nitrate	mgN/L	< 0.05
Nitrite	mgN/L	< 0.03
Sodium	mg/L	18.4
Sulphate	mg/L	80
Sulphide	mg/L	< 0.05
Total Oxidized Nitrogen	mgN/L	< 0.05

Analysis by ALSWater.

The dissolved oxygen concentration in the water column is shown in Fig. 4a. It shows 2 profiles, obtained about one month apart. Fig. 4 also shows the vertical location of the rings of coupons as held inside the cylinders (Fig. 2). Fig. 4b shows the water temperature profiles. Using the locations of the coupons shown in Fig. 4a, the results shown in Fig. 3 can be readily interpreted, as shown in Fig. 5 for the 0.06%wt C steel. In this case mass losses for coupons exposed to similar DO concentrations have been grouped. For the main data set, with DO around 2.5mg/L, the interpreted trend show has been drawn through the data. It is based on generally similar trends observed in data for generally similar DO concentrations and similar water temperature (Melchers & Chernov, 2010). The trends for the lower DO concentrations are schematic, based, again on the earlier work. Although not shown, the data and the possible interpreted trends are almost identical for the lower carbon steels.

It is seen (Fig. 5) that when dissolved oxygen concentration is considered the results in Fig. 3 take on a more consistent pattern. The coupons exposed in the height range 900-1150mm from the base of the tank can be considered to form one cohort and the trend of corrosion loss with time shows a distinct bi-modal characteristic, as expected. The coupons with the lower corrosion losses at longer exposure periods are all associated with lower levels of water DO concentrations.

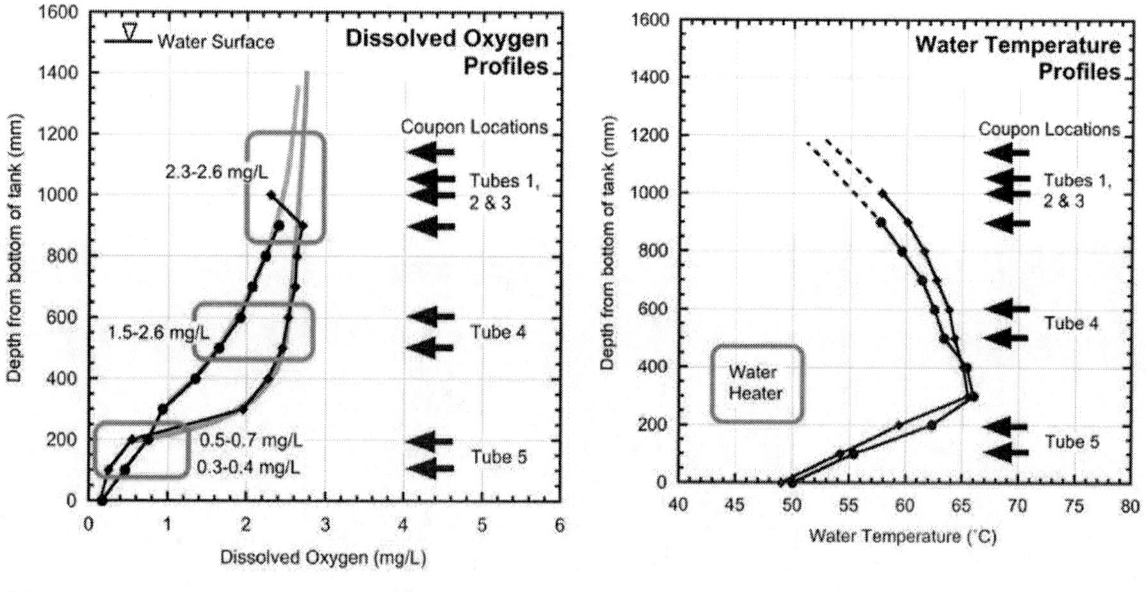

(a) (b)

Figure 4. Profiles of (a) dissolved oxygen concentration and (b) water temperature.

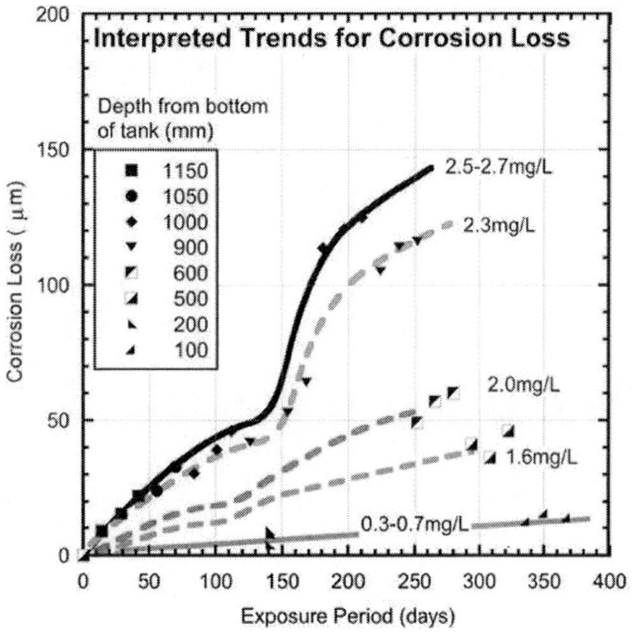

Figure 5. Corrosion loss as a function of exposure period for all mild steel (0.06%wt C) coupons grouped by DO concentration, together with corresponding interpreted trends. The upper two trends correspond to an average water temperature of around 55°C and the lower ones to around 60-65°C (Fig. 4b).

4. OTHER OBSERVATIONS

4.1 Rusts on coupons and in the tank

Unlike conventional red-brown corrosion products that form on steels under immersion conditions, the only rust formed on the steel coupons, throughout the experiment, was black in colour. XRD analysis showed that it was almost pure magnetite (Fig. 6). This is not inconsistent with corrosion of steel in the reduced DO in the water. It was noted that the coupons with the greatest corrosion loss also had the greatest amount of magnetite and that this was difficult to remove. Conversely, coupons with little corrosion loss exhibited much less magnetite and this was easy to remove by washing with some light brushing under water.

Shortly after the commencement of the experiment it was noted that there was some evidence of light brown-red colouring on the interior surface of the tank and also on the white plastic tubes holding the coupons and their polyethylene ropes. XRD analysis of these deposits showed them to be predominantly hematite and some goethite (Fig. 7). The build-up of these deposits continued throughout the experiment. The results of the quantitative analyses of the rusts are shown in Table 4. Fig. 8 shows the inside of the one of the plastic coupon holders.

Table 4. Composition of the iron oxides formed during the experiment.

			Red-brown Rust	Black Rust
Hematite	iron(III) oxide	α Fe_2O_3	~ 83 %	
Goethite	iron oxyhydroxide	α-FeO(OH)	~ 17 %	
Magnetite	iron(II,III) oxide	$Fe^{2+}Fe^{3+}{}_2O_4$		> 97.0%
Jacobsite	iron(II,III) manganese oxide	Mn Fe_2O_4		< 3.0%

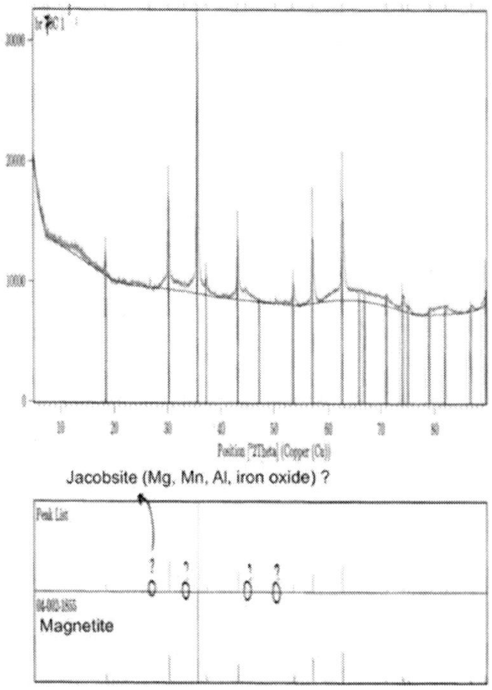

Figure 6. XRD profile for the black rusts showing predominantly magnetite.

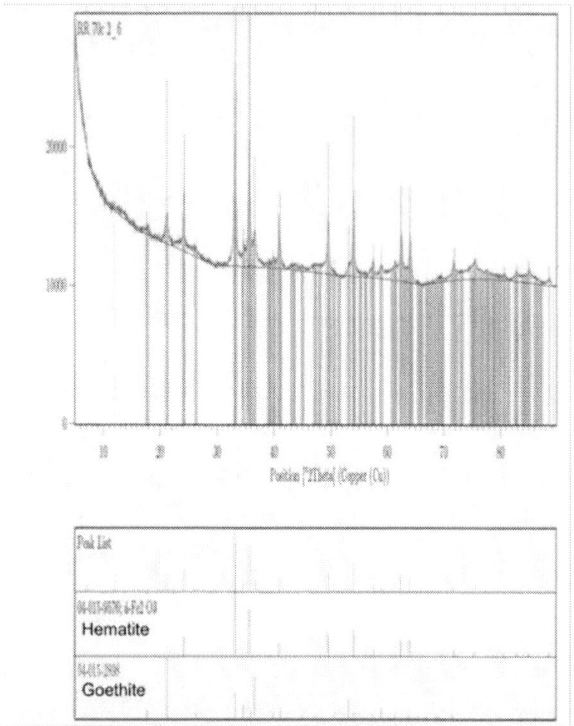

Figure 7. XRD profile for the red-brown rusts showing predominantly hematite and some goethite.

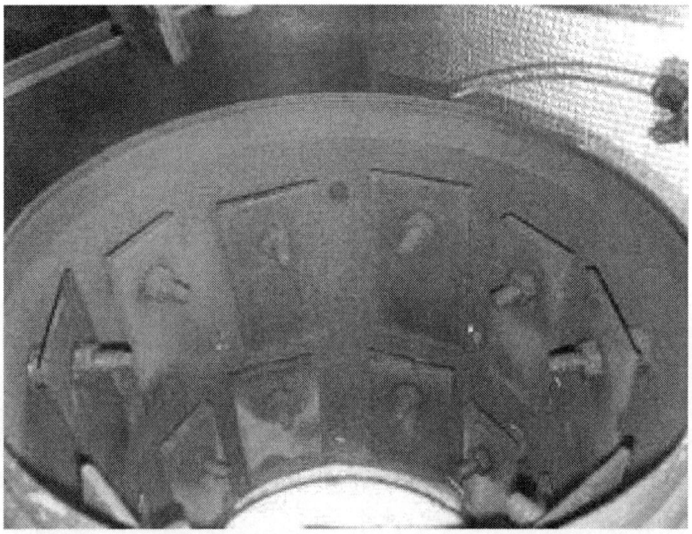

Figure 8. Inside view of circular tube (originally) white plastic coupon holder showing widespread presence of brown-red rusts but black rust on the steel coupons.

4.2 Corrosion patterns on coupons

During the washing and cleaning of the coupons after recovery it was noticed that a lighter area, roughly circular, was very clearly visible around the hole that was used to suspend the coupon using a nylon bolt and nut (Fig. 8). Initially it appeared that this lighter area had undergone less corrosion. For convenience that area will be termed a 'corona' herein, partly in view of the fact that for the earlier coupons in particular the corona tended to disappear from view. For the coupons exposed for longer periods the corona tended to remain visible. They were not all the same size and were not always obvious, although in all cases the coupons showed what appeared to be a ring of differential aeration in the region under the nylon nut. Two examples are shown in Figs. 9 and 10.

Figure 9. Coronas on one side (from left to right) of the 0.04%wtC, 0.01%wt C and 0.06%wt C (mild) steels after 42 days exposure. In each case the shape of the hexangular nylon nut is clearly visible and the coronas appear to represent differential aeration almost uniformly around the edge of the nut. Note also a smaller corona on the left coupon around the identification hole. In this case the mild steel shows the largest corona.

Figure 10. Coronas on one side of 0.06%wt C, 0.04%wt C and 0.01%wt C steels after 181 days exposure. In these cases, too, the shape of the hexangular nylon nut is clearly visible. In this case the low carbon coupon has the largest corona. They were not observed on the other side of the coupons.

5. DISCUSSION

What was started as a simple but detailed confirmation of corrosion loss trends (and development of pitting corrosion not reported here), revealed some rather unexpected observations. Working backwards, to try to understand better the corona phenomena, microscope observations of the metal surfaces were made to establish changes in coupon corroded surface topography in moving away from the hole for the nylon bolts. Also, micrometer observations were made of the changes in coupon thickness with location on the coupons.

In addition to the coupons shown in Figs. 9 and 10, the surfaces of coupons from 14 days exposure onwards were examined. They also showed some degree of development of coronas, on one side (the side where the nut was located, and thus as shown in Fig. 2) and in general the coronas were larger for longer exposures. Also, the coronas were much more prominent on the side of the coupon that was immediately under the nylon nut (on one side). In all cases this was the thickest part of the cross-sectional profile. This relatively narrow, approximately annular, zone extended radially outwards from the edge of the fixing hole to a distance less than the width of the nut, typically 1-2mm outwards. The surface of that zone in all cases was pitted. These pits were almost exclusively isolated, circular shaped pits, similar to those reported by Butler et al. (1972) for steels in neutral solutions, remarkably uniform in size, about 100 microns in diameter and typically about 300-350 microns deep. This was irrespective of steel composition. On the same side as the nut, beyond the zone under the nut the surface was corroded in what appeared to be to a rather uniform plateau that had corroded some 100-120 microns inwards. This zone corresponded to the corona, and was distinctly different in appearance under the microscope than the annular zone that has been under the nut. It appeared as generally corroded but with many circular and near circular pits, and some irregular shaped localised corrosion. Again, the typical depth of these was remarkably uniform over this zone, with depths 300-350 microns, but just occasionally (estimated at <5% of all pits) deeper - 600-700 microns deep in total. Many of these showed some degree of internal plateauing.

There was no evidence for any of the coupons examined, of a ring of localised corrosion (a groove) aligned with the edge of the nylon nut, as might be expected for classical differential aeration at an edge. Under the microscope there appeared to be just a gradual transition from the thicker part of the coupon to the slightly thinner parts elsewhere.

Beyond the approximately annular region associated with the corona, the corroded surface was of similar depth overall, extending over the whole of the rest of the coupon, almost right to the edges. However, although the depth of corrosion overall was similar, there was an obvious difference in surface texture when viewed under the microscope. Compared to the surface in the corona zone, that of the outer surface was distinctly 'rougher' in appearance, with very few circular pits and much localised corrosion. The depths of such localised and pitting corrosion were around 300-350 microns in most cases, with very occasional much deeper pits, generally similar to that observed in the corona zone. The surface also was slightly darker in appearance (cf. Figs. 9 and 10). This and the different 'roughness' may account for their different overall appearance. It does not explain the mechanism(s) responsible. However, some speculation is possible.

Under immersion conditions the nylon nut against the steel surface will act to set up classical differential aeration (DA) along the edges of the nut (Evans, 1960). This can be seen in the distinctive hexagonal shape of the surface of the coupon immediately adjacent to the fixing hole. Similarly, the identification holes drilled through the coupons all can act as initiators of differential aeration (Fig. 9 lower left). Typically, the DA mechanism will draw ferrous ions Fe^{2+} from under the nylon nut, causing localised material loss. The mechanism involved is essentially the same as pitting corrosion and requires a cathodic region, typically radially around it. At the same time, since the whole coupon is in the exposure environment, there also will be generalized corrosion over the rest of the surface, and some interaction can be expected between them. It is considered likely that the relative rates of occurrence of these two independent processes, corrosion driven by differential aeration from around the nut, and general corrosion elsewhere, will determine the extent of the corona. This also may explain the differences between the considerably 'smoother' or less pitted corona zone compared with the corroded surfaces further away. In this regard, the differences in pit behaviour for steels with different carbon contents is relevant (Melchers et al. 2016). It is clear that further explanation is required.

Regarding the unexpected deposition of hematite on the plastic surfaces, ropes and epoxy coating (Fig. 8) and the almost exclusive deposition of magnetite on the corroding coupons, it may be speculated that this is the result of the use of the deoxidizer sodium hydrogen sulphite ($NaHSO_3$). Although a commercial deoxidizer widely used in industry, there has been a lessening of its use, apparently because of side-effects (personal correspondence). The rust deposition shown in Fig. 8 is likely to be one these. Under normal immersion corrosion conditions the rust products form initially as $Fe(OH)_2$ some of which deposit on the cathodic parts of a steel surface, with electron transfer from anodic zones largely through the metal. Typically, the layers of $Fe(OH)_2$ closest to the metal surface would transform to Fe_3O_4 (magnetite) but the external surface is then oxidized to $Fe(OH)_2$ under the availability of O_2 (i.e. DO) and H_2O (Evans & Taylor, 1972; Wranglen, 1974). However, this process is likely to be slowed under depressed DO conditions, or, more likely, with the presence of $NaHSO_3$, there will be a reaction with the developing $Fe(OH)_2$ external to the magnetite layer, thereby producing hematite, which is water-insoluble material and thus will tend to deposit on surfaces (particularly those that are, relatively, alkaline):

$$2Fe(OH)_2 + 2NaHSO_3 + 1.5O_2 \rightarrow Fe_2O_3 + 2NaHSO_4 + 2H_2O \quad (1)$$

Other reactions also are possible, such as the direct oxidation of $NaHSO_3$ with the dissolved oxygen (DO) in the water to form sodium hydrogen sulphate ($NaHSO_4$) (acidic in solution), leaving less oxygen in solution for the corrosion process. Note that these reactions are consistent with the observed high Fe, sodium and sulphate contents in the water, and its low pH at the end of the experiment (Table 3).

Irrespective of the precise mechanisms and reactions involved, the net effect is the depression of the amount of DO in the water and reduction in the amount of corrosion of the steels, as indeed is the intended use of $NaHSO_3$. From the present perspective, however, it also means that the corrosion product on the surface of the metal is almost entirely magnetite. This, in turn, has interesting implications.

Rust deposition has long been considered the cause of the early decline of the rate of corrosion as a function of exposure time and thus as a function of increased rust deposition (Tammann, 1923; Evans 1960). In fact, Evans and Taylor (1972) noted that freshly deposited magnetite has little in common, structurally and physically, with conventional magnetite mineral, noting also that after deposition as a soft material it gradually becomes more structured and less permeable with time. Further, the build-up of the rust layers and the reduction in their oxygen permeability properties has been proposed as the driver for the bi-modal behaviour (Fig. 5) (Melchers, 2014). Typically, in addition to magnetite, the rust layers are composed of $Fe(OH)_2$, $Fe(OH)_3$ and FeOOH but these are all known to be relatively much more permeable than layered magnetite. It is therefore of interest that the present experimental results show that the bi-modal behaviour is attained with magnetite as the only (significant) rust product in the rust layer. The corollary of this is that, at least for ferrous metals, magnetite is an important part of the development of bi-modal behaviour.

Comparison of the trends in Fig. 5 with other experimental results for elevated temperatures and/or low DO conditions (Mercer & Lumbard, 1995; Melchers & Chernov, 2010; Melchers 2006) shows a high degree of consistency, but direct comparison is rendered difficult because of the use, in the present case, of the deoxider $NaHSO_3$ which also lowers the water pH. This has an effect on the development of corrosion with time (Melchers 2006). The present experimental results add to the range of information available but it is clear that further efforts are required to correlate the various observations and to derive the parameters for the bi-model model as functions of DO, temperature and pH.

Finally, while the use $NaHSO_3$ as a deoxider might, in retrospect, be considered to have been a mistake, it did reveal two features not, apparently, previously noted in the literature - coronas and magnetite as the only corrosion product on the corroding steel.

6. CONCLUSION

The following conclusions may be drawn:

1. The bi-modal trend for corrosion loss was observed for 3 low alloy steels with carbon content of 0.01%wt, 0.04%wt and 0.06%wt in ultra-pure water dosed only with $NaHSO_3$ to achieve a DO concentration of around 2.3mg/L waters and heated to around 60°C water temperature,

2. The bi-modal trend was observed with magnetite as the only significant rust product on the corroding steel,

3. The occurrence of the magnetite on the coupon and hematite in all other surfaces in the exposure tank is attributed to the deoxiding effect of the added $NaHSO_3$,

4. Irrespective of carbon content, after about 4 weeks exposure all coupons revealed the presence of coronas around the location of the fixing hole and also around some identification holes, and this is attributed to a development of differential aeration starting under the nylon fixing nut.

7. ACKNOWLEDGMENTS

The present work is one component of a research project dealing with the development of the bi-modal characteristic of immersion and other corrosion, funded by the Australian Research Council under a DORA award. The authors much appreciate that support, as well as that of the Civil Engineering laboratories at The University of Newcastle, and in particular that of senior electronics technical officer Ross Gibson.

8. REFERENCES

Butler, G., Stretton, P. and Beynon, J.G. (1972) Initiation and growth of pits on high-purity iron and its alloys with chromium and copper in neutral chloride solutions, British Corrosion Journal 7(7) 168-173.

Evans U.R. (1960) The corrosion and oxidation of metals: scientific principles and practical applications, London, Edward Arnold.

Evans U.R. and Taylor, C.A.J. (1972) Mechanism of atmospheric rusting, Corrosion Science, 12: 227-246.

Forgeson, B.W., Southwell, C.R. & Alexander, A.L. (1960) Corrosion of metals in tropical environments - Part 3 - Underwater corrosion of ten structural steels, Corrosion, 16: (3)105t-114t.

Kreysa, G, Eckermann, R. (1992) Dechema Corrosion Handbook, Vol. 11, New York, VCH Publishers.

Melchers, R.E. (2006) Modelling immersion corrosion of structural steels in natural fresh and brackish waters, Corrosion Science, 48(12): 4174-4201.

Melchers, R.E. (2014) Microbiological and abiotic processes in modelling longer-term marine corrosion of steel, Bioelectrochemistry, 9; 89-96.

Melchers, R.E. & and Chernov, B.B. (2010), Corrosion loss of mild steel in high temperature hard freshwater, Corrosion Sci., 52, 449-454.

Melchers, R.E. Chaves, I.A. Jeffrey, R. (2016) A conceptual model for the interaction between carbon content and manganese sulphide inclusions in the short-term seawater corrosion of low carbon steel, Metals 6(6): 132. doi: 10.3390/met6060132.

Mercer, A.D. and Lumbard, E.A. (1995) Corrosion of mild steel in water, British Corrosion Journal 30: 43-55.

Schumacher, M. (Ed) 1979. Seawater Corrosion Handbook, Noyes Data Corporation, New Jersey.

Tammann, G. (1923) Lehrbuch der Metallographie, 2nd Edn., Leipzig, Leopold Voss.

Uhlig, H.H. & Revie, R.W. (1985) Corrosion and Corrosion Control, 3rd Ed, John Wiley and Sons, New York.

Wranglen, G. (1974) Pitting and sulphide inclusions in steel, Corrosion Science, 14: 331-349.

9. AUTHOR DETAILS

Robert E. Melchers is Professor of Civil Engineering at The University of Newcastle, Australia. He holds a BE and MEngSc from Monash University and a PhD from the University of Cambridge, UK. He is a Fellow of the Australian Academy of Technological Sciences and Engineering and an Honorary Fellow of The Institution of Engineers Australia.

His most recent awards are the 2009 ACA Corrosion Medal, the 2012 Jin S Chung Award (International Society of Offshore and Polar Engineers) and the 2013 John Connell Gold Medal (The Institution of Engineers, Australia). He was the Institution of Engineers, Australia Structural College Eminent Speaker for 2014.

Robert Jeffrey is a research scientist at The University of Newcastle, Australia where he has been investigating corrosion in marine, tidal and atmospheric conditions for many years. He holds a BSc (Chem) and PhD from Newcastle University. Robert is also principal consultant for Pacific Testing Pty Ltd, a company specializing in corrosion problems.

He is an honorary life member of the ACA(2011), been awarded the ACA Corrosion Medal (2012), a past president of the ACA and has been on the committee of the Newcastle branch for over twenty years. He co-authored a paper that won the prestigious T P Hoar Prize (Institute of Corrosion, UK) and has twice been presented with the ACA's Marshal Fordham award, the AC Kennet award.

THE ENTRANCE BRIDGE: DETERIORATION AND REMEDIATION CHALLENGES

M Karlaftis [1], B McGuiness [1] & M Ali [2]
Marine & Civil Maintenance Pty Ltd [1], Sydney, Australia,
GHD Pty Ltd [2], Sydney, Australia

SUMMARY: The Entrance Bridge, on the Central NSW coast, was built circa 1967 and consists of 18 x 26m spans of cast in situ hollow box concrete super structure supported on pairs of 900mm diameter reinforced concrete columns.

The condition of the bridge was severely affected by its exposure to the marine environment with concrete spalling and delamination identified on the majority of the pier columns and in the deck soffit.

NSW Roads & Maritime Services (RMS) engaged GHD in 2006 to undertake a condition assessment and design of a cathodic protection system for the bridge. GHD further undertook a strength and stability assessment for the proposed remediation works proposed. Removal of damaged concrete and drilling holes in the columns for cathodic protection were assessed against detailed analysis of the structure. Subsequently, a staged concrete repair methodology was proposed with the utilization of strain gauge monitoring of the structure to analysis its response to the remediation work.

Marine and Civil Maintenance (MCM) was awarded the repair and cathodic protection contract in 2014. The parties to the contract devised innovative access and repair methodologies. The challenges included: working in a sensitive environment over an endangered seagrass, working on a bridge with load restriction and staging the works in order not to undermine its structural integrity, carrying out concrete demolition work around sensitive strain gauges, and responding to significant increases in work scope by adapting the access system and methodologies proactively through the project.

This paper details the challenges faced in refurbishing this unique bridge and the innovation employed in overcoming the challenges.

1. INTRODUCTION

1.1 The Structure

The Entrance Bridge (numbered BN1712 by RMS Newcastle Asset Management) is situated over the entrance to Tuggerah Lakes in the Wyong local Government area. Refer Photo 1. It was built circa 1968 and consists of eighteen (18) spans with seventeen piers of two single columns within the waterway. It is a continuous hollow box girder reinforced concrete structure. Several piers are connected directly into the deck, with others through fixed pit bearings and piers on the approaches of the bridge having glide restraint bearings.

Photo 1. General View of The Entrance Bridge

1.2 History of the Bridge

Opened in 1934, the inaugural Entrance Bridge was originally a wooden single lane bridge with bypass bays at intervals that allowed traffic travelling in the opposite direction to pass [1]. The original bridge was adequate until 1965 as it only served traffic travelling as far as The Entrance North.

Around 1955, the Number 11 bus operated by Red Bus Services crashed through a wooden retaining wall and landed in The Entrance Channel after its brakes failed while turning onto The Entrance Bridge. The driver and three passengers were uninjured, although the boat that the bus landed on was heavily damaged.

In 1965, Wilfred Barret Drive was opened by the Deputy Premier, the Hon. Pat Hills. The road was named after Councillor Wilfred Barret who was a member of the first Wyong Shire Council. The road continues to be an important link between The Entrance and Toukley, passing through Wyrrabalong National Park.

By the end of 1965 the bridge was considered dangerous and too costly to maintain. The traffic flow over the old bridge at The Entrance increased to such an extent the structure was under constant repair. Work began on the current two lane, concrete structure on 17 July 1967 after a successful tender by Transbridge Pty Ltd. The bridge was opened to traffic on 20 December 1968 and officially opened by the NSW Premier, the Hon. Robert Askin on 18 April 1969.

2. REMEDIATION METHODOLOGY DEVELOPMENT

2.1 Condition Assessment

RMS observed significant deterioration on the concrete piers in mid-2000 and in 2007,commissioned GHD to undertake a condition investigation (consisting of both durability and structural assessments) of The Entrance Bridge. The investigation was undertaken in November 2007.

The bridge deck soffit was found to be suffering from spalling and delamination. The pier column was mainly suffering from significant vertical cracks in addition to isolated concrete spalling.

Photo 2. Concrete spalling leaving exposed reinforcement on the deck soffit

Photo 3. Typical cracks observed in the pier columns

In addition to visual inspection, GHD undertook the following diagnostic testing on a selected number of pier columns:

- Covermeter survey;
- Chloride content analysis;
- Electrode potential mapping;
- Carbonation depth measurement;
- Concrete electrical resistivity measurements; and
- Reinforcement electrical continuity testing

The investigation finding summary included

- The Pier Columns, Deck Soffits and Deck Sides were deteriorating due to corrosion of reinforcement. In 2007, total damaged area was estimated at approximately ~60 m^2
- The dominant corrosion mechanism for the Pier Columns, Deck Soffits and Deck Sides was chloride ingress. Carbonation was not considered to be the mode of deterioration present to these elements. When the GHD's 2007 investigation results were compared with the results in earlier investigations undertaken by others, it became clear that the damage had progressed significantly during 2000-2007.
- For the deck soffit, the damage was mainly around the Pier columns where reinforcements were too congested and concrete cover to reinforcement was less.
- For the pier columns, the damage was mainly in the bottom 2m section above the high water level (HWL) mark. The chloride content was also found to be high for the bottom 2m above HWL, exceeded the industry accepted corrosion initiation threshold value of 0.06% (by weight of concrete) at the reinforcement level.
- The reinforcement was generally found to be electrically continuous that made any electrochemical repair such as cathodic protection (CP) feasible.

After the investigation/assessment finding, three repair options were presented and discussed with RMS:

- Option 1: Conventional patch repair to damaged areas and coating application to all elements.
- Option 2: Combination of CP patch repair, CP to the lower 3m of Pier Columns, Deck Soffits and Deck Sides, and coating application to Footpath Slab. Decorative coating to all.
- Option 3: Combination of CP patch repair, CP to the lower 3m of Pier Columns, CP to the deck soffit around the pier columns (3mx3m), and concrete patch repair and coating to the remaining areas. Decorative coating to all.

A risk based analysis and workshop was undertaken for these options considering the cost, risk and technical benefits with Option 3 being the preferred option and adopted by the client.

2.2 Concept Design

A concept CP design was prepared in August 2012 that recommended application of a CP system to the following areas:

- 4m section of the pier columns: (1m below the MHWS and 3m above MHWS) using discrete anodes;
- Deck Soffit (~3 m x ~3 m) around the piers, also using discrete anodes.
- Two Transformer Rectifier (TR) units, one at the south end and the other at the North end.

2.3 Detailed CP Design

The detailed CP design was a continuation of the concept CP design with the additional design features:

- Utilisation of AS2832.5 guidelines and experience gained by the designer in CP to similar structures
- Incorporation of vertical CP zoning to account for tidal, splash and atmospheric areas
- Incorporation of horizontal CP zoning to (1) limit the Zone current, (2) minimise attenuation (voltage drop), and (3) minimise use of very large sized cables.
- Monitoring of the systems via four permanent reference electrodes per CP Zone (typical)
- A workshop with RMS Engineers and the CP designer determined that due to dense reinforcement present in the pier columns splash and atmospheric area, discrete anode installation may not be feasible. Ribbon anodes with high cover (40mm) from concrete surface was specified for the splash and atmospheric areas of the pier columns. Additionally, a waterproofing coating was specified on the top 4mm of the ribbon anode slots to minimise anode backfill deterioration
- Each CP Zone current was limited to 5A.

Table below provides CP zoning details.

Table 1: Electrical Zoning Detail of the CP System

T/R UNIT	Zone	ANODE TYPE	ELEMENTS COVERED
TRU 1 –South TR Section A	1	DISCRETE	Tidal Region of Pier Columns 3 - 5.
	2	RIBBON	Splash Region of Pier Columns 2 - 5
	3	RIBBON	Atmospheric Region of Pier Columns 2 - 5
	4	DISCRETE	Deck soffit of Pier 2 - 5
TRU 1–South TR Section B	5	DISCRETE	Tidal Region of Pier Columns 6 – 8
	6	RIBBON	Splash Region of Pier Columns 6 - 9
	7	RIBBON	Atmospheric Region of Pier Columns 6 - 9
	8	DISCRETE	Deck soffit of Pier 6 - 9
TRU 2–North TR Section C	9	DISCRETE	Tidal Region of Pier Columns 10, 12 & 13
	10	RIBBON	Splash Region of Pier Columns 10 - 13
	11	RIBBON	Atmospheric Region of Pier Columns 10 - 13
	12	DISCRETE	Deck soffit of Pier 10 - 13
TRU 2 –North TR Section D	13	DISCRETE	Tidal Region of Pier Columns 14 – 16
	14	RIBBON	Splash Region of Pier Columns 14 - 17
	15	RIBBON	Atmospheric Region of Pier Columns 14 - 17
	16	DISCRETE	Deck soffit of Pier 14 - 17

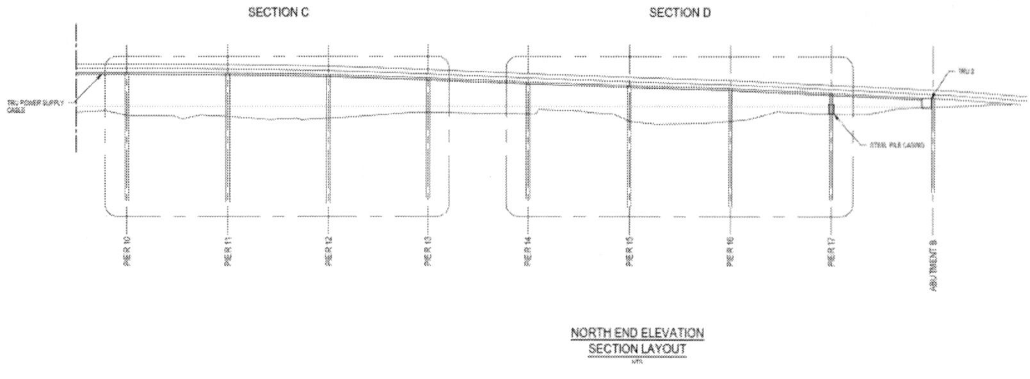

Figure 1. Section identifiers and pile numbers for the northern section of the bridge

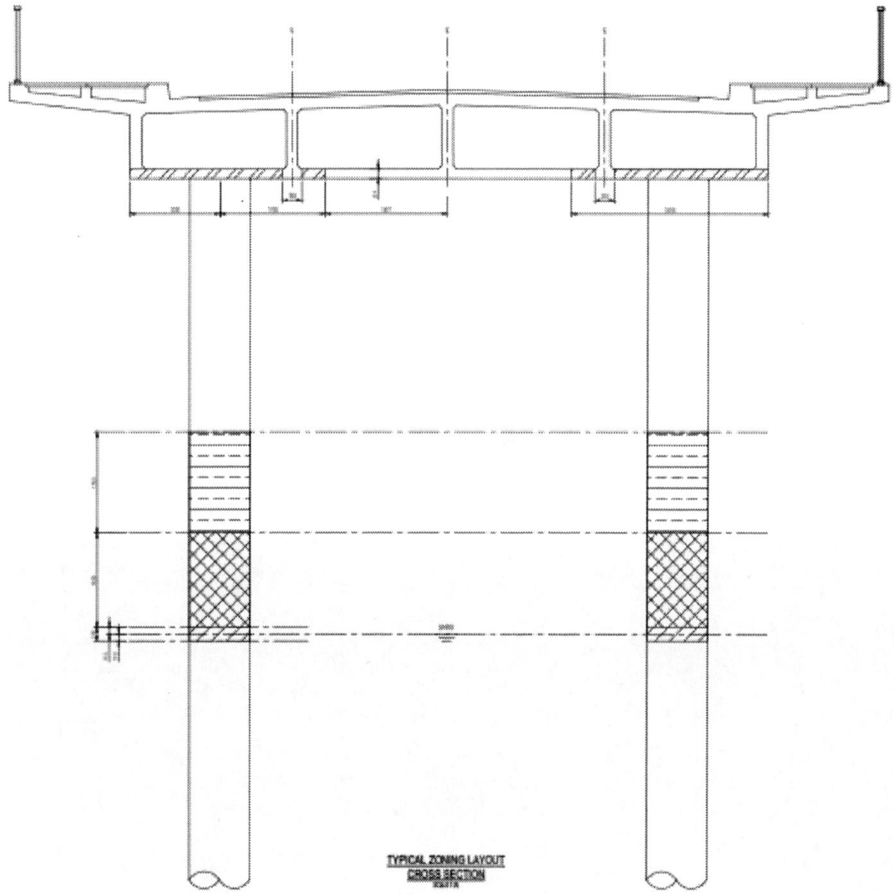

Figure 2. CP Zoning (vertical) to the piers

3. REMEDIATION CONTRACT SCOPE

3.1 Contract Particulars

In November 2013, The Roads & Maritime Services NSW (RMS) invited tenders for the repair and protection works to be completed over an approximate 12-month period.

The contract was based on New South Wales Government GC21 (Ed. 2) General Conditions of Contract and incorporated a payment schedule consisting of both lump sum and contract rate items.

The tender contract period (based on initial scope) was 40 weeks from the date of award, with a defects liability period of 12 months following practical completion. The contract period was extended post-tender (and progressively throughout the contract) to approximately 48months, to allow for increased repair quantities, design amendments and seasonal weather events.

The major items of work identified in the tender were:

- Concrete repairs to piers, bridge soffit generally, bridge soffit 3mx3m around pier columns, bridge sides, bridge footpath soffit and sides and stairs.
- Application of cathodic protection system to pier columns in accordance with specification BN361 and drawings.
- Application of cathodic protection system to bridge deck soffit (3m x 3m) around pier column tops in accordance with specification BN361 and drawings.
- Installing sacrificial anodes to stair soffit.
- Application of silane treatment to non-CP sections of the bridge including the remainder of pier columns, bridge deck soffit, bridge sides, footpath soffit and sides, stairs and abutment A and B faces.
- Application of decorative coating (coloured) system to pier columns, deck soffit, bridge sides, footpath soffit and sides, stairs and abutment A and B faces.
- Demolish existing CP trial area and remove thermal monitoring box.
- Scupper Extensions

The contract was awarded to Marine & Civil Maintenance Pty Ltd (MCM) in March 2014.

3.2 Practical Constraints

Due to the slender nature of the columns, RMS considered it prudent to undertake a structural assessment of the deterioration and the proposed repair actions. A structural assessment of the repair was undertaken that suggested the following safety measures for the repair process:

- No heavy vehicle more than T44 loading allowed on the bridge.
- Refurbishment of one column of the pier to be carried out at a time. Work on subsequent column to be carried out after 14 days of completion of the first column.
- Works not to be undertaken on two adjacent piers concurrently
- Vehicular speed restricted to 50 km/hr.
- No pedestrian crowd loading allowed on the footpath.
- No drilling on the new patched grout until it reaches 40MPa strength or higher.
- Strain gauges shall be provided to monitor increase in stress in column section when concrete removal is undertaken for the repair.
- Identification, isolation and protection of sensitive areas including Indigenous Heritage sites and areas of protected marine seagrass

3.3 Staging of Works

The works were staged to ensure protection of the structure during the works in accordance with the practical restraints defined in the project documentation.

Throughout the project, various access systems were installed to ensure at least one work location was accessible at any one time within each compartment, defined as follows:

- Compartment 1 (Abutment A – Pier 6)
- Compartment 2 (Pier 7 – Pier 12)
- Compartment 3 (Pier 13 – Abutment B)

4. ACCESS

Access was provided to all areas of the bridge soffit, piers, abutments and bridge sides, using an innovative rail mounted, aluminium system developed collaboratively between MCM, rope access subcontractor Townview Australia and structural engineers, Core Project Consulting. This system placed significantly less load on the structure than standard scaffolding, and imposed a reduced shadow on the seagrasses.

Throughout the project, as the scope of work increased other access systems were developed and used to ensure suitably accessible works areas and continued productivity of an expanding labour force.

The following is a high-level summary of the access systems implemented across the project:

- A twin-level aluminium work platform was used to provide access to the piers. The platform also gave access to the pier soffits above the pier columns. All concrete repairs and CP installation to these areas was via this access system. The system was installed and maintained by qualified Industrial Rope Access personnel subcontracted to MCM

- A rail mounted gantry system was originally developed to provide access to the span decks for the Non-CP concrete repairs and the coating application works. Similarly, this system was installed and maintained by Industrial Rope Access personnel. This system was generally used to provide access to span 17, 12 and 2 only. As the gantry was typically 1m wide it became inefficient in locations where the Non-CP concrete repair areas on the deck soffits increased significantly in quantity and size

- Subsequently, a custom designed barge and fixed scaffolding access system was used to replace the gantry access system in the more deteriorated areas. The barge consisted of 2 pontoons (12m x 3m) with a fixed scaffolding system installed by a subcontractor after a stability assessment was carried out by MCM's project marine engineer. The system was designed to allow its overall height to be adjusted for different spans

- Due to the present of an identified and protected Aboriginal Midden at the north abutment, a mobile scaffold was used to provide access to the operations in this area whilst minimising the impact on sensitive areas

- For works to span 1 over the public access road at the south abutment, fixed scaffolding with appropriate traffic control measures was used to provide access to these areas. The work to span 1 was done in two consecutive sections in order to maintain the operation of the road throughout the works

- The works to the stairs at the north end of the bridge was carried out via chain suspended steel access decks

- Cabling works for the CP system was undertaken via mobile elevated work platforms mounted on purpose designed barges, designed and certified by project marine engineers

- Industrial Rope Access as per AS/NZS 4488 was regularly used to gain access to various areas unreachable from the fixed and suspended stage sections and formed the basis of the projects emergency access response plan

Photograph 4. Access Systems

5. REPAIR METHODOLOGY

5.1 Concrete Repairs – General

Much of the concrete breakout was undertaken initially with handheld jackhammers due to the relatively small patch sizes. Hydrodemoliton was employed progressively as the Non-CP concrete repair quantities increased significantly.

The repairs to the piles were generally carried out using form and pour methods, with dry sprayed gunite used for the soffit areas. Smaller patches in the piles and soffit were backfilled by hand rendering. The repair materials were bagged proprietary cementitious products designed for the two processes.

Due to structural integrity concerns for the concrete repair to the pier columns, stain gauges were installed prior to any breakout work commencing. The base line results were obtained and the strain gauges remained in place to the end of the repair work on each column to monitor any structural changes and to identify potential hazards.

5.2 Concrete Repairs - Scope

The extent of repairs required increased significantly over expectations as the works progressed. The table following indicates the difference between the contract provisional quantities initially nominated and the final breakout areas.

Table 2: Concrete Repair Quantities

Concrete Repairs	Estimated Qty	Actual Qty	Difference
P4. CP Concrete Repairs within pier column areas	30m2	54.46m2	24.46m2
P5. CP Concrete Repairs within deck soffit areas	54m2	89.14m2	35.14m2
P6. Non-CP Conventional concrete repair within column areas	8m2	1.41m2	-6.59m2
P7. Non-CP Conventional concrete repair within deck soffit areas	23m2	163.68m2	140.68m2
P8. Non-CP Conventional concrete repair within bridge sides and footpath areas	8m2	20.90m2	12.90m2

5.3 Steel Reinforcement Preparation & Replacement

All exposed reinforcement was cleaned to the standard noted in the technical specification. In CP repairs the exposed steel was prepared equivalent to SA 2 as defined in AS1627.4, while in non-CP repairs the exposed steel was prepared equivalent to SA2.5.

After breakout of defective concrete, all exposed reinforcement was visually inspected and any reinforcement that had lost more than10% of its cross-section area was augmented with new reinforcing.

5.4 Concrete Repairs

The concrete repairs on the piles were generally undertaken using a formed and poured application. In total 54.46m2 at an average of 100mm deep was applied using a high-performance microconcrete, with high strength and low shrinkage.

For repairs within CP areas and larger non-CP repairs in the deck soffits, the concrete was reinstated with dry-sprayed gunite, also with high strength and low shrinkage.

All other small patch repairs were hand patched with a high-build render of similar strength and shrinkage.

5.5 Rebar Continuity

Reinforcement continuity was tested at all negative connections and at locations where steel was exposed.

For all discontinuities on any part of the structure, 6mm diameter continuity bars were welded in place and the steel retested.

5.6 Sacrificial Anodes

In accordance with the technical specification, in all non-CP repairs, sacrificial zinc anodes were fixed to the reinforcing steel all round the perimeter of each repair, the spacing being dependent on the density of reinforcement at each location.

Because the size of such repairs was smaller than expected but their number was much greater, the number of anodes was considerably higher that estimated. The table below quantifies this difference.

Table 3: Sacrificial Anode Quantities

Sacrificial Anodes	Tender Qty	Actual Qty	Difference
P2. Establishment of continuity Reo & ancillary steel	200No	2401No	2201No

5.7 Impressed-Current Cathodic Protection

An ICCP system was installed on the bridge pier columns up to 3m from the mid tide level in 3 zones (tidal, splash and atmospheric). A fourth zone for the ICCP system was also installed in the bridge deck soffit (to a 3m x 3m area) around the pier column tops..

5.7.1 Design

The ICCP system was designed for minimal maintenance over a 50-year design life.

The performance criteria for the ICCP installation was:

- A minimum 150mV negative shift from the "Base" potential to the "Instant Off" potential
- A minimum 100mV potential decay from the "Instant Off" potential and the potential measured after the CP system has been turned off for a period of up to 24 hours
- A potential that is consistently less negative than -150mV with respect to Ag/AgCl/0.5M KCl after the CP system has been turned off for a maximum period of 72 hours
- An "Instant Off" potential more negative than -720mV with respect to Ag/AgCl/0.5M KCl

The design current density of steel reinforcement and ancillary embedded steelwork was $20mA/m^2$.

The ICCP system design is based on 4 subzones per pier with the tidal zone consisting of 2 rows of 9 vented discrete anodes to each pier column. For this section of the installation, proprietary discrete anodes were provided with a current rating of 5.1mA per anode.

In the splash zone, the anode system consisted of 9 rows of ribbon anode installed on both pier columns at 150mm centres. The anodes utilised were 19mm wide mixed-metal oxide-coated titanium mesh with a maximum allowable current density for the anode ribbon of $110mA/m^2$, or 5.28mA per metre of ribbon.

For the atmospheric zone in the pier columns, the anode system consisted of 9 rows of ribbon anode installed on each column at 180mm centres. Similarly, the anodes utilised were 19mm wide mixed-metal oxide-coated titanium mesh with a maximum allowable current density for the anode ribbon of $110mA/m^2$, or 5.28mA per metre of ribbon.

The deck soffit zone consisted of 98 vented anodes installed in the deck soffit at the top of both pier columns. In the soffit areas the discrete anodes had a current rating of 1.9mA for a 100mm long anode.

Zoning of the system was established as per the details provided in Table 1.

For each subzone, two positives and one negative connection were installed with a total of 256 positive and 128 negative connections installed across the project.

Additionally, five reference electrodes and five reference returns were installed to monitor the performance of the system with a total of 80 reference electrodes and 80 reference returns being installed.

All cable tails running from the structure CP connections and references electrodes were embedded in chases cut into the concrete and terminated in junction boxes, with one junction box per pier.

A stainless-steel cable tray was installed on the entire length of the bridge to direct the CP cables to one of the two transformer-rectifiers (TRU's) installed on each bridge abatement. Each TRU has the capability to monitor and control 8 independent electrical zones. The TRU at the south abutment controls and monitors the zones of Sections A and B, while the TRU at the North abutment controls and monitors Sections C and D.

5.7.2 Installation

The ribbon anodes for each respective zone was installed parallel to the concrete surface in chases 45mm deep and 30mm wide with the discrete anodes being installed in 22mm diameter holes 130mm deep from the concrete surface.

For the ribbon anodes, after marking out the locations of the anodes, the depth of cover to the reinforcing steel was confirmed with cover meters and ground penetrating radar (GPR).

All concrete saw cutting was carried out with electrical tools and the chases pre-soaked before installation of the ribbon and backfill material. As an additional durability measure, a cementitious waterproofing coating was applied over the ribbon chases in the splash and atmospheric zone.

Before final installation of the anode ribbon, the ribbon was cut to length and then temporarily removed. The chase was washed out and soaked, and a 5mm layer of grout was applied to the base of the anode chase. The anode ribbon was reinserted and pushed firmly into the mortar and welded in place with a hand-held spot welder. Plastic dowels were wedged in the chase to secure the ribbon in place and the mortar backfill was completed before the initial layer had set. The backfill was stopped short 4mm from the concrete column surface to allow for the coating to be applied in the chase.

Where the anodes were to be installed in a section of pier column that required extensive repairs, the anodes were fixed to the reinforcing steel stirrups using insulating spacers after the concrete had been broken out and before backfilling the repair. Proprietary plastic chairs were fixed to the exposed 10mm stirrups or main bars at nominal 300-400mm spacing, and the ribbon was fixed to the chairs with plastic cable ties.

Once the ribbons were fixed in place, and no more than one week after concrete breakout, the concrete repair was formed and poured (or hand patched) into place as described above.

The anode ribbons within the same zone were connected by 12mm conductor bar. The conductor bar was installed in the concrete in a chase 'on the flat' 20mm wide and 50mm deep. Each conductor bar was shielded from the other CP zones with double layer mastic lined heat shrink sleeve. Each anode system was tested against the rebar and the other CP anodes zones for short circuits.

For the discrete anode installation in the deck soffits, GPR surveys were used to map the position of reinforcing bars, to assist in drilling the anode holes.

Once the hole was drilled, its proximity to steel reinforcement was checked with a down the hole cover meter probe. If the internal side cover to the steel reinforcement was less than 10mm and the drilled hole could not be easily relocated, the section of the hole identified as being close to the steel reinforcement was coated with a layer of epoxy resin. Additionally, if the internal concrete cover at the base of the hole was less than 10mm between the steel reinforcement and the anode, the base was treated with a single coat of epoxy to ensure isolation. This was done in order to pre-emptively eliminate the possibility of short circuits in the system.

In the tidal areas, the discrete anodes were installed perpendicular to the column concrete surface. To begin with, a maximum of four of the 18 anodes could be installed at one time, in order to avoid compromising the structural integrity of the column. Subsequently, the installation procedure was changed and all anodes were installed vertically into the concrete cover, rather than between the bars. This permitted all the anodes to be installed at the same time, and thus expedite the work.

As the discrete anodes were vented to release gases built up at the anode surface, the plastic venting tubes were connected with T tube sections to those of adjacent anodes and a single common tube was directed, within chases cut in the concrete, to discharge well above the water line.

5.7.3 Testing

Extensive testing was carried out throughout the installation process to ensure the durability and performance of the system and demonstrate conformance with the technical specification and associated standards.

Prior to full scale installation of the system, a trial installation was undertaken to ensure sufficient capacity within the system design and adequacy of the installation and testing methodologies.

During the concrete breakout, the electrical continuity of the steel reinforcement was checked at all accessible locations. In areas of discontinuity, electrical continuity was established by welding a 6mm diameter bar between continuous and non-continuous bars.

The resistivity of the backfill mortar was checked to provide reassurance that the product was compatible with an ICCP system and would not impede long term performance of the system over the desired design life.

Once the installation of the CP system commenced, adequacy of the installed CP components was checked as each component was being installed. This included continuity testing for the following:

- Anode ribbons to conductor bar/wire
- Positive connections to conductor bar
- Negative connections & reference returns to steel reinforcement

Short circuit testing was carried out between each anode and cathode and between anodes of adjacent zones to confirm the absence of short circuits.

The reference electrodes had the additional testing requirements:

1. Surface potential mapping using portable half-cell reference electrodes was carried out to determine the desired location of a reference electrode in both active and passive regions
2. On installation, the high and low impedance potentials were taken in both wet and dry concrete in order to confirm encapsulation of the permanently installed references
3. Before commissioning the CP system, the natural potential of all reference electrodes was taken as a base line for future monitoring and assessment

5.7.4 Commissioning & Monitoring

Two TRU's, able to be manually controlled and operated on site, were connected to 240VAC power supplies at either bridge abutment. The TRU frames were bolted onto a concrete plinth and installed at the following locations:

- North TRU – At the North abutment on the east side
- South TRU – At the South abutment adjacent to the West column of pier 1

Commissioning of the ICCP system was undertaken on 29th September 2016 with an initial performance assessment being undertaken two weeks following.

Tables 2 & 3 following shows the initial commissioning results of the system installed on the Entrance Bridge. Subsequent monitoring of the system continues to demonstrate that all zones are achieving full protection in accordance with Australian Standard criteria.

Table 4: System Commissioning and Initial Performance Assessment (South TRU)

ZONE	PIER	References	Natural Potentials (mV) (A)	Set Zone Currnet (and voltage)	Initial ON Potentials	Potentials after 3hrs operation	Current & Voltage as Found	ON POTNETIALS	IINSTANT OFF (B)	3HRS OFF POTENTIALS	Potential Shift (B-A)	24hr OFF	Potnetial Decay Criterion (C-B)
			No tidal										
TIDAL - A1	3	A1BW	-388	0.25A	-562	-636	0.25A	-838	-665	-583	-277	-512	153
	4	A1CE	-493	3.1V	-543	-642	5.2V	-775	-723	-725	-230	-651	72
	5	A1DW	-353		-500	-602		-697	-642	-473	-289	-379	263
SPLASH - A2	2	A2AW	-278		-527	-745		-778	-459	-325	-181	-274	185
	3	A2BE	-293	0.8A	-464	-608	0.8A	-638	-426	-331	-133	-305	121
	4	A2CW	-244	1.7V	-317	-555	2.8V	-462	-435	-317	-191	-252	183
	5	A2DE	-216		-452	-693		-718	-532	-350	-316	-251	281
ATMO - A3	2	A3AE	-63		-407	-505		-527	-416	-226	-353	-116	300
	3	A3BW	-78	0.55A	-353	-488	0.55A	-511	-383	-218	-305	-105	278
	4	A3CE	-161	2.0V	-439	-531	2.7V	-558	-471	-273	-310	-169	302
	5	A3DW	-144		-419	-531		-553	-399	-219	-255	-142	257
DECK - A4	2	A4AW	19		-385	-490		-547	-202	-65	-221	-19	183
	3	A4BE	-85	0.8A	-410	-537	0.8A	-598	-360	-135	-275	-91	269
	4	A4CW	-130	4.0V	-240	-328	**7.2V**	-314	-287	-146	-157	-117	170
	5	A4DE	-74		-582	-694		-753	-289	-122	-215	-78	211
TIDAL - B5	6	B5AE	-471		-660	-872		-913	-620	-586	-149	-527	93
	7	B5BW	-529	0.25A	-570	-713	0.25A	-750	-747	-680	-218	-577	170
	8	B5CE	-533	1.9V	-678	-879	3.5V	-793	-657	-590	-124	-530	127
	9	B5DW	no tidal										
SPLASH - B6	6	B6AW	-166		-1173	-1210		-1406	-547	-268	-381	-187	360
	7	B6BE	-426	0.84A	-451	-512	0.84A	-530	-523	-460	-97	-408	115
	8	B6CW	-182	2.5V	-601	-652	3.3V	-691	-467	-229	-285	-172	295
	9	B6DE	-253		-372	-473		-477	-436	-328	-183	-254	182
ATMO - B7	6	B7AE	-193		-404	-455		-466	-433	-253	-240	-178	255
	7	B7BW	-96	0.55A	-329	-385	0.55A	-380	-326	-167	-230	-100	226
	8	B7CE	-94	2.3V	-390	-429	2.9V	-403	-369	-196	-275	-113	256
	9	B7DW	-96		-315	-382		-408	-369	-205	-273	-119	250
DECK - B8	6	B8AW	-103		-532	-590		-630	-227	-170	-124	-118	109
	7	B8BE	-41	0.84A	-257	-336	0.84A	-300	-253	-103	-212	-53	200
	8	B8CW	-72	4.7V	-662	-714	**8.7V**	-822	-407	-145	-335	-73	334
	9	B8DE	-214		-240	-298		-296	-285	-231	-71	-207	78

Table 5: System Commissioning and Initial Performance Assessment (North TRU)

ZONE	PIER	References	Natural Potentials (mV) (A)	Set Zone Currnet (and voltage)	Initial ON Potentials	Potentials after 3hrs operation	Current & Voltage as Found	ON POTNETIALS	IINSTANT OFF (B)	3HRS OFF POTENTIALS	Potential Shift (B-A)	24hr OFF	Potnetial Decay Criterion (C-B)
TIDAL - C9	10	C9AE	-363		-721	-723		-743	-557	-449	-194	-364	193
	11	C9BW	-355	0.35A	-435	-510	0.35A	-492	-466	-370	-111	-362	104
	12	C9CE	-241	3.1V	-341	-472	5.7V	-464	-400	-315	-159	-263	137
	13	C9DW	-376		-654	-930		-1200	-603	-454	-227	-390	213
SPLASH - C10	10	C10AW	-258		-552	-660		-707	-558	-351	-300	-255	303
	11	C10BE	-244	0.84A	-658	-706	0.84A	-750	-544	-287	-300	-239	305
	12	C10CW	-156	2.3V	-249	-346	3.1V	-359	-340	-207	-184	-155	185
	13	C10DE	-314		-812	-894		-988	-584	-419	-270	-337	247
ATMO - C11	10	C11AE	-235		-638	-750		-756	-524	-318	-289	-222	302
	11	C11BW	-134	0.55A	-239	-334	0.55A	-340	-314	-146	-180	-119	195
	12	C11CE	-127	2.0V	-315	-421	2.7V	-449	-427	-182	-300	-123	304
	13	C11DW	-127		-559	-656		-699	-512	-253	-385	-138	374
DECK - C12	10	C12AW	-54		-226	-273		-252	-169	-101	-115	-56	113
	11	C12BE	-46	0.8A	-66	-113	0.8A	-135	-132	-96	-86	-66	66
	12	C12CW	-67	4V	-152	-203	6.4V	-204	-190	-118	-123	-75	115
	13	C12DE	-345		-351	-361		-363	-323	-352	22	-344	-21
TIDAL - D13	14	D13AE	-411		-1198	-1276		-1634	-873	-577	-462	-527	346
	15	D13BW	-325	0.25A	-397	-490	0.25A	-504	-455	-417	-130	-367	88
	16	D13CE	-283	3.4V	-354	-777	**7.3V**	-793	-458	-336	-175	-300	158
	17	D13DW	No tidal										
SPLASH - D14	14	D14AW	-225		-363	-524		-519	-439	-297	-214	-235	204
	15	D14BE	-192	0.84A	-355	-503	0.84A	-538	-459	-321	-267	-235	224
	16	D14CW	-136	2.5V	-363	-522	3.6V	-548	-432	-271	-296	-164	268
	17	D14DE	-217		-527	-591		-602	-470	-245	-253	-209	261
ATMO - D15	14	D15AE	-76		-506	-674		-692	-404	-134	-328	-76	328
	15	D15BW	-90	0.56A	-360	-555	0.55A	-595	-513	-332	-423	-173	340
	16	D15CE	-122	2.1V	-519	-655	3.1V	-651	-468	-250	-346	-140	328
	17	D15DW	-119		-259	-390		-396	-363	-181	-244	-119	244
DECK - D16	14	D16AW	-110		-271	-414		-429	-319	-168	-209	-105	214
	15	D16BE	-81	0.8A	-485	-636	0.8A	-720	-425	-178	-344	-88	337
	16	D16CW	-33	4.3V	-70	-161	**9.2V**	-181	-147	-85	-114	-44	103
	17	D16DE	8		-1	-51		-90	-86	-58	-94	-23	63

5.8 Concrete Coatings

5.8.1 Silane Impregnation

A number of trials were undertaken to maximise durability and efficiency of the installed remediation methodologies and systems.

As part of these trials, it was established that silane impregnation could be achieved via the application of a single component, penetrating silane-siloxane primer that formed part of the decorative coating system, rather than by use of a dedicated silane
impregnation.

The primer was applied in two coats following completion of the concrete repairs and following preparation of the concrete surface by water-blasting to ensure adequate penetration. This treatment not only provided savings to the client, but also increased labour force productivity and efficiency of the protective coating system due to the alleviation of any application compatibility issues.

5.8.2 Decorative Coating

An elastomeric waterproofing and decorative product was applied as the decorative coating for the project with a specified colour of ANS N22 Cloud Grey.

The coating was applied by both roller and airless spray, a minimum of 2 hours after the application of the second coat of the primer.

5.8.3 Anti-Graffiti Coating

Due to the bridge location and at the request of the client, an anti-graffiti coating was applied to the abutment and deck soffit and sides between pier 17 and Abutment B and the walkway and soffits at span 18 (easily accessible locations on the bridge structure).

5.9 Stair Remediation

The repair of the Stairs at the Northern end of the bridge was treated as a variation part way through the contract.

The concrete repairs to these areas were carried out as non-CP conventional concrete repairs and included sacrificial anodes. Depending on the size of the repair, it was undertaken with form and pour, gunite spray and hand patch methodologies.

As a means of providing long term durability to the stair structure, sacrificial block anodes were installed on the piles and the lower landing to provide protection to the steel reinforcement.

A summary of the work on the stairs is shown below.

Table 6: Concrete Repair Quantities – Variation Works

Location	Concrete Repairs (m2)	Sacrificial Anodes Installed (No)	Block Anodes (No)
South Pile	3.22	4	2
North Pile	3.85	4	2
Lower Landing	0.65		2
Lower Soffit	5.56	22	
Middle Landing	1.35	12	
Upper Soffit	3.19	13	
Upper Landing	0.11		

5.10 Bearing Plates

Inspection of the bearing plates was also undertaken as a variation to the contract, in order to take advantage of the labour and access already established onsite. The work included the removal of the existing petrolatum protective tape from the bearing to allow the client's representative to inspect the existing condition of the bearing plate. Once the inspection was completed the bearing plate was coated with a new petrolatum protective tape. Each bearing plate was coated with petrolatum primer, mastic and finally a petrolatum tape.

The overall inspection indicated that the plates were in sound condition, and that the protective coating system is providing adequate protection and durability to these critical elements.

6. CONCLUSIONS

This project is a typical example of current practice in the repair and protection of concrete bridge substructures in Australia. The particular site conditions required flexibility and continual planning in order to carry out complex work on a variety of workfronts situated in a well-patronised boating and fishing spot, with road traffic continuing above and a range of structural and environmental restraints applying to all tasks.

The logistics of getting staff, material, plant and equipment safely to and from the work face proved critical to ensuring work force efficiency with a high standard of quality. Safety and environmental protection was a major focus throughout the project and compliance with all regulations and standards was undertaken cooperatively in collaboration between the parties.

7. ACKNOWLEDGEMENTS

The authors wish to thank owner and client Roads & Maritime Services NSW, designer GHD Pty Ltd and contractor Marine and Civil Maintenance Pty Ltd for permission to publish this paper.

8. REFERENCES

1. The Entrance Bridge, accessed May 2017, https://en.wikipedia.org/wiki/The_Entrance_Bridge

9. AUTHOR DETAILS

MICHAEL KARLAFTIS, BEng-Civil, MIEAust.
Michael Karlaftis is the Engineering Manager for Marine & Civil Maintenance Pty Ltd.

Michael is a Civil Engineer with experience in Geotechnical investigations, footing designs and Structural Design before specialising in 2001 in remedial engineering, including design of impressed current cathodic protection and prevention systems, and project management. He also holds a Certificate in Residential Construction and an Advanced Certificate in Electronic Communication Systems. Since 2006 Michael has been responsible for project management of a wide range of significant infrastructure cathodic protection projects. Michael was the MCM project manager for this project.

BLANE MCGUINESS, BEng (Hons), MBA, AFAIM
Blane McGuiness is the Strategy & Development Manage for Marine & Civil Maintenance Pty Ltd
Blane is a specialised corrosion engineer with over 10 years' international experience in Cathodic Protection system design, installation, commissioning and monitoring, as well as concrete & steel structures condition assessments and durability consultation. He provides engineering and business management services nationally across MCM's projects, tenders and business development activities.
Blane is recognised in the industry for his technical knowledge and his capabilities in extending the life of an asset, evident from the range of engineering and management services provided across multiple major projects.

MOHAMMAD ALI, BEng, PhD

Mohammad Ali has a PhD in Civil Engineering with CP as Specialisation. He has over 30 years' of experience, both in Australia and overseas, in investigation of deteriorated steel and concrete structures, CP and durability. He has designed numerous CP systems in Australia and overseas during his 20+ years working life in Australia. He is currently a Principal with GHD and was the Past President of the ACA

THE INFLUENCE OF SOIL MOISTURE ON THE CORROSION OF MILD STEEL IN CLAYS

R.B. Petersen and R.E. Melchers
Centre for Infrastructure Performance and Reliability, The University of Newcastle, Australia 2308

SUMMARY: This paper presents the results of an experimental study on the influence of both soil moisture and compaction on the corrosion of mild steel buried in clay soils. Mild steel samples were buried in soils prepared at different moisture contents and compacted to different levels, and were left to corrode for 3 months. Mass loss due to corrosion in the soils was measured after 3 months exposure and plotted against both moisture content and degree of saturation. The results were compared to results from similar tests on sand and loam soils.

Keywords: Mild steel, Clay soils, Corrosion, Moisture, Mass loss, Experimental investigation.

1. INTRODUCTION

Soil moisture is a critical component in the corrosion of buried ferrous metals. Previous studies have highlighted the importance of soil moisture on the long-term corrosion of both cast iron and steel pipes buried in soil (Cole and Marney 2012, Melchers and Petersen 2016, Petersen and Melchers 2016).

For short-term exposures it is known that the relationship between corrosion and soil moisture is a highly non-linear function. This is because of the wetting properties of the metal surface relative to the moisture content of the soil. Tests over 6 months exposure by Gupta & Gupta in 1979 showed that little corrosion occurs for low moisture contents but that it increases very quickly for just a small further increase in moisture content (see Figure 1). They found that the corrosion loss reached a maximum value at a critical moisture content, which varied depending on the soil texture. In their short-term tests there was a subsequent drop in corrosion for moisture contents higher than this critical point. They considered this drop was related to a lack of oxygen in these short term tests. Gupta and Gupta also plotted mass loss versus the degree of saturation of the soil and they found that the plots for the different soils were similar (see Figure 2). From these graphs they concluded that corrosion in soil becomes significant once the degree of saturation of the soil was greater than 0.50, and that the critical level of moisture (where the mass loss was greatest) was at a degree of saturation of 0.65.

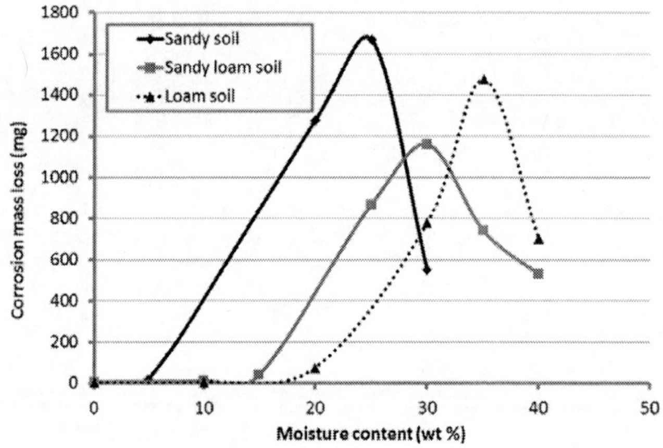

Figure 1 Corrosion mass loss versus moisture content for sandy and loamy soils (adapted from Gupta and Gupta, 1979)

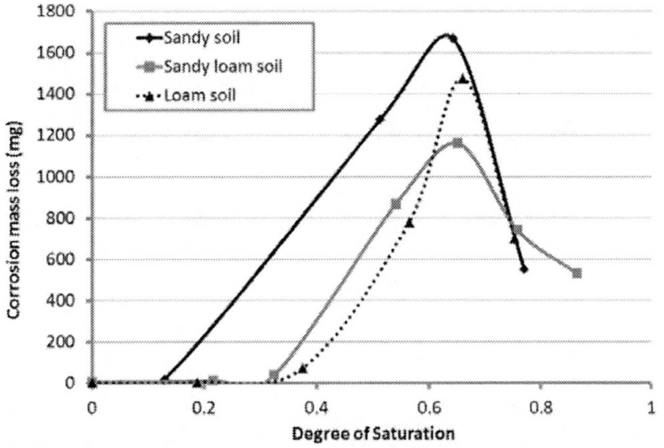

Figure 2 Corrosion mass loss versus degree of saturation for sandy and loamy soils (adapted from Gupta and Gupta, 1979)

The Gupta and Gupta test results above relate only to sandy soils and sandy loams. Clay soils were not considered. However, clay backfill is predominant in many Australian water utility networks (Petersen and Melchers 2016). To the authors knowledge the type of study performed by Gupta and Gupta on sand and loam soils has not been done on clay soils. It is unknown if the relationship between corrosion and soil moisture in a clay soil follows a similar trend, hence there is a need to conduct similar experiments on clays to determine this relationship.

For highly impermeable soils like clays, the effect of compaction on corrosion could be important. Poor compaction leads to the possible inclusion of air voids next to the surface of the metal potentially increasing corrosion, whereas greater compaction effort would likely reduce corrosion (Romanoff 1957, Melchers and Petersen 2016). It follows that the effect of compaction of clay soils on corrosion losses requires investigation.

A laboratory program is currently underway at The University of Newcastle to replicate and extend the Gupta and Gupta type of study to clay soils. It involves determining the corrosivity of mild steel in clay soil for various soil moisture contents and compaction levels, for exposures ranging from 3 to 12 months. This paper presents the results of samples exposed for 3 months.

2. EXPERIMENTAL METHOD

The clay soil used in this investigation was collected from the native backfill surrounding a 90 year old cast iron water main in Strathfield, NSW. This pipeline was selected as the location for soil sampling because it has been the focus of an intensive corrosion study over the past 5 years, and a large amount of data on the corrosion of this pipe and the surrounding soil conditions have been collected (Petersen and Melchers 2016). The soil was collected in plastic buckets and delivered to the Civil Engineering Laboratories at The University or Newcastle, where the corrosion tests took place. In its original state the raw soil was wet and was stuck together in large clods roughly the size of two clenched fists. To homogenise the soil in preparation for the experiments, the soil was first broken down into small clods and laid out on a plastic sheet and allowed to air dry for at least 24 hours. The air-dried soil was then broken down further using a mortar and pestle until it could pass through a 4 mm sieve. The raw soil contained a very small percentage of small rocks and pebbles and fine grass roots, and these were removed from the soil at this stage by hand. All of the air-dried soil passing the 4 mm sieve was then thoroughly mixed to obtain a homogeneous mixture for the corrosion experiments.

The homogenised batch of clay was then divided into nine separate batches, and these batches were then prepared to different moisture contents, ranging between air-dry and fully saturated. The batches were prepared to moisture contents of 4.8, 8.4, 10.6, 13.0, 15.5, 18.0, 20.7, 23.9, and 25.9 (% g/g of dry soil). The clay soils were prepared to the desired moisture content by adding measured amounts of water to the soil and mixing it into the clay by hand. The wetted clays were then stored in sealed plastic containers and left to cure for a week to allow the moisture to mix more evenly throughout the clay. After this curing period, the moisture content of each batch was checked using an oven-drying approach (Standards Australia 2005).

Mild steel coupons 60 mm x 50 mm x 3 mm were used for the corrosion tests. The coupons were cut from sheet metal, and a unique set of holes were drilled into each sample following a numbered guide for identification purposes. A total of 81 coupons were cut and numbered. The coupons were cleaned of any mill scale prior to exposure by submerging them for 1 minute in an inhibited hydrochloric acid solution, in accordance with ASTM G1-03 (ASTM International 2011). The coupons were then washed in water and stored in an oven at 60 degrees for an hour, and following this drying period the coupons were weighed to determine their pre-exposure mass. The coupons were then stored in the oven until ready to be buried in the soil samples for testing.

Thick-walled PVC pipe sections plus end caps were used to contain the soil samples and the buried coupon for the corrosion tests. Each soil corrosion test container had a 100 mm internal diameter and a height of 120 mm, and was sized in accordance with ASTM G162 – 99 (ASTM International 2010).

The prepared soil was added to the container (PVC pipe section with bottom end cap) in four approximately equal layers. The mild steel coupon was laid horizontally in the middle of the third layer, as shown in Figure 3. The orientation of the coupon surfaces in relation to the top and bottom of the container was also noted. Each soil layer was compacted equally by forcing a cylindrical steel plunger down onto the surface of the soil in the container. The cylindrical steel plunger fit closely into the PVC container so that each soil layer was compacted uniformly across its entire surface. A Tinius Olsen machine was used to apply the downward vertical force on the cylindrical steel plunger. The test container and the cylindrical steel plunger are shown in the Tinius Olsen machine in Figure 4.

Figure 3 Mild steel coupon placement in the third soil layer. Note that soil is still added on top of the coupon shown in this figure to fill the container.

Figure 4 Compaction setup showing test container and cylindrical plunger in the Tinius Olsen machine.

Three different compaction forces were used. They were 2 kN, 3 kN and 4 kN. The lowest compaction level of 2 kN was chosen to represent a standard level of compaction in the field. This force was estimated by considering the load applied to an upper layer of soil by a standard heavy earthworks machine, with a mass of 1000 kg, and a track area of 0.4 m^2. Initial density checks on samples compacted in this way showed densities that were similar to those calculated from undisturbed samples taken from the field. The other two levels of compaction were chosen to investigate the influence of additional compaction effort on corrosion in soils.

During the compaction procedure, the load was applied at a rate of approximately 0.2 kN per second, and the load was removed once the load reached the target value (of 2 kN, 3 kN or 4 kN). During compaction of the fourth and final soil layer a steel collar was used to help guide the compacting steel plunger (this is shown in Figure 4). After compaction of the final layer the compacted clay soil was struck flush with the top of the container, and the top cap was then fitted to the top of the container. Any air gaps between the end caps and the PVC pipe section were sealed with electrical adhesive tape to limit moisture loss from the container over the duration of the tests.

Three corrosion tests were prepared at each level of soil moisture and compaction to provide enough tests so that results for coupon corrosion could be obtained at three different exposure periods. This gave 27 corrosion tests at each exposure period, for a total of 81 corrosion tests. The selected exposure periods were 3 months, 6 months and 12 months.

To date only the first set of 27 coupons has been retrieved (the 3 month exposure tests). To remove the mild steel coupons the containers were opened and the soil from around the coupon was removed using a hand spade. The coupon was removed from the soil sample and any soil loosely adhered to the surface of the mild steel coupon was brushed or wiped off by hand. The coupons were then stored in a desiccator to limit any further corrosion until they were cleaned. At the time of coupon removal a sub-sample of the soil was also taken and its moisture content was checked using an oven drying method to see if any moisture was lost over the duration of the test. In general, very little to no moisture was lost from the containers over the three month test period.

The coupons were cleaned using electrolysis and an ultrasonic technique in accordance with ASTM G1-03 (ASTM International 2011). Coupons were cleaned using electrolysis in rounds of 2 hours in duration. After each round of electrolysis the surfaces of the coupons were scrubbed with a lightly abrasive cloth to remove any of the rust products that were loosened by the electrolysis process. The coupons were then dried thoroughly using a paper towel and then placing them in an oven at 60 degrees Celsius for at least an hour. Once the coupon was dry its mass was measured and this mass was compared to the original mass of the coupon to determine the mass loss over the exposure period. The cleaning procedure was repeated until the change in mass loss between successive cleaning rounds was less than 5 %. Finally, the samples were cleaned in an ultrasonic bath for 10-30 minutes to remove small amounts of rusts deep in surface corrosion pits. Following cleaning in the ultrasonic bath the final remaining mass of the coupon was measured and the final mass loss was determined. This final mass loss was then taken as the mass loss caused by soil corrosion over the exposure period.

3. OBSERVATIONS AND RESULTS

As noted, the first set of coupons was retrieved after 3 months of exposure in the soil samples. When removing coupons from the soil it was found that for soils with moisture content of 13.0 % and below the soil fell away easily from the coupons. In these cases, only light brushing of the soil was required to remove the relatively dry soil from the surface of the coupons. For soils with moisture content of 15.5 % and above the soil was found to adhere better to the surface of the coupon (as shown in Figures 5a and 5b), and required more effort to remove it.

In very wet soils with moisture contents above 20.7 % it was noticed that when removing the coupon from the soil a thin layer of dark coloured rusts usually was left behind. This rust layer covered both sides the entire surface of the coupon. This is shown in the photograph in Figure 5c. For drier soils this was not the case. Instead, rust products stayed stuck to the surface of the coupon and did not completely cover the coupon. Some soil appeared also to be stuck into these dry rusts.

(a) Moist soil adhered to test coupon (mc = 15.5 %)

(b) Very wet soil adhered to test coupon (mc =25.9 %)

(c) Soft rust products left behind in very wet soil (mc = 20.7 %)

Figure 5 Some observations from coupon recovery.

Figure 6 shows photographs of coupons after they were retrieved from the soil samples, at the 9 different moisture contents, after 3 months of exposure. These samples were retrieved from soil samples compacted with a force of 2 kN. These results are typical of both sides of the coupon, and they are also typical for coupons buried in the soils compacted with the higher compaction levels. The photographs show typical rust products on the surface of the coupons. As noted above the rust products on the surface of the coupons buried in the drier soils tended to stick to the surface of the coupons, contained some soil particles, and did not cover the entire surface of the coupon. The amount of the rusts appeared to increase with the moisture content in the soil, at least until a moisture content of around 15.5 - 18 % was reached. Beyond these moisture contents (i.e. at mc = 20.7 % and higher) the rusts formed were different, as noted above and shown in Figures 7 g-i. These rusts were darker and easier to remove, and (as noted above) in some cases the rust products came off the surface of the coupon completely (Figure 5c). In these cases, the rusts covered the entire surface of the coupon.

(a) mc = 4.8%.

(b) mc = 8.4 %.

(c) mc = 10.6 %.

(d) mc = 13.0 %.

(e) mc = 15.5 %.

(f) mc = 18.0 %.

(g) mc = 20.7 %.

(h) mc = 23.9 %.

(i) mc = 25.9 %.

Figure 6 Photographs of coupons after they were retrieved from soil samples after 3 months of exposure.

Figure 7 shows photographs of the same coupons after they were cleaned using electrolysis and ultrasonics as outlined above. These photographs show the corrosion damage on each coupon and how it changed as a function of the soil moisture content. The corrosion damage on coupons buried in the drier soils was typically of the form of groups of small pits in distinct areas distributed over the surface of the coupon. The corrosion on coupons buried in moderately wet soils typically was of the same form as those in the drier soils; however the depth and extent of corrosion was greater. Once the moisture content of the soil reached 20.7 % it appeared that the corrosion became more uniform over the surface of the coupon, and the overall corrosion damage had reduced (Figure 7 g). At very high soil moisture contents very little corrosion damage was observed (Figure 7h & i).

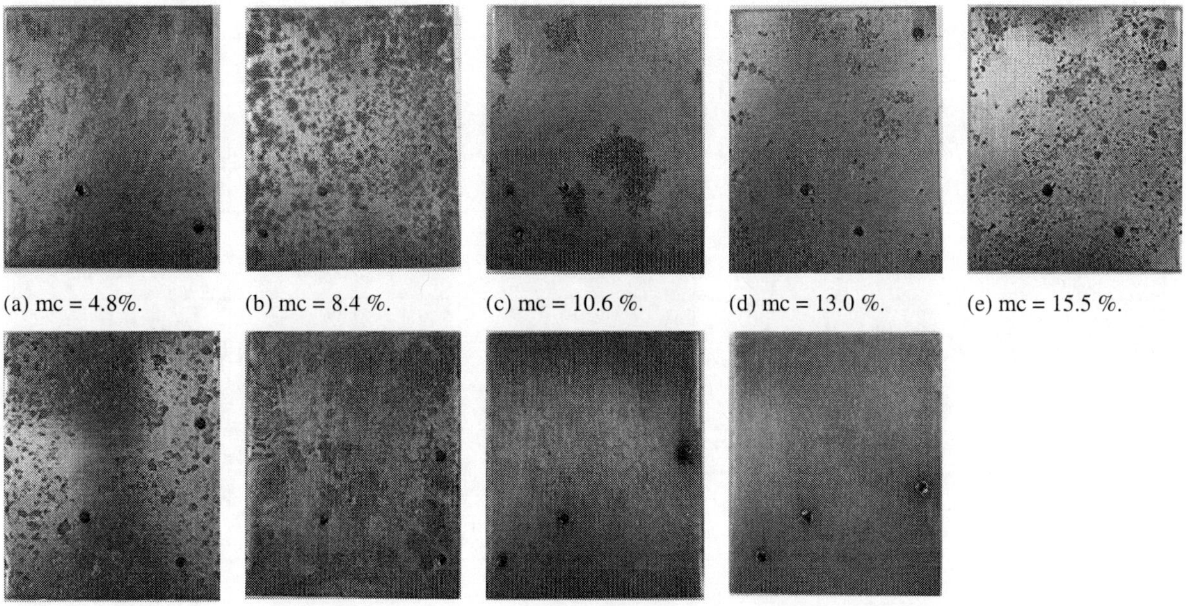

(a) mc = 4.8%. (b) mc = 8.4 %. (c) mc = 10.6 %. (d) mc = 13.0 %. (e) mc = 15.5 %.

(f) mc = 18.0 %. (g) mc = 20.7 %. (h) mc = 23.9 %. (i) mc = 25.9 %.

Figure 7. Photographs of coupons exposed for three months after they were cleaned using electrolysis.

Table 1 presents the corrosion mass loss results for all samples extracted after 3 months of exposure. The mass loss results are expressed as a function of the original mass of the coupon. Also tabulated are the moisture content of the soil (as a percentage weight of dry soil), and a calculated value of the degree of saturation of the soil. The degree of saturation was calculated for each test container using the moisture content of the soil, the mass of the soil in the container, the volume of the soil in the container, and an estimate of the soil particle density of 2900 kg/m^3. Figure 8 shows the corrosion losses plotted against moisture content for the three different levels of compaction. Figure 9 shows the same corrosion losses plotted against the degree of saturation of the soils for the three different levels of compaction.

Table 1. Mass loss results for soil corrosion tests exposed for 3 months

Test ID	Compaction level (kN)	Moisture content mc (% g/g dry soil)	Estimated Degree of Saturation Sw	Corrosion mass loss (% of original mass)
1	2	4.8	0.18	0.134
10	2	8.4	0.25	0.559
19	2	10.6	0.30	0.675
28	2	13.0	0.37	0.982
37	2	15.5	0.44	1.881
46	2	18.0	0.63	2.000
55	2	20.7	0.89	1.152
64	2	23.9	1.01	0.188
73	2	25.9	1.01	0.179
4	3	4.8	0.17	0.053
13	3	8.4	0.27	0.513
22	3	10.6	0.31	0.541
31	3	13.0	0.39	0.932
40	3	15.5	0.48	1.650
49	3	18.0	0.71	2.140
58	3	20.7	0.93	0.926
67	3	23.9	1.04	0.151
76	3	25.9	1.02	0.157
7	4	4.8	0.17	0.097
16	4	8.4	0.27	0.389
25	4	10.6	0.33	0.770
34	4	13.0	0.38	1.034
43	4	15.5	0.53	1.607
52	4	18.0	0.76	2.003
61	4	20.7	1.00	0.407
70	4	23.9	1.05	0.077
79	4	25.9	1.05	0.070

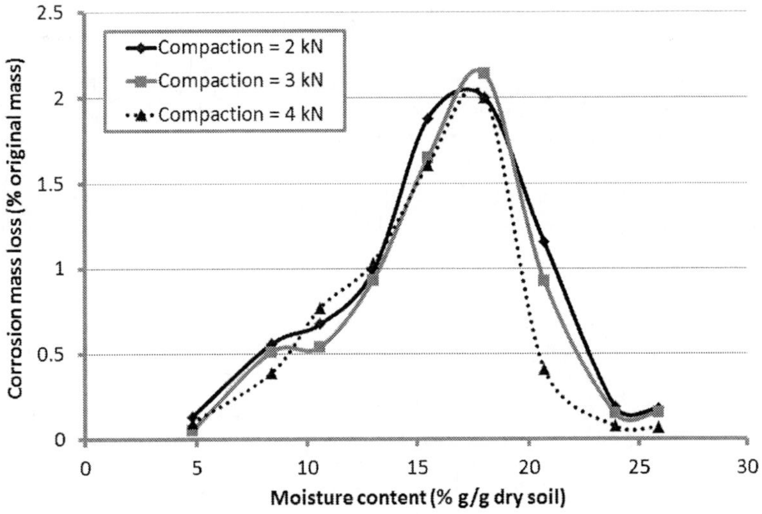

Figure 8 Corrosion mass losses, expressed as a percentage of the original coupon mass, versus soil moisture content for different levels of soil compaction.

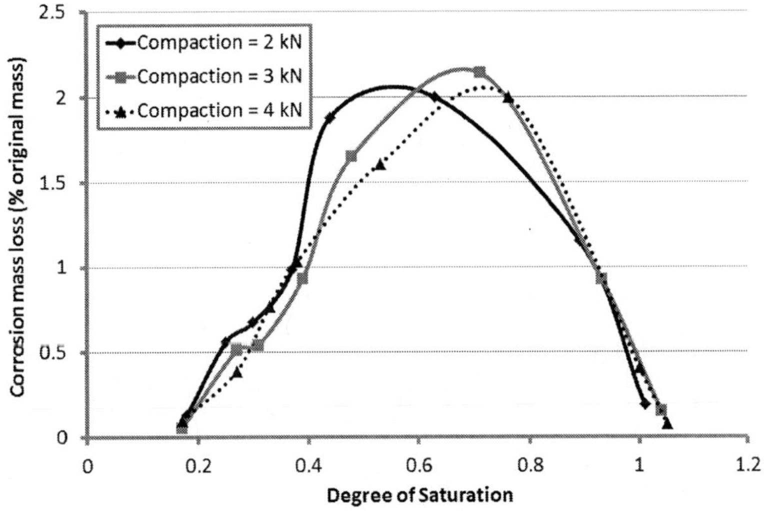

Figure 9 Corrosion mass losses, expressed as a percentage of the original coupon mass, versus the calculated degree of saturation of the soil for different levels of soil compaction.

4. DISCUSSION

The trend of mass loss versus moisture content for clay soils (Figure 8) is similar to the trends already observed for sand and loam soils (Figure 1). For the clay soil studied herein there is low corrosion at low moisture contents, but the amount of corrosion increases quickly for a small increase of moisture content beyond a certain value. The mass loss becomes significant when the moisture content is approximately 15 %, and the mass loss reaches a maximum value at a moisture content of approximately 17.5 %. These values are slightly less than those observed for both the sand and sandy loam soils. Further increases in moisture beyond 17.5 % lead to a decrease in mass loss and this is likely a result of a lack of oxygen in the test containers.

Similarly, the trend of mass loss versus degree of saturation for clay soils (Figure 9) is similar to the trends already observed for sand and loam soils (Figure 2). For the clay soil used herein it appears that corrosion becomes significant when the degree of saturation is at least 0.50, and the mass loss reaches a maximum value at a degree of saturation between 0.5 and 0.7. In these tests the corrosion remains significant until a degree of saturation of approximately 0.8. These results are similar to those from the sand and loam soils, indicating that the mass loss versus degree of saturation trend is independent of soil texture.

Figures 8 and 9 both show that changing the compaction effort did not significantly change the corrosion trends. This may be because the range of compaction effort was simply not large enough to have any significant effect on the conditions at the corroding interface. It is a matter that could require further investigation. Also, the present tests considered corrosion of coupons buried in soils that were compacted under very controlled conditions. It is likely that setting up an experiment with poor, non-uniform or very little compaction effort, which would give larger air voids at the surface of the coupon, would give significantly different results, and may shed some more light on the influence of compaction on corrosion in the field. This also is a matter for further investigation.

This paper presents only the results for coupons exposed for three months. Coupons still need to be retrieved for exposures of 6 months and 12 months to ascertain whether and how the corrosion versus moisture trends change over time. Future work also involves comparing the results of these relatively short-term and well-controlled tests to observations of long-term corrosion of ferrous metals buried in the field.

5. CONCLUSIONS

Mild steel samples were buried in clay soil prepared to different moisture contents, compacted to different levels, and exposed for 3 months. It was found that trend of corrosion mass loss versus moisture content for clay follows the same trends observed for sand and loam soils. For a clay soil corrosion becomes significant (i.e. there is enough water for corrosion to occur) when the degree of saturation of the clay is at least 0.50, and in these short-term tests corrosion remains significant until a degree of saturation of 0.80. Further study is required to see whether this is also true as time progresses. In these tests compaction effort did not significantly influence corrosion mass loss.

6. ACKNOWLEDGMENTS

The authors thank the staff of Sydney Water for their assistance in collecting the clay soil from field sites used in this experimental program. The authors would also like to thank staff at the University of Newcastle laboratories and undergraduate student Mohammad Rafiq for their help in preparing the samples for testing, and Igor Chaves for his advice and guidance in cleaning the mild steel specimens before and after soil exposure.

The research reported herein was supported financially by the Critical Pipes Project, a research consortium of Australian Water utilities Sydney Water Corporation, Melbourne Water, Water Corporation (WA), South Australia Water Corporation, South East Water, Hunter Water Corporation and City West Water, together with UK Water Industry Research Ltd. and the Water Research Foundation of the USA. The research partners are Monash University, University of Technology, Sydney and The University of Newcastle. The second author also acknowledges with gratitude the continued support of the Australian Research Council.

7. REFERENCES

ASTM International (2010) "Standard Practice for Conducting and Evaluating Laboratory Corrosion Tests in Soils", ASTM G162-99, ASTM International, West Conshohocken, PA.

ASTM International (2011) "Standard Practice for Preparing, Cleaning and Evaluating Corrosion Test Specimens", ASTM G1-03, ASTM International, West Conshohocken, PA.

Cole IS and Marney D (2012) The science of pipe corrosion: A review of the literature on the corrosion of ferrous metals in soils, Corrosion Science 56 5-16.

Gupta SK and Gupta BK (1979) The critical soil moisture content in the underground corrosion of mild steel, Corros. Sci. 19: 171-178.

Melchers, R.E. and Petersen, R.B. (2016) Romanoff's soil corrosion data deconstructed, (In) Proc. Conf. Corrosion and Prevention, Auckland, New Zealand, Nov 13-16, Aust. Corrn. Assn., Melbourne. CD ROM, Paper 053.

Petersen, R.B. and Melchers, R.E. (2016) Factors involved in the long-term corrosion of buried cast iron, (In) Proc. Conf. Corrosion and Prevention, Auckland, New Zealand, Nov 13-16, Aust. Corrn. Assn. Melbourne. Paper 067.

Romanoff M (1957) Underground Corrosion, National Bureau of Standards Circular 579, US Government Printing Office, Washington, DC.

Standards Australia (2005), Soil moisture content tests – Determination of the moisture content of a soil – Oven drying method (standard method), AS 1289.2.1.1 – 2005. Standards Australia, NSW.

8. AUTHOR DETAILS

Robert Petersen is a Research Associate in the Centre for Infrastructure Performance and Reliability at the University of Newcastle, Australia. He holds a BE (Civil Engineering, 2005) and a PhD (2009), both from the University of Newcastle. His PhD research was in the area of mathematical modeling of the in-plane shear behavior of structural masonry strengthened with fibre reinforced polymers. His current research interest is the corrosion of cast iron and steel water mains buried in soil.

Robert E Melchers is Professor of Civil Engineering at The University of Newcastle, Australia. He holds a BE and M Eng Sc from Monash University and a PhD from the University of Cambridge, UK. He is a Fellow of the Australian Academy of Technological Sciences and Engineering.

His most recent awards are the 2009 ACA Corrosion Medal, the 2012 Jin S Chung Award (International Society of Offshore and Polar Engineers) and the 2013 John Connell Gold Medal (The Institution of Engineers, Australia).

THE EFFECT OF COASTAL SEAWATER TEMPERATURE ON CORROSION OF STEEL WITH DIFFERENT CARBON CONTENTS

R. Jeffrey & R.E. Melchers
Centre for Infrastructure Performance and Reliability, The University of Newcastle, Australia

SUMMARY: This paper considers the combined effect of water temperature and the carbon content of low alloys steels on their immersion corrosion loss. Experimental results for exposures over 1.75 years are reported for low alloy steels with carbon contents of 0.01%wt, 0.04 %wt and 0.06 %wt, recovered at intervals of about 4 weeks. One set of each of these three types of steel coupons was in a tank with periodic renewal of natural coastal seawater and another immersed in a tank with seawater from the same source but heated to a constant 30°C. In both tanks the water was changed slowly, taking about 5 minutes to replace the 200L in each of the respective drums, at 12 hour intervals. The water in that tank was preheated to avoid sudden temperature changes. In both cases the water was essentially stagnant throughout the experiment. For the first four months the 0.01% C corroded slightly more than the 0.04%C which was in turn slightly higher than the 0.06% C. In all cases, there was a bimodal trend in the corrosion mass loss. There was a clear difference in corrosion mass loss as a function of water temperature but this was greater for the longer exposures. A slight difference in corrosion loss was observed between the steels with different carbon contents.

Keywords: Corrosion, Steel, Carbon content, Temperature.

1. INTRODUCTION

A key factor in the immersion corrosion of steel is the water temperature and also the dissolved oxygen content of the water (Mercer & Lumbard, 1995; Jones, 1996). Temperature influences the dissolved oxygen (DO) content of the water, but DO also tends to decrease with water depth (Svedrup et al. 1942). While both influences have been much studied in laboratory experiments for short-term exposures [less than a year] (Mercer & Lumbard, 1995), there is much less information available for longer term effects, such as for 3 or more years exposure. Experiments in fresh water tanks show that corrosion loss is lower for lower DO content (Melchers & Chernov, 2010), and this also applies to marine exposures. However, the effect is not a simple linear function, in part because the trend in corrosion loss with time is not linear. All the available evidence shows it has a bi-modal function. Fig. 1 shows a schematic view of the bi-modal behaviour for corrosion loss derived form coupon mass loss data (Melchers, 2003a).

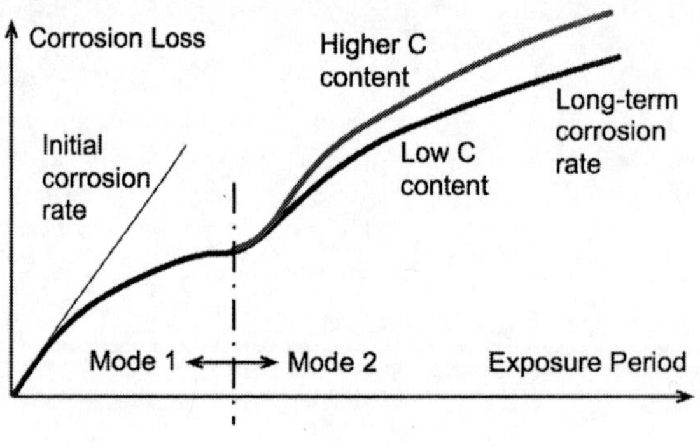

Figure 1. Schematic diagram of bi-modal behaviour with increased corrosion as a result of carbon content, as derived from various data sources (Melchers 2003a).

Steel composition traditionally has been claimed to have little influence on the corrosion of steels in immersion conditions (Evans, 1960) but this observation largely was derived from relatively short-term observations. Using the bi-modal model to interpret a wide variety of data from many different sources, it was shown that some constituents of low alloy steels affect the early behaviour and others are more influential later (Melchers, 2004). Further, there remains a degree of uncertainty about the influence of carbon content. Using data published in the literature it was found that carbon content had no discernible effect for shorter term exposures, but that after several years a higher carbon content did appear to be associated with increased corrosion loss (Melchers 2003b), as shown in Fig. 1. This led to the hypothesis that carbon content in the steel might have some role in the development of the bi-modal behaviour.

The present paper reports an investigation of the effect of carbon content on the short and longer term immersion corrosion of low carbon steels, and also the effect of water temperature on any such effect. As described below, for logistic reasons the immersion testing was carried out at ambient temperature (20˚C nominal) and separately at a (nominally) constant elevated temperature (30˚C). This will affect the DO content of the water. There were no other DO effects as the experiments were carried out at similar water depths. The next section describes the experiments, followed by a summary of the results obtained and their interpretation.

2. EXPERIMENTAL PROGRAMME

2.1 Test Coupons

Steels of three different low carbon contents were used in the experiments. They were sourced from different suppliers and ranged from nominally 'pure' iron, which was actually a steel with 0.01% carbon content, to a mild steel with nominally 0.06% carbon content (Table 1). The 'pure' iron was very difficult to obtain and only in 1m long strips 38mm wide by 6.4mm thick. For the other two steels the coupons were cut from standard strip width of 50mm and 2 and 4mm thicknesses to assist distinguishing between the steel compositions. These combinations have slightly different edge to surface ratios but, as noted below, this was considered through using the total surface area for deriving equivalent corrosion losses from the mass losses. Also, to total length of edges is very similar for all three sample types. Further, for immersion corrosion of coupons of these sizes there is no a significant edge ratio effect. This is unlike the well-known edge ratio effect for atmospheric corrosion coupons (Roberge, et al. 2002).

Table 1 Size of coupons (100mm long) and their composition (%wt)

	Cross-section (mm)	C	Mn	P	S	Si	Ni	Cr	Mo	Cu
'Pure' iron	38 x 6.4	0.01	0.05	0.007	0.004	0.00	0.02	0.02	0.00	0.01
Low carbon steel	50 x 4	0.04	0.19	0.005	0.013	0.00	0.04	0.11	0.02	0.04
'Mild' steel	50 x 2	0.06	0.20	0.009	0.009	0.00	0.01	0.01	0.00	0.02

The three steels were also analysed for V, Al, B, Nb, Zr, Ti, Sn, W, N, Co and As, all of which were found to be below 0.001% or below limits of detection.

2.2 Test environments

Two test environments were used. Both used natural coastal seawater drawn from the test site at Taylors Beach (about 40 km north of Newcastle NSW in the Port Stephens waterway). The water quality for this site has been described previously (Melchers & Jeffrey, 2014) and can be considered to be clean coastal seawater with minimal nutrient loadings.

The first test environments used the natural seawater directly, and was thus always at ambient temperature. For this experiment the coupons were exposed in a 200 L sloping base fibreglass seawater tank, as also used for previous experiments (Jeffrey and Melchers, 2014) (Fig. 1).

The second test environment used the natural seawater from the same source but heated nominally to 30 ± 0.5 ˚C throughout the duration of the experimental programme. For this a 200 L polythene exposure vessel was used (Fig. 2). To ensure the water remained within the desired temperature range a 200 L header tank (Fig. 2) was used for pre-heating the water prior to discharge into the exposure tank. This was to heat the water slowly and in stages to ensure microorganisms were not exposed to sudden temperature changes. The pre-heating tank was equipped with 3 x 300W aquarium heaters to heat the seawater to 30 °C and the sample tank had 1x 300W aquarium heater to maintain the temperature. The heaters were supported horizontally close to the bottom of the tanks and were controlled by an Omron (E5CZ-R2) digital temperature controller (each tank) with thermocouple sensors in the tanks. A PLC controlled the valves, pumps and disabled the heaters if water level was low (Figures 3).

The seawater temperature was recorded in both exposure tank during the 2016 exposure period, using Tinytag® data loggers.

Figure 1. Coupons as suspended in the 200 L sloping-bottom natural seawater exposure tank.

Figure 2. Heated seawater installation with heating drum on stand and exposure drum at right.

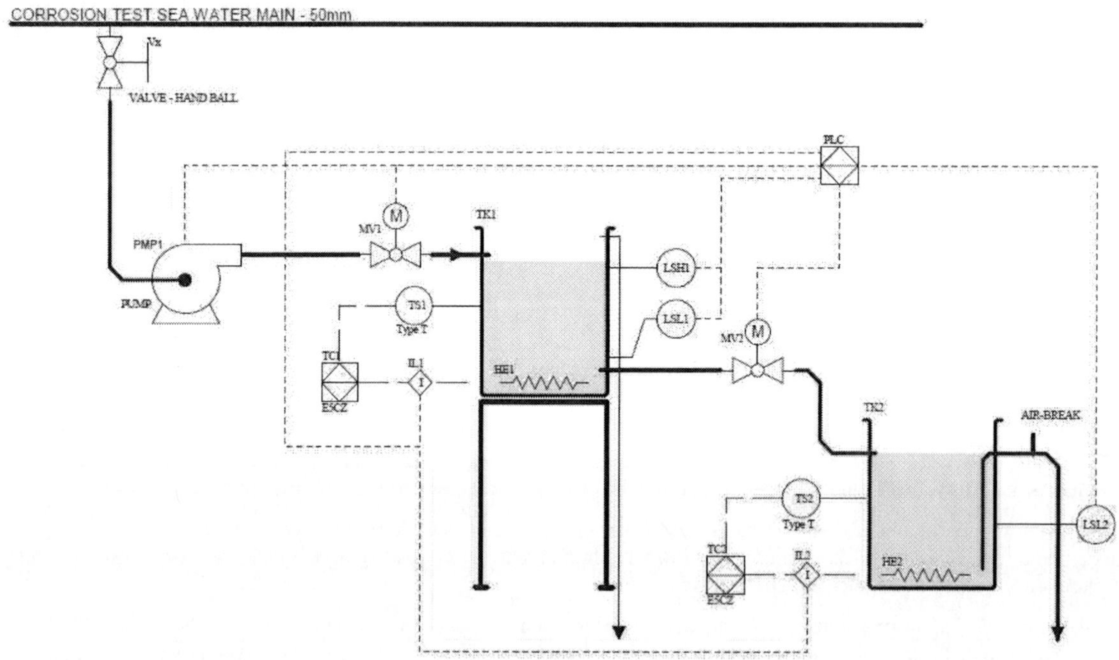

Figure 3. Layout of piping and electronics for the experimental system for heated exposure environment.

For both exposure environments the water was changed every 12 hours. In both cases this was controlled by means of a timing valve that allowed fresh water to displace the water from the tanks. The velocity involved was less than 0.01 m/s. As a result, the water could be considered close to stagnant throughout the experiment.

In both environments the coupons were suspended at a depth of 100mm from the water surface using simple plastic suspension hangers and coloured cable ties to distinguish between the different grades of steel (Fig. 1).

2.3 Test program

The test programme consisted of exposing multiple coupons to each environment. All coupons were drill-coded for unique identification, cleaned from mill-scale using inhibited 50% HCl solution. This method has been used before and is consistent with achieving surfaces of steel similar to practical infrastructure. All coupons were weighed to the nearest 0.1 mg prior to exposure.

For each exposure environment, and for each of the 3 steel types, coupons were recovered at approximately 4 week intervals after the date of first immersion (17 Sept. 2015), that is at 28, 56, 84, 112, 143, 168, 223, 256, 306, 336, 364, 421, 446, 476, and 511 days exposure, with multiple coupons at 446 days to obtain an estimate of variability (not discussed herein). At any one time only one coupon was recovered for each environment and for each steel type, working on the basis that with the close recovery intervals estimates of variability could easily be made. This approach has been used previously (Melchers & Jeffrey, 2014).

2.4 Recovery and processing

Upon recovery, it was observed that in most cases the corrosion product on the coupon surface appeared as made-up of a mixture of various shades of orange-brown hydroxides. When this external layer was carefully removed at discrete locations it was observed that there was a black magnetite layer under the orange-brown layer. The magnetite also was immediately adjacent to the metal surface. Similar observations have been made before for steels exposed to seawater immersion conditions (e.g. Jeffrey & Melchers, 2009). In the early recoveries, the black layer was soft and only poorly adherent to the steel substrate. This allowed the corrosion products to be removed readily by washing and scrubbing with a soft plastic brush. Later in the exposure program typically some remnant rust had to be cleaned with inhibited 50 % HCl solution. After cleaning and thorough drying the coupons were weighed to the nearest 0.1 mg and mass loss calculated relative to the initial mass of the coupon. This allowed corrosion loss to be estimated from the mass loss using the surface area of the coupons.

3. RESULTS

3.1 Overall results

The mass losses obtained throughout the experiments are plotted as a function of exposure time and average seawater temperature in Figs. 4-6 for each of the 3 steels. It is immediately obvious that, irrespective of carbon content, corrosion is greater at higher temperature, as expected. It also is clear that the effect of water temperature is comparatively greater at longer exposures.

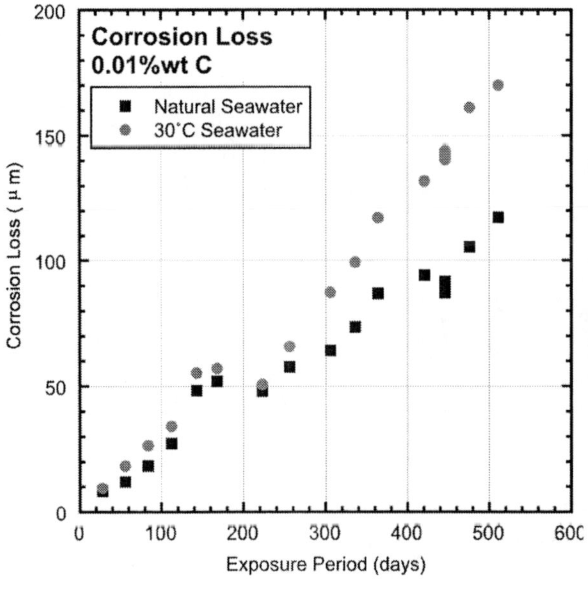

Figure 4. Corrosion loss as a function of time and average water temperature for 0.01%wt C steel in 20°C (av.) and 30°C seawaters.

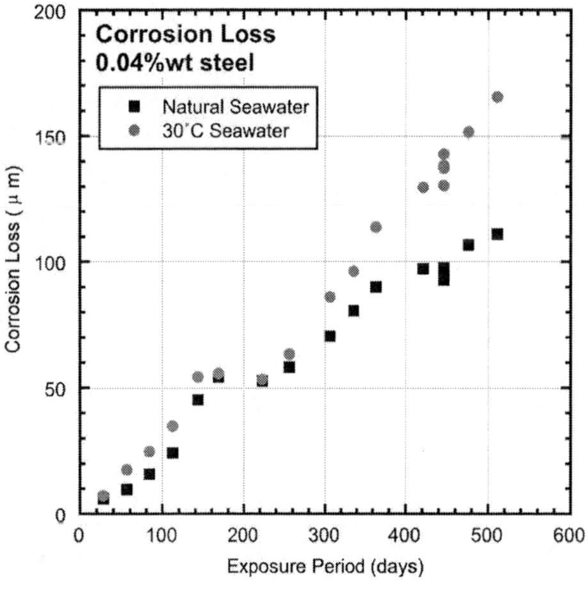

Figure 5. Corrosion loss as a function of time and average water temperature for 0.04%wt C steel in 20°C (av.) and 30°C seawaters.

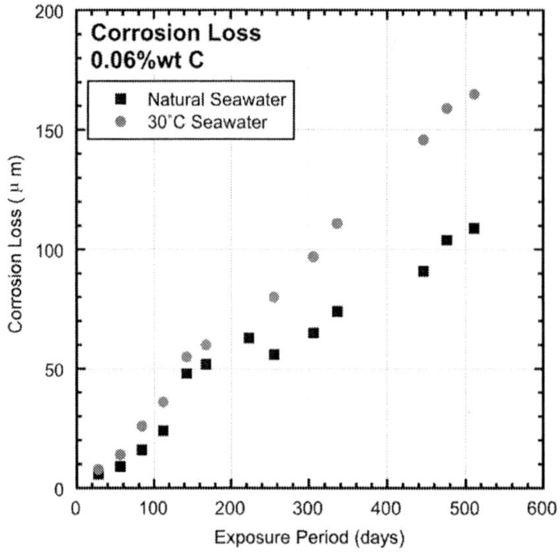

Figure 6. Corrosion loss as a function of time and average water temperature for 0.06%wt C steel in 20˚C (av.) and 30˚C seawaters.

3.2 Variation in water temperature

The recorded variation in water temperature for the 20˚C nominal natural seawater stream recorded over the last 15 months is shown in Fig. 7. It is seen that the actual temperature varied between 11.0 to 29.6˚C with an average of 21.0˚C. This corresponds to 70 to 511 days in Figs 4-6, that is, over the bulk of the data. The highest temperature was in the period corresponding to 100 to 260 days when the average water temperature was about 25˚C. The lowest water temperature range was from about 320 to 420 days with an average water temperature of about 16˚C.

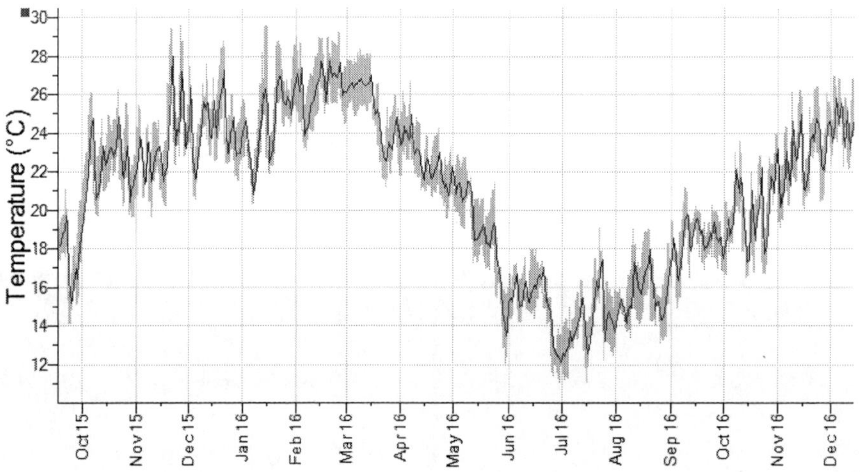

Figure 7. Temperature variation for the natural seawater in the unheated tank (20˚C nominal), corresponding to the period 70-511 days of exposure.

The recorded temperature variation in the nominally 30˚C tank is shown in Fig. 8. It shows almost no variation, although the maximum temperature reached 31.9°C and the minimum dropped to 29.3˚C. The average water temperature for the 15 month period of record was 30.4°C.

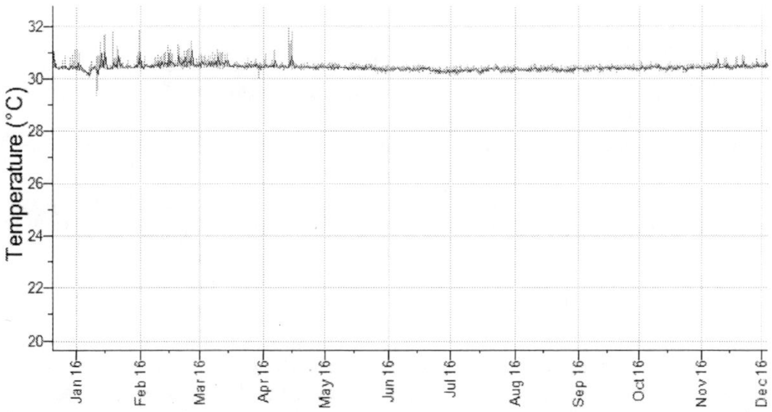

Figure 8. Temperature variation for the exposure tank nominally at 30˚C.

3.3 Comparison of corrosion loss with water temperatures

Direct comparison of the corrosion loss trends for the nominally 20˚C exposures and the nominally 30˚C exposures is complicated by the variation shown in Fig. 7. It is clear that in the period to about 200 days exposure the water temperature of the natural seawater was roughly 4-5˚C lower than that in the 30˚C tank, even though the corrosion losses are only slightly different. On the other hand, after about 200 days exposure, the corrosion losses in the natural seawater are significantly less than those in the nominally 30˚C tank, even though the temperature difference is only about 3-4˚C. One reason for this is the Arrhenius behaviour of the chemical reactions governing corrosion as a function of water temperature but it cannot be the only reason. As noted previously, the mechanisms corresponding to corrosion behaviour in Mode 1 (Fig. 1) is different to that in Mode 2, and thus different temperature sensitivities must be expected.

3.4 Effect of carbon content

The effect of carbon content of the steels on corrosion loss as a function of exposure period is shown in Fig. 9 for exposure in the natural seawater (20˚C nominal) and in Fig. 10 for exposure in the seawater heated to 30˚C. It is evident that there is minimal difference in the corrosion losses and in the trends with time. Superficially this applies both for the early period, corresponding to Mode 1 in Fig. 1. However, closer examination shows that for exposures in the 30˚C seawater there is a slight increase in corrosion loss with higher carbon content. This is consistent with earlier findings based on literature date (Melchers, 2003b). A similar effect of carbon content is evident, but much less so, for the 20˚C (nominal) natural seawater exposures.

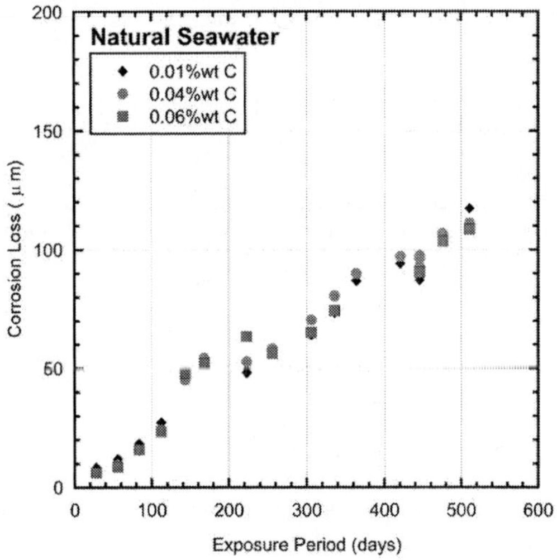

Figure 9. Corrosion losses for the 3 steels with different carbon contents in unheated natural seawater.

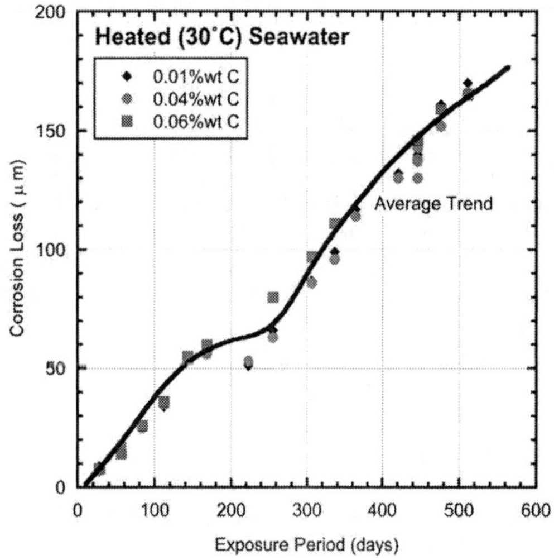

Figure 10. Corrosion losses for the 3 steels with different carbon contents in 30˚C (nominal) seawater. Average trend shown also.

4. DISCUSSION

After allowing for the variability that inevitably is present in data from exposure trails, it is clear that the data in Fig. 10 for the constant 30˚C exposure conditions shows a bi-modal trend for all three steels. It is less clear (Fig. 9) for the coupons exposed to the natural seawater that varied in temperature as shown in Fig. 7. A simple correction to the corrosion losses in Fig. 9 can be made by reducing the corrosion losses for the period when the water temperature was above 20˚C and increasing them for the period when the water temperature was below 20˚C, in both cases using a simple linearization of Arrhenius law at 20˚C. It is known from previous work that under diffusion control the corrosion rate approximately doubles for a 30˚C increase in temperature (IJsseling, 1989). Diffusion control applies in phases 1 onwards (Fig. 1) and thus it will affect mainly the early exposure period of 10-100 days when the water temperature was around 18-19˚C (Fig. 7). As shown on Fig. 9, the trend in corrosion loss for this period is less than subsequently, and this as would be expected. A small correction is therefore appropriate to allow for the lower water temperature in this early period. It is shown schematically in Fig. 11. Subsequently, in the period about 120-260 days the water temperature was around 25˚C and for this period the corrosion losses have been decreased slightly, again as shown schematically in Fig. 11. Thereafter the water temperature was only slightly less than 20˚C and the original data should be increased slightly. Irrespective of the precise adjustments, it is clear that the resulting data clearly shows a bi-modal trend (Fig. 11).

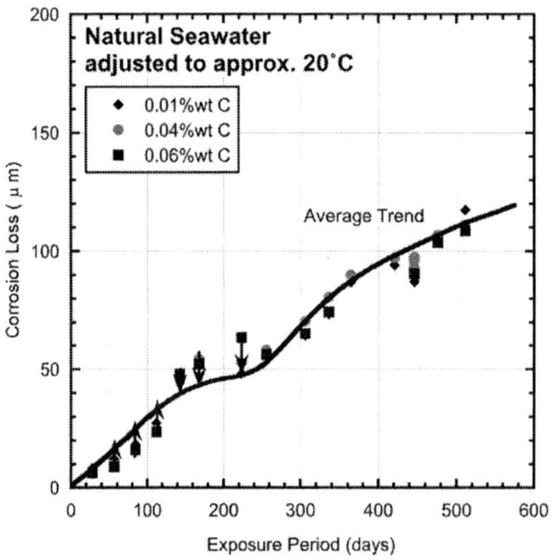

Figure 11. Corrosion loss data (Fig. 9) after adjustment of early corrosion losses for water temperature (shown schematically with arrows) together with interpreted average trend.

Overall, the above shows the importance of allowing for actual variations in water temperature in interpreting trends in data for corrosion losses, particularly for relatively short-term exposures. However, it should be clear that annual averages are adequate for exposures extending over years and decades.

The present results support the earlier notion that carbon content is of negligible importance in shorter-term (<100 days) exposures in seawater for seawater temperatures around 15 to 30˚C. However, a preliminary investigation of the depth of pitting for the 3 different steels has shown a difference in maximum pit depth and in the number of pits per unit area (Melchers, et al. 2016). This will be reported in more detail in due course.

5. CONCLUSION

The test results show that:

1. The carbon content in the range 0.01%wt – 0.06%wt of low alloys steels has little influence on the corrosion loss in seawater in the range 15-30˚C for exposures less than about 100 days and a very small effect between about 150-350 days with higher carbon steels showing slightly more corrosion loss, after which the effect becomes negligible. However, for practical purposes the differences are negligible, and

2. The effect of seawater temperature was as expected with higher water temperatures elevating the corrosion loss but retaining the bi-modal characteristic of the corrosion trend.

6. ACKNOWLEDGMENTS

The present work is one component of a research project dealing with the development of the bi-modal characteristic of immersion and other corrosion, funded under a DORA grant by the Australian Research Council. The authors appreciate that support, as well as that of the Civil Engineering laboratories at The University of Newcastle, and in particular Ross Gibson, and also the continued support of NSW Fisheries at Taylors Beach.

7. REFERENCES

Evans U.R. (1960) The corrosion and oxidation of metals: scientific principles and practical applications, London, Edward Arnold.

Jones, D. (1996) Principles and Prevention of Corrosion, Second Edition, Upper Saddle River, NJ, Prentice-Hall.

Jeffrey, R. & Melchers, R.E. (2009) Corrosion of vertical mild steel strips in seawater, Corros. Sci., 51(10) 2291-2297.

Melchers, R.E. (2003a) Modelling of marine immersion corrosion for mild and low alloy steels - Part 1: Phenomenological model, Corrosion, 59(4) 319-334.

Melchers, R.E. (2003b) Effect on marine immersion corrosion of carbon content of low alloy steels, Corros. Sci., 45(11) 2609-2625.

Melchers, R.E. (2004) Effect of small compositional changes on marine immersion corrosion of low alloy steel, Corros. Sci., 46(7) 1669-1691.

Melchers, R.E. & Chernov, B.B. (2010) Corrosion loss of mild steel in high temperature hard freshwater, Corros. Sci., 52: 449-454.

Melchers, R.E. & Jeffrey, R.J. (2014) Long-term corrosion of mild steel in natural and UV-treated coastal seawater, Corrosion, 70 (8) 804- 818.

Melchers, R.E. Chaves, I.A. & Jeffrey, R. (2016) A conceptual model for the interaction between carbon content and manganese sulphide inclusions in the short-term seawater corrosion of low carbon steel, Metals 6(6) 132. doi: 10.3390/met6060132.

Mercer, A.D. & Lumbard, E.A. (1995) Corrosion of mild steel in water, British Corrosion Journal, 30: 43-55.

Roberge, P.R., Klassen, R.D. & Haberecht, P.W. (2002) Atmospheric corrosion modelling - A review. Materials & Design, 23; 321-330.

Svedrup, H.U., Johnson, M.W. & Fleming, R.H. (1942) The Oceans: Their Physics, Chemistry and General Biology, McGraw-Hill, New York.

IJsseling, F.W. (1989) Guidelines for corrosion testing for marine applications, British Corrosion Journal, 24; 55-78.

8. AUTHOR DETAILS

Robert Jeffrey is a research scientist at The University of Newcastle, Australia where he has been investigating corrosion in marine, tidal and atmospheric conditions for many years. He holds a BSc (Chem) and PhD from The University of Newcastle. Robert is also principal consultant for Pacific Testing Pty Ltd, a company specializing in corrosion problems.

He is an honorary life member of the ACA (2011), been awarded the ACA Corrosion Medal (2012), is a past president of the ACA and has been on the committee of the Newcastle branch for over twenty years. He co-authored a paper that won the prestigious T P Hoar Prize (Institute of Corrosion, UK) and has twice been presented with the ACA's Marshal Fordham award and the AC Kennet award.

Robert E. Melchers is Professor of Civil Engineering at The University of Newcastle, Australia. He holds a BE and MEngSc from Monash University and a PhD from the University of Cambridge, UK. He is a Fellow of the Australian Academy of Technological Sciences and Engineering and an Honorary Fellow of The Institution of Engineers Australia.

His most recent awards are the 2009 ACA Corrosion Medal, the 2012 Jin S Chung Award (International Society of Offshore and Polar Engineers) and the 2013 John Connell Gold Medal (The Institution of Engineers, Australia). He was the Institution of Engineers, Australia Structural College Eminent Speaker for 2014.

EFFECT OF WATER VAPOUR ON OXIDATION BEHAVIOUR OF Ni-Cr ALLOYS IN CO_2 ATMOSPHERE AT 650°C

Y. Xie, J. Zhang, D. J. Young
School of Materials Science and Engineering, University of New South Wales, Sydney, NSW 2052, Australia

SUMMARY: Binary Ni-Cr alloys containing 5-30 wt% Cr were oxidized in Ar-20CO_2-20H_2O (vol%) gas mixtures at 650°C. Weight gain kinetics for alloys reacted with electropolished surfaces generally followed the parabolic rate law, and reaction rates increased with increasing Cr content of the alloy up to 25 wt%. All alloys underwent both internal oxidation and external oxide scaling. Scattered Ni nodules covered by a discrete thin film of NiO were observed on Ni-5Cr and Ni-10Cr, but a large number of loosely connected NiO particles were formed on the other alloys. The depth of the internal oxidation zone (IOZ) in the various alloys increased with increasing Cr concentration from 5 to 25 wt%, but decreased in Ni-30Cr which developed a continuous chromia band at the IOZ-alloy interface in some areas. Scale buckling was found in the alloys containing 20-30 wt% Cr. In addition, carbides were formed in these high Cr level alloys, along the grain boundaries where narrow chromia bands were formed at the intersection of grain boundaries with the surface. Compared with the results obtained in dry Ar-20CO_2 gas, the weight gain kinetics were slowed, but the thickening kinetics of IOZ were unchanged by the presence of water vapour.

Keywords: Oxidation, Carburization, Ni-based alloys.

1. INTRODUCTION

Oxyfuel process has been considered as one of the most promising technologies for CO_2 capture in fossil fuel fired power plants. In this process, the exhaust gas enriched in CO_2 and H_2O is highly corrosive. In addition, higher operation temperatures, e.g. above 700°C are required to achieve a higher thermal efficiency of the oxyfuel power plant. The heat exchanging components in boilers are exposed to such a harsh environment that the constructional iron-based chromia-foming alloys have been reported to undergo accelerated oxidation and severe carburization [1-5]. Therefore, nickel-based alloys should be used due to their superior creep and corrosion resistance. However, there is very limited information on the oxidation behaviour of Ni-based alloys exposed to CO_2/H_2O-rich gases relevant to the oxyfuel environment.

The oxidation of binary Ni-Cr alloys in air and oxygen has been well studied, which mainly focused on the formation of external chromia scale and its temperature dependence [6, 7]. A preliminary study has demonstrated the significant effect of temperature on the formation of protective chromia scale on Ni-Cr alloys in dry CO_2 atmosphere [8]. However, the influence of water vapour on their oxidation behaviour was found to be varied with the oxidizing environments, e.g. gases containing high amounts of free oxygen (wet air) [9, 10] or gases with low oxygen partial pressure (H_2/H_2O or pure steam) [9, 11, 12]. These investigations were usually carried out in either CO_2-rich gases or H_2O-containing environments. However, there is limited information about the oxidation behaviour of Ni-Cr alloys exposed to CO_2/H_2O-rich gases.

In the present study, the oxidation behaviour of model Ni-Cr alloys with different Cr contents in Ar-20CO_2-20H_2O (vol%) environment was studied at 650°C and compared with the behaviour in Ar-20CO_2 atmosphere. The main purpose is to elucidate the effect of water vapour present in CO_2 atmosphere in terms of the resultant oxidation kinetics and scale morphologies.

2. EXPERIMENTAL DETAILS

Six model alloys, Ni-5, 10, 15, 20, 25 and 30 Cr (nominal composition, all in wt%) were prepared by arc melting pure metals Ni (99.99%) and Cr (99.995%) under a protective Ar-5%H_2 gas atmosphere, using a non-consumable electrode. The formed alloy buttons were annealed at 1100°C for 50 h in a flowing Ar-5%H_2 gas for homogenisation and then slowly cooled in the furnace. The as-annealed alloys had coarse and irregular grains, generally with an average grain size of about 1.0 ~ 1.5 mm for Ni-5Cr and Ni-10Cr, 0.5 ~ 1.0 mm for Ni-15Cr and Ni-20Cr, 0.2 ~ 0.5 mm for Ni-25Cr and Ni-30Cr alloys. Analysis by X-ray diffraction confirmed that all alloys were single-phase austenitic. Rectangular alloy specimens with the approximate size of (12-15) mm× (6-8) mm× (0.8-1.2) mm were ground to a 1200 grit finish, polished down to a 3 μm finish and then electropolished in 15% concentrated hydrochloric acid to remove the work hardened surface region. Finally, all the specimens were ultrasonically cleaned with ethanol prior to exposure.

The isothermal corrosion experiments were carried out in a vertical furnace under atmospheric pressure, and the test specimens were exposed to a flowing Ar-20CO_2-20H_2O mixture with a linear gas flow rate of 2 cm/s at 650°C for up to 500 h. Each exposure experiment was continuous till the predetermined time. Wet gases of controlled water vapor pressure were generated by passing a mixture of argon and carbon dioxide through a thermostatted water saturator. The demineralized water in contact with the gas mixture was set at such a temperature that an excess of water vapor was produced. This excess of water was subsequently condensed by cooling the wet gas in a distillation column at the temperature corresponding to the required water content. Before reaction, a quartz reactor was purged with pure argon for 1 h to expel atmospheric oxygen, and then the furnace started to heat at a rate of 70°C/min and maintained at the required temperature. After that, the furnace was lifted to position the samples into the hot zone and the reaction gas was introduced subsequently. After exposure, the furnace was lowered to leave the samples out of the furnace for rapid cooling to room temperature.

Weight changes of oxidized samples were measured by an analytical balance (Mettler Toledo XP205) with an accuracy of 0.01 mg. Each point in the weight gain curves shown in this work represents a separate sample. Corrosion products after exposure were characterized using X-ray diffraction (XRD; PANanalytical Xpert MPD) with Cu-Kα radiation, optical microscopy (OM), scanning electron microscopy (SEM; Hitachi S3400) equipped with energy dispersive X-ray spectroscopy (EDS; Bruker). Etchant solutions of hydrochloric acid and nitric acid at a volume ratio of 5:1 were used to reveal the carbides in the alloy matrix. For cross section analysis, the selected oxidized samples were mounted in epoxy resin before sectioning, and then metallographically ground and polished to 0.1 μm.

3. RESULTS

3.1 Oxidation kinetics

Figure 1 shows the weight gain kinetics of the studied alloys in wet CO_2 gas. It is indicated that the weight uptakes of alloys increased approximately parabolically with time. Reaction rates of alloys with Cr contents of 5 and 10 wt% were very low, while Ni-15Cr alloy reacted obviously faster. With the Cr concentrations increasing to 20 wt% and above, the weight gains were relatively irregular with some fluctuations, but consistently higher than those of alloys with lower Cr contents. Parabolic rate constants, k_w, were calculated with the equation:

$$(\frac{\Delta W}{A})^2 = 2k_w t \tag{1}$$

where ΔW is the weight gain, A the surface area of the tested specimen, k_w the parabolic rate constant, and t the oxidation time. The values of k_w under wet conditions obtained by regression of the data shown in Fig. 1 were summarized in Table 1, together with those obtained under dry condition [8] for comparison. It is obvious that, in wet CO_2 gas, the values of k_w for alloys containing 15 to 30 wt% Cr were in the same magnitudes, which were about two magnitudes larger than those of Ni-5Cr and Ni-10Cr alloys.

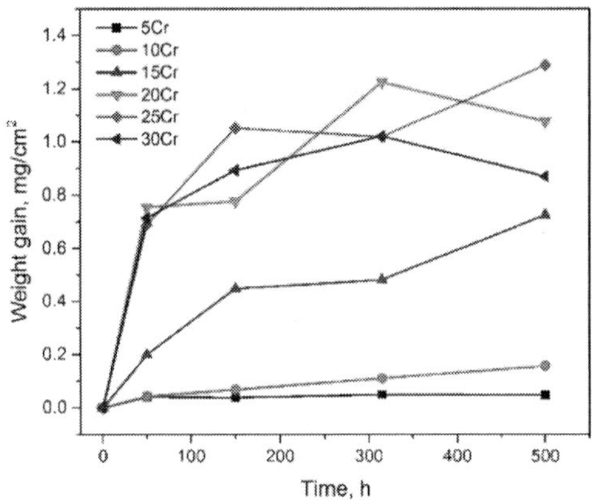

Figure 1 Weight gain kinetics of various alloys exposed to wet CO_2 gas at 650°C

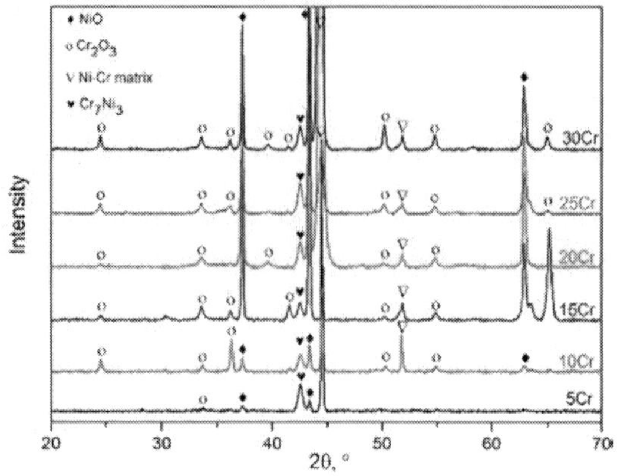

Figure 2 XRD patters of the alloys after exposure in wet CO_2 gas at 650°C for 500 h

Table 1. Parabolic oxidation rate constant, k_w ($10^{-15}g^2cm^{-4}s^{-1}$), for alloys reacted in wet and dry CO_2 at 650°C

Alloy	Ni-5Cr	Ni-10Cr	Ni-15Cr	Ni-20Cr	Ni-25Cr	Ni-30Cr
k_w for wet CO_2	0.8	6.2	136.3	434.7	498.6	315.3
k_w for dry CO_2	910.8	764.3	1069.7	1162.9	979.7	374.9

3.2 Scale morphology

Figure 2 shows the surface XRD patterns of alloys oxidized for 500 h in wet condition, indicating the oxidation products of all the alloys principally consisted of NiO and Cr_2O_3. BSE-SEM cross-section images of alloys after reaction in wet CO_2 gas for 500 h are shown in Fig. 3. A shallow internal oxidation zone (IOZ) mostly composed of spherical precipitates was formed on Ni-5Cr, with sparse dendritic precipitates penetrating deep into the matrix. Some small nodules were observed locally above the IOZ of Ni-5Cr, and more nodules appeared on Ni-10Cr. These nodules were confirmed by EDS to be Ni metallic particles covered partially by a thin layer of NiO which was too thin to be distinguished on Ni-5Cr but became apparent on Ni-10Cr. The expulsion of metallic Ni from the underlying alloy substrate was attributed to relieving stress induced by internal oxidation of Cr [13]. The Cr-rich oxide precipitates within the IOZ was determined to be Cr_2O_3 based on the XRD spectrums shown in Fig. 2.

The scales developed on Ni-15Cr consisted of two parts: the outer part comprised of discrete islands randomly scattered on the alloy surface; the inner part appeared to be a uniform and continuous layer which actually consisted of very fine precipitates densely distributed in alloy substrate. In order to determine the phases present in different parts, oxidized sample surface underwent a series of consecutive grinding and XRD analyses. Figure 4 shows the XRD patterns corresponding to the approximate locations indicated in Fig. 3c. Combined with the EDS analyses (not shown here), it is indicated that the outer part was mainly composed of NiO and the inner part was a mixture of Cr_2O_3 and alloy matrix. Thus, the inner part can be regarded as IOZ. It is worth noting that marked Ni nodules were also found in the outer part and surrounded by thick NiO which exhibiting an elongated shape.

Alloys with higher Cr levels retained similar IOZ morphology as Ni-15Cr but the amount of outer loosely connected NiO granules increased. A large number of needle-shape NiO oxides, somewhat morphologically similar to the whiskers frequently reported in H_2O-containing environment [14, 15], were observed in the outermost scales, especially for Ni-20Cr. Besides, scale spallation also occurred in the alloys containing ≥ 15 wt% Cr to variable extents, which may be correlated to the poor adhesion of the outer NiO particles. It should be noted that the continuous Cr-rich oxide band (shown to be Cr_2O_3) formation at the IOZ-alloy interface within some grains of Ni-30Cr (Fig. 3f) was not necessarily the case for all the grains, which would be mentioned in the following section. In general, the depth of IOZ increased with alloy Cr content up to 25 wt% and then decreased at 30 wt%, which is in general agreement with the gravimetric data shown in Fig. 1.

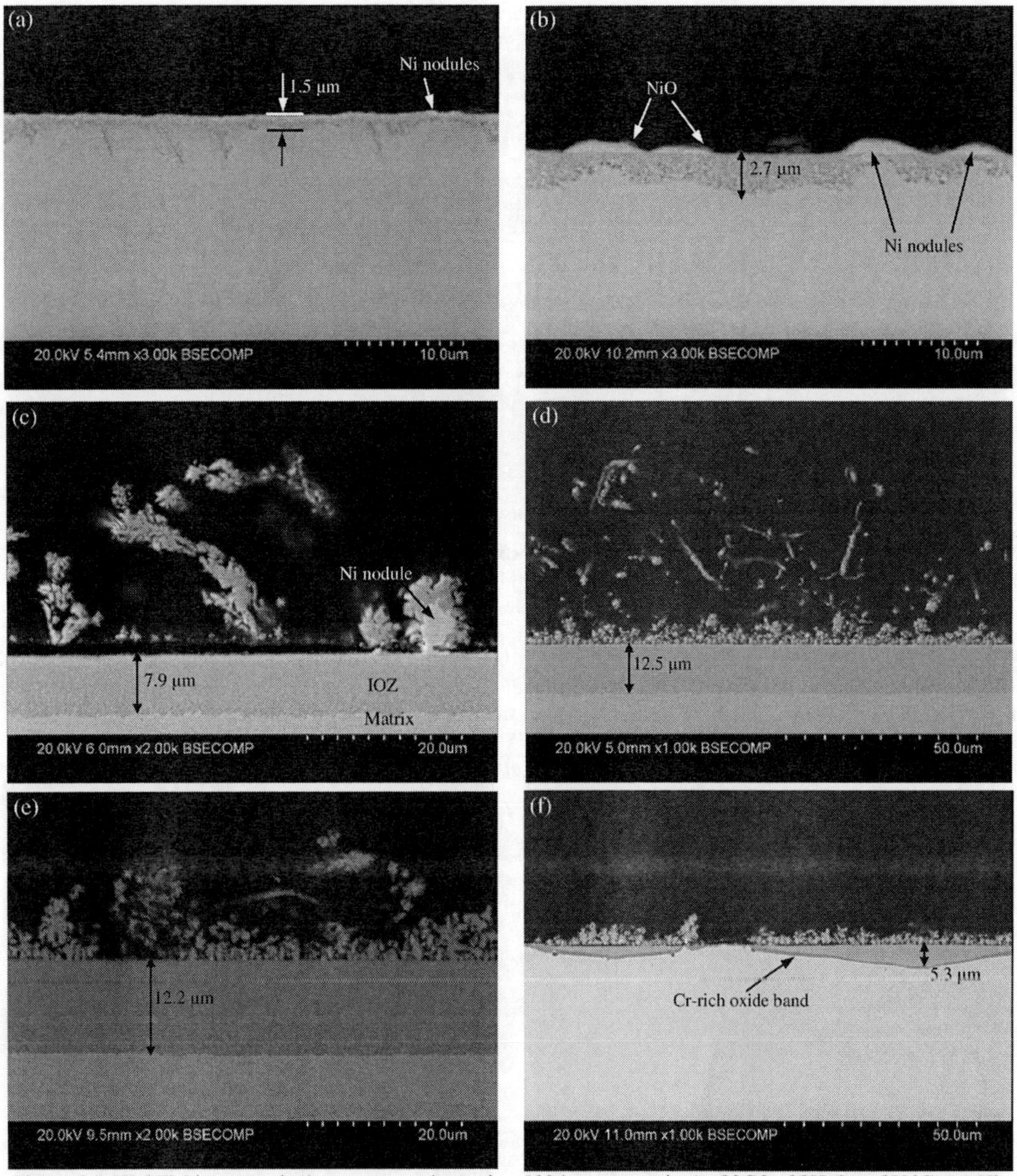

Figure 3 BSE-SEM images of alloy cross-sections after 500 h exposure in Ar-20CO_2-20H_2O at 650°C: (a) Ni-5Cr, (b) Ni-10Cr, (c) Ni-15Cr, (d) Ni-20Cr, (e) Ni-25Cr, and (f) Ni-30Cr

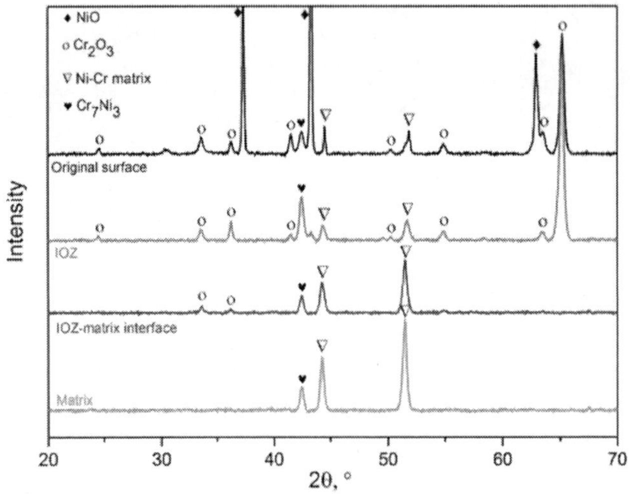

Figure 4 XRD patters obtained from the consecutive grinding of scale formed on Ni-15Cr

Except the predominant morphologies shown above, the oxide scales formed on the alloys with 20-30 wt% Cr also presented some local buckling as shown in Figure 5, but this phenomenon was not visible on the alloys with lower Cr contents. The continuity of IOZ generated on these alloys was interrupted by this local bulge. The IOZ above and directly adjacent to the scale buckling was split into two layers with alloy matrix enclosed. Some cracks were found at the top or root of bulged IOZ. EDS line profile shown in Fig. 5(d) demonstrated there was no apparent difference in chemical compositions between the two separated IOZ layers on Ni-25Cr, but the alloy matrix enclosed was Cr depleted to a large extent. For Ni-30Cr, a Cr-rich oxide band was observed at the interface between IOZ and metal matrix but ceased at the bottom of the regions where scale buckling was generated.

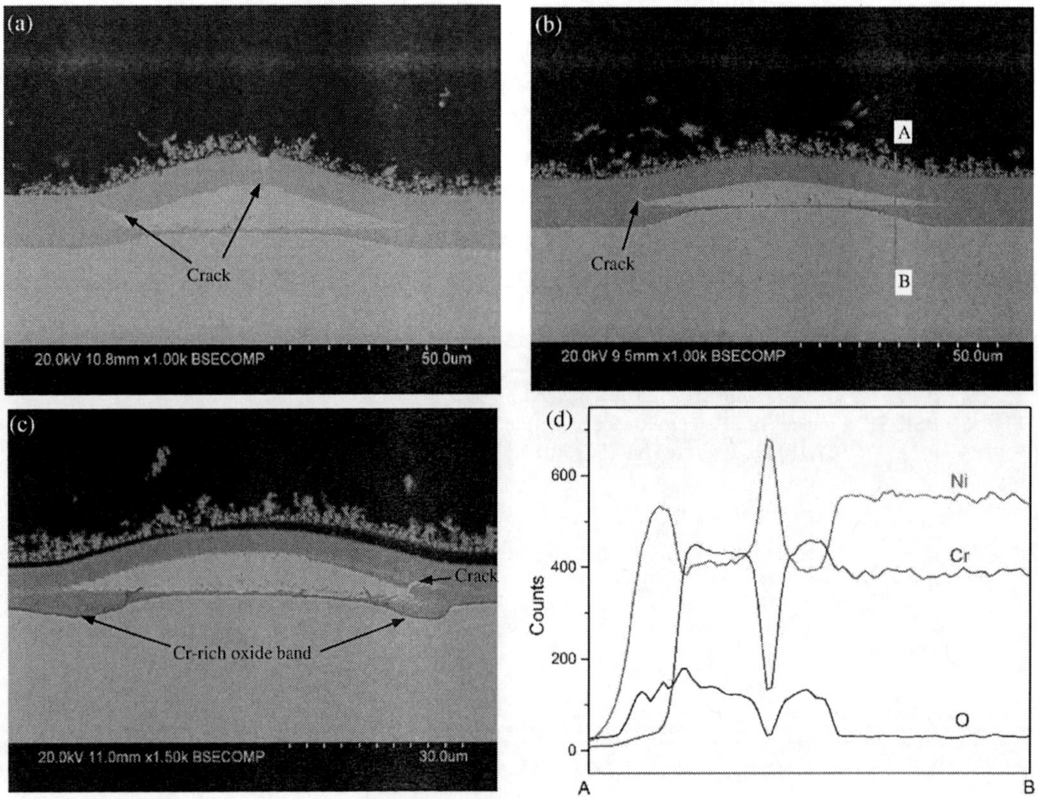

Figure 5 Scale buckling formed on (a) Ni-20Cr, (b) Ni-25Cr, (c) Ni-30Cr; (d) EDS line profiles along A-B shown in (b)

3.3 Carburization

Intergranular carbide precipitates were observed by etching in the alloys containing 20-30 wt% Cr after exposure in wet CO_2 gas. Figure 6 presents the typical structures of carbides distributed along the grain boundaries. A high magnification SE-SEM image of carbides in Fig. 6a and its corresponding EDS line profile analysis confirmed the formation of Cr-rich carbide precipitates. No carbide was observed in the lower Cr content alloys under the same condition.

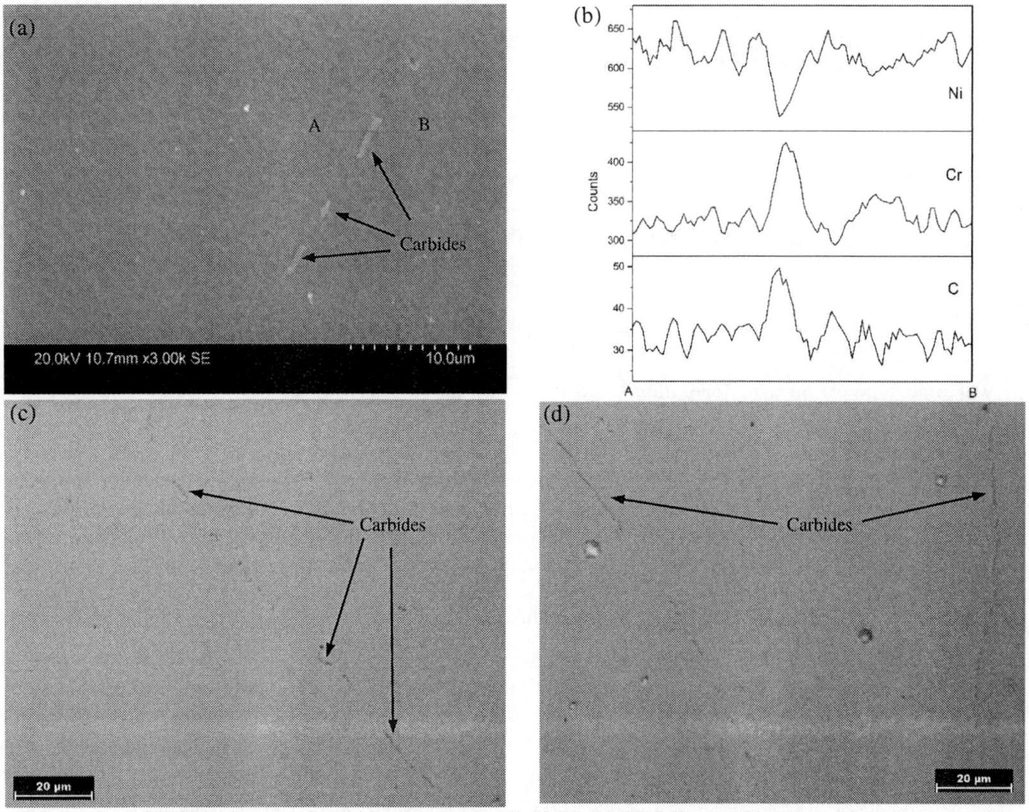

Figure 6 (a) SE-SEM image of carbides formed in Ni-20Cr, (b) EDS line profiles shown in (a), (c) and (d) optical image of carbides formed in Ni-25Cr and Ni-30Cr, respectively.

4. DISCUSSION

The oxide scale structures observed in the present study are summarized in Table 2, where the results from dry CO_2 condition [8] are also listed for comparison. For dry and wet CO_2 gases, the equilibrium oxygen partial pressure (p_{O_2}) and carbon activity (a_c) of the atmosphere can be determined by the following reactions:

$$CO_2 = 2CO + \frac{1}{2}O_2 \quad (2)$$

$$2CO = CO_2 + C \quad (3)$$

$$H_2O = H_2 + \frac{1}{2}O_2 \quad (4)$$

$$CO + H_2 = H_2O + C \quad (5)$$

At equilibrium, wet CO_2 gas yields a p_{O_2} of 1.1×10^{-9} atm and a_c of 7.7×10^{-16}, while for dry CO_2 gas p_{O_2} = 5.1×10^{-9} atm and a_c = 1.6×10^{-15}. Under both exposure conditions, the corresponding values of p_{O_2} and a_c are sufficient to incur oxidation of Ni and Cr but insufficient for carburization. Therefore, dry and wet CO_2 oxidizing conditions are similar with regard to p_{O_2} and a_c. However, Table 2 clearly suggests that oxide scale morphologies developed in wet CO_2 gas differ from those formed in dry CO_2 gas. This difference should be attributable to the presence of H_2O gaseous species in the atmosphere, which will be discussed in the following section.

Table 2. Oxide scales developed after exposure for 500 h in different environments

Alloy	Wet CO_2 gas		Dry CO_2 gas	
	Outer scale	Inner scale	Outer scale	Inner scale
Ni-5Cr	Thin NiO film and small Ni nodules	Shallow IOZ	Continuous NiO layer	Intermediate layer (NiO + Cr_2O_3) + deep dendrite IOZ
Ni-10Cr	Thin NiO film and Ni nodules	IOZ	Continuous NiO layer	Intermediate layer (NiO + Cr_2O_3) + IOZ
Ni-15Cr	Scattered NiO islands and Ni nodules	IOZ	Continuous NiO layer	Discrete intermediate layer (NiO + Cr_2O_3) + IOZ
Ni-20Cr	Loosely connected NiO granules	IOZ	Continuous NiO layer	IOZ
Ni-25Cr	Loosely connected NiO granules	IOZ	Continuous NiO layer	IOZ + semicontinuous chromia band
Ni-30Cr	Loosely connected NiO granules	IOZ + chromia band	Continuous NiO layer	Intermediate layer (NiO + Cr_2O_3) + continuous chromia band

4.1 Effect of water vapour on scale formation

Compared with the results obtained under dry CO_2 condition [8], it is found that the weight gains of Ni-5Cr and Ni-10Cr decreased significantly from about 1.5-1.8 to 0.05-0.15 mg/cm^2 when water vapour was added into the atmosphere. For the alloys with 15-25 wt% Cr, however, this decrease was less pronounced, from approximately 1.0-2.0 to 0.7-1.3 mg/cm^2. Further increasing Cr content to 30 wt%, Ni-30Cr exhibited only slightly reduced weight gains after exposed to wet CO_2 gas relative to dry CO_2 gas. It is demonstrated in Table 1 that the parabolic rate constants of various alloys oxidized in wet condition were consistently lower than those in dry condition, but the gap between them gradually decreased with increasing Cr concentration of the alloys. This difference in oxidation kinetics can be manifested to some extent by the scale morphologies developed under different reaction conditions.

For all the alloys reacted with dry CO_2 atmosphere, a continuous and relatively compact NiO layer of appreciable thickness was formed above their original alloy surfaces. Underneath the external NiO layer, an intermediate layer (NiO + Cr_2O_3) and an innermost IOZ were developed in the alloys with 5-15 wt% Cr. When the Cr concentration of the alloy was increased to 20 wt%, the intermediate layer disappeared. A chromia band appeared at the reaction front of IOZ in Ni-25Cr and Ni-30Cr. However, there was no continuous NiO layer developed on the alloys oxidized in wet CO_2 atmosphere. As shown in Fig. 3, NiO presented discrete thin films on Ni-5Cr and Ni-10Cr, but showed numerous islands on Ni-15Cr. NiO particles continued to increase in number and were more closely spaced as the Cr concentration of the alloy increased to above 20 wt%, but they still failed to form a continuous layer. In addition, it is found that the depth of IOZ also increased with increasing alloy Cr level up to 25 wt%. Correspondingly, the compressive stress resulted from the internal oxide precipitation could also increase. It was reported that external NiO layer and Ni nodules helped to relieve the stress [16]. Therefore, the increase in Ni nodule size and NiO particle population density observed on the alloys was explainable. The scale buckling shown in Fig. 5 may be ascribed to the excessive compressive stresses which resulted from the deep IOZ but could not be fully accommodated by the formation of external NiO particles. In the case of the alloys oxidized in dry CO_2 atmosphere, although even deeper IOZ was developed compared with in wet CO_2 gas, the continuous external NiO layer combined with numerous pores distributed in IOZ may accommodate the stresses. Therefore, no visible scale buckling existed on the alloys.

The reason for discrete Ni nodules accompanied with few NiO on low Cr level alloys but a large amount of NiO particles on high Cr level alloys could be related to an increased amount of internal oxides which lead to more Ni nodule formation and therefore NiO. Despite this, the question remains why NiO exhibited scattered particle structure rather than a continuous layer in the presence of water vapour, although the oxygen partial pressure of the atmosphere was not greatly changed. Raynaud and Rapp [15] reported that NiO whisker formation was enhanced when water vapour was added to oxygen, and they proposed that the easier breakup of H_2O molecule compared with oxygen accelerated the whisker growth kinetics. This mechanism could also be applicable for the present study, as the outermost NiO particles found on Ni-20Cr were somewhat similar to whiskers in morphology. Although the difference in oxidation kinetics under wet and dry exposure conditions can be qualitatively explained by the corresponding different oxide scale structures developed on the alloys, some uncertainties still exist and more work is required to ascertain them.

According to the classical theory of oxidation [17], the Cr concentration of the alloy should reach a critical value, N_{Cr}, to prevent internal oxidation and form an external chromia scale, which can be estimated by the equation:

$$N_{Cr} = \left(\frac{\pi g^*}{2v} \frac{V_m}{V_{CrO_v}} \frac{N_O^{(s)} D_O}{\tilde{D}_{Cr}} \right)^{\frac{1}{2}} \quad (6)$$

where g^* is a critical volume fraction of oxide precipitates, generally approximated as about 0.3 [18], v the stoichiometric coefficient of CrO_v (1.5 for $CrO_{1.5}$), V_m the molar volume of alloy, V_{CrO_v} the molar volume of the oxide, D_O the diffusion coefficient of oxygen in the alloy, $\tilde{D}_{Cr}$ the alloy inter-diffusion coefficient, $N_O^{(s)}$ the mole fraction of oxygen dissolved in the alloy surface.

As has been discussed in dry CO_2 gas [8], N_{Cr} equals to 0.31 at 650°C. Since the Cr concentration of Ni-30Cr is slightly higher than this value, the observed chromia bands in both dry gas and wet gas at the scale-alloy interface of this alloy are explainable, although the morphologies of oxide scales are different as indicated in Table 2. This result is also consistent with the report [19] that water vapour does not significantly change oxygen permeability in nickel.

It is generally accepted that when an internal oxidation process is controlled by inward oxygen diffusion, the reaction kinetics can be expressed by parabolic rate law:

$$X_i^2 = 2k_p^{(i)} t \quad (7)$$

where X_i is the penetration depth of internal oxides reached in the exposure time, t, and $k_p^{(i)}$ the parabolic rate constant. In this paper, X_i was measured from the original alloy surface or inner layer-IOZ interface to the internal oxide precipitation front, depending on whether inner layer existed. Based on Wagner's analysis [17], the parabolic rate constant can be related to oxygen permeability:

$$k_p^{(i)} = \frac{\varepsilon N_O^{(s)} D_O}{v N_{Cr}^{(0)}} \quad (8)$$

where ε is a diffusion blocking factor and $N_{Cr}^{(0)}$, the initial mole fraction of Cr in the alloy. It should be noted that the prerequisite for the validity of Eqn (8) is that the oxygen permeability is much larger than the product $N_{Cr}^{(0)} \tilde{D}_{Cr}$, i.e. $N_O^{(s)} D_O >> N_{Cr}^{(0)} \tilde{D}_{Cr}$.

To ascertain the effect of water vapour on oxygen permeability in the various alloys investigated, the penetration depth of IOZ were plotted in Fig. 7a and 7b as a function of reaction time after alloys were exposed to wet and dry CO_2 gases, respectively. Error bars on the plots represented the standard deviations in depth caused by depth measurement uncertainties. Alloys of Ni-5Cr and Ni-30Cr were excluded from Fig. 7, because it was manifested that the accelerated diffusion of oxygen along the acicular precipitate-matrix interface in Ni-5Cr and the continuous chromia band established at the scale-alloy interface in Ni-30Cr could apparently influence the nature of internal oxidation kinetics [8]. It is clear that, as expected from the above discussion, the rate of IOZ advancement roughly conformed to a parabolic law under both wet and dry conditions. Table 3 summarize the corresponding values of $k_p^{(i)}$ calculated by regression on Eqn (7).

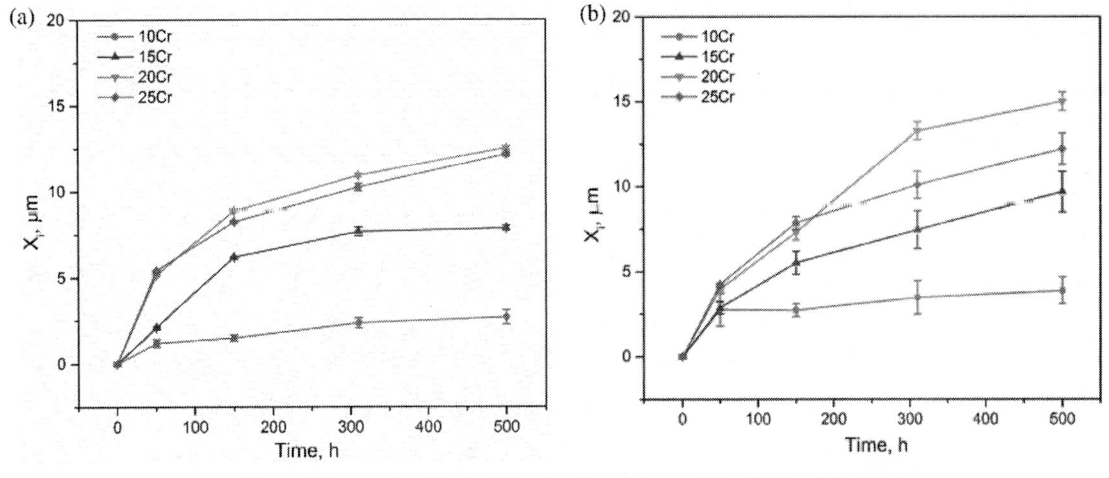

Figure 7 Internal oxidation kinetic plots: (a) wet CO_2, (b) dry CO_2 [8]

Table 3. Parabolic rate constant $k_p^{(i)}$ (10^{-3} μm^2/h) obtained after exposure in different conditions

Alloy	Ni-10Cr	Ni-15Cr	Ni-20Cr	Ni-25Cr
Wet CO_2 gas	7.8 ± 0.5	74.5 ± 9.9	173.6 ± 14.3	159.5 ± 11.3
Dry CO_2 gas	17.1 ± 2.8	92.9 ± 1.5	237 ± 15.3	156.6 ± 7.1

The value of inter-diffusion coefficient, $\tilde{D}_{Cr}$, at 650°C was extrapolated from higher temperatures [20]. Assuming ε =1 and v =1.5 in Eqn (8), $N_O^{(s)}D_O$ was calculated individually for each alloy as listed in Table 4. Comparing the corresponding values of $N_O^{(s)}D_O$ and $N_{Cr}^{(0)}\tilde{D}_{Cr}$ confirmed the validity of Eqn (8). Table 4 shows that the permeability values obtained in wet CO_2 gas were similar to those found in dry CO_2 gas, which is again confirmed that oxygen permeability in nickel is not affected by the presence of water vapour.

Table 4. $N_{Cr}^{(0)}\tilde{D}_{Cr}$ and $N_O^{(s)}D_O$ (10^{-14} cm^{2}s^{-1}) in alloys oxidized in different environments

Alloy	Ni-10Cr	Ni-15Cr	Ni-20Cr	Ni-25Cr
$N_{Cr}^{(0)}\tilde{D}_{Cr}$	0.03	0.04	0.06	0.07
$N_O^{(s)}D_O$ in wet CO_2 gas	0.36	5.15	1.59	1.81
$N_O^{(s)}D_O$ in dry CO_2 gas	0.79	6.43	2.17	1.78

4.2 Carburization

For carburization taking place in Ni-Cr alloys, the reaction can be expressed by:

$$\underline{Cr} + v\underline{C} = CrC_v,\ \Delta G \quad (9)$$

the free energy change can be obtained from the reactions:

$$Cr + vC = CrC_v,\ \Delta G^o \quad (10)$$

$$Cr = \underline{Cr},\ \Delta\overline{G}_{Cr} \quad (11)$$

$$C = \underline{C},\ \Delta\overline{G}_{C} \quad (12)$$

where underlining indicates a solute species. Thus, the equilibrium condition for carbide precipitation is written as:

$$K_{10} = \frac{a_{CrCv}}{a_{Cr}a_C^v} = \exp\left(-\frac{\Delta G^o - \Delta\overline{G}_{Cr} - v\Delta\overline{G}_C}{RT}\right) \quad (13)$$

where carbide activity is assumed to be unity, ΔG^o, $\Delta\overline{G}_{Cr}$ and $\Delta\overline{G}_C$ the standard free energy changes for the formation of CrC_v and the partial molar free energies of dissolution of Cr and C in nickel, respectively [21, 22]. The value of chromium activity in Ni-30Cr at 650°C is approximated to that calculated at 800°C [8], which yields a_{Cr} = 0.46. Then, the corresponding carbon activity is calculated to be 3.7×10^{-5} for $Cr_{23}C_6$ precipitation in Ni-30Cr.

Under wet CO_2 test condition, intergranular carbides were observed in alloys containing 20-30 wt% Cr. The typical structure of the oxide scale developed on Ni-30Cr alloy after etching was shown in Fig. 8, and the similar oxide morphology was also observed on Ni-20Cr and Ni-25Cr. Within the grain, internal oxidation zone was covered with discrete NiO particles, which means oxidizing gas molecules could access to the alloy surface readily. Accordingly, the value of a_C and p_{O_2} at the NiO-alloy interface was likely to be the same as the bulk gas. As calculated in the previous section, the carbon activity of the bulk gas was clearly too low to form carbides in these alloys. However, not within the grains, but along the grain boundaries did these alloys form carbides. As a result, it is reasonably inferred that the local carbon activity of these two places should be different. Examining the scales closely adjacent to the grain boundary shown in Fig. 8b, it is found that the place where the grain boundary intersected the alloy surface was covered by a narrow continuous band of oxide which cut into the alloy matrix obliquely at a shallow angle. The phase of this oxide band was determined to be Cr_2O_3 by EDS analysis (not shown here), and it should be attributed to fast chromium diffusion to the surface along the grain boundaries. Similar phenomena were also reported elsewhere [23, 24]. Fig. 8c indicate that discrete carbides were formed along the grain boundary right below the chromia band.

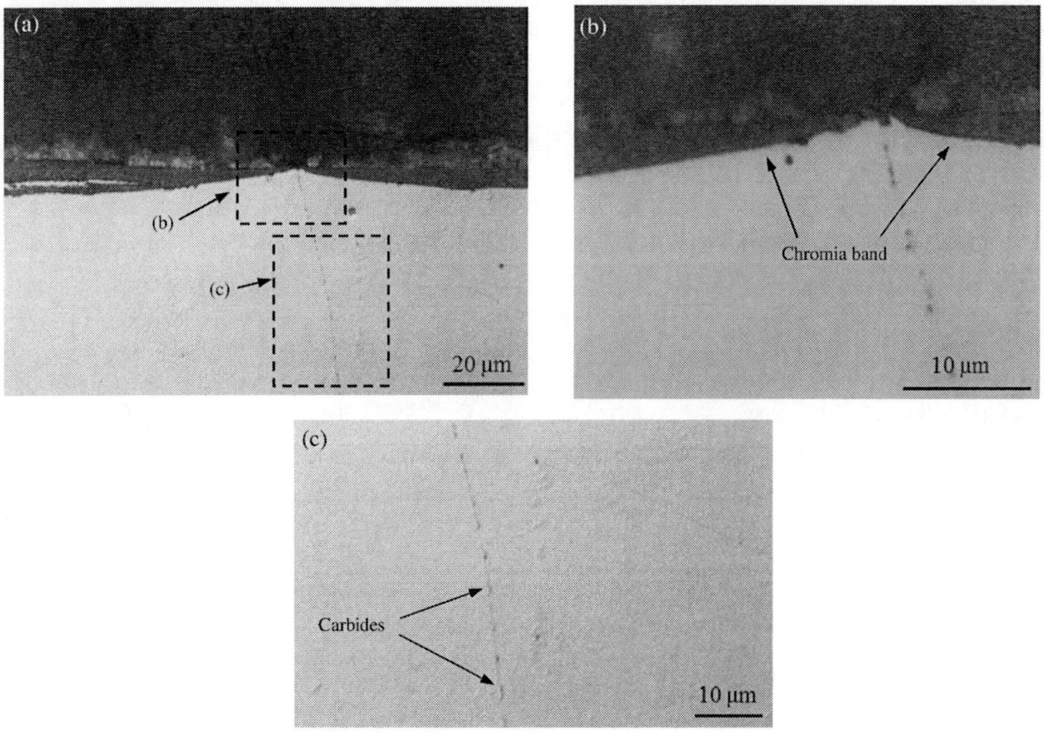

Figure 8 Cross-sections of Ni-30Cr oxidized for 310h in wet CO_2 gas, after etching, (b) and (c) enlarged areas indicated in (a)

The interfacial carbon activity in places where the chromia band is in contact with the alloy matrix is inversely proportional to the oxygen partial pressure at this location via the following equation [8]:

$$a_C = K_3 K_2^2 \frac{p_{CO_2}}{p_{O_2}} \tag{14}$$

where K_2 and K_3 are the equilibrium constant of Eqn (2) and (3), respectively. The value of p_{CO_2} is unknown, but the maximum value is assumed to be the partial pressure of CO_2 in the bulk gas. The local oxygen partial pressure is equal to 1.5×10^{-33} atm, based on chromia-alloy equilibrium. Substituting this value into Eqn (14) yields $a_C = 2.6\times10^9$ which is presumably 14 orders of magnitude larger than what is required to form $Cr_{23}C_6$ carbides in Ni-30Cr at 650°C, indicating the thermodynamic possibility of carbide formation. In addition, the facilitated carbon diffusion and carbide nucleation at grain boundaries also favour the occurrence of carbide precipitation in terms of the reaction kinetics.

5. CONCLUSIONS

The oxidation behaviour of Ni-Cr alloys in wet CO_2 gas at 650°C has been studied in this paper. The weight gain kinetics of the tested alloys generally obeyed parabolic rate law, and the change of corresponding rate constants correlated well with their distinct scale morphologies. A discrete thin NiO film overlying Ni nodules were observed on the surfaces of Ni-5Cr and Ni-10Cr, which was replaced by numerous loosely connected NiO granules or islands on the other higher Cr-containing alloys. The thickness of IOZ increased progressively from Ni-5Cr to Ni-25Cr and then decreased at Ni-30Cr. Compared with the results obtained in dry CO_2 gas, the internal oxidation kinetics were not promoted in the presence of water vapour, which indicated that water vapour addition did not enhance the oxygen permeability in the alloys. In wet CO_2 gas, the origin of scale buckling existed in higher Cr level alloys (Cr $\geq$ 20 wt%) is considered to be related with the internal stress created by internal oxide precipitation.

Intergranular carbides were formed in the alloys containing 20-30 wt% Cr exposed to wet CO_2 gas. Instead of prevailing IOZ produced in the grain, a narrow chromia band was developed over the intersections of grain boundaries with the surface of these alloys. The greatly enhanced carbon activity beneath the chromia band indicated the thermodynamic possibility of carbide formation.

6. ACKNOWLEDGMENTS

Financial support from the Australian Research Council's Discovery program is gratefully acknowledged.

7. REFERENCES

1. Abellán JP, Olszewski T, Meier GH, Singheiser L and Quadakkers WJ, The oxidation behaviour of the 9% Cr steel P92 in CO_2- and H_2O-rich gases relevant to oxyfuel environments, Int J Mat Res (2010) 287-299.

2. Abellán JP, Olszewski T, Penkalla HJ, Meier GH, Singheiser L and Quadakkers WJ, Scale formation mechanisms of martensitic steels in high CO_2/H_2O-containing gases simulating oxyfuel environments, Mat High Temp 26 (2009) 63-72.

3. Nguyen TD, Zhang J and Young DJ, Water vapour effects on corrosion of Fe-Cr and Fe-Cr-Ni alloys containing cerium and manganese in CO_2 gas at 818°C, Corros Sci 89 (2014) 220-235.

4. Gheno T, Monceau D, Zhang J and Young DJ, Carburisation of ferritic Fe-Cr alloys by low carbon activity gases, Corros Sci 53 (2011) 2767-2777.

5. Essuman E, Meier GH, Żurek J, Hänsel M and Quadakkers WJ, The effect of water vapour on selective oxidation of Fe-Cr alloys, Oxid Met 69 (2008) 143-162.

6. Wood GC and Hodgkiess T, Characteristic scales on pure Nickel-chromium alloys at 800°-1200°C, J Electrochem Soc 113 (1966) 319-327.

7. Giggins CS and Pettit FS, Oxidation of Ni-Cr alloys between 800° and 1200°C, Trans Met Soc AIME 245 (1969) 2495-2507.

8. Xie Y, Zhang J and Young DJ, Temperature effect on oxidation behaviour of Ni-Cr alloys in CO_2 gas atmosphere, J Electrochem Soc 164 (2017) C285-C293.

9. Michalik M, Hänsel M, Zurek J, Singheiser L and Quadakkers WJ, Effect of water vapour on growth and adherence of chromia scales formed on Cr in high and low pO_2-environmetns at 1000 and 1050°C, Mater High Temp 22 (2005) 213-221.

10. Essuman E, Meier GH, Zurek J, Hänsel M, Norby T, Singheiser L and Quadakkers WJ, Protective and non-protective scale formation of NiCr alloys in water vapour containing high- and low pO_2 gases, Corros Sci 50 (2008) 1753-1760.

11. Simon D, Gorr B, Hänsel M, Shemet V, Christ H-J and Quadakkers WJ, Effect of in-situ gas changes on thermally grown chromia scales formed on Ni-25Cr alloy at 1000°C in atmospheres with and without water vapour, Mater High Temp 32 (2015) 238-247.

12. Hänsel M, Garcia-Feresnillo L, Tobing SL and Shemet V, Effect of H_2/H_2O ratio on thermally grown chromia scales formed on Ni25Cr alloys in Ar-H_2-H_2O atmospheres at 1000°C, Mater High Temp 29 (2012) 187-192.

13. Newman RC and Scenini F, Another way to think about the critical oxide volume fraction for the internal-to-external oxidation transition? Corrosion 64 (2008) 721-726.

14. Hänsel M, Quadakkers WJ and Young DJ, Role of water vapor in chromia-scale growth at low oxygen partial pressure, Oxid Met 59 (2003) 285-301.

15. Raynaud GM and Rapp RA, In situ observation of whiskers, pyramids and pits during the high-temperature oxidation of metals, Oxid Met 21 (1984) 89-102.

16. Bradley L, Wood GC and Stott FH, The effects of stress on the internal oxidation of Ni-Cr and Ni-Al alloys, Mater Sci Forum 251-254 (1997) 341-348.

17. Wagner C, Reaktionstypen bei der oxydation von legierungen, Z Elektronchem 63 (1959) 772-782.

18. Rapp RA, The transition from internal to external oxidation and the formation of interruption bands in silver-indium alloys, Acta Metall Mater 9 (1961) 730-741.

19. Guo P, Zhang J and Young DJ, Oxygen permeability measurements in Ni using H_2-H_2O, CO/CO_2 and Ni/NiO rhines pack atmospheres, Oxid Met 83 (2015) 223-235.

20. Askill J, Tracer diffusion in the chromium-nickel system, Phys Stat Sol, 8 (1971) 587-596.

21. Shatynski SR, The thermochemistry of transition metal carbides, Oxid Met 13 (1979) 105-118.

22. Allen AT and Douglass DL, Internal-carburizing behaviour of Ni-V, Ni-Cr, and Ni-3Nb alloys, Oxid Met 51 (1999) 199-219.

23. Wood GC, Stott FH, Whittle DP, Shida Y and Bastow BD, The high-temperature internal oxidation and intergranular oxidation of nickle-chromium alloys, Corros Sci 23 (1983) 9-25.

24. Calvarin G, Molins R and Huntz AM, Oxidation mechanism of Ni-20Cr foils and its relation to the oxide-scale microstructure, Oxid Met 53 (2000) 25-48.

8. AUTHOR DETAILS

Yun Xie is a PhD candidate at the School of Materials Science and Engineering, UNSW. His work mainly focuses on the high temperature oxidation behaviour of Ni-based alloys in CO_2/H_2O-rich atmosphere.

Dr Jianqiang Zhang is Associate Professor at the School of Materials Science and Engineering, UNSW. His research is mainly on high temperature corrosion of metals and alloys in oxidation and carburisation gas atmospheres.

David Young is Emeritus Professor at the School of Materials Science and Engineering, UNSW. He has worked in the field of high temperature corrosion for more than 40 years, publishing extensively in the research literature.

AN INVESTIGATION INTO THE CORRELATION BETWEEN HALF CELL, CURRENT INJECTION TESTING AND THE CURRENT REQUIRED FOR PROTECTION

P. Nicholls[1], J. Gastal[2], B. Ward[2], J. Mclean[2], B McGuinness [3]
[1] Select Solutions, Melbourne, [2] Freyssinet Australia, [3] SRG.

SUMMARY: Half-cell testing is a central part of the site of testing which is commonly used to gauge the risk of corrosion of steel in concrete. The mostly non-destructive, quick and portable nature of this test method make it ideal for use onsite and contributes significantly to its popularity. Another test methodology which is becoming more commonly used during the installation of impressed current cathodic protection systems (ICCP) is the current injection test. Whilst this test method is not currently standardised, it has become a useful way to ascertain if an individual anode is short circuited to steel reinforcement as well as the degree of anode encapsulation. Anecdotally it has been noted that, when considered globally, the amount of current injected into a concrete structure scales inversely to the half-cell test results. This paper exams the half cell and current injection results from 120 concrete beams within a single wharf structure in southern Victoria subject to impressed current cathodic protection using discrete anodes. These results are then compared against the current required to protect the concrete beams in line with the criteria defined in AS 2832.5-2008 Cathodic protection of metals Part 5 Steel in concrete structures Section 2.2.

Keywords: Impressed Current Cathodic Protection, Condition assessment, Half cell

1. INTRODUCTION

Steel reinforced concrete is one of, if not the most commonly used building material for civil infrastructure. Due to its prolific use, its degradation represents a significant cost to asset owners the world over. This cost is controlled via the use of regular inspections, monitoring, maintenance and rehabilitation.

A central tool used as part of the regular monitoring regime is half-cell testing. The theoretical basis of half-cell testing is the mixed potential theory of corrosion, which states that the corrosion current flowing through steel reinforcement is linked to the potential of the steel measured relative to a known standard [1]. Whilst there are other methods which may be used to estimate the corrosion current within a steel element [2], few are as simple or as practical. Given that access to concrete elements of interest is commonly restricted or limited, test regimes which require complicated apparatus or require long term analysis are not always applicable, making half-cell testing a preferred option.

When conducting half-cell surveys, to ensure accurate readings and interpretation it is important to account for the following:

- The standard potential of the particular species of reference electrode used. When measuring the potential of steel in a site environment it is essential that the difference between the reference species used and that of the standard being compared to is accounted for in order to ensure that the results are interpreted correctly [3].
- Local conditions of the concrete. Half-cell testing relies on the concrete as an electrolyte. The resistivity of concrete varies significantly with the local conditions which will alter the recorded readings [3]. It is common practice to saturate the surface of the concrete using potable water prior to taking the reading. Despite this the test is still susceptible to changes in the moisture content of the concrete which can change dependant on the exposure

of the zone in question (splash zone, tidal zone, submerged zone and atmospheric zone) and changes in the relative humidity of the atmosphere.

- The location of the reference electrode relative to the steel. Whilst the resistivity of concrete varies from structure to structure and within any given structure with the prevailing conditions, it is generally quite high (resistivity can range from 10^4 Ωm to 10^9 Ωm [4]). This means that relatively small increase in the distance between the steel and the external reference can correspond to significant increases in the resistance, which in turn effects the recorded results. Additionally areas where the steel is dense may result in the average reading for all bars being recorded, not the reading for a particular piece of steel.
- Continuity of the steel and integrity of the steel connection. If the steel is not continuous, or the resistance of the connection to the reinforcement is not low, the results captured will not be representative of the potential of the steel reinforcement.
- The exposure zone of the concrete can significantly alter its potential. Concrete which is partially or completely submerged can become saturated. As a result the potential of the steel reinforcement in these areas can be negatively polarised.
- Measurements should be conducted across a regularly spaced grid. This enables areas where the potential changes to be easily identified and increases the repeatability of the test.

Noting the limitations discussed above, it is commonly accepted in the industry that a mixed potential relative to Silver-silver chloride (Ag-AgCl) more negative than -200mV is indicative of corrosion [2].

One of the commonly used methods for arresting corrosion of the streel reinforcement in concrete is (ICCP). Of principal importance to the operation of an ICCP system is the correct installation of the anodes. In order to be effective, the anodes need to be electrically isolated from the steel that is to be protected, fully encapsulated within the repair mortar and ideally, apply a uniform current density. It is possible to test for these variables using the current injection test. This test involves connecting the reinforcement and anode together via a power source of known voltage and recording the resultant current injected in a specified time frame. This is then compared to the surrounding anodes to ensure that each anode is operating in a similar manner, and to identify any which appear to be incorrectly installed. This test is sensitive to many of the same factors as the half-cell test which include:

- The time at which the recording is noted. Once connected to the battery, the steel will begin to polarise, reducing the current drawn. As a result in order to generate comparable results the recording must be taken at the same time after the initial connection.
- The resistivity of the concrete [5]. The current in question will flow through the concrete electrolyte. Given that the voltage is constant, changes in the resistance will affect the recorded current. As the resistivity of the concrete changes with the age of the concrete, the exposure environment and relative humidity, so too will the injected current.
- The resistance between the steel and the anode will vary as a function of the distance between them, with the magnitude of the variation dependent on the resistivity of the concrete. As the resistance varies, so too will the magnitude of the current injected.
- Corrosion activity of the steel. Steel which is actively corroding will draw more current than steel which is completely or partly passive.

The purpose of this paper is to compare the results for these two test methodologies across a structure with the current required to meet the Australian Standard Protection Criteria[6].

2. METHODOLOGY

The testing described below was undertaken as part of the quality control of the installation of an ICCP system on a wharf structure located in southern Victoria. The wharf structure consisted of 135 concrete beams, with each beam being further divided into four zones based on an exposure classification as shown below in Figure 2. Zone 1 is situated within the tidal zone, zone 2 is within the splash/tidal zone, zone 3 is within the atmospheric zone, and zone 4 is within the atmospheric/splash zone. The anode layout, consisting of anode number, size and position, is identical between beams 14 to 61. For the remainder of the paper this will be known as the 14-61 cohort. A different anode layout to that of the 14-61 cohort exists between anodes 64 to 135 however within beams 64 to 135 the anode layout is identical. For the remainder of the paper this will be known as the 64-135 cohort. The remaining 15 beams are not considered within this study. The decision to exclude these beams was taken in order to ensure that the sample size of a given anode layout was large enough to ensure that the local variations are accounted for. Testing was carried out concordantly with construction activities between June 2015 and December 2016 after which the ICCP system was energised.

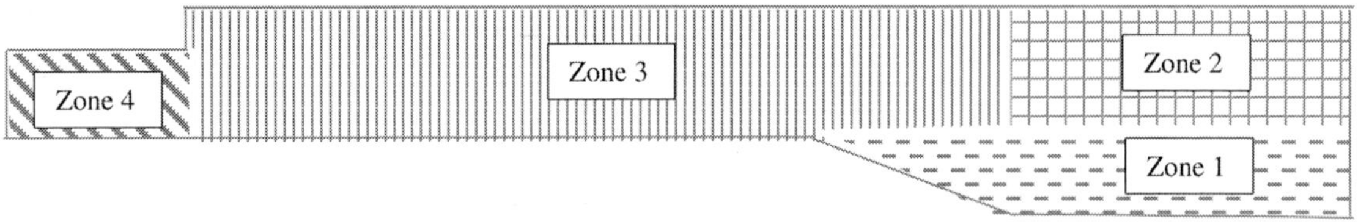

Figure 1. Schematic showing a typical beam and zone layout.

2.1 Half Cell Testing

2.1 Half Cell Testing

Half-cell testing was conducted across each of the four zones within the concrete beams using a Ag-AgCl reference

1. Test grid was marked on beam at 500mm centres. Test grid locations beam to beam were maintained as uniform as possible.
2. A spray bottle was used to saturate the concrete at each test location locally prior to testing.
3. Negative connection was made to the steel reinforcement either via a welded connection or a low resistance clip.
4. Readings were taken at each location using a portable Ag-AgCl reference electrode and a high impedance multimeter.

Please note that in some instances testing was only conducted along one of the beam faces.

Testing was conducted across a period of months during a range of tidal and weather conditions.

2.2 Current Injection Test

After each discrete anode was installed it was left for a period of no less than 48 hours to allow the encasement mortar to set and the moisture level to stabilise before a current injection test was performed. The test apparatus consisted of a sealed six volt battery which was connected to the reinforcement and anode. A one ohm resistor was installed within this circuit to allow the potential difference either side of the resistor to be recorded using a high impedance multimeter. As the resistance was known to be one ohm the noted potential difference can be considered to be equivalent to the current. A schematic of this is shown in the Figure 2 below.

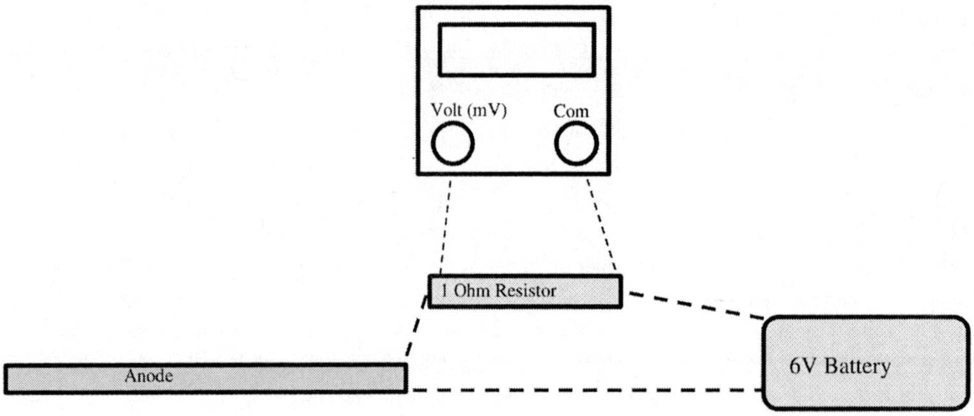

Figure 2. Schematic of Current Injection Test Apparatus

In addition to the resistor a 500mΩ fuse was placed in the circuit to protect the battery in the event of a direct short circuit between the steel and anode. For each anode the current injection test was conducted in the following manner:

1. A cable was connected to the steel reinforcement either via a welded connection or a low resistance clip.
2. A cable was connected to the anode tail via an alligator clip completing the circuit and injecting current into the anode.
3. The technician conducting the test allowed the current to run for five seconds before taking a reading.
4. Where individual anode readings were fell outside of the range allowed by the Technical Specification (plus/minus 50% of the beams zonal mean) they were replaced and the test repeated on the newly installed anode.

Current injection testing was conducted between two days and several weeks after the installation of the anodes across a range of tidal and weather conditions, however all anodes in a specific zone were tested at the same time.

2.3 Current Required For Protection

The final variable examined in this paper is the amount of current required to achieve protection as Australian Standard 2832.5. It is assumed that achieving protection criteria represents that corrosion has been arrested within the beam. It should be noted that since the system was commissioned the current outputs of a number of elements have been reduced whilst maintaining the protection criteria. It is anticipated that overtime it will be possible to continue to refine the current whilst continuing to meet the protection criteria.

3. RESULTS AND DISCUSSION

3.1 Half Cell Testing

The results of the half-cell tests within each of the 500mm by 500mm test grids has been averaged and the averaged data per zone is presented in Figure 3 and Figure 4 as well as the tabulated zonal average presented in Table 1 and Table 2 below.

North Face vs $Cu/CuSO_4$

	Z	Y	X	W	V	U	T	S	R	Q	P	O	N	M	L	K	J	I	H	G	F	E	D	C	B	A	
1	-257	-228	-208	-203	-191	-189	-189	-198	-203	-209	-215	-208	-217	-233	-246	-247	-237	-262	-271	-277	-286						1
2	-330	-297	-260	-212	-207	-204	-224	-272	-300	-288	-271	-249	-253	-274	-324	-350	-356	-372	-363	-354	-383	-400	-410	-416	-412	-451	2
3																			-479	-492	-491	-507	-499	-493	-506	-542	3
	Z	Y	X	W	V	U	T	S	R	Q	P	O	N	M	L	K	J	I	H	G	F	E	D	C	B	A	

South Face vs $Cu/CuSO_4$

	A	B	C	D	E	F	G	H	I	J	K	L	M	N	O	P	Q	R	S	T	U	V	W	X	Y	Z	
1						-302	-289	-277	-266	-255	-250	-242	-230	-222	-218	-211	-213	-203	-200	-185	-188	-178	-174	-215	-240	-238	1
2	-460	-423	-395	-382	-382	-394	-390	-390	-397	-346	-320	-307	-280	-276	-274	-256	-295	-270	-251	-232	-212	-204	-202	-272	-294	-301	2
3	-542	-513	-500	-506	-519	-511	-530	-481																			3
	A	B	C	D	E	F	G	H	I	J	K	L	M	N	O	P	Q	R	S	T	U	V	W	X	Y	Z	

Figure 3. Averaged half-cell results beams 14 to 61. All results are presented as the potential difference (mV) with reference to a Copper - copper sulphate (Cu/CuSO4) reference electrode.

North Face vs $Cu/CuSO_4$

	Z	Y	X	W	V	U	T	S	R	Q	P	O	N	M	L	K	J	I	H	G	F	E	D	C	B	A	
1	-178	-160	-148	-165	-140	-124	-121	-127	-123	-128	-121	-101	-96	-117	-128	-136	-166	-182	-200	-220	-225						1
2	-237	-234	-233	-210	-184	-181	-187	-199	-220	-219	-211	-193	-200	-202	-222	-242	-257	-257	-280	-291	-282	-279	-286	-290	-293	-314	2
3																			-338	-361	-375	-370	-372	-400	-376	-390	3
	Z	Y	X	W	V	U	T	S	R	Q	P	O	N	M	L	K	J	I	H	G	F	E	D	C	B	A	

South Face vs $Cu/CuSO_4$

	A	B	C	D	E	F	G	H	I	J	K	L	M	N	O	P	Q	R	S	T	U	V	W	X	Y	Z	
1						-244	-238	-219	-203	-190	-161	-141	-130	-115	-119	-131	-130	-129	-118	-115	-118	-128	-135	-156	-164	-174	1
2	-345	-334	-317	-305	-295	-291	-289	-285	-279	-251	-241	-227	-217	-197	-205	-208	-222	-223	-195	-176	-179	-189	-196	-222	-232	-252	2
3	-409	-400	-400	-389	-370	-364	-360	-347																			3
	A	B	C	D	E	F	G	H	I	J	K	L	M	N	O	P	Q	R	S	T	U	V	W	X	Y	Z	

Figure 4. Averaged half cell results beams 64 to 135. All results are presented as the potential difference (mV) with reference to a Copper - copper sulphate (Cu/CuSO4) reference electrode.

Table 1. Average half cell half cell and standard deviation data for beams 14 to 61

	Average Potential Relative to Cu/CuSO4 (mV)	Standard Deviation (mV)
Zone 1	-507	88
Zone 2	-398	79
Zone 3	-245	79
Zone 4	-262	81

Table 2. Average half cell half cell and standard deviation data for beams 64 to 135

	Average Potential Relative to Cu/CuSO4 (mV)	Standard Deviation (mV)
Zone 1	-377	93
Zone 2	-295	72
Zone 3	-176	72
Zone 4	-199	72

For both the 14 to 61 cohort and the 64 to 135 cohort the order of zones from most to least negative was zone 1, zone 2, zone 4 and zone 3. This order mirrors the degree of exposure with zone 1 being tidally affected twice daily, zone 2 being semi-submerged and exposed to splash, zone 4 being exposed to splash generated by waves crashing onto the sheet pile wall at the rear of the wharf and zone 3 being the least exposed to tidal or splash condition's.

It is interesting to note that in general the potentials found within beams 14 to 61 were more negative than those recorded in beams 64 to 135. This is consistent with observations from site which found that beams 14 to 61 were in a worse condition than beams 61 to 135. This may be due to this part of the structure being older as well as different construction techniques.

The standard deviation data was fairly consistent across the zones and between the 14 to 61 cohort and the 64 to 135 cohort.

3.2 Current Injection Test

The results of the current injection testing has been averaged and is represented in Figure 5, Figure 6, Figure 7 and Figure 8 and the averaged data per zone is presented in Table 3 and Table 4 below.

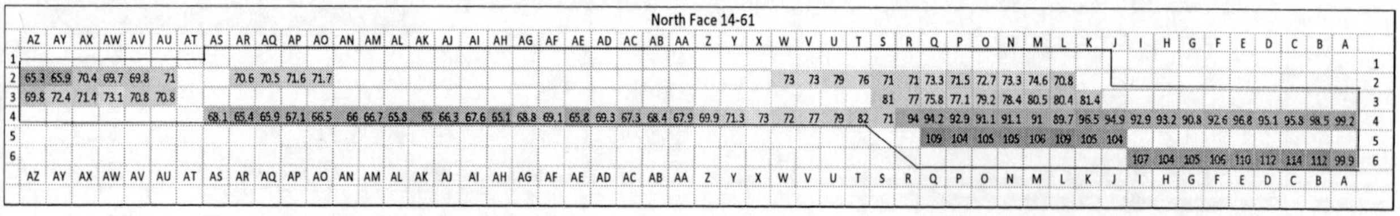

Figure 5. Current Injection Data (mA) for the North Face of Beams 14 to 61

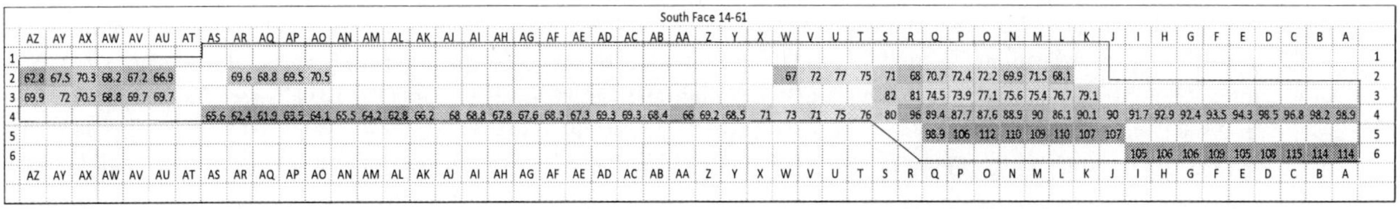

Figure 6. Current Injection Data (mA) for the South Face of Beams 14 to 61

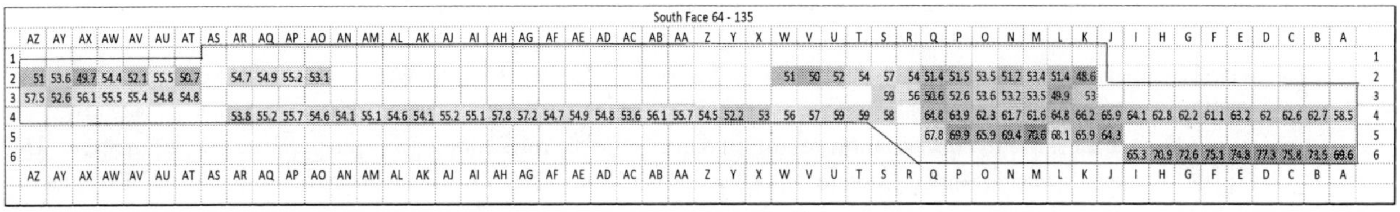

Figure 7. Current Injection Data (mA) for the North Face of Beams 64 to 135

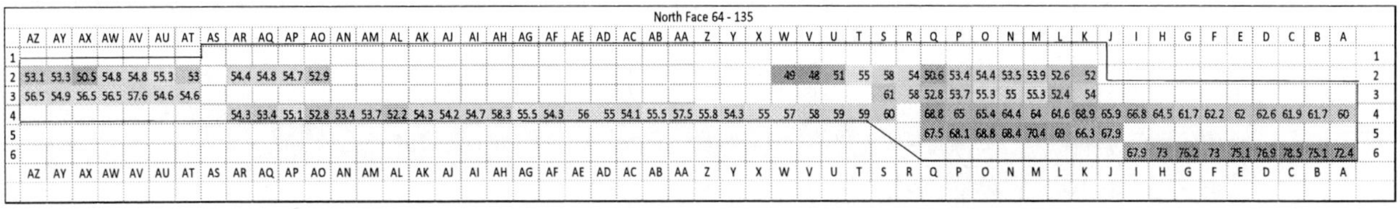

Figure 8. Current Injection Data (mA) for the South Face of Beams 64 to 135

Table 3. Average Current Injection Results for Beams 14 to 61

	Average Current Injected (mA)	Standard Deviation (mA)
Zone 1	108	20
Zone 2	93	16
Zone 3	71	13
Zone 4	69	14

Table 4. Average Current Injection Results for Beams 64 to 135

	Average Current Injected (mA)	Standard Deviation (mA)
Zone 1	71	26
Zone 2	64	22
Zone 3	55	18
Zone 4	55	19

The current injection test results were found to vary in a similar manner to the half-cell tests. This similarity is seen in both the zone averages presented in the tables and also in the heat maps presented in the images above. Areas which were found to have negative potentials in the half cell testing were shown to draw a higher current from the six volt battery. Whilst it is known that the amount of the current drawn will vary with both the distance between the steel and the anode as well as the degree of grout encapsulation, the similarity of the current injection and half-cell results indicates that this is not the only factor driving the amount of current which is drawn. Additionally, it is considered that the large number of test locations (over 23,000 current injection data points collected) will mean that the influence of these two factors will be averaged out. Given that all elements of the current injection test regime were kept the same, the driving force behind the amount of current drawn must be intrinsic to the test sites themselves. This may be due to the difference in resistivity of the concrete itself. Where zones are more exposed to sea water the resistivity is expected to be lower, meaning that for a given driving voltage, a higher current will flow.

There is a strong similarity between the results of all of the tests conducted in sections 3.1 and 3.2. This underlines the interconnection between the factors which drive and affect corrosion within the marine environment. The pattern of negative potentials within the half-cell results (zone 1, 2 ,4 and 3) is driven by a combination of the lower resistivity in the more exposed areas and the higher chloride concentration due to the resultant sea water ingress, and finally a negative polarisation of the steel due to concentration polarisation driven by the restriction of oxygen to the surface of the steel. This same pattern is seen in the current injection results however in this instance it is likely due solely to the lowered resistance caused by the (or saturation by) presence of the seawater.

3.3 Current Required to Achieve Protection

The amount of current required to achieve protection within each of the zones, normalised to current per square meter of steel surface area, is presented in Figure 9, Figure 10, while the passing criteria achieved presented by zone is presented in Table 5.

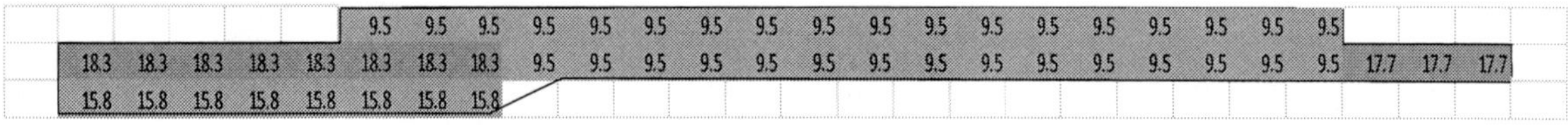

Figure 9. Beams 14-61 Current (mA) per m^2 of steel reinforcement

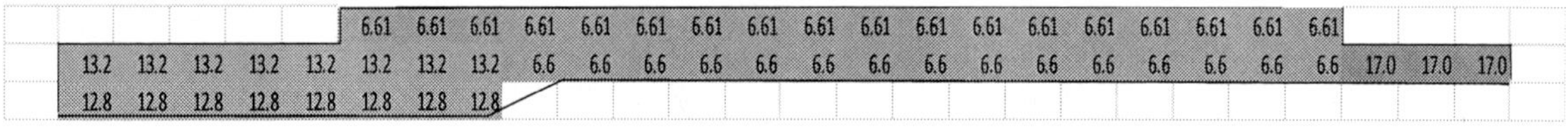

Figure 10. Beams 64-135 Current (mA) per m^2 of steel reinforcement

Table 5. Criteria Achieved by Zone

	100 mV Depolarisation	Absolute Passive	Absolute Potential	Absolute Passive & Absolute Potential	100 mV Depolarisation & Absolute Passive	100 mV Depolarisation & Absolute Potential	All Criteria	Failed
Zone 1	69%	13%	17%	0%	10%	0%	0%	2%
Zone 2	83%	31%	1%	0%	28%	0%	0%	3%
Zone 3	97%	88%	0%	0%	87%	0%	0%	2%
Zone 4	100%	93%	0%	0%	93%	0%	0%	0%

Between the two cohorts a lower amount of current per square metre of steel was required within the 64-135 cohort to achieve protection compared to the 14-61 cohort. This is in agreement with the general observation that the beams in the 14-61 cohort were in worse condition as well as the pattern seen in sections 3.1 and 3.2 which also show the 14-61 cohort to be more corrosive.

The results presented above show that according to the current per meter squared of steel surface area zone 3 required the least amount of current to achieve protection as per the Australian Standard. This relationship was seen in both the 14-61 and 64-135 cohorts and conforms to the results of sections 3.1 and 3.2 which show that zone 3 is generally the most passive of the zones.

The other zones however do not adhere to the zone order which appeared to be formed from sections 3.1 and 3.2. As previously stated sections 3.1 and 3.2 would indicate that zone 1 may have been the most corrosive (and therefore require the highest level of current to achieve protection), followed by zone 2 and 4 in that order. This is not what is seen. Figure 9 shows that for cohort 14-61 zone 2 required the most current per square meter of steel, followed by zone 4 and 2 respectively. Within Figure 10, which uses the same metric to compare the 64-135 cohort, zone 4 required the highest level of current to achieve protection followed by zones 2 and 1. This implies that zones 2 and 4, which are exposed to sea water splash but not immersion, are more corrosive than zone 1, which is periodically immersed. Should this be the case this is likely to be due to a restriction in the amount of oxygen available at the steel surface in zone 1 as oxygen transport through the aqueous phase is less than that through the gas phase. This is consistent with the results of sections 3.1 and 3.2 in that oxygen restriction will cause concentration polarisation, decreasing the potential of the steel in zone 1 relative to the other zones (section 3.1) whilst the higher water content will lower the resistivity of the concrete compared to the other zones (sections 3.2). The results also indicate that the relative corrosivity of zones 2 and 4 changes between the cohorts, this is likely due to a change in the geometry of the beams and environmental exposure of the elements between the cohorts.

It is important to note at this time that the current required to achieve protection as per the Australian Standard and the corrosivity of a zone is not necessarily equivalent. It is assumed that achieving protection as per the Australian Standard is indicative that corrosion has been arrested within the element, and that the current required to achieve this is the minimal amount possible. This will not always because the case as firstly, whilst the Australian Standard is conservative, the only way to be sure that corrosion is arrested is to conduct direct measurements (for example via LPR) and secondly given the infancy of the ICCP system it is likely that the applied current will continue to be refined during the next several years.

Table 5 shows the breakdown of which criteria was achieved by zone. For simplicity during this study within the 100mV decay criteria there was no discrimination between a 24 and 72 hour period. As would be expected no individual element achieved all criteria. Additionally no individual element achieved the absolute passive and absolute potential criteria. Zone 3, demonstrated to be quite passive in the above tests, showed a high portion of zones passing the 100mV and absolute passive criteria and no beams passing the absolute potential criteria. A similar relationship was seen in zone 4 where a significant number of beams were found to be passing the absolute passive criteria, which may be indicative that in general the relatively high current supplied to this zone may be decreased. The number of beams where zone 2 passed the absolute passive criteria was lower than that of zone 4, which would indicate that the conditions in this zone are less passive than those of zone 4.

Within zone 1, 13% or 15 beams tested achieved the absolute passive criteria. Upon investigation 12 of these 15 were located in the 64-135 cohort which has previously been seen to be less corrosive, with the remaining 3 located in the 14-61 cohort. This number is higher than might be expected given that this zone was demonstrated to be more negative than other zones.

Finally the only zones to achieve the absolute potential criteria were zones 1 and 2, where 17% and 1% respectively of elements achieved this value. This due to the regular immersion of zone 1. It is likely that the reference in zone 2 which achieved this criteria was located at a lower extremity of the zone.

Please note that a total of 8 elements failed to pass the criteria during the study. These are believed to be specific and isolated incidents by the authors and not indicative of a pattern.

4. CONCLUSIONS

In general, all tests showed that the 14-61 cohort was in worst condition than the 64-135 cohort. Once the installation specific variation is averaged out of the current injection test, the results were found to vary in a similar manner to the half-cell tests with areas shown to have negative potentials drawing a higher current. This resulted in the zones ordered from most to least negative aligning with the zone order from most to least current drawn, zones 3, 4, 2, 1. It is believed that this pattern is due to the exposure of the zones, with the lower higher content of zone 1 driving a lower resistivity and potential, which decreases as the zones are less exposed to the splash from the sea. When the amount of current required to achieve protection is included this pattern breaks down, with zone 1 notably requiring less current to achieve sustained protection than zone 2 and in one instance zone 2 and 4. This may be due to the restriction of oxygen to the steel surface. Surprisingly zone 4 appeared to require a higher amount of current to achieve a protection criteria than would be expected from the half cell and current injection testing, however given the high number of beams where zone 4 passed the absolute passive criteria it may be that the current to this zone is able to be decreased.

5. ACKNOWLEDGMENTS

Thanks to Jeff Pritchard, Nick Critchley and Tom Wenzel of Freyssinet Australia for their assistance and guidance in putting this study together. Thank you to Alexander Certoma, Omar Dennaoui, Ahmed Dahash, Mustafa Altaweel, Essam Al-Jebory, David Lim and Issuru Perera for all of their help undertaking the numerous test measurements. A giant thank you to Eugène Moucadeau and Kate Humphreys whose dedication made certain that we were able to capture the data.

6. REFERENCES

1. G.K Glass, A.M Hassanein, N.R Buenfeld, "Monitoring the passivation of steel in concrete induced by cathodic protection", Corrosion Science, Volume 39, Number 8, pp 1548, 1997
2. Australian Standard Handbook HB 84 "Guide to concrete repair and protection", 2006.
3. ASTM C-876-87(91) *Standard Test Method for Half Cell Potentials of Uncoated Reinforcing Steel in Concrete*, American Society for Testing and Materials, 1999.
4. Neville, Adam M. Properties of concrete / A.M. Neville. -- 5th ed. p. cm. ISBN 978-0-273-75580-7 (pbk.) 1. Concrete. I. Title. TA439.N48 2011. 620.1'36--dc23.
5. Morris, W., E. Moreno, and A. Sagüés, *Practical evaluation of resistivity of concrete in test cylinders using a Wenner array probe*. Cement and Concrete Research, 1996. **26**(12): p. 1779-1787.
6. AS 2832.5-2008 Cathodic protection of metals Part 5 Steel in concrete structures, Australian Standards, 2005.

7. AUTHOR DETAILS

Peter is a project engineer with Select Solutions. He has several years' experience working as a project engineer primarily focused on the assessment, maintenance and remediation of civil infrastructure. Peter has experience in the design review, and assessment of both steel and concrete and steel cathodic protection systems.

No Photo Available

Joe is a project engineer with Freyssinet International Major Projects. He has significant years' experience working as a project engineer, primarily working on the delivery of significant, steel, concrete and timber infrastructure. Joe has experience in the design review, and assessment of both steel and concrete and steel cathodic protection systems.

Ben is a project engineer with Freyssinet Australia. He has a wealth of hands-on experience in relation to corrosion and Cathodic Protection systems, both as a technician and in design and commissioning. He has also been responsible for steel asset monitoring and condition assessments.

Jack McLean works as a Project Engineer with New South Wales' growing remedial division. Bringing a number of years of project management and technical expertise in both a contracting and consultancy role, Jack is responsible for the successful delivery and management of remedial and cathodic protection projects across Victoria and New South Wales. As part of this role, Jacks responsibilities include procurement, labour supervision, contracts administration, financial management, subcontract management, programming, quality assurance, HSE management as well as technical supervision in design.

With his background in corrosion control and cathodic protection. Jack can provide the "glue" in the project team between technical expertise and project management.

Brendan is a project engineer with SRG. He has a wealth of Cathodic Protection experience through investigation, design and installation, and expansive knowledge of cathodic protection systems such as ICCP, galvanic and novel. Over the past 5 years Brendan has gained experience in many high risk work environment supervising and advising during Cathodic Protection installations.

EFFECTS OF Ti AND Al ON CORROSION OF Ni-20Cr AND Ni-30Cr ALLOYS IN DRY CO_2 GAS

T. D. Nguyen, J. Zhang, D. J. Young
School of Materials Science and Engineering, University of New South Wales, Sydney, NSW 2052, Australia

SUMMARY: Model alloys Ni-20Cr and Ni-30Cr with and without 2% Ti and 1.5% Al (wt. %) were exposed to Ar-20CO_2 gas at 700 °C. Samples were ground and then electropolished before reaction to avoid cold working effects on alloy diffusion. The undoped Ni-20Cr alloy formed a thick NiO scale and a deep internal oxidation zone containing Cr-rich oxide precipitates. The additions of Ti and Al to Ni-(20, 30)Cr dramatically improved corrosion resistance by promoting chromia formation. All alloys were carburised beneath their scales, forming intergranular Cr-rich carbide precipitates.

Keywords: Alloy, Titanium, Aluminium, High temperature corrosion.

1. INTRODUCTION

Oxy-fuel combustion is a feasible solution being developed and tested to reduce and capture CO_2 emission from coal fired power plants. In this process, coal is burned in a mixture of oxygen and recirculated flue gas instead of air. The resulting flue gas is rich in CO_2 and H_2O. Corrosion resistance of stainless steels used in heat exchangers is degraded rapidly in this highly corrosive gas.

Nickel-base alloys are usually used in many applications requiring high mechanical strength and corrosion resistance. It has been reported [1] that electropolished Ni-20Cr and Ni-30Cr failed to form protective Cr_2O_3 scales in dry CO_2 at 700 °C. The additions of Ti [2-3] and Al [3-5] to Ni-base alloys have been shown to improve oxidation resistance significantly. The aim of this paper was to study the effects of Ti and Al on corrosion of Ni-(20, 30)Cr at 700 °C in Ar-20CO_2.

2. MATERIALS AND EXPERIMENTS

Six model alloys, Ni-20Cr-(2Ti, 1.5Al) and Ni-30Cr-(2Ti, 1.5Al) (weight %), were prepared by arc melting pure metals Ni (99.95%), Cr (99.995%), Ti (99.5 %) and Al (99.95 %) in a flowing gas mixture of Ar-5%H_2. The arc-melted buttons were annealed at 1000 °C in a flowing Ar-5%H_2 gas for homogenization. Rectangular alloy samples were surface ground to a 1200-grit finish, electropolished to remove deformed surface material, and ultrasonically cleaned in alcohol prior to reaction. Etchants used to reveal carbides in reacted samples were HNO_3 : HCl = 2 ml : 10 ml solution for Ni-20Cr-(Si, Al, Mn) alloys and Murakami's solution (2 g $K_3Fe(CN)_6$ + 2 g KOH + 20 ml H_2O) for Ni-30Cr-(Si, Al, Mn) alloys.

All samples were reacted for up to 500 h at 700 °C in a flowing Ar-20%CO_2 mixture with a linear flow rate of 2 cm/s at a total pressure of 1 atm. Weight changes of reacted samples were measured using an analytical balance with an accuracy of 0.01 mg. In weight change curves, each point corresponds to a separate sample. Reacted samples were characterized by X-ray diffractometry (XRD), scanning electron microscopy (SEM), transmission electron microscopy (TEM), energy dispersive analysis (EDX) and selected area electron diffraction (SAED).

3. RESULTS

3.1 Corrosion of Ni-20Cr-(Ti, Al) alloys

Weight gain kinetics of Ni-20Cr-(2Ti, 1.5Al) alloys are shown in Figure 1. The weight uptakes were small and increased with time, but those of Ni-20Cr alloys containing Ti or Al were reduced dramatically. No scale spallation was observed after cooling.

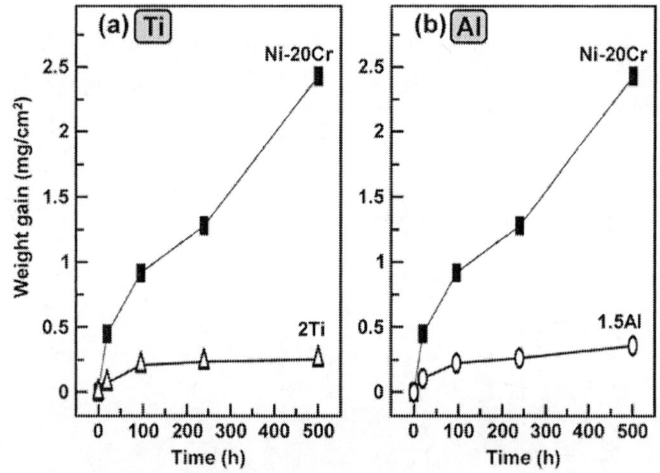

Figure 1 Weight gain kinetics of Ni-20Cr alloys containing (a) Ti and (b) Al

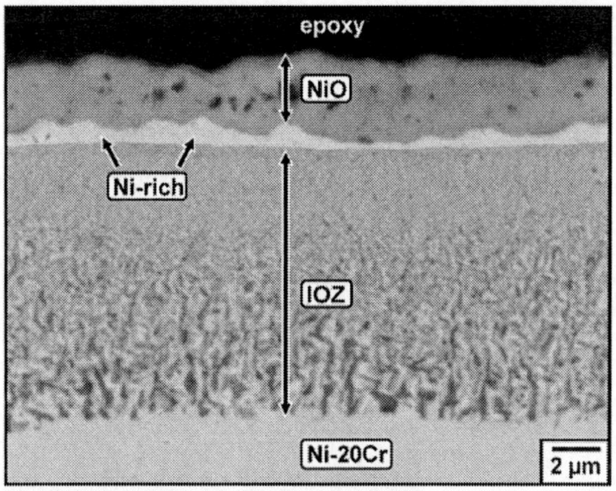

Figure 2 BSE-SEM cross-section of Ni-20Cr reacted for 240 h

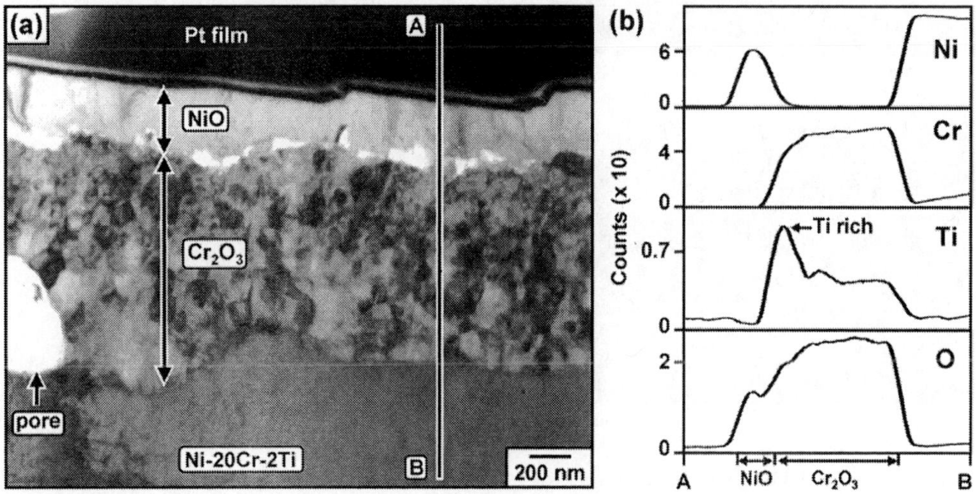

Figure 3 Ni-20Cr-2Ti reacted for 500 h: (a) BF-TEM image and (b) EDX line profiles along A-B in (a)

A cross-section of the undoped Ni-20Cr alloy after reaction for 240 h is shown in Figure 2. Analysis by XRD and EDX showed that the alloy formed a NiO scale (about 2.6 µm thick), an intermediate Ni-rich layer and an underlying internal oxidation zone (IOZ) containing small Cr-rich oxide precipitates. The NiO scale and IOZ were still observed after 500 h reaction but not the Ni-rich layer at the scale-IOZ interface.

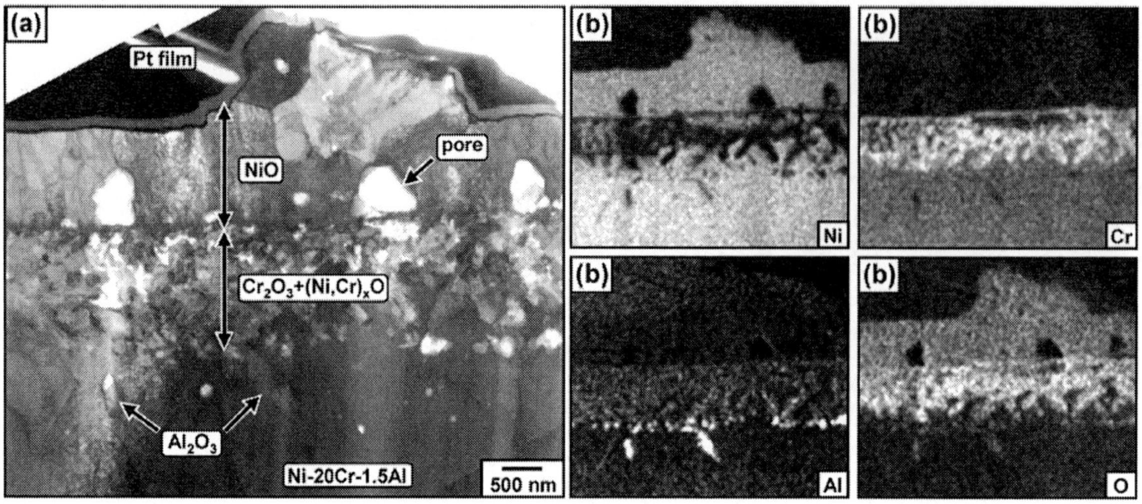

Figure 4 Ni-20Cr-1.5Al reacted for 500 h: (a) BF-TEM image and (b) EDX mapping of (a)

A bright-field (BF) TEM image of a scale on Ni-20Cr-2Ti reacted for 500 h is shown in Figure 3. The alloy developed a thin duplex scale of an outer NiO layer (about 0.2 μm thick) and an inner Cr_2O_3 layer containing a small amount of Ti (about 0.9 μm thick) (Figure 3). Figure 3b also shows an enrichment of Ti at the interface between NiO and Cr_2O_3. The Ni-20Cr-1.5Al alloy during reaction for 500 h developed a thick duplex scale of an outer NiO layer (about 1.2 μm thick) and an inner layer containing mixtures of Cr_2O_3 and (Ni, Cr) oxides (Figure 4). An IOZ containing Al_2O_3 particles was observed beneath the scale.

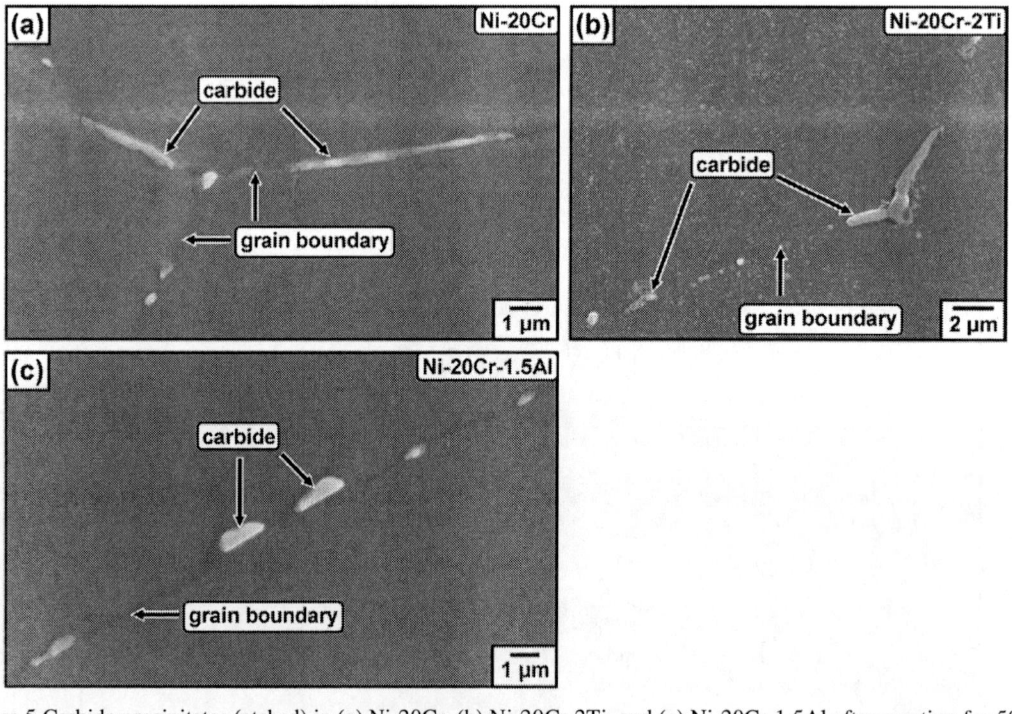

Figure 5 Carbide precipitates (etched) in (a) Ni-20Cr, (b) Ni-20Cr-2Ti, and (c) Ni-20Cr-1.5Al after reaction for 500 h

All alloys undertook carburisation beneath their scales, resulting in small intergranular precipitates (Figure 5). Analysis by EDX (not shown) confirmed the precipitates to be Cr-rich carbides. The carbides in Ni-20Cr were identified by selected area diffraction (SAD)/TEM (not shown) to be Cr_7C_3.

3.2 Corrosion of Ni-30Cr-(Ti, Al) alloys

Weight gain kinetics of Ni-30Cr-(2Ti, 1.5Al) alloys are shown in Figure 6. Weight gains were very small and increased with reaction time. The Ni-30Cr-1.5Al alloy had very small weight uptakes in comparison with Ni-30Cr and Ni-30Cr-2Ti. No scale spallation was observed after cooling.

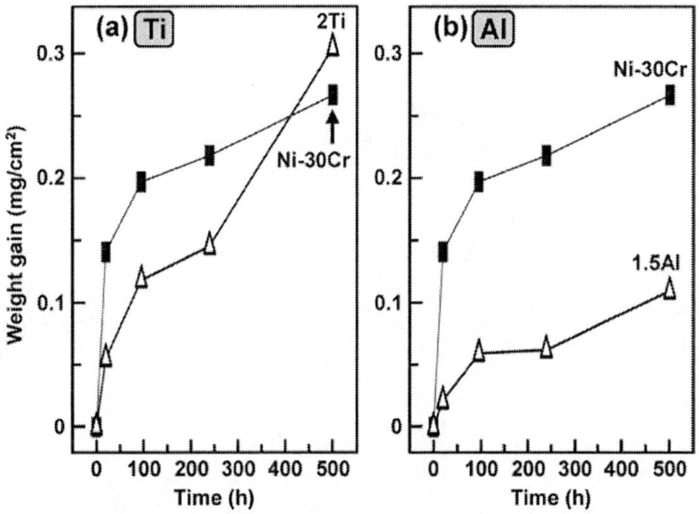

Figure 6 Weight gain kinetics of Ni-30Cr alloys containing (a) Ti and (b) Al

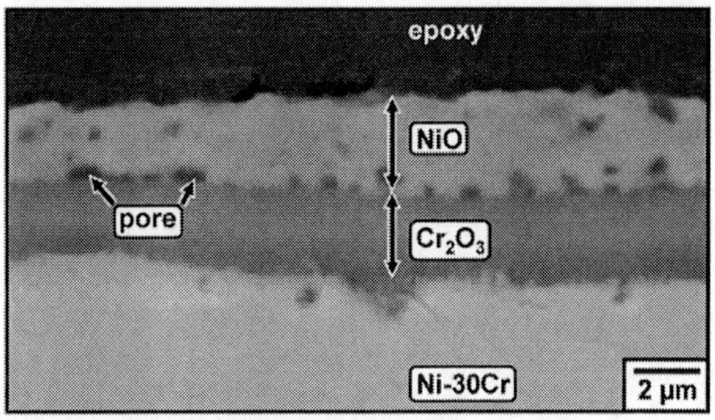

Figure 7 BSE-SEM cross-section of Ni-30Cr reacted for 500 h

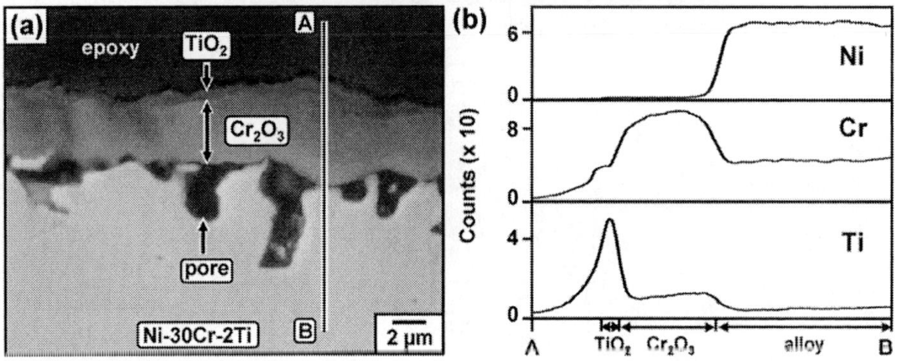

Figure 8 Ni-30Cr-2Ti reacted for 500 h: (a) BF-SEM image and (b) EDX line profiles along A-B in (a)

The pure Ni-30Cr alloy after 500 h formed a scale consisting of an outer NiO layer (about 2.4 μm thick) and an inner Cr_2O_3 layer (about 2.3 μm thick) (Figure 7), and only a Cr_2O_3 scale on other areas, with a fraction of less than 1% of surface area (not shown). The additions of Ti and Al suppressed growth of the NiO scale. A cross-section of Ni-30Cr-2Ti reacted for 500 h is shown in Figure 8. The scale was analysed by XRD (not shown) and EDX (Figure 8b) and shown to consist of an outer TiO_2 layer (about 0.6 μm thick) and an inner Cr_2O_3 layer (about 3.3 μm thick) with a small amount of Ti dissolved (2.7 at%). The Ni-30Cr-1.5Al alloy formed a very thin scale, as shown in Figure 9a. The corresponding EDX (Figure 9b) and XRD (not shown) indicated the scale was made up of an outer, thin NiO layer and an inner Cr_2O_3 layer. Internal Al_2O_3 precipitates were formed beneath the scale (Figure 9b).

All Ni-30Cr-(Ti, Al) alloys were carburised beneath their scales, forming intergranular Cr-rich carbide precipitates (Figure 10). Analysis by SAD (Selected Area Diffraction) /TEM (not shown) indicated the carbides formed in the undoped Ni-30Cr alloy were $Cr_{23}C_6$.

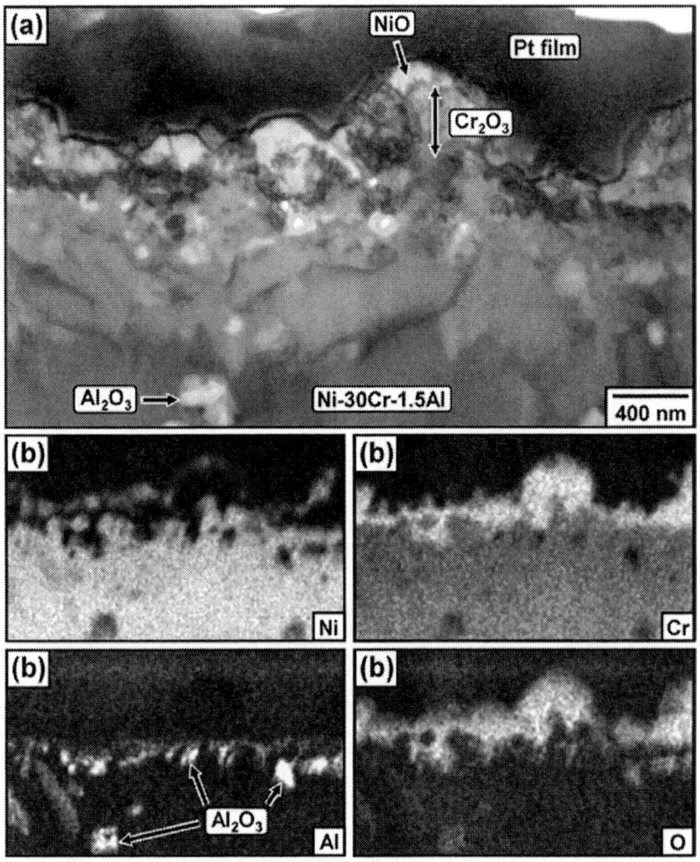

Figure 9 Ni-30Cr-1.5Al reacted for 500 h: (a) BF-TEM image and (b) EDX mapping of (a)

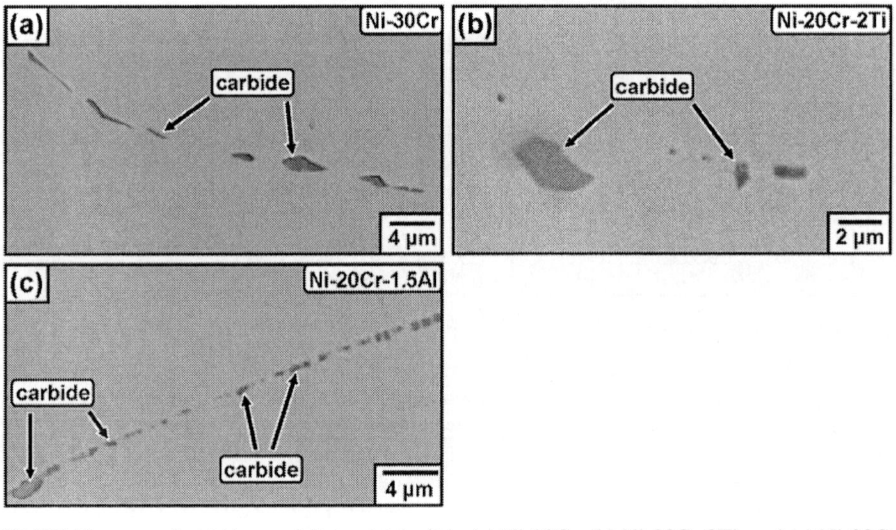

Figure 10 BSE-SEM images of carbide precipitates (etched) in (a) Ni-30Cr, (b) Ni-30Cr-2Ti, and (c) Ni-30Cr-1.5Al after reaction for 500 h

4. DISCUSSION

4.1 Effects of Ti and Al on oxidation

All alloys in this study underwent both oxidation and carburization when exposed to Ar-20CO_2. The oxygen partial pressure and carbon activity of the reaction gas were controlled by the dissociation reactions:

$$CO_2 = CO + (1/2)\ O_2 \tag{1}$$

$$2CO = CO_2 + C \tag{2}$$

The undoped Ni-20Cr alloy formed a NiO scale and an underlying Cr-rich IOZ (Figure 2), while Ni-30Cr formed a duplex scale of NiO and Cr_2O_3 layers (Figure 7) in some surface areas and a single Cr_2O_3 scale on some other regions (not shown). It is clear that Ni-30Cr had higher Cr supply to form the dense Cr_2O_3 scale layer and prevented the formation of an IOZ. This result suggests that 30% Cr is a critical concentration for the conversion of internal oxidation to external oxidation at this temperature.

The additions of Ti and Al to Ni-20Cr significantly reduced weight uptakes (Figure 1) and resulted in different corrosion products. The Ti and Al-bearing alloys formed an outer NiO scale layer and an inner Cr_2O_3-rich layer (Figures 3,4). This observation implies that addition of Ti or Al reduces the critical Cr concentration for the transition from internal to external chromium oxidation, from 30% to 20%. The improved oxidation resistance of Ti and Al containing alloys is thought due to the formation of additional stable Ti and Al-rich oxide particles on the sample surface in the early stage, reducing Ni-rich oxide formation and oxygen inward diffusion. Consequently, internal oxidation of chromium is significantly reduced and a continuous Cr_2O_3-rich layer was formed. However, the Cr concentration in the undoped Ni-20Cr alloy was not sufficient to form an exclusive single Cr_2O_3 scale, and an outer NiO scale layer was still formed on Ni-20Cr-2Ti and Ni-20Cr-1.5Al.

The benefit of Ti and Al additions on promoting formation protective scales was clearly observed for Ni-30Cr with higher Cr alloy concentration. The Ni-30Cr-2Ti alloy formed a duplex scale of TiO_2 and Cr_2O_3 layers (Figure 8). The Ni-30Cr-1.5Al alloy formed a very thin scale of an outer NiO layer and an inner Cr_2O_3 layer, with internal Al_2O_3 particles beneath the scale (Figure 9). Similar Ti and Al effect was also observed in the oxidation of Ni-19Cr-10Co-6W-4Mo-2Al-2.5Ti in steam at 800 °C which resulted in a complex scale of TiO_2/$MnCr_2O_4$/Cr_2O_3 layers and internal oxide particles of Ti and Al [2]. In this study, a small amount of titanium was detected throughout the Cr_2O_3 layers on Ni-(20,30)Cr-2Ti (Figures 3b,8b) and a separate TiO_2 layer over a Cr_2O_3 layer (Figure. 8b). It is concluded that Ti can penetrate the chromia layer.

4.2 Effects of Ti and Al on carburisation

All Ni base alloys in this study formed intergranular Cr-rich carbide precipitates beneath their scales (Figures 5,10). At first sight, carburisation of the alloys by Ar-20CO_2 were impossible because the equilibrium carbon activity of the reaction gas at 700 °C is very small, 6.4×10^{-15} [9]. However, it has been shown that carbon activity set by the scale-alloy interfacial equilibrium was high enough to carburise the underlying alloys [6]. Carbon was shown to penetrate Cr_2O_3 scales along grain boundaries [7]. Therefore, carburisation products were formed in the underlying Fe-base [6,8] and the Ni-base alloys (Figures 5,10) after reaction in CO_2. The addition of Ti and Al did not affect the carburisation too much, indicating capability of carbon penetration through TiO_2 thin layer and internal oxidation zone with Al_2O_3 precipitates.

5. CONCLUSIONS

The undoped Ni-20Cr and Ni-30Cr alloys formed non-protective, Ni-rich oxide scales. The additions of Ti and Al to Ni-(20, 30)Cr significantly improved the oxidation resistance by forming more protective scales. Carbon from the reaction gas penetrated the scales and carburised the underlying alloys, forming Cr-rich carbide precipitates in all alloys.

6. ACKNOWLEDGMENTS

Financial support from the Australian Research Council's Discovery program is gratefully acknowledged.

7. REFERENCES

[1] Nguyen TD, Xie Y, Ding S, Zhang J, Young DJ, Oxidation behaviour of Ni-Cr alloys in CO_2 at 700 °C, Oxid. Met. 87 (2017) 605-616.

[2] Abe F, Araki H, Yoshida H, Okada M, The role of aluminum and titanium on the oxidation process of a nickel-base superalloy in steam at 800 °C, Oxi. Met. 27 (1987) 21-36.

[3] Nagai H, Okabayashi M, Mitani H, The effects of rare earths and reactive elements on the oxidation resistance of Ni-20Cr alloy, Trans. JIM 21 (1980) 341-348.

[4] Giggins CS, Pettit FS, Oxidation of Ni-Cr-AI alloys between 1000 ° and 1200 °C, J. Electrochem. Soc. 118 (1971) 1782-1790.

[5] Stott FH, Wood GC, The mechanism of oxidation of Ni-Cr-Al alloys at 1000°–1200°C, Corros. Sci. 11 (1971) 799-812.

[6] Gheno T, Monceau D, Zhang J, Young DJ, Carburisation of ferritic Fe–Cr alloys by low carbon activity gases, Corros. Sci. 53 (2011) 2767–2777.

[7] Young DJ, Nguyen TD, Felfer P, Zhang J, Cairney JM, Penetration of protective chromia scales by carbon, Scripta Mater. 77 (2014) 29-32.

[8] Nguyen TD, Zhang J, Young DJ, Martensite formation in Fe-9Cr alloys exposed to low carbon activity gas, Scripta Mater. 69 (2013) 9-12.

[9] Xie Y, Zhang J, Young DJ, Temperature effect on oxidation behavior of Ni-Cr Alloys in CO_2 gas atmosphere, J. Electrochem. Soc. 164 (2017) 285-293.

8. AUTHOR DETAILS

Thuan Dinh Nguyen is a postdoctoral fellow at the School of Materials Science and Engineering, UNSW.

Dr Jianqiang Zhang is Associate Professor at the School of Materials Science and Engineering, UNSW. His research is mainly on high temperature corrosion of metals and alloys in oxidation and carburisation gas atmospheres.

David Young is Emeritus Professor at the School of Materials Science and Engineering, UNSW. He has worked in the field of high temperature corrosion for more than 40 years, publishing extensively in the research literature.

ELECTROCHEMICAL AND STATIC HPHT STUDY OF PORPHYRIN DERIVATIVES AS EFFECTIVE CORROSION INHIBITOR FOR N80 STEEL IN SWEET CORROSION ENVIRONMENT

Ambrish Singh[1, 2], Yuanhua Lin[1, 2]
[1] State Key Laboratory of Oil and Gas Reservoir Geology and Exploitation (Southwest Petroleum University), Chengdu, Sichuan 610500, China.
[2] School of Material Science and Engineering (Southwest Petroleum University), Chengdu, 610500, China.

SUMMARY : Five Porphyrins 5,10,15,20-Tetra(4-pyridyl)-21H,23H-porphine (P1), 5,10,15,20-Tetraphenyl-21H,23H-porphine (P2), 5,10,15,20-Tetrakis(4-hydroxyphenyl)-21H,23H-porphine (P3), 4,5,10,15,20-Tetrakis(pentafluorophenyl)-21H,23H-porphinepalladium(II) (P4), 4,4′,4″,4‴-(Porphine-5,10,15,20-tetrayl)tetrakis(benzoic acid) (P5) were studied using electrochemical impedance spectroscopy (EIS), potentiodynamic polarization, static high pressure high temperature (HPHT), contact angle measurement, scanning electrochemical microscopy (SECM), and atomic force microscopy (AFM) in 3.5% NaCl solution saturated with CO_2 for corrosion of N80 steel. The inhibition efficiency was found to increase with increase in the concentration of inhibitor. The adsorbed film on N80 steel surface containing inhibitor was confirmed by the SECM, AFM, and contact angle results. All the results obtained were in good agreement with each other.

Keywords: SECM, Porphyrins, AFM, Contact Angle, N80 steel

1. INTRODUCTION

The Porphines molecule has planar structure and four pyrrole subunits with the nitrogen atoms as the bonding sites. The metal is at the centre bonded to the four nitrogen atoms of the pyrrole subunits [1]. The whole molecule structure shows conjugated double bonds due to the –CH= bridging groups and thus, a very mobile electronic system. These properties are very important in the adsorption process of these compounds on diverse metals surfaces. Porphines are capable to develop supramolecular structures on metal surfaces as a result of metal-porphyrin and porphyrin-porphyrin interactions. The molecules of Porphines can reconfigure the electron distribution of the aromatic ring and change their properties in order create ordered molecular layers [2]. Such structures are supposed to act as effective corrosion inhibitors, the electrode surface being protected by the film barrier avoiding the diffusion of electroactive species towards the electrode surface and the accumulation of such species in its proximity. The building of the protective barrier by adsorption of porphyrin molecules and the corrosion inhibiting effect are affected by the properties of such molecules, a result of some factors such as: peripheral functional groups, steric influence, electron density at donor centers [3]. In the present paper the effect of five Porphines on the corrosion of N80 steel in 3.5% NaCl solution saturated with CO_2 is investigated.

2. EXPERIMENTAL DETAILS

Porphines of analytical grade were bought from the chemical shop and were used in the corrosion studies. Figure 1 represents the molecular structure of all Porphines. Corrosion tests were also performed on a N80 steel of the following percentage composition (wt.%): C 0.31; Si 0.19; Mn 0.92; P 0.010; S 0.008; Cr 0.2; Fe balance, which were metallographically abraded according to ASTM A262, with fine grade emery papers from 600 to 1200 grade. N80 steel coupons were used for the weight loss and electrochemical study.

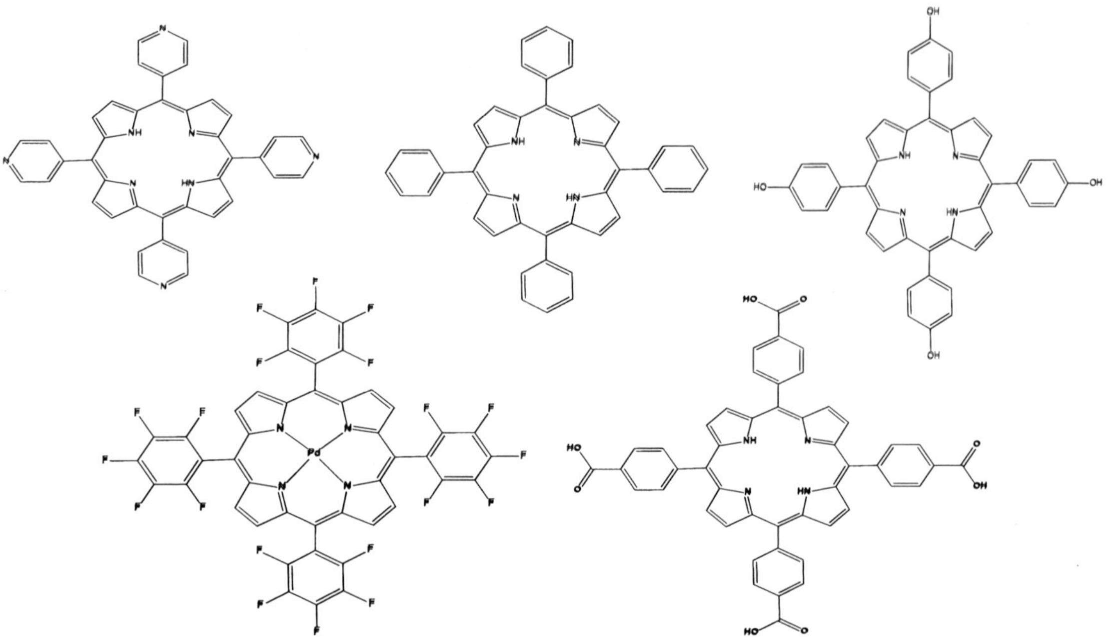

Figure 1: Structure of Porphines

2.1 Weight loss experiments

Weight loss experiments were performed at different temperatures to determine various thermodynamic parameters. The temperature was increased to find out the performance of the Porphyrins at high temperature. All the experiments were performed using a static high pressure high temperature (HPHT) Autoclave in presence of CO_2.

2.2 Electrochemical measurements

The electrochemical studies were performed using an Autolab three electrode cell assembly at room temperature. The N80 steel were the working electrode, platinum electrode was used as an auxiliary electrode, and standard calomel electrode (SCE) was used as a reference electrode. All electrochemical measurements were carried out using Autolab Potentiostat/Galvanostat (Model GSTAT302N) with EIS software Autolab Instruments Inc., Netherland. Autolab applications include software FRC and GPES for EIS measurements, polarization curves, and data fitting.

Tafel curves were obtained by changing the electrode potential automatically from −300 to +300 mV versus corrosion potential (E_{corr}) at a scan rate of 1 mV s^{-1}. EIS measurements were carried out in a frequency range from 100 kHz to 0.00001 kHz under potentiodynamic conditions, with amplitude of 10 mV peak-to-peak. The linear Tafel segments of anodic and cathodic curves were extrapolated to corrosion potential to obtain corrosion current densities (I_{corr}) [5]. The inhibition efficiency was evaluated from the measured I_{corr} values using the relationship:

$$\eta\% = \frac{I^0_{corr} - I^{'}_{corr}}{I^0_{corr}} \times 100 \quad (1)$$

where, I^0_{corr} and $I^{'}_{corr}$ are the corrosion currents in absence and in the presence of an inhibitor, respectively. The charge transfer resistance values were obtained from the diameter of the semi circles of the Nyquist plots. The inhibition efficiency of the inhibitor has been found out from the charge transfer resistance values using the following equation:

$$\eta\% = \frac{R^{'}_{ct} - R^0_{ct}}{R^{'}_{ct}} \times 100 \quad (2)$$

where, $R^{'}_{ct}$ and R^0_{ct} are the charge transfer resistance in presence and in the absence of an inhibitor, respectively.

2.3 Surface Characterization

Contact angle measurements were done using sessile drop technique on DSA100 Kruss instrument made in Germany. All the samples were carefully cleaned to avoid the surface contaminations. SECM studies were carried out using an electrochemical work station of CHI900C model consisting of the three electrode assembly with N80 steel, platinum electrode, and Ag/AgCl/KCl (saturated) electrode as working, counter and reference electrodes, respectively [6,7]. AFM studies were performed using NT-MDT SOLVER Next AFM/STM instrument. The scan size of each sample used in AFM is 10μm × 10μm.

3. RESULTS AND DISCUSSION

3.1 Weight loss tests

Weight loss tests were performed at varying temperatures from 40°C to 100°C. The inhibition efficiency was found to decrease for all the inhibitors. The inhibitor concentration of 400 ppm was used for all inhibitors. After the preliminary test it was observed that there was no significant change in the inhibition efficiency after increasing the inhibitor concentration to 500, 600 and 700 ppm.

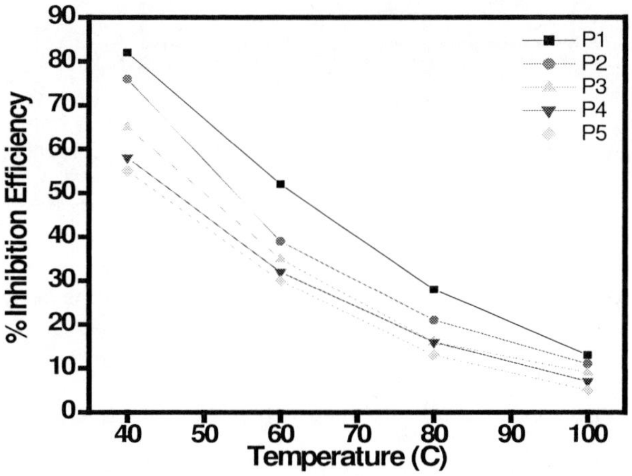

Figure 2: Weight loss data for N80 steel in 3.5% NaCl solution in the absence and presence of Porphines at various temperatures.

3.2 EIS Measurement

Impedance spectra for N80 steel in CO_2 saturated 3.5% NaCl in absence and presence of different concentrations of porphyrins is shown as Nyquist plots (Figure 3). The impedance diagram (Nyquist) contains depressed semicircles with the center under the real axis with one capacitive loop in the high frequency (HF) zone, and one inductive loop in the lower frequency (LF) zone. Such behavior is characteristic for solid electrodes and often referred to as frequency dispersion and has been attributed to roughness and other inhomogeneities of solid surface [8,9]. The LF inductive loop may be a consequence of the layer stabilization by-products of the corrosion reaction at the electrode involving inhibitor molecules and their reactive products. Considering the dependence of inhibition efficiency on inhibitor concentration as represented in Figure 3, it seems likely that the inhibitor molecules act by adsorbing and blocking the available active centres at the steel surface [10]. R_S represents the solution resistance, $Y°$ is the Constant phase element (CPE), R_{ct} is the Charge transfer resistance, n represents the phase shift or the roughness, and L represents the inductance as shown in Table 1.

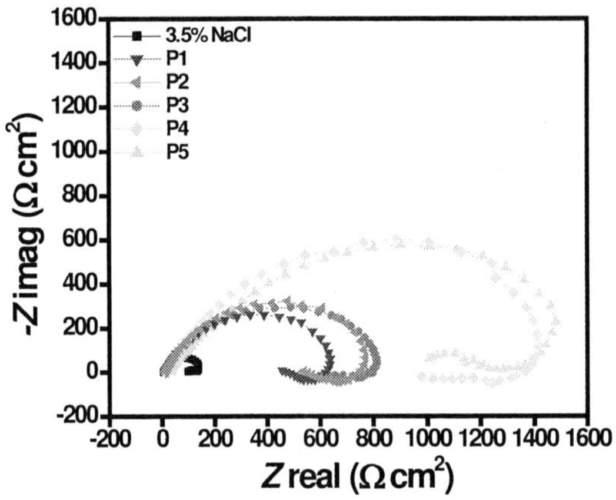

Figure 3: Nyquist plot at an amplitude of 10 mV for N80 steel in 3.5% NaCl solution in the absence and presence of Porphines.

It is clear from table 1 that by increasing the inhibitor concentration the inhibition efficiency increases. This situation was the result from an increase in the surface coverage by the inhibitor, which led to an increase in the inhibition efficiency [11].

Table 1: Electrochemical impedance parameters at an amplitude of 10 mV for N80 steel in 3.5% NaCl solution in the absence and presence of Porphines.

Solution	R_s	R_{ct}	n	$Y_\circ$	L	η	Surf. coverage
	(Ω cm^2)	(Ω cm^2)		(Ω^{-1}s^n/cm^2)	(H cm^2)	%	θ
3.5% NaCl	9.6	128	0.755	298	-	-	-
P1	3.3	590	0.757	121	113	78	0.78
P2	5.9	610	0.771	103	124	79	0.79
P3	2.7	775	0.787	92	99	83	0.83
P4	3.9	1015	0.792	87	52	87	0.87
P5	3.2	1402	0.812	65	37	91	0.91

It is apparent from Nyquist plots that the impedance response of N80 steel in inhibited 3.5% NaCl solution saturated with CO_2 has significantly changed after the addition of Porphines in solution and that the impedance of inhibited substrate increases with increasing concentration of the inhibitor [12].

3.3 Polarization Measurements

Polarization curves for N80 steel at various concentrations of PFs in CO_2 saturated 3.5% NaCl are shown in Figure 4. The model we use for corrosion process assumes that the rates of both the anodic and cathodic processes are controlled by the kinetics of the electron transfer reaction on the metal surface. This is generally the case for corrosion reactions [13]. The extrapolation of Tafel straight line allows the calculation of the corrosion current density (I_{corr}) as shown in Figure 4. The values of I_{corr}, the corrosion potential (E_{corr}), cathodic and anodic Tafel slopes (b_c, b_a) and inhibition efficiency (η%) are given in table 2.

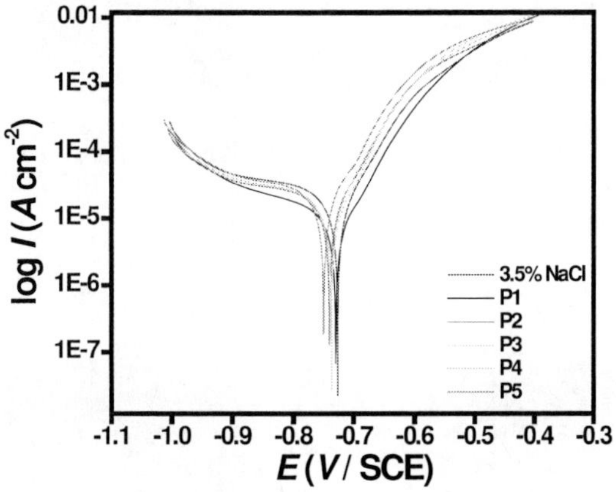

Figure 4: Tafel polarization plots for N80 steel in 3.5% NaCl solution in the absence and presence of Porphines.

Table 2: Polarization parameters for N80 steel in 3.5% NaCl in the absence and presence of different concentrations of Porphines.

Tafel data						
Solution	E_{corr} **(V vs. SCE)**	I_{corr} **(mA cm^{-2})**	b_a **(mV d^{-1})**	$-b_c$ **(mV d^{-1})**	η **(%)**	**Surface coverage** θ
3.5% NaCl	-0.739	0.91	82	214	-	-
P1	-0.729	0.23	67	171	75	0.75
P2	-0.733	0.18	78	119	80	0.80
P3	-0.727	0.12	65	145	87	0.87
P4	-0.740	0.09	72	133	90	0.90
P5	-0.746	0.08	58	98	91	0.91

In literatures, it has been reported that (i) if the displacement in E_{corr} is >85 mV with respect to E_{corr}, the inhibitor can be seen as a cathodic or anodic type and (ii) if displacement in E_{corr} is <85, the inhibitor can be seen as mixed type. In the present study, shift in E_{corr} values is less than 85 mV, suggesting that Porphines act as mixed type of inhibitor [14-18].

3.4 Scanning Electrochemical Microscopy (SECM)

In general, feedback modes require an oxidizable/reducible species to be added in the solution phase for SECM operation. This species is known as the redox mediator. The faradaic current related to the redox transformation of the mediator is thus measured at the tip, usually by setting the tip potential at a value sufficiently far from its corresponding standard potential to achieve a diffusion-limited regime [19]. In this mode, the SECM tip is rastered in the X-Y plane of the surface of the sample inside an electrochemical cell. In SECM, the proximity of the tip to the substrate is the perturbation in the measurement that constitutes the SECM response. The important advantage of the SECM technique is that it operates on both insulating (coated / films) and conducting (non-coated) surfaces, thus allowing one to easily distinguish between the coated and corroded surfaces [20]. Figure 5 presents the morphology of the specimens visualized by scanning electrochemical microscope.

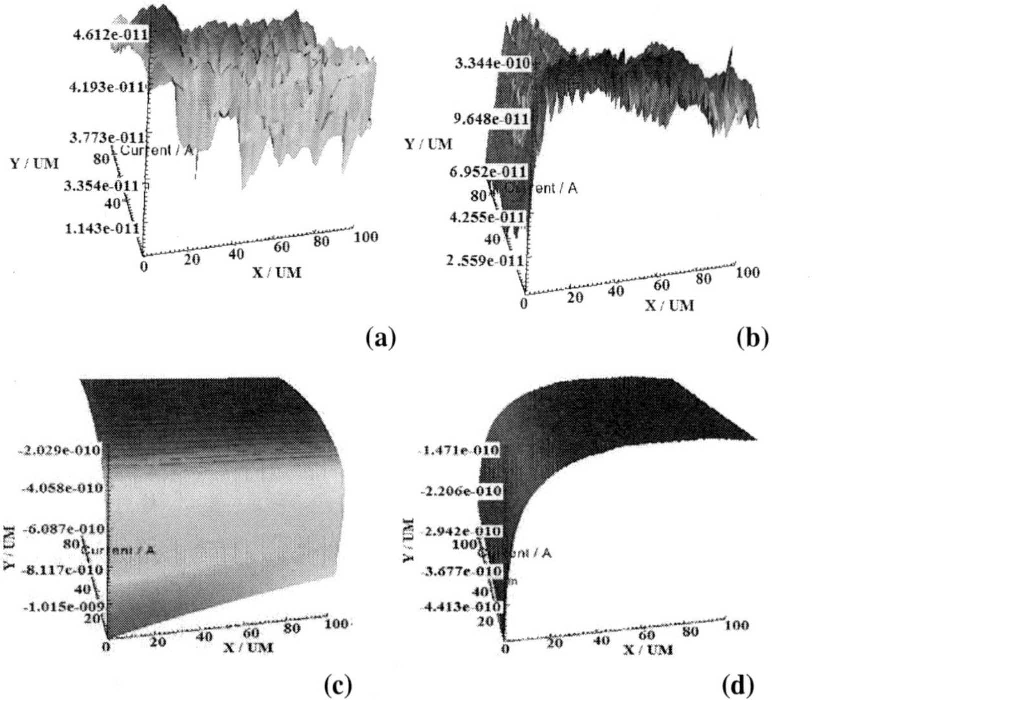

Figure 5: SECM figures for (a) 3.5% NaCl *x* axis 3D N80 steel (b) 3.5% NaCl *y* axis 3D N80 steel (c) 3.5% NaCl *x* axis 3D P5 inh N80 steel (d) 3.5% NaCl *y* axis 3D P5 inh N80 steel.

The current at the tip becomes smaller when the tip is closer to the substrate, and tends to zero when the distance between tip and substrate, approaches zero. Changes in the faradaic current measured at the tip are then exclusively related to variations in the distance between the tip and the substrate underneath, which originates from the scanned surface either progressing in the solution towards the tip (resulting in smaller faradaic currents being measured at the tip), or the surface showing a depression. This is due to the free diffusion of oxygen to the tip. This may be hindered by the proximity of the specimen with respect to the time; the sample was exposed in the test environment [21-25]. A negative feedback effect is observed if the probe is brought to the vicinity of an insulating (inhibitor film), the steady-state current that flows through the tip, tends to be smaller. This is a result of the insulating film partially blocking the diffusion of O towards the tip as shown in Figure 5. Also, a redox competition effect could be expected to occur within the metal. This may be due to the result from a competition between the tip and the steel surface, which is directly exposed to the electrolyte. An increase in the current is observed when a conductor, *i.e.* 3.5% NaCl solution without inhibitor is approached, because the redox mediator is regenerated at the surface as shown in Figure 4. In the absence of inhibitor, the N80 steel surface remains conductive, which is evidenced by an increase in the current as the surface, is approached [26].

3.5 Contact Angle

The contact angle measurements were carried out at room temperature. A baseline test without corrosion inhibitor was carried out first after which the corrosion inhibitors were injected on the N80 steel surface. Droplets were placed on the steel surface by a syringe. For each concentration of corrosion inhibitor the contact angle measurements were repeated 3 times. The syringe used was cleaned with acetone each time before the tests. The contact angle of steel surface without inhibitor was measured as 14.7° meaning that the surface behaved as hydrophilic (favors water) [27]. The contact angle increased with the addition of inhibitor and the steel surface became hydrophobic (does not favors water). At high concentration of the Porphines, the droplet formed a very small contact angle as is evident from Figure 6. All the Porphines showed hydrophobic nature but PF-5 inhibitor showed the best result with an angle of 124.3°. These results confirm the formation of a water repellant or hydrophobic layer on the N80 steel surface when Porphines are present.

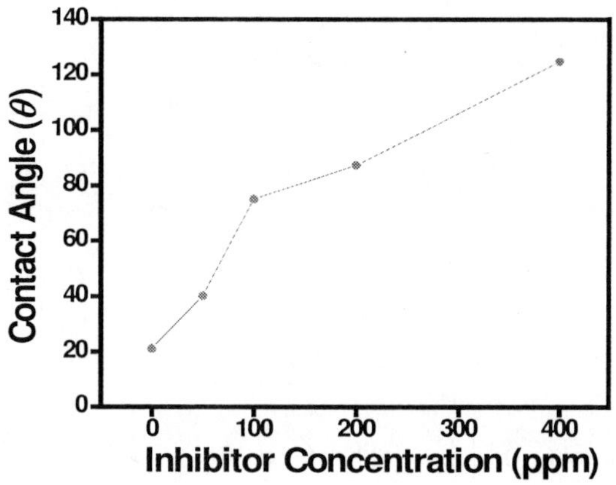

Figure 6 : Contact angle of P5 inhibitor for N80 steel

3.6 Atomic Force Microscopy (AFM)

The atomic force microscope gives a powerful means of characterizing the microstructure. The three-dimensional AFM images of steel surface without and with the presence of inhibitors are shown in Figure 7. In uninhibited solution, N80 steel surface was damaged strongly attributed to the dissolution of oxide film. When inhibitor is added in that solution the surface appears more flat, homogeneous and uniform and height scale of surface started to decrease [28].

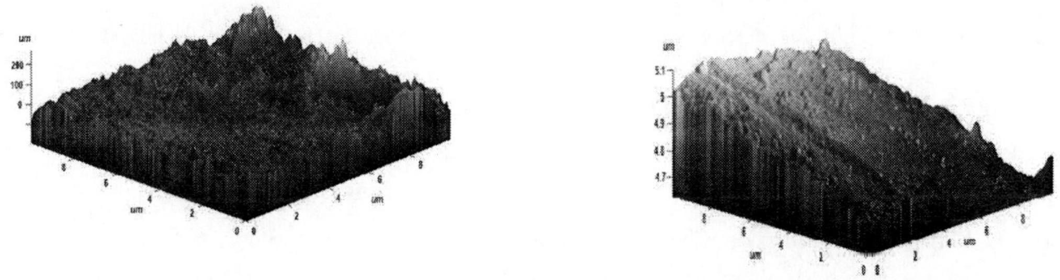

Figure 7: AFM images for (a) 3.5% NaCl solution + N80 steel (b) 3.5% NaCl solution + N80 steel with P5 inhibitor.

4. CONCLUSIONS

- Porphines are good inhibitor for N80 steel corrosion in CO_2 saturated 3.5% NaCl solution. Inhibition efficiency obtained from different methods employed are in reasonable agreement with each other.
- Polarization curves measurement (< 85 mV) indicated that Porphines acted as mixed type inhibitor.
- The SECM and AFM analyses showed that the inhibition of N80 steel corrosion occurred due to the formation of a protective film on the metal surface through adsorption of the constituents of Porphines.

5. ACKNOWLEDGMENTS

Authors are thankful for the Sichuan 1000 Talent fund, Sichuan Province project and Open Fund (PLN1411) of State Key Laboratory of Oil and Gas Reservoir Geology and Exploitation, Southwest Petroleum University.

6. REFERENCES

[1] T Ahamad, V Kumar, N Nishat, Polymer International **55** (2006)1398–1406.

[2] B Evgenij, JR Macdonald, Impedance Spectroscopy Theory, Experiment, and Applications, John Wiley & Sons, New Jersey, (2005).

[3] FRF Fan, J Kwak, O Lev, Analytical Chemistry **61** (1989) 132-138.

[4] AJ Bard, G Denuault, C Lee, D Mandler, DO Wipf, Accounts Chemical Research **23** (1990) 357-363.

[5] C Lee, AJ Bard, Analytical Chemistry **62** (1990) 1906 -1913.

[6] MA Quraishi, A Singh, VK Singh, DK Yadav, AK Singh, Materials Chemistry and Physics **122** (2010) 114-122.

[7] AO Yuce, R Solmaz, G Kardas, Materials Chemistry and Physics **131** (2012) 615-620.

[8] AY Musa, RTT Jalgham, AB Mohamad, Electrochemica Acta **56** (2012) 176-183.

[9] QQ Liao, ZW Yue, D Yang, ZH Wang, ZH Li, HH Ge, YJ Li, Thin Solid Films **519** (2011) 6492-6498.

[10] I Ahamad, R Prasad, MA Quraishi, Corrosion Science **52** (2010) 1472-1481.

[11] A Singh, I Ahamad, VK Singh, MA Quraishi, Journal of Solid State Electrochemistry **15** (2011) 1087-1097.

[12] DK Yadav, MA Quraishi, Industrial Engineering and Chemical Research **51** (2012) 14966-14979.

[13] KF Khaled, MA Amin, Corrosion Science **51** (2009) 1964-1975.

[14] A Khamis, MM Saleh, MI Awad, Corrosion Science **66** (2013) 343-349.

[15] DK Yadav, DS Chauhan, I Ahamad, MA Quraishi, RSC Advances **3** (2013) 632-646.

[16] KM Ismail, Electrochemica Acta **53** (2008) 5953-5960.

[17] Ambrish Singh, Y. Lin, W. Liu, D. Kuanhai, J.Pan, B. Huang, C. Ren, D. Zeng, J. Tai. Inst. Chem. Eng. **45** (2014) 1918.

[18] DK Yadav, MA Quraishi, B Maiti, Corrosion Science **55** (2012) 254-266.

[19] Ambrish Singh, Y. Lin, W.Liu, S. Yu, J.Pan, C. Ren, D. Kuanhai, J. Ind. Eng. Chem. **20** (2014) 4276.

[20] KM Ismail, Electrochemica Acta **53** (2008) 5953-5960.

[21] Ambrish Singh, Yuanhua Lin, Eno. E. Ebenso, Wanying Liu, Jie Pan, Bo Huang, J. Ind. Eng. Chem. **24** (2015) 219.

[22] H Keles, M Keles, I Dehri, O Serindag, Colloid Surface A **320** (2008) 138-145.

[23] J Izquierdo, L Nagy, JJ Santana, G Nagy, RM Souto, Electrochemica Acta **58** (2011) 707-716.

[24] BM Quinn, I Prieto, SK Haram, AJ Bard, Journal of Physical Chemistry B **105** (2001) 7474-7476.

[25] M Tsionsky, AJ Bard, D Dini, F Decker, Chemistry of Materials **10** (1998) 2120-2126.

[26] A Maljusch, C Senoz, M Rohwerder, W Schuhmann, Electrochemica Acta **82** (2012) 339-348.

[27] Yuanhua Lin, Ambrish Singh, E. E. Ebenso, Y. Wu, C. Zhu, H. Zhu, J. Tai. Inst. Chem. Eng. **46** (2015) 214.

[28] Ambrish Singh, Yuanhua Lin, I. B. Obot, E. E. Ebenso, K. R. Ansari, M. A. Quraishi, Appl. Surf. Sci. **356** (2015) 341.

IMPROVING CORROSION RESISTANCE OF MONEL 400 ALLOY BY CVD GRAPHENE COATING

A. Sanjid[1], R.K. Singh Raman [1, 2]
[1]Department of Chemical Engineering, Monash University, Clayton, Australia
[2]Department of Mechanical & Aerospace Engineering, Monash University, Clayton, Australia

SUMMARY: Corrosion resistance of Monel 400 alloy due to graphene coating was investigated. An ultra-thin graphene layer was developed over Monel 400 alloy by chemical vapour deposition (CVD) technique, and the presence of multilayer graphene was confirmed by Raman spectroscopy. Potentiodynamic polarization tests showed the graphene coating to improve corrosion resistance of the bare Monel 400 alloy by nearly two orders of magnitude. The improvement which is contrary to those reported in a few other studies is attributed to the effective barrier properties due to less defective nature and the multi-layer nature of the CVD graphene developed in this study.

Keywords: Graphene coating; chemical vapour deposition; Monel 400 alloy; Corrosion resistance

1. INTRODUCTION

Monel 400 alloy is widely employed in marine applications because of its high strength and corrosion resistance. However, the corrosion resistance of the alloy can be impaired under aggressive oceanic condition. Ultra-thin graphene has been investigated in recent times for its unique properties as corrosion barrier coatings (1). However, graphene as corrosion barrier coatings for Monel-400 alloy has not been explored yet.

Cavitation, marine biofouling and stress corrosion cracking are common modes of degradation and failure of Monel 400 in marine environment (2). Other factor that can cause degradation is the shear stress due to sea water that proportionally increases with the velocity of sea vessel which can eventually disrupt the protective oxide film on the alloy surface that provides protection against corrosion, thereby exposing the bare alloy to aggressive chloride ion containing sea-water. It is well known that besides its properties of remarkable impermeability for fluids and inertness to aggressive chemicals (including those that cause corrosion), graphene also possesses remarkable toughness, which is likely to resist the shear stress due to sea water. There are recent reports suggesting a multi-layer graphene coating to be more effective in providing durable corrosion resistance to copper and nickel (3). On the basis of these attractive properties of graphene, this research aims to produce high quality multilayer CVD graphene on Monel-400 alloy with minimal defects, with a view to improving the corrosion resistance in aggressive sea water.

2. EXPERIMENTAL DETAILS

2.1 Sample preparation

A plate of Monel 400 alloy (175mm×100mm×2mm) was procured from VDM Metals Australia Pty Ltd. (composition of the samples is given in Table 1). The plate was machined into coupons (15mm×15mm×2mm). These coupons were ground with successively finer grinding papers, i.e., 180, 320, 800, 1200 and 2500 grade. The ground coupons were polished with pads that were charged with diamond polishing paste (3 μm), and washed with acetone for 5 min in ultrasonic bath, and then dried under compressed air.

Table 1. Composition of the Monel-400 alloy used in this study

Elements	Carbon	Nickel	Manganese	Silicon	Copper	Iron
Composition	0.1%	63.8%	1.0%	0.1%	32.8%	2.0%

2.2 CVD growth of graphene on coated samples

CVD growth of graphene on ground and polished specimens was performed using a set-up shown in Figure 1. The specimens were loaded inside the quartz tube of the furnace which was evacuated to 5 mtorr through a vacuum pump (Edwards, nXDS10i, Germany). Ar and H2 mixture was introduced into the quartz tube at a flowrate of 250 sccm at 1.8 torr pressure. The samples were held for 40 minutes at 1060 °C under the reducing atmosphere of Ar/H2 flow. After that, hexane was passed for 60 minutes at 1060 °C at a flow rate of 1 sccm.

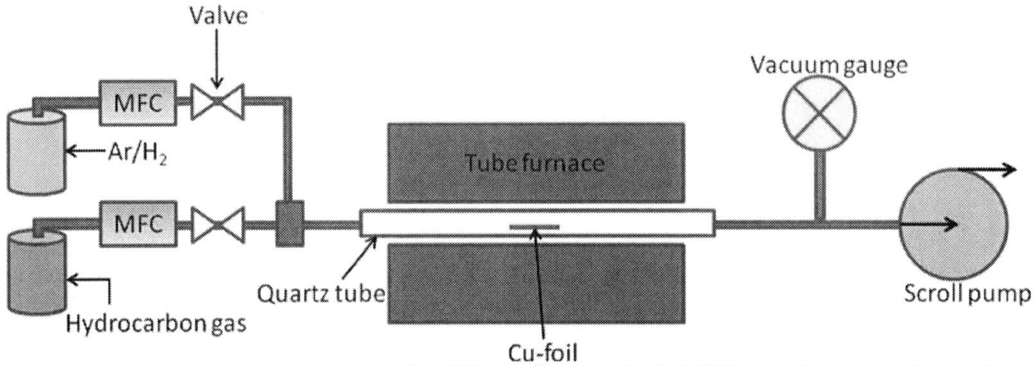

Figure 1: Schematic diagram of typical set-up for CVD graphene synthesis (MFC: mass flow controller (4). Note: In this study, Monel 400 coupons were used in place of Cu-foil.

2.3 Characterization of graphene

The samples subjected to CVD treatment as described at 2.2 above were characterized by Renisaw Invia Raman Spectrometer that is equipped with 514 nm wavelength green laser with 1 μm spot size. With a scan range of 100-3200 wavenumbers. Raman spectra were recorded at 10-12 different spots for a given sample.

2.4 Electrochemical test

Electrochemical tests were performed on the graphene coated and bare Monel 400 specimens in 0.1 M NaCl solution, using a Princeton applied research potentiostat (Model 2273), using a conventional three electrode electrochemical cell. Saturated calomel electrode was used as the reference electrode platinum mesh as counter electrode. The exposed area of the working electrode (Monel 400) was 0.785 cm2. Specimens were held at the open circuit potential (OCP) for 1 h prior to starting the electrochemical measurements, for stabilizing the surface potential of the specimen.

3. RESULTS

Figure 2 is the representative Raman spectrum of graphene coated sample, where G peak at 1584 cm-1 and 2D peak at 2682 cm-1 indicate the presence of graphitic carbon (5). The ratio of G and 2D peak intensities (IG/I2D) indicates the number of layers of graphene present at that point. The nature of the Raman spectra recorded at 10-12 different locations on the surface of the coated samples was similar (i.e., G and 2D peak intensities at different locations were respectively similar). The IG/I2D value of ~1.8 in the Raman spectrum suggests the presence of a multilayer graphene. The relatively small D peak at 1332 cm-1 suggests a low content of defects or disorders of the graphene plane; hence, less conducting pathways through these coating defects, which is beneficial for corrosion resistance. No peaks were found at around 300-600 cm-1, indicating the absence of metal oxides or hydroxides (6). So, the overall quality of graphene was remarkable.

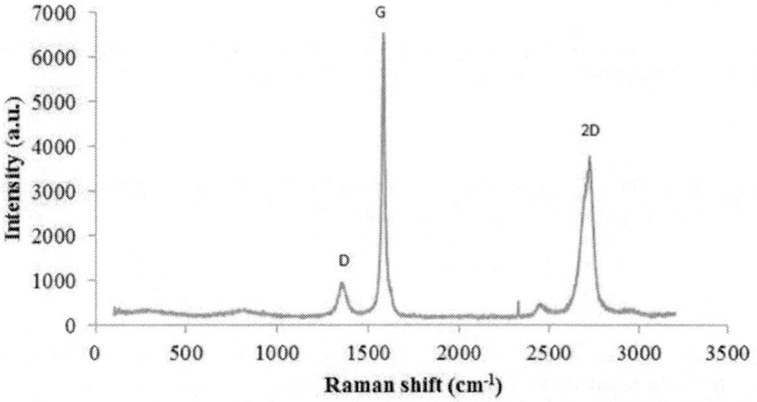

Figure 2. Raman spectrum of graphene coated Monel 400 alloy

Potentiodynamic polarization (PDP) tests were carried out to compare the anodic dissolution rate of graphene coated and uncoated Monel-400 alloy. A comparison of the intercepts of the anodic and cathodic arms of the coated and uncoated alloy suggests the corrosion resistance of the graphene coated alloy to be almost two orders of magnitude lower than the bare samples. Also, the E_{corr} of the graphene coated alloy is 1~50 mV more positive than that of the bare alloy. The polarisation plots in Fig 3 are the representatives of triplicate runs for both graphene coated and bare samples. These results suggest the graphene coating to have considerably improved the corrosion resistance of Monel 400 in 0.1 M NaCl solution.

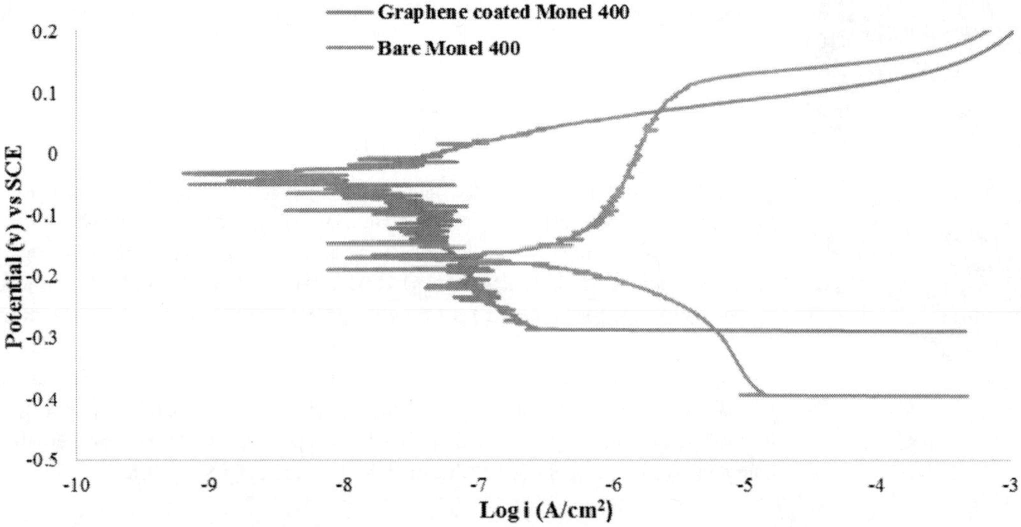

Figure 3. Potentiodynamic polarization plot of the graphene coated and bare Monel 400 alloy in 0.1 M NaCl solution.

The observed results of graphene coating improving corrosion resistance of Monel 400 alloy by two orders of magnitude (Fig.3) have also been confirmed by the results of electrochemical impedance spectroscopy (EIS). Two main features bode plot and Nyquist plot of EIS data supported the findings of PDP test. The broader nature of the phase angle vs frequency plot and the increased impedance in impedance vs frequency plot clearly showed significant improvement in corrosion resistance for the graphene coated samples.

4. CONCLUSIONS

Multilayer CVD graphene was successfully developed over Monel 400 alloy. This coating improved corrosion resistance of the alloy by nearly two orders of magnitude in 0.1 M NaCl solution.

5. ACKNOWLEDGMENTS

The authors would like to acknowledge Department of Chemical Engineering, Monash University, MIGR, MCEM and MCATM for providing required facilities and funding for this project.

6. REFERENCES

1. Raman RS, Banerjee PC, Lobo DE, Gullapalli H, Sumandasa M, Kumar A, et al. Protecting copper from electrochemical degradation by graphene coating. Carbon. 2012;50(11):4040-5.
2. Tuthill A. Guidelines for the use of copper alloys in seawater. NiDI Publication. 1988;12003:12-5.
3. Raman RS, Tiwari A. Graphene: The thinnest known coating for corrosion protection. Jom. 2014;66(4):637-42.
4. Tiwari A, Chakraborty Banerjee P, Singh Raman RK, Majumder M, editors. CVD Graphene on Metals for Remarkable Corrosion Resistance. Corrosion & Prevention 2012, Melbourne, Australia; 2012: Australasian Corrosion Association.
5. Ferrari AC. Raman spectroscopy of graphene and graphite: disorder, electron–phonon coupling, doping and nonadiabatic effects. Solid state communications. 2007;143(1):47-57.
6. Raman RS, Gleeson B, Young D. Laser Raman spectroscopy: a technique for rapid characterisation of oxide scale layers. Materials science and technology. 1998;14(5):373-6.

7. AUTHOR DETAILS

A. Sanjid received his B.Sc. degree in Mechanical Engineering from Bangladesh University of Engineering and Technology (BUET) in 2012. Right after his graduation he joined Centre for Energy Sciences in University of Malaya (UM) in and completed his research based masters in 2015. In the same year, he has joined Department of Chemical Engineering at Monash University, Australia to pursue his PhD. Currently, he is working on graphene coatings for corrosion resistance. . He has published over 20 peer-reviewed international journal publications and 6 reviewed international conference publications.

Professor Raman Singh's expertise includes: Alloy Nano/Microstructure Corrosion Relationship, Stress Corrosion Cracking (SCC), Corrosion/SCC of Biomaterials, Corrosion Mitigation by Novel Material (Graphene), Advanced and Environmentally Friendly Coatings, High Temperature Corrosion (HTC), Microbiologically Influence Corrosion (MIC). Resources in Prof Singh's group include three separate laboratories (one each for research in SCC and HTC, MIC and electrochemical corrosion and coatings) and ~25 researchers. He has published over 125 peer-reviewed international journal publications, 14 book chapters and over 90 reviewed conference publications. His professional responsibilities include leadership (as co-chairman) of a few international conferences and membership of editorial/review boards of a few journals.

ACCURATE ASSESSMENT OF INSTANTANEOUS CORROSION RATES

A J Betts[1], J F Cassidy[2], D Culliton[1]

[1]Applied Electrochemistry Group, FOCAS Institute, [2]School of Chemical and Pharmaceutical Sciences, Dublin Institute of Technology, Kevin St, Dublin 8, Ireland

SUMMARY: Slow potentiodynamic polarisation curves are often utilised for determination of corrosion rates, through identification of corrosion current densities. As this electrochemical approach is relatively easy to perform, this classic technique can be carried out at various stages of a corrosion investigation (albeit on identical samples) to ascertain instantaneous corrosion rates occurring during different stages of a corrosion process. Assuming that the corrosion is uniform, (which is in fact most often not the case, especially during the initial stages of corrosion) the corrosion current obtained from such scans is then typically converted to a corrosion rate. Most often the Tafel extrapolation method is used to estimate Tafel slopes in conjunction with a Stern-Geary treatment to ascertain corrosion current densities and hence evaluate an alloy's corrosion rate in a selected environment. Problems encountered with this commonly used method, which is strictly only applicable to irreversible reactions for corrosion rate determination (using estimates of Tafel slopes) are discussed. This approach however is not suitable for many alloy/environment combinations, such as aluminium alloys in neutral saline environments where it is extremely difficult to determine accurate Tafel slope values.

Following discussion of electrode kinetics in electrochemistry and corrosion studies, an alternative but little used method based on the Allen-Hickling plot approach to elucidate corrosion rates is outlined. After showing its derivation, potentiodynamic polarisation curves of two cast aluminium alloys, LM24/A380 and LM25/A356, that are not amenable to the standard Tafel extrapolation method are presented and discussed. The Allen-Hickling method overcomes many of the inherent issues associated with the classic Tafel extrapolation and Stern-Geary polarisation resistance treatments. Allen-Hickling plots minimise problems associated with contributions from concentration polarisation and the inability of the standard electrochemical methods to deal with corrosion processes that involve significant contributions from back reactions, commonly seen in the corrosion of aluminium and other metal alloys. Finally a discussion of other approaches based on polarisation curve analysis is given which indicates that the Allen-Hickling methodology deserves more attention from the corrosion science community.

Keywords: Electrode kinetics, Allen-Hickling plot, cast aluminium corrosion

1. INTRODUCTION

It is often necessary to know the instantaneous corrosion rate of a metallic alloy in a corrosive service environment. Examples of such situations include the development of new alloys, assessing the performance of novel surface coatings or treatments and the investigation of the efficacy of new corrosion inhibitors. Corrosion rates are seldom, if ever constant and invariably are time-dependent. In practise they can be found by using a range of techniques; both electrochemical and non-electrochemical. Examples of the former include direct current techniques such as potentiodynamic polarisation curves (together with Tafel Extrapolation and use of the Stern-Geary polarisation resistance technique) and electrochemical noise measurements [1]. Alternating current electrochemical techniques such as Electrochemical Impedance Spectroscopy, Electrochemical Frequency Modulation (EFM), faradaic rectification and harmonic analysis are also used [2]. Non-electrochemical methods including gravimetric measurements (weight loss/time), atomic absorption spectrophotometry, colorimetric and ion couple plasma ICP analysis of dissolved metal ions and determination of gas evolution (most often hydrogen) [3]. In practice it is necessary to use both electrochemical and non-electrochemical methods in order to validate corrosion rates obtained from experimental studies, as chemical processes may sometimes be involved.

Potentiodynamic polarisation curves are a mainstay of most corrosion laboratories. This electrochemical approach is relatively straightforward to perform enabling the assessment of a corrosion current density i_{corr} resulting from a mixed potential [4]. However it is important to understand the issues involved in the conduction of such measurements. First and foremost it is necessary to attempt to obtain a pseudo steady-state during the polarisation scan by using slow potential scan rates (often less than 0.5 mV/s). This is done in order to prevent a significant non-faradaic contribution resulting from capacitance effects (such as double layer charging). For this reason many researchers utilise potential step measurements instead to allow for capacitive discharge (chronamperometry) to take place before measuring resultant currents.

Another major concern is the occurrence of concentration polarisation [5]. Arising from limitations of the mass transport of reactive species to an electrode surface, it may cause significant changes in surface concentrations. Thus reactive species' surface concentrations may differ substantially from those in the bulk electrolyte (ie. the corrosive environment). Hydrodynamic methods such a rotating disk electrode, rotating cylinder electrode or a wall-jet or channel flow electrode may help overcome this. Finally an ohmic resistance due to iR drop from a low conductivity test electrolyte (solution) can occur between the Luggin capillary containing a reference electrode and the corroding electrode surface. The presence of adsorbed surface species (salt films or other surface films) can also influence the currents measured in polarisation curves. High conductivity electrolytes including the 3.5% NaCl (0.56 M NaCl) commonly employed in corrosion investigations also assist. Furthermore a Luggin capillary placed in close proximity to the working electrode can diminish the iR contribution from poorly conducting electrolytes. Potentiostat feedback and current interrupt methods are used to account for iR effects as can a numerical adjustment following a separate EIS measurement and analysis.

In this paper it is intended to critically examine the use of the Tafel Extrapolation (TE) method often used in conjunction with the Stern-Geary polarisation resistance method. Potentiodynamic studies are presented on the corrosion of two cast aluminium alloys, LM24 /A380 and LM25/A356 in 0.1 M NaCl. Another approach which has not been widely adopted by the corrosion research community is examined, the Allen-Hickling plot method [6]. Although it has been used in other areas of electrochemistry, including lithium battery research [7-10], sulphur electrochemistry [6, 11-12], fuel cell research [13,14] and studies of hydrogen reactions [15,16], oxygen reduction reactions [17-19] and aluminium electrodeposition [20,21] and other electrochemical research [22-28], with few exceptions [29-32] it has seldom been used for corrosion rate determination.

1.1 The Tafel Equation and its Role in Electrochemical Corrosion.

The Tafel Equation (TE) was originally proposed by Julius Tafel in a seminal paper in 1905 [33]. It is interesting to note that the TE was derived empirically by Tafel, from studies of the hydrogen evolution reaction (HER) on a range of metals (Ni, Cu, Au, Bi, Sn and Hg). Tafel's main research interest was the use of electrochemical methods for the reduction of organic compounds. Only much later was the equation adapted for use in corrosion science and engineering investigations, largely through the efforts of Wagner and Traud [34] who insightfully developed mixed potential theory and later Stern and Geary [35] who applied it to the corrosion of ferrous metal alloys in strong acid electrolytes. The latter developed the polarisation resistance analysis technique which requires knowledge of Tafel slopes. It is important to differentiate between the two applications of the TE; in electrochemistry it involves redox reactions whereas in corrosion research a mixed potential is studied. Unfortunately the literature often uses nomenclature that is slightly different in each area which can lead to some confusion. In its most simplistic form for an electrochemical reaction involving the transfer of electrons (termed activation or charge transfer control) the Tafel Equation is given by the expression:

$$\eta = a \pm b\log(i) \quad (1)$$

where η, denotes the overpotential, which is simply the difference between the applied potential (E) and the formal or equilibrium (reversible) potential E_{eq} from the thermodynamic Nernst Equation ($\eta = E - E_{eq}$). Also in equation 1, the first term *a* (equation 2) is a constant (the exchange current density) and b is the Tafel slope (in mV/decade).

$$a = -2.303 b \log i_0 \quad (2)$$

Values of the exchange current density i_0, can be determined directly from the intercept of the ln i (or log i) axis through a η value of 0 in a semilogarithmic plot. Often k_0 the standard rate constant is used in the electrochemical literature which is derived from i_0 Hence a semi-logarithmic plot of η versus log(i) will yield a Tafel slope of b [5]. Under oxidising potential values b_a denotes an anodic Tafel slope, whereas under reducing potential values b_c is a cathodic Tafel slope (which has a negative value). The TE can also be used to assist the elucidation of corrosion reaction mechanisms. However it is important to understand the theoretical framework upon which these assessments are made and to be aware of the assumptions and limitations inherent in the TE approach.

Equation 1 is only applicable at high overpotential η values which are far removed from the equilibrium potential E_{eq} , where it is assumed the reverse reaction contributes <1% of the current [5]. Thus the reaction under investigation is essentially irreversible. The TE is derived from the more general Butler-Volmer (BV) equation, one of the fundamental equations of electrochemical kinetics [3]. Another way of expressing equation 1 is

$$\eta = \frac{RT}{\alpha_a F}\ln\left(\frac{i}{i_0}\right) = 2.303\frac{RT}{\alpha_a F}\log\left(\frac{i}{i_0}\right) \qquad (3)$$

The TE is extensively used by electrochemists and corrosion scientists and engineers and other materials-oriented researchers. It is fundamentally important in studies of electrochemical (electrode) kinetics, also sometimes known as electrodics [36]. Unfortunately though, in many instances the TE is not used correctly, as is often the case of analysis of polarisation curves. Its constraints and limitations are insufficiently recognised, as described above or are even completely ignored. This is especially noticeable when the Stern-Geary approach for the determination of corrosion rates is used erroneously through use of incorrect values of the Tafel slopes, b_a and b_c. In the Tafel extrapolation methodology linear Tafel slopes are extrapolated to the corrosion potential E_{corr} in order to ascertain a corrosion current density i_{corr}. Directly proportional to the corrosion rate (CR), it forms part of the following equation [1-3].

$$CR = \frac{M i_{corr}}{nF\rho} \qquad (4)$$

where M is the metal's or alloy's molar mass, n the number of electrons transferred, F is 96485 $Cmol^{-1}$ and ρ is the density of the metal or alloy under investigation. According to this expression CR is measured in units of cm/s, which can readily be converted to other units such as mm/y.

1.2 Development of the Wagner-Traud Equation in Corrosion Science

In the mid 20th century two developments thrust the TE into the forefront of corrosion research. The first was the development of the Wagner-Traud mixed potential theory, which applied a modified Butler-Volmer (BV) equation to a new application of electrochemistry [34]. This theory proposed that two separate, but coupled electron transfer reactions occur simultaneously on a corroding metal surface which could be described by an equation that was similar in appearance to the BV equation. As in the fundamental electrochemical redox case, one reaction is an oxidation reaction (anodic metal dissolution), the other a cathodic reduction reaction. It proved to be a significant advance in the theoretical description of electrochemical corrosion. The two most common cathodic reactions are hydrogen ion reduction (hydrogen evolution reaction HER), typically occurring in deaerated acid solutions and the oxygen reduction reaction, ORR (either a 2 e^- or a 4e^- process) which takes place in the presence of oxygen (ie. in an aerated electrolyte). Another cathodic reaction is metal electrodeposition from ions present in the electrolyte (ie electroplating).The Wagner-Traud modified form of the fundamental BV equation is given by [1,2]

$$i = i_{corr}(\exp(\alpha_a(\Delta E)F/RT) - (\exp(-\alpha_c(\Delta E)F/RT)) \qquad (5)$$

$$\text{with } \Delta E = E - E_{corr} \qquad (6)$$

where E is the applied potential (V) E_{corr} is the corrosion potential (which can differ from the Open Circuit Potential OCP as a result of the polarisation process), i_{corr} the corrosion current density in mA/cm^2, R is 8.314 $JK^{-1}mol^{-1}$ and T is temperature (K). The symbols α_a and α_c represent the anodic and cathodic transfer coefficients respectively which are discussed in the next section.

Just as in the electrochemical literature describing redox processes (used in the BV equation), an αn product term is often erroneously used in the modified BV equation. Another way of expressing this current-overpotential relationship is

$$i = i_0\left(\exp\left(\frac{2.303\Delta E}{b_a}\right) - \exp\left(\frac{-2.303\Delta E}{b_c}\right)\right) \qquad (7)$$

Frankel [37] suggested that equations 5 and 7 be named Wagner-Traud (WT) equations, which clearly differentiates it from the electrochemical BV equation. Even though it appears similar to the BV equation and is indeed often erroneously referred to as the BV equation in the corrosion literature, it should more correctly be regarded as an extension of the BV equation. Whereas the BV equation refers to redox processes, the WT equation is applicable to a corrosion process normally involving two different, but coupled electrochemical reactions occurring on a metal surface immersed in a corrosive solution. This concept is embodied in the Stern description of the origin of polarisation curves [35].

For low values of η , the exponential term in the BV equation is normally approximated by a simple two term Taylor series expansion yielding the charge transfer resistance R_{ct} given by the following equation [2]

$$i = i_0\left(\frac{F\eta}{RT}\right) \qquad (8)$$

Where $\eta / i = R_{ct}$ (9) and thus $R_{ct} = \frac{RT}{Fi_0}$ (10)

However by using a 3 term Taylor series expansion it can be shown [38] that the transfer coefficient α can have a significant influence on the current measured in the low overpotential region, Consequently a non-linear function which is more accurate at values of α significantly different from 0.5, is obtained (equation 11 below). If α is 0.5, then equation 11 simply reduces to the familiar linear expression given by equation 10 above.

$$i = i_0\left(\frac{F\eta}{RT}\right)\left(1+\left((\alpha-0.5)\left(\frac{F\eta}{RT}\right)\right)\right) \qquad (11)$$

Recently the experimentally determined quantity α (transfer coefficient) has been defined in the context of the BV equation in an IUPAC Technical Report [39]. An accompanying IUPAC Recommendation defines the transfer coefficients according to the following expressions:

$$\alpha_a = (RT/F)(d\ln i_a/dE) \quad and \quad \alpha_c = -(RT/F)(d\ln|i_c|/dE) \qquad (12) \quad \text{or}$$

$$\alpha_a = 2.303(RT/F)(d\log i_a/dE) \quad and \quad \alpha_c = -2.303(RT/F)(d\log|i_c|/dE) \quad (13)$$

where i_a and i_c are the anodic and cathodic current densities, corrected for any changes in reactant concentration at the electrode surface with respect to the bulk value. In practice such a correction can be carried out by imposing hydrodynamic conditions to ensure efficient mass transfer. From these expressions it is evident that the transfer coefficient is also the product of the reciprocal of the Tafel slope (1/b) with the term (RT/F). The IUPAC report also highlights a major misconception involving use of a transfer coefficient-electron number product, αn in the BV formulation. Incorrect use of the product αn is often found in the literature [5].

1.3 Electrochemical Assessment of Corrosion Rate

A major advance in corrosion research occurred in 1957, when M Stern published the first of a series of papers using the WT formulation of the BV to predict the shape of Tafel plots, essentially polarisation curves obtained over a range of a few hundred mV either side of E_{corr}. The first paper with A L Geary [35], is probably one of the most cited papers in corrosion science. It gave an expression enabling use of electrochemical methods such as potentiodynamic polarisation to directly determine corrosion rates. Although the initial paper described a theoretical model, two following papers confirmed its applicability [40,41], albeit for corrosion of ferrous metals in deaerated acids. It assumed that the predominant process was charge transfer and that a limiting current due to concentration (diffusion-limited) polarisation occurred only at high potential values far away from E_{corr}. Unfortunately it used a nomenclature that has since resulted in a lot of confusion with β_a and β_c used to represent the anodic (b_a) and cathodic (b_c) Tafel slopes respectively [36, 42] as opposed to their other descriptive use in electrochemistry as symmetry factors in the single electron transfer BV equation.

The most famous result from the Stern paper published in 1958 [41] is the polarisation resistance equation, which requires knowledge of b_a and b_c to enable estimation of i_{corr} through measurement of the polarisation resistance R_p (Ωcm^2).

$$i_{corr} = \frac{b_a b_c}{2.303(b_a + b_c)R_p} \qquad (14)$$

According to this expression the i_{corr} is inversely proportional to R_p and most importantly depends upon the values of the Tafel slopes b_a and b_c.. In practice this can be simplified if the b_c value is very much larger than the b_a value to yield

$$i_{corr} \approx \frac{b_a}{2.303R_p} \qquad (15)$$

Apart from the case of non-passive ferrous metals (eg iron, mild steel) in deaerated acid solutions, linear anodic and cathodic Tafel plots are seldom observed. In fact it is far more common for non-Tafel behaviour (ie curved semi-logarithmic Tafel plots) to be obtained. It is normally recommended that linear Tafel slopes be evaluated only if they occur over at least one decade of logarithmic current density. Some even recommend two decades (eg [43]) Several examples of such curved Tafel plots (polarisation curves) displaying a variety of linear (activation polarisation) and non-linear sections are displayed in the book by Thompson and Payer [44].

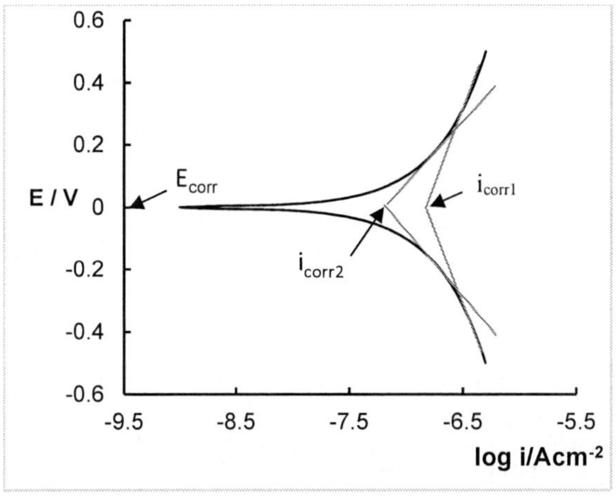

Figure 1. Polarisation curve (Tafel plot) representative of a highly resistive surface film showing erroneous extrapolation of pseudo Tafel slopes back to 2 different values of i_{corr}, labelled i_{corr1} and i_{corr2} at E_{corr} of 0.0V.

Figure 1 (left) illustrates a common problem with use of the Tafel plot extrapolation of apparently linear sections of semi-logarithmic plots back to yield values of i_{corr}.. Although there may appear to be linear sections, this semi-logarithmic plot was in fact derived from a simple ohmic resistance. Therefore both i_{corr} values depicted along the line with E=0 are incorrect representations of the true (polarisation) resistance value, as has been noted by others [45]. Such extrapolations are often incorporated in software accompanying electrochemical equipment supplied by many leading potentiostat manufacturers and may lead to incorrect analysis of instantaneous corrosion rates.

Considerable care should always be taken when carrying out such Tafel extrapolations as they are often not linear and the overpotential at which the linear region commences should be larger than several tens of mV (in order to minimise the significance of any back reaction).

1.4 The Allen-Hickling Formulation of Electrode Kinetics

Also originally published in 1957, [6] the same year as the more famous Stern-Geary paper, this approach offers a different method to estimate i_{corr} (and hence evaluation of corrosion rates). In addition it can often provide mechanistic insights into electrochemical reactions such as those involving corrosion, through estimation of transfer coefficients. Developed for studying the kinetics and mechanism of the reduction of sodium sulphide in solution, it utilises the BV equation to derive a different expression from that normally given by the conventional Tafel equation approach. Although little known in corrosion research, the Allen-Hickling equation may be re-formulated to yield a different expression for corrosion utilising the WT formulation of the BV equation.

In electrode kinetics it is extremely improbable that two electrons are ever transferred in an electron transfer process simultaneously. Instead it is generally considered that redox and corrosion reactions take place in a sequence of one electron transfer steps. One of these is an irreversible, rate-determining step (rds). All other steps are assumed to be relatively fast and in-equilibrium [5]. This is termed the quasi-equilibrium approach and assumes that the exchange current density of the rds is at least 100 times less than that of any other step. It also assumes that the rds of the forward reaction is the same as the backward reaction rds (the principle of microscopic reversibility) and there is no electrode blockage by adsorbed species. Another assumption is that the reaction is in a steady state. This serves as the basis of an alternative, but much more complex steady state analysis which is also used in electrode kinetics. [30 A T Marshall]

Using this quasi-equilibrium approach for a sequence of electron transfer (multistep) reactions Fletcher [46] derived expressions for the transfer coefficients α_a and α_c given in terms of the symmetry factor β (generally ~0.5) for a single electron transfer process. In the general case in which the rds occurs vs times in the electrochemical (redox) reaction sequence then [39]

$$\alpha_a + \alpha_c = n / v \qquad (16)$$

This is important, as the assumption that the sum of the transfer coefficients ($\alpha_a + \alpha_c$) is always 1.0 is not always strictly correct. It applies when the symmetry factor β is used for a single electron transfer reaction, although sometimes with an n/v ratio of 1 it is also appropriate. Therefore individual transfer coefficients α_a and α_c can have values ranging from less than 0.5 to more than 1, giving rise to Tafel slopes from as low as 15 mV/decade to more than 240 mV/decade [43]

Now for an oxidation (corrosion) process (and by noting that α_c is equivalent to (n/v - α_a) in equation 16) above it can be shown that

$$\log\left(\frac{i}{\left[1-\exp\left(\frac{-nF\Delta E}{vRT}\right)\right]}\right) = \frac{\alpha_a F\Delta E}{2.303RT} - \log i_{corr} \qquad (17)$$

Therefore a plot of the left hand side of this equation against the overpotential ΔE, will yield an intercept of log i_{corr} and a slope of (α_aF/2.303RT). A similar derivation yields the following expression for the concomitant cathodic reaction which is given by the following expression

$$\log\left(\frac{i}{\left(exp\left(\frac{nF\Delta E}{vRT}\right)-1\right)}\right) = \log i_{corr} - \frac{\alpha_c F\Delta E}{2.303RT} \qquad (18)$$

In both of these equations the n/v ratio term represents the sum of the anodic and cathodic transfer coefficients (equation 16) and can range in value, depending upon the species being oxidised (eg Al to Al^{3+}) or reduced (eg the ORR).

2. ASSESSMENT OF THE CORROSION RATE OF TWO CAST ALUMINIUM ALLOYS

2.1 Use of Polarisation Curves to Assess the Corrosion Rate of Cast Aluminium Alloys.

In order to investigate different approaches to the electrochemical assessment of the corrosion rate of metallic alloys two commonly used cast aluminium alloys were selected and subjected to potentiodynamic polarisation testing. Due to their light weight aluminium alloys are widely used in the aerospace, automotive as well as many other manufacturing industries. Although the addition of small amounts of other elements results in enhanced mechanical properties, (strength) this is often achieved at the expense of their corrosion resistance. Elements such as silicon, copper, magnesium and iron in relatively small amounts may exist as components of intermetallic (IM) particles in aluminium alloys [47,48]. Such IMs can act as defect sites for the initiation of localised corrosion processes, such as pitting when exposed to corrosive solutions such as those containing chloride ions [49].

Two cast alloys were selected; LM24/A380 (AlSi8Cu3Fe) and LM25/A356 (AlSi7Mg0.5) as very little information on their corrosion characteristics is available in the literature. The nominal composition of each alloy [50] is presented in Table 1 and was confirmed by spark spectrometric analysis, conducted at CMA Ltd in TCD in Dublin. Both alloys were supplied by Alucast Ltd formerly of Co. Wexford, Ireland.

Table 1 Nominal Composition of Cast Aluminium Alloys LM24/A380 and LM25/A356. (All figures are in weight percent and the balance is aluminium). [50]

Alloy	Si	Mg	Fe	Cu	Mn	Ni	Zn	Pb	Cr
LM 24/A380 (AlSi8Cu3Fe) EN 1706 AC-46500	8.0-11	0.05-0.55	0.6-1.2	2.0-4.0	<0.55	<0.55	<3.0	<0.35	<0.15
LM25/A356 (AlSi7Mg0.5) EN 1706 AC-42000	6.5-7.5	0.2-0.65	<0.5	<0.2	<0.35	<0.15	<0.15	<0.15	-

Potentiodynamic polarisation curves (Tafel plots) of both alloys involved the same procedure. Following grinding with progressively finer grades of SiC paper (down to a 1200 grit finish), cleaning in distilled water and an acetone rinse, samples were immersed in 0.1 M NaCl solution at room temperature and their Open Circuit Potential (OCP) monitored during the course of a 24 hour period. Duplicate samples were tested and were found to be closely matched. Once a stable OCP value was established (at 24 hours), polarisation tests were then conducted using a Solartron 1288 electrochemical interface/potentiostat, controlled by Corr View software (Scribner and Associates). A Saturated Calomel Reference Electrode (SCE) was employed, together with a Pt mesh counter electrode. In both alloys steady OCPs were obtained only after many hours exposure, most probably due to the initial occurrence of pitting, probably involving the IMs. The presence of these IMs was confirmed by both metallographic and SEM/EDS analyses. Hydrogen bubbles were observed at various stages of the test period. However the presence of a stable OCP suggests that a reasonably steady state corrosion process was achieved at longer times of exposure for both alloys (> several hours) in the test solution.

In an attempt to cause minimal surface damage during the polarisation test procedure, separate anodic and cathodic scans were conducted on two different samples of each alloy. Both polarisation scans commenced at ±10 mV from the OCP, depending upon scan direction, starting at low current densities, moving past E_{corr} towards higher current density regions. Upon completion of each separate scan, they were then concatenated (after slight potential adjustment, typically a few mV) to produce a complete polarisation curve (Tafel plot), such as those displayed in Figures 2 and 5. Naturally aerated quiescent solutions at room temperature (approximately 18°C) were used throughout. A potential scan rate of 0.5 mV/s was employed in an attempt to minimise the influence of capacitive charging. The plots are uncorrected for iR as this was very small up until high current densities.

Figure 2 (right) shows the combined anodic and cathodic scans obtained on an LM24/A380 alloy sample. The cathodic region clearly shows a concentration polarisation with current density essentially independent of applied potential. No distinct linear Tafel behaviour is evident in this cathodic section, although extrapolation from the highest overpotential region (most negative potential values) leads to an estimated i_{corr} value of approximately 3.4 μAcm^{-2}.

In contrast a linear Tafel section is evident in the upper anodic portion of the Tafel plot, very near the E_{corr} enabling estimation of a Tafel slope to be carried out; 28 mV/decade. Extrapolation of the cathodic section of the curve yields an i_{corr} value of approximately 0.33 μAcm^{-2} which indicates a relatively low corrosion rate, assuming that the corrosion process has transitioned from an initial localised corrosion process to a more general or uniform type of corrosion behaviour over an extended period of time (24 hours). Thus there appears to be a ten-fold difference between the two i_{corr} values. This illustrates the difficulty in accurately assessing the i_{corr} and/or corrosion rate from Tafel extrapolation using both b_a and b_c values.

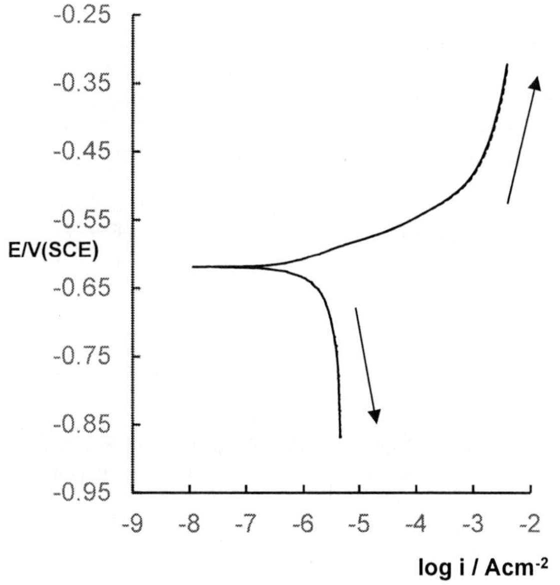

Figure 2 Tafel Plot showing combined cathodic/anodic polarisation curve of LM24/A380 in aerated quiescent 0.1 M NaCl after 24 hours exposure. Scan Rate of 0.5 mV/s. Arrows indicate the scan directions.

The overall shape of this polarisation curve is similar to many others reported for aluminium alloys recorded in neutral NaCl solutions, with a low Tafel slope in the anodic portion [47-49] and a concentration polarisation occurring at high potentials. In addition under such quiescent conditions in aerated solutions, used in this work cathodic corrosion can ensue. This is a chemical (as opposed to electrochemical) corrosion process resulting from surface alkalisation due to the onset of the ORR, producing OH^- [49]. This ionic species, confined to the Nernst diffusion layer can attack the alloy matrix.

Examination of the low potential (ie low ΔE) region enables a plot of the polarisation resistance behaviour to be constructed from both the separate anodic and cathodic portions of the Tafel plots. Figures 3 and 4 both show essentially linear polarisation resistance behaviour, with similar R_p values of approximately 14,000 Ωcm^2 in this LM24/A380 alloy.

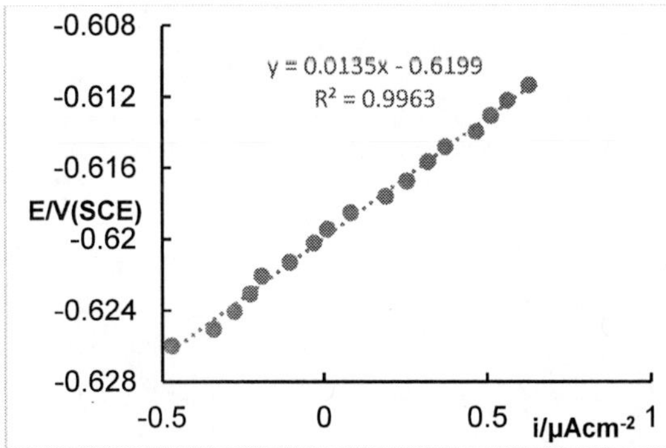

Figure 3. Polarisation resistance plot obtained from anodic section of LM24/A380 polarisation curve in 0.1M NaCl solution.

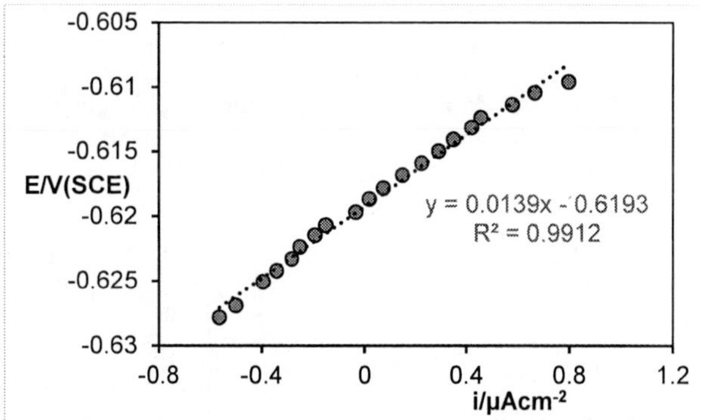

Figure 4. Polarisation resistance plot obtained from cathodic section of LM24/A380 polarisation curve in 0.1M NaCl solution.

In contrast the polarisation curve of the LM25/A356 alloy differs from its LM24/A380 counterpart. Here the cathodic region displays a linear section with a very high (in terms of magnitude) Tafel slope of approximately -440 mV/decade, indicating a contribution from another cathodic process in addition to diffusion-limited oxygen reduction. This could be hydrogen evolution (HER) in addition to oxygen reduction (ORR), but a re-deposition process may also be responsible. Extrapolation of this linear section of the cathodic Tafel line to E_{corr} of -0.688V(SCE) yields an i_{corr} value of approximately 0.17 μAcm^{-2} which is ~20 times less than that of its LM24/A380 counterpart, which contains significantly higher levels of Si, Cu and Fe. This confirms that the LM25/356 is a more corrosion-resistant alloy.

In the upper section of the Tafel plot (Figure5) the anodic polarisation curve also shows different behaviour from its LM24/A380 counterpart. Starting from the E_{corr} at about -688mV(SCE), there is a region indicating the presence of a semi-protective (passive) surface film with current density values ranging from <0.1 μAcm^{-2} up to about 0.4 μAcm^{-2}. Some small current spikes are also evident in this region, possibly due to pit initiation events (metastable pitting). At about -580 mV(SCE) the current increased rapidly, probably due to the onset of an activation controlled oxidation reaction.

Figures 6 and 7 below show polarisation resistance plots of both anodic and cathodic sections of the LM25/A356 alloy samples.

These are clearly non-linear and it is difficult to assign a R_p value the plots possibly due to the occurrence of competitive back (reverse) reactions at lower potentials and the surface film at potentials slightly above E_{corr}.

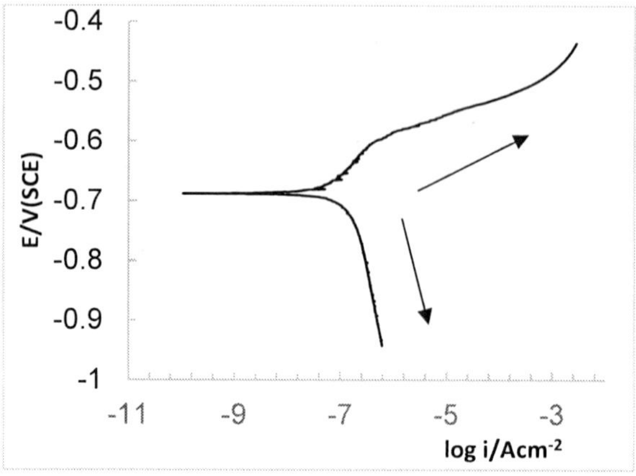

Figure 5 Tafel Plot showing polarisation curve of LM25/A356 alloy in aerated quiescent 0.1M NaCl solution following 24 hours exposure. Scan rate of 0.5 mV/s. Arrows indicate the scan directions.

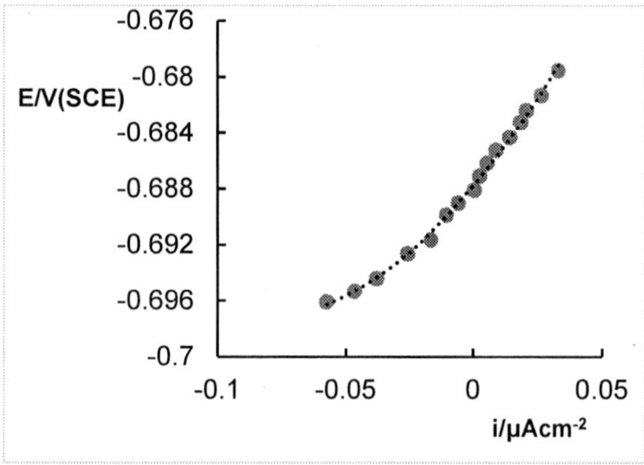

Figure 6. Polarisation resistance plot of anodic section of LM25/A356 sample in 0.1M NaCl solution.

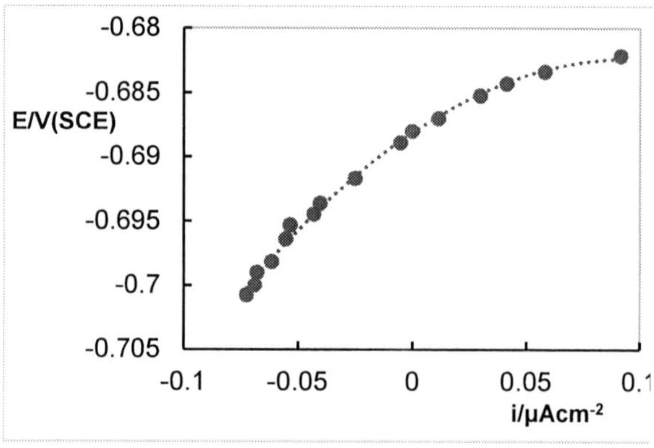

Figure 7 Polarisation resistance plot of cathodic section of LM25/A356 sample in 0.1 M NaCl solution.

Using the Allen-Hickling approach [6], plots of the logarithmic current function versus the overpotential (ΔE) were obtained producing an intercept from which i_{corr} can be estimated. Two plots emanating from the LM24/A380 sample polarisation data are presented in Figures 8 and 9. The anodic plot (Figure 9) is clearly influenced by surface processes and deviates significantly from linearity at very low ΔE values. In these plots an n/v ratio of 1 was used.

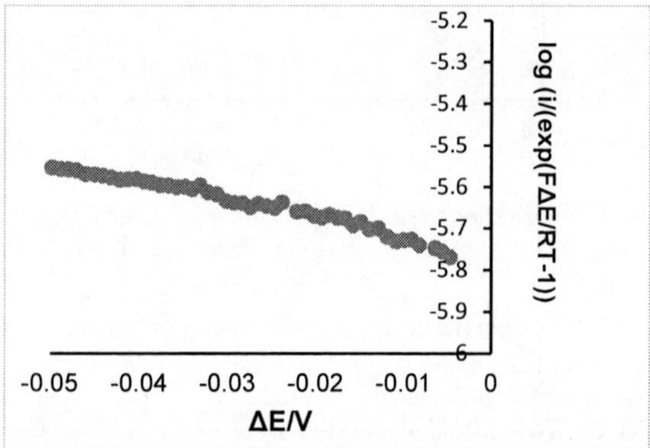

Figure 8. Allen-Hickling plot of ΔE versus logarithmic current function for cathodic section of LM24/A380 alloy polarisation curve in aerated 0.1M NaCl solution (n/v=1)

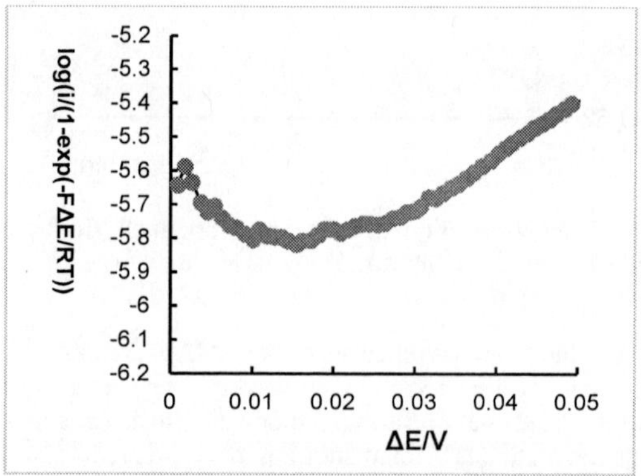

Figure 9. Allen-Hickling plot of ΔE versus logarithmic current function for anodic section of LM24/A380 alloy polarisation curve in aerated 0.1M NaCl solution (n/v=1)

Extrapolation of the linear portions of these plots to the ΔE value of 0 (E_{corr}) gives values of i_{corr} of 1.6 μAcm^{-2} for the cathodic and 0.6 μAcm^{-2} for the anodic sections respectively, corresponding to corrosion rates of about 0.002 µm/h and 0.00075 µm/h using equation 1. Although obtained at an n/v ratio of 1 and they could well converge at different n/v ratios. The anodic section displays an upward curve at ΔE of less than about 20mV, possibly due to the contribution of a cathodic back reaction as suggested by the curved polarisation resistance plots. Some have accounted for this back reaction by using a graphical approach noting that linear Tafel line data from one side (half-cell reaction) can be extended and added to the other non-linear section, and then adjusted. This method is described by Jones for steel in deaerated H_2SO_4 [3] and by McCafferty for titanium in boiling 1 M H_2SO_4 [51], but in practice is seldom applied to Tafel plot analysis.

3. DISCUSSION

Determination of the instantaneous corrosion rate of most metals through the application of slow potentiodynamic electrochemical sweeps is potentially an attractive, although tricky procedure to undertake without unduly influencing the surface reactions. For example care must be taken to ensure that the OCP is reasonably steady before commencing the polarisation sweep. In the case of the LM24/A380 and LM25/A356 alloys this took 24 hours before sufficient stability was attained. Another issue is the choice of starting potential. Slow potentiodynamic sweeps commencing at large negative potentials risk artificially changing the surface chemistry of the electrode through driving the ORR and/or HER cathodic reactions. This can lead to excessive hydroxide ion formation in the former and production of large amounts of adsorbed species (both atomic and molecular hydrogen as bubbles) in the latter case. Lack of hydrodynamic control could lead to build-up of these species. For this reason commencing sweeps at the E_{corr} was undertaken in this work.

As illustrated above it can be is difficult to determine the intersection point of the i_{corr} from Tafel extrapolation back to E_{corr} for both anodic and cathodic branches of polarisation curves. It is generally considered that the cathodic polarisation section, during which the metal surface remains largely unoxidised is perhaps more representative and consequently this is often used to estimate corrosion rates [47]. It is however not a very accurate estimate. Anodic regions are even more prone to surface modifications such as surface roughening (due to metal dissolution) and the formation of deposits and films as a result of the polarisation procedure. Such effects may interfere with the charge transfer (activation-controlled) oxidation process upon which the Tafel extrapolation procedure depends.

It is invariably good practice to measure accurate corrosion rates through the use of additional techniques to validate polarisation curve results. For example this approach was used to study the corrosion behaviour of a Zn27Al1.5Cu0.02Mg alloy immersed in 3.5wt% NaCl solution (pH 6.7) [52] The polarisation curves appear very similar to the LM24/A380 alloy plot shown in Figure 2, with an activation controlled anodic portion (b_a of 40mV/decade) and a diffusion-controlled cathodic reaction (ORR) displaying a very high b_c value (several hundreds of mV/decade). Extrapolation of the cathodic section to E_{corr} gave excellent agreement with both weight loss and polarisation resistance test values. Anodic Tafel extrapolation was however much more problematic, as selection of an appropriate Tafel slope section was difficult.

As demonstrated in Figure 1 selection of the right section of the Tafel slope section for extrapolation back to E_{corr} is critically important for accurate corrosion rate assessment. It has been suggested that overpotential values greater than 40-100mV away from the E_{eq} or E_{corr} be selected, but this led to significant over-estimation of i_{corr} values reported by Poorsqemi et al [53] in their study. Extrapolation of their curves approximately 30 mV from E_{corr} yielded much better correlation with corrosion rates measured independently by solution analysis and weight loss methods.

As an alternative approach the Allen-Hickling plot method was investigated in this work. It gave good results and shows much promise, especially in cases where back reactions can be significant, such as the anodic hydrogen evolution reaction (or negative difference effect) [54]. In addition to giving values of the appropriate transfer coefficient, α_a or α_c it provides an alternative method for the determination of the corrosion rate (through evaluation of i_{corr}). It has several advantages over the traditional approach utilising lower overpotentials approximately from 20-60mV thereby making it less susceptible to concentration polarisation effects which can be significant at high ΔE values such as those utilised in normal Tafel extrapolation studies. In addition use of lower potential ranges means that there is likely to be less iR drop.

Another useful method to ascertain accurate corrosion rates is mathematical modelling utilising the fundamental BV or WT [55] equations to describe the electrochemical reactions. Care has to be taken to account for iR contributions, concentration polarisation and surface film (resistance) effects in order to simulate polarisation curves. In a series of papers Flitt and Schweinsberg carried out a comprehensive synthesis of complete polarisation curves of ferrous alloys [56, 57] and zinc [58]. Using a proprietary software SYMADEC, they successfully deconvoluted a number of component anodic and cathodic reactions through variation of reversible potentials (E_{eq}), Tafel slopes, limiting current densities and i_0 values in order to match experimental curves. A similar approach for an electrochemical reaction was reported using the Solver Excel program [59].

Curved Tafel plots are most often observed in practice, with often only a very limited linear range, far less than the suggested 1-2 decades. This is especially true when high resistance passive films are present (in the anodic section of for example stainless steel alloys) and instances where the concentration polarisation effect dominates, particularly in the cathodic region, as a result of mass transport limitations. The latter can be accounted for by using the limiting current (i_l) correction [60]

$$i = i_0(1-\frac{i}{i_l})^m \exp(-2.303\left(\frac{\Delta E}{b_c}\right) \quad (19)$$

where m is the order of the ORR on a sold surface ($0.5 \leq m \leq 1$) and i_l is the limiting current density. A similar expression was used in the mathematical simulations of Flitt and Schweinsberg. However Nagy and Thomas [61] pointed out that for corrosion processes a better description is given by a modified WT equation containing an additional term, λ which modifies the cathodic process component.

$$i = i_{corr}(\exp(2.303\Delta E/b_a) - \lambda \exp(-2.303\Delta E/b_c)) \quad (20)$$

$$\lambda = 1/(1-\frac{i_{corr}}{i_l}+\frac{i_{corr}}{i_l}\left(\frac{-2.303\Delta E}{b_c}\right)) \quad (21)$$

In the case of diffusion (mass transport) control $i_l = i_{corr}$, for mixed control $i_l > i_{corr}$ and for charge transfer control $i_l = \infty$, $b_a = 2.303RT/\alpha_a F$ and $b_c = -2.303FT/\alpha_c F$. The Allen-Hickling approach is less susceptible to this concentration polarisation effect and certainly warrants further study. Another equation describing the total cathodic current density accounting for both charge transfer and concentration polarisation was developed by Flitt and Schweinsberg for a redox process [55, 56].

$$i = (i_0 \exp(-\alpha F\eta/RT)/[1+(i_0 \exp(-\alpha F\eta/RT)/i_l] \quad (22)$$

There is a fundamental difference however between electrochemical redox reactions where the anodic and cathodic reactions are the reverse of each other (the Principle of Microscopic Reversibility) and the mixed potentials of the corrosion case [61]. In the latter, the two partial reactions are not identical. Whereas the net external current is zero at equilibrium and the material fluxes are also zero to/from the surface in an electrochemical redox reaction. Mixed potential theory means this does not hold for corrosion. Instead the net external current is not zero at E_{corr}, as there is a net overall reaction taking place at the surface i_{corr}. Hence the polarisation equation is necessarily different from the normal mass transport corrected electrochemical kinetic equation most often applied (such as equation 19).

As noted earlier, the choice of potentiodynamic scan rate is important when conducting polarisation experiments. It is actually preferable to use potentiostatic methods to overcome capacitive charge contribution to current density, but they are time-consuming experiments. Hence potentiodynamic polarisation scans are far more often used. To overcome any uncertainty over choice of scan rate Revie has tabulated [62], maximum permissible scan rates (in mV/s) for assessment of instantaneous corrosion rates of metals. In addition Zhang et al [63] proposed a correction routine in an attempt to remove the capacitive current contribution from use of higher scan rates, leaving only the faradaic (charge transfer) current component for analysis.

Increased surface roughness and adsorption of corrosion reactants, intermediates and corrosion products can greatly affect Tafel slopes. For this reason various isotherms (eg. Langmuir, Temkin, Frumkin) have been used [36, 64]. Adsorbates can lead to increased pseudocapacitance, which can contribute to the current density on corroding metal electrodes. Burstein and Wright [16] carried out a study of the oxidation of adsorbed hydrogen on nickel and adopted a modified Allen-Hickling plotting procedure involving use of a Langmuir isotherm to account for the presence of surface H_{ad} species. The best fit was obtained using an effective n/v ratio of 1, thereby identifying the reaction's rate determining step. This nicely illustrates another benefit of the Allen-Hickling approach, the ability to ascertain mechanistic information.

4. CONCLUSIONS

The aim of this paper was to critically review use of the Tafel extrapolation, which is widely used in corrosion investigations and to explore a little-used alternative approach, the Allen-Hickling plot method. A modified expression was developed for both anodic and cathodic processes occurring during corrosion processes such as those on aluminium alloys LM24/A380 and LM25/A356 in 0.1 M NaCl solution which were not readily amenable to Tafel extrapolation or Stern-Geary polarisation resistance analysis. Various effects such as the existence of iR drop, capacitive charging leading to non-faradaic currents, concentration polarisation, surface roughening, the existence of adsorbed species and surface film formation should be taken into account when measuring potentiodynamic polarisation curves (Tafel plots). Consequently when performing such experiments potentiostat iR compensation, use of a relatively conducting (supporting) electrolyte and imposition of good hydrodynamic conditions (in order to prevent concentration polarisation) should be taken into account during the design and performance stages.

Although often used in literature dealing with electrode kinetics, the transfer coefficient product term αn, often presented in literature Butler-Volmer equations is incorrect. It is now recommended that single transfer coefficient terms (α_a and α_c) be used in conjunction with an n/v ratio (assuming a quasi-equilibrium electrode kinetics approach) to account for rate determining steps in electrochemical and corrosion mechanisms.

Linear semi-logarithmic Tafel plots are only strictly appropriate in limited cases; for example non-passive ferrous metals in deaerated acid solutions under well-controlled hydrodynamic conditions (free from concentration polarisation effects). For a variety of reasons most metals and alloys do not display such linear behaviour. It is recommended that a variety of complementary approaches, including use of Allen-Hickling plots can help validate corrosion rates of metals and alloys along with other methods to overcome this issue.

5. ACKNOWLEDGEMENTS

The authors would like to acknowledge discussions with participants in COST Action MP1407 and the support of Prof. M Devereux and Prof. S Jerrams.

6. REFERENCES

1. Nagy Z, DC Electrochemical Techniques for the Measurement of Corrosion Rates in Modern Aspects of Electrochemistry , J. O'M. Bockris, B. E. Conway B E, White R E (eds) 1993 (MAOE, volume 25) 135-190
2. Kelly R G, Scully J R, Shoesmith D W and Buchheit R G, Electrochemical Techniques in Corrosion Science and Engineering, CRC Press, 2003
3. Jones D A, Principles and Prevention of Corrosion, 2nd ed, Prentice Hall, London 1996
4. E J Calvo, Fundamentals; The basics of electrode reactions, in Comprehensive Chemical Kinetics Vol 26 Electrode Kinetics: Principles and Methodology, C H Bamford and R G Compton (eds) Elesevier 1986
5. Bard A J and Faulkner L R, Electrochemical Methods: Principles and Applications Wiley and Sons, New York 2001
6. Allen P L and Hickling A, Electrochemistry of Sulphur. Part 1. Overpotential in the discharge of the sulphide ion Transactions of the Faraday Soc., 53 (1957) 1626-163
7. Xu J and Farrington G C, A Microelectrode Study of Lithium Electrokinetics in Poly(ethylene glycol dimethyl ether) and 1,2-Dimethoxyethan, Journal of the Electrochemical Society 142 (10) (1995) 3303-3309
8. Chiku M. Tsujiwaki W, Higuchi E. and Inoue H. Determination of the rate-determining step in the electrochemical oxidation of Li metal at the Li negative electrode/Li_2S–P_2S_5 solid electrolyte interface, Journal of Power Sources, 244, (2013), 675-678
9. Scarr R F, Kinetics of the Solid Lithium Electrode in Propylene Carbonate, Journal of the Electrochemical Society 117(3) (1970) 295-299
10. Hedges W M, Pletcher D and Gosden C, Microelectrode Studies of the Li/Li^+ Couple in $SOCl_2/LiAlCl_4$ Journal of the Electrochemical Society 134 (6) (1987) 1334-1340
11. Pujare N U, Tsai K J and Sammells A F An Electrochemical Claus Process for Sulfur Recovery, Journal of the Electrochemical Society 136 (12), (1989), 3662-3678
12. Baltrūnas G. Valiūnienė A Margarian Z..Viselgienė G. Popkirov G, The electroreduction kinetics of silver sulfite complexes Electrochimica Acta 53(22), (2008), 6513-6520
13. Lu S, Investigation of Lithium Ferrite as an Alternate Anode Material in Molten Carbonate Fuel Cells, Journal of the Electrochemical Soc. 142 (11), (1995), 3812-3815
14. Nishina T, Takahashi M and Uchida I, Gas Electrode Reactions in Molten Carbonate Media IV . Electrode Kinetics and Mechanism of Hydrogen Oxidation in $(Li + K)CO_3$ Eutectic, Journal of the Electrochemical Society, 137(4), (1990) 1112-1121
15. Prosini P P, Pozio A. Botti S. Ciardi R Electrochemical studies of hydrogen evolution, storage and oxidation on carbon nanotube electrodes Journal of Power Sources 118 (1-2) (2003), 265-269
16. Burstein G T and Wright G A Oxidation of atomic hydrogen adsorbed on nickel, Journal of Electroanalytical Chemistry and Interfacial Electrochemistry, 50(3), (1974), 399-401.
17. Co A C and Birss V, Mechanistic Analysis of the Oxygen Reduction Reaction at (La,Sr)MnO3 Cathodes in Solid Oxide Fuel Cells, Journal of Physical Chemistry B 110, (2006) 11299-11309
18. Soderberg J N, Co A C,. Sirk A H C and Birss V I, Impact of Porous Electrode Properties on the Electrochemical Transfer Coefficient, Journal of Physical Chemistry B, 110 (21), (2006) 10401–10410
19. Taylor R J and Humffray A A, Electrochemical Studies on Glassy Carbon Electrodes II, Oxygen Reduction in Solutions of High pH (>10), Journal of Electroanalytical Chemistry, 64 (1975) 63-84
20. Schulze K and Hoff H Electrode Kinetics of Aluminium in Chloride Melts with Respect to Electrocrystallization, Electrochmica Acta, 17 (1972) 119-131
21. Eckert J and Galova M, Electrode Kinetics of the Aluminium Deposition from Tetrahydrofuran Electrolytes, Electrochimica Acta, 26 (8), (1981), 1169-1175
22. Bishop E, Hitchcock P H, Mass and Charge Transfer Kinetics and Coulometric Current Efficiencies, The Analyst, 98, (1973), 465-474
23. Orata D and Okong'o E, Allen–Hickling equation applied to the quasi-reversible polyaniline redox system Reactive and Functional Polymers, 45 (3) (2000), 211-216

24. B I Bindra P and Roldan J Mechanisms of Electroless Metal Plating II Formaldehyde Oxidation, Journal of the Electrochemical Society, 132 (11), (1985) 2581-2589
25. Airey P L and Sutton H C Electrochemistry of the HO_2 and O^-_2 radicals under steady state conditions. Part 1. The electrochemistry of O^-_2 at neutral pH Journal of the chemical Society, Faraday Trans. 1, 72, (1976) 2441-2451
26. Kuhn A T and Randle T H, Kinetics and mechanism of the cerium(IV)/cerium(III) redox reaction on a platinum electrode Journal of the Chemical Society, Faraday Trans. 1, 79 (1983) 1741-1756
27. Kuhn A T and Randle T H, Effect of oxide thickness on the rates of some redox reactions on a platinum electrode Journal of the Chemical Society, Faraday Trans. 1, 81 (1985), 403-419
28. Randle T H and Kuhn A T, The influence of platinum (phase) oxide on the electrode kinetics of the manganese(III)/manganese(II) and cerium(IV)/cerium(III) redox couples in sulphuric acid, Electrochimica Acta, 31(7), (1986), 739-744.
29. Ghosh S, Vandarkuzhali S. Gogoi N. Venkatesh P, .Seenivasan G. Reddy B P Nagarajan K Anodic dissolution of U, Zr and U–Zr alloy and convolution voltammetry of $Zr^{4+}|Zr^{2+}$ couple in molten LiCl–KCl eutectic Electrochimica Acta 56 (2011) 8204-8218
30. Kublanovsky V, Bersirova O, Dikusar A, Bobanova Z, Cesiulis H, Sinkeviciute J and Prosyceva I, Electrodeposition and Corrosion Properties of Nanocrystalline Fe-W alloys, Problems in Corrosion and Protection of Materials, 7, (2008), 308-314
31. Vernickaite E, Antar Z, Nicolenco A, Kreivaitis R, Tsyntsaru N, Cesiulis H Tribological and Corrosion Properties of Iron-Based Alloys 8th International Conference BALTTRIB'2015 26–27 November 2015 , Lithuania
32. Cesiulis H and Sinkeviciute, Problems in Corrosion and Protection of Materials, 4, (2004), 128-134
33. Burstein G T, A Century of Tafel's Equation: 1905–2005, Corrosion Science 47 (12) (2005) 2858–2870
34. Wagner C and Traud W (and Mansfeld F), On the interpretation of Corrosion Processes Through the Superposition of Electrochemical Partial Processes and on the Potential of Mixed Electrodes, Corrosion 62(10) (2006) 843
35. Stern M and Geary A L Electrochemical polarization I. A theoretical analysis of the shape of polarization curves Journal of the electrochemical society, 104 (1) (1957) 56-63 and Scully. J R, Polarisation Resistance Method for Determination of Instantaneous Corrosion Rates, Corrosion, 56(2), (2000) 199-218
36. Bockris J O'M and Reddy A K N, Modern Electrochemistry 2B Electrodics in Chemistry Engineering, Biology and Environmental Science, Springer 2001
37. Frankel G S, Wagner-Traud to Stern-Geary Development of Corrosion Kinetics, in Corrosion Science a Restrospective, Electrochemical Society Proceedings 13 (2002), 33-41
38. Mann R F, Amphlett J C, Peppley B A and Thurgood C P, Application of Butler-Volmer Equations in the Modelling of Activation Polarization for PEM Fuel Cells, Journal of Power Sources, 161 (2006) 775-781
39. Guidelli R, Compton R G, Feliu J M, Gileadi E, Lipkowski J, Schmickler W and Trasatti S, Defining the Transfer Coefficient in Electrochemistry: An Assessment (IUPAC Technical Report), Pure and Applied Chemistry, 86(2), (2014), 245-258 and 259-262
40. Stern, M Electrochemical Polarization iii Further Aspects of the Shape of Polarization Curves Journal of the Electrochemical Society 105(11) (1957), 645-650
41. Stern M A Method For Determining Corrosion Rates From Linear Polarization Data, Corrosion 14(9)(1958)60-64.
42. Bockris J O'M and Nagy Z, Symmetry Factor and Transfer Coefficient. A Source of Confusion in Electrode Kinetics, Journal of Chemical Education 50(12) (1973) 839-842
43. Fletcher S, Tafel Slopes From First Principles, Journal of Solid State Electrochemistry, 13, (2009) 537-549
44. Thompson N G and Payer J H, DC Electrochemical Test Methods, NACE International 1998
45. Kirkland N T, Birbilis N and Staiger M P Assessing the Corrosion of Biodegradable Magnesium Implants; A Critical Review of Current Methodologies and Their Limitations, Acta Biomaterialia, 8 (3), (2012) 925-936
46. Fletcher S, The Theory of Electron Transfer, Journal of Solid State Electrochemistry 14 (2010), 705-739
47. Birbilis N and Buchheit R G, Electrochemical Characteristics of Intermetallic Phases in Aluminium Alloys: An Experimental Survey and Discussion, Journal of The Electrochemical Society, 152 (2005) B140-B15
48. Birbilis N and Buchheit R G, Investigation and Discussion of Characteristics for Intermetallic Phases Common to Aluminium Alloys as a Function of Solution pH, Journal of the Electrochemical Society, 155 (3) (2008) C117-126
49. Sukiman NL, Zhou X, Birbilis N, Hughes A E, Mol, J M C Garcia, S J, Zhou X, Thompson, G E, Durability and Corrosion of Aluminium and its Alloys: Overview, Property Space, Techniques and Developments, INTECH 2 2012
50. http://www.kayepresteigne.co.uk/ Accessed 30th May 2016
51. McCafferty E, Validation of Corrosion Rates Measured by the Tafel Extrapolation Method, Corrosion Science, 47 (2005) 3202-3215
52. Bobic B, Bajat J, Acimovic-Pavlovic Z, Rakin M and Bobic I, The Effect of T4 Heat Treatment on the Microsctucture and Corrosion Behaviour of Zn27Al1.5Cu0.02Mg, Corrosion Science 53(1) (2011) 409-417
53. Poorqasemi E, Abootalebi O, PeikariM, and Haqdar F, Investigating the Accuracy of the Tafel Extrapolation Method in HCl Solutions, Corrosion Science, 51 (5), (2009) 1043-1054
54. Frankel G S, Fajardo S and Lynch B M, Introductory lecture on corrosion chemistry: a focus on anodic hydrogen evolution on Al and Mg, RSC Faraday Discussions., 180 (2015), 11-33
55. Frankel G S, Electrochemical Techniques in Corrosion: Status, Limitations and Needs, Journal of ASTM International 5(2) (2007), 1-27

56. Flitt H J and Schweinsberg, D P, A guide to polarisation curve interpretation: deconstruction of experimental curves typical of the $Fe/H_2O/H^+/O_2$ corrosion system, Corrosion Science, 47 (2005), 2125-2156
57. Flitt H J and Schweinsberg D P, Evaluation of corrosion rate from polarisation curves not exhibiting a Tafel region Corrosion Science, 47 (2005), 3034-3052
58. Flitt H J and Schweinsberg D P, Synthesis, matching and deconstruction of polarization curves for the active corrosion of zinc in aerated near-neutral NaCl solutions, Corrosion Science, 52, (2010), 1905–1914
59. Howard E and Cassidy J F, Use of Excel to Obtain Voltammetric Information on Microelectrodes, Journal of Chemical Education, 77, (2000), 409-411
60. Cáceres L, Vargas T and Herrara L, Determination of electrochemical parameters and corrosion rate for carbon steel in un-bufffered sodium chloride solutions using a superposition model, Corrosion Science, 49 (2007), 3168-3184
61. Nagy Z and Thomas D A, Effect of Mass Transport on the Determination of Corrosion Rates from Polarisation Measurements Journal of the Electrochemical Society 133(10), (1986) 2013-2017
62. Revie W R , Uhlig's Corrosion Handbook, Wiley New York 2011
63. Zhang W L, Jiang Zh H, Yao Zh P, Song Y amd Wu Zh D, Effects of Scan Rate on the Potentiodynamic Polarization Curve Obtained to Determine the Tafel Slopes and Corrosion Current Density, Corrosion Science, 51 (2009) 581-587
64. Aljourani J, .Raeissi K and Golozar A, Benzimidazole and its derivatives as corrosion inhibitors for mild steel in 1M HCl solution Corrosion Science, 51 (8) (2009) 1836-1843

7. AUTHOR DETAILS

Dr Tony Betts is Head of Research in the College of Sciences and Health in the Dublin Institute of Technology, Dublin Ireland. He cofounded the Institute's Applied Electrochemistry Group with Professor John Cassidy and has been in DIT since 2004. Especially interested in materials-oriented electrochemistry, he was previously a corrosion scientist in Forbairt, now Enterprise Ireland and before his move to Ireland undertook the same role in MPT of Industrial Research Ltd (formerly DSIR). Formerly the New Zealand President of the ACA he was the PF Thompson Lecturer in the ACA Conference held in Auckland in 2000.

Professor Cassidy was awarded a diploma in Applied Science by Dublin Institute of Technology, his BSc (Applied Sciences) by the University of Dublin, and completed his PhD at the University of Utah, USA. He has since lectured in Analytical Chemistry in DIT, Kevin Street. He was appointed Assistant Head of School in 2001 and awarded Professorship of DIT in 2009. His research interests include Analytical Chemistry and Instrumentation. This involves the theory and operation associated with Modern Analytical Instruments in the area of spectroscopy, electrochemistry and chromatography.

Dr. David Cullion is an Early Stage Researcher and Project Manager in Mechanical Engineering. His main areas of research include surface degradation phenomena and treatments, adapted electrochemical technologies, light metal alloys, renewable energies and mechanistic water-transfer systems. In his Ph.D he developed a novel conformal anti-tribocorrosion surface technology. The technology is the subject of a number of industrial projects in the wastewater and electronics industries and a patent has been filed. Dr. Culliton combines this with extensive experience in Project Management covering the IT, Construction and Research environs. He is currently involved in the development of pan-European and Global research projects.

NATURAL FRESH AND SALT WATER IMMERSION CORROSION TRIALS OF COMPOSITE ALUMINIUM SYNTACTIC FOAMS

I. Chaves[1], T.Fiedler[2]
[1]Centre for Infrastructure Performance and Reliability, The University of Newcastle, Australia,
[2]School of Engineering, The University of Newcastle, Australia

SUMMARY: Lightweight composite materials such as syntactic perlite aluminium foams have been gaining much attention from the automotive, aerospace and nautical industries due to their superior mechanical and energy absorbing properties. Yet, there is still no empirical evidence of its mechanical performance under extreme corrosive marine environments. This study reports on 18 months corrosion field trials of such composite foam immersed in natural temperate marine waters off the southeast
coast of Australia. Further, natural fresh water corrosion trials were also performed on the foams in order to quantify potential long-term phenomenological factors. Unloading Young's modulus, 1% offset yield stress as well as plateau stress after exposure fall within the scatter of previously reported unexposed samples. The results indicate no short-term loss of mechanical performance for natural exposure conditions. Reasons and implications are discussed.

Keywords: Aluminium, Syntactic Foam, Corrosion, Compression, Strength.

1. INTRODUCTION

In design, structural capacity and serviceability are governed by material mechanical and geometrical properties. Corrosion poses a risk and jeopardises structural integrity and capacity by diminishing the available cross-section. Further, it may cause localised perforation through pitting or crevice corrosion and therefore affect containment or buoyancy capacity. Despite their excellent general corrosion resistance [1] passive metals such as aluminium alloys are, in particular, quite susceptible to such localised attacks [2].

Composite materials, such as the perlite-aluminium syntactic foams, have been studied extensively because of their multi- functional properties and economic advantages [3]. A typical perlite aluminium foam consists of a metallic matrix containing porous perlite particles. It has very low-density (0.7-1.05 g/cm3). This is significant when aluminium and its alloys are the metals of choice for applications in the aerospace, defence, and marine industries and increasingly are being used in structural and other commercial infrastructure. With a total porosity of 61%, which is much higher than that of
other syntactic foam materials [3], it is important to empirically quantify and characterise the mechanical properties of such composite material under corrosive conditions, especially since positive correlations between higher levels of porosity and higher corrosion rates have been postulated.

For marine immersion environments this becomes even more of an issue as complex bi-modal corrosion mechanisms take place over increased exposure periods [1,4].

In order to initiate studies of such material subject to aggressive corrosion environment this paper focuses on the short-term aerobic corrosion characterisation of the composite perlite-AA601.3 syntactic foam immersed in natural marine waters. Since the main mechanism for aluminium corrosion involves pitting and crevice corrosion under localised areas that are in an anoxic condition this study is important and has implications for the long-term phenomenological modelling. These are discussed in further sections.

2. MATERIAL AND METHODS

2.1 Composite foam mechanical, chemical and metallurgical characterisation

Porous expanded perlite (EP) particles, a silicate base volcanic glass [3], were used to produce relatively low-cost and low-density aluminium syntactic foam. EP is obtained by heating raw perlite rock above 870°C which then expands to 15-20 times of its original volume generating voids within the volcanic glass. These porous particles are then combined with melt aluminium in a simple casting process to create the syntactic foam. Chemical characterisation of both the aluminium alloy as well as the perlite is presented in Table 1 according to manufacturer product specification. EP particles were obtained from Australia Perlite Pty [5] ranging from 3-4mm in diameter. The aluminium alloy was sourced from Rio Tinto Aluminium and complied with the chemical requirements of the Alloy AA601.3.

Table 1. Chemical characterisation of aluminium alloy and expanded perlite syntactic foam

AA601.3 Aluminium (%wt)											
Si	Fe	Cu	Mn	Mg	Sr	Zn	Ti	Ni	V	Ga	Ca
7.3	0.10	< 0.01	< 0.01	0.39	0.003	< 0.01	0.12	0.01	0.01	0.01	Trace

Expanded Perlite Chemical Analysis (%wt) [5]			
Silica	74.0	Potassium Oxide	4.0
Aluminium Oxide	14.0	Titanium Oxide	0.1
Ferric Oxide	1.0	Heavy metals	Trace
Calcium Oxide	1.3	Sulphate	Trace
Magnesium Oxide	0.3	Moisture	0.3
Sodium Oxide	3.0	Organic Content	< 0.1
Expanded Perlite Physical Properties [5]			
Thermal Conductivity	0.0361 @ 0°C	Fusion Point	1260-1340°C
Specific Gravity	0.055-0.30 g/ml	Softening Point	871-1093°C
Colour	Off white		

A more detailed description of the manufacturing of the syntactic foam is readily available [3]. However, in summary EP particles were furnace dried for 30min at 250°C, encased within a graphite mould and pushed downwards into a 25mm diameter graphite crucible containing molten AA601.3 alloy at 720°C all within a counter-gravity pressure infiltration set-up. The entire assembly was then left to cool to room temperature at which point the samples could be manually removed from the mould.

Due to the likely difficulty in controlling sample density, longer 150mm samples were prepared and then cut into smaller samples. Because of slight differences in sample density, the samples were organised so that at each recovery a sample of low density, medium and higher density were retrieved. Prior to exposure sample length, diameter and mass were recorded. For each dimension an average of 5 measurements was used and an average of 3 readings was taken for the initial mass.

2.2 Exposure site

The test site is located in the temperate waters at Taylor's Beach, on the southeast coast of Australia. Securely enclosed within the premises of the NSW Port Stephens Fisheries Research Institute the site has temperate annual water temperatures (14~26°C) with low average annual pollution levels (<0.06 mgN/L of DIN) and low average annual water velocities (<0.1 m/s) [4]. The Taylor's Beach test site has been utilised as a corrosion exposure testing site previously [6]. It is well documented and allows results of the present project to be compared with previous reported results [7].

The experiment was performed in accordance with the standard practice for preparing, cleaning and evaluating corrosion test specimens G1-03 [8]. For each foam density, and for each exposure period, triplicate samples were used in accordance to established corrosion testing procedures [9]. Hence, a total of 3 samples were retrieved per recovery (Fig. 1). UV resistant identification cable ties were used (Fig. 1). Samples were recovered after one week, 1 month and after 6 months of exposure.

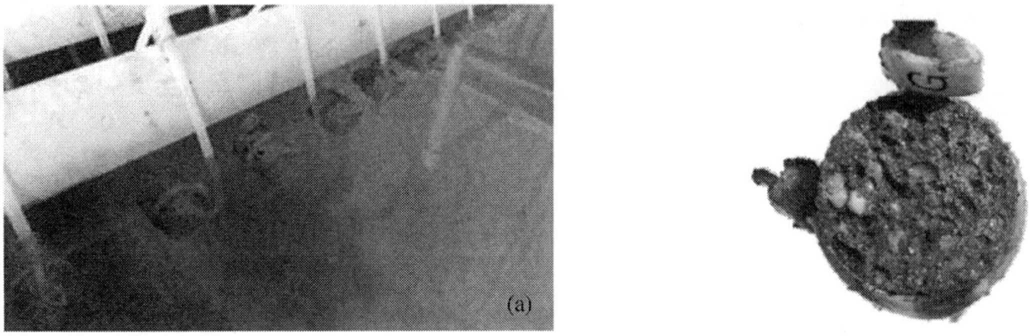

(b)

Figure 1. Samples exposed at temperate waters test site (a) and upon recovery (b)

2.3 Sample cleaning and data obtention

Upon retrieval samples were carefully hosed-down on site to remove mud and marine debris. Following transportation back to the University of Newcastle Civil Engineering Laboratories, the samples were subjected to standardised mechanical and ultrasonic cleaning [8]. A photographic record was made throughout, including the state of each sample prior to mechanical cleaning, post-mechanical cleaning, and post-ultrasonic cleaning (Fig. 2). Ultrasonic cleaning was employed to avoid any potential additional loss of base metal [10]. Further, it facilitates analysis of corrosion product phases, and observations of any peculiar localised corrosion [11].

Mechanical cleaning was performed by scrubbing with a rigid nylon brush under running tap water. Samples were then submerged for 20 minutes in ultrasonic deionised water (approx. 500ml) bath [8]. Following mechanical and ultrasonic cleaning, samples were oven dried for 24 hours. They were then weighed, measured and stored in a desiccator. The results are presented in the next section.

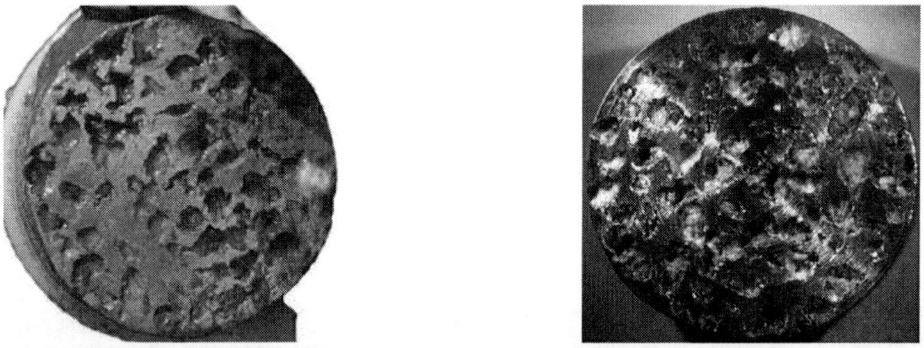

(a) (b)

Figure 2. Samples prior to mechanical cleaning (a) and after mechanical and ultrasonic cleaning (b)

Samples were tested following ISO13314 [12] for the mechanical characterisation of porous metals. All compressions tests were conducted on a uni-axial Shimadzu testing machine with a 50kN load cell and constant crosshead displacement speed of 1mm/min. In order to determine the onset of the required unloading cycle a single sample of average density was compressed and its plateau stress determined to be 60 MPa. All subsequent samples have thus been unloaded at 70% of this value (42 MPa) before reloading after reaching 20% of the plateau stress (12 MPa). Samples were then compressed until the load limit (50 kN) of the testing machine was reached. The unloading modulus is the gradient of the secant line of the hysteresis loop created by the unloading cycle. Furthermore, the 1% offset yield stress and plateau stress (average stress between 20-40% macroscopic compression) have been evaluated.

3. RESULTS

The results for the mechanical properties of the tested samples are presented in Figures 3 to 5. Is it well established that density has a strong impact on the mechanical properties of cellular metals [5,13,14,15,16]. Accordingly, the results are plotted versus density to allow a better comparison of the samples (Fig. 3-5).

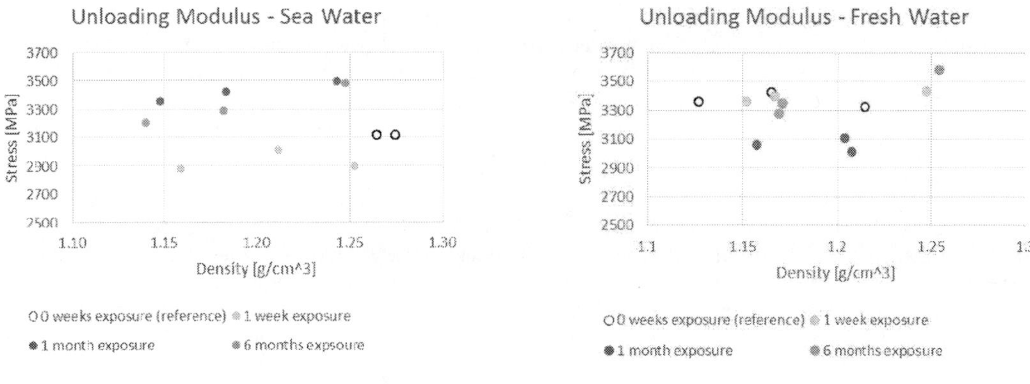

(a) (b)

Figure 3. Young's Modulus (unloading-modulus) versus density for each exposure period of the sea water trials (a) and fresh water trials (b).

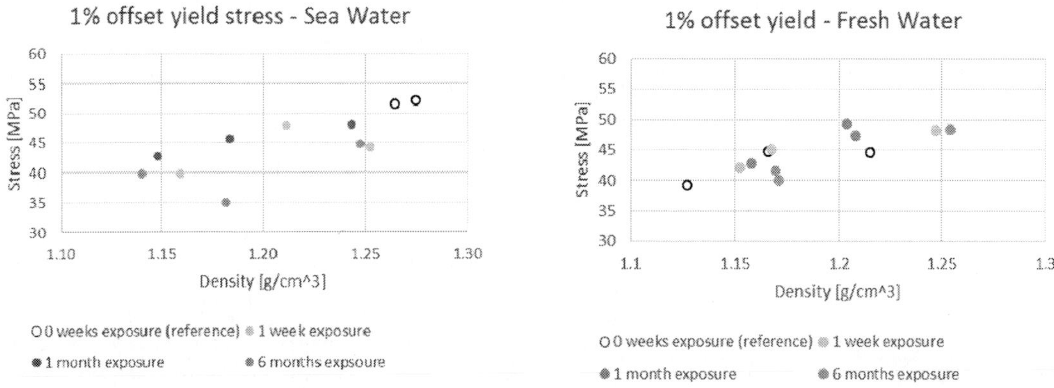

(a) (b)

Figure 4. 1% offset yield stress versus density for each exposure period of the sea water trials (a) and fresh water trials (b).

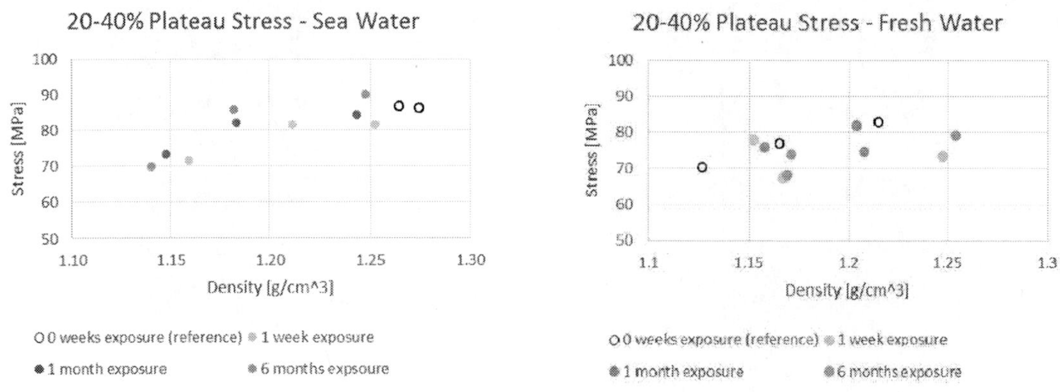

(a) (b)

Figure 5. Plateau stress versus density for each exposure period of sea water trials (a) and fresh water trials (b).

4. DISCUSSION

The investigated syntactic foam consisted of a metallic matrix containing a large number of small expanded perlite particles. The shape, position and size of these particles vary randomly within certain boundaries. Particles mostly resemble ellipsoids [3] and exhibit minimum half axis lengths between 3 and 4 mm (due to sieving of the particles) [3]. However, the surface of these particles is often locally non-convex and exhibits a large number of small indentations. The position of the particles is a result of their shape and filling sequence into the mould prior to casting. The geometry (meso- structure) of the load-bearing aluminium is now determined by the inverse space of the perlite particles, i.e. the volume which is not occupied by expanded perlite. As a result, even samples with a similar density (i.e. volume fraction aluminium) exhibit different meso-structures and thus dissimilar mechanical properties. This scattering of properties is well documented in the literature [5,17,18,19]. The challenge of this study is the isolation of the impact of corrosion from this intrinsic variation.

Figure 3 shows the dependence of the elastic unloading modulus on density and exposure time. A distinct scatter of the data points can be observed. Minimum values are observed for samples of the 1 week exposure group; however, this deviation of elastic properties most likely can be attributed to random variation of the material meso-structure. The data could further be interpreted to show an increase of the unloading modulus with exposure time. Theoretically, sedimentation of suspended solids and or salt particles within the perlite could occur and thus increase the stiffness of the material following extended marine exposure. Observations for the long-time exposure samples currently in progress may confirm this trend.

The 1% offset yield stress (Fig. 4) shows a systematic dependence on the initial density. It can be observed that the material strength increases steadily with the initial density. The explanation is a larger volume fraction of aluminium (conversely a smaller volume fraction of expanded perlite particles) of these samples. Unlike the weak perlite particles, the aluminium is able to withstand a significant external load, thus increasing the overall strength of a sample. The same trend is visible for the plateau stress (Fig. 5). In the case of the plateau stress, a lower scatter of data points was observed. This can be explained by the averaging of stresses over large deformations (20-40%) for which initial geometrical differences of the samples have limited significance. Examination of the exposure time for the 1% offset yield against the plateau stresses reveals no systematic correlation. Thus, the data suggests that short term exposures up to 6 months do not change the mechanical strength of perlite-metal syntactic foam.

4.1 Mass-gain effect

Following the retrieval, cleaning and drying of each of the seawater exposed samples a mass gain was recorded for samples exposed in sea waters; i.e. the final dried samples weighed more than pre-experimental mass. Theoretically, the mass gain could be attributed to sedimentation of dissolved salts and or other suspended solids within the pores of the expanded
perlite pores [3]. As the natural sea water flows through the suspended EP-aluminium foam samples in the experimental apparatus, salt dissolved in the water also passes through the pores. When the samples are retrieved, any water trapped within the samples is removed by the oven drying process. While the water evaporates in this process, it is likely that some of the dissolved salt remains in the perlite pores and contribute to the final mass of the sample. Attempts to manually remove the crystallised salt sediment during the mechanical cleaning process were unsuccessful, presumably due to the intricate geometrical structure of the foam samples (Fig 6).

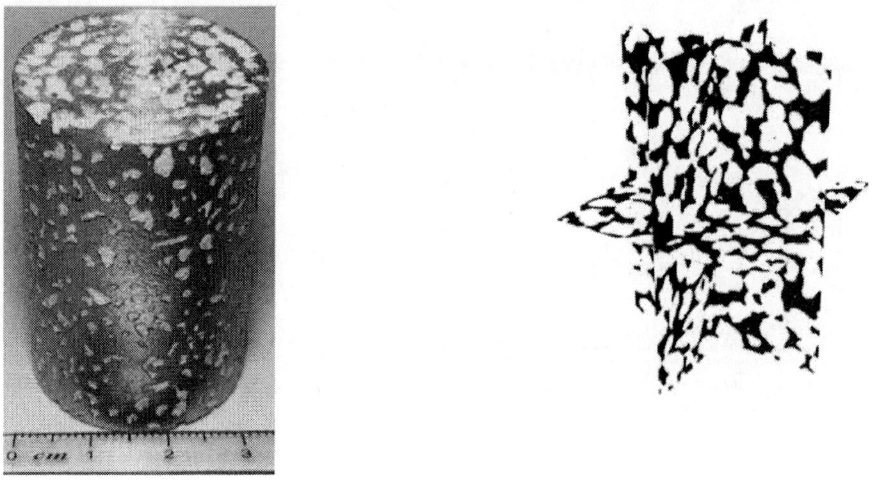

(a) (b)

Figure 6. (a) Aluminium syntactic foam, (b) orthogonal slices of micro-tomography data [5].

The mass gain in the seawater samples is considered likely to also contribute to the scatter in the total mass results. Specifically, the distinct increase in unloading modulus with exposure period could be attributed to the deposited salt increasing the stiffness and hence the elastic modulus of the material. Furthermore, due to the intricate geometrical structure of EP-aluminium foam samples mass loss is one of the few methods available to quantify the impact of corrosion of the material. It is therefore desirable to find a way to normalise the change in sample mass back to a mass loss by removing the additional mass (i.e. the mass gain). This is desired to enable an assessment of the corrosion behaviour of EP-aluminium foam. To investigate this further, two hypothesises were considered; the first was the theory that the salt mass comes from instantaneous dissolved salts only, and the second that trapped salt builds up and increases with exposure duration.

The mass of instantaneous dissolved salts in the sample was estimated using the following method. The porosity of the sample was estimated using a ratio of the measured density of the cleaned and dried sample and the known density of the aluminium matrix. This enabled an estimation of the volume of pores in each EP-aluminium matrix sample. It was assumed that at the time of retrieval, all of these pores were filled with seawater. The mass of salt contained in this seawater was estimated according to the average salt content of seawater of 35 g/L. This value for salt mass gain was subtracted from the final dried mass of each sample with the intention of reducing the final mass to a mass loss. The results of this calculation returned the 1-week and 1-month samples to a final mass loss value, however the 6-month samples still exhibited a mass gain. This result indicates that mass gain does not only depend on the instantaneous volume of seawater contained in the sample but also is a function of the duration of sample exposure.

This led to the investigation of the second hypothesis; the samples of longer exposure duration experience a build-up of trapped salt sediments in the pores resulting in a final salt mass at a greater concentration than the salinity of seawater. This theory is also supported in considering a possible influence of mass gain on the mechanical properties of the seawater samples. It appears that the mass gain variable contributes to the distinct increase in unloading modulus with exposure duration; over time the increasing sedimentation of crystallised salt increases the stiffness of the EP-aluminium foam. In order to further analyse the relationship between mass gain and exposure duration the experimental data is presented in Figure 7.

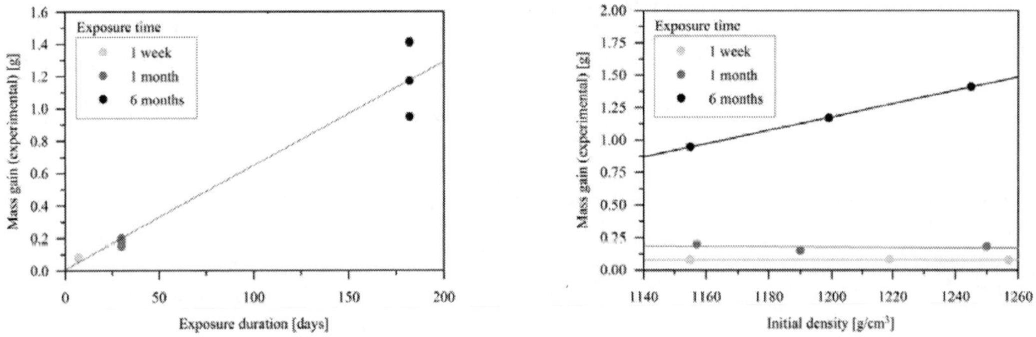

(a) (b)

Figure 7. Mass gain versus exposure duration (a) and mass gain versus initial density (b).

An analysis of Figure 7 indicates that there is a linear relationship between initial density and quantity of mass gain; across all exposure durations the samples of higher density exhibit a greater mass gain and this relationship becomes more profound with longer periods of exposure to seawater. The likely explanation for this trend is for the EP-aluminium samples of higher density there is a greater volume fraction of aluminium metal and a lower volume of expanded perlite pores. When salt sediments flow into the samples of higher density there are fewer interconnections between the pores that would allow the particles to be washed out again by residual water flow during the experiment or water flow during post- experimental cleaning.

4.2 Long-term corrosion effects

The data collected in this experiment were also evaluated with the purpose of predicting the long term corrosion behaviour of EP-aluminium foam. However the scatter in the results and the identification of mass gain mean that any long term predictions using this data are based on the relationships identified in the previous section and will require a much larger data set for validation.

Using the observed relationships between mass gain and both exposure duration and initial density, the corrosion mass gain behaviour of EP-aluminium foam was modelled. A sample of average density across those considered in this study was assumed for the evaluation of long term behaviour; initial density of 1220 kg/m^3. Using the linear exposure duration trend in Figure 7(a), obtained from the difference between a sample's mass before and after exposure, the average mass gain of this sample was calculated for each exposure period. Although Figure 7(a) exhibits vertical scatter due to initial density, the hypothetical sample was allocated an average density in order to increase the accuracy of this estimation. Using the relationship in Figure 7(b), the mass gain of this sample for each exposure duration was calculated based on initial density. A comparison of the mass gain values from each relationship indicated the mass gain based on initial density was the higher value for each exposure duration (Fig. 7b). The difference between this value and the average mass gain from Figure 7(b) was assumed to be the corrosion mass loss quantity which has been replaced by salt particle mass gain. This value was recorded as the mass loss of an average density EP-aluminium foam sample for each exposure duration.

The results of the above calculations were compared to the long term corrosion behaviour of aluminium reported in the literature [1]. This comparison indicated that the EP-aluminium foam experiences a greater aluminium corrosion mass loss than solid aluminium coupons. This is consistent with some of related previous research; metallic foam materials corrode at a faster rate compared to a bulk metal due to the high interfacial area within the pore structure contributing to the overall exposed area of metal [20]. In comparing the two trends, the Southwell data [21] was exposed to fully immersed natural seawater conditions not unlike the EP-aluminium seawater samples. The base aluminium alloy in both experiments is an alloy of similar chemical composition. Therefore the increase in corrosion mass loss from aluminium coupon to EP-aluminium foam should be proportional to the increase in aluminium surface area between the datasets.

5. CONCLUSIONS

The present study investigated the impact of corrosion of EP-aluminium syntactic foam on the mechanical properties of the material. There is no previous corrosion research on this specific material and this study also serves as a baseline prototype for future research. This study was conducted by exposing foam samples in natural seawater tanks and retrieving sample sets after designated durations for compression tests. The mechanical properties of plateau stress and elastic modulus were obtained for each sample and compared with other durations in order to predict trends. Overall, instead of enabling the identification of experimental trends, the data indicated that the plateau stress and elastic unloading modulus depend on multiple variables including initial density, foam geometrical variation and mass gain in the seawater samples. The scatter in experimental data due to these variables required calculation assumptions to draw trends and to predict the long term corrosion behaviour of EP-aluminium foam. As a result, the trends and predictions require additional experimental data for
validation. Due to the prototype nature of the reported experiments, some unforeseen variables were expected but could not be isolated from the results. Improvements to experimental methodology to control these variables in future research are recommended.

6. ACKNOWLEDGMENTS

The authors acknowledge the continued financial support provided by the Australian Research Council. The continued support of the Fisheries Research Station at Taylors Beach (Department of Primary Industry, NSW) is very much appreciated. The technical assistance of Ms. Mengxia Liang is also appreciated.

7. REFERENCES

1. R.E. Melchers, Trends in long-term corrosion of aluminium alloys in marine, industrial and urban environments, Corrosion and Materials, 40 (2015): 48-53.

2. Z.S. Smialowska, Pitting corrosion of aluminium, Corrosion Science, 51 (1999): 719-727.

3. M. Taherishargh, I.V. Belova, G.E. Murch, T. Fiedler, Low-density expanded pearlite-aluminium syntactic foam, Material Science and Engineering, 604 (2014): 127-134.

4. R.E. Melchers, Long-term immersion corrosion of steels in seawaters with elevated nutrient concentration, Corrosion Science, 81 (2014): 110-116.

5. Ausperl, Australian Perlite in: http://www.ausperl.com/perlite/ (accessed on 22 Nov 2016).

6. R.E. Melchers, R. Jeffrey, Corrosion of long vertical steel strips in the marine tidal zone and implications for ALWC, Corrosion Science, 65 (2012): 26-36.

7. R.E. Melchers, Modelling of marine immersion corrosion for mild and low-alloy steels-Part 2: uncertainty estimation, Corrosion, 59 (2003): 335-344.

8. ASTM G103, Standard practice for preparing, cleaning and evaluating corrosion test specimens, ASTM International, Pennsylvania, USA, (2011).

9. F.A. Champion, Corrosion testing procedures, 2nd ed., Chapman and Hall, (1964): 448 pages.

10. I.A. Chaves, R. Jeffrey, R.E. Melchers, Rust removal from steel coupons after short-term marine immersion, Corrosion, 71 (2015): 811-820.

11. I.A. Chaves, R.E. Melchers, Pitting corrosion in pipeline steel weld zones, Corrosion Science, 53 (2011): 4026-4032.

12. ISO 13314, Mechanical testing of metals: ductility testing, compression test for porous and cellular metals, 1st ed., International Standards, 13 pages.

13. C. Gaillard, J.F. Despois, A. Mortensen, Processing of NaCl powders of controlled size and shape for the microstructural tailoring of aluminium foams, Material Science Engineering, 374 (2004): 250-262.

14. R.A. Palmer, K. Gao, T.M. Doan, L. Green, G. Cavallaro, Pressure infiltrated syntactic foams – Process development and mechanical properties, Material Science Engineering, 464 (2007): 85-92.

15. I.A. Orbulov, J. Dobransky, A. Nemeth, Microstructural characterisation of syntactic foams, Material Science, 44 (2009): 4013-4019.

16. D.P. Mondal, S. Das, N. Ramakrishnan, K.U. Bhasker, Cenosphere filled aluminium syntactic foam made through stir- casting technique, Applied Science Manual, 40 (2009): 279-288.

17. Q. Zhang, P.D. Lee, R. Singh, G. Wu, T.C. Lindley, Micro-CT characterization of structural features and deformation behaviour of fly ash/aluminium syntactic foam, Acta Materials, 57 (2009): 3003-3011.

18. M. Alizadeh, M.A. Mirzaei, Compressive properties and energy absorption behaviour of Al–Al2O3 composite foam synthesized by space-holder technique, Materials Design, 35 (2012): 419-424.

19. M.S. Goel, M. Peroni, G. Solomos, D.P. Mondal, V.A. Matsagar, A.K. Gupta, M. Larcher, S. Marburg, Dynamic compression behaviour of cenosphere aluminium alloy syntactic foam in: Materials Design, 42 (2012): 418-423.

20. I.B. Singh, D.P. Mondal, M. Singh, A. Bhadkul, N. Jha, Corrosion behaviour of AA2014 aluminium alloy-cenospheres syntactic foam in 3.5% NaCl solution, Chemical technology, 21 (2014): 168-175.

21. C.R. Southwell, A.L. Alexander, C.W. Hummer, Corrosion of metals in tropical environments-Aluminium and magnesium, Material Protection, 12 (1965): 30-35.

8. AUTHOR DETAILS

Igor A Chaves is a Research Academic at the University of Newcastle, Australia. He is a Civil Engineer, holds a Master's degree in structural engineering and has been awarded a PhD for his studies in corrosion analysis and prediction. His research interests also include structural reliability and risk assessment of infrastructure.

He is the recipient of the ACA's 2011Young Corrosion Professional Award, 2012 ACA best paper Marshal Fordham Award, and 2016 Brian Cherry International Scholarship Award.

Thomas Fiedler is an Associate Professor at the University of Newcastle, Australia He is a Mechanical Engineer, holds a Master's degree in mechanical and material science engineering and a PhD in cellular materials (metals and ceramics) and composite materials.

His most recent awards are the 2013 Pro-Vice Chancellor award for excellence in research performance, and the prestigious 2016 excellence in innovation award as a direct recognition from industry partners of his practically applied research solutions.

LONG-TERM MARINE IMMERSION CORROSION FIELD TRIALS OF SIMULATED STEEL WELD HAZ

S. Krismer[1], I. Chaves[2]
[1]Bureau Veritas Asset Integrity and Reliability Services Pty Ltd, Newcastle, Australia; [2]Centre for Infrastructure Performance and Reliability, The University of Newcastle, Australia

SUMMARY: Corrosion deterioration will cause the actual structural safety of an asset to decline, slowly, with operational use and with time. Moreover, welded areas of such assets have been shown to corrode at an even higher rate compared to adjacent base metal. Yet, controversy still remains as to a more appropriate and accurate representation of the phenomena. This paper reports on a long-term investigation of the effect of weld heat affected zone microstructural variation on corrosion rates. Commercial normalized fine grained steel was heat treated in order to produce samples with a homogeneous microstructure representative of a specific region of a Heat Affected Zone. Immersed for a total of 12 months in temperate natural seawater, corrosion trend loss results for the samples reinforce earlier observations that microstructural changes induced within the heat affected zone do result in relative differences in corrosion rate. Reasons and implications are discussed within.

Keywords: Mild Steel, Weld, Microstructure, Heat Treatment, Seawater, Corrosion.

1. INTRODUCTION

Following from recently published results regarding the effect of microstructure in relation to HAZ corrosion of mild steel welds exposed to marine environment [1], representable long-term information is crucial for an appropriate, and therefore, more accurate corrosion damage predictions. It has been demonstrated how severe localised corrosion can be considered in design to produce a sufficiently safe asset with a more reliable long-term service life [2]. Also, the underlying cause of severe pitting corrosion of low-carbon mild steels located at or near the weld has been considered many times [3-9] but some discussions remains regarding the specific mechanism involved.

In the previous study [1], a better understanding of the dominant factors that drive the preferential corrosion effect within the weld zones were discussed. By simulating the various microstructural regions within weld head affected zone (HAZ) and independently assessing their corrosion rates when immersed in marine waters the early short-term observed results provided support for the notion that the microstructural changes within the HAZ of a weld do indeed contribute to increased corrosion rates in this region. However, further long-term research was recommended in order to conclusively establish the mechanisms, and their relative dominance compared to other potential causative factors such as residual stress, mill scale, and weld metal composition. Hence, this paper reports long-term exposure corrosion losses for the various microstructural variations of mild steel welds inherently produced in the HAZ. Results are presented and implications of the long-term consistency of such microstructural influence on corrosion rates are discussed.

2. EXPERIMENTAL METHODOLOGY

As per the previous study [1], samples from a normalized fine grained mild steel pipe (ASTM A333/A333M – 15) were heat treated to produce a supercritical coarse grain microstructure, a subcritical spheroidised microstructure, a quenched microstructure, and a quenched and tempered microstructure. The various samples groups and the recovery periods are shown in Table 1.

Table 1. Experimental plan.

Group	Heat Treatment	Exposed Surface Microstructure
1	None (Control)	As received normalised fine grain with decarb layer
2	None	Normalised fine grain; No decarb layer
3	Spheroidised	Held at 650°C for 1h; Oven cooled; No decarb layer
4	Coarse Grain	Held at 1050°C for 1h; Oven cooled; No decarb layer
5	Quenched & Tempered	Held at 950°C for 1h; Quenched in 17°C tap water; Held at 400°C for 1h; Oven cooled; No decarb layer
6	Quenched	Held at 950°C for 1h; Quenched in 17°C tap water; No decarb layer

NOTE: Each group had a triplicate set (e.g. 1A, 1B, 1C) recovered at 3 months, 6 months and 12 months.

The original material (Groups 1 and 2) microstructure consisted of as received normalised fine grain structure, as shown in Figure 1a, with some banding of the blocky pearlite. The spheroidised samples (Group 3) had the pearlitic lamella partially spheroidised, still clustered in the pearlite colonies, with some grain boundary carbide precipitation as shown in Figure 1b and 1c. The grain coarsened (Group 4) samples had double the original grain size, with a blocky ferrite/pearlite microstructure as shown in Figure 1d. It was noted that there was considerable variation in the ferrite grain size, ranging from <10μm to 200μm, which is consistent with selective grain growth effects. Also, there was extensive grain boundary carbide precipitation, forming a continuous layer at some grain boundaries. The samples for Group 5 had a quenched and tempered microstructure, consisting of type low carbon martensite / bainite / ferrite as shown in Figure 1e. Group 6 was very similar, but was left in the as-quenched condition.

The samples were prepared such that the face from the outer diameter of the pipe was linished to 120 grit, while the side faces were left as cut (for the control samples) or with the oxides from heat treatment still present. The back face (internal surface of pipe) was not prepared, leaving whatever oxide scales were present. For the original material, this was in the form of mill scale, while for the heat treated samples the oxide layer was thicker and less coherent. This fact is considered to be highly relevant to the interpretation of the corrosion phenomena observed.

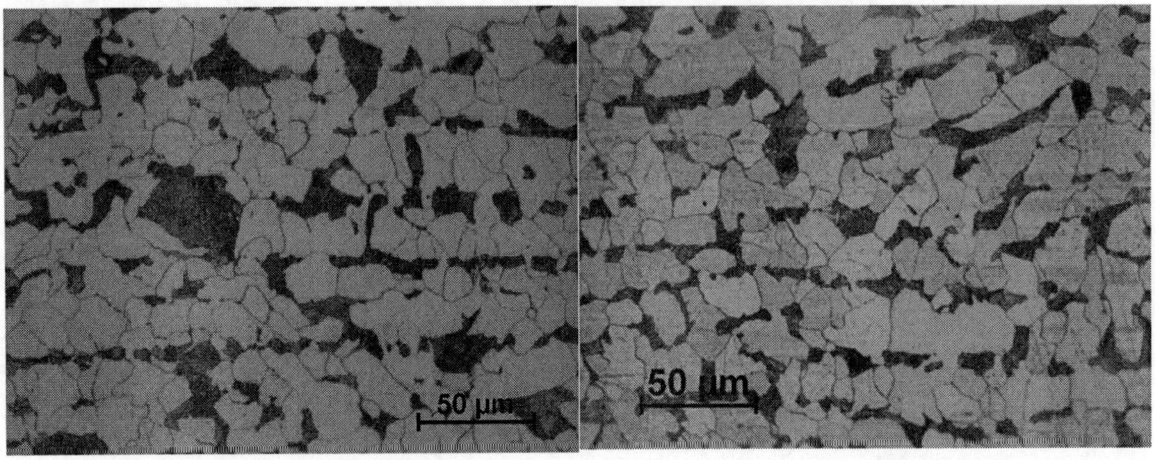

(a) Group 1 – As-received (500x objective) (b) Group 3 – Spheroidised (500x objective)

Figure 1. Core longitudinal microstructure of each group (2% Nital Etch). (Continued on next page).

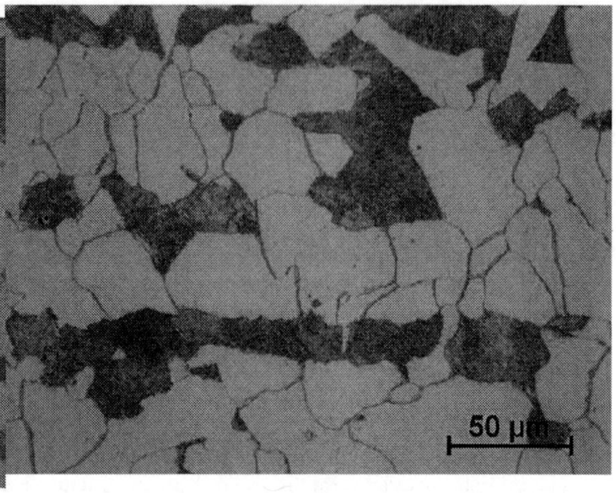

(c) Group 3 – Spheroidised (1000x objective)

(d) Group 4 - Coarse Grain (500x objective)

(e) Group 5 – Quenched and Tempered (500x objective)

(f) Group 6 – Quenched (500x objective)

Figure 1. Core longitudinal microstructure of each group (2% Nital Etch).

Samples of the parent and heat treated microstructures were immersed in natural seawater at the NSW Port Stephens Fisheries Research Institute, at Taylors Beach NSW. All samples were submerged using plastic CTMP1000 Smart Tube UV resistant cable ties and coded for identification at a constant 100mm depth within a sheltered fibre glass tank (Fig. 5) which is slowly but constantly flushed with fresh seawater at a rate of 10 litres per minute. Seawater chemistry is presented in Table 2. Such condition ensured that no external environmental factors such as storm runoff or stagnant water conditions would interfere with the experiment.

Table 2. Typical seawater chemistry at Taylors Beach test site.

Parameter	Units	Typical value	Parameter	Units	Typical value
Alkalinity($CaCO_3$)	ppm	409 – 419	Nitrite	ppm	< 0.003 – 0.011
Ammonia	ppm	0.017 – 0.090	pH		8.1 – 8.3
Calcium	ppm	374 – 409	Salinity	ppt	25.7 – 31.8
Chloride	ppm	21,000	Sulphate	ppm	1600 – 2750
Diss. Oxygen	%	90	Total Phosphorus	ppm	0.003 – 0.07
Nitrate	ppm	0.017 – 0.050	Water Temp.(annual mean)	^{o}C	20

Once recovered, samples were subjected to standardised cleaning procedure consisting of mechanical brushing using nylon brush under tap water followed by 2 hour electrolytic cleaning [10]. Once removed from the electrolysis bath, samples were rinsed under tap water, submerged for 20min in an ultrasonic bath (Branson 40 kHz SONOGEN Ultrasonic Cleaner), oven dried and weighed for mass loss evaluation. Subsequently, the surface condition of each sample was methodically inspected and imaged under an upright bright light optical Zeiss AXIO Imager A1 microscope. In addition, a subjective visual evaluation was made to compare the appearance of the corrosion severity and morphology.

3. RESULTS

A visual assessment and comparison was made of all of the test samples following cleaning. Representative examples of the prepared face of the 12 month exposure samples are shown in Figure 2. In general, there was very little difference between the appearances of most of the samples, except for Group 4. Group 4 had a rougher surface appearance with more pitting evident than the other samples. The other samples had some localised pitting, with vermicular shaped corrosion regions. The texture from the linishing during preparation was still evident on most of the samples, indicating uniform type corrosion was also occurring. Although the prepared face of the samples did not exhibit significant variation, the back face, which had not been linished, showed the most corrosion. The appearance of the back face of the 12 month exposure group samples is presented in Figure 3. The appearance of Group 1 and 2 is very similar, which makes sense since they both had the original mill scale remnant. Group 3 has slightly more corrosion, and would have had slightly thicker and different oxide due to the spheroidising heat treatment. Group 4 exhibited severe corrosion pitting, even more than the other heat treated samples of Groups 5 and 6. Groups 5 and 6 were quite similar in appearance of the corrosion of the back face, with deep hemispherical type pits of varying sizes. It was noted that the mill scale present on Groups 1 and 2 provided partial corrosion protection up to at least 6 months, with patchy areas of corrosion developing as shown in Figure 4. In fact, isolated regions of the oxide layer were still present in a number of areas on the samples even after 12 months exposure.

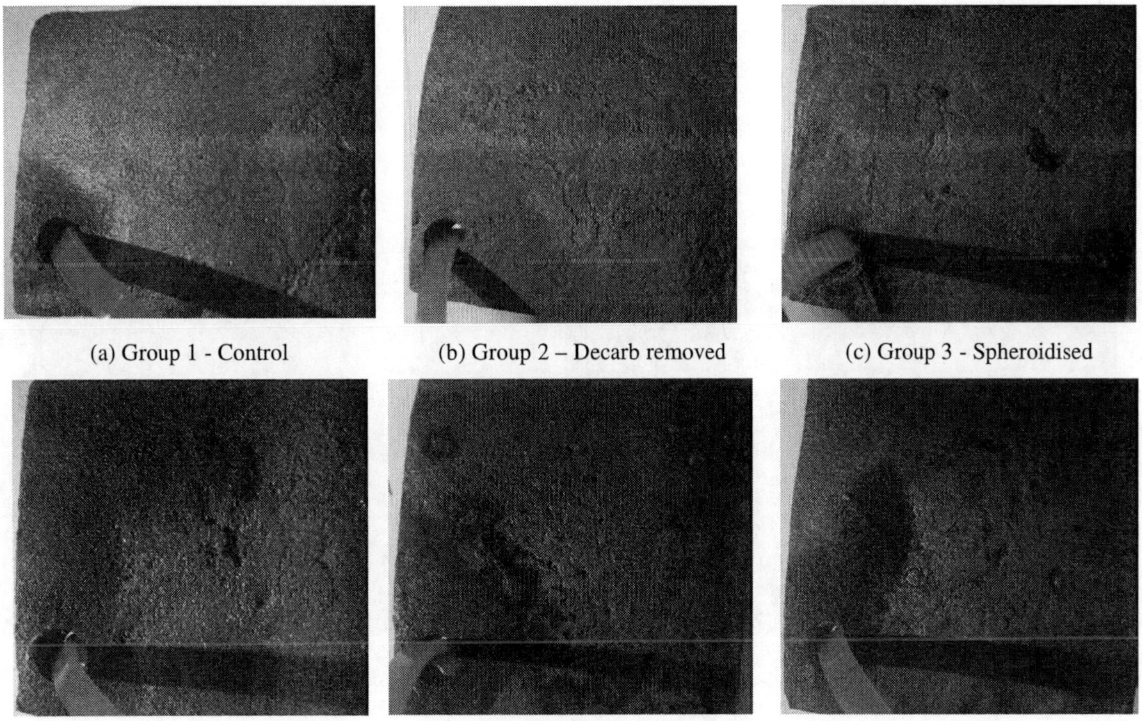

(a) Group 1 - Control

(b) Group 2 – Decarb removed

(c) Group 3 - Spheroidised

(d) Group 4 – Grain Coarsened

(e) Group 5 – Quenched and Tempered

(f) Group 6 - Quenched

Figure 2. Photos showing appearance of prepared face for 12 month samples

(a) Group 1 - Control

(b) Group 2 – Decarb Removed

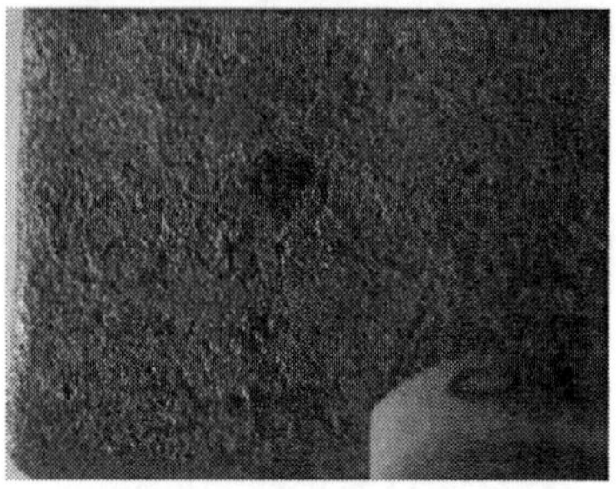

(c) Group 3 - Spheroidised

(d) Group 4 – Grain Coarsened

(e) Group 5 – Quenched and Tempered

(f) Group 6 – Quenched

Figure 3. Photos showing appearance of back face (unprepared) for 12 month samples.

(a)

(b)

Figure 4. Photos demonstrating the protective effect of mill-scale on the back face of the control samples after 6 months exposure.

A subjective visual assessment was made based on apparent severity and nature of the corrosion for the samples from each test cohort and exposure period. The samples were then ranked in comparative order from the most corrosion to the least corrosion, as shown in Table 3. In the process of doing this comparative assessment, it was noted that the samples for each period could be grouped. These groups have been visually colour coded in Table 3.

Table 3. Subjective Visual Assessment and ranking in order of severity

Most Corrosion ↓ Least Corrosion	3 Months	6 Months	12 Months
	Group 4	Group 4	Group 4
	Group 5	Group 5	Group 5
	Group 6	Group 6	Group 6
	Group 3	Group 3	Group 3
	Group 2	Group 2	Group 2
	Group 1	Group 1	Group 1

Mass losses were obtained by weighing each sample individually before exposure and after cleaning off the rust products as described in previous section. Table 4 summarizes the average percentage mass loss for each group after 3, 6 and 12 months marine immersion exposure. This data is also presented graphically in Figure 5. Once graphed, it becomes clear that all samples exhibit a distinct corrosion rate change over the 12 month exposure period, with the rate appearing to accelerate between 6 and 12 months of exposure. The recorded average percentage mass loss results were then ranked from most to least, as shown in Table 5. It was noted that the results for the samples could be grouped, similar to the findings for the visual assessment as previously shown. Although the cohorts for the relative groups remained the same, the individual relative ranking between the samples in the groups was noted to be different for mass loss and visual severity. This is examined in more detail in the discussion.

Table 4. Percentage Mass loss

Group	Heat Treatment	3 month	6 months	12 months
1	None (C)	-1.45%	-2.60%	-6.35%
2	None (D)	-1.38%	-2.50%	-7.35%
3	Spheroidised (S)	-1.65%	-3.02%	-7.29%
4	Coarse Grain (CG)	-2.27%	-4.22%	-10.03%
5	Quenched & Tempered (QT)	-1.63%	-2.74%	-7.94%
6	Quenched (Q)	-1.70%	-3.35%	-7.69%

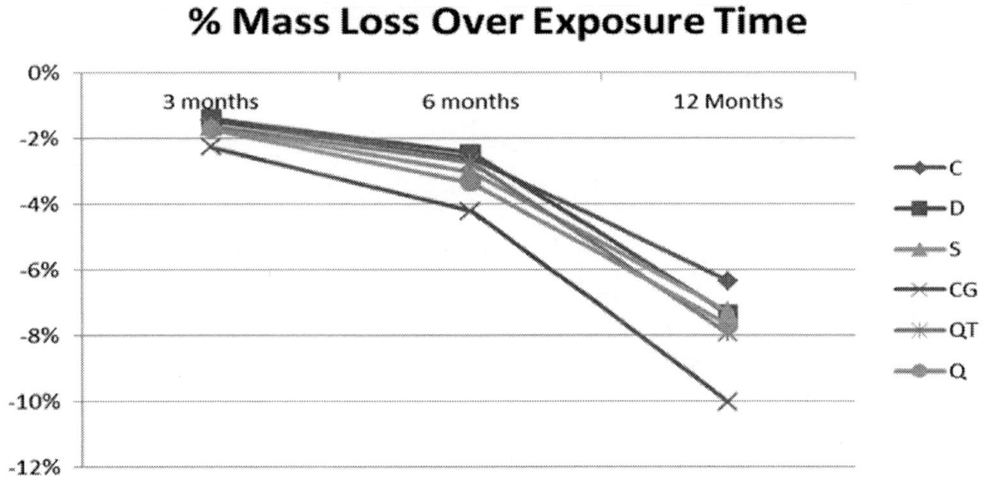

Figure 5. Graph showing the change in % mass loss over time for each sample.

Table 5. Percentage Mass Loss

Most Corrosion	3 Months	6 Months	12 Months
↓	Group 4	Group 4	Group 4
	Group 6	Group 6	Group 5
	Group 3	Group 3	Group 6
	Group 5	Group 5	Group 2
	Group 1	Group 2	Group 3
Least Corrosion	Group 2	Group 1	Group 1

The depth of pits was measured using a microscope focussed successively on the pit base and on the surrounding surface, thereby obtaining relative pit depth. Measurements were recorded by means of digital Linear Variable Differential Transformer (LVDT) of 1 micron sensitivity. Table 6 summarises the average pit depths as measured from the six deepest pits observed on the prepared face of each sample. The trends over time for each sample are shown graphically in Figure 6. The parent metal samples (Groups 1 and 2) demonstrate similar increased corrosion rate changes from the 6 to 12 month period as observed with percentage mass loss. The spheroidised samples (Group 3) showed no obvious change over the exposure periods. The heat treated samples all appeared to have an inverse inflection, with a decrease in apparent corrosion rate based on the average pit depth.

Table 6. Average Pit Depth (Microns)

Group	Heat Treatment	3 month	6 months	12 months
1	None (C)	49.7	66.0	140.4
2	None (D)	53.8	60.7	130.8
3	Spheroidised (S)	111	75.7	91.6
4	Coarse Grain (CG)	139.8	212.8	190.6
5	Quenched & Tempered (QT)	78.5	138.1	119.9
6	Quenched (Q)	61.7	124.4	100.2

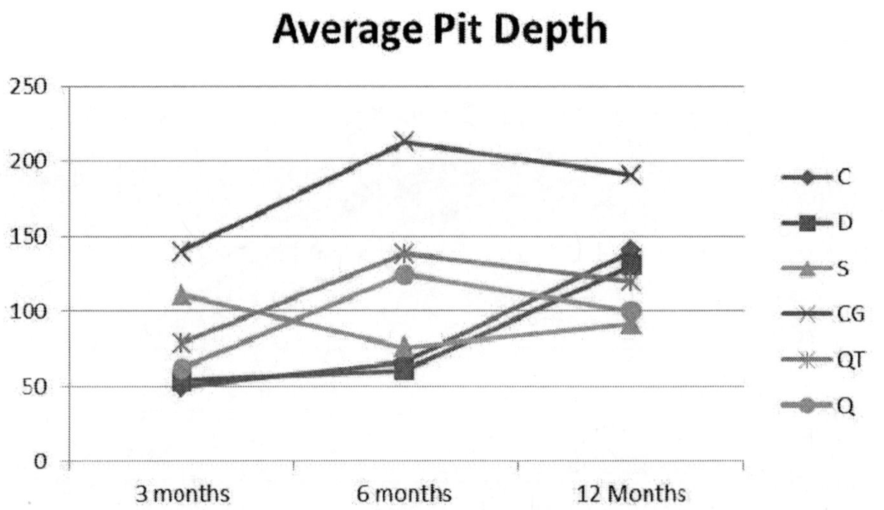

Figure 6. Graph showing the change in average pit depth over time for each sample.

4. DISCUSSION

One of the key observations made from the 12 month exposure samples was that the back face, which is in fact the internal surface of the pipe material used to manufacture the test coupons, had much more severe corrosion for all of the test groups than the prepared surface. However, it was also noted that at the 3 month exposure period, there was very little corrosion activity on the back face of the samples. This indicates that there is a definite short term protective effect afforded by the presence of an oxide layer – particularly mill scale. Corrosion is evident after 6 months, but it would appear that the corrosion rate is significantly higher for the back face compared to the prepared face for the last 6 months of the exposure period. It was noted that the mill scale had been partially lost in some areas for Groups 1 and 2, with pitting occurring of the exposed areas. However, after 12 month exposure period, the oxide layers appeared to have been completely lost and severe pitting was present on all samples. Although the reasons for this are not entirely clear, it is suspected that the more severe corrosion on the back face of the samples is due to preferential colonisation of biological and microbial activity in the crevices within the oxide scale on the surface, and without the negative effects on ongoing growth due to corrosion products. In this theory, the more established colonies then result in more rapid corrosion rates once the tight, adherent high temperature oxide layer is finally lost. This may be creating a false increase in the corrosion rate, and additional studies would be required to assess whether the observed corrosion rate changes occur for samples that have no surface oxide films.

It had previously been noted [1] that after 3 months exposure the nascent pit morphology was different between the sample types, and the same general morphological features were still evident after 12 months, although much more pronounced. Rounded shallow crater type pitting was evident on the samples in Groups 5 and 6. The pitting in the original material (Groups 1 and 2) was much smaller in size and depth than for Groups 5 and 6. The samples from Group 3 tended to be more similar to Groups 5 and 6 initially, but were found to correlate best to Groups 1 and 2 after long term exposure. This is consistent with the spheroidised microstructure being more similar to the ferrite/pearlite microstructure of Groups 1 and 2, than the martensitic microstructure of Groups 5 and 6. The pitting in Group 4 was less uniform in shape and had visually rougher edges. On this basis, it would seem that the pit morphology is strongly driven by microstructure. However, this may be the extent to which the pitting data may be used based on this experimental methodology. Of particular concern was that the pit depths exhibited an inverse inflection (i.e. decrease in apparent corrosion rate) with increasing exposure time for the heat treated samples. The reasons for this effect were not clearly determined in this experiment, but it would suggest that uniform corrosion is overtaking the pitting corrosion on the prepared face of the samples.

It was previously noted in the 3 month exposure trials that the results may be significantly influenced by the variability introduced by microbiological corrosion effects associated with exposure in sea water. It was observed that there were regions with much more severe localised pitting than the bulk surface of the samples. The regions of pitting were generally around the edges of the samples and were obviously associated with microbial biofilm effects. Therefore, measurements of the deepest pits were considered likely to not be indicative of the general corrosion rate and significantly affected by localised pitting phenomena. Measurements of average pit depth for the deepest pits will there also be similarly affected, and this distortion of pit depth results appears to be exacerbated with increased exposure time. This makes simple interpretation of localised phenomena such as pit depths and even mass loss difficult to interpret. It was previously suspected that the localised pitting effects may also influence the mass loss. Therefore, a subjective assessment of corrosion morphology, excluding obvious highly localised corrosion due to microbiological activity, was considered to be more indicative of corrosion rates at the 3 month period. However, based on the findings of this experiment, it would appear that the localised pitting effects from microbiological activity do not skew the data over longer exposure times. When averaged for the three coupons for each sample group, it was found that the overall percentage mass loss displayed the same general trends as the subjective visual assessment. This suggests that simple mass loss is in fact a viable means for assessing corrosion rates even in small sample sizes when exposed in marine environments.

In comparing the visual severity ranking with the percentage mass loss ranking, it is clear that there are very strong similarities and correlations. In all cases, there were 3 distinct categories of samples. In all instances, Group 4 had the most severe corrosion. The next category typically consisted of Groups 5 and 6, which is expected due to the similarity in microstructure between these groups. Group 3 appeared to transition from being initially with the Group 5 and 6 samples, to unique at 6 months exposure, to being with the Group 1 and 2 samples after 12 months exposure from the visual assessment. However, the percentage mass loss suggested that it remained with the Group 5 and 6 category. An anomaly was evident in that the Group 2 samples had more mass loss than Group 3 after 12 months exposure. All 3 samples showed similar values, all higher than Group 3. The reasons for this are not clear. However, after 12 months there was a general cluster of results between Groups 2, 3, 5 and 6, with Group 1 much lower, and Group 4 much higher. It is difficult to interpret why Group 2 is different from Group 1, and the authors are reluctant to suggest that there is an appreciable affect from having the decarburised layer present on the surface (group 1) which promotes a slower corrosion rate. In reality, it is quite possible that this variation was induced by an unknown factor in the test set-up arrangement under natural marine exposure conditions, which promoted more severe corrosion of this particular cohort of samples from Group 2. Further insights may be made though, based on the clustering of the results. It would appear that quenching, quenching and tempering, and spheroidising have only a minor effect on the corrosion rates, but the grain coarsened heat treatment in this case promotes severe corrosion. The reasons for this were expounded upon in the previous study [1], and no further insight was gained by longer exposure times.

Spheroidising seems to have little effect on the corrosion compared to the parent material. Martensitic type microstructures exhibit more severe corrosion. This validates the theory posited that the deformed higher latent energy crystal lattice type for the bainitic/martensitic microstructural phases does influence the corrosion rate. The most severe corrosion of the grain coarsened sample was previously theorised to be due to the presence of grain boundary carbide films in the sample coupons. The increased exposure period did not provide any further insight into the nature of this effect, but did confirm that this type of microstructural feature is severely deleterious to corrosion resistance.

5. CONCLUSIONS

The results of the 12 month exposure trials on the homogeneous microstructure samples clearly indicate that there are differences in corrosion rate and morphology clearly attributable to the microstructure type and features, which are evident over both short and longer term exposure periods.

Overall, the general trends in corrosion morphology and rates after 12 months are consistent with the initial results after 3 months exposure. Spheroidising seems to have little effect on the corrosion compared to the parent material. Martensitic type microstructures exhibit more severe corrosion than a normalised microstructure. Although the grain coarsened microstructure produced for this experiment was found not to be representative of what is typically found in a Heat Affected Zone, this experiment clearly showed that there was a deleterious effect caused by the grain coarsened microstructure with grain boundary carbide films. Additional testing is required to determine the separate effects of grain size and grain boundary carbides. In terms of experimental methodology, it was clear that improvements can be made in further studies. The overall degree to which microstructure influences the preferential corrosion at the Heat Affected Zone of weldments is not clear. At present, it is still strongly suspected that there is some contribution to corrosion rates from the residual stresses inherently formed by arc-welding processes, and further experiments are required to assess the influence of this factor. Also, further trials would need to be carried out on samples that better represent all zones of a weldment, particularly the weld metal and grain coarsened area.

6. ACKNOWLEDGMENTS

The authors acknowledge the continued financial support provided by the Australian Research Council and the valuable support of the EDX team at the Newcastle University's Central Scientific Services. Expert consult from Professor Eric Kisi and Dr. Heber Sugo – Newcastle University Mechanical Engineering Department is acknowledged. The continued support of the Fisheries Research Station at Taylors Beach (Department of Primary Industry, NSW) is very much appreciated.

7. REFERENCES

1. Chaves IA and Krismer S (2016) Effect of microstructure in relation to HAZ corrosion of mild steel welds in marine environment, Proc. of Corrosion & Prevention 2016, Paper 34, November, Auckland, New Zealand.

2. Chaves IA and Melchers RE (2014) Extreme value analysis for assessing structural reliability of welded offshore steel structures, Structural Safety (50) 9-15.

3. Eid NMA (1989a) Localized corrosion at welds in structural steels under desalination plant conditions - Part 1: Effect of surface roughness and type of welding electrode, Desalination 78(1): 397-406.

4. Eid NMA (1989b) Localized corrosion at welds in structural steel under desalination plant conditions - Part 2: Effect of heat treatment, test temperature and test media, Desalination 73(1): 407-415.

5. Lee C, Bond S and Woollin P (2005) Preferential weld corrosion: Effects of weldment microstructure and composition, (In) Proc. of NACE International Corrosion Congress, April, Seoul, South Korea.

6. Davis JR (2006) Corrosion of Weldments, American Society for Metals International (eds), Ohio, USA.

7. Matsushima I (2011a) Carbon steel corrosion in freshwaters, (In) Uhlig's Corrosion Handbook (3rd ed.) page 599, John Wiley & Sons Ltd. (eds), New Jersey, USA.

8. Matsushima I. (2011b) Localized corrosion of iron and steel, (In) Uhlig's Corrosion Handbook (3rd ed.) page 616, John Wiley and Sons (eds), New Jersey, USA.

9. Wang LW, Liu ZY, Cui ZY, Du CW, Wang XH and Li XG (2014) In Situ Corrosion Characterisation of Simulated Heat Affected Zone on API X80 Pipeline Steel, Corrosion Science 85(1): 401-410.

10. ASTM G1-03 (2011) Standard practice for preparing, cleaning and evaluating corrosion test specimens, ASTM International, Pennsylvania, USA.

8. AUTHOR DETAILS

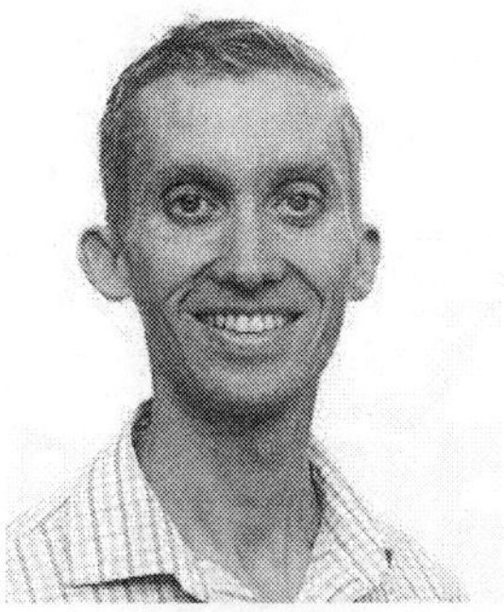

Simon Krismer is a Senior Materials Engineer and welding expert at Bureau Veritas Asset Integrity and Reliability Services Pty Ltd. Simon has more than 16 years of materials performance and root cause failure analysis consulting experience. Simon also is the current ACA Newcastle branch president and has been involved with promoting the ACA and its values for more than 10 years.

Igor A Chaves is a Research Academic at the University of Newcastle, Australia. He is a Civil Engineer, holds a Master's degree in structural engineering and has been awarded a PhD for his studies in corrosion analysis and prediction. His research interests also include structural reliability and risk assessment of infrastructure.

He is the recipient of the ACA's 2011Young Corrosion Professional Award, 2012 ACA best paper Marshal Fordham Award, and 2016 Brian Cherry International Scholarship Award.

COMPARATIVE MARINE IMMERSION CORROSION OF SHIP CONSTRUCTION STEELS EXPOSED IN ARCTIC AND TEMPERATE WATERS

B.B. Chernov[1], I.A. Chaves[2], A.M. Nugmanov[1], R.E. Melchers[2]
[1]G.I. Nevelskoy Maritime State University, Vladivostok, Russia,
[2]Centre for Infrastructure Performance and Reliability, The University of Newcastle, Australia

SUMMARY: Arctic waters present significant challenges for ship constructional steels. Mostly they are alloyed to meet strength and ductility conditions sufficient to ensure adequate structural safety conditions for low temperature conditions. However, their corrosion resistance characteristics are not well understood and may be important for periods when ice conditions are sufficient to severely damage or remove protective coatings. Although cathodic protection usually would be expected to be sufficient to maintain the steels in adequate condition, the unprotected corrosion behaviour under relatively short-term (months) exposure conditions is of interest to industry. The present paper reports interim progressive corrosion mass loss results for 5 different specialized low alloy structural steels used in ship construction. The steels for the samples were supplied by Pusan National University and were manufactured in Korean steel mills. Corrosion loss results are given for progressive periods up to 2-year immersion exposures in unpolluted seawater near Vladivostok (Russia) and, for comparison, in temperate Pacific Ocean waters at Taylors Beach, Australia (16-24°C). The Russian waters are subject to ice conditions for 5-8 months in winter and relatively warm seawater conditions in summer (around 24-30°C).

Keywords: Steel, Marine, Corrosion, Arctic, Temperate, Mass Loss.

1. INTRODUCTION

Marine corrosion is a significant proportion of the estimated annual cost of corrosion that is somewhere between 3% and 4% of the gross domestic product for industrialised countries [1]. Experience shows that many coastal and maritime infrastructures are approaching the end of their estimated design life, following the boom in urbanisation and construction in the 1960s and 1970s, including in shipping [2]. Increasingly this is requiring investment by asset owners for repair, rehabilitation or replacement. Decisions about such options should be based on quantitative estimates of likely future corrosion or deterioration. To assist such decisions, this paper reports data and trends for corrosion mass loss over several years obtained in field trials for five different structural steels relevant to the maritime industry and certified by one or more classification societies such as the American Bureau of Shipping [3].

For the testing programme two natural seawater testing locations were used and an appropriate field testing regime adopted [4]. Even though field testing may involve a loss of some degree of experimental control, trails in natural conditions are preferable to laboratory-based studies to obtain realistic quantitative results for seawater corrosion [5]. If properly done, they also obviate the need to translate laboratory results, that usually are only short term, to what might be expected under field conditions. Further, commercially available realistic specimens (i.e. normal material composition(s), reasonably sized samples, and realistic external surface finishes) can be employed. The procedures involved are well developed from previously established field trials [6,7].

The next section describes the experimental program. It was conducted in both temperate and sub-arctic conditions. Also described are the material composition and metallurgical observations. This is followed by a description of the trends in the results for corrosion loss and the evolution with time, separated into ordinary and hardened steel grades.

2. EXPERIMENTAL DETAILS

2.1 Exposure Sites

The experiments were performed in accordance with the standard practice for preparing, cleaning and evaluating corrosion test specimens G1-03 [8]. Exposure racks are constructed of permissible materials with UV resistant nylon bolts and nuts used to hold the test specimens. The racks were suspended vertically in constant immersed condition and arranged in separate groups to facilitate periodic sample removal without disturbing other samples (Fig 1a). Care was taken to ensure that samples did not come in contact with one another or any of the other racks during exposure.

One test site was in the temperate seawater test site located at Taylor's Beach, on the southeast coast of Australia. The site is securely enclosed within the premises of the NSW Port Stephens Fisheries Research Institute. The seawater has a temperature range between 14~26°C, has low average annual pollution levels (0.065 mgN/L DIN) and is subject to low average annual water velocities (<0.1 m/s) [9]. Taylor's Beach test site has been utilised as a corrosion exposure test site previously [10]. It is well documented and allows results of the present project to be compared with previous reported results [6, 9, 10].

Figure 1 (a) Location of samples at the Taylor's Beach site

The second test site is located in Sub-Arctic waters at Slavyanka Bay, just outside the port of Vladivostok, Russia, chosen for convenience and for its relatively high security (Fig. 1b). The surface water temperatures at the site test site ranges form -1.7°C in mid winter to 23°C in mid-summer. The tidal range is around 0.5m and water visibility is variable, typically between 10 and 35m. As expected, the water quality when determined for the present experimental work, as not as good as that at Taylors Beach.

Figure 1 (b) Test site at Slavyanka Bay.

2.2 Test samples

The steel samples used for the present study are standard grades as defined in International Standard ASTM A131 [11].

They included the ordinary strength grades A, B, D as well as the higher strength hardened grades AH and DH. Their compositions are shown in Table 1. Tensile testing was performed for diagonal, transversal and longitudinal properties and all samples satisfied minimum yield point, elongation and modulus requirements, according to ASTM A131, of 235MPa for ordinary strength steel and 390MPa for higher strength steel [11]. Specifically, grades A, B, D showed average yield points at 281MPa, 329MPa and 394MPa respectively and hardened grades AH and DH showed average yield points at 419MPa and 399MPa respectively.

Table 1. Chemical composition of steel samples (%wt).

Grade	C	Si	Mn	P	S	Cr	Ni	Cu	Mo	Al	Ti	Nb	V
A	0.15	0.01	0.88	0.015	0.007	0.02	0.01	0.02	trace	-	-	-	-
B	0.14	0.19	0.90	0.019	0.010	0.02	0.02	0.02	0.002	0.05	-	-	-
D	0.15	0.24	0.89	0.018	0.012	0.02	0.02	0.05	0.001	0.04	-	-	-
AH	0.16	0.37	1.45	0.016	0.004	0.01	0.00	0.01	0.002	0.03	0.2	0.001	0.015
DH	0.16	0.39	1.43	0.016	0.002	0.01	0.02	0.01	0.001	0.03	0.002	0.002	0.002

For the present experiments, the steels were made available by Pusan National University, South Korea, and had been cut from steels manufactured in Korean steel mills. The steel plates as supplied were cut to coupon size (100 x 100 x 6mm nominal thickness) using an automated guillotine to ensure uniformity in size throughout. A total 200 samples were cut. Each sample was marked with notch identification marks were used (Fig. 2a). Prior to exposure each sample was chemically pre-cleaned as per standardised procedure [8] and weighed to determine mass loss to 0.01 gm.

At the Taylors Beach site, for each steel grade, and for each exposure period, triplicate samples were exposed (and recovered) in accordance with established corrosion testing procedures [4]. Three racks containing one sample of each steel grade (Fig. 2b) were recovered at each time point in the programme. For the Vladivostok test site, due to extreme conditions, only one sample per grade was able to be recovered per exposure period. At both sites, for the first year, samples were collected on a quarterly basis and then at (approximately) 6 month intervals for the second and third year of exposure.

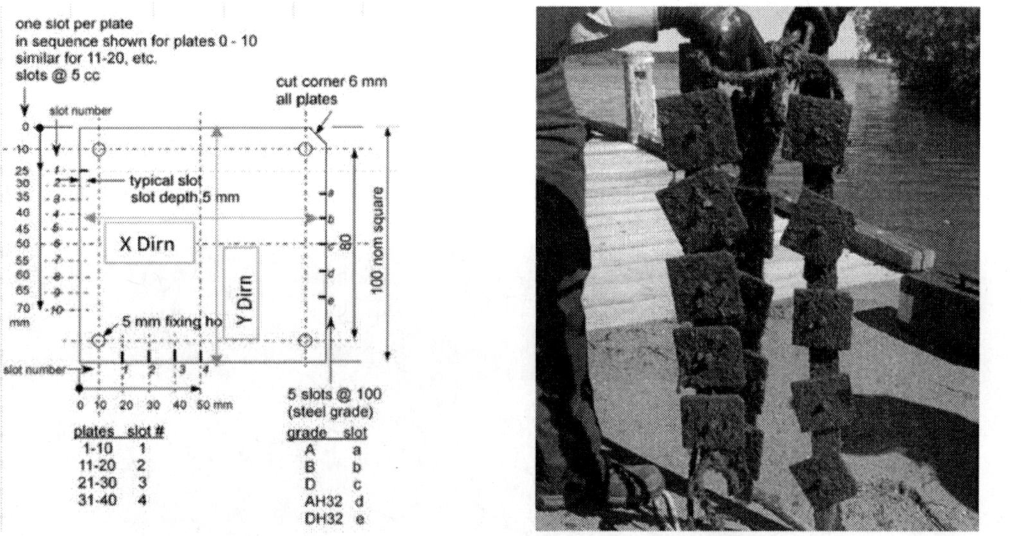

(a) (b)

Figure 2. (a) Test sample with identification coding (b) Typical samples immediately after recovery.

2.3 Sample Cleaning and Data Recovery

After removal of the scheduled number of test racks from the water, the samples on it were hosed-down on-site to remove mud and marine debris. The racks and the samples were then transported immediately to the University of Newcastle Civil Engineering Laboratories. There the samples were removed from the test racks and subjected to standardised mechanical and chemical cleaning [8]. A photographic record was made of the state of each sample prior to mechanical cleaning, post-mechanical cleaning, and post chemical cleaning (Fig. 3). Recording of all steps facilitates analysis of observations of any peculiar localised corrosion [12].

(a) (b)

Figure 3. Typical samples (a) prior to cleaning, (b) after chemical cleaning.

Mechanical cleaning took form of scrubbing with a rigid nylon brush under running tap water. Samples were then submerged for a min of 10 minutes in a solution of 500mL hydrochloric acid (32% HCL, sp gr 1.19), mixed with 3.5 g of inhibitor hexamethylene tetramine and reagent water to make 100 mL [8]. Appropriate safety measures were taken for the handling, neutralisation and removal of the chemical solution upon completion of the cleaning process. The correct choice of the cleaning process is important to avoid any potential further loss of base metal [13]. Following mechanical and chemical cleaning, samples were thoroughly rinsed under tap water and further scrubbed with a rigid nylon brush in order to remove any remaining trace chemicals [13]. Properly sorted, samples were then acetone and paper towel dried, weighted, measured and stored.

After cleaning each sample was weighed to determine mass to 0.01 gm. This was then compared to the initial mass to determine mass losses.

2.4 Material Mechanical, chemical and Metallurgical Characterisation

For the metallurgical microstructural characterisation, representative sample offcuts were obtained. To cut the small samples needed for detailed examination, a slow cutting wet drop saw was used to ensure that no further heating occurred that might interfere with the representativeness of the offcuts. The offcuts were then cold mounted into epoxy resin (Struers SpeciFix-20), ground and polished to a 1 micron finish according to appropriate standards [14]. A 2% Nital (2mL of Nitric Acid and 100mL of ethanol) was used to etch the offcuts resulting in a clear visualisation of the steel grade microstructure (Fig. 4 and 5) under a 200x objective magnification upright white light AXIO Imager A1 Zeiss Microscope.

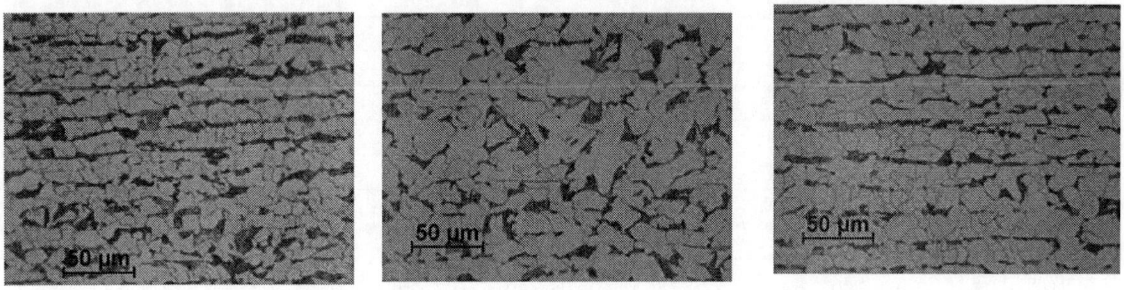

(a) Steel A (b) Steel B (c) Steel D

Figure 4. Micrograph of ordinary ABS steel grades.

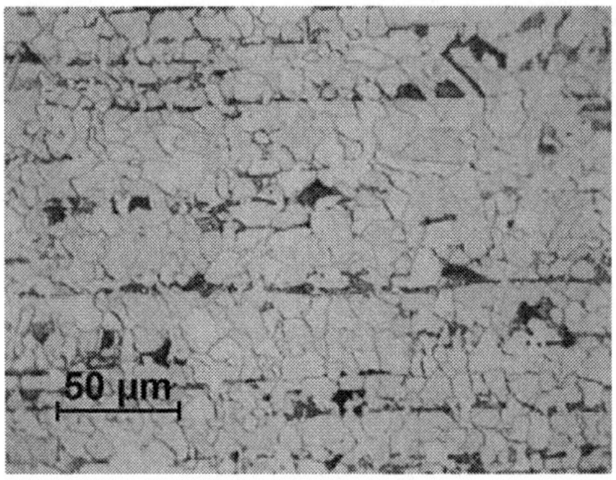

(a) Steel AH

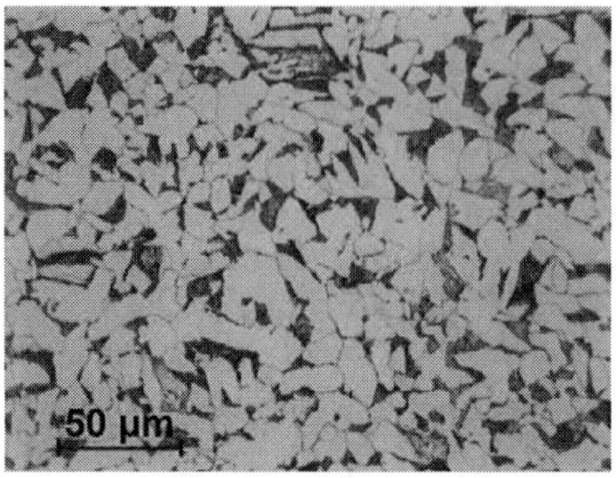

(b) Steel DH

Figure 5. Micrograph of Hardened high strength ABS steel grades (a) AH and (b) DH.

Examination of the microstructure showed the following features for each of the grades as follows:

A – Mix of banded plus blocky pearlite in a ferrite matrix. Pearlite slightly spheroidised. Some grain boundary carbide precipitates. Greatest proportion of pearlite in all samples.

B – Mix of blocky ferrite and pearlite, with some banding of pearlite. Suggests that this has been processed at a higher temperature, with partial re-austenitizing to break up the banded structure.

D – Highly banded pearlite in a matrix of ferrite, some blocky pearlite present. Pearlite lamellae unresolvable. Very fine MnS stringers were noted, which were not seen in other samples. These were only in the centre region of the plate.

AH – Mostly ferrite with less pearlite. Pearlite was banded with some blocky areas. The pearlite morphology was significantly different from all other samples, with coarse lamella evident as well as some spheroidisation. Ferrite grain size was also mixed, with some quite large ferrite grains indicative of grain growth due to holding at a high temperature during manufacture.

DH – banded plus blocky pearlite in a ferrite matrix. Pearlite was unresolvable and partially spheroidised. Some grain boundary carbide precipitates were evident.

3. EXPERIMENTAL RESULTS

3.1 Environmental Conditions

Table 2 summarizes the seawater temperature records and the average DIN values for both test sites.

Table 2. Seawater temperatures and Dissolved Inorganic Nitrogen concentration values for each site

Taylors Beach			Vladivostok		
Recovery	Exposure (years)	Temp. (°C)	Recovery	Exposure (years)	Temp. (°C)
Jul 2013	0.25	18	Dec 2012	0.5	7
Oct 2013	0.5	21	Jun 2013	1	9
Jan 2014	0.75	26	Nov 2013	1.5	8
Apr 2014	1	23	Jul 2014	2	17
Oct 2014	1.5	21	Nov 2014	1.5	11
Apr 2015	2	23	Jul 2015	3	17
Oct 2015	2.5	20	Nov 2015	3.5	8
Apr 2016	3	24	Jul 2016	1	15
Oct 2016	3.5	22	Oct 2016	4.5	8
Apr 2017	4	22	Jun 2017	5	13
Annual mean		22			12

The water quality at Taylors Beach has been characterized several times [10]. The water quality at the Vladivostok site measured for the project during the summer is summarized in Table 3. The DIN concentration is dominated by the nitrate concentration. It is known in general to be much less in winter months.

Table 3. Summary of summer water quality at the Vladivostok site

Parameter	Units	Typical value
Salinity	%	26.8
pH		8.0
Conductivity	mS	54.6
Dissolved Oxygen	mg/L	5.3
Nitrate	mgN/L	12.1
Nitrite	mgN/L	< 0.001
Ammonia	mgN/L	< 0.001
Calcium	mg/L	312
Total Phosphorous	mgP/L	0.22

3.2 Corrosion Losses

The mass losses computed from the final mass and the initial mass of each coupon were converted to equivalent corrosion loss on one side of a sample. This facilitates comparison of the present experimental results to those presented in the literature. The conversion of measured mass loss to sample depth loss was calculated by dividing the mass loss ($m_0 - m_1$), where m_0 is initial mass and m_1 is corroded sample mass, by the product between the sample surface area (α) and density (γ), taken for all coupons as $7.847e^{-3}$ g/mm^3, as follows:

$$\text{Depth loss (mm)} = (m_0\text{-}m_1) \bullet \gamma \bullet \alpha \quad (1)$$

Mass loss was considered to occur on all faces of the sample (i.e. uniformly as well as localised pitting). Thus, the calculated depth is scaled by factor α to obtain corrosion depth on just the face-plate of the sample. The factor is calculated by subtracting the percentage of surface area made up by sample edges from unity. An expression for factor α is shown in the following equation (Eq. 2), where x_o refers to the side length of the sample before exposure, y_o refers to the height of the sample before exposure, with to being sample thickness before exposure and x_{cut} refers to the edge length of each sample's cut identification corner (Fig. 3b), $x_{cut} = 6$ mm.

$$\alpha = \{1 - [(2\, x_o\, t_o + 2\, y_o\, t_o) / (2\, x_o\, t_o + 2\, y_o\, t_o + 2\, x_o\, y_o - 0.5\, x_{cut})]\} \quad (2)$$

Corrosion depth loss (mm) calculated for each exposure period (in years) is shown in Figures 6 to 8. Of these, Figure 6 shows the combined corrosion loss data for all grades of steel, Figure 7 shows the corrosion losses for ordinary steel grades A, B and D, and Figure 8 shows the corrosion loss data for hardened higher strength steel grades AH and DH. As described further in the Discussion section, each Figure also shows what is considered the most appropriate interpreted trend consistent with the data, consistent with the requirement that corrosion cannot decrease with increased exposure time.

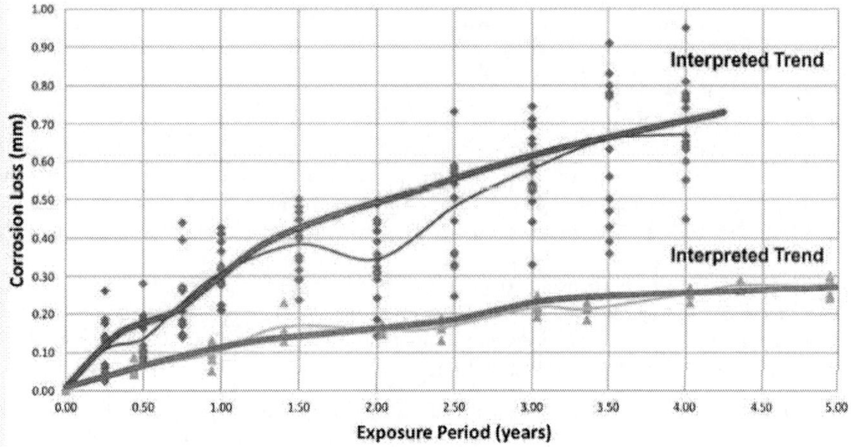

Figure 6. Taylors Beach (blue diamonds) and Vladivostok (green triangles) corrosion loss data and mean and interpreted trends for all steel grades combined.

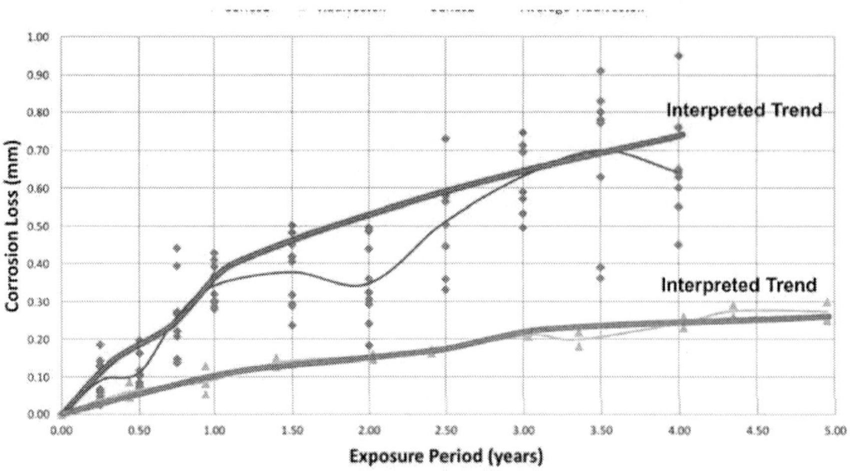

Figure 7. Taylors Beach (blue diamonds) and Vladivostok (green triangles) corrosion loss data and mean trends for ordinary steel grades A, B and D combined.

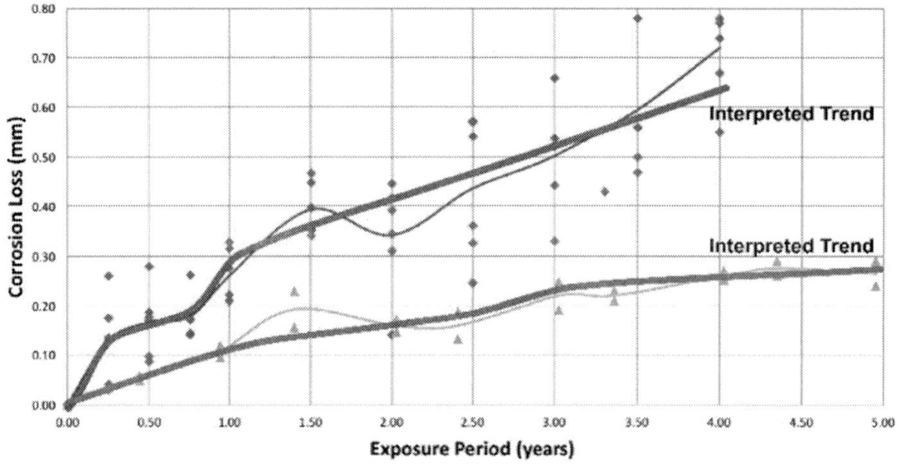

Figure 8. Taylors Beach (blue diamonds) and Vladivostok (green triangles) corrosion loss data and mean trends for hardened steel grades AH and DH combined.

4. DISCUSSION

4.1 Temperature and Nutrient Effects

Following on from the earlier work [6] that established the bi-modal characteristic it was expected that the trends in the data for corrosion loss for both sites would be consistent with that pattern. It can be seen from Figs. 6-8 that there is considerable variability in the data, and also some inconsistencies, particularly the latter, with some data presenting average values that decrease with increased exposure time. This cannot be correct, theoretically, and suggests that conditions at the sites were not quite as uniform as had been supposed previously. This appears to apply particularly for those coupons recovered at Taylors Beach at 2 years exposure. Despite these observations, it is also clear that at a more general level each of the data sets can be considered to be consistent, with some variability, with the bi-modal trends shown as heavy lines marked 'interpreted trends' on Figs. 6-8. Irrespective of steel grade, Figs. 6-8 show very clearly the effect of annual average water temperature (Table 2) on corrosion loss. This is not new and has been demonstrated both in field [6, 9, 10] and laboratory studies [15].

Comparing Figs. 7 and 8 for normal and for hardened steels respectively shows that at both sites the early parts of the trends, corresponding to Mode 1 of the bi-modal model [6] has somewhat less corrosion overall for the hardened steels compared with the normal steels. Also, the initial corrosion rates between the two sites are, roughly, three-four times for the Taylors Beach site compared with the Vladivostok site, and this is broadly consistent with the effect of water temperature as shown previously [6, 16]. However, this does not extend through to the longer term corrosion losses, for which the Vladivostok site shows a much greater relative loss when considered in terms of water temperature alone. As noted previously [9, 17], the longer term corrosion is directly affected by the concentration of dissolved inorganic nitrogen (DIN) in the water. As shown in Table 2, the annual average concentration of DIN at Vladivostok site is much higher than that at the Taylors Beach site, and this must be considered in any comparisons. However, it is noted that water temperature plays a part in this relationship [9].

4.2 Material Metallurgical Effect

It is also possible that microstructural influences have an effect, particularly in the early stages of corrosion. As noted, the corrosion and the rate of corrosion for the hardened steels is higher for the first few months of the first. A possible explanation for this is the microstructural effect of small crystal sizes, and hence longer overall length in grain boundaries enhancing kinetic corrosion reactions.

As can be seen in Figs. 4 and 5, typically low carbon steel ferrite (white) and pearlite (black) microstructure is present. It is interesting to note the differences in crystal concentration and distribution for the different steel grades. The micrographs of the coupons without hardening (A, B, D) show the presence of nearly 80% of ferrite-rich areas (i.e. Figs. 4a, 4b and 4c) while highly clustered ferrite crystals can be observed in steel coupons that were hardened (AH, DH) (e.g. Figs. 5a and 5b). It should be clear that a more homogeneous and finer distribution of ferrite and pearlite is associated with a longer total length of grain boundaries. Both conditions, grain boundary length and micro-segregation of elements, are considered to be linked to elevated severity of localised corrosion arrack [18]. Usually this leads to more numerous and deeper pits and hence higher corrosion mass losses [18].

From the microstructures of the sample offcuts investigated, it appears that the layout of lamellar cementite (pearlite) and ferrite regions has an influence on the early corrosion rates. As shown in Fig. 8, the experimentally observed mass loss of hardened steel grades AH and DH was higher compared to the ordinary steels. Based on the mechanistic theory of preferential grain boundary attack and influence of micro-alloying elements as noted by Butler et al. [18] the higher mass losses are likely to be attributable to a higher rate of localised corrosion across the entire surface, with areas of localized corrosion then amalgamating into what is perceived as a general corrosion. This assumed mechanism becomes plausible when taken into consideration the reported experimental observations of numerous shallow secondary pitting following localised pitting [19]. This is another area for further research, as currently on-going.

5. CONCLUSIONS

1. Average corrosion loss trends follow similar bi-modal pattern in the sub-Arctic region as well as in more moderate seawater climates.
2. The data for the Sub-Arctic observations indicate a much slower rate of development of the corrosion loss trends.
3. The hardened steels grades exhibited a noticeably higher rate of very early corrosion loss compared with normal steel grades, that is, during the first year of exposure.
4. The observed results provide support for the notion that increase in hardness promotes reduced crystal size, increases the total length of grain boundaries, and therefore increasing the susceptibility of hardened steel to corrosion during the aerobic phase of the bi-modal phenomenological model.

6. ACKNOWLEDGMENTS

The authors acknowledge the financial support provided by the Australian Research Council, the supply of the shipping grade steels for the coupons by Prof Jeom-Kee Paik, Pusan National University, Korea, and the valuable support of the EDX team at The University of Newcastle's Central Scientific Services. The continued support of the Fisheries Research Station at Taylors Beach (Department of Primary Industry, NSW) is very much appreciated.

7. REFERENCES

1. Schmitt G (2009) Global needs for knowledge dissemination, research, and development in materials deterioration and corrosion control, World Corrosion Organisation, New York

2. Val DV and Stewart MG (2003) Life-cycle cost analysis of reinforced concrete structures in marine environment, Structural safety (25) 343-362.

3. ABS (2014) Rules for Materials and Welding- Part 2, American Bureau of Shipping, 439 pages

4. Champion FA (1964) Corrosion testing procedures 2nd Ed., Chapman and Hall, 448 pages.

5. Lee JS, Ray RI and Little BJ (2010) The influence of experimental conditions on the outcome of laboratory investigations using natural coastal seawaters, Corrosion (66) 1–6.

6. Melchers RE (2003) Modelling of marine immersion corrosion for mild and low-alloy steels-Part 2: uncertainty estimation, Corrosion 59: 335-344.

7. Melchers RE and Chernov B (2010) Corrosion loss of mild steel in high temperature hard freshwater, Corrosion Science (52) 449-454.

8. ASTM G1 (2011a) Standard practice for preparing, cleaning and evaluating corrosion test specimens, ASTM International, Pennsylvania, USA.

9. Melchers RE (2014a) Long-term immersion corrosion of steels in seawaters with elevated nutrient concentration, Corrosion Science 81:110-116.

10. Melchers RE and Jeffrey R (2012) Corrosion of long vertical steel strips in the marine tidal zone and implications for ALWC, Corrosion Science 65: 26-36.

11. ASTM G52-00 (2013) Standard Practice for Exposing and Evaluation Metals and Alloys in Surface Seawater.

12. Chaves IA and Melchers RE (2011) Pitting corrosion in pipeline steel weld zones, Corrosion Science (53) 4026-4032.

13. Chaves IA, Jeffrey R and Melchers RE (2015) Rust removal from steel coupons after short-term marine immersion, Corrosion (70) 804-818.

14. ASTM E3 (2011) Standard guide for preparation of metallographic specimens, ASTM International, Pennsylvania, USA.

15. Mercer AD and Lumbard EA (1995) Corrosion of mild steel in water, British Corrosion Journal, 30: 43-55.

16. Melchers RE (2002) Effect of temperature on the marine immersion corrosion of carbon steels, Corrosion (58) 768-782.

17. Melchers RE and Jeffrey R (2013) Accelerated low water corrosion of steel pilling in harbours, Corrosion Engineering Science and Technology (48) 496-505.

18. Butler G, Stretton P and Beynon JG (1972) Initiation and growth of pits on high-purity iron and its alloys with chromium and copper in neutral chloride solutions, British Corrosion Journal (7) 168-173.

19. Jeffrey R and Melchers RE (2007) The changing topography of corroding mild steel surfaces in seawater, Corrosion Science (49) 2270-2288.

8. AUTHOR DETAILS

Boris B Chernov is a Professor in Chemical Engineering at the G.I. Nevelskoy Maritime State University, Vladivostok, Russia. His research interests include corrosion loss of mild steel in high temperature hard freshwaters. He is the author of numerous conference and journal papers.

Igor A Chaves is a Research Academic at the University of Newcastle, Australia. He is a Civil Engineer, holds a Master's degree in structural engineering and has been awarded a PhD for his studies in corrosion analysis and prediction. His research interests also include structural reliability and risk assessment of infrastructure.

He is the recipient of the ACA's 2011Young Corrosion Professional Award, 2012 ACA best paper Marshal Fordham Award, and 2016 Brian Cherry International Scholarship Award.

Anas M. Nugmanov is a Scientific employee at the G.I. Nevelskoy Maritime State University, Vladivostok, Russia.

Robert E Melchers is Professor of Civil Engineering at The University of Newcastle, Australia. He holds a BE and MEngSc from Monash University and a PhD from the University of Cambridge, UK. He is a Fellow of the Australian Academy of Technological Sciences and Engineering and an Honorary Fellow of The Institution of Engineers Australia.

His most recent awards are the 2009 ACA Corrosion Medal, the 2012 Jin S Chung Award (International Society of Offshore and Polar Engineers) and the 2013 John Connell Gold Medal (The Institution of Engineers, Australia). He was the Institution of Engineers, Australia Structural College Eminent Speaker for 2014.

LONG-TERM MARINE PITTING CORROSION OF ALMGSI ALUMINIUM ALLOYS

M. Liang[1], I. A. Chaves[1], R. E. Melchers[1]
[1]Centre for Infrastructure Performance and Reliability, The University of Newcastle, NSW2308, Australia

SUMMARY: AlMgSi aluminium alloys (i.e. 6000 series) are one of the so-called 'marine alloys' with good performance in aggressive marine environments. Advantages include high strength-to-weight ratio and good corrosion resistance. They are commonly used for modern small to medium size ship construction. There remains a need for more explicit long-term corrosion information for accurate lifetime prediction. This paper reports 2-year field exposure data for 6060-grade aluminium in temperate immersion, tidal and splash zone exposures. Duplicate sample strips were recovered from the tidal and atmospheric splash zones respectively. Corrosion losses were estimated and the surface morphology and chemistry of the corroded surfaces were investigated by scanning electron microscopy and energy dispersive spectroscopy. Extensive pit depth measurements were made using a bright light optical microscope and differential focus technique. The results show that the 6060-grade aluminium alloy suffered pitting corrosion and intergranular corrosion simultaneously in the marine environments. Crystallographic corrosion morphology was observed in some of the extreme pits. The AlMgSi alloy samples were most resistant to corrosion under tidal condition but less resistant in the splash zone. The most aggressive environment was seawater immersion at 0.64m depth. A statistical analysis using extreme value theory indicates that after 2 years' exposure the pitting corrosion mechanism of the maximum pit may be different from that for shallower pits. The reasons for this and the implications are discussed.

Keywords: Aluminium; Pitting corrosion; Seawater; Splash zone.

1. INTRODUCTION

AlMgSi aluminium alloys are extensively used as extrusion profiles in various applications, e.g. architecture, automobile, aerospace and other structural constructions (Davis 1999). They are one of the so-called 'marine alloys' with good performance in aggressive marine environments. Advantages include high strength-to-weight ratio and good corrosion resistance (Kaufman 2000, Vargel 2004). The excellent resistance to general corrosion in most environments for aluminium alloys benefits from a protective oxide film formed immediately after exposure to the air (Szklarska-Smialowska 1986, de la Fuente et al. 2007). This micron-scale film usually has very low solubility in the pH range of 4~9. Thus, for the typical surface seawater pH of around 8.2 (Gimenez et al. 1981), the aluminium alloy is in a passive state.

In marine environments, AlMgSi alloys may be susceptible to intergranular corrosion under one of the following conditions: 1) the Si/Mg ratio exceeds the stoichiometric ratio for the intermetallic compound $Mg_{\#}Si$ (Zeng et al. 2011); 2) unfavourable heat treatment is introduced during the progress of precipitation strengthening (Svenningsen et al. 2006); 3) high copper content (generally over 0.2wt.%) in the alloys (Zhan et al. 2008). Intergranular corrosion occurs only at the grain boundaries or closely adjacent regions which selectively dissolve due to the anodic characteristic caused by precipitates depletion. Pitting, however, is a very common form of corrosion attack for aluminium alloys in marine environment because of the high chloride concentration and high conductivity of seawater (Vargel 2004). Pitting may cause perforation and degradation of strength and thus pose a threat to structural components. It is of great interest for both researchers and engineers due to the difficulty of inspection and therefore prediction. Potential mechanisms of pitting initiation and propagation as well as the corresponding various influencing factors are well-documented (Szklarska-Smialowska 1986, Vargel 2004). They are mostly based on electrochemical experiments conducted in simulated seawater. Further, the exposure periods were very short, commonly expressed in hours, days or weeks (Melchers 2014). It is therefore of interest to explore the corrosion behaviour and in particular pitting corrosion for long-term field exposures (measured in years).

Very little data for longer-term marine exposures of aluminium alloys is reported in the open literature. Examples include the up to 16 years' exposure in five natural tropical environments reported by Southwell et al. (1965), some 10 years of seawater exposure of eight types of pure aluminium and 5000 series alloys reported by Alior (1974), 10 years' coastal atmospheric exposure at Alicante (Spain) documented by Otero et al. (1978), a 5 years' exposure program of testing seawater corrosivity around the world presented by Phull et al. (1997), and more recently a study of up to 20 years exposure in urban, coastal and industrial environments reported by Sun et al (2009).

For aluminium alloys in practice, usually only the deepest pit depths are of interest since they are the most probable ones to cause the first perforation of plates (Melchers & Jeffrey 2008). Extreme Value (EV) statistics has been used widely for aluminium alloy pitting, following Aziz (1956). For practical data sets, EV statistics also have been applied to pitting of oil well tubing (Eldredge 1957), pitting corrosion of carbon steels (Melchers 2005, Chaves & Melchers 2011, Chaves 2013, Wang & Melchers 2017) and cast ions (Asadi & Melchers 2017) and also corrosion induced cracking (Kowaka & Tsuge 1994). In each case, the Gumbel EV distribution was used to describe the uncertainty of pit depths. This can then be used to estimate the probability of plate perforation (e.g. Melchers 2008).

This paper reports 2-year field exposure data for 6060 aluminium exposed in temperate immersion, tidal and splash zones. Extreme pit depth data was analysed by using statistical extreme value theory. Corrosion morphologies and microstructures were also investigated. These investigations are described below, followed by a discussion of their implications.

2. EXPERIMENT AND PROCEDURES

2.1 Materials

The chemical composition of the grade 6060 T5 temper aluminium alloy test material is shown in Table 1 (obtained from a certified testing laboratory). No surface preparation was applied and all surfaces were exposed 'as-fabricated'.

Table 1. Composition of AlMgSi alloy (% by weight)

Si	Mg	Fe	Cu	Mn	Pb	Ni	Zn	Ti	Sn	Cr	Al
0.44	0.4	0.18	0.01	0.06	<0.01	<0.01	<0.01	0.01	<0.01	<0.01	Balance

2.2 Experiments

For the field exposures, duplicate aluminium sample strips (40mm wide and 6mm thick) were mounted horizontally on a 0.7m ×0.8m×3m fiberglass reinforced plastic frame and exposed to several natural seawater conditions. Two strips (T1 and T2) were mounted at tidal zone (0.2m from low tide water line), and four strips (S1, S2, S3 and S4) were attached at splash zone (1.4m from low tide water line). Fig. 1 shows a schematic view of coupon deployment.

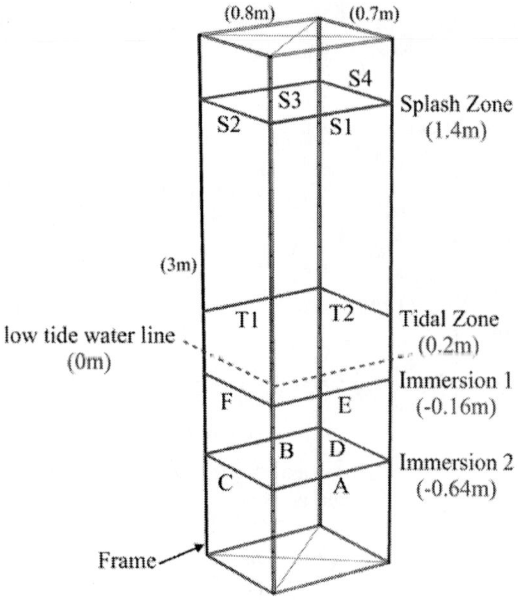

Figure 1 Schematic view of 6060 aluminium sample strips and their location.

The strips were exposed at the NSW Fisheries Research Station at Taylors Beach, Port Stephens, NSW, Australia (32.33S, 151.03E). Located 17 km from the open sea, the inlet continues a further 2 km inland and disperses onto a region of mangrove covered mud flats (Jeffrey & Melchers 2002).

All strips were recovered after 2 years. Light tap water and gentle nylon brush pre-cleaning stripped off fouling, corrosion products and sand sediments loosely attached to their surfaces. Each strip was then cut into six 60mm×40mm×6mm coupons and indentation-coded for identification. The coupons were cleaned following standard ASTM G1-03 (2011), i.e. immersed in the ultrasonic bath of a 90°C solution with 5 vol.% phosphoric acid and 2 wt.% chromium trioxides for 10 minutes, rinsed abundantly with distilled water and finally stored in a 56°C oven.

2.3 Measurements

Pit depths were measured by an upright white light optical microscope (Zeiss AXIO Imager A1) using the change in focal distance between the bottom and surrounding surface of a pit. Depth sensitivity was about 0.1 μm. This method is particularly suitable for small pits with narrow openings which are difficult to measure with a gauge probe type instrument (ASTM 2013). A surface profiler was not used in this case because of the large areas of the coupons of interest.

To avoid bias induced by edge effects, only the pits located in the central area (around 2mm distance from each edge) of the coupon were measured. A considerable number of pits (hundreds) were examined and their depths were recorded, on both sides of each coupon, from which the 5 deepest pit depths per coupon were extracted. Each depth reading was obtained from an average of three consecutive readings for the same pit. Scanning Electron Microscopy (Philips XL30 SEM) and Energy Dispersive Spectroscopy (Zeiss Sigma VP FESEM) were also utilised, to observe topography of and within the pits and to analyse the composition of different parts and materials of the coupons.

3. INTERPRETATION OF RESULTS

3.1 Pit Depth Data

As expected, for an aluminium alloy, pitting was observed as the dominant form of corrosion and general corrosion was observed, visually, to be negligible. There was considerable scatter in the pit depth data as shown in Fig. 2. It shows that each data set has a left-skewed distribution, corresponding to pits with lower depths, whereas the deeper pit depths on the right are spread out. For strips exposed at the same level, the data sets also show some degree of variability. This can be seen, for example, for immersion 2 for which the pit depths of strip D show a comparatively uniform increase from left to right. Data of strip A, however, has the highest scatter among the strips, with the smallest depth being 140μm but the deepest one being as high as over 2mm.

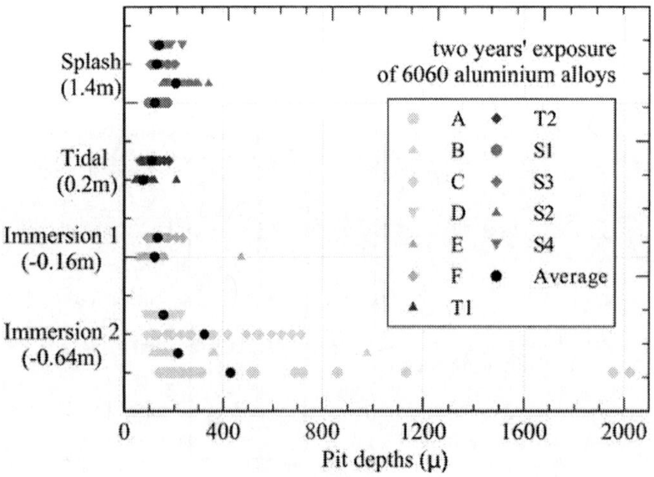

Figure 2 Extreme pit depths distribution.

Despite the scatter, it is clear that there are differences in pitting between the different exposure conditions. The average extreme pit depths (black dots in Fig. 2) are summarised in Table 2. It shows that the strips constantly immersed at 0.64m seawater depth have the largest average pit depth as well as the deepest pit. The strips exposed in tidal zone (0.2m) showed the smallest average pit depth. Generally similar results have been obtained for steels exposed at the same test site (Melchers & Jeffrey 2012).

Table 2. Average of deepest 30 extreme pit depths on 6060 aluminium sample strips

Average extreme pit depths (µm)	Strip #1	Strip #2	Strip #3	Strip #4	Total average
Splash	122	204	129	138	148
Tidal	74	110	~	~	92
Immersion 1	120	132	~	~	126
Immersion 2	429	215	323	156	281

To analyse the pit depth data, it is usual to select a range of maximum pit depths and sort them in ascending order with each data point assigned a cumulative probability. This is so-called 'rank order method' (Galambos 1987). The cumulative probability of ith pit depth $F_i(y)$ can be estimated from:

$$F_i(y) = \frac{i}{N+1} \quad (1)$$

where N represents the total number of pit depths considered and i is the rank order of the data. The pit depths (y) were then plotted on a Gumbel plot that has the vertical axis representing the standardised variable (w) obtained from (Galambos 1987):

$$w = -\ln(-\ln F_i(y)) \quad (2)$$

If the maximum pit depths (y) are indeed Gumbel-distributed, the data set plotted on a probability plot would present itself in a linear fashion (Galambos 1987). Following the method outlined above, the data for tidal and splash zones in Fig. 1 are plotted on Gumbel plots in Fig. 3 and Fig. 4 separately. The 'best-fit' lines for each data set ('T' for Tidal and 'S' for Splash) are also shown.

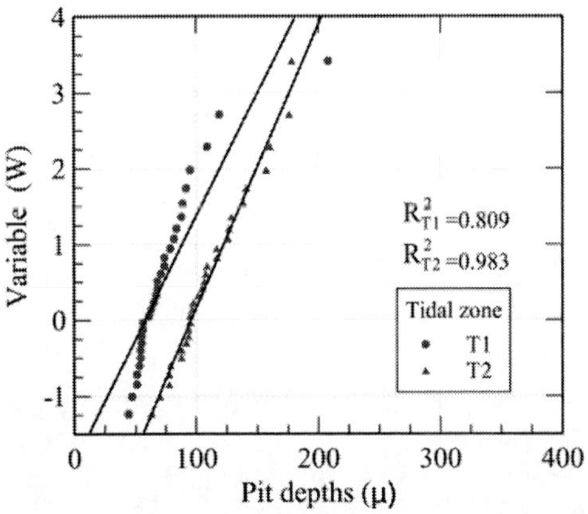

Figure 3 Gumbel probability plot showing maximum pit depths for 6060 aluminium sample strips exposed two years in tidal zone (0.2m).

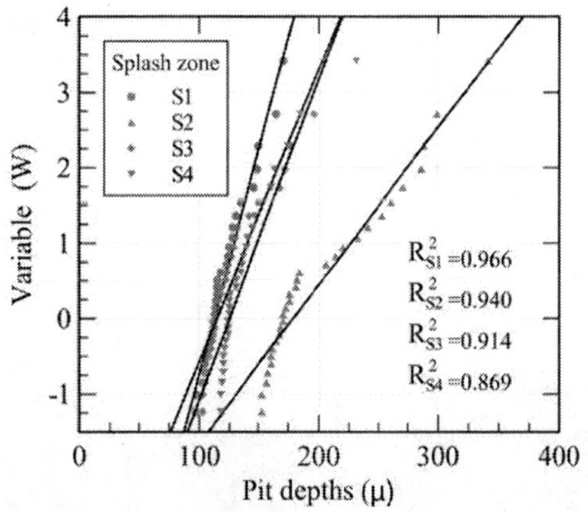

Figure 4 Gumbel probability plot showing maximum pit depths for 6060 aluminium sample strips exposed two years in splash zone (1.4m).

The remained question is to what degree does the maximum pit depth data follow the Gumbel probability distribution. The correlation coefficient of the best-fit line, based on the overall data deviation can be quantified by least squares method ($R^\#$). In the tidal zone (Fig. 3), data set T2 is well fitted by a single straight line with a high $R^\# = 0.983$. But the other data set T1 starts to deviate, and has a slightly lower $R^\# = 0.809$. As to the data in the splash zone (Fig. 4), comparatively high values of correlation coefficient were obtained for strips S1 to S3. However, the value of strip S4 is lower, being 0.869. It could be considered, based on these interpretations, that the Gumbel plot is not a good fit for the complete data sets in the tidal and splash zones. It is noted here that previously (Liang et al. 2016), pit depth data for immersion conditions, plotted on a Gumbel plot, showed a generally similar non-linear trend. This aspect is addressed further in the discussion section below.

3.2 Corrosion Morphology

Fig. 5 and Fig. 6 show the typical corrosion morphologies of the tidal zone and splash zone after exposure. The surfaces have been thoroughly cleaned by standardised procedures (ASTM 2011).

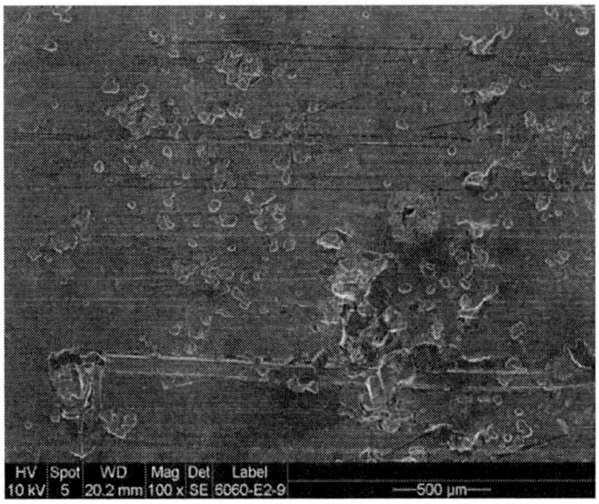

Figure 5 SEM image of typical corroded surfaces exposed in tidal zone (0.2m).

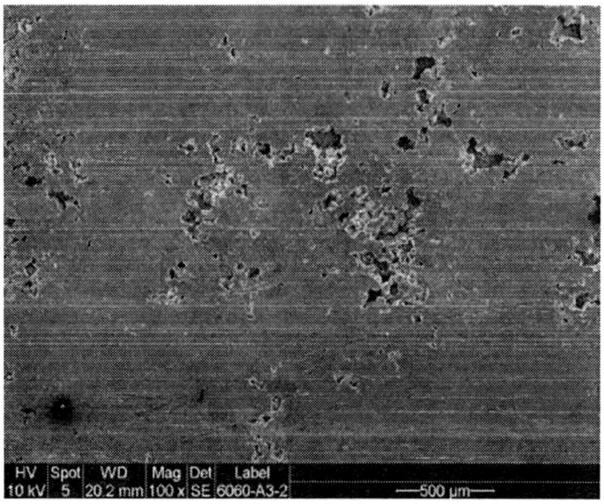

Figure 6 SEM image of typical corroded surfaces exposed in splash zone (1.4m).

In the present experiments, a large portion of the surfaces remained unattacked, with some clusters of pits apparently randomly distributed on the coupon surfaces. However, the pitting corrosion morphology was quite different between the tidal zone and the splash zone. Compared to coupons in the splash zone, the pits on the surface of coupons retrieved from tidal exposure generally have wide near-round openings and very shallow depths, often no deeper than 10 µm. Among those 'average' pits, some deeper pits were observed (as shown by arrows in Fig. 5). The magnified view in Fig. 7 shows an example of such deeper pit with irregular pit openings compared to the 'average' surrounding pits. This topography is unlike previously reported pits for aluminium alloys that were isolated and hemispherical (Ezuber et al. 2008). It is possible that such irregular pits can propagate along paths related to the grain boundaries of the alloy (Jones & Hoeppner 2009). Also, a corroded grain with distinct boundary edges can be observed in the centre of the pit (shown by red arrow in Fig. 7). This suggests that grain boundary attacks (or intergranular corrosion) potentially is possible in conjunction with pitting, at least for AlMgSi aluminium alloy exposed in the tidal zone.

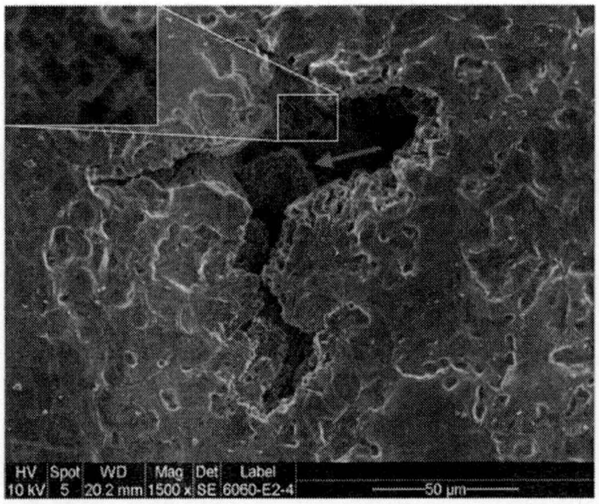

Figure 7 Magnified SEM image of typical extreme deep pits showing by arrows in Fig. 5.

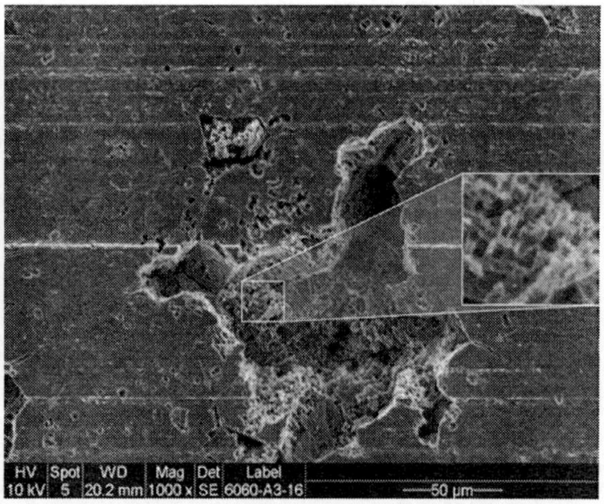

Figure 8 SEM image of magnified pits in splash zone.

For coupons exposed in the splash zone, the number of pits was much greater compared to the number of pits on coupons in the tidal zone. Pit clustering areas also appear more prevalent. Fig. 6 shows one of the typical pit clustering areas. Again, compared to the tidal zone, the pits in the splash zone are much deeper and appear more irregular in shape. The magnified view of pits in Fig. 8 suggests that the corrosion is actually related to intergranular corrosion (Davis 1999, Zeng et al. 2006, Svenningsen et al. 2006a). This is supported by what appear to be dendritic trenches at the edges of pits (shown by red arrows in Fig. 8). The sidewalls of pit present a feature suggesting whole grains had been peeled off and pulled out of the pit cavity. Taking these observations into consideration, it can be postulated that, the corrosion of aluminium alloy in the splash zone is the result of a synergy of different mechanisms including both pitting and intergranular corrosion.

Furthermore, it is interesting to note that crystallographic facets were observed in some deeper pits in the tidal zone and almost in all pits in the splash zone (see the zoom-in images in Fig. 7 and Fig. 8). Similar geometric facets have been found previously (Gimenez et al. 1981, Zaid et al. 2008), and this observation has been explained as likely the result of the attack occurring according to well-defined crystallographic directions in pure aluminium and aluminium alloys (Gimenez et al. 1981). It is likely that the corrosion mechanism of these pits with crystallographic features is different from that of the shallow broad pits. This is a matter for further investigation.

4. DISCUSSION

4.1 Influence of Microstructure

The microstructure of the surface (such as micro-defects and intermetallic particles) have some influence on the pitting corrosion of aluminium (Frankel 1998, Meng et al. 2009), but this has been considered mainly in terms of pitting susceptibility rather than pitting progression (Frankel 1998). For example, Fig. 9 shows the bulk microstructure of the 6060 aluminium alloy considered here, and for this the sizes of intermetallic particles (white spots) are slightly non-uniform, with diameters varying from hundreds of nanometres to several microns (blue arrows). This variation in particle size is likely to pose different potentials to the aluminium substrate during electrochemical reactions and thus may cause different rates of local corrosion. The grain sizes also are non-uniform, varying from tens of microns to more than one hundred microns. It has been reported that the corrosion rate of aluminium alloys had a tendency to change as grain size varies (Ralston et al. 2011). In addition, in this research, pitting in tidal and splash zones seems to be associated with intergranular corrosion and would appear to occur predominantly at the grain boundaries. As a result, the shape and size possibly effect the rate of corrosion propagation (Davis 1999). This could explain why there are differences between different strips data sets exposed to the same conditions.

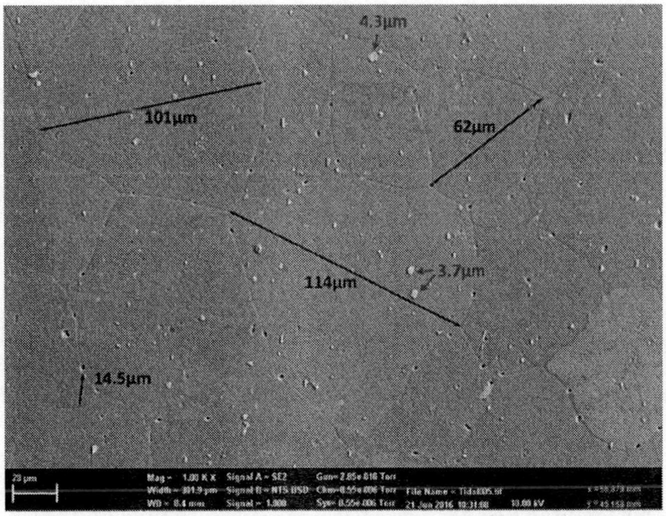

Figure 9 SEM image of the microstructure in the bulk material revealing different sizes of grains and intermetallic particles (scale bar=20μm).

Another possible factor leading to the data scatter between duplicate strips located at the same level is the orientation of strips. During the long-term exposure, the prevailing wind driven corrosive salt spray might reach the surface of strips. On the other hand, the rain may wash away the contaminants brought by the wind (Walton & King 1955). Overall, the effect of prevailing wind is considered to be minor for the present results (Jeffrey & Melchers 2008).

4.2 Extreme Value Analysis

What also interesting is, despite some linearity (e.g. high $R^{\#}$ values for data set of strip T2 in Fig. 3 and for strips S1 and S2 in Fig. 4), not all data sets can be well fitted by Gumbel plot. Note that the maximum pit depth data point on strip T1 (top right in Fig. 3) clearly influences the overall deviation. It such extreme maximum pit is considered as an outlier and therefore omitted, the goodness-of-fitting improved significantly (the value of $R^{\#}$ increased from 0.809 up to 0.975) as shown in Fig. 10.

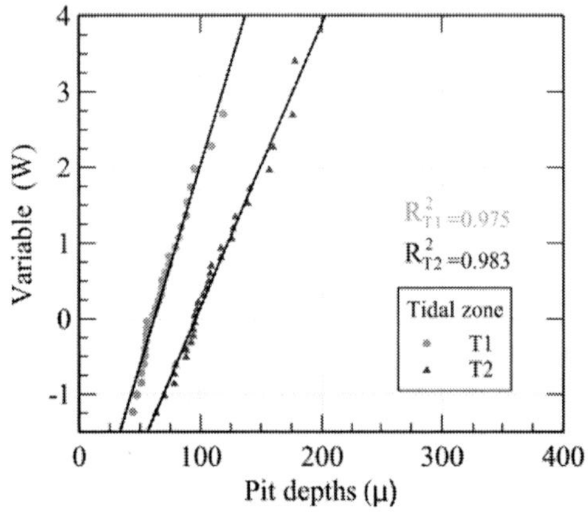

Figure 10 Replotted data in Fig. 3 with the maximum pit depth data being deleted.

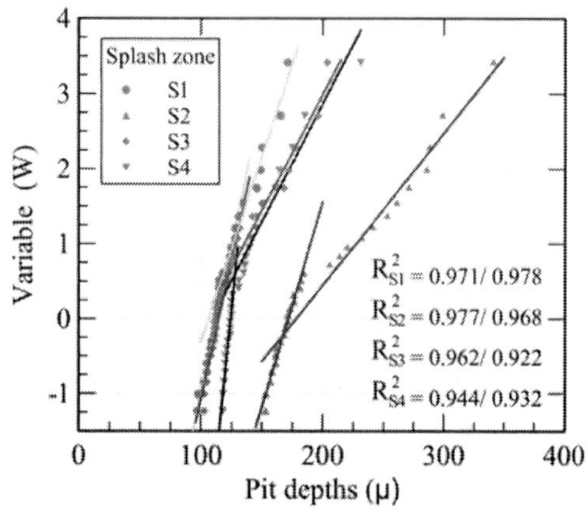

Figure 11 Gumbel probability plot with interpreted trend lines showing the same data in Fig. 4.

As for splash zone, each data set can be subdivided into two subsets and best-fitted by two straight lines (Fig. 11). The corresponding correlation coefficient values for each data set increased compared to that of the single straight line. For example, when the data set for strip S4 is fitted by a single straight line, the value of $R^{\#}$ is 0.869 (see Fig. 4). But if the data set is divided into two groups (i.e. lower extreme pit depths and deeper extreme pit depths) and each group is fitted by the same method, the values of $R^{\#}$ became 0.944 and 0.932, suggesting a better goodness-of-fitting. The lower pit depth data (lower left of Fig. 11) show little variation about linearity, and suggest the data are consistent with the Gumbel probability distribution. The pit depth data for deeper pits, however, show more deviations and the trend line is somewhat flatter compared to the trend line for the shallower pits. This means that it is likely that the population of the deeper extreme pits is different from that for the shallower extreme pits. This phenomenon also has been observed for pitting of other aluminium alloys exposed in wet environments (Melchers 2005).

Following from the above, it can be noted that two main issues should be considered when applying statistical analysis: the homogeneity of the pit depth data and the representativeness of sampling (Melchers 2005). They are important for developing an appropriate probabilistic model for the deepest pits.

The pattern shown in Fig. 11 for maximum pit depth probability distribution may be related to earlier suggestions (Melchers 2005), that the deeper extreme pits belong to the population that initiates at first exposure and then grows stably into extreme pits, while the pits with smaller depths constitute another population that repassivates shortly after first exposure and that pits tend to grow somewhat randomly during later exposure. This may explain the similar fitting pattern in Fig. 11, each data set can be divided into two statistical populations, i.e. the smaller extreme pits and the deeper extreme pits.

In addition, Fig. 12 shows the cross-section of the pit with maximum depth observed in the splash zone. It is interesting to note that crystallographic facets were not observed in this pit. Instead, the internal topography shows dimple-like smooth walls. Further, the space in the pit is very irregular, suggesting a complex pitting corrosion progress. There appears no evidence of intergranular corrosion in the pit. Such difference in pitting morphology compared to that of pits in Fig. 7 ~ 8 may be a result of different pitting mechanisms, and thus contribute to the deviation of deeper extreme pits from the single straight line in Gumbel plot.

In this context, it may be speculated that the maximum depth pit on coupons in the tidal zone is the result of an accelerated corrosion rate, controlled under a different dominant pitting mechanism compared to the mechanism for the less deep pits. A possible reason is the randomly distributed nature of intermetallic particles as these facilitate pitting corrosion (Frankel 1998). The AlFeSi type (Fig. 13(b)) intermetallic particles (spots 1-4 in Fig. 13(a)) appear to be randomly distributed in the material (see also Fig. 9). These particles are likely to be cathodic to the matrix and thus promote anodic dissolution (i.e. pitting) of aluminium substrate during corrosion (Yasakau et al. 2007). There are some fine white intermetallics dispersed in the dark grey corrosion products (spots 5-6) in the centre of Fig. 13(a). It is feasible that the corrosion process, when it reaches the area where impurity particles have accumulated, will allow the propagation rate to increase.

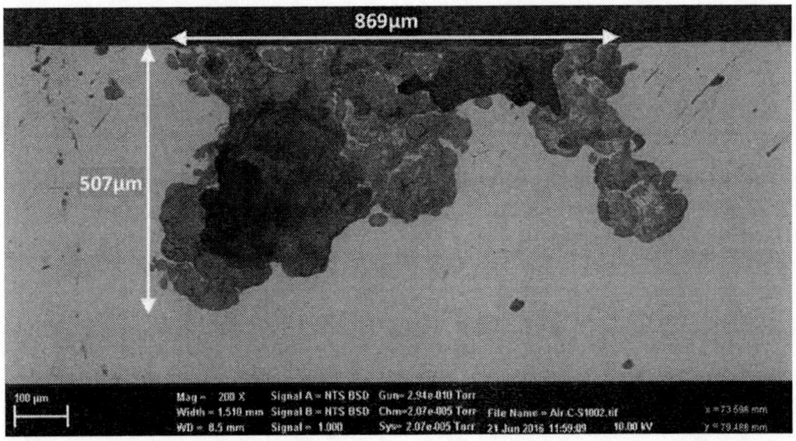

Figure 12 SEM image of the cross-section for maximum pit on coupon exposed in the splash zone (scale bar=100μm).

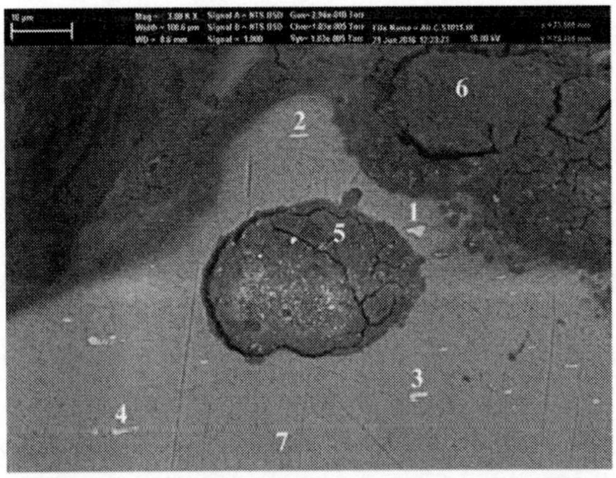

(a)

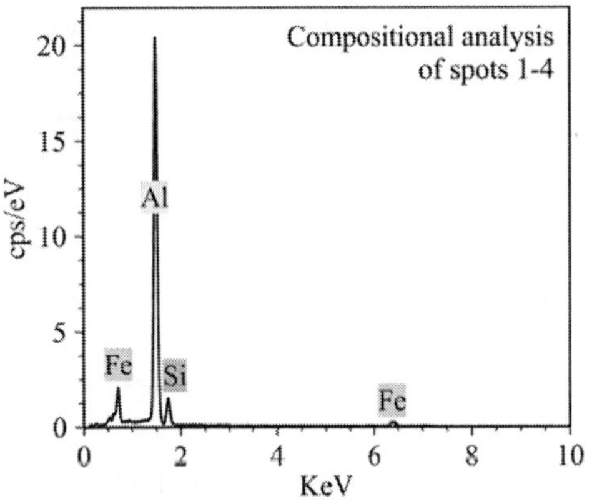

(b)

Figure 13 (a) Cross-section image of maximum pit on coupon exposed in the splash zone and (b) EDS analysis result of intermetallic particles in the material (scale bar=10μm).

Due to the limited number of sample strips, the probability of obtaining pits of certain large depths decreases (Aziz & Godard 1955). Only a limited number of extreme depth pits can be expected to form in a given area (Melchers 2005). So, it is likely that the area of exposure in the tidal zone is not large enough to capture additional deep pits formed with the same pitting mechanism as the extreme pit shown in Fig. 12. This also is an area for further study.

5. CONCLUSIONS

1) Grade 6060 T5 aluminium alloys showed the best resistance to pitting corrosion under marine tidal condition but they were less resistant in the splash zone. The most aggressive environment for AlMgSi alloys in this research was seawater immersion at 0.64m depth.

2) Under marine conditions, 6060 aluminium alloys showed both pitting corrosion and intergranular corrosion simultaneously. The more severe intergranular corrosion was considered to contribute to the increased penetration rate under splash zone and immersion conditions.

3) The complete sets of extreme pit depths data could not be fitted to the conventional Gumbel Extreme Value probability distribution function. The data for the splash zone could, however, be or better fitted by two distinct straight lines on a Gumbel plot.

4) Consistent with some other observations in the literature, crystallographic facets were observed in the pits for AlMgSi alloys in marine environments. Why this was not observed for the deepest pit in the splash zone remains a matter for further research.

6. ACKNOWLEDGMENTS

The financial support provided by the University of Newcastle and the Australian Research Council is gratefully acknowledged. The authors also acknowledge the technic support from the Central Scientific Service of The University of Newcastle.

7. REFERENCES

Ailor WH (1974) Ten-year seawater tests on aluminum, Corrosion in Natural Environments 558 117-134.

Asadi ZS and Melchers RE (2017) Extreme value statistics for pitting corrosion of old underground cast iron pipes, Reliability Engineering & System Safety 162 64-71.

ASTM (2011) Standard practice for preparing, cleaning, and evaluating corrosion test, G1-03: 1-11.

ASTM (2013) Standard guide for examination and evaluation of pitting corrosion, G46-94: 1-8.

Aziz PM (1956) Application of the statistical theory of extreme values to the analysis of maximum pit depth data for aluminum, Corrosion 12(10) 35-46.

Aziz PM and Godard HP (1955) Influence of specimen area on the pitting probability of aluminum, Journal of The Electrochemical Society 102(10) 577-579.

Chaves IA (2013) Development of behaviour models for pitting corrosion of mild steel pipeline welds in marine, PhD Thesis 65-77.

Chaves IA and Melchers RE (2011) Pitting corrosion in pipeline steel weld zones, Corrosion Science 53 4026-4032.

Davis JR (1999) Corrosion of aluminum and aluminum alloys, ASM International.

Eldredge GG (1957) Analysis of corrosion pitting by extreme-value statistics and its application to oil well tubing caliper surveys, Corrosion 13 67-76.

Ezuber H, El-Houd A and El-Shawesh F (2008) A study on the corrosion behavior of aluminum alloys in seawater, Materials & Design 29(4) 801-805.

Frankel GS (1998) Pitting corrosion of metals a review of the critical factors, Journal of The Electrochemical Society 145(6) 2186-2198.

Galambos J (1987) The asymptotic theory of extreme order statistics, Second Edition, Kluwer, New York.

Gimenez P, Rameau JJ and Reboul MC (1981) Experimental pH potential diagram of alumnium for sea water, Corrosion 37(12): 673-682.

Jeffrey R and Melchers RE (2002) Shape and size effects for marine immersion coupons, British Corrosion Journal 37(2) 99-104.

Jeffrey R and Melchers RE (2008) Three year observations of corrosion losses for steels at a severe marine atmospheric site, (In) Corrosion and prevention 2008, Wellington, New Zeland, paper 039.

Jones K and Hoeppner DW (2009) The interaction between pitting corrosion, grain boundaries, and constituent particles during corrosion fatigue of 7075-T6 aluminum alloy, International Journal of Fatigue 31(4) 686-692.

Kaufman JG (2000) Introduction to aluminium alloys and tempers, ASM Internatial.

Kowaka M and Tsuge H (1994) Introduction to life prediction of industrial plant materials: application of the extreme value statistical method for corrosion analysis, New York: Allerton Press.

De la Fuente D, Otero Huerta E and Morcillo M (2007) Studies of long-term weathering of aluminium in the atmosphere, Corrosion Science 49(7) 3134-3148.

Liang MX, Chaves IA and Melchers RE (2016) Long-term pitting corrosion of 6060 aluminium alloy in natural seawater, (In) Corrosion and prevention 2016, Aukland, New Zeland, paper 70.

Melchers RE (2014) Bi-modal trend in the long-term corrosion of aluminium alloys, Corrosion Science 82 239-247.

Melchers RE (2008) Extreme value statistics and long-term marine pitting corrosion of steel, Probabilistic Engineering Mechanics 23(4) 482-488.

Melchers RE (2005) Statistical characterization of pitting corrosion-Part 1: Data analysis, Corrosion 61(7) 655-664.

Melchers RE and Jeffrey R (2012) Corrosion of long vertical steel strips in the marine tidal zone and implications for ALWC, Corrosion Science 65 26-36.

Melchers RE and Jeffrey R (2008) Probabilistic models for steel corrosion loss and pitting of marine infrastructure, Reliability Engineering and System Safety 93(3) 423-432.

Meng G, Wei L, Zhang T, Shao Y, Wang F and Dong C et al. (2009) Effect of microcrystallization on pitting corrosion of pure aluminium, Corrosion Science 51(9) 2151-2157.

Otero E, Lizarbe R and Feliu S (1978) Behaviour of aluminium during 10 years' exposure at a coastal test site at alicante, British Corrosion Journal 13(2) 82-84.

Phull BS, Pikul SJ and Kain RM (1997) Seawater corrosivity around the world: results from five years of testing, Corrosion Testing in Natural Waters ASTM STP 1300: 34-73.

Ralston KD, Fabijanic D and Birbilis N (2011) Effect of grain size on corrosion of high purity aluminium, Electrochimica Acta 56(4) 1729-1736.

Southwell CR, Alexander AL and Hummer CW (1965) Corrosion of metals in tropical environments- aluminum and magnesium, Materials Protection 4(12) 30-35.

Sun S, Zheng Q, Li D and Wen J (2009) Long-term atmospheric corrosion behaviour of aluminium alloys 2024 and 7075 in urban, coastal and industrial environments, Corrosion Science 51(4) 719-727.

Svenningsen G, Larsen MH, Nordlien JH and Nisancioglu K (2006a) Effect of high temperature heat treatment on intergranular corrosion of AlMgSi(Cu) model alloy, Corrosion Science 48(1) 258-272.

Svenningsen G, Larsen MH, Nordlien JH and Nisancioglu K (2006b) Effect of thermomechanical history on intergranular corrosion of extruded AlMgSi(Cu) model alloy, Corrosion Science 48(12) 3969-3987.

Svenningsen G, Larsen MH, Nordlien JH and Nisancioglu K (2006c) Effect of artificial aging on intergranular corrosion of extruded AlMgSi alloy with small Cu content, Corrosion Science 48(12) 1528-1543.

Szklarska-Smialowska Z (1986) Pitting Corrosion of Metals, NACE, Houston.

Vargel C (2004) Corrosion of Aluminium, Elsevier, London.

Walton CJ and King W (1955) Symposium on atmospheric corrosion of non-ferrous metals/resistance of alumnium-base alloys to 20-year atmospheric exposure, (In) 50th annual meeting of American Society for Testing Materials: 1-158.

Wang X and Melchers RE (2017) Long-term under-deposit pitting corrosion of carbon steel pipes, Ocean Engineering 133 231-243.

Yasakau KA, Zheludkevich ML, Lamaka SV and Ferreira GS (2007) Role of intermetallic phases in localized corrosion of AA5083, Electrochimica Acta 52 7651-7659.

Zaid B, Saidi D, Benzaid A and Hadji S (2008) Effects of pH and chloride concentration on pitting corrosion of AA6061 aluminum alloy, Corrosion Science 50(7) 1841-1847.

Zeng FL, Wei ZL, Li JF, Li CX, Tan X and Zhang Z et al. (2011) Corrosion mechanism associated with Mg_2Si and Si particles in Al-Mg-Si alloys, Transactions of Nonferrous Metals Society of China (English Edition) 21(12) 2559-2567.

Zhan H, Mol JMC, Hannour F, Zhuang L, Terryn H and De Wit JHW (2008) The influence of copper content on intergranular corrosion of model AlMgSi(Cu) alloys, Materials and Corrosion 59(8) 670-675.

8. AUTHOR DETAILS

Mengxia Liang is a PhD student in The University of Newcastle, Australia. She holds a Master's degree in Material Processing Engineering from Sichuan University, China. Her research topic is the long-term pitting corrosion behaviour of aluminium alloys.

Igor A Chaves is a research academic at the University of Newcastle, Australia. He is the recipient of the 2012 Marshall Fordham Award (Australasian Corrosion Association) and the 2011 Young Corrosion Professional Award. He is a civil engineer and holds a Master's degree in structural engineering and a PhD in his studies in corrosion analysis and prediction. His research interests also include structural reliability and risk assessment of infrastructure.

Robert E Melchers is Professor of Civil Engineering at The University of Newcastle, Australia. He holds a BE and M Eng Sc from Monash University and a PhD from the University of Cambridge, UK. He is a Fellow of the Australian Academy of Technological Sciences and Engineering and an Honorary Fellow of The Institution of Engineers Australia.

His most recent awards are the 2009 ACA Corrosion Medal, the 2012 Jin S Chung Award (International Society of Offshore and Polar Engineers) and the 2013 John Connell Gold Medal (The Institution of Engineers, Australia). He was the 2014 Eminent Speaker for the College of Structural Engineers, The Institution of Engineers, Australia.

ASSESSING THE ORIGINS OF CORROSION ON STAINLESS STEELS

G. Sander[1,2], V. Cruz[1,2], M. Brameld[2,3], N. Birbilis[1,2]
[1]Department of Materials Science and Engineering, Monash University, Clayton, Victoria, 3800, Australia
[2]Woodside Innovation Centre, Monash University, Clayton, Victoria, 3800, Australia
[3]Woodside Energy, 240 St Georges Terrace, Perth WA 6000, Australia

SUMMARY: High resolution optical imaging was carried out during the *in-situ* imaging of 316L stainless steel subject to potentiostatic polarisation. Experiments were performed in a custom open-top 3D printed electrochemical cell. The tested specimens were maintained at a fixed potential of 50 mV below their respective pitting potential in order to produce, and subsequently allow the quantification, of metastable pit formation. Testing was carried out in 0.1 M sodium chloride (NaCl) solution, whilst the electrochemical data and images were simultaneously recorded and analysed to reveal metastable pit characteristics and behaviour. The physical and chemical mechanisms of metastable pit nucleation, growth and repassivation are also elaborated herein. The specimens studied include 316L stainless steel additively manufactured (i.e. 3D printed) by Selective Laser Melting (SLM), allowing a comparison with wrought 316L.

1. INTRODUCTION

The pitting corrosion of stainless steels (SS) in saline or acidic environments has been a constant concern and topic of investigation for scientists and engineers. This localised form of attack can be severe, leading to premature failures in a diverse range of engineering structures [1-4]. The formation of pits is still considered to be unpredictable in regards to where (on the material surface) and when (during the material lifetime) they will occur. A general assessment of pitting corrosion can be performed through electrochemical polarisation tests such as cyclic-potentiodynamic polarisation (CPP) [5] and exposure tests such as immersion in $FeCl_3$ solution [6]. However, such methods, and most available testing methods for pitting have limitations with respect to providing information required to predict and compare pitting propensity between similar alloys. It has been hypothesised elsewhere [7-8] that pit devolvement occurs in three stages: pit nucleation, metastable pitting and stable pit propagation. The nucleation of pits is usually associated to micro electrochemical cells close to heterogeneities on an alloy surface [2] which may lead to the dissolution of the protective oxide layer, bare metal exposure and consequently, localised attack. The nucleated pit will experience a metastable stage where it could either spontaneously undergo repassivation or grow further into the metal [9]. The role of chromium in SS allows the repair of damage on its oxide film, which can retard metastable pit transition to stability [2]. The formation and propagation of stable pits on stainless steels will occur if the bare metal surface is in contact with an aggressive electrolyte, and if the chromium concentration in the alloy (locally or overall) cannot successfully regenerate a passive oxide layer [10]. The pitting potential (E_{pit}) defines the potential in which a passive layer for an alloy cannot be sustained, with stable pitting occurring freely for potentials more noble to E_{pit} [11]. The detailed mechanisms of pit nucleation and metastable pitting behaviour are not yet fully understood.

Selective Laser Melting (SLM) is an additive manufacturing (also known as 3D printing) process that consists of local melting using a laser beam focused on a powder bed. Components are built-up layer by layer until a final part is completed [12-13]. Complex designs may be produced using the SLM process, employing computer aided design (CAD) [14]. The production of alloys through additive manufacturing has been reported to influence mechanical properties [15-18], microstructure [19-21] and durability [22-25]. The visual and electrochemical investigation of 316L SS specimens produced conventionally and produced using SLM is investigated in this work.

2. EXPERIMENTAL METHODS

2.1. Sample Preparation and Experimental set-up

Wrought and additive manufactured (AM) AISI 316L stainless steel specimens were employed to perform a comparative characterisation of metastable pitting in this study. The AM specimens were produced by SLM using 316L powder (Sandvik-Osprey, UK) with a particle size distribution ranging from 10 to 53 µm. The SLM process was performed using an EOS M280 instrument, with a Yb-fibre laser (1070nm) of 100 µm diameter, a 165 W laser power and 960 mm/s laser scan speed in a nitrogen-purged atmosphere. Wrought 316L rods were obtained from a distributor in Victoria, Australia. The test specimens were cut into small parallelepipeds and cold mounted in epoxy resin, with an electrical connection to the specimen through the resin. Specimens were metallographically prepared by grinding the surface using silicon carbide (SiC) paper to a 2000 grit finish. Some specimens which were to be used for optical imaging were also polished using diamond paste suspensions to a 1 µm mirror finish. All specimens were ultra-sonically cleaned prior to electrochemical testing and dried using compressed air. Enamel laquer was used to electrically isolate the metal/epoxy interfaces, and to eliminate the possibility of crevice corrosion. The electrolyte used in all electrochemical testing was 0.1 M NaCl.

2.2. Electrochemical tests

Specimens were tested using an electrochemical flat cell (Princeton Applied Research, K0235) employing a saturated calomel (SCE) reference electrode, a platinum mesh as the counter electrode). Cyclic potentiodynamic polarisation (CPP) tests were performed using a scan rate of 1 mV/s from – 25 mV_{SCE} below the OCP to + 1000 mV_{SCE} reversing the scan direction when the anodic current density reached 100 $\mu A/cm^2$. The typical average pitting potentials of specimens were determined using cyclic potentiodynamic polarisation. The CPP test was performed 5 times for each specimen to confirm reproducibility of the collected data. Metastable pitting was characterised using chronoamperometry (CA) where a fixed potential of 50 mV below the respective E_{pit} value, was applied to each specimen. In-situ imaging was performed simultaneously with the CA experiment, using the set-up presented in Figure 1. The electrochemical cell was designed using CAD software and 3D printed by fused deposition modelling using Acrylonitrile-Butadiene-Styrene (ABS) as the build material. The cell was designed to hold the specimen being analysed for optical microscopy observation, in-situ during the electrochemical testing. The same of reference and counter electrodes used for the CPP testing were used for the CA testing. In addition, a high-definition video imaging of the specimen surface was recorded during each CA testing.

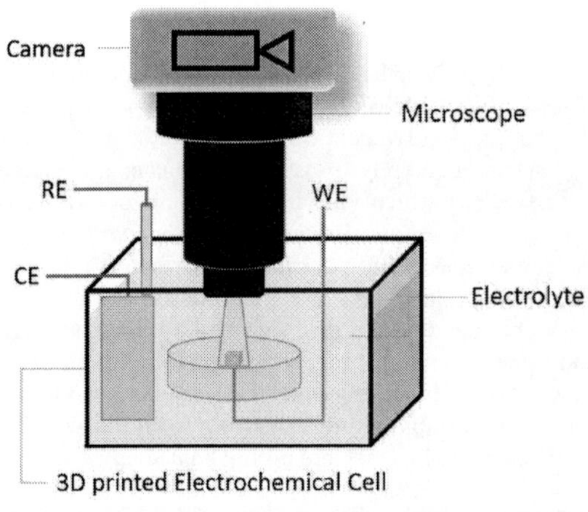

Figure 1: Custom electrochemical cell 3D printed using fused deposition modelling, and experimental set-up schematic of the chronoamperometry testing.

3. RESULTS AND DISCUSSION

The E_{pit} of the wrought and SLM 316L SS specimens were measured and found to be in the range 30 mV_{SCE} to 95 mV_{SCE} (Figure 2) having average potentials of 32.5 mV_{SCE} and 90 mV_{SCE} respectively. Sander *et al.* [22] have suggested that this distinction of E_{pit} values is caused by different inclusion compositions, sizes and porosities in the microstructures of AM 316L stainless steels. The average of the measured E_{pit} values were used to define the anodic polarisation potentials for the CA testing; resulting in the wrought 316L being CA tested at + 27.5 mV_{SCE} and the SLM 316L CA tested at + 85 mV_{SCE}.

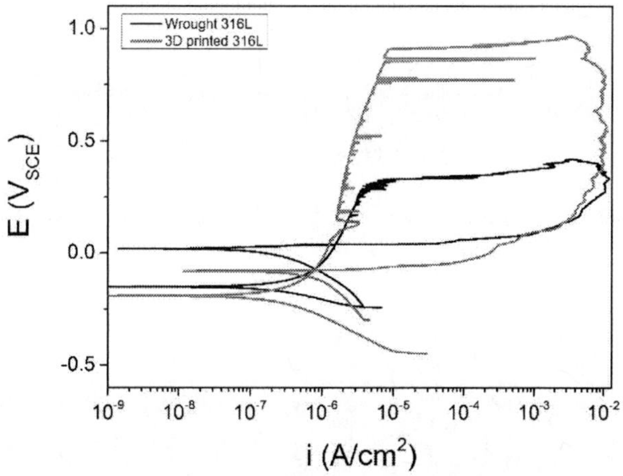

Figure 2: Cyclic-potentiodynamic polarisation urves for the wrought, and the selective laser melting, 316L SS in 0.1 M NaCl.

The potentials immediately below E_{pit} increase the probability to develop metastable pits. These events can be detected and quantified by current density peaks during electrochemical polarisation tests [11]. The metastable pit frequency (Mpf) of the specimens were calculated based on the number of peaks (N) detected in the current density vs time plot. The exposed surface area (A) and the test duration (t) were substituted in Eqn (1) to obtain the Mpf values for each specimen.

$$Mpf = \frac{N}{A\,.\,t} \quad (cm^{-2}.s^{-1}) \qquad (1)$$

The SLM 316L specimen had an Mpf of 0.29 $cm^{-2}.s^{-1}$ whilst the wrought sample had a higher Mpf (0.41 $cm^{-2}.s^{-1}$). This would suggest that the metastable pitting events, whilst different in Mpf, were overall however not too dissimilar for both specimens. However, the intensity of peaks was visibly higher on the wrought 316L as a function of time (Figure 3). The charge passed through the metastable pits (calculated by peak integration) was significantly different, being 21 mC/cm^2 for SLM 316L and 116.3 mC/cm^2 for wrought 316L. This indicates that the material damage arising from metastable pitting was higher for wrought 316L compared to SLM 316L.

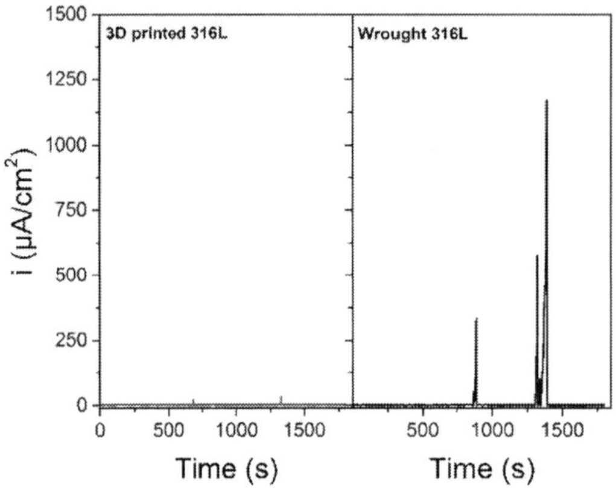

Figure 3: Chronoamperometry results measured from potentiostatic testing of wrought, and the selective laser melting, 316L SS.

The simultaneous in-situ imaging allowed observation and quantification of damage arising from metastable pitting events. The SLM 316L had a number of pores on its surface pre-exposure (Figure 4.a) while the wrought 316L specimen was free of porosity (Figure 5.a).

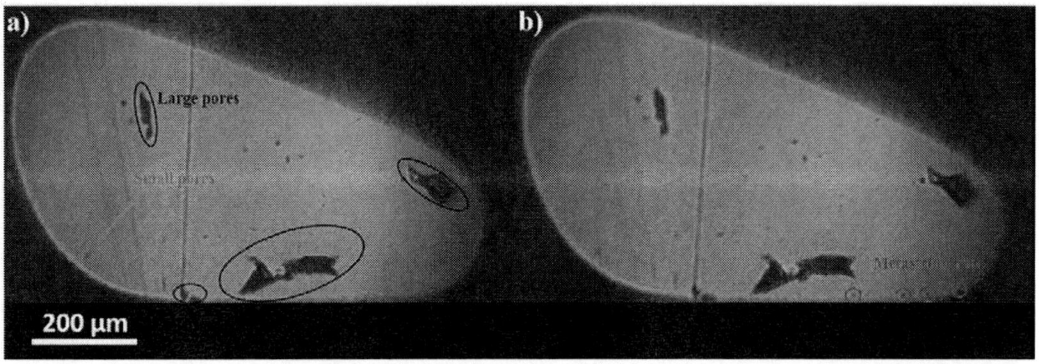

Figure 4: Snapshots from video imaging during CA testing of SLM 316L at 85 mV_{SCE} in 0.1 M NaCl: a) at thecommencement of the testing highliting the sample's pores and b) at the end of the testing highliting the metastable pits formed.

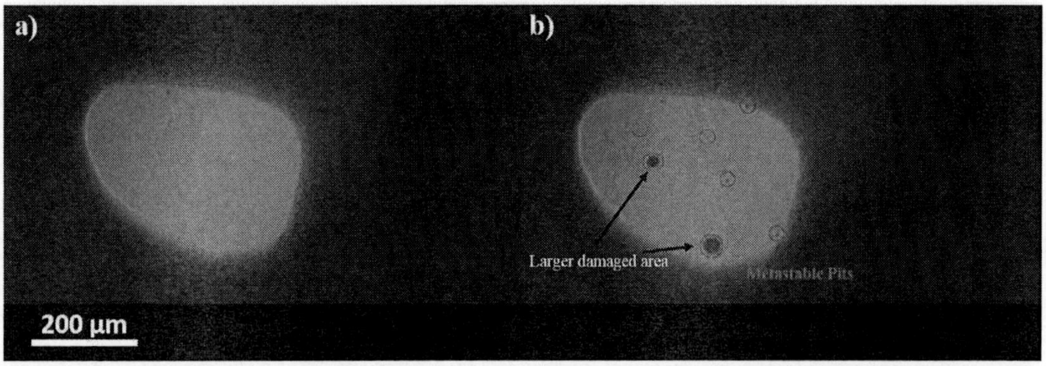

Figure 5: Snapshots from video imaging during CA testing wrought 316L at $27.5mV_{SCE}$ in 0.1 M NaCl: a) at the commencemnt of the testing and b) at the end of the test highliting the metastable pits formed.

Following the potentiostatic polarisation (CA testing), the two analysed specimens revealed evidence of damage on their respective surfaces. The SLM sample suffered less damage on its surface (Figure 4.b) than the wrought 316L (Figure 5.b). The metastable pits presented as circular shapes and were not nucleated near any residual pores from the SLM production process, which can be observed from Figures 4.a and 4.b. Pores may be potential pit nucleation sites if their volumes were reduced to micro crevice dimensions, however this was not studied or observed in this study. The presence of such porous features on the surface could potentially allow an oxygen concentration gradient and high concentration of anionic species (i.e. chloride) increasing the acidity inside the cavities [2, 10]. Acidic environments (especially rich in Cl^- anions) allow the chromium to undergo hydrolysis which has been shown to facilitate oxide layer breakdown [10]. The structural effects on corrosion for SLM prepared specimens, although they appear to have enhanced corrosion resistance from the tests herein, do merit specific future works. The microstructures of SLM 316L have been found to differ from the conventional produced specimens mostly due to faster cooling rates when compared to conventional manufacturing processes [26-28]. This difference in the solidification process can potentially affect inclusions' characteristics. The alloying elements that form inclusions in these alloys wouldn't have enough time to diffuse, nucleate and grow to detrimental sizes without post-printing heat treatment. Therefore, it is expected that the SLM 316L would have a reduced number of (chemically distinct) pit nucleation sites, justifying the increase in pitting corrosion resistance as observed in the present work. The damage on the wrought SS was more drastic and larger pits were formed as shown in Figure 5.b. The presence of MnS inclusions in conventionally produced austenitic SS' microstructures has been associated to decreasing the materials resistance to pitting corrosion [3, 8-9]. This association supports the hypotheses that the SLM 316L may develop inclusions with different compositions or smaller sizes when compared to the wrought 316L. The study presented herein aimed to be a preliminary investigation of the pitting corrosion on SS utilising two 316L specimens produced by different manufacturing methods.

4. CONCLUSIONS

Stainless steel 316L specimens were investigated using potentiodynamic and potentiostatic polarisation tests in order to assess pitting corrosion. The samples were produced by two different manufacturing methods: wrought and additive manufacturing (Selective Laser Melting). The respective pitting potentials (E_{pit}) were determined in order to quantify the electrochemical potential range in which metastable pits would be evidence (and capable of subsequent quantification). Metastable pits events were quantified while in-situ video imaging was performed. The work herein allowed the following summary:

- The SLM 316L was more resistant to pitting corrosion than wrought 316L in 0.1 M NaCl.
- The average E_{pit} of the SLM sample was 62.5 mV_{SCE} more noble than the average E_{pit} of the wrought 316L
- Both (wrought and SLM) 316L samples developed metastable pits when polarised 50 mV_{SCE} below their respective E_{pit}. The Mpf was similar for the analysed specimens whilst the charge passed through the metastable pits were significantly different, the charge (representing the amount of alloy dissolution) was notably higher for the wrought 316L in comparison to the SLM 316L.
  The pores in the SLM 316L specimens, do not appear to serve as pit nucleation sites.
- Works to date have suggested that inclusions may be the main pit nucleation sites in stainless steels, and herein we suggest that SLM production can have a beneficial influence on inclusion annihilation.
- Significant future work remains in order to holistically assess the corrosion of SLM prepared alloys, incorporating chemical analysis, surface analysis and electron microscopy.

5. AKNOWLEDGEMENTS

The authors are grateful to Woodside Energy and the Monash Centre of Additive Manufacturing. We also thank Mr. Marten Jurg for technical assistance.

6. REFERENCES

1 – Frankel GS, Sridhar N. Understanding localized corrosion. Materials Today. 2008 Oct 31;11(10):38-44.

2 – Frankel GS. Pitting corrosion of metals a review of the critical factors. Journal of the Electrochemical Society. 1998 Jun 1;145(6):2186-98.

3 – Ryan MP, Williams DE, Chater RJ, Hutton BM, McPhail DS. Why stainless steel corrodes. Nature. 2002 Feb 14;415(6873):770-4.

4 – Sedriks AJ. Corrosion of stainless steel, 2. 1996

5 – Szklarska-Smialowska Z. Review of literature on pitting corrosion published since 1960. Corrosion. 1971 Jun;27(6):223-33.

6 – ASTM G48-11(2015), Standard Test Methods for Pitting and Crevice Corrosion Resistance of Stainless Steels and Related Alloys by Use of Ferric Chloride Solution, ASTM International, West Conshohocken, PA, 2015, www.astm.org

7 – Burstein GT, Pistorius PC, Mattin SP. The nucleation and growth of corrosion pits on stainless steel. Corrosion Science. 1993 Jan 1;35(1-4):57-62.

8 – Pardo A, Merino MC, Coy AE, Viejo F, Arrabal R, Matykina E. Pitting corrosion behaviour of austenitic stainless steels–combining effects of Mn and Mo additions. Corrosion Science. 2008 Jun 30;50(6):1796-806.

9 – Stewart J, Williams DE. The initiation of pitting corrosion on austenitic stainless steel: on the role and importance of sulphide inclusions. Corrosion Science. 1992 Mar 1;33(3):457465-63474.

10 – Jones, D. (1992). Principles and prevention of corrosion. New York: Macmillan, pp.119-221.

11 – Scully JR, Budiansky ND, Tiwary Y, Mikhailov AS, Hudson JL. An alternate explanation for the abrupt current increase at the pitting potential. Corrosion Science. 2008 Feb 29;50(2):316-24.

12 – Tolosa I, Garciandía F, Zubiri F, Zapirain F, Esnaola A. Study of mechanical properties of AISI 316 stainless steel processed by "selective laser melting", following different manufacturing strategies. The International Journal of Advanced Manufacturing Technology. 2010 Nov 1;51(5):639-47.

13 – Röttger A, Geenen K, Windmann M, Binner F, Theisen W. Comparison of microstructure and mechanical properties of 316L austenitic steel processed by selective laser melting with hot-isostatic pressed and cast material. Materials Science and Engineering: A. 2016 Dec 15;678:365-76.

14 – Gibson I, Rosen D, Stucker B. Additive manufacturing technologies: 3D printing, rapid prototyping, and direct digital manufacturing. Springer; 2014 Nov 26. P. 22 – 25.

15 – Frazier WE. Metal additive manufacturing: a review. Journal of Materials Engineering and Performance. 2014 Jun 1;23(6):1917-28.

16 – Riemer A, Leuders S, Thöne M, Richard HA, Tröster T, Niendorf T. On the fatigue crack growth behavior in 316L stainless steel manufactured by selective laser melting. Engineering Fracture Mechanics. 2014 Apr 30;120:15-25.

17 – Herzog D, Seyda V, Wycisk E, Emmelmann C. Additive manufacturing of metals. Acta Materialia. 2016 Sep 15;117:371-92.

18 – Mower TM, Long MJ. Mechanical behavior of additive manufactured, powder-bed laser-fused materials. Materials Science and Engineering: A. 2016 Jan 10;651:198-213.

19 – Baufeld B, Brandl E, Van der Biest O. Wire based additive layer manufacturing: Comparison of microstructure and mechanical properties of Ti–6Al–4V components fabricated by laser-beam deposition and shaped metal deposition. Journal of Materials Processing Technology. 2011 Jun 1;211(6):1146-58.

20 – Jia Q, Gu D. Selective laser melting additive manufacturing of Inconel 718 superalloy parts: Densification, microstructure and properties. Journal of Alloys and Compounds. 2014 Feb 5;585:713-21.

21 – Gu DD, Meiners W, Wissenbach K, Poprawe R. Laser additive manufacturing of metallic components: materials, processes and mechanisms. International materials reviews. 2012 May 1;57(3):133-64.

22 – Sander G, Thomas S, Cruz V, Jurg M, Birbilis N, Gao X, Brameld M, Hutchinson CR. On The Corrosion and Metastable Pitting Characteristics of 316L Stainless Steel Produced by Selective Laser Melting. Journal of The Electrochemical Society. 2017 Jan 1;164(6):C250-7.

23 – Dai N, Zhang LC, Zhang J, Chen Q, Wu M. Corrosion behaviour of selective laser melted Ti-6Al-4V alloy in NaCl solution. Corrosion Science. 2016 Jan 31;102:484-9.

24 – Sun Y, Moroz A, Alrbaey K. Sliding wear characteristics and corrosion behaviour of selective laser melted 316L stainless steel. Journal of materials engineering and performance. 2014 Feb 1;23(2):518-26.

25 – Prashanth KG, Debalina B, Wang Z, Gostin PF, Gebert A, Calin M, Kühn U, Kamaraj M, Scudino S, Eckert J. Tribological and corrosion properties of Al–12Si produced by selective laser melting. Journal of Materials Research. 2014 Sep;29(17):2044-54.

26 – Casati R, Lemke J, Vedani M. Microstructure and fracture behavior of 316L austenitic stainless steel produced by selective laser melting. Journal of Materials Science & Technology. 2016 Aug 31;32(8):738-44.

27 – Zhong Y, Liu L, Wikman S, Cui D, Shen Z. Intragranular cellular segregation network structure strengthening 316L stainless steel prepared by selective laser melting. Journal of Nuclear Materials. 2016 Mar 31;470:170-8.
28 – Sun Z, Tan X, Tor SB, Yeong WY. Selective laser melting of stainless steel 316L with low porosity and high build rates. Materials & Design. 2016 Aug 15;104:197-204.

7. AUTHOR DETAILS

Guilherme Sander is a PhD candidate at Monash University and studies the corrosion of 3D printed stainless steels. In addition, Guilherme is experience in additive manufacturing of metals and polymers, engineering design, software development and materials chemistry. Guilherme's PhD is focused on the printing parameter optimisation of different stainless steels and the correlation of these parameters to the materials properties in corrosive environments. Guilherme is passionate about materials durability and experienced in many areas of corrosion engineering.

ADDRESSING THE CHALLENGES OF ACCELERATED LOW WATER CORROSION AT THE PORT OF MELBOURNE

J McLean[1] N Critchley[1], M De Silva[2].
1. Freyssinet Australia, Melbourne, Australia
2. Port of Melbourne Corporation, Melbourne, Australia.

SUMMARY: Accelerated Low Water Corrosion (ALWC) is a well-known phenomenon and International codes cover the treatment, control and prevention of ALWC. In 2015 Port of Melbourne surveys detected the influx of ALWC causing bacteria into Port waters. If left untreated ALWC could cause a reduction in the service life of some assets.

To address the risk of ALWC, the Port of Melbourne undertook a program for the design, installation and commissioning of impressed current cathodic protection systems to a number of wharf assets at risk of ALWC beginning in 2015. This was undertaken with a focus on having the ICCP systems operational within 18 months. The systems were designed with tailored operational polarisation criteria over and above that which is used to arrest more "typical" corrosion of steel in seawater.

This paper looks at the challenges of designing and installing these systems on operational wharf terminals, taking into account the conservative protection criterion of -900 mV instant OFF (Ag/AgCl), the demanding working environment and the criticality of achieving protection in an accelerated timeframe.

Keywords: Steel, Seawater, ALWC, ICCP.

1. INTRODUCTION

The Port of Melbourne is the largest container port in Australasia [1]. With over 3000 ship visits annually, it is one of the most heavily utilised shipping facilities in the Southern Hemisphere. Over the last five years, routine diving inspections identified the presence of Accelerated Low Water Corrosion (ALWC) inducing bacteria on a number of Port of Melbourne assets in the Port Philip Bay and Yarra Entrance. As an identified risk to the structural capacity and ongoing use of steel wharf elements, a program was put in place to arrest this particularly aggressive form of corrosion.

To achieve this, a two staged design, installation and commissioning program was undertaken for the delivery of Impressed Current Cathodic Protection (ICCP) systems on these critical wharf sites. In order to minimise the risk of disruption to the ongoing use of the Port facilities due to structural failure, this program of works was undertaken within an accelerated time frame, with a focus on delivering a tailored, turnkey solution for each unique site.

This paper discusses the challenges that were faced in the design and installation of these systems as well as the solutions arrived at to overcome these. As part of this discussion, a review of the following items will be undertaken:

1. Designing to achieve the conservative -900mV (Instant OFF) criteria,
2. Material selection and specification to achieve longevity and suitability of climate,
3. Consideration of operational use of port facilities,
4. Monitoring and control systems,
5. Staging of the works,
6. Installation challenges (access, climate, coactivity)

2. BACKGROUND

Corrosion of steel structures in a marine environment is a commonly accepted risk that has been controlled and managed by asset holders for many decades via sacrificial steel thicknesses, protective surface treatments and cathodic protection. The predictability of common steel corrosion (or 'rusting') tends to allow programmed repairs and monitoring systems to control these risks within regular maintenance programs. According to the Maritime Navigation Commission however [2], within the past 20 years there has been a growing awareness of an accelerated form of corrosion localised at the low-water, tidal zone of maritime structures. This Accelerated Low Water Corrosion (ALWC), is a microbiologically induced corrosion (MIC) that results in rapid pitting of steel structures. This highly aggressive type of corrosion is less predictable and its rapid onset can create significant challenges for an asset owner.

In 2014, the presence of ALWC inducing bacteria was detected at Port of Melbourne, in Victoria. To address the risk of this bacteria and its potential effects on the structural capacity of these wharf assets; the asset owner, Port of Melbourne, decided to implement the use of Impressed Current Cathodic Protection (ICCP) systems on these assets. These works would be undertaken over a two-staged, 18-month delivery process.

To ensure successful delivery of this program of works, the four key success criterion for the Ports Delivery Strategy were:

- Deploy functional systems that address the ALWC whilst not in any way compromising normal operations of the terminals
- Implement state-of-the-art intelligent systems so that remote control and monitoring is possible
- Incorporate safety-in-design throughout the concept and detailed design process and,
- Develop installation methodologies that cater for uninterrupted terminal operations during the installation and commissioning process.

In order to achieve this program within the time frame, the works were staged in the following order:

1. **Stage 1**: 2 sites (30 and 33 South Wharf)
2. **Stage 2**: 7 sites (Ann Street Pier, Gellibrand Pier, 32 South Wharf, 2 Webb Dock East, 3 Webb Dock East, Swanson Dock East Approach and Swanson Dock West Approach)

Design and installation works for these projects began in late 2015 with project completion mid-2017.

3. DESIGN CONSIDERATIONS

Design of impressed current cathodic protection systems for marine structures presents its own set of challenges. Items such as the conservative ALWC protection criteria, lifecycle of system componentry, integration with port activities and access limitations to site all had to be considered at the design stage of the works. This section will discuss these challenges and the methodologies and systems used to overcome them.

3.1 Meeting -900mV design criteria

One of the principal challenges faced by a design team protecting steel under the effects of ALWC is the additional current requirement to halt corrosion. According to the report of working group 44 of the Maritime Navigation Commission [2], in aerobic conditions, protection of steel is achieved at a potential of -800 mV (with respect to an Ag/Ag Cl reference electrode). This value is derived from Pourbaix diagrams and is quoted in many international standards [3]. In anaerobic conditions however, as is the case with ALWC, an electrical potential of -900 mV relative to the reference electrode is required to suitably combat this aggressive form of corrosion.

From a design perspective, this -100 mV difference requires significantly more current. For the ALWC works, this was achieved by selecting conservative design criteria, adopting an initial, average and final current density requirement 50% greater than that specified in AS 2832.3 (as seen in Table 1 below). Although this increase in current has significant implications for the system design and installation, the criticality of addressing this particularly virulent form of corrosion must be considered.

Table 1. Current Density Requirement – Aerobic VS Anaerobic (ALWC) Conditions

Current Requirement (AS 2832.3)	Bare Steel in Water	Bare Steel in Sea Mud
Initial	120 mA/m^2	20mA/m^2
Average	90mA/m^2	20mA/m^2
Final	90mA/m^2	20mA/m^2
Adopted Current Requirement (ALWC)	**Bare Steel in Water**	**Bare Steel in Sea Mud**
Initial	180 mA/m^2	30mA/m^2
Average	135mA/m^2	30mA/m^2
Final	135mA/m^2	30mA/m^2

3.2 Design for Longevity and Climate

In order to achieve a serviceable system life in a marine environment longevity of design must be considered. In a particularly aggressive marine climate, system componentry is exposed to tidal movements and splash zones which can seriously affect their usable service life. In order to ensure that the design for the ALWC program allowed for these conditions appropriate material selection and installation techniques was integral. To achieve this the following requirements were specified as part of the design:

3.2.1 Cable Tray

Outside of hazardous areas, a UV stabilised glass reinforced PVC cable tray was used. The advantage of adopting this material is the ease of workability and the resistance to corrosion. Minimising the use of metallic componentry also eliminates the need for electrical isolations and earthing.

Inside hazardous areas, such as petrochemical wharf areas, cable tray was made from marine grade aluminium to allow for electrical continuity and earthing. The advantages of using aluminium (as compared to stainless steel) is its workability whilst also being lightweight and durable. The disadvantage was that electrical isolation was required between the aluminium tray and stainless steel brackets to eliminate the risk of dissimilar metal corrosion.

Cable tray was hung from wharf soffits and mounted to the edge of capping beams where appropriate, care was taken to ensure that tray was installed as far from the splash zone as practicable. Additionally, locations for cable tray were selected where the risk of mechanical damage from day to day operations was also minimised. Both the PVC and aluminium cable tray can be seen in Figure 1 and Figure 2 below.

Figure 1 - PVC Cable Tray

Figure 2 – Aluminium Cable Tray

3.2.2 Metallic Fixings/Brackets

Metallic items above the low tide level such as cable tray brackets, conduit brackets, fixings and fasteners were all made of Marine Grade 316 Stainless Steel. This ensures added durability from atmospheric and crevice corrosion. Stainless Steel brackets that were welded to mild steel elements were coated with epoxy mastic at the weld location to prevent the onset of dissimilar metal corrosion and ensure protection of the welds.

Metallic items installed below the low tide level, such as anode brackets were designed to be under the influence of the cathodic protection system. As such the use of 250L mild steel was adopted. The low carbon content of this steel reduces the risk of hydrogen embrittlement whilst welding, ensuring a durable and higher quality fixing.

3.2.3 Cables

All cables outside of hazardous areas were a minimum, standard double insulated PVC/PVC (SDI) or XLPE/PVC. This was selected as double insulated cables have a much greater resistance to mechanical abrasion and a longer design life than single insulated cables.

Inside hazardous areas, the use of Steel Wire Armoured (SWA) cable was used. This provided additional mechanical protection and durability. However, this introduced an additional challenge to the installation works as the cable was significantly more cumbersome, heavier and less flexible than regular SDI cable. It also must be noted that special, earthed hazardous area junction boxes were required where cables had to be joined. This added another element of complexity and required the use of specialist electrical personnel.

3.3 Suitability for Port Operations

The Port of Melbourne is the most heavily utilised shipping facility in Australia. When designing for port applications, practicality is one of the key considerations and challenges faced by all designers. For the ALWC works a number of elements had to be considered to ensure that the systems were suitable for the ongoing and future use of port operations. Some of the key challenges faced at the design stage included:

3.3.1 Ship Berthing and Stevedoring activities

Allowances had to be factored into the design to ensure that ship berthing activities would not be affected by the cathodic protection system. Location and placement of cable tray, conduits and other hardware had to be adjusted to ensure that rope activities and line boats could access bollards and tie off points when required. Some examples of where this was achieved included the use of a rope guide rail to protect cable tray installed on capping beam as well as raising suspended cable tray to allow for lines boat access, as shown in Figure 3 and Figure 4 below.

Figure 3 – Rope Guide Rail to protect cable tray

Figure 4 – Raised Cable Tray to allow Line Boat Access

3.3.2 Underwater Protrusions

When installing anode brackets on sheet pile wall in berthing areas and shipping channels, consideration must be made of the risk of introducing underwater hazards. For the ALWC Works, elements such as vessel draught and safety for diving activities had to be considered. This was planned in collaboration with key stakeholders and control of these risks were managed by lowering anodes to safe depths as well as using suitable signage and demarcations as shown in Figure 5 and Figure 6 below.

Figure 5 – Caution Signage for Diving Operations

Figure 6 – Port Lateral Mark: Demarcating Submerged Hazard

3.3.3 Electrical Interference and Isolation of Foreign Structures

An additional consideration during the design was the risk of stray current interference with foreign structures. Although this is often viewed as a minimal risk in a marine environment (due to the relatively low resistivity of the electrolyte) it must be considered at design stage. One of the key considerations in the design for the ALWC project was to ensure no stray current effects were impressed upon berthed vessels.

This risk was overcome by ensuring anodes were designed such that their placement, size and output current had minimal influence on any surrounding metallic structures, such as vessels, barges and other floating plant.

3.4 Designing for Monitoring and Control

Due to the rapid deterioration of steel under the influence of ALWC, regular monitoring and maintenance of the cathodic protection system must be undertaken to ensure its ongoing effectiveness. Although this is the case with any Cathodic Protection (CP) system, the deleterious effects caused by this particularly aggressive form of corrosion justifies a monitoring schedule above and beyond what may be required for a traditional cathodic protection system.

Planning for this monitoring and maintenance schedule across multiple sites presented a significant logistical design issue for the project team. To accomplish this, the Port of Melbourne specified that each individual Transformer Rectifier Unit be furnished with remote monitoring and adjustment capabilities. This afforded the benefit of real-time data systems checks, monitoring of protection levels and adjustments to system outputs as required.

Through system design, these online monitoring and adjustment capabilities allowed for safe and regular system checks at an increased frequency, whilst minimising the requirement to physically access isolated works areas.

4. INSTALLATION WORKS AND CHALLENGES

The installation works across multiple sites within an operational port environment, introduced a number of staging, planning, access, construction and monitoring challenges. This section will explore these key issues encountered and the avenues that were adopted to address them.

4.1 Staging the Works (Dealing with Project Complexity)

To address the inherently complexity of the works, whilst prioritising critical sites, a flexible approach was adopted for the programming and scheduling of works. Instead of concurrently completing the installations across all nine sites and delivering them as a full 'package of works', they were split up into smaller, individual packages, completed and handed over to the client as they were finished. This was in the form of two separate stages, breaking down the nine sites across the project into these two stages, as shown in Figure 7 below.

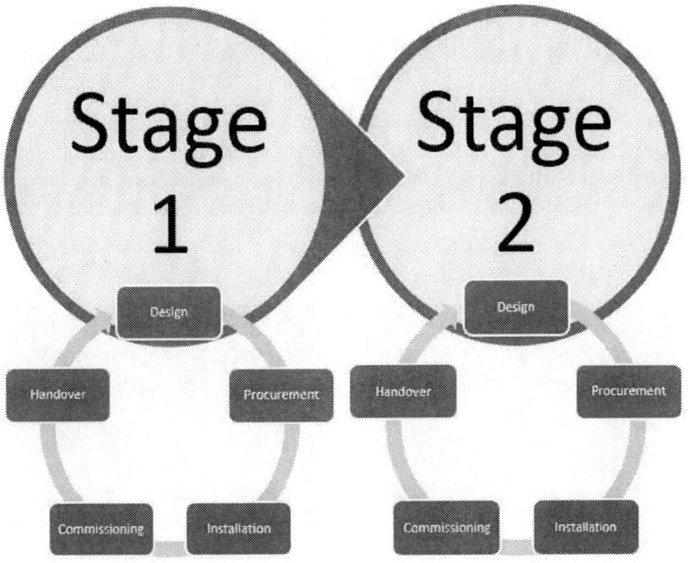

Figure 7 –Staged Works

Adopting a staged rollout of the ALWC protection program allowed for more stakeholder engagement in the delivery of the final product. Lessons learnt could be implemented over the subsequent stages and asset risk criticality could be prioritised and staged accordingly.

The added benefit of adopting staged work packages was that economies of scale could still be achieved by doing the works concurrently. This allowed the project team to reduce overhead costs and increase efficiency of project management resources.

4.2 Working around Operational Wharves

A key requirement for the installation works at each site was that there be minimal impact to the operations of the Port of Melbourne's tenants. Working across nine individual wharf structures, over a two-stage project portfolio involved significant stakeholder engagement and planning. Of these nine sites, six of them were tenanted for differing operational uses including two container terminals, one Roll On - Roll Off (RORO) berths, one tug boat operator, a swing basin and a petrochemical terminal. Each of these different operational uses presents their own unique set of challenges that needed to be managed accordingly.

Working around live shipping movements is inherently risky, aside from the substantial Occupational Health and Safety considerations, the hidden costs of potential disruptions to shipping operations are significant. To facilitate this project requirement a systematic communication and stakeholder engagement plan was formulated. This plan detailed a number of important aspects, including:

- Estimated Duration of the works
- Design intention and concept drawings
- Working Hours and Locations
- Access system and emergency response plan
- Vessel Isolation "Lock-Out" Procedure (where required)

By engaging the tenants at the planning stage of the works, a clear line of communication could be maintained between all stakeholders, allowing for installation works to be undertaken with minimal disruptions to port operations.

4.3 Access

Access for carrying out under wharf works and installation of cathodic protection systems is a serious challenge. Splash zones are intermittently accessible due to tidal movements and both wet and dry works must be planned and staged accordingly.

For a complex program of works like that undertaken across multiple sites, a contractor must be innovative and flexible with their use of access systems. The use of floating platforms, semi-submersible platforms, suspended gantries and scaffolding are all used for dry works, with selection criteria depending upon such variables as overhead room/height of the deck soffit, pile spacing, shipping and tidal movements. Some examples of different access systems can be seen in Figure 8 and Figure 9 below.

Figure 8 – Floating Pontoons to Access Tidal Areas

Figure 9 – Scaffold Access

Undertaking concurrent installations across multiple sites requires the use of all these different types of systems. There is no one size fits all rule for safe and efficient access to under-wharf areas. In order to ensure this was achieved for the works, each separate works area had to be individually considered and suitable access systems designed accordingly.

Installation and welding works below the low water line are undertaken by specialist diving subcontractors who must consider rescue plans, laydown for a dive vessel/vehicle as well as access for umbilical's and equipment. Similarly, with above water works, these access limitations were unique to each works area and site-specific dive plans had to be formulated to ensure that the works were undertaken safely.

4.4 Weather and Climate

Planning of works around environmental conditions was crucial to the success of the project. Each of the nine individual sites was exposed to its own unique set of environmental challenges, further impacted by factors including access limitations under wharves, strength and direction of winds, tidal movements and rain events.

To minimise the potential impact of delays caused by adverse weather conditions, systematic forecasting, monitoring and planning of environmental conditions was undertaken. This was crucial to ensure that downtime was kept to a minimum and production rates were maintained.

One of the advantages of working across multiple sites concurrently is that adverse weather effects in one area may not be present at different sites in the Port Philip Bay. For example, when NE Winds prevented works at Gellibrand Pier or Ann Street, the work crew were able to swiftly relocate to the more sheltered Webb Dock area, where works were unaffected. Working across multiple sites allowed for an agile workforce that could be flexible to deal with adverse weather conditions, as they arose.

5. CONCLUSIONS

The use of ICCP systems for ALWC applications is a widely accepted practice that, when designed installed and commissioned correctly, can increase the usable life of a steel asset immensely. With the discovery of ALWC in the Port of

Melbourne, the application of this type of system was deemed necessary to arrest this particularly aggressive type of corrosion.

To undertake the scope of works required for the delivery of these systems, a number of design and installation challenges had to be overcome. In order to ensure continued success for future works, the following key findings and lessons learnt must be considered:

- Adoption of a conservative design criteria to allow for the additional current requirement to address ALWC. For the purpose of this program, a current density requirement 50% greater than a "typical" design was adopted.
- Design for durability and longevity of system materials and their componentry. Consideration has to be made for the fit-for use purpose of these materials over the systems intended design life.
- Innovative thinking and flexible use of access systems must be adopted to allow for variable conditions across multiple sites. Consideration must be made of environmental factors, productivity rates and health and safety of personnel.
- Sufficient lines of communication must be maintained at all times to ensure that key stakeholders are aware of all aspects of the works and that significant issues are identified and controlled as they arise.
- Programming and scheduling of works should be flexible and agile; ensuring that risk criticality can be addressed, early value can be delivered and economies of scale can be achieved.

Overall the delivery of this complex program of works was undertaken successfully with minimal disruptions to the Port of Melbourne and their tenants. Whilst considering the vast scope of works, design difficulties and the challenges involved for a safe and efficient installation; the project was able to achieve all four key criterion of the projects delivery strategy and was a major achievement to all of those involved.

6. ACKNOWLEDGMENTS

The author would like to acknowledge the contributions made by specialist subcontractors and consultants in their assistance with the delivery of these works.

7. REFERENCES

1 Port of Melbourne, "Quick Facts", http://www.portofmelbourne.com/about-us/quick-facts, Accessed 16/06/2017

2 International Navigation Association (PIANC), *Accelerated Low Water Corrosion (ALWC), Final Report of PIANC Working Group 44,* , PIANC General Secretariat, Brussels, Belgium, 2005.

3 AS 2832.5 – *Cathodic Protection of Metals – Fixed Immersed Structures*

8. AUTHOR DETAILS

Jack McLean works as a Project Engineer with New South Wales' growing remedial division. Bringing a number of years of project management and technical expertise in both a contracting and consultancy role, Jack is responsible for the successful delivery and management of remedial and cathodic protection projects across Victoria and New South Wales. As part of this role, Jacks responsibilities include procurement, labour supervision, contracts administration, financial management, subcontract management, programming, quality assurance, HSE management as well as technical supervision in design.

With his background in corrosion control and cathodic protection. Jack can provide the "glue" in the project team between technical expertise and project management.

Nicholas Critchley works with Freyssinet Australia as The Victorian Remedial Manager. He has over 15 years' experience in the infrastructure asset maintenance, rehabilitation & protection industry. Specializing in concrete and steel structure inspection and investigation, design strategies for refurbishment and protection, bid and project management and business development and strategy. During his career Nicholas has worked in a variety of fields and positions throughout Australia and China. Following his graduation (Bachelor of Civil Engineering, RMIT, Melbourne, 1999) Nicholas worked his first professional 13 years with Savcor. Initially employed as a Project Engineer in Melbourne, Nicholas has since fulfilled Project, Operations, State and Business Development Management roles Melbourne, Brisbane and Guangzhou, China.

Michael De Silva is a Project Manager with a combined 15 years of experience working at the intersection of Emerging Technology, Strategy and Engineering. Michael has a proven track record of managing and delivering multi-disciplinary projects of up to $20M in value, and in combination his leadership skills and commercial nous, has consistently exceeded corporate strategic goals. In addition to managing and delivering multiple projects across an annual $50M capital works programme, Michael also leads an innovation group tasked with the implementation of emerging technology into various business streams across the corporation.

CONDITION AND RISK ASSESSMENT OF GRIFFITH G4 SEWER RISING MAIN

A Hackett
Kellogg Brown and Root, Sydney, Australia

SUMMARY: Griffith City Council (GCC), in the NSW Riverina, has centralised their sewage treatment and has an extended network of series pumping and rising mains. The G4 SPS rising main is approximately 7.1 km long and was reportedly constructed in 2-3 phases from 1974 and 1991 from cast-iron and ductile Iron. The main travels alongside the main irrigation canal for a large proportion of its length, which also provides drinking water to the town. Previous failures, current low flow rates and high detention times, and high levels of hydrogen sulphide measured at the treatment plant inlet channel led GCC to initiate detailed condition and risk assessment of the main.

KBR were engaged to review available operational and performance data, previous investigation reports, and undertake condition and risk assessment. External physical testing of the main including Broadband Electromagnetic scanning (BEM), physical inspection and spot ultrasonic testing at 30 representative and expected worst case locations. Bulk soil resistivity and transient pressure monitoring were also undertaken. Quantitative failure risk assessment and economic life modelling was undertaken by specialists WISER analysis.

The findings validated the approach of point sampling with detailed scanning at regular intervals and high risk locations. The general rate of corrosion was found to be substantially lower than feared, but several critical points with significantly elevated risk or deterioration were identified due to local design features. A staged strategy was recommended involving condition monitoring, operational changes, and air valve relocation. The estimated cost of these measures over 5 years of approximately $100,000, providing a cost effective means of managing risk while deferring major rehabilitation or renewal works for a substantial period.

Keywords: Pipeline corrosion, sewage rising main, BEM, risk assessment.

1. INTRODUCTION

Griffith City Council (GCC) G4 Sewer Pumping Station (SPS) Rising main is approximately 7.1 km long and was reportedly constructed in 2-3 phases from 1974 and 1991 from Cast-iron and Ductile Iron. The main travels alongside the main irrigation canal for a large proportion of its length, which also provides drinking water to the town. Previous failures, current low flow rates and high detention times, and high levels of hydrogen sulphide measured at the wastewater treatment plant (WWTP) inlet channel led GCC to initiate detailed condition and risk assessment of the main.

KBR were engaged to review available operational and performance data, previous investigation reports, and undertake condition and risk assessment. KBR conducted extensive physical testing of the main including Broadband Electromagnetic scanning (BEM), physical inspection and spot ultrasonic testing at 30 locations. Quantitative failure risk assessment and economic life modelling was undertaken by WISER analysis.

The evaluation was initially expected to be a fairly straight-forward assessment of internal corrosion rates, pipe materials and load conditions to determine the timeframe to renewal or relining. The major risks turned out to be very localised and related to design details. Previous failures may have been due to pipe manufacturing defects in a batch of pipes.

1.1 General Rising Main Issues and Risks

Sewer rising mains (AKA "Force mains") are a category of pressure pipeline which have some specific properties that can lead to very high corrosion rates, premature failure and high failure consequences. The issues were not necessarily well understood at the time that many rising mains in Australia were being designed and constructed in the 1960's and 1970's. In addition, there are a number of design assumptions including sewage composition and future flow projections, as well as operational changes that can substantially affect the conditions.

The primary deterioration process for steel, concrete, and cement lined pipes is biogenic corrosion caused by the generation of hydrogen sulphide (H_2S) that is produced when sewage becomes depleted of oxygen or "septic". Corrosion of metallic surfaces can occur directly from the liquid phase. Very severe degradation of cementitious surfaces occurs in gas pockets when sulphur oxidising bacteria on the surface metabolise H_2S and produce sulphuric acid. Typically the crown of the pipe suffers the worst internal corrosion in rising mains.

The factors that influence corrosion and failure rates include the composition of the sewage including trade wastes, the length of the upstream system, detention time in the pump station and rising main, the velocity of flow in the rising main, the presence of flat grades, crests and gravity sections, the location of air-release valves, and the nature of the pipe surfaces.

External pipe corrosion may also be significant, with low lying swampy areas, creek crossings, salinity and acid-sulphate soils in coastal areas being possible risk factors. External corrosion areas typically do not align with internal corrosion risk areas. Low points are typically closer to the sewage pump station, and prone to the highest static, dynamic, and surge pressures associated with pump operation. Sewer rising mains also have a failure consequence profile strongly weighted towards public health and amenity, and environmental degradation, with 3rd party and infrastructure damage less significant. Determining the risk profile of a rising main therefore involves understanding the asset as designed and constructed in its physical environment, consequence zones, as well as the past, present and future operating context.

2. G4 RISING MAIN

2.1 Route

Griffith is a regional centre in the heart of the Murrumbidgee Irrigation area in the NSW Riverina. The population of the town is approximately 19,000. The topography is very flat with the route following an irrigation canal for the majority of its length, but climbs Benerembah Hill in the centre of the town. The G4 rising main crosses through the town to the WWTP; a total distance of approximately 7km.

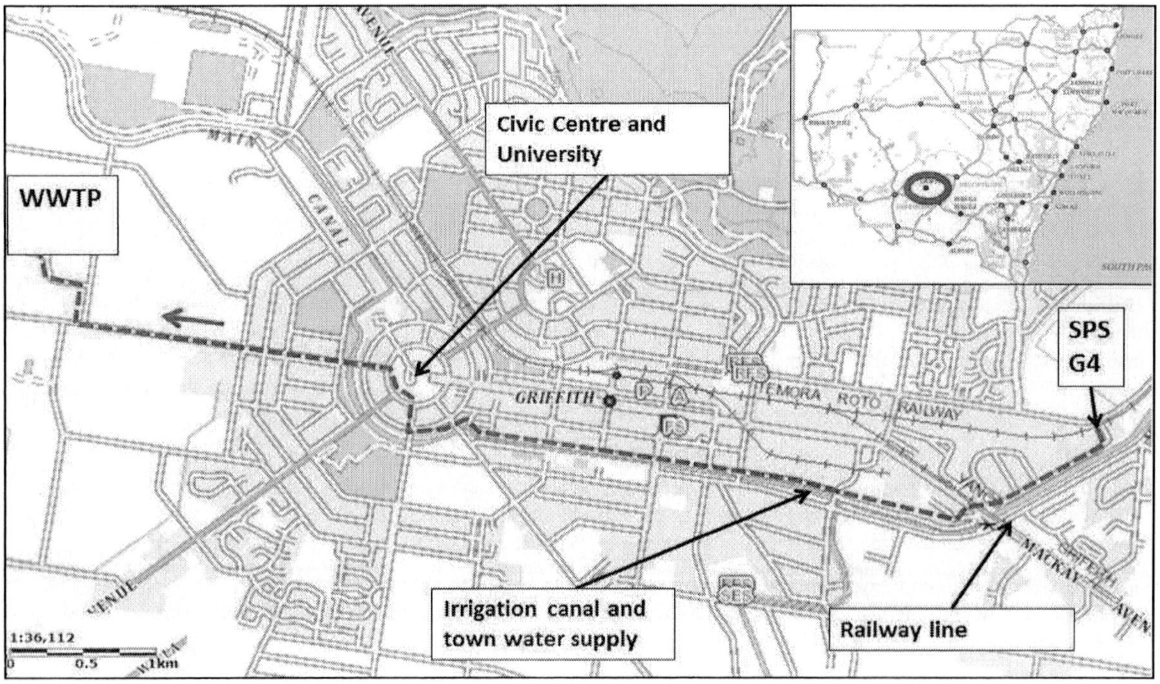

Figure 1. Rising Main Route

2.2 Design and Construction

The G4 pipeline as defined is 7.2 km long and consists of five segments differentiated by size, material or age. The original cast iron cement lined (CICL) main was reportedly laid in 1975-76 from the G4 pump station and interconnected to the twin DN300 CICL sewer rising mains near G1 pump station.

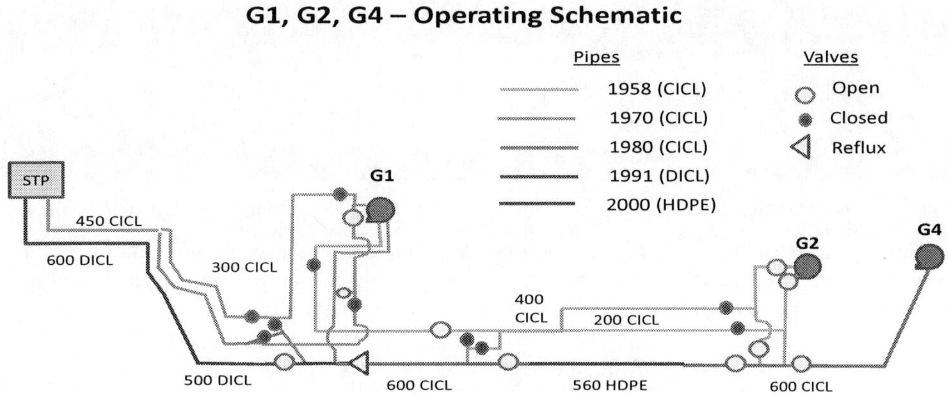

Figure 2. Rising main schematic - current

This section of DN600 CICL main was not used until G4 pump station was constructed around 1980. The twin DN300 CICL rising main continued onto Benerembah Hill where it was connected to the DN450 CICL G1 syphon/gravity main to the WWTP until a new DN500/600 ductile iron cement lined (DICL) pipe was laid in 1991. All the flows from G1 and G2 Sewage Pump Stations (SPS's) were re-directed to the G4 common rising main in 2007.

Table 1. G4 Rising Main Asset Description (upstream to downstream)

Asset ID	Length (m)	Nominal Diameter (mm)	Material	Year Laid	External protection	Comment
SRM00012	1,887	600	CICL - Class B	1975	Nil	Not used until 1980
SRM00087	845	560	HDPE	2000	N/A	Slip-line rehabilitation
SRM00027	629	600	CICL - Class B	1975	Nil	Not used until 1980
SRM001222/18	1,303	500	DICL - Class K9	1990	PE Sleeved	
SRM00028	2,447	600	DICL - Class K9	1990	PE Sleeved	

The rubber ring jointed pipes were laid in soil trenches and typically backfilled with native soil consisting of a sandy clay. The original depth of cover was between 300mm and 2m (avg 1.0m). The 1990 pipes were laid in a protective polyethylene "Greensleeve". No soil data was available from the original construction, but published maps and tests in the area recorded in the eSPADE database pointed to a classification similar to Marah Clay loam or Willbriggie clay loam. These are red clay or sandy clay of calcareous origin. pH data, where available, is between 6.5 and 8.5. Combined with generally low moisture, the conductivity and external corrosion risk was expected to be low for the majority of the alignment.

Although the 1975 pipes were recorded as CICL, a dimpled exterior observed at some sites is generally considered characteristic of DICL pipes. When aligning with ultrasonic point measurements, the Broadband Electromagnetic (BEM) scanned thickness matched better if the material is processed as DICL, however electron microscopy of a failed pipe sample was more characteristic of grey cast-iron. 1975 is reported to be the time of transition from grey cast-iron to ductile cast-iron in Australia. It appears likely that the pipes are a mixture of late CICL and early DICL.

Figure 3. 1990 - 600mm DICL pipe

Figure 4. 1975 - 600mm dimpled "CICL" pipe

The long section profile shows that the gradient is almost flat, with a total change in elevation of only 12m, with 9m occurring over a 500m section in the centre. In fact, the pipeline is only technically a rising main for 60% of its length. After the high point, it is a gravity pipe or inverted syphon until the treatment plant.

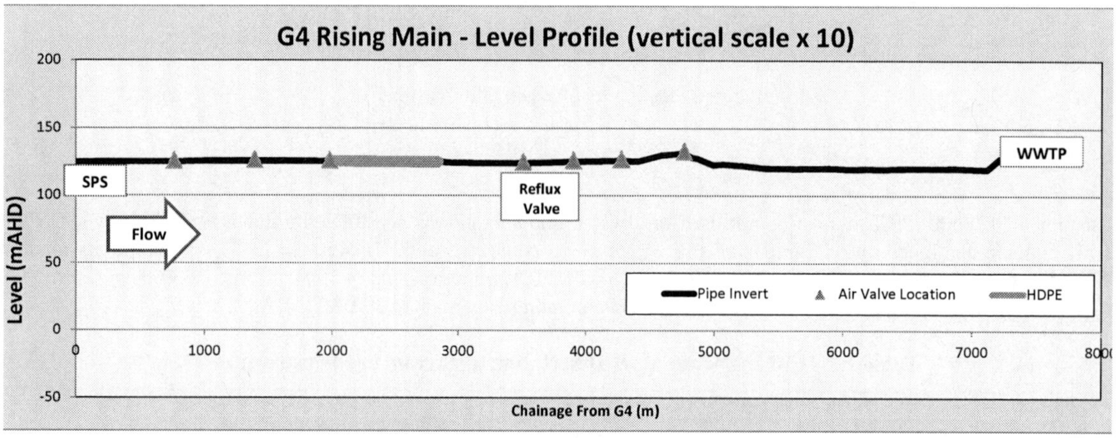

Figure 5. Rising main long section

2.3 Operating context and History

The G4 Rising Main carries flow from residential, light industrial and commercial areas, generally representative of Griffith as a whole. Figure 7 shows the overall network schematic of the G4 rising main. There is therefore significant detention overnight with the result that the sewage is often septic in the G4 SPS wet well before pumping as reflected in observed concrete degradation.

The G4 pumping station wet-well, pumps and rising main were designed for significant catchment growth, which has not eventuated. Minor future population growth will probably be offset by ongoing water efficiencies for the foreseeable future.

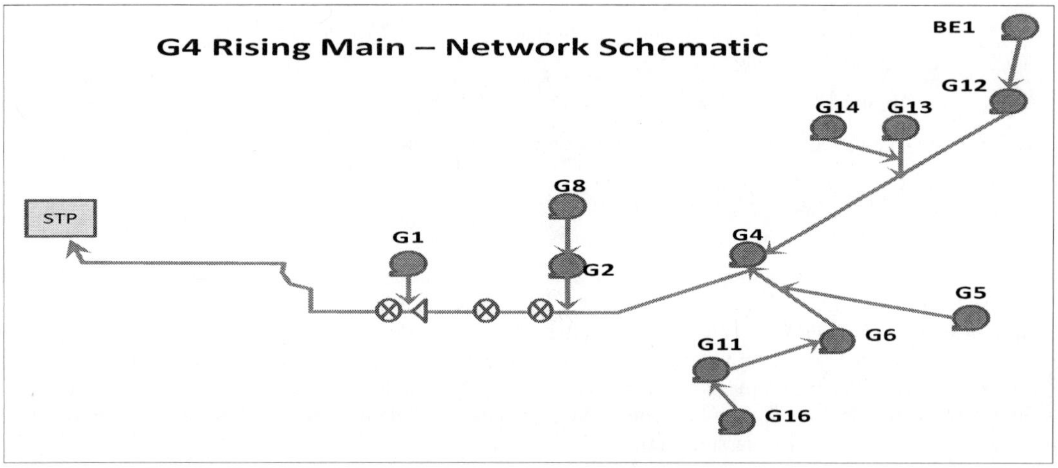

Figure 6. Rising main long section

A significant change to the operating environment of the G4 rising main is the contribution of trade wastes from food production and wineries in the catchment. Much of the planned and early flows came from these factories. Increasing controls on the quality of discharges as well as economic factors led to a movement of production out of town and an 80% drop in trade waste discharges. The overall effect of this on pipe corrosion is unclear as the decrease in Biological Oxygen Demand (BOD) is offset by increased detention times. High BOD is indicative of high sulphate loads and the tendency of the sewage to become septic, generating dissolved sulphides and hydrogen sulphide gas.

Table 2. Trade Waste Discharge Changes

Trade Waste Type	Avg Concentration - 1999			1999	2014
	BOD avg mg/L	SS avg mg/L	pH avg	Effluent Outflow avg kL/day	Effluent Outflow avg kL/day
Food Processing	1450.0	320.0	8.2	530.0	0.0
Winery	3384.4	520.0	7.8	370.0	110.0
Total				**900.0**	**110.0**

Gas phase H_2S and Temperature monitoring data from October 2014 to February 2015 showed levels in the G4 wet-well vary between 2 and 20 ppm (average 5ppm) with occasional spikes to 100ppm. G4 receives flow from several small tributaries, and the source of sulphide release may be primarily related to one or more influent flows rather than the main wet-well volume.

Monitoring in the Griffith Sewage Treatment Plant inlet works shows that H_2S concentration typically varies between 5 and 100 ppm (average 30-40ppm). This may not be exclusively related to the G4 discharge. Overall, hydrogen sulphide (H_2S) monitoring indicates that the G4 flow is probably septic for some of the pumping time. STP monitoring shows extended high levels for the morning peak indicating septic flows from overnight detention.

Liquid phase monitoring had been undertaken in association with trials of ferrous chloride and magnesium hydroxide dosing at various pump stations. This had a substantial impact at the discharge end, and was implemented at several SPSs including G4 from 2005 to 2012. Figure 16 shows the Hydraulic profile of the main along with modelled and measured pressures.

A desktop surge analysis was undertaken in 2003 which predicted surge pressures from -10m to +33m without air valves. The normal operating range was predicted to be 5m to 15m. The worst case surge pressures were considered unlikely to have been responsible for the observed failures. No transient monitoring had been undertaken.

Various operational and structural risk reduction measures were put in place after the last significant failure in 1999

- 2000 - Slip lining of 850m of the G4 Rising Main from the current G2 discharge to the G2 cross connection at Jondaryan Avenue
- 2003 - Installation of six air valves with carbon filters on the G4 Rising main between Whybrow St and Murrumbidgee Avenue

- 2007 - Removal of the surge chamber, modifications to the bend, and installation of an air-valve and carbon filter at the crest at Benerembah St
- 2007 - Installation of a reflux valve upstream of the G1 discharge
- 2007 - Re-direction of the G2 and G1 discharges to the common G4 rising main
- June 2005-June 2012 - Ferrous Chloride dosing at various SPS including G1, G3, G4.

2.4 Failure History

A significant failure in 1999 was investigated in detail, and refers to multiple similar failures occurring in a 500m section of the rising main between commissioning in 1980 and 1999. The failure mechanism involved a full-length longitudinal split. By co-incidence, Rock Solid have a sample of the specific 1999 pipe that they use for testing and calibration. The detailed material investigations could not determine a cause of the failure, as the section of failure had no significant surface load, and pressures are well within the corroded pipe capacity. It was noted that the pipe sprung apart, and remained 20mm apart at the split. The pipe corrosion and loss of cement lining is characteristic of a gravity section, even though the failure location was on a low, permanently pressurised section.

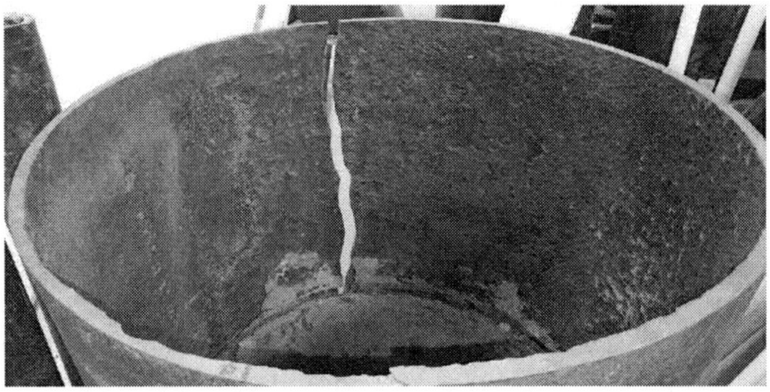

Figure 7. Failed pipe segment from G4 rising main (Courtesy of Rock Solid)

3. TESTING AND ANALYSIS

3.1 Investigation rationale

Based on the material, measured sewage quality, geometry and failure history of the long flat rising main, the overwhelming risk and deterioration process was considered to be H_2S attack of cement lining and steel due to sewage septicity. The highest risk sections were considered to be the gravity sections downstream of the crest at Benerembah St, but static or moving air pockets and corrosion are possible along the entire length. Subsequent to the methodology being developed and planned, additional drawings showed that the gravity section is largely an inverted syphon due to an elevated STP outlet. This reduced the degradation risk profile significantly, with the the pipe full for the majority of the length. The general methodology remained applicable, with a focus on high-point sampling.

The main was laid at a depth of 300mm–1,500mm below the surface, not under paved or improved surfaces, with few other buried services. Access was therefore available to expose and test the top half of the pipe without shoring for most of the length for an acceptable cost.

The likely failure mechanisms therefore involve internal corrosion due to biogenic sulphuric acid and hydrogen sulphide generated within the pipe, degradation of the cement lining, and steel section loss above the flow level leading to:

- pipe perforation and leakage of sewage into the environment
- soil scouring and formation of sinkholes, siltation of the sewer and odour problems
- Pipe collapse under soil or external loading

3.2 Field testing

The best method to quantify deterioration and failure risk is to physically measure the remaining steel section above the flow line in sufficient locations to determine the longitudinal corrosion profile. External BEM scanning from 9 o'clock to 4 o'clock at approximately 250m spacing was proposed to provide this profile. Exposure and inspection of the pipe at these locations also facilitated an external inspection of pipe which will give a good picture of the coating condition and a direct assessment of external corrosion to add to the risk model. The locations included the significant high points, and utilised air valve, reflux valve and gauging pits where possible.

Bulk in-situ soil resistivity testing was proposed in the vicinity of a selection of the excavations to provide a picture of the general soil corrosivity profile and variability. Testing via 4-pole Wenner method was proposed at seven locations adjacent to test pits.

General corrosion and graphitisation of the pipe due to internal and external factors is also assessed by pipe exposure and BEM scanning. Broadband Electromagnetic (BEM) Scanning was proposed as the most suitable method for pipe metal thickness estimation due to tolerance of surface roughness and coatings, resolution, speed, cost and robustness of the equipment and method. Point ultrasonic spot tests were undertaken at each site and referenced to the BEM test cell. The surface was smoothed with a file as required to achieve full contact.

Figure 8 - Wenner 4-pole soil resistivity testing

Figure 9 - BEM Scanning in excavation with HSK unit

Concerns over transient pressures arising from past failure and surge modelling were addressed by installing a surge capable monitor and simulating the worst case event of multiple pumps stopping. A Telog LPR-31i Line Pressure Recorder was used, capable of measuring transient pressures to 0.1s intervals. The logger resolution was set to 20s. Pressures variation of 20% outside the normal recording resolution triggers a short interval transient recording period. The gauge was placed on air-valve risers at 2 locations near the G4 pump station, and downstream of the reflux valve. The worst case was simulated with multiple pumps stopping at the tributary pump stations, and all three pump stations operating together,

3.3 Condition Risk Modelling

KBR engaged WISER Analysis to undertake physical probabilistic failure (PPM) and economic life modelling (PELA) of the rising main, based on the asset, environment and condition data.

Physical probabilistic modelling was undertaken on 10 condition zones, 5 each on the nominal CICL and DICL sections. No assessment was undertaken on the HDPE section. The number of zones is determined as an appropriate level of segmentation given the variability in degradation, environment and loading conditions, and the resolution and extent of testing undertaken. Internal pressure parameters were based on the measured figures, factored up at the direction of GCC to allow for future operating scenarios.

The loading classification is the worst case for each section based on the surface access. Most of the main is accessible for parking on by trucks or equipment and the W80 wheel load is applied. Road and highway crossings may also be susceptible to M1600 loads. The railway crossing has the 300LA classification applied.

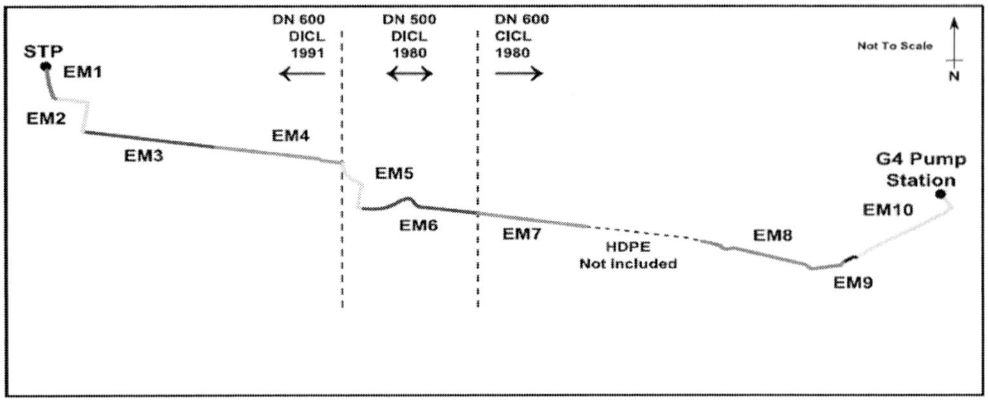

Figure 10. Likelihood and economic modelling zones defined for quantitative risk modelling

A qualitative consequence assessment was undertaken to support the risk and preliminary options assessment process. The consequence of failure is generally uniform and fairly low. Several critical points were identified relating to risk of water supply contamination, a children's playground, and very high profile locations.

Table 3. Failure Consequence Assessment

Economic Modelling Zone	Public Amenity	Direct Repair Cost	Infra-structure damage	Public Health	Environ-mental damage	Traffic Impact	GCC Reputation
EM1	Low	Low	Low	Low	Low	Low	Low
EM2	Low	Low	Low	Low	Low	Medium	Low
EM3	Low	Low	Medium	Low	Low	Low	Medium
EM4	Medium	Medium	Medium	Medium	Low	Medium	Medium
EM5	Medium	Low	Medium	Low	Low	Medium	High
EM6	High	Low	Low	High	Low	Low	High
EM7	Medium	Low	Low	Low	Low	Low	Medium
EM8	Low	Low	Low	High	Low	Low	Low
EM9	Low	Low	Low	Low	Low	Low	Low
EM10	Low	Low	Low	Low	Low	Low	Low

4. TESTING RESULTS

4.1 External Corrosion

External corrosion was assessed visually based on presence of corrosion products. Pitting was not significant enough to warrant measurement. All external corrosion was general, with no significant pitting, and therefore no reference surface was available for section loss measurement. BEM scanning takes into account external section loss. The pipe surface observations align with the general soil resistivity testing data in finding minimal external corrosion. Surface rust was evident associated with moisture trapped under the sleeving, but no significant section loss was found on the sleeved DICL sections. Where the CICL or DICL sections were backfilled with native soil without sleeving, no significant external corrosion was observed.

Minor pitting was observed in buried pipes at sites 18, 23A, 24 and 27. At each of these sites, imported sand or gravel backfill was observed. Site 23A had a substantial hard surface build-up which was difficult to remove. It is not possible to accurately estimate actual section loss at this site without high pressure water or abrasive cleaning. Overall the rate of external corrosion of buried pipes is sufficiently low that it may be neglected as a possible failure cause, and further detailed investigation was not recommended

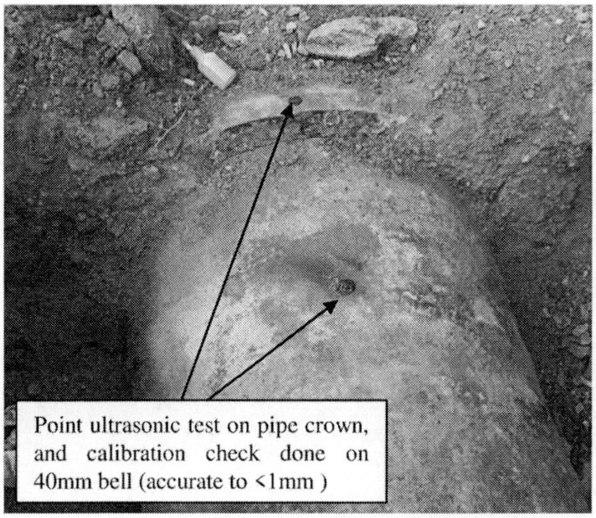

Figure 11. 600mm CICL pipe with fine gravel bedding

Figure 12 - CICL pipe in air valve pit

Corrosion in air valve pits is more significant, where the moist environment, occasional H_2S, and effect of the tapping bands and wall penetrations is causing general corrosion and minor section loss. The rate of corrosion is estimated at 0.1-0.2 mm per year. The extent is not sufficient to cause concern for the pipe load capacity, but problems may be experienced with the tapping band seals.

Depending on the extent of internal corrosion, the external corrosion rate may lead to perforations and failure of the DICL pipes in 20-30 years. The external corrosion of the CICL pipes in the air-valve pits is unlikely to be a determining factor in the life of the pipeline.

Soil resistivity testing generated raw results in the range of 370-4000 ohm.m. All samples fall well into the range of "Essentially non-corrosive". The soils were generally red sandy clays with little moisture. This aligns with the observation of no significant corrosion for pipes buried in native soil.

Imported sand or gravel bedding was associated with slightly higher external corrosion than native backfill. Significant soil moisture was only observed at one site, which was the lowest point where the pipe crosses underneath the canal. External corrosion was no more severe here than at other sites in native soil. (I.e. minimal)

4.2 Pipe thickness measurement

The BEM scanning was undertaken at 31 sites including 23 excavations, 6 air-valve pits and one reflux valve pit. The scanning of the 500mm and 600mm sections showed random thickness distribution with possible crown thinning at only one site which was a local high point. Full circumference scanning was undertaken at two locations to assess the risk of invert corrosion.

Scanning of the 600mm CICL sections identified possible crown thinning at 9 of 12 sites, with the greatest effect noted upstream of the air-valve at site 24 and the reflux valve at site 19.

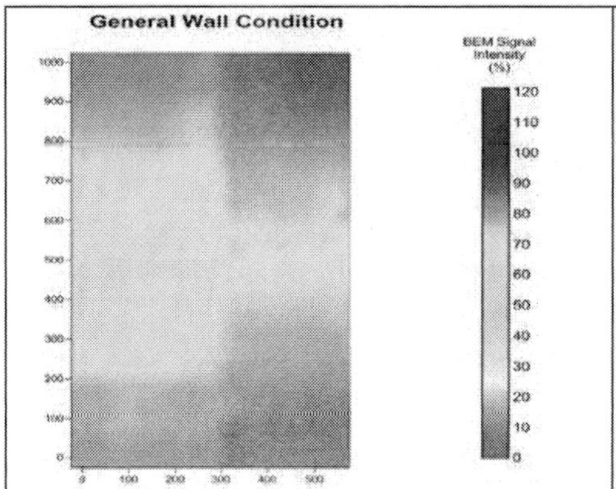

Figure 13. Site 24 - Air valve site pipe thickness

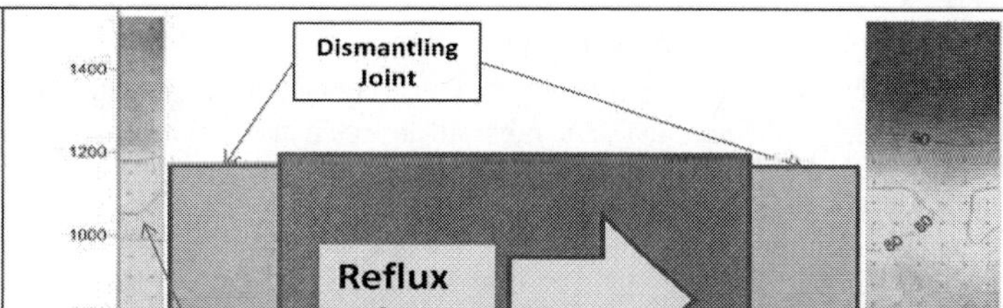

Figure 14. Reflux valve pipe thickness

The overall pipe thickness profile along the pipeline is shown in figure 15 below. The chart shows design pipe thickness tolerances along with BEM the inferred thickness range at each site (lines for clarity), as well as a point ultrasonic measurement typically aligned with the minimum BEM cell (50 x 50 mm) found on the crown. The local high points are identified, as well as significant fittings.

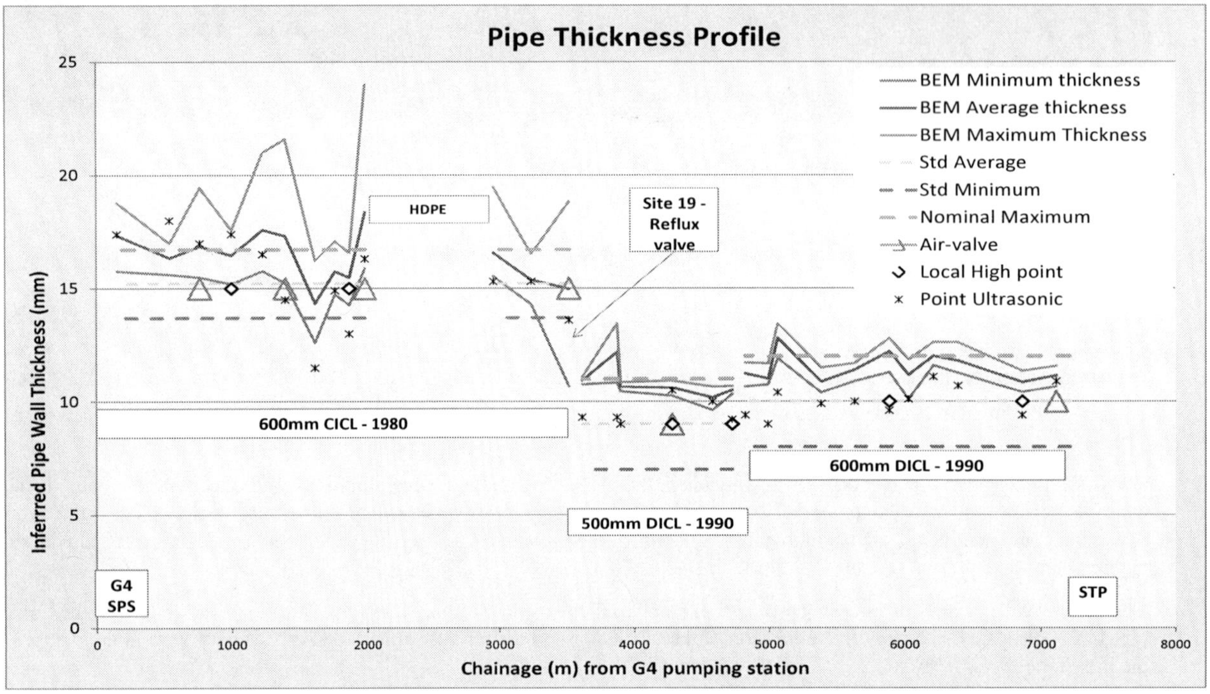

Figure 15. Reflux valve pipe thickness

Several features may be observed from the chart.

- The pipe thicknesses are typically above average based on published standards
- The potential high points are no more prone to corrosion, suggesting that standing air-bubbles do not remain, but are pushed through.
- CICL pipe thicknesses are quite variable, suggesting inconsistent manufacturing. This accords with the presence of dimpling observed on some pipes, but not others.
- The ultrasonic check on DICL is consistently less than the inferred BEM. thickness.

The UT - BEM discrepancy was discussed with Rock Solid Group. RSG preferred to maintain their standard DICL calibration curve for DICL, rather than make a special adjustment to align with spot ultrasonic.

4.3 Pressure and Transient Monitoring

Modest surge pressures are identified by gauging associated with pump operation on the G4 SPS. Pressures monitored up to 2m below the nominal static pressure expected at each site upstream of the reflux valve. Occasional pressure drops to 0m measured by GCC are longer term changes associated with leakage through the SPS reflux valves exceeding leakage through the Canal St reflux valve.

The surges occur over a relatively long period, as no "transient event" recording was triggered by KBR's gauge at either site. No vacuum pressures were identified during the monitoring period. Peak pressures of 13m were identified in the 600mm CICL and the 500mm DICL sections.

KBR's gauge at Whybrow St read lower than expected from the calculated hydraulic profile. At Canal Rd, KBR's gauge read higher than GCC's gauge, and higher than expected from the calculated hydraulic profile.

Lower than expected peak pressures can arise from a smaller installed pump than assumed, or pump problems such as ragging or impellor wear. Higher than expected pressures can arise from surges, or pipe constrictions, from partially closed valves, air-pockets, or silt accumulation.

The pressure at Canal St / City Park were observed to drop below the minimum static pressure for a significant period. This may have been due to development of a syphon over the crest, or due to scouring to repair a break on the G1 tributary main. The minimum pressure is slightly low for syphoning from the STP level.

The monitoring showed that the pressure regime is substantially less severe than the modelled waterhammer that was undertaken before air-valve and reflux valve installation. Transients will not be significant in the effective life of the pipeline under the current configuration.

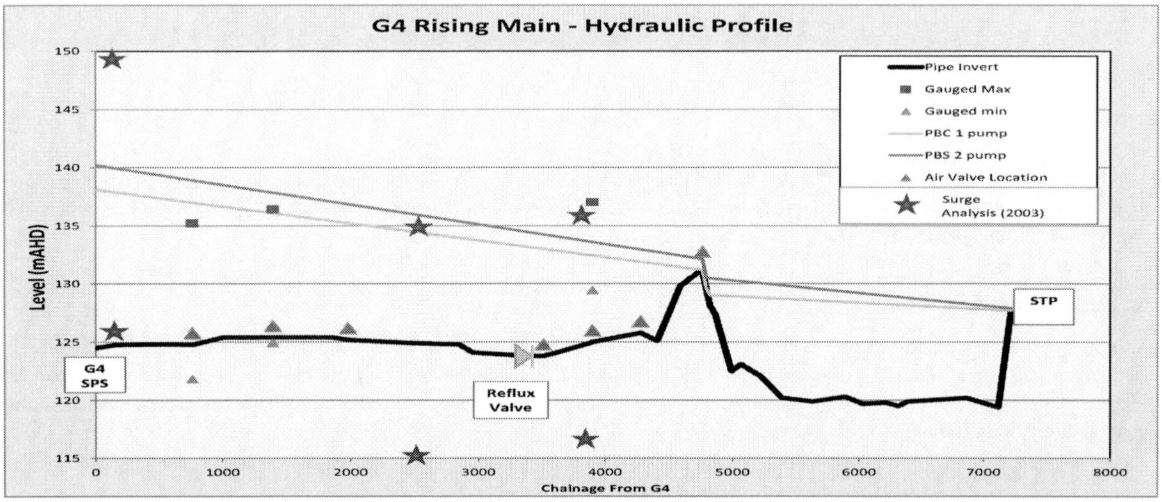

Figure 16. Pressure monitoring

5. CONDITION RISK AND ECONOMIC MODELLING

The analysis has estimated a corrosion rate for each likelihood zone based on the testing undertaken and the probable pipe properties from all available sources. Based on these distributions and other variables such as depth, pressure and surface loading, a physical probabilistic failure model was prepared to determine the probable time to failure of each zone. Due to the physical material property and failure mode differences, separate models were used for DICL and CICL. The modelling was undertaken by WISER Analysis.

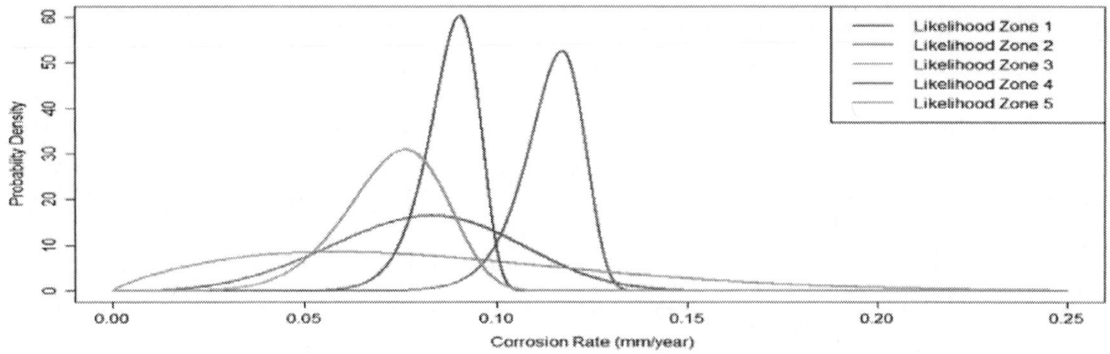

Figure 17. Weibull probability density for maximum corrosion rates in DICL zones

Significant variability arises from the limited number of samples in each section, and particularly the variability and uncertainty associated with the pipe thickness of the 600mm CICL section. This is reflected in the significant difference between the "Pessimistic" and "Expected" failure frequency outcomes determined.

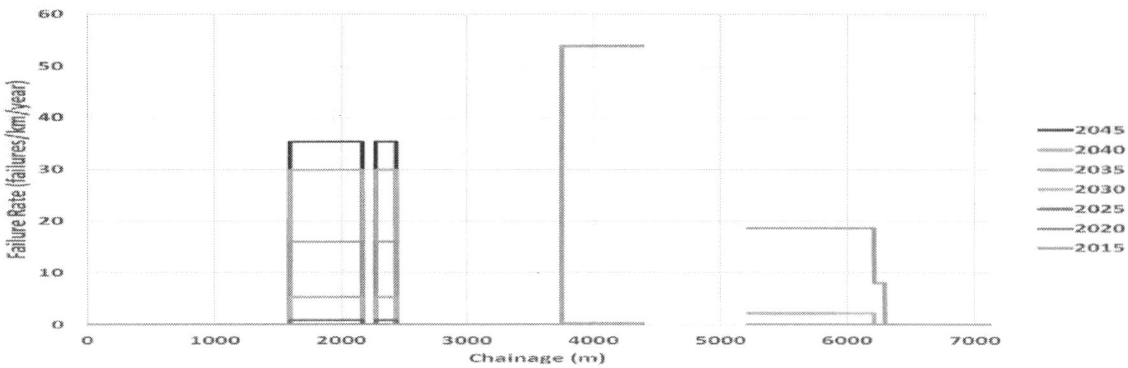

Figure 18. Cumulative pipe failures modelled on G4 main

The analysis highlighted that the majority of the pipe had a remaining life of greater than 30 years. Some sections may experience failures in 10-15 years.

6. DISCUSSION

Spot testing was undertaken at representative locations along the main, with a focus on local high points. There is evidence of crown corrosion at a number of sites, but not all locations. The evidence of corrosion does not necessarily correspond to identified plan high points. This suggests that permanent air-bubbles are not remaining in minor crests, but are being pushed along and discharged at air valves.

The analysis does not take into account the system modifications undertaken since 2003, the installation of air-valves in 2003 and the reflux valve in 2007. The installation of air valves is likely to have a significant beneficial effect on corrosion of the 600mm CICL section between McKay and Lenehan St (corresponding to the critical economic modelling Zone EM8). The analysis is therefore conservative. Consequently the corrosion rate assumptions should be viewed cautiously, and condition monitoring was recommended to provide a better post-2003 corrosion rate assessment before scheduling renewals.

6.1 High risk condition locations

Several locations were identified where the spot testing regime measured elevated localised corrosion, or where there is a very high risk of very localised corrosion that may not be captured by the point testing undertaken. These specific sites are described below.

Benerembah St Crest

The high point at Benerembah St is a particularly complex location where there are a number of specific locations where H_2S may be released. Corrosion can be very localised as H_2S is heavier than air. Although the two sites chosen should be in the gravity (vapour) zone, they are not guaranteed to identify the worst case. Figure 19 identifies the normal situation and possible critical locations.

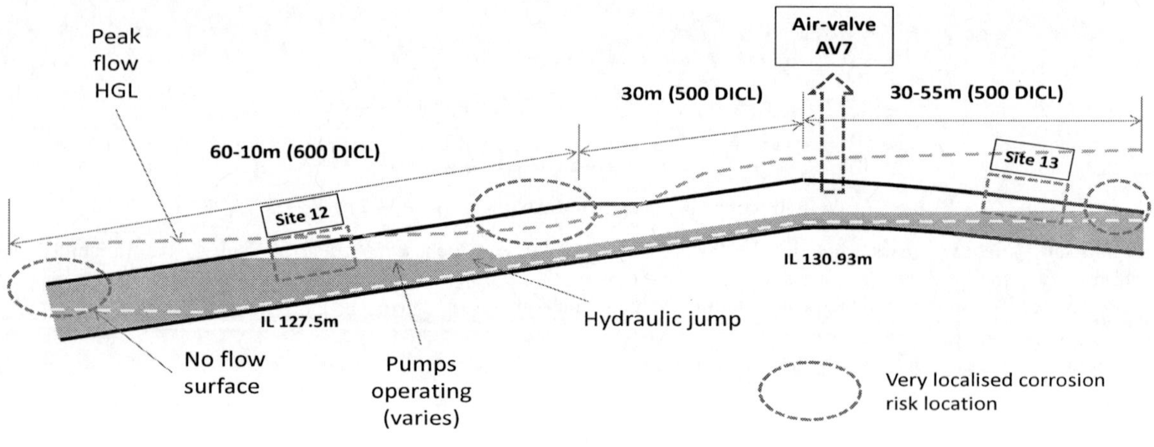

Figure 19. Localised Corrosion Risk Area - Benerembah St

Canal St Reflux Valve

The reflux valve has elevated corrosion immediately upstream. This probably results from a local air pocket trapped by the valve. Similarly the elevated corrosion identified downstream may be caused by turbulence associated with the reflux valve.

The concern is that if this level of corrosion has occurred since the reflux valve was installed in 2009, then the rate of corrosion is such that pipe perforation may be expected within 10 years. The corrosion may also pose a risk to the reflux valve.

Monitoring of section loss in 2 years' time was recommended to confirm that the section loss is associated with the valve installation. Inspection may be possible through the top of the reflux valve during maintenance.

Sites 23-24 - Crown Thinning

Site 24 exhibits clear signs of crown thinning with thickness down to approximately 65% of the maximum thickness. Site 24 is in the middle of the long zero grade section from Mckay Avenue to Lenehan St. This is the section considered most likely to experience crown corrosion due to gas accumulation with gas movement halted by the 0.17% grade down to the G2 connection. The installation of AV17 may be expected to reduce air accumulation significantly. Some gas may still be generated and cause corrosion downstream of the AV as far as Lenehan St. Site 23A is a very shallow site that lends itself well to monitoring thickness.

6.2 High risk consequence locations

Four locations were identified as particularly high consequence locations including the public exposure associated with Benerembah Hill, the main public playground at City Park, and two locations where a pipe leak or rupture may directly contaminate the canal and town water source. Figure 20 shows the location where the high risk material (CICL), greatest and general corrosion, coincides with the highest consequence risk. Although the risk of failure is still low, and monitoring is easily accomplished, the consequence can be mitigated via surface works at a relatively low cost.

Figure 20. Localised high consequence location - Public health / Reputation

6.3 Previous Failure

The conclusion reached is that the previous pipe failures by longitudinal splitting were potentially the result of a manufacturing defect where a batch of pipes was produced with residual stresses causing fast fracture under minor corrosion, and pressures well below their theoretical capacity.

7. RECOMMENDED RISK MANAGEMENT STRATEGY

The recommended risk management strategy consists of a mixture of condition monitoring, air-valve installation, and surface drainage modifications, most of which can be executed by Council without further design or specialist input. The upfront /annual cost of $64,000/$6,000 compares favourably with minimum rehabilitation cost of approximately $600,000 for the key sections.

Table 4. Risk Management Options Summary

Option	Scope	Estimated cost	Annual benefit of deferral (at 4%)	Commence[2]
General Pipe Corrosion				
Condition Monitoring	5- yearly BEM – 8 sites (Air valve pits plus Site 23a)	$3,700 /yr	N/A	2020
Liquid Phase Interventions	Oxygen, chemical or biological dosing	Detailed analysis required		
Operational Diversions	Divert to G1 rising main	$10,000	N/A	2016
Surface drainage modifications	Grade surfaces away from canal	$12,000	N/A	2016
New 250mm pipe	1980m beside existing	$1,174,311	$46,972	>2025
275mm slip-line	Nominal 1km slip-line PE	$265,000	$10,580	>2025
560mm slip-line	Nominal 1km slip-line PE	$792,000	31,700	>2025
Localised Corrosion upstream of Reflux Valve				
Condition monitoring	Point ultrasonic testS – 2-yearly	$1,000 / yr	N/A	2017
Install Air Valve	Offset pit and valve	$40,000	$1,600	2017
Upstream lining	100m CIPP	$108,400	$4,340	2020
Benerembah St Crest and gravity section				
Condition monitoring	Install Tee and laser scan 5-yearly	$12,000 install tee $1,000 scan/yr	N/A	2016
Slip-lining or CIPP exposed section	180m slip-line or CIPP	$199,400	$7,980	2017
Eliminate section	500m x 500DICL	$696,000	$28,000	2017
Barometric loop at STP	6m elevated outlet pipe riser and steel frame	$194,000	$7,800	2017

8. CONCLUSIONS

The sampling and investigation process provided a good understanding of the general pipeline condition, that turned out to be better than expected. Quantitative probabilistic failure and economic modelling indicated that the majority of the pipeline should last over 15 years before corrosion related failures would commence.

A number of specific problem corrosion or high consequence sites were identified as part of the testing in conjunction with evaluation of materials and construction, current and historical operational aspects, topography, environmental and social factors. Condition monitoring, minor operational changes, and measures to mitigate high consequence failures were identified enabling significant capital expenditure to be deferred with substantial cost saving.

9. ACKNOWLEDGMENTS

I would like to acknowledge the support of staff from Griffith City Council, Bellatos Plumbing, Rock Solid Group, and Scott Gould and David Marlow from WISER analysis for their professional input into the successful execution of the project.

10. REFERENCES

Condition Assessment and Risk Management Guidelines: Rising Main Sub-systems, Water Services Association of Australia, September 2013

Physical Properties of Soils in the Murrumbidgee and Coleambally Irrigation Areas Survey, CSIRO Land and Water Science Report 23/08, January 2013

AS 1723-1975 - Centrifugally Cast Grey Iron Pressure Pipes

AS 2280 1999 - Ductile Iron Pressure Pipes and Fittings

ASTM G57-06 - Standard Test Method for Field Measurement of Soil Resistivity Using the Wenner Four-Electrode Method

11. AUTHOR DETAILS

Andrew Hackett (CPEng) is a Principal Civil Engineer with KBR specialising in condition assessment and risk management of water and wastewater infrastructure.

TRACTION STRAY CURRENT MITIGATION SYSTEMS

Richard A Brodribb
M. Brodribb Pty Ltd, Melbourne

SUMMARY: Traction Stray current mitigation systems are installed on metallic structures to reduce the adverse effects of stray current from, primarily, DC traction systems causing accelerated corrosion on pipes. This review paper examines the history and usage and durability of current mitigation systems including TRADs and conductance bonds drawing from experience both in Australia and overseas. The electrical circuits of such devices are assessed and a new low forward loss control scheme is proposed.

Keywords: Traction systems, stray current mitigation, drainage bond.

1. INTRODUCTION

Trains, subway and trams often use electric current, transmitted over wires or rails, to drive the electric motors that move the vehicle. The current is generated at a substation, passes through one feed conductor, through the motor and then returns via another conductor to the substation.

When the current passes through rails which are not isolated from ground, leakage current may leave the rail, pass into the ground and return to a substation via parallel electrical paths as well as that of the rail return to the substation.

The parallel paths do not have to have as low a resistance as the rail for current to be returned through them and a high resistance parallel path will still allow some current to flow through it. If the rail return path resistance increases, more current will flow through the parallel paths.

These parallel paths are commonly through metallic structures such as buried pipes and the flow of such currents may cause corrosion on the rail or rail support where the current leaves the rail and on the pipe where the current leaves the parallel path formed by the pipe to return via some path to the substation. This effect is commonly referred to as "stray current corrosion"

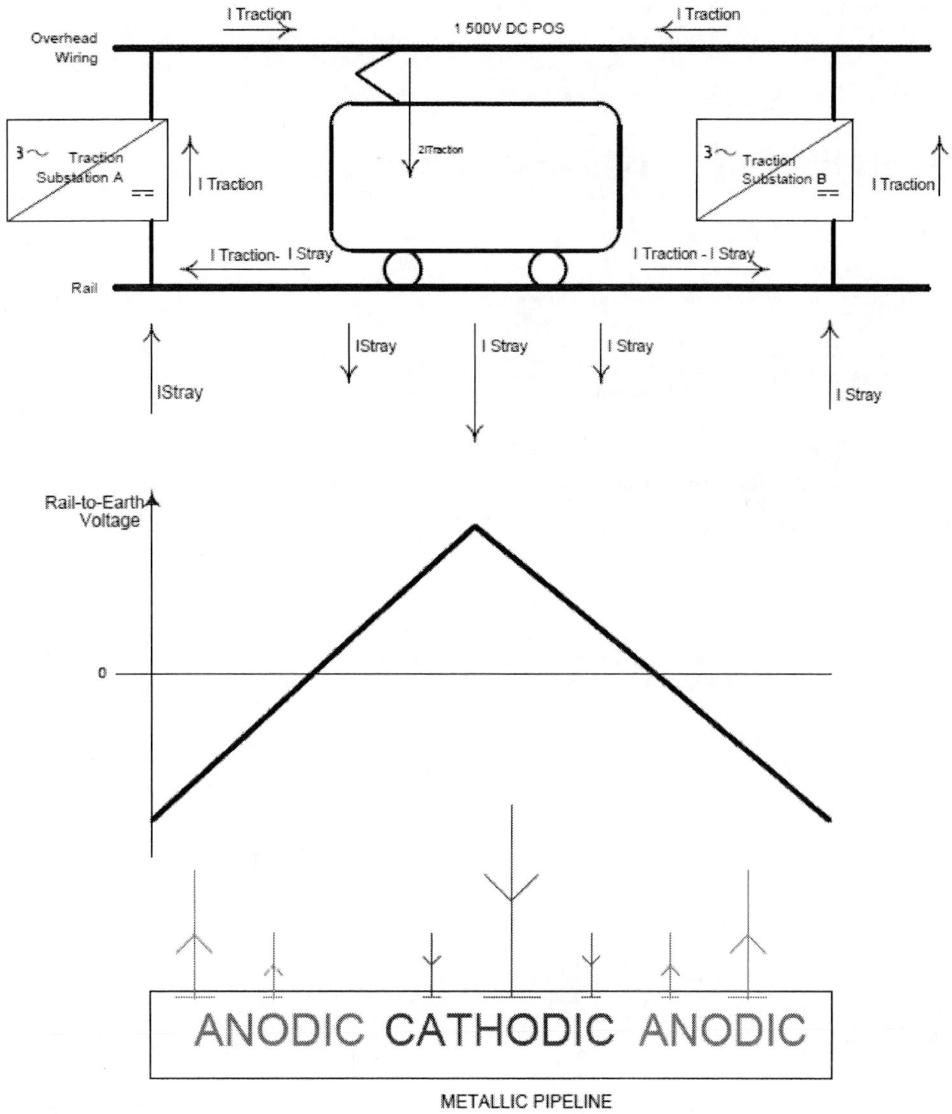

Figure 1 Stray current generation (after (1))

The rail become positive with respect to earth when a train travels over it and negative with respect to earth at the traction substations where the current returns to the DC power supply.

An adjacent metallic structure such as a pipe, may become a path of current return if it is not isolated from the surrounding earth. Current enters the pipe at an area the rail is positive with respect to earth, marked cathodic in Figure 1. At this point the track and track supports will corrode and the pipe receives cathodic protection although the level of cathodic protection may exceed that which is desirable.

Current discharges from the pipe where the rail becomes more negative at the substation return. Current leaves the pipe at the sections shown anodic and the pipe corrodes.

The harmful effects of stray current corrosion from electrical traction systems on adjacent metallic infrastructure were noted within a few years of the installation of the first such systems. In the UK a court judgement (National Telephone Company v. Graff Baker 1893) ruled that tramway companies were entitled to use the earth as a return as long as they did not exceed reasonable limits. (2). In the USA and UK (3), and Australia (4), stray current corrosion effects from the newly installed DC train and tram systems on pipeline, telecommunications and electricity utility companies also resulted in early legal actions against rail and tram companies. (from (5), refs quoted 14 – 21).

FIGURE 2. Examples of Stray-Current Corrosion (Circled Areas) of Cast-Iron Water Mains 1918 (Photocopy courtesy of Chicago Transit Authority).

Figure 3 Stray current damage to cast iron pipes, Chicago 1918 (from (5))

Estimates have been made that the cost of stray current corrosion is approximately 5% of the total cost of corrosion in the US (6) , which has been itself estimated to be 3.1% of US GDP. (7). US GDP was approximately 74 T$ in 2016 (8), so with the above estimation stray current corrosion cost the US economy $115 000 000 dollars per annum.

Barlo (9) and others also note that the cost of stray current corrosion to the transport utilities is generally unknown and the cost of stray current damage is often borne by the foreign structure owner.

2. MITIGATING STRAY CURRENT CORROSION.

2.1 Isolation of the rail.

If the current could be constrained to the rail return by electrical isolation from the ground then stray currents would not occur and subsequent stray current corrosion in a foreign metallic structure could not take place.

Rail to ground resistance can be increased by use of concrete rather than wooden sleepers, insulating pads under the rail fastening clips which secure the rail to the sleeper, ballast cleaning to increase bulk resistance to ground and finally by complete isolation of the rail. It is likely that this last alternative is only suitable for new track construction. (10)

The levels of acceptable resistance to ground for track in NSW are given as 10 Ω/km for rail pairs in NSW (11) and a minimum of 2Ω/km to permit the signaling system to operate (12).
A track and ballast in very good condition will exhibit a resistance of >50Ω/km to earth. This is a low level of isolation in comparison to electrical equipment which typically might have a resistance of 10 000 000Ω to earth at the point of connection and thus parallel paths of current cannot completely be eliminated from the rail to a foreign structure.

2.2 Bonding to the rail.

If current flow to and from the foreign structure cannot be stopped then the corrosion on the foreign structure could be reduced by making the anodic current leaving the pipe to the rail come from one point on the structure and be directly bonded to the rail. The bond would then "drain" the stray current from the foreign structure to the rail. The current, instead of leaving the surface of the pipe into the earth and thence to the rail, causing corrosion on the pipe surface, would pass directly through this metallic bond to the rail and no corrosion would occur on the pipe surface.

Corrosion would still occur on the rail where the current leaves the rail for the foreign structure.

This arrangement is shown in figure 3.

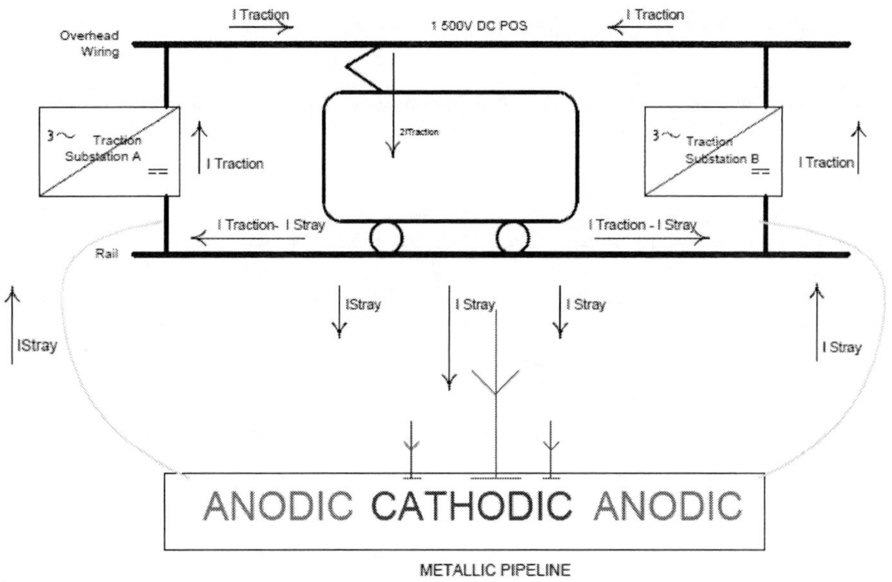

Figure 3 Metallic bond to rail

Bonding such like this was installed on early traction systems where the effect of stray current had been observed. However, a direct connection such as this will cause the pipeline to become substantially negative with respect to earth and to other structures, thus widening the source of the stray current being from the rail to the rail and this bonded foreign structure with adverse effects on neighboring utilities (13) and so the practice was generally discontinued.

2.3 Controlling drainage current.

The simple metallic bond of Figure 3 has a number of disadvantages including the potential of the pipe moving with the rail attracting current from adjacent structures and the possibility of high cathodic currents entering the pipe from the positive rail section. High cathodic currents can lead to overprotection and hydrogen embrittlement (14).
The drainage current can be limited by replacing the metallic bond with a resistor and the direction of current flow can be controlled by the addition of a diode. The diode only allows current to flow from the structure to be drained when this structure is more positive (anodic) than the collection point at the substation. Preventing reverse current flow to the foreign structure reduces stray current in the situation where the pipe is more anodic than the rail at a distance from the substation as may occurring when the train or tram uses regenerative braking.

A typical diode biased drainage bond is shown in figure 4

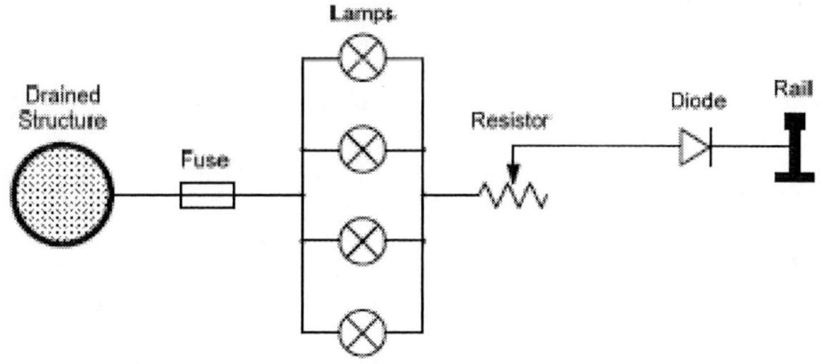

Figure 4 Drainage bond with diode and resistor (ref TH EL 1200)

2.4 Practical considerations of bond design.

The current (I) that flows through the bond is DC and the voltage difference (V) between the bond and the rail is also DC. Thus, the bond, in operation, must dissipate a power (W) in heat loss of the drainage current x the voltage difference, following W = V x I.

The maximum voltage the rail can move with respect to ground is typically limited to +/- 60V DC for safety reasons (8), so the maximum voltage across the bond is limited to approximately 60V.

The devices must be limited to withstand both the worst-case voltage and the worst-case current without failure while still providing a low resistance current path to the rail from the structure. With the example shown above the fuse is rated at 25A so the maximum power dissipated before fuse failure would be 25A x 60V or 1 500W, which may result in cabinet overheating if sustained.

Impressed Current Cathodic protection systems.
Drainage bonds protect the foreign structure by "draining" the stray current from the rail but an alternate protection strategy is also used, that of impressed current cathodic protection (ICCP).

An ICCP system mitigates the effect stray current corrosion operates by holding the foreign metallic structure to negative earth, independently of the rail, at the level required to stop corrosion.

Several types of ICCP system are in use around Australia

1. **Thyristor Drainage unit (TDU).** The TDU comprises a controllable DC rectifier which controls the drainage current from the foreign structure by altering the voltage of the foreign structure with respect to earth. The return current at the substation is measured and a current of some small, linear proportion of the return current, is drawn from the foreign structure by altering the foreign structure voltage. The ratio of current for drainage to substation return current required for protection is obtained from field testing. (15).
   The sense parameter may also be rail to earth voltage

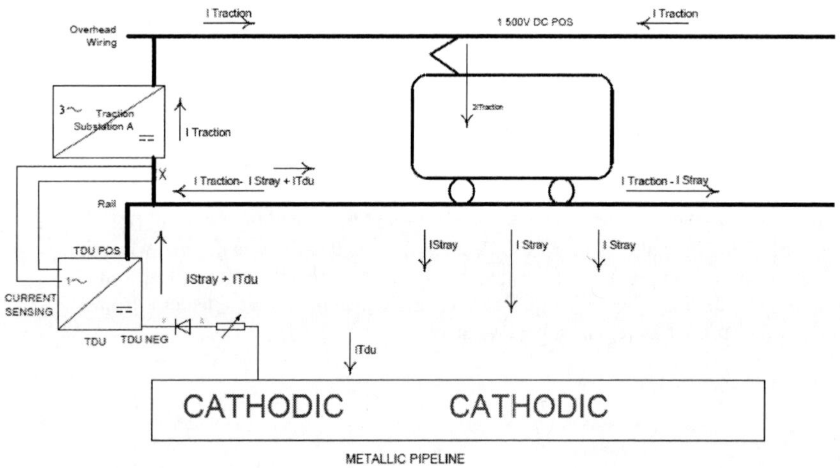

Figure 5 TDU and Drainage Bond

The power conversion system of the TDU comprises an isolation transformer and thyristor controlled rectifier bridge giving a variable DC output. The thyristor bridge can be precisely controlled by an electronic regulator circuit, allowing the output of the TDU to be set via the required parameters of measured current or rail voltage.

The drainage bond is still used for the return path but the TDU increases or decreases the pipe potential in proportion to the expected stray current.

As the TDU is a source of voltage, the power loss in the drainage bond section is reduced as well as providing more precise control of the stray current.

Referring to figure 6, the voltage drop across the bond resistor is now Vtdu – Vs, rather than Vr – Vs. As Vtdu can be up to 30V the maximum voltage the resistor in the bond sees is now 30V for a 60V rail maximum. Halving the voltage across the resistors of the reduces the power dissipated to 25% of the value without the TDU, in the worst case.

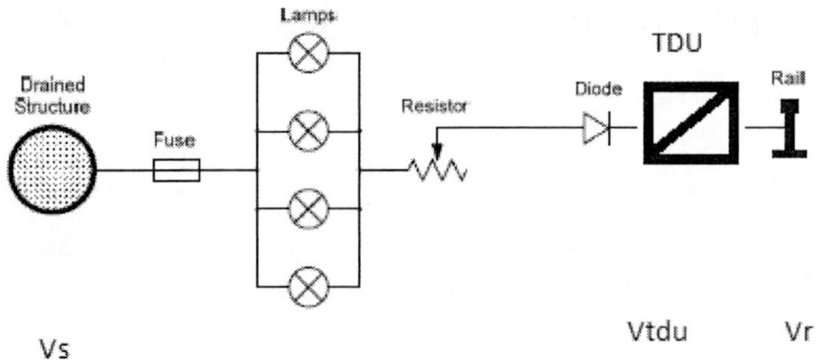

Figure 6 Voltage drop with Bond and TDU

The rating of the TDU in Victoria, is typically 30V at currents of up to 100A.

2. **TRAD - Transformer Rectifier Assisted Drainage system.** The TRAD is used in NSW rail systems and is an autopotential controlled thyristor rectifier variant of the current controlled TDU connected in the same manner as the TDU but without the external diode. In normal operation, the TRAD uses the pipeline structure voltage compared against a stable reference to earth as the control parameter.

   The TRAD adjusts its voltage output so that this value, structure to reference voltage, is held at a level which indicates cathodic protection of the foreign structure. The TRAD is therefore, like the TDU, an ICCP system using the rail as the anode and protecting the foreign pipeline by cathodic protection.

   Unlike the TDU, however, the TRAD will continue to operate to hold the pipe voltage negative to earth, even when no stray current flows from the traction system. A resistor in the pipe circuit, like that of the bond, limits the current that could flow if the pipe became more positive than the rail.

   TRAD ratings for a recent project were 0 – 50V 0 – 30A.

   Both the TDU and the TRAD can only operate when the pipeline is more cathodic than the rail to collect stray current and more anodic with respect to the substation so that current can be returned. If the pipe becomes more positive (anodic) than the rail uncontrolled stray current may flow back from the pipe to the rail in the vicinity of the braking train, collected from other structures. This situation can occur with regenerative braking.

3. REGENERATIVE BRAKING AND STRAY CURRENT MITIGATION.

DC powered trains and trams assist their braking by reversing the action of the DC motor into a DC generator. When the train or tram brakes the DC field supply from the overhead bus to the motor is removed and the motor becomes a generator of a voltage proportional to its shaft speed. If the generator output is then connected to a resistive load bank the kinetic energy, proportional to the velocity squared, of the moving train is efficiently transformed into heat in the load bank, slowing the train. Additional mechanical braking brings the train to a halt.
Regenerative braking has been used on trains since the early 1900's.(16)

The use of internal resistive load banks does not alter the direction of current flow from overhead rail to return rail and so does not alter the return rail potential.

More recent advances in DC motor control technology allowed for the DC braking current to be returned to the overhead positive rail, saving energy and reducing running costs.

If the direction of DC current is reversed, as in braking the local effect at the train is to make the rail more negative than the protected structure, so current might flow from the pipe to the rail, corroding the pipe at the point adjacent to the rail, the reverse of the usual stray current situation.

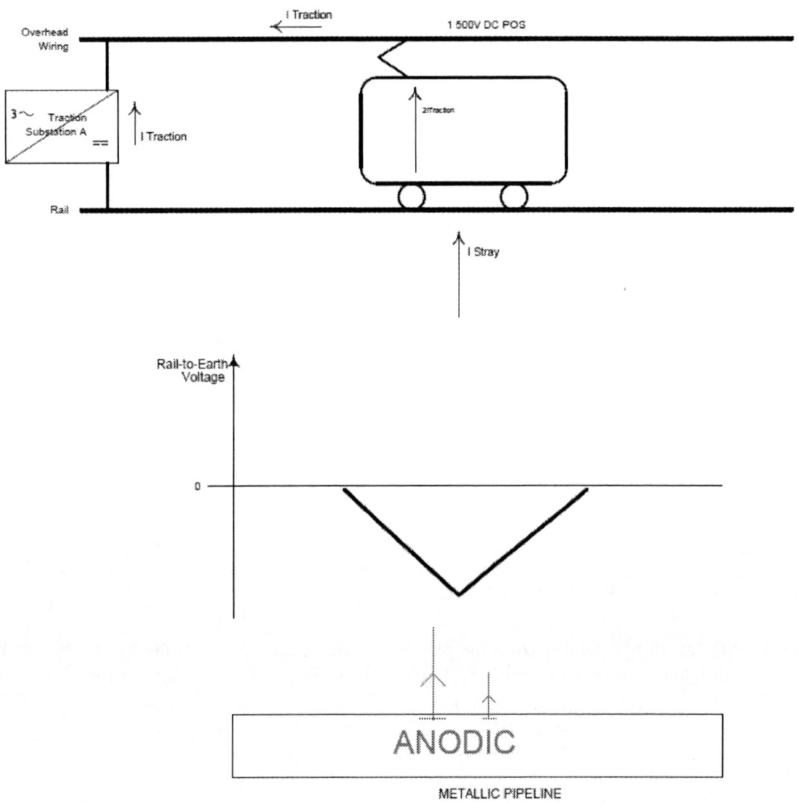

Figure 7 Regenerative braking

4. VARIABLE CONDUCTANCE DRAINAGE BOND (VCDB) AND DC SUPPLY.

The introduction of regenerative braking trams resulted in anodic conditions on structures which could not be correlated to substation load and could not therefore be protected using the proportional current injection system of the TDU (17)

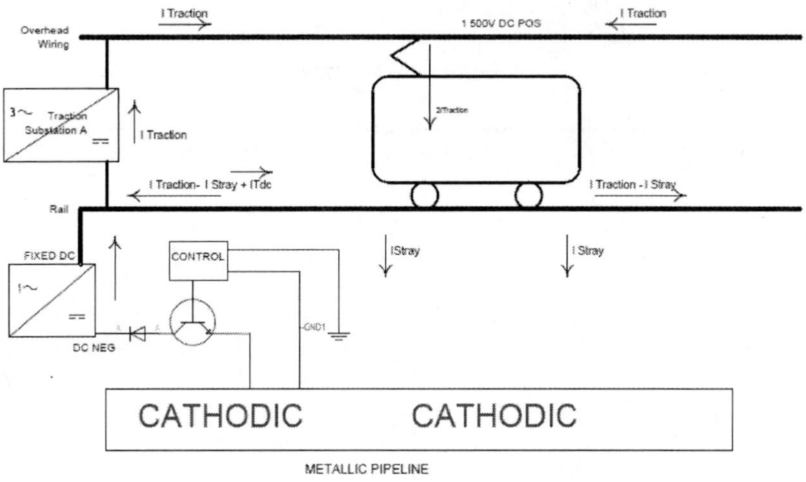

Figure 8 Variable conductance drainage bond

The VCDB combines a fixed voltage DC power supply with an electronically variable resistor, usually a large transistor or MOSFET. The VCDB senses potential to earth via the same pipeline structure to reference system of the TRAD, but rather than increasing or reducing the voltage of the pipe with respect to ground, as does the TRAD, it adjusts the value of the bond resistance so that when no protection is required for the pipe the bond resistance is high and when protection of the pipe is required, the bond resistance is reduced, allowing current to flow.

The resistance of the VCDB is subject to the same load conditions as the TDU in normal operation and the semiconductor device must be operated in the linear region, that is, not switched completely on or completely off. To dissipate the heat the device must be mounted on a large heatsink. A semiconductor device operating in the linear region is also power limited so a sudden increase in DC voltage across the device may cause it to fail.

5. A LOW LOSS SOLUTION FOR REGENERATIVE STRAY CURRENT TDUS.

Regenerative braking returns high currents to the rail in the reverse direction to normal current flow. This local high current can make the pipe positive with respect to the rail, overcoming the voltage offset supplied by the rectifier. TDU's and TRAD's cannot operate correctly in this case because voltage because once the pipe is positive with respect to rail uncontrolled current flows through the freewheeling diode that is part of such thyristor rectifier circuits. The blocking diode that is part of the conventional bond is also ineffective as it and the freewheel diode, are in series so and the current must be limited by an external resistor, which must be suitably sized to operate under regeneration conditions.

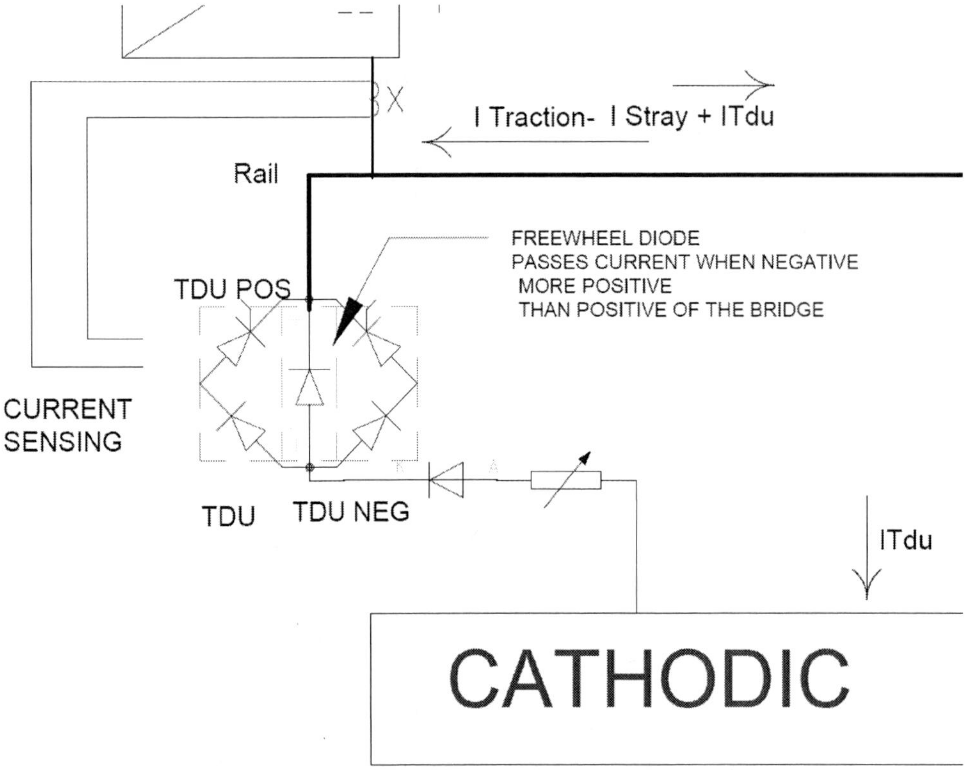

Figure 9 Freewheeling diode in TDU

Replacement of the half bridge with a full bridge, comprised only of thyristors and without a freewheeling didoe, could allow TDU type device to generate either positive voltage, as in the usual case, or negative voltage, as required for the control of currents from regenerative braking. The full bridge operates, in fact, in the same regenerative manner as the regenerative brake on the train, returning the extra energy back to the AC mains rather than dissipating this energy in a resistor.

Such full bridge control, suitably filtered, could replace both the diode and the heat dissipating dropping resistor without requiring any semiconductor device to operate in the more susceptible linear mode.

6. CONCLUSION

Stray current corrosion problems from DC traction systems have been present since the first DC traction systems were installed adjacent to buried pipes. Various solutions for stray current traction mitigation have been and are used and these and their limitations are discussed.

Current railway practices standards and guides expect that rail construction will minimize stray current, and standards note "minimize stray current by good design" (18), "Drainage bonds and TRAD shall only be used as a last resort…" (19). However, it would appear that, with a legacy of a railway system over 100 years old, and still going strong, that stray current corrosion mitigation systems still have a future.

7. REFERENCES

1. T HR EL 120002 GU Electrolysis from Stray DC Current Version 1.0 01 May 2014. State of NSW, Transport NSW.

2. Design Standards Stray Current Management. Tramway Technical Guidance Note 3. Office of Rail Regulation. UK.

3. D Paul, "DC Stray Current in Rail Transit Systems and Cathodic Protection" IEEE Industry application Magazine Jan 2016, NY, NY.

4. VEC Resource Manual, EnergySafe Victoria, July 2014. Melbourne, Victoria.

5. Barlo, T, Zdunek, A, "Stray Current Corrosion in Electrified Rail Systems – Final Report" May, 1995.

6. Barlo, T, Zdunek, A, *op.cit. p2.*

7. Corrosion Costs and Preventive Strategies in the US. NACE Publication FHWA-RD-01-156.

8. Bureau of Economic Analysis, US. Department of Commerce, Table 1.1.5 Gross Domestic Product 2016.

9. Barlo, T, Zdunek, A, *op. cit.,* p.2.

10. Barlo, T, Zdunek, A, *op. cit.,* p. 16.

11. T LR 00001 ST Traction Power System Requirements Version 1.0 25 May 2017. State of NSW, Transport NSW.

12. T HR EL 120002 GU Electrolysis from Stray DC Current Version *op. cit.,* p.7

13. D Paul, *op. cit.,* p.11.

14. T HR EL 120002 GU Electrolysis from Stray DC Current Version, *op. cit.,* p. 12

15. VEC Resource Manual, *op.cit.,* p. 14.

16. Raworth, Alfred. "Regenerative control of electric tramcars and locomotives". Proceedings of the Institution of Electrical Engineers 1906–1907. 38: 374–398.

17. VEC Resource Manual, *op.cit.,* p. 16.

18. Design Standards Stray Current Management, *op.cit.*, p.7.

19. T LR 00001 ST Traction Power System Requirements, *op.cit.*, p.21.

8. AUTHOR DETAILS

Richard Brodribb is the Managing Director of M. Brodribb Pty Ltd.
He has been involved with the design and manufacture of a wide range of DC power supplies for cathodic protection, battery charging and other DC applications for over 40 years.
He has a B.E from Monash (Elec Eng) and a M.B.A from RMIT. He is an MIEEE member and a member of NACE.

USING THERMOPLASTICS FOR INFRASTRUCTURE PROTECTION IN MARINE AND OTHER CHALLENGING ENVIRONMENTS

T. J. Davison
Enviropeel Systems, UK

SUMMARY: Corrosion Inhibiting Thermoplastics offer the ability to deal with a multitude of corrosion problems on whole assemblies with just a single application and minimum surface preparation. This would obviously be of enormous benefit to owners and operators, particularly in the mining and offshore industries where cost-effective remedies can be hard to find, but can such a claim be justified? This paper examines the mechanisms that provide the protection, how the material is used and what the past fifteen years of experience applying the material in a wide range of different environments can tell us about its efficacy, longevity and potential for the future.

Keywords: Thermoplastics, Corrosion, Infrastructure, Marine, Mining, Offshore

1. INTRODUCTION

Figure 1 Rusty valve deck on offshore infrastructure

Figure 2 Rusting bolt from DNV testing

Steel structures in coastal and offshore environments are characterized by high levels of corrosion and the need for constant vigilance in order to maintain function and prevent failure of containment and/or delivery systems. Environmental and safety considerations also demand effective long-term remedies where structural integrity may be at risk. In the UK, safety authorities have identified corrosion prevention on North Sea structures as a key area requiring improved performance from operators.

Solutions include prevention through the use of design, material selection and advanced coating systems yet, despite everything, rust abounds.

Figure 2 shows corrosion in galvanized PTFE bolts after only 18 months exposure a splash zone environment in long-term DNV testing.

In the most intractable areas, galvanic and pitting corrosion may be visible within months - and service-life extensions mean that even the best-maintained structures will have corrosion problems in areas which are difficult to resolve.

Figure 3 shows well heads from an unmanned platform in the North Sea. Many platforms in the North Sea and elsewhere were constructed with an expected service life of a maximum thirty years but, in 2010 when these photographs were taken, they were already ten years past their sell by date and becoming unfit for purpose.

Figure 3 Ageing infrastructure in the North Sea

Figure 4 Remedial application to North Sea Wellhead

In some cases, infrastructure was replaced where metal loss threatened integrity but, with only a few years left of production in a stagnant market, wholesale renewal of such costly facilities could not be justified and operators wanted to find a safe and effective remedy for their problems. Accordingly, in 2010, a trial application was requested from a UK company offering a sprayable thermoplastic anti-corrosion system and a series of extremely corroded wellheads were protected using the system. At the time, the company applying the system indicated to the customer that it would stop the existing corrosion and prevent any further corrosion for at least 10 years.

As often happens in these cases, personnel and budget priorities changed and, despite the success of the initial application, the operators didn't follow through and it wasn't until 2016 that an inspector, who happened to be going to the platform decided to inspect the well heads to see what had happened. His report showed that everything the company had promised about the levels of protection had been true. Of the sample substrates from which he removed material he concluded:

"... the 2010 application ... has arrested the corrosion of the flanges. The encapsulation properties of the product have performed as expected in preventing moisture ingress and further corrosion."

Before going any further, we should look at the system and how it works.

2. HOW THE SYSTEM WORKS

Thermoplastics have a long track record of corrosion and impact protection through their use as dipping agents for precision parts and machined tools in workshop or factory environments, as shown in Figure 5. More recently, advances in delivery systems and material development have produced durable, UV-resistant coatings with enhanced anti-corrosion properties that can be spray applied on to any substrate, in any location.

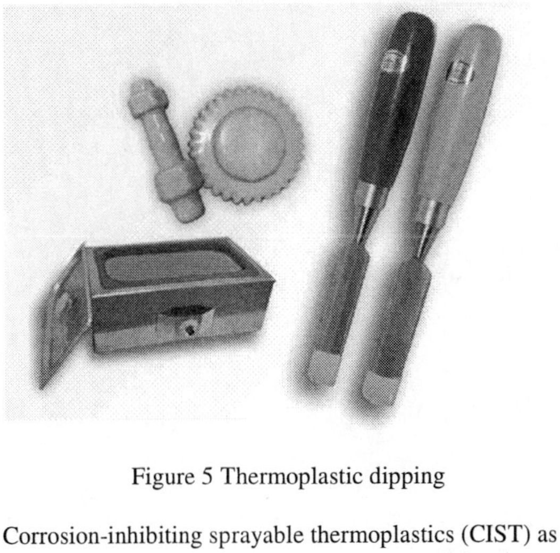

Figure 5 Thermoplastic dipping

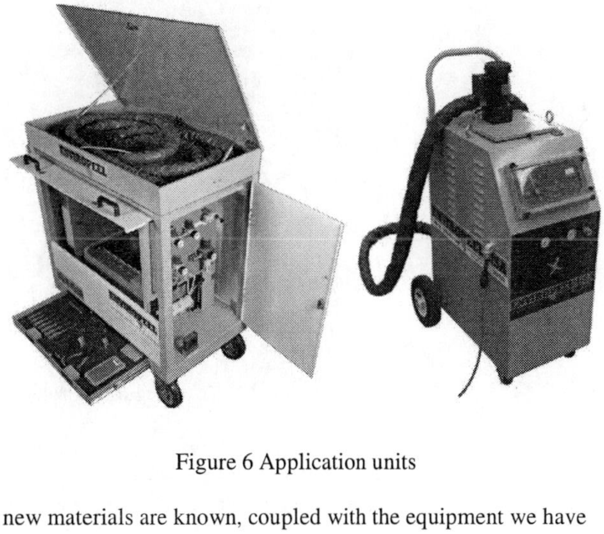

Figure 6 Application units

Corrosion-inhibiting sprayable thermoplastics (CIST) as the new materials are known, coupled with the equipment we have seen, claim a number of advantages over traditional coating systems:

- Adaptability – a 'one size fits all' strategy means the same material and equipment can be used on the smallest component as well as a large complex assembly
- Portability – the equipment is mobile and can be taken to site where it can be easily moved around and has the ability to deliver material up to 19 metres away from the application unit using hoses and spray guns
- Easy to prepare - the material requires very little preparation – once loose surface materials are removed, the material can be directly applied to rusty surfaces
- Active protection – the material contains corrosion-inhibiting oils which are circulated on to the substrate throughout the lifetime of the application, arresting existing corrosion and preventing any further deterioration
- Passive protection – the application forms an impermeable close-fitting barrier to ingress of moisture and contaminants
- Highly resistant to UV and salty, wet marine environments
- Low environmental impact – re-usable, recyclable and non-toxic, the system is non-hazardous and can be virtually waste free.
- Can be applied to complete assemblies – no need to separate parts or shut down plant

The system works by combining a small percentage of inhibiting oil into a UV-resistant polymer and creating a resilient hard rubbery thermoplastic that is solid at ambient temperatures.

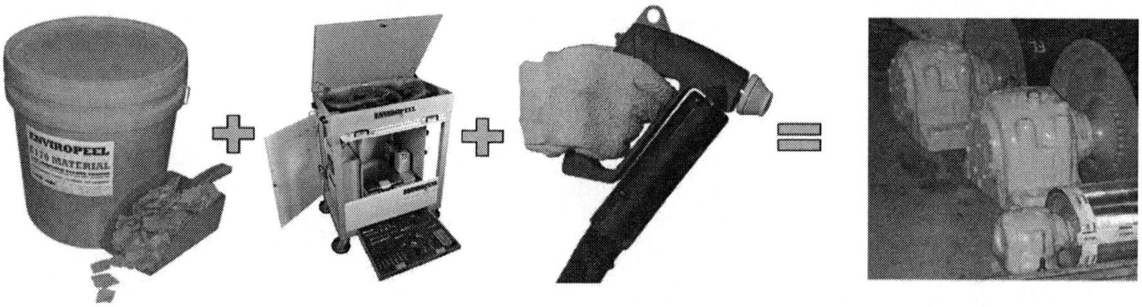

Figure 7 Stages of the system

The material is chipped for easy handling and fast melting. It is entered into the heating tank of specially designed application units, where it is heated to form a liquid which is then delivered on demand through heated hoses to the spray gun which applies the material to the substrate. Rapid cooling of the material causes the application to tighten around the substrate to form a close-fitting cocoon. The protection mechanisms of the material are represented in Figure 8..

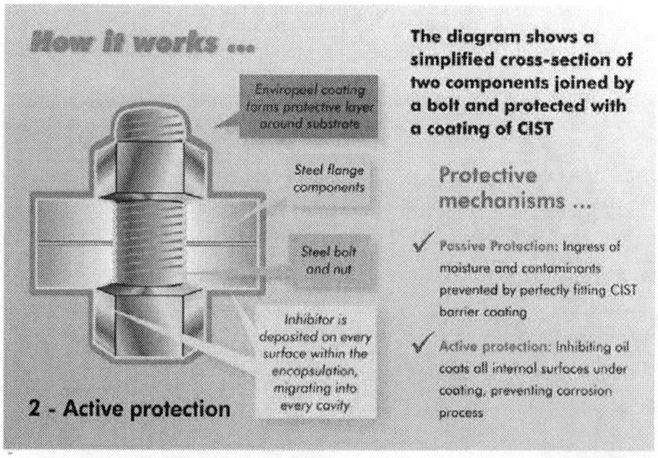

Figure 8 Protective mechanisms

The outer barrier prevents ingress of moisture and oxygen while, internally, inhibiting oils volatilized by the heating process are released on to all surfaces inside the encapsulation. Minute quantities of oil within the material continue to circulate on to the steel surface for as long as the material remains in place.

The combination of ingress prevention, excluding moisture and oxygen, and prevention of the galvanic process through an active oil coating means the system is effective in providing both remedial and pro-active corrosion prevention to complete assemblies such as standby equipment, valves and flanges as well as small components or substrates.

So, this is the theory, how does the product actually perform? Four frequently asked questions are:

1. How does the product perform in marine and high UV environments?
2. How long will it last?
3. How long before the inhibitor runs out?
4. How impervious to water is the material?

First of all we will look at testing in the light of these questions.

3. TESTING

3.1 Accelerated Weathering

A common weathering test is ASTM G154, in which samples are exposed to 42 cycles of 8 hours of UVA-340 ultra-violet light at a temperature of 60°C followed by 4 hours of condensation at 50°C for a total of 21 days of exposure. Like many other standard tests, this may provide a benchmark for comparison but does it has very little to say about what will happen in the real world.

Other tests such as ASTM D4587 and G 155 provide for a similar combination of UV irradiation, heat and water while ASTM B117 focuses on a combination of water salt and heat. All these tests are widely accepted in assessing suitability of materials for outdoor use but, for example, a 1000 hrs of UVA exposure is only the equivalent of 1 year in South Florida and a bolt that has passed a 1000 hrs hot salt fog test may be completely unserviceable within 18 months in some environments, as we have seen.

So, in order to provide some real-world equivalency, the scale of the testing for CIST was revised to be significantly longer and under more extreme conditions.

HOUR COUNTER	MONTH	WEEK 1	WEEK 2	WEEK 3	WEEK 4
672	1	UV	UV	HSF	UV
1344	2	UV	HSF	UV	UV
2016	3	HSF	UV	UV	HSF
2688	4	UV	UV	HSF	UV
3360	5	UV	HSF	UV	UV
4032	6	HSF	UV	UV	HSF
4704	7	UV	UV	HSF	UV

Indicates the samples in UV Cabinet

Indicates the samples in HSF Cabinet

Figure 9 Exposure schedule

For the weathering test, it was decided that the rapid half-daily cycles of wet and dry specified under ASTM G154 were not representative of natural weather cycles and a regime was implemented that allowed much longer periods of exposure to each element of the test. The cycles were between UV exposure and hot salt fog (HSF) exposure and the test was to run for 6 months.

HSF PARAMETER			
Salinity, IN	5%	Salinity, OUT	5%
pH		Fog Collectors Data	
pH in	7.2	Fog Collector 1	2.0 ml/hr/80cm²
pH out	6.5	Fog Collector 2	1.8 ml/hr/80cm²
Temperature		Pressure	
Saturated Air Temp	47°C	Chamber Pressure	1 bar
Chamber Temp	35°C	Air Intake Pressure	0.9 MPa
Hygrometer		Downtime	
Wet Bulb	35°C	From	-
Dry Bulb	36°C	To	-
UV PARAMETER			
UV LAMP NO	4	PLACEMENT	360 DEGREE
WAVELENGTH		15W/ 450nm	
HEATER NO	2	AVERAGE TEMPERATIRE	46 DEGREE C

Figure 10 Exposure parameters

The UV exposure was via 4 x 15w 450nm UVA tubes arranged to give continuous 360° exposure while the HSF element employed standard ASTM B117 parameters with 5% salinity, fog temperature 47°C and chamber temperature 35°C.

The samples were prepared using carbon steel test pieces, blasted and coated with a single coat of epoxy and fastened using standard nickel-plated B7 bolts. They were then coated with CIST using 2 coats at 2mm per coat, with edges sealed using standard cable ties and the pipe ends left unprotected.

Figure 11 Samples at start of test

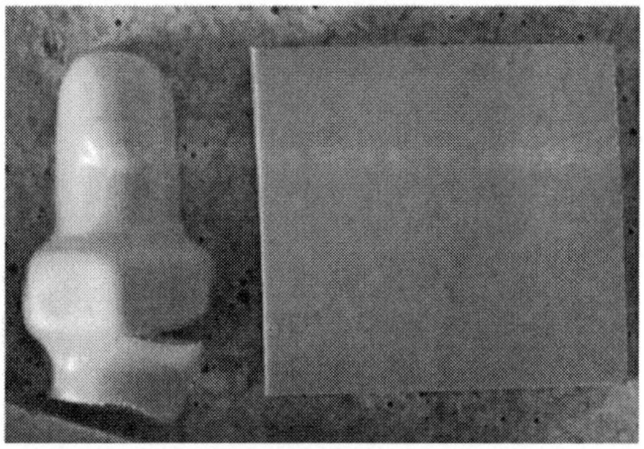

Figure 12 Slight fading after 4500 hrs

The test pieces were then entered into the UV cabinet to initiate their exposure (Fig 11). Samples were exposed to the weathering process in different orientations to expose all areas, testing for ingress as well as weathering. A number of small buff-coloured CIST coated bolts were included to allow interim removals as required.

As the number of weeks increased, it became increasingly obvious that a longer period was needed as there was no discernible effect on the CIST. It was decided to leave the samples in for another six months and remove one of the small bolts to assess the changes so far. The material was cut from the bolt to see if there had been any changes either to the bolt or the CIST coating.

The bolt itself displayed no signs of deterioration, and although the exterior of the CIST coating was found to be relatively free of inhibiting oil, inside the encapsulation inhibitors were evident on all surfaces. Visual examination of the surface showed slight fading, about 5% compared with a reference sample (Figure 12) and microscopic analysis of the surface of the coating showed no deterioration of the surface structure (Figure 13). Analysis of a section through the coating confirmed a slight oil loss at the surface but showed that oil was uniformly present throughout the coating (Figure 14).

Figure 13 No crazing or cracking

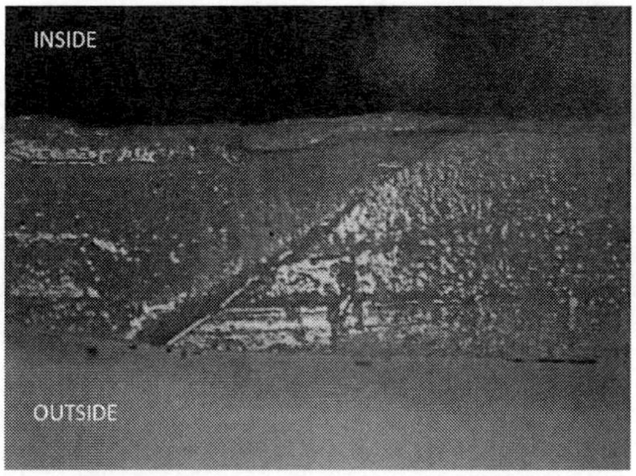

Figure 14 Slight drying on outer edge

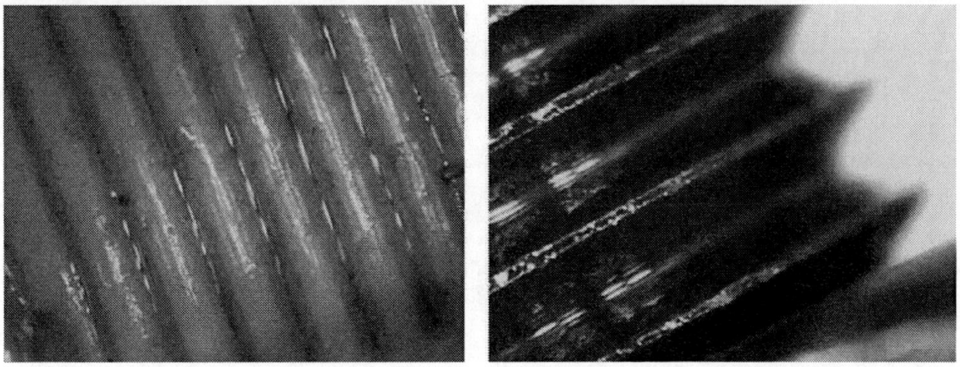

Figure 15 Bolt thread and removed coating show optimum oil deposition

It was concluded that the accelerated weathering cycle over 6 months had no detrimental effect on the CIST encapsulation and that all performance criteria remained at 100%.

3.2 ASTM B117 Hot Salt Fog Testing

Most authorities agree that there can be no extrapolation of an HSF test to correlate with a defined number of years - but they will agree that the longer the test, the more likely it is that a successful outcome will be achieved in the real world. So it was decided that an unprecedented 10,000 hours (nearly 14 months) of continuous HFS testing would go a long way to showing what CIST could achieve.

Figure 16 Control plus one sample removed at 4,500 hrs and final sample at 10,000 hrs

After 10,000 hours the uncoated control piece displayed significant levels of deterioration but the CIST samples were in excellent condition. No corrosion or deterioration of any kind could be detected within the CIST encapsulation, bolts, nuts and flange were in perfect condition.

3.3 Cyclic Inhibitor Loss Test

CIST is an active coating, the deposition of the inhibiting oil on to the substrate and the ability of this inhibition process to prevent corrosion are key elements in the overall corrosion protection package so, if this deposition is a continuous process, how long can the coating maintain its performance?

In fact, the process by which inhibiting oil is maintained on the encapsulated substrate is more akin to the water cycle that governs the world's rainfall than it is to a constantly pouring tap. On application, the hot material deposits a thin film of oil between the coating of CIST and the substrate. On cooling, some of this is reabsorbed into the CIST and, as the substrate surface and ambient temperatures go through their natural cycle, a continuous cycle of deposition and absorption is maintained. Loss through the surface of the CIST encapsulation is negligible, it soon becomes almost dry to touch but a 'dew point' effect is maintained between the substrate and the CIST producing sufficient inhibitors to fill the pits and crevices within the bolted structure, coating all internal surfaces without escaping into the environment.

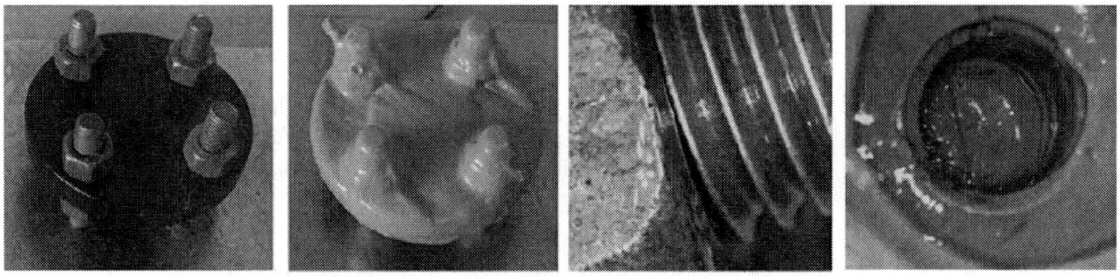

Figure 17 Cyclic test piece before and after coating plus thread and coating interior after test

For this test, an encapsulated substrate was subjected to a continuous cycle of temperature variations from 45°C to 4°C on a 24-hour cycle for six months. The weight of the substrate and the applied CIST material was recorded before and after encapsulation. The sample was suspended over a tray to collect any potentially escaping liquids.

A laboratory oven was used to raise the substrate to 45°C during the day, overnight the substrate was transferred to a refrigerator monitored at 4°C. The substrate was removed from cooling at 9 am and kept at ambient temperature until 11.00 when it was replaced in the oven. The oven was set to 45°C and switched on for 3 hours. After 3 hours the oven was switched off and the substrate allowed to cool in ambient temperatures for 2 hours before repeating the cycle, five days on, two days off. At weekends the substrate was left alternately at 45°C or ambient temperatures.

Over the six-month period there was no significant change in weight nor any escape of oil on to the tray. Initially weekly measurements were taken but, with no changes found, the interval was increased to monthly. At the end of six months, the material was removed and examined. Good oil deposits were found within the encapsulation, on bolt threads, metal and CIST interior surfaces. Other than some soiling from handling, no surface effects were found on the CIST exterior.

Date	Weight uncoated	Weight coated
17-8-12	1427.5 g	1565.0g
21-9-12	N/A	1565.0
18-10-12	N/A	1565.1
23-11-12	N/A	1565.0
2-1-13	N/A	1565.0
31-1-13	N/A	1564.9
28-2-13	1427.6	1565.0

Figure 18 Cyclic test results

3.4 SUSTAINED HIGH TEMPERATURE TEST

A measured and weighed sample of CIST material was placed in a laboratory oven for 3 months at 80°C in order to establish the rate of inhibitor loss at elevated temperatures. At the end of each week the material was removed from the oven, measured, weighed and immediately returned to the oven to maintain its temperature. Weight loss was consistent with an average loss of 0.08% per week. Weekly measurements showed no measurable shrinkage over the period.

	Date	Measured Weight	Weekly Loss (gm)
Mar	11	2778.5	
	18	2775.9	2.6
	26	2774.2	1.7
Apr	2	2772.1	2.1
	8	2769.8	2.3
	15	2767.6	2.2
	22	2765.3	2.3
	29	2763.2	2.1
May	8	2761.5	1.7
	14	2758.6	2.9
	20	2756.7	1.9
	28	2754.2	2.5
Jun	4	2752.9	1.3
	11	2749.8	3.1
	18	2747.8	2
	25	2745.7	2.1
July	2	2743.5	2.2
	8	2741	2.5
	16	2738.9	2.1
	23	2736.2	2.7
	30	2734.7	1.5
Aug	6	2732.6	2.1
	13	2730.3	2.3
	21	2728.5	1.8
	28	2726	2.5
Sep	4	2723.8	2.2
	12	2721.5	2.3
Loss over 26 wks		57	57

ELEVATED TEMP WEIGHT LOSS TEST

Total Loss Over 26 Weeks (gm)	57
Percentage loss over 26 weeks	2.05%
Average loss per week	0.08%
Annual loss	4.10%

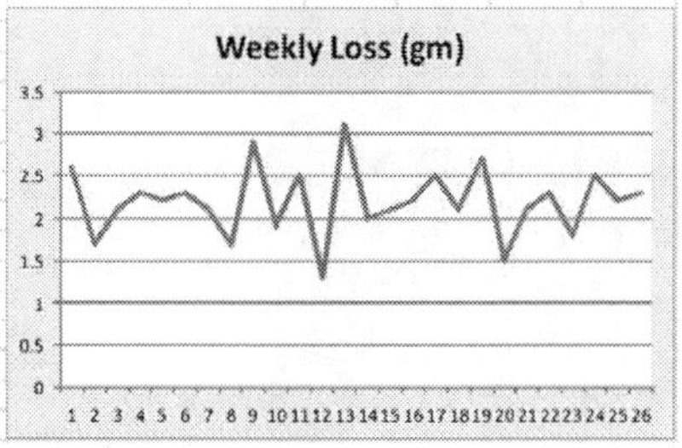

Figure 19 Test results

3.5 Impermeability – Sustained Deluge Test

Not satisfied that standard tests were sufficiently demanding, a deluge test was devised to run on a CIST-coated flange and pipe. A 20% saline solution was cycled at approximately 400 litres an hour for 11 hours a day, 5 days out of 7 for 6 weeks. Ambient temperatures were between 24°C and 38°C. The steel pipe substrate and flange were new and uncoated low-carbon steel, a section along the length of the pipe was abraded prior to coating with CIST to provide the best potential conditions for corrosion. The CIST encapsulation was sealed using a standard tie-wrap technique.

At the end of the test the CIST was cut away to reveal the substrate. No corrosion of any kind was found within the CIST encapsulation, whereas significant corrosion occurred along the length of the exposed, unprotected pipe.

Figure 20 Test substrate at 7 days, 28 days and 38 days

Figure 21 Completely unblemished interior after 6 weeks

So, if testing provides such encouraging results, how has the material performed in the real world?

4. REAL WORLD EXAMPLES

In the fifteen years since the first introduction of CIST there has been a wide range of applications in a variety of industries, particularly in the Offshore, Mining, Gas and Power Generation industries. We have already seen the results for application results for a major oil company on one of its unmanned platforms in the North Sea, but these images are from other platforms in the area, where CIST applications were first introduced in 2003.

Figure 22 Failing carbon steel bolts in stainless flanges

Figure 23 Mixed carbon and stainless substrate after 7 years

First applications were introduced to remedy potential bolt failure on small flanges but soon spread to include a range of other substrates. Particularly vulnerable are carbon-steel bolts in stainless steel joints as Fig 22 shows. Significantly, seven years after receiving CIST protection, similar substrates in the North Sea were uncovered for maintenance purposes and no corrosion was found to have taken place.

In 2008, a platform that had significant corrosion even before it was commissioned, had more than 8000 flanges coated to stabilize rust across the structure.

Experience on these platforms has led to successful applications on offshore facilities from California to the Congo for Total, Nobel, ConocoPhillips, Shell and a number of smaller companies.

Figure 24 More than 8000 flanges coated

Figure 25 A group of small coated flanges

5. AUSTRALIA

CIST is scheduled to be used on offshore facilities in Australia for the first time this year but Australia was one of the first places to recognize the value of CIST, introducing standby protection for the mining industry in 2000. Although most mining in Australia takes place a long way from the sea, it all ends up in ships and is often transported using long conveyor systems which are very exposed to environmental conditions.

Figure 26 Severe rusting from high salt levels

Figure 27 No corrosion on test bearing

These two conveyor bearings illustrate both the problem and the solution. The Fig 26 is from a salt conveyor in Port Hedland. Bearings on these conveyors were failing in as little as nine months from exposure to highly saline water and contaminants from the transport of mineral salt. Following the success of CIST for standby protection, trials were undertaken for the introduction of CIST protection on live bearings. Slide 27 shows removal of material from test application revealing perfectly preserved substrate.

The trials were an immediate success, bearings protected with CIST continued to perform without any deterioration for more than 4 years. Data collected for BHP Billiton and Rio Tinto in 2005 showed a 500% increase in equipment lifespans and a significant reduction in accidents because of the reduced maintenance requirements. CIST is now used across the mining industry in Australia for conveyor and standby protection.

Recently, an engineer who was a pioneer for CIST in Australia, sent us photos of applications his company had made up to 12 years earlier. Following removal, the substrates were found to be in excellent condition despite having had no attention whatsoever in the intervening years. He was ecstatic.

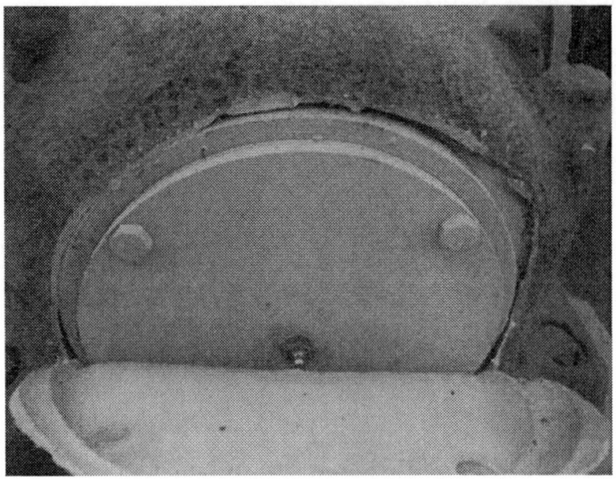

Figure 28 After 12 years

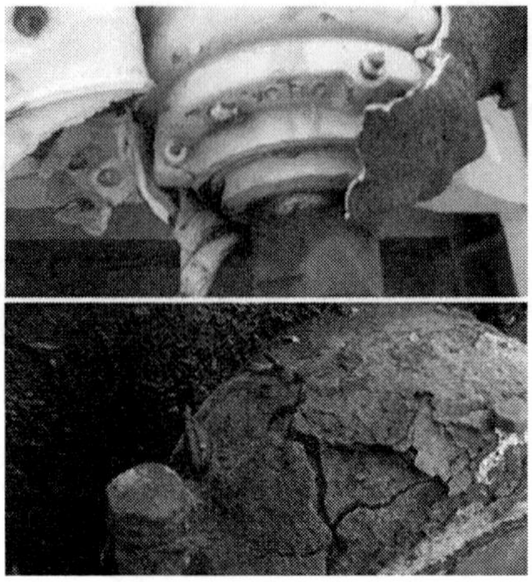

Figure 29 12 years with CIST
And 12 years without

6. CONCLUSION

Many other applications in other industries have benefited from the unique qualities of CIST. In the US, results from applications in the mining industry have duplicated the savings and bearing replacement reductions found in Australia, with applications to underground conveyors in Canadian potash mines commencing in June this year. Gas production in Oman, high voltage switchgear in the UK and Europe, bolt protection on wind-turbine towers in the US and Europe – the range of applications has been enormous.

Figure 30 CIST coated bolts on a wind turbine tower

Figure 31 CIST being applied on to high voltage switchgear in the UK

Nothing is perfect, but the results seem to show that CIST has some very effective answers for corrosion and ingress problems in marine and other challenging environments - particularly on complex and vulnerable assemblies where standard coating systems cannot be used.

STRAY CURRENT CORROSION DETECTION - THE USE OF PENETRATION PROBES

B. Sandberg[1], B. Lindell[2]
[1]Swerea KIMAB, Stockholm, Sweden [2]3CCC, Billeberga, Sweden

SUMMARY: Pipelines are protected against corrosion by a combination of a high quality coating and cathodic protection systems. They are in many cases exposed to interference from dc and/or ac currents. If the current is transferred to the pipeline in a conductive way it can be an advantage not to have a coating with very high resistance. This enables mitigations as local groundbeds or introducing a voltage drop in the pipe. To detect ac corrosion the local corrosion rate must be monitored. The local corrosion rate can be evaluated by excavating test coupons but penetration probes give an economic benefit and results are received without delay, giving the most cost effective way to monitor a highly influenced section of the pipe network.

Keywords: Stray current, Soil, Pipeline, Protective coating, Probe.

1. INTRODUCTION

High quality coatings on buried pipelines decrease the current density needed for cathodic protection. It also lowers the risk for interference between different cathodic protection systems. However, the sensitivity for the pipeline itself to be influenced by stray current, both ac and dc, increases when the coating resistance increases. Both measurements and mitigations are in many cases complicated to perform. This paper will try to identify the most important parameters to consider when dealing with stray current interference on well coated pipelines. A strategy how to protect pipelines in areas of complex stray current interference, including the use of coupons or special probes as an early warning of corrosion, is presented.

2. STRAY CURRENT INTERFERENCE

When dealing with stray current interference many factors have to be considered. The interfering source can have a fix location or move with time. The current output from a fix source may also vary in magnitude as well as in current direction.

Historically dc currents have been the main concern. Today, with high quality coatings, also ac corrosion takes place as a consequence of high current densities in small coating damages caused by increased ac potentials of the pipe. Performed research has prevailed ac corrosion to be dependent on a combination of ac interference and the cathodic protection level (1, 2).

dc interference is caused by conductive transfer of currents. To be able to fully evaluate an interference situation it is essential to understand that the influence is caused by changes in soil potentials, see figure 1.

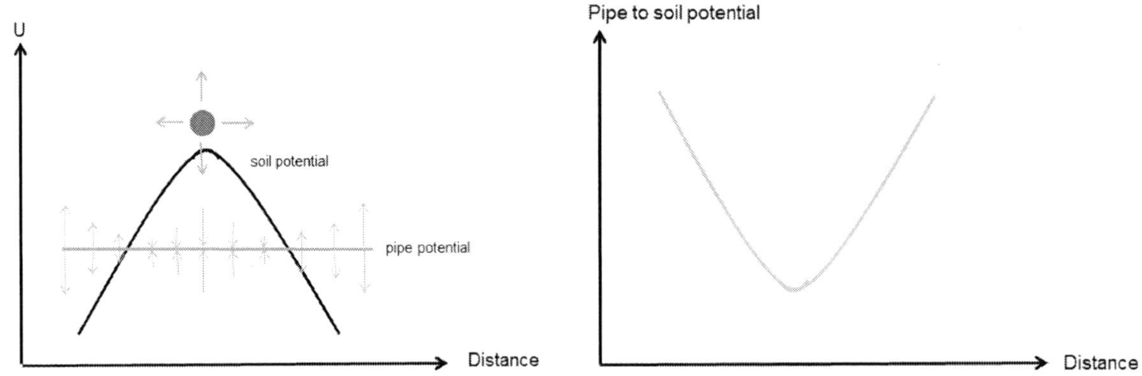

Figure 1. Changes in "pipe to soil potentials" due to changes in soil potentials caused by dc interference.

If the interfered construction is poorly coated the inleaking as well as the outleaking current creates a secondary electrical field, lowering the strength of the original field. If the current flow in the pipe is high enough also the potential of the pipe itself will change, see figure 2. In the case of a very well coated pipeline only small currents are picked up, resulting in no secondary field and no potential drop in the pipeline. The pipe acts as an "antenna" and all differences in soil potentials are transferred to changes in cathodic protection levels, "pipe to soil potentials", in accordance to figure 1.

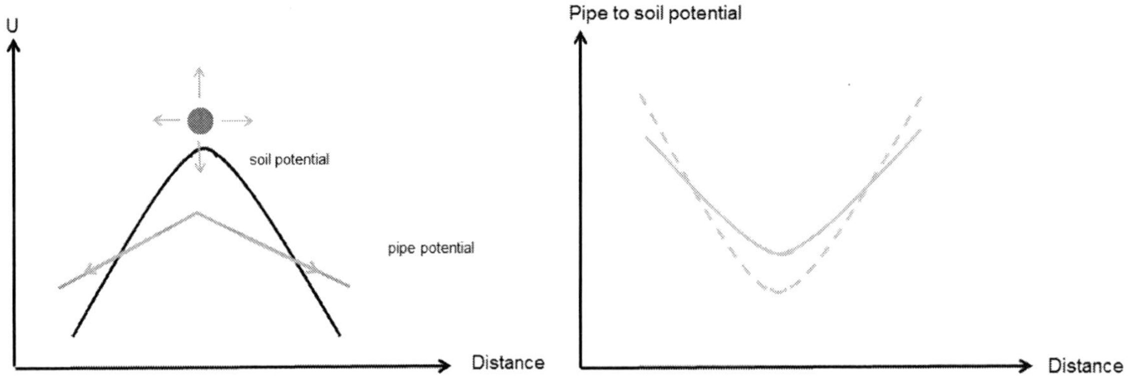

Figure 2. Current flow in the pipe decrease the difference in "pipe to soil potentials" due to changes in pipe potentials.

As the origin of the interference is differences in soil potentials it is evident that the soil resistivity and how it varies is of great importance. In central Europe the bed rock is located some 2 km below the ground surface. Often the resistivity decreases with depth. A typical profile is shown in figure 3. In the same figure we can see the electrical field from an hemispheric electrode, diameter 1 m, having an applied voltage of 2 V. The low resistivity in deeper soil layers results in a relatively high leakage current but despite this the electrical field drops quickly. 95 m from the electrode the soil potential has fallen from 2,0 to 0,7 V.

In northern Europe, Scandinavia, the bed rock goes all the way up to the ground surface. In Sweden the top soil mainly consists of moraine or sand. How this shallow bed rock influence the electrical gradient in the soil is shown in figure 4. For the sake of comparison we use the same resistivity in the top soil as in figure 3.

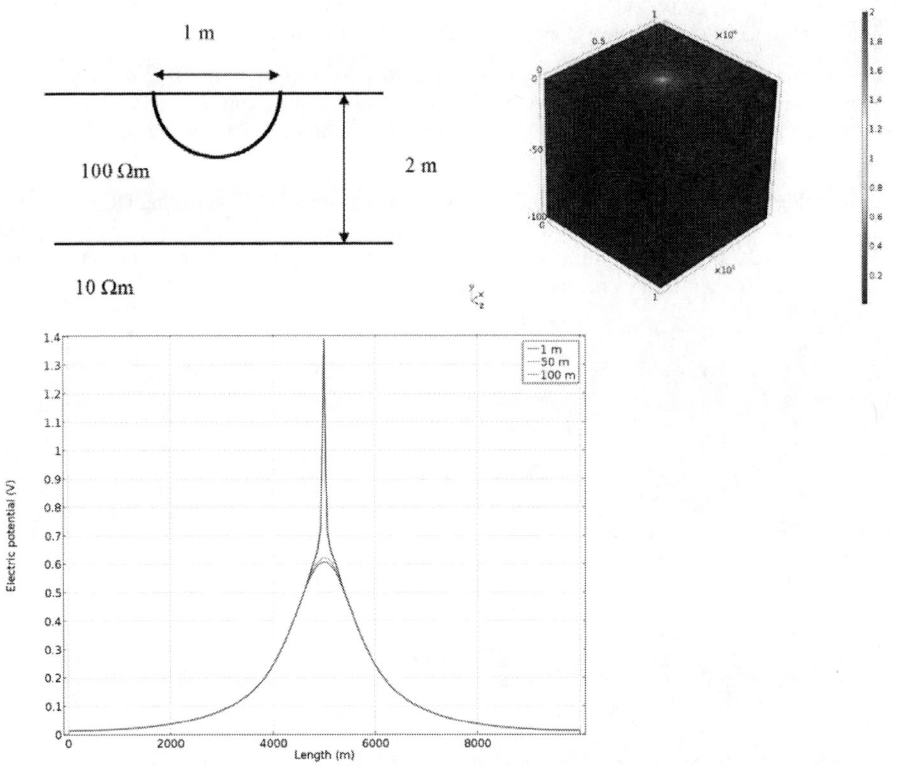

Figure 3. Electrical field in the ground close to a hemispheric electrode having a radius of 0,5 m and a voltage of 2V. Bed rock deep down.

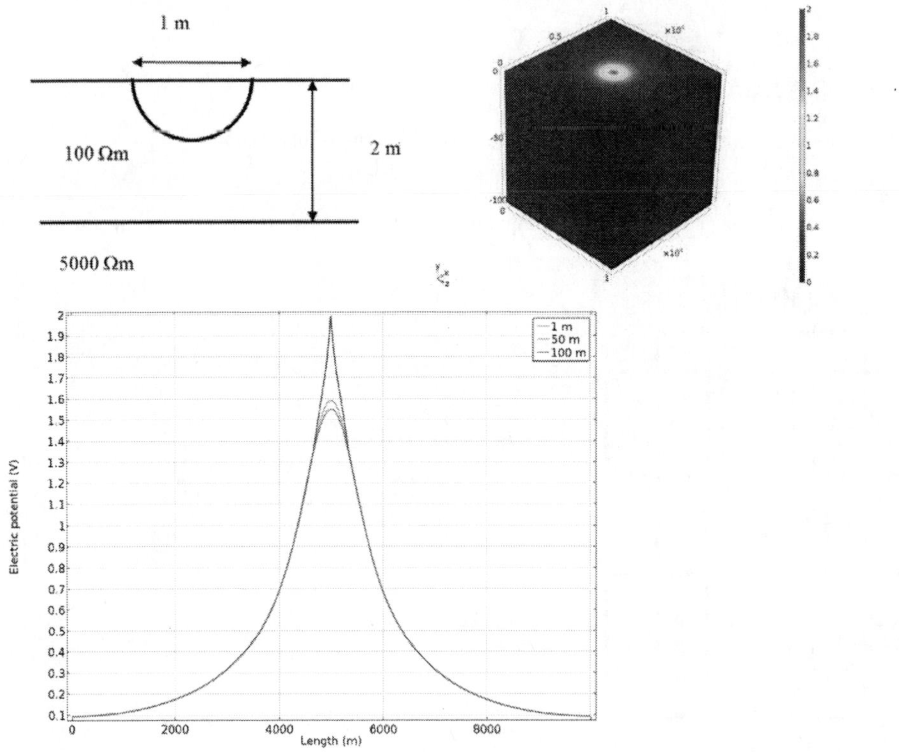

Figure 4. Electrical field in the ground close to a hemispheric electrode having a radius of 0,5 m and a voltage of 2 V. Shallow bed rock.

In this case the current is only 3 % of that in the previous case with a low resistivity in the deeper soil. Despite the low current, the soil potential 0,7 V is reached as far away as 960 m away from the interfering source.

The mostly populated areas in Scandinavia are however situated in areas with clayey deposits of marine origin on top of the bed rock. These clays have a very low resistivity. The combination of very low resistivity in the top soil and high resistivity in the underlying shallow bed rock results in higher current leakages from interfering sources and extensive electrical fields in the ground.

One could think that, under these geological circumstances, we would try to reduce the number of interfering DC systems to a minimum. This is however not the case. Instead the DC trams and DC trains are situated in these densely populated areas with clay on top of the shallow bed rock. In addition we also have a large number of High Voltage Direct Current transmission systems (HVDC) connecting Sweden with its neighboring countries, see figure 5.

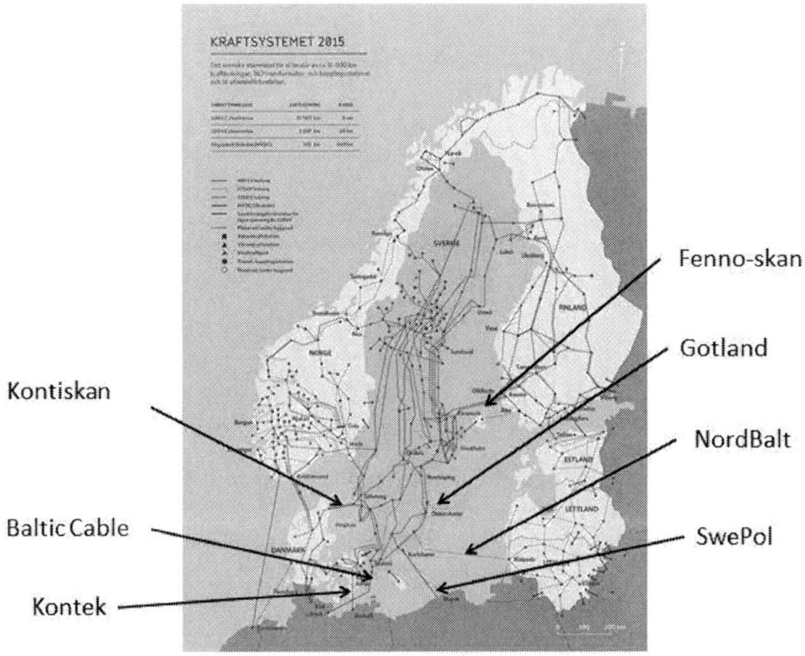

Figure 5. 400 kV power network in the Nordic countries. Names given for HVDC systems in Sweden.

Only one is a true monopolar system. The other systems are not bipolar but consists of two monopolar systems using the same electrodes, see figure 6.

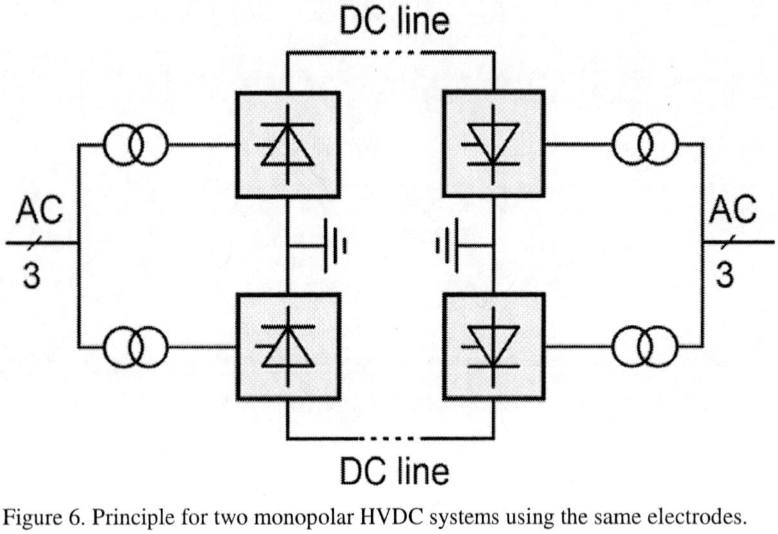

Figure 6. Principle for two monopolar HVDC systems using the same electrodes.

The two systems are seldom in balance and during long periods of maintenance they are operated as monopoles. The areas of interference can be large, see figure 7. To add to the complexity the polarity of the electrodes change depending on which system transferring the highest current.

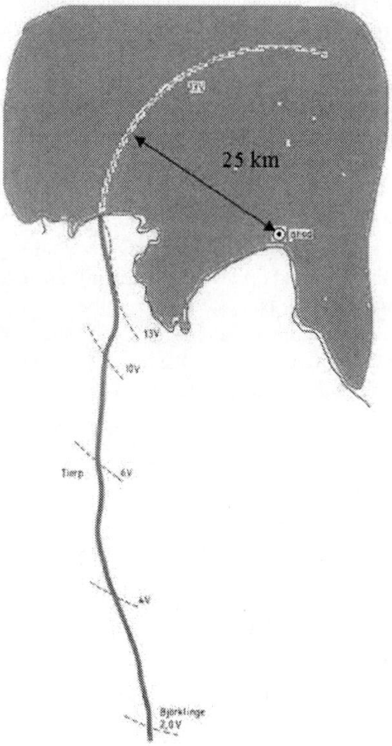

Figure 7. Electrical field around the Fennoskan electrode, situated north of Stockholm, during period with full unbalance current (1250 A). Red line represents a planned but never installed pipeline between Sweden and Finland.

ac interference on well coated pipelines is in most cases a result of induction caused by parallel high voltage power lines. In Sweden we also have ac interference due to conductive interference. The electric field measured around a power line tower crossing a gas pipeline is presented in figure 8 (3).

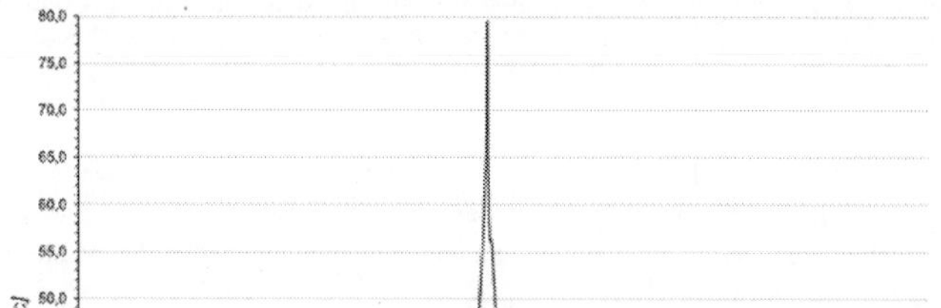

Figure 8. Conductive ac gradient around a power line tower.

10 km away the potential of the soil is above 10 V against remote ground. In some areas in Sweden remote earth does not exist.

3. COUNTER MEASURES

Available methods for mitigating DC and AC interference are given in EN 50262 (DC) and EN 15280 (AC). Mitigations can be applied on both the interfering and the interfered construction.

In the case of DC interference it is important to distinguish between permanent interference and interference varying in time (trams and trains).

In the past, DC interference caused by trams or trains was often counteracted by installing drains (direct, forced or polarized). These measures increased the stray current flow in the ground causing complex interference situations. Today efforts are being made in order to increase the isolation between the rail and the ground as much as possible, installing the substations closer and lowering the resistance of the rail system by using welded rails and cross bondings rail-rail. Interference on pipes and cables are counteracted by separate cp systems. On well coated pipes the protection potential will vary from place to place and from time to time, due to changes in soil potential. By logging the protection potential for 24 hours in many different places the most exposed place can be identified. Installing a local cp at this spot is not an option. The total current demand for the whole pipeline is so small that the current output from the cp system will not be able create the necessary electric field around the groundbed. The location or design of the groundbed cannot alter the interference situation. Instead the current flow to and from coupons have to be registered, see EN 50162 (4), and the current output from the cp system has to be adjusted to a level giving minimum outleaking current in the most exposed places, see figure 9.

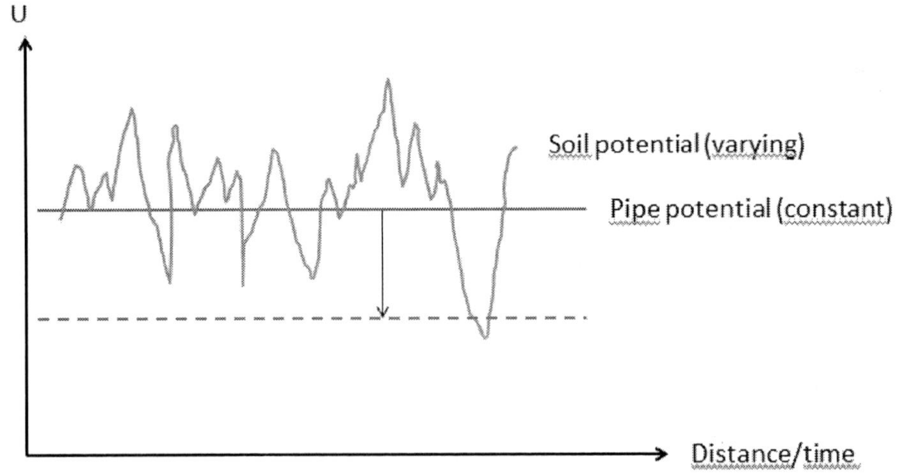

Figure 9. Interference on well coated pipeline. The pipe to soil potential is totally controlled by the soil potential. Protection by adjusting the pipe potential to a level which avoids “longer” periods of outleaking current from coupons.

Short anodic peaks are not hazardous. The reason for this is that it is becoming more and more clear that the protection caused by cp in soil is not a consequence of lowering the potential to immunity but by creating passivity, see figure 10 (2). To break down this passivity, the pH in the electrolyte in the direct vicinity of steel has to be lowered. This is not achieved during a short anodic peak. The steel will still be passive but on a more positive potential.

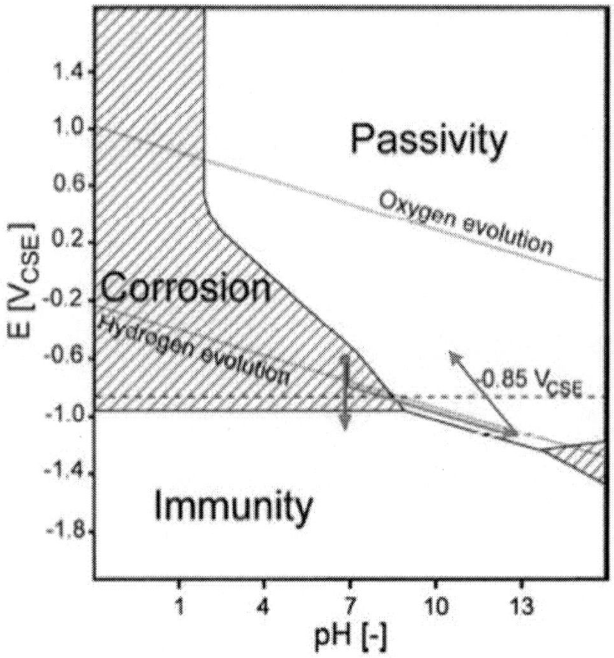

Figure 10. Polarization of steel with cathodic current. Green: Concentration polarization of a coating defect in soil. Red: Activation polarization of a coating defect in soft streaming water. Blue: Effect of short anodic peak.

This has been beautifully shown by Bette (7). By exposing steel samples in tap water for a repeated anodic current using different length of the exposure time he could show that short interference periods are less harmful, see figure 11.

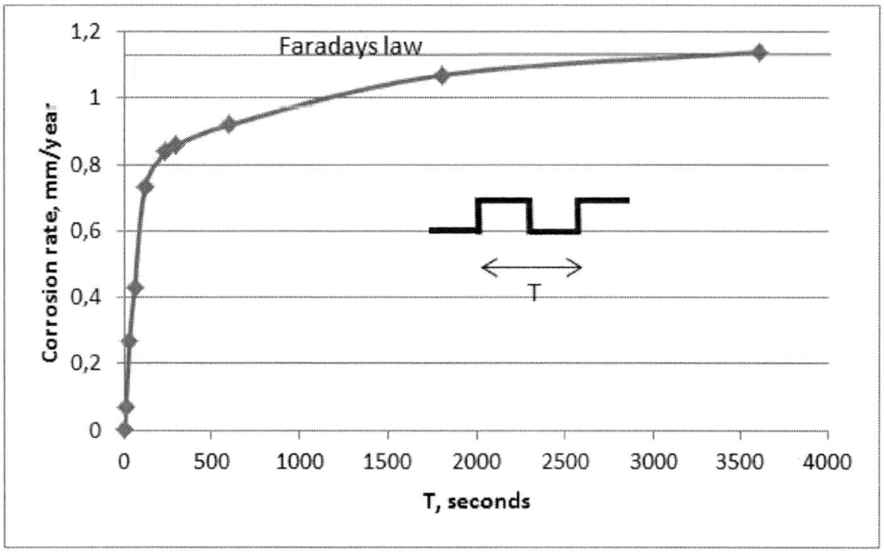

Figure 11. Corrosion rate as a function of length of interference time (7).

There are many different methods to mitigate permanent DC interference on pipelines as drainage, groundings, local cathodic protection or changes in the electrical continuity of the pipeline (isolation flanges). If overprotection is not accepted the tool-box is limited for pipelines having a high quality coating. All mitigation methods, except isolating flanges will just lower the potential level of the whole pipeline not being able to decrease differences in protection levels, see figure 12-14.

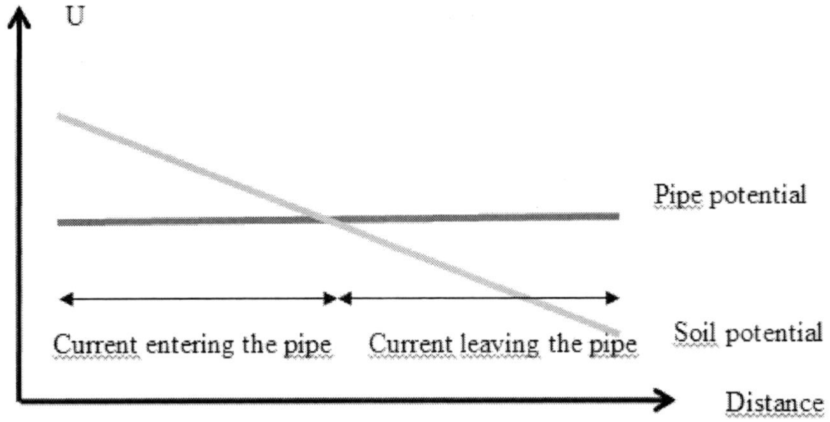

Figure 12. Pipeline with high quality coating in an interfering electrical field.

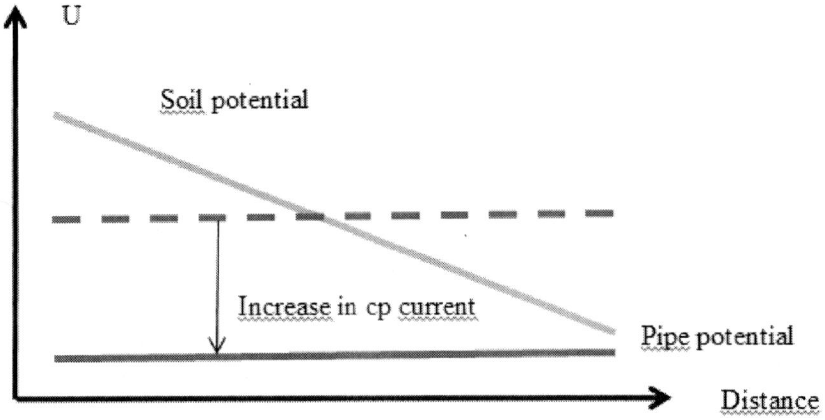

Figure 13. The effect of mitigations as drainages or increased cp current.

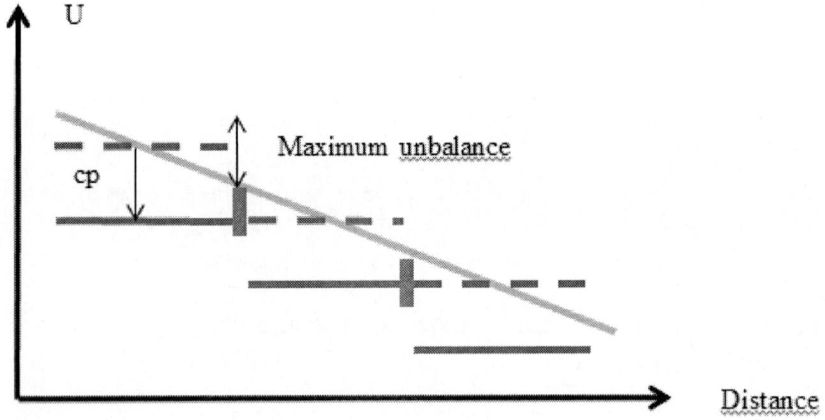

Figure 14. The effect of installing isolation flanges.

Without coating or a coating with low resistance to ground it is more easy to mitigate the unbalance by introducing a current flow in the pipe, see figure 15, or influence the electric field by a local cp installation, se figure 16.

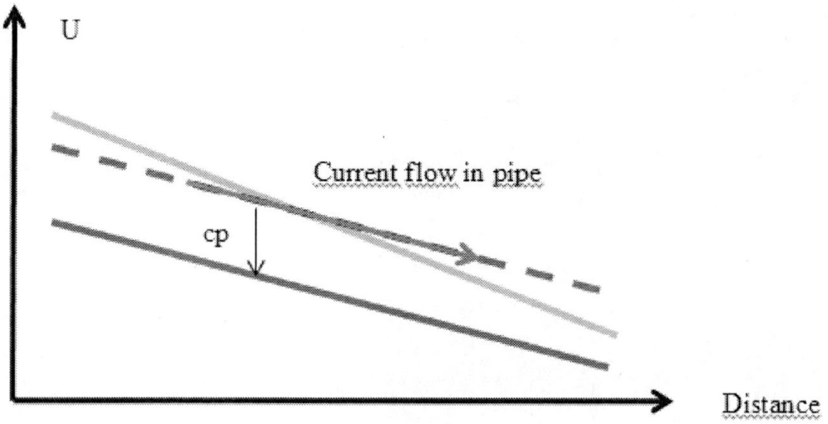

Figure 15. Effect of current flow in interfered pipeline

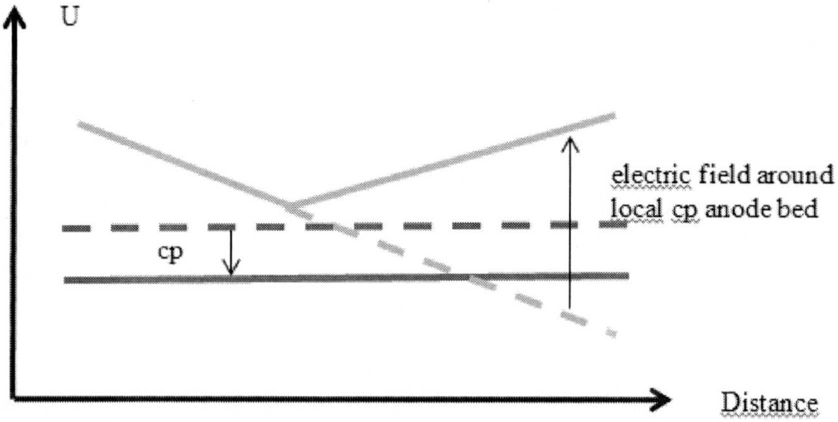

Figure 16. Effect of local cp on poorly coated pipeline in stray current area.

ac corrosion can be mitigated by using construction or operation measures (6). Construction measures can either be performed on the high voltage power line or on the pipeline. Increasing the separation between pipeline and power line reduces capacitance, inductive and conductive coupling. The method is especially suitable for conductive coupling in the vicinity of power line towers. Other measures to reduce the long term conductive coupling are to isolate the top wire from the towers or increase the bedding material around the tower footings. The phase conductors can also be transpositioned at suitable positions. The efficiency of this measure is however limited for underground pipelines because the characteristic length of the pipeline is generally shorter than the distance between the transposition points. On the pipeline, installation of isolating joints is the only practical construction measure. Changing bedding material is an uncertain method as the bedding material can be drained, eluted or mixed. To install an insulated wire in close proximity but not connected to the pipeline is not so effective in reducing long term interference.

Operation measures can only be performed on the pipeline. Groundings can be installed, directly or via DC decoupling devices. AC compensation methods have been tried. The cp protection level can be adjusted to a low or more than normally high level.

Approximately 1 % of all lightning's has a current higher than 200 kA and a duration longer than 2 ms which is the condition needed for melting steel. During the period 1971-2002 the whole European gas net has been hit 20 times resulting in gas leakage. In 2008 the first leak occurred on a high pressure gas pipeline in Sweden and it was caused by lightning, see figure 17. The pipe had a polyethylene coating and also an external layer of fibrous concrete. Fibrous concrete increase the risk as the attack is concentrated to one spot. Without concrete lightning results in many diffuse coating damages over a 10-15 m distance.

Figure 17. Hole in high pressure gas pipeline caused by lightning having melted the steel.

Higher positioned areas are often more exposed to lightning. Groundings, installed to mitigate ac interference, are on the other hand normally located in lower located agricultural areas. Lightning are drawn to the pipe in the highlands due to the coupling with the groundings in lower areas. Also in this case, with a combination of ac interference and frequent lightning, a less high resistance coating could be a better solution than polyethylene covered with concrete.

4. MEASUREMENT TECHNIQUES

The normal way to assess the effectiveness of the cathodic protection is dc potential measurements. In the case of constant dc interference, potential measurements are normally satisfying. Valuable information is also received by performing gradient measurements between two reference electrodes. If the dc interference varies with time it is necessary to measure the current flow between the pipe and a coupon buried in the soil during a period of 24 hours (4, 5).

There are several different criteria's used for ac interference involving ac voltage, dc potential, ac and dc current density, ac/dc current density ratio as well as soil resistivity (6). ac voltage should be measured against remote earth, which however in many cases is difficult if the interference is conductive, see figure 9. Current densities are measured using coupons being connected to the pipeline and installed in the same bedding material as the pipeline.

In areas with incomplete protection or where the protection level is hard to determine it is advisable to perform some kind of corrosion rate measurements. In principle three different techniques are available:

- Weight loss measurements
- Electrical resistance measurements
- Perforation measurements

Weight loss measurements are performed by installation of pre-weighed coupons. After some time the coupon is excavated, brought to laboratory for cleaning, inspection and weighing. The primary advantage is that the inspection gives detailed information of the maximum as well as the average corrosion rate. The primary disadvantage is that the coupon gives no information until it is excavated.

Electrical resistance measurements require the installation of ER probes. Corrosion is detected by increase of the electrical resistance of the coupon when corrosion progressively decreases the thickness of the coupon. The primary advantage of the technique is that the average corrosion rate can be followed continuously and used to optimize cathodic protection levels. The primary disadvantage is that localized corrosion is not detected.

Perforation measurements are made with special perforation probes. As on normal coupons, the ac and dc interference and the current density can be measured. Additionally the time when a critical predetermined corrosion depth is reached can readily be determined. Hence, the efficiency of the measures taken to decrease the ac corrosion rate can be verified in the field.

There are two different designs today. In both cases, the probe consists of a thin steel plate which external side is exposed to the soil. When the corrosion process has perforated the steel plate a signal is given. In design one, water entering the hole increases the humidity inside the probe and is measured as a decrease in resistance. In design two, the probe is pressurized and the perforation detected by a fall in pressure. By using a pressure switch the probe can readily be integrated into a monitoring system. The primary advantage is the simple handling and especially the information about the corrosion depth. Corrosion caused by ac interference mostly results in local attacks. The current density is uneven due to a thick chalk layer being formatted on the steel surface. Therefore the perforation probes provide an additional safety level by giving an alarm value. It is the only coupon that provides information on corrosion depth even in the case of very local corrosion attack. The primary disadvantage is that the information of ongoing localized corrosion is not available until the coupon is perforated. This can however be dealt with by chosing a suitable wall thickness of the probe.

5. DISCUSSION

Interference caused by dc currents can be difficult to evaluate and also to mitigate. Adding ac interference does not make the situation less complex. It is recommended to apply a low level of cathodic protection or an extra high protective current density in the case of ac interference. To succeed to avoid a normal cp level when also dc interference is on hand is close to impossible when protecting pipelines with a high quality coating. In some areas the corrosion protection is totally dependent on the coating being free from holidays. In this situation it is better to use a coating with lower resistance making it possible to install local cathodic protection which has the possibility to influence the electrical field in the soil as well as the potential drop in the pipe itself.

Existing polyethylene coated pipelines in these complex interference areas are exposed to an increased risk for premature perforation due to localized corrosion. Locating, excavating and repair of coating holidays decrease the risk of damages. It is however recommended to install perforation probes in strategic places as road crossings.

To measure current densities and/or voltages is only a way to estimate the corrosion risk. Evaluation of the uniform corrosion rate is not enough. To get an early warning if severe pitting corrosion is taking place is however of great interest. This can be accomplished with test coupons. The same information can also be collected using a perforation probe but in a more economical way and the warning is received without delay.

6. CONCLUSIONS AND RECOMMENDATIONS

The following conclusions and recommendations can be drawn:

- Pipelines having a high quality coating are more easily influenced by interference, from both ac and dc sources.
- Mitigations are hard to perform and it is recommended to separate areas with simultaneous ac and dc interference by isolating flanges and to use a poorer coating on the exposed pipe section.
- To detect ac corrosion the local corrosion rate must be monitored.
- The local corrosion rate can be evaluated by excavating test coupons but penetration probes give an economic benefit and results are received without delay, giving the most cost effective way to monitor a highly influenced section of the pipe network.

7. REFERENCES

1. U. Angst, M. Buchler, B. Martin, H.-G. Schoneich, G. Haynes, S. Leeds and F. Kajiyama: Cathodic protection of soil buried steel pipelines – a critical discussion of protection criteria and threshold values. Materials and Corrosion 2016, 67, No. 11

2. M. Büchler, B. Ackland and U. Angst: The historic evolution of cathodic protection criteria. Ceocor 2016.

3. Edwall, H-E & Malmborn, K: Practical case of measures to reduce the stray current impact on a 40 km long natural gas pipeline caused by intersecting High Voltage transmission lines. Ceocor 2011.

4. EN 50162:2005. Protection against corrosion by stray current from direct current systems.

5. EN 13509:2003. Cathodic protection measurement techniques.

6. EN 15280:2013. Evaluation of ac corrosion likelihood of buried pipelines applicable to cathodically protected pipelines.

7. Bette, U: Reduction of stray currents and stray current measurements. Ceocor 2017.

8. AUTHOR DETAILS

B Sandberg is Senior Scientific Advisor at the Research Institute Swerea KIMAB, a company formed through the merge between the Swedish Corrosion Institute (SCI) and the Metals Research Institute in 2006. At that time, he was the managing director of SCI. He has been active in the field of corrosion since 1981.

Bjorn Lindell is the founder and CEO of 3C Corrosion Control Company in Sweden. He holds a BSc degree in physical chemistry and has more than 40 years' experience of cathodic protection. It covers design and implementation for land based and submerged structures as well as protection of concrete structures.

CORROSION INHIBITION OF HY80 WITH LA(*TRANS*-4OHCIN)$_3$ & IMIDAZOLINIUM-*TRANS*-4OHCIN

R. A. Catubig[1], A. E. Somers[1], W. C. Neil[2], G. M[c]Adam[2], R. Yunis[1] & M. Forsyth[1]
[1]Institute for Frontier Materials, Deakin University, Burwood 3125 VIC Australia
[2]Defence Science & Technology Group, Fishermans Bend 3207 VIC Australia

SUMMARY: HY80 steel has been commercially used as a material for naval hull construction since the 1950s. Repair of damaged HY80 steel due to corrosion attack is both costly and time consuming. The presence of microbial species in seawater may also lead to microbiologically induced corrosion (MIC). A cost effective method for protecting HY80 from corrosion is by the incorporation of inhibitors into coatings that have anti-microbial activity and can stifle the effects of corrosion. This work examines the synergistic corrosion inhibition afforded by mixing lanthanum *trans*-4-hydroxy cinnamate (La(4OHCin)$_3$), known to be an effective corrosion inhibitor on steel, and imidazolinium *trans*-4-hydroxy cinnamate (IMI-4OHCin), a potential MIC inhibitor, on HY80 immersed in artificial seawater.

Pitting characteristics are explored with optical profilometry. Compared to the Uninhibited HY80 steel surface, the average number of pits was reduced in the presence of both La(4OHCin)$_3$ and IMI-4OHCin after a 24 hour immersion period. Furthermore, the corrosion inhibition mechanisms of each inhibitor tested separately and when mixed together is discussed with potentiodynamic polarisation.

Keywords: Steel, seawater, corrosion, inhibitor, rare-earth, profilometry

1. INTRODUCTION

The high salt content in seawater (approximately 0.55 M of Cl^- ions present) is one of the most naturally corrosive environments to steel.[1] The steel alloy HY80 has been commonly used as part of hulls on ships since the 1950s.[2]

In addition to the high salt content of seawater, the presence of bacteria and other organisms may also result in microbiologically induced corrosion (MIC). In some cases, the occurrence of MIC can significantly increase the severity of corrosion attack. This results in severe corrosion damage and extended down time for performing expensive repairs.[3] Therefore it would be beneficial to utilise an inhibitor that can stifle both microbial attachment and corrosion.

It is well known that hexavalent chromate-based inhibitors are incredibly efficient in limiting corrosion on steel and it is similarly well known that they are toxic and carcinogenic.[4] Numerous "greener" alternative inhibitor systems have been explored with rare-earth based inhibitors showing promise for steel and aluminium.[5-9] Additionally, the incorporation of organic ligands with the rare-earth inhibitors seemingly provides added functionality and improved inhibition, which is commonly referred to as synergistic inhibition.[8]

With the aim of developing a multi-functional inhibitor system that can stifle MIC by inhibiting both microbial attachment and corrosion this work evaluates the corrosion inhibiting performance on the marine alloy HY80 of two inhibitor compounds in artificial seawater. One of the most effective rare-earth based corrosion inhibitors for steel, lanthanum 4-hydroxy cinnamate (La(4OHCin)$_3$) is mixed with a potential microbial inhibitor, imidazolinium 4-hydroxy cinnamate (IMI-4OHCin), and the presence of synergistic inhibition is explored. The work focuses on the corrosion inhibition afforded by the inhibitors. Experimentation on the efficiency to stifle microbial attachment will be part of a future work program.

2. EXPERIMENTAL DETAILS

2.1 HY80 alloy composition

Commercially available HY80 steel alloy, taken from only a single batch of material, was used in all experiments. The composition of the HY80 steel, determined by Inductively Coupled Plasma Atomic Emission Spectroscopy, is presented in Table 1.

Table 1: Typical composition (%) for the major elemental additions to the HY80 alloy.

Element	Fe	C	Mn	Si	S	P	Ni	Cr	Mo	Cu	Al
Nominal Composition (%)	Bal	0.14	0.33	0.26	0.01	0.01	3.10	1.53	0.31	0.11	0.015

All sample preparation and experiments were conducted under standard laboratory conditions (23±1°C), neutral pH and in artificial seawater. The steel samples were initially surface treated by immersing in concentrated hydrochloric acid 32% for 15 minutes, rinsed with deionised water and dried with nitrogen gas to remove any debris/impurities on the alloy surface. The treated samples were then encased in cold mounted epoxy. The alloy surface was polished to P1200 grit silicon carbide polishing paper with deionised water as the lubricant. Subsequent to polishing, the alloy surfaces were rinsed with deionised water, dried with nitrogen gas and stored in a desiccator for approximately 1 hour prior to testing. Figure 1 shows an optical profilometry image of the as-polished HY80 surface prior to testing.

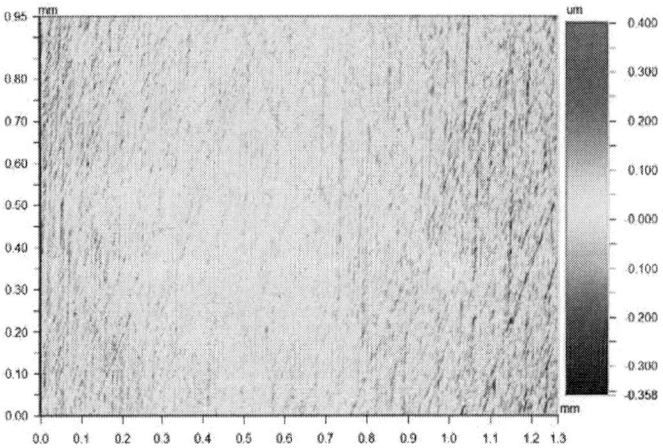

Figure 1: Optical profilometry image of the HY80 surface after exposure to HCl for 15 minutes and polishing to P1200 grit SiC polishing paper. The height profile is represented by the colour scale on the right.

2.2 Immersion corrosion testing

For each test, immersion samples with a surface area of 20 mm x 20 mm were placed in separate 80 ml solutions. The artificial seawater was prepared with analytical grade NaCl (Chem-Supply, ≥99.7%), $MgCl_2$ (Sigma, ≥98.0%), $MgSO_4$ (Merck, ≥96.0%), $CaCl_2$ (Sigma, ≥96.0%) and KCl (Univar, ≥99.8%) as shown in Table 2. Artificial seawater was used so that the composition of the electrolyte was known and constant throughout all testing. Since microbial inhibition studies will be part of a future work program, no microbes were tested in this work.

Table 2: Composition of artificial seawater in 1L of deionised water.[10]

Salt	NaCl	$MgCl_2$	$MgSO_4$	$CaCl_2$	KCl
Mass (g)	26.524	2.441	3.301	1.140	0.726
Concentration (M)	0.454	0.026	0.027	0.010	0.010
Concentration (ppm)	26524.0	2440.7	3300.6	1140.4	725.5

Synthesis of $La(4OHCin)_3$ and IMI-4OHCin were conducted as per the published work by Deacon *et al.* and Chong *et al.* respectively.[11, 12] Each inhibitor was dissolved into analytical grade ethanol (Chem-Supply, ≥96.5%), sonicated to assist in dissolution and finally mixed into the artificial seawater solution. The final solution was heated and stirred at approximately 50 °C to allow for the ethanol to evaporate. The concentrations of the individual inhibitors (Table 3) were tested as they were close to saturation levels in the artificial seawater and therefore an initial starting point in quantifying the possible maximum level of inhibition.

Polished samples were immersed for 24 hours. After the samples were removed from their respective solutions, they were rinsed gently with deionised water, dried with nitrogen gas and stored in a desiccator.

Table 3: List of inhibitors tested with their abbreviations concentration and chemical structure.

Inhibitor	Abbreviation	Concentration (mM)	Structure
No Inhibitor (artificial seawater only)	Uninhibited	-	-
Lanthanum 4-hydroxy cinnamate	$La(4OHCin)_3$	1.1 (800 ppm)	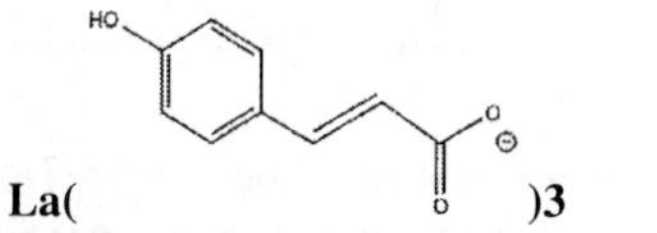
Imidazolinium 4-hydroxy cinnamate	IMI-4OHCin	10.0 (2475 ppm)	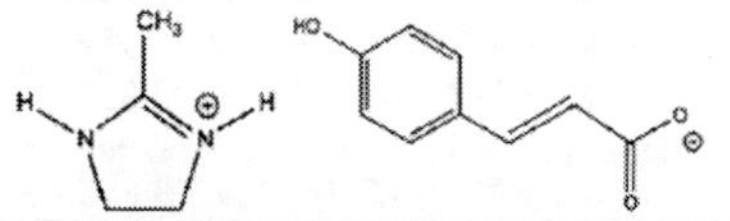
Lanthanum 4-hydroxy cinnamate + Imidazolinium 4-hydroxy cinnamate	La+IMI	$La(4OHCin)_3$: IMI-4OHCin = 1.1 : 10.0 (800 ppm : 2475 ppm)	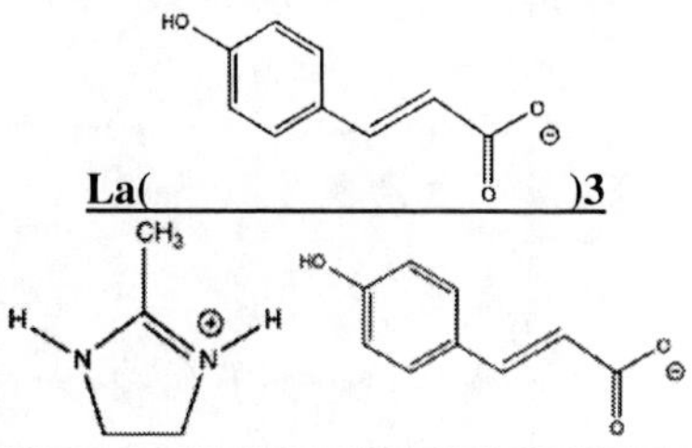

2.3 Optical Microscopy

Subsequent to the 24 hour immersion period, each sample was photographed with a Leica MZ6 optical microscope and a Leica DFC320 digital camera and processed with the Leica Application Suite version 4.0.0 in visually differentiate the corrosion damage on each sample.

2.4 Optical profilometry

The corrosion products on the immersion samples were removed by immersion for at least 3 minutes in a solution containing 25 ml analytical grade hydrochloric acid (Chem-Supply ≥31.5%w/w), 25 ml deionised water and 0.175 g of analytical grade hexamethylene tetramine (Sigma, ≥99.0%). The samples were rinsed with deionised water, dried with nitrogen gas and stored in a desiccator.

3D surface optical profilometry was conducted with a Bruker Contour GT-K1 3D optical profiler. Each sample was analysed as quickly as possible (within 30 minutes) after removal of the corrosion product.

2.5 Potentiodynamic polarisation

A standard three-electrode cell was used for all polarisation tests. The counter electrode and reference electrode were titanium mesh and a saturated calomel electrode (SCE) respectively. In order to minimise IR drop and bring the reference electrode closer to the working electrode (the polished HY80 steel), the reference electrode was housed in a glass luggin capillary containing the test solution. The tip of the luggin was approximately 2D in distance away from the working electrode, whereby D is the diameter of the luggin tip (2.5 mm).[13] The surface area of the working electrode was 10 mm x 10 mm.

The potentiodynamic polarisation scans for all inhibited samples were made from -150 to 250 mV relative to the open circuit (OCP) at a scan rate of 0.167 mV/s. The Uninhibited sample was scanned from -60 to 250 mV instead. This was because polarising at more negative potentials without the presence of an inhibitor resulted in significant changes to the condition of the steel, shifting E_{corr}, and therefore limiting consistency of the results. The polarisation scans were performed after an OCP period of 30 minutes with triplicate measurements. A 300 ml solution volume was tested open to air. Electrochemical tests were conducted using a 16 Channel-VMP3 (variable multichannel potentiostat) with the EC-lab software version 11.03.

3. RESULTS

3.1 Optical microscopy 24 hours

After the 24 hour immersion period, optical images were obtained prior to removal of corrosion product and presented in Figure 2 to Figure 5. From visual inspection, the Uninhibited sample showed significant corrosion attack on about 50% of the total surface area of the steel (a representative area shown in Figure 2). The other area appeared to be free from corrosion attack. Using the optical microscope, the seemingly corrosion-free surface did indeed have corrosion attack and pits but this corrosion attack was more concentrated on the rest of the sample.

A surface film over the entire surface was observed on the sample immersed in solution containing $La(4OHCin)_3$ (Figure 3). Some sites showed caps of corrosion product (yellow arrow). The whole IMI-4OHCin sample surface was covered with orange coloured corrosion product/deposits. Some pitted sites were also observed as indicated by the yellow arrow in Figure 4.

The sample immersed into the La+IMI mixture also showed the presence of a surface film and locally pitted sites. The film is distinctly different to both the $La(4OHCin)_3$ and IMI-4OHCin inhibitors tested separately. The colourful bands such as the band passing from the top left to the middle bottom of Figure 5 is present on approximately a quarter of the sample surface and is visible to the naked eye. This reflects the deposited film was inherently different for all inhibited surfaces.

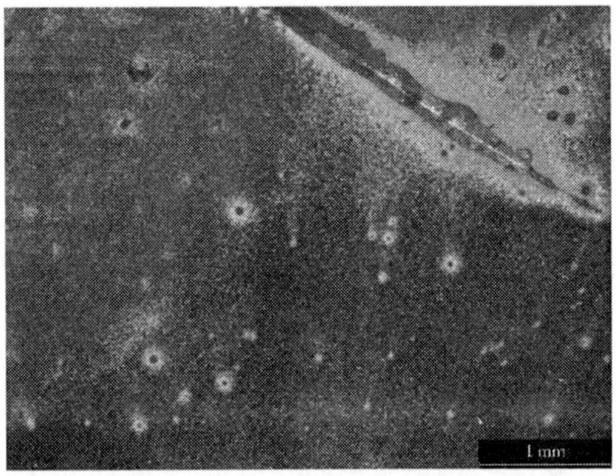

Figure 2: Optical image at 2.5x magnification of the Uninhibited sample prior corrosion product removal. A scratch on the sample is visible on the top right. The scale bar is 1 mm.

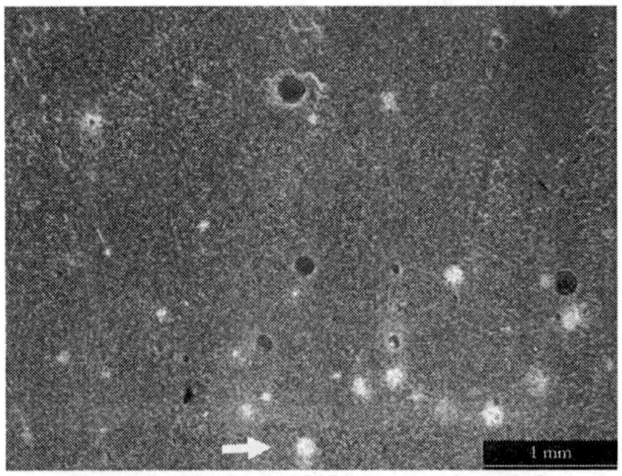

Figure 3: The $La(4OHCin)_3$ sample at 2.5x magnification prior corrosion product removal. A surface film and pits were present. The yellow arrow shows an example cap of corrosion product. The scale bar is 1 mm.

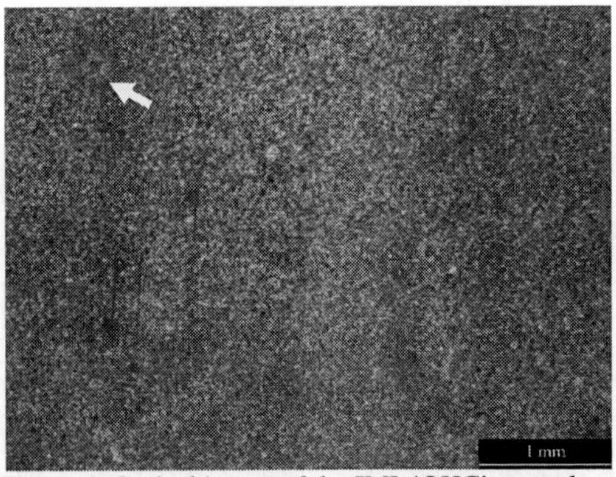

Figure 4: Optical image of the IMI-4OHCin sample at 2.5x magnification prior corrosion product removal. The whole surface was covered in an orange deposit. The yellow arrow shows an example of a pitted site. The scale bar is 1 mm.

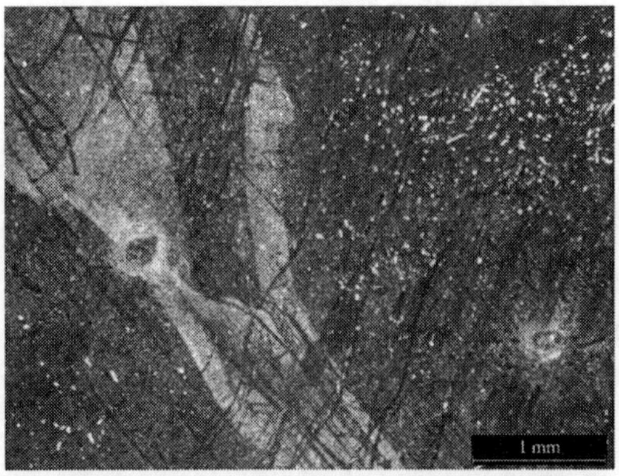

Figure 5: Optical image of the La+IMI sample at 2.5x magnification prior corrosion product removal. A film is present but appears thinner than when each inhibitor was tested separately. The scale bar is 1 mm.

3.2 Optical profilometry 24 hours

To analyse the influence of each inhibitor on the resulting pitting characteristics after a 24 hour immersion period, the corrosion product on each sample was removed and the underlying surface analysed with optical profilometry. Figure 6 presents the maximum pit depth for each sample, which is calculated from the average of the ten deepest pits on each sample. The maximum depth instead of the average depth is presented to illustrate the worst cases of pitting present on each sample, therefore providing a more accurate estimation of the severity of damage on the HY80 steel. Compared to the Uninhibited sample, the presence of all inhibitors resulted in the deepening of all pits, suggesting increased local corrosion attack. Although within one standard deviation, the La+IMI mixture yielded the shallowest of all inhibited samples.

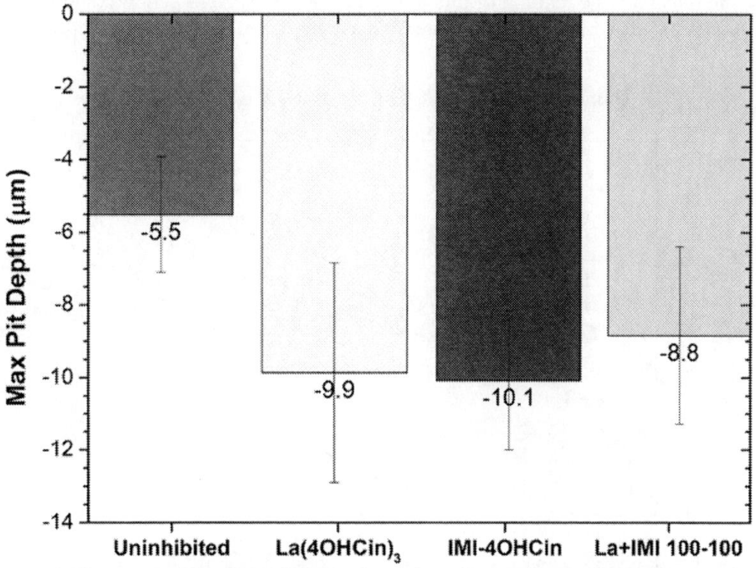

Figure 6: The maximum pit depth (calculated from the average of the 10 deepest pits).

Similarly the maximum pit diameter was determined for all samples and presented in Figure 7. The maximum pit diameter was calculated from the average of the 10 widest pit diameters. The $La(4OHCin)_3$ and IMI-4OHCin inhibitors tested separately yielded wider maximum pit diameters than the Uninhibited sample. However, given that they were all within one standard deviation of each other as well as the relatively large standard deviation, this suggests the pit diameters were relatively similar between the Uninhibited, $La(4OHCin)_3$ and IMI-4OHCin inhibitors.

In contrast, the La+IMI mixture resulted in maximum pit diameters that were 33% of the Uninhibited sample as well as a comparatively smaller standard deviation. Also the maximum pit depth value of 17.6 μm was found to be outside the standard deviations of the Uninhibited, $La(4OHCin)_3$ and the IMI-4OHCin, indicating a significant reduction in pit diameter widening with respect to the other samples.

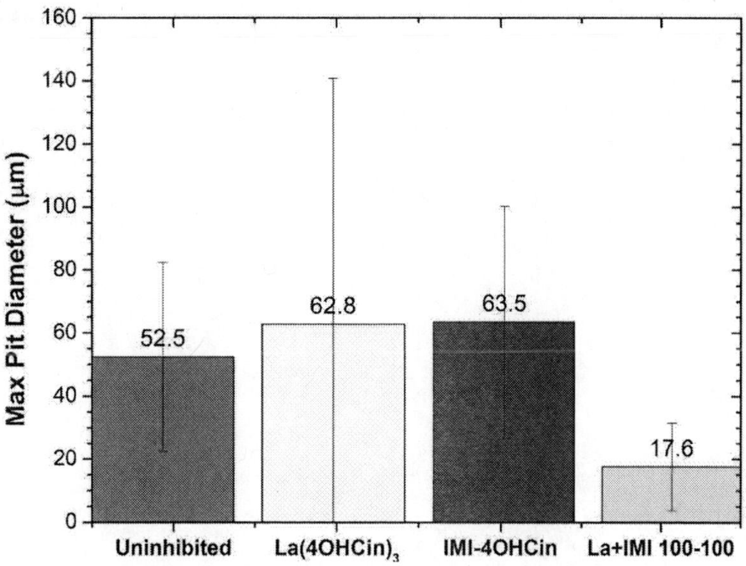

Figure 7: The maximum pit diameter (calculated from the average of the 10 widest pits).

In addition to the dimensions of pit depth and diameter, the average number of pits per scanning area (1.21 mm^2) was also determined (Figure 8). It must be remembered that the initiation, growth and proximity of pits are some of the factors that can change the size and shape of pits. Differentiating between these types of pits could not be done by the Optical profilometry software and thus only the number of pits at the 24 hour immersion period could be measured. All inhibitors reduced the average number of pits compared to the Uninhibited sample and is ordered as follows from largest to smallest average number of pits:

Uninhibited > IMI-4OHCin > La(4OHCin)$_3$ > La+IMI

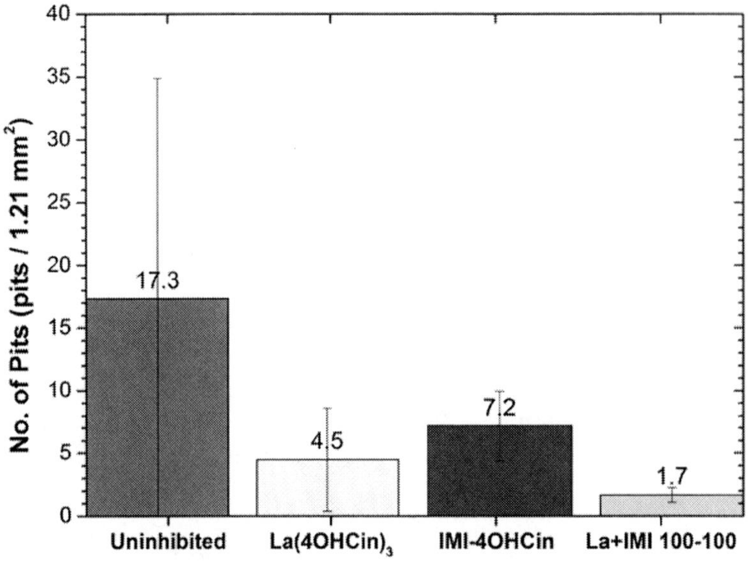

Figure 8: The average number of pits within the scanning area of the optical profilometry instrument (1.21 mm^2).

3.3 Potentiodynamic polarisation

The potentiodynamic polarisation curves for each test solution are presented in Figure 9. The La(4OHCin)$_3$ inhibitor yielded a reduction in cathodic current densities relative to the Uninhibited solution. Anodic current densities were also reduced by up to 2 orders of magnitude. This suggests a mixed mode of inhibition.

In the presence of the IMI-4OHCin inhibitor, there is no noticeable reduction in cathodic current densities but up to 3 orders of magnitude reduction in anodic current densities. This indicates a predominantly anodic mode of inhibition afforded by the IMI-4OHCin inhibitor.

The La+IMI mixture demonstrated reductions in both cathodic and anodic current densities, the latter having a reduction of up to 4 orders of magnitude. This demonstrates a mixed mode of inhibition while effectively stifling anodic reactions.

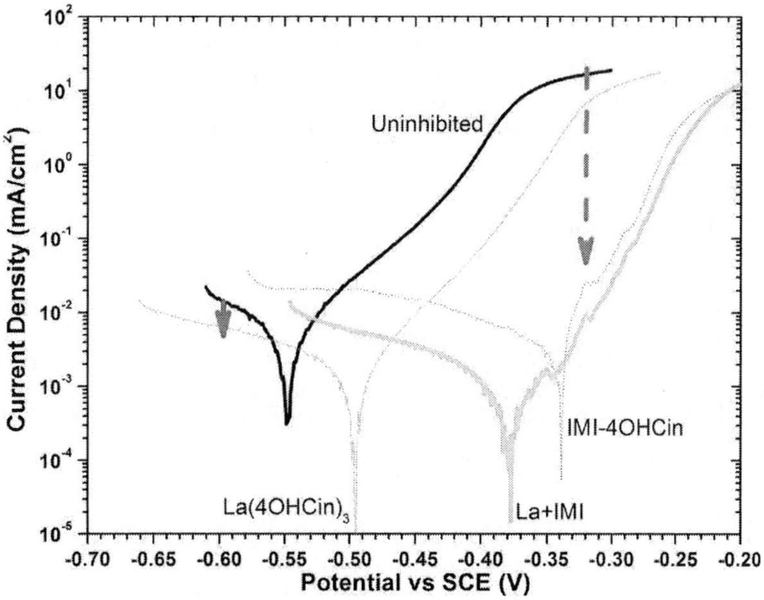

Figure 9: Potentiodynamic polarisation curves for the Uninhibited (black solid), La(4OHCin)$_3$ (orange dotted), IMI-4OHCin (blue dotted) and the La+IMI mixture (green solid).

Figure 10 shows the average i_{corr} values for each inhibitor with the IMI-4OHCin inhibitor having a similar average to the Uninhibited solution indicative of insignificant levels of inhibition. The La(4OHCin)$_3$ inhibitor was less than half the i_{corr} value of the Uninhibited solution suggesting significant inhibition of corrosion processes. The La+IMI mixture yielded just over half the value of the La(4OHCin)$_3$ and approximately ¼ of the Uninhibited sample suggesting a greater level of inhibition was achieved by the mixture compared to using La(4OHCin)$_3$ separately.

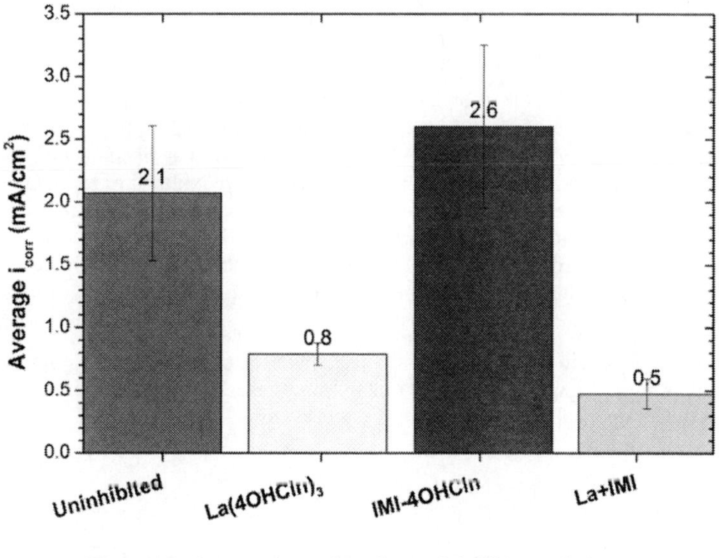

Figure 10: Average i_{corr} values for each inhibitor solution.

4. DISCUSSION

4.1 Corrosion inhibition of La(4OHCin)$_3$ and IMI-4OHCin tested separately

The IMI-4OHCin inhibitor at a concentration of 10 mM was only successful in reducing the number of pitted sites relative to the steel surface with no inhibitor present. The average i_{corr} for the IMI-4OHCin inhibitor was greater (although still within one standard deviation). Since the average i_{corr} value is assumed to be an estimate of the corrosion rate, the higher i_{corr} value suggests that the severity of corrosion in the presence of the IMI-4OHCin inhibitor likely exacerbated the corrosion attack relative to the Uninhibited sample. The IMI-4OHCin sample also had the most visible corrosion product (Figure 4), which coincides with the higher average i_{corr} value. The IMI-4OHCin inhibitor has been shown to provide significant inhibition on mild steel (AS 1020) at a concentration of 4 mM in a 0.01 M NaCl solution at neutral pH 2 and pH 8.[14] In the current work, initial observations would suggest that corrosion of the HY80 steel in the artificial seawater appears to be too aggressive since the IMI-4OHCin inhibitor could not reduce the corrosion rate less than the Uninhibited sample. However, its ability to limit corrosion at pH 2 as observed by Chong *et al.*[14] may have provided some level of inhibition locally within the highly acidic conditions expected at the tip of pits. This resulted in reducing the occurrence of pit initiation and therefore a reduced number of pits.

The pitting characteristics in the presence of $La(4OHCin)_3$ 1.1 mM was similar to the IMI-4OHCin with regards to the maximum pit depth and diameter, however, a reduced number of pit sites and a smaller average i_{corr} was achieved with the $La(4OHCin)_3$ inhibitor. This indicates that a larger percentage of the exposed steel surface was protected from pitting corrosion than with the IMI-4OHCin inhibitor. The pits that do initiate are not stifled significantly resulting in similar maximum pit depth to the IMI-4OHCin inhibitor. It can be speculated that by reducing the number of actively pitting sites, both inhibitors have effectively reduced the anodic surface area. The cathodic surface area would be relatively large compared to the anodic surface area, resulting in corrosion being driven into the small number of pits.

These observations indicate that in the presence of the $La(4OHCin)_3$ inhibitor, pitting initiation is stifled resulting in the reduced number of pits. Corrosion is therefore more localised in the remaining pits. Given that only one immersion time (24 hours) was explored in this work, it is not clear whether $La(4OHCin)_3$ provides limited inhibition towards pit growth or pit growth was only inhibited on some pits but not all. If the latter was taking place, it may explain the relatively large standard deviations in maximum pit depth (Figure 6). Nonetheless, the reduced corrosion rate (average i_{corr} in Figure 10) indicates that corrosion over the steel surface was stifled in the presence of the $La(4OHCin)_3$ inhibitor compared to when no inhibitor was present.

4.2 Corrosion inhibition of the La+IMI mixture

The reduction in the maximum pit diameter, average number of pits and the average i_{corr} value by a greater degree than the individual inhibitors separately, all indicate that the La+IMI mixture was providing synergistic inhibition. When tested as separate salts (ie. Na(4OHCin) and IMI-Cl), Chong *et al.* has shown that the $4OHCin^-$ anion provided significant inhibition whereas the imidazolinium cation consistently yielded the smallest reductions in the average i_{corr} value.[15] Similarly, Blin *et al.* showed that the $4OHCin^-$ anion achieved higher inhibition percentages compared to the lanthanum cation.[16] This indicates that a significant amount of inhibition is due to the presence and behaviour of the 4OHCin anion in solution. Since the anion for both the $La(4OHCin)_3$ and IMI-4OHCin inhibitors was $4OHCin^-$, there will be a higher concentration of the $4OHCin^-$ anion in the La+IMI mixture compared to when the inhibitors were tested separately and will likely contribute a significant portion of the overall corrosion inhibition of steel.

It has been demonstrated by Seter *et al.* that an increase in the $La(4OHCin)_3$ concentration in solution increased the likelihood of forming larger and more complicated complexes such as $La(4OHCin)_4$ (as opposed to $LaCl_3$), which was also correlated with improved corrosion inhibition.[15] It is possible that a similar phenomenon took place in the La+IMI mixture whereby the increased $4OHCin^-$ anion concentration increased the likelihood of forming large complexes that were more effective in stifling corrosion attack. The presence of the IMI^+ cation should not be forgotten and likely also contributing to the improved synergistic inhibition afforded by the La+IMI mixture. While the extra inhibition is welcome in any inhibitor, the majority of corrosion inhibition will rely on the $La(4OHCin)_3$ inhibitor and the IMI-4OHCin is expected to provide significant MIC inhibition properties. Further work on the speciation of these inhibitors is needed to confirm this hypothesis.

The significant reduction in maximum pit diameter, average number of pits, average i_{corr} and even a minor reduction in maximum pit depth all illustrate the improved corrosion inhibition by the La+IMI mixture in artificial seawater relative to the individual inhibitors tested separately or the absence of any inhibitor.

5. CONCLUSIONS

The effectiveness to stifle corrosion by mixing lanthanum 4-hydroxy cinnamate (a known effective corrosion inhibitor on steel) and imidazolinium 4-hydroxy cinnamate (a candidate for a microbial inhibitor) was investigated. The La+IMI mixture achieved greater levels of corrosion inhibition compared to the individual inhibitors tested separately suggesting a synergistic type of inhibition. This was demonstrated by the smaller pit diameters, lower average number of pits and reduced average i_{corr} values.

6. ACKNOWLEDGMENTS

The authors acknowledge the Defence Science and Technology Group for funding the project and collaboration with the Institute for Frontier Materials, Deakin University.

7. REFERENCES

1. Kester DR, Duedall IW, Connors DN and Pytkowicz RM, Preparation of artificial seawater. Limnology and Oceanography, 12(1) (1967) 176-9.
2. Montemarano TW, Sack BP, Gudas JP, Vassilaros MG and Vanderveldt HH, High strength low alloy steels in naval construction. Journal of Ship Production, 2(3) (1986) 145-62.
3. Qian PY, Lau SCK, Dahms HU, Dobretsov S and Harder T, Marine biofilms as mediators of colonization by marine macroorganisms: Implications for antifouling and aquaculture. Marine Biotechnology, 9(4) (2007) 399-410.
4. Twite RL and Bierwagen GP, Review of Alternatives to Chromate for Corrosion Protection of Aluminum Aerospace Alloys. Prog Org Coat, 33(2) (1998) 91-100.
5. Hinton BRW, Corrosion inhibition with rare-earth-metal salts. J Alloy Compd, 180 (1992) 15-25.
6. Isaacs HS, Davenport AJ and Shipley A, The Electrochemical Response Of Steel To The Presence Of Dissolved Cerium. J Electrochem Soc, 138(2) (1991) 390-3.
7. Blin F, Leary SG, Deacon GB, Junk PC and Forsyth M, The nature of the surface film on steel treated with cerium and lanthanum cinnamate based corrosion inhibitors. Corrosion Sci, 48(2) (2006) 404-19.
8. Forsyth M, Seter M, Hinton B, Deacon GB and Junk P, New 'Green' Corrosion Inhibitors Based on Rare Earth Compounds. Australian Journal of Chemistry, 64(6) (2011) 812-9.
9. Forsyth M, Wilson K, Behrsing T, Forsyth C, Deacon GB and Phanasgoankar A, Effectiveness of rare-earth metal compounds as corrosion inhibitors for steel. Corrosion, 58(11) (2002) 953-60.
10. Subow NN, Oceanographical tables, Institute O (Hydrometeorological Commision), Moscow, 1931.
11. Deacon GB, Forsyth M, Junk PC, Leary SG and Lee WW, Synthesis and Characterisation of Rare Earth Complexes Supported by para-Substituted Cinnamate Ligands. Z Anorg Allg Chem, 635(6-7) (2009) 833-9.
12. Chong AL, Forsyth M and Macfarlane DR, Novel imidazolinium ionic liquids and organic salts. Electrochim Acta, 159 (2015) 219-26.
13. Pletcher D, Greff R, Peat R, Peter LM and Robinson J, Instrumental Methods in Electrochemistry, (Elsevier Science), 2001.
14. Chong AL, Mardel JI, MacFarlane DR, Forsyth M and Somers AE, Synergistic Corrosion Inhibition of Mild Steel in Aqueous Chloride Solutions by an Imidazolinium Carboxylate Salt. Acs Sustainable Chemistry & Engineering, 4(3) (2016) 1746-55.
15. Seter M, Hinton B and Forsyth M, Understanding speciation of lanthanum 4-hydroxy cinnamate and its impact on the corrosion mechanism for AS1020 steel. J Electrochem Soc, 159(4) (2012) C181-C9.
16. Blin F, Leary SG, Wilson K, Deacon GB, Junk PC and Forsyth M, Corrosion mitigation of mild steel by new rare earth cinnamate compounds. J Appl Electrochem, 34(6) (2004) 591-9.

8. AUTHOR DETAILS

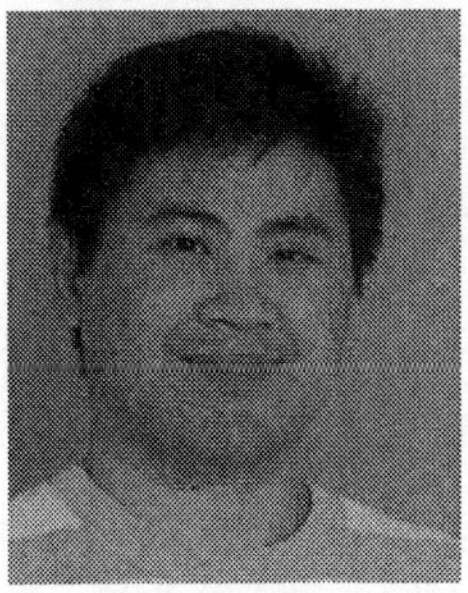

Dr Rainier Catubig is an Associate Research Fellow at IFM, Deakin University. He completed his PhD in corrosion inhibition mechanisms of rare earth mercaptoacetate inhibitors for the aerospace alloy, AA2024-T3, in 2015 at Deakin University. He has experience in inspecting the effect of corrosion on multilevel apartment buildings and conducting concrete core surveys in aged Olympic sized swimming pools to estimate the remaining service life. He currently investigates rare earth corrosion inhibitors for naval applications.

Dr Anthony Somers is a Research Fellow in the Institute for Frontier Materials (IFM) at Deakin University where he completed his PhD in 2014. His current research areas are in oil miscible ionic liquid lubricants and corrosion inhibitors for steels. He is particularly interested in the surface interactions of these salts with metal surfaces in the areas of lubrication and corrosion inhibition.

Dr Wayne Neil is a corrosion scientist in the Defence Science and Technology Group. His research interests include microbiologically influenced corrosion; corrosion of marine alloys including marine grade aluminium, copper nickel, stainless steels and high strength steels; environmentally friendly corrosion inhibitors; and corrosion of magnesium alloys.

Dr Grant McAdam is currently a Senior Research Scientist in Defence Science and Technology Group. His research interests include corrosion protection of magnesium and aluminium aerospace alloys, environmentally friendly corrosion inhibitors and corrosion protection schemes, fibre optic corrosion sensors, environment characterisation, and corrosion of marine alloys.

Dr Ruhamah Yunis is working as an Associate Research fellow at Deakin University, Burwood, Australia since 2015. Her current research focuses on the synthesis, purification and characterization of ionic liquids, rare-earth corrosion inhibitors, lubricants, organic ionic plastic crystals and polymer electrolytes.

Professor Maria Forsyth is the Associate Director of ARC Centre of Excellence for Electromaterials Science, an Australian Laureate Fellow and an Alfred Deakin Professorial Fellow at Deakin University. Her work is directed towards development and understanding of charge transport at metal/electrolyte interfaces within electrolyte materials. These include ionic liquids, polymer electrolytes, plastic crystals and rare earth inhibitor complexes. Understanding these materials is used to design new materials and processes to control these phenomena in two key areas: corrosion inhibition and electrochemical devices. Professor Forsyth is a co-author of over 360 journal publications. She has delivered more than 25 invited and plenary talks in the past 5 years and currently has more than 11000 citations.

AUTHOR INDEX

Annual Conference of the Australasian Corrosion Association 2017

Corrosion and Prevention 2017

Sydney, Australia
12 - 15 November 2017

Volume 2 of 2

ISBN: 978-1-5108-5699-8

Printed from e-media with permission by:

Curran Associates, Inc.
57 Morehouse Lane
Red Hook, NY 12571

Some format issues inherent in the e-media version may also appear in this print version.

Printed by Curran Associates, Inc. (2018)

For permission requests, please contact Australasian Corrosion Association
at the address below.

Australasian Corrosion Association
P.O. Box 112 Kerrimuir
Victoria 3129
Australia

Phone: 61 3 9890 4833
Fax: 61 3 9890 7866

aca@corrosion.com.au

Additional copies of this publication are available from:

Curran Associates, Inc.
57 Morehouse Lane
Red Hook, NY 12571 USA
Phone: 845-758-0400
Fax: 845-758-2633
Email: curran@proceedings.com
Web: www.proceedings.com

TABLE OF CONTENTS

VOLUME 1

VOLUME 2

STUDIES OF THE MICROBIOLOGICALLY INFLUENCED CORROSION OF STAINLESS STEEL ALLOYS BY SULPHATE REDUCING BACTERIA

M.A. Javed[1], W.C. Neil[2], G. McAdam[2] and S.A. Wade[1]
[1]Faculty of Science, Engineering and Technology, Swinburne University of Technology, Australia
[2]Defence Science and Technology Group, Melbourne, Australia

SUMMARY: Microbiologically influenced corrosion (MIC) of carbon steel by sulphate reducing bacteria (SRB) is well documented. However, contradictory reports can been found in the literature on the subject of SRB influenced corrosion of passive alloys, particularly stainless steels. In this paper we will present results of laboratory tests looking at the MIC of stainless steels. The first study that will be presented investigated the initial bacterial attachment ($\leq$ 18 h) and subsequent longer-term (3 months) corrosion behaviour of three different stainless steel alloys (316, 904 L and 2507) using a relatively standard MIC test setup. Polished metal coupons, including carbon steel samples as a control, were exposed to modified Baar's medium with SRB (*Desulfovibrio desulfuricans*) under anaerobic conditions. While increased corrosion of the carbon steel samples was observed compared to abiotic (i.e. no SRB present) controls, no evidence of increased corrosion of any of the stainless steel coupons was found during these tests. A second study was undertaken to look at the whether adding NaCl to the modified Barr's medium would affect the outcome of MIC studies using stainless steels. Again, no evidence of increased corrosion of the stainless steel coupons due to the presence of SRB was found for the test arrangement used.

Keywords: carbon steel, microbiologically influenced corrosion, stainless steels, sulphate reducing bacteria

1. INTRODUCTION

Microbiologically influenced corrosion (MIC) is a phrase describing corrosion that is affected by the presence and activity of microorganisms [1, 2]. MIC can occur for wide variety of materials exposed to aqueous environments [3, 4], as microorganisms tend to colonise and form biofilms on metal surfaces in these environments. The presence of biofilms on a metal surface can influence the corrosion process through a number of different processes including (but not limited to) changes in the local concentration of electrolyte constituents, redox potential, pH and/or oxygen concentration at the metal/solution interface [2, 4]. MIC is commonly associated with extremely high corrosion rates, much greater than one would normally expect for a particular environment in the absence of microorganisms [5]. MIC has been reported to account for a considerable fraction of overall corrosion costs [6, 7], and due to its damaging effects in different industries, worldwide costs in the range of billions of dollars per annum appear realistic [8]. The industries most affected by MIC include maritime, oil and gas production, nuclear power, waste water treatment, chemical processing and fire sprinkler water systems [1].

Sulphate reducing bacteria (SRB) are anaerobic microorganisms that use sulphate as a terminal electron acceptor for their respiratory metabolism and produce sulphide as a metabolic by-product [9, 10]. They are one of the most widely reported causative groups of bacteria associated with MIC [6, 11, 12]. SRB have been isolated from wide variety of environments including seawater [13], natural costal clay soils [14] and off-shore oil extraction plants [6]. They have been shown to cause localised corrosion attack on the surface of metallic materials, especially ferrous alloys, when present in the biofilm formed on these surfaces [15-17]. Several corrosion mechanisms for SRB have been documented in the literature including cathodic depolarisation [18], anodic depolarisation [19], formation of iron sulphides (FeS) [20], production of exopolymers capable of binding metal ions [21], sulphide induced stress corrosion cracking [22] and/or hydrogen embrittlement [23].

Extensive research has been conducted on the initial bacterial attachment and subsequent MIC of carbon steel by SRB. Beech and Gaylarde [24] studied the initial attachment of *Pseudomonas fluorescens* and *Desulfovibrio desulfuricans* to mild steel and stainless steel surfaces. They observed that the number of bacterial cells attached to mild steel was greater than stainless steel, irrespective of the incubation time and medium used. Javed et al. [15] found high localised corrosion attack on carbon steel samples exposed to test media with SRB compared to the same test media without SRB. King et al. [25, 26] studied the corrosion of mild steel in the presence of anaerobic SRB bacteria and proposed that it was due to the formation of FeS film on the metal surface. The FeS film formed acts as a cathode relative to bare steel surface which acts as an anode. Beech and co-workers [21, 27] suggested that a galvanic couple can be formed between the bacterial exopolymer capable of binding metal ions and steel, resulted in accelerated corrosion of metal substrate.

The action of SRB corrosion, however, is not limited to just carbon steels. Several studies have also explored the role of SRBs on the corrosion of stainless steel alloys. Compared to carbon steels, stainless steels generally tend to offer better corrosion resistance. The greater corrosion resistance of stainless steels is primarily due to the formation of a stable passive layer on the surface [28]. However, despite their good general corrosion resistance, stainless steels are still vulnerable to MIC. Localised pitting corrosion attack on stainless steels due to MIC has been reported by different industries such as maritime, oil and gas, pulp and paper, power generation and water supply [29-32]. SRB have been reported to be associated as contributing factors in most of the field reports of MIC failures in stainless steels [33-36]. Laboratory attempts have been made to replicate the rapid sustained SRB influenced corrosion of stainless steels that occur in the field. However, most of these studies have either produced insignificant rates of corrosion [37-39] or used unrealistic potentials for electrochemical measurements [40-42].

Interestingly, a number of contradictory reports can be found in the scientific literature on laboratory based studies of the SRB corrosion of stainless steels. Sheng et al. [37] for example studied the MIC of austenitic stainless steel (AISI 316) in the presence of SRB (*D. desulfuricans*) and a local marine isolate in enriched artificial seawater at different time intervals up to 44 days. They found localised pitting corrosion attack on the surface of 316 stainless steel after 4 days of immersion in the presence of both bacterial strains. The maximum pit depth determined was reported to be ~100 nm with SRB and ~40 nm with marine isolate at the end of 44 days of immersion. In contrast, Lutterbach and Franca [43] exposed the relatively less corrosion resistant grade stainless steel AISI 304 to cooling seawater for 30 days. Despite a high concentration of SRB present in the biofilm formed on the stainless steel, no signs of corrosion were detected on its surface at the end of 30 days of immersion. Similarly, Ortiz et al. [44] studied the corrosion behaviour of mild steel and AISI 304 stainless steel samples exposed to two different bacterial strains i.e. *Pseudomonas fluorescence* and a SRB for 7 days. They found significant corrosion attack on mild steel samples, however, no MIC attack was observed on any of the stainless steel samples under similar exposure conditions. These conflicting reports suggest that the SRB corrosion of stainless steel is a complex issue and further studies are required to help elucidate their role in MIC of these alloys.

In the present study, the initial bacterial attachment and subsequent corrosion of different grades of stainless steel alloys (i.e. austenitic 316, super austenitic 904 L and duplex 2507) has been systematically investigated in the presence of *D. desulfuricans* as a model system. Marine grade carbon steel (DH36) coupons were used as a control and exposed to similar testing conditions under an anaerobic environment. A range of surface characterisation techniques, including light microscopy, 3D profilometery and scanning electron microscopy (SEM) were used to determine the initial bacterial attachment, biofilm formation and the resulting corrosion after a 3 months immersion period. The influence of the chemical composition of test media, specifically NaCl levels, on the SRB corrosion of the stainless steel alloys was also studied.

2. EXPERIMENTAL PROCEDURE

2.1. Metal coupon preparation

The metal coupons used in this study were cut from plate samples into squares with dimensions of 25 mm × 25 mm using an automatic abrasive cutting machine with water-cooling. The chemical composition of the metal coupons (Table 1) was determined using inductive couple plasma atomic emission spectroscopy. All the metal coupons were finely ground using an automatic grinding and polishing machine, through series of silicon carbide paper with decreasing grit size finishing at 4000 grit. After grinding, the coupons were ultrasonically cleaned in an acetone for 15 min followed by rinsing with water and ethanol. Finally, the coupons were dried under warm air and stored in a desiccator before use.

Table 1. Elemental composition in weight per cent of the different alloys used in this work.

Alloy Type	Fe	C	Mn	Si	S	P	Ni	Cr	Mo	Cu	Ti	Al
DH36	**Bal**	0.16	1.28	0.25	0.01	0.01	0.01	0.02	<0.01	<0.01	0.02	0.045
316	**Bal**	0.04	1.52	0.60	0.01	0.03	10.0	16.9	2.10	0.29	0.01	-
904 L	**Bal**	0.02	0.52	0.46	0.01	0.02	17.5	20.2	5.90	0.68	<0.01	0.04
2507	**Bal**	0.03	0.41	0.29	0.01	0.02	6.75	25.0	3.62	0.07	<0.01	-

2.2. Surface roughness measurements

A 3D optical profilometer was used to measure the surface roughness of the prepared metal coupons, as explained in more detail previously [45, 46]. Briefly, the coupons were first scanned to take an overview of sample surface features and then detailed topographical measurements were taken at 10 random locations on each sample. The raw data obtained from the profilometer was filtered and processed according to the ISO 4288:1996 [47] standard using VISION 64^{TM} software.

2.3. Preparation of test medium and bacterial culture

Bacterial attachment and longer-term (3 month) corrosion tests were conducted in ATCC 1249 modified Baar's (MB) medium, a medium commonly used in laboratory MIC tests with SRB. The chemical composition (g/L) of the MB medium is given Table 2.

Table 2. Chemical composition (g/L) of modified Baar's (MB) medium used in this study.

K_2HPO_4	NH_4Cl	$MgSO_4.7H_2O$	$CaSO_4.2H_2O$	Sodium lactate	Sodium citrate	Yeast extract	5% $Fe(NH_4)_2(SO_4)_2$
0.5	1.0	4.1	1.3	5.7	5.0	1.0	1.0

Tests were also conducted in MB medium with different NaCl concentrations (i.e. no added NaCl, a relatively low chloride of 5 g/L and a chloride level used to simulate seawater 35 g/L) for 15 days. The test media was prepared by mixing the salts in MilliQ water. The pH of all the test media was then adjusted to 7.5 ± 0.1 using a 3.5 M KOH solution. The media was afterwards sterilised for 20 min by autoclaving at 121 °C and 103 kPa. After cooling, 0.1 ml of 5% $(NH_4)_2Fe(SO_4)_2$ solution was added to 5.0 ml of the previously autoclaved medium using syringe filters (0.02 μm in pore size).

The SRB strain used in this study was *D. desulfuricans* (ATCC 27774). To prepare for testing the bacteria were first taken from -80 °C storage and grown in MB medium at 37 °C for four days under anaerobic conditions. Compact W-Zip Seal Pouches with AnaeroGen Sachets were used to maintain anaerobic conditions during the tests, which were verified using the anaerobic redox indicator resazurin. Prior to use in corrosion test, a 1 mL aliquot of the four day old *D. desulfuricans* culture was inoculated into 400 mL of fresh test media of interest and incubated for three days at room temperature within a level 2 physical containment (PC2) cabinet under anaerobic conditions. All the experiments were conducted at room temperature (~21 °C), with the aim of better matching real life MIC conditions compared to the elevated temperatures typically used to grow SRBs. The number of bacterial cells was determined in the standard MB medium via the spread plate method using tryptic soy agar after every two weeks during longer-term (3 months) immersion tests.

2.4. Initial bacterial attachment studies

For the initial bacterial attachment studies, the prepared metal coupons were immersed in 45 ml of MB medium inoculated with *D. desulfuricans* in a sterile polypropylene container. Duplicate metal coupons were used for each exposure time, after which the test was repeated to give n = 4 for each time interval. The tests were carried out under anaerobic conditions and in static batch mode at 21°C. During tests, the coupons were fully immersed with a horizontal orientation. To observe the early stages of bacterial attachment, the coupons were removed from the test media after 1, 4 and 18 h of immersion. Coupons were then gently rinsed with sterile saline solution followed by a sterile deionized (DI) water rinse and left to dry in air in a PC2 cabinet at room temperature [48, 49]. A SEM was used to image the bacterial cells attached to the coupon surfaces. Images were taken at 10 different locations on each coupon and bacterial cells were counted manually using Image J software [46].

2.5. Longer-term corrosion studies

For longer-term corrosion studies, metal coupons (n = 5) were vertically hung on Nylon string and immersed in ~400 ml of MB medium inoculated with *D. desulfuricans*. Each metal coupon was immersed in an individual 500 ml test bottle. The coupons were fully immersed in the test media for 3 months. The test media was replenished (75 % of the total volume) with fresh sterile medium after every two weeks with the aim of maintaining the bacterial density and the resulting metabolic activity in the steady state growth phase. For comparison, control experiments (n = 5) were performed by immersing metallic coupons in MB medium without the SRB. The bulk pH and redox potential (E_h) of the test media was measured periodically during the longer-term immersion tests. The number of viable bacterial cells in the test media was monitored every month using the spread plate method as described earlier. The experiments were carried out under anaerobic conditions and in static batch mode at 21 °C.

2.6. MIC test with different chloride levels

Additional immersion studies were conducted to investigate the effect of chloride levels on MIC of 316 stainless steel by SRB. Stainless steel coupons (n = 3) hung vertically on Nylon string were exposed to MB medium with different chloride (NaCl) levels for 15 days. Tests were performed under both abiotic and biotic conditions in an anaerobic environment using AnaeroGen pouches in static batch mode at 21 °C as described previously.

2.7. SEM and EDS observations

A SEM (Zeiss SUPRA model 40VP-25-38) at 3 kV operating voltage and magnifications between 500× and 15000× was used to analyse the biofilm and/or corrosion product morphology on the coupons retrieved after immersion testing. The coupons taken from the biotic test media were washed twice with sterile PBS and underwent fixation and dehydration before imaging with the SEM as explained elsewhere in detail [15]. Briefly, the coupons were fixed with 2.5 vol. % solution of glutaraldehyde in PBS for 30 min at 21 °C, the coupons were then washed twice with DI water followed by stepwise dehydration with 25, 50, 75, 90 and 100 vol. % ethanol for 10 min each. Finally, the coupons were dried in air in a PC2 cabinet prior to SEM imaging. Chemical analysis of the biofilm and/or corrosion products on the surface of the coupons was carried out using SEM coupled with energy dispersive X-ray spectroscopy (EDS).

To allow assessment of the extent of corrosion damage under biofilm/corrosion products, ultrasonic cleaning of the coupons was performed. The coupons were immersed in beaker with 100 ml of passive cleaning solution depending on the alloy type, and then ultrasonically cleaned in an ultrasonic bath for 1-2 min to remove the biofilm and/or corrosion products. Clarke's solution was used for cleaning of carbon steel coupons [49] and a 25 vol. % solution of nitric acid in water was used for cleaning of stainless steel coupons [37]. Finally, the coupons were rinsed with DI water followed by an ethanol rinse and dried under a nitrogen gas flow. The dried coupons were weighed and stored in desiccator before being analysed via optical microscope, 3D optical profilometer and SEM.

2.8. Mass loss and corrosion rate

The initial and final mass of the metal coupons, i.e. before and after corrosion, were measured using a high accuracy mass balance. The mass loss was then calculated for each alloy type and corrosion rates were determined, as per ASTM G1-03 standard, as shown below [50]:

$$\text{Corrosion rate } (\mu\text{m/y}) = \frac{8.76 \times 10^7 \times \text{mass loss (g)}}{\text{density (g/cm}^3) \times \text{area (cm}^2) \times \text{time (h)}}$$

2.9. Statistical analysis

Statistical analysis (ANOVA) was performed on the numerical data obtained in this study, with the help of the statistical functions available in the MINITAB 16 and Origin Pro 8.6 software. The p values less than or equal to 0.05 were considered significant.

3. RESULTS

3.1. Surface roughness analysis

The results of surface roughness analysis of the ground coupons are shown in Table 3. The results show that the surfaces of all alloys had average surface roughness values within the nanometre range. Statistical analysis showed no significant difference ($p > 0.05$) between the surface roughness parameters of the different stainless steel alloys. A small but statistically significant difference ($p < 0.05$) was found in the surface roughness parameters between the DH36 and all other stainless steels alloys.

Table 3. Surface roughness analysis of the different alloy substrates, obtained using 3D profilometer.

Alloy Type	Surface roughness (nm ± SE)		
	Sa	Sq	Sz
DH-36	6.4 ± 0.1	8.3 ± 0.3	169 ± 11
316	2.8 ± 0.1	3.9 ± 0.1	107 ± 9
904 L	3.2 ± 0.1	4.3 ± 0.1	86 ± 7
2507	2.6 ± 0.1	3.5 ± 0.1	83 ± 5

3.2. Initial bacterial attachment (≤ 18 h)

The analysis of bacterial cell attachment on different alloys was performed for 1, 4 and 18 h. For all alloys tested, the number of bacterial cells attached to the surface was found to increase with immersion times (Figure 1). However, the rate and magnitude of bacterial attachment depended on the type of alloy substrate. A variation in bacterial cells attachment was also observed across different alloys at any fixed time interval. It is believed however that difference in magnitude of the numbers of bacteria attached observed for the various alloys is not likely to be significant in terms of bacterial attachment.

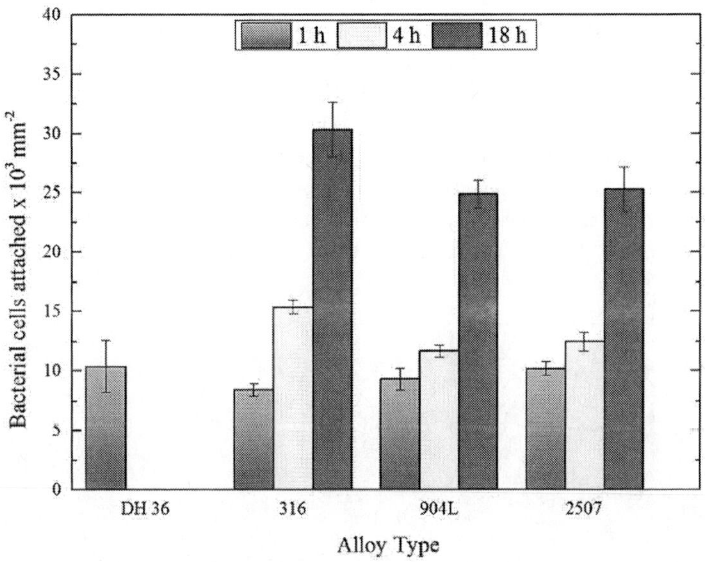

Figure 1. Number of bacterial cells attached to different alloys as a function of time in MB medium inoculated with *D. desulfuricans*. Data shown: Mean ± SE, n = 40.

For DH36, the analysis of bacterial cell attachment could only be performed for 1 h, as by 4 h the bacteria had produced sufficient amounts of extracellular polymeric substances (EPS), creating a biofilm which prohibited the counting of individual bacteria (see Figure 2). No obvious EPS production and/or biofilm formed on any of the stainless steel alloys tested. The spatial distribution of the initial bacterial attachment on 316 stainless steel after 1, 4 and 18 h of immersion is shown in Figure 2, indicating a relatively uniform distribution of bacterial cells. Similar distribution of bacterial cells was found on the surface of 904 L and 2507 stainless steel alloys (data not shown).

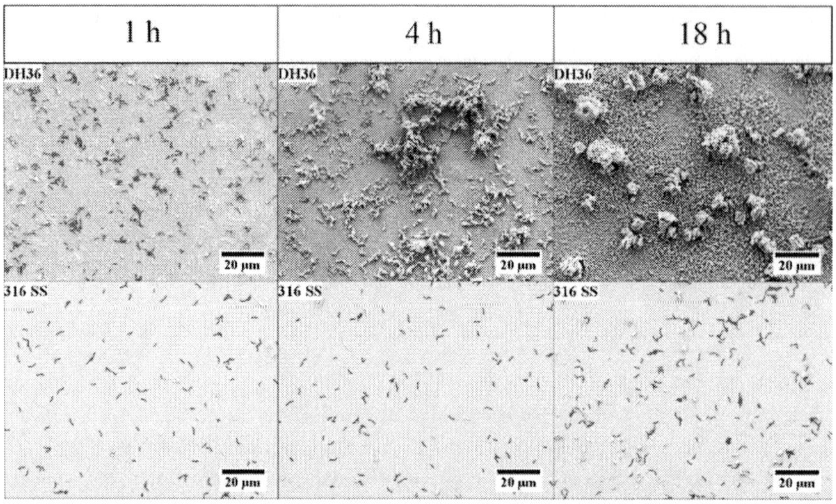

Figure 2. SEM images showing examples of initial bacterial cell (black) attachment on different alloy surfaces in MB medium inoculated with *D. desulfuricans* after different times of immersion.

3.3. Longer term immersion tests (3 months)

SEM images of different alloy surfaces after immersion in MB medium for 3 months, but before cleaning, are shown in Figure 3. SEM images of alloy surfaces immersed in abiotic MB medium are shown in Figures 3 a-d. A thick film with cracks formed on DH 36 (Figure 3a). No such film/corrosion products were formed on any of the stainless steel alloys (Figures 3 b-d). Figures 3 (e-h) show SEM images of alloy surfaces immersed in biotic MB medium for 3 months. All of the alloy surfaces were covered with a thick biofilm containing bacteria, extracellular polymeric substances and FeS by-products. The FeS is produced by the metabolic activity of SRB in media containing soluble ferrous ions [10, 15]. An EDS analysis of the corrosion products/biofilm (as discussed later) showed high content of elemental iron and sulphur, indicating the presence of FeS.

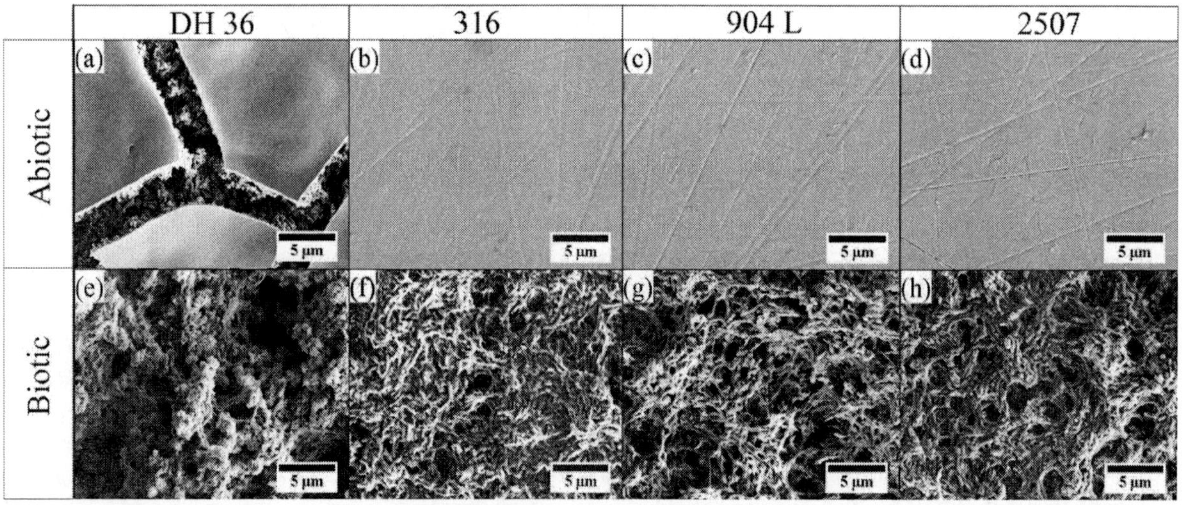

Figure 3. SEM images showing corrosion products and biofilms formed on different alloy surfaces immersed in MB medium under both abiotic (a-d) and biotic (e-h) conditions for 3 months.

SEM images of different alloy surfaces following immersion in MB medium for 3 months, after the removal of biofilm/corrosion products, are shown in Figure 4. Figures 4 a-d show the alloy surfaces which had been immersed in the abiotic MB medium. Clear evidence of corrosion attack on the surface of DH 36 under both abiotic and biotic conditions can be seen. No obvious corrosion, however, was observed on any of the stainless steel alloys immersed in either abiotic or biotic MB medium.

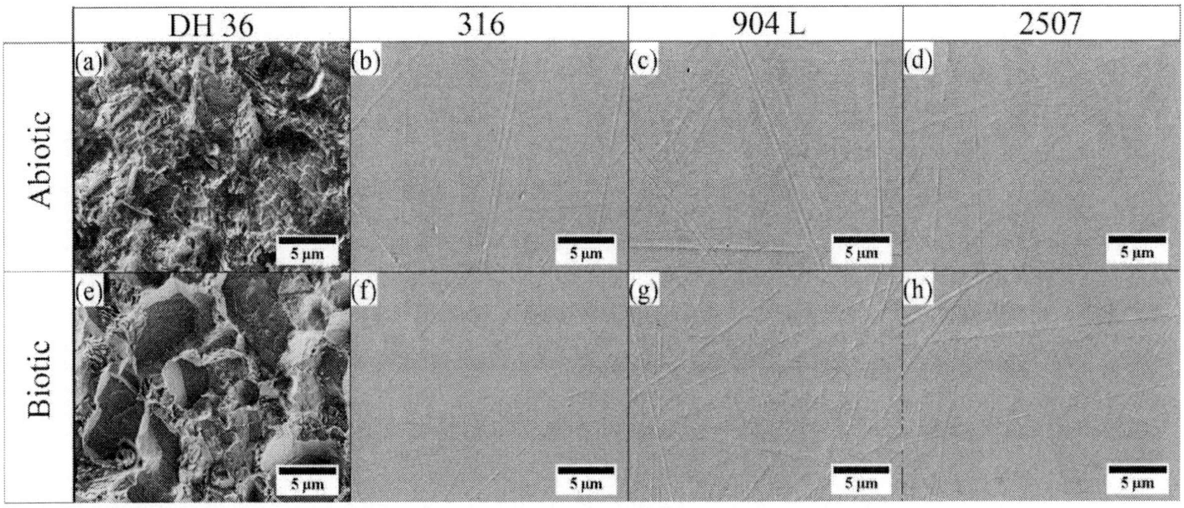

Figure 4. SEM images showing the surface morphology of different alloys in MB medium under both abiotic (a-d) and biotic (e-h) conditions for 3 months, after removal of corrosion products / biofilms formed.

EDS spectra of the biofilm/corrosion products formed on the different alloy surfaces after 3 months of immersion in MB medium containing SRB are shown in Figure 5. Each of the EDS spectra show the presence of strong peaks of Fe and S, which is likely to be due to the formation of FeS by SRB. It was also found that the Fe-S ratio of the biofilm/corrosion products formed on different alloy surfaces at the end of 3 months was approximately similar i.e. within the range of ~1.4 – ~1.8. The Fe/S ratio has been shown to provide information on the mineralogical nature of FeS film formed (for example, mackinawite, pyrrhotite and/or pyrite) which in turn affects the corrosion process [51-53]. Previous studies have shown that a non-protective FeS film formed when the concentration of iron ions exceeded the concentration of sulphide ions in the test media [8, 15]. In the present study, a non-protective FeS film formed on all of the alloys tested as indicated by the EDS data (i.e. the relative concentration of Fe ions was ~1.5 times higher than S ions). While it was beyond the scope of the current work it may be interesting in future studies to identify the exact changes in the type(s) of FeS produced during the formation of biofilm using techniques such as X-ray diffraction and/or Raman spectroscopy.

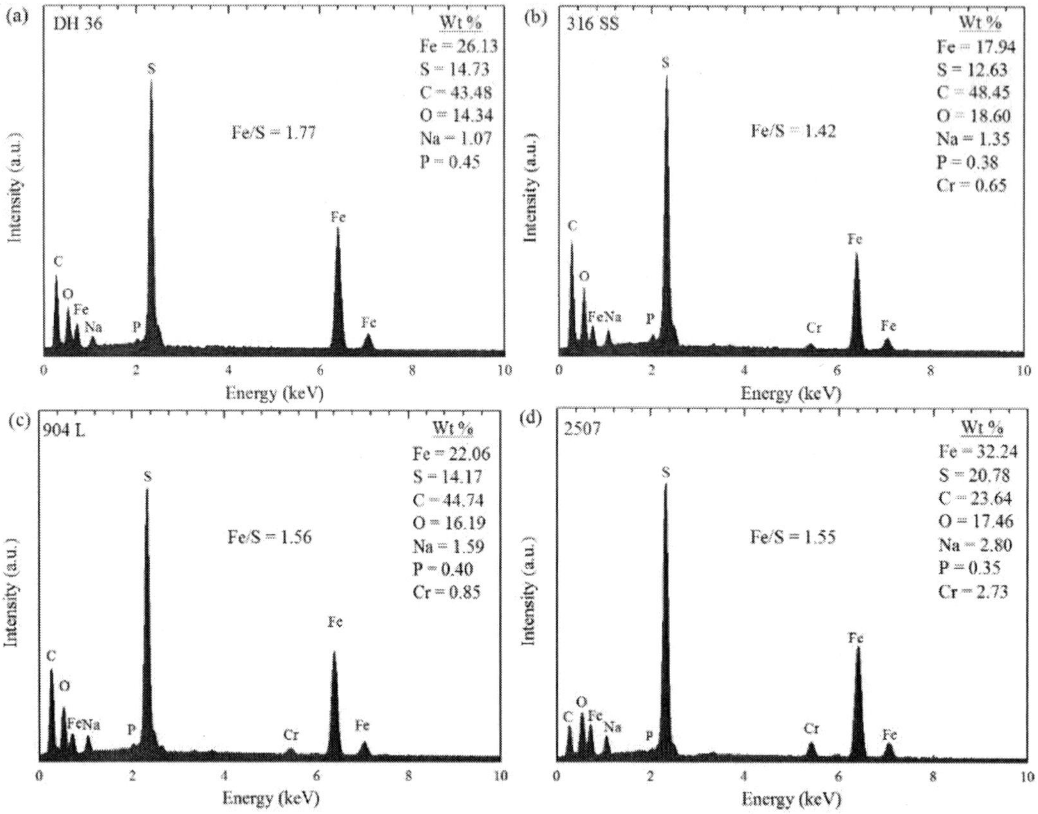

Figure 5. EDS spectra of biofilm/corrosion products formed on different alloys immersed in MB medium inoculated with *D. desulfuricans.*

Longer term immersion tests showed significantly increased corrosion of DH 36 alloy immersed in the MB medium under biotic conditions, compared to abiotic conditions (Figure 6). Compared to the DH36 alloy, however, the stainless steel alloys tested showed very low corrosion rates (≤ 0.15 µm/y) under both abiotic and biotic conditions (Figure 6). Statistical analysis showed no significant difference ($p > 0.05$) between the corrosion rates of any of the stainless steel alloys tested under both abiotic and biotic conditions.

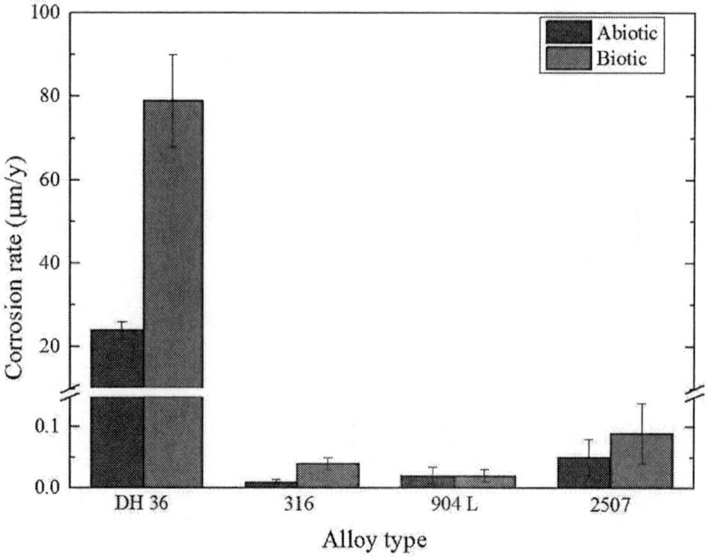

Figure 6. Corrosion rates of different alloys tested under abiotic and biotic conditions in MB medium after 3 months of immersion. Data Shown: Mean ± SE, n = 6.

3.4. NaCl levels and MIC of stainless steel

SEM analysis was performed on 316 stainless steel alloy immersed in MB medium with (i) no added chlorides (0 g/L NaCl), (ii) relatively low chlorides (5 g/L NaCl) and (iii) a chloride level to simulate seawater (35 g/L NaCl) for 15 days. No corrosion attack was observed on the surface of 316 stainless steel alloy under abiotic conditions in any of the test media (data not shown). Under biotic conditions, a SRB biofilm was formed on the surface of 316 stainless steel alloy immersed in standard MB medium (Figure 7 a) and the MB medium with a relatively low NaCl level i.e. 5 g/L (Figure 7 b). However, no biofilm was seen on the surface of 316 stainless steel alloy immersed in MB medium with a chloride level of 35 g/L (Figure 7 c) at the end of 15 days of immersion. No corrosion attack was observed on the surface of stainless steel alloy for any of the chloride levels tested after the removal of biofilm and/or corrosion products (Fig. 7 d-f).

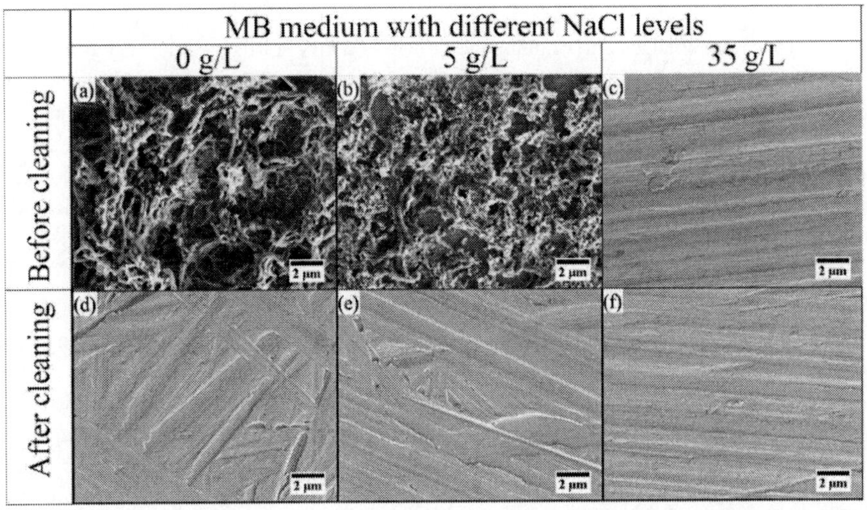

Figure 7. SEM images of 316 stainless steel (a-c) before and (d-f) after cleaning immersed in biotic MB medium for 15 days with different NaCl levels.

3.5. pH, Redox potential and viable bacterial counts

The pH, redox potential (E_h) and viable bacterial counts in the test media was measured at different time intervals during the 3 month immersion studies (Figure 8). For long-term immersion studies performed under abiotic conditions no significant change was observed in the pH and E_h values of the bulk test media over the duration of the testing. The average pH and E_h values were found to be equivalent to ~6.5 and -20 mV vs SCE, respectively. Under biotic conditions, after two weeks of immersion, the pH values slightly increased to more alkaline values (~7.7) with a decrease in the E_h values (~-100 mV vs SCE) which then remained relatively constant for the rest of the experiment. Similar trends of pH and E_h values was observed for the MB medium with 0 and 5 g/L NaCl levels after two weeks of immersion under both abiotic and biotic conditions. However, no change was observed in the pH and E_h values of MB medium with high NaCl levels (i.e. 35 g/L) under biotic conditions.

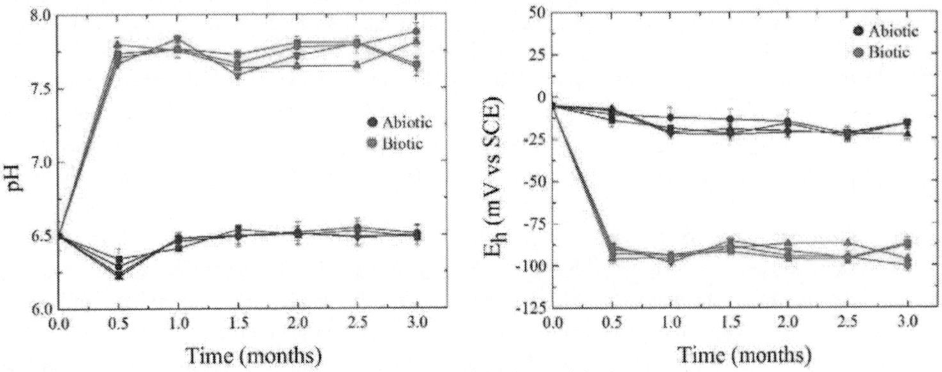

Figure 8. pH and redox potential (E_h) of MB medium under abiotic and biotic conditions measured after every two weeks during long term corrosion tests.

The viable bacterial counts were determined at different time intervals to provide an insight into possible population size effects on the corrosion behaviour of the different alloys used in this study. The results showed no significant change in the number of viable bacterial cells in the test medium with different alloy types over the course of the 3 months immersion studies. The average value of viable bacterial counts in SRB inoculated MB medium for the different alloy types was found to be ~2×10^8 cells/ml. For the tests investigating the effect of NaCl levels the number of SRB in the MB medium was only monitored visually and indicated by the blackening of test medium. The blackening of the test medium occurred due to the formation of FeS by-products (black in colour) during the metabolic activity and growth of SRBs. After 3 days of inoculation, the MB media with 0 and 5 g/L NaCl had both turned black indicating the growth of SRBs. However, the colour of the MB medium with 35 g/L NaCl remain unchanged throughout the 15 days of testing suggesting that SRBs growth was possibly delayed due to the high salinity of the test medium.

Additional short-term tests were conducted to examine the growth of SRBs in MB medium with different NaCl level (0, 5, 10, 15 and 35 g/L). The SRB growth was monitored both visually and using TSA plates. After two weeks of incubation, visual analysis showed blackening of the MB medium with 0, 5, 10 and 15 g/L NaCl. No blackening of the MB medium was observed with 35 g/L. On the other hand, the TSA plate counts showed growth of SRB in all of the test media. However, estimated number of SRBs found in the MB medium with 35 g/L were ~100 times less compared to rest of the MB medium (i.e. with 0, 5, 10 and 15 g/L NaCl). These findings showed that growth of SRB cells in MB medium slowed with an increase in the concentration of NaCl present in the test medium.

4. DISCUSSION

In this study, we have investigated the initial bacterial attachment (≤ 18 h) and relatively longer term (3 months) corrosion of stainless steel alloys immersed in MB medium with and without SRB for different time intervals. Carbon steel (DH 36) coupons were used as a control for comparison. Additional experiments were performed on stainless steel samples in MB medium with different NaCl levels with and without SRB. The results obtained showed accelerated corrosion of carbon steel coupons in the presence of SRB, however no MIC attack was found on any of the stainless steel coupons in any of the test media under both abiotic and biotic conditions.

The surface roughness of the substrate has been reported to potentially effect initial bacterial attachment to a surface and subsequent biofilm formation [54, 55]. In order to minimise the influence of the surface roughness, coupons were polished to a 4000 grit finish and a detailed analysis of surface roughness parameters was carried out. The surface roughness analysis results show that the surfaces of all alloys used had average surface roughness values within the nanometre range. No significant difference was found in surface roughness between the polished surfaces of the different stainless steel alloys. A small variation in bacterial cell attachment was observed across different stainless steel alloys at any fixed time interval (Figure 1). However, this small difference in magnitude of the numbers of bacteria attached to different stainless steel alloys is not likely to be significant, in terms of bacterial attachment. In contrast to stainless steels alloys tested, the DH36 samples showed relatively more attachment of bacterial cells and greater EPS production after 4 h of immersion. After 18 h, the DH36 surface was fully covered with SRB biofilm, whereas individual bacterial cells were present on the surface of stainless steel alloys. This difference in bacterial attachment and biofilm formation might be due to the difference found in the average surface roughness parameters between the DH36 and stainless steel alloys. The slight difference in surface roughness between these alloys after polishing at same 4000 grit finish is likely due to the difference in the hardness of these material (e.g. DH36 is relatively softer than stainless steel alloys).

Another possible driver for the rapid colonisation of SRB on DH36 might be the availability of iron ions at the metal-liquid interface. In case of stainless steel alloys, a protective passive oxide layer is present which prevents dissolution of metal species such as iron and therefore reduces their concentration at the metal-liquid interface. No such passive film covers the surface of carbon steel where relatively soluble iron oxides are present [56]. The concentration of metal ions would therefore be greater at the carbon steel surface compared with stainless steel surface. *D. desulfuricans* are known to have a relatively high requirement for inorganic iron in culture media for their metabolic activity and growth [10, 57]. Studies have shown that the interaction between the EPS and availability of metal ions at the surface could be an important factor influencing not only the initial attachment of bacterial cells but also the quality and quantity of biofilm formed on these surfaces [58, 59]. Microbial chemotaxis towards metal ions is a well-known phenomenon. Beech and Gaylarde [60] found that the interaction between EPS and metal ions could play an important role both in the corrosive action of SRB and in the attachment of SRB cells to steel surfaces. The other potential factors affecting the attachment of bacterial cells to metal substrates include the presence of physical forces, surface electrostatic charge and hydrophobic interactions between bacterial cells and metal surfaces [61, 62]. It would be interesting in future studies to investigate the effect of these factors on the attachment of bacterial cells to the metal substrates.

Longer-term (3 month) immersion studies showed significantly high corrosion of DH36 in biotic conditions compared to abiotic conditions. These results are consistent with previous studies which showed increased corrosion of steel in the presence of SRB [15, 25, 63, 64]. Two well-known mechanisms reported in literature to explain an accelerated corrosion of steel by SRB include (i) FeS film formed by SRB [20, 26] and (ii) binding of metal ions by EPS produced by SRB [21]. No evidence of general or localised corrosion attack was found on any of the stainless steel alloys tested under same test conditions. Conflicting reports have been found in literature on SRB corrosion of stainless steel as summarised in the introduction section of this paper. The reason for the conflicting results of MIC of stainless steels may be due to varying laboratory conditions and test media used [48, 65]. Webster and Newman [66] examined the impact of media constituents on SRB corrosion of stainless steel alloy. They concluded that the composition of SRB growth medium used in the laboratory studies of MIC of stainless steels is very important. Two critical factors are (i) the presence of chloride and (ii) relative concentration of chlorides to inhibiting anions in the test medium. They suggested that chloride must be present in a concentration at least comparable to that of all other anions combined, otherwise corrosion was inhibited even at high sulphide concentrations up to 100 ppm. Anions, including sulphate, hydroxide, phosphate, acetate, carbonate and nitrate can stabilise passive film and inhibit pitting corrosion attack on stainless steel alloys.

To check the effect of chloride levels and SRB attack on stainless steels, additional short term (15 days) experiments were conducted by exposing 316 stainless steel samples to SRB inoculated test medium with different NaCl levels. The chlorides to anions ratios (Cl^-:anions) of MB medium with 0, 5 and 35 g/L NaCl were calculated and found to be equivalent to 0.2:1, 1:1 and 6.4:1. Again, no MIC attack was found on stainless steel coupons in any of the test media under both abiotic and biotic conditions after 15 days of immersion studies. One of the argument raised for no MIC attack seen in this case could be the small test duration i.e. 15 days. Further longer-term immersion studies on different chlorides levels and MIC attack on stainless steels are currently in progress and might give some useful insight into the relationship between them. In addition to chloride ions effects, we are also investigating some other potential factors which could affect the SRB corrosion of stainless steels such as:

- *Removal of iron ions from the test media*: The iron ions are usually added in the test media to support the rapid growth of SRB [10, 57]. The removal of iron ions from the test media would permit high build up of hydrogen sulphide (H_2S) which would otherwise react with iron ions and form FeS by-products. Studies have shown that the presence of high concentration of H_2S reduces the pitting potential of stainless steel and makes it susceptible to localised corrosion [67-69].

- *Purging of test media with air/oxygen*: Newman et al. [70] showed that SRB influenced corrosion in stainless steel is unlikely to occur if oxygen is not present in the bulk electrolyte. The authors concluded that cathodic current provided by the oxygen reduction reaction at discrete cathodic sites was required for high rates of anodic dissolution. Purging of oxygen in the SRB inoculated test media would also results in the oxidation of sulphides (H_2S, HS^-) produced by the SRB to thiosulphate ($S_2O_3^{-2}$) ions. Thiosulphate ions have been shown to cause severe pitting corrosion of stainless steel in newspaper machines [71, 72].

- *Effect of test medium:* The test medium used for MIC studies have been shown to significantly affect the outcome of these studies. For example, Javed et al. [48] showed that the corrosion of steel could either accelerate and/or inhibit in the presence of same bacterial strain depending on the growth medium used for MIC tests [48]. Similarly, some components of the test medium either act as corrosion inhibitors and/or interfere with the measurement techniques and thus can affect the corrosion process, for example; yeast extract can interfere with electrochemical measurements [66, 73]. Further studies of different test media compositions (e.g. without yeast extract) are of interest.

The results of pH (~7.7) and E_h (~-100 mV vs SCE) of SRB inoculated test media showed that SRB growth occur at near neutral pH and negative redox potential values [6, 10]. The increase in the bulk pH of the test media is reported to be due to the oxidation of lactate present in the test media into bicarbonate by the SRB [74, 75]. Under biotic conditions, no difference in the bulk pH/E_h values was observed in the test media with different alloys tested however a significant difference was seen in the nature and extent of SRB corrosion attack on DH36 and stainless steel alloys. This shows that, in the presence of bacterial biofilm, measuring bulk pH/E_h values of test media is not reliable to predict the corrosion of metal substrate and might be different from the pH/E_h values at the metal-biofilm interface [76].

5. CONCLUSION

The following conclusion can be drawn from this investigation of initial SRB attachment and subsequent corrosion of different stainless steel alloys:

1. In the initial attachment tests ($\leq$ 18 h) the numbers of SRB cells attached to all of the alloys increased with time; however, the rate and magnitude of bacterial attachment depended on the type of alloy substrate e.g. DH36 and stainless steel alloys.
2. After 18 h of immersion, a thick biofilm formed on the surface of DH36 whereas individual SRB cells were present on the surface of the different stainless steel alloys tested. However, after 3 months of immersion, the SRB cells formed a thick biofilm layer on the surface of all of the stainless steel alloys.
3. Longer term (3 months) immersion studies showed significantly higher general corrosion rates for DH36 in biotic conditions compared to abiotic conditions. No evidence of general or localised corrosion was found on any of the stainless steels alloys either the abiotic or biotic conditions.
4. No corrosion attack was found on stainless steel alloy tested in MB medium with different NaCl levels in both abiotic and/or biotic conditions after 15 days of immersion.

Further studies aimed at improving the understanding of the effect of different laboratory based testing parameters (e.g. chloride levels, immersion temperatures, oxic/anoxic conditions, iron ions levels etc) on the corrosion of stainless steel alloys when exposed to SRB are currently in progress. It is envisaged that the information obtained from these laboratory-based studies will provide support towards the eventual development of detection and mitigation strategies for MIC attack in stainless steel alloys.

6. ACKNOWLEDGEMENTS

The Commonwealth of Australia through the Defence Science and Technology Group financially supported this research project. The authors would also like to thank Dr Melanie Thomson for providing the SRB strain used in this work.

7. REFERENCES

1. Little BJ, Lee JS. Microbiologically Influenced Corrosion: John Wiley & Sons, Inc. 2007.
2. Borenstein SW. Microbiologically Influenced Corrosion Handbook: Cambridge, Woodhead.1994.
3. Little BJ, Wagner P, Mansfeld F. Microbiologically influenced corrosion of metals and alloys. International Materials Reviews. 1991; 36(6): 253-272.
4. Videla HA. Manual of Biocorrosion: CRC Press, Inc. 1996.
5. Wade SA, Mart PL, Trueman AR. Microbiologically influenced corrosion in maritime vessels. Corrosion and Materials. 2011; 36(5): 68-79.
6. Hamilton WA. Sulphate-reducing bacteria and anaerobic corrosion. Annual Reviews Microbiology. 1985; 39: 195-217.
7. Flemming HC. Economical and technical overview. In Microbially influenced corrosion of materials: scientific and engineering aspects: Springer, Berlin Heidelberg. 1996.
8. Enning D, Garrelfs J. Corrosion of iron by sulfate-reducing bacteria: New views of an old problem. Applied and Environmental Microbiology. 2014; 80(4): 1226-1236.
9. Hamilton WA. Biocorrosion: the action of sulphate-reducing bacteria. In Biochemistry of microbial degradation: Springer, Netherlands. 1994.
10. Postgate JR. The sulphate reducing bacteria: Cambridge University Press, New York. 1984.
11. Booth GH. Sulphur bacteria in relation to corrosion. Journal of Applied Bacteriology. 1964; 27(1): 174-181.
12. Beech IB. Biocorrosion: role of sulfate reducing bacteria. In Encyclopedia of environmental microbiology: John Wiley & Sons, Inc. 2003.
13. Hardy JA. The enumeration, isolation and characterization of sulphate-reducing bacteria from north sea waters. Journal of Applied Bacteriology. 1981; 51(3): 505-516.
14. Salvarezza RC, Videla HA. Passivity breakdown of mild steel in sea water in the presence of sulfate reducing bacteria. Corrosion. 1980; 36(10): 550-554.
15. Javed MA, Stoddart PR, Wade SA. Corrosion of carbon steel by sulphate reducing bacteria: Initial attachment and the role of ferrous ions. Corrosion Science. 2015; 93: 48-57.

16. King RA, Miller JDA. Corrosion by the sulphate-reducing bacteria. Nature. 1971; 233: 491-492.
17. AlAbbas FM, Williamson C, Bhola SM, Spear JR, Olson DL, Mishra B, Kakpovbia AE. Influence of sulfate reducing bacterial biofilm on corrosion behavior of low-alloy, high-strength steel (API-5L X80). International Biodeterioration and Biodegradation. 2013; 78: 34-42.
18. Von Wolzogen Kühr CAH, Van Der Vlugt LS. The graphitization of cast iron as an electrochemical process in anaerobic soils. Water. 1934; 18: 147-165.
19. Crolet J. From biology and corrosion to biocorrosion. Oceanologica acta. 1992; 15(1): 87-94.
20. Booth GH, Elford L, Wakerley DS. Corrosion of mild steel by sulphate-reducing bacteria: An alternative mechanism. British Corrosion Journal. 1968; 3(5): 242-245.
21. Beech IB, Sunner J. Biocorrosion: Towards understanding interactions between biofilms and metals. Current Opinion in Biotechnology. 2004; 15(3): 181-186.
22. Javaherdashti R, Raman RS, Panter C, Pereloma EV. Microbiologically assisted stress corrosion cracking of carbon steel in mixed and pure cultures of sulfate reducing bacteria. International Biodeterioration and biodegradation. 2006; 58(1): 27-35.
23. Biezma MV. The role of hydrogen in microbiologically influenced corrosion and stress corrosion cracking. International Journal of Hydrogen Energy. 2001; 26(5): 515-520.
24. Beech IB, Gaylarde CC. Attachment of *Pseudomonas fluorescens* and *Desulfovibrio desulfuricans* to mild and stainless steel - First step in biofilm formation. Journal of Applied Bacteriology. 1989; 67(2): 201-207.
25. King RA, Miller JDA, Wakerley DS. Corrosion of mild steel in cultures of sulphate-reducing bacteria: Effect of changing the soluble iron concentration during growth. British Corrosion Journal. 1973; 8(2): 89-93.
26. King RA, Wakerley DS. Corrosion of mild steel by ferrous sulphide. British Corrosion Journal. 1973; 8(1): 41-45.
27. Beech IB, Cheung CWS. Interactions of exopolymers produced by sulphate-reducing bacteria with metal ions. International Biodeterioration and Biodegradation. 1995; 35: 59-72.
28. Blair M. Cast stainless steels. Properties and selection: irons, steels, and high performance alloys, Vol 1, ASM Handbook, ASM International, 2005, pp. 1409-1442
29. Flemming HC. Biofouling in water systems - cases, causes and countermeasures. Applied Microbiology and Biotechnology. 2002; 59(6): 629-40.
30. Little BJ, Lee JS, Ray RI. Microbiologically influenced corrosion: Global phenomena, local mechanisms. Corrosion and Materials. 2011; 36(1): 46-51.
31. Cord-Ruwisch R, Kleinitz W, Widdel F. Sulfate-reducing bacteria and their activities in oil production. Journal of Petroleum Technology. 1987; 39: 97-106.
32. Flemming HC, Meier M, Schild T. Mini-review: microbial problems in paper production. Biofouling. 2013; 29(6): 683-696.
33. Naik NJ, Desai PB. Sulfate reducing bacteria — an important agent for bio-fouling in cooling water system of fertilizer industries. In Microbes in applied research: Current advances and challenges: World scientific Publishing Co. 2012.
34. Liu W. Rapid MIC attack on 2205 duplex stainless steel pipe in a yacht. Engineering Failure Analysis. 2014; 42: 109-120.
35. Brennenstuhl AM, Gendron TS, Cleland R. Mechanisms of underdeposit corrosion in freshwater cooled austenitic alloy heat exchangers. Corrosion Science. 1993; 35(1-4): 699707-705711.
36. Scott PJB, Davies M. Microbiologically influenced corrosion of high molybdenum austenitic stainless steel. CORROSION; April 17-21; New Orleans (LA): NACE International 1989. Paper 89186.
37. Sheng X, Ting YP, Pehkonen SO. The influence of sulphate-reducing bacteria biofilm on the corrosion of stainless steel AISI 316. Corrosion Science. 2007; 49(5): 2159-2176.
38. Ringas C, Robinson FPA. Corrosion of stainless steel by sulfate reducing bacteria - Total immerson test results. Corrosion. 1988; 44(9): 671-678.
39. Alfaro-Cuevas-Villanueva R, Cortes-Martinez R, García-Díaz JJ, Galvan-Martinez R, Torres-Sanchez R. Microbiologically influenced corrosion of steels by thermophilic and mesophilic bacteria. Materials and Corrosion. 2006; 57(7): 543-548.
40. Ringas C, Robinson FPA. Corrosion of stainless steel by sulfate reducing bacteria - Electrochemical techniques. Corrosion. 1988; 44(6): 386-396.
41. Zhang Q, Wang P, Zhang D. Stainless steel electrochemical corrosion behaviors induced by sulphate-reducing bacteria in different aerated conditions. International Journal of Electrochemical Science. 2012; 7: 11528-11539.
42. Ismail KM, et al. The influence of bacteria on the passive film stability of 304 stainless steel. Electrochimica Acta. 1999; 44(26): 4685-4692.

43. Lutterbach MTS, De França FP. Biofilm formation in water cooling systems. World Journal of Microbiology and Biotechnology. 1996; 12(4): 391-394.
44. Ortiz C, Guiamet PS, Videla HA. Relationship between biofilms and corrosion of steel by microbial contaminants of cutting-oil emulsions. International Biodeterioration. 1990; 26(5): 315-326.
45. Javed MA, Stoddart PR, McArthur SL, Wade SA. The effect of metal microstructure on the initial attachment of *Escherichia coli* to 1010 carbon steel. Biofouling. 2013; 29(8): 939-952.
46. Javed MA, McArthur SL, Stoddart PR, Wade SA. Techniques for studying initial bacterial attachment and subsequent corrosion. Corrosion and Prevention; November 2013; Brisbane: Australian Corrosion Association (ACA). Paper 64.
47. ISO. 1996. Standard 4288. Geometrical Product Specifications (GPS) – surface texture: profile method – rules and procedures for the assessment of surface texture. Available from: http://www.iso.org/iso/home.html
48. Javed MA, Stoddart PR, Palombo EA, McArthur SL, Wade SA. Inhibition or acceleration: Bacterial test media can determine the course of microbiologically influenced corrosion. Corrosion Science. 2014; 86: 149-158.
49. Javed MA, Neil WC, Stoddart PR, Wade SA. Influence of carbon steel grade on the initial attachment of bacteria and microbiologically influenced corrosion. Biofouling. 2016; 32(1): 109-122.
50. ASTM, Standard practice for preparing, cleaning, and evaluating corrosion test specimens [standard G1-03], in: ASTM Int. ASTM Handbook 3.02 Corrosion of Metals, Wear and Erosion, West Conshohocken, PA, 2011, pp. 20–28.
51. El Mendili Y, Abdelouas A, El Hajj H, Bardeau JF. Phase transitions of iron sulphides formed by steel microbial corrosion. RSC Advances. 2013; 3(48): 26343-26351.
52. Zhao X, Duan J, Hou B, Wu S. Effect of sulfate-reducing bacteria on corrosion behavior of mild steel in sea mud. Journal of Material Science and Technology. 2007; 23: 323–327.
53. Newman RC, Rumash K, Webster BJ. The effect of pre-corrosion on the corrosion rate of steel in neutral solutions containing sulphide: relevance to microbially influenced corrosion. Corrosion Science. 1992; 33(12): 1877-1884.
54. Crawford RJ, et al. Surface topographical factors influencing bacterial attachment. Advances in Colloid and Interface Science. 2012; 179-182: 142-149.
55. Hilbert LR, Bagge-Ravn D, Kold J, Gram L. Influence of surface roughness of stainless steel on microbial adhesion and corrosion resistance. International Biodeterioration and Biodegradation. 2003; 52(3): 175-185.
56. Grubb JF, DeBold T, Fritz JD. Corrosion of wrought stainless steels. Corrosion: Materials, Vol 13B, ASM Handbook, ASM International, 2005, pp. 54–77.
57. Hauser JY, Holder GA. Iron availability in mixed cultures of sulfate-reducing bacteria. Biotechnology and Bioengineering. 1986; 28(1): 101-106.
58. Allison DG, Sutherland IW. The role of exopolysaccharides in adhesion of freshwater bacteria. Journal of General Microbiology. 1987; 133(5): 1319-1327.
59. Beech IB, Sunner JA, Hiraoka K. Microbe-surface interactions in biofouling and biocorrosion processes. International Microbiology. 2005; 8(3): 157-168.
60. Beech IB, Gaylarde CC. Adhesion of Desulfovibrio desulfuricans and Pseudomonas fluorescens to mild steel surfaces. Journal of Applied Bacteriology. 1989; 67: 201-207.
61. Katsikogianni M, Missirlis YF. Concise review of mechanisms of bacterial adhesion to biomaterials and of techniques used in estimating bacteria-material interactions. European Cells and Materials. 2004; 8: 37-57.
62. Garrett TR, Manmohan B, Zhibing Z. Bacterial adhesion and biofilms on surfaces. Progress in Natural Science. 2008; 18 (9): 1049-1056.
63. Hardy JA, Bown JL. The corrosion of mild steel by biogenic sulfide films exposed to air. Corrosion. 1984; 40(12): 650-654.
64. El Hajj H, Abdelouas A, El Mendili Y, Karakurt G, Grambow B, Martin C. Corrosion of carbon steel under sequential aerobic–anaerobic environmental conditions. Corrosion Science. 2013; 76: 432-440.
65. Wade SA, Javed MA, Palombo EA, McArthur SL, Stoddart PR. On the need for more realistic experimental conditions in laboratory-based microbiologically influenced corrosion testing. International Biodeterioration and Biodegradation. 2017; 121: 97-106.
66. Webster BJ, Newman RC. Producing rapid sulfate-reducing bacteria (SRB) influenced corrosion in the laboratory. In Microbiologically influenced corrosion testing: ASTM International. 1994.
67. Werner SE, Johnson CA, Laycock NJ, Wilson PT, Webster BJ. Pitting of type 304 stainless steel in the presence of a biofilm containing sulphate reducing bacteria. Corrosion Science. 1998; 40: 465-480.
68. Davoodi A, Pakshir M, Babaiee M, Ebrahimi GR. A comparative H2S corrosion study of 304L and 316L stainless steels in acidic media. Corrosion Science. 2011; 53: 399-408.

69. Newman RC, Wong WP, Garner A. Technical note: A mechanism of microbial pitting in Stainless Steel. Corrosion. 1986; 42: 489-491.
70. Newman R, Webster B, Kelly R. The electrochemistry of SRB corrosion and related inorganic phenomena. ISIJ International. 1991; 31(2): 201-209. CORROSION; April 17-21; New Orleans (LA): NACE International 1989. Paper 89186.
71. Newman RC, Wong WP, Ezuber H, Garner A. Pitting of stainless steels by thiosulfate ions. Corrosion. 1989; 45: 282-287.
72. Laitinen TMJ. Comparison of stainless steels in simulated paper machine environments. Corrosion. 1999; 55: 858-869.
73. Lee JS, Little BJ, 2015. Technical note: electrochemical and chemical complications resulting from yeast extract addition to stimulate microbial growth. Corrosion; 71: 1434-1440.
74. Baumgartner LK, Pamela RR, Christophe D, Alan WD, Buckley DH, Spear JR, Przekop KM, Visscher PT. Sulfate reducing bacteria in microbial mats: changing paradigms, new discoveries. Sedimentary Geology. 2006; 185(3): 131-145.
75. Teclu D, Tivchev G, Laing M, Wallis M. Determination of the elemental composition of molasses and its suitability as carbon source for growth of sulphate-reducing bacteria. Journal of Hazardous Materials. 2009; 161(2): 1157-1165.
76. Dexter SC, Chandrasekaran P. Direct measurement of pH within marine biofilms on passive metals. Biofouling. 2000; 15(4): 313-325.

8. AUTHOR DETAILS

Dr Muhammad Awais Javed is currently working as a research fellow at Swinburne University of Technology. He has won several awards at different forums including best presentation award at Brian Cherry Forum in 2014 and the ACA Young Corrosionist Conference Scholarship in 2015. Recently, he was nominated and selected for Fresh Science Victoria 2016, a national competition helping early-career researchers find, and then share, their stories of discovery. His research interests include microbiologically influenced corrosion, marine corrosion, metallic coatings, materials microstructure characterization and its effect on corrosion.

Dr Wayne Neil is a corrosion scientist in the Defence Science and Technology Group. His research interests include microbiologically influenced corrosion; corrosion of marine alloys including marine grade aluminium, copper nickel, stainless steels and high strength steels; environmentally friendly corrosion inhibitors; and corrosion of magnesium alloys.

Dr Grant McAdam is currently a Senior Research Scientist in Defence Science and Technology Group. His research interests include corrosion protection of magnesium and aluminium aerospace alloys, environmentally friendly corrosion inhibitors and corrosion protection schemes, fibre optic corrosion sensors, environment characterisation, and corrosion of marine alloys.

Assoc. Prof. Scott Wade joined Swinburne University of Technology in 2009 and is the Program Coordinator of the Biomedical Engineering undergraduate degree in the Faculty of Science, Engineering and Technology. He has worked on various aspects of fundamental corrosion research as well as collaborating with leading industry and research organisations to provide information and guidance on applied corrosion problems. His current research interests include microbiologically influenced corrosion, marine corrosion, accelerated low water corrosion, biofouling, and the development of corrosion/biofouling resistant coatings.

NOVEL SMART-RELEASE CHROME-FREE INHIBITING PIGMENTS FROM LAB INVESTIGATION TO INDUSTRIAL COATINGS

P. Dodds[1,2], G.Williams[2], C. Griffiths[1], P. Ansell[2] and A. Nielsen[2]
[1]Hexigone Inhibitors Ltd, Swansea, UK. [2]Swansea University, Swansea, UK.

SUMMARY: Smart-release chrome free inhibiting pigments were investigated as novel replacements for hexavalent chromate within primer systems of multi layer organic coating systems. The novel pigments move research away from sparingly soluble salts towards micro-reservoirs of inhibitors, that are released in response to environmental triggers. The inhibiting pigments were assessed using an in-situ scanning kelvin probe technique for their effectiveness at preventing cathodic delamination, on hot dipped galvanised steel substrates. The use of a model coating within a delamination cell experiment, allowed assessment of the novel pigments within 24 hours, with a spatially resolved map of the corrosion potential (E_{corr}) profiles providing delamination rate kinetics (K_{del}). The highest inhibition of delamination was shown by the benzotriazolate and magnesium (II) ion containing pigments. When used in synergy, the amount of each inhibitor required to give the high level of inhibition, was halved. The high through-put screening using the scanning Kelvin probe technique, led to the inclusion of the highly efficient pigments being formulated into architectural coil coated steel primers and assessed using ASTM-B117 and humidity testing. The results of 1000 hours humidity and neutral salt spray testing showed no sign of creep from the scribe or edge creep.

The current study suggests that a benzotriazolate ion-containing pigment systems are strong, environmentally acceptable contenders to replace hexavalent chromate, for the inhibition of corrosion on hot dip galvanised steel.

Keywords: Hot Dipped Galvanised Steel, Chrome-free, Smart-release, Protective Coating, Novel

1. INTRODUCTION

There has been substantial academic research to find a replacement for hexavalent chromium pigments within protective primer systems, which has recently become increasingly important, due to the restriction of use by the European Chemicals Agency (ECHA). The highly effective hexavalent chromium compounds are excellent corrosion inhibitors but are also carcinogenic, mutagenic, toxic and harmful to the environment and aquatic ecosystems[1].

The use of smart-release reservoirs within protective coatings has allowed the research to move away from sparingly soluble salts, to include chemicals that are not normally compatible with full organic coating systems[2]. The reservoirs hold onto the corrosion inhibitors until a trigger in the coating environment causes the release of the inhibitor, which then prevents further corrosion.

One of the current industry standards for chrome free pigments is zinc phosphate, with a new more soluble family of polyphosphates purported to be the best replacement. The protection offered by the phosphates and polyphosphates is thought to be via anodic passivation, where the phosphates/polyphosphates form a layer over the metal defect, preventing metal dissolution. Polyphosphate based corrosion inhibitor pigments are the third generation of phosphate based corrosion inhibitor systems. Strontium aluminium polyphosphate (SAPP) and zinc aluminium polyphosphate (ZAPP) have been studied to determine their corrosion inhibition compared to zinc phosphate[3]. It was realised, when using electrochemical impedance spectroscopy that both inhibitors performed better than zinc phosphate(ZP) [3]. The improved inhibition is thought to be due to a higher concentration of phosphate, as the polyphosphates will break down into orthophosphate. The effect of phosphate on reducing the corrosion rate is likely to be it reaction with zinc, to produce and insoluble precipitate of zinc phosphate on the metal surface[4]. There is also thought to be a protection effect of the aluminium, due to its increase in solubility at high pH. This increase in solubility would mean it is more readily available at the cathodic front, forming insoluble precipitates of aluminium hydroxide or as zinc aluminate, blocking the mass transport of sodium ions from the defect to the cathodic front.

The failure mechanism of coated hot dip galvanised steel, that this piece of work focusses on, is cathodic disbondment. The cause of cathodic disbondment is suggested to be caused by; base-catalysed polymer degradation, base-catalysed hydrolysis of interfacial bonds, dissolution of amphoteric metal oxide film (zinc) and attack on the metal-polymer bond by the reactive intermediates and hydrogen peroxide, produced by the cathodic oxygen reduction reaction. Assessment of the effectiveness of pigments at inhibiting cathodic disbondment of zinc and iron, can be performed using a scanning Kelvin probe technique and a delamination cell. The technique was pioneered by Stratman et. al.[5] and has been used as an experimental technique for a substantial number of investigations[6–15].

Previous work, investigating ion exchange resins loaded within corosion inhibitors, focused on metallic cations as the inhibitors, firstly in a cross-linked sulphonated polystyrene (Dowex®)[6], and improving on this work, using a resin composed of a co-polymer of divynl benzene/styrene with a sulphonic acid group (Amberlite®) [7]. The exchange capacity of the latter was three times that of the Dowex® product, and is thought to have increased the high efficacy of the Mg^{2+}/Ca^{2+} loaded pigments.

To investigate anion exchange resins, a divynl benzene/styrene co-polymer was used, with a quaternary ammonium functional group (Amberjet®). The resins are commonly found within water treatment and chemical synthesis, due to their ability to sequester anions from solution. The characteristic of the resin, to sequester anions from solution has been manipulated, so that the resins uptake and store the corrosion inhibiting anions.

2. EXPERIMENTAL DETAILS

2.1 Materials

Hot dip galvanised (HDG) steel samples consisting of 0.7 mm gauge mild steel coated on both sides with a 20 µm zinc layer containing 0.15 wt.% aluminium were supplied by Tata Steel UK. EDTA (ACS reagent, 99.4-100.6%), phenyl phosphonic acid (98%), sodium benzoate (reagent plus), sodium phosphate, sodium metavanadate, sodium chromate, sodium chloride, benzotriazole, Amberjet® 4200 Cl. All chemicals were obtained from the Sigma Aldrich Company.

2.2 Methods

A 0.25M stock solution of each of the organic anions and was made using deionised water. To dissolve the benzotriazole, drops of 0.1M NaOH were used to increase the pH to around 8.2. The solutions were added to the resin beads in the ratio of 100ml per 10g of Amberjet® 4200Cl resin beads. Each solution and bead mixture was agitated via stirring for 2 hours, after which the solution was decanted and replaced with fresh stock solution in the same 100ml per original 10g weight ratio. The solutions were stirred for a further 2 hours and again the solution was replaced for fresh stock of 0.25M target anion solution. The beads were left overnight under stirring to ensure maximum exchange had taken place. The beads were filtered using a Buchner filtration system and washed with 500ml of deionised water, to ensure any chloride solution was removed. The beads were dried for 12 hours at 40°C and then milled using a Retsch Planetary Ball Mill at 350 RPM for 1 hour and sieved through a 20µm mesh.

$$Mpig = \frac{\emptyset \cdot M_{pol} \cdot \rho_{pig}}{(1-\emptyset) \cdot \rho_{pol}} \quad (1)$$

Where Mpig is mass of pigment, Mpol is mass of polymer, ρpig is the density of pigment, ρpol is the density of polymer and $\emptyset$ is the required volume fraction of pigment.

The sample was prepared as described by Stratmann et al[16] and is detailed within [7].

The SKP reference probe was positioned 12mm away from the defect and was scanned towards and perpendicular to the coating-defect boundary. The SKP reference probe was a gold wire with a 125µm diameter, vibrated at 280Hz with an amplitude of 40µm. Scans were taken at 0hrs and every 60 minutes up to 72 hours. A 100µm gap between the reference probe and sample was used, with E_{KP} points measured at 20 per mm. The environmental chamber of the SKP allows a constant humidity of 95% R.H. with the temperature held at 25°C. Full details of the SKP equipment and calibration have been reported previously[17].

3. RESULTS

The combination of Amberjet® and inorganic anions proved to be either; detrimental to the coatings adhesion (figure 2), or offered no increased protection against cathodic delamination of HDG, against an unpigmented system (Figure 1). The combination of the resin and an effective corrosion inhibitor such as chromate (Figure 4, v), proved that the system is viable for protecting against cathodic delamination, and that it was the presence of certain inhibitor anions that had the deleterious effect. Organic anions were then explored as the anion to be captured with the matrix, offering corrosion protection when required.

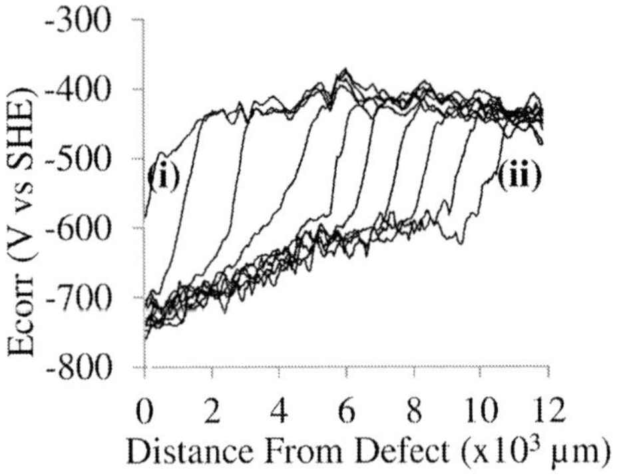

Figure 1 Ecorr Vs Distance from Defect Profile for PVB coated HDG. i = 240mins then every hour up until ii = 780 mins

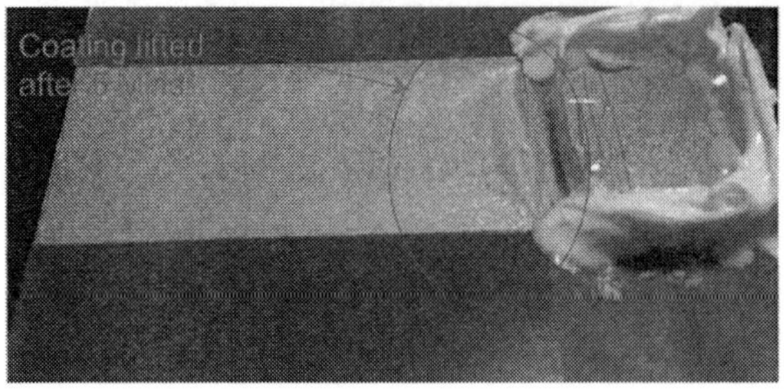

Figure 2 The delamination cell coated with PVB containing 0.1 PVF of Amberjet® with a phosphate counter-ion

The four anions chosen were EDTA, phenyl phosphonate, benzoate and benzotriazole. It has been previously demonstrated that when included directly in a coating, phenyl phosphonate shows excellent inhibition of cathodic delamination of coated hot dip galvanised steel [18], and that benzotriazole, when included in a anion exchange pigment, is highly efficient at inhibiting filiform corrosion of coated aluminium[19]. The initial screening of the organic anion containing pigments at 0.1 PVF, highlighted that the EDTA-Amberjet® had the same effect on adhesion as the phosphate-Amberjet® pigment. The inclusion of the pigment within the model coating led to rapid electrolyte ingression beneath the coating.

The highest performing inhibiting pigment was the combination of benzotriazole and Amberjet® 4200 Cl. The E_{corr} Vs distance profile in Figure 3 shows that there is an inhibition of the delamination cell establishing itself until 2520 minutes, after the electrolyte was added to the defect. The delamination rate of the cathodic front, once the cell has established itself, is sluggish. Figure 4 demonstrates the low rate of k_{del} for the delamination of coating that contains the benzotriazole/resin pigment, which is half that of the chromate containing pigment, even after inhibiting the cell from establishing itself, up to 2520 minutes.

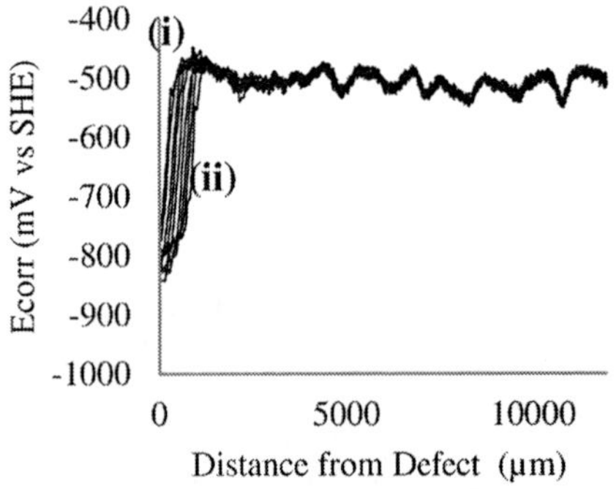

Figure 3 E_{corr} vs Distance profiles for 0.1 PVF benzotriazole in Amberjet® in PVB on HDG. (i) is 2520 mins then every two hours up until (ii)3960 mins

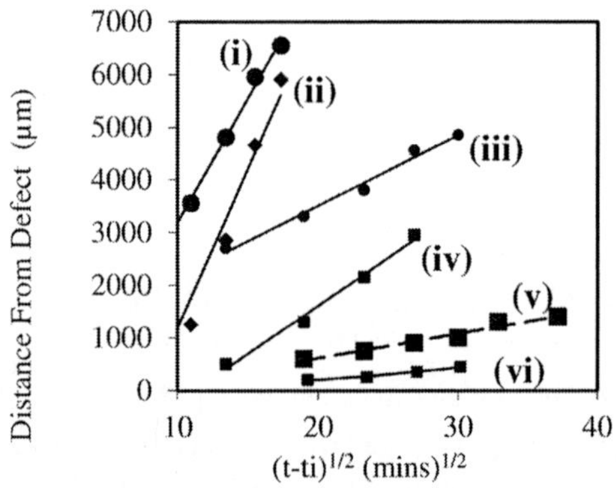

Figure 4 Delamination distance vs $t^{1/2}$ plots for PVB coatings containing (i) 0.1 PVF Amberjet®, (ii) unpigmented, (iii) Amberjet®- PhPhos, (iv) 0.1 PVF Amberjet®-benzoate, (v) Amberjet®-CrO_4^{2-}, (vi) Amberjet®-BTA

4. DISCUSSION

The high inhibition effects of the benzotriazole/resin pigments led to the inclusion of the pigments into full industrial coating systems. The pigment was included as a sole inhibitor into polyurethane (PU), polyester (PE) and acrylic primer formulations. The primers were applied to hot dipped galvanised steel with a chrome-free pre-treatment. The PE and PU primers were coated with their respective resin top coats and the acrylic primer coated with a 200µm plastisol topcoat. The multilayer organic coated HDG panels were scribed and subjected to 1000 hours of neutral salt spray testsing (ASTM-B117).

The images in Figure 5 show there is no progression of cathodic delamination of the coated HDG from 250-1000 hours. There is some white staining, which is thought to be the benzotriazole protection.

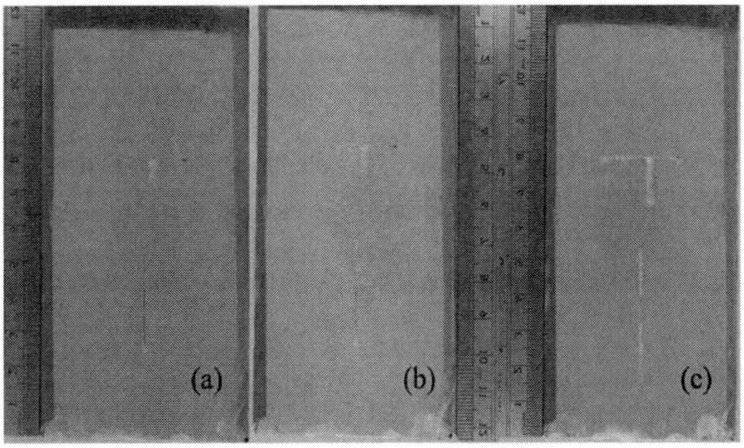

Figure 5 HDG panels with chrome-free pre-treatment, acrylic primer containing resin/benzotriazole pigment, top-coated with a 200μm plastisol. Images at (a)250 hours (b) 500 hours (c) 1000 hours of ASTM B117 testing

The images in Figures 5 and 6 demonstrate the compatibility of the exchange resin/benzotriazole pigment, with PE and PU resin systems. The work was completed by a customer within their independent UKAS accredited testing facilities. There is very little progression from the scribe, with the resin system inhibiting cathodic delamination. The cut-edge corrosion does not progress from the edge of the test panels and this is usually due to another failure mechanism, anodic dissolution. It is thought that the benzotriazole reduces the potential of the zinc on the cut-edge so that it no longer cathodically protects the steel, resulting in reduced anodic undermining of the coating.

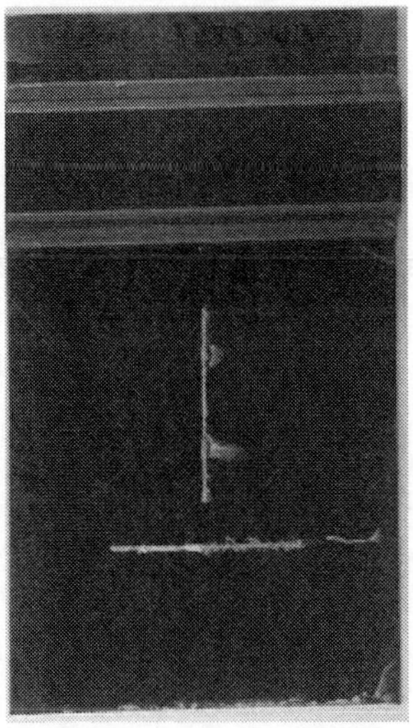

Figure 6 HDG panels with chrome-free pre-treatment, PE primer containing resin/benzotriazole pigment, top-coated with a 25μm PE topcoat. Image after 1000 hours UKAS accredited ASTM-B117 testing.

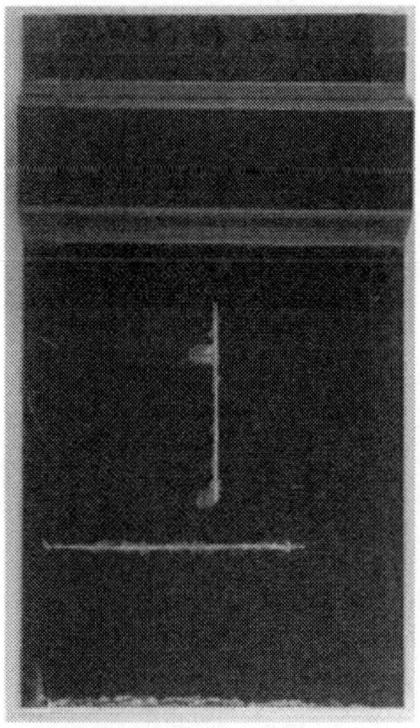

Figure 7 HDG panels with chrome-free pre-treatment, PU primer containing resin/benzotriazole pigment, top-coated with a 25μm PU topcoat. Image after 1000 hours UKAS accredited ASTM-B117 testing.

5. CONCLUSIONS

The work presented here demonstrates the scanning Kelvin probe as a rapid through-put experimental technique, enabling the researcher to screen potential new inhibitors within 24 hours. Once an excellent corrosion inhibitor is realised, it can be included into full industrial coating systems for traditional accelerated testing. The anion exchange resin was shown to be effective at inhibiting cathodic disbondment, when it is used in-conjunction with organic anions, the most effective anion being benzotriazole. It is envisaged that the exchange resin/benzotriazole combination can be used as a pigment to replace the toxic and carcinogenic strontium/zinc chromate pigments currently used.

6. ACKNOWLEDGMENTS

This work was made possible by the European Social Fund.

7. REFERENCES

1. Wise SS, Xie H, Fukuda T, et al (2014) Hexavalent chromium is cytotoxic and genotoxic to hawksbill sea turtle cells. Toxicol Appl Pharmacol 279:113–8.

2. Abdullayev E, Price R, Shchukin D, Lvov Y (2009) Halloysite tubes as nanocontainers for anticorrosion coating with benzotriazole. ACS Appl Mater Interfaces 1:1437–43.

3. Naderi R, Attar MM (2008) Electrochemical assessing corrosion inhibiting effects of zinc aluminum polyphosphate (ZAPP) as a modified zinc phosphate pigment. Electrochim Acta 53:5692–5696.

4. Naderi R, Attar MM (2009) The inhibitive performance of polyphosphate-based anticorrosion pigments using electrochemical techniques. Dye Pigment 80:349–354.

5. Stratmann M, Leng A, Fürbeth W, et al (1996) The scanning Kelvin probe; a new technique for the in situ analysis of the delamination of organic coatings. Prog Org Coatings 27:261–267.

6. Williams G, Geary S, McMurray HN (2012) Smart release corrosion inhibitor pigments based on organic ion-exchange resins. Corros Sci 57:139–147.

7. Dodds PC, Williams G, Radcliffe J (2016) Chromate-free smart release corrosion inhibitive pigments containing cations. Prog Org Coatings.

8. Williams G, McMurray HN (2001) Chromate Inhibition of Corrosion-Driven Organic Coating Delamination Studied Using a Scanning Kelvin Probe Technique. J Electrochem Soc 148:B377.

9. Williams G, McMurray HN, Worsley DA (2002) Cerium(III) Inhibition of Corrosion-Driven Organic Coating Delamination Studied Using a Scanning Kelvin Probe Technique. J Electrochem Soc 149:B154.

10. Williams G, McMurray HN, Bennett A (2014) Inhibition of corrosion-driven organic coating delamination from a zinc surface using polyaniline pigments. Mater Corros 65:401–409.

11. Williams G, McMurray H. (2016) Inhibition of corrosion driven delamination on iron by smart-release bentonite cation-exchange pigments studied using a scanning Kelvin probe technique. Prog Org Coatings.

12. McMurray HN, Coleman A, Watson T, Williams G (2007) SKP and Micrography Used to Determine the Coating Disbondment Mechanism during FFC on Iron. Meet Abstr MA2007-02:882.

13. Glover CF, Williams G (2013) Inhibition of Corrosion-Driven Organic Coating Delamination on Hot Dip Galvanized Steel by Phenyl Phosphonic Acid. ECS Trans 50:391–403.

14. Mabbett I, Elvins J, Gowenlock C, et al (2014) Addition of carbon black NIR absorber to galvanised steel primer systems: Influence on NIR cure of polyester melamine topcoats and corrosion protection characteristics. Prog Org Coatings 77:494–501.

15. Williams G, Holness RJ, Worsley DA, McMurray HN (2004) Inhibition of corrosion-driven organic coating delamination on zinc by polyaniline. Electrochem commun 6:549–555.

16. Fürbeth W, Stratmann M (2001) The delamination of polymeric coatings from electrogalvanised steel – a mechanistic approach. Corros Sci 43:207–227.

17. Williams G, Gabriel A, Cook A, McMurray HN (2006) Dopant Effects in Polyaniline Inhibition of Corrosion-Driven Organic Coating Cathodic Delamination on Iron. J Electrochem Soc 153:B425.

18. Glover CF, Subramanian R, Williams G (2015) In-Coating Phenyl Phosphonic Acid as an Etch-Primer Corrosion Inhibitor System for Hot Dip Galvanized Steel. J Electrochem Soc 162:C433–C441.

19. Williams G, McMurray HN (2012) Inhibition of filiform corrosion on organic-coated AA2024-T3 by smart-release cation and anion-exchange pigments. Electrochim Acta 69:287–294.

8. AUTHOR DETAILS

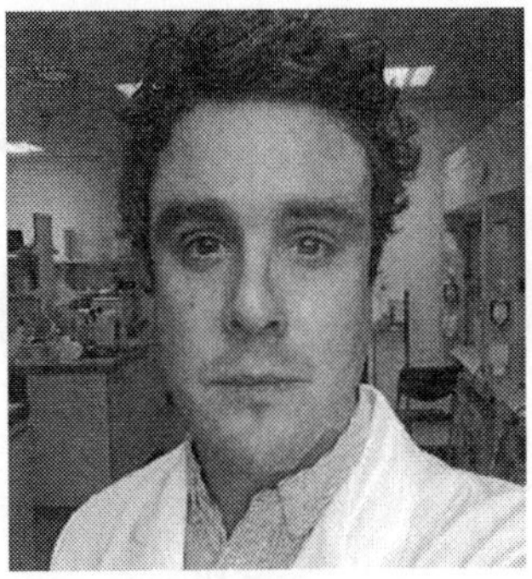

Dr. P. Dodds is the Chief Technical Officer/CEO at Hexigone Inhibitors Ltd. and a Royal Academy of Engineering Enterprise Fellow at Swansea University.

Prof. Geraint Williams is the Head of Materials Research Centre at Swansea University and is the Chief Scientific Advisor to Hexigone Inhibitors Ltd.

AN INVESTIGATION INTO THE ATMOSPHERIC CORROSION OF BRASS USING SODIUM CHLORIDE-CONTAINING WATER DROPLETS

A. Nielsen, G. Williams, N. Cooze, P.C. Dodds, P. Ansell & E. Strutt
Swansea University, Swansea, UK

SUMMARY: A combination of time-lapse photography and height-regulated scanning kelvin probe (SKP) potentiometry is used to study a cathodic secondary spreading phenomenon on brass, following exposure to $NaCl_{(aq)}$ containing water droplets. The rates at which these highly alkaline electrolytic films proceed from the droplet edge are measured and are shown to follow parabolic kinetics indicating that the process is controlled by the migration of Na^+. A solid Cl^- sequestering corrosion product is identified under the droplet. This, coupled with Na^+ migration from the droplet bulk into the secondary spread zone, lowers the solute concentration causing progressive evaporation under constant humidity; shown to be most extreme at lower $NaCl_{(aq)}$ droplet concentrations.

1. INTRODUCTION

Imitating gold at a fraction of the cost is an attribute of brass that has driven watchmakers and minters to utilise this alloy, providing their products with desirable aesthetics while being commercially feasible. Unfortunately, tarnishing under atmospheric conditions reveals the imitator all too soon. This study aims to simulate and measure a fundamental corrosion mechanism on brass that arises through exposure to $NaCl_{(aq)}$ containing water droplets.

As a $NaCl_{(aq)}$ water droplet becomes deficient in oxygen a differential aeration cell is established whereby cathodic oxygen reduction becomes preferable at the droplet circumference. This region draws sodium cations (Na^+) and associated water to counteract the electronegativity induced by an accumulation of hydroxide anions (OH^-) produced by the cathode. A resulting secondary spreading phenomenon is observed comprising dilute sodium hydroxide (NaOH). The thin, highly alkaline film originates at the droplet perimeter and proceeds radially from the droplet periphery.

The secondary spreading phenomenon has been analysed on metals such as iron and zinc whereby a time frame of the chemical composition inside and adjacent to the droplet is well documented (1–3). This work aims to combine height-regulated SKP potentiometry and time-lapse photography to measure the progression of secondary spreading after exposure to $NaCl_{(aq)}$ containing water droplets on brass; specifically, to visually identify the $NaCl_{(aq)}$ concentrations associated with secondary spreading, along with analysis of the droplet profiles and how they vary as a function of time.

2. EXPERIMENTAL PROCEDURE

2.1 Materials & Preparation

50×50×1mm brass coupons comprising 63 wt.% Cu and 37 wt.% Zn were purchased from Goodfellow Metals Ltd. Samples were ground using P1200-2500 SiC paper and polished with a 5μm alumina suspension, prior to being rinsed with distilled water and dried.

2.2 Time-lapse photography

To investigate the atmospheric corrosion using time-lapse photography solutions of 0.1 – 5 mol/dm^3 $NaCl_{(aq)}$ were used and a 1μl droplet of solution was applied to the brass surface using an Eppendorf repeater pipette. A reservoir of identical solution was included around the raised sample to provide a correct and constant humidity, thus preventing droplet evaporation. A Canon EOS 600D digital SLR camera was used to take close up images of the droplet at regular intervals; the spreading phenomenon is observable with the naked eye so time-lapse photography using a high-resolution lens served sufficiently to record the displacement of the spreading relative to the droplet edge. In order to do this a 0.5mm increment ruler was positioned next to the sample at equal height. The high reflectivity of the polished brass coupons made it necessary to angle the camera slightly to eliminate the reflection of the lens. A spirit level was used to ensure this angle was only made in the plane parallel to the ruler; it is for this reason that spreading effect measurements were only recorded in the plane perpendicular to that of the ruler.

2.3 Height-regulated SKP potentiometry

Similar investigation was carried out using an in-house designed SKP that takes its measurement using a 100μm diameter gold vibrating reference probe (4). Details of the SKP calibration method are described elsewhere (5). The humidity was maintained using electrolyte reservoirs positioned inside the SKP chamber of identical concentration to the droplet under study. The relative humidity (RH) was monitored using a data logging RH probe providing the humidity necessary to equilibrate the droplet for a given $NaCl_{(aq)}$ concentration. The probe was maintained at 100μm proud of the sample using a height regulated method that works by utilising the ratio of Kelvin currents at primary and secondary harmonic frequencies; used as a quantity to define the probe-to-sample distance. The principles of harmonic distortion and its association to topographic measurements are described fully elsewhere (6, 7). This approach enabled the mapping of the hemispherical geometry associated with a water droplet wetting a metal surface accurate to within 10μm.

3. RESULTS AND DISCUSSION

3.1 Time-lapse photography analysis

The secondary spreading phenomenon is studied on brass following exposure to $NaCl_{(aq)}$ containing water droplets of 1μl volume. Reasonably attributing the phenomenon to electrochemical effects relies on ensuring the droplet is stable prior to data capture. This is done using electrolytic reservoirs, of identical composition to the droplets, which fix the relative humidity such that the rate of evaporation and condensation are in equilibrium. Figure 3.1 illustrates a timeline of the secondary spreading over 60 hours.

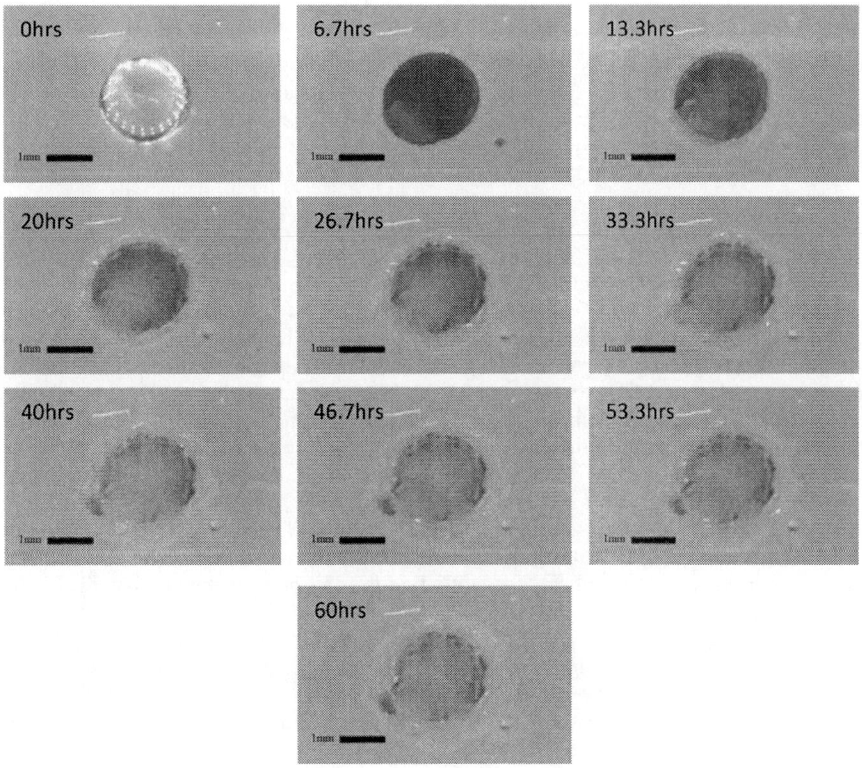

Figure 3.1 – Secondary spreading initiated and observed over 60 hours using a 1μl, 0.1mol/dm^3 $NaCl_{(aq)}$ droplet on a brass surface.

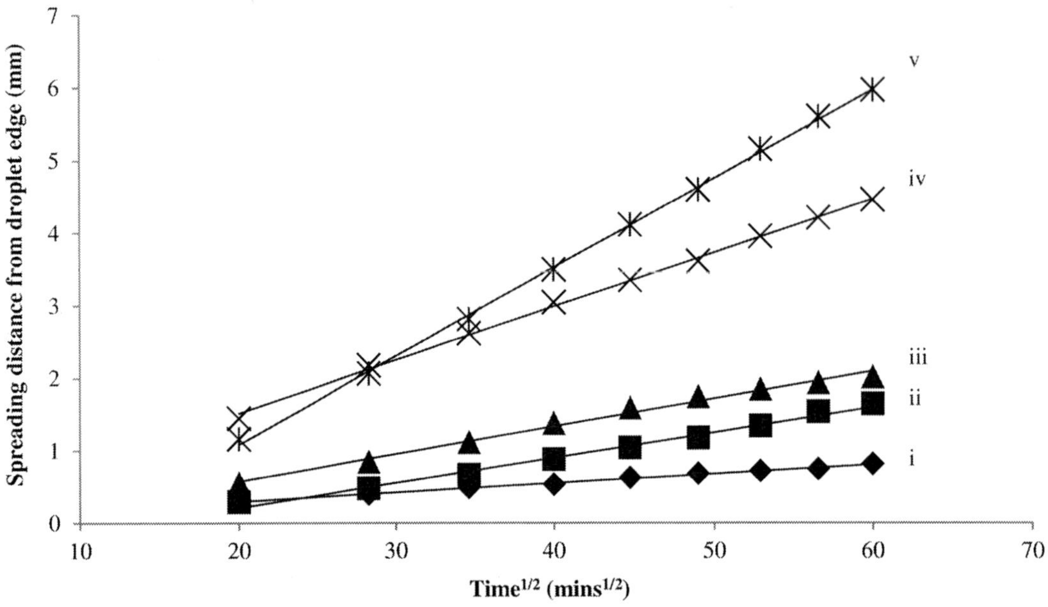

Figure 3.2 – Plots of secondary spreading distance measured from the droplet edge vs the square root of holding time for $NaCl_{(aq)}$ concentrations of i) 0.1, ii) 0.5, iii) 1, iv) 2 and v) 5mol/dm^3. The relative humidity is maintained in each case using an electrolyte reservoir of identical composition to the studied droplet. 1μl droplets are applied to the brass surface in each instance.

Plots of the time dependent secondary spreading from the droplet edge, recorded using close-up time-lapse photography are shown in Figure 3.2 The rates of secondary spreading observed at $NaCl_{(aq)}$ concentrations ≤0.1mol/dm^3 are negligible, while the highest recorded rates of spreading are associated with 5 mol/dm^3 $NaCl_{(aq)}$ droplets. The radial growth of spreading is shown to be linear vs square root of holding time, suggesting that the process is most likely controlled by the migration of Na^+ species from the droplet bulk to the edge of the spreading zone.

3.2 Height-regulated SKP droplet profile analysis

Using height-regulated SKP potentiometry it is possible to measure the droplet topography accurate to within 10μm. A drying effect has been captured using this non-contact height profiling technique whereby the topography of the $NaCl_{(aq)}$ containing water droplet has been characterised as a function of time. For each studied concentration, the droplet maintains a stable profile within 3 hours. Significant drying has been measured by periodically height profiling a 0.1 mol/dm^3 $NaCl_{(aq)}$ droplet. The change in volume is significant compared to that observed in higher $NaCl_{(aq)}$ concentrations. Figure 3.3 demonstrates the changing droplet geometry over the course of 3 hours for droplets with varying $NaCl_{(aq)}$ concentrations. It is shown that a 0.1 mol/dm^3 droplet essentially fully dries over the course of 3 hours. However, as the $NaCl_{(aq)}$ concentration is raised less drying is observed, such that negligible drying occurs at $NaCl_{(aq)}$ concentrations ≥ 2 mol/dm^3. This information is shown graphically in Figure 3.4 by plotting the change in maximum recorded height for each droplet concentration over a 3-hour exposure time to atmospheric conditions vs $NaCl_{(aq)}$ droplet concentration. It is shown that when the droplet concentration is 0.1 mol/dm^3, a maximum height change of 495μm is measured over 3 hours. Conversely, when studying droplet concentrations of 5 mol/dm^3 a maximum height change of 22.5μm is measured over 3 hours. It is proposed that the drying effect (observed over the $NaCl_{(aq)}$ concentrations investigated) is a direct response to the secondary spreading along with formation of Cl^- sequestering corrosion product inside the confines of the droplet.

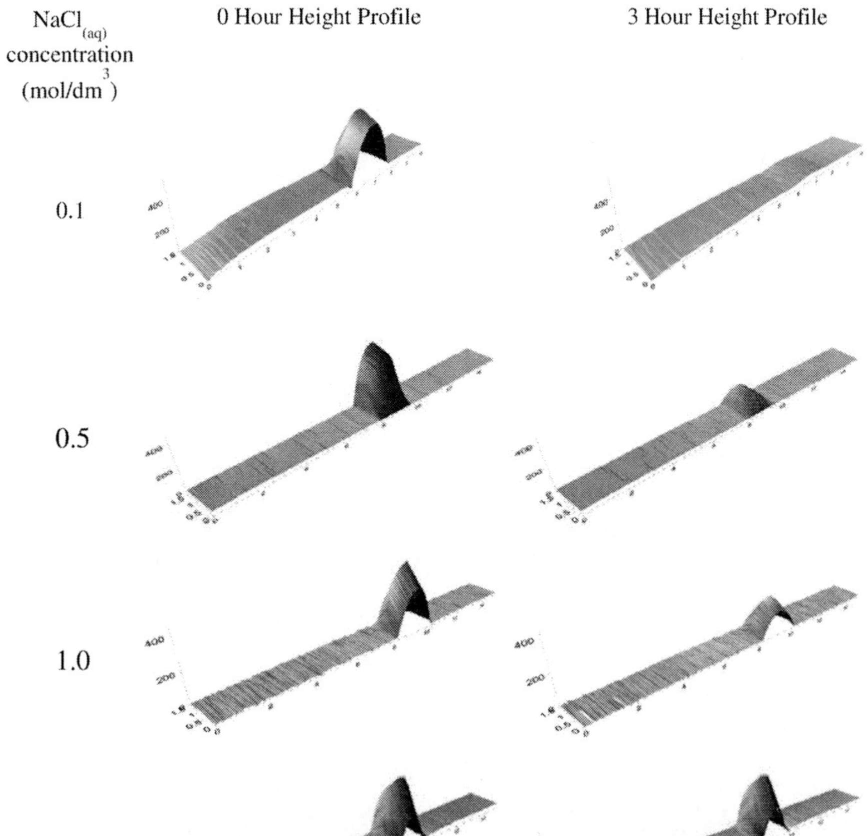

Figure 3.3 - SKP non-contact height profiles for 1μl droplets measured at 0 and 3 hours after being applied to the surface of brass, studied over a 0.1 – 5 mol/dm^3 concentration range.

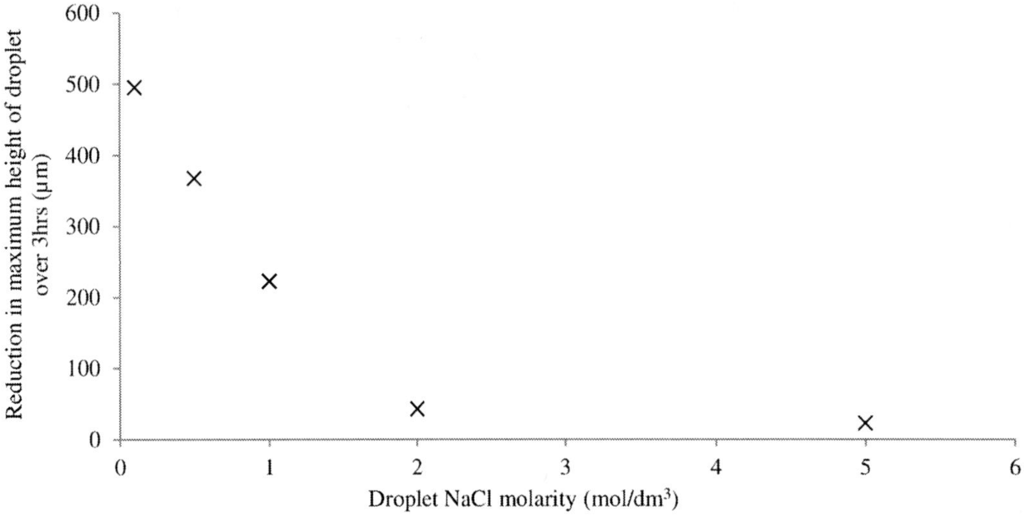

Figure 3.4 – A chart describing the variation in maximum droplet height from t = 0 → 3 hours as a function of droplet $NaCl_{(aq)}$ concentration.

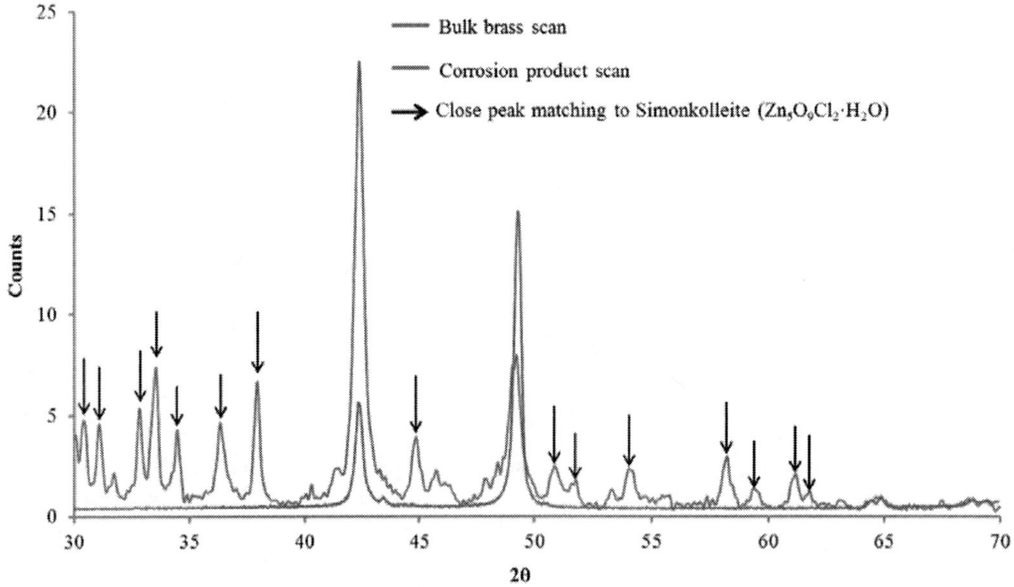

Figure 3.5 – GI-XRD spectra for bulk brass (shown in red) and residual corrosion product underneath droplet (shown in blue).

3.3 Mechanism causing droplet drying

Grazing-incidence X-ray diffraction (GI-XRD) has been used to identify the crystalline structure of the corrosion product found underneath the droplet at the brass surface. The red spectrum displayed in Figure 3.5 is measured from the polished bulk of the brass specimen. Peaks are observed at approximately 47° and 49° and have been closely related to a brass phase and pure copper respectively. The blue spectrum is measured from the corrosion product that forms underneath the droplet on the surface of the brass. Lots of close peak matching to a chloride-containing zinc corrosion product, Simonkolleite ($Zn_5O_9Cl_2 \cdot H_2O$), is observed. It has been shown in Figure 3.4 that a significant reduction in maximum height is observed over 3 hours when investigating 0.1 mol/dm^3 $NaCl_{(aq)}$ water droplets compared to a negligible variation in maximum height over 3 hours associated with highly concentrated droplets. It is proposed that the measured height change is a consequence of droplet evaporation – consider a $NaCl_{(aq)}$ containing water droplet positioned on the surface of a brass specimen at t = 0 in equilibrium with its humidified environment:

- Initially a differential aeration cell is setup whereby the edges of the droplet become preferential sites for oxygen reduction because the O_2 diffusion path length is least arduous.
- Consequently, an accumulation of OH- ions collect at the perimeter.
- Na^+ ions from the droplet bulk move to counteract the electronegativity pulling associated H_2O (due to its dipole moment) towards the droplet perimeter.
- A very thin, highly alkaline electrolytic film proceeds from the droplet edge – the growth of which is dictated by the net migration of Na^+ from the droplet bulk.
- The formation of a Cl^- sequestering corrosion product (Simonkolleite) occurs in parallel to the secondary spreading.
- Both processes lower the solute content in the droplet bulk and the water activity increases.
- The droplet is no longer in thermodynamic equilibrium with the humidity provided via reservoir of identical $NaCl_{(aq)}$ concentration to that of the droplet at t = 0.
- The droplet must therefore evaporate to reduce its water activity and re-establish equilibrium with its environment.

At low $NaCl_{(aq)}$ concentrations the relative effect that the electrochemical processes have with regards to the reduction in droplet bulk solute content is far more significant than that of highly concentrated droplets, and therefore significant drying is observed at low $NaCl_{(aq)}$ droplet concentrations.

The schematic shown in Figure 3.6 illustrates how the secondary spreading draws Na^+ from the droplet bulk, forming a thin film of dilute NaOH along with precipitation of Na_2CO_3. The formation of a zinc corrosion product (Simonkolleite) and its action to sequester Cl^- from the droplet bulk is also shown.

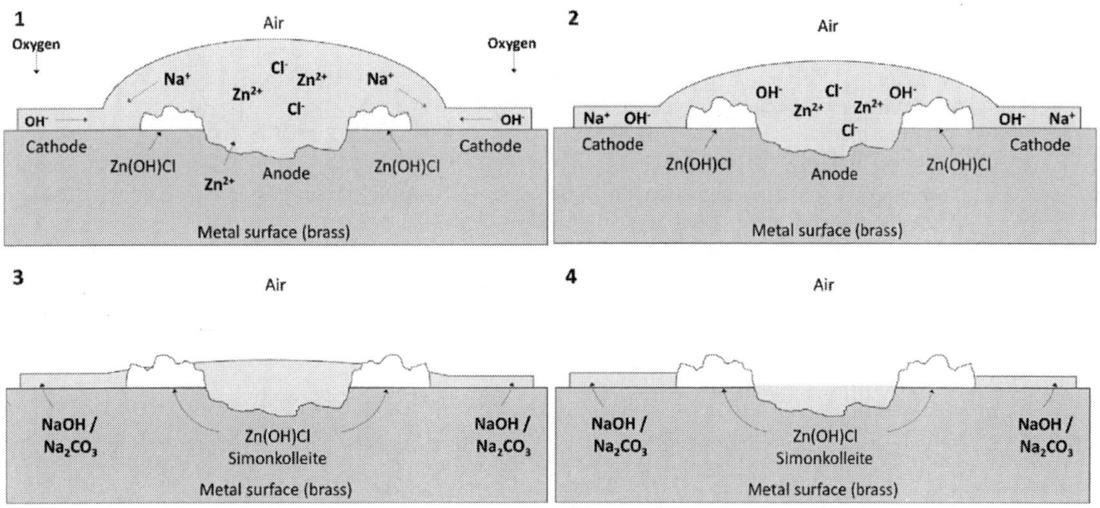

Figure 3.6 - Schematic describing the mechanism of the secondary spreading process on brass after exposure to a low $NaCl_{(aq)}$ concentration electrolyte droplet under atmospheric conditions.

4. CONCLUSIONS

This investigation details the corrosion induced mechanistic behaviour of a brass surface following exposure to sodium chloride containing water droplets. A time dependent visual analysis has provided kinetics associated with the radial spreading of a thin, highly alkaline electrolytic film at various $NaCl_{(aq)}$ concentrations. It can be deduced that when the spreading is plotted vs the square root of holding time, a linear trend is observed. The parabolic kinetics indicates that the secondary spreading process is controlled by the net Na^+ migration from the droplet bulk to the extremities of the secondary spread zone. A height regulated scanning Kelvin probe study enabled side profiles of the droplets to be generated as a function of time. A 0.1 mol/dm^3 $NaCl_{(aq)}$ droplet under equilibrated humidity experiences a significant reduction in height over 3 hours; this is attributed to droplet evaporation. If the droplet concentration is increased to 5 mol/dm^3 significantly less drying is observed over the same time period. A grazing incidence X-ray diffraction analysis found that a considerable portion of the corrosion product forming underneath the main droplet comprised Simonkolleite. It has been proposed that the secondary spreading acts to drive Na^+ from the droplet bulk while the formation of Simonkolleite entraps aqueous Cl^- within a solid corrosion product. Both processes cause a reduction in droplet solute concentration, leading to consequential evaporation of the droplet most observable at lower $NaCl_{(aq)}$ concentrations. This is done so that a thermodynamic equilibrium is met once again between the humidified environment and the droplet.

REFERENCES

1. Neufeld a. K, Cole IS, Bond a. M, Furman S a. The initiation mechanism of corrosion of zinc by sodium chloride particle deposition. Corros Sci. 2002;44(3):555–72.

2. Zhang J, Wang J, Wang Y. Micro-droplets formation during the deliquescence of salt particles in atmosphere. Corrosion. 2005;61(12):1167–72.

3. Tsuru T, Tamiya KI, Nishikata A. Formation and growth of micro-droplets during the initial stage of atmospheric corrosion. Electrochim Acta. 2004;49(17-18):2709–15.

4. Williams G, McMurray N. Latent fingermark visualisation using a scanning Kelvin probe. Forensic Sci Int. 2007;167:102–9.

5. Rossi S, Fedel M, Deflorian F, del Carmen Vadillo M. Localized electrochemical techniques: Theory and practical examples in corrosion studies. Comptes Rendus Chim. 2008;11(9):984–94.

6. Cheran LE, McGovern ME, Thompson M. Surface immobilized biochemical macromolecules studied by scanning Kelvin microprobe. Faraday Discuss [Internet]. 2000;(116):23–75. Available from: http://www.ncbi.nlm.nih.gov/pubmed/11197481

7. Searle JR, Glover CF, Khan K, Wilson BP, Williams G, Worsley D a. Electrochemical Scanning Techniques for the Examination of Bi-Metallic Coins. ECS Trans [Internet]. 2015;64(28):47–59. Available from: http://ecst.ecsdl.org/cgi/doi/10.1149/06428.0047ecst

CHALLENGES AND OPPORTUNITIES FOR CHROMIUM FREE INHIBITORS FOR GALVANISED STEEL

O. Gharbi[1], C. Smith[2], R. Donahue[1], N. Birbilis[1]
[1]Department of Materials Science and Engineering, Monash University, Victoria, Australia
[2] PPG, McNaughton Rd, Clayton, Australia

SUMMARY: In this work, the corrosion inhibition mechanisms of several inhibitors applied to galvanised steel were investigated. Herein, electrochemical measurements such as potentiodynamic polarisation testing were used to evaluate the level of inhibition for each system. In addition, scanning electron microscopy and EDX mapping were performed on the test samples to provide a mechanistic insight into the performance of inhibitor systems.

1. INTRODUCTION

Galvanised steels are extensively used in various applications such as structural exteriors, roofing, cladding, etc. [1]. The galvanised layer, nowadays evolved from pure Zn to nominally be a Zn-Al or Al-Zn alloy, to more contemporary Al-Zn-Mg alloy, acts to sacrificially protect the steel substrate. Where improved corrosion resistance, aesthetics and longevity of the galvanised layer are required, a primer and topcoat are also applied to galvanised steel [1,2].

To date, primers typically involve functional additions of hexavalent (or trivalent) chromium containing compounds to enhance the corrosion resistance provided by the coating system [3,4]. The so called self-healing properties provided by such chromium-containing coatings are known to halt further surface corrosion, as they are purported to form a protective film upon the alloy surface [4,5]. However, hexavalent chromium has a detrimental impact on health [6] and the environment, and therefore needs to be replaced by suitable alternatives [4]. To this end, the substitution of the hexavalent chromium with the less toxic trivalent chromium, became rather widespread in the past decade [7–9]. However, recent work has indicated that, during the coating growth, trivalent chromium was oxidized by hydrogen peroxide (generated by the oxygen reduction), forming hexavalent chromium in the coating [8]

The corrosion protection afforded by chromate inhibition is the benchmark for next generation chromium-free systems to match or exceed. Herein, we present some results related to the ability of Cr-free inhibitors to impart protection of galvanised steel. In this context, the inhibitors are used to retard the dissolution (or passivate) the galvanised coating. In spite of the galvanising being sacrificial, increasing the longevity of galvanising is industrially important (and routine). Furthermore, most coated commercial steels are in fact coated galvanised steels; and to this end the primer coating is nominally inclusive of a corrosion inhibitor for life extension of the coating system. The testing of inhibitors herein (in aqueous electrolytes) is intended to serve as a proxy for inhibitor performance when the inhibiting compound is incorporated into a primer coating. Attention is given to the selection of several Cr-free systems, such as a rare earth based inhibitor [10], molybdenum (molybdate) [11], permanganate, and the organic compound benzotriazole (BTA) [12,13].

2. EXPERIMENTAL

Commercially produced Al-Zn-Mg based (galvanised) coatings upon mild steel (coil) sheets were used in this study. Prior to experiments / testing, specimens were ultrasonically cleaned in ethanol and dried under compressed air. Separate anodic and cathodic potentiodynamic polarisation testing was conducted in the range of -0.02V vs E_{corr} to a potential corresponding to 1 mA cm^{-2} for the anodic scan and 0.02V vs E_{corr} to -2V vs. E_{ref} for the cathodic scan. A Bio-Logic VMP-3 potentiostat was used herein, with testing conducted at a scan rate of 1 mV s^{-1}. Polarisation tests were conducted using a PAR K0235 flatcell with a test electrolyte of 10 mM NaCl. All reagents used for the experiments herein were reagent grade.

Surface characterisation was conducted upon the galvanised coatings prior to, and after, polarisation testing. An FEI Quanta-FEG scanning electron microscope (SEM) was used, in conjunction with energy dispersive X-ray analysis (EDX) mapping in order to study surface composition and elemental distribution.

3. RESULTS AND DISCUSSION

The first portion of this study was devoted to the assessment of the selected inhibiting strategies. Galvanised steel sheets were exposed to 10 mM NaCl in the presence of several compounds including benzotriazole (BTA), Cerium-based, Molybdenum-based and Manganese-based inhibitors at 10 mM for a period of 30 mins. Each of these systems was previously purported in studies to serve as promising inhibitors for Mg, Al or steel applications [12,14,15], and were selected for trialling on this basis.

Potentiodynamic polarisation scans that were performed after 30 mins immersions to assess the inhibition characteristics of each system are presented in **Fig 1.**

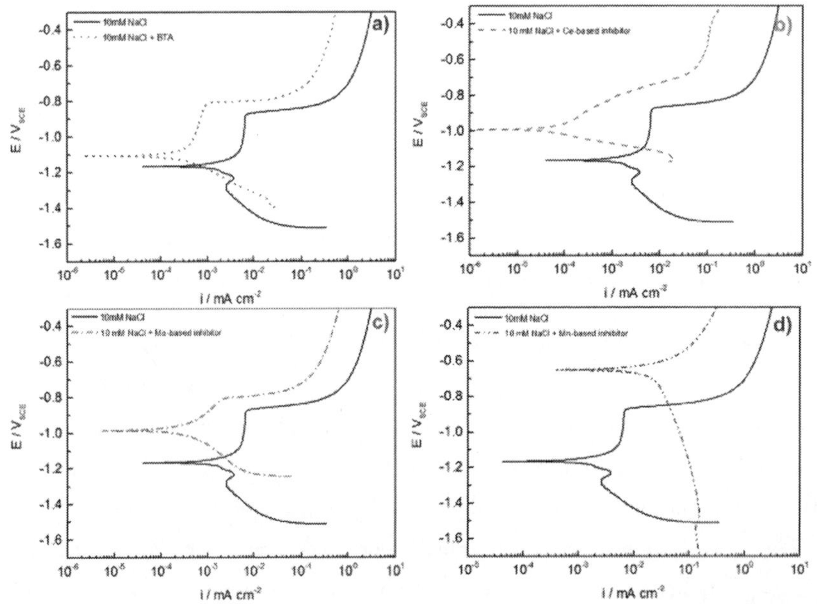

Figure 1: Potentiodynamic polarisation results of the Al-Zn-Mg coated (galvanised) steel in the presence of 10 mM NaCl and a) Benzotriazole (BTA), b) Ce-based inhibitor, c) Mo-based inhibitor and d) Mn-based inhibitor.

The inhibiting influence of BTA is clearly discernible from the potentiodynamic polarisation data following 30 mins exposure to the inhibitor containing electrolyte; whereby the anodic current over the range of potentials from ~-1.1 to -

$0.8V_{SCE}$ was reduced by approximately one order of magnitude (**Fig. 1 a**). However, there was no evidence to suggest that BTA can inhibit the cathodic reaction; noting that cathodic inhibition is also a hallmark of the performance of chromate based systems [3,4]. It has been recently reported that BTA was able to impart significant corrosion resistance for Mg-alloys, where BTA could suppress Mg dissolution – by promoting the nucleation of a highly crystalline $Mg(OH)_2$ [12]. The potentiodynamic polarisation testing revealed that the Ce compound was capable of reducing both the cathodic and anodic kinetics by approximately 2 orders of magnitude (**Fig. 1 b**). Prior works have indicated that a number of Ce-based compounds (including Ce-nitrate [16], Ce-dibutyl phosphate [17], and Ce-chloride [18,19]), can act as effective cathodic inhibitor for Al alloys via selectively precipitating upon on the cathodic sites. Testing herein also revealed that the molybdate inhibitor was capable of reducing both the cathodic and anodic current densities by approximately 1 order of magnitude (**Fig. 1 c**). Indeed, several researchers have confirmed the beneficial effect of Mo-based compounds such as sodium molybdate [20–22] on corrosion suppression of Zn and Al alloys [11,22]. The observations herein can be rationalised by such prior studies, as the galvanised layer primarily consists of Zn and Al.

In contrast, the permanganate-based inhibitor revealed a detrimental effect, concomitant with an increase of the cathodic and anodic current densities observed from potentiodynamic polarisation testing (**Fig. 1 d**). To summarise the results obtained from potentiodynamic polarisation testing **Fig. 2** reveals the relative inhibiting performance as estimated from the comparison of corrosion current densities determined by Tafel extrapolation [23].

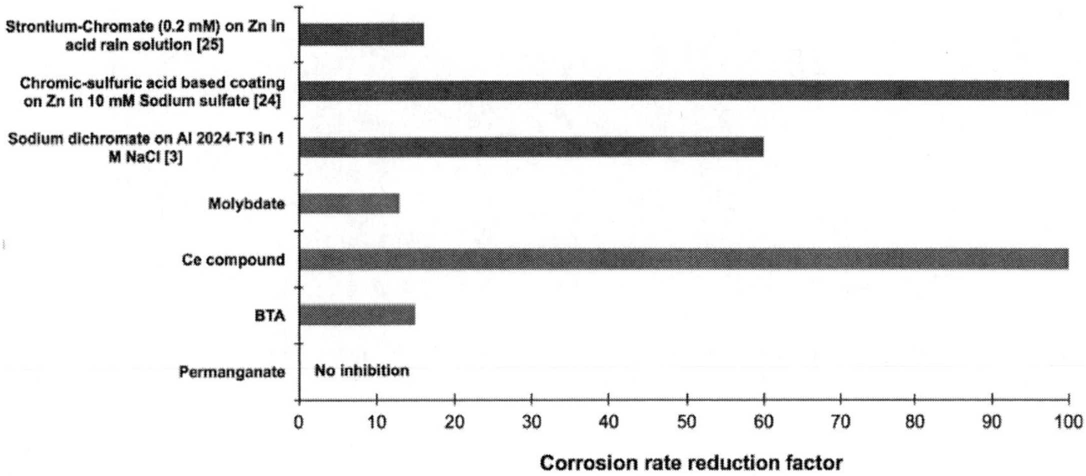

Figure 2: Summary of corrosion rate reduction of the inhibited systems relative to uninhibited systems, information regarding chromates is a range that is taken from [4,24,25]**. It is noted that the chromate values represent a variety of like substrates.**

Scanning electron microscopy in addition to energy dispersive X-ray spectroscopy (EDX) mapping was carried out on specimens prior to, and following exposure to the electrolytes described above. A selection of such images are presented in **Figs. 3 to 6,** and aid in the revelation of surface morphology and chemical distribution which serves to reveal the mechanistic function associated with the inhibiting systems.

The SEM micrograph and EDX mapping of the surface prior to testing (taken in backscattered electron mode) reveals a highly heterogeneous microstructure, where three distinct areas can be identified: bright tunnels which is attributed to Zn, surrounded by an Al based matrix and finally, dark areas that are periodically observed across the surface and are found to be Mg-Si based phases (**Fig. 3**). From the microstructural observations, it is expected that the individual electrochemical reactivity of each of these phases would inevitably yield to localized corrosion.

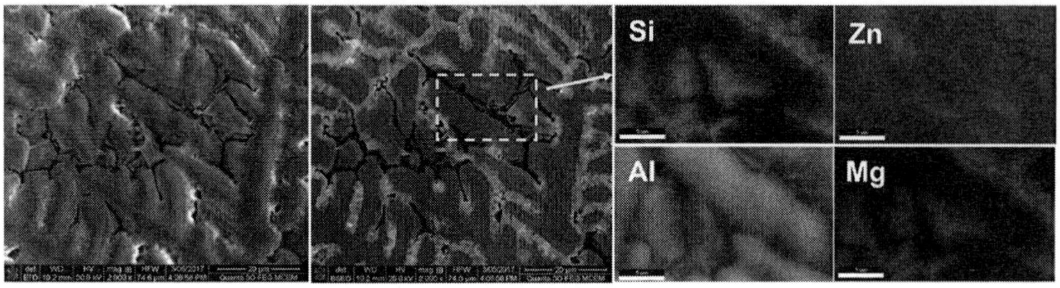

Figure 3: Scanning electron micrograph of the galvanized layer in backscattered electron mode and EDX map of the selected zone emphasizing the heterogeneous structure of the galvanised layer.

The SEM image of the galvanised layer after polarisation testing in 1 mM NaCl is displayed in **Fig 4**. As readily seen in the EDX mappings and the BSE image, the surface shows clear evidence of localised corrosion: the EDX mapping reveals that the Mg-Si was heavily attacked during the polarisation test. On the other hand, with the Al signal, the EDX map does not highlight any kind of preferential attack of this phase, which is believed to remain intact after the corrosion test. The shape of the Zn rich tunnels is considerably narrower, suggesting that those areas underwent active dissolution during the test. Interestingly, the O signal correlates with Zn and seems, to a lesser extent, distributed around the Mg-Si phase. These observations indicate the formation of a Zn and Mg/Si oxide on the surface.

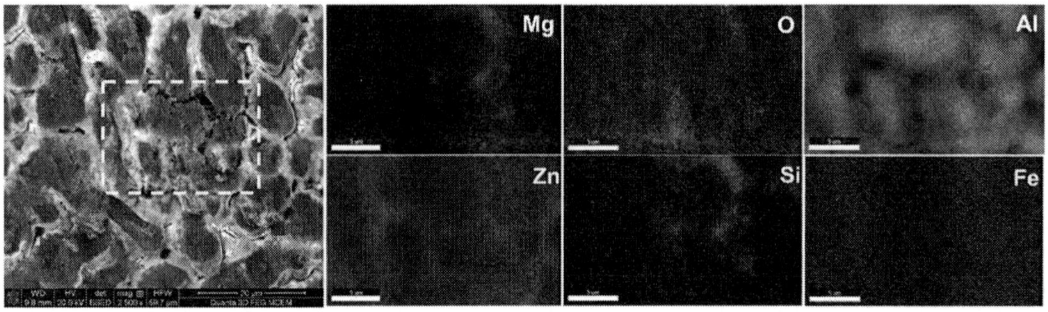

Figure 4: Scanning electron micrograph of the galvanised layer taken in BSE mode and a EDX map of the selected zone showing localised corrosion around the Mg-Si zones and Zn areas.

The SEM micrograph in BSE mode after polarisation testing with the presence of Manganese–based inhibitor and a EDX map of the surface are displayed in **Fig. 5**. Interestingly, O and Mn signals correlate and are predominantly found around the Zn/Mg-Si interface. Following the test, Al signal remains significant and does not coincide with O signal, strongly suggesting that Al rich areas were not oxidized during the test. According to these results, Mn-based oxides seem to preferentially precipitate on the Mg-Si and Zn rich areas which are expected to be the most reactive zones of the layer.

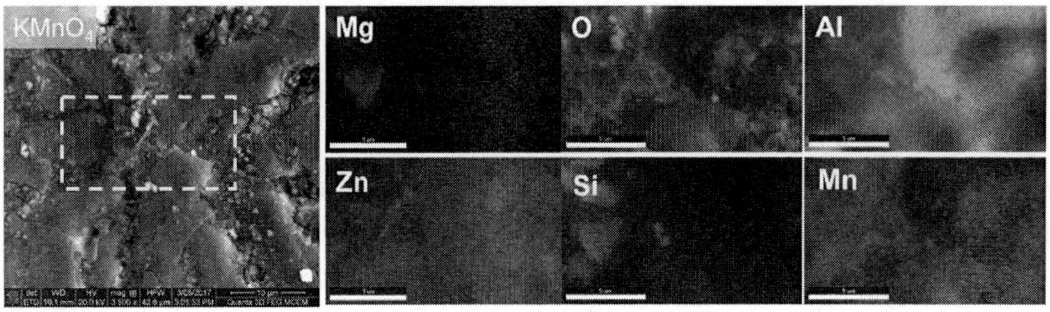

Figure 5: SEM micrographs and EDX mapping of the surface after polarisation testing in the presence of an Mn-based inhibitor.

The surface after polarisation with the presence of BTA and the corresponding EDX map are presented in **Fig. 6**. The surface morphology and topography is shown by the SEM micrographs taken in secondary electron mode, which reveals significant changes to the surface morphology, most likely influenced by the addition of BTA. The formation of a thin layer, covering principally the active Zn phases (and to some extent Al phases) is revealed by the secondary electron image. This result is in good agreement with a prior studies [26,27] where it was suggested that the chemisorption of BTA^- anions layer on Zn, Fe or Cu was the predominant inhibition mechanism encountered.

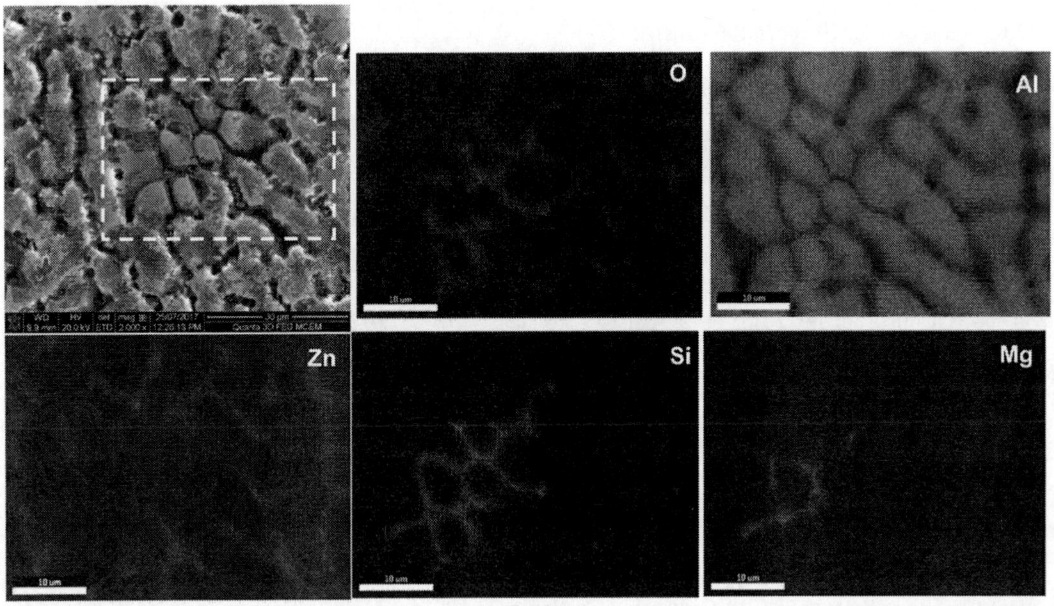

Figure 6: Secondary electron image of the galvanised steel substrate after polarisation test with the presence of BTA.

On the other hand, the EDX map shows partial dissolution of Mg indicating that BTA did not sufficiently promote the formation of a protective Mg oxide. These observations could be explained by the pH of the electrolyte (pH = 6.5) creating *a priori*, a non-favorable environment for Mg oxide precipitation at the first stages of the exposure. However, a certain amount of Mg appears to remain on the surface and is in fact strongly correlated to O and Si signal highlighting the formation of a significant Mg-Si oxide layer on the surface.

4. CONCLUSIONS

The reactivity of Al-Zn-Mg based (galvanised) coatings on steel in the presence of several inhibitors was studied using potentiodynamic polarisation and scanning electron microscopy. Results revealed that it was possible to evaluate the propensity of inhibition for the inhibiting compounds tested via the change in corrosion current for each inhibitor system relative to the uninhibited scenario. Such electrochemical tests are, granted, instantaneous tests and may not reveal long term effects, however electrochemical testing is an important methodology on the basis that testing of coated steels is not readily possible by longer term methods such as mass loss. Results herein were also supplemented by with microstructural characterisation and EDX mapping. It was determined that Ce-based compounds as well as Mo-based compounds were shown to act as both cathodic and anodic inhibitors (i.e. mixed inhibitors). Conversely, benzotriazole (BTA) appeared to principally reduce anodic kinetics by one order of magnitude potentially by promoting Zn- or Mg-oxide (or hydroxide) formation. The Mn-based compound tested (potassium permanganate) however, was shown to dramatically accelerate the corrosion of the substrate used herein (the Al-Zn-Mg based (galvanised) coating). Of the inhibitors explored, the performance relative to chromate based systems was promising, which was discussed in the context of future work.

5. REFERENCES

[1] D. Lindström, I. Odnevall, Long-term use of galvanized steel in external applications . Aspects of patina formation , zinc runoff , barrier properties of surface treatments , and coatings and environmental fate, Env. Monit Assess. 173 (2011) 139–140. doi:10.1007/s10661-010-1377-8.

[2] P.R. Roberge, R. Pierre, Handbook of Corrosion Engineering, McGraw-Hil, 1999.

[3] I.. Zin, R.. Howard, S.. Badger, J.. Scantlebury, S.. Lyon, The mode of action of chromate inhibitor in epoxy primer on galvanized steel, Prog. Org. Coatings. 33 (1998) 203–210. doi:10.1016/S0300-9440(98)00056-3.

[4] M.W. Kendig, R.G. Buchheit, Corrosion inhibition of aluminum and aluminum alloys by soluble chromates, chromate coatings, and chromate-free coatings, Corrosion. 59 (2003) 379–400. doi:10.5006/1.3277570.

[5] S. Thomas, N. Birbilis, M.S. Venkatraman, I.S. Cole, Self-repairing oxides to protect zinc : Review , discussion and prospects, Corros. Sci. 69 (2013) 11–22. doi:10.1016/j.corsci.2013.01.011.

[6] I. MONOGRAPHS, Evaluation of Carcinogenic Risks to Humans - Chromium, Nickel and Welding, 1990.

[7] K. Cho, V. Shankar Rao, H. Kwon, Microstructure and electrochemical characterization of trivalent chromium based conversion coating on zinc, Electrochim. Acta. 52 (2007) 4449–4456. doi:10.1016/j.electacta.2006.12.032.

[8] J.T. Qi, T. Hashimoto, J.R. Walton, X. Zhou, P. Skeldon, G.E. Thompson, Trivalent chromium conversion coating formation on aluminium, Surf. Coatings Technol. 280 (2015) 317–329. doi:10.1016/j.surfcoat.2015.09.024.

[9] L. Li, A.L. Desouza, G.M. Swain, Effect of Deoxidation Pretreatment on the Corrosion Inhibition Provided by a Trivalent Chromium Process (TCP) Conversion Coating on AA2024-T3, J. Electrochem. Soc. 161 (2014) 246–253. doi:10.1149/2.031405jes.

[10] J. Mardel, S.J. Garcia, P.A. Corrigan, T. Markley, A.E. Hughes, T.H. Muster, et al., Progress in Organic Coatings The characterisation and performance of Ce (dbp) 3 -inhibited epoxy coatings, Prog. Org. Coatings 70. 70 (2011) 91–101. doi:10.1016/j.porgcoat.2010.10.009.

[11] G.M. Treacy, G.D. Wilcox, M.O.W. Richardson, Behaviour of molybdate-passivated zinc coated steel exposed to corrosive chloride environments, J. Appl. Electrochem. 29 (1999) 647–654. doi:10.1023/A:1026461924663.

[12] J.-L. Wang, C. Ke, K. Pohl, N. Birbilis, X.B. Chen, The Unexpected Role of Benzotriazole in Mitigating Magnesium Alloy Corrosion: A Nucleating Agent for Crystalline Nanostructured Magnesium Hydroxide Film, J. Electrochem. Soc. 162 (2015) C403–C411. doi:10.1149/2.0781508jes.

[13] C. Dodds, G. Williams, P. Ansell, C. Glover, A Novel, Highly Effective, Chrome-Free, Smart-Release, Economically Viable Corrosion Inhibitor- from Lab Kinetics through to Industrial Coating Incorporation, in: Act. Self-Healing Org. Coatings, 2016: p. 1.

[14] S.B. Madden, J.R. Scully, Inhibition of AA2024-T351 Corrosion Using Permanganate, J. Electrochem. Soc. 161 (2014) 162–175. doi:10.1149/2.075403jes.

[15] M.F. Montemor, D. V Snihirova, M.G. Taryba, S. V Lamaka, I.A. Kartsonakis, A.C. Balaskas, et al., Evaluation of self-healing ability in protective coatings modified with combinations of layered double hydroxides and cerium molibdate nanocontainers filled with corrosion inhibitors, Electrochim. Acta. 60 (2012) 31–40. doi:10.1016/j.electacta.2011.10.078.

[16] M.L. Zheludkevich, R. Serra, M.F. Montemor, K.A. Yasakau, I.M.M. Salvado, M.G.S. Ferreira, Nanostructured sol-gel coatings doped with cerium nitrate as pre-treatments for AA2024-T3 Corrosion protection performance, Electrochim. Acta. 51 (2005) 208–217. doi:10.1016/j.electacta.2005.04.021.

[17] N. Birbilis, R.G. Buchheit, D.L. Ho, M. Forsyth, Inhibition of AA2024-T3 on a Phase-by-Phase Basis Using an Environmentally Benign Inhibitor, Cerium Dibutyl Phosphate, Electrochem. Solid-State Lett. 8 (2005) C180–C183. doi:10.1149/1.2073672.

[18] S. Joshi, W.G. Fahrenholtz, M.J. O'Keefe, Alkaline activation of Al 7075-T6 for deposition of cerium-based conversion coatings, Surf. Coatings Technol. 205 (2011) 4312–4319. doi:10.1016/j.surfcoat.2011.03.073.

[19] a. E. Hughes, J.D. Gorman, P.R. Miller, B. a. Sexton, P.J.K. Paterson, R.J. Taylor, Development of cerium-based conversion coatings on 2024-T3 Al alloy after rare-earth desmutting, Surf. Interface Anal. 36 (2004) 290–303. doi:10.1002/sia.1652.

[20] O. Lopez-Garrity, G.S. Frankel, Synergistic Corrosion Inhibition of AA2024-T3 by Sodium Silicate and Sodium Molybdate, ECS Electrochem. Lett. 3 (2014). doi:10.1149/2.0021410eel.

[21] M. García-Rubio, P. Ocón, a. Climent-Font, R.W. Smith, M. Curioni, G.E. Thompson, et al., Influence of molybdate species on the tartaric acid/sulphuric acid anodic films grown on AA2024 T3 aerospace alloy, Corros. Sci. 51 (2009) 2034–2042. doi:10.1016/j.corsci.2009.05.034.

[22] V. Moutarlier, M.P. Gigandet, J. Pagetti, L. Ricq, Molybdate/sulfuric acid anodising of 2024-aluminium alloy: influence of inhibitor concentration on film growth and on corrosion resistance, Surf. Coatings Technol. 173 (2003) 87–95. doi:10.1016/S0257-8972(03)00511-5.

[23] M. Stern, A.L. Geary, Electrochemical Polarization, J. Electrochem. Soc. 104 (1957) 559. doi:10.1149/1.2428653.

[24] A.A.O. Magalhaes, I.C.P. Margarit, O.. Mattos, Electrochemical characterization of chromate coatings on galvanized steel, Electrochim. Acta. 44 (1999) 4281–4287. doi:10.1590/S1516-14392013005000179.

[25] I.M. Baghni, S.B. Lyon, B. Ding, The effect of strontium and chromate ions on the inhibition of zinc, Surf. Coatings Technol. 185 (2004) 194–198. doi:10.1016/S0257-8972(03)01316-1.

[26] S. Da Costa, J. Rubim, S. Agostinho, Spectroelectrochemical study of the corrosion of a copper electrode in deaerated 1.0 M HCl solutions containing Fe(III): Effect of the corrosion inhibitor benzotriazole, J. Electroanal. Chem. Interfacial Electrochem. 221 (1987) 1–21.

[27] P.G. Cao, J.L. Yao, J.W. Zheng, R.A. Gu, Z.Q. Tian, Comparative Study of Inhibition Effects of Benzotriazole for Metals in Neutral Solutions As Observed with Surface-Enhanced Raman Spectroscopy, Langmuir. (2002) 100–104.

6. AUTHOR DETAILS

Dr. Oumaïma Gharbi is a postdoctoral research fellow, at the Department of Materials Science and Engineering, Monash University.

DETERMINATION OF THE LONG-TERM CORROSION RESISTANCE OF PEO-COATED MAGNESIUM ALLOYS: A REVIEW

Riyad O. Hussein[1], Vahid Dehnavi[2], Derek O. Northwood[3]

[1]Faculty of Engineering, Technology and Science, Higher Colleges of Technology, Al Ain, UAE,
[2]Department of Chemistry, Western University, London, Ontario, Canada,
[3]Department of Mechanical, Automotive & Materials Engineering, University of Windsor, Ontario, Canada

SUMMARY: Although magnesium alloys have found application in the automotive, aerospace, defense, communications and computer hardware industries, their more widespread application has been restricted by a susceptibility to corrosion, especially in acidic and salt-water environments. Several surface treatments have been used to enhance the corrosion resistance. One of those treatments, plasma electrolytic oxidation (PEO), also known as micro-arc oxidation (MAO), has proven successful in improving the corrosion and wear resistance of magnesium and its alloys. The corrosion protection afforded by the PEO coatings is usually evaluated using potentiodynamic polarization or EIS techniques after shorter immersion times. Such data can be misleading when evaluating the long-term performance. In this paper we review corrosion data for PEO-coated magnesium alloys for longer immersion times and relate this to the microstructure and chemistry of the PEO coatings. As well as potentiodynamic polarization and EIS results, we present results for the drop tests a new technique that is being developed to quickly rank the corrosion resistance of PEO coatings. This test would seem to have its limitations for more corrosion resistant coatings.

1. INTRODUCTION

Magnesium is a strong, lightweight, and easily machined material which offers weight savings in the aerospace, automotive, consumer electronics, biomaterials and defense industries [1]. However, magnesium is highly chemically reactive which can restrict its use in certain applications where corrosion and wear resistance are of importance, e.g. automotive [2]. Much effort has been directed to understanding and improving the corrosion behavior of Mg [3-6]. There are generally two possible ways to improve the corrosion behavior of Mg and Mg alloys; modify the composition and microstructure, which is a challenging and would be a long-term goal [7-9], and secondly by surface treatments or by forming coatings [1,6], which produce protective ceramic, polymer or composite layers.

Even with adjustment of alloy composition to improve corrosion resistance, magnesium alloys may remain very reactive in various environments. Consequently, one of the major methods for improving the corrosion resistance of magnesium alloys is an appropriate surface treatment [1,5,6,9]. Plasma electrolytic oxidation (PEO), also referred as microarc oxidation (MAO) method has demonstrated a particular effectiveness in producing coatings with not only corrosion protection but also enhanced wear properties, hardness and toughness with better thermal stability and dielectric properties [1,10]. Plasma Electrolytic Oxidation (PEO) is a high voltage plasma-assisted anodic oxidation process that transforms the magnesium metal surface, as well as other lightweight metals Al and Ti, into thick, ultra-hard ceramic oxides by a plasma discharge in an electrolytic bath [1,5, 6]. It has been shown that PEO technology has a number of unique advantages: including technological simplicity and the possibility to coat any size and/or complex structure without any pretreatments or post treatments. The PEO process is capable of producing uniform and very thick coatings on the inner and outer surfaces of the substrate materials and the PEO oxide coatings is integral with the metal substrate [5].

2. PEO COATINGS

The PEO process involves the creation of plasma discharges around a sample immersed in an alkaline electrolyte causes partial instant melting of the oxide layer and promotes the formation of a highly adherent ceramic oxide coating. The formation mechanisms of the coating layer by PEO involve electro-, thermal-, and plasma- chemical reactions in the electrolyte [5, 6]. A schematic diagram of the setup for PEO coating is shown in Fig. 1, and is described in more detail in ref [11]. The electrodes are connected to one or two pulsed DC power supplies depending on the current mode, unipolar or bipolar. These power supplies operate under a constant current or voltage control function.

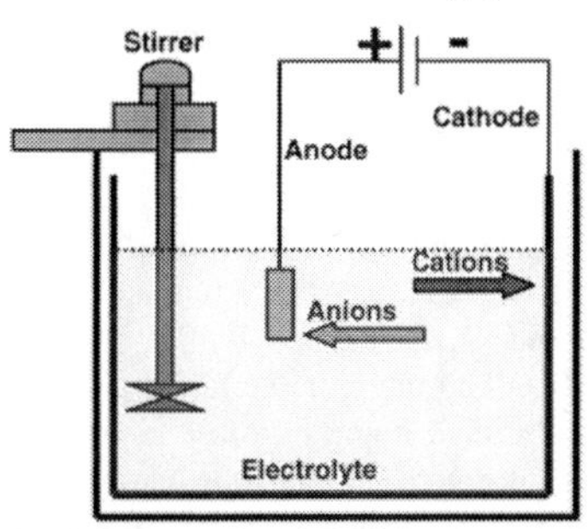

Fig. 1 Schematic of PEO Process [11]

Generally, the corrosion resistance of PEO-coated magnesium alloys depends on coating structures, mainly coating compactness in terms of porosity and other defect levels, relative thicknesses of the coating's three layers, and chemical composition of the coating. PEO coatings consist of three layers [12]: a porous outer layer with several large-sized, deep pores/cavities, an intermediate layer that is relatively dense and a thin inner dense layer (barrier layer) that is well adhered to the substrate (Fig. 2(b)). It is this the dense inner coating/substrate interface layer that ultimately provides the best corrosion performance [5].

Examination of the surface morphology PEO coatings, Fig. 2(a), reveals the complex nature of these coatings and defects in them. The surface of the coatings shows a 'pancake' structure where the center of each pancake is a discharge channel through which the Mg and Mg2+ from the substrate were likely surged out of the channel. It is then combined with the electrolyte anion including OH-1 and reacted with O2 generated due to electrolysis, and finally sintered and quickly solidified by the relatively cool electrolyte, leaving distinct boundaries that define each pancake [13,14]. Fig. 2(a), shows micropores and cracks of different size, which helps to release residual stress of the coating: the size and density reflecting the strength of the discharges that produced the pores. Since PEO coatings contain those unavoidable pores, although most of them are discontinuous and the number of through thickness pores is very small, the overall corrosion resistance of PEO coated magnesium alloy depends on the retarding effect of the porosity level of outer layer and on the substrate/coating interface corrosion resistance. The coating/substrate interface layer seems to have an important effect on the corrosion resistance; it acts as a barrier layer from the aggressive corrosion electrolyte. Therefore, the more compact and thicker the coating, and the presence of more corrosion resistance phases, the slower and more difficult is the access of the corrosive species.

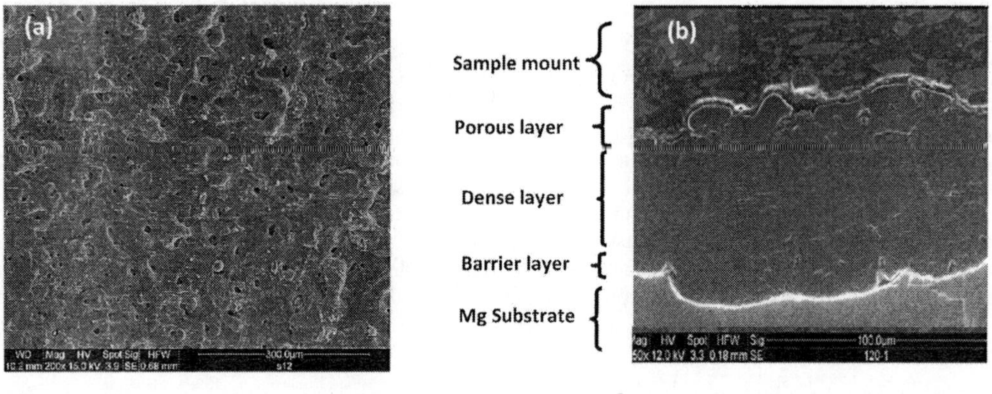

Fig. 2 SEM micrographs of PEO coating on AM60B alloy showing (a) the surface morphology and (b) cross-section.

3. CORROSION CHARACTERIZATION OF COATED and UNCOATED Mg ALLOYS: GENERAL COMMENTS

Magnesium corrosion reaction in aqueous environments is generally an electrochemical reaction with water generating magnesium hydroxide compounds on the surface and a release of hydrogen gas [5,6]:

$Mg + 2H_2O \rightarrow Mg(OH)_2 + H_2$ **(1)**

This overall reaction may be divided into the following anodic and cathodic partial reactions:

$Mg \rightarrow Mg^{2+} + 2e^-$ (anodic/oxidation reaction) **(2)**

$2H_2O + O_2 + 4e^- \rightarrow 4OH^-$ (cathodic/reduction reaction) **(3)**

$Mg^{2+} + 2OH^- \rightarrow Mg(OH)_2$ (product formation) **(4)**

Song et al [15] have shown that the above overall reaction for pure Mg is also applicable for the corrosion processes of AZ91D Mg alloy. However the existence of the alloying elements (Al, Zn) will have some effect on the reactions.

3.1 Potentiodynamic Polarization

A number of excellent reviews describe the existing techniques to measure the corrosion rate for coated and uncoated Mg alloys and provide instructions on their proper use including the advantages and drawbacks of each method [16,17]. Potentiodynamic polarization studies and impedance spectroscopy (EIS) test methods are the most commonly used to determine corrosion rates of the PEO coatings [18]. The electrochemical technique of Tafel extrapolation of polarization curves is widely used. This technique uses data obtained mainly from cathodic polarization measurements, because it is experimentally simple, fast and provides a measurement of the corrosion rate at a particular time over which the measurement was carried out. The studies of Mg corrosion by Song and Atrens [19] and Shi et al [20] indicated that the Mg corrosion rate evaluated by Tafel extrapolation from polarization curves deviates from corrosion rates evaluated from weight loss and hydrogen evolution by 48% to 96%. Furthermore, the short-term corrosion tests for Mg alloys did not agree with long-term tests [20].They suggest that the Tafel extrapolation test for Mg should be supplemented by use of other simple measurements (weight loss, hydrogen evolution rate, rate of Mg^{2+} leaving surface). There is also the question of non-linearity of the Tafel plots.

3.2 Electrochemical Impendence Spectroscopy (EIS)

Electrochemical impedance spectroscopy (EIS) has been also used to study corrosion of Mg, through the application of a time varying voltage and measurement of the current response, the ratio of two (voltage and current) gives the frequency-dependent impedance. The term impedance refers to the frequency dependent resistance to current flow of a circuit element including resistor, capacitor, and inductor. EIS can provide valuable information about surface treatment layers on PEO-coated magnesium, and the interfaces between electrolyte/coating/substrate [21,22], identify diffusion and provides information on the capacitive behavior of the system. It also allows the kinetics of heterogeneous electron-transfer reactions, coupled chemical reactions, or adsorption processes to be studied [23], and can provide information about pitting and crevice corrosion [24]. EIS is theoretically complex and depends on the proper selection of an equivalent circuit; such equivalent circuit should be based on both the physical structure of the PEO coatings and their impedance responses. The choice of the circuit was a balance between a reasonable fitting of the experimental values and a good description of the electrochemical system by keeping the number of circuit elements at a minimum. Each element in the equivalent circuit should correspond to some specific activity in the electrochemical cell [25]. This technique is preferred for coated metals and is sensitive to the early stages of coating degradation. EIS may be able to distinguish between two or more electrochemical reactions taking place. However, as noted by Frankel [26] this technique "is prone to artifact and misinterpretation". This is particularly true with respect to the choice of equivalent circuit.

3.3 Drop Test

The drop test is a simple method to evaluate the corrosion resistance of PEO coatings on magnesium alloys. In this method, a solution consisting of an oxidizing compound, an acid, and water is prepared and a drop of the solution is dripped on the surface of the sample. After penetrating through the coating and upon contact with magnesium substrate, the colour of the droplet changes because of the reaction that occurs. The time it takes for the colour to change is a measure of the corrosion resistance of the coating.

A typical drop solution composition consists of 0.05 g of $KMnO_4$, 5 mL of HNO_3, and 95 mL of water [27-29]. The nitric acid in the solution reacts with the substrate upon contact and generates reducing compounds such as NO_x which reduce Mn(VII) into Mn(II) and the colour of the solution changes from purple to colourless. Li et al. [30] used a drop solution composed of 3 g/L $K_2Cr_2O_7$ and 250 mL/L HCl and recorded the time at which the colour of the solution transformed form orange to green as a measure of the corrosion resistance of anodized coatings on an AZ91D Mg alloy. Li et al [30] describe the drop test as "a fast corrosion test".

4. CORROSION CHARACTERIZATION OF COATED and UNCOATED Mg ALLOYS: EFFECT OF IMMERSION TIME

4.1 Potentiodynamic Polarization and EIS

The effect of increasing the immersion time on the electrochemical corrosion behavior of all Mg-alloys, regardless of compositions causes an increase in corrosion as confirmed by the potentiodynamic polarization and EIS studies. Increasing the immersion time increases the dissolution of the alloy through the corrosion of Mg as a result of the continuous interaction with the corrosive media, which prevents the alloy from forming any oxide layer, and/or corrosion inhibitors that can minimize the corrosion. This is reflected on the increased values of i_{Corr} and the decreased values of R_p for the Mg alloys with immersion time. This effect was also observed from EIS impedance and Nyquist plots. Due to the variation of the experimental test parameters including solution composition, and concentration and in addition to the test time and other parameters, comparison of the results is very difficult. However, it is illustrative to compare the corrosion protection properties afforded by the PEO-coatings for different immersion times for select Mg-alloys.

Cheng et al. [31] have made measurements of the corrosion rate of PEO coatings on AM60 Mg alloys using both a concentrated and dilute NaOH based electrolyte for 50min treatment time. EIS as function of immersion time from 1 to 24 h and potentiodynamic polarization studies using 3.5% NaCl were used to evaluate the corrosion resistance of these coatings. As shown in Fig. 3, and based on the i_{corr} value, the corrosion rate of coatings prepared from the concentrated electrolyte (2.643×10^{-7} A/cm^2) was less than those prepared using the dilute electrolyte (1.042×10^{-5} A/cm^2). EIS studies also verify the results of the polarization studies where, after 24 h of immersion in 3.5% NaCl, the coatings prepared using the concentrated electrolyte have a |Z| value of 1000 Ω cm^2 whereas it is 100 Ω cm^2 for coatings obtained using the dilute electrolyte. The EIS measurements, as shown in Fig 4(a&b) for |Z| values for concentrated and dilute electrolytes respectively indicate that increasing the immersion time from 1 to 24 h will increase the corrosion rates.

Lim et al. [35] reported the corrosion resistance measurements of AZ31 Mg alloy PEO coated with and without the addition of CeO_2 particles, using potentiodynamic polarization and EIS tests (up to 40 h) in 3.5% NaCl. The impedance (|Z|) value of PEO coatings prepared without CeO_2 particles is reduced from ~1.5×10^5 to 4×10^3 Ω cm^2 by increasing the immersion time from 10 min to 40 h. However, after 40 h the values of |Z| approaches the values of the uncoated Mg alloy (see Fig. 5). The potentiodynamic and EIS tests indicated that the corrosion resistance of AZ31 Mg alloy is significantly improved by CeO_2 incorporation. The high impedance of CeO_2-containing PEO coated Mg alloy was still maintained after 40 h immersion in 3.5 wt% NaCl solution. Coatings prepared with the addition of 30 g/l of CeO_2 particles, exhibit |Z| value ~ 1.5×10^5 Ω cm^2 after 40 h of immersion.

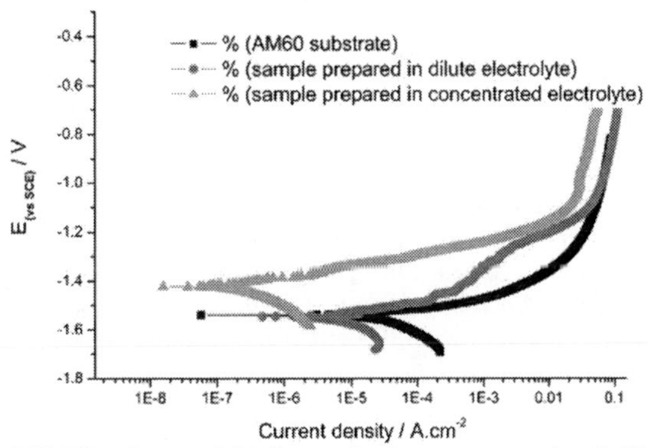

Fig. 3. Potentiodynamic curves of AM60 substrate, MAO samples in concentrated and dilute electrolytes in 3·5 wt-% NaCl solution [31].

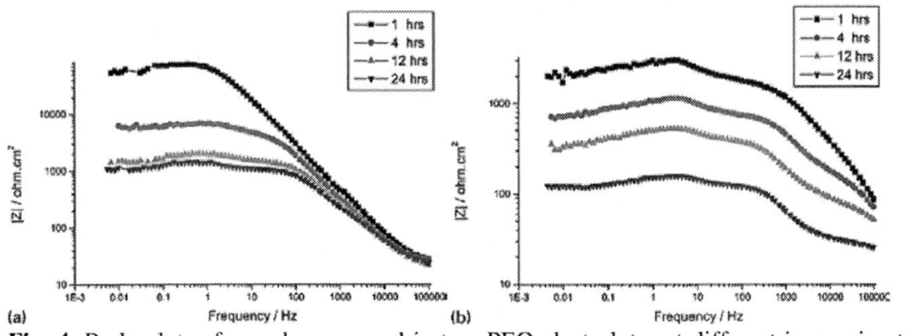

Fig. 4. Bode plots of samples prepared in two PEO electrolytes at different immersion times in 3·5 wt-% NaCl solution [31].

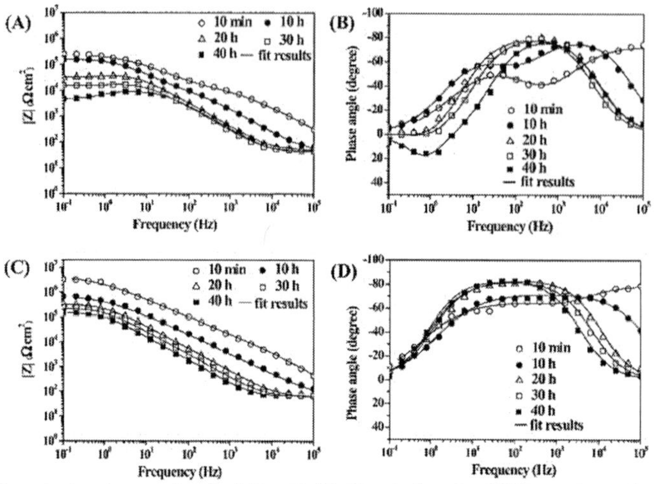

Fig. 5 Variation of Bode plots during immersion in 3.5 wt% NaCl solution for PEO coatings obtained in (A),(B) CeO2-free and (C),(D) 30 g/l CeO_2-containing electrolytes [32].

Hussein et al. [33] have studied the effect of current density using bipolar current modes on the electrochemical corrosion protection performance of uncoated and the PEO coated AZ91 Mg alloy in a 3.5% NaCl solution for different immersion times up to 24 h using different experimental measurements. The potentiodynamic polarization test was performed and the change in resistance as a function of immersion time served as a basis for the comparison of corrosion performance of these coatings. The potentiodynamic polarization curves of the uncoated (S0) AZ91D Mg alloy and the PEO coated specimens (S1, S2 and S3 coated with current density of 0.05 , 0.07 and 0.11A/cm^2 respectively) for different times up to 24 h are shown in Fig. 6. Compared to the uncoated AZ91D, all PEO-coated samples exhibited a higher polarization resistance, a lower corrosion current density and a higher (more noble) corrosion potential. The first point to note is that the variation of corrosion resistance with current density is not linear. Some of the polarization curves eg. S3 at 0.5h, and S1 at 24h, exhibit evidence of a "passive behaviour". However, this passive behaviour does not lead to significant reductions in corrosion current densities, particularly at longer immersion times. Compared to uncoated alloy, the corrosion rate from potentiodynamic polarization was reduced by at least three orders of magnitude (Table 1). According to this study the amount of porosity and other coating defects rather than the coating thickness, are the main factor that controls the corrosion protection performance of the PEO coatings this porosity is indirectly proportional to the current density. Hence, coating S2, having relatively thick and dense oxide layers, has a higher polarization resistance and lower corrosion current densities than the other PEO coatings for 0.5h and 6h immersion times (Table 1). For a 24h immersion time the ranking is sample S1 which shows a relatively low corrosion rates compared with S2 and S3 This may be attributed to the lower coating thickness of S1coating that may be partially dissolved during the immersion period and during the penetration of liquid into the porous layer causes a thinning of the MgO dense layer.

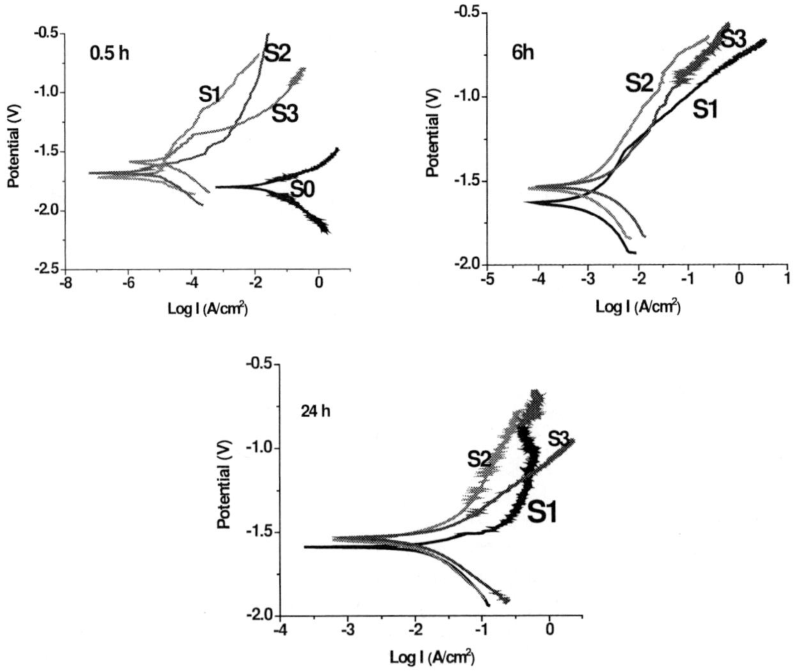

Fig. 6. Potentiodynamic polarization curves of the uncoated (S0) and PEO-coated samples using different current densities (S1-S3) after three different immersion times [33].

Table 1. Potentiodynamic polarization corrosion test results in3.5%wt NaCl solution [33]

Immersion time	S	E_{corr} (V)	i_{corr} $\mu A/cm^2$	R_p ($K\Omega/cm^2$)
0.0 h	S0	-1.797	22.43	1.29
0.5 h	S1	-1.717	0.005	5214.0
	S2	-1.673	0.004	5947.7
	S3	-1.586	0.011	3270.2
6 h	S1	-1.632	1.05	76.52
	S2	-1.541	0.801	88.12
	S3	-1.525	2.12	37.93
24 h	S1	-1.585	14.39	2.47
	S2	-1.547	10.93	5.80
	S3	-1.536	10.85	4.19

The Nyquist and the Bode diagrams from the EIS analysis of experimental and fitted curves of the untreated and PEO-treated AZ91D Mg alloy at three different current densities for three different immersion times were also reported [33]. The fitted Nyquist and Bode plots results shown in Fig. 7 are for AZ91D alloy without and with PEO treatment after immersion in 3.5 % NaCl solution for 0.5h . Nyquist plots of untreated AZ91D alloy were characterized by an inductive loop in the low frequency (LF) region (enlarged part of Fig. 7). The values of the circuit elements calculated from the fitting are summarized in Table 2. The equivalent circuits used for curve fitting are given in Figs. 10 and listed in the Tables 3-4.

Table 2. Fitting results of EIS diagrams of PEO treated AZ91D Mg alloy using different current densities [33].

Imm. time	S	Rs $\Omega \cdot cm^2$	CPE1-Q $\mu F/cm^2\ s^{1-n}$	CPE1-n	R1 ($\Omega \cdot cm^2$)	CPE2-Q $\mu F/cm^2\ s^{1-n}$	CPE2-n	R2 ($\Omega \cdot cm^2$)
0.5h	S1	70	0.14	0.68	1.67×10^5	0.70	0.55	2.5×10^{23}
	S2	70	0.20	0.63	2.30×10^5	0.70	0.60	2.2×10^7
	S3	67	0.24	0.68	2.00×10^5	0.74	0.74	2.0×10^6
6h	S1	70	1.1	0.67	3.97×10^4	9.2	0.78	1.7×10^4
	S2	70	0.75	0.61	5.81×10^4	25.7	0.99	2.0×10^3
	S3	70	1.1	0.62	7.87×10^2	0.0059	0.93	2.8×10^3

EIS studies reveal that the data obtained for the PEO-coated AZ91D coating immersed for o.5h in the solution, exhibits a much higher corrosion rate than those for 6 and 24h immersion time, and the magnitudes of the resistance R_1 for the coated samples are in the range (0.167-0.23 M Ω cm^2), and Q_1 values increases from 0.14 for S1 to 0.24 $\mu F/cm^2\ s^{n-1}$ for S3. The |Z| of these coatings vary from $10^{5.5}$ Ω cm^2 for S3 to $10^{6.5}$ Ω cm^2 for S1and S2. The Nyquist curves for the uncoated AZ91D Mg alloy presented in Fig. 7 exhibit negative values in the low frequency range forming a curve known as an inductive loop [34]. This loop may be attributed to decomposition of metal to ions leading to the formation of corrosion products. The impedance modulus at the LF range for uncoated sample is $10^{3.1}$ Ω cm^2 (Fig. 7(b)).

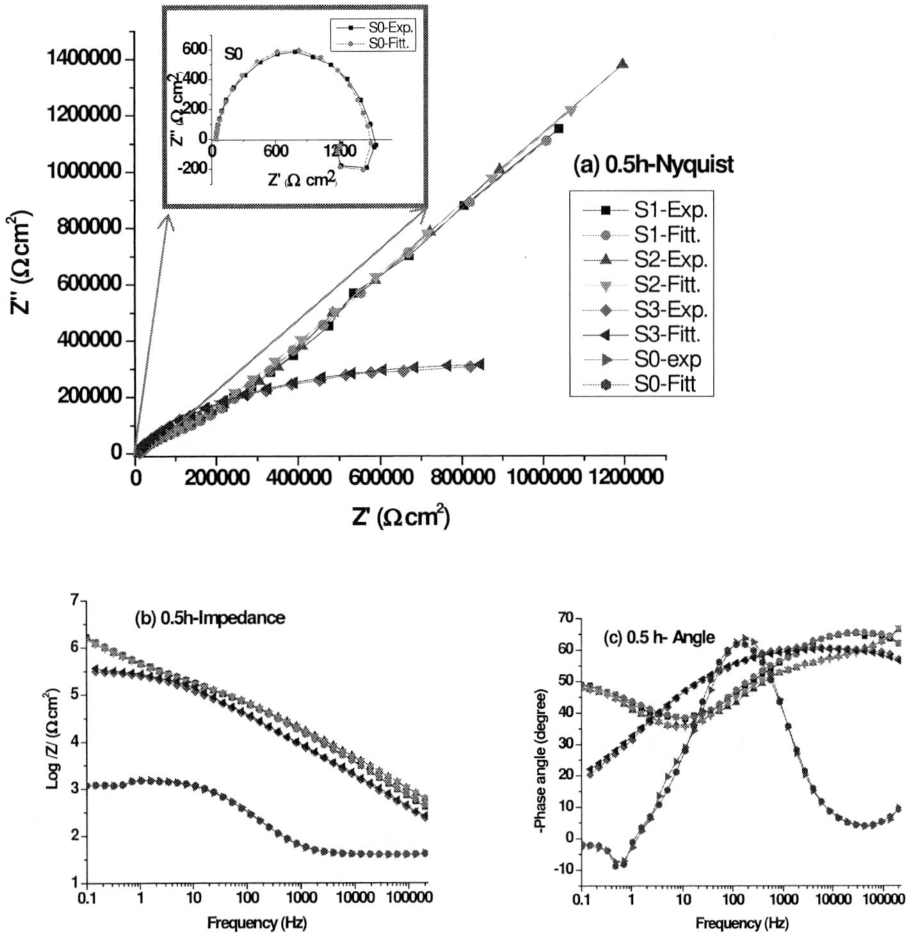

Fig. 7 EIS plots of coated and uncoated AZ91DMg alloy at different current densities for 0.5 h immersion time ((a) Nyquist, (b) Impedance and (c) angles plots) [33].

The Nyquist and Bode plots of experimental and fitting curves as a function of current density of coated AZ91D for 6h immersion time are shown in Fig. 8. Although, the same equivalent circuit was used as for the 0.5h immersion time, PEO-coated Mg alloy immersed for 6h shows different EIS behavior, and the difference was present for all of the frequency range of the EIS plots. The experimental Nyquist diagram recorded shows a semicircular trend with diameters in the range of 29-60 kΩ cm^2. This loop may be attributed to decomposition of metal to ions leading to the formation of corrosion products, and then to the adsorption of electro-active species of the electrolyte [34]. In the LF range, the phase angle is shifted to lower angle around 0 to -7° for S3 and (S1&S2) samples respectively. The |Z| of these coatings ranges from $10^{4.5}$ to $10^{4.75}$ Ω cm^2. As shown in Table 2, the values of R1 ranging from 5.8×10^2 Ω cm^2 for S3 to 7.9×10^4 M Ω cm^2 for S1 sample.

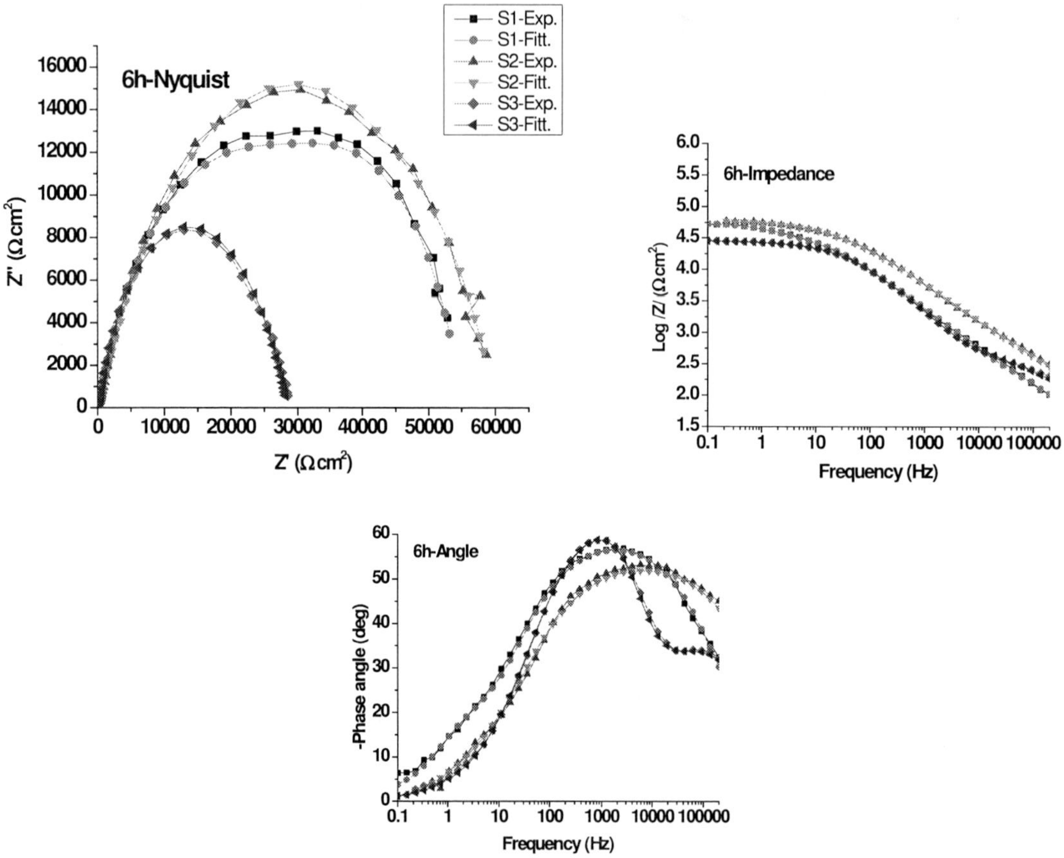

Fig. 8. EIS plots of coated and uncoated AZ91DMg alloy at different current densities for 6 h immersion time ((a) Nyquist, (b) Impedance and (c) angles plots) [33].

According to the change of the EIS behavior of the coated AZ91D Mg alloys immersed for 24h from that of the 0.5h and 6h immersion samples where the difference is due to the existence of both a kinetic and diffusion processes in the circuit. As the immersion time prior to corrosion testing reaches 24 hours, the corrosion rates for the PEO-coated samples were comparable to those for the uncoated alloy with no immersion before corrosion testing. The longer immersion time allows the electrolyte to pass through the coating easily and reach the substrate. The diffusion of electrolyte species into the pores will be easier at higher frequencies thereby resulting in adsorption at the inner barrier layer. Due to this adsorption, the Nyquist curve shows an inductive behavior in the LF range [33]. Fig. 9 (b) and Table 3 show the equivalent circuit giving the best fit for the impedance data of the 24h immersion time.

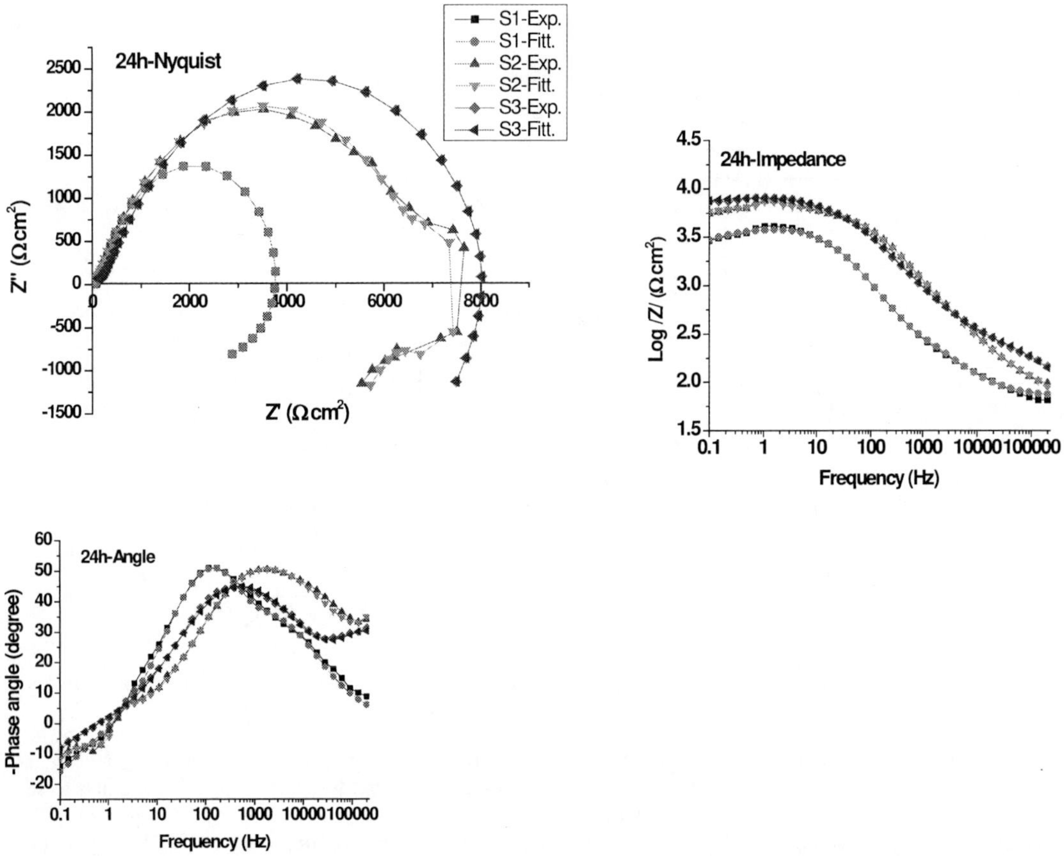

Fig. 9. EIS plots of coated and uncoated AZ91DMg alloy at different current densities for 24 h immersion time ((a) Nyquist, (b) Impedance and (c) angles plots) [33]

Table 3. Fitting results of EIS diagrams of PEO treated AZ91D Mg alloy using different current densities for 24h immersion time [33].

S	Rs $\Omega\cdot cm^2$	CPE1-Q $F/cm^2\ s^{1-n}$	CPE 1-n	R1 $\Omega\cdot cm^2$	CPE2-Q $\mu F/cm^2\ s^{1-n}$	CPE2 -n	R2 $\Omega\cdot cm^2$	CPE3-Q $\mu F/cm^2\ s^{1-n}$	CPE3-n	R3 $\Omega\cdot cm^2$
S1	70	1.5×10^{-5}	0.63	1.56×10^4	17.4	1	40	-26.8	0.22	-21740
S2	70	2.3×10^{-6}	0.67	-95	1.8	0.4	5537	300	0.002	-11.81
S3	70	4.5×10^{-8}	0.86	185	0.82	1	-5.6	3.8	0.64	8563

Table 4. Fitting results of EIS diagrams of untreated AZ91D Mg alloy (S0) for 0h immersion time [33].

S	Rs $\Omega\cdot cm^2$	CPE1-Q $F/cm^2\ s^{1-n}$	CPE 1-n	R1 $\Omega\cdot cm^2$	CPE2-Q $\mu F/cm^2\ s^{1-n}$	CPE2 -n	R2 $\Omega\cdot cm^2$	CPE3-Q $F/cm^2\ s^{1-n}$	CPE3 -n	R3 $\Omega\cdot cm^2$
S0	70	6.0×10^{-9}	1	-30	10.16	0.88	1433	0.057	0.55	0.98

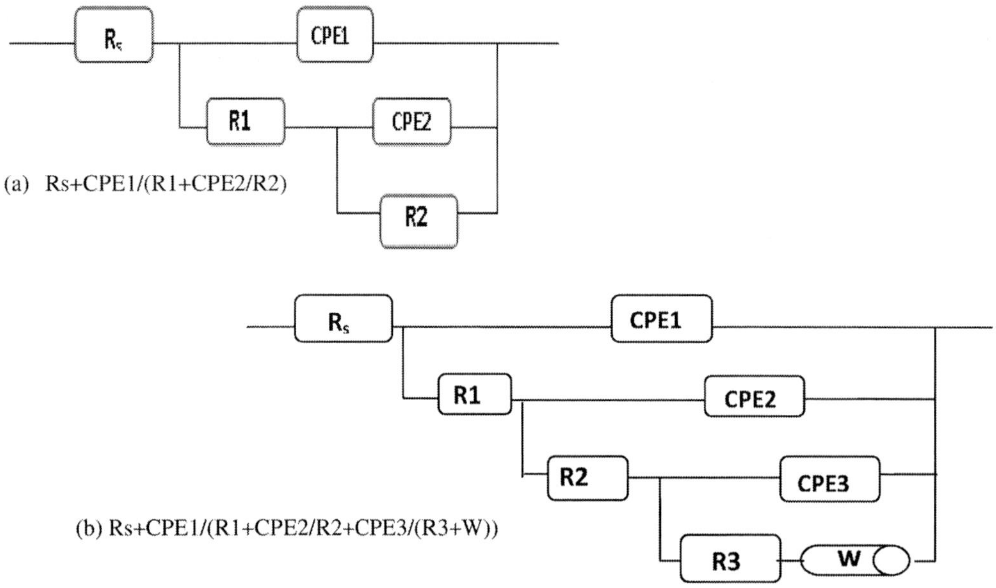

Fig. 10 The equivalent circuit model for (a) coated AZ91DMg alloy for 0.5, 6h immersion time, (b) coated Mg alloy for 24h immersion time and uncoated Mg alloys.

Sealing and post-treatment the coating, influence the porosity and corrosion resistance of PEO coatings on pure Mg and Mg alloys. Duan et al. [35] have explored the use of an organic sealing agent coated by multi-immersion method to seal the pores and microcracks of the outer layer and integrated with the PEO coatings. Guo et al. [36] and Lu et al. [37] have studied the effect of poly-L-Lactic acid (PLLA) to seal the pores of the PEO coatings deposited on WE42 Mg alloys using an alkaline electrolyte solution containing NaOH and Na2SiO3. Sealing the pores and cracks of the PEO coating with PLLA has enabled an improvement of the corrosion protection of the Mg alloys.

4.2 Drop Test

The drop test is a relatively new test and there are limited results in the literature. Our own preliminary studies have shown that the drop test has some advantages and some limitations. It is a simple method to quickly rank the relative corrosion resistance of PEO coatings on Mg alloys but provides no mechanistic information. Determining the exact time it takes for the drop solution to change colour can be challenging. For less resistant coatings, the colour of the whole droplet changes very rapidly and is easy to judge.

Figs. 11, 12 and 13 illustrate drop test results for a AM50 Mg-alloy PEO-coated. The alloy was PEO coated using a DC power mode at a constant current density of 40 mA/cm^2. The electrolyte used consisted of 20g/l NaOH, 80 g/l $Na_2SiO_3.5H_2O$, and 80 g/l $Na_2B_4O_7.10H_2O$ and the temperature was maintained at kept at 25±3 °C . The drop test solution was of 0.05 g $KMnO_4$, 5mL HNO_3, and 95 mL water. Fig. 11 shows the colour transformation during a drop test for a sample coated for 1 minute. After only 7 seconds the droplet turned completely colourless. However, for more resistant coatings, the colour change is gradual and makes it difficult to judge. Another issue is that in coatings with good corrosion resistance, the droplet dries out before it changes colour. Fig. 12 illustrates the colour change of the droplet on a sample coated for 20 min. After 60 min the droplet almost dried out but no colour change was observed. The surface of the coating showed no sign of damage after the droplet was rinsed (Fig. 12-d).

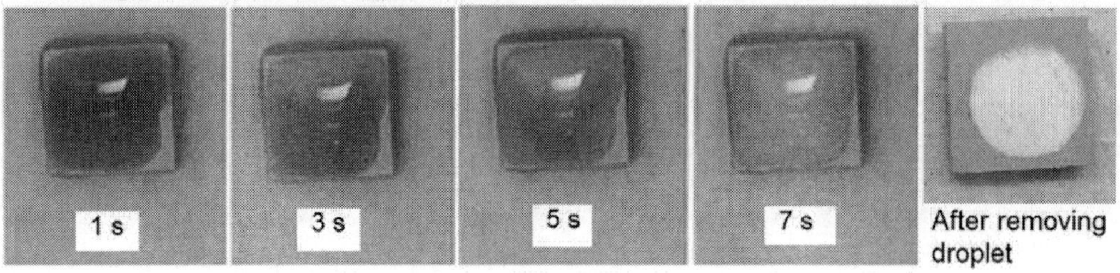

Fig. 11. Transformation of droplet colour with time on an AM50 die-cast Mg alloy PEO coated for 1 min. After removing the droplet the substrate is exposed.

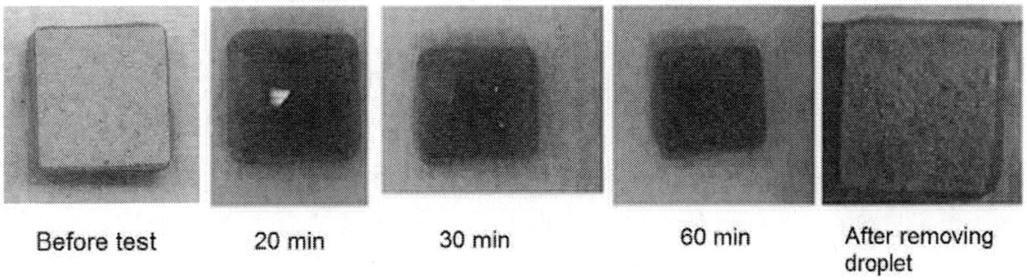

Fig. 12. Transformation of droplet colour with time on an AM50 die-cast Mg alloy PEO coated for 20 min.

The corrosion resistance, as measured by the drop test time (time taken for droplet colour to change after is applied on the sample) is plotted as a function of PEO processing time in Fig. 13. Longer coating times produced more corrosion resistant coatings. For samples coated for 15 minutes, or longer, the droplet dried out after 75 minutes without any colour change. In order to make the drop test more useful in testing thicker, and higher quality, coatings it is suggested that a higher concentration of HNO_3 in the droplet be investigated.

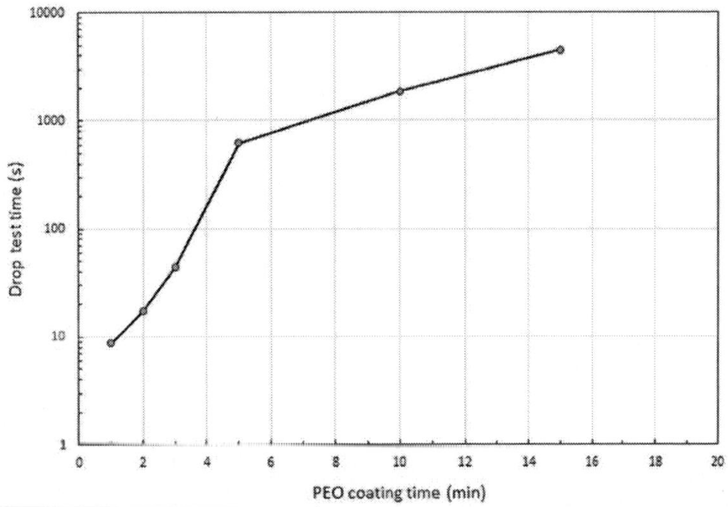

Fig. 13. Corrosion resistance of PEO coated AM50 in drop test.

5. CONCLUSIONS

1. Potentiodynamic polarization and EIS results show that PEO coatings offer significant corrosion protection to Mg alloys for short immersion times prior to corrosion testing. The high impedance and R2 values in the EIS results demonstrated that the PEO coating did not undergo any significant degradation and that the coating at the metal-substrate interface was providing significant protection.
2. Potentiodynamic polarization and EIS studies, however, show that corrosion rates for the same coated materials increased rapidly with increase in immersion time.
3. Sealing the coating would appear to be necessary to protect against long term exposure in aggressive environments.
4. The drop test was found to be a quick method to rank the corrosion resistance of PEO coated Mg alloys. The results showed that as the corrosion resistance of the coating increased, it took a longer time for the droplet colour to change. However, for coatings with higher corrosion resistance, the droplet dried out before any change in colour. To solve the issue, a solution with a higher concentration of acid could be used.

6. REFERENCES

[1] Montisci S., Surface technologies show the way for magnesium. Materials World, 24 (12) (2016) 27-29.

[2] Hussein R.O., Northwood D. O., Nie, X., Plasma electrolytic oxidation (PEO) coatings on magnesium alloy AZ91D for improved corrosion and wear resistance. Materials Forum, 38 (2014) 100-114.

[3] Song G. L. and Atrens A., "Corrosion mechanisms of magnesium alloys," Advanced Engineering Materials, 1(1999) 11-33.

[4] Zeng R-C, Zhang J., Huang W-J, Dietzel W., Kainer K.U., Blawert C., Wei K.E., Review of studies on corrosion of magnesium alloys,Trans. Nonferrous Met. Soc. China, 16 (2006) s763-s771

[5] Hussein R. O., Nie X. and Northwood D. O., The application of Plasma Electrolytic Oxidation (PEO) to the production of corrosion resistant coatings on magnesium alloys: a Review, Corrosion and Materials, 38 (1) (2013) 53-63.

[6] Hussein R.O., Northwood D. O., Production of anti-corrosion coatings on light alloys (Al, Mg, Ti) by Plasma Electrolytic Oxidation (PEO), 'In: Developments in Corrosion Protection, (M. Aliofkhazraei, Ed.) InTech Open Publishers, ISBN 980-953-307-1100-2. 2013

[7] Kaesel VT, Tai PT, Bach FW, Haferkamp H, Witte F, Windhagen H. Approach to control the corrosion of magnesium by alloying. In: Proceedings of the Sixth International Conference on Magnesium Alloys and their Applications. New York: Wiley VCH, 2004. p. 534–539.

[8] Song G, Atrens A, Recent insights into the mechanism of magnesium corrosion and research suggestions. Advanced Engineering Materials, 9(3) (2007) 177–183.

[9] Gray JE, Luan B. Protective coatings on magnesium and its alloys – a critical review. J. Alloys and Compounds, 336 (2002) 88–113.

[10] Hussein R.O., Zhang P., Northwood D.O. and Nie X., Improving the corrosion resistance of magnesium alloy AJ62 by a plasma electrolytic oxidation (PEO) coating process, Corrosion and Materials, 36 (3) (2011) 38-49.

[11] Hussein R. O., Northwood D. O., and Nie X., The effect of processing parameters and substrate on the corrosion resistance of plasma electrolytic oxidation (PEO) coated magnesium alloys, Surface & Coatings Technology, 237 (2013) 357–368.

[12] Hussein R. O., Nie X. and Northwood D. O., An investigation of ceramic coating growth mechanisms in Plasma Electrolytic Oxidation (PEO) processing, Electrochimica Acta, 112 (2013) 111-119.

[13] Arabal R, Matykina E, Hashimoto T, Skeldon P, Thompson GE. Characterization of AC PEO coatings on magnesium alloys. Surf Coat Technol,, 203 (2009) 2207–20.

[14] Dunleavy CS, Golosnoy IO, Curran JA, Clyne TW. Characterisation of discharge events during plasma electrolytic oxidation. Surf Coat Technol, 203 (2009) 3410–9.

[15] Song G, Atrens A, Dargusch M. Corrosion behaviour of AZ21, AZ501 and AZ91 in sodium chloride, Corrosion Science, 41 (2) (1998) 249-273.

[16] Sankara Narayanan T S N , Song Park , Min Ho Lee, Strategies to improve the corrosion resistance of microarc oxidation (MAO) coated magnesium alloys for degradable implants: Prospects and challenges , Progress in Materials Science, 60 (2014) 1–71

[17] Gray JE, Luan B. Protective coatings on magnesium and its alloys—a critical review. J. Alloys Compd, 336 (2002) 88–113.

[18] Gu XN, Li N, Zhou WR, Zheng YF, Zhao X, Cai QZ, Corrosion resistance and surface biocompatibility of a microarc oxidation coating on a Mg–Ca alloy. Acta Biomaterialia, 7 (2011) 1880–1889.

[19] Song G., Atrens A., Understanding magnesium corrosion – a framework for improved alloy performance, Advanced Engineering Materials, 5 (2003) 837–858.

[20] Shi Z, Liu M, Atrens A, Measurement of the corrosion rate of magnesium alloys using Tafel extrapolation, Corrosion Science, 52 (2010) 579-588).

[21] Zhang Y, Yan C, Wang F, Li W, Electrochemical behavior of anodized Mg alloy AZ91D in chloride containing aqueous solution, Corrosion Science, 47 (2005) 2816–2831

[22] Arrabal R, Mota JM, Criado A, Pardo A, Mohedano M, Matykina E, Assessment of duplex coating combining plasma electrolytic oxidation and polymer layer on AZ31 magnesium alloy. Surface & Coatings Technology, 206 (2012) 4692–4703.

[23] Huang VM, Wua S-L, Orazema ME, Pébèreb N, Tribollet B, Vivier V, Local electrochemical impedance spectroscopy: A review and some recent developments. Electrochimica Acta, 56 (2011) 8048– 8057.

[24] Mansfeld F, Lin S, Kim K, Shih H, Pitting and surface modification of SiC/Al. Corrosion Science, 27 (1987) 997-1000.

[25] Cottis R, Turgoose S, Electrochemical Impedance and Noise, NACE International, 1999. ISBN 1-57590-093-9.

[26] Frankel G. S., Electrochemical techniques in corrosion: status, limitations, and needs, Journal of ASTM International, Vol. 5, No. 2, Paper ID JAI101241,1-27,(2008).

[27] Zhao L., Cui C., Wang Q., Bu S., Growth characteristics and corrosion resistance of micro-arc oxidation coating on pure magnesium for biomedical applications, Corrosion Science 52 (2010) 2228–2234. doi:10.1016/j.corsci.2010.03.008.

[28] Lin X., Tan L., Zhang Q., Yang K., Hu Z., Qiu J., et al., The in vitro degradation process and biocompatibility of a ZK60 magnesium alloy with a forsterite-containing micro-arc oxidation coating., Acta Biomater. 9 (2013) 8631–42. doi:10.1016/j.actbio.2012.12.016.

[29] Han J., Wan P., Sun Y., Liu Z., Fan X., Tan L., et al., Fabrication and Evaluation of a Bioactive Sr–Ca–P Contained Micro-Arc Oxidation Coating on Magnesium Strontium Alloy for Bone Repair Application, J. Mater. Sci. Technol., 32 (2016) 233–244. doi:10.1016/j.jmst.2015.11.012.

[30] Li W., Zhu L., Li E., Zhao B., Growth characterization of anodic film on AZ91D magnesium alloy in an electrolyte of Na,SiO, and KF, J. Univ. Sci. Technol. Beijing, 13 (2006) 450-455.

[31] Cheng YL, Qin TW, Li LL, Wang HM, Zhang Z. Comparison of corrosion resistance of microarc oxidation coatings prepared with different electrolyte concentrations on AM60 magnesium alloy. Corros Eng Sci Technol, 46 (2011) 17–23.

[32] Lim TS, Ryu HS, Hong SH. Electrochemical corrosion properties of CeO2-containing coatings on AZ31 magnesium alloys prepared by plasma electrolytic oxidation. Corrosion Science, 62 (2012) 104–111.

[33] Hussein R. O., Nie X. and Northwood D. O. (2013) Effect of cathodic current density on the corrosion protection of the oxide coatings formed on AZ91D magnesium alloy by plasma electrolytic oxidation. Corrosion and Prevention 2013, Australia 2013, paper no. 112.

[34] Ghasemi A, Raja VS, Blawert C, Dietzel W, Kainer KU, The role of anions in the formation and corrosion resistance of the plasma electrolytic oxidation coatings. Surface & Coatings Technology, 204 (2010) 1469-1478.

[35] Duan H, Du K, Yan C, Wang F. Electrochemical corrosion behavior of composite coatings of sealed MAO film on magnesium alloy AZ91D. Electrochimica Acta, 51 (2006) 2898–908.

[36] Guo M, Cao L, Lu P, Liu Y, Xu X. Anticorrosion and cytocompatibility behavior of MAO/PLLA modified magnesium alloy WE42. J Mater Sci Mater Med, 22 (2011)1735–40.

[37] Lu P, Cao L, Lin Y, Xu X, Wu X. Evaluation of magnesium ions release, biocorrosion, and hemocompatibility of MAO/PLLAmodified magnesium alloy WE42. J Biomed Mater Res B, 96B (2011) 101–109.

7. AUTHOR DETAILS

Dr. Riyad O. Hussein received his first PhD in physics and applied physics from The University of Strathclyde, U.K., Glasgow in 1990, and his second PhD in Engineering Materials at University of Windsor, Windsor, Canada. Dr. Hussein is currently Associate Professor in the Faculty of Engineering, Technology and Science, Higher Colleges of Technology, Al Ain, UAE. He worked in research and academia for about 27 years before joining his current position. Current research areas includes: plasma electrolyte Oxidation (PEO) coating of Al and Mg for corrosion and wear resistance, plasma discharge analysis of the PEO process.

Dr. Vahid Dehnavi is a postdoctoral fellow in the Department of Chemistry at Western University, London, Ontario, Canada. He received his Ph.D. in Engineering Materials at Western University in 2014. Before getting his Ph.D., he was a research associate and lecturer in the field of Materials Science and Engineering for 5 years. His current research areas include improving plasma electrolytic oxidation coatings properties for applications in automotive and biomedical industries, corrosion of magnesium, copper and their alloys, and the effect of microstructure on the properties and corrosion behaviour of metals.

Professor Derek O. Northwood is currently Distinguished University Professor, and Professor of Engineering Materials, in the Department of Mechanical, Automotive and Materials Engineering, University of Windsor, Windsor, Ontario, Canada. He is licensed Professional Engineer in Ontario (P. Eng.) and holds Chartered Professional Engineer (CPEng) registration in Australia (1990). He has 40+ years of experience in academia. Current corrosion-related research includes: protective coatings for light metal alloys (Mg, Al, Ti); surface engineering of steel and cast iron for automotive applications; and surface patterning for improved corrosion resistance. In November 2016, Professor Northwood was inducted as a Fellow of the Royal Society of Canada (FRSC).

A REVIEW ON THE PASSIVATION ABILITY OF NANOSTRUCTURED STEEL SURFACES CREATED BY NANOCRYSTALLIZATION TREATMENT

M. Laleh [1], Mike Y. Tan [1,2], Anthony E. Hughes [2,3], R. Parvizi [1,2]

[1]Deakin University, School of Engineering, Faculty of Science, Engineering and Built Environment, Geelong Waurn Ponds Campus, Waurn Ponds, Victoria 3216, Australia

[2]Deakin University, Institute for Frontier Materials, Geelong Waurn Ponds Campus, Waurn Ponds, Victoria 3216, Australia

[3]Commonwealth Scientific and Industrial Research Organization (CSIRO), Mineral Resources, Private Bag 33, Clayton South, Victoria 3169, Australia

SUMMARY: Passivation is a process through which metals are protected from corrosion due to the formation of a nanometer-thick oxide layer on their surfaces. The great practical significance of passivation is that it protects metals and alloys from corrosion that occurs in the atmosphere, water and aggressive chemical environments. Studies on passive film formation and surface passivity of metals and alloys are not new. Substantial efforts have been made in recent decades to understand the thermodynamics and kinetics of the formation, growth and also breakdown of the passive films under various environmental conditions. It has been reported that the electrochemical stability of the passive film is mainly dependent on its chemical composition and structure (e.g. thickness, compactness, etc.) which are in turn influenced by the chemical composition and microstructure of the substrate as well as the conditions of the surrounding environment (i.e. pH, temperature, etc.). So far, various methodologies including surface treatments, surface coatings, inhibitors, etc. have been used to enhance the passivation ability and passive film stability upon the metals and alloys. The aim of this review paper is to investigate works that have been undertaken to improve the electrochemical corrosion behavior of the passive film formed on a steel surface, with particular focus on surface nanocrystallization treatments to understand the effect of grain size on the passivity behavior. It has been shown that currently there is no general agreement in the literature on the effects of grain refinement on the passivation ability and corrosion performance of stainless steels. The main reason for this disagreement is the complex contribution of various factors resulted from different manufacturing processes and surface treatments which could pose complex effects on corrosion resistance of a material. Each one of these factors has its own effect on the passivity and corrosion behavior and they are also highly interdependent on each other. One of the significant gaps in corrosion-related literature pointed out in this paper is that when studying the passivity and corrosion resistance of nanocrystalline stainless steels, all the possible determining factors like grain size, grain boundary character, texture, chemical composition, phase composition, surface roughness, residual stresses and inclusions should be considered.

Keywords: Passivation; Corrosion; Nano-crystallization; Steel

1. INTRODUCTION

Stainless steels have been used extensively in various engineering applications such as steam power plants, chemical, oil and gas industries owing to their excellent corrosion resistance. The superior corrosion resistance of the stainless steel arises from a nanometer-thick passive film which naturally forms on its surface and limits susceptibility to attack in corrosive environments. However, localized breakdown of passive films leads to accelerated dissolution of the underlying material, eventually results in the failure of stainless steels [1-3].

Some metals and alloys, like iron and steels, are inherently reactive and so can react with oxygen to form passive films on their surfaces, when exposed to corrosive environments. The passive film then acts as a barrier between the metallic substrate and the environment. The chemical composition of the passive films depends on many factors such as the chemical composition of the substrate, pH, temperature, applied potential, etc. The structure reported in the most of the investigations for the passive film on stainless steels is composed of an inner Cr-rich and an outer Fe-rich layer. The protective efficiency of this passive film is found to be dependent on its chemical composition, microstructure, and electrical properties [4-7]. Therefore, many investigations have been done over the past decades to examine and improve the properties of the passive films formed on stainless steels and to understand and prevent the localized corrosion [8-14], in order to achieve a longer working life of the components made from stainless steels.

The phenomena of passivity and localized corrosion have been studied extensively. For example, the physico-electrochemical aspects of passivity and mechanism of growth and breakdown of passive films have been reviewed by Macdonald [15] and recently by Maurice and Marcus [16] on the nanoscale growth, structure, and local properties of passive films. Different aspects of pitting corrosion with a focus on the initiation and propagation stages of pit formation have been reviewed by Soltis [17], who stated that although there are a considerable number of theories and mechanisms for passive film breakdown, it remains still least understood. Three most relevant mechanisms for passivity breakdown that have been discussed in literature are: (1) the penetration mechanism, (2) the film breaking mechanism and (3) the adsorption mechanism. Surface analytical methods that have been used for studying the passivity of metals were reviewed by Strehblow [18]. Despite the large number of studies have been carried out on passive films on stainless steel, the critical factors controlling the growth and stability of passive films are still not fully understood [19].

In recent years, nanostructured materials have attracted a growing interest in academia and industry, not only because of their good mechanical and physical properties, e.g. hardness, wear, and Young's modulus, but also because of their possibly enhanced electrochemical properties compared to the coarse-grained counterparts. It is generally shown that nanostructure may change the thickness and compactness of passive film, mainly due to this fact that high number of grain boundaries in the nanostructured metals provide more pathways for passivating elements to diffuse to the surface to form a passive film. A vast number of studies on the relationship between the grain size and corrosion behavior have been performed for different metals and alloys systems by using different treatments. It has been shown that nanostructure can also impact intergranular corrosion resistance. For example, surface nanocrystallization of an austenitic stainless steel using surface mechanical attrition treatment (SMAT) resulted in a very low degree of sensitization compared to the un-treated alloy, which was mainly attributed to the formation of twin boundaries in the microstructure of the SMAT-ed alloy. These twin boundaries were not susceptible to carbide precipitation because of their more regular and coherent atomic structure and lower grain boundary energy than those of other grain boundaries [20]. Some of the past studies on how grain size refinement affects the electrochemical corrosion behavior of nanocrystalline materials have been summarized in a series of literature reviews by Liu et al. [21], Ralston and

Birbilis [22] and more recently Gupta and Birbilis [23]. However, the effect of grain refinement on the corrosion behavior is still a controversial issue and both decrease and increase of corrosion resistance have been reported after reducing grain size [24,25].

Most previous research in this field has focused mainly on the effect of grain size on the passivity and corrosion behavior, while the effects of other factors such as inclusions size and distribution, and texture have rarely been identified. This remains as an open research area and needs more investigations to clarify the changes induced by different techniques and subsequently their effects on the passive film formation and stability. The main objective of this paper is to review the various nanocrystallization processes that have already been used to manipulate the passivation ability and pitting corrosion resistance of stainless steels. Particular focus is on sputtering and severe plastic deformation (SPD) techniques that have previously been used to create bulk or surface-nanostructured stainless steels.

2. MAHNETRON SPUTTERING

Magnetron sputtering is a type of physical vapor deposition (PVD) coating technology by which a thin nanocrystalline layer with the exact chemical composition of the target is produced. The corrosion behavior of nanocrystalline stainless steel films produced by magnetron sputtering has widely been investigated during the last two decades [26-36]. A comprehensive summary of the electrochemical corrosion resistance of nanocrystalline stainless steels fabricated using magnetron sputtering is presented in Table 1. In most cases, the sputtered nanosize-grained samples have shown a wider passive range and a higher pitting corrosion resistance compared to coarse-grained counterparts, in spite of the same chemical compositions. This is believed to be due to the availability of much more diffusion paths for chromium to enrich in the passive film for nanocrystalline materials [31]. It has been verified that the passive film nucleates preferentially at the active sites like dislocations, grain boundaries, triple junctions, etc. and then gradually spread over the rest of the surface [33]. Passive film grows progressively on coarse grained surfaces, while it grows instantaneously on the nanocrystalline surfaces. Because of the smaller densities of these active sites on the surface of coarse-grained sample, the nucleation sites for passive film are limited which then resulted in uneven growth of the passive film. For the nanocrystalline sample, a large fraction of grain boundaries provides more nucleation sites for passive film growth. In the latter case, the distance between the neighboring nucleation sites was short which helped to rapid formation of a uniform passive film. A similar explanation has been also proposed by Pan et al. [34-36]. Another reason for the enhanced corrosion resistance is the homogeneous chemical composition of the sputtered films.

Ye and coworkers [27] showed that nanocrystallization using magnetron sputtering has different effects on corrosion resistance of 309 SS depending on the electrolyte used in electrochemical measurements. In Na_2SO_4 solution, there is not any significant difference between the passivity behavior of nanocrystalline (NC) and coarsecrystalline (CC) samples which can be attributed to the formation of compact passive films with the same electronic characteristics on both samples; however, a more homogeneous, compact and stable passive film is formed on the NC sample in NaCl solution. Corrosion behavior investigation of a nanocrystalline 309 SS coating prepared by magnetron sputtering in the transpassive region showed a lower corrosion current density and dissolution rate for the passive film formed on the nano-crystalline sample compared with the bulk stainless steel in Na_2SO_4 solution, which was attributed to the more homogeneous distribution of Cr at the surface of the sputtered coating [30]. Similar results obtained by Li and coworkers [32] who achieved a continuous and uniform thin passive film on sputtered nanocrystalline stainless steel.

All in all, most of the literature has reported that the surface nanocrystallization using magnetron sputtering caused an increase in the corrosion resistance of stainless steels. This can be explained in two ways: first, the nanostructured nature of the sputter deposited films help to provide more diffusion pathways for passivating elements (e.g. Cr) to diffuse to the surface to form a Cr-enriched passive film. Second, films produced by magnetron sputtering generally have more homogeneous chemical composition than the bulk counterparts. Inhomogeneous microstructure of the bulk stainless steels, especially presence of inclusions, has been known as one of the main reasons for localized corrosion initiation.

Table 1: Summary of the electrochemical corrosion parameters of nanocrystalline stainless steels produced by magnetron sputtering; E_{corr}: corrosion potential, i_{corr}: corrosion current density, E_b: breakdown potential. NC and CC stand up for nano-crystalline and coarse-crystalline samples, respectively.

Material	Environment	Grain size (nm)	E_{corr}^{NC}-E_{corr}^{CC} (mV)	$i_{corr}^{NC}/i_{corr}^{CC}$	E_b^{NC}-E_b^{CC} (mV)	Passive range (NC relative to CC)	Reference
Austenitic SS	3.5 wt.% NaCl	50-100	~90	~1 (similar)	~510	wider	[26,32]
309 SS	Na_2SO_4 + NaCl	50	-	-	0-800	wider	[27]
Fe-10Cr	H_2SO_4+Na_2SO_4+NaCl	20-30	61	~2.4	similar	-	[28]
304 SS	0.3 wt.% NaCl	~25	107	-	~850	-	[29]
304 SS	0.3 wt.% NaCl	50	-	~1 (similar)	~700	wider	[34]
304 SS	0.05 MH_2SO_4 + 0.2MNaCl	50	-	-	~800	wider	[35,36]

3. SEVERE PLASTIC DEFORMATION

One of the most used techniques for microstructure refinement of metals and alloys is severe plastic deformation (SPD). Reviews on the principles of various SPD processing techniques have been summarized by Valiev et al. [37], and a latter by Estrin and Vinogradov [38] on the properties achievable by SPD processing. So far different SPD methods have been developed for fabrication of nanostructures in the bulk or at the surface of metals and alloys. The plastic deformation induced by shot peening [39], ultrasonic peening [40], ball milling [41], equal channel angular pressing (ECAP) [7,42,43], cold rolling [44-47], laser shock peening [48,49], sandblasting [50], surface mechanical attrition treatment (SMAT) [51-53] and high pressure torsion (HPT) [54] has been reported to have a significant influence on the corrosion performance of stainless steels. The extent of the improvement or deterioration effect of the SPD processes on the corrosion performance are mainly dependent upon the degree of grain refinement, volume fraction of the grain boundaries, residual stresses and density of dislocations induced by each treatment. However, in most cases, the enhanced corrosion properties of the nanocrystalline materials has commonly related to the more diffusion paths for ions and atoms to migrate through grain boundaries and high density of dislocations [55-57], similar to that explained for nanostructured sputtered films in the previous section. Although it is fundamentally accepted that the type and extent of residual stresses can alter the free energy state and corrosion resistance of materials, the effects of residual stresses on passivity and pitting corrosion behavior have not fully been understood to date and different experimental results are reported in the literature. Some reports indicate a possible beneficial influence of the compressive residual stresses on the corrosion resistance [58-61].

In the following sections, some examples showing the effects of grain refinement induced using different types of SPD techniques on the passivity are provided and discussed.

The improved corrosion resistance of ultrafine grained AISI 2205 duplex stainless steel fabricated by cold rolling and reversion annealing could be mainly attributed to the formation of a thicker passive film on its surface, as indicated by Jinlong et al. [45]. Moreover, according to the results of XPS analysis, they suggested that grain refinement promoted the enrichment of Cr in the passive film, which can be considered as another important factor in improving corrosion resistance. Zheng and coworkers [42] stated that according to the XPS measurements, the thickness and composition of the passive films formed on the ECAPed and un-treated alloys were very similar. They claimed that this is probably due to the un-changed Cr diffusion properties for the nanocrystalline sample at room temperature. However, the results of the electrochemical measurements indicated a nobler open circuit potential, a lower corrosion current density and a higher corrosion resistance for the ECAPed samples than the un-treated counterpart. So, the authors suggested that compactness and stability rather than the thickness and composition change of the passive film have resulted in the high corrosion resistant 304 stainless steel prepared by ECAP [42]. Studies on the electrochemical corrosion behavior of bulk ultrafine-grained mild steel fabricated by ECAP showed a lower passive current density and an enhanced formation ability of a continuous dense passive film compared to the as-cast sample, which was attributed to the high number of preferential nucleation sites in the nanocrystalline sample and also to the refinement of the iron carbide particles during the ECAP processing [62]. Hasegawa and Osawa [63] observed a higher general corrosion rate but an improved pitting corrosion resistance for ultrafine grained AISI 304 SS and related this behavior to an increase of grain boundary density and formation of stable passive film on the surface. Similar results were obtained by Di Schino and Kenny [64] on the improved localized corrosion resistance of ultrafine grained AISI 304 stainless steel in 3.5 wt.% NaCl solution. The enhanced pitting corrosion resistance of bulk nanocrystalline stainless steel produced by a severe rolling technique indicated formation of a more compact passive film resulting in a lower corrosion rate and pitting susceptibility [65].

Electrical properties of the passive films have also been reported to have a significant effect on the corrosion resistance of the stainless steels. For example, Jinlong and Hongyun [66] reported that in spite of the disappearance of the most densely packed crystallographic plane (111) γ in the nano/ultrafine-grained 321 stainless produced by cold rolling, the corrosion resistance was increased in 0.5M H_2SO_4 solution probably due to the significant lower donor density (measured by Mott-Schottky method) compared to the coarse grained sample. This is in a good agreement with the findings of Ref. [45].

It is generally believed that phase transformation occur during some SPD treatments, which can subsequently affect the corrosion behavior of the material. For example, Peguet and coworkers [46] confirmed a complex effect of cold working on the pitting phenomena of stainless steel; volume fraction of strain-induced martensite and dislocation density were introduced as controlling factors during different stages of cold working. Grain refinement of 2025 duplex stainless steel by rolling caused a more positive corrosion potential, open circuit potential and higher corrosion resistance than the coarse grained sample in 0.1 M NaCl solution [67]. Corrosion assessment of ultrafine grained 316L stainless steel showed an improved pitting corrosion resistance and larger charge transfer resistance due to an increased density of grain boundaries and also the homogeneous distribution of non-metallic phases and pit-forming impurities after a warm multi-axially forging treatment [68]. Surface chromium enriched 316L and AISI 304 stainless steels showed improved corrosion resistance in borate buffer solution, the positive effects of nanocrystallization and more chromium content in the passive film overcame the negative effect of strain-induced α'-martensite [69,70]. The authors claimed this behavior is mainly due to stronger O_2 adsorption and chemical activity of chloride ions on the surface of the passive film formed on nanocrystalline sample. A summary of the data for electrochemical corrosion behavior of nanocrystalline stainless steels fabricated using SPD techniques is presented in Table 2.

Based on the results that have been reported in the literature, the effects of severe plastic deformation on passive film formation and stability are highly inconsistent. This inconsistency is not surprising, considering this fact that different SPD techniques causes various types of microstructural and compositional changes (such as grain refinement, texture, phase transformation, etc.), which can be evaluated carefully during electrochemical investigations.

4. SUMMARY

This review paper summarizes information on major nanocrystallization techniques that have been used as potential methods for improving the passivity of stainless steel and the enhancement of the corrosion resistance of stainless steels. Currently there is no general agreement in the literature on the effects of grain refinement on the passivation ability and corrosion performance of stainless steels. The main reason for this disagreement is the complex contribution of various factors resulted from different manufacturing processes and surface treatments which could pose complex effects on the corrosion resistance of a material. Each one of these factors has its own effect on the passivity and corrosion behavior and they are also highly interdependent on each other. One of the significant gaps in corrosion-related literature pointed out in this paper is that when studying the passivity and corrosion resistance of nanocrystalline stainless steels, all the possible determining factors like grain size, grain boundary character, texture, chemical composition, phase composition, surface roughness, residual stresses and inclusions should be considered.

Table 2: Summary of the electrochemical corrosion behavior of nanocrystalline stainless steels produced by different severe plastic deformation methods; E_{corr}: corrosion potential, i_{corr}: corrosion current density, E_b: breakdown potential. NC and CC stand up for nano-crystalline and coarse-crystalline samples, respectively.

Material	Environment	Treatment	E^{NC}_{corr}-E^{CC}_{corr} (mV)	$i^{NC}_{corr}/i^{CC}_{corr}$	E^{NC}_{b}-E^{CC}_{b} (mV)	Corrosion resistance	Comments	Reference
AISI 4145	Therminol	Laser shock peening	0-7	0.1-0.004	2-5	improved	Existence of high-level compressive residual stresses and uniform microstructural distribution	[48]
304 SS	Borate buffer solution	Cold rolling + HT	similar	~0.9	similar	improved	Fewer strain induced α'-martensitic and more nanotwins in nano/ultrafine grained sample	[44]
AISI 321	3.5 wt.% NaCl	Ultrasonic peening	(-90)-(+131)	similar	-	improved or decreased	Both the grain refinement and the volume fraction of the strain-induced martensite affect the corrosion	[40]
AISI 2205	Borate buffer solution	Cold rolling	-	-	~100	improved	Thicker passive film and significant decreased donor and acceptor densities in ultrafine grained sample	[45]
Iron	0.5 M H_2SO_4	Cold rolling	-11	1.34	-	decreased	Nanocrystallization accelerated the kinetics of the anodic metal dissolution process	[71]
304 SS	3.5 wt.% NaCl	Sandblasting annealing	-	-	-	improved or decreased	More diffusion of chromium to the surface, good adherence of passive film to annealed sample	[50]
Ferrite steel	0.01-1 M NaCl 0.09 M NaOH	ECAP	(-36)-(+36)	0.32-1.46	-	improved or decreased	Improvement in corrosion resistance of Nanostructured material is sensitive to Cl^- ions concentration	[43]
430 SS	0.1 M NaOH	Cold rolling + HT	~(-50)	-	-	decreased	Higher point defect density of the fine-grained sample than that of coarse-grained one	[72]
1Cr18Ni9 Ti SS	3.5 wt.% NaCl	Shot peening	-	-	-	improved	The passive film on the surface of the shot-peened surface is more easier to form and is more stable	[39]
316L SS	Saliva solution	Cold rolling + nitriding	70	-	-	improved	Grain refinement led to lower passive current density and a thicker or compact passive	[47]
Fe-10Cr Fe-20Cr	0.5 M H_2SO_4	High energy ball milling	15 and 5	similar	similar	improved	Higher chromium enrichment of the passive film formed on nanocrystalline sample	[41]
AISI 409 and 304	0.6 M NaCl	SMAT	(-540)-83	0.02-4.3	-	improved or decreased	It depends on the surface roughness, microstrain and defect density of the samples induced by SMAT	[51, 52]

5. ACKNOWLEDGEMENTS

Financial support from Deakin University Postgraduate Research Scholarship (DUPR) is greatly appreciated.

6. REFERENCES

[1] Ryan MP, Williams DE, Chater RJ, Hutton BM, McPhail DS, Why stainless steel corrodes, Nature 415 (2002) 770-774.

[2] Punckt C, Bolscher M, Rotemund HH, Mikhailov AS, Organ L, Budiansky N, Scully JR, Hudson JL, Sudden onset of pitting corrosion on stainless steel as a critical phenomenon, Science 305 (2004) 1133-1136.

[3] Marcus P, Maurice V, Strehblow HH, Localized corrosion (pitting): A model of passivity breakdown including the role of the oxide layer nanostructure, Corrosion Science 50 (2008) 2698-2704.

[4] Olsson COA, Landolt D, Passive films on stainless steels-chemistry, structure and growth, Electrochimica Acta 48 (2003) 1093-1104.

[5] Maurice V, Yang WP, Marcus P, XPS and STM investigation of the passive film formed on Cr(110) single-crystal surfaces, Journal of The Electrochemical Society 141 (1994) 3016-3027.

[6] Maurice V, Yang WP, Marcus P, X-ray photoelectron spectroscopy and scanning tunneling microscopy study of passive films formed on (100) Fe-18Cr-13Ni single-crystal surfaces, Journal of The Electrochemical Society 145 (1998) 909-920.

[7] Zheng ZJ, Gao Y, Gui Y, Zhu M, Studying the fine microstructure of the passive film on nanocrystalline 304 stainless steel by EIS, XPS, and AFM, Journal of The Solid State Electrochemistry 18 (2014) 2201-2210.

[8] Frankel GS, Pitting corrosion of metals: a review of the critical factors, Journal of The Electrochemical Society 145 (1998) 2186-2198.

[9] Frankel GS, Sridhar N, Understanding localized corrosion, Materials Today 11 (2008) 38-44.

[10] Maurice V, Yang WP, Marcus P, XPS and STM study of passive films formed on Fe-22Cr (110) single-crystal surfaces, Journal of The Electrochemical Society 143 (1996) 1182-1200.

[11] Ryan MP, Newman RC, Thompson GE, Atomically resolved STM of oxide film structures on Fe-Cr alloys during passivation in sulphuric acid solution, Journal of The Electrochemical Society 141 (1994) L164-L165.

[12] Nanjo H, Newman RC, Sanada N, Atomic images of 304SS surface after electrochemical treatments, Applied Surface Science 121/122 (1997) 253-256.

[13] Gods P, Isgor OB, Carpenter GJC, Li J, McRae GA, Gu GP, Nano-scale study of passive films and chloride-induced depassivation of carbon steel rebar in simulated concrete pore solutions using FIB/TEM, Cement and Concrete Research 47 (2013) 55-68.

[14] Oh K, Ahn S, Eom K, Jung K, Kwon H, Observation of passive films on Fe-20Cr-xNi (x=0,10,20 wt.%) alloys using TEM and Cs-corrected STEM-EELS, Corrosion Science 79 (2014) 34-40.

[15] Macdonald DD, Passivity-the key to our metals-based civilization, Pure Applied Chemistry 71 (1999) 951-978.

[16] Maurice V, Marcus P, Passive films at the nanoscale, Electrochimica Acta 84 (2012) 129-138.

[17] Soltis J, Passivity breakdown, pit initiation and propagation of pits in metallic materials-review, Corrosion Science 90 (2015) 5-22.

[18] Strehblow HH, Passivity of metals studied by surface analytical methods, a review, Electrochimica Acta 212 (2016) 630-648.

[19] Frankel GS, Li T, Scully JR, Localized corrosion: passive film breakdown vs pit growth stability, Journal of The Electrochemical Society 164(4) (2017) C180-C181.

[20] Laleh M, Kargar F, Suppression of chromium depletion and sensitization in austenitic stainless steel by surface mechanical attrition treatment, Materials Letters 65 (2011) 1935-1937.

[21] Liu L, Li Y, Wang F, Electrochemical corrosion behavior of nanocrystalline materials-a review, Journal of Materials Science and Technology 26(1) 2010 1-14.

[22] Ralston KD, Birbilis N, Effect of grain size on corrosion: a review, Corrosion 66 (2010) 075005-01-075005-13.

[23] Gupta RK, Birbilis N, The influence of nanocrystalline structure and processing route on corrosion of stainless steel: a review, Corrosion Science 92 (2015) 1-15.

[24] Ralston KD, Birbilis N, Davies CHJ, Revealing the relationship between grain size and corrosion rate of metals, Scripta Materialia 63 (2010) 1201-1204.

[25] Ralston KD, Fabijanic D, Birbilis N, Effect of grain size on corrosion of high purity aluminium, Electrochimica Acta 56 (2011) 1729-1736.

[26] Li T, Liu L, Zhang B, Li Y, Wang F, Growth of single corrosion pit in sputtered nanocrystalline stainless steel film, Corrosion Science 111 (2016) 186-198.

[27] Ye W, Li Y, Wang F, Effect of nanocrystallization on the corrosion behavior of 309 stainless steel, Electrochimica Acta 51 (2006) 4426-4432.

[28] Meng G, Li Y, Wang F, The corrosion behavior of Fe-10Cr nanocrystalline coatings, Electrochimica Acta 51 (2006) 4277-4284.

[29] Inturi RB, Szklarska-Smialowska Z, Localized corrosion of nanocrystalline 304 type stainless steel films, Corrosion 48 (1992) 398-403.

[30] Ye W, Li Y, Wang F, The improvement of the corrosion resistance of 309 stainless steel in the transpassive region by nano-crystallization, Electrochimica Acta 54 (2009) 1339-1349.

[31] Ryan MP, Laycock NJ, Isaacs HS, Newman RC, Corrosion pits in thin films of stainless steels, Journal of The Electrochemical Society 146 (1999) 91-97.

[32] Li T, Liu L, Zhang B, Li Y, Wang F, An investigation on the continuous and uniform thin membrane passive film formed on sputtered nanocrystalline stainless steel, Corrosion Science 104 (2016) 71-83.

[33] Gaberscek M, Pejovnik S, Impedance spectroscopy as a technique for studying the spontaneous passivation of metals in electrolytes, Electrochimica Acta 41 (1996) 1137-1142.

[34] Pan C, Liu L, Li Y, Wang F, Pitting corrosion of 304 SS nanocrystalline thin film, Corrosion Science 73 (2013) 32-43.

[35] Pan C, Liu L, Li Y, Wang S, Wang F, Passive film growth mechanism of nanocrystalline 304 stainless steel prepared by magnetron sputtering and deep rolling techniques, Electrochimica Acta 56 (2011) 7740-7748.

[36] Pan C, Liu L, Li Y, Zhang B, Wang F, The electrochemical corrosion behavior of nanocrystalline 304 stainless steel prepared by magnetron sputtering, Journal of The Electrochemical Society 159 (2012) C453-C460.

[37] Valiev RZ, Estrin Y, Horita Z, Langdon TG, Zehetbauer MJ, Zhu YT, Producing bulk ultrafine-grained materials by severe plastic deformation, JOM 58 (2006) 33-39.

[38] Estrin Y, Vinogradov A, Extreme grain refinement by severe plastic deformation: a wealth of challenging science, Acta Materialia 61 (2013) 782-817.

[39] Wang T, Yu J, Dong B, Surface nanocrystallization induced by shot peening and its effect on corrosion resistance of 1Cr18Ni9Ti stainless steel, Surface & Coatings Technology 200 (2006) 4777-4781.

[40] Mordyuk BN, Prokopenko GI, Vasylyev MA, Iefimov MO, Effect of structure evolution induced by ultrasonic peening on the corrosion behavior of AISI-321 stainless steel, Materials Science and Engineering A 458 (2007) 253-261.

[41] Gupta RK, Singh Raman RK, Koch CC, Electrochemical characteristics of nano and microcrystalline Fe-Cr alloys, Journal of Materials Science 47 (2012) 6118-6124.

[42] Zheng ZJ, Gao Y, Gui Y, Zhu M, corrosion behavior of nanocrystalline 304 stainless steel prepared by equal channel angular pressing, Corrosion Science 54 (2012) 60-67.

[43] Hadzima B, Janecek M, Estrin Y, Kim HS, Microstructural and corrosion properties of ultrafine-grained interstitial free steel, Materials Science and Engineering A 462 (2007) 243-247.

[44] Jinlong L, Hongyun L, Tongxiang L, Wenli G, The effects of grain refinement and deformation on corrosion resistance of passive film formed on the surface of 304 stainless steels, Materials Research Bulletin 70 (2015) 896-907.

[45] Jinlong L, Tongxiang L, Chen W, Limin D, Comparison of corrosion properties of passive films formed on coarse grained and ultra fine grained AISI 2205 duplex stainless steel, Journal of Electroanalytical Chemistry 757 (2015) 263-269.

[46] Peguet L, Malki B, Baroux B, Influence of cold working on the pitting corrosion resistance of stainless steels, Corrosion Science 49 (2007) 1933-1948.

[47] Jinlong L, Tongxiang L, Improved corrosion resistance of 316L stainless steel by nanocrystalline and electrochemical nitridation in artificial salvia solution, Applied Surface Science 359 (2015) 158-165.

[48] Lu JZ, Han B, Cui CY, Li CJ, Luo KY, Electrochemical and pitting corrosion resistance of AISI 4145 steel subjected to massive laser shock peening treatment with different coverage layers, Optics & Laser Technology 88 (2017) 250-262.

[49] Lim H, Kim P, Jeong H, Jeong S, Enhancement of abrasion and corrosion resistance of duplex stainless steel by laser shock peening, Journal of Materials Processing Technology 212 (2012) 1347-1354.

[50] Wang XY, Li DY, Mechanical and electrochemical behavior of nanocrystalline surface of 304 stainless steel, Electrochimica Acta 47 (2002) 3939-3947.

[51] Balusamy T, Kumar S, Sankara Narayanan TSN, Effect of surface nanocrystallization on the corrosion behavior of AISI 409 stainless steel, Corrosion Science 52 (2010) 3826-3834.

[52] Balusamy T, Sankara Narayanan TSN, Ravichandran K, Song Park Il, Ho Lee M, Influence of surface mechanical attrition treatment (SMAT) on the corrosion behavior of AISI 304 stainless steel, Corrosion Science 74 (2013) 332-344.

[53] Lu AQ, Zhang Y, Li Y, Liu G, Zang QH, Liu CM, Effect of nanocrystalline and twin boundaries on corrosion behavior of 316L stainless steel using SMAT, Acta Metallurgica Sinica 19 (2006) 183-189.

[54] Hug E, Prasath Babu R, Monnet I, Etienne A, Moisy F, Pralong V, Enikeev N, Abramova M, Sauvage X, Radiguet B, Impact of the nanostructuration on the corrosion resistance and hardness of irradiated 316 austenitic stainless steel, Applied Surface Science 392 (2017) 1026-1035.

[55] Laleh M, Kargar F, Velashjerdi M, Low-temperature nitriding of nanocrystalline stainless steel and its effect on improving wear and corrosion resistance, Journal of Materials Engineering and Performance 22 (2013) 1304-1310.

[56] Tong WP, Tao NR, Wang ZB, Lu J, Lu K, Nitriding iron at lower temperatures, Science 299 (2003) 686-688.

[57] Yu X, Jiang Z, Zhao J, Wei D, Zhou C, Huang Q, Effect of grain-refined microalloyed steel on the formation mechanism of a tight oxide scale, Corrosion Science 85 (2014) 115-125.

[58] Lu JZ, Luo KY, Yang DK, Cheng XN, Hu JL, Dai FZ, Qi H, Zhang L, Zhong JS, Wang QW, Zhang YK, Effect of laser peening on stress corrosion cracking (SCC) of ANSI 304 austenitic stainless steel, Corrosion Science 60 (2012) 145-152.

[59] Peyre P, Carboni C, Forget P, Beranger G, Lemaitre C, Stuart D, Influence of thermal and mechanical surface modifications induced by laser shock processing on the initiation of corrosion pits in 316L stainless steel, Journal of Materials Science 42 (2007) 6866-6877.

[60] Rhouma AB, Braham C, Fitzpatrick ME, Ledion J, Sidhom H, Effects of surface preparation on pitting resistance, residual stress, and stress corrosion cracking in austenitic stainless steels, Journal of Materials Engineering and Performance 10 (2001) 507-514.

[61] Badran O, Kloub N, Al-Tal M, The effect of shot peening and polishing on the pitting corrosion resistance of stainless steel, American Journal of Applied Sciences 5 (2008) 1397-1402.

[62] Zhang L, Ma A, Jiang J, Yang D, Song D, Chen J, Sulphuric acid corrosion of ultrafine-grained mild steel processed by equal-channel angular pressing, Corrosion Science 75 (2013) 434-442.

[63] Hasegawa M, Osawa M, Corrosion behavior of ultrafine grained austenitic stainless steel, Corrosion-NACE 40 (1984) 371-374.

[64] Di Schino A, Kenny JM, Effects of grain size on the corrosion behavior of refined AISI 304 austenitic stainless steel, Journal of Materials Science Letters 21 (2002) 1631-1634.

[65] Wang SG, Sun M, Long K, The enhanced even and pitting corrosion resistances of bulk nanocrystalline steel in HCl solution, Steel Research International 83 (2012) 800-897.

[66] Jinlong L, Hongyun L, Comparison of corrosion properties of passive films formed on phase reversion induced nano/ultrafine-grained 321 stainless steel, Applied Surface Science 280 (2013) 124-131.

[67] Jinlong L, Tongxiang L, Chen W, Limin D, Effect of ultrafine grain on tensile behavior and corrosion resistance of the duplex stainless steel, Materials Science and Engineering C 62 (2016) 558-563.

[68] Muley SV, Vidvans AN, Chaudhari GP, Udainiya S, An assessment of ultra fine grained 316L stainless steel for implant applications, Acta Biomaterialia 30 (2016) 408-419.

[69] Jinlong L, Meng Y, Miura H, Tongxiang L, The effect of surface enriched chromium and grain refinement by ball milling on corrosion resistance of 316L stainless steel, Materials Research Bulletin 91 (2017) 91-97.

[70] Jinlong L, Guo W, Liang T, Yang M, The effect of ball milling time and surface enriched chromium on microstructures and corrosion resistance of AISI 304 stainless steel, Materials Chemistry and Physics 197 (2017) 79-86.

[71] Oguzie EE, Wang SG, Li Y, Wang FH, Corrosion and corrosion inhibition characteristics of bulk nanocrystalline ingot iron in sulphuric acid, Journal of The Solid State Electrochemistry 12 (2008) 721-728.

[72] Fattah-alhosseini A, Vafaeian S, Influence of grain refinement on the electrochemical behavior of AISI 430 ferritic stainless steel in an alkaline solution, Applied Surface Science 360 (2016) 921-928.

CHALLENGES WITH DESIGNING RETROFIT CATHODIC PROTECTION SYSTEMS IN PETROCHEMICAL COMPLEXES

C. Chan
SMEC, Sydney, Australia

SUMMARY: Corrosion of pipeline services are of particular concern as unplanned shutdowns and remediation of contaminated soil can be extremely expensive for petrochemical complexes.

As a plant's production capacity expand over time, pipeline extensions are tied in to the existing pipelines to support the increased production capacity. Different section of the pipelines added to the plant have their own dedicated ICCP system. Sometimes these systems are designed and installed by different CP companies. Each section of the pipelines is electrically isolated with insulating flange kits.

To highlight these challenges, a previous project carried out by the author is used as a case study.

The pipelines were segregated into 50 areas within the plant. Ten existing ICCP systems were installed at locations within the plant intended to provide protection to the entire pipelines and storage tanks within the plant. The designed total current output was in the order of 250 Amps.

The pipelines within the plant are electrically bonded together via copper cables installed across the flanges of the pipelines to mitigate the risk of any stray current effects from the ICCP systems. The cables are directly bonded across the block valves which are buried underground.

The Plant is approximately 20 years old and four out of ten of the existing ICCP systems were found to be failing with very low current output even though maximum 50V driving voltage had been applied.

This paper discusses some challenges of the retrofit and discovery of existing CP design faults, the incorrect installation of bonding cables and the effect of current distribution due to electrical isolation at various parts of the pipelines.

Many of the faults found are commonly experienced in the Petrochemical complexes and could be easily overlooked.

Keywords: Cathodic Protection, Protection Criteria, Retrofit ICCP System, Electrical Continuity, High Resistance Bond, Pipeline Corrosion

1. INTRODUCTION

1.1 Plant Description

The Plant is situated in a dedicated heavy industrial zone where all the petrochemical, oil and gas refineries are located. The Plant was constructed approximately 20 years ago with production capacity of approximately 600,000 metric tons/year (mty) of ethylene and 300,000 mty of propylene. The Plant consists of complicated network of underground pipelines with eight aboveground storage tanks. The tanks are installed with HDPE liner with anode grid system installed between the tank bottom and the liner.

The hydrant pipelines and cooling water pipelines are electrically bonded together via single core copper cables buried underground.

The hydrant pipelines are made of carbon steel and interconnected with cast iron valve buried underground. Single core copper cables are welded across the cast iron valve to electrically bond the steel carbon pipes together underground.

Ten (10) impressed current CP systems were installed to protect the aboveground storage tanks, Cooling Water Pipelines and hydrant pipelines during construction in 1997.

Past CP system surveys had shown reduced current outputs from some individual ICCP systems and insufficient protection levels on the pipelines which suggested that various CP systems are failing to maintain adequate levels of protection.

This paper discusses the challenges with designing retrofit ICCP systems, the limited current distribution within the Plant and the root cause of these constraints.

In order to achieve even levels of protection on all the buried services within the Plant. As a minimum, the following requirements must be fulfilled:

- The system must have sufficient output capacity to provide adequate protective current.

  Additional current capacity must be provided to accommodate for losses to other steel services

  within the Plant.
- The current must be evenly distributed amongst the buried steel services within the Plant.
- Shielding effects must be considered.
- All steel services must be electrically bonded together that are intended to be protected by the CP systems.
- All steel services must be electrically isolated that are not intended to be protected by the CP systems.

In order to fulfil the above requirements, especially where congested steel services are installed closely to each other within a Plant environment, the design of a retrofit ICCP system must adopt a conservative approach when determining the design parameters.

Figure 1 below show the general arrangement of the Plant's services and the route of the buried

Hydrant pipelines.

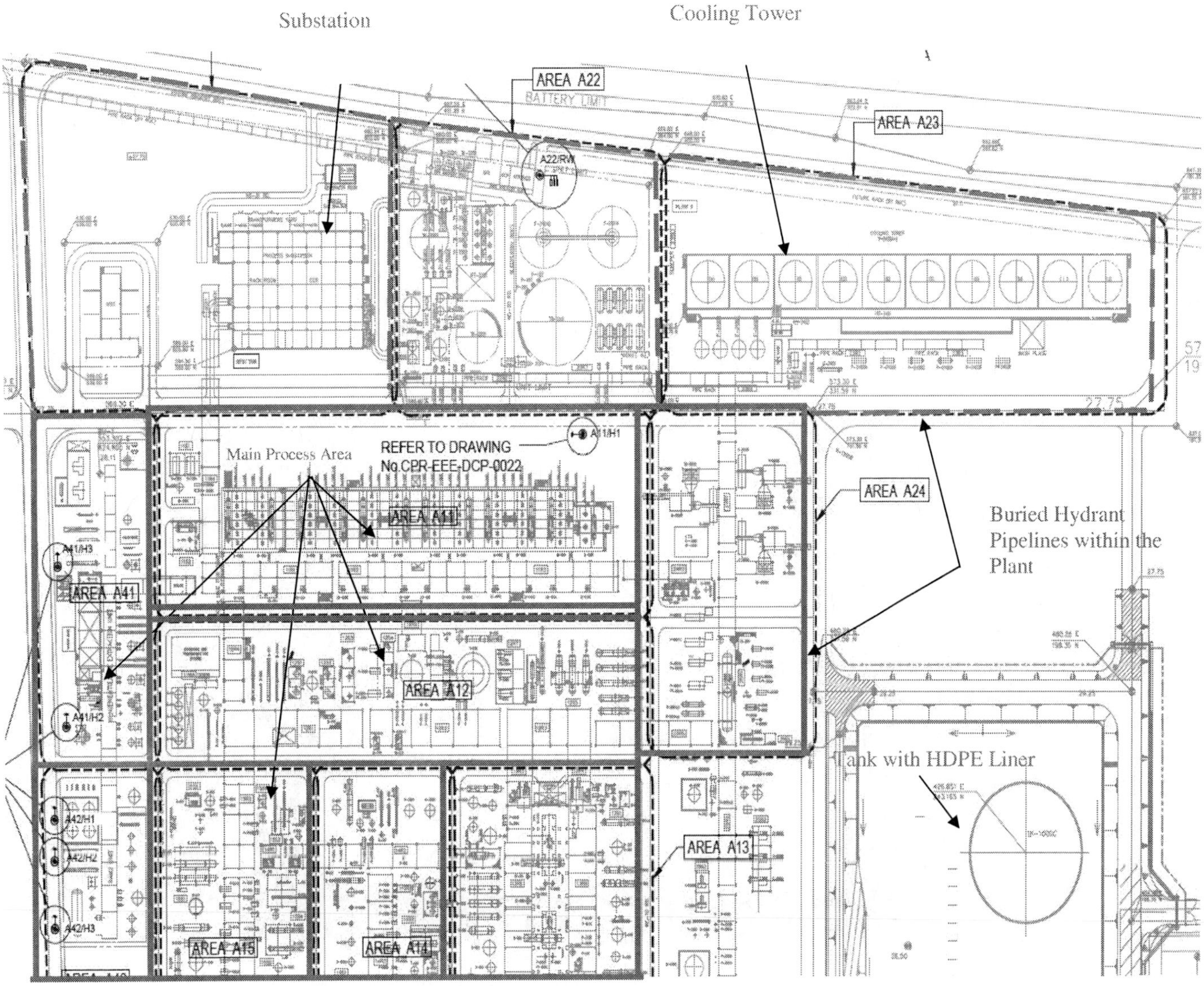

Figure 1: Plant General Arrangement and Underground Pipeline Route

2. RETROFIT ICCP SYSTEM

Prior to commencing the design of the new ICCP system to boost the protection levels of the buried pipelines, a desktop review of as-built drawings and site visit were undertaken for site familiarisation.

The as-built drawings showed that the buried section of the pipelines were electrically bonded via copper cable. Electrical continuity testing at area of the pipelines nearest to the proposed groundbed location were undertaken to confirm continuity.

Continuity testing was undertaken by measuring the potentials of the pipelines at dedicated test hole where the portable copper/copper sulphate reference electrode was placed at a fixed location. The positive test lead was then connected to the underground side of the flanges. Depending on the distance separating the pipelines, typically a difference in potential of 10mV or less during current interruption was adopted to confirm electrical continuity of buried section of the pipelines as depicted in Figure 2 below.

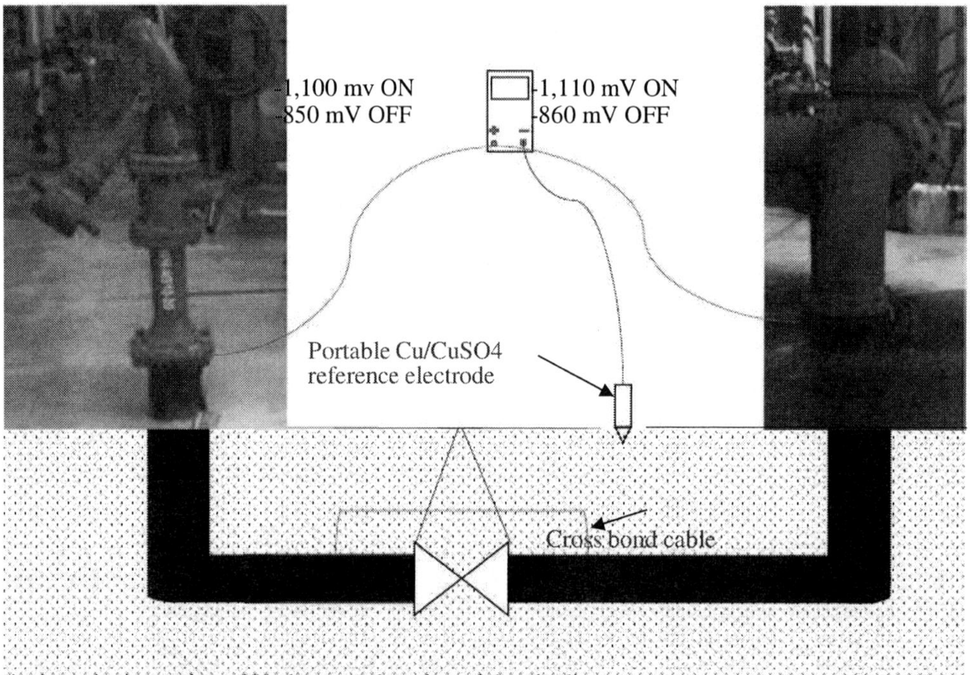

Figure 2: Electrical continuity testing of Underground section of the pipeline

As part of the retrofit work, two high current output anode groundbeds was installed along the pipeline route at location where the pipelines are receiving the lowest amount of CP protection, but furthest away from the plant services to avoid any current shielding from the copper earthing rods.

The combined maximum output current for the retrofit ICCP system were designed to have the capacity to output current of 200 Amps DC.

A total of 45 test points were nominated and installed across the plant to capture the current distribution profile of the new ICCP system.

Prior to commissioning the new ICCP system, the existing ICCP systems were switched off for approximately 6 weeks to allow the pipelines to depolarise in order to measure the native state potentials.

The commissioning consisted of energising the new ICCP system at 10% out the maximum current output for approximately 2 weeks (whilst all the existing TRUs were switched off) and then adjusting the current output to reflect the rate of polarisation. Potentials measurement were conducted on weekly basis to monitor the polarisation levels.

During the 8 weeks energisation, approximately 60% of the test points either had experienced low polarisation shift or no significant shift as highlighted in Table 1 below. Most of the test points that did not achieve the protection criteria were mostly located at the inner section of the plant. This suggested that there was some shielding preventing the CP current from flowing to the inner section of the plant. Subsequently, the existing ICCP systems were also switched ON to assess whether the polarisation will improve.

Areas	Test Points	CP "OFF" Potential (mV)				1st Polarization Shift (mV)	2nd Polarization Shift (mV)	3rd Polarization Shift (mV)
		Completed Depolarization	Polarization after 24 hrs	Polarization after 10 days	Polarization after 60 days			
		03-05-13	13-05-13	20-05-13	17-07-13			
[redacted]								
	H3	-440	-488	-484	-495	48	44	55
12	H2	-307	-331	-330	-328	24	23	21
	H3	-299	-328	-327	-325	29	28	26
[redacted]								
15	H3	-354	-412	-424	-430	58	70	76
	H4	-294	-384	-387	-398	90	93	104
21	H2	-415	-481	-523	-464	66	108	49
	H3	-287	-303	-307	-312	16	20	25
[redacted]								
	RW	-447	-458	-459	-458	11	12	11
23	RCW2	-543	-573	-558	-572	30	15	29
	CW3	-398	-472	-430	-446	74	32	48
[redacted]								
27	H3	-558	-560	-570	-573	2	12	15
	H4	-547	-610	-611	-623	63	64	76
[redacted]								
41	H3	-374	-364	-362	-379	-10	-12	5
	H5	-324	-333	-335	-332	9	11	8
42	H2	-443	-423	-443	-461	-20	0	18
	H3	-359	-358	-351	-361	-1	-8	2
51	H2	-332	-352	-335	-359	20	3	27
	H5	-444	-463	-443	-458	19	-1	14

Table 1: Polarisation shift potentials

Following the energisation of all the ICCP systems for approximately 2 weeks, there was some improvement in the potential levels, however the potentials were still considered low and did not meet the protection criteria.

Subsequently, site investigation works were undertaken in stages to find the root cause of the limited current being distributed to the inner section of the plant.

2. SITE INVESTIGATION

2.1 Stage 1: Pipe to Soil potential measurements

Pipe to soil potentials was undertaken from the pipelines closest to the groundbed to the remote location where there were no potential shifts from ON/OFF. It was observed that the soil potential gradient was limited to approximately 30 metres from the anode groundbed to the pipelines within the plant as indicated in Figure 3 below. There were no potential shifts observed beyond the 30 metre radius towards the inner part of the plant.

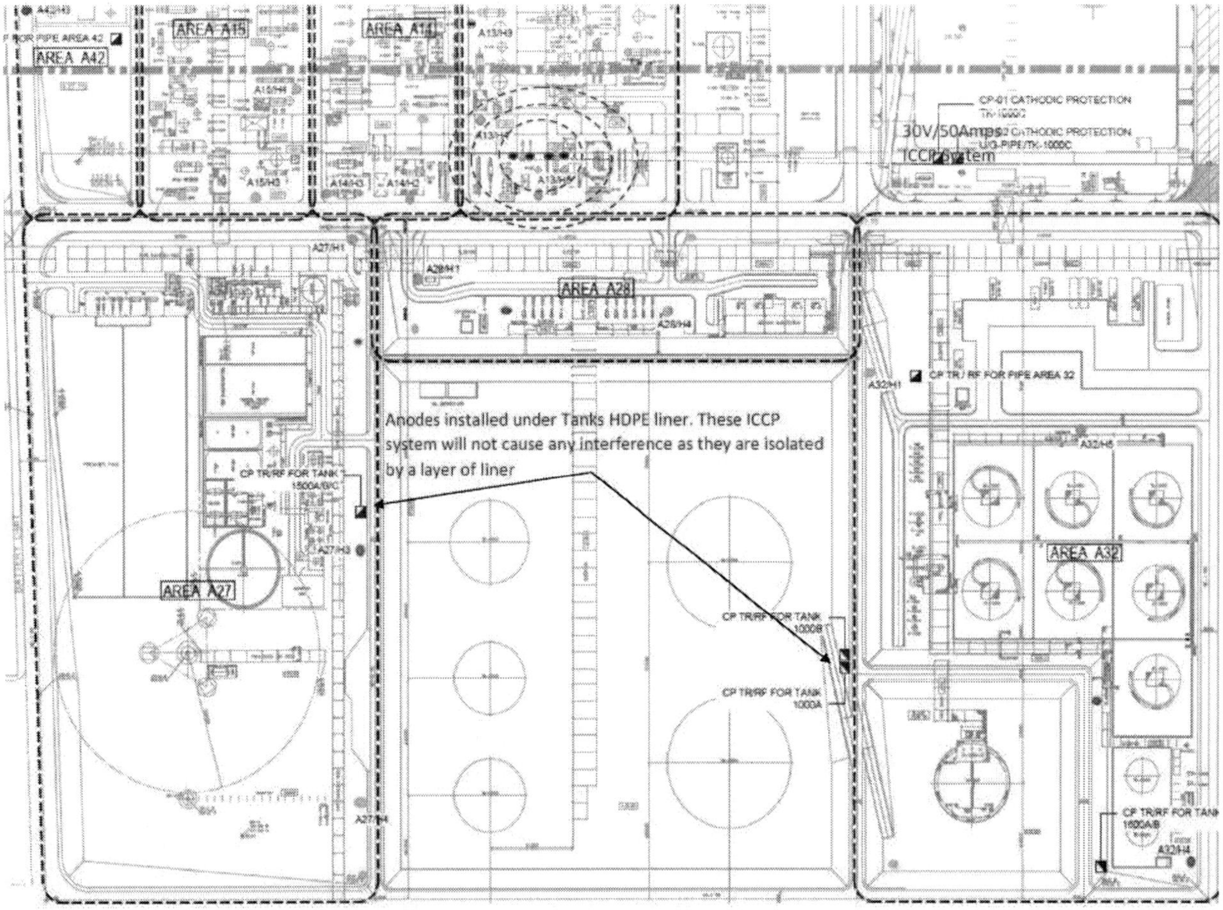

Figure 3: Soil potential gradient

Most of the pipes beyond 30 metres from the groundbed were not being protected as very small shift of potential were observed whilst the TRU was interrupted.

Current measurement at individual earth stakes were undertaken using a portable clamp meter. It was observed that the earth stakes were experiencing high current measurements indicating that the CP current had been taken up by these earth stakes, even though this can only give an indication as there are other variables that could influence the current measurements using a portable current clamp meter.

2.2 Stage 2: Electrical Continuity Test

Electrical continuity tests were undertaken in areas following the pipeline route. During the continuity test, it was observed that the existing insulating flange kits have failed as the underground section of the pipe was shorted to the aboveground section. This has a major impact on current demand as the earthing systems are now electrically shorted with the pipelines.

Furthermore, it was observed that the hydrant pipelines located at the inner part of the plant were electrically isolated to the rest of the pipelines. This was due to the fact that cross bonding cables originally installed to provide electrical continuity along the various tie-in points had experienced high resistance due to the age and possible disconnection of some of the cables. These high resistance bonds have prevented the current from flowing to the inner sections of the pipelines thereby causing very low potential shifts depicted in Figure 4 below.

The plot plan below shows the location of low potential levels due to the high resistance bond.

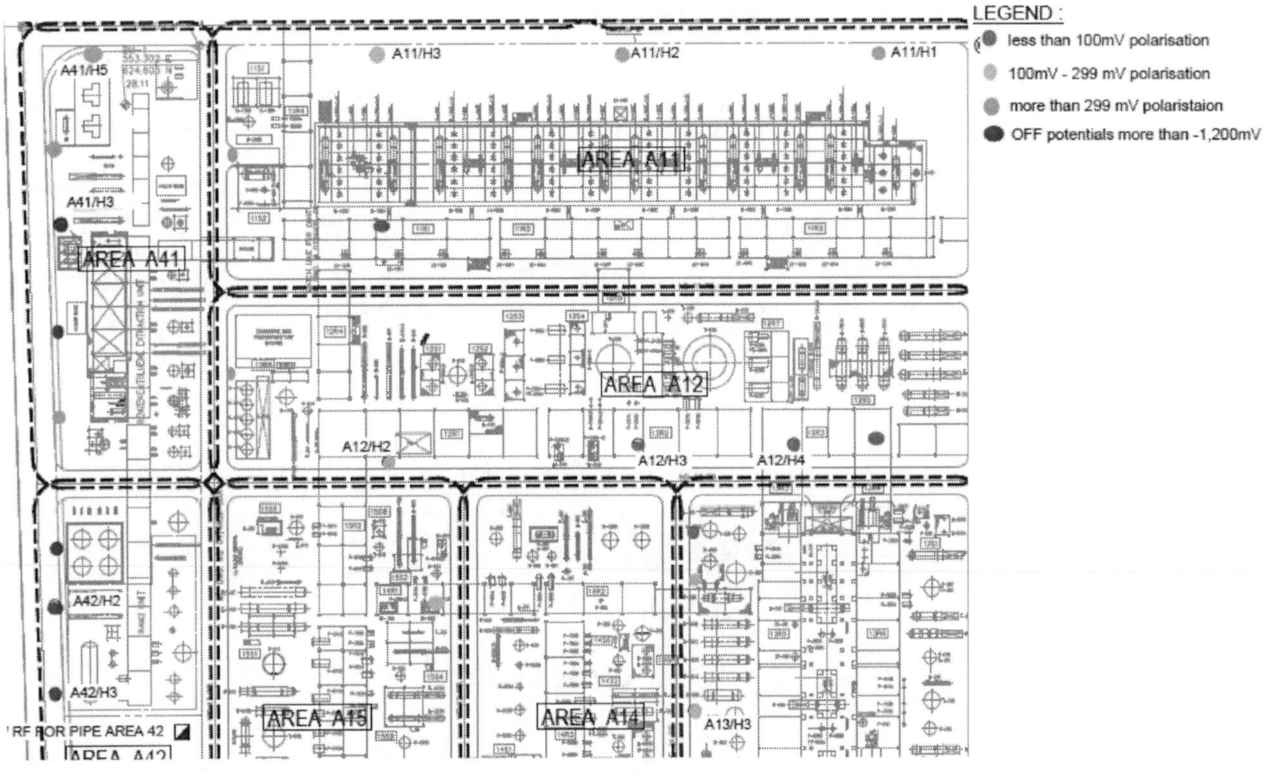

Figure 4: Colour coded potential shift measurements

3. HIGH RESISTANCE BOND

During construction, it was intended for each section of the pipelines where installed with block valves to be cross bonded with a single core copper cable thermite welded directly to the buried pipelines as per Figure 5 below. As time progressed, the copper cables were either broken or had developed high resistive bonds between sections of the pipelines that were installed with block valves.

This had resulted in various sections of the pipelines being electrically isolated, hence blocking the CP current from flowing to the inner section of the pipelines located in the middle of the plant. Diagram below shows the configuration of the bonding cables across the block valves.

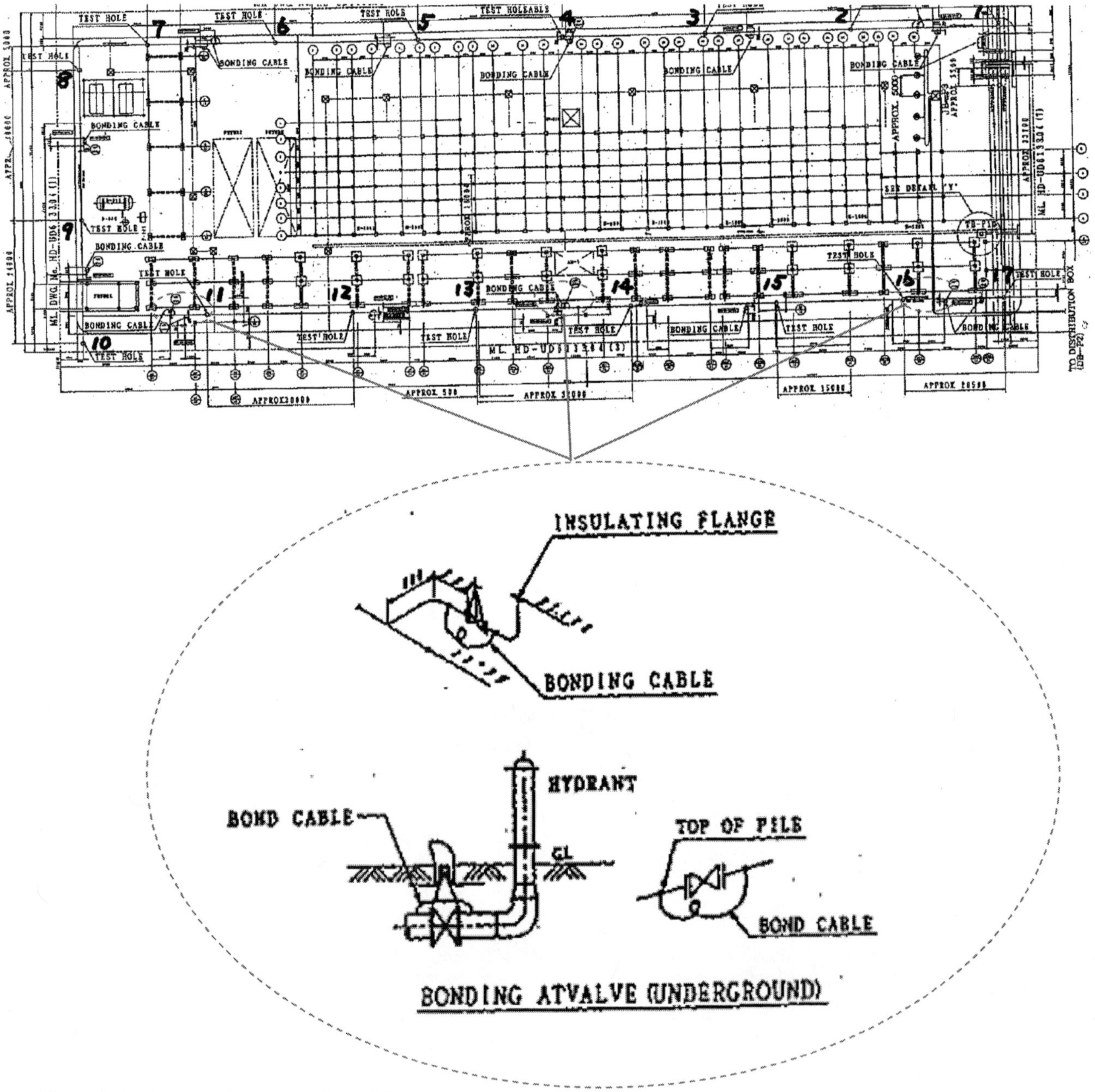

Figure 5: Underground cross bonding cable

4. EARTHING SYSTEM CP CURRENT SINK

The Plant was installed with different types of earthing systems with a common earthing grid system and individual earth stakes within the Plant.

The Plant's common earth grid system comprised of bare copper stranded wires installed within the Plant's earthing grid which provided a low resistance to earth. Any exposed metallic structures such as pipelines, tanks, supports, fences, steel reinforcement, electrical services within the plant area were earthed in the event of power surges and lightning strikes.

In a congested plant configuration, these earthing grid is likely to be electrically continuous with nearby electrical earthing systems due to interconnection via pipes.

The original design of impressed current cathodic protection systems (ICCP systems) to protect the entire surface area of buried steel services may not have considered drainage of CP current to these copper wires.

Whilst it is normal practice to electrically isolate the aboveground section of the pipes to the buried section via insulating flange kits, maintaining an effective isolation is often very difficult, especially for hydrant and cooling water pipelines in plant environments.

When an electrical short occurs, the affected pipelines will be electrically bonded to the entire copper earthing system which draws most of the CP current to the earth's copper wires.

In addition, some of the existing anode groundbeds installed are at close proximity to the copper wires, where most of the CP current is being drawn to these earthing wires with only limited CP current going to the pipelines.

5. ORIGINAL CATHODIC PROTECTION SYSTEM DESIGN

There are two types of ICCP systems installed within the plant. The horizontal anodes are designed to protect the above ground storage tanks and vertically installed anodes for the hydrant pipelines. These anodes are treated as separate ICCP systems to other steel structures as the HDPE liner containment will act as a barrier for CP current to flow beyond the tank's bottom plate.

The first type of ICCP system comprised of horizontal anodes installed between the tank bottom plate and a layer of HDPE liner. These anodes were providing low output current due to the nature of anode grid configuration and the close proximity from the anodes to the tank base.

The second type of ICCP system comprised of vertically installed MMO canister anodes installed at approximately 5m depth. Five anode wells are designed to provide current output of around 30 Amps which were intended to protect approximately 1,500 square metres of buried pipelines. The anodes are installed in close proximity to steel vessels. These steel vessels were installed with individual copper earthing rods and some anode groundbeds are in close proximity to these earthing rods as shown in Figure 6 below. Due to the close proximity, the current from the ICCP system is being drawn to the earthing rods thereby acting as a current sink.

Refer below for location of steel vessels with earthing rod stakes and anode grounded.

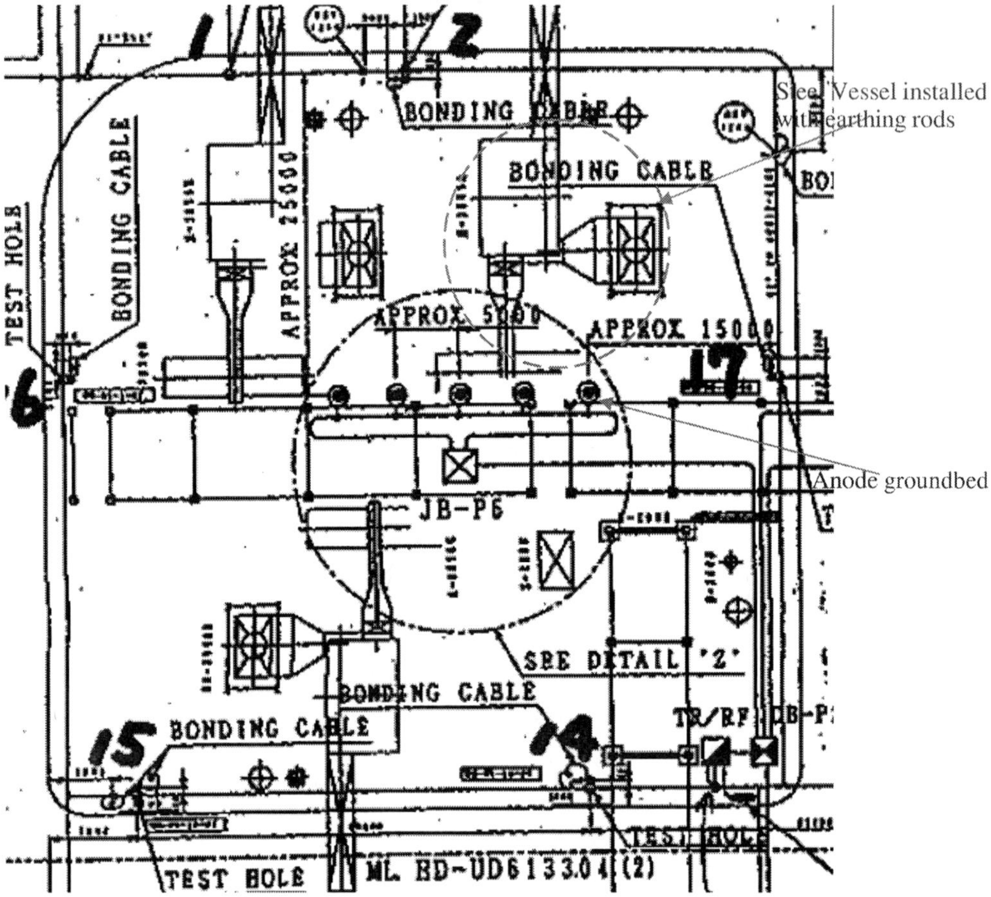

Figure 6: Original anode groundbed locality plan

6. PROPOSED SOLUTION

Due to the various challenges, the approach adopted was to provide better CP current distribution across the plant by undertaking the following actions:

- Replace all failed anodes and re-connect to existing TRU and junction boxes. Adjust current output as required to ensure that the pipelines are not over-protected.
- Cross bond selected sections of the hydrant pipelines to ensure that the entire length is electrically continuous.
- Cross bond cooling water and hydrant pipelines within the plant.
- Replace aboveground insulating flange kits to electrically isolate earthing rods and other metallic services exiting the plant.
- Install new anode groundbed as far away from the Plant's buried services as possible. A remote grounbed location is preferred if Plant configuration allows.

This approach was considered to present a practical and cost effective solution to the challenges outlined above.

7. CONCLUSIONS

The design of retrofit ICCP systems comes with challenges which are mainly due to:

- Incorrect as-built drawings provided for underground metallic services
- Electrical shorts plant's ICCP systems to copper earthing and metallic services external to the Plant
- Failure of above ground insulating flange kits
- Failure of underground cross bond cables
- Incorrect placement of anode groundbeds

In order to minimise any unforeseen challenges, it is imperative that a detailed site investigation is undertaken prior to commencing the design of any retrofit CP systems.

Depending on the unique Plant configuration and constraints, detailed site investigation should consist of, but not limited to the following works:

- Undertake electrical continuity test on representative amount of steel services intended to be protected by the CP systems
- Undertake electrical isolation test on any steel services exiting from the Plant
- Determine location of steel services in comparison to as-built drawings. Any discrepancies with actual site condition from the as-built drawings should be marked up and convey to plant owners.
- Assess the functionality of existing (original) ICCP systems
- Undertake a current drainage test wherever possible

- Apply a conservative amount of contingencies for current losses to other services for designing retrofit CP systems

8. REFERENCES

1. Australian Standard AS2832.3 (2003), "Cathodic Protection of Metals Part 2: Compact Buried Structures".

2. NACE SP 0169-2007, "Control of External Corrosion on Underground or Submerged Metallic Piping Systems".

9. AUTHOR DETAILS

C Chan is a Senior Corrosion Protection practitioner at SMEC Australia Pty Ltd.

He has more than 10 years' experience in cathodic protection related works has developed a wide range of experience covering condition assessment of reinforced concrete, buried and immersed steel structures, cathodic protection (CP) design for buried and immersed steel structures, monitoring of existing CP systems, site investigation to identify CP system deficiencies, stray current investigation on pipelines and tanks, site survey, troubleshooting of CP systems, project management and construction management during installation and commissioning of CP systems.

IMPORTANCE OF ACCURATE PIPELINE CORROSION DIAGNOSIS FOR OPTIMIZED CORROSION MANAGEMENT

S. Loftus, D. Sandana, I. Laing
ROSEN UK, Newcastle-upon-Tyne, UK

SUMMARY: Corrosion engineers are frequently faced with the challenging task of diagnosing pipeline corrosion with only minimal data available relating to the system conditions. The extent and reliability of such a corrosion diagnosis can be heavily influenced by the quality and quantity of data available.

A corrosion diagnosis methodology integrating the organization's long-term experience in corrosion engineering, inspection technologies and inspection signal analysis has been developed. To ensure that all probable corrosion conditions are critically addressed for a pipeline system the methodology relies on (i) 'Pattern Recognition', drawing on knowledge of pipeline 'symptoms' (*e.g.* direct inspection results, elevated bacterial levels in deposits) experienced in similar operational/environmental conditions; and (ii) 'Differential Diagnosis' to determine the most likely corrosion mechanism(s). Key pipeline information is obtained *via* the collaborative management of a Request Form of Information, appraising the key data required from the pipeline operator. Well-founded corrosion engineering judgements are agreed for use as data inputs under circumstances where only minimal operational and environmental data is available. For internal corrosion diagnosis specifically, corrosion models such as NORSOK and ECE are used alongside flow assurance analysis (OLGA/PIPESIM) as an integral part of the evaluation process to identify key corrosion 'hot-spots'. Corrosion engineering expertise is critical in understanding the limitations of modelling, evaluating the reliability of the modelling outputs and the overall likelihood of realization of the identified threats. Under the proposed diagnosis methodology, a detailed evaluation of corrosion 'symptoms' and assessment of pipeline conditions can be used to pinpoint a single candidate corrosion condition, targeting specific remedial actions and mitigation strategies (treatment plan) for an optimized pipeline corrosion management strategy.

The corrosion diagnosis methodology is presented in this work with a case study of an operational pipeline for which the methodology has been applied to, and the corresponding improvements and optimizations recommended for the overall pipeline corrosion management strategy.

Keywords: Corrosion diagnosis, corrosion management, integrity.

1. INTRODUCTION

It is well-known that corrosion is an expensive problem, with the global cost of corrosion estimated to be US$2.5 trillion (equivalent to 3.4% of the global Gross Domestic Product (GDP) (2013)). However if available corrosion control practices are used effectively, it is estimated that savings of between 15 and 35% of the cost of corrosion could be realised, *i.e.* between US$375 and $875 billion annually on a global basis [1]. To achieve the full extent of these savings, corrosion management and its integration into an organisation's integrity management system must be accomplished by implementing a Corrosion Management Strategy (CMS).

For pipeline systems in particular, a CMS should be an integral part of a Pipeline Integrity Management System (PIMS). However a CMS and the overall process of managing corrosion is of such importance that it can also be run independently as a parallel process to PIMS. Once established, effective management of corrosion involves the continuous review, maintenance, and improvement of a tailor-made CMS to ensure the safe operation of a pipeline system throughout its intended service or production life.

An example of an integrity management cycle is illustrated in Figure 1. It is now well-established that the first key stage in developing a PIMS is to conduct an Integrity Review Process (IRP) [2 - 4], in which all potential internal and external corrosion threats are determined and evaluated, with corrosion hot spots identified for a given system. The IRP can be conducted using a variety of different assessment methods; examples include Corrosion Risk Assessments (CRA), Risk Based Inspections (RBI) and Internal/External Corrosion Direct Assessment (ICDA/ECDA) methodologies (Figure 1).

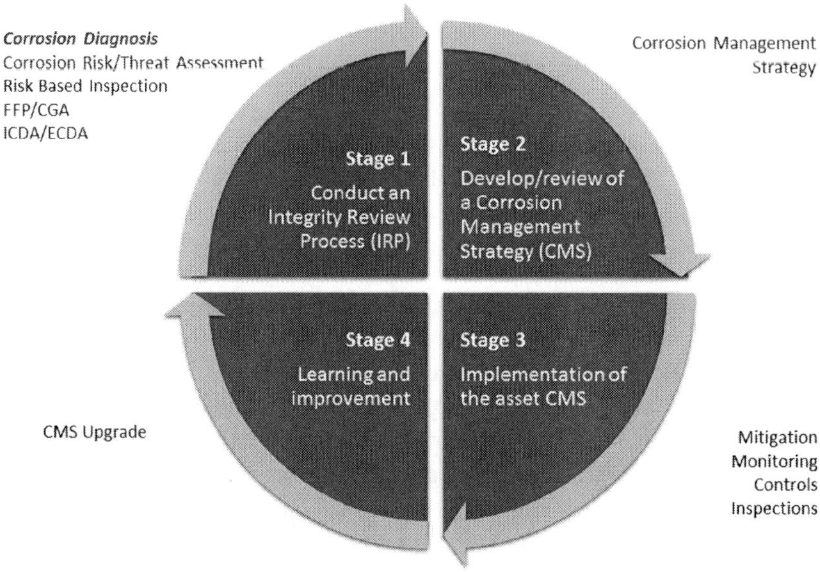

Figure 1: Integrity management cycle.

Only once an understanding of the corrosion threats is gained and a risk prioritisation conducted as part of the IRP, can cost-effective mitigation strategies (Stage 2) and inspection/maintenance regimes (Stage 3) be appropriately developed and optimised. Such an approach is essential in the current climate where integrity management budgets have become slim or in the case of ageing asset operations, when an operator faces pressures of an increasing maintenance burden against a decline in production [5]. It is important to have a comprehensive and structured methodology in place for evaluating corrosion, with corrosion diagnosis a key element underpinning all of the IRP assessment methods. Corrosion engineers are frequently faced with the challenging task of diagnosing pipeline corrosion with only minimal data available relating to the system conditions. The extent and reliability of such a corrosion diagnosis can be heavily influenced by the quality and quantity of data available. In order to address this, a corrosion diagnosis methodology integrating the organization's long-term experience in corrosion engineering, inspection technologies and inspection signal analysis has been developed to overcome issues associated with lack of available data, to optimise pipeline corrosion management. Improvements at all stages in the integrity management cycle will ultimately result in the optimisation of corrosion management, whether it be in terms of cost-effectiveness, resource optimisation or reduced risk of incidents and pipeline safety. Refining the accuracy and reliability of the corrosion diagnosis is therefore a critical first step in the early stages of optimising a CMS.

2. CORROSION DIAGNOSIS METHODOLOGY

Typically pipeline corrosion control measures are determined by the composition of the fluids transported (internal) and the soil conditions in which the pipe is laid (external). For example, corrosion inhibitor may be dosed into pipeline systems known to have a high CO_2 content to lower internal corrosion rates, or suitable coatings and cathodic protection (CP) systems can be applied to the pipeline to prevent external corrosion. Once the pipeline is in service, it often turns out that other additional factors, usually of an operational nature, pose a serious threat to the integrity of the pipeline [6]. An example is bacterial contamination carried over to an export pipeline from upstream processing operations. Under these circumstances it becomes necessary to conduct an IRP for the in-service pipeline and comprehensively diagnose any potentially new operational corrosion condition.

The term 'diagnosis' is typically associated with the field of medicine, where it is defined as 'the process of determining by examination the nature and circumstances of a diseased condition' [7]. Corrosion diagnosis can be defined in more general terms as a means of determining or analysis of the cause and nature of a corrosion problem or condition. Like a disease, corrosion is a problematic condition that requires accurate diagnosis in order to provide the correct and effective treatment.

The key objectives of the corrosion diagnosis methodology developed are outlined as follows:

1. Identify 'symptoms' *i.e.* indications of a corrosion condition, from a comprehensive review of pipeline inspection and monitoring data.
2. Assess and evaluate the symptoms alongside key operating conditions and environmental conditions of the pipeline, either internally or externally depending on the nature of corrosion.
3. Based on a review of the symptoms and pipeline conditions, determine the most probable corrosion/degradation mechanisms and their causes.

2.1 Identification of Symptoms

For the purposes of the corrosion diagnosis methodology, a corrosion symptom is defined as an observable sign (or indication) of a potential corrosion condition, typically representative of a departure from the expected pipeline condition under normal operation. The identification of corrosion symptoms involves the review of pipeline inspection and monitoring data relevant to the pipeline system. Symptoms indicating a corrosion condition are determined from a comprehensive review of the following types of data and documentation:

- Results of direct and/or indirect inspections and assessments *e.g.* in-line inspection (ILI), CP surveys, *etc.*
- Results of any monitoring or sampling conducted *e.g.* corrosion rate monitoring, fluid sampling, *etc.*
- Previous failures related to corrosion.

From these types of data sources, examples of corrosion symptoms could include:

- Corrosion concentrations in distribution patterns from ILI data.
- Areas of over/under-protected CP potentials reported from CP surveys.
- Coating defects reported from surveys such as a direct current voltage gradient (DCVG) survey.
- Abnormal/elevated corrosion rates reported from in-line corrosion rate monitoring (*e.g.* linear polarization resistance probe) or external corrosion coupons.
- High iron counts in sampled process fluids.
- Elevated bacterial numbers in deposits from pig trash analysis.
- Historic widespread internal corrosion failures.

2.2 Evaluation of Symptoms and Assessment of Pipeline Conditions

Once the symptoms of a corrosion condition have been identified, an assessment of key pipeline operational and environmental conditions is required in order to identify and evaluate any conditions that could be conducive to a corrosion condition and aligned with the identified symptoms. The performance of any corrosion control measures in place should be audited as part of the root cause evaluation, as they too may affect the symptoms or corrosion condition present.

A comprehensive Request Form of Information (RFI) is used to manage the data requirements for this evaluation stage. The RFI aims to collect as much information as possible for the pipeline system, with data requested under the following general categories:

- Pipeline general data, *e.g.* service type, age, diameter, length, *etc.*
- Design, construction and commissioning data, *e.g.* pipeline location and route including environmental data, basis of design documentation, *etc.*
- Operational parameters, *e.g.* operating temperature, pressure, fluid composition, *etc.*
- Corrosion mitigation data, *e.g.* internal and external corrosion protection details.
- Pipeline history, *e.g.* previous inspections, reported incidents or failures, rehabilitation procedures/actions.

In the absence of any essential data, well-founded corrosion engineering judgements are agreed for use with the client/parties involved, with the potential impact on the assessment discussed and evaluated. Note that the quality and confidence in such an assessment/evaluation process is dependent upon the quantity and quality of operational and environmental data available.

2.3 Principles of Diagnosis

Two diagnosis methodologies that are commonly recognized in the field of medicine are pattern recognition and differential diagnosis. Once an evaluation of corrosion symptoms and assessment of pipeline conditions has been conducted, the principles behind these diagnostic methodologies can be applied effectively in the diagnosis of corrosion, as described below:

Pattern Recognition

This method uses experience to recognize a pattern of characteristics. It is mainly based on certain symptoms or signs being associated with certain corrosion conditions, not necessarily involving the more cognitive processing involved in a differential diagnosis.

Differential Diagnosis

This method is based on considering as many candidate corrosion mechanisms or conditions as possible that can possibly cause the signs or symptoms, followed by a process of elimination or at least of rendering the entries more or less probable by further processing and evaluation, aiming to reach the point where only one candidate condition remains probable. The final result may also remain a list of possible (or interacting) conditions, ranked in order of probability or severity.

2.4 Determination of Probable Corrosion Mechanisms

For an initial screening diagnosis of the corrosion condition, pattern recognition would typically be used first to draw on knowledge and experience of the identified symptoms and/or corrosion conditions under known similar operational/environmental conditions in other systems. To avoid a biased diagnosis both the pattern recognition and differential diagnosis methodologies are used to ensure that by considering all symptoms and available pipeline data, all potential corrosion conditions have been considered systematically and carefully, and eliminated or rendered less probable as necessary.

Usually the pipeline service type (*i.e.* oil production, dry gas, wet gas, *etc.*) will dictate which corrosion conditions are most probable in a given pipeline system. Each different pipeline service type may be exposed to varying internal and external environments meaning that only certain corrosion/degradation mechanisms are likely in each case. For example, microbiologically influenced corrosion (MIC) would not typically be considered probable in a nominally dry gas pipeline system; the conditions conducive to MIC, including the presence of water (and contaminated fluids) and accumulation of deposits would not normally be expected in such a system.

Any changes in operational or environmental conditions along the length of a pipeline may result in localized corrosion conditions, meaning that only certain corrosion mechanisms are probable in particular pipeline sections.

2.5 Evaluation of the Corrosion Condition and Prognosis

After the corrosion problem has been effectively diagnosed and an understanding of the most probable corrosion threats is gained, the corrosion condition is evaluated to understand the severity of the corrosion threat to the system. In effect a CRA is conducted for the pipeline to assess the severity of the corrosion threat.

Evaluating the corrosion condition further can also be achieved by a prognosis; for the purposes of the corrosion diagnosis methodology, a prognosis is defined as an assessment of how the corrosion problem has *and* potentially will affect the pipeline condition. The prognosis for the pipeline system considers the effects of the corrosion/degradation mechanisms on the pipeline integrity, evaluating corrosion rates determined from direct pipeline inspections and/or corrosion rates determined from corrosion modelling associated with the relevant corrosion condition. Additional assessment methods such as a Fitness-for-Purpose (FFP) assessment or an estimated remaining life assessment of the pipeline can be conducted to also ascertain how the corrosion condition has affected *and* potentially will affect the pipeline in future operations if no remedial treatment plan is implemented.

2.6 Development of a Treatment Plan

Once the corrosion problem has been effectively diagnosed and evaluated, a treatment plan can be developed and optimised for the pipeline system. The treatment plan for the pipeline comprises of appropriate recommendations on mitigation, inspection and monitoring, and intervention strategies required in order to address and effectively manage the corrosion condition (and specific mechanisms) determined by the corrosion diagnosis. The formulation of such recommendations involves the following key steps:

- Identify mitigation and control strategies for the identified corrosion mechanism(s).
- Analyze, comment on and provide recommendations on, for example:
    - Chemical treatments, *e.g.* method, injection locations, frequency, dosing rates *etc.*
    - Physical and chemical monitoring, *e.g.* type of analysis, sampling locations, frequency, calibration requirements.
    - Monitoring controls, including type of probes and coupons used, their locations and frequency of monitoring.
    - Inspection activities, including type, location and frequency.
    - Operational pigging activities, including type of tools (pigs), frequency and recommendations on pig trash analysis.

- Define the type, timetable and frequency for future inspections.

3. CASE STUDY

A case study is presented for a 14 inch gas pipeline, 9 km in length, commissioned in 2001 and subject to an ILI in 2015 [8]. The pipeline had historically operated in a wet condition but is now considered to be operated dry. As the pipeline had operated under two distinct gas compositions (wet and dry), neither of the ICDA methodologies for wet or dry gas were applicable. As a result a hybrid methodology combining these two methodologies had to be considered. Historical pipeline operation data was not available, so assumptions based on well-founded corrosion engineering judgements were made in the assessment.

The IRP conducted for this pipeline system focussed on internal corrosion using ICDA methodologies, with a corrosion diagnosis and CRA carried out as part of the process to identify the most probable internal corrosion threats and evaluate the overall corrosion condition respectively. In addition OLGA multiphase flow modelling software was used as part of the corrosion evaluation to identify key internal corrosion 'hot-spots' along the length of the pipeline.

3.1 Corrosion Diagnosis: Corrosion Symptoms and Assessment of Pipeline Conditions

The inlet gas transported by the pipeline is from two separate sources. The first source has no drying facilities and is considered to be untreated, whereas the second is dehydrated to a target maximum dew point of 0°C under operating conditions. Since gas dehydration facilities were installed the target dew point was only met 41% of the operational time between the two gas streams. Therefore water condensation, drop-out or hold-up, was likely to have been a common occurrence in the pipeline since commissioning.

There is no direct corrosion inhibitor or chemical injection into the pipeline, hence corrosion control is solely reliant upon the absence of water. However, pig trash analysis data indicated the presence of liquids, confirming that free water is present within the pipeline. Additionally the pig trash analysis indicated sludge's and solids; as a consequence it is possible that microbial activity and hence MIC was a probable threat to the pipeline. This was further justified based on the high levels of acid-producing bacteria (APB) and sulphate-reducing bacteria (SRB) measured within the recovered sludge and liquids. The presence of bacteria within the pipeline is considered unusual; typically the threat of MIC is considered to be low in gas pipeline systems due to the absence or lack of nutrients required to sustain microbial growth in the condensed water present [9]. To add further confusion to this unusual finding, the bacterial counts within the retrieved sludge and fluids were found to be highly variable with little consistency. The cell measurements only considered planktonic bacteria however, with no sessile cell measurements being recorded. The highest planktonic levels were reported to be 4.7×10^3 cells/ml.

The level of planktonic bacteria that would typically be associated with a microbial problem are those in excess of 10^3 cells/ml. However, this is a planktonic bacteria measurement and it is ultimately sessile bacteria (those attached to the metal surface beneath biofilms) that would pose a corrosion problem. Planktonic numbers can only practically be used as a microbial activity indicator [10]. The presence of microbes in the pipeline is likely to be explained by contamination from non-sterile conditions during sample retrieval; this was ultimately determined to be the likely source of the bacteria.

It was confirmed through gas analysis at the outlet that no H_2S was present in the gas but 1.39 mol.% of CO_2 was present. With the presence of water confirmed in the pipeline, sweet corrosion was expected to be the predominant corrosion threat.

An ILI run using a Magnetic Flux Leakage (MFL) tool was conducted in 2015. The distribution of internal corrosion features can be seen in Figure 2. The internal corrosion concentrations exhibited in the pipeline are further indications (symptoms) of an existing corrosion condition.

To summarise from the assessment of pipeline conditions and review of available inspection and monitoring data, the following key corrosion symptoms were identified for the pipeline system:

- Presence of water indicated from pig trash analysis data and operational history.
- Presence of sludge and solids with elevated levels of planktonic bacteria from pig trash analysis data.
- Presence of CO_2 from outlet gas analysis.
- Internal corrosion concentrations in distribution patterns from ILI data.

The corrosion diagnosis determined CO_2 (sweet) corrosion to be the predominant corrosion threat, due to the presence of water and CO_2 in the pipeline and without any chemical inhibition. Top of Line (TOL) CO_2 corrosion is mainly dependent upon the rate of condensation and the replenishment of liquids. In gas pipelines the condensation rate below which the TOL corrosion rate is considered negligible depends on the organic acid concentration in the condensed water. As the presence of organic acids was not expected, the limit where TOL corrosion will become apparent is at water condensations rates in the range of 0.15 - 0.25 g/m^2s [11]. Based on the available information, at this stage of the corrosion diagnosis TOL CO_2 corrosion was also considered a probable threat to the pipeline system.

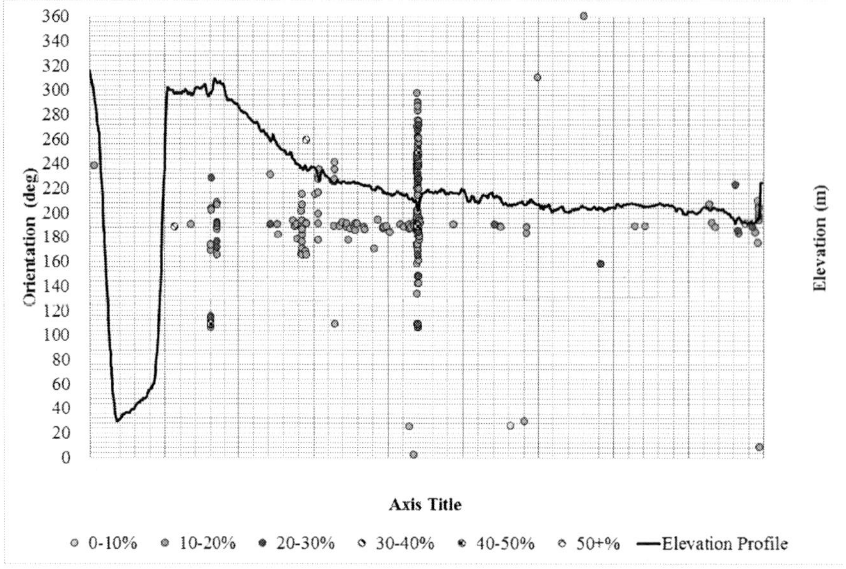

Figure 2: ILI results.

MIC was considered a threat but ruled out by differential diagnosis on the basis of a lack of substantial evidence to confirm that MIC was active in the pipeline; there was inconsistent microbial data, with no sessile bacteria reported to be indicative of active MIC. It was deemed that contamination from non-sterile conditions during sample retrieval was the likely source of the planktonic bacteria measured.

3.2 Corrosion Risk Assessment and Flow Modelling: Evaluation of the Corrosion Condition

The corrosion diagnosis determined that CO_2 corrosion and TOL CO_2 corrosion were the predominant threats in the gas pipeline system. High risk locations that were considered most susceptible to CO_2 corrosion within the pipeline were those where liquid hold-up and water condensation was most likely to occur, in addition to areas subjected to high wall shear stresses. These locations are likely to be located where there are changes in flow regime, especially if slugging is experienced in the pipeline.

The NOROSK M-506 CO_2 corrosion rate calculation model [12] was used initially to estimate potential corrosion rates in the pipeline system. In the calculation of the sweet corrosion rate the following parameters were used:

- 18.3°C (worst case temperature at which the dew point is exceeded).
- 26.9 barg.
- 1.39 mol.% CO_2.
- pH 4.1 (unsaturated) and 5.6 (saturated).

For such conditions at the pipeline inlet, and considering occasions where the dew point exceeds the outlet temperature of the pipeline, the calculated CO_2 corrosion rates were 1.9 mm/yr. (pH = 4.1) and 0.82 mm/yr. (pH = 5.6), Figure 3. Although depending on the rate of water replenishment, these corrosion rates would be transient, only occurring when water was present in the pipeline.

Corrosion will only occur where water is present; as part of the corrosion diagnosis and ICDA methodologies it is important to identify the locations where water either condenses from the pipeline or hold-up locations are identified. The locations where this corrosion would likely occur can be determined from flow modelling.

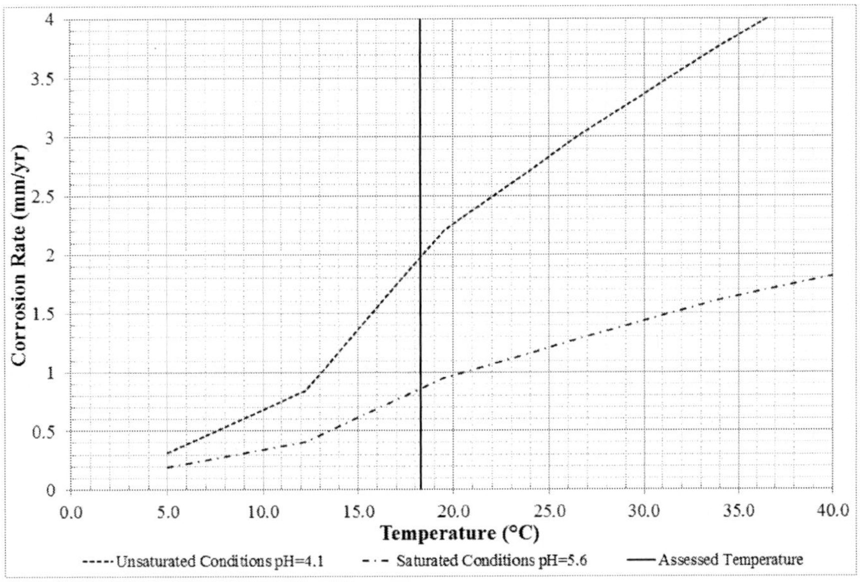

Figure 3: NORSOK M-506 calculated corrosion rate and profile.

The indirect inspection stage associated with the ICDA methodology was conducted using OLGA multiphase flow modelling software, taking into account the expected pipeline conditions. As was determined by the CRA the main risk to the pipeline is considered to be from CO_2 corrosion and TOL CO_2 corrosion, therefore identification of critical locations was based on the location of the key parameters needed for CO_2 corrosion to occur. These parameters are considered to be locations associated with;

- High condensation or water hold-up.
- High wall shear stresses.
- Regions of flow slugging.

Operational data indicated that the condensation of free water was a likely scenario. This was confirmed by retrieved liquids during pigging, but was also confirmed by the Pressure-Temperature (PT) profile calculated by the OLGA dynamic flow simulator, against the expected dew point curve calculated by Multiflash PVT Software, Figure 4. The PT profile of the pipeline lies to the left of the water dew point curve, indicating that condensation will occur throughout the entire pipeline. The water dew point was calculated using operational measurements of the typical water dew point, 18.3°C at 8.6 barg.

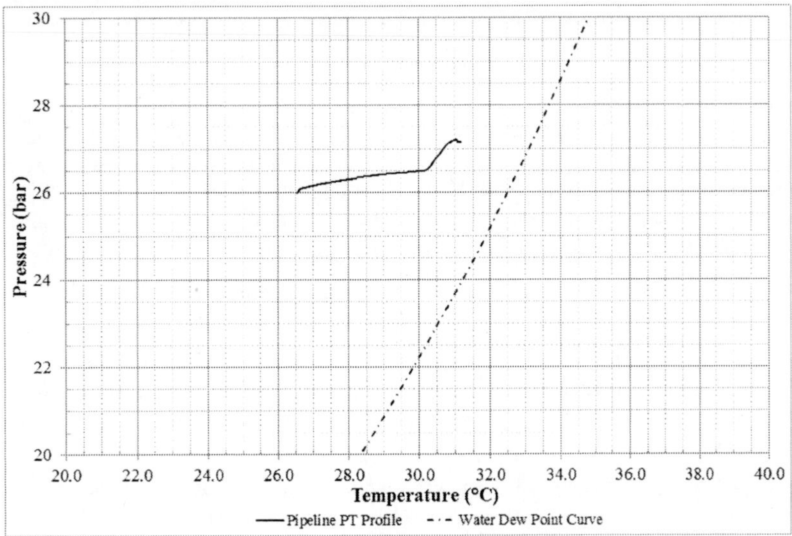

Figure 4: Pipeline PT profile *vs.* water dew point curve.

The condensation is ultimately the source of free water in the pipeline. Although gas drying facilities exist, the untreated wet gas from the second source is ultimately supplying the gas with water that is capable of condensing under the typical operational conditions.

Pipelines with a low condensation rate typically exhibit bottom of the line corrosion as the liquid water film naturally gravitates to the bottom of the pipe. Under these conditions and due to the rate of iron saturation, the rate of water replenishment is not sufficient to support the corrosion reaction at the top of the pipeline. Where the condensation rate is sufficient, TOL corrosion will be observed.

All of the peaks in the condensation rate identified through modelling were directly correlated with pipeline inclinations and the rate of cooling at those locations. Under the operational conditions the maximum condensation rate was observed to be 0.00296 g/m^{2}s, significantly below that needed to sustain TOL corrosion. As a result, corrosion would be expected to occur at liquid hold-up locations and present at the bottom of the line. Based on the results of the flow modelling, the threat of TOL CO_2 corrosion is ruled out by differential diagnosis as the condensation rate is found to be too low to sustain TOL corrosion. CO_2 corrosion, realised predominantly at the bottom of the pipeline, remains the primary candidate corrosion condition for the pipeline system determined from the corrosion diagnosis and evaluation of the corrosion condition.

The flow regime within the pipeline also plays a crucial role in the identification of critical locations for corrosion. As can be seen from Figure 5, of the expected flow regimes the majority of the pipeline exhibits stratified flow. However, there are some locations where slug flow is predicted to occur under steady-state operation. As expected, these events align with locations where liquid hold-up and high shear stresses take place. These again coincide with low points, most notably at the major river crossing at the beginning of the pipeline and again at a second river crossing towards the middle of the pipeline.

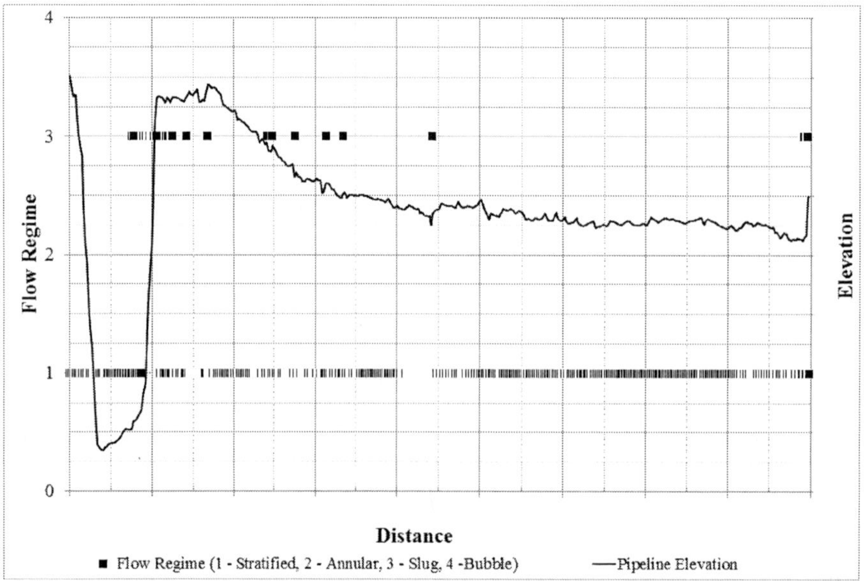

Figure 5: Pipeline flow regime profile.

Areas where liquids accumulate present ideal conditions for corrosion to occur. Where water stagnates or holds up, corrosion will occur at the liquid gas interface. Fourteen such locations were identified along the pipeline and as expected, are associated with steep uphill sections of the pipeline where liquids would accumulate. These locations correspond to points where the critical inclination angle of the pipeline, as specified in the dry gas ICDA (DG-ICDA) methodology, is exceeded. All of the locations where this critical angle is exceeded exhibited liquid hold-up, accounting for up to 25% of the pipeline cross-sectional area.

Wall shear stresses act as a catalyst to the corrosion reaction, typically by promoting the removal of protective corrosion scales or films. As would be expected the locations of shear stress corresponds to the locations where liquid hold-up occurs. This is due to the movement of the liquid film against the pipe wall and also transient slugging events at these locations.

The calculated water phase shears stress also identified 14 locations close to or exceeding the critical threshold of 20 Pa, thereby increasing the likelihood of the onset of corrosion at these locations. The maximum shear stress predicted was found to be 32.1 Pa.

In addition to flow modelling, corrosion rate modelling was conducted for the length of the pipeline, taking into account all of these influencing factors within the pipeline. The corrosion rates were estimated using both the NORSOK M-506 and de Waard 95 corrosion model, the standard corrosion package built into OLGA.

As can clearly be seen from the corrosion rate plot, Figure 6, both corrosion models predict not only different corrosion rates in general, but also where corrosion rate peaks occur. The highest corrosion rate prediction of NORSOK was 4.16 mm/year, located at distance "X". Comparatively the de Waard model predicts the maximum corrosion rate of 0.82 mm/year, to occur at location "Y".

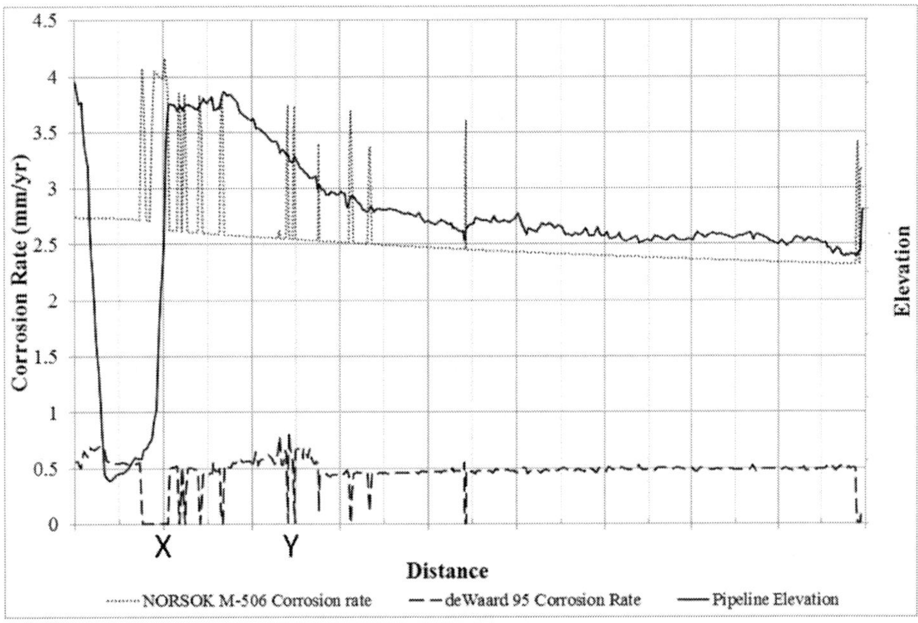

Figure 6: Estimated corrosion rates within pipeline from corrosion rate modelling in OLGA.

Both corrosion models take into account various parameters in different ways, such as the effect of shear stress, which is mainly influenced by the liquid film velocity. This effect is much more pronounced within the NORSOK model, hence why there is such a variance in corrosion level along the pipeline compared to that predicted by de Waard. The major differences between corrosion models and the reasons why different models predict different results has been discussed in great detail [13,14] and is beyond the scope of this work. However, the key fact to draw from these works is different corrosion models will always differ in the corrosion rate prediction, even under a given set of conditions. Consequently, corrosion models should not be used in isolation to identify inspection locations or hotspots. Instead the raw inputs into these models, can provide a much better indication as to where the most severe corrosion will occur.

Corrosion models are nevertheless useful to give an indication of the corrosion activity. Additionally, in the absence of actual corrosion rate measurements, they provide a valuable input into remaining life calculations (or prognosis), for example. Therefore, in this case, rather than considering the numerical number of the corrosion rate, the magnitude of this number should be compared.

The flow modelling of the pipeline indicated 14 locations with the potential for accelerated CO_2 corrosion to occur. It was observed that some of these locations exhibit similar characteristics, with shear stress and water hold-up relatively similar between each site. The major variable between each location is the temperature and pressure of the system. In order to identify the more critical locations or 'hot spots', a function of the total shear stress was considered together with the total liquid hold-up to highlight where the combination of these parameters was the highest, designated 'Inspection magnitude', shown in Figure 7. In summary, six of these locations were found at the uphill section of the first river crossing, five at inclines between the first and second river crossings; another at the incline of the second river crossing itself; whilst the final two were at the pipeline outlet. Based on the tight grouping of the six locations around the first river crossing this location was deemed to be the most critical.

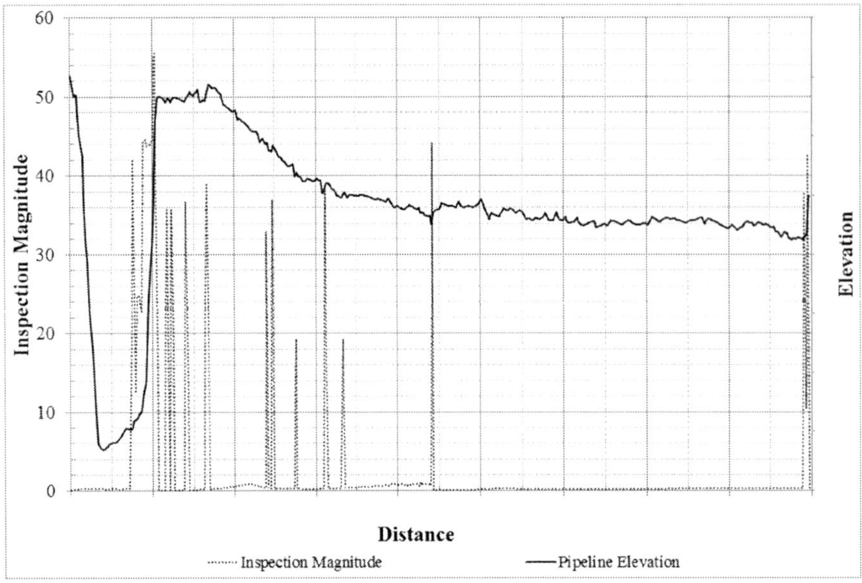

Figure 7: Inspection magnitude.

Many of these locations show clear commonalities in water hold-up, the shear stresses and flow regime changes. Taking these parameters into account, along with the temperature and pressure at each location, led to identification of the most critical locations. From the initial 14 locations, seven were identified that would be considered as representative of any point within the pipeline, labelled as A to G in Figure 8. Again, using the results of the flow modelling and corrosion assessment together, these seven locations were then ranked in order of criticality. In summary, locations B and F were considered to be the two most critical locations.

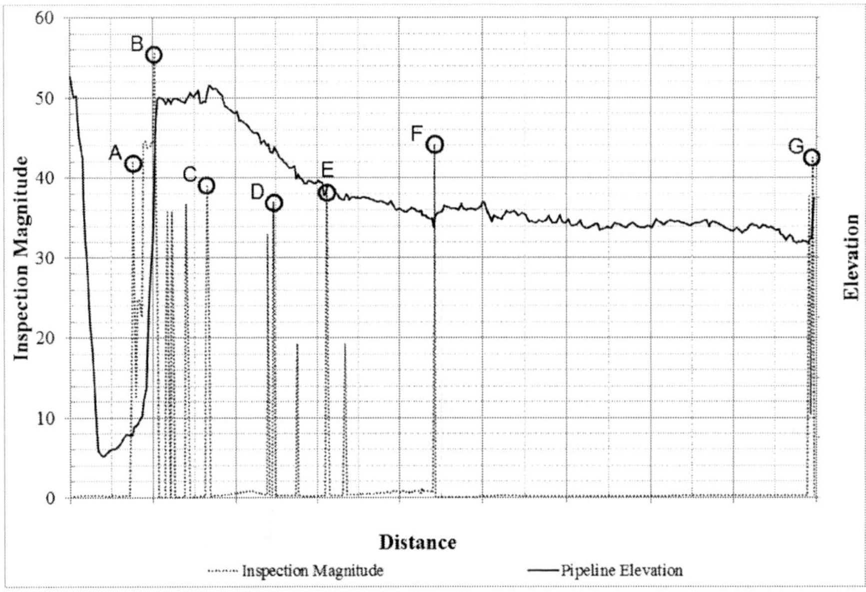

Figure 8: Critical inspection locations.

All the corrosion locations identified by indirect inspection showed very good correlation with the ILI inspection results (Figure 2) except for locations A and B. Discounting locations A and B the overall percentage error of the identified locations was 0.68 %.

Although the hot spot locations from the indirect inspection stages correlate well to the actual corrosion clusters, corrosion engineering experience and the OLGA multiphase model strongly suggests that corrosion should be occurring at locations A and B due to the presence of water accumulation, high shear stresses and flow slugging. The fact that no corrosion was found would indicate either a problem with the indirect inspection or direct inspection results. As the indirect inspection correctly identified 5 of the 7 critical corrosion locations to a <1% degree of accuracy, it is deemed unlikely it was due to an error in modelling. In order to identify the possible cause for this discrepancy, a further review of the ILI report and results was conducted.

Analysis of the ILI inspection report indicated that there were multiple factors that could have led to the reason no corrosion was found at location A or B. The speed tolerance of the tool was exceeded numerous times during the ILI run. As a result, the normal tool sizing tolerances were not applicable. In addition to this, the inspection tool was only designed to work effectively up to a maximum wall thickness of 12.70 mm, whereas within the first river crossing the pipeline has a maximum wall thickness of 14.27 mm. In this case the normal detection and sizing tolerances will not apply. Finally, although not fully quantifiable, debris is known to exist and was observed in the line which will further affect the sizing and detection tolerances. Given the steep gradient of the first river crossing it would be expected the majority of debris would exist there. Again, this is coincident with the locations A and B.

Review of the speed excursion plot alone presents a strong case that corrosion features have not been identified, as corrosion features stop at the exact location where the tool speed tolerance is exceeded, and are not identified again until it falls below the tolerance limit. This comparison can be seen in Figure 9. Given the commonality in corrosion initiation factors at both the first and second river crossings identified by OLGA (shear stress, liquid hold-up) this would present a strong case that similar corrosion will also exist in this section. However given the higher temperature and pressure at the pipeline inlet, corrosion could possibly be more pronounced in this area. Therefore based on the increased wall thickness, a corrosion feature of 39% Wall Thickness (WT) feature would be equivalent to at least a 17.5% WT feature in this section, not accounting for the increased temperature or pressure. Through comparison of the estimated corrosion rate between the first and second river crossing this increases the potential total wall loss by a further 13%. Increasing the potential wall loss at this location to approximately 20% WT at locations A and B. However, it must also be considered that other factors such as the pipeline debris could exacerbate corrosion in this section.

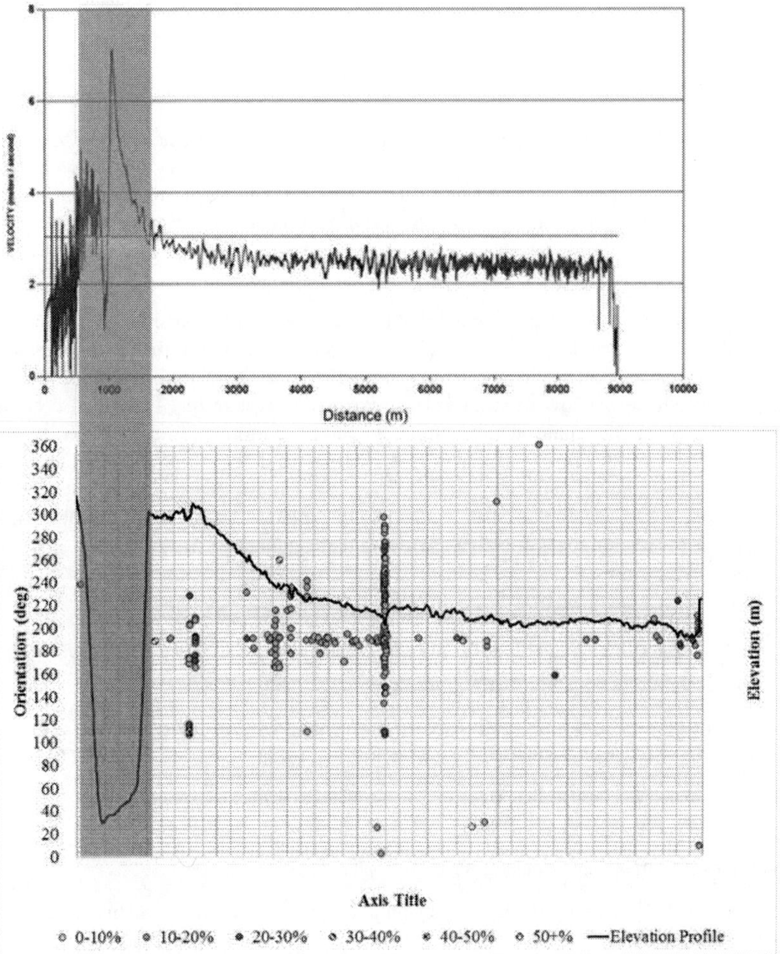

Figure 9: Direct inspection speed excursion to feature detection comparison.

From the ILI, the deepest corrosion feature within the pipeline was found within the second river crossing and was measured as 39% WT or 2.5 mm in depth. Based simply on the length of time the pipeline has been in service, the average, maximum corrosion growth rate (based on the deepest corrosion feature of 2.5 mm over the length of service) may be considered as 0.18 mm/yr. However, this growth rate should be treated with caution as it does not consider periods where corrosion has been active or dormant. The actual corrosion growth rate during active periods could be up to those calculated by the NORSOK model, before it returns to a dormant condition. The resulting corrosion rate at these locations was determined to be 0.21 mm/yr. again based on the actual measured corrosion rate at location F.

3.3 Treatment Plan

Following the corrosion diagnosis and evaluation of the corrosion condition, CO_2 corrosion realised at the bottom of the pipeline was determined to be the primary corrosion threat to the pipeline system. In order to effectively manage this threat of internal corrosion in the gas pipeline system it was recommended, as part of a treatment plan, to primarily minimise the entry of water into the gas stream. This would involve reviewing the requirements for gas drying operations at the second gas source in order to minimise subsequent condensation of free water in the gas stream. As part of such measures it would be recommended to continue to monitor dew point control and report on its effectiveness. Currently there are no chemicals injected into the pipeline to mitigate internal corrosion. If dew point control is lost for a significant period of time it would be recommended to inject chemicals to inhibit internal corrosion; these could include corrosion inhibitor, methanol or glycol for gas systems. If the entry of water into the gas stream cannot be managed effectively through implementation of gas drying operations, the continuous use of chemical inhibition should be considered to lower the rates of internal CO_2 corrosion.

An additional recommended corrosion control measure would be to increase the frequency of operational pigging in order to reduce the amount of liquids hold-up in the pipeline and therefore minimise the effects of internal corrosion at the bottom of the line.

4. CONCLUSIONS

A corrosion diagnosis methodology integrating the organization's long-term experience in corrosion engineering, inspection technologies and inspection signal analysis has been developed to overcome issues associated with lack of available data, to optimise pipeline corrosion management. Improvements at all stages in the integrity management cycle will ultimately result in the optimisation of corrosion management, whether it be in terms of cost-effectiveness, resource optimisation or reduced risk of incidents and pipeline safety. Refining the accuracy and reliability of the corrosion diagnosis is therefore a critical first step in the early stages of optimising a CMS.

In the specific case study presented it was demonstrated that through the use of corrosion assessments and flow modelling in the evaluation of the internal corrosion condition, improvements to the accuracy of the corrosion diagnosis could be made. TOL corrosion was initially considered a probable threat until flow rate modelling indicated that the condensation rate expected in the pipeline would be too low to sustain TOL corrosion; as a result, TOL CO_2 corrosion was ruled out by differential diagnosis. CO_2 corrosion, realised predominantly at the bottom of the pipeline, then became the primary candidate corrosion condition for the pipeline system determined from the corrosion diagnosis and evaluation of the corrosion condition. As such, targeted remedial actions and mitigation strategies (treatment plan) were recommended, aimed at controlling the entry and accumulation of water in the pipeline system and control of internal corrosion rates through chemical inhibition, for an optimized pipeline corrosion management strategy.

5. REFERENCES

[1] G. Jacobson (Editor), NACE International IMPACT – International Measures of Prevention, Application, and Economics of Corrosion Technologies Study, NACE International, 1st March 2016.

[2] Pipeline Performance in Alberta, 1990-2005. In: Alberta Energy and Utilities Board Report 2007-A, Alberta Energy and Utilities Board, Calgary, Alberta, Canada, April 2007.

[3] A. Morshed, 2007. Offshore assets: from corrosion engineering to corrosion management. Materials Performance, 46, 10, p34 (2007).

[4] Idem, 2008. Asset corrosion management using KPIs. Ibid., 47, 5, p50.

[5] D. Sandana, J. Soltis. Effective integrity management of pipeline through corrosion management and visualisation tools, J. of Pipe. Eng., 14(4), 269 (2015).

[6] M.J.J. Simon Thomas, L.H. Prager, C.V.M. Voermans, B.P. Miglin, B.F.M. Pots, I.J. Rippon, "Deterministic Pipeline Integrity Assessment to Optimize Corrosion Control and Reduce Cost", CORROSION 2002, paper no. 02075, Houston, TX: NACE International, 2002.

[7] Dictionary.com: http://www.dictionary.com/browse/diagnosis. Last accessed 10th July 2017.

[8] L. Barton, A. Pinto, I. Laing, R. Ladwa, "ILI Validation of Internal Corrosion Direct Assessment for Corrosion Identification and Inspection Prioritisation", Proc. ASME 2017 Int. India Oil & Gas Pipeline Conf., IOGPC2017-2448, April 2017.

[9] R. Sooknah, S. Papavinasam, R.W. Revie, M. De Romero (2007) Modelling the Occurrence of Microbiologically Influenced Corrosion, CORROSION/2007 Conference, National Association of Corrosion Engineers (NACE), Paper no. 07515.

[10] J.F.D. Stott (2010). 2.20 Corrosion in Microbial Environments. In: Shreir's Corrosion - Volume II: Corrosion in Liquids, Corrosion Evaluation. Edited by Cottis, B., Graham, M., Lindsay, R., Lyon, S., Richardson, T., Scantlebury, D. & Stott, H. 4th ed. Amsterdam: Elsevier. p1169-1190.

[11] Nyborg, R. & Dugstad, A. (2007) Top of Line Corrosion and Water Condensation Rates in Wet Gas Pipelines, NACE International, Paper 07555, NACE Corrosion 2007.

[12] NORSOK Standard M-506 CO_2 Corrosion Rate Calculation Model, March 2005.

[13] R. Nyborg, "Overview of CO_2 Corrosion Models for Wells and Pipelines", NACE International, CORROSION 2002, Paper No. 02233, Houston, TX (2002).

[14] R. Nyborg, "Field Data Collection, Evaluation and Use for Corrosivity Prediction and Validation of Models, Part II: Evaluation of Field Data and Comparison of Prediction Models", NACE International, CORROSION 2006, Paper No. 06118, Houston, TX (2006).

6. AUTHOR DETAILS

Steven Loftus is a Senior Corrosion Engineer at ROSEN UK, and has over 5 years of experience in corrosion consultancy and corrosion and materials engineering work in the oil and gas industry. He acts as an experienced corrosion professional with particular strengths in corrosion diagnosis, corrosion risk assessment reporting for onshore/offshore pipelines, in-depth understanding of corrosion science and mechanisms, and corrosion management strategies. He has thorough knowledge of industry best-practice and regulatory requirements, and has project experience in the UK and Europe, Australia, Middle East and Asia Pacific regions. Before joining ROSEN UK, Steven completed his MSc in Corrosion Control Engineering at The University of Manchester, UK, and is a certified NACE Senior Internal Corrosion Technologist.

A NOVEL FLOW LOOP FOR TESTING THE SUSCEPTIBILITY OF MATERIALS TO STRESS CORROSION CRACKING

R Rihan
Petroleum Research Center, Kuwait Institute for Scientific Research, Al-Ahmadi, Kuwait

SUMMARY: A novel stress corrosion cracking (SCC) testing rig has been designed and constructed at the Kuwait Institute for Scientific Research (KISR) for investigating the susceptibility of materials to SCC in flowing fluid using the fracture mechanics circumferential notched tensile (CNT) specimen. The SCC testing rig consists mainly of a flow loop that has test sections, autoclave, flowmeter, and pump. The flow loop can load more than one CNT specimen at the same time at different stress intensity factor (K_I) values, has the ability to remove the failed specimen and load another one while the experiment is still running, and runs tests at various flow velocities at the same time. The SCC testing flow loop is constructed from a corrosion resistant material such as Hastelloy C276.

The SCC testing flow loop can be used for determining the threshold stress intensity factor (K_{Iscc}) and crack growth rate in flowing fluid for different materials in different corrosive environments at different temperatures. The CNT specimen technique simplifies and speeds up the testing procedure as well as provides considerable cost advantages over the conventional fracture mechanics-based test techniques.

The SCC testing flow loop can be used to determine the susceptibility of the construction materials of major industries, such as oil and gas, petrochemical, desalination plants, and power plants to SCC. The new mechanical SCC testing flow loop was registered as a patent as a recognition of its novelty and practical applications in several industries for the simple and cost-effective determination of the design and prediction of life data. This paper presents a detailed description of the novel SCC testing flow loop.

Keywords: Stress corrosion cracking, Circumferential notched tensile specimen, Threshold stress intensity factor (K_{Iscc}), Crack growth rate, Stress corrosion cracking testing flow loop.

1. INTRODUCTION

The experience of various industries calls for a fresh approach to investigate the failed components and develop a simple monitoring technique. The accurate monitoring of the stress corrosion crack growth and life prediction can be carried out by the fracture mechanics approach. However, the traditional fracture mechanics techniques such as; compact tension, double cantilever beam, cantilever bend (CB), and center cracked plate (Sedriks, 1990; Turnbull, 1992) are prohibitively expensive and time-consuming, which restrict their use as a commonly used technique for the stress corrosion cracking (SCC) monitoring. This paper aims to use a recently introduced SCC fracture mechanics circumferential notched tensile (CNT) specimen (shown in Figure 1) technique (Rihan, 2005; Rihan et al., 2004a,b; 2005a,b,c; 2006 a,b,c,d,e; 2008; Singh Raman et al., 2006a,b; 2007a,b,c; Ibrahim et al., 2008; Pal and Singh Raman, 2010), which is a simple and cost-effective technique for evaluating the SCC susceptibility of engineering materials.

The CNT specimen has been tested successfully by Rihan (2005), Rihan et al. (Rihan et al., 2004a,b; 2005a,b,c; 2006 a,b,c,d,e; 2008; Singh Raman et al., 2006a,b; 2007a,b,c; Ibrahim et al., 2008), and Pal and Singh Raman (2010) for determining the susceptibility of different ferrous alloys to SCC in different corrosive environments at different temperatures. The obtained threshold stress intensity factor (K_{Iscc}) value using the CNT specimen was found very close to

K_{Iscc} values obtained by other researchers using different standard specimens (Rihan et al., 2006a; Singh Raman et al., 2007a; Ibrahim et al., 2008; Pal and Singh Raman, 2010).

The CNT specimen has been recently used to generate the K_{Iscc} of solution-annealed austenitic AISI 304 stainless steel and sensitized austenitic AISI 304 stainless steel in 42% $MgCl_2$ environment at 154°C (Pal and Singh Raman, 2010). The generated K_{Iscc} data using the CNT technique have been compared with those generated using the traditional techniques, such as compact tension and double cantilever beam specimens, reported in the literature. The generated K_{Iscc} for the solution-annealed austenitic AISI 304 stainless steel using the CNT specimens was found to be ~14 MPa $m^{1/2}$, which was similar to the one generated using conventional double cantilever beam specimens by Speidle (1981) and Denhard (1960) (K_{Iscc} = 10 -12 MPa $m^{1/2}$) and Mariano and Spinelli (2004) (K_{Iscc} = 15 MPa $m^{1/2}$). The generated K_{Iscc} for the sensitized austenitic AISI 304 stainless steel using the CNT specimens was found to be 8 MPa $m^{1/2}$, which was similar to the one generated using the conventional double cantilever beam specimens by Mariano and Spinelli (2004) (K_{Iscc} = 7.8 MPa $m^{1/2}$).

The results of this study confirm the reliability of the CNT specimens and support the findings of the earlier studies by the author with others (Rihan et al., 2006a; Singh Raman et al., 2007a; Ibrahim et al., 2008), which were performed to generate the K_{Iscc} of heat treated 4340 steel in sodium chloride (NaCl) solutions at room temperature. The K_{Iscc} has been determined to be 15 MPa $m^{1/2}$. The obtained K_{Iscc} value using the CNT specimen was found to be very close to K_{Iscc} values, approximately 14 MPa $m^{1/2}$, obtained by other researchers (Brown and Beachem, 1965; Smith et al., 1968) using standard specimens such center cracked, surface flawed, and cantilever bend specimens.

The author with others have invented two novel CNT test rigs to test the susceptibility of engineering materials to SCC in stagnant (not flowing) fluid. The two CNT rigs are small enough that it can be placed on desks, tables, or shelves. These rigs were registered as patents (Rihan et al., 2006d; Rihan et al., 2013).

The first CNT testing rig (shown in Figure 2a) is sealed by Polytetrafluoroethylene (PTFE) O-rings and ferrules, which can handle temperatures up to 230°C and pressures up to 85 bar and has the capability of performing electrochemical measurements. The second CNT testing rig (shown in Figure 2b) is a modified version of the first CNT testing rig, which can be operated at a relatively higher temperature (~360°C) and pressure (~200 bar). The second CNT testing rig is sealed using PTFE Vee Packings. As PTFE is not recommended to be used at temperatures above 230°C, the sealing point was placed away from the autoclave and was fitted with a cooling system with circulating water. The corrosion cell is sealed by threading the PTFE Vee Packing holder and the specimen holding/sealing nut to the body of the corrosion cell (autoclave) using National Pipe Taper (NPT) thread, as shown in Figure 2b. It is equipped with a circulating system to allow the test solution movement in the notch region of the CNT specimens, as shown in Figure 2b. The solution moves through the circulating tube by natural thermal convection as a result of the significant difference in the temperature between the autoclave and the circulating tube.

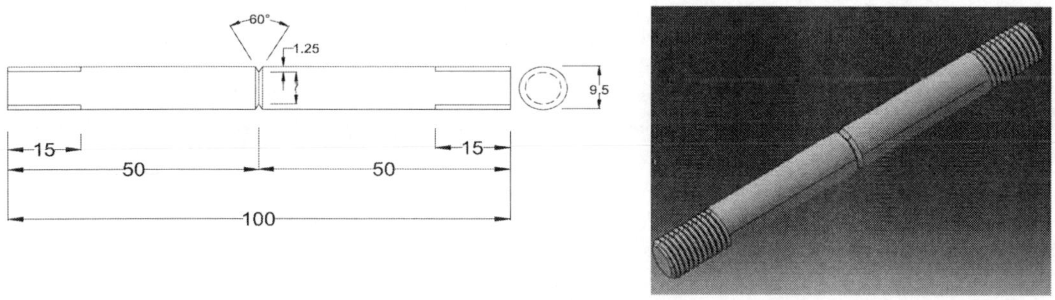

Figure 1 The CNT specimen (all dimensions in mm)

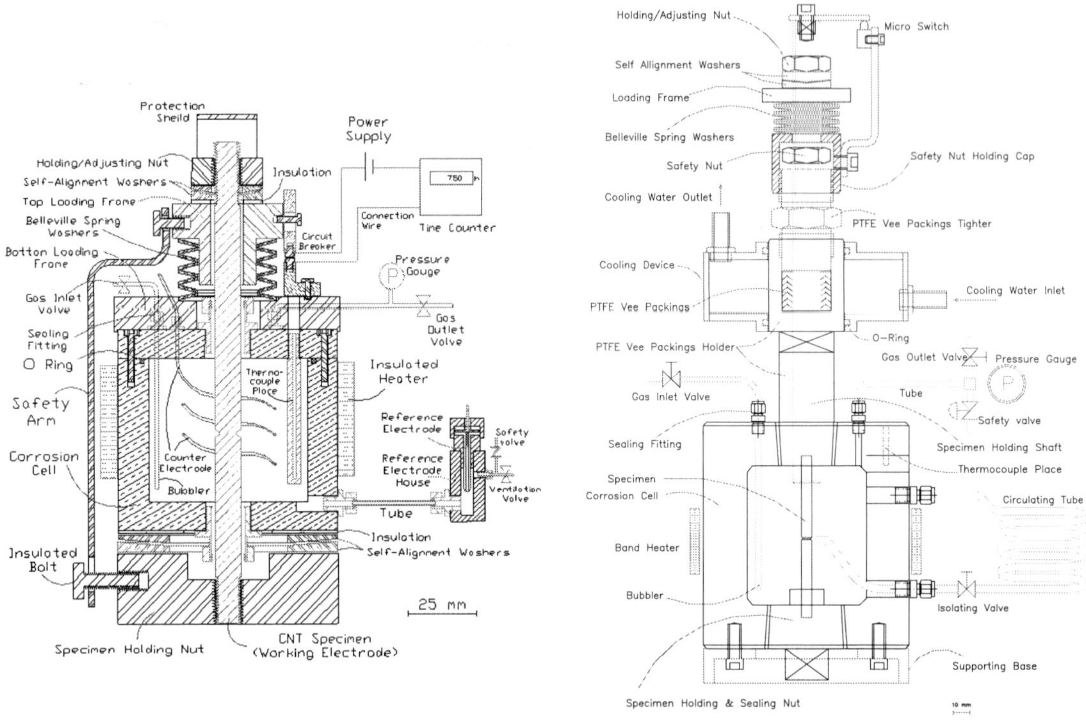

(a) The first CNT testing rig (Rihan et al., 2006d)

(b) The second CNT testing rig (Rihan et al., 2013)

Figure 2 The two CNT testing rigs' diagrams

To the best of the author's knowledge, a few experimental studies (Choi et al., 1982; Fong, 2007; Fuller and Macdonald, 1984; Kwon et al., 2000; Shim and Szklarska-Smailowska, 1987a,b) were performed on the effect of fluid flow velocity on SCC. Only two of these studies (Fong, 2007; Kwon et al., 2000) were conducted using the fracture mechanics compact tension standard specimen, while the other studies (Choi et al., 1982; Fuller and Macdonald, 1984; Shim and Szklarska-Smailowska, 1987a,b) used the non-fracture mechanics slow strain rate test specimen. The results of these studies, generally, indicate that the SCC susceptibility decreases in the presence of flowing fluid and decreases with increasing the fluid velocity.

In these studies, the flowing conditions within the autoclave were provided by a rotating stirrer, and the exact orientation of the flow and flow rate past the specimen surfaces (including the crack mouth) was not measured. Instead, the flow rate was specified by the rotation rate of the stirrer. Therefore, it can be concluded that no fracture mechanics study was conducted at a known flow rate or fluid velocity.

The aim of this work was to design and construct a novel SCC testing rig for monitoring and investigating the susceptibility of metals and alloys to SCC in flowing fluid using the fracture mechanics CNT specimen. The new SCC testing rig can be used for determining the K_{Iscc} and crack growth rate in the flowing solution of different materials in different corrosive environments at different temperatures using the CNT specimen. To the best of the author's knowledge, K_{Iscc} has never been determined in the flowing fluid environment.

2. RESULTS

The design and construction of the novel SCC testing rig is an integral part of this work. The initial design of the new SCC testing rig consists mainly of a flow loop that has test sections, autoclave, flowmeter, and cooler, as shown in Figure 3. The fluid moves through the circulating tube by natural thermal convection as a result of the significant difference in temperatures between the hot autoclave and the cooler, which creates a natural circulation. This has been used in some boilers, reactors, and solar water heating where the required flow rate is not high (Teir, 2002; Norton, 2011; IAEA, 2005). The fluid velocity needs to be determined at the crack tip since it is the most important part of the fracture mechanics specimens where the crack initiates. The test sections are placed just after the autoclave, as shown in Figure 3, in order to expose the notched region of the CNT specimens, the crack tip, to hot flowing fluid. Then, the hot fluid will pass through the cooler before returning to the autoclave.

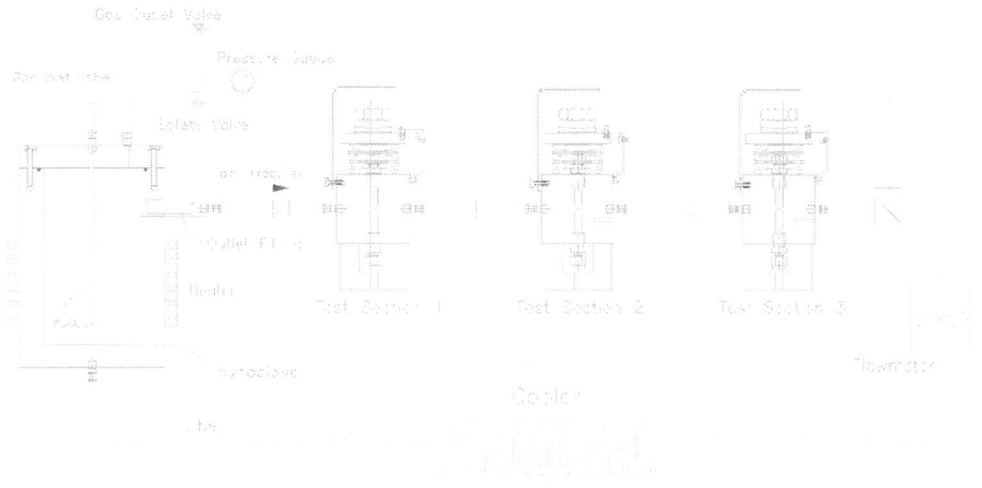

Figure 3 The initial design of the novel SCC testing flow loop (the high-temperature test version)

In cases where the experimental work has to be performed at room temperature, the test sections' line and the cooler line are exchanged, as shown in Figure 4. This allows the fluid to be cooled to room temperature before entering the test section. Also, the experimental work can be performed at a high temperature and room temperature at the same time by installing some of the test sections before the cooler and some after the cooler, as shown in Figure 5.

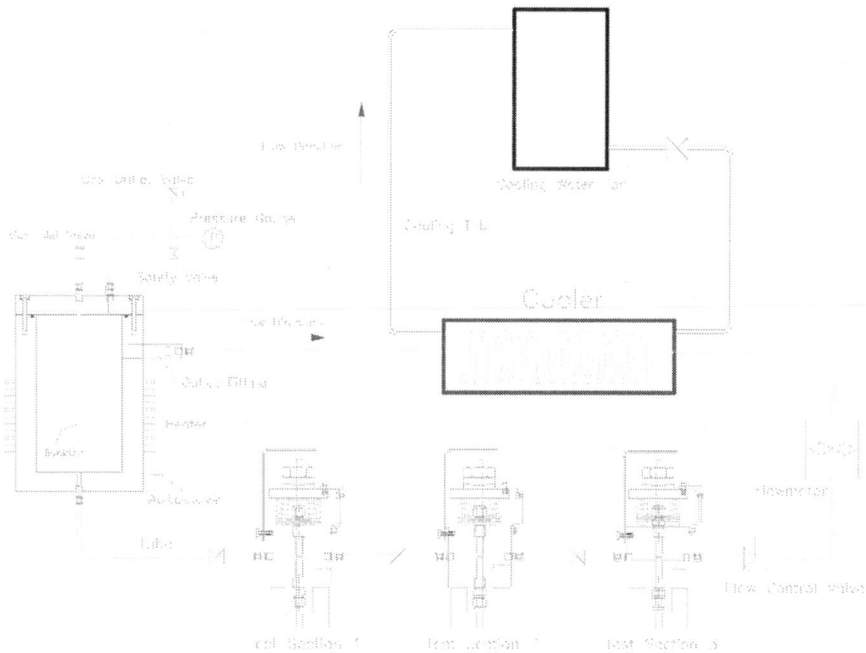

Figure 4 The initial design of the novel SCC testing flow loop (the room temperature test version)

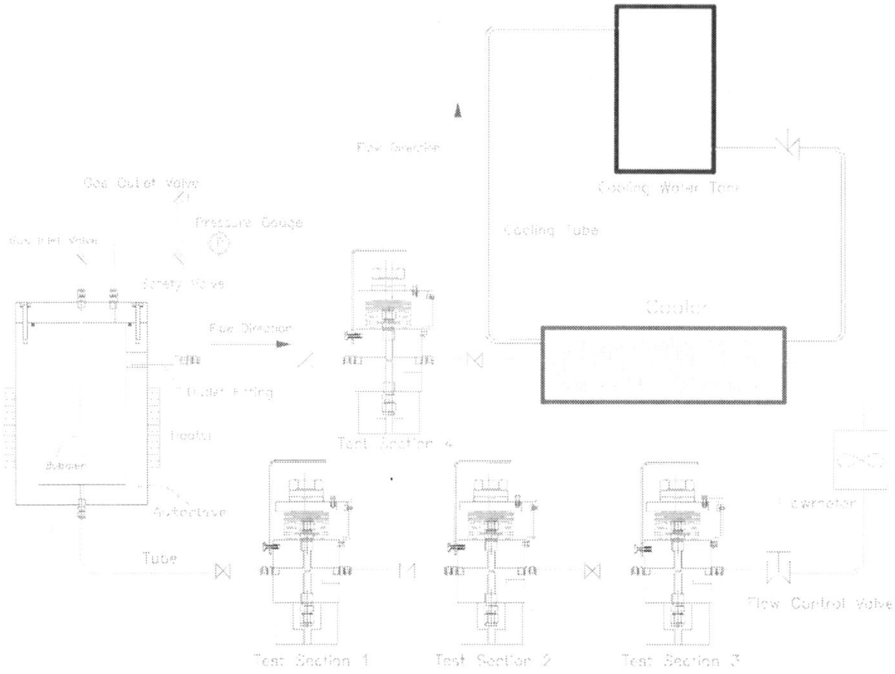

Figure 5 The initial design of the novel SCC testing flow loop (the hot and room temperature test version)

The initial model was tested successfully, but the flow rate was low in which the flowmeter was not able to detect such a low flow rate. Therefore, the initial design (Figure 3) of the flow loop was modified and a variable speed pump for circulating the fluid was added to the flow loop. The new design of the SCC testing flow loop is shown in Figure 6. The main advantage of adding a pump is to allow the control of the velocity of the fluid through the system and to generate high flow rates, which simulate the actual operating conditions in industrial plants. The SCC testing flow loop is constructed from a corrosion resistant material such as high-nickel alloys, namely, Hastelloy C276. The SCC testing flow loop is provided with test sections, autoclave (pressure vessel acting as a reservoir for the solution), heating system, flowmeter, safety valve, valves, and fittings. The SCC testing flow loop was registered as a patent (Rihan, 2017) in recognition of its novelty and practical applications in several industries.

The system has a flow line (by–pass line) parallel to the test section line. The by–pass line is used to pass the solution through the system during the heating process (while the test section line is closed). Once the required temperature is reached, the test section line is opened (then the by–pass line is closed) in order to expose the specimens directly to the solution at the required temperature. The nominal diameter of the piping system generally is ½ inch, while the nominal diameter of the test section and the by-pass lines is 3/8 inch. The piping and fittings are threaded according to the NPT standards. The system has a filling and draining valve for filling the system with the test fluid at the start of the experiment and draining the system at the end of the experiment.

The flow loop can have more than one test section installed in a series, as shown in Figure 6. This allows researchers to test more than one CNT specimen at various flow velocities and K_I values. Each test section can be isolated and removed from the loop and another test section can be installed while the experiment is still running by closing the two valves that connect the test section with the loop. When the CNT specimen of one of the test sections fails, the failed CNT specimen can be replaced by opening the valve of the by–pass line followed by closing the two valves that connect that test section with the loop. Then, a connection tube is positioned in the test section's place followed by opening the two isolating valves and closing the valve of the by–pass line. Once the failed CNT specimen is replaced with a new one, the test section is returned to the flow loop. For example, if the CNT specimen of the Test Section 4 fails, the test sections' line after replacing Test Section 4 with a connection tube will look like the one as shown in Figure 7.

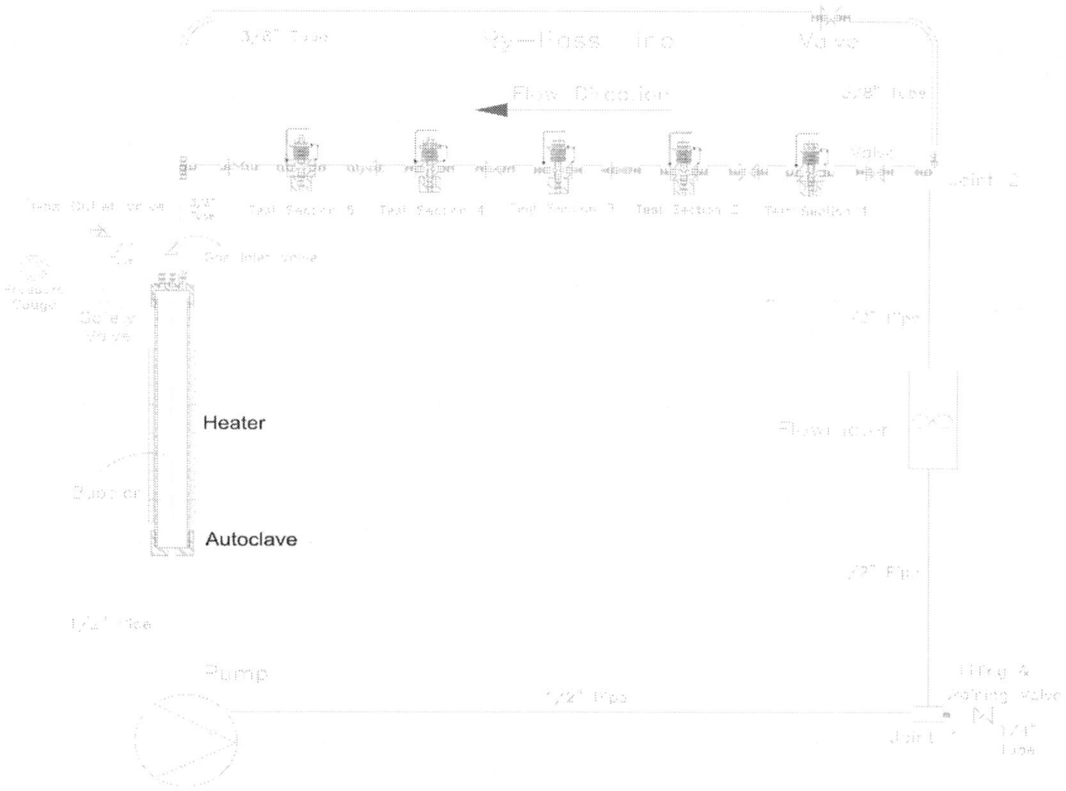

Figure 6 The new design of the SCC testing flow loop

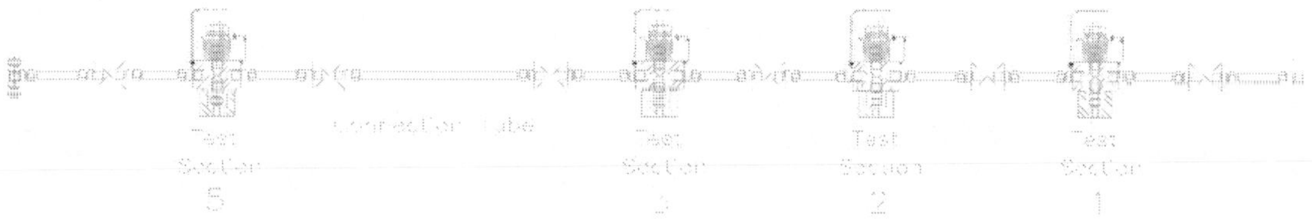

Figure 7 A connection tube is positioned in the place of Test Section 4

The most important part of the flow loop is the test section (shown in detail in Figure 8). The test section is designed to simulate the corrosive environment present in industry and to house the specimen. Five to six test sections are required to be built in order to test more specimens at the same time. Each test section can have different flow velocities by fabricating the test sections with different flow channel diameters.

The test section is equipped with the following:

- Working electrode (CNT specimen). There is only one working electrode inserted through the center of the test section (as shown in Figure 8). The working electrode is constructed from the test material. In cases when some tests are required to be performed under the applied potential, the test section can be equipped with counter and reference electrodes.

  The advantages of the CNT specimen are as follows:

  - The small cross section of the specimen makes it possible to load the specimen to quite high-stress levels using moderate loads. This means that either small load fram all capacity machines are required for applying the loads.

- ❑ Reducing material requirements. In many cases, materials over 20 mm or even 10 mm thick may not be available. This is a characteristic of many pressure vessels or structural steels, especially expensive materials such as stainless steel or duplex stainless steel. If post-exposure or aged samples are to be taken out of old structures, there may be quite severe size restrictions; for example, the wing of Boeing 747 has no material thick enough to manufacture a standard test specimen (Ibrahim, 1989).
- ❑ The CNT specimen is cheap to produce because of its small cylindrical shape, which reduces the cost of each specimen by a factor of 10.
- ❑ The CNT specimen has no edge effect since the notch of the CNT specimen has no end of crack in the plane stress region. This is an important fact that enables a small sized specimen, such as the CNT specimen achieve the plane strain condition (Ibrahim, 1989; Ibrahim and Stark, 1987, 1990).
- ❑ In cases where there is insufficient material available from a failed component to machine a long CNT specimen, short CNT specimens are machined at the available length and then elongated by joining extension rods through welding or threading since the region of interest is the notch region.

- Belleville spring washers. The Belleville spring washers are used to exert a certain tensile load on the CNT specimen through the loading frame.
- Loading frame. The loading frame is used for exerting a certain tensile load on the specimen. The CNT specimen is loaded by applying a compressive load on the Belleville spring washers through a loading machine. Then, the holding/adjusting nut is adjusted and the applied machine load is removed. The stored energy in the springs is transmitted to the CNT specimen as a tensile load.
- Safety arms. The safety arm is used for preventing the specimen from being ejected as a projectile when the specimen breaks.
- Microswitch. The microswitch is connected to a time counter through an electrical circuit. During the experiment and before the specimen breaks, the electrical circuit will be closed. Once the specimen breaks, the loading frame moves up and opens the circuit at the microswitch point and the counter stops counting. Thus, the time to failure is determined.

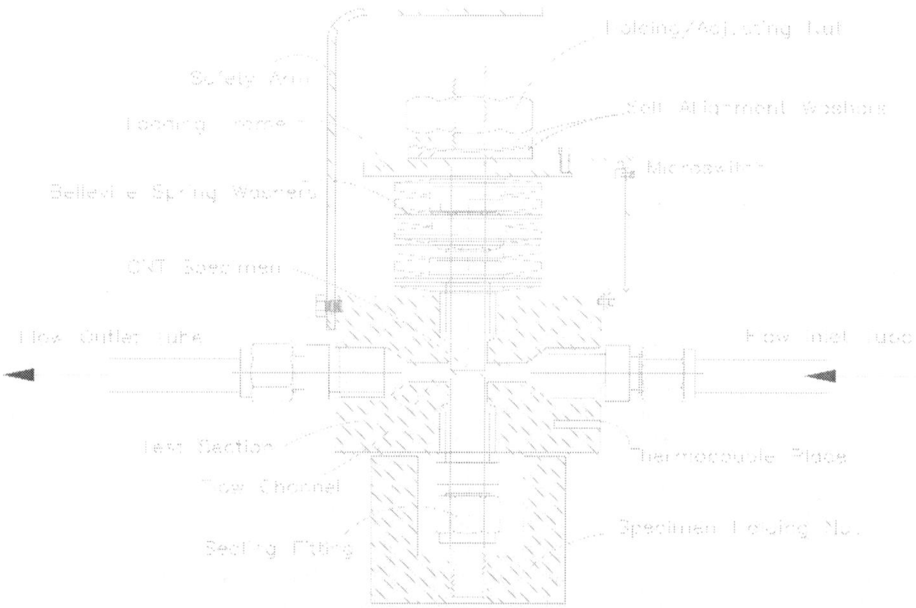

Figure 8 Detailed drawing of a test section of the flow loop

The advantages of the SCC testing flow loop are as follows:

- Has the ability to load more than one CNT specimen at the same time.
- Has the ability to load more than one CNT specimen at different K_I values.
- Has the ability to remove the failed specimen and load another one without stopping the experiment.

- Has the ability to run tests at various flow velocities at the same time by fabricating test sections with different flow channel diameters.
- Has the ability to run tests at a high temperature and room temperature at the same time.
- Allows researchers to reproduce the SCC problems of flowing corrosive liquids in a laboratory setting and to research on alternative operating conditions, which can eliminate or reduce the SCC problems.

The CNT specimen is notched in order to simulate an ideal plane crack with an essentially zero root radius to satisfy the assumption of K_I analysis (Griffith, 1920; Inglis, 1913; Irwin, 1948; Orowan, 1948). The machined notch is extended by fatigue since the fatigue crack is the sharpest crack that can be reproduced in the laboratory. The machined notch is extended to eliminate any effect of the machined notch geometry. The pre-cracking (fatigue crack) of the CNT specimen was achieved by mounting the CNT specimen on a rotating lathe machine, as shown in Figure 9, in order to fatigue crack the CNT specimen, as shown in Figure 10.

Figure 9 A CNT specimen is mounted on a rotating lathe machine

Figure 10 The fatigue crack of a CNT specimen

3. CONCLUSIONS

1. A novel SCC testing flow loop has been designed and constructed at the Kuwait Institute for Scientific Research (KISR) for investigating the susceptibility of materials to SCC in flowing fluid using the fracture mechanics CNT specimen.
2. The construction of the SCC testing flow loop allows researchers to reproduce the SCC problems of flowing corrosive liquids in a laboratory setting and to research on alternative operating conditions, which can eliminate or reduce the SCC problems.

4. ACKNOWLEDGMENTS

The author would like to acknowledge the support received from the management of the Kuwait Institute for Scientific Research (KISR) by funding this work through Project No. PP053K.

5. REFERENCES

Brown, B.F.; and C.D. Beachem. 1965. Study of stress factor in corrosion cracking by use of pre-cracked cantilever beam specimen. *Corrosion Science* **5**:745–750.

Choi, H.; F.H. Beck; Z. Szklarska-Smialowska; and D.D. Macdonald. 1982. The effect of fluid flow on the stress corrosion cracking of ASTM A508 CI 2 steel and AISI type 304 stainless steel in high temperature water. *Corrosion* **38**:76–85.

Denhard, E.E. 1960. Effect of composition and heat treatment on the stress corrosion cracking of austenitic stainless steels. *Corrosion* **16**:359–370.

Fong, C. 2007. Flow effects on the stress corrosion cracking of sensitized stainless steels under BWR environments. Presented at the Canadian Nuclear Society - 13th International Conference on Environmental Degradation of Materials in Nuclear Power Systems, Canada, April 19–23.

Fuller, G.A.; and D.D. Macdonald. 1984. The effect of fluid flow on the stress corrosion cracking of AISI 304 stainless steel in 0.01 M Na_2SO_4 solution at 280 C. *Corrosion* **40**:474–477.

Griffith, A.A. 1920. The phenomena of rupture and flow in solids. *Philosophical of the Transactions Royal Society of London* **221**:163–197.

IAEA. 2005. Natural circulation in water cooled nuclear power plants. International Atomic Energy Agency, Austria.

Ibrahim, R.N. 1989. The development of a small K_{Ic} specimen and its application to sustained load cracking in aluminum pressure vessels. PhD thesis, University of New South Wales, Australia.

Ibrahim, R.N.; and H.L. Stark. 1987. Validity requirements for fracture toughness measurements from small circumferentially notched cylindrical specimens. *Engineering Fracture Mechanics* **28**:455–460.

Ibrahim, R.N.; and H.L. Stark. 1990. Establishing K_{Ic} from eccentrically fatigue cracked small circumferentially grooved cylindrical specimens. *International Journal of Fracture* **44**:179–188.

Ibrahim, R.N.; R. Rihan; and R.K. Singh Raman. 2008. Validity of a new fracture mechanics technique for the determination of the threshold stress intensity factor for stress corrosion cracking (K_{Iscc}) and crack growth rate of engineering materials. *Engineering Fracture Mechanics* **75**:1623–1634.

Inglis, C.E. 1913. Stress in a plate due to the presence of cracks and sharp corners. Proceedings, Institute of Naval Architects. March 14.

Irwin, G.R. 1948. Fracture dynamics. in *Fracturing of Metals*. Cleveland: American Society of Metals, pp. 147–166.

Kwon, H.S.; A. Wuensche; and D.D. Macdonald. 2000. Effects of flow rate on crack growth in sensitized type 304 stainless steel in high-temperature aqueous solutions. *Corrosion* **56**:482–491.

Mariano, N.A.; and D. Spinelli. 2004. Stress corrosion cracking of stainless steel used in drill collars. *Materials Science and Engineering A* **385**:212–219.

Norton, B. 2011. Solar water heaters: A review of systems research and design innovation. *Green* **1**: 189–207.

Orowan, E. 1948. Fundamentals of Brittle Behaviour in Metals. Fatigue and Fracture of Metals. New York: Wiley.

Pal, S.; and R.K. Singh Raman. 2010. Determination of threshold stress intensity for chloride stress corrosion cracking of solution-annealed and sensitized austenitic stainless steel by circumferential notch tensile technique. *Corrosion Science* **52**:1985–1991.

Rihan, R.; R.N. Ibrahim; and R.K. Singh Raman. 2004a. The determination of K_{Iscc} of mild steel in hot caustic by using small circumferential notched tensile (CNT) fracture toughness specimens. Presented at the International Conference on Structural Integrity and Fracture, Brisbane, Australia, Sept. 26–29.

Rihan, R.; R.K. Singh Raman; and R.N. Ibrahim. 2004b. Circumferential notch tensile (CNT) tests for determination of K_{Iscc} using small fracture mechanics specimens. Presented at the Environment – Induced Cracking of Metals Conference, Banff, Canada, Sept. 19–23.

Rihan, R.O. 2005. A novel fracture mechanics approach for the determination of susceptibility of engineering materials to stress corrosion cracking. PhD thesis, Monash University, Melbourne, Australia.

Rihan, R.; R.K. Singh Raman; and R.N. Ibrahim. 2005a. Circumferential notched tensile (CNT) testing of cast iron for determination of threshold (K_{Iscc}) for caustic crack propagation. *Materials Science and Engineering* A **407**:207–212.

Rihan, R.; R.K. Singh Raman; and R.N. Ibrahim. 2005b. Circumferential notched tensile (CNT) tests for generating K_{Iscc} data for cast iron vessels used in hot caustic solutions. Presented at the 8th International Conference & Exhibition, Melbourne, Australia, Apr. 20–22.

Rihan, R.; R.K. Singh Raman; and R.N. Ibrahim. 2005c. Determination of stress corrosion cracking (SCC) susceptibility (K_{Iscc}) at imposed potentials using a novel circumferential notch tensile (CNT) technique. Proceedings, Corrosion and Prevention, Gold Coast, Australia, Nov. 20–23.

Rihan, R.; R.K. Singh Raman; R.N. Ibrahim; and D. Gerrard. 2006a. Determination of the threshold stress intensity factor (K_{Iscc}) of 4340 steel using small circumferential notched tensile (CNT) specimens. Proceedings, Corrosion and Prevention 2006 "Steel and Concrete - Nothing Lasts Forever", Hobart, Australia, Nov. 19–22.

Rihan, R.; R.K. Singh Raman; and R.N. Ibrahim. 2006b. Determination of crack growth rate and threshold for caustic cracking (K_{Iscc}) of a cast iron using small circumferential notched tensile (CNT) specimens. *Materials Science and Engineering* A **425**:272–277.

Rihan, R.; R.K. Singh Raman; and R.N. Ibrahim. 2006c. Circumferential notched tensile (CNT) tests for generating K_{Iscc} data for cast iron vessels used in hot caustic solutions. *International Journal of Pressure Vessels and Piping* **83**:388–393.

Rihan, R.; R.N. Ibrahim; and R.K. Singh Raman. 2006d. Stress corrosion cracking testing rig and method. Provisional Patent. Registered at Watermark Patent & Trade Mark Attorneys. Australia. 26th May 2006.

Rihan, R.; R.K. Singh Raman; and R.N. Ibrahim. 2006e. A novel fracture mechanics approach for determination of crack growth rate and threshold for caustic cracking (K_{Iscc}) of ferrous alloys using small circumferential notched tensile (CNT) specimens. International Conference and Exhibition on Pressure Vessels and Piping, Chennai, India, Feb. 7–9.

Rihan, R.; R.K. Singh Raman; and R.N. Ibrahim. 2008. Circumferential notched tensile testing for correlation of the stress intensity factor (K_I) and stress corrosion crack growth rate. *Metallurgical and Materials Transaction* A **39**:1475–1478.

Rihan, R.; M. Qubbaj; M. Basha; and L. Al-Hadhrami. 2013. Stress corrosion cracking testing device. U.S. Patent Number 8474324 B2.

Rihan, R. 2017. System for testing stress corrosion cracking. U.S. Patent Number 9541485 B1.

Shim, S.H.; and Z. Szklarska-Smailowska. 1987a. Effect of fluid flow on the stress corrosion cracking of AISI 304 stainless steel in pure water and 0.01 M Na_2SO_4 solutions differing in pH. *Corrosion* **43**:286–90.

Shim, S.H.; and Z. Szklarska-Smailowska. 1987b. Effect of fluid flow rate on the intergranular stress corrosion cracking of AISI 304 stainless steel. *Corrosion* **43**:280–286.

Singh Raman, R.K.; R. Rihan; and R.N. Ibrahim. 2006a. Determination of threshold for stress corrosion cracking (K_{Iscc}) by a novel approach using round tensile specimens. *Metallurgical and Materials Transaction: A* **37A**:2963–2973.

Singh Raman, R.K.; R. Rihan; and S.S. Pal. 2006b. The determination of the susceptibility of AISI 1020 mild steel welds to stress corrosion cracking (SCC) in hot caustic solutions using circumferential notched tensile (CNT) specimens. WTIA-Monash Symposium, Monash University, Melbourne, Australia.

Singh Raman, R.K.; R. Rihan; and R.N. Ibrahim. 2007a. Validation of a novel approach to determination of threshold for stress corrosion cracking (K_{Iscc}). *Materials Science and Engineering: A* **452-453**:652–656.

Singh Raman, R.K.; R. Rihan; and R.N. Ibrahim. 2007b. Circumferential notch tensile testing, role of imposed electrochemical potentials in susceptibility of steel to caustic cracking. *Journal of the Electrochemical Society* **154**:C658–C662.

Singh Raman, R.K.; R. Rihan; and R.N. Ibrahim. 2007c. Role of imposed potentials in threshold for caustic cracking susceptibility (K_{ISCC}): investigations using circumferential notch tensile (CNT) testing. *Corrosion Science* **49**:4386–4395.

Smith, H.R.; D.E. Piper; and F.K. Downey. 1968. Study of stress-corrosion cracking by edge-force loading. *Engineering Fracture Mechanics* **1**:123–128.

Sedriks J. 1990. Stress Corrosion Cracking Test Methods, Publications of NACE, National Association of Corrosion Engineers, US.

Speidel, M.O. 1981. Stress corrosion cracking of stainless steel in NaCl solutions. *Metallurgical Transactions A* **12A**:779–789.

Teir, S.; and A. Kulla. 2002. *Steam/Water Circulation Design*. Helsinki University of Technology. Department of Mechanical Engineering. Energy Engineering and Environmental Protection Publications. Steam Boiler Technology eBook.

Turnbull, A. 1992. Test methods for environment assisted cracking. *British Corrosion Journal* **27**:271–289.

6. AUTHOR DETAILS

R. Rihan is a Research Scientist in the Petroleum Research Center of the Kuwait Institute for Scientific Research. He received his Ph.D. in Corrosion Engineering from Monash University, Australia, in 2005. He has extensive experience in the corrosion and materials' performance. Dr. Rihan has numerous publications in international journals and conference proceedings, covering wide areas of corrosion such as stress corrosion cracking, H_2S/CO_2 corrosion in oil and gas industry, erosion-corrosion, mechanical design and construction of corrosion testing rigs, emulsion and multiphase flow in petroleum pipelines, corrosion inhibitors, and galvanic corrosion. He has also patents dealing with stress corrosion cracking devices.

JUDICIOUSLY CLASSIFYING EXPOSURE ENVIRONMENTS FOR CONCRETE AND STEEL

N. Tripathi, H. Fatemi
Roads and Maritime Services

SUMMARY: When designing concrete and steel elements, it is of particular importance to consider the severity or corrosivity of the environment it will be exposed to during its service life. Atmospheric corrosivity for the two main materials used in bridges, concrete and steel, is not classified uniquely. Whereas durability of concrete elements is determined based on exposure classification in AS 5100.5, corrosion protection of steel elements is based on corrosivity categories in AS 4312. It is the authors' observation that structural designers lack adequate understanding of the difference in the two approaches for classifying the severity of the environment. This often results in the designer nominating the exposure classification based only on AS 5100.5. Whereas this may be adequate for durability design of concrete elements, it does not permit selection of an appropriate protective coating system from AS 2312.1 that is based on corrosivity categories of AS 4312.

This paper briefly describes the difference in the approaches adopted in the above two reference standards for classifying the atmospheric corrosivity for reinforced concrete and structural steel elements. It also demonstrates the importance of having an adequate understanding of the difference in the two approaches using case studies and how a gap in this knowledge can result in higher technical risk. The paper uses a case study to demonstrate it is essential for the designer to know the severity of the environment to which a product may get exposed during the various stages of its procurement e.g. manufacturing, storage and transport.

Keywords: Exposure classification, Corrosivity category, Durability design.

1. INTRODUCTION

One of the most critical parameters required for selection of an appropriate corrosion protection method is the severity or corrosivity of the environment to which the element is or will be exposed during its service life. Determination of corrosivity of the environment is critical to durability design of bridges and other structures. However, it has been observed by the authors that the effect of atmospheric corrosivity has not been dealt with the same approach for steel and concrete – the two main materials used in bridges. Where a bridge has both reinforced concrete and structural steel elements, the structural designer, usually with very limited exposure to corrosion behaviour of steel, may struggle to select an appropriate protection mechanism for steel elements. This particularly applies for selection of a paint coating system from AS 2312, using the exposure classification from AS 5100.5 (AS, 2014a & 2017a). Since there is no industry standard establishing correlation between the exposure classifications of AS 5100.5 and the corrosivity categories of AS 2312 or AS 4312, finding equivalency between the two poses even bigger challenges (AS, 2017a, 2014a & 2008).

While the natural environment existing around a structure containing concrete and steel elements is the same, the durability design needs to encompass different approaches adopted in the aforementioned standards. A steel rebar in concrete is protected from corrosion due to high alkalinity of the concrete and the thickness of concrete available between the rebar and the environment, i.e. concrete cover. However, the rebar may lose its protection as a result of degradation of concrete and ingress of corrosive species from the environment. On the other hand, protection of structural steel exposed to the atmosphere is usually provided by protective coatings.

This paper briefly describes the difference in approach adopted in the above two reference standards for classifying the atmospheric corrosivity for reinforced concrete and structural steel elements. It also demonstrates the importance of having an adequate understanding of the difference in the two approaches using case studies highlighting how a gap in this knowledge can increase technical risk. The paper also uses a case study to demonstrate that it is essential for the designer to not only know severity of the service environment but also the environment to which a product may get exposed during the various stages of its procurement e.g. manufacturing, storage and transport.

The objective of the paper is to bring to the attention of the designers, specifiers, durability consultants and other relevant industry professionals, the existing lack of appropriate guide on this matter.

2. ENVIRONMENTAL CORROSIVITY IN THE STANDARDS

Australian Standard AS 4312:2008, is used for determination of corrosivity categories for structural steel and has classified the Australian environment based on the categories described in ISO 9223:2002 that, in turn, are based on corrosion rate of mild steel obtained from a large number of data collected by exposure of test coupons at different locations. The categories in AS 4312:2008 were adopted from ISO 9223:2002 with the exception that the most severe category, C5, has been subdivided into C5-I and C5-M and an additional category (T) was included to represent inland tropical zone (relevant to organic coatings only). This concept is in line with the concept adopted in ISO 12944-2:1998.

The current version of ISO 9223 (ISO, 2012a), has added one more corrosivity category, CX, that is classified as 'Extreme' and has steel corrosion rate of 200 to 700 micron per year. However, this category is not included in AS 4312:2008 simply because it was not present in the earlier version of ISO 9223 that formed the basis (ISO, 2002). The recently revised version of AS/NZS 2312 (AS, 2014a) has adopted the categories in ISO 9223:2012 and includes this new category, CX. These categories are also used in ISO 14713 for selection of metallic coatings, though the corrosion rate of steel for the first two categories, C1 and C2, are altered slightly (ISO, 2009). The ISO 9223:2002 categories have been modified slightly in ISO 12944-2:1998 for selection of protective paint coatings. In ISO 12944-2, the C5 (very high) category is divided into C5-M Marine and C5-I Industrial to account for the differing performance of protective coatings in these two environments, even though the corrosion rate of steel may be the same (ISO, 1998). This approach has also been adopted in AS 4312:2008 and AS/NZS 2312:2014. Furthermore, a Tropical (T) category has been introduced for these standards to account for the fact that tropical environments degrade organic coatings, such as paint, at a greater rate than would be inferred from steel corrosion rate. Whereas AS/NZS 2312 was revised in 2014 and has incorporated the latest changes of ISO 9223:2012, AS 4312 has not been revised since 2008.

Protection of structural steel from corrosion in AS/NZS 5100.6 has been approached on the basis of AS 2312 (part 1and 2). The standard clearly states durability of a coating system can be estimated from AS 2312. For protection of reinforcing steel embedded in concrete elements, AS 5100.5 for bridges and AS 3600 for buildings, provide the basis (AS, 2017a-b & 2009a). The standards use the term exposure classification instead of corrosivity category. Even though, 5100.5:2017 supplement notes definitions of environmental conditions have been derived from the general concepts given in AS 2312:2002, the exposure classifications in AS 5100.5 (A, B1, B2, C1, C2 and U) are not only different in nomenclature from AS 2312:2002 (A, B, C, D, E and tropical) but also in several other terms. The exposure classifications in Table 4.3 of AS 5100.5:2017 have been shown to be dependent not only on distance from the shore line but also on three climatic zones in Australia, depicted in Fig. 4.3 of the standard.

Even though the actual corrosion mechanism for rebars in concrete is same as to structural steel exposed to atmosphere, initiation of rebar corrosion requires removal of a passivation layer, created due to alkalinity of concrete. Therefore, for the rebar corrosion to occur there must be deterioration in concrete and, hence, corrosivity of the environment in AS 5100.5 has not been chosen on the basis of mild steel corrosion rate, contrary to the approach adopted in AS 2312 or AS 4312. However, there are common factors among both approaches such as the distance from coastline. The approach adopted in As 5100.5 identifies the exposure classification as B1 as long as the concrete element is located more than 50 kilometres from the coast in nonindustrial tropical climate zone. The same environment also exists for any climatic zone for the concrete element located one kilometre to 50 kilometres from a coastline. Severity of the zone has been escalated to B2 for the elements located one kilometre from a coastline. Contrary to this, AS2312 or AS 4312 classify all tropical environments with the same severity irrespective of the distance from the coast as this environment is uniquely associated to the Australian environment and is specific to the degradation of organic coatings (AS, 2017a, 2014a & 2008).

Exposure classifications A, B1, B2, C1 and C2 in AS 5100.5:2017 represent increasing degrees of corrosivity of environment, while classification U represents an environment not given in the standard and for which a degree of severity of corrosion should be appropriately assessed. Determination of corrosivity category is essential for selection of a protection system for structural steel. Assessment of exposure classification is the first step in selection of a suitable surface protection system for concrete elements.

Table 1 shows an amalgamation of the requirements in different standards by representing common features of the two approaches, which is the distance from coastline. However, there are many other factors that affect the corrosivity categories and exposure classification. Therefore, the table should not be used on its own, without referring to the respective standard, to determine the environmental corrosivity for one material based on the other.

Table 1. Environmental corrosivity in steel and concrete

ISO 9223: 2002 Category	ISO 12944 Category	Former AS/NZS 2312 Category	ISO 9223 : 2012 Category	AS/NZS 2312.1 :2014 and AS/NZS 5100.6 Category	AS/NZS 2312.2 2014, ISO 9224:2012 and AS 4312 Category	AS 5100.5 Exposure Classification	Typical exterior environment and distance from coastline (meters/kilometres)
C1	C1	A	C1	C1	C1	—	Few Alpine regions
—	—	—	—	—	—	A	>50 km from coastline
C2	C2	B	C2	C2	C2	A	>50 km from coastline
C3	C3	C	C3	C3	C3	—	50m to 50km from coastline
—	—	—	—	—	—	B1	1 km to 50 km from coastline
C4	C4	D	C4	C4	C4	—	On the coast line and up to 50m to 1km from coastline
—	—	—	—	—	—	B2	<1 km from coastline
C5	C5-I	E-I	C5-I	C5-I	C5-I	—	offshore and on the beachfront and acidic environment with a pH < 5
—	—	—	—	—	—	C1*	In spray zone
						C2*	In tidal/splash zone
C5	C5-M	E-M	C5-M	C5-M	C5-M	—	offshore and on the beachfront
—	—	—	—	—	—	C1	In spray zone
						C2	In tidal/splash zone
—	—	—	CX	CX	CX	—	Surf beach coastline regions with very high salt deposition and in severe acidic industrial environments
—	—	—	—	—	—	U	Exposure environment not specified in the Standard
	—	F	—	T	—	—	Inland tropical
—	—	—	—	—	—	U	Exposure environment not specified in the Standard

*Exposure classification of C1 and C2 is allocated in this table based on the definition of corrosivity category C5 of AS/NZS 2312, i.e. offshore and on the beachfront in regions of rough seas and surf beaches. However, in accordance with AS 5100.5 industrial refers to areas that are within three kilometres of industries that discharge atmospheric pollutants and an exposure classification of B1 is stipulated in the standard.

3. CASE STUDIES

Steel, concrete and composite bridge designs are based on prescriptive normative provisions included in the aforementioned standards. However, based on the authors' experience in durability design review of bridge projects, structural designers are usually conversant with either steel or concrete durability provisions. This has often been found to have resulted in project documents having adequate durability provisions for one material that the designer is conversant while lacking such provisions for the other material. It would be in the interest of the project to have the design and verification of both steel and concrete durability undertaken by specialists in both materials. This is of particular importance due to the differences in durability design approaches for steel and concrete bridges, as contained in the applicable standards. The different approach adopted by the two standards in dealing with the corrosivity of the environment to which a steel and concrete element is exposed adds to the complications. The following case studies demonstrate the issue in bridge project structures with at least 100 years of design life.

3.1 Bridge bearing plates

A bridge was constructed as part of a project located approximately 500 kilometres north of Sydney. Exposure classifications of A, B1, B2 and C were assigned for different concrete elements in accordance with AS 5100.5 (AS, 2017a). Also, the atmospheric corrosivity category of C3 was assigned to minor steel elements in accordance with AS 4312 and, hence, was considered to be applicable for the bearing plates (AS, 2008). This was consistent with B1 exposure classification assigned to the abutment headstocks, approach slabs, wing walls, concrete box girder end blocks and in general for elements not expected to be in contact with ground water. As a result, carbon steel was selected for bearing plates, at all piers except for the one which was over the river. On the other hand, an exposure classification of B2 was assigned to the concrete pier columns as these were near, or just above, splash zone. According to Table 1, this exposure classification (B2) is consistent with C4 corrosivity category and, therefore, is applicable to the bearing plates attached to the top of the column, in contrary to the category C3 determined on the basis of the exposure classification of head stocks (B1). A different material with greater corrosion resistance, say stainless steel, may have been required based on category C4 for the bearing plates (PDRS, 2015).

Corrosion of bearing plates has been observed even in the presence of surface protection e.g. galvanising or painting. This would jeopardise the function of the bearing and, in turn, be detrimental to the structural integrity of a bridge. The cause of the observed corrosion problem on bearing plates was identified as an inconsistent assignment of corrosivity category of minor steel elements with exposure classification of the concrete bridge. This had resulted in selection of materials and surface protections that would have not performed as per design requirements over the expected service life and required early intervention for maintenance. Roads and Maritime peer reviews the durability designs from both steel and concrete perspectives by the subject matter experts to identify any inconsistency in design and prevent the problem from reoccurring.

It is apparent from the above that even though the durability design of both steel and concrete have been consistent with the relevant standards, they were not fully matching when they were put together with an analogy that is given in Table 1. This is of particular importance as elements such as bearing plates cannot be easily replaced after construction and therefore the corrosivity category and, in turn, construction materials should be selected with due diligence to minimise whole of life cycle cost of the asset. This requires a holistic approach to durability design of steel and concrete when they are designed to last for 100 years in an environment as part of one structure such as a bridge. Not only should the corrosivity categories and exposure classifications be assigned in accordance with their respective standards, but there should also be a consistency between durability design of steel and concrete. This highlights the importance of a guide to put different pieces of the puzzle together similar to the analogy of Table 1.

3.2 Steel posts of retaining walls

An exposure classification of B1 was assigned for retaining wall precast concrete panels for a project in Sydney in accordance with AS 5100.5 (AS, 2017a). On the other hand, corrosivity category of retaining wall steel posts was considered as C3 in accordance with AS 4312 (AS, 2008). To achieve the 100 years of design life for steel posts without maintenance, they were deigned to be protected with an 85 micron layer of hot dip galvanised coating (HDG600) in accordance with AS/NZS 2312.2 and be subsequently painted with coating system PRN C34 in accordance with AS/NZS 2312.1 (AS, 2014a-b). It was expected that the combination of HDG600 and PRN C34 coating could provide protection for the first 40 years. Then the steel post was designed with an annual rate of corrosion of 0.02 mm/face/year of buried faces for the next 60 years, i.e. total of 1.2mm sacrificial thickness on each face, in accordance with medium exposure classification of AS 2159 (AS, 2009b). However, the corrosion problem was that the surface preparation of the stipulated coating system, PRN C34, required a blasted profile of 75-100 microns that would remove the whole HDG coating from the substrate. Therefore, the assumption of 40 years of no corrosion would not have been valid and an unprotected surface would have caused section losses more than the design assumptions. This could have been disastrous for the retaining walls. Roads and

Maritime did not accept the assumption of 40 years of no corrosion and instead accepted only 10 years of no corrosion with HDG without painting. The rest of service life was covered by assigning sacrificial thickness for 90 years (PDRS, 2017).

This highlights the importance of having in-depth knowledge of not only the corrosivity of environment in accordance with the relevant standards, but also the integrities of the designed protection systems including the implications of each protection system on the others, when they are used in combination such as painting over a galvanised layer.

3.3 Galvanised steel straps of reinforced soil walls

This case study relates to a project located approximately 150 kilometres north of Sydney. The project involved construction of several concrete bridges in which the abutments consisted of reinforced soil walls. These walls were made of precast concrete panels held in position by galvanised steel reinforcement anchored to the panels at one end with the other end extending into the retained engineered fill. The length extending into the fill varied depending upon the frictional force required to hold a particular panel in positon.

Corrosion of steel in soil was estimated in accordance with AS 2159. An exposure classification of nonaggressive was assigned for the engineered fill with pH > 5; chloride > 0.5%, resistivity > 5,000 ohm cm. The expected corrosion rate of mild steel in this environment was 0.01 mm/year in accordance with AS 2159 (AS, 2009b). Since the reinforcing bars were Hot Dip Galvanised (HDG), an average life of 10 to 15 years was also assumed for HDG in this environment in accordance with AS/NZS 2041 (AS, 2001). These assumptions resulted in selection of HDG steel reinforcements with a minimum sacrificial corrosion thickness of the mild steel substrate of 0.85 mm per face for 100 years of design life without maintenance, as require by the project specification. However, within one and half years of installation in ground, some of the bars were observed to have developed brown rust. This prompted the project team to visually examine a number of bars, several of which were found to have developed ferrous corrosion products, indicating the galvanised layer was consumed and the corrosion has extended to the steel substrate (PDRS, 2001).

The ensuing investigation involved a number of experts from the corrosion industry working in an effort to establish the cause of the accelerated corrosion of the galvanised steel straps buried in ground. These experts were commissioned separately by each of the suppliers/contractors in the supply chain. These included the manufacturer and supplier of the straps, construction contractor, main project contractor and the client. Even though the consultants were engaged by different entities, they shared the knowledge and outcome for the investigation to be able to establish the cause of corrosion. All possibilities were examined. The different hypothesises put forward looked into the possibilities of contamination of soil and water used in the construction through different sources, storage conditions, contamination of the product during transport and so on. Presence of excess sulphur in any form, chloride and other salts, microbiologically induced corrosion, effect due to leaching of fertilisers from the nearby farming and presence of excess carbon in the soil due to burning of timber were all examined. Even though there was no strong evidence to support a particular cause, the most accepted and practical possible reason was considered to be the product contamination with saline water during shipment from overseas which resulted in dramatic increase in corrosion rate and drastic reduction of galvanised coating. This was supported to some extent by presence of chloride on the tested samples. Some of the challenges faced during the transport of overseas procured products include appropriate protection, handling and storage.

The premature corrosion of the reinforcing bars led to remedial work on the infrastructure project with significant cost. This highlights that a structural designer should be conversant not only with the mainstream design standards for environmental corrosivity of steel and concrete, e.g. AS 4312 and AS 5100.5 (AS, 2008 & 2017a), but also with other relevant standards, e.g. AS/NZS 2312 and AS 2159 (AS 2014a & 2009b). Moreover, designers may need to specify all required boundary conditions that have been assumed as part of their design to achieve the expected design, such as delivery condition of the product. Exposure of the product to an environment, different to the service condition, during any stage of its procurement, e.g. transport, has the potential to affect its durability, as demonstrated in this case study.

4. CONCLUSIONS

Based on the aforementioned discussion and case studies, the following conclusions can be drawn:

4.1 A more consistent approach is required for both concrete and steel relevant standards for environmental corrosivities. This will enable the establishment of a better correlation between the two, based on common factors such as distance from the coastline.

4.2 There is evidence to support that more awareness is required among structural designers about the different approaches adopted in the standards for classifying the corrosivity or severity of the environments. This may necessitate development of guidelines/technical notes with a holistic approach in durability design of steel in conjunction with concrete materials.

4.3 The product needs to be designed for the different environments to which it may get exposed during different stages of procurement. This may require considering contractual provisions, regardless of the ratio of steel to concrete in materials used, for design and verification of durability by both steel and concrete durability and materials specialists.

4.4 Durability design requires a holistic approach considering not only the environmental corrosivities but all other parameters that can influence the design life of the structures. Therefore, it is prudent to ensure consistency of assigned corrosivity between different construction materials as well as the delivery conditions and fabrication processes.

5. ACKNOWLEDGMENTS

The authors thank the chief executive of Roads and Maritime for permission to publish this paper. The opinions and conclusions expressed in this paper are those of the authors and do not necessarily reflect the views of Roads and Maritime.

6. REFERENCES

International Standardisation Organisation (ISO, 2002). "ISO 9223 Corrosion of metals and alloys-Corrosivity of atmospheres-Classification, determination and estimation", Geneva, Switzerland.

International Standardisation Organisation (ISO, 2012a). "ISO 9223 Corrosion of metals and alloys-Corrosivity of atmospheres-Classification, determination and estimation", Geneva, Switzerland.

International Standardisation Organisation (ISO, 2012b). "ISO 9224 Corrosion of metals and alloys - Corrosivity of atmospheres - Guiding values for the corrosivity categories", Geneva, Switzerland.

International Standardisation Organisation (ISO, 1998). "ISO 12944-2 Paints and varnishes - Corrosion protection of steel structures by protective paint systems - Part 2: Classification of environments" Geneva, Switzerland.

International Standardisation Organisation (ISO, 2009). "ISO 14713-1 Zinc coatings - Guidelines and recommendations for the protection against corrosion of iron and steel in structures - Part 1: General principles of design and corrosion resistance"

Standards Australia (AS, 2002). "Joint Australian/New Zealand Standard AS/NZS 2312 Guide to the protection of structural steel against atmospheric corrosion by the use of protective coatings", Sydney, NSW, Australia.

Standards Australia (AS, 2008). "Australian Standard AS 4312 Atmospheric corrosivity zones in Australia", Sydney, NSW, Australia.

Standards Australia (AS, 2009a). "Australian Standard AS 3600 Concrete Structures", Sydney, NSW, Australia.

Standards Australia (AS, 2009b). "Australian Standard AS 2159 Piling, design and installation", Sydney, NSW, Australia.

Standards Australia (AS, 2011). "Australian/New Zealand Standard AS/NZS 2041 Buried corrugated metal structures", Sydney, NSW, Australia.

Standards Australia (AS, 2014a). "AS/NZS 2312.1Guide to the protection of structural steel against atmospheric corrosion by the use of protective coatings - Part 1: Paint coatings", Sydney, NSW, Australia.

Standards Australia (AS, 2014b). "AS/NZS 2312. "Guide to the protection of structural steel against atmospheric corrosion by the use of protective coatings - Part 2: Hot dip galvanizing", Sydney, NSW, Australia.

Standards Australia (AS, 2017a). "Australian Standard AS 5100.5 Bridge design-Part 5 Concrete", Sydney, NSW, Australia.

Standards Australia (AS, 2017b). “Australian Standard AS 5100.6 Bridge design-Part 6 Steel and composite construction”, Sydney, NSW, Australia.

Project Durability Reports (PDRS, 2011). Sydney, NSW, Australia.

Project Durability Reports (PDRS, 2015). Sydney, NSW, Australia.

Project Durability Reports (PDRS, 2017). Sydney, NSW, Australia.

7. AUTHOR DETAILS

Narendra Tripathi is Bridge Engineer with Roads and Maritime Services, Transport for NSW. He is a post-graduate from the University of Sydney and has extensive professional experience, obtained in private and public sector organisations. He is a Chartered Professional Engineer (structural and civil), certified NACE CIP L3 and SSPC Concrete Coating Inspector, NACE and ACA Corrosion Technologist and WTIA Quality Control Welding Coordinator (QCWC). He represents AUSTROADS on Standards Australia technical committee on corrosion and also on Australian Paint Approval Scheme.

Hamid Fatemi is the concrete durability technologist in Bridge and Structural Engineering, Roads and Maritime Services. He holds a PhD in materials engineering and a master’s degree in metallurgical engineering. Hamid has professional experience in excess of 20 years in materials selection, testing and characterisation, concrete durability condition assessment, cathodic protection and surface protection systems. He is a fellow member of the Institution of Engineers Australia (FIEAust) and a charted professional engineer (CPEng) in the Structural College of IEAust. Hamid became certified as a corrosion technologist and a NACE CIP L2 coating inspector by the Australasian Corrosion Association (ACA). He also represents AUSTROADS on Standards Australia technical committee MT/007- non-destructive testing of metals and materials.

ARTIFICIAL INTELLIGENCE ASSISTED CONDITION ASSESSMENT

W Nash[1], L Holloway[1,2], T. Drummond[3], N Birbilis[1]
[1]Department of Materials Science and Engineering, Monash University, Clayton, Australia
[2]MEnD Consulting Pty Ltd, West Perth, Australia
[3]Department of Electrical and Computer Systems Engineering, Monash University, Clayton, Australia

SUMMARY: The confluence of aging infrastructure and commodity prices has created a scenario where maintenance costs are often prohibitive and inspection has become the primary means of ensuring the integrity of critical infrastructure. This inspection comes at a considerable cost due to the need provide access to equipment that is often elevated and operating, for example, in the oil and gas environment. Deep learning and artificial intelligence (A.I.) computer vision systems are revolutionising many fields, and condition assessment can take advantage of this progress to provide expert opinion systems on site, and direct inspection resources to areas of high risk. In this work, the capability of semi-supervised deep learning networks to conduct rapid condition assessment of complex steel structures is presented. Currently this technology is applied in post-processing of site photos, although this provides for some time saving, a roadmap is provided to deploying this software for on-site assessment in real time, for example with cameras mounted on drones, or augmented reality headsets, and explore the challenges remaining for field deployment. Some pitfalls of artificial intelligence systems and how to avoid them are discussed, whilst a discussion of the extensibility of these systems is also be presented to illustrate how A.I. can be used to tackle problems in other areas.

Keywords: Corrosion, Inspection, Artificial Intelligence, Deep Learning

1. INTRODUCTION

Artificial Intelligence (A.I.) has become ubiquitous in our modern lives. From automatic detection of faces in applications such as Snapchat, to Google suggested search results from your recent browsing history, the results of machine learning are reaching further into our lives every day. In the context of A.I., highly publicised results that the broader public can relate to include when the IBM supercomputer (known as Watson) beat the world champions at Jeopardy, or more recently when the AlphaGo A.I. from Google's DeepMind beat the world champion at the ancient Chinese strategy game 'Go'. Other impacts like the electioneering of Brexit and the Trump 2016 campaign by data mining company Cambridge Analytica (https://motherboard.vice.com/en_us/article/mg9vvn/how-our-likes-helped-trump-win) receive less coverage, although arguably represents both a more insidious and meaningful impact of A.I.

A.I. sophistication has accelerated thanks to increases in Graphic Processing Unit (GPU) performance and the creation of very large datasets collected via the Internet. Indeed, if you've ever completed a CAPTCHA (http://www.captcha.net/) when buying AFL tickets, or taken a Facebook quiz, you've contributed to training a machine. At the same time, developments in the field of A.I. have led to deeper network architectures with complex structures to improve performance in training and application. Up until the 1990s Artificial Neural Networks (ANNs) were largely constrained to one input layer, one hidden layer and one output layer (1). Today's Deep Learning networks utilise multiple hidden layers, which leads to very compute intensive weight matrices, for instance the VGG16 ImageNet-2014 winning network from University of Oxford has 138M parameters. (VGG stands for Visual Geometry Group, 16 refers to the number of layers in the network)

Researchers in A.I. have in part been driven by machine learning contests. In the field of computer vision, the present benchmark competition is known as ImageNet (http://www.image-net.org/); where the dataset for 1000 classes is provided to competitors for training automated classification systems. The aim of the research herein is to retrain an image recognition network to identify a less well-defined category: corrosion. Corrosion may occur in many shapes and forms, depending on the substrate and environment. Initially the network presented was trained to detect atmospheric corrosion of mild steel.

It is intended that an automatic corrosion detector will supplement visual inspection of assets, leading to faster inspections, reduced inspection costs and mitigate many of the safety precautions presented to inspectors. The end-goal is to equip a robot with corrosion detecting cameras for autonomous inspection. The manuscript herein takes a high-level look at the methods utilised, the challenges presented, and presents a framework for overcoming these.

2. EXPERIMENTAL DETAILS

A Fully Convolutional Network (FCN) was selected for the corrosion detector Deep Learning network architecture because it can identify where in an image it detects corrosion, and can handle variable image input resolutions. The FCN architecture primarily consists of convolutional layers, so called because they convolve the preceding layer to produce feature maps, thus the trained FCN performs as a "nonlinear filter" (2) of the input image. The input to the corrosion detector FCN consists of the individual pixel intensities in each of the red, green and blue channels of an image. The FCN outputs a densely-labelled image, where each pixel is coloured according to its detected class – within computer vision this is known as semantic segmentation of the image (3).

The initial step for machine learning is to build the dataset for training of the network. The building of a dataset for training in the context of condition assessment includes compiling and labelling images of corrosion in different settings, on various asset types, and from diverse angles. Herein, the training dataset was produced for strongly supervised training with individual pixels labelled by colour as corrosion, no corrosion, or ignore (for areas that researchers were unable to differentiate, e.g., in shadow or areas of heavy staining). Strongly supervised training is defined as training where the level of labelling detail is the same as the required output of the network (4). A total of 166 high resolution images were processed to generate their associated "ground truth" label images. An example of an original image is presented in **Figure 1** and the corresponding ground truth image is presented in **Figure 2.** The training set comprised 80% of the data set images, and 20% of the images were retained in the validation set for network performance evaluation.

Figure 1 Example of Original Image for Training

Figure 2 Groundtruth Image, Corrosion labelled as magenta (R,G,B = [255,0,255]), background labelled as red (R,G,B = [255,0,0])

The corrosion detector network utilises the FCN implementation of VGG16 in TensorFlow code from (5) modified for the corrosion dataset.

The machine learning network utilised herein for corrosion condition detection was trained on an NVidia TitanX 1080 Graphical Processing Units (GPUs) for 60,000 iterations using ADAM optimisation (6) with an initial learning rate of 10^{-5} to minimise cross-entropy loss (7). The minimisation of cross entropy loss, defined as the difference between the predicted probability distribution and the true probability distribution, provides a strong training signal even when the true probability is small and as such is a good loss measure for image recognition problems.

3. RESULTS

Herein, a corrosion detection network is described that incorporated 166 training images, and was deployed in the analysis of images of corrosion in various settings. Following network 'training', the validation set was evaluated by the network, cross-entropy loss (previously defined in the Experimental section) was determined to be 0.11. The value of 0.11 is near to the global minimum for the network as monitoring indicates that the loss approaches an asymptote, and prior network iterations at the beginning of training had values of cross-entropy loss in the range of 0.93 – 0.98.

Figure 3 provides an example output of the application of the trained network, where green pixels are labelled positively as 'corrosion'. **Figure 4** presents the same image as in **Figure 3**, however the network out is presented in heatmap contrast for 'corrosion' activation scores for each pixel. The original image is 5760 × 3840 pixels, of a visually complicated structural support and piping structure. The inference was performed on the same GTX 1080 GPU as used for training, downsizing the image to 3000 × 2000 pixels, and analysis of such an image was executed in approximately 0.9 seconds. The downsizing was carried out due to memory limitations of the GPUs.

Figure 3 Example Output of Corrosion Detection Network, Pixels the network has labelled as corrosion are coloured Green

Figure 4 Example Output of Corrosion Detection Network, presented as a heatmap

4. DISCUSSION

After training on a dataset of just 166 examples (each example being an image of corrosion that had been manually labelled for corrosion) the network presented herein was able to identify the majority of corroded areas within subsequent (computer only) analysis of images. However, when the network is assessed by an expert (i.e. human, corrosion engineer) there were instances that false positives could be present for some features, in particular the brown coloured sheet metal to the top left of the image (in Figure 3). False negatives are also present where corrosion is not labelled by the network, e.g., on the handrail of Figure 3. The impact of utilising a network that may produce false positives and false negatives will depend on the nature of the application, which is not the purpose of the present study. However, a network with as high fidelity as possible is required in cases when engineering decisions may follow. It is anticipated that further training will reduce these errors, particularly if the training set is expanded to include more examples that include the problematic features.

In terms of deployment to a drone or crawler robot there are several remaining barriers to implementation of a corrosion detection network that are being worked on at present, including the following:

4.1 Computer power required to run A.I. network

Mounting any detection system on a robot introduces weight and (mobile) power limits. In order to overcome these limitations, the computer network should be optimised to be able to run on a single low power chip. The corrosion detection network presented herein comprises some 138 million neurons as adapted from the VGG16 architecture. We expect that this is vastly more neurons than is required to accurately detect corrosion, and that many the neurons are not contributing to the detection in a significant way. The advantage of utilising such a large network architecture is that it can be adapted to other tasks, and in the future the same network could be trained to detect multiple features of interest, for examples cracks in concrete, paint defects, and of course corrosion of other alloys such as aluminium or galvanising. Meanwhile, for deployment of the corrosion detection network it should be compressed into a smaller form, without significant loss of performance. This can be achieved by network distillation as detailed in (8). The minimum architecture size for acceptable results can be found by incrementally reducing the depth of the network and measuring performance.

4.2 Access for inspection

Often a key consideration for visual inspection crews is gaining access, and gaining it safely adds considerable expense to the task. Although we anticipate that automated inspection robots will reduce the cost of safe inspection, there are still some important considerations and limitations. Drones are a great choice for quick surveying, however they present their own safety hazards with spinning rotors and radio frequency controls that restrict their use in the presence of critical infrastructure, such as oil and gas facilities, or above crowded areas. Furthermore, the ability to fly drones is also restricted by several laws that require adherence to airspace controls, defined in Australia under *Civil Aviation Safety Regulations Part 101*. For assets that require frequent inspections a specialist cable or rail mounted system may be appropriate, and a modified spider-cam could be envisioned for offshore structures. More portable options include a pole cam, and a robotic boat for wharf inspections. It is also anticipated that Augmented Reality implementations may be popular to assist inspectors alongside other asset data.

4.3 Choosing an appropriate imaging resolution for detection

The resolution of input images is directly related to the compute time of inference, with each pixel value needing to be loaded into the RAM. For training, very high-resolution images were used, in the order of 35 Mega-Pixels (MP) that needed to be sliced into smaller images to prevent 'out of memory' errors when the GPUs run out of memory. For deployment, it will be necessary to adopt an imaging resolution that satisfies the detail required by asset owners, but that is able to be processed in adequate time with limited compute resources. One option is to compress the images for inference, however it has been found that the network will misinterpret finer details under this scenario. Full resolution semantic segmentation has been achieved in real time using a cascading network balancing network depth and resolution in (9). Using a similar approach, it is expected that the final automated imaging system will be able to handle 1980 × 1080 pixels or more in real time, the final image resolution will depend on GPU memory available.

4.4 Locating detected corrosion

When deploying an accurate corrosion detection network, it may not be as meaningful if a drone surveys a site and returns many images of corroded assets without identifying the defect locations. Therefore, it is necessary to assign identifying features to the images captured. In the future, it may be possible to locate defects through site wide geographical information systems. However, in the near-term location information is likely to be obtained by pairing Geographical Positioning Systems (GPS) with camera pose direction metadata. It may also be possible to equip the robot with a paint marker to allow repair crews to quickly locate defects, and eventually a robot that can undertake minor repairs in-situ is entirely possible.

5. CONCLUSIONS

- The research presented herein provides an overview of the demonstration that so-called Deep Learning networks can be used successfully as a corrosion detector, although more work is needed to reduce false positives (i.e. erroneous detection of corrosion) and false negatives (i.e. missed detection of corrosion).
- The accuracy of the corrosion detection network will depend heavily on the training dataset. This is akin to the efficacy of any training (or education) based exercise. For example, to disambiguate features that are under-represented in training images, the network presented struggles. An example of features underrepresented in the dataset includes rust coloured items such as, for example, red bricks. As a result, a wide training dataset is required for a general or universal corrosion detection network; however either specific or minimalist training datasets could be utilised in specialised cases when unique infrastructure is being assessed exclusively.
- Limitations in the practicality of deploying automated inspection that are yet to be overcome include a reduction in the computing power required for image labelling, provision of access for crawlers and drones, selecting appropriate resolution equipment for image capturing, and attaching location data to the information collected.

6. ACKNOWLEDGMENTS

We gratefully acknowledge support from Woodside Energy and MEnD Consulting.

7. REFERENCES

1. Schmidhuber J. Deep Learning in Neural Networks : An Overview. Neural Networks. 2015;61(Jan):85–117.
2. Long J, Shelhamer E, Darrell T. Fully Convolutional Networks for Semantic Segmentation. Proc IEEE Conf Comput Vis Pattern Recognit. 2015;3431–40.
3. Borenstein E, Ullman S. Class-Specific, Top-Down Segmentation. In: Heyden A., Sparr G., Nielsen M. JP, editor. ECCV 2002 Lecture Notes in Computer Science [Internet]. Heidelberg: Springer, Berlin; 2002. p. 109–22. Available from: http://link.springer.com/10.1007/3-540-47967-8_8

4. Branson S, Perona P, Belongie S. Strong Supervision From Weak Annotation : Interactive Training of Deformable Part Networks.

5. Teichmann M, Weber M, Zoellner M, Cipolla R, Urtasun R. MultiNet: Real-time Joint Semantic Reasoning for Autonomous Driving. 2016; Available from: http://arxiv.org/abs/1612.07695

6. Kingma DP, Ba J. Adam: A Method for Stochastic Optimization. 2014; Available from: http://arxiv.org/abs/1412.6980

7. Boer PDE. A Tutorial on the Cross-Entropy Method. 2005;19–67.

8. Hinton G, Vinyals O, Dean J. Distilling the Knowledge in a Neural Network. NIPS 2014 Deep Learn Work [Internet]. 2015;1–9. Available from: http://arxiv.org/abs/1503.02531

9. Zhao H, Qi X, Shen X, Shi J, Jia J. ICNet for Real-Time Semantic Segmentation on High-Resolution Images. 2017; Available from: http://arxiv.org/abs/1704.08545

8. AUTHOR DETAILS

Will Nash is undertaking a Research Masters at Monash University, his thesis title is Artificial Intelligence in Materials Engineering. Prior to undertaking postgraduate study, he worked as a consulting engineer for roughly 9 years in Australia and Canada. He has spent many days working from precarious perches in inclement weather inspecting bridges, wharves, reservoirs and everything in between.

Liam Holloway is Managing Director and Principal Engineer at MEnD Consulting. Liam has over 15 years' experience working as an engineer in the field of concrete durability, inspection and repair. He completed is PhD by research on the application of organic inhibitors for the mitigation of reinforcement corrosion in marine structures in 2001 at Monash University. Since that time Liam has worked as a consulting engineer and spent 8 years working as National Engineering Manager for a large remedial contractor. In 2014 Liam established MEnD Consulting, a small practice specialising in the investigation and remedial engineering for existing infrastructure.

Tom Drummond is a Professor of Electrical and Computer Systems Engineering. His research areas include Computer Vision, Augmented Reality, Robotics and Machine Learning.

Nick Birbilis is a Professor, and Head of Materials Science and Engineering at Monash University. His research spans the broad areas of corrosion and corrosion control.

REMAINING LIFE ESTIMATES USING MONTE CARLO SIMULATION & WEIBULL SURVIVAL ANALYSIS

J-A. Latham, A. Spark & S. Furman
AECOM Australia, Melbourne, Australia

SUMMARY: Concerns over potential corrosion issues with copper and mild steel heating hot water and chilled water HVAC pipes in a number of buildings constructed in 1970-80s prompted a condition survey to determine the remaining life of and identify pipework requiring immediate replacement. Ultrasonic thickness (UT) measurements were taken on site at representative areas and used as a basis to determine the level of corrosion currently sustained. However, due to the age of the buildings, original information concerning the type of pipe and the original wall thicknesses was unavailable. Subjective assessments and deterministic models commonly used to determine the remaining life of assets will typically result in a range of years, e.g. a remaining life of 10-15 years. However, with no previous wall thickness measurements available an in-situ corrosion rate could not be applied to estimate remaining life of the surveyed pipework in this way. Furthermore, using deterministic methods there is no indication of how likely it is to fail at 10 years versus 15 years. That is, is it a 100% chance of failure in 15 years or less; is there still a relatively high chance of failure in 10 years? This information can be critical in assessing the risk of an asset to allow for budget planning by an asset owner. The use of survival analysis techniques allows for remaining life to be represented as a probability that can then be tied directly to risk.

To determine the estimate of remaining life a number of assumptions on the internal and external corrosion rates and failure condition were made. Using Monte Carlo simulations the variance in corrosion rates as well as initial wall thickness could be taken into account. Using Weibull Survival Analysis an estimate of life was able to be presented as a probability of failure. This approach was able to take into account the uncertainty of the assumptions used, as well as provide a profile that could be presented as risk to the client.

Keywords: Pipe, Weibull, Monte Carlo, Survival Analysis, Reliability Analysis, Remaining Life.

1. INTRODUCTION

The aim of this work was to estimate the remaining life of HVAC (heating, ventilation and air conditioning) heating hot water and chilled water pipes in two buildings and undertake a risk assessment to provide an indication of risk of failure of the pipes. No historical information, including past failure of pipes, original wall thickness at time of installation, or any past wall thickness measurements were available for use during this assessment. In the absence of such information, assumptions on the corrosion behaviour introduce uncertainty that is best taken into account using probabilistic models.

Survival analysis, also called "time-to-event analysis", is the statistical method of analysing a set of data where the outcome is a unit of time to an event of a single variable, termed "survival time" [1] [2]. In engineering the event is typically defined by failure and therefore survival analysis is similar to reliability engineering analysis, which describes the ability for a system to function for a set period of time. The difference is that for survival analysis the focus is on the probability that failure will occur by a certain time, whereas reliability analysis focuses on the probability that failure will not occur by a certain time [3]. That is if at time point T the probability of failure, where failure occurs at time t, is:

$$S(t) = \Pr(t < T) \quad (1)$$

Thus the reliability will be:

$$R(t) = 1 - S(t) = \Pr(t > T) \quad (2)$$

Therefore if the probability of failure is 70% the reliability is 30%. In the context of HVAC water pipes, failure can be defined as a leak or burst in the pipe.

parametric (can predict all future values from its known finite parameters, e.g. all values in a normal distribution can be predicted knowing the mean and standard deviation), nonparametric (parameters are infinite and prediction of future values must be based on not only the parameters, but the current state of the values) and semiparametric (a mix between parametric and nonparametric) methods for the analysis of survival data. Non-parametric methods include the Kaplan Meier Product Limit methods, semi-parametric models include the Cox-Proportional Hazard method and parametric methods involve Weibull and Exponential methods [2]. Parametric models, such as the Weibull method used in the assessment of the pipes discussed in this paper, assume the survival time follows a distribution [4]. The probability of failure is therefore obtained by fitting a distribution to the data and calculating the cumulative distribution function (CDF).

For the purpose of survival analysis of assets, data used is ideally lifetime data collected on the asset or a range of similar assets that can provide time scales for failure events. When using lifetime data a common issue will be that not all assets may have failed within the observation time. This results in right censoring, as the failure time is beyond the measurement period and therefore cannot be known or reported [1] [2]. Assuming the observation time is the same for all assets a probability of failure within the observation time can be easily assessed but the probability of failure beyond this time is unknown. If the observation times vary due to differing asset age / installation dates or when observation began this leads to left truncated data. The presence of censoring means that standard statistical methods cannot be applied to meaningfully assess the data; which is a strength of survival analysis [1] [2].

In the absence of lifetime failure data, where a deterministic model can be used to estimate remaining life, randomly generated variables can be used instead to produce a distribution of survival time. Monte Carlo simulation is a method of using randomly generated numbers to assess the statistics of a problem [5]. In order to run a Monte Carlo simulation the inputs must first be represented as a number of probability density functions (PDFs). The Monte Carlo method then involves repeatedly sampling these PDFs to produce a large quantity of variables that are then tallied, or computed to provide a distribution of possible outcomes [5].

2. METHODOLOGY

2.1 Pipes Investigated

Twelve HVAC heating and chilled pipes were investigated in this work. These pipes transport hot (80-85°C) and chilled (6°C) water for the purpose of heating and cooling of two buildings built in the 1970s and 1980s in the Melbourne CBD area. Both are closed systems using a sodium nitrite ($NaNO_2$) corrosion inhibitor with a sodium hydroxide (NaOH) pH buffer to maintain pH above 9.

The pipes selected for inspection and testing were chosen in order to obtain a representative sample of varying corrosion conditions in the chilled and hot water systems of each building, including:

- Pipe diameter
- Hot or chilled water
- Location in building (i.e. floor level)
- Pipe type (e.g. main riser, air handling unit (AHU), branch run, etc.).

Table 1 lists the pipes selected for inspection and testing.

Limited design documentation from the time of installation meant that the material type and grade utilised for some pipes was unclear. Based on photographs taken on site it appears that the pipes are primarily mild steel. We were unable to obtain a copy of the historical Australian Standards for the specification of copper pipes at the time of construction, we are confident that there is no significant difference in the copper alloys used to what is available today (copper alloy C1220). Without material data sheets from the time of construction, the specific composition of the steel pipes cannot be confirmed.

Table 1. Summary of Pipes Inspected

Ref	Material[1]	Building	Level	System[2]	Pipe Type	Diameter (mm)	Age[3] (years)
1	Mild Steel	1	2	HHW	Main Riser	100	40
3	Mild Steel	1	6	HHW	Branch Run	38	40
4	Mild Steel	1	7	HHW	Branch Run	38	40
6	Mild Steel	1	Plant	HHW	AHU Branch	50	40
7	Copper	1	Plant	CHW	AHU Coil Branch	28	?
8	Mild Steel	2	1	HHW	Branch Run	32	33
10	Mild Steel	2	2	HHW	Riser Offset	50	33
11	Mild Steel	2	2	CHW	Riser Offset	40	33
12	Mild Steel	2	5	HHW	Branch Run	15	33
13	Mild Steel	2	5	CHW	Branch Run	25	33
14	Mild Steel	2	7	HHW	AHU Branch	65	33
15	Mild Steel	2	7	CHW	AHU Branch	80	33

Table Notes:

1. Assumed based on visual appearance in photos from visual inspection
2. CHW = Chilled Water; HHW = Heating Hot Water
3. The age of pipework has been assumed to be from the date of construction. The age of the copper pipe is not known

2.2 Wall Thickness Measurements

Ultrasonic thickness (UT) measurements were taken at four axial locations around the circumference of each pipe.

2.3 Corrosion and Corrosion Rates

2.3.1 Assumptions and Limitations

Due to the age of the buildings and the limited amount of design information available, there were several limitations to the investigation. The remaining life of the pipes was defined, for the purpose of the investigation, as the time to plastic deformation of the pipe. The deformation results from internal and external corrosion causing a reduction in pipe wall thickness and subsequent reduction in pipe hoop strength. The internal corrosion in the chilled and hot water heating systems was assumed to be controlled by the use of corrosion inhibitors. As these systems are closed loop systems, the water chemistry can be closely controlled and monitored to achieve acceptable corrosion protection. External corrosion of the pipes is expected to be very low, especially as the pipes are primarily located in building cavities and covered by insulating material. However, condensation of water on the external surface of the chilled water pipe may occur, and where it occurs it was expected to lead to increased external corrosion of this pipework.

The use of corrosive fire retardants (chloride compounds) in insulation can also exacerbate corrosion; however, the effect of this cannot be quantified and it is not known whether they have been used in any of the insulation present and therefore has not been considered.

For the purposes of our remaining life estimates, we have made the following general assumptions:

- Both hot and chilled water pipes experience internal and external corrosion
- External corrosion is atmospheric and follows the equations in ISO 9224
- External corrosion is linear and general in nature, which is based on the age being greater than 20 years in accordance with the methodology in ISO 9224

- Internal corrosion is considered linear and general in nature due to the presence of inhibitors in a closed system that actively prevent / minimise the effect of localised corrosion sites
- Chilled water (CHW) is no less than 4°C
- Heating hot water (HHW) is a maximum of 85°C
- The internal corrosion rate increases by a factor of two for every 18°C rise in temperature [6]
- Appropriate inhibitor levels will be maintained in the water over the lifetime of the pipe (based on recent water chemistry analysis results)
- There is no galvanic corrosion occurring between the mild steel and copper pipes (based on recent water chemistry results, which did not show elevated Fe or Cu ions in the water).

Further assumptions and limitations are outlined in the following sections.

2.3.2 Internal Corrosion

Both chilled and hot water systems use a $NaNO_2$ corrosion inhibitor. pH is held slightly alkaline using pH buffer NaOH. A summary of the pH and inhibitor levels are listed in Table 2.

Table 2. Recent Water Quality Summary. HHW: Heating Hot water; CHW: Chilled water.

Item	Building 1		Building 2	
	CHW	HHW	CHW	HHW
pH	9.2	9.8	9.2	9.2-9.6
[$NaNO_2$] (ppm)	1100	1540-1860	1120	1400-1890

$NaNO_2$ is a widely used corrosion inhibitor for closed cooling systems, and inhibitor concentrations of 600-1200 ppm will provide suitable corrosion protection to steel at water pH > 7 [7]. Greater concentrations of inhibitor (often 5000-7000 ppm) are needed for adequate inhibition in systems containing both steel and copper [7]. Review of the literature shows that at levels of nitrite inhibitor of over 1000 ppm, the corrosion rate in chilled water should be no more than 0.5 mpy [7] [8] (equivalent to 0.013 mm/year), even in a closed system with mixed metallurgy [7]. A corrosion rate of 0.5 mpy or below for mild steel is considered to be a "good" level of corrosion inhibition in closed systems [9].

Information on the actual internal corrosion rates of the pipes investigated as part of this project was not available. Initial thickness measurements (that would allow for calculation of an average corrosion rate over the life of the pipe) or coupon corrosion tests were also not available. As a result, literature values of expected corrosion rates based on the inhibitor levels in these pipes has been used to estimate the remaining life.

While information on likely corrosion rates could be found for mild steel pipe systems, similar information for copper piping was not available. Based on the outcome of "good" [9] levels of corrosion inhibition on the steel, it has been assumed that similar levels of inhibition are being achieved on the copper pipes. A summary of the internal corrosion rates used to estimate remaining life are provided in Table 3.

It has been assumed that the level of inhibition of Cu is sufficient and therefore any galvanic corrosion effects due to deposition of Cu particles on mild steel pipe sections is negligible and thus has not been considered.

Table 3. Corrosion Rates Used for the Internal Corrosion of Pipes containing chilled water [9]

Material	Internal Corrosion Rate, r_{corr} (mm/year)
Mild Steel	0.005-0.013
Cu	0.003-0.006

The internal corrosion has been assumed to be linear in nature, where the loss of material can be calculated from Equation (1).

$$\Delta T = r_{corr} \cdot t \qquad (3)$$

Where T is the material thickness

r_{corr} is the internal corrosion rate

t is time.

2.3.3 External Corrosion

The corrosion has been assumed to be atmospheric and no additional influence from the insulation composition has been taken into account. External corrosion of the pipes is expected to be low as the pipes are primarily located in building cavities and covered by insulating material.

Based on the definitions of corrosivity in ISO 9223, we have assumed a corrosivity of C2 for chilled pipes (for a minor frequency of condensation), and C1 for hot water pipes (see Table 4).

Table 4. ISO 9223 C1 and C2 First year Corrosion Rates Used for the External Corrosion of Pipes

Material	**External Corrosion Rate, R_{corr} (mm/year)**	
	C1 Classification	**C2 Classification**
Mild Steel	0-0.0013	0.0013-0.025
Copper	0-0.0001	0.0001-0.0006

The external loss of material has been calculated using the equations in ISO 9224. As the age of the pipes is greater than 20 years, we have assumed that the atmospheric corrosion is linear, where the loss of material can be calculated in accordance with Equation (2).

$$\Delta T = R_{corr}[20^b + b(20^{b-1})(t - 20)] \quad (4)$$

Where T is the material thickness
R_{corr} is the first year corrosion rate
b is the time exponent
t is time.

2.4 Minimum Wall Thickness

Remaining life was estimated based on the limited information available to give confidence to the client of when the pipes need to be replaced as a whole. Ongoing monitoring will be required to ensure localised corrosion has not resulted in an isolated section of pipe failing, however that is outside the scope of this estimation. The estimate of remaining life has been based on the time calculated for the wall thickness to reduce to a level that will cause failure. How thin the walls can be is dependent on the pipe pressure, the diameter, and the tensile strength of the material.

The minimum allowable wall thickness can be calculated according to Barlow's equation for hoop stress in Equation (3):

$$P = \frac{2ST}{OD \cdot SF} \quad (5)$$

Where: P is fluid pressure
T is wall thickness
OD is outside pipe diameter
SF is safety factor
S is Material Strength (Ultimate tensile or yield).

Our calculations for each pipe were made on the following assumptions taking into account the length of time under consideration:

- An operating fluid pressure of 5 bar (0.5 MPa)
- Corrosion will be uniform (see assumptions in Section 2.3.1)
- Failure is considered to be when plastic deformation of the pipe occurs. Therefore the yield strength has been used to calculate the minimum wall thickness.
- A safety factor of 1.

2.5 Monte Carlo Modelling and Weibull Survival Analysis

A combination of Monte Carlo simulation and Weibull survival analysis was used to calculate the remaining life of each pipe in terms of probability of failure.

Below is a summary of the process:

- UT site measurement data was divided into frequency bins to allow for comparison to standard distribution types (e.g. normal distribution). It was found that the data was generally skewed, and a standard normal distribution was not applicable. To account for the skew in the measurements, a Weibull distribution was used. An example of this fit is shown for Pipe Reference 4 in Figure 1. 2000 random pipe wall thicknesses were generated based on this distribution.
- Corrosion rate information obtained from literature for both internal and external corrosion is listed as a range. It was assumed that the range of corrosion rates exists as a normal distribution, where the upper and lower limits are three standard deviations from the mean (accounting for 99.73% of possible corrosion rates). Based on this, a set of 2000 random corrosion rates were generated for each corrosion type (internal chilled, internal hot & external – see Section 2.3).
- 2000 Monte Carlo simulations were performed by calculating the years to failure based on Equation (4) with random inputs for each of the variables of initial wall thickness, internal and external corrosion rate.
- The resulting distribution of time to failure from the Monte Carlo simulations was fit to a Weibull distribution and the CDF (cumulative distribution function) was calculated to provide the probability of failure at any given year.

$$t = \left(\frac{\frac{T_i - T_f}{R_{corr}}}{b \cdot 20^{b-1}} \right) \Big/ \left(1 + \frac{r_{corr}}{R_{corr} \cdot b \cdot 20^{b-1}} \right) \qquad (6)$$

Where t is time

T_f is the thickness at failure (see Section 2.4)

T_i is the initial thickness

r_{corr} is the internal rate of corrosion (see Section 2.3.2)

R_{corr} is the external rate of corrosion (see Section 2.3.3)

b is the time exponent (ISO 9224).

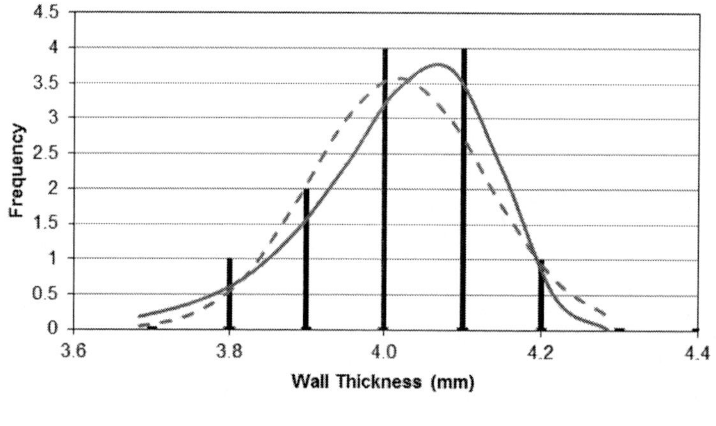

Figure 1. Normal and Weibull Distribution of wall thickness measurements on Pipe Reference 4.

2.6 Selection of Consequence

The consequence rating for each pipe was selected based on our understanding of the impact of a pipe failure on the operation the two buildings and selection was based on the following considerations:

- Location of the pipe:
  - Potential to cause damage prior to detection
  - Ease of access for repair.
- Result of injury due to leak:
  - Slips
  - Scalding (hot water only).
- Relative cost of repairing pipework.

All pipes inspected (and hence assessed as part of this report) were relatively easy to access (located in ceilings) and leaks are expected to be easily detected. Potential for injury is still present, and the consequence of injury with hot water is considered to be greater than for chilled water. For this reason, all chilled water pipes have been considered to have a minor consequence, whereas hot water pipes have been considered to have a moderate consequence.

3. RESULTS

3.1 Ultrasonic Thickness Measurements

The ultrasonic thickness (UT) measurements for each pipe inspected are summarised in Table 8.

Table 5. Summary of UT measurements

Ref	Material*	Building	Level	System	Pipe Diameter (mm)	Average Remaining Wall Thickness (mm)			
						Top of pipe	Bottom of pipe	Side 1	Side 2
1	Mild Steel	1	2	HHW	100	4.8	4.7	4.7	4.7
3	Mild Steel	1	6	HHW	38	3.2	3.2		
4	Mild Steel	1	7	HHW	38	4.1	4.0		
6	Mild Steel	1	Plant	HHW	50	3.5	3.4		
7	Copper	1	Plant	CHW	28	1.4	1.4	1.4	1.4
8	Mild Steel	2	1	HHW	32	2.9	3.0		
10	Mild Steel	2	2	HHW	50	3.5	3.6	3.7	3.6
12	Mild Steel	2	5	HHW	15	2.7	2.8		
13	Mild Steel	2	5	CHW	25	3.1	3.1		
14	Mild Steel	2	7	HHW	65	3.6	3.6	3.6	3.8
15	Mild Steel	2	7	CHW	80	3.8	3.8	3.8	3.9

3.2 Remaining Life Outcomes

The outcome of the survival analysis is shown in Table 9. The results indicate that there is very little probability of failure in the next 20 years (2037) for the pipes tested. The pipes which show an increased probability of failure over the next 40 years are the heating hot water pipes. This can be attributed to the influence of increasing temperature with increasing corrosion rate [6]. These outcomes assume that the chemical inhibition levels are maintained in the water, and internal and external corrosion rates remain at the levels noted above.

Table 6. Probability of Failure of Inspected Pipes

	Pipe Reference										
	1	3	4	6	7	8	10	12	13	14	15
	MS	MS	MS	MS	Cu	MS	MS	MS	MS	MS	MS
Year	**HHW**	**HHW**	**HHW**	**HHW**	**CHW**	**HHW**	**HHW**	**HHW**	**CHW**	**HHW**	**CHW**
2017	0.0%	0.0%	0.0%	0.0%	0.0%	0.0%	0.0%	0.0%	0.0%	0.0%	0.0%
2021	0.0%	0.0%	0.0%	0.0%	0.0%	0.0%	0.0%	0.0%	0.0%	0.0%	0.0%
2027	0.0%	0.0%	0.0%	0.0%	0.0%	0.0%	0.0%	0.0%	0.0%	0.0%	0.0%
2037	0.0%	0.1%	0.0%	0.4%	0.0%	0.3%	0.0%	0.5%	0.0%	0.1%	0.0%
2038	0.0%	0.2%	0.0%	0.6%	0.0%	0.4%	0.1%	0.8%	0.0%	0.1%	0.0%
2039	0.0%	0.3%	0.0%	0.8%	0.0%	0.6%	0.1%	1.1%	0.0%	0.1%	0.0%
2040	0.0%	0.5%	0.1%	1.1%	0.0%	0.9%	0.2%	1.7%	0.0%	0.2%	0.0%
2041	0.0%	0.7%	0.1%	1.4%	0.0%	1.4%	0.2%	2.5%	0.0%	0.3%	0.0%
2042	0.0%	1.0%	0.1%	1.9%	0.0%	2.0%	0.3%	3.6%	0.0%	0.4%	0.0%
2043	0.0%	1.4%	0.2%	2.5%	0.0%	2.7%	0.5%	5.0%	0.0%	0.5%	0.0%
2044	0.1%	1.9%	0.2%	3.3%	0.0%	3.8%	0.7%	7.0%	0.0%	0.8%	0.0%
2045	0.1%	2.6%	0.3%	4.2%	0.0%	5.2%	0.9%	9.5%	0.0%	1.0%	0.0%
2046	0.1%	3.6%	0.5%	5.3%	0.0%	7.1%	1.3%	12.8%	0.0%	1.4%	0.0%
2047	0.1%	4.8%	0.6%	6.7%	0.0%	9.4%	1.7%	17.0%	0.0%	1.9%	0.0%
2048	0.2%	6.4%	0.8%	8.4%	0.0%	12.3%	2.3%	22.1%	0.0%	2.5%	0.0%
2049	0.3%	8.4%	1.1%	10.4%	0.0%	16.0%	3.1%	28.3%	0.0%	3.3%	0.0%
2050	0.3%	10.9%	1.4%	12.8%	0.0%	20.5%	4.0%	35.4%	0.0%	4.2%	0.0%
2051	0.5%	14.0%1	1.9%	15.6%	0.0%	25.8%	5.2%	43.6%	0.0%	5.4%	0.0%
2052	0.6%	17.7%	2.4%	18.8%	0.0%	32.0%	6.7%	52.4%	0.0%	6.9%	0.0%
2053	0.8%	22.2%	3.1%	22.5%	0.0%	39.0%	8.6%	61.5%	0.0%	8.7%	0.0%
2054	1.0%	27.4%	4.0%	26.6%	0.0%	46.7%	10.8%	70.5%	0.0%	10.9%	0.0%
2055	1.2%	33.4%	5.0%	31.2%	0.0%	54.9%	13.5%	78.8%	0.0%	13.6%	0.0%
2056	1.6%	40.1%	6.2%	36.2%	0.0%	63.3%	16.8%	85.9%	0.0%	16.7%	0.0%
2057	2.0%	47.4%	7.7%	41.6%	0.0%	71.4%	20.6%	91.4%	0.0%	20.3%	0.0%

Table Notes:

1. "MS" is for mild steel
2. "Cu" is for copper.

3.3 Evolution of Risk

Likelihood and consequence ratings were cross referenced to the risk definitions in Table 5, which allowed the evolution of risk of failure over the next 40 years to be tabulated in Table 10. For all pipes, the risk is predicted to remain low for the next 20 years. The risk has been determined based on the corrosion rates noted in Section 3 and the continuation of effective internal inhibition of the recirculating water.

Table 7. Evolution of Risk of Failure

	Pipe Reference										
	1	3	4	6	7	8	10	12	13	14	15
	Mild Steel	Mild Steel	Mild Steel	Mild Steel	Copper	Mild Steel	Mild Steel	Mild Steel	Mild Steel	Mild Steel	Mild Steel
Year	HHW	HHW	HHW	HHW	CHW	HHW	HHW	HHW	CHW	HHW	CHW
2017	Low	Low	Low	Low	Low	Low	Low	Low	Low	Low	Low
2021	Low	Low	Low	Low	Low	Low	Low	Low	Low	Low	Low
2027	Low	Low	Low	Low	Low	Low	Low	Low	Low	Low	Low
2037	Low	Low	Low	Low	Low	Low	Low	Low	Low	Low	Low
2038	Low	Low	Low	Low	Low	Low	Low	Low	Low	Low	Low
2039	Low	Low	Low	Low	Low	Low	Low	Moderate	Low	Low	Low
2040	Low	Low	Low	Low	Low	Low	Low	Moderate	Low	Low	Low
2041	Low	Low	Low	Low	Low	Moderate	Low	Moderate	Low	Low	Low
2042	Low	Low	Low	Low	Low	Moderate	Low	Moderate	Low	Low	Low
2043	Low	Moderate	Low	Low	Low	Moderate	Low	Moderate	Low	Low	Low
2044	Low	Moderate	Low	Low	Low	Moderate	Low	Moderate	Low	Low	Low
2045	Low	Moderate	Low	Low	Low	Moderate	Low	Moderate	Low	Moderate	Low
2046	Low	Moderate	Low	Low	Low	Moderate	Moderate	Moderate	Low	Moderate	Low
2047	Low	Moderate	Low	Low	Low	Moderate	Moderate	Moderate	Low	Moderate	Low
2048	Low	Moderate	Low	Low	Low	Moderate	Moderate	Moderate	Low	Moderate	Low
2049	Low	Moderate	Moderate	Low	Low	Moderate	Moderate	Moderate	Low	Moderate	Low
2050	Low	Moderate	Moderate	Low	Low	Moderate	Moderate	Moderate	Low	Moderate	Low
2051	Low	Moderate	Moderate	Low	Low	Moderate	Moderate	Moderate	Low	Moderate	Low
2052	Low	Moderate	Moderate	Low	Low	Moderate	Moderate	High	Low	Moderate	Low
2053	Low	Moderate	Moderate	Low	Low	Moderate	Moderate	High	Low	Moderate	Low
2054	Low	Moderate	Moderate	Moderate	Low	Moderate	Moderate	High	Low	Moderate	Low
2055	Moderate	Moderate	Moderate	Moderate	Low	High	Moderate	High	Low	Moderate	Low
2056	Moderate	Moderate	Moderate	Moderate	Low	High	Moderate	High	Low	Moderate	Low
2057	Moderate	Moderate	Moderate	Moderate	Low	High	Moderate	High	Low	Moderate	Low

4. DISCUSSION

Using a deterministic approach to solving for remaining life using a single set of wall thicknesses would lead to an outcome of either a single remaining life for each pipe (i.e. end of life at year t) or a range of remaining lives (i.e. *x-y* years) based on the range of thickness measurements and/or literature corrosion rates. The limitation of this approach is that without the probability tied to the outcome it is not clear which value should be used to make business decisions, and what the implication of this choice may be now and in the future. The use of probability also allows the risk to be presented as a cost. For example, if the failure of a pipe is expected to incur a cost $10,000 and there is a 15% probability of failure in year *t*, the risk cost (or risk exposure) would be equal to $1,500. These figures can be used to prioritise maintenance and repair works across an asset portfolio. This is especially critical as risk appetites may vary.

Our assessment of the remaining life of various hot and chilled HVAC water pipes tested in Building 1 and 2 found that they are all at low risk of failure in the short, medium and long term, based on the corrosion rates used and the continuation of effective internal inhibition of the recirculating water. This outcome is the result of a combination of low operating fluid pressure and effective corrosion inhibition within the pipes. At a fluid pressure of 0.5 MPa, the uniform section loss required for failure (material yield) is considerable. Effective corrosion inhibition ensures that rates of loss are low, particularly in chilled pipe where corrosion rates are slower.

Additional modelling for a scenario of "poor" inhibition predicts probabilities of failure of 100% for numerous HHW pipes within 20 years. This emphasises the importance of monitoring and maintaining water inhibition levels, but also the need to have a better understanding of the actual corrosion rates within the systems.

It is important to note that the results presented here are reliant on the selected test locations being representative of pipework conditions within the buildings and on the use of literature to determine probable corrosion rates. It is important to note that this modelling does not take into account the possibility of data beyond the distributions that are used.

5. CONCLUSIONS

The Monte Carlo simulation method is a useful tool for producing statistical data that can be analysed where none, or an insufficiently small number, is available. The use of a statistical approach allows for the variance in corrosion rates and initial wall thickness to be taken into account; and therefore accounts for the uncertainty of the assumptions used. Survival analysis can then be used to assign a probability of failure over time, allowing the data to be presented as a risk, which is critical for prioritising or seeking future funding for maintenance and repair works on assets by asset owners.

6. REFERENCES

[1] T. G. Clark, M. J. Bradburn, S. B. Love and D. G. Altman, "Survival Analysis Part I: Basic conceots and first analyses," *British Journal of Cancer,* vol. 89, pp. 232-238, 2003.

[2] S. Prinja, N. Gupta and R. Verma, "Censoring in Clinical Trials: Review of Survival Analysis Techniques," *Indian J Community Med,* vol. 35, no. 2, pp. 217-221, 2010.

[3] S. Zacks, Introduction to Reliability Analysis: Probability Models and Statistical Methods, Binghampton: Springer-Verlag, 2012.

[4] B. George, S. Seals and I. Aban, "Survival analysis and regression models," *J Nucl Cardiol,* vol. 21, no. 4, pp. 686-694, 2014.

[5] R. L. Harrison, "Introduction to Monte Carlo Simulation," *AIP Conf Proc,* pp. 17-21, 2010.

[6] F. L. T. Charng, "TDA Progress Report 42-69; Review of Corrosion Causes and Corrosion Control in a Technical Facility," NASA, 1982.

[7] General Electric, "Chapter 32 - Closed Recirculating Cooling Systems," 2012. [Online]. Available: https://www.gewater.com.handbook/cooling_water_systems/ch_32_closed.jsp. [Accessed 3 11 2016].

[8] C. M. M. H. M.R. Ali, "Effect of Molybdate, Nitrite and Zinc Ions on the Corrosion Inhibition of Mild Steel in Aqueous Chloride Media Containing Cupric Ions," *Journal of Scientific Research,* vol. 1, no. 1, pp. 82-91, 2009.

[9] Association of Water Technologies, "Molybdate and Non-Molybdate Options for Closed Systems," Association of Water Technologies, 2005.

7. AUTHOR DETAILS

Julie-Anne is a Senior Materials Engineer within the Strategic Asset Management and Advanced Materials team at AECOM in Melbourne, Australia.

Julie-Anne has five years industry experience coordinating and undertaking condition inspections, non-destructive and destructive testing and analysis of detailed testing of numerous structures. She has prepared deterioration models for estimation of remaining life, condition assessment reports and technical repair specifications. Julie-Anne has also reviewed and prepared asset management plans and strategies for water and civil infrastructure clients. Prior to joining AECOM Julie-Anne completed a PhD in corrosion engineering and electrochemistry.

Amy Spark is a Senior Engineer within the Strategic Asset Management and Advanced Materials team at AECOM in Melbourne, Australia.

Amy has eight years of experience as an engineer in a variety of environments and industries and holds a PhD in corrosion engineering. She has worked on numerous projects in durability / materials engineering and asset management. Amy was awarded her PhD in 2016 having studied Microbiologically Influenced Corrosion (MIC) on the buried surface of steel water pipelines. She enjoys using her engineering expertise to improve public health and infrastructure outcomes in line with government regulations and international standards.

Sarah is an Associate Director in the Strategic Asset Management and Advanced Materials Group at AECOM.

She is a graduate of the Corrosion Science and Engineering Masters' program at UMIST in England and has approximately 25 years' experience in corrosion related consulting and research. She has specific experience in durability planning for new infrastructure, condition assessment and rehabilitation of aged infrastructure, performance assessments of materials and protective coatings, materials selection, specification of protective coatings, failure analysis and cathodic protection design.

EFFECT OF NI-RE DIFFUSION BARRIER ON NI-AL COATING STRUCTURE AND ITS OXIDATION IN DRY CO_2 GAS AT 650 °C

M Li[1, 2], C Zhou[2], J Zhang[1] & DJ Young[1]

[1]School of Materials Science and Engineering, University of New South Wales, Sydney, NSW 2052, Australia

[2]School of Materials Science and Engineering, Beijing University of Aeronautics and Astronautics, Beijing 100191, China

SUMMARY: Nickel aluminide coating has been widely applied on the low-corrosion resistant alloys to resist oxidation, metal dusting and carburization at moderate and high temperatures. However, interdiffusion between the Ni-Al coating and the alloy matrix will alter both the coating and matrix chemistry, and as a result degrade the protective coating effect and the mechanical properties of the substrate. In this work, Ni-Re electroplated interlayer was prepared as a diffusion barrier to inhibit interdiffusion between Ni-Al coating and an iron alloy substrate during both the Al-pack cementation and the oxidation processes. For comparison purposes, a sample coated with Ni-Al only was also prepared. This Ni-Al coating created a rather thick IDZ (interdiffusion zone) with some detrimental intermetallic precipitates during Al-pack cementation. However, almost complete blocking of interdiffusion was observed in the Ni-Al/Ni-Re coating system, with no IDZ and no detrimental intermetallic phase precipitation. An oxidation experiment at 650 °C was carried out in dry CO_2 gas for 660 h. Severe internal oxidation and interdiffusion occurred in the Ni-Al coating, while only slight internal oxidation was found in the Ni-Al/Ni-Re coating. The high thermodynamic stability of the Ni-Re diffusion barrier effectively restrains interdiffusion and oxidation during dry CO_2 exposure.

Keywords: Ni-Al coating; Ni-Re interlayer; Interdiffusion; CO_2 corrosion.

1. INTRODUCTION

Global utilization of clean energy has been highly promoted to mitigate the CO_2 emission. As heat-resisting steels for serving at moderate and high temperatures, Fe-Cr-Ni alloys have been widely used in various hot components, such as the venturi of gas turbine engines and heat exchangers of petroleum chemical equipment which all saturate with CO_2-rich atmospheres at moderate and high temperatures [1-3]. These commercial Fe-Cr-Ni alloys are capable of forming a dense and continuous chromia scale in air. However, the protective chromia scale has been shown to suffer breakaway oxidation in CO_2 [4-6]. An effective way to protect these alloys is by coating with nickel aluminides.

Recent studies indicated that nickel aluminide coatings showed high resistance to oxidation, metal dusting and carburization attributed to the alumina protective scale formed at moderate and high temperatures [7-9]. Even in very low oxygen partial pressures such as in metal dusting atmospheres, aluminides form dense oxide scales and thus separate the base material from the corrosive carburizing environment. However, a deleterious problem of this coating is that intensive interdiffusion occurs between nickel aluminide coatings and the alloy substrates [10-13], and that rapidly drops the concentration of Al in the coating even at a relatively low temperature (650 °C) [14]. Since the protective Ni-Al layer cannot be sustained when Al concentration decreases to a certain level, the coating fractures rapidly under high temperature service environment [15-17]. Moreover, due to the outward diffusion of alloy elements such as Fe, Cr, etc. and inward diffusion of Al and Ni, the thick interdiffusion zone (IDZ) and enormous detrimental brittle phases of (Fe, Al) and (Cr, Al) intermetallic compounds are formed [18-20]. These detrimental brittle phases in IDZ are vulnerable to carburization or oxidation, which dramatically leads to the deterioration of mechanical properties of the coating and the alloy substrate. Therefore, it is essential to restrain the interdiffusion and to achieve a long life of Ni-Al coating for protection of the iron alloy substrate from high temperature carburization.

It is reported that several types of diffusion barriers have been developed for inhibiting the interdiffuison between the aluminide coatings and the alloy substrates. Ceramic interlayers (Al-Al_2O_3, Al_2O_3) [21, 22], metallic interlayer Ni-W [23] and NiCrO [24] diffusion barriers were prepared by pre-deposited, sputtering or electroplating onto the alloy substrates for fabricating multi-aluminizing coatings. However, these coatings just retard rather than fully block interdiffusion between the aluminide coatings and the alloy substrates. Owing to the high melting point and thermodynamic stability at high temperatures and a desired compatibility with both coating and alloy stability, refractory element Rhenium (Re) was found out to be a possible candidate as a diffusion barrier to inhibit the elemental interdiffusion [25-27]. In this work we designed a multi-aluminizing coating with Ni-Re electroplated interlayer as diffusion barrier between the Ni-Al coating and the iron alloy substrate. By an integrated approach of experimental characterization, the effectivity and mechanism of interdiffusion inhibiting of the Al-coatings were revealed. The exposure experiment to CO_2 was also conducted on the Ni-Al/Ni-Re coating to evaluate the corrosion resistance performance.

2. EXPERIMENTAL DETAILS

Samples with dimensions of 20 mm x 5 mm x 2 mm were cut from steel alloy (the chemical composition is shown in Table1), then abraded to a final 800 grit SiC paper and ultrasonically cleaned in acetone. After that, the samples were electrodeposited with a Ni-Re interlayer from a nickel sulfate bath (200 g/L $NiSO_4 \cdot 6H_2O$, 0.2 g/L $C_{12}H_{25}SO_4Na$, 10 g/L $NiCl_2$, 40 g/L H_3BO_3) with addition of 1.68 g/L KReO4. Then an electroplated layer of Ni was deposited in the nickel sulfate bath. Subsequently, the deposited samples were aluminized at 650 °C for 5 h using a conventional pack-cementation in a powder mixture of Al (25 wt.%) + Al_2O_3 (70 wt.%) + NH4Cl (5 wt.%) in an Ar atmosphere.

The exposure surfaces of all the coated samples were placed parallel to the gas flow of dry CO_2 (80 ml/min) in a ceramic tube furnace at 650 °C and a total pressure of 1 atm. The weight changes of the corroded samples were measured using an analytic balance with the accuracy of 0.01 mg.

The surface phases and morphologies were identified by the X-ray diffraction (XRD) and scanning electron microscope (SEM). The samples were mounted in epoxy resins for the cross-section characterization by the back scattered electron images (BSE) and focused ion beam (FIB) analysis.

Table 1. Chemical composition (wt %) of the steel alloy substrate.

Elements	Cr	Ni	Mn	Si	P	C	S	Fe
Contents	23.73	12.04	1.27	0.54	0.025	0.03	0.004	Bal.

3. RESULTS AND DISCUSSION

3.1 Coating microstructure

The surface composition of the raw Ni-Al and Ni-Al/Ni-Re coatings are investigated by XRD patterns shown in Figure 1. It indicates that the outer layer of Ni-Al coating mainly consists of the δ-Ni_2Al_3 phase, which supports the study that the Ni film can convert into a coating of aluminide in its δ-phase reported by Tan [28]. While for the Ni-Al/Ni-Re coating, the characteristic peaks corresponding to δ-Ni_2Al_3 phase have nearly disappeared and replaced by almost all the $NiAl_3$ characteristic peaks.

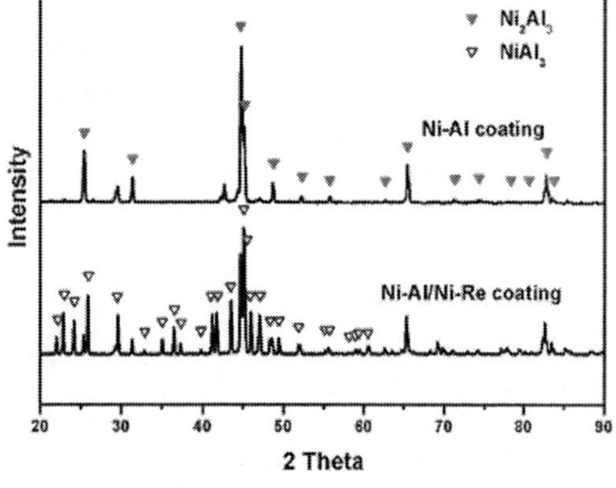

Figure 1 XRD patterns of the Ni-Al and Ni-Al/Ni-Re coatings

The cross-sectional BSE images of the as-prepared coatings are shown in Figure 2. It can be seen from Figure 2a that the thickness of the Ni_2Al_3 outer layer in the Ni-Al coating is 20 μm, which is only one-third of the thickness of IDZ. The thick IDZ formed due to interdiffusion by consuming massive amounts of Al element from protective Ni-Al layer. Meanwhile, a lot of detrimental phases of brittle intermetallic compounds of (Fe, Al) and (Cr, Al) formed in IDZ. Combined with the analysis of XRD patterns in Figure 1 and BSE images in Figure 2b, the outer layer (10 μm) and inner layer (60 μm) of the as-prepared Ni-Al/Ni-Re coating can be determined to be the $NiAl_3$ and the Ni_2Al_3 phase, respectively. According to the Ni-Al phase diagram [29], the $NiAl_3$ phase delays the transformation to the δ-Ni_2Al_3 when the concentration of Al are high at temperature below 854 °C, indicating the diffusion of Al is depressed in the Ni-Al/Ni-Re coating compared with the conventional Ni-Al coating [10, 14].

Noticeably, no interdiffusion zone was found in the Ni-Al/Ni-Re coating and the Ni-Re barrier (5 μm) remained stable underneath the Ni_2Al_3 inner layer (Figure 2b). Apparently, the Ni-Re barrier effectively blocks the interdiffusion between the Ni-Al coating and the iron alloy substrate during the Al packing cementation process.

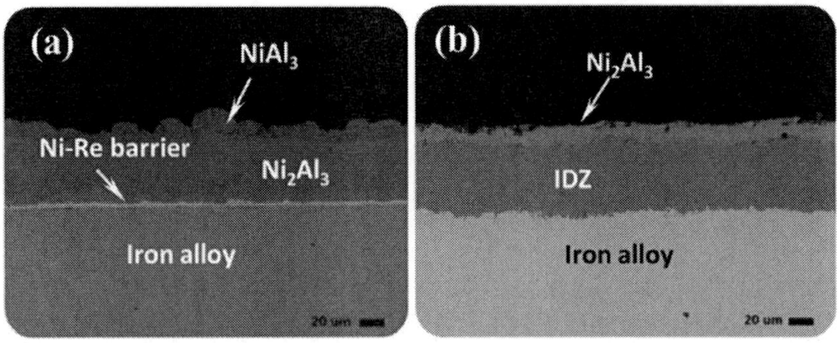

Figure 2 BSE cross section images of (a) the Ni-Al coating and (b) the Ni-Al/Ni-Re coating, respectively.

3.2 Corrosion exposure to CO_2

After being exposed to CO_2 at 650 °C for 660 h, the oxidation behaviours of the Ni-Al coating and the Ni-Al/Ni-Re coating were investigated. The XRD patterns in Figure 3 show that the Ni-Al coating remains the outer layer of Ni_2Al_3 phase, and the Ni-Al/Ni-Re coating shows an outer layer of mostly Ni_2Al_3 phase and a small amount of $NiAl_3$ phase. It is also noticed that the phase of $NiAl_3$ in Ni-Al/Ni-Re coating decreased a lot after the CO_2 exposure experiment. No obvious oxides were detected directly by XRD measurement on the surface of both coatings.

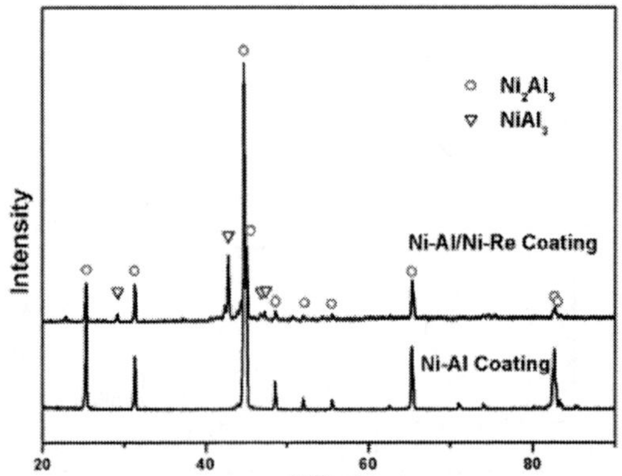

Figure 3 XRD patterns of the Ni-Al coating and the Ni-Al/Ni-Re coating after being exposed to CO_2 at 650 °C for 660 h

The FIB cross-sectional images of the exposed Ni-Al coating and Ni-Al/Ni-Re coating are shown in Figure 4. It is noticed that an alumina scale (determined by TEM in Figure 5) forms on both coatings after being exposed to CO_2 at 650 °C for 660 h. It may be too thin to be detected by XRD. The external alumina scale on Ni-Al coating presented a discontinuous and rough structure (Figure 4a). Meanwhile, small holes were also observed in the Ni_2Al_3 outer layer. For the Ni-Al/Ni-Re coating, a dense outer layer of Ni_2Al_3 + $NiAl_3$ is observed in Figure 4b. The morphology of the alumina scale formed on the

Ni-Al/Ni-Re coating is shown in the TEM image (Figure 5a), which presents a quite dense and continuous structure. The EDS maps in Figure 7 confirm that the oxide scale is Al_2O_3.

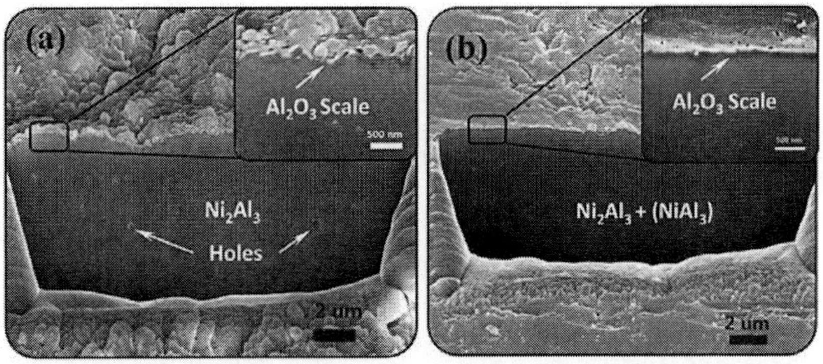

Figure 4 Cross-sectional FIB view of (a) the Ni-Al coating (b) the Ni-Al/Ni-Re coating after being exposed to CO2 at 650 °C for 660 h.

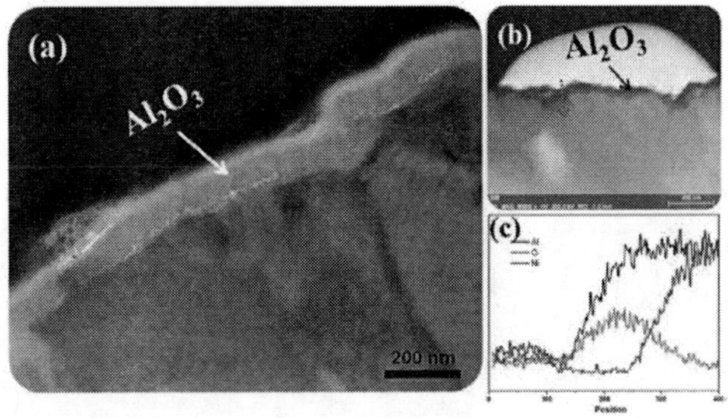

Figure 5 TEM image of the Ni-Al/Ni-Re coating (a), SEM image (b) and the corresponding EDS line scan (c).

In order to analyze the internal diffusion and reaction between the coating and steel alloy substrate after the CO_2 exposure, the metallographic cross-sectional images of the exposed Ni-Al coating and Ni-Al/Ni-Re coating are shown in Figure 6. The Ni-Al coating without a Ni-Re barrier (Figure 6a) shows large holes in the interdiffusion zone (IDZ). According to the Kirkendall effect [30], elements of Al, Cr and Fe from the coating and steel substrate react with O to form the corresponding oxides, which leads to the outward diffusion of these elements and voids formed along the diffusion path. In terms of the Ni-Al/Ni-Re coating (Figure 6b), no voids were observed in both of the outer layer and the Ni-Re diffusion barrier after being exposed to CO_2 at 650 °C for 660 h. Obviously, the interdiffusing behaviour was well inhibited by the Ni-Re layer during the exposure experiment.

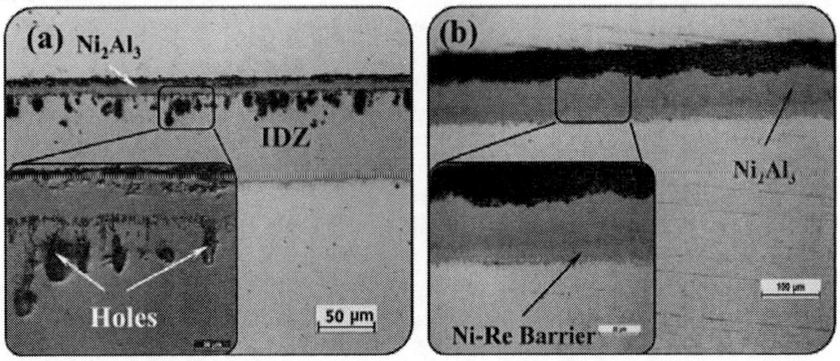

Figure 6 Metallographic Cross sections of (a) the Ni-Al coating and (b) the Ni-Al/Ni-Re coating after being exposed to CO_2 for 660 h at 650 °C.

The further analysis was investigated by BSE images and EDS mappings. Figure 7 shows the corresponding images of the Ni-Al coating with the Ni_2Al_3 outer layer peeled off during the grinding process. A thin layer (indicated by yellow arrow in

SEM image in Figure 7) with high concentration of O and Al was observed, implying that a severe internal oxidation of Al occurs between the Ni_2Al_3 outer layer and IDZ, which leads to numbers of large voids in IDZ.

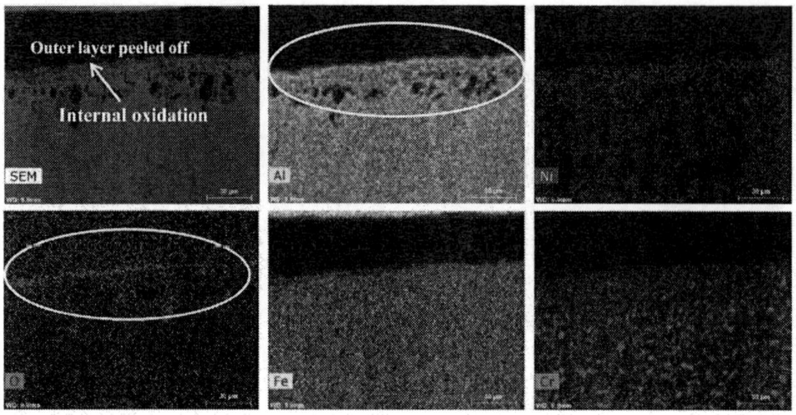

Figure 7 SEM image and EDS mapping of the Ni-Al coating after the CO_2 exposure at 650 °C for 660 h.

Element mapping images in Figure 8 show that most of Re remains in the difussion barrier after being exposed to CO_2 at 650 °C for 660 h. No further IDZ formed in the Ni-Al/Ni-Re coating during this period. The Ni-Re diffusion barrier slightly increased in width (20 μm) due to the interdiffusion between the coating and the steel substrate. Slightly high concentrations of O and Al were detected in the Ni-Al diffusion barrier (indicated by yellow circles), suggesting that minor internal oxidation occurs in the exposed Ni-Al/Ni-Re coating.

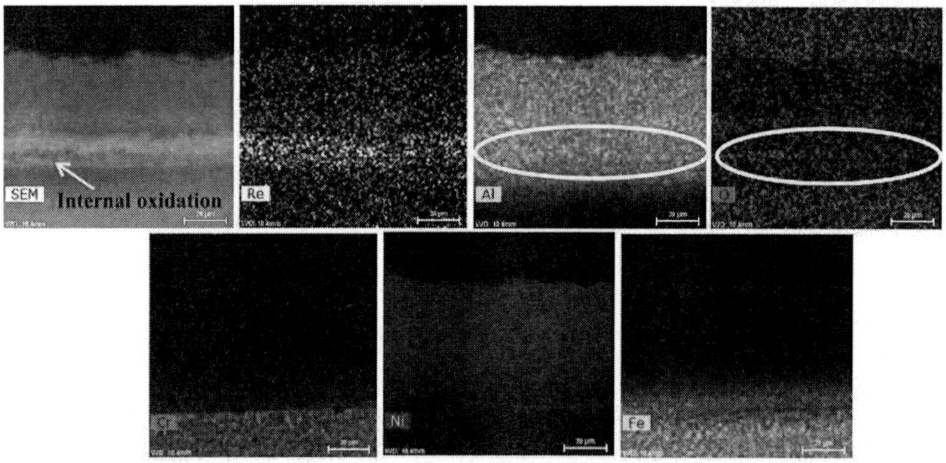

Figure 8 SEM image and EDS mapping of the Ni-Al/Ni-Re coating after the CO_2 exposure at 650 °C for 660 h.

4. CONCLUSIONS

The Ni-Al coatings with and without a Ni-Re diffusion barrier were prepared to protect the iron alloy substrate from CO_2 corrosion. The Ni-Al/Ni-Re coating showed an excellent retardation of interdiffusion without the formation of IDZ and detrimental phase precipitation during the aluminization process which however, are main problems for the Ni-Al coating without a diffusion barrier. A dense and continuous Al_2O_3 scale formed on the surface after being exposed to CO_2 at 650 °C for 660 h in the coated sample with the diffusion barrier but a non-continuous Al_2O_3 for that without the diffusion barrier. The superior barrier effect of Ni-Re on interdiffusion was found to be still effective after the CO_2 exposure for 660 h, and only minor internal oxidation occurred in the Ni-Al/Ni-Re coating. However, a significant number of large voids was observed in a rather thick IDZ in the Ni-Al coating without the diffusion barrier.

5. ACKNOWLEDGMENTS

Financial support from the National Natural Science Foundation of China is gratefully acknowledged.

6. REFERENCES

[1] S. Sridhar, P. Rozzelle, B. Morreale, D. Alman, Materials challenges for advanced combustion and gasification fossil energy systems, Metall. Mater. Trans. A 42A (2011) 871–877.

[2] C.S. Giggins, F.S. Pettit, Corrosion of metals and alloys in mixed gas environments at elevated temperatures, Oxid. Met. 14 (1980) 363–413.

[3] J. Piron Abellan, T. Olszewski, H.J. Penkalla, G.H. Meier, L. Singheiser, W.J. Quadakkers, Scale formation mechanisms of martensitic steels in high CO_2/ H_2O-containing gases simulating oxyfuel environments, Mater. High Temp. 26 (2009) 63–72.

[4] I. Kvernes, M. Oliveira, P. Kofstad, High-temperature oxidation of Fe-13Cr-xAl alloys in air-H2O vapor mixtures, Corros. Sci. 17 (1977) 237–252.

[5] T. Gheno, D. Monceau, D. Young, Mechanism of breakaway oxidation of Fe–Cr and Fe-Cr-Ni alloys in dry and wet carbon dioxide, Corros. Sci. 64 (2012) 222–233.

[6] Y. Ikeda, K. Nii, Mechanism of accelerated oxidation of Fe–Cr alloys in watervapor containing atmosphere, Trans. Natl. Res. Inst. Met. 26 (1984) 52–62.

[7] D.B. Miracle, Overview No. 104 The physical and mechanical properties of NiAl, Acta Mater. 41 (1993) 649-684.

[8] Z.D. Xiang, S. R. Rose, P.K. Datta, Low-temperature formation and oxidation resistance of nickel aluminide/nickel hybrid coatings on alloy steels, Scr. Mater. 59 (2008) 99-102.

[9] W. Wang, B. Yang, L. Du, W. Zhang, Diffusion research between Ni_3Al coating and titanium alloy produced by plasma spraying process. Appl. Surf. Sci. 10(2010) 3342-3345.

[10] E. Basuki, A. Crosky, B. Gleeson, Interdiffusion behaviour in aluminide-coated René 80H at 1150 °C, Mater. Sci. Eng. A, 224 (1997) 27.

[11] A. Paul, A. A. Kodentsov, F. J. J. Van Loo, Bifurcation of the Kirkendall plane during interdiffusion in the intermetallic compound β-NiAl, Acta Mater. 52 (2004) 4041-4048.

[12] H. Wei, W. Gong, S. Huang, C. Zhou, Single-phase interdiffusion in the Ni_3Al-Mo ternary system, Scr. Mater. 64(2011) 53-56.

[13] D. Yao, J. Yang, W. Gong, C. Zhou, Interdiffusion behavior between Nb and $MoSi_2$ intermetallic compound, Mater. Sci. Eng. A, 527(2010) 6787-6793.

[14] Z.D. Xiang, D. Zeng, C.Y. Zhu, D.J. Wu, P.K. Datta, Degradation kinetics at 650 °C and lifetime prediction of Ni2Al3/Ni hybrid coating for protection against high temperature oxidation of creep resistant ferritic steels, Corros. Sci. 53(2011) 3426-3434.

[15] O. knotek, E. Lugscheider, F. Löffler, W. Beele, Diffusion barrier coating with active bonding designed for gas turbine blades, Surf. Coat. Technol. 68-69 (1994) 22-26.

[16] R. Braun, C. Leyens, Protective coatings on orthorhombic Ti_2AlNb alloys, Mater. High Temp. 22 (2005) 437-447.

[17] K. Yuan, R. Eriksson, R. L. Peng, X. H. Li, S. Johansson, Y. D. Wang, Modeling of microstructural evolution and lifetime prediction of MCrAlY coatings on nickel based superalloys during high temperature oxidation, Surf. Coat. Technol. 232 (2013) 204-215.

[18] X.P. Tan, H.U. Hong, B.G. Choi, I.S. Kim, C.Y. Jo, T. Jin, Z. Q. Hu, Characterization of topologically close-packed phases in secondary reaction zone in a coated CMSX-4 single crystal Ni-based superalloy, J. Mater. Sci. 48 (2013) 1085-1089.

[19] C.M.F. Rae, M.S. Hook, R.C. Reed, The effect of TCP morphology on the development of aluminide coated superalloys, Mater. Sci. Eng. A, 396 (2005) 231.

[20] D.K. Das, K.S. Murphy, S.W. Ma, T.M. Pollock, Formation of secondary reaction zones in diffusion aluminide-coated Ni-base single-crystal superalloys containing ruthenium, Metall. Mater. Trans. A, 39A (2008) 1647-1657.

[21] H.Q. Li, Q.M. Wang, S.M. Jiang, J. Gong, C. Sun, Ion-plated Al-Al2O3 films as diffusion barriers between NiCrAlY coating and orthorhombic-Ti2AlNb alloy, Corros. Sci. 52 (2010) 1668-1674.

[22] H.Q. Li, Q.M. Wang, S.M. Jiang, J. Gong, C. Sun, Oxidation and interfacial fracture behaviour of NiCrAlY/Al2O3 coatings on an orthorhombic-Ti_2AlNb alloy, Corros. Sci. 53 (2011) 1097-1106.

[23] E. Cavaletti, S. Naveos, S. Mercier, P. Josso, M.P. Bacos, D. Monceau, Ni-W diffusion barrier: Its influence on the oxidation behaviour of a β-(Ni, Pt)Al coated fourth generation nickel-base superalloy, Surf. Coat. Technol. 204 (2009) 761-765.

[24] H. Yao H, Z. Bao, M. Shen, S. Zhu, F. Wang, A magnetron sputtered microcrystalline β-NiAl coating for SC superalloys. Part II. Effects of a NiCrO diffusion barrier on oxidation behavior at 1100 °C, Appl. Surf. Sci. (2017).

[25] S. Prins, R. Arroyave, Z.K. Liu. Defect structures and ternary lattice site preference of the B2 phase in the Al-Ni-Ru system, Acta Mater. 55 (2007) 478-4787.

[26] Y. Matsumura, M. Fukumoto, S. Hayashi, A. Kasama, I. Iwanaga, R. Tanaka, T. Narita, Oxidation Behavior of a Re-Base Diffusion-Barrier/β-NiAl Coating on Nb-5Mo-15W at High Temperatures, Oxid. Met. 61 (2004) 105-124.

[27] F.H. Latief, K. Kakehi, Effects of Re content and crystallographic orientation on creep behavior of aluminized Ni-base single crystal superalloys. Mater. Des. 49 (2013) 485-492.

[28] X. Tan, X. Peng, F. Wang, The mechanism for self-formation of a CeO_2 diffusion barrier layer in an aluminide coating at high temperature, Surf. Coat. Technol. 224 (2013) 62-70.

[29] I. M. Robertson, C. M. Wayman. Ni5Al3 and the nickel-aluminum binary phase diagram, Metallography, 17(1984) 43-55.

[30] M.M.P. Janssen, G.D. Rieck, Reaction diffusion and Kirkendall-effect in the nickel-aluminum system. Transactions of the Metallurgical Society of AIME, 239(1967) 1372-1385

7. AUTHOR DETAILS

Meifeng Li is a joint-PhD from Beihang University and the University of New South Wales. Her PhD project is about investigating and improving the resistance performance of Ni-Al coatings in various environment.

Dr Chungen Zhou is a Professor at the School of Materials Science and Engineering, BUAA. His research is mainly on high temperature oxidation and the surface coating technology

Dr Jianqiang Zhang is Associate Professor at the School of Materials Science and Engineering, UNSW. His research is mainly on high temperature corrosion of metals and alloys in oxidation and carburisation gas atmospheres.

David Young is Emeritus Professor at the School of Materials Science and Engineering, UNSW. He has worked in the field of high temperature corrosion for more than 40 years, publishing extensively in the research literature.

RESEARCH INTEGRITY IN CORROSION RESEARCH

A. J. Betts[1], T.A. Green[2]
[1]Applied Electrochemistry Group, FOCAS, Dublin Institute of Technology, Dublin, Ireland, [2] Department of Chemical and Process Engineering, University of Strathclyde, UK

SUMMARY: Research Integrity is becoming increasingly important throughout the scientific community. Issues covered by this all-encompassing term can include good data management and security, use of appropriate research methodology, analysis of data based on sound theoretical foundations and correct use of statistics. Ethical issues such as publication authorship, intellectual property rights, conflict of interest and research misconduct including plagiarism, fabrication and falsification of data are also covered.

One of the most significant types of research misconduct is that of scientific fraud. Many well-publicised cases have been reported over the years, most notably in the bioscience, and health/life science fields. Several historical examples are presented which illustrate the nature of such transgressions. A variety of causes have been identified, the misconduct most often being uncovered at a later stage after the findings have been disseminated. Surprisingly much of the research work conducted in some of these areas has proven difficult to replicate resulting in demands for a re-evaluation of all original data leading to the call for open access to all data (Open Science). Although few cases have been reported in the corrosion science area, it has not proven entirely immune from this problem. Some examples reported in the open literature in corrosion science and engineering are presented which illustrate that it is not immune to research misconduct.

As scientists and engineers it behooves us to recognize that rigorous standards of ethical behaviour are always required in the conduct of research. However many of the pressures that corrosion researchers face are similar to those of other scientists, particularly research funding and career advancement. Sometimes this can compromise ethics concerning research in the field with damaging consequences to both the individual involved and the institution's reputations. Sound experimental design with evidence of repeatability of results, high quality journal paper peer-review and editorial processes and verification practices and the need to carefully balance research/scientific needs versus those of industry and commercial enterprises are some of the most important issues of concern.

Keywords: Research Integrity, Scientific Misconduct, Corrosion Research

1. INTRODUCTION

1.1 Description of Research Integrity

Just as Quality Assurance rose to prominence in the manufacturing and service industries at the end of the 20th century, the concept of Research Integrity is becoming increasingly important in 21st century scientific research. But what exactly is

Research Integrity? In fact there is no commonly accepted definition worldwide and neither is there for the associated issues of research misconduct, conflict of interest or other key terms that describe unacceptable research practices. Generally research integrity encompasses a range of issues dealing with both the performance of the research activity itself and the dissemination or reporting of the results and their interpretation. In principle scientific research should always be designed and conducted in accordance with ethical principles and with appropriate critical review processes to ensure this. Since 2007 four World Conferences on Research Integrity have been held promoting the exchange of information and with the intention of developing ways to promote research integrity and to harmonise efforts to foster responsible research practices.

Such efforts became necessary because research regulations and accepted research practices can often vary significantly from country to country and even among professional organisations themselves in different fields. For example in Europe one description is found in the *European Code of Conduct for Research Integrity* [1] which identifies eight key principles governing good research practice which are integral to the notion of research integrity. They are:

- objectivity
- reliability in performing research
- impartiality and independence
- duty of care
- openness and accessibility
- honesty in communication
- fairness in providing references and apportioning credit and
- responsibility for the scientists and researchers of the future.

Many European universities have adopted this code and in Ireland there is now a *National Policy Statement on Ensuring Research Integrity* launched in early 2017 [2]. In the UK there is a similar code [3], whereas in Australia [4] and in New Zealand [5] slightly different approaches have been adopted.

One of the most important issues arising from Research Integrity is that of Research Misconduct which describes a range of activities which may compromise research performance or outputs, the three most common types being:

- *Data Fabrication* – defined as the fraudulent creation of results and their recording and/or reporting
- *Data Falsification* – manipulation of research results, equipment, materials or processes, or changing or omitting data or results, such that the research is not accurately represented in the final research output, most often a journal or a conference paper
- *Plagiarism* – the wilful appropriation of other people's ideas, processes, results, or written records without correct attribution. Information or ideas obtained through confidential review of others' research proposals and manuscripts is included in this definition.

Each of these may compromise the integrity of the research. Data fabrication and falsification are probably the most serious offences which may ultimately undermine the development of knowledge itself and can have far reaching consequences outside the research community itself. Although plagiarism (and self-plagiarism) may appear slightly less serious, the corrupting effect on the principle of open communication and sharing of knowledge for wider benefit means that repeated or significant plagiarism is problematic.

Whilst fabrication, falsification, and plagiarism represent the most serious examples of misconduct, there are other types of poor practices which are probably more widespread and therefore potentially more damaging to the reputation of research and the research community's integrity. Such poor practices include:

- *Data-related poor practice* - not preserving primary data, poor data management and/or storage and backup;
- *Publication-related practice* - claiming undeserved authorship, denying authorship to contributors, artificially proliferating publications;
- *Personal behaviours* - inadequate leadership/mentoring of the next generation of researchers and scholars, inappropriate personal behaviour;
- *Financial and other malpractice* - peer review abuse, non-disclosure of a conflict of interest, misrepresenting credentials;
- *Poor research procedures* - harmful, dangerous or unethical research methods

Research integrity is becoming more important in both academic institutional and industrial research programmes, where basic principles such as experimental replication and repeatability have come under threat in recent years. It has even given rise to a new field of academic study and a journal devoted to the topic was launched in 2015 [6].

1.2 Aims and Objectives

The aim of this paper is to bring this important issue to the attention of the corrosion research community, including scientists, engineers and technologists, thereby raising awareness of its growing importance and potential future impact. Following a brief historical overview of scientific misconduct cases, mainly in the biological and health/life sciences, some examples of misconduct in corrosion research are presented and discussed. Possible reasons for the lack of research integrity are given, as well as suggestions for the prevention of such problems in future research endeavours.

1.3 Historical Examples of Research Misconduct

In principle every scientist should strive to ensure that their research is carried out with the requisite rigour and strictly to the highest ethical standards. Unfortunately in practice this ideal is not always achieved for a variety of reasons, as is illustrated by notable historical examples especially in the bioscience and health/life sciences. Famous examples include the Piltdown Man hoax, Lysenko and his rigid adherence to the Lamarkian view of evolution, Mark Spector and his cell biology results and more recently examples in the field of stem cell research. Even though the majority of these have arisen in the biosciences and health sciences, it is not solely the preserve of these fields and some other cases in the physical sciences are also well-known.

One of the best known cases in the physical sciences was that of "polywater", an example of what the American Chemist Irving Langmuir termed "Pathological Science" [7]. He described behaviour where "people are tricked into false results....by subjective efforts, wishful thinking or threshold interactions". In the case of polywater a Russian scientist Nikolai Fedyakin in 1961 reported water present in glass capillaries with a range of strange properties; an elevated boiling point, depressed freezing point and a much higher viscosity in comparison to normal water. Careful investigations conducted by Boris Derjaguin confirmed these findings, which aroused much interest. However later investigations discovered that the water was contaminated with compounds that gave rise to these unusual properties (as a result of the water's colligative properties) [8].

A more recent case of misconduct in the physical sciences was that of Hendrik Schön a young German physicist who rose to prominence after making a series of astonishing breakthroughs in experiments involving organic-based semiconductor materials. These were later discovered to be fraudulent, but only after he had been awarded a series of prestigious prizes (later to be rescinded) including the 2002 Young Investigator Award of the Materials Research Society [9]

Perhaps the best-known case of pathological science which occurred in a field closely allied to Corrosion Science was the "cold fusion" claims of M Fleischmann and S Pons made in Utah USA. Announced in a tumult of publicity in 1989 they claimed to have produced excess heat along with helium, neutrons and tritium by-products when heavy water (D_2O) was electrolysed on the surface of a palladium electrode [10]. Efforts to duplicate their results met with varied success. Finally panels assembled by the US Department of Energy in November 1989 and again in 2004 concluded that there was no convincing evidence for the occurrence of fusion at room temperature. Work has since continued on the topic now known as either: Low-Energy Nuclear Reactions (LENR), Condensed Matter Nuclear Science (CMNS), or CANR (Chemically Assisted Nuclear Reaction) [11].

Other examples of pathological science include the N-rays of Blondlot (1903) and Benveniste's Water Memory or homeopathy report published in Nature in 1988. The above historical examples (amongst others) show a wide range of questionable behaviour, ranging from Langmuir's pathological science to outright fraud and seem to have occurred for a wide variety of reasons [12].

2. EXAMPLES OF RESEARCH MISCONDUCT IN CORROSION SCIENCE

Although examples of such misbehaviour in corrosion science are comparatively rare, the authors have identified several cases which are presented in this section. Their impact varied widely. Whilst some were only potentially of minor interest to researchers in their own particular narrow research field, others were of far more widespread concern, with high potential impact and serious implications for population health and wellbeing. By no means a comprehensive list, it does however indicate that the problem of research misconduct is not solely the preserve of the biosciences and social sciences. There may be other cases in the corrosion science realm which remain yet to be revealed.

2.1 Alleged Corrosion Research Misconduct Example 1: Galvanic Corrosion in the Water Supply Industry

The first example deals with research conducted on copper-lead pipe galvanic (dissimilar or bimetallic) corrosion in Washington D.C. drinking water that was later questioned by two well-respected corrosion scientists. It is well-known that, depending upon a number of different factors, lead (in old pipe installations) and copper can enter drinking water through corrosion of metal plumbing systems. As a result of drinking water from these pipes health problems from both metals (copper and lead) may arise. They can range from stomach illness to irreversible brain and nervous system damage, with children being especially vulnerable. In 1991 the EPA in the USA published a regulation to control lead and copper release in drinking water known as the Lead and Copper Rule (LCR) [13]. The rule requires water utilities to monitor drinking water at their customer's taps. Accordingly, if the lead concentrations exceed 15 ppb or copper concentrations are above 1.3 ppm in more than 10% of samples, the utility company must then take remedial action to control the corrosion. If the 15 ppb level for lead is exceeded, replacement of the lead service lines (LSL) may have to be undertaken. Copper is the preferred metal for this approach, with partial replacement most often carried out.

Partial replacement of the lead pipe with one made from copper is likely to result in galvanic corrosion as the dissimilar metal pipes are always electrically connected by an earth strap. A further complication arose when the water treatment was changed from one involving chlorination to one using chloramine, whose affect was not as well-documented. This resulted in a "DC water crisis" with almost 15,000 homes having partial LSL replacement between 2004 and 2008 at an estimated cost of over $100 million [14]. This approach was called into question by the US Centers for Disease Control and Prevention (CDC) who issued a report in 2010 stating that "...children living in housing where a lead line was partially replaced after 2003 were more likely to have (elevated blood lead levels) than children living in housing without a lead service line" [15]. In addition "partial lead service line replacement was not effective in decreasing risk for higher blood lead levels", with the risk similar to that for people who never had their lead lines replaced.

Depending upon a number of factors such as water chemistry, water flow rates and the existence of passivated surfaces, most corrosion scientists consider that galvanic corrosion between lead and copper pipes in water systems may occur. For example the BSI Commentary on Corrosion at Bimetallic Contacts and its Alleviation PD 6484:1979 [16] predicts that lead when connected to copper in fresh water will suffer a range of potential corrosion outcomes; from "no additional corrosion or only very slight additional corrosion to slight or moderate additional corrosion which may be tolerable in certain circumstances". The effect is variable though, with bicarbonates, sulfates and silicates mitigating against corrosion, whilst aeration and elevated temperatures will promote it. More recent studies have confirmed the enhancement of lead corrosion by copper, which may explain the elevated blood lead levels mentioned in the CDC Report. Consequently it is normally recommended that when carrying out a partial LSL replacement procedure utilising a copper pipe joined to an existing lead pipe, that an insulating or dielectric spacer (such as a PVC pipe or flange) be inserted between the two metals, which are then usually electrically joined together by an earth strap.

However one study carried out by a team of researchers reported in the peer-reviewed *Journal of the American Water Works Association* (JAWWA) seemed to disagree with this commonly held view [17]. Contrary to expectations, this research concluded that galvanic corrosion was "limited to the immediate vicinity (~5 mm) of the lead-copper connection" (a distinctly localized region) and that, most importantly, "accelerated metal release with this type of galvanic coupling may be minimal". This is a significant finding as it would suggest there was no need to insert a spacer between the dissimilar metal pipes, nor was the lead level in the drinking water likely to increase significantly over time once the new installation was commissioned. In addition, the authors of the article also stated that if the lead and copper pipe were separated by 1 cm to 15 cm and an external wire (such as an earth strap) used to maintain electrical contact, the measured potential of the "entire lead coupon shifts in an anodic direction" and thus "galvanic coupling has likely accelerated lead release by up to ten-fold". This was supported in their paper by measurements of corrosion potential (E_{corr}) at different locations over the galvanically connected metal pipe surfaces. Significantly no galvanic current measurements were taken nor were any total lead concentrations in the recirculated water determined by commonly used methods such as AA or ICP.

Corrosion experts were surprised at this finding, as it is often considered that pipe galvanic corrosion is most often limited to a region of about 1.5 times the pipe diameter and increased separation of the bimetallic couple generally reduces galvanic corrosion through the operation of an ohmic resistance or "distance effect" [18]. The highest galvanic corrosion rate is obtained in the absence of any ohmic resistance (*iR*). However, increasing the solution *iR* by increasing the separation of the anode and cathode elements will reduce the extent of galvanic corrosion. Shortly after publication two well-respected corrosion scientists, Marc Edwards and Gerald Frankel, questioned the findings of the JAWWA study [19]. In a Discussion document (entitled "Effect of Changing Water Quality on Galvanic Coupling") published in the same journal in 2012 [20] Edwards pointed out a number of problems with the original JAWWA paper. In addition to the lack of recognition of decades of perceived corrosion wisdom backed up by literature evidence, and after contacting the original paper's authors for answers, Edwards pointed out a number of other issues with the paper. These included discrepancies between data presented in some of the figures and that discussed in the text, problems with graphics depicted in some of the figures (especially figures 9 and 10) which were performed by a graphic artist with mis-labelled axes. Some data was missing and errors were apparent in data presentation such as combining several separate curves to create a single curve unrepresentative of the actual time-dependent corrosion behaviour. Furthermore, Edwards mentions other concerns such as text being written before completion of experiments, only short term results (< 5 days) being obtained, errors in total lead determination (i.e. failure to ascertain particulate lead) and providing supporting evidence that was not able to be verified. As a consequence of this Edwards asked for the paper to be retracted by the journal.

After corresponding with the journal editors and seeking information from the authors, in addition to using a freedom of information request, Edwards wrote a letter to the journal editor which was published in January 2014 [19]. Published in the same issue was another letter by Gerald Frankel. Both highly respected corrosion scientists expressed severe reservations about the science described in the original JAWWA article which appears to contradict both experience obtained in the field and laboratory studies conducted on galvanic corrosion over many years. Results from the original study which purport to show that use of partial LSL replacement will only have a minimal effect on lead release in potable water (and thus lowered entry into the bloodstream of consumers) could have immense public health significance. In response to these criticisms, the journal editors however did not retract the paper. Instead they published an "Expression of Concern" [21] along with the recommendation that readers refer to a report (public project paper) [22] which summarises information from both the original study and from a separate one performed by Marc Edwards in the same locality.

In his Discussion document Edwards ascribes the errors published in the original paper to Langmuir's "pathological science" where "well-intentioned scientists have been tricked into false results by wishful thinking and other factors". It does seem strange that the three reviewers of the original paper did not detect important issues concerning aspects of the corrosion study which was later called into question by Edwards and Frankel. The reviewers may well not have understood the fundamentals of corrosion science at a sufficient level which highlights one of the essential problems in the reviewing process; namely reviewers must have sufficient knowledge of the area to make valid judgements. Edwards later confirmed flaws in the original study by conducting a carefully controlled investigation which was published in the NACE journal *Corrosion* in 2012 [23]. Contrary to the original paper's results, Edwards et al confirmed that direct coupling of 1.9 cm diameter lead and copper pipes caused a sudden large increase in lead release. In addition increasing lead-copper separation distance through insertion of PVC pipe sections resulted in diminished levels of lead with lower levels occurring with longer PVC lengths. It is also worth noting that Edwards used a more suitable methodology in his study. Rather than relying upon measurements of E_{corr} (which does not give any information on corrosion rates) he determined the corrosion current, the level of total lead in the water using ICP-MS and also conducted a visual inspection of scale build-up. The study was conducted over a much longer time period (4 months), instead of only a few days. The Edwards study confirmed that, as expected, directly connecting lead and copper pipes together will increase the rate of lead release and the use of an insulating material to effect pipe separation will diminish it.

2.2 Alleged Corrosion Research Misconduct Example 2: Metallic Corrosion Inhibition

The second example concerns a number of papers published over a period of many years in high-impact corrosion, electrochemical and materials science journals. Curiously, it was one of the other members of the research group who highlighted the problem. He published a series of five papers in 2013 and 2014 [24-28] that raised concerns about data presented in a large number of journal articles (over 40) on inhibition and corrosion studies on a variety of metals published by other members of the group from 2001 until 2011. Apart from one paper which uses the term "Scientific Integrity" in its title, the other four use "Scientific Fraud" as part of their titles which demonstrates his view of the work's authenticity.

The first, published in 2013, entitled "Scientific Fraud in the Digital Age" [24, 39] employed graph digitising software (UN-SCAN-IT) in conjunction with commercial graphing programs (Origin and Sigma Plot) to compare results presented in two PhD student theses with those published ostensibly from the same work in reputable peer-reviewed journals. The journals cited in this review included: *Applied Surface Science*, *Corrosion Science*, *Journal of Applied Electrochemistry*, *Electrochimica Acta* and the open access journal *International Journal of Electrochemical Science*. Comparable figures identified in the two theses which had virtually the same captions in the journal articles often had a very different appearance which is evident by eye. Plots presented in the review paper as evidence of the alleged fraud consisted of Nyquist EIS plots, inhibition efficiency versus concentration plots, potentiodynamic polarisation curves and plots of anodising potentials as a function of time. They appeared different in most but not all cases (e.g. the Tafel polarisation plots looked very similar). Also it should be noted that the scales on the EIS Nyquist plot axes were different in both cases, being in ohm in the thesis and ohm cm^2 (or sometimes incorrectly ohm cm^{-2}) in the papers. Of most concern was the finding that another thesis from 2007 contained results which served as the basis for three papers. The data showed no sign of any random (or systematic) error and through digital analysis it was claimed that the weight loss, Nyquist EIS and Tafel polarization plots were all adjusted by use of a multiplication or scaling factor. The review stated that "we proved that these results are fabricated and never been experimentally obtained". Furthermore it was also concluded that "the same fabrication technique was used in falsifying the data presented in other articles" (of which 26 were listed).

A similar conclusion was reached in the next article in the series entitled "Scientific Fraud and its Implications on Electrochemical and Corrosion Science Research" [25] in which data from inhibition of low carbon steel was examined. The paper in question "The inhibition of low carbon steel corrosion in hydrochloric acid solutions by succinic acid" was published in *Electrochimica Acta* in 2007 and which has been cited more than 280 times. Weight loss data as a function of time (gravimetric/time graphs) show perfectly straight lines and values at different inhibitor concentrations again seem to have been generated by multiplying by a constant factor. Digitisation of the polarisation (Tafel) curves indicated that they appeared to have been shifted along the potential axis which suggested that the data had simply been manipulated by multiplication by different constants. Further evidence for this manipulation is that the anodic and cathodic Tafel slopes reported are remarkably consistent ($\pm$ 1.5 mV dec^{-1}) over an extremely wide range of pH and inhibitor concentrations. EIS Nyquist plots also seemed to have been generated by applying such scaling factors in the real and imaginary axes. In addition, the appearance of an inductive loop was thought by the reviewer to be unusual under such conditions (1.0 M HCl). It was also recommended, that in future publications that the journal editors and reviewers use a similar approach in order to identify any suspicious data that had simply been scaled by such a process. The present authors have noticed some further anomalies in the paper. Figure 13, for example, purports to show the EDX spectra of the steel substrate but, rather than two narrow peaks (K-alpha and K-beta) characteristic of iron an unusually broad single peak occurs between 6 and 8 keV.

Similarly the polarisation resistance values, R_p, in Table 2 does not reflect the values expected from the quoted Tafel slopes (b_a and b_c) and the value of i_{corr} obtained by Tafel analysis.

In the third review paper entitled "Scientific Fraud in Corrosion Science Research: A Review" [26] published in a different journal in 2014, the digitisation approach was again used to investigate the veracity of weight loss and iron dissolution data published in a number of papers by the same authors. Eight cases were presented which were alleged to show evidence of data fabrication. The work was all published in reputable journals with relatively high impact factors. In the first case published in *Corrosion Science* [27] 162 points said to be obtained from steel dissolution in 4.0 M H_2SO_4 which all align as completely straight lines are shown. This is surprising as the acid solution was stagnant and such perfect alignment would not be expected. Digitisation analysis seemed to indicate that all the lines with various amounts of a glycine derivative inhibitor were scaled by different multiplication factors. The next example using similar glycine derivative inhibitors was also published in 2010 in the same journal [28]. This time dissolved aluminium was measured in 0.5 M KSCN as a function of time. As before, remarkably consistent linear plots were obtained of dissolved aluminium against time; all 126 points lying exactly on the straight lines, which again appeared to be possibly obtained by a scaling process. Another paper [29] published originally in 2009 in *Corrosion Science* also dealt with the corrosion of aluminium in weakly alkaline solutions inhibited by polyacrylic acids. Once again plots obtained over time (this time representing weight loss) were extraordinarily linear and all described as having "zero percent error". Like the previous examples they also appeared to be directly scaled as were the results in the next two studies examined, which both involved measurements of weight loss corrosion of steels published in 2007 in *Electrochimica Acta* [30] and in *International Journal of Electrochemical Science* in 2008 [31].

Another study on the inhibition of copper corrosion published in the *Journal of Applied Electrochemistry* in 2006 presented remarkably linear weight/time data (all 110 data points lying exactly on one of the 11 lines and appearing to pass directly through the origin) as well as polarisation and impedance data obtained from 25° C up to 65° C [32]. There appears to be virtually perfect alignment with remarkably consistent agreement between results obtained from all three experimental methods (gravimetric and electrochemical) which is presented in a figure showing inhibition efficiency versus logarithm of the inhibitor concentration. It was suggested that this too could have been obtained by a scaling procedure. The final two examples presented in the Review paper were both published in the journal *Materials Chemistry and Physics* in 2001 and 2002 [26]. They present data that was supposedly obtained from aluminium corrosion in different inhibitors and once again present perfectly linear weight loss vs time plots that may have been scaled.

The next paper in the series which was published in the same journal as the Review article (also in 2014) was entitled "Scientific Integrity in the Digital Age: Data Fabrication" [33]. Although it listed 30 articles from 2001 to 2011 which utilised "the same fabrication technique....to fabricate the data presented in other articles" it discussed in some depth with two articles by the same individuals as before. The first published in the *Journal of the Portuguese Electrochemical Society* (*Portugaliae Electrochimica Acta*) investigated "The inhibition of Uniform and Pitting Corrosion Processes of Al Induced by SCN^- anions: Part 1 The Effect of Glycine" [34]. Visual inspection of their weight loss data (obtained through use of ICP-AES) again revealed that all 84 data points were located exactly "on straight lines with zero percent error". It then described this as "experimentally impossible and has never been reported in the literature except for articles published (by) the same authors", citing twelve of their publications. Once more it was claimed that use of a simple multiplication or scaling procedure achieved this experimentally unlikely result. Less evident but just as disconcerting was the author's assertion that the electrochemical cyclic polarisation plots were also fabricated. Through use of the digitizing software it was possible to replicate all the curves by multiplication (or division) by constant factors to scale the current density data and thus obtain an exact match of the published polarisation curves.

A similar analysis was performed on cyclic voltammetry (electrochemical) data reported in *Electrochimica Acta* [35] entitled "On the Role of NO_2^- ions in Passivity Breakdown of Zn in deaerated Neutral Sodium Nitrate Solutions and the Effect of Some Inorganic Inhibitors". A digitisation analysis again seemed to indicate that the data was simply scaled and the reported current peaks although diminished with each successive scan, did not appear to change their position (on the potential axis) possibly due to a change in surface ohmic current contribution.

The final paper in this series by the same author "Scientific Fraud and the Power Structure of Science" was also published in *Research in Chemical Intermediates* in 2014 [36]. Although this paper discusses aluminium corrosion results in a paper published in *Corrosion Science* [29], it also questions the results presented in more than other 40 articles by the same authors, all published in high profile journals. It even includes a link to supplementary material which is available on a YouTube channel [37].

The experimental work mentioned on this critical review was performed as part of a Masters thesis from 2005. It presented weight loss data showing seven perfectly straight lines of weight loss as function of immersion time, with all 96 data points lying exactly on one of the 7 straight lines. Not a single point lies off any of the lines. The use of a scalar (multiplier) to obtain the data is again suggested. Furthermore in this example careful digitization analysis also seems to suggest that polarization curves (Tafel plots) and also Nyquist EIS plots were obtained using a similar multiplier (scaling) approach.

This series of review papers raises serious concerns about the authenticity of the work reported in over 40 peer-reviewed papers, as well as the standard of the editorial and journal review processes. In multiple cases the published weight loss/time and metal dissolution/time the data seems uniformly (and suspiciously) perfect without any outliers. Nor are any error bars apparent in any of the plots which seem to be single value results. It is not clear if replicate samples (even duplicate or triplicates) were ever tested and there is no evidence of any random (or systematic) errors occurring in any of the data reported. As the critical author commented they were "zero error plots" and it is surprising that questions had not been raised earlier, given that the articles went back to 2001. A digitisation analysis appeared to confirm that a simple scaling procedure could have been used to obtain such graphs. Consequently it was suggested that journal editors and/or reviewers use such an approach to ascertain the likely provenance of results presented for review. The use of a similar multiplier or scaling technique to obtain electrochemical data (e.g. Tafel plots and EIS plots) and other results (such as inhibition efficiency/inhibitor concentration plots) is somewhat surprising. It is also more disturbing as this is less obvious upon visual inspection, although corroding electrodes are not usually dimensionally stable and normally their surfaces change from scan to scan.

A response to these claims was published in the same journal also in 2014 [38]. Although it was suggested that it would be the first in a series whose intended aim was to "respond to the Paper Scientific Integrity in the Digital Age- Data Fabrication" it appears to be the only paper to have been produced thus far, at least according to a check carried out using Google Scholar. This response article is perplexing, as it concentrates largely on the fact that the numerous weight loss/immersion time and comparable metal dissolution concentration/immersion time plots all go through the origin. Several examples of data going through the origin are presented in this paper. None of the plots however presented as supporting evidence in Figures 2-4 are of weight loss/time or metal concentration/time. Instead they are plots of I-V in a CdS thin film (photo-electrochemical), a polarisation resistance plot (which traverses the origin at an overpotential of zero) and an erosion/time plot of a wear experiment. Such plots are not comparable to the weight/dissolution figures presented in the offending papers. However two plots (figures 5 and 6) of weight loss versus time are presented which do appear to show similar behaviour to those in the papers. Derived from studies conducted on inhibition of aluminium corrosion, both reports which emanate from another group in the same country are published in a very low impact journal (IF < 0.5). Another report which also appears to support their contention that weight loss/time plots do indeed traverse the origin was published in a high impact journal, *Corrosion Science* (5 year IF>5). However this journal also published many of the articles questioned in the series of five critical reviews [24-28] previously discussed.

More important though than the question of whether such weight loss/time (and metal dissolution/time) plots do indeed go through the origin, is the failure of the author of the response paper to mention the perfect alignment ("zero error") of his weight loss (gravimetric) corrosion data. Such perfect straight lines of what amounts to thousands of data points over all the papers examined in the critical review series require a very good explanation, which is not provided. The response paper does, however, point out that 18 papers do not contain plots with figures with straight lines going through the origin. Although this assertion may be correct, an inspection of seven of these papers published in *Electrochimica Acta* revealed that they did not contain weight loss data at all. Many of them presented results of pitting corrosion studies for which weight loss/time plots would have been totally inappropriate. The author does not address in his rebuttal nor other issues associated with rescaling of polarisation, EIS, and other data sets, but does note that he has co-authored three papers with the person accusing him of fraud. Many of these papers contain similar deficiencies (e.g. perfectly linear weight loss plots).

Crucially **none** of the papers mentioned in the critical series appear to have been retracted and are still available for use by corrosion science researchers. At least ten of the suspect papers have been cited more than 100 times (Scopus) so that the corrosion community in general appears unaware of these issues.

2.3 Alleged Corrosion Research Misconduct Example 3: Author Issues, Plagiarism and Retracted Papers in Corrosion Science

The third corrosion science example involves issues of authorship and plagiarism/self-plagiarism during dissemination of research results. In the first case reported on the web site Retraction Watch, [40] two papers published in 2014 in the *Journal of Dispersion Science and Technology* [41, 42] seem to be both substantially similar to another paper, published in the same journal in the previous year [43]. Both papers were retracted and also appeared to have used data published in a 2013 article from another journal, *Industrial & Engineering Chemistry Research*, "without proper citation." [44]. Like many of the papers mentioned in example 2 above, these publications dealt with corrosion inhibitors, in this case for prevention of mild steel corrosion in hydrochloric acid. The *Industrial & Engineering Chemistry Research* paper shares several authors with the 2013 *Journal of Dispersion Science and Technology* paper.

According to the publishers of the *Journal of Dispersion Science and Technology*, "these actions constitute a breach of warranties made by the authors with respect to originality. We note that we received, peer-reviewed, accepted, and published the article in good faith based on these warranties, and censure this action. The retracted article will remain online to maintain the scholarly record, but it will be digitally watermarked on each page as RETRACTED". A similar notice was also issued for another paper in the same journal [45]. The issue largely centres on self-plagiarism; that is using identical results for more than one publication without acknowledgement of the source of the data and attempting to present this as original material when it clearly is not.

.

Several other instances of similar misbehaviour were also found in other retracted papers in the corrosion science area. A paper entitled "Mechanism Responsible for Synergy Between Fretting and Corrosion for Three Biomaterials in Saline Solution" published in the journal *Wear* in 2001 was retracted as the work was "substantially based without the author's permission, on material submitted" for another person's PhD "and that it was submitted for publication without "either their or the third author's knowledge [46]. Consequently the "Editor-in-Chief of *Wear* (and the publisher), who published the article in good faith, now wish to treat this paper as retracted and express their disappointment at the inconvenience caused to" the former PhD student and the article's third author. This appears to be a clear case of plagiarism of another researcher's work. It also implicates another author who did not appear to be aware of the paper before submission which is unacceptable behaviour.

Another retracted paper, was published in *Materials Science and Engineering; A* in 2005 [47]. Entitled "Mechanical Properties and Corrosion resistance of Low-Alloy Steels in Atmospheric Conditions Containing Chloride" the Editor requested that the article be retracted as it "duplicates significant parts of a paper that had already appeared in *Corrosion Science* 47(2005) 1001-1021". It was pointed out that "one of the conditions of submission of a paper for publication is that authors declare explicitly that their work is original and has not appeared in a publication elsewhere". Evidently this was not the case in this instance resulting in a situation where data was not appropriately cited. The retraction statement concluded that "the scientific community takes a very strong view on this matter" and furthermore "apologies are offered to readers of the journal that this was not detected during the submission process". This is not an uncommon practice. Indeed once when serving as technical convenor of a Conference, one of the authors of this paper highlighted such an issue with a submitted conference paper that was substantially the same as another paper albeit with a different title, that had just been published in a reputable journal.

A paper "Structure adhesion and corrosion resistance study of tungsten bisulfide doped with titanium deposited by DC magnetron co-sputtering" published in *Applied Surface Science* [48] was retracted at the request of the Editor-in-Chief "because this paper had been written and submitted to the journal by the corresponding author, without the consent of (some of) the co-authors". According to the retraction notice "one of the conditions of submission of a paper for publication is that all authors have approved the manuscript and have agreed to submission to a journal. Since this condition has been violated, this paper will be retracted". Just as in the previous example some authors had not approved submission of the manuscript before submission to the journal.

The final example involves self-plagiarism in which images taken from other publications were inserted into a paper published recently (2016) in the *Journal of Materials Science: Materials in Medicine* [49]. The retraction note from the journal stated that "following a complaint about the republishing of unreferenced data in several journals including (this journal) the following issues were detected: there are textual similarities and duplication of images in the....paper and in "The effect of electropulsing induced gradient topographic oxide coating of Ti-Al-V alloy strips on the fibroblast adhesion and growth" authored by Xiaoxin Ye, Zion Tse, Guoyi Tang and Guolin Song, and published in *Surface and Coatings Technology*, notably Figures 2 and 3(a) in the latter article appear as Figures 5 and 7 in (this journal) paper without reference to the original publication. In addition, similarities in other Figures (for example, Figures 8 and 9) have been noted with a number of other publications....". Two Materials science journals were named. It was also noted that (the paper) was "hereby retracted at the request of the Editor-in-Chief and with the agreement of the authors".

3. DISCUSSION

In the first example considered there does not appear to be any direct evidence of data fabrication, falsification or plagiarism which are the normal indicators of research misconduct. Although an "Expression of Concern" was raised, the paper in question was never formally retracted. The major issue raised by Edwards and Frankel was that that the findings are at odds with conventional corrosion theory and many previous (and later) studies of galvanic corrosion in copper-lead piping. Edwards does, however, find some examples of poor research practices relating to the collection, storage, analysis and presentation of data. The authors were not able to explain adequately how the visual appearance of graphs changed over time, how and when the data was collected, nor did they supply copies of the original data for inspection as requested. The journal reviewers may also have been either negligent or unfamiliar with fundamental aspects of corrosion science, especially the theory of galvanic corrosion mechanisms.

The second example appears to be a blatant case of data fabrication and falsification involving a large number of papers, over an extended time period and published in reputable electrochemistry and corrosion journals. The review analyses seems to show clear evidence of systematic data fabrication, and the rebuttal hardly addresses the most significant issues. In many instances, the authors of the original papers seem to have taken corrosion data and simply rescaled it to generate additional data sets that nominally reflect changes in the experimental conditions. This data fabrication technique has been applied variously to weight loss measurement, polarisation and EIS plots. Digitisation of the data sets typically shows that they are exact ratios of each other – even though there is no plausible physical explanation for this to be the case. For linear plots these manipulations are often obvious, but for more complex curves (such as polarisation and EIS plots) these are often much more difficult to detect without a detailed analysis. Finally, there are many examples of extremely precise and highly linear data (e.g. weight loss measurements) presented in the papers, where even a cursory review of the corrosion literature will indicate that this is exceedingly unusual and, given the nature of corrosion processes, extremely unlikely.

It is useful to relate example 2 to the case of Hendrik Schön [9] who used very similar techniques to fabricate results on a wide range of semiconductor devices. The authors of the report investigating allegations of Schön's misconduct asked: "Is there clear evidence that the data does not come from the measurements described?" The first evidence they identify is *data substitution* "...in which data sets for distinct experimental conditions show unreasonable similarity to each other, in some cases after multiplying one data set by a constant factor". In the publications of Schön there are many examples where results are multiplied or rescaled by a constant factor (often an integer) to generate additional data. This is precisely the kind of data fabrication identified in the second example. Additionally, the authors of the report discuss the concept of *unreasonable precision*, as an indicator that data does not come from direct measurements. It is clear that the weight loss data reported in example 2 could be considered an example of this. Similarly, the Tafel parameters (b_a and b_c) reported often show remarkable consistency (± 1.5 mV dec^{-1}) over a wide range of conditions, and this could also be considered an example of suspicious and/or unreasonable measurement precision.

The third and final case presents several examples showing a range of misbehaviours in publications that are clearly in breach of good scientific practice. Ranging from appropriation of another person's data, self-plagiarism (re-use of the same figures and data in different journal articles) to improper attribution of authors without their consent, these are all unacceptable practices. However, unlike the papers discussed in the second example, all of these papers were retracted and the issues clearly made apparent for the discernment of other researchers.

In practise, research integrity issues exist which are of more widespread concern than the narrow interests of the researchers themselves. Other stakeholders in the research sphere include funding agencies, industry partners, conference organisers, journal publishers, editors and reviewers and ultimately the consumers or public who may be affected by the research findings. All need to be aware of such issues which can not only damage the reputations of individuals but of institutions as well and can be viewed as a betrayal of trust. A 2016 paper [50] highlights many of the reasons for scientific misconduct which has become more prevalent in recent times. For example in life sciences and biomedical research the percentage of articles retracted has increased by 1000% since 1975, with 67% being due to misconduct, including fraud and suspected fraud (43.4%), duplicate publication (14.2%) and plagiarism (9.8%). The costs of scientific research misconduct are significant. The direct costs of handling a research misconduct case is US$525,000 with over US$110 million incurred annually at the institutional level in the USA. Such results led the journal *Nature* in 2015 to state that "pretending research misconduct does not happen is no longer an option".

Several reasons for such practices have been advanced including the notion of "perverse incentives" arising from the need to achieve targets through quantitative performance metrics, sometimes at the expense of research quality and the "hypercompetitive funding environment" [50]. In the latter case more researchers are now applying for a diminishing pool of funding as has happened in the EU with its latest H2020 funding scheme claiming a success rate of only 14% (less than 1 in 7). Together with institutional requirements to improve global rankings and the concept of academic freedom to pursue risky ideas diminishing, (the "trampling of the scientific ethos") enormous pressure has been brought to bear on researchers. Consequently there is a strong temptation to engage in dubious research practices. One survey reported in Edwards and Roy [50] conducted in 1989 noted that 34% of researchers had engaged in practices such as "dropping data points on a gut feeling" and "changing the design methodology and results of a study in response to pressures from a funding source". It would be surprising if the percentage had not increased since then.

Research integrity issue can be tackled on many fronts. As indicated by the examples in given in section 2 science is still essentially a self-policing activity. It has largely been other scientists or domain experts, who have highlighted research misconduct issues in published articles. In conjunction with the pressures faced by researchers highlighted above and with a rapid increase in journal numbers, including open access journals where researchers pay for publication, there is a need for more journal editors and reviewers. However, many of the reviewers lack experience or expertise meaning that it is likely that more reports of research misconduct will appear. Publishers are well aware of this and have set up a forum for editors and publishers of peer reviewed journals to discuss all aspects of publication ethics (Committee on Publication Ethics, COPE) [51]. Established in 1997 by a small group of journal editors it now has over 10,000 members worldwide from all academic fields. Although COPE does not investigate individual cases, it actively encourages editors to ensure that cases are thoroughly investigated by an employer or research institution such as a university. Use of computer tools such as digitisation of data and data analytics and other methods can assist this work through identification of unexpected trends.

Another publication issue is the re-use of digital images in multiple publications, which is a problem especially in the biomedical field and is more prevalent in higher impact journals.

Research funders themselves are also highly aware of research integrity issues. Following such misconduct cases there is a strong move for original research data to be available for researchers and others to interrogate. Notwithstanding difficulties such intellectual property rights and commercial constraints, the concept of Open Science has been promulgated as a way to peer-review and authenticate data. However this has been difficult to carry out and is still under development.

A variety of approaches aimed at tackling this important issue are being adopted ranging from training of PhD students to monitoring of research outputs and validation of results. More difficult to achieve is the need to reduce "perverse incentives" such as the favouring of quantity of outputs over quality, especially in the very competitive funding environment that currently exists and the need to actively discourage unethical behaviour in favour of the pursuit of the practice of science in the service of humanity. Such challenges apply to all scientific fields including Corrosion Science.

4. CONCLUSIONS

Research integrity which covers more than just ethical behaviour is becoming more important in science, especially in the biosciences and health/life sciences. Several cases of alleged research misconduct reported in the open literature indicate that although it is not yet a big problem in corrosion research, it has occurred. This makes it extremely important that various research stakeholders such as funding organisations, journal editors and reviewers and above all researchers themselves are fully aware of this issue and its ramifications.

The three examples of alleged research misconduct presented demonstrate different levels ranging from plagiarism to serious data fabrication. In the first example it could well lead to serious public health damage. Critical judgments on the quality of research work presented in peer-reviewed papers is necessary to prevent dissemination of potentially incorrect and/or possibly harmful results and it is up to all researchers to contribute, both as researchers and as reviewers and evaluators of research proposals and peer-reviewed papers.

Real corrosion data will be inherently noisy, as unlike a normal electrochemical working electrode (e.g. Pt, Au) which remains dimensionally stable (apart from adsorption), a corroding working electrode undergoes spatial and temporal changes. Processes such as film formation and removal, dissolution, intermetallic particle detachment and metal detachment will take place on a corroding electrode over a period of time. In practice this may well lead to significant amounts of random errors, especially in gravimetric and electrochemical data. Such inherent variability is well understood by most corrosion researchers. Therefore it is both surprising and disappointing that abnormally invariant data reported in many papers identified in the second case has not been questioned more vigorously by journal editors, referees/reviewers or indeed by other researchers.

It is important that journal editors and reviewers possess the requisite expertise and experience to identify possible fraudulent or fabricated data and plagiarism issues. The use of a simple digitisation analysis is one method of detection as was performed in the second example presented. Image processing techniques are routinely used by scientific publishers to detect manipulation of images, so it not unreasonable that such approaches could be applied to other forms of data. However it is important to note that, in many cases, the fabrication of data appeared to be so obvious that simple visual inspection should have been sufficient to arouse suspicion. Quality testing of journal reviewer's competence may also be a helpful initiative to help prevent such occurrences.

There are now many on-line resources for dealing with scientific misconduct. For example, the blog "Retraction Watch" includes an analysis and commentary on publications that have been recently retracted. Similarly, the website "Pub-Peer" allows researchers to anonymously review and discuss published research. In many cases, such discussions have led to the identification of scientific misconduct and the retraction of papers. Most scientific publishers have developed standardised procedures for handling ethical allegations relating to issues such as plagiarism, authorship, conflicts of interest and data fraud or falsification as illustrated in the third example. With such resources and procedures now available, there are no real impediments for scientists and researchers to raise legitimate concerns about suspected research misconduct.

The reasons for such unscrupulous behaviour (research misconduct) can range from Langmuir's "pathological science" and willingness to suspend objective judgement of the work's quality in order to gain attention and advance one's career, to more sinister motives, sometimes for financial benefit. This has become a major problem throughout science over the past few decades. In an increasingly competitive funding environment this has caused severe pressure on researchers leading occasionally to unethical behavior.

Vigilance and adherence to the requisite high standards for research will prevent instances of willful research misconduct and save science's reputation for being trustworthy and reliable. This is vitally important as each case undermines its integrity and future prospects and diminishes its status in the public domain.

5. ACKNOWLEDGMENTS

The authors wish to acknowledge Prof. Marc Edwards for supplying published documents and helpful discussions with Prof Sudipta Roy and other partners in COST Action MP1407, Prof. M Devereux and Prof. S Jerrams.

6. REFERENCES

1. http://www.allea.org/wp-content/uploads/2017/05/ALLEA-European-Code-of-Conduct-for-Research-Integrity-2017.pdf Accessed 20 June 2017
2. http://www.iua.ie/wp-content/uploads/2014/06/National-Policy-Statement-on-Ensuring-Research-Integrity-in-Ireland-2014.pdf Accessed 14 May 2017
3. http://ukrio.org/publications/code-of-practice-for-research/ Accessed 12 June 2017
4. https://www.nhmrc.gov.au/research/responsible-conduct-research/australian-research-integrity-committee Accessed 14 May 2017
5. https://royalsociety.org.nz/what-we-do/research-practice/ Accessed 16 May 2017
6. https://researchintegrityjournal.biomedcentral.com/ Accessed 12 June 2017
7. Langmuir I, Pathological Science, Physics Today, 42 (10) (1989) 36–48
8. Franks, F, Polywater. The MIT Press. p. 140 1981 ISBN 0-262-06073-6.
9. https://media-bell-labs-com.s3.amazonaws.com/pages/20170403_1709/misconduct-revew-report-lucent.pdf Accessed 29 June 2017
10. Fleischmann M, Pons S Electrochemically induced nuclear fusion of deuterium, Journal of Electroanalytical Chemistry, **261** (2A) (1989) 301–308,
11. http://cen.acs.org/articles/94/i44/Cold-fusion-died-25-years.html Accessed 22 June 2017
12. https://www.irishtimes.com/news/science/william-reville-fraud-is-now-the-biggest-enemy-of-science-1.2661922 Accessed 2 June 2017
13. https://www.epa.gov/dwreginfo/lead-and-copper-rule Accessed 19 June 2017
14. http://www.infrafunding.wa.gov/downloads/2016_Conference_Presentations/S50.pdf Accessed 12 June 2017
15. https://www.cdc.gov/nceh/lead/publications/ Accessed 10 June 2017
16. BSI Commentary on Corrosion at Bimetallic Contacts and its Alleviation PD 6484:1979
17. Boyd, G.R., Reiber, S.H., McFadden, M.S. and Korshin, G.V. Effects of changing water quality on galvanic coupling, Journal AWWA (American Water Works Association) 104(3) 2012 E136-E149 and Boyd, G.R.; Reiber, S.H.; McFadden, M.S.; & Korshin, G., Expanded Summary: Effect of Changing Water Quality on Galvanic Coupling. Journal AWWA, 104(3) (2013) 45
18. Landolt D, Corrosion and Surface Chemistry of Metals, EPFL Press 2007 281
19. Edwards M Letters to the Editor, Galvanic Corrosion Article Draws Comment. Journal AWWA, 104(12) 2012 8. and Edwards M and Frankel G Letters to the Editor, Concerns Continue Regarding Galvanic Coupling Article. Journal AWWA, 106(1) (2014) 18.
20. Edwards, M. Discussion: Effect of Changing Water Quality on Galvanic Coupling. Journal AWWA, (2012) 104:12:65. http://dx.doi.org/10.5942/jawwa.2012.104.0151.
21. JAWWA Expression of Concern in http://retractionwatch.com/2015/04/15/expression-of-concern-opens-floodgates-of-controversy-over-lead-in-water-supply/ Accessed 22 June 2017
22. Giammar D E, Welter G J, Cantor A Review of Previous Water Research Foundation Projects on Galvanic Corrosion in http://www.waterrf.org/resources/Lists/PublicProjectPapers/Attachments/3/4349_LiteratureReview.pdf Accessed 28 June 2017
23. St. Clair J, Stamopoulos C and M. Edwards, Technical Note: Increased Distance Between Galvanic Lead:Copper Pipe Connections Decreases Lead Release, Corrosion, 68(9) (2012) 779-783
24. Khaled K F, Scientific fraud in the digital age. Der Pharma Chemica. 5 (2013) 28-38
25. Khaled K F , Scientific fraud and its implications on electrochemical and corrosion science research. Der Pharma Chemica. 5 (2013) 256–263
26. Khaled K F, Scientific fraud in corrosion science research: a review, Research on Chemical Intermediates 40 (2014) 1735-1752

27. Amin M A Ibrahim M M Corrosion and corrosion control of mild steel in concentrated H2SO4 solutions by a newly synthesized glycine derivative Corrosion Science 53 (2011) 873-885
28. Amin M A, A newly synthesized glycine derivative to control uniform and pitting corrosion processes of Al induced by SCN^- anions – Chemical, electrochemical and morphological studies, Corrosion Science 52 (2010) 3243-3257
29. Amin M A, Abd El-Rehim S S, El-Sherbini E E F, Hazzazi O A, Abbas M N, Polyacrylic acid as a corrosion inhibitor for aluminium in weakly alkaline solutions. Part I: Weight loss, polarization, impedance EFM and EDX studies, Corrosion Science 51, (2009) 658-667
30. Amin M A, Abd El-Rehim S S, El-Sherbini E E F, Bayoumi R S, The inhibition of low carbon steel corrosion in hydrochloric acid solutions by succinic acid Part I. Weight loss, polarization, EIS, PZC, EDX and SEM studies Electrochimica Acta 52 (2007) 3588-3600
31. Amin M A, Abd El-Rehim S S, El-Sherbini E E F, Bayoumi R S, Chemical and Electrochemical (AC and DC) Studies on the Corrosion Inhibition of Low Carbon Steel in 1.0 M HCl Solution by Succinic Acid - Temperature Effect, Activation Energies and Thermodynamics of Adsorption, International Journal of Electrochemical Science, 3, (2008) 199-215
32. Amin A, Weight loss, polarization, electrochemical impedance spectroscopy, SEM and EDX studies of the corrosion inhibition of copper in aerated NaCl solutions, Journal of Applied Electrochemistry, 36 (2006) 215-226
33. Khaled K F, Scientific integrity in the digital age: data fabrication, Research in Chemical Intermediates 5(1) (2013) 28-38
34. Amin M A, Mohsen Q, Mersal G A M , Inhibition of Uniform and Pitting Corrosion Processes of Al Induced by SCN- anions – Part I. Effect of Glycine Portugaliae Electrochimica Acta 28 (2010) 95-112
35. Amin M A, Hassan H H, Abd El Rehim S S On the role of NO2 – ions in passivity breakdown of Zn in deaerated neutral sodium nitrite solutions and the effect of some inorganic inhibitors Potentiodynamic polarization, cyclic voltammetry, SEM and EDX studies Electrochimica Acta 53 (2008) 2600-2609
36. Khaled K F, Scientific fraud and the power structure of science, Research on Chemical Intermediates 40 (2014) 2785-2798
37. Amin M A, Experimental findings with typical straight lines passing through the origin are not necessarily referring to data fabrication: response (I) on the paper titled "Scientific integrity in the digital age—data fabrication" Research on Chemical Intermediates 41 (2015) 4423-4254
38. Baykoucheva S, Managing Scientific Information and Research Data Chandos Publishing 2015 22
39. https://www.youtube.com/user/fassadsci Accessed 24 June 2017
40. http://retractionwatch.com/2015/02/18/acid-studies-burned-duplication-test-corrosion-papers/ Accessed 27 June 2017
41. Ghasemi O, Ghadimi M R , and Ghasemi V, Adsorption and Inhibition Effect of Benzaldehyde Schiff Bases on Mild Steel Corrosion in 1 M HCl Medium, *Journal of Dispersion Science and Technology* 35(8) (2014) 1143–1154
42. Ghasemi O, Ghanbariadivi H and Ghasemi V, Inhibition Effect of Hydroxybenzaldehyde Schiff Bases on the Corrosion of Mild Steel in Hydrochloric Acid Solution, *Journal of Dispersion Science and Technology* 35(5) (2014): 706–716
43. Ghasemi O, Danaee I, Rashed G R, Rashvand Avei M, and Maddahy M H, Adsorption and Inhibition Effect of Synthesized N,N′-bis(3-Hydroxybenzaldehyde)-1,3-Propandiimine on the Corrosion of Mild Steel in 1 M HCl, *Journal of Dispersion Science and Technology* 34(7) (2013): 985–995
44. Gholami M, Danaee I, Maddahy M H, and Rashvand Avei M, Correlated ab Initio and Electroanalytical Study on Inhibition Behavior of 2-Mercaptobenzothiazole and its Thiole-Thione Tautomerism Effect for the Corrosion of Steel (API 5 L X52) in Sulphuric Acid Solution, *Industrial & Engineering Chemistry Research* 52(42) (2013)
45. Gholami, M, Danaee I, Maddahy M H, and Rashvand Avei M, Correlated ab Initio and Electroanalytical Study on Inhibition Behavior of 2-Mercaptobenzothiazole and its Thiole-Thione Tautomerism Effect for the Corrosion of Steel (API 5 L X52) in Sulphuric Acid Solution, *Industrial & Engineering Chemistry Research* 52 (42) (2013)
46. He D, Zhang T, Wu Y, Mechanisms Responsible for Synergy Between Fretting and Corrosion for Three Biomaterials in Saline Solution. Wear 250 (1-12) (2001) 180-187
47. Chen Y Y, Tzeng H J, Wei L I, Shih H C, Mechanical properties and corrosion resistance of low-alloy steels in atmospheric conditions containing chloride, Materials Science and Engineering A 398 (1-2) (2005) 47-59
48. De La Roche J, González J M, Restrepo-Parraa E, Sequeda F, Alleh V, Scharf T W, Structure adhesion and corrosion resistance study of tungsten bisulfide doped with titanium deposited by DC magnetron co-sputtering, Applied Surface Science, 320 (2014) 538-543 and 349 (2015) 997
49. Ye X, Tang G, Effect of coupling asynchronous acoustoelectric effects on the corrosion behavior, microhardness and biocompatibility of biomedical titanium alloy strips, Journal of Materials Science: Materials in Medicine 26 (2015) 53 and 27 (2016) 173
50. Edwards M A and Roy S, Academic Research in the 21st Century: Maintaining Scientific Integrity in a Climate of Perverse Incentives and Hypercompetition, Environmental Engineering Science 34(1) (2017) 51-61
51. https://publicationethics.org/ Accessed 19 June 2017

7. AUTHOR DETAILS

Dr Tony Betts is Head of Research in the College of Sciences and Health in the Dublin Institute of Technology, Ireland. He co-founded the Institute's Applied Electrochemistry Group and has been in DIT since 2004. Especially interested in materials-oriented electrochemistry, he was previously a corrosion scientist in Forbairt, now Enterprise Ireland. Before moving to Ireland he undertook a similar role in MPT of IRL (formerly DSIR). A former President of the New Zealand ACA, he was the P F Thompson Lecturer in ACA's Auckland Conference in 2000 and has been a Member of ACA for over 25 years.

Dr. Todd Green is Knowledge Exchange Associate at the Department of Chemical and Process Engineering, University of Strathclyde. He obtained a BSc(Hons) and PhD in Physical Chemistry from the University of Western Australia and performed post-doctoral research in electrochemistry at Aarhus University, Denmark (1995) and electrochemical engineering at the University of Newcastle, UK (1996 - 1998). After working as a Process Engineer in the microfabrication of optoelectronic, semiconductor and microsystem devices, he moved to Strathclyde. Dr Green is a member of the Institute of Engineering and Technology (IET) and NACE International. His research interests encompass electrochemical engineering, corrosion and microfabrication.

AN OVERVIEW OF CORROSION PROTECTION PROVIDED BY CARBON FIBRE REINFORCED POLYMER BONDED ON REINFORCED CONCRETE

A Wei[1], R Al-Ameri[1], M Tan[1] & Y.C. Koay[2]
[1]School of Engineering, Deakin University, Geelong, Australia,
[2]VicRoads Asset Services (Structures), Melbourne, Australia

SUMMARY: Reinforced concrete (RC) has been widely used in the construction for many years because of its stability, economy, and resiliently in extreme conditions. However, steel corrosion is regarded as one of the most important issues affecting the durability and stiffness/strength of RC structures. Carbon fibre reinforced polymer (CFRP) has been used as repairing and strengthening method via bonding onto RC. CFRP is now widely used due to its high corrosion resistance, high mechanical strength and ease of application on site. Meanwhile, CFRP has also been investigated recently as an anode material in the impressed current cathodic protection (ICCP) system due to its electrically conductive properties.

This paper presents an overview of steel corrosion affected by parameters of CFRP bonded on RC surface and the ICCP system using CFRP as anode. Different parameters such as epoxy types, CFRP layers, exposed environment and corrosion measurements are discussed and evaluated. The possibility of CFRP using as the dual function of both anode material in ICCP system and strengthening material on RC against corrosion is examined.

The current results indicate that CFRP has advantages of strengthening RC structures suffering from corrosion and improving corrosion resistivity. However, limited literature could be found with regard to steel corrosion investigation of RC structures with bonded CFRP under sustained loading. Further working on steel corrosion affected by CFRP simulated in the actual RC structures are expected and will be studied in a PhD project using electrochemical methods such as: Half-cell potential, Linear polarised resistance (LPR), and electrochemical impedance spectroscopy (EIS). The ICCP system using CFRP as the anode material is also investigated using simulations of a marine environment subject to wet-dry cycle, sustained loading, and high temperatures.

Keywords: Steel corrosion, Carbon fibre reinforced polymers, Reinforced concrete, Impressed current cathodic protection.

1. INTRODUCTION

Reinforced concrete (RC) has been widely used in the construction field because of its stability, economy, and resilience in extreme conditions [1, 2]. As concrete has high compressive strength but low tensile strength, steel bars are embedded in concrete so as to improve the tensile capacity of RC structures. However, corrosion of steel bars is regarded as one of the most important issue for the durability of the structures, especially for bridges and buildings [3, 4]. The features of failures induced by corrosion in RC structures are usually considered as the reduction of the cross section in steel bars, the decrease of bonding between steel bars and concrete, the degradation of steel properties and concrete cover cracking and ultimately spalling [5]. In order to prevent structures failure and destruction induced by steel corrosion, a huge cost of repairing and maintenance is spent worldwide [6]. Therefore, corrosion of steel bars should be understood and managed for the durability of RC structures.

Methods of corrosion repair and prevention have been used to delay and increase corrosion resistance in the RC. A typical technique for repairing is concrete patching and enlarge the cross section to improve the strength, in which deteriorated concrete is removed corroded steel bars cleaned and areas reinstated with more steel bars and patching mortar. However, this method may add further dead load to the structure that was not initially designed and lead to the structures being in the unsafe state when adding more steel bars and concrete on the existing structures [7]. Additionally, corrosion rate could be increased after concrete patching due to the electrochemical reaction between the newly repaired areas and surrounding areas. Surface coatings are another preventive method in creating a barrier to stop corrosion activity by restricting moisture, oxygen and chloride contamination, an example of such coatings is epoxy-coated reinforcement and galvanised reinforcement [8]. However, the coating of steel bars is costly, and the protection will be reduced when the epoxy layer is damaged during the process of transport or installation. . Corrosion inhibitors added to RC can also slow or prevent chloride ion diffusion and increase corrosion resistance [9]. Additionally, adequate cover of RC surface is another pathway in the long-term durability protection from steel corrosion.

An innovative technique to the traditional repair method is the application of carbon fibre reinforced polymers (CFRP), as seen in Figure 1. CFRP has been applied in RC structures for strengthening and also has advantages in terms of high chemical resistance, minimal impact of structures, high strength to weight ratio, high tensile strength [1, 10] and convenient installation [11, 12]. CFRP has been used for shear, flexural or axial loading enhancement in columns, slabs and walls for many types of structures. Strengthening, bonding between CFRP and concrete, stiffness, and ductility, mechanical and electrical performance for CFRP on RC structures has been studied by others [13]. Heffernan and Erki [14] and Bonfiglioli et al. [15] investigated the fatigue life of RC beams and indicated that CFRP could enhance the structures through the externally bonding strength. Ye et al. [16] studied the shear strength of RC columns strengthened by CFRP and concluded that shear strength of RC columns was significantly increased with external strengthening. White et al. [17] performed experimental and modelling study in strain rate of RC beams after CFRP strengthening, and the results indicated that CFRP could increase capacity, stiffness and energy absorption by approximately 5% in. Ye et al. [18] indicated that the ductility of RC columns could be substantially improved with CFRP strengthening according to study the seismic performance of RC columns under constant axial load and lateral cyclic load. Therefore, CFRP strengthens the mechanical performance and loading capacities of RC structures.

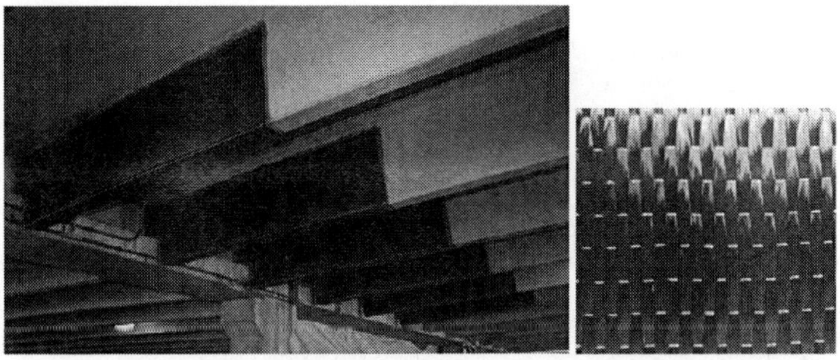

Figure 1 Structural strengthening using CFRP

(Available at: http://truesdellcorp.com/index.php/northeast-wisconsin-tech-structural-failure-repair)

Due to the property of high resistance, CFRP has also been studied for the delay of aggressive ions diffusion and whether it can decrease corrosion rate in RC structures. Different parameters of fibres such as types, number of layers and application directions have been studied [19, 20]. Corrosion induced by chloride ions is considered as the most destructive mechanism of corrosion in RC structures, as chloride diffusion results in the breakdown of passive layers and then leads to pitting corrosion occurring in the electrochemical reactions [21]. CFRP with epoxy has been successfully proved in improving the strength and stiffness of RC structures exposed to various environment based on limiting the solution absorption in RC structures [22]. Additionally, CFRP is used by asset owners due to its desirable properties for repair strategies due to its corrosion resistance. In recent times, efforts have been made to use CFRP as an anode in an impressed current cathodic protection (ICCP) system. In an ICCP system, a cathodic current is applied to the steel reinforcement, and then the steel potential is shifted towards the lower corrosion rate material [23]. The potential dual function of CFRP to provide both strengthening and to act as the ICCP anode has significant practical benefits for structure repairs.

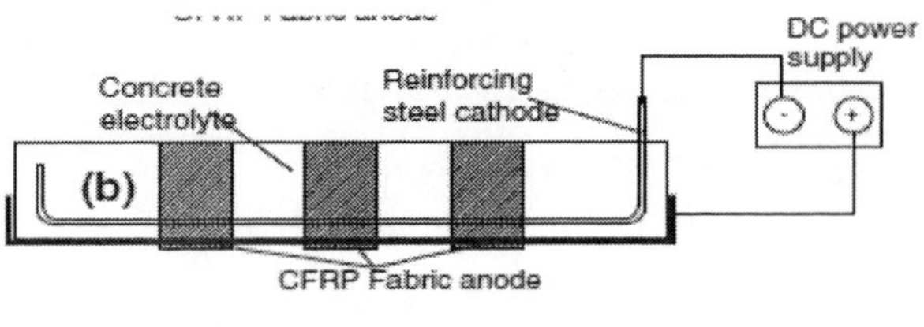

Figure 2 Schematic ICCP application to corroded reinforced concrete beams [24]

On the basis of these perceived benefits, a number of studies were commenced to investigate the use of CFRP as both anode material and strengthening material in ICCP system [25-27]. Literature about steel corrosion behaviour affected by CFRP strengthening on actual RC structures is unclear. Therefore, in order to promote the research of CFRP for dual function, this paper will develop an overview of steel corrosion performance affected by CFRP bonded on RC and the CFRP anode using in the ICCP system.

2. CORROSION OF STEEL REINFORCEMENT EMBEDDED IN CONCRETE STRUCTURES ADHERED BY CFRP

Limited studies have been carried out on the effect of corrosion using external CFRP adhered on RC structures. Studies have begun to investigate corrosion of steel reinforcement embedded in concrete structures with CFRP adhered. Subhani et al. [28] investigated the bond strength using different kinds of epoxy between CFRP and concrete in retrofitted beams in the marine environment. A proposed long-term model in predicting the bond strength during flexural tests was developed in a wet – dry cycle environment. El-Hacha et al. [29] also studied the effect of severe environmental conditions on CFRP wrapped RC columns. Thirty-six plain concrete cylinders (150 x 300mm) were cast and tested, in which nine of them were unstrengthened as the reference and the rest of the cylinders were strengthened by two layers of CFRP. Specimens were exposed to high temperatures (45 °C), or heating–cooling cycles (23 to 45°C). The results indicated that the concrete strength increased by 43% and 74% in unconfined and heating-cooling cycles, respectively using two layers CFRP wrapping. The axial strain performance in confining specimens was improved by up to 4 times. Additionally, corrosion behaviours have been investigated on various parameters, such as CFRP wrapping materials, numbers of layers, exposure environment, bonding behaviour between CFRP and concrete, and corrosion measurement techniques have been compared, as seen in Table 1.

2.1 The functions of CFRP in corrosion of RC

The authors could find very little literature on the CFRP corrosion behaviour in RC structures. Badawi and Soudki [25] studied the effect of CFRP confinement on cracking induced by corrosion in RC beams. Uniform and shear span corrosion was demonstrated at three different degrees of corrosion. In this case, impressed current was applied to accelerate the corrosion process. The results indicated that CFRP could reduce corrosion expansion by up to 70% and slow down the corrosion rate by up to 35% based on the calculation of the corrosion mass loss.

Gadve et al. [30] studied the corrosion rate of reinforcement on corroded cylinders in salt water. Pull out strength, mass loss, half-cell potential of steel bars, and cell voltage were measured. The conclusions indicated that CFRP could delay the corrosion rate and improve corrosion resistance in RC structures.

Zhuang et al. [31] investigated the effectiveness of CFRP against steel corrosion on RC piles. Twelve RC piles strengthened or unstrengthened by CFRP were studied in a simulated tidal marine environment for 6 hours from 25°C to 40°C. The impressed current technique was performed to accelerate the corrosion process. Half-cell potential measurement, steel weight loss and strains of piles were measured at various corrosion degrees. The results illustrated that CFRP could reduce corrosion activity by up to 45% compared to 52% of unstrengthened piles using the half-cell potential measurement, and reduce by 20% the corrosion rate from mass loss test.

Masoud et al. [32] studied the structural behaviour of RC beams strengthened by CFRP in the marine environment under monotonic and fatigue loading. Unstrengthening, wrapping using CFRP and flexural strengthening were investigated. The experimental results indicated that CFRP application on RC was effective in repairing damages and protecting structures against corrosion. Ultimate monotonic strength was increased by about 37% - 87%, and the fatigue of strengthened samples was increased by from 2.5 and 6.0 times compared to the unstrengthened similarly corroded specimens.

In summary, RC specimens in the above literature were generally studied using one layer CFRP for strengthening and exposed to a simulated marine environment where chloride ions can lead to localised corrosion due to passive film destruction. Most of the research studie corrosion behaviour using impressed currents to accelerate corrosion processes for pre-corroded RC specimens. The methods of half-cell potential and mass loss were widely used for corrosion measurement. Chloride ion diffusion and penetration can be slowed and oxygen in the corrosion process can be restricted due to the external CFRP and epoxy [33] acting as a barrier. It can be concluded that CFRP could improve corrosion resistance and reduce the corrosion rate by up to 45% when strengthened on RC structures.

Table 1. Corrosion protection using CFRP adhered on RC structures

Reference	CFRP adhered types	CFRP layers	Pre-corroded	Exposed environment	Long Term Sustained Loading	Duration	Corrosion measurement methods	Key conclusions
Badawi and Soudki [25]	U shape wrap	One	Yes	Salted concrete	NA	145 days	Crack width; Strain; Mass loss	CFRP reduce corrosion rate by 35%.
Gadve et al. [30]	Wrapping cylinders	One	Yes	3.5% NaCl solution;	NA	25 days	Half-cell potential; Pull-out strength test; Mass loss; Cell voltage	CFRP slow down corrosion rate.
Zhuang et al. [31]	Wrapping piles	One	Yes/NA	Marine; Wet-dry cycle; 25-40°C temperature	NA	45 days	Half-cell potential; Mass loss; Strains test	CFRP reduce corrosion rate by 20% - 45%.
Masoud et al. [32]	U-shape on beams	One	Yes	3% NaCl in concrete mixture	NA	44 days	Monotonic and fatigue loading test	CFRP is efficient in reducing steel corrosion.
Wootton et al. [6]	Wrapping cylinders	One to three	Yes	Saltwater	NA	50 days	Mass loss	CFRP increase corrosion resistance.
Debaiky et al. [7]	Wrapping Columns	One/Two	Yes/NA	Total chloride solution	NA	100/200 /300 days	Half-cell potential; LPR (Current density I_{corr}); Crack mapping; Mass loss; Strain/stress test	CFRP significantly decrease corrosion rate.
Suh et al. [34]	Wrapping pre-stressed piles	One to Four	NA	Wet-dry cycle; marine environment	NA	1200 days	Half-cell potential; LPR (Corrosion rate); Mass loss; Bond test	CFRP with epoxy against cracks; slow down corrosion rate.
Soudki et al. [35]	Strips/U shape on beams	One/Four	NA	Wet- dry cycle; 3% NaCl	NA	600 days	Half-cell potential; LPR (Corrosion resistance & Current density I_{corr}); Mass loss; Failure test	CFRP enhance loading capacity by 11 to 28 % and reduce corrosion rate.

2.2 Corrosion affected by different various external factors in RC with CFRP

Different parameters can influence the corrosion process for RC structures with CFRP applied, such as epoxy types, CFRP wrapping layers, exposed environment and sustained loadings. Wootton et al. [6] investigated the corrosion performance of steel bars embedded in concrete cylinders with CFRP external wrapping. An impressed current was applied to accelerate the corrosion process. Measurement of corrosion potentials, impressed current flow levels and mass loss were performed. The results indicated that CFRP could decrease steel mass loss, reduce chloride content of concrete and slow down cracking and spalling. However, the results were different dependant on the various parameters of wrapping fibres orientation and wrap layer numbers. The epoxy type used to bond the CFRP also significantly affects the corrosion resistance in the RC.

Debaiky et al. [7] discussed the corrosion activities at initiation and propagation stages when using CFRP as a wrapped repair material corroded RC columns. CFRP wrapping patterns of one layer or two layers, fully or partially were compared during the experiment. Results were analysed by using the methods of half-cell potential, LPR, crack mapping and the weight loss of steels. The experimental results indicated that CFRP wrapping delayed corrosion occurring before corrosion propagation, and significantly decreased the corrosion rate especially when RC columns were fully covered.

Suh et al. [34] investigated the effectiveness of repairing techniques for corrosion control using CFRP over three years' of experiments under marine environment wet-dry cycle conditions. Two types of prewrapped substrate were prepared: full repair types where the delaminated concrete was removed and reformed, and another type was sealed with epoxy where cracks were performed and then wrapped by CFRP. The experimental results concluded that CFRP with epoxy sealing cracks was effective and the corrosion rate was dramatically decreased even under severe corrosion damage conditions.

The behaviours of CFRP strengthened RC beams in a corrosive environment were examined by Soudki et al. [35]. Eight beams were pre - cracked and then repaired using CFRP and subject to a wet-dry cycle environment. Three uncracked beams were prepared as the reference specimens at room temperature. Two types of fabrics and strips were compared. A failure test and an electrochemical corrosion rate test were practiced. The results indicated that CFRP and resin system could enhance the loading capacity by 11% to 28%, reduce the corrosion rate, and decrease the chloride diffusion. The stiffness and yield load of the RC beams were not significantly impacted when exposed to this harsh environment.

In conclusion, although CFRP could significantly reduce steel corrosion rate and improve corrosion resistance in RC structures, the various external factors in RC with CFRP could also play an important role in affecting corrosion, such as epoxy types, CFRP wrapping layers, exposed environment and sustained loadings. The increasing wrapping layers can increase corrosion resistance from one to two layers, however, three wrapping layers will not significantly effect on corrosion resistance when compared to two layers [6]. The marine environment and associated wet-dry cycle conditions were well studied in the above literature. Corrosion measurement via Half-cell potential, LPR and mass loss were widely used to assess the corrosion process conditions. However, a larger number of specimens are expected to further simulate structures conditions so as to identify the detailed influence results. As the corrosion process is an electrochemical reaction process, it is reasonable to detect the corrosion process of reinforcement in concrete using electrochemical methods.

2.3 Corrosion measurements in RC with external CFRP adhered

In order to determine corrosion rates and patterns in RC structures, non-destructive methods have been necessarily applied in assessing the corrosion conditions without destroying the current service structures. There are no direct methods available for corrosion rate measurement during the whole progress, therefore, the parameters that indirectly measure the corrosion process have to be studied [30]. Electrochemical monitoring methods have advantages of non-destructive and fast testing of RC structures [36]. Therefore, the corrosion process of reinforcement in concrete is generally detected using electrochemical methods as the corrosion process is an electrochemical reaction. Existing monitoring methods for steel corrosion in RC structures have been reviewed in Table 1, which mainly includes the half-cell potential measurement; concrete resistivity measurement; linear polarised resistance (LPR); mass loss, and bond strength test method. Among of these methods, half-cell potential, LPR and mass loss measurements are the predominantly used test methods in these studies. Half – cell potential method is broadly used in RC structures and is standardized by ASTM C876-15 (1999) [37, 38]. LPR method has been successfully evaluated for steel corrosion testing and monitoring in highly – resistive concrete structures [39, 40].

3. CFRP AS ANODE IN IMPRESSED CURRENT CATHODIC PROTECTION SYSTEMS

Some researchers have studied CFRP as the anode in impressed current cathodic protection (ICCP) system in recent times. The dual function of CFRP for both strengthening and an ICCP anode due to the CFRP being electrical conductive [24, 41].

3.1 CFRP sheet applications for ICCP system on RC cylinders

Research has mainly focused on the use of CFRP as an ICCP on simple cylinders with one steel bar as the cathode or a single piece of CFRP in simulated corrosion solutions. Zhu et al. [42] studied the mechanical and electrochemical performance of CFRP in oxygen evolution environment by conducting accelerated polarization. Different amounts of current density were applied under various test durations. The driving voltage, potential and tensile strength testsfair were carried out. The experimental results concluded that CFRP could be used as both an anode in ICCP system and the passive material for strengthening RC structures.

Gadve et al. [41] studied the corrosion protection provided to steel bars embedded in concrete using CFRP adhesively wrapped outside the surface of cylinders. CFRP was used as the conductive anode without using any external anode, while the steel bars were the cathodes in the pre-corroded specimens. External DC power supply was applied to impress the constant current of 50 mA (current density 870 mA/cm^2) between CFRP and the steel bar. The specimens were exposed to the 3.5% NaCl solution at 50°C temperature environment. The specimens were tested for pull-out strength, mass loss, potentiodynamic scans and the half-cell potential of steel. The experimental results indicated that CFRP could be active in protecting steel bars from corrosion for the CP function.

Zhu et al. [43] studied the deterioration of RC with CFRP anode after polarization. The steel bars were accelerated corrosion at first, and then the polarization test was performed between the anode of CFRP and the cathode of steel bar. Conductive carbon paste and carbon fibre reinforced mortar were used to connect between CFRP and concrete. The current density of 1244 mA/m^2 and 2488 mA/m^2 was applied for 25 days. Scanning electron microscopy (SEM) and X-ray diffraction (XRD) techniques were used to study the deterioration mechanism on the CFRP interface with concrete. The feeding voltage was applied during the polarization period. The experimental results illustrated the deterioration of the interface between the CFRP and the contact material may occur due to the chemical changes.

Electrical and mechanical behaviour testing of CFRP were carried out by Zhu et al. [13] in a different solution environments. Behaviour of CFRP as an anode in the ICCP system was investigated in RC structures. The results indicated that the corrosion potential was acceptable according to the "instant off" potential tests and met requirements [44]. The mechanical and electrical performance of CFRP was not adversely affected with a current density up to 2 A/m^2.

CFRP has been also applied as a novel external anode to study the electrochemical chloride extraction in RC specimens by Zhu et al. [45]. Three different specimens with a water-cement (w/c) ratio of 0.32, 0.4, and 0.5 and three different constant current densities of 1, 2, and 3 A/m^2 were applied. The results concluded that the bond loss decreased when the current density increased and with lower of w/c ratio's. There was no significant degradation of CFRP anode during the treatment system.

The literature above mainly studies the CFRP as anode using very simple cylinders with one steel bar. The possibility of CFRP used as the active anode in ICCP systems was investigated. Different impressed current densities from the power source and different concrete properties have been investigated. Up to 2488mA/m^2 could be applied in the ICCP system. However, real RC structures should be further studied with CFRP stirrups connected with the longitudinal steel bars for ICCP system.

3.2 CFRP applications on RC in ICCP system

Traditional CP and CFRP anode for ICCP have been compared by Lambert et al. [24]. The dual function of CFRP for both strengthening and acting as the ICCP anode in RC structures have been discussed. U-shaped wrapping of CFRP was applied and strips on the tensile side of RC beams were extended. Different current densities impressed for ICCP and potential decays were measured. Meanwhile, load-deflection was tested and the relationship of load-deflection was developed. The experimental results illustrated that CFRP could increase the stiffness and reduce the ultimate deflection of beams. U-shape wrapping of CFRP was 13.5% larger than the without U-shape wrapping. The current density could be applied up to 128 mA/m^2 for ICCP, and CFRP was suitable for the dual function of strengthening and ICCP in RC structures.

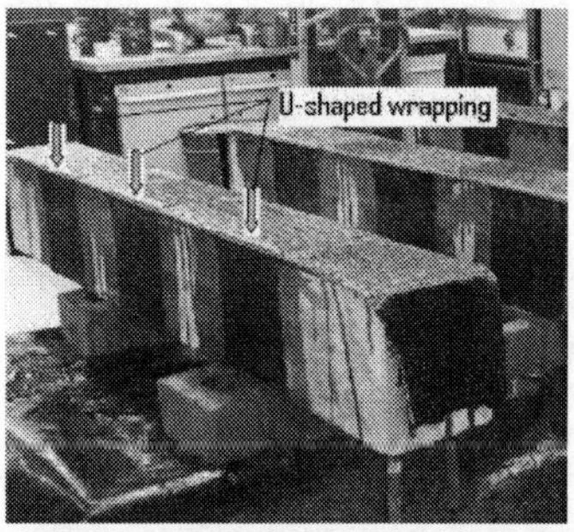

Figure 3 Schematic ICCP application to corroded reinforced concrete beams [24]

Van Nguyen et al. [46] investigated the performance of CFRP as ICCP anode in RC beams for up to 1100 hours. CFRP anode has been assessed in both concrete and calcium hydroxide solution. A maximum current density of 128 mA/m^2 of steel area was recommended in this experiment. The current was checked and the "on" and "instant-off" potential values of steel were recorded daily. The results concluded that CFRP can be employed effectively as the anode for ICCP system. However, CFRP was extended out of the edge of both RC beams, which is not suitable in real structures.

Gadve et al. [47] also investigated both the passive and active protection of steel corrosion using CFRP adhered to the concrete surface under highly corrosive environment. About 2 to 20% of conductive particulates by weight of epoxy were mixed into the epoxy for uniform conductivity of CFRP. An external DC power supplier was applied to impress the constant current of 150 mA for 1500 hours. Four parameters of flexure strength, mass loss, half-cell potential and cell voltages were studied. The corrosion potential measurement was obtained from the potentiodynamic scan on cylindrical specimens daily. The experimental results indicated the effective performance of CFRP as the strengthening material and ICCP anode material.

In summary, CFRP in ICCP system could act as anodes adhered on RC and no other external anode is needed, which is cost saving opportunity. However, graphite particles were widely used in the conductive epoxy mixture, which is only suitable for the new CFRP application on RC. Normal structures with CFRP for strengthening used the non-conductive epoxy resin for bonding. Therefore, new connection technique could be a challenge for removing the current cured epoxy with CFRP, and then be connected with an electrical wire for impressing current supply.

4. SUMMARY

CFRP could significantly decrease the corrosion rate of RC structures. Current investigations mainly focused on steel corrosion under marine or wet-dry cycle environment without sustained loading. Limited literature could be found in steel corrosion investigation of RC structures external bonded by CFRP under sustained loading. Further working on steel corrosion affected by CFRP simulated in the actual RC structures are expected to be deeply and comprehensively studied.

Active protection using CFRP as an anode has been started to research for the dual function of both strengthening and ICCP anode system. The dual function of CFRP application could provide cost savings and easier maintenance. However, current investigations mainly focus on experiments using simple specimens, which are questionable. New connection techniques are a challenge for removing the current dry epoxy in CFRP, in which CFRP conductively connected with an electrical wire for impressing current supply.

A number of techniques have been applied to allow for steel corrosion measurement. Non-destructive methods are recommended in order to assess the corrosion activity without destroying existing service structures. As there is no single method can be analysed in the whole corrosion process, several methods could be applied in RC structures with CFRP. Corrosion measurements of half-cell potential, LPR, and mass loss were the major used techniques in RC with CFRP.

5. REFERENCES

1. Mehta, K.P., *Reducing the environmental impact of concrete.* Concrete international, 2001. **23**(10): p. 61-66.
2. Michel, A., et al., *Experimental investigation of the relation between damage at the concrete-steel interface and initiation of reinforcement corrosion in plain and fibre reinforced concrete.* Corrosion Science, 2013. **77**: p. 308-321.
3. El Maaddawy, T., A. Chahrour, and K. Soudki, *Effect of fiber-reinforced polymer wraps on corrosion activity and concrete cracking in chloride-contaminated concrete cylinders.* Journal of Composites for Construction, 2006. **10**(2): p. 139-147.
4. Hariche, L., et al., *Effects of reinforcement configuration and sustained load on the behaviour of reinforced concrete beams affected by reinforcing steel corrosion.* Cement and Concrete Composites, 2012. **34**(10): p. 1202-1209.
5. Andrade, C. and C. Alonso, *On-site measurements of corrosion rate of reinforcements.* Construction and building materials, 2001. **15**(2): p. 141-145.
6. Wootton, I.A., L.K. Spainhour, and N. Yazdani, *Corrosion of steel reinforcement in carbon fiber-reinforced polymer wrapped concrete cylinders.* Journal of Composites for Construction, 2003. **7**(4): p. 339-347.
7. Debaiky, A.S., M.F. Green, and B.B. Hope, *Carbon fiber-reinforced polymer wraps for corrosion control and rehabilitation of reinforced concrete columns.* Materials Journal, 2002. **99**(2): p. 129-137.
8. Tang, F., et al., *Corrosion resistance and mechanism of steel rebar coated with three types of enamel.* Corrosion Science, 2012. **59**: p. 157-168.
9. Pandiarajan, M., et al., *Corrosion inhibitor for concrete corrosion–An overview.* Eur. Chem. Bull, 2013. **2**(1): p. 1-8.
10. Nossoni, G., *Modeling the corrosion rate of steel reinforcement in FRP-wrapped concrete.* Journal of Composites for Construction, 2015. **20**(3): p. 04015068.
11. Robert, M. and B. Benmokrane, *Behavior of GFRP reinforcing bars subjected to extreme temperatures.* Journal of Composites for Construction, 2009. **14**(4): p. 353-360.
12. Al-Salloum, Y., *Experimental and analytical investigation of compressive strength of FRP-confined concrete.* Final Rep. King Saud University. Kingdom of Saudi Arabia, 2007.
13. Zhu, J., et al., *Behavior of CFRP plate in simulated ICCP system of concrete structures.* 2014.
14. Heffernan, P. and M. Erki, *Fatigue behavior of reinforced concrete beams strengthened with carbon fiber reinforced plastic laminates.* Journal of composites for construction, 2004. **8**(2): p. 132-140.
15. Bonfiglioli, B., G. Pascale, and S. Martinez de Mingo, *Dynamic testing of reinforced concrete beams damaged and repaired with fiber reinforced polymer sheets.* Journal of materials in civil engineering, 2004. **16**(5): p. 400-406.
16. Ye, L., et al., *Shear strength of reinforced concrete columns strengthened with carbon-fiber-reinforced plastic sheet.* Journal of Structural Engineering, 2002. **128**(12): p. 1527-1534.
17. White, T.W., K.A. Soudki, and M.-A. Erki, *Response of RC beams strengthened with CFRP laminates and subjected to a high rate of loading.* Journal of Composites for Construction, 2001. **5**(3): p. 153-162.
18. Ye, L., et al., *Experimental study on seismic strengthening of RC columns with wrapped CFRP sheets.* Construction and Building Materials, 2003. **17**(6): p. 499-506.

19. da Fonseca, B.S., et al., *Influence of GFRP Confinement of Reinforced Concrete Columns on the Corrosion of Reinforcing Steel in a Salt Water Environment.* Journal of Materials in Civil Engineering, 2014. **27**(1): p. 04014107.
20. Khoe, C., R. Sen, and V.R. Bhethanabotla, *Oxygen permeability of fiber-reinforced polymers.* Journal of Composites for Construction, 2010. **15**(4): p. 513-521.
21. Khoe, C., R. Sen, and V.R. Bhethanabotla, *Oxygen permeability of FRP-concrete repair systems.* Journal of Composites for Construction, 2011. **16**(3): p. 277-285.
22. Gholami, M., et al., *A review on steel/CFRP strengthening systems focusing environmental performance.* Construction and Building Materials, 2013. **47**: p. 301-310.
23. Pedeferri, P., *Cathodic protection and cathodic prevention.* Construction and Building Materials, 1996. **10**(5): p. 391-402.
24. Lambert, P., et al., *Dual function carbon fibre fabric strengthening and impressed current cathodic protection (ICCP) anode for reinforced concrete structures.* Materials and structures, 2015. **48**(7): p. 2157-2167.
25. Badawi, M. and K. Soudki, *Control of corrosion-induced damage in reinforced concrete beams using carbon fiber-reinforced polymer laminates.* Journal of composites for construction, 2005. **9**(2): p. 195-201.
26. Soudki, K.A. and T.G. Sherwood, *Behaviour of reinforced concrete beams strengthened with carbon fibre reinforced polymer laminates subjected to corrosion damage.* Canadian journal of civil engineering, 2000. **27**(5): p. 1005-1010.
27. Pantazopoulou, S., et al., *Repair of corrosion-damaged columns with FRP wraps.* Journal of composites for construction, 2001. **5**(1): p. 3-11.
28. Subhani, M., R. Al-Ameri, and A. Al-Tamimi, *Assessment of bond strength in CFRP retrofitted beams under marine environment.* Composite Structures, 2016. **140**: p. 463-472.
29. El-Hacha, R., M.F. Green, and G.R. Wight, *Effect of severe environmental exposures on CFRP wrapped concrete columns.* Journal of composites for construction, 2010. **14**(1): p. 83-93.
30. Gadve, S., A. Mukherjee, and S. Malhotra, *Corrosion of steel reinforcements embedded in FRP wrapped concrete.* Construction and Building Materials, 2009. **23**(1): p. 153-161.
31. Zhuang, N., et al., *Corrosion Activity on CFRP-Strengthened RC Piles of High-Pile Wharf in a Simulated Marine Environment.* Advances in Materials Science and Engineering, 2017. **2017**.
32. Masoud, S., K. Soudki, and T. Topper, *CFRP-strengthened and corroded RC beams under monotonic and fatigue loads.* Journal of composites for construction, 2001. **5**(4): p. 228-236.
33. Spainhour, L.K. and I. Thompson. *Effect of carbon fiber jackets on reinforced concrete columns exposed to a simulated tidal zone.* in *Second International Conference on Fiber Composites in Infrastructure*. 1998.
34. Suh, K., et al., *Effectiveness of fiber-reinforced polymer in reducing corrosion in marine environment.* ACI structural journal, 2007. **104**(1): p. 76.
35. Soudki, K., E. El-Salakawy, and B. Craig, *Behavior of CFRP strengthened reinforced concrete beams in corrosive environment.* Journal of composites for construction, 2007. **11**(3): p. 291-298.
36. Ribeiro, D. and J. Abrantes, *Application of electrochemical impedance spectroscopy (EIS) to monitor the corrosion of reinforced concrete: A new approach.* Construction and Building Materials, 2016. **111**: p. 98-104.
37. Standard, A., *C879: Standard Test Method for Corrosion Potentials of Uncoated Reinforcing Steel in Concrete.* ASTM International, West Conshohocken, 2009.
38. Ahmad, S., *Reinforcement corrosion in concrete structures, its monitoring and service life prediction—a review.* Cement and concrete composites, 2003. **25**(4): p. 459-471.
39. Gonzalez, J., et al., *Errors in the electrochemical evaluation of very small corrosion rates—I. polarization resistance method applied to corrosion of steel in concrete.* Corrosion Science, 1985. **25**(10): p. 917-930.
40. Al-Amoudi, O.S.B. and M. Maslehuddin, *The effect of chloride and sulfate ions on reinforcement corrosion.* Cement and Concrete Research, 1993. **23**(1): p. 139-146.
41. Gadve, S., A. Mukherjee, and S. Malhotra, *Active protection of fiber-reinforced polymer-wrapped reinforced concrete structures against corrosion.* Corrosion, 2011. **67**(2): p. 025002-1-025002-11.
42. Zhu, J.-H., et al., *Mechanical and Electrochemical Performance of Carbon Fiber Reinforced Polymer in Oxygen Evolution Environment.* Polymers, 2016. **8**(11): p. 393.
43. Zhu, J.-H., et al., *Polarization induced deterioration of reinforced concrete with CFRP anode.* Materials, 2015. **8**(7): p. 4316-4331.
44. Takewaka, K., *Cathodic protection for reinforced-concrete and prestressed-concrete structures.* Corrosion science, 1993. **35**(5-8): p. 16171625-16231626.
45. Zhu, J.-H., et al., *Application of carbon-fiber-reinforced polymer anode in electrochemical chloride extraction of steel-reinforced concrete.* Construction and Building Materials, 2016. **120**: p. 275-283.
46. Van Nguyen, C., et al., *The performance of carbon fibre composites as iccp anodes for reinforced concrete structures.* ISRN Corrosion, 2012. **2012**.
47. Gadve, S., A. Mukherjee, and S. Malhotra, *Corrosion protection of fiber-reinforced polymer-wrapped reinforced concrete.* ACI Materials Journal, 2010. **107**(4): p. 349-356.

6. AUTHOR DETAILS

Aifang Wei, also known as Amy, is a PhD student in the School of Civil Engineering at Deakin University. She is working on improving corrosion resistance using advanced material of Carbon Fibres in reinforced concrete structures. Her PhD project mainly focuses on electrochemical methods for steel corrosion testing, monitoring and protecting embedded in concrete using CFRP adhered and epoxy treated chopped carbon fibre reinforced concrete.

Dr Riyadh Al-Ameri is a senior lecturer in Structural Engineering at the School of Engineering, Deakin University since 2010. He received his MSc & PhD in Structural Engineering from Cardiff University (UK, 1990). He has more than 25 years of mixed industrial and academic experience. Before joining Deakin, took a senior role with the construction industry in design and consultation space. Visiting Academic and Honorary Associate at Monash University (2005). Received awards for Excellence in Industry Collaborations, Endeavour Research Fellowship, and Teaching Excellence. Published more than sixty papers. Keynote speaker and member of the International Scientific committee of several international conferences and current reviewer for many international journals in his field. Al-Ameri is a Fellow of the Institution of Engineers Australia, chartered Structural Engineer, and on the National Engineering register in Civil and Structural colleges. Member of the National Executive Committee of the Australian Association of Steel Concrete Composite Structures (AASCCS), the International Association for Bridge Maintenance and Safety (IABMS), the International Institute for FRP in Construction (IIFC), and the American Society of Civil Engineers (ASCE).

Mike Yongjun Tan is a Professor in Applied Electrochemistry and Corrosion Technologies at Deakin University in Australia. He is also a Research Program Leader of the Energy Pipelines Cooperative Research Centre. Dr Tan's principal teaching and research interests are in corrosion science and engineering and their applications for enhancing the reliability and durability of civil and industrial infrastructures. He contributed to electrochemical methods for corrosion testing, monitoring and prediction and corrosion inhibitor and anti-corrosion coating research. He is the author of some 150 publications and a book entitled 'Heterogeneous Electrode Processes and Localised Corrosion' (2012 John Wiley & Sons).

Dr. Yew-Chin Koay is a Team Leader Structural Technology and Assets at VicRoads. He completed his PhD degree from RMIT University and has almost 20 years experience as civil, structural and infrastructure engineer; postdoctoral fellow and lecturer in Australia, China, Malaysia and South Korea. Dr. Koay received several prestigious fellowships such as Endeavour Research Fellowship, Korea Sceince and Engineering Foundation, VicRoads Kerry Burke Memorial Scholarship, etc. He is a founding member of Young Researcher and Graduate Symposium (YRGS), an advisory board member of Australian Network of Structural Health Monitoring (ANSHM) and committee member of Standards Australia. He was a Chair Scientific Committee of Austraods Bridge Conference (ABC) 2017.

APPROACHES TO OVERCOMING ONGOING PIPELINE CORROSION MONITORING CHALLENGES

K. Wang[1], F. Varela[1], and M. YJ. Tan[1,2]
[1]Institute for Frontier Materials, [2]School of Engineering
Deakin University, Waurn Ponds, Australia

SUMMARY: Steel pipelines are susceptible to corrosion when installed underground, especially when they are under the effect of factors such as stray currents or disbonded coatings. These factors generally increase the risk of corrosion because they could locally nullify cathodic protection (CP). Corrosion monitoring, if performed properly, can help enhance the safe and economical operation of pipelines, by providing a means of inexpensively collecting site-specific and in-situ corrosion information at any required time and frequency. The capability of corrosion monitoring to complement corrosion inspection techniques is well accepted, however, its practical application in complex industrial conditions, especially in heterogeneous systems such as underground pipelines, has been limited. Unlike conventional inspection techniques, corrosion monitoring measures corrosion happening on the pipeline surface indirectly. It measures what is occurring on the surfaces of the corrosion probes, based on the principle that corrosion occurring on a large structure such as a pipeline would also occur on a smaller probe made of the same material and exposed to the same environmental condition. Nevertheless, ensuring that the key elements of the corrosion processes affecting a large and complex structure are simulated by strategically designed and located discrete probes is currently an unsurpassed challenge. A rational strategy must be developed to ensure the reliability of the collected information using corrosion probes. In this paper, the application of corrosion monitoring in heterogeneous conditions and several factors affecting the design and location of corrosion monitoring probes for early warning of external corrosion on buried pipelines have been discussed. It is concluded that in order to ensure the reliability and accuracy of corrosion monitoring as an early warning tool, three general rules should be observed in order to ensure that the probes are capable of simulating the worst possible corrosion situation along the monitored structure, and collecting information regarding the maximum corrosion rate on their own surface.

Keywords: Pipeline, Cathodic protection, Corrosion monitoring, Stray current, Disbonded coating

1. INTRODUCTION

Steel pipelines are extensively applied as safe and economical means to transport gas, oil and other fluids to locations where they are consumed or where they are refined. In 2013, there were 34,612 Km of steel pipelines in Australia, which are an essential piece of infrastructure for the national energy industry. Due to the low corrosion resistance of pipeline steel, the external surface of an underground pipeline is protected by organic coatings and cathodic protection (CP)[1-3]. Insulating polymer coatings are the first line of defence against external corrosion. However, some coating defects would inevitably exist exposing the pipe's bare metal surface directly to the environment[4]. When the coatings fail, CP works as a backup to prevent corrosion by polarizing the metal to a CP potential accepted as safe[5]. However, both protection systems could fail, especially when disbonded coatings and stray current are present, resulting in corrosion of the pipeline. Recent corrosion failures, such as the explosions in Qingdao and Kaohsiung[6], have shown the importance of making efforts to ensure the structural integrity of this infrastructure assets, especially in high consequence areas. Inspection and structural health monitoring[7] are practical approaches to asset corrosion management, which is particularly critical when the operational life of aged pipelines is extended beyond their initial design life.

Corrosion inspection is commonly applied for pipeline corrosion management. A large number of inspection methodologies, including potential surveys[8], current attenuation[9], ultrasonic testing and in-line inspection (ILI) tools[10], have been used to detect and inspect defects which could threat the integrity of pipelines. Inspection is able to detect corrosion by measuring the accumulation of metal loss, which could be induced by corrosion reactions or physical stresses. However, they are often not sensitive enough to detect corrosion damage at its initial stage. In some extreme cases, for example, for pipelines buried deeply underground, under rocky terrain and paved roads, potential survey methods are even incapable to detect defects regardless of the size of defects[11]. On the other hand, conventional ILI tools are not employable for all pipelines since they require launchers, receivers and a relatively obstacle free path. However, even for the cases which are suitable for ILI, the high costs involved in using these tools limit the frequency of inspection (usually between 5 and 15 years). As a consequence, inspection alone is insufficient to keep track of dynamically environmental conditions affecting the initiation and propagation of pipeline corrosion.

Corrosion monitoring technology could be used to complement current inspection techniques. Corrosion monitoring aims to provide early warning of corrosion that could cause failure of a pipeline system, so that effective corrosion mitigation procedures could be evaluated and implemented based on the obtained parameters, such as corrosion rate, corrosion pattern and corrosion mechanism information[12-14]. Corrosion monitoring is generally based on the application of specially designed probes[10], which are commonly made by sensing elements of the same material as the monitored infrastructure. The benefits of corrosion monitoring as an early warning system, however, are only achievable if corrosion probes are able to monitor corrosion reliably and accurately. The meaning of reliability of corrosion monitoring could be defined as that corrosion probes should be capable of simulating the possibly worst-case scenario corrosion situation along the pipeline. The meaning of accuracy of corrosion monitoring could be defined as that the information collected by the probe, including maximum corrosion rate, should reflect the corrosion processes occurring on the probe surface.

A common issue in current industry application of corrosion probes is that intuitively, in some cases, a corrosion probe is used with an expectation of monitoring corrosion in a similar manner as using a thermometer to measure temperature, which assumes a homogeneous and continuous environment. Inevitably, these practices could lead to confusing and misleading results, because localised corrosion is typically predominant in such environments. Therefore, the main challenge of corrosion monitoring comes from localised corrosion.

2. WORKING PRINCIPLE OF MONITORING PROBES

In order to effectively monitor localised corrosion, probe measurement techniques should be sensitive to measure parameters related to the controlling corrosion processes/mechanism. The premise for achieving this is to understand the working principles and limitations of different techniques for corrosion probe measurement. Over the past decades, many corrosion probes and associated measurement techniques have been reported in the historic literature for laboratory and field corrosion testing and monitoring applications. They can be generally divided into two distinct groups: non-electrochemical and electrochemical probes. Non-electrochemical probes are based on the measurement of physical parameters for corrosion assessment. They monitor various physical parameters such as strain[15], resistance[16] or reluctance[17] to reflect the structural degradation, which is induced by accumulated metal loss. However, non-electrochemical monitoring methods are not sensitive to the initial stages of corrosion, especially localised corrosion. In addition, non-electrochemical methods provide little insight on the performance of the CP system and mechanisms leading to corrosion.

Electrochemical probes, which measure electrochemical parameters for corrosion assessment, are designed based on the electrochemical nature of corrosion. The electrochemical parameters that are fundamentally related to the thermodynamics and kinetics of corrosion reactions could be utilised to indicate the intrinsic characteristics of corrosion processes. Corrosion potential is the most common electrochemical parameter monitored from buried pipelines, however, the accuracy of the value of corrosion potential is often unsatisfactory due to measurement errors introduced by the IR-drop[18]. The polarization resistance technique is not suitable for underground infrastructures because of the high resistivity of the media and the heterogeneous nature of the soil[19]. On the other hand, the conventional one single electrode based electrochemical corrosion probes are not capable of reflecting the localised corrosion distribution[20]. Multi-electrode array, also known as Wire Beam Electrode (WBE), has been successfully employed in mapping the localised corrosion distribution, for example the localised corrosion of carbon steel in aqueous solution[21], crevice[22], soil[23], water/gas interface[24] and even under coatings[25].

3. CORROSION MECHANISMS AFFECTING PROBE DESIGN

Corrosion probes should also be capable of simulating the controlling corrosion mechanisms to monitor localised corrosion in order to provide early warning. What is more, the corrosion probe should be designed to reflect the possibly worst corrosion situation under a certain corrosion mechanism in any site-specific location. Corrosion under coating disdondment and stray current corrosion are typical worst-cases of corrosion on buried pipelines. The following sections discuss the mechanisms of corrosion under disdondment and stray current corrosion and their impact on the probe's design.

3.1 Corrosion under disbonded coatings and probe design requirements

When a CP potential is applied on a coated pipeline, the adhesion between the coatings and pipeline could degrade from where defects (pinholes, holidays and ruptures) exist, which is known as cathodic disbondment[26,27]. Coating disbondment allows the formation of a crevice between pipeline surface and the coating. Crevice corrosion is difficult to be mitigated by

CP because of the inability of CP current to reach areas under disbondment (cathodic shielding)[28]. The CP current has to flow through the holiday to the crevice bottom, and the ohmic potential drop across the solution trapped under disbondment reduces the applied CP potential to less negative values[26]. The potential at the tip of the crevice is less negative than that of the opening, suggesting the crevice tip could not be properly protected[29]. Many types of corrosion, such as pitting[30], microbiologically induced corrosion (MIC)[31] and stress corrosion cracking (SCC)[32] have been observed under disbonded coatings. The different types of corrosion processes occurring under disbonded coatings are usually investigated in aqueous solution. However, in soil, the actual corrosion process occurring under disbonded coatings can be different and more complex, because the trapped solution under disbondment could be absorbed by the unsaturated soil particles, leaving an air gap of zero current under CP. It has been noted that this air gap can isolate sections within the crevice from the area outside the crevice where CP is effective[33]. Within these isolated areas, the system is at open circuit potential and a combination of cathodes and anodes is developed. After a transitory period, the cathodes tend to locate close to the air gap, where oxygen is constantly replenished and the anode tends to be stable at the crevice tip.

When disbonded coating is responsible for the ineffectiveness of CP, a corrosion monitoring probe should be designed to simulate the crevice corrosion under disbonded coatings. Varela *et al.*[34] firstly developed an electrochemical probe based on WBE to reproduce corrosion under disbondment. The probe surface is covered with PMMA cover to simulate disbonded coatings. The results show that the probe is capable of evaluating the actual polarization level achieved and cathodic reaction locations both outside and inside the crevice. Moreover, coating type and crevice dimension, which could influence the corrosion under disbondment areas, contribute to probe design. The first factor is that the type of coating covering on the probe surface to simulate the disbonded coating. Corrosion under disbondment was usually investigated by simulated crevice setup, which utilised polymethyl methacrylate (PMMA)[35] or Lucite sheet[36] instead of real coating to simulate 'ideally shielding coating' impermeable to CP current and chemical species. However, thickness[37,38] and aging of real coatings[39] could influence the oxygen permeation, pH retention and CP current permeation. Therefore, the effect of real coatings on the localised corrosion behaviour under disbondment coatings should be investigated. The second factor is the dimension (e.g. gap size and crevice length) of the crevice formed on the probe surface. Larger gap size would lead to an enhanced penetration of the cathodic currents. Consequently, even the deeper region of the crevice could be polarised[40]. However, it is found that the larger gap size would allow the dilution of the pH generated inside the crevice due to more effective ions transportation[35]. Furthermore, the available data of crevice geometry effect is carried out in aqueous test, however, when the test is conducted in the soil, more soil would be trapped under the disbonded coating with a larger gap size, which reduces the solution volume trapped in the crevice and blocks oxygen diffusion and CP current. In addition, in the unsaturated soil, the continuity of the air gap formed in the crevice would be disrupted when the air gap is large due to sand distributing on the crevice surface[28]. Therefore, it is a pending issue to understand the effect of gap size on the localised corrosion behaviour of the probe in the soil. These issues suggest the challenges of designing a corrosion probe for simulating and monitoring corrosion under disbonded coatings on pipelines.

3.2 Stray current corrosion and probe design requirements

Another typical worst-cases of corrosion on buried pipelines is caused by stray currents that are undesirable currents deviated from the intended path[41]. Due to metal's low electrical resistance properties, metallic infrastructures, such as pipelines, tend to be the alternative path of such stray currents. Depending on their sources, there are three types: direct current (DC) stray current, alternating current (AC) stray current and telluric current[11]. In recent decades, electric transit systems have been developing rapidly and serving as the biggest and the best known source of DC stay current. Even if the running rails are isolated from ground, current could leak to the ground, depending on the rail-to-earth conductance term for each running rail[41,42]. Stray current corrosion is location-dependent. At the position at which stray current enters the pipe the steel is catholically polarised, while at the position where stray current exits from the pipe the potential is shifted anodically[43]. At the stray current pick up point, in some cases, hydrogen embrittlement and accelerated coating damage could be induced as a result of overprotection, whereas at the discharge point insufficient protection could accelerate the corrosion of pipeline[10]. In addition, the attack is extremely localised and can have dramatic consequence. For these reasons, the location of corrosion monitoring device is important for detecting stray current corrosion.

In terms of corrosion caused by stray current, corrosion failure prefers to occur at uncoated surfaces, therefore a corrosion probe for simulating and monitoring stray current corrosion can be an uncoated metal surface. Huo *et al.*[44] has applied WBE probe to monitor the evolution of dynamic anodic transient effect on the corrosion occurring on the buried pipeline surface[44]. It has been shown that WBE probe is able to measure local anodic currents and visualise current distributions over the probe surface under the effects of CP and dynamic anodic transients. However, the optimum probe dimension needs to be clearly understood. Historically, efforts have been made to evaluate the effect of specimen size on pitting corrosion exposed to aggressive species. Aziz et al.[45] found that the pitting probability of 2S aluminium specimen in tap water increased with the increase of exposed area, until remaining constant as the value of specimen area reaching 60cm^2. In addition, the authors pointed out that the pitting probability became zero at some small value of the exposed area. Burstein et al.[46] found that the decrease of 316 stainless steel specimen size increased the pitting potential measured in acidic chloride solutions using potentiodynamic technique. Likewise, the results of Li[47] presented that the average pitting potential decreased significantly with increasing specimen size. In addition, the pitting probability was quantitatively correlated with specimen surface area ratio. Furthermore, the maximum pitting depth, which is associated with pitting damage estimation, is a function of exposed specimen area[48]. However, localised corrosion induced by stray current would be different. Stray current is generally dynamic (fluctuation with time). When stray current is absent, different size defects have different current under CP, leading to different concentration of hydroxyl generated on the defect surface[49]. When metal surface is under attack of discharged stray current, different size of defects would experience different initial states. Especially, smaller defect tends to have more positive polarization potential[50]. Thus, it requires more understanding how will the defect dimension influence the localised corrosion distribution induced by stray current.

Assuming that there is an 'ideal probe', which could simulate the controlling corrosion mechanism of pipeline defects and provide precise measurements of parameters regarding the controlling corrosion processes/mechanisms, the selection of the most adequate location for its installation is still not trivial. Therefore, the second challenge of corrosion monitoring results from probe installation.

4. FACTORS INFLUENCING PROBE INSTALLATION

In terms of practical applications, we should consider not only the reliability and accuracy of corrosion monitoring system but also its economic cost. If the quantity of monitoring points is above the optimum number, the costs of monitoring would exceed any savings generated[51], then the application of probes would be counterproductive. The installation of corrosion probes would be critical to obtain as much corrosion related information as possible from the optimized number of probes. Therefore, the corrosion probes should be installed at the 'worst-case scenario' locations among a certain length of pipeline, representing the possibly maximum corrosion rates in such a segment. The following sections present how to identify the possibly worst scenario concerning corrosion under disbonded coatings and corrosion induced by stray current.

4.1 Environmental factors influencing probe installation

Soil is typically a multiphase corrosion system, discontinuous and heterogeneous, consisting of gas phase, liquid water phase and organic solid phase, which changes the pH retention, oxygen concentration, electrolyte available could contribute to localised corrosion and affect corrosion monitoring[52]. The effect of soil moisture on corrosion under disbondment has been evaluated by Varela *et al*[33]. An air gap will form in the crevice when the soil is unsaturated or when the soil undergoing wet/dry cycle. However, more soil factors need to be investigated to understand the effect of different environmental factors, such as, soil particle[53-55] and carbon dioxide[56,57], on the corrosion behaviour under disbondment.

4.2 Corrosion mechanism affecting probe installation

The probe installation for stray current corrosion monitoring should be based on the analysis and understanding of corrosion mechanisms. Two simple and specific cases are discussed below to show where to install probes for stray current corrosion monitoring. In the first example, the stray current is caused by the impressed current from a CP system when the pipeline crosses a foreign protected pipeline (Fig.1a). If the probe is installed at location A and B, where the stray currents are picked up, more negative current could be possibly recorded. However, if the probe is installed at the intersection between two pipelines (i.e. location C), where the stray currents are discharged, the probe is capable of catching up the maximum corrosion rate. Therefore, location C in this case with possibly maximum corrosion rate is more reliable for probe installation. In the other example, the stray current is caused by the DC transit system when the pipeline is parallel or cross the tracks (Fig.1b). Likewise, the location near the DC substation, where stray current is discharged, tending to form the worst corrosion, is the best location for installing the probe to monitor the stray current corrosion reliably. In addition, at the stray current pick up point, in some cases, accelerated coating damage could be induced as a result of overprotection, therefore, the pickup points could be the possible installation locations for monitoring corrosion under disbondment. Currently, as mentioned before, except potential measurement, the results of ER probe[58] and galvanic probe[59] are not capable of indicating the occurring of pick-up and discharge. However, the potential measurement is subjected to the effect of IR drop. In addition, these different probes could not reflect the localised corrosion nature of stray current at the discharge point and could not provide the actual polarization at the pick-up point which is associated with cathodic disbondment. Although WBE probe has demonstrated potential capability for monitoring localised corrosion induced by simulated anodic transients, more experiments are required to understand the effect of monitoring location on stray current corrosion monitoring.

Furthermore, how to identify the installation location for dynamic and more complex stray current corrosion is still unclear. For example, if the discharge point is dynamically changing in locating, such as in the case of a tram with regenerative braking, there is no single location where the probe should be installed. Therefore, further research exploring the possibility to track the location of the discharge point using several probes should be carried out.

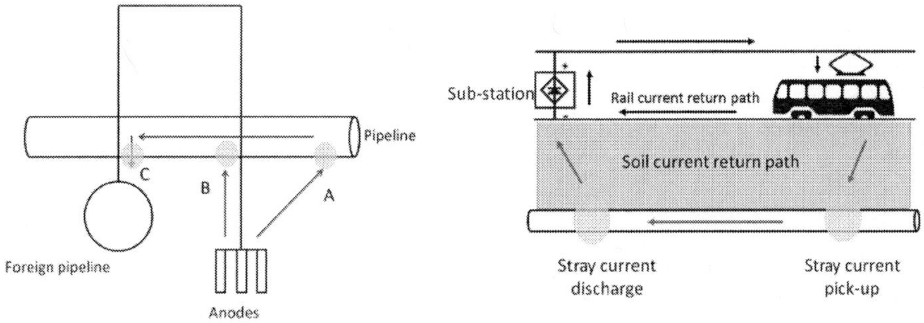

(a) (b)

Fig.1 Stray current caused by an impressed current CP system (a) and DC transit system (b)

4.3 Probe installation orientation issues

Interestingly, different orientation of the bare metal defects, for example upwards and downwards, could produce different water dispense patterns because of gravity, which could also bring about different corrosion patterns of defects. Therefore, it is necessary to study the effect of orientation on corrosion behaviour under the interference of stray current.

In terms of corrosion under disbondment, the possible defects on the pipeline surface could have different orientations, such as facing downwards, upwards, sideways or lying flat. The effect of probe orientation has also been investigated in saturated sandy soil at hydrogen evolution potential[33]. If they are overprotected (such as -1250mV$_{CSE}$), the accumulated gas would occupy different positions of the crevice, for example, if the opening crevice faces downwards the hydrogen bubbles would accumulate at the whole area of the crevice, while if the opening crevice faces sideways the hydrogen bubbles would accumulate at the side of the crevice. The accumulation of hydrogen would shield CP current, which generates low current density area under disbonded coatings. More corrosion is expected at these low current areas. Therefore, in above same environment of the saturated sandy soil at hydrogen evolution potential, the corrosion probes should be installed with faces downwards to simulate the worst corrosion situation. However, in unsaturated soil, different volume of solution pattern by virtue of gravity and absorption by soil particles would be generated on pipeline crevice surface under disbonded coatings. Therefore, it is necessary to study the effect of orientation on corrosion under disbondment.

5. FINAL REMARKS

In order to use corrosion monitoring probes as an effective early warning tool, corrosion probes should be representative of the worst possible corrosion situation along the monitored pipeline section. In order to achieve this purpose, three important aspects should be considered: probe design, measurement technique and installation. Based on the preceding literature review and analysis of industry case studies, three rules are proposed:

1. Probe design/selection: Corrosion probes should simulate the controlling corrosion mechanism affecting the structure and recreate the worst possible condition for that specific corrosion mechanism.

2. Probe measurement: The electrochemical technique on which the probe operates should be sensitive to measure parameters related to the controlling corrosion processes/mechanism.

3. Probe installation: Corrosion probes should be strategically installed at worst-case scenario locations.

If corrosion probes are designed and applied according to these 'rules', the results of the probes would be reliable and accurate enough to provide early warning of corrosion in buried pipelines.

Although this hypothesis might seem straightforward, the implementation of these basic concepts to a heterogeneous and dynamic system such as an energy pipeline is deceptively challenging to fulfil these requirements because of the insufficient understanding of the pipeline corrosion process. As discussed on the preceding discussions, in addition to the complex effect of environmental factors on corrosion, the effect of the probe's geometrical dimensions, orientation and location are not well understood, denying the possibility of accurate and reliable corrosion monitoring.

6. ACKNOWLEDGMENTS

This work was funded by the Energy Pipelines CRC, supported through the Australian Government's Cooperative Research Centres Program. The funding and in-kind support from the APGA RSC is gratefully acknowledged.

7. REFERENCES

1. Yan, M., *et al.*, *Corrosion Science* (2008) **50** (5), 1331
2. Yan, M., *et al.*, *Corrosion engineering, science and technology* (2007) **42** (1), 42
3. Song, F., and Sridhar, N., *Corrosion Science* (2008) **50** (1), 70
4. Papavinasam, S., and Doiron, A., *CORROSION 2009* (2009)
5. Vandelinder, L., *L. S. Vandelinder, Ed. 364 pages, 8. 5 x 11 in.(22 x 28 cm), hard cloth. NACE, Houston, Texas, 1984. Item* (1984) (51020)
6. Li, X., *et al.*, *Nature* (2015) **527** (7579), 441
7. Hodges, S., *et al.*, *NACE International, Corrosion 2010* (2010)
8. Matocha, G., *et al.*, Analysis of ILI, ACVG and DCVG Indications in Conjunction With CIS: A Case Study. In *2012 9th International Pipeline Conference*, American Society of Mechanical Engineers(2012), pp 9
9. Pope, R., *Corrosion* (1946) **2** (6), 307
10. Varela, F., *et al.*, *Corrosion Engineering, Science and Technology* (2015) **50** (3), 226
11. Papavinasam, S., *Corrosion control in the oil and gas industry*. Elsevier: 2013
12. Huang, Y., and Ji, D., *Sensors and Actuators B: Chemical* (2008) **135** (1), 375
13. Agarwala, V. S., *et al.*, Corrosion detection and monitoring-A review. In *CORROSION 2000*, NACE International(2000)
14. Song, H.-W., and Saraswathy, V., *Int. J. Electrochem. Sci* (2007) **2**, 1
15. Ren, L., *et al.*, *Optical Fiber Technology* (2014) **20** (1), 15
16. Legat, A., *Electrochimica Acta* (2007) **52** (27), 7590
17. Song, S., *et al.*, *JOURNAL OF CHEMICAL INDUSTRY AND ENGINEERING-CHINA-* (2001) **52** (7), 622
18. Żakowski, K., and Sokólski, W., *Corrosion Science* (1999) **41** (11), 2213
19. Callow, L., *et al.*, *British Corrosion Journal* (1976) **11** (3), 132
20. Tan, Y. M., and Revie, R. W., *Heterogeneous electrode processes and localized corrosion*. John Wiley & Sons: 2012
21. Tan, Y., *Journal of the Electrochemical Society* (2009) **156** (6), C195
22. Tan, Y.-J., *et al.*, *Corrosion science* (2001) **43** (10), 1919
23. Naing Aung, N., and Tan, Y.-J., *Corrosion Science* (2004) **46** (12), 3057
24. Tan, Y.-J., *et al.*, *Corrosion science* (2001) **43** (10), 1931
25. Wang, T., and Tan, Y.-J., *Corrosion Science* (2006) **48** (8), 2274
26. F. Gan, Z.-W. S., G. Sabde, and D.-T. Chin, *Corrosion* (1994) **50** (10), 13
27. Sabde, D.-T. C. a. G. M., *Corrosion* (1999) **55** (3), 9
28. Wang, W., *et al.*, *Journal of Loss Prevention in the Process Industries* (2014) **29**, 163
29. R. N. PARKINS, A. J. M., J. H. HOLBROOK, and R. R. FESSLER, *corrosion* (1985) **41** (7), 9
30. Chen, X., *et al.*, *Journal of Applied Electrochemistry* (2008) **39** (5), 697
31. Xu, J., *et al.*, *Materials Chemistry and Physics* (2013) **142** (2), 692
32. Maocheng, Y., *et al.*, *Corrosion Science* (2016)
33. Varela, F. N., Electrochemical methods for monitoring localized corrosion under cathodic protection; Deakin University2015
34. Varela, F., *et al.*, *Journal of The Electrochemical Society* (2015) **162** (10), C515
35. Varela, F., *et al.*, *Electrochimica Acta* (2015) **186**, 377
36. Chen, X., *et al.*, *Corrosion Science* (2015) **101**, 1
37. Kuang, D., and Cheng, Y. F., *Corrosion Science* (2015) **99**, 249
38. Fu, A., and Cheng, Y., *Progress in Organic Coatings* (2011) **72** (3), 423
39. Pei, Y.-m., *et al.*, *Polymer degradation and Stability* (2011) **96** (7), 1179
40. Kuang, D., and Cheng, Y., *MATERIALS PERFORMANCE* (2015) **54** (5), 40
41. Chen, Z., *et al.*, *Journal of Natural Gas Science and Engineering* (2013) **15**, 76
42. Allahkaram, S. R., *et al.*, *Journal of Natural Gas Science and Engineering* (2015) **26**, 453
43. von Baeckmann, W., *et al.*, *Handbook of cathodic corrosion protection*. Gulf Professional Publishing: 1997
44. Huo, Y., *et al.*, *Electrochemistry Communications* (2016) **66**, 21
45. Aziz, P., and Godard, H. P., *Journal of the Electrochemical Society* (1955) **102** (10), 577
46. Burstein, G., and Ilevbare, G., *Corrosion Science* (1996) **38** (12), 2257
47. Li, L., and Sagues, A., *Corrosion* (2001) **57** (1), 19
48. Fontana, M. G., *Corrosion engineering*. Tata McGraw-Hill Education: 2005
49. Orazem, M., *et al.*, *Corrosion* (1997) **53** (4), 264
50. Freiman, L., *Protection of metals* (2001) **37** (6), 605
51. Webster, S., and Woollam, R., *BP Sunbury report no. ESR* **95**
52. Cole, I. S., and Marney, D., *Corrosion Science* (2012) **56**, 5
53. He, B., *et al.*, *Engineering Failure Analysis* (2015) **58**, 19
54. Smith, D. C. A., *Anti-Corrosion Methods and Materials* (1981) **28** (2), 4
55. Smith, C., *Anti-Corrosion Methods and Materials* (1981) **28** (2), 4
56. Liu, Z., *et al.*, *Corrosion Science* (2012) **55**, 54
57. Xu, L., and Cheng, Y., *Corrosion Science* (2012) **59**, 103

58. Li, S., *et al.*, *Sensors and Actuators B: Chemical* (2007) **120** (2), 368
59. Choi, Y., *et al.*, *Corrosion* (2007) **63** (10), 951

8. AUTHOR DETAILS

Ke Wang, is a PhD student at the Institute for Frontier Materials, Deakin University. He finished his master in Tianjin University, China. Now he is working on an industry project, which attempts to improve the reliability and accuracy of corrosion monitoring.

Facundo Varela, also known as Bob, is a research fellow at the Institute for Frontier Materials at Deakin University. He has recently completed his PhD project that focused on new electrochemical methods for monitoring localized corrosion under cathodic protection. As a research fellow, he is working on an Energy Pipeline CRC sponsored project aiming to perform field trials of the sensors developed during his PhD.

Technologies at Deakin University in Australia. He is also a Research Program Leader of the Energy Pipelines Cooperative Research Centre. Dr Tan's principal teaching and research interests are in corrosion science and engineering and their applications for enhancing the reliability and durability of civil and industrial infrastructures. He contributed to electrochemical methods for corrosion testing, monitoring and prediction and corrosion inhibitor and anti-corrosion coating research. He is the author of some 150 publications and a book entitled 'Heterogeneous Electrode Processes and Localised Corrosion' (2012 John Wiley & Sons).

INFLUENCE OF BIODIESEL ON CORROSION BY *BACILLUS MEGATERIUM*

Y. M. Pusparizkita, T. Setiadi, A. Harimawan
Department of Chemical Engineering, Institut Teknologi Bandung, Indonesia

SUMMARY: Biofuels such as biodiesel have preliminary been used as an alternative source of energy due to the diminishing of petroleum reserves. It has been currently mixed with diesel oil until 30% v/v. However, the hygroscopic nature of biodiesel makes it more susceptible to contamination by microorganisms. The commonest type of microorganism is SRB. However, recent studies have found other types of bacteria, namely, *Bacillus sp.,* which may contribute biocorrosion in tanks containing hydrocarbon product. In this work, the presence of B. *megaterium* could promote the corrosion rate up to 10 times compared to electrochemical corrosion rate. The higher concentration of biodiesel in diesel oil made increasing the growth of bacteria in the biofilm and the rate of corrosion. Variations of biodiesel concentration did not give significant effect on the biofilm composition and the corrosion products. The biofilm composition formed group containing alcohol (O-H), fatty acids (C-H), proteins (N-H), amides (C-N), carbonyl and carboxylates, while corrosion products formed by *B. megaterium* containing Fe_2O_3 and Fe_3O_4. Accordingly, this bacteria formed pitting corrosion.

Keywords: Corrosion, Biofilm, Petroleum Products, Biofuels, Fatty Acid Methyl Esters, Bacteria.

1. INTRODUCTION

Metal is the main material for a component of various applications in an industrial process equipment because of its important mechanical property compared to other materials. Carbon steel is generally used because of its strength and inexpensive material (1). However, carbon steel can be easily corroded. Corrosion is a major cause of damage to pipe and tool process components affecting operating and maintenance costs. Damage caused by biocorrosion itself is accounted for about US $ 2 billion in the oil and gas industry in 2001 (2). Moreover, 40 % of all corrosion of the internal gas and petroleum industry pipelines can be attributed to biocorrosion (3). In the case of corrosion in storage tanks, the presence of microbial induced corrosion (MIC) caused damage by forming pits on the inner surface of the bottoms AST containing crude oil (4).

Biocorrosion or MIC is a corrosion process due to the presence and activity of microorganisms that affect corrosion on metal surfaces (5). Products produced through MIC metabolism of extracellular polymeric substance (EPS), organic and inorganic acids will form biofilms that can change the condition of the surrounding environment and trigger corrosion. The microorganisms involved in the corrosion in the petroleum production system generally come from the anaerobic sulphate reducing bacteria (SRB) (6). However, recent studies have demonstrated the involvement of other types of bacteria such as iron oxidizing bacteria, manganese, acid and methanogens in this biocorrosion process (7-9). For some species that play a role, Bacillus sp. are known to dominate the biodegradation of petroleum hydrocarbons (10). *Bacillus megaterium* is one of the species identified in diesel and naphtha transport pipelines and may potentially cause biocorrosion (11).

Due to the decline in the reserves of petroleum, which is still the main energy source, alternative energy sources should be found. Biofuels are expected to reduce the petroleum consumption. Correspondingly, the Indonesian government has established a policy to increase the utilization of biodiesel by mixing it into certain types of fuel oil. Nevertheless, the hygroscopic nature of biodiesel makes it more susceptible to contamination of microorganisms and thus accelerating the corrosion in fuel storage tank systems. Thus, it is significant to study the issue of biodiesel percentage in diesel oil on corrosion of carbon steel by Bacillus sp. in particular *Bacillus megaterium.*

2. EXPERIMENTAL DETAILS

This biocorrosion research was conducted using *Bacillus megaterium* PI12 obtained from National University of Singapore (NUS). The metal specimen selected was carbon steel of ST 37. The incubation medium of diesel oil was obtained from

PERTAMINA gas station and subsequently mixed with concentrations of 15 %, 20 % and 30 % v/v biodiesel produced from the laboratory.

Furthermore, the experimental study investigated the corrosion characteristics of carbon steel immersed in hydrocarbons (diesel oil and biodiesel) containing *B. megaterium* cultures. Characteristics to be studied involved the number of bacterial colonies on metal surfaces, biofilm composition and morphology, surface metal morphology, corrosion rate and corrosion products. At certain time intervals of corrosion process, surface properties of metal and biofilms were analyzed. This research consisted of three main stages: (1) preparation, (2) experimental and (3) corrosion analysis.

2.1 Sample preparation

The preparation stage consisted of the culture stage, preparation of the test solution, the metal preparation and the reactor.

The medium used to increase the number of *B. megaterium* was Bushnell Haas medium. The composition of Bushnell Haas medium is presented in Table 1. *B. megaterium* was cultured in a sterilized medium with a working volume of 100 mL until it entered a stationary growth stage. The acclimatization was carried out by taking 2 mL inoculated bacteria into an Erlenmeyer containing 300 mL of Bushnell Haas medium and 1 g of diesel oil as a carbon source. Then they were incubated in a rotary shaker being operated at a temperature of 27 ^{0}C and 200 rpm for 30 days. The hydrocarbon to be used as the medium was sterilized with a pore size membrane of 0.45 µm. Specimens preparation followed several steps. First, the metal was cut into 1 cm x 1 cm based on ASTM G 31-72 standard. After that, the specimens were polished using 240-1200 grid abrasive paper (based on ASTM G 1-81) and washed using ethanol to remove fat and other impurities. Once clean, the specimens were stored in a desiccator.

Table 1. Bushnell Haas composition (g/L)

$MgSO_4$	0,2
$CaCl_2$	0,02
KH_2PO_4	1
K_2HPO_4	1
NH_4NO_3	1
$FeCl_3$	0,05

2.2 Experimental Procedure

Types of hydrocarbons medium were diesel oil mixed with 15%, 20% and 30% v/v concentrations of biodiesel. Immersion was done by hanging the metal specimens in a submersion reactor containing 10% v/v culture of *B. megaterium.* The volume of immersion reactor is 800 mL. All submersion reactors contained the sterilized steel and hydrocarbons. The reactor was then stored at 25°C under anaerobic condition for 10 days (Figures 1 and 2).

Figure 1. Submersion reactor

Figure 2. Submersion process

2.3 Corrosion analysis

Experimental analysis of the specimens included bacterial colonies on biofilm by total plate count (TPC) method, biofilm morphology with scanning electron microscopy (SEM), corrosion product with X-ray diffraction (XRD), corrosion rate with the gravimetric method, and biofilm composition with Fourier transform infrared (FTIR). Two coupon replicates were assessed for each analysis method. Moreover, several sample preparation steps were done before the analysis.

2.3.1 Total Plate Count (TPC)

The biofilm attached to the metal surface was taken using a spatula then dissolved it in 10 mL sterile Aqua dm. Furthermore, samples were diluted by Aqua at 100, 1000, and 10000 times. From each dilution, the yield took 0.1 mL and uniformly scratched using the drizzle suds on the nutrient agar (NA) surface in the petri dish.

2.3.2 Scanning Electron Microscopy (SEM)

The sample preparation for biofilm morphology analysis using SEM was done by fixing the biofilm on the metal surface using formaldehyde in a phosphate buffer with a pH of 7.3 for 24 hours. After that, the sample was rinsed with Aqua dm and followed with ethanol 20%, 50%, 75%, and 100%. Next samples were then dried and stored in a desiccator.

2.3.3 X-ray Diffraction (XRD)

The corroded samples were subsequently subjected to XRD analysis. In this way, the sample preparation procedure and XRD measurement refer to ASTM A751.

2.3.4 Gravimetri

The corrosion rate of the samples was determined by the Gravimetri method. The sample preparation and experimental corrosion procedure refer to ASTM G1.

2.3.5 Fourier Transform Infrared (FTIR)

The biofilms formed on carbon steel are released from carbon steel by spatula and dried in the desiccator. The composition of biofilms was examined by FTIR.

3. RESULTS

Bacterial growth in the biofuels was examined by counting the number of bacterial colonies in the biofilm formed. The number of bacterial colonies in biofilms was analyzed to determine the ability of *B. megaterium* to live in biofilms on the metal surface. The number of *B. megaterium* colonies in the biofilm is presented in Figure 3. Generally, the number of *B. megaterium* colonies has a similar profile at different concentrations of biodiesel during the first five days of immersion. The number of bacterial colonies shows to growth in 15% biodiesel concentration from the first day to the fifth day. Likewise, at 20% and 30% biodiesel concentrations, the number of bacterial colonies grown increasingly on the biofilm until the fifth day. Then the number of bacterial colonies growing on the biofilm tended to decrease from seventh to the final day of immersion. Also, it shows that the different concentration of biodiesel has a significant effect on the number of bacterial colonies growing on the biofilm on the metal surface. There are more bacteria colonies growing and living at 30% biodiesel concentrations compared to 15% and 20% biodiesel.

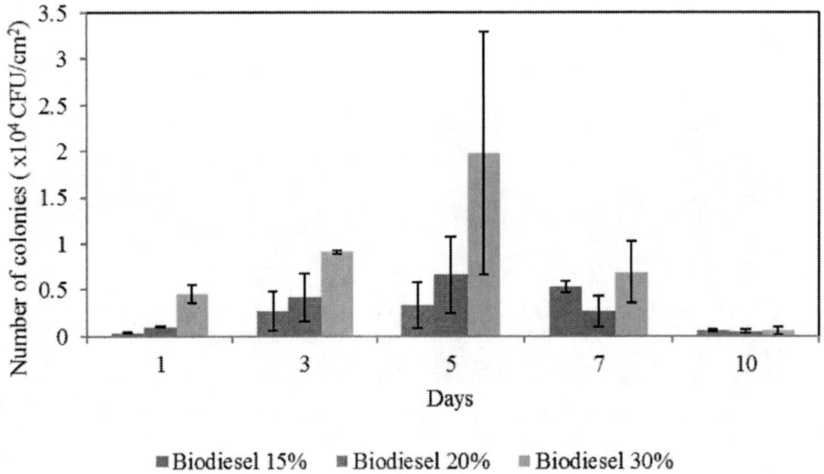

Figure 3. The growth of *B. megaterium* in biofilm

The observation of biofilm morphology was done by using SEM. SEM observations were performed for samples that had been immersed for one, five and ten days. The biofilm morphology formed over ten days due to the activity of bacteria can be seen in Figures 4, 5, and 6 for various biodiesel concentrations. Figure 4 shows that at the first day of incubation, biofilm morphology could not been seen on the metal surface with a soaking medium of 15% and 20% biodiesel. The presence of spots on the metal surface shows that single cell bacteria may begin to interact and attach to the metal surface. However, the biofilm growth has occurred on the surface of the metal with the medium of 30% biodiesel.

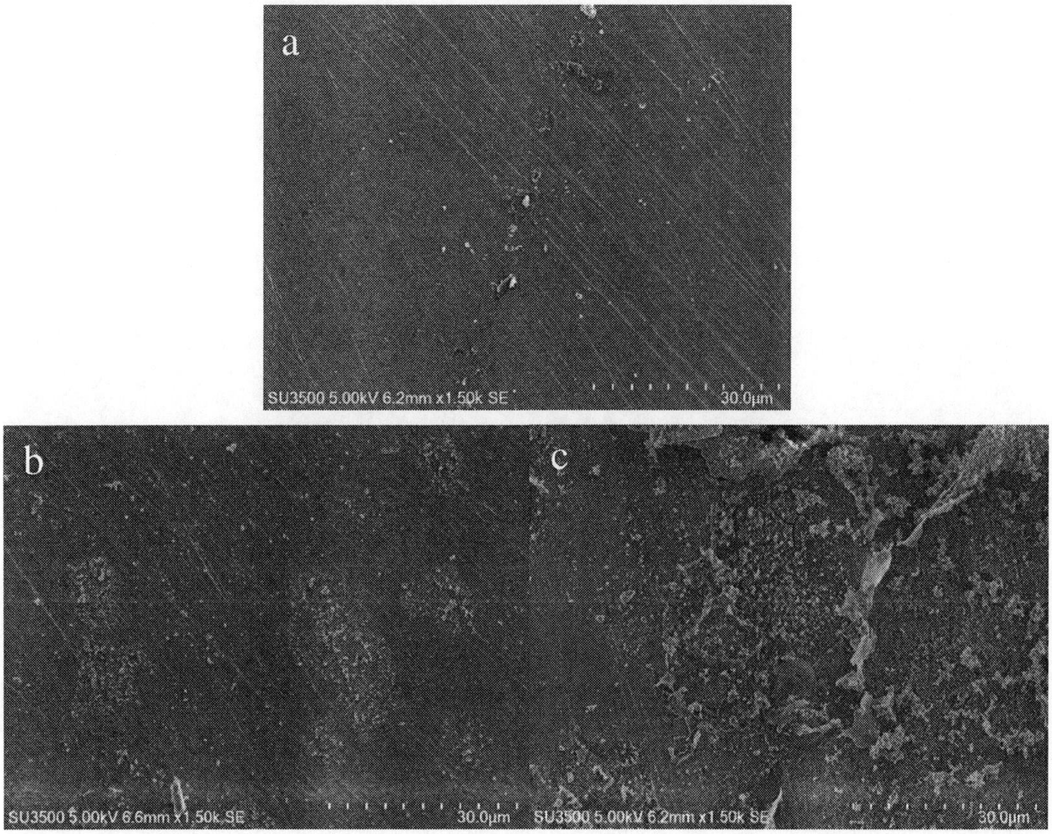

Figure 4. Morphologies of biofilm formed in 1 day incubation at various biodiesel concentration: (a) 15%, (b) 20%, and (c) 30%

In contrast, Figure 5 presents results for the obtained morphology on the metal surface after five days of immersion with the medium of 15% and 20% biodiesel. The biofilms formed on the metal surface with the 30% immersion of biodiesel appear thickened with the cover of metal surfaces covered by the biofilm layer. The thickening of the biofilms formed can indicate that the process of biofilm formation enters the colonization stage. At this stage, the bonds between microorganisms are getting stronger and the biofilm thickness is increasing as biofilm compiler such as the EPS rises along with the growth of bacterial colonies in the biofilm (12).

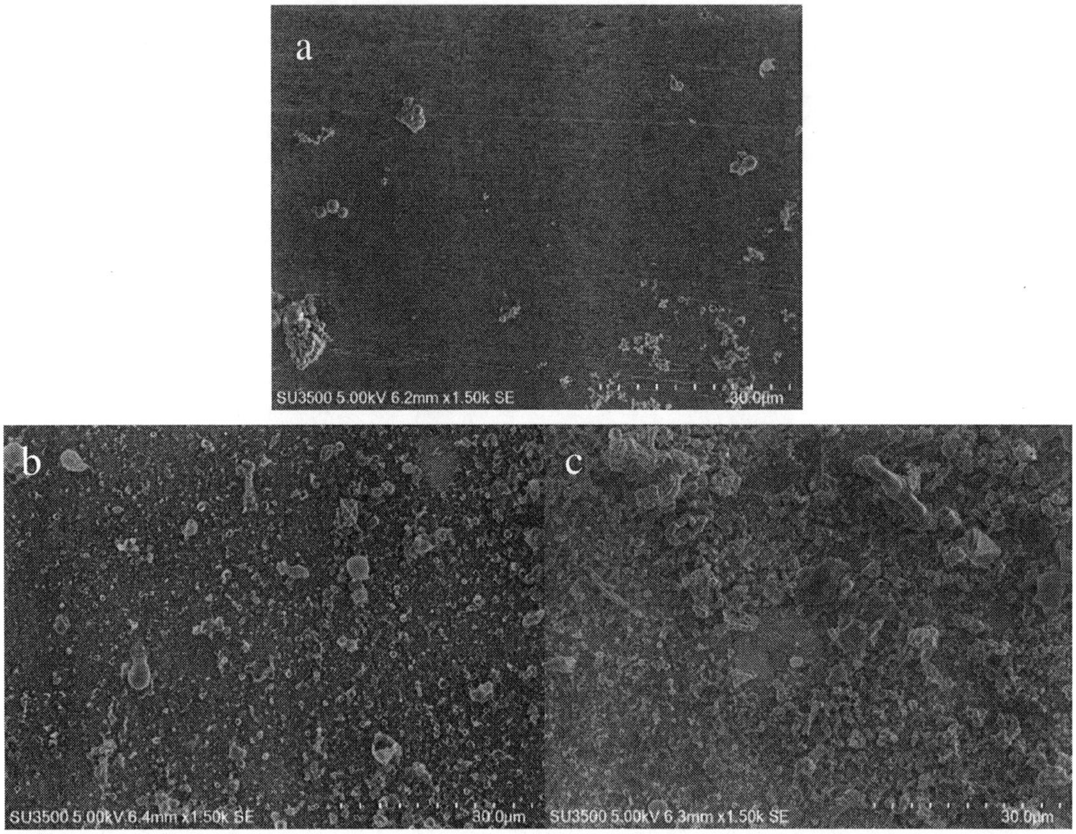

Figure 5. Morphologies of biofilm formed in 5 days incubation at various biodiesel concentration: (a) 15%, (b) 20%, and (c) 30%

At the tenth day of the observation, it can be seen in Fig. 6, that the biofilm formed on the metal surface has undertaken the maturation phase and entered into the final phase for all variations of biodiesel concentration. It is very obvious where biofilms run into lysis so that bacteria spread and allows colonization elsewhere (12). The lysis causes a decrease in the number of colonies on the metal surface as shown in Figure 3. After the tenth days, an increase in the number of colonies can occur because bacteria begin to colonize and form biofilms elsewhere.

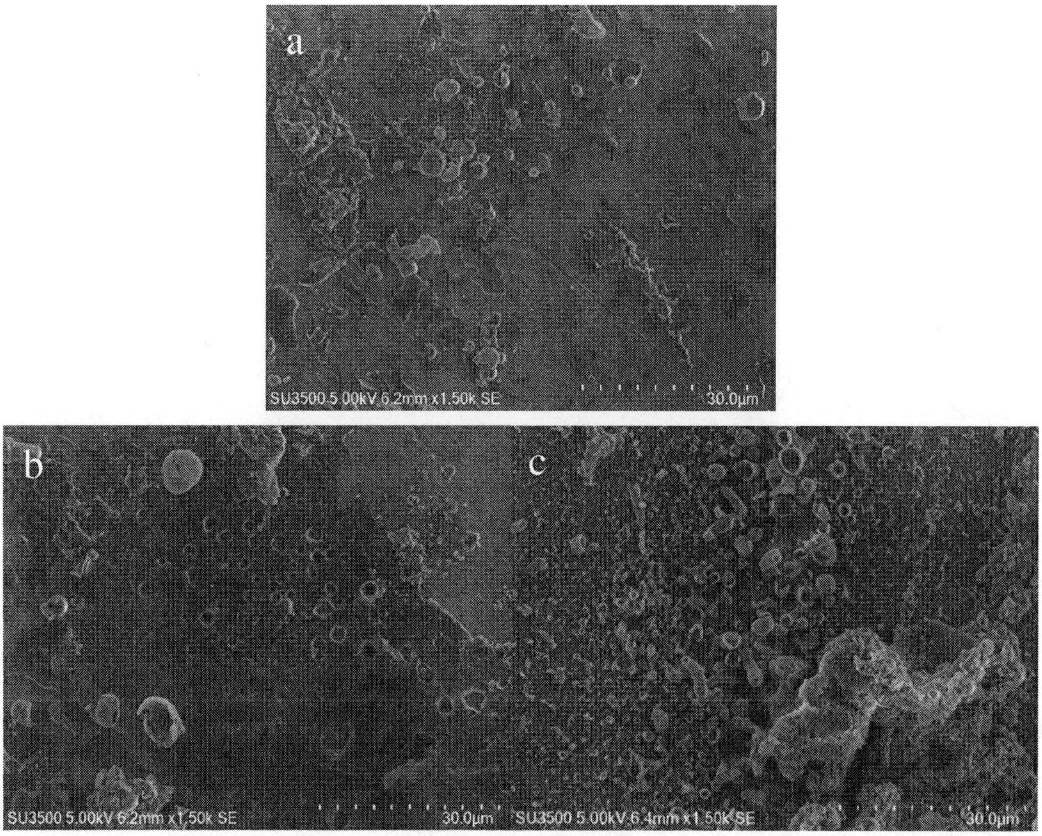

Figure 6. Morphologies of biofilm formed in 10 days incubation at various biodiesel concentration: (a) 15%, (b) 20%, and (c) 30%

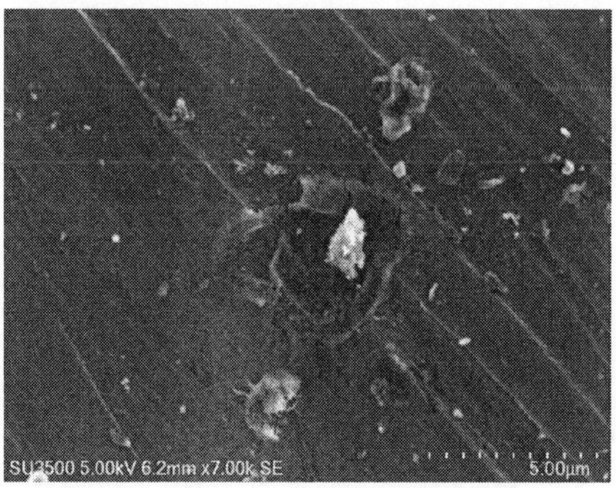

Figure 7. Corrosion attack form of *B. megaterium*

The damage caused by the activity of these bacteria is corrosion attack in the form of pitting corrosion. The occurrence of corrosion on metal surfaces can be triggered by metabolite products produced by the activity of *B. megaterium* in biofilms. The metabolic products that constitute the biofuels can be predicted by analyzing the biofilm composition.

Observations of biofilm composition with FTIR were performed on samples that had been immersed for ten days. The FTIR spectrum of biofilms formed at each biodiesel concentration by *B. megaterium* can be seen in Figure 8. The biofilm composition produced at each biodiesel concentration did not give a significant difference. The type of microorganism used to also produce a biofilm composition that is not much different. The composition of the biofilm formed comprises the presence of a hydroxyl (OH) group at a wavelength of 3400 cm^{-1}, a methylene group (= CH_2) at the wavelengths of 2920, 2850 and 1460 cm^{-1} and the - (CH_2) n- 720 cm^{-1}. The products of oxidation of biodiesel in the form of carboxylic acids, aldehydes, and alcohols in the biofilm can be corrosive, thus affecting the corrosion rate in metals (13).

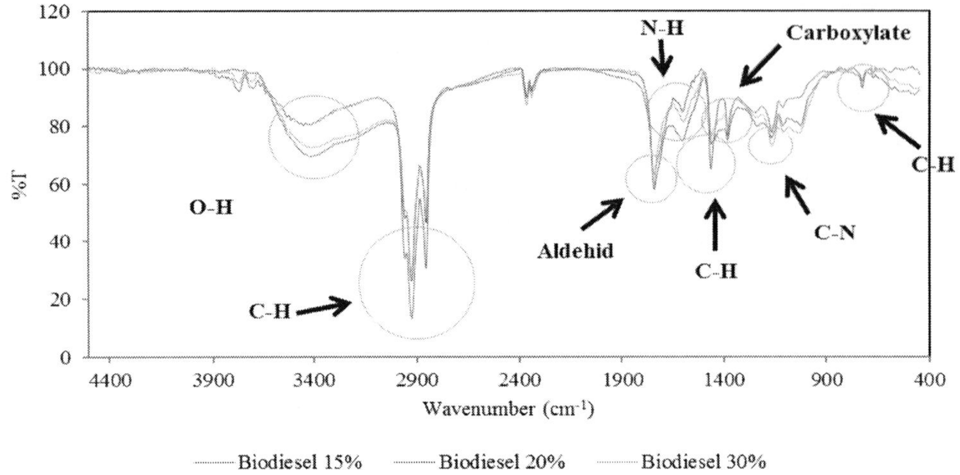

Figure 8. FTIR spectrum of *B. megaterium* biofilm at various biodiesel concentration

The corrosion rate is a corrosion parameter that can describe the corrosion reaction kinetics occurring in a metal in a corrosive environment. The corrosion rate can be calculated by determining the loss of mass occurring in the metal. In this variation of biodiesel concentration, the result of mass loss is presented in Table 1.

In general, loss mass value of the carbon steel in experiments for the variation of biodiesel concentration with *B. megaterium* bacteria provide a mass loss value below 1%. In the variation of biodiesel concentration, the mean loss value of the mean mass of the variation sample increases proportional to the biodiesel concentration in the medium. Based on the mass loss value in Table 1, the corrosion rate of each variation of biodiesel concentration can be determined. The corrosion rate of each variation of biodiesel concentration can be seen in Figure 9.

Table 1 Weight loss of specimens at various biodiesel concentration during incubation

Variations of Biodiesel	15%					
Days to-	1	3	5	7	10	Average
Mass loss of specimen variation (%)	0,04±2E-4	0,13±2E-4	0,19±3E-4	0,08±2E-4	0,7±5E-3	0.23±3E-3
Mass loss of control specimen (%)	0,01±4E-5	0,06±2E-4	0,02±4E-6	0,06±8E-5	0,27±1E-4	0.08±1E-3
Variations of Biodiesel	20%					
Days to-	1	3	5	7	10	Average
Mass loss of specimen variation (%)	0,04±6E-5	0,14±7E-4	0,45±1E-3	0,58±4E-3	0,4±2E-3	0,32±3E-3
Mass loss of control specimen (%)	0,01±5E-6	0,06±2E-4	0,02±6E-5	0,06±1E-5	0,29±2E-4	0.09±1E-3
Variations of Biodiesel	30%					
Days to-	1	3	5	7	10	Average
Mass loss of specimen variation (%)	0,05±3E-4	0,1±3E-4	0,14±5E-4	0,66±4E-3	0,83±3E-3	0,36±4E-3
Mass loss of control specimen (%)	0,02±5E-5	0,08±6E-4	0,02±5E-5	0,08±7E-5	0,45±2E-3	0.13±2E-3

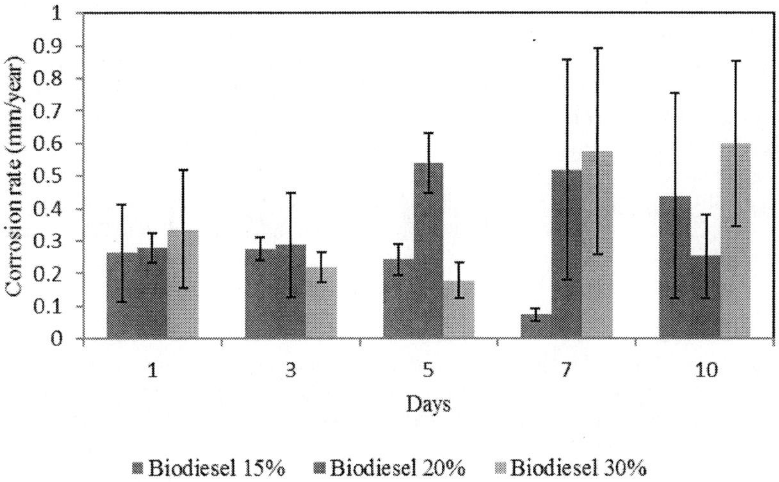

Figure 9. The corrosion rate of carbon steel at various biodiesel concentration

The presence of *B. megaterium* effects electrochemical reactions in the corrosion process in order to form certain corrosion products. The corrosion product was analyzed to predict the corrosion mechanism that occurs due to the activity of *B. megaterium*. Corrosion product of the metals due to the activity of *B. megaterium* bacteria was analysed by XRD for mineral oxide composition determination.

XRD measurement results with 15% and 20% biodiesel concentration differences provided three peaks at 2θ position of 44°, 82°, and 98°. The peak formed in the 2θ 44° and 82° corresponding to ferrous metal (Fe). At position 2θ 78° and 95°, they belong to Fe_2O_3. Conversely, the peak at position 2θ 73° corresponds to Fe_3O_4. At the concentration of biodiesel 30%, three peaks were formed at 2θ position of 44°, 80°, and 86°. The peak at position 2θ 44° relates to iron metal (Fe). At position 2θ 80° Fe_2O_3 was identified and the peak at 2θ 86° positions related to Fe_3O_4. The peaks formed at 2θ position of 73° 78° 80° 86° and 95° for this variation of biodiesel concentration indicate the corrosion product produced by *B. megaterium* activity.

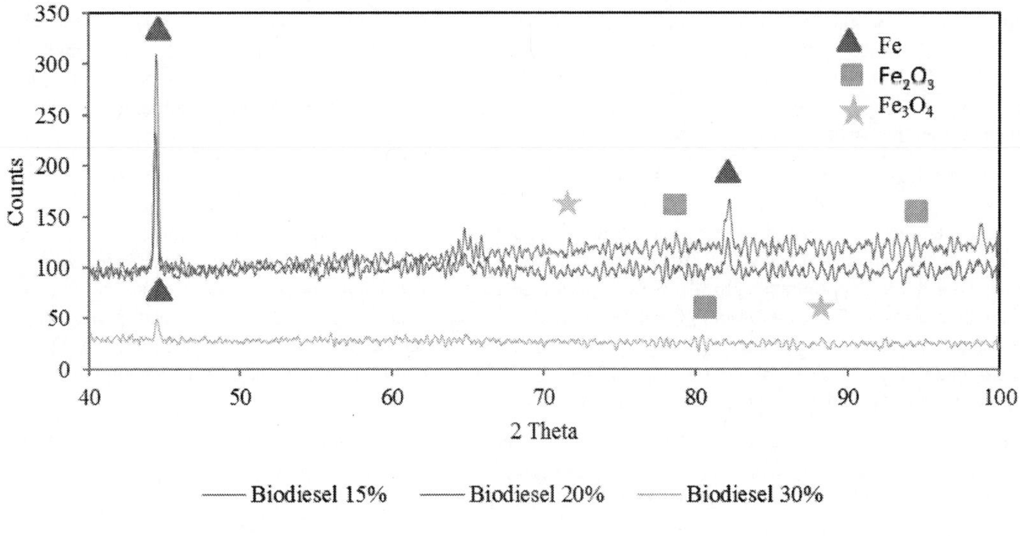

Figure 10. XRD spectra of carbon steel surface at various biodiesel concentration

4. DISCUSSION

4.1 Bacteria growth on the Biofilm

The growth of *B. megaterium* in biofilm is influenced by the concentration of biodiesel in diesel fuel. With increasing concentrations of biodiesel than the number of bacteria that grow and live will also increase. Biodiesel is utilized by *B.*

megaterium as a source of nutrients. The biodiesel properties that are readily biodiesel in aerobic and anaerobic conditions compared to diesel fuel provide benefits for microorganisms to grow and multiply (14-17). This is supported by research conducted by Stolz et al.(18) and Sendzikiene et al.(19) that the components of fatty acid methyl esters (FAMEs) are more rapidly degraded up to four times than diesel. In addition, the high percentage of unsaturated esters such as 40.2% palmitate (16: 0) and 42.4% oleate (18: 1) in biodiesel made from palm oil makes it easier to degrade so it can be utilized by microorganisms in the metabolic process (20). In the anaerobic system, the microorganisms hydrogenate the components of unsaturated fatty acids into components of saturated fatty acids that can poison the microorganism itself, resulting in a decrease in the number of living colonies (15, 21). The decrease in the number of colonies of microorganisms can also be caused by the lysis process in the biofilm, so that microorganisms have spread into the immersion medium. This is supported by the observation of biofilm morphology on the metal surfaces shown in Figure 6.

4.2 Chemical Composition of Biofilm

According to Busalmen et al. (22), the presence of biofilm compounding groups because *B. megaterium* produce peroxidase enzymes in the form of hydrogen peroxide (H_2O_2) to degrade hydrocarbons and use them as carbon sources. In addition, these microorganisms also produce catalase that overcomes the toxic properties of H_2O_2 and then breaks it down into water and oxygen. Oxygen can be exploited by microorganisms in the process of respiration and produce water. H^+ ions are obtained from the degradation of hydrocarbons and electrons provided by the enzyme cytochrome oxidase (23).

$$\frac{1}{2}O_2 + 2H^+ + 2e^- \rightleftarrows H_2O \qquad (1)$$

Hydrocarbon degradation products in the form of hydrogen and carbon can be utilized both microorganisms form the group (-O-CH_2-). In the iron metals *B. megaterium* are able to convert iron into iron oxide compounds by utilizing the existing oxygen and eventually formed (-CH_2-CH_2-) (23). In addition, the C-H group at wavelengths below 3000 cm-1 can show saturated hydrolysis of unsaturated hydrolysis hydrocarbons that contain lots of biodiesel (21, 24). The presence of biodiesel oxidation events also allows the formation of oxidation products in the form of carboxylic groups at 1370 cm-1 wavelengths, aldehydes at 1700 cm-1 wavelength and alcohol (O-H) at 3400 cm-1 wavelengths (25). The N-H group at the 1600 cm-1 wavelength supported by the C-N group at the wavelength of 1160 cm-1 shows the amide II protein compound which is the constituent of EPS.

4.3 Corrosion Rate

Increased corrosion rate occurring in the first three days for variation of 15% biodiesel concentration and the first five days for variation of 20% concentration caused by environmental influence and biofilm forming activity by *B. megaterium*. The biofilms formed may contribute to corrosion processes (26). From SEM analysis seen single *B. megaterium* cell began to stick with metal surface since the first day. However, the absence of a complete biofilm causes the environment to be more influential on the increase of corrosion rate. While at 30% biodiesel concentration besides environmental influence, biofilm formation by *B. megaterium* seen from SEM analysis resulted in higher corrosion rate compared to 15% and 20% concentration on the first day.

The biofilms thickened up to the fifth day of observation as the number of bacteria increased. The maximum number of colonies can indicate that the bacteria have entered the stationary phase so that bacteria can not grow and develop further because of the source of the nutrients that are running out. This can decrease the rate of corrosion (27). The decrease in corrosion rate can also be caused by biofilms that have been formed by bacteria protecting the metallic layer from a potentially corrosive environment (28-30). However, the large number of bacteria can also increase the rate of corrosion as in the concentration of 20% biodiesel. This is because *B. megaterium* that are still productive, can oxidize biodiesel and produce acidic products such as carboxylic acids, aldehydes, alcohols and saturated fatty acids (15, 21, 25). The metabolite that accumulate in the biofilm can decrease in the pH value of the medium and cause an acidic environment which affects the amount of mass loss of metal specimens and corrosion rate.

Increasing the mass loss of metal specimens and the corrosion rate occurring on the tenth day by *B. megaterium* can be due to the lysis process occurring from the observed biofilm morphology. The lysis events cause biofilm damage so that the carbon steel surface will be more easily corroded by the environment.

Table 2 Effect of biodiesel concentration on corrosion rate control and *B. megaterium*

Variation	Diesel Oil	Control		
		15%	20%	30%
Corrosion rate (mm/year)	0,039	0,085±0,056	0,087±0,060	0,130±0,113
Increased corrosion rate compared to diesel oil		2,1x	2,2x	3,3x
Variation	**Diesel Oil**	***B. megaterium***		
		15%	**20%**	**30%**
Corrosion rate (mm/year)	0,039	0.259±0.196	0.367±0.222	0.381±0.2673
Increased corrosion rate compared to diesel oil		6.6x	9.4x	9.8x

Based on research conducted Rajasekar et al. (10) the average corrosion rate of carbon steel metal in diesel with 10 days of submersion was 0.039 mm/year. *B. megaterium* can accelerate corrosion rate up to 10 times. In control, the concentration of biodiesel on diesel can lead to increased corrosion rate up to 3 times. This is because the concentration of biodiesel in the diesel fuel influences the final pH of immersion medium. Biodiesel is essentially acidic, so increasing the concentration of biodiesel in diesel causes a decrease in pH (31, 32). This may affect the increase in corrosion rate that occurs.

4.4 Corrosion Mechanism by *B. megaterium*

The corrosion mechanism by *Bacillus megaterium* on carbon steel metal at various concentrations of biodiesel can be estimated from the resulting corrosion product. The corrosion mechanism of both bacteria on the carbon steel metal can be estimated from the resulting corrosion product. Based on the results of XRD analysis, the resulting corrosion products are mostly Fe_2O_3 and Fe_3O_4.

Corrosion by microorganisms or biocorrosion begins when biofilms on the surface of the metal begin to form and there has been corrosion on the surface of carbon steel metal. The biofilms formed on the surface of the carbon steel metal may limit the diffusion of the dissolved compounds on the medium into the biofilm and vice versa. *B. megaterium* produce acidic metabolite products resulting in changes in environmental conditions in the form of changes in acidity levels in biofilms.

Both corrosion products are the result of the reaction between Fe^{2+} ions with water. The corrosion mechanism begins with the corrosion process of carbon steel metal by the dissolved compound in the biofilm environment. The soluble compound in the biofilm environment that is thought to be the trigger for corrosion is the acidic compound of metabolite and oxygen. Oxygen is a water-soluble compound contained in biodiesel as well as the degradation of the biodiesel itself.

The oxygen will react with H^+ from methylation products of *B. megaterium* which are acidic and produce water. The water then reacts with ferrous ions from reduced iron to form Fe_2O_3 and Fe_3O_4. Based on the mechanism described, the corrosion reaction occurring in the carbon steel metal by *B. megaterium* in the experiment is attempted to be described in Figure 11 as well as the following reactions.

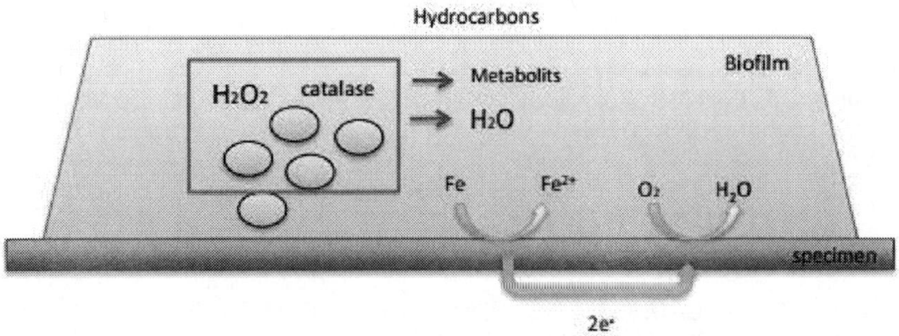

Figure 11. Corrosion mechanism of carbon steel by *B. megaterium* in the medium of hydrocarbons

The reaction that takes place in the solution, is a cathodic reaction or a reduction reaction. The dominant reduction reaction occurs as an oxygen reduction reaction in an acidic environment (Eqn. 2). The reaction to the metal is an anodic reaction in the form of an iron oxidation reaction into ferro (Fe^{2+}) and ferric (Fe^{3+}) (Eqn. 3 and 4) ions. The cathodic and anodic reactions occur on the surface of the carbon steel metal. The reaction of *B. megaterium* is a hydrolysis reaction of H_2O_2 to oxygen and water (Eqn. 5). In addition, oxygen will be utilized *B. megaterium* for respiration to produce water (Eqn. 6). A series of such reactions produce corrosion products of Fe_2O_3 and Fe_3O_4 (Eqn. 7 and 8).

Solution	: $\frac{1}{2}O_2 + 2H^+ + 2e^- \rightleftarrows H_2O$	(2)
Metal	: $Fe \rightleftarrows Fe^{2+} + 2e^-$	(3)
	: $Fe^{2+} \rightleftarrows Fe^{3+} + 1e^-$	(4)
Mikroorganism	: $H_2O_2 \rightleftarrows \frac{1}{2}O_2 + H_2O$	(5)
	: $O_2 + 4H^+ + 4e^- \rightleftarrows 2H_2O$	(6)
Corrosion product	: $3Fe^{2+} + 4H_2O + 6e^- \rightleftarrows Fe_3O_4$	(7)
	: $2Fe^{2+} + 3H_2O + 4e^- \rightleftarrows Fe_2O_3$	(8)

5. CONCLUSIONS

From the research it can be concluded that:

1. Bacterial growth in biofilms is influenced by the concentration of biodiesel in diesel. The higher the concentration of biodiesel in the diesel fuel, the growth of bacteria in biofilm is also increasing. The biodiesel properties that are easily degraded can be utilized by bacteria as a source of nutrients for growth.
2. Variations in biodiesel concentrations in diesel do not have a significant effect on the biofilm composition formed on the surface of carbon steel. The biofilm composition formed consists of hydroxyl (O-H), fatty acid (C-H), protein (N-H), carbonyl and carboxylic groups.
3. In the absence of microorganisms, the addition of biodiesel to diesel can increase the corrosion rate up to three times as much as diesel because the biodiesel is acidic and can decrease the environmental pH value so it affects the corrosion rate.
4. The corrosion rate of mixed diesel and biodiesel medium increased up to ten times due to the activity of *B. megaterium*.
5. Variance of biodiesel concentration does not affect product and corrosion form on metal surfaces. The corrosion products produced by *B. megaterium* are Fe_2O_3 and Fe_3O_4 with corrosion form of corrosion of wells.

6. ACKNOWLEDGMENTS

The authors would like to express their special thanks to PMDSU Scholarship which has provided research fund, Chemical Engineering Program of Institut Teknologi Bandung which has provided facilities and infrastructure in the implementation of this research, staff of Microbiology and Technology Laboratory of Bioprocess for the assistance that has been given in making research.

7. REFERENCES

1. Panossian Z, Almeida NL, Sousa RM, Pimenta Gd, Marques LB. Corrosion of carbon steel pipes and tanks by concentrated sulfuric acid: A review. Corrosion Science. 2012:1-11.
2. Koch G, Brongers M, Thompson N, Virmani Y, Payer J. Corrosion costs and preservative strategies in the United States. Washington, DC: Federal Highway Administration; 2001.
3. Graves J, Sullivan E. Internal Corrosion in gas gathering system and transmission lines. Mater Prot. 1996;5:33-77.
4. Groysman A. Corrosion in Systems for Storage and Transportation of Petroleum Products and Biofuels: Springer; 2014.
5. Beech IB, Gaylarde CC. Recent advances in the study of biocorrosion-an overview. Rev Microbiol. 1999;30:177-90.
6. Von Wolzogen Kuehr CAH, van der Vlugt LS. GRAPHITIZATION OF CAST IRON AS AN ELECTROBIOCHEMICAL PROCESS IN ANAEROBIC SOILS. 2006.
7. Zhu XY, Lubeck J, Kilbane JJ. Characterization of Microbial Communities in Gas Industry Pipelines. Applied and Environmental Microbiology. 2003;69(9):5354-63.
8. Muthukumar N, Mohanan S, Maruthamuthu S, Subramanian P, Palaniswamy N, Raghavan M. Role of Brucella sp. and Gallionella sp. in oil degradation and corrosion. Electrochem Commun 2003;5:412-26.
9. Jan-Roblero J, Posadas A, Zavala-Dıaz de la Serna F, Garcıa R. Phylogenetic characterization of bacterial consortia obtained of corroding gas pipelines in Mexico. World J Microbiol Biotechnol 2008;24:1775-84.
10. Rajasekar A, Ganesh Babu T, Maruthamuthu S, Karutha Pandian ST, Mohanan S, Palaniswamy N. Biodegradation and corrosion behaviour of Serratia marcescens ACE2 isolated from an Indian diesel-transporting pipeline. World Journal of Microbiology and Biotechnology. 2007;23(8):1065-74.

11. Rajasekar A, Anandkumar B, Maruthamuthu S, Ting Y-P, Rahman PKSM. Characterization of corrosive bacterial consortia isolated from petroleum-product-transporting pipelines. Applied Microbiology and Biotechnology. 2009;85(4):1175-88.
12. Heyer A, D'Souza F, Morales CFL, Ferrari G, Mol JMC, Wit JHWd. Ship ballast tanks a review from microbial corrosion and electrochemical point of view. Ocean Engineering. 2013;70:188-200.
13. Fazal MA, Haseed ASMA, Masjuki HH. Degradation of automotive materials in palm biodiesel. Energy. 2012;40:76-83.
14. Zhang X, Peterson C, Reece D, Haws R, Moller G. Biodegradability of biodiesel in the aquatic environment. Trans ASAE. 1998;41:1423-30.
15. Lee JS, Ray RI, Little BJ. Microbiological and corrosivity characterizations of biodiesels and advanced diesel fuels DTIC Document. 2009.
16. Aktas DF, Lee JS, Little BJ, Duncan KE, Perez-Ibarra BM, Suflita JM. Effects of oxygen on biodegradation of fuels in a corroding environment. International Biodeterioration & Biodegradation. 2013;81:114-26.
17. Lee JS, Ray RI, Little BJ, Duncan KE, Oldham AL, Davidova IA, et al. Sulphide production and corrosion in seawaters during exposure to FAME diesel. Biofouling. 2012;28:465-78.
18. Stolz JF, Follis P, Floro G, Donofrio R, Buzzelli J, Griffin WM. Aerobic and anaerobic biodegration of the methyl esterified fatty acids of soy diesel in freshwater and soil environments. 1995.
19. Sendzikienea E, Makarevicienea V, Janulisa P, Makareviciuteb D. Biodegradability of biodiesel fuel of animal and vegetable origin. Eur J Lipid Sci Technol. 2007;109:493-7.
20. El-Araby R, Amin A, El Morsi AK, El-Ibiari NN, El-Diwani GI. Study on the characteristics of palm oil–biodiesel–diesel fuel blend. Egyptian Journal of Petroleum. 2017.
21. Sousa DZ, Smidt H, Alves MM, Stams AJ. Ecophysiology of syntrophic communities that degrade saturated and unsaturated long-chain fatty acids. FEMS Microbiol Ecol. 2009;68:257-72.
22. Busalmen JP, Vazquez M, de Sanchez SR. New evidence on the catalse mechanism of microbial corrosion. Electrochem Acta. 2002;47:1857-65.
23. Rajasekar A, Babu TG, Pandian STK, Maruthamuthu S, Palaniswamy N, Rajendran A. Role of Serratia marcescens ACE2 on diesel degradation and its influence on corrosion. Journal of Industrial Microbiology & Biotechnology. 2007;34(9):589-98.
24. Coates J. Interpretation of Infrared Spectra, A Practical Approach. R.a. Meyers ed. Chichester: John Wiley & Sons Ltd; 2000.
25. Jakeria MR, Fazal MA, Haseeb ASMA. Influence of different factors on the stability of biodiesel: a review. Renew Sust Energ Rev. 2014;30:154-63.
26. Achten WMJ, Almeida J, Fobelets V, Bolle E, Mathijs E, Singh VP, et al. Life cycle assessment of Jatropha biodiesel as transportation fuel in rural India. Appl Energy. 2010;87:3652-60.
27. Pedersen A, Hermansson M. The effects on metal corrosion by Serratia marcescens and Pseudomonas sp. Biofouling. 1989;1(4):313-22.
28. Saravia SGGd, Mele MFLd, Videla HA. Biofilms and Inorganic Passive Layers in the Corrosion of Cu/Ni Alloys in Chloride Environments. Corrosion Science. 1990;46:302-6.
29. Lewandowski Z, Dickinson W, Lee W. Electrochemical interactions of biofilms with metal surfaces. Water Sci Technol. 1997;36:295-302.
30. Gu JD, Roman M, Esselman T, Mitchell R. The role of microbial biofilms in deterioration of space station candidate materials. International Biodeteriora- tion & Biodegradation. 1998;41:25-33.
31. Fazal MA, Haseeb ASMA, Masjuki HH. Corrosion mechanism of copper in palm biodiesel. Corrosion Science. 2013;67:50-9.
32. Eryilmaz T, Yesilyurt MK. Influence of blending ratio on the physicochemical properties of safflower oil methyl ester-safflower oil, safflower oil methyl ester-diesel and safflower oil-diesel. Renewable Energy. 2016;95:233-47.

8. AUTHOR DETAILS

Yustina Metanoia Pusparizkita is a Chemical Engineering PhD student at the Bandung Institute of Technology, Indonesia where she also received her MT (2017) degree. She got a full scholarship from the government of Indonesia for her master and doctoral programs. Her general research area is biocorrosion in biofuel tanks and is currently investigating the role of concentration variation of biodiesel in diesel oil and types of microorganisms influencing the biocorrosion. Yustina expects to improve the biocorrosion resistance of the storage material and complete her PhD dissertation in 2019.

Tjandra Setiadi is a Professor in Bioprocess Engineering, Department of Chemical Engineering, Faculty of Industrial Technology, Institut Teknologi Bandung (ITB)

Ardiyan Harimawan is Assistant Professor in Bioprocess Engineering, Department of Chemical Engineering, Faculty of Industrial Technology, Institut Teknologi Bandung (ITB) His research division is Chemical Engineering Product Design and Development

IN SITU SCANNING KELVIN PROBE STUDY OF CATHODIC DISBONDMENT ON TRIVALENT CHROMIUM COATED STEEL

J. Edy[1], N. McMurray[1], A. deVooys[2]
[1]Swansea University, Swansea, Wales, U.K., [2]TATA Steel Europe, The Netherlands

SUMMARY: European REACH regulations have prompted the need to replace the use of hexavalent chromium in the electroplating process for Electro Chromium Coated Steel (ECCS). Trivalent Chromium Coated Technology steel has been developed by TATA Steel to replace ECCS. ECCS is lacquered in its final application and has excellent resistance to underfilm corrosion. Therefore, prototype Cr^{3+} coated steel materials were investigated for their resistance to cathodic disbondment, in-situ, using the Scanning Kelvin Probe technique (SKP).

Keywords: cathodic disbondment, scanning kelvin probe, chromate-free, steel passivation, packaging steels

1. INTRODUCTION

Coatings for packaging steels, mainly Electro Chromium Coated Steel (ECCS) are traditionally produced from electrolyte baths containing hexavalent chromium (Cr^{6+}). Such packaging steels consist of a low gauge (0.13-0.49mm) low carbon steel substrate with a thin coating that is composed of a chromium metal base layer and a top layer of chromium oxide, deposited from chromic acid.

Due to the international concern regarding the environmental [1]–[4] and health implications [2], [4]–[8] surrounding use of hexavalent chromium in production processes, there is a substantial interest in the development of Cr^{6+}-free coatings. The replacement system must be equivalent or better than the existing system, non-toxic and suitable for the food industry. One such replacement system has been developed based upon the electrodeposition of a chromium metal-oxide-carbide coating from an electrolyte containing trivalent (Cr^{3+}) chromium salts, to produce a Cr^{3+}coated steel [9], [10].

Electrodeposited coatings on packaging steels are present mainly to provide corrosion resistance. However, it has been found for ECCS that iron can become exposed after can formation, providing initiation sites for corrosion [11], [12]. Due to this, ECCS is always coated with an organic lacquer or laminate. ECCS has excellent resistance to corrosion driven disbondment of such coatings and any replacement passivation system must also perform the same.

This piece of research aims to study the relationship between Cr^{3+} coated steel coating thickness in resisting corrosion driven coating disbondment of organic coatings at penetrative defect sites. The well understood cathodic disbondment mechanism, in which delamination of the organic coating is driven by a cathodic reaction, such as oxygen reduction, is studied here.

The cathodic disbondment corrosion cell consists of an anode located near the defect site, which is coupled to a cathodic delamination front via a thin (<5μm) gel like electrolyte, which ingresses beneath the coating. Cathodic disbondment has been shown previously on iron [13], [14] and on Cr3+ coated steel but not on conventional ECCS electrodeposited from Cr^{6+}bath [15].

Cathodic delamination was studied on Iron, ECCS, and Cr^{3+} coated steels overcoated with a poly(vinyl) butyral coating (PVB). Samples were studied by using a Stratmann-type cell [4], [6] and using 0.86M aqueous sodium chloride initiating electrolyte to accelerate corrosion conditions. The time dependent advancement of cathodic disbondment was determined by repeated in situ scanning of the surface using the scanning kelvin probe (SKP) apparatus. The ability of the SKP to spatially visualise the distribution of localised free corrosion potential has been demonstrated elsewhere [14]–[16], [18]–[21].

2. EXPERIMENTAL DETAILS

2.1 Materials

Iron samples of 99.5% purity were purchased from Goodfellow Cambridge Ltd. All samples of chromium coated packaging steels were donated by TATA Steel Packaging Europe. Cr^{3+} coated steels studied here had chromium coatings of varying thicknesses, as measured by X-ray Photoelectron Spectroscopy (XPS) internally at TATA Steel Research and Development Centre, Ijmuiden, The Netherlands.

Polyvinyl Butyral (PVB) and all other chemicals used were obtained from Aldrich Chemical Co. and were of analytical grade purity. All samples were cleaned and degreased using ethanol and distilled water and air dried before experimentation.

2.2 Surface characterisation

X-ray Photoelectron Spectroscopy (XPS) spectra and depth profiles were recorded on a Kratos Axis Ultra using monochromated Al Kα X-rays of 1486.8eV. The measured spot size was 700 μm x 300 μm. Depth profiles were recorded using 4 KeV Ar^+ ions to create a sputter crater of 3mm x 3mm. Sputter rate was calibrated against XRF results and was 1.59nm min^{-1}.

The surface morphology of samples studied here was characterised using either a Zeiss Evo LS25 SEM or JEOL 7800 FEG SEM (Field Emission Gun – Scanning Electron Microscope). Unless otherwise stated in the SEM images shown, overview images were obtained using an accelerating voltage of 7kV, and working distance of 8.1mm.

2.3 Cathodic disbondment study using Scanning Kelvin Probe (SKP)

All samples (50x50mm coupons) were coated with 15.5% (w/w) Poly(Vinyl) Butyral (PVB) of molecular weight 70,000 – 100,000, dissolved in ethanol, using insulating tape height guides to give an air-dried thickness of 30μm. Following bar casting, tape was pulled back to create the electrolyte reservoir (defect site) of dimensions approximately 20x30mm. A schematic illustration is shown in Figure 1.

The Scanning Kelvin Probe (SKP) reference probe was positioned at 12mm away from the defect and programmed to scan towards and perpendicular to the defect site, as illustrated in Figure 2. The SKP reference probe was a gold wire of diameter 125μm and was vibrated at 280Hz with an amplitude of 40μm. The gap between the reference probe and sample was 100μm, with E_{KP} points measured at 20 per mm. The SKP chamber allows a constant humidity of 95% relative humidity to be obtained at 25°C and was calibrated according to a procedure described elsewhere [13].

Changes in E_{corr} between the intact metal-polymer interface and delamination region, where ingress of electrolyte has occurred, was indirectly measured by the kelvin probe. As cathodic delamination proceeds, away from the defect site, distinct E_{corr} profiles develop, a typical example is shown in Figure 3 .

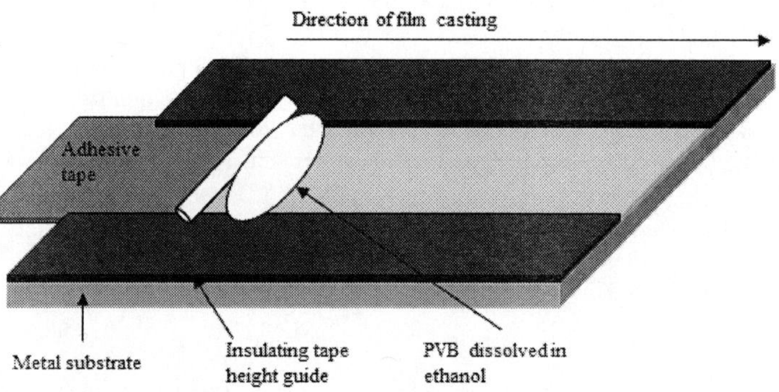

Figure 1 Schematic of Stratmann Cell prepared for study of cathodic disbondment using the Scanning Kelvin Probe (SKP)

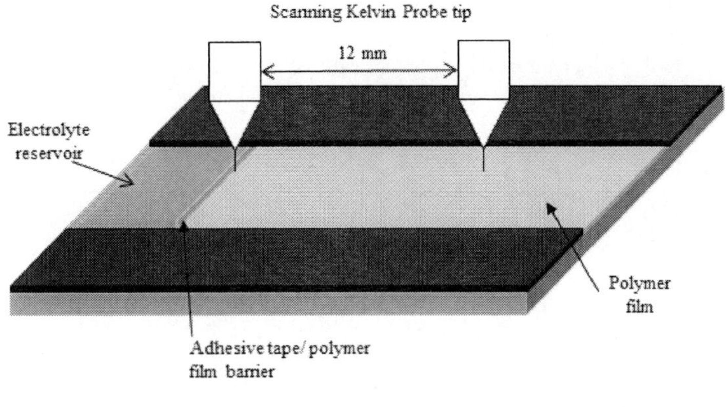

Figure 2 Schematic showing experimental procedure for a typical SKP cathodic delamination experiment

2.4 XPS characterisation of chromium coatings

Table 1 Chromium layer thickness of ECCS and Cr^{3+} coated steel

Chromium coating type	($mg.m^{-2}$)			
	Total Chromium	Chromium metal	Chromium Oxide	Chromium Carbide
Cr^{6+} (ECCS)	93.4	82.1	9.3	2.0
Cr^{3+} coated Steel	56.9	32	18.2	6.7
	58.4	36.3	4.2	17.9
	62.2	45	5	12.6
	67.0	44.5	5.5	17
	80.5	57.9	6.8	15.8
	94.0	69.6	13.3	11.1
	133.9	102	8.2	23.7

3. RESULTS AND DISCUSSION

3.1 Cathodic delamination of PVB coated Iron

Cathodic disbondment kinetics of the PVB coating on pure iron was determined, as described elsewhere [15]. This was conducted to obtain baseline characteristics without the presence of any chromium passivation coating. Following an initiation time for corrosion, time dependent $E_{corr}(x)$ profiles, where x is distance from the defect site, were established and are displayed in Figure 3. Delamination of PVB coated iron was consistently initiated within 1 hour of the 0.86 Mol dm^{-3} NaCl electrolyte being added to the defect site.

The E_{corr} measured over the intact portion of the coating remained high and consistent at ca. 0.30V vs. SHE, after the calibration has been applied, in agreement with values reported previously [13], [15]. E_{corr} values at the delamination front sharply fell to values close to that of a bare metal in contact with electrolyte in the defect site, here measured ca. -0.10V vs SHE. Anodically active iron is expected as -0.44V vs SHE [14], [16], and so values here are more anodic than expected.

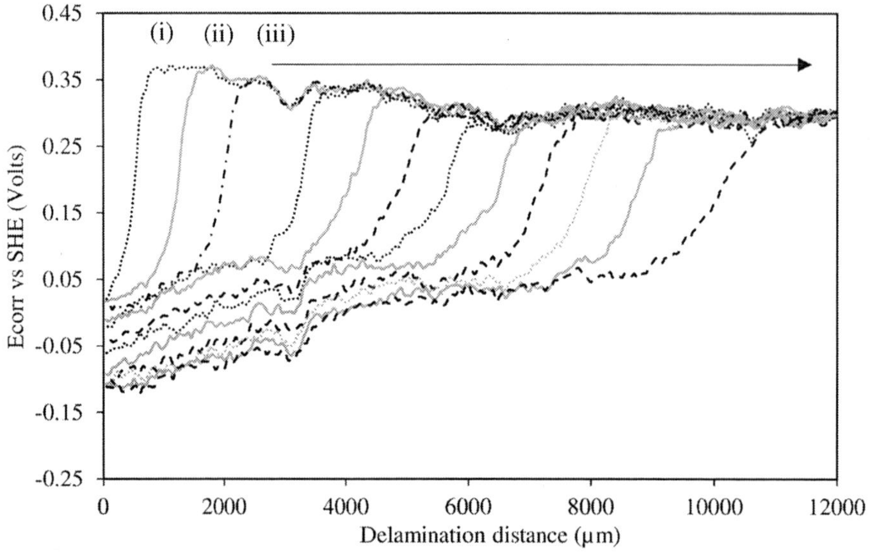

Scan time key (left to right): i.) 60 minutes, (ii.) 120 minutes, (iii.) 180 minutes and two-hour intervals thereafter.
Figure 3 SKP derived E_{corr} vs. distance (x) profiles for iron overcoated with a 30_{um} PVB film

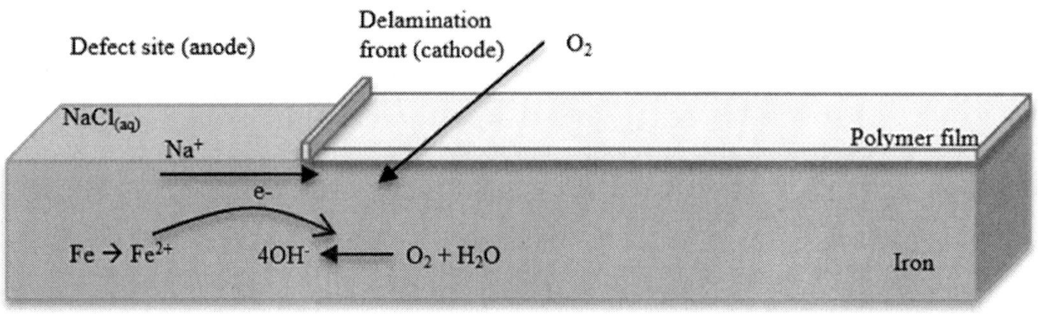

Figure 4 Schematic diagram showing the mechanism of cathodic disbondment for organically coated iron

During coating disbondment anodic activity is constrained to the external coating defect site [16]. A local cathode exists at the delamination front, where the oxygen reduction reaction (ORR) occurs creating a zone of high pH according to:

$$O_2+2H_2O+4e^- \leftrightarrows 4OH^- \quad (1)$$

This results in hydrolysis of interfacial bonds between polymer and metal substrate, leading to coating disbondment. Electrolyte ingresses underneath the coating from the defect site. The underfilm electrolyte links the ionic transport between the anodic defect and cathodic site. As coating delamination proceeds, the underfilm electrolyte ingresses further under the coating. A schematic of this mechanism is shown in Figure 4.

The linear E_{corr} gradient between the delaminated and intact regions is explained from the presence of ionic conductivity in the underfilm electrolyte [14], [15]. A semi empirical method of determining the delamination front has been developed as the maximum gradient in the time dependent $E_{corr}(x)$ profiles [13], [14], [16]. As delamination proceeds with time, the linear gradient moves further in distance away from the defect site.

On iron the rate at which this progression increases with time decreases as it moves further away from the defect site. This is typical of a delamination rate controlled by the migration of sodium ions from the external defect site electrolyte to the cathodic delamination front [14], [15]. The transport of sodium ions along the coating-metal interface has been confirmed as the rate determining step for cathodic disbondment on steel by Sorensen et al.[22].

For a system where delamination is controlled by the migration of sodium ions from the defect site to the cathodic front, delamination distance (x_{del}) is related to the time since electrolyte contact (t_{del}) by Equation 2, where k_d is the delamination rate constant and t_i is the initiation period.

$$x_{del} = k_d(t_{del} - t_i)^{1/2} \quad (2)$$

The kinetic plot for delamination on iron is displayed in Figure 6. Kinetics are obtained from time dependent E_{corr} profiles for iron and were plotted as x_{del} vs ($t_{del} - t_i$) for easy comparison against Cr^{3+} coated steel. Delamination kinetics for iron show a parabolic curve consistent with delamination being limited by the migration of Na^+ ions. The K_d value obtained for iron from the initial linear slope of the plot was (503 ± 9.2) μm min $^{-1/2}$. Confidence limits displayed here correspond to one standard deviation of the mean.

3.2 Cathodic delamination of Electro Chromium Coated Steels (ECCS)

Cathodic delamination on conventional ECCS, where the coating is deposited from a Cr^{6+} electrolyte, has been studied elsewhere and was not observed over 96 hours [15]. A potential plateau was observed, and it was suggested that there is no coupling at the defect site and so E_{corr} values remain higher than that of bare metal in contact with electrolyte, so delamination does not occur [15].

3.3 Resistance to cathodic delamination on Trivalent Chromium Coated steel

3.3.1 Introduction

In contrast to ECCS, cathodic delamination was observed on Cr^{3+} coated steel, as found previously [15]. Figure 5 shows E_{corr}(x) profiles that are separated by a constant distance as time proceeds, suggesting that the system has linear kinetics and is thus controlled by electron transfer.

The presence of electron transfer controlled kinetics has been verified on this Cr^{3+} system previously by Wint et al. [15], using methodology from Stratmann et al. [14]. It was found by Stratmann et al. [14] that a linear relationship exists between delamination distance and square root of time, showing that diffusion of cations from the anode to the cathode is the rate determining step for systems under parabolic controlled kinetics [14], [15]. It was also found that a slower rate would be found for smaller cation sizes, due to the dependence of diffusion coefficient in aqueous electrolytes [14].

However, this rate will not be influenced in a system under electron transfer controlled kinetics. Given this, Wint et al. [15] produced E_{corr}(x) profiles on a Cr^{3+} coated steel sample where corrosion was initiated, in separate experiments, using NaCl and CsCl. Delamination rates were found, within experimental error, to be similar and thus confirming presence of electron transfer kinetics for Cr^{3+} coated steel [15]. The E_{corr}(x) profiles and kinetic plots for Cr3+ coated steel presented here are in agreement with these findings by Wint et al. [15].

For a system where the rate determining step for cathodic delamination is controlled by the rate of electron transfer between the defect site and cathodic front, delamination distance (x_{del}) is related to the time since electrolyte contact (t_{del}) by Equation 3, where k_d is the delamination rate constant and t_i is the initiation period.

$$x_{del} = k_d(t_{del} - t_i) \quad (3)$$

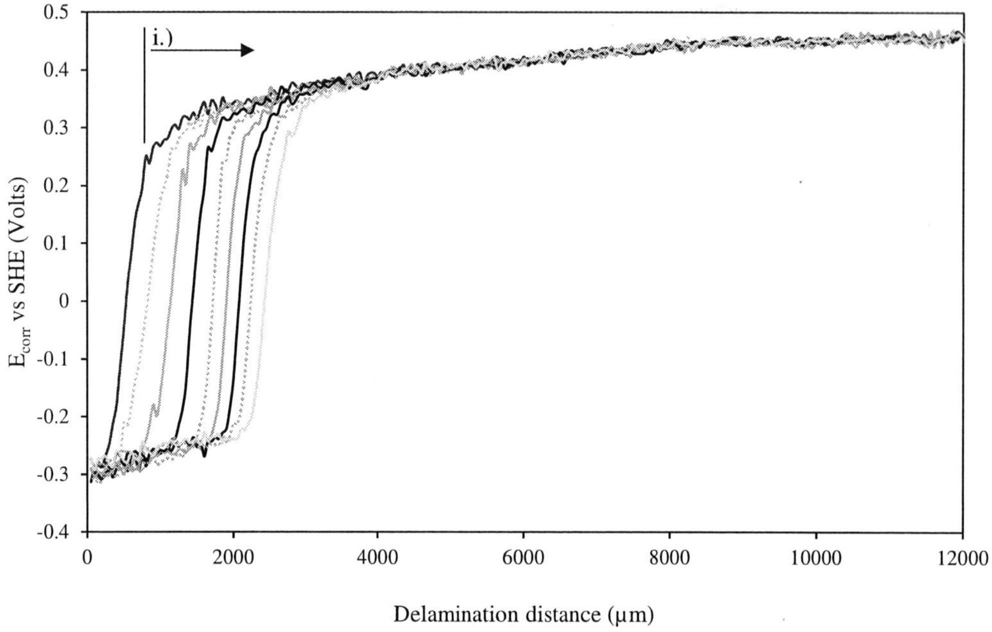

Scan time key (left to right): i.) 2460 minutes after initiation and four-hour intervals thereafter.

Figure 5 A Typical SKP derived E_{corr} vs. distance (x) profile for Cr^{3+} coated steel overcoated with a 30_{um} PVB film.

3.3.2 Effect of Cr^{3+} coated steel coating composition on rate of cathodic delamination

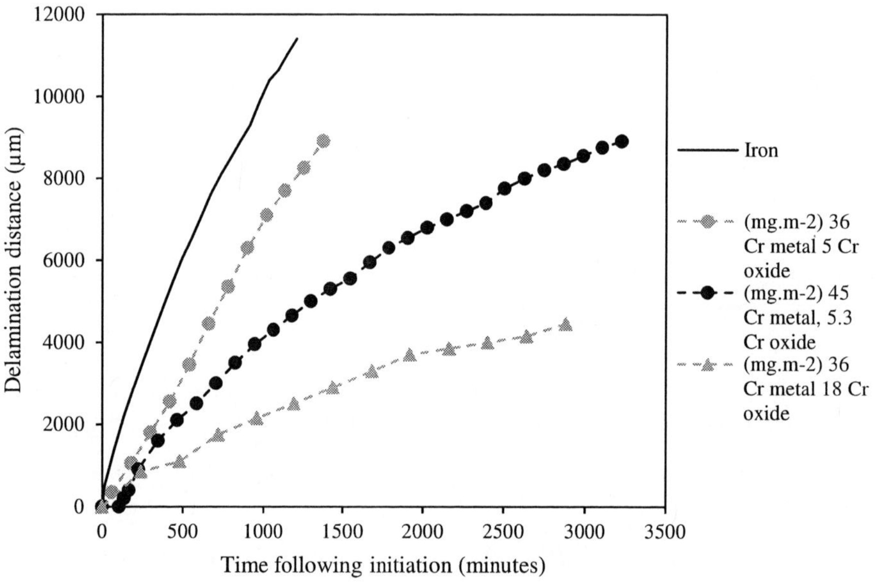

Figure 6 Delamination distance plotted against time since initiation of delamination for Iron and Cr^{3+} coated steel of varying coating weights (mg.m-2) which showed predominantly **parabolic delamination kinetics.**

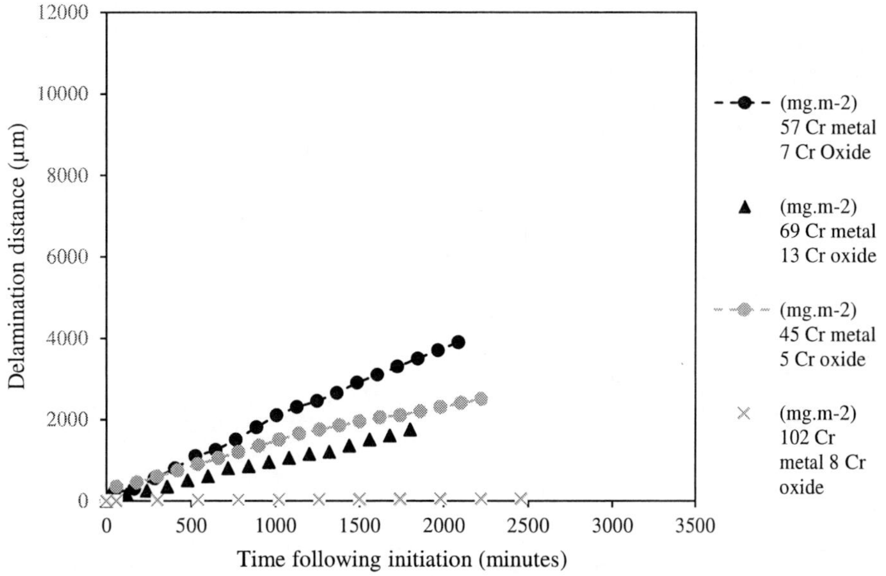

Figure 7 Delamination distance plotted against time since initiation of delamination for Cr^{3+} coated steel variants which exhibited predominantly **linear delamination kinetics**.

Delamination rates for Iron and Cr^{3+} coated steel is summarised in Figures 6 and 7. Figure 6 shows the delamination rate for Iron and low coating weight Cr^{3+} coated steel where the delamination kinetics observed was primarily parabolic, indicating mass transport control. Figure 7 shows the delamination rates for high coating weight Cr^{3+} coated steel, which displayed primarily linear kinetics, indicating that above a certain coating thickness electron transfer is sufficiently slow enough to be the rate limiting step. The influence of coating composition on cathodic disbondment rate will be described by discussing each coating parameter in turn in the following sub-sections 3.3.3 – 3.3.5.

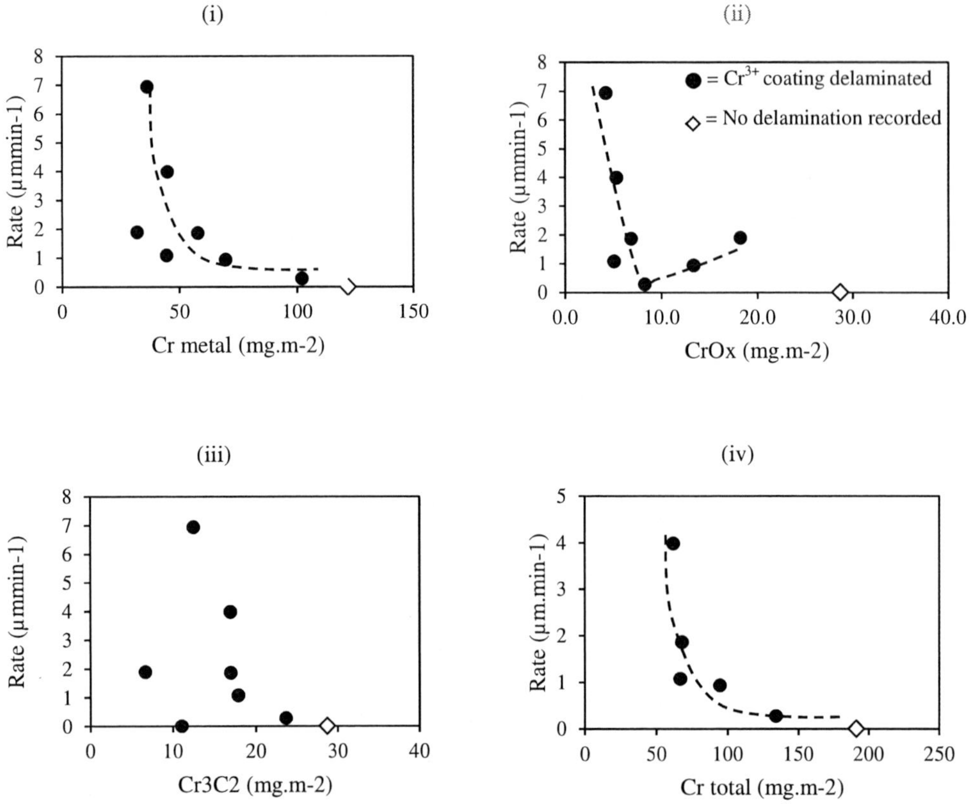

Figure 8 Cathodic delamination rate (µm. min^{-1}) plotted against the coating parameters of Cr^{3+} coatings ($mg.m^{-2}$): (i) Chromium metal (ii) chromium oxide (iii) chromium carbide (iv) Total chromium. The diamond marker point indicates a Cr^{3+}coating variant that did not exhibit cathodic disbondment over 96 hours.

3.3.3 Influence of total chromium and chromium metal coating weight on rate of cathodic disbondment

Cr^{3+} coated steel samples of varying coating thicknesses, detailed in Table 1, were studied. Three repeat scans for each variant were carried out. The delamination rates for Cr^{3+} coated steel is reported against total chromium coating weight ($mg.m^{-2}$) in Figure 8(iv). Unlike the crystalline chromium metal-oxide layer on ECCS, the chromium layer on the Cr^{3+} coated steel under study here is an amorphous mixed chromium metal-oxide-carbide layer [10]. Therefore, delamination rates are also plotted against all coating components, shown in Figure 8 (i-iv).

Deposition of the trivalent chromium coating onto the steel strip is accompanied by a pH change at the strip surface associated with the evolution of hydrogen gas [9], [10]. Deposition of the Cr^{3+} coated steel coating is suggested to be blocked where bubbles of hydrogen are being evolved, thus creating a level of porosity in the Cr^{3+} coated steel coating. This porosity is proposed to provide sites of iron exposure that can support cathodic oxygen reduction and thus support cathodic disbondment [15].

The rate of cathodic disbondment was seen to sharply decrease with increasing overall Cr coating thickness up to ~120 $mg.m^{-2}$. Furthermore, delamination rate also decreased with increasing Cr metal coating thickness between c.a. 50 and 75 $mg.m^{-2}$, above 100$mg.m^{-2}$ performance of the coating was very good.

It is proposed that this is due to greater surface coverage of the steel substrate and less porosity in the chromium electrodeposit making the surface less able to support cathodic disbondment. SEM characterisation of thin and thick Cr^{3+} coatings suggest that thicker Cr^{3+} coatings have reduced porosity. Illustrative examples are shown in Figure 9 and 10.

SEM images show that thicker Cr^{3+} coatings have fewer and smaller pores. In addition, the pores in thicker coatings appear to be either clustered together and not evenly spread out across the surface. The cathodic delamination front may only proceed where there is a consistent path of sites, spatially close to one another, capable of supporting electrochemical cathodic activity across a surface. If a sufficient inert matrix exists between these sites, advancement of the cathodic front may be prevented.

The effect of chromium metal and chromium oxide levels on the performance of polymer coated packaging steel cans in saline sterilisation tests has been previously studied by Boelen et al. [11]. The authors reported that at low levels of Cr metal, iron became exposed after can formation but this exposure was reduced at high Cr metal levels [11]. The authors reported improved performance for cans with higher Cr metal thicknesses and attributed this to reduction in iron exposure, consistent with findings presented here [11].

3.3.4 The role of chromium carbide content in coating on resisting cathodic delamination

The presence of carbide in the Cr^{3+} coated steel coating is coincidental because of the nature of the deposition process. During the electrodeposition of Cr^{3+} coated steel coating Cr^{3+} exists in the electrolyte solution as the very stable $Cr(H_2O)_6^{3+}$ complex [9], [10]. Therefore, Formate ($HCOO^-$) is also included in the electrolyte to destabilise this $Cr(H_2O)_6^{3+}$ complex to enable electrodeposition of chromium onto the steel substrate [9], [10]. It has been found that chromium metal is co-deposited alongside chromium carbide, and generally more carbide is present as the amount of metal in the coating increases [10].

No correlation between chromium carbide coating content versus delamination rate was found as shown in Figure 8(iii). Although they may contribute to increased surface coverage of the steel substrate, it cannot be confirmed from the data here what the role of chromium carbide in the Cr^{3+} coating has with regards to cathodic disbondment.

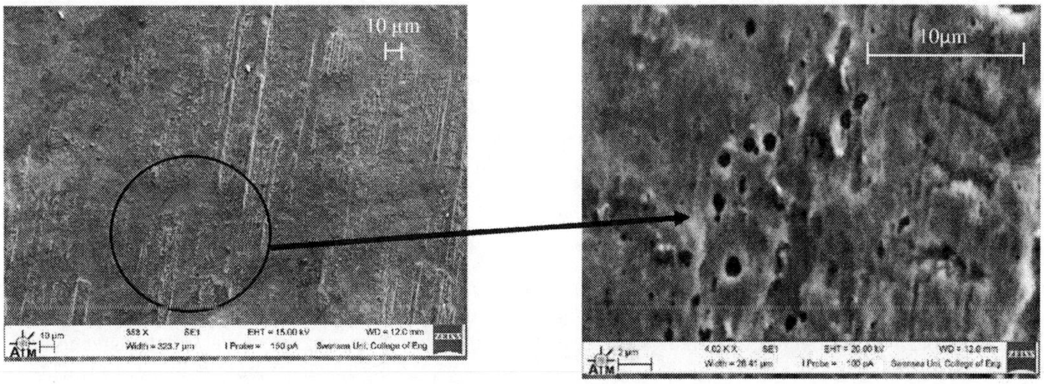

Figure 9 Surface overview of Cr^{3+} coated steel with low coating weight 59 $mg.m^{-2}$ Cr total - insert shows proposed porosity in coating thought to be initiating sites for corrosion.

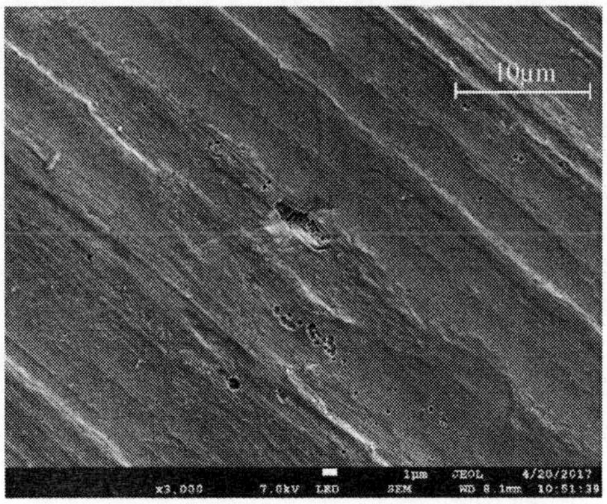

Figure 10 SEM image of Cr^{3+} coated steel surface of high coating weight (164 $mg.m^{-2}$ total-Cr) showing reduced porosity compared to low coating weight.

3.3.5 The Influence of outer oxide levels on the rate of cathodic disbondment

A reduction in delamination rate was also observed for Cr^{3+}coatings with increasing chromium oxide content. A sharp drop in delamination rate is observed as chromium oxide levels increase up to 10 $mg.m^{-2}$ as shown in Figure 8(ii). Above 10 $mg.m^{-2}$ an increase in delamination rate is observed, before dropping to zero at 28 $mg.m^{-2}$ chromium oxide.

At low Cr^{3+} coating weights delamination kinetics were predominantly parabolic indicating that the delamination rate is limited by the migration of sodium ions underneath the polymer coating from the external defect site to the delamination front. However, for thicker Cr^{3+} coatings delamination kinetics were observed to be predominantly linear indicating that the rate of cathodic disbondment on thicker coatings is limited by the rate of electron transfer across the surface, and is in agreement with findings on this substrate previously by Wint et al. [15].

Chromium oxide is a wide-band gap semiconductor and therefore behaves as an electrical insulator. It is proposed here that chromium oxide may slow the rate of electron transfer across the surface such that it becomes the rate determining step for cathodic disbondment. It is suggested from the data presented here that elevated levels may be more effective blockers of electron transfer and through this cause a reduction in cathodic disbondment rate.

These findings are also in line with those reported by Boelen et al. [11], who saw that Electro Chromium coated steel with higher oxide levels had better adhesion to the coating and thus had better resistance to the propagation of underfilm corrosion. Boelen et al., [11] also reported an improvement in corrosion performance, in terms of polymer film blistering, for ECCS cans with higher chromium oxide levels.

The reduction in performance above 10$mg.m^{-2}$ chromium oxide may be explained by cracking of the Cr^{3+} coating that occurs as the coating shrinks during the drying process following electrodeposition. Thicker coatings may encounter higher residual stresses as they dry compared to thinner coatings, causing them to crack and create exposure sites to the underlying steel.

Despite this, a further increase in chromium oxide led to increased performance. At c.a. 28 $mg.m^{-2}$ chromium oxide no delamination was recorded over 96 hours, equal to that of ECCS. However, this variant also had high coating thickness of all other parameters, as shown by the diamond marker point in Figure 8(i-iv). Therefore, this may infer that if there is sufficient coverage of the steel, cracking in the oxide layer does not affect resistance to cathodic disbondment.

4. CONCLUSIONS

The coating thickness of Cr^{3+} coated steel was investigated for its resistance to corrosion driven cathodic disbondment of a PVB coating. Cathodic disbondment was observed on Cr^{3+} coated steel and the kinetics of cathodic delamination was observed to be predominantly linear on thicker coatings, consistent with delamination being under electron transfer control. It was shown that overall coating thickness has a strong negative impact on the rate of cathodic disbondment. This is proposed here to be due to a reduction in porosity in thicker coatings, thus reducing areas of steel exposure that could support cathodic oxygen reduction and enable a continuous cathodic front to propagate. In addition, levels of chromium oxide were also strongly negatively correlated with delamination rate. It is proposed that chromium oxide in the coating impairs electron transfer and thus reduces delamination. Cathodic delamination was not observed on a Cr^{3+} coated steel variant of high overall thickness, where it is proposed that porosity was sufficiently reduced, and where adequate oxide was present in the coating. No correlation was found between delamination rate and the amount of chromium carbide in the coating.

5. ACKNOWLEDGEMENTS

The authors would like to thank TATA Steel Europe and the Engineering and Physical Science Research Council (EPSRC) for their financial support of the Materials Advances Through Training Engineering Researchers (MATTER). Finally, the authors would like to thank TATA Steel Packaging Europe for donating prototype samples of chromium coated steels and for giving resources to this study.

6. REFERENCES

[1] H. J. Gibb, P. S. J. Lees, P. F. Pinsky, and B. C. Rooney, "Lung cancer among workers in chromium chemical production," *Am. J. Ind. Med.*, vol. 38, no. March, pp. 115–126, 2000.

[2] B. C. Schneider, S. L. Constant, S. R. Patierno, R. A. Jurjus, and S. M. Ceryak, "Exposure to particulate hexavalent chromium exacerbates allergic asthma pathology," *Toxicol. Appl. Pharmacol.*, vol. 259, no. 1, pp. 38–44, 2012.

[3] S. S. Wise, C. Wise, H. Xie, L. J. Guillette, C. Zhu, J. P. Wise, and J. P. Wise, "Hexavalent chromium is cytotoxic and genotoxic to American alligator cells," *Aquat. Toxicol.*, vol. 171, pp. 30–36, 2016.

[4] K. L. Witt, M. D. Stout, R. A. Herbert, G. S. Travlos, G. E. Kissling, B. J. Collins, and M. J. Hooth, "Mechanistic insights from the NTP studies of chromium.," *Toxicol. Pathol.*, vol. 41, no. 2, pp. 326–42, 2013.

[5] M. Peter and M. Neil, "Factors affecting the reduction and absorption of hexavalent chromium in the gastrointestinal (GI) tract: Potential impact on evaluating the carcinogenicity of ingested hexavalent chromium," Washington, D.C., 2012.

[6] W. F. Kelly, P. Ackrill, J. P. Day, M. O. Hara, C. T. Tye, I. Burton, C. Orton, and M. Harris, "Cutaneous absorption of trivalent chromium : tissue levels and treatment by exchange transfusion," pp. 397–400, 1982.

[7] H. W. Kuo, "Chromium (VI) induced oxidative damage to DNA: increase of urinary 8-hydroxydeoxyguanosine concentrations (8-OHdG) among electroplating workers," *Occup. Environ. Med.*, vol. 60, no. 8, pp. 590–594, 2003.

[8] R. S. Luippold, K. A. Mundt, R. P. Austin, E. Liebig, J. Panko, C. Crump, K. Crump, and D. Proctor, "Lung cancer mortality among chromate production workers," *Occup. Environ. Med.*, vol. 60, no. 6, pp. 451–457, 2003.

[9] M. Steegh and J. Wijenberg, "Patent WO 2014 079911 A2.pdf," 2014.

[10] J. H. O. J. H. O. J. Wijenberg, M. Steegh, M. P. P. Aarnts, K. R. R. Lammers, and J. M. C. M. C. Mol, "Electrodeposition of mixed chromium metal-carbide-oxide coatings from a trivalent chromium-formate electrolyte without a buffering agent," *Electrochim. Acta*, vol. 173, pp. 819–826, 2015.

[11] B. Boelen, H. Den Hartog, and H. Van der Weijde, "Product performance of polymer coated packaging steel, study of the mechanism of defect growth in cans," *Prog. Org. Coatings*, vol. 50, no. 1, pp. 40–46, 2004.

[12] X. Zhang, B. Boelen, P. Beentjes, J. M. C. Mol, H. Terryn, and J. H. W. de Wit, "Influence of uniaxial deformation on the corrosion performance of pre-coated packaging steel," *Prog. Org. Coatings*, vol. 60, pp. 335–342, 2007.

[13] G. Williams, a. Gabriel, a. Cook, and H. N. McMurray, "Dopant Effects in Polyaniline Inhibition of Corrosion-Driven Organic Coating Cathodic Delamination on Iron," *J. Electrochem. Soc.*, vol. 153, no. 10, p. B425, 2006.

[14] A. Leng, H. Streckel, and M. Stratmann, "The delamination of polymeric coatings from steel. Part 2: First stage of delamination, effect of type and concentration of cations on delamination, chemical analysis of the interface," *Corros. Sci.*, vol. 41, pp. 579–597, 1999.

[15] N. Wint, A. C. A. de Vooys, and H. N. McMurray, "The corrosion of chromium based coatings for packaging steel," *Electrochim. Acta*, no. 2015, pp. 1–11, 2016.

[16] A. Leng, H. Streckel, and M. Stratmann, "The delamination of polymeric coatings from steel. Part 1: Calibration of the Kelvinprobe and basic delamination mechanism," *Corros. Sci.*, vol. 41, pp. 547–578, 1999.

[17] P. Delahay, "A Polarographic Method for the Indirect Determination of Polarization Curves for Oxygen Reduction on Various Metals," *J. Electrochem. Soc.*, vol. 97, p. 205, 1950.

[18] G. Grundmeier, W. Schmidt, and M. Stratmann, "Corrosion protection by organic coatings: electrochemical mechanism and novel methods of investigation," *Electrochim. Acta*, vol. 45, no. 15–16, pp. 2515–2533, 2000.

[19] P. C. Dodds, G. Williams, and J. Radcliffe, "Chromate-free smart release corrosion inhibitive pigments containing cations," *Prog. Org. Coatings*, vol. 102, p. , 2016.

[20] G. Williams, S. Geary, and H. N. McMurray, "Smart release corrosion inhibitor pigments based on organic ion-exchange resins," *Corros. Sci.*, vol. 57, pp. 139–147, 2012.

[21] M. Rohwerder and F. Turcu, "High-resolution Kelvin probe microscopy in corrosion science: Scanning Kelvin probe force microscopy (SKPFM) versus classical scanning Kelvin probe (SKP)," *Electrochim. Acta*, vol. 53, pp. 290–299, 2007.

[22] P. A. Sørensen, K. Dam-Johansen, C. E. Weinell, and S. Kiil, "Cathodic delamination: Quantification of ionic transport rates along coating-steel interfaces," *Prog. Org. Coatings*, vol. 68, no. 1–2, pp. 70–78, 2010.

7. AUTHOR DETAILS

James Edy is a 3rd Year PhD student, part of the Corrosion and Coating Group at Swansea University. He is sponsored by TATA Steel Packaging and his project aims to characterise the underfilm corrosion performance of a novel trivalent chromium passivation system for packaging steel cans. The work presented today discusses the resistance of these trivalent chromium coatings to cathodic disbondment, as studied using the Scanning Kelvin Probe (SKP).

Professor Neil McMurray, alongside Professor Dave Worsley, is a group leader of the Corrosion and Coating Group within the Materials Research Centre (MRC), College of Engineering at Swansea University. He has over 20 years' experience in corrosion science research including development of scanning electrochemical techniques such as Scanning Vibrating Electrode Technique (SVET) and Scanning Kelvin Probe (SKP). He has over 50 publications in the field of corrosion science with regards to underfilm corrosion mechanisms and corrosion inhibitor pigments. He is the primary academic supervisor for James Edy.

Dr. Arnoud de Vooys (PhD) is an electrochemist who is the principle researcher at TATA Steel R&D Ijmuiden, The Netherlands. He has many publications investigating corrosion of coated steel products using the Scanning Vibrating Electrode Technique (SVET) and Electrochemical Impedance Spectroscopy. He directly supervises James' research project giving guidance and input from both an academic and industrial viewpoint.

EVALUATION OF THE SURFACE TOPOGRAPHY OF CORRODED OLD CAST IRON PIPES

Z. Soltani Asadi, R.E. Melchers
Centre for Infrastructure Performance and Reliability, The University of Newcastle, Australia

SUMMARY: Pitting corrosion is among the main reasons of the failure of cast iron water mains world-wide. Estimating the growth and the progression of pits along the surfaces of pipes would assist in predicting the remaining service life of the pipes. In this study, topographies of the surfaces of old cast iron pipes (up to 129 years old) are investigated by computing the contour maps of the surfaces. The analysis of the digital contour maps from real pipes supports previous observations that pitting corrosion is not a continuous process. It appears that a sequence of events occurs, in stages, which changes the topographies of the surfaces. Soon after first exposure, micro pits are generated, and these grow in both size and depth. A number of pits may grow faster and sideways and eventually form larger pits, termed broad or macro pits. A new population of pits then tends to initiate on the inside surfaces of broad pits, which results in the pits having a bench like shape. Since other pits nucleate and commence to grow after a time, or stop growing, it was seen that after a long exposure period there were still a significant number of isolated pits on the surfaces.

Keywords: Pitting corrosion, Cast iron pipes, Surface topography, Pit populations

1. INTRODUCTION

Pitting corrosion of cast iron pipes is among the main reasons of the deterioration and the perforation of pipes after many years exposure in soil environments. Since in Australia and the UK the majority of the pipes are internally cement lined, the most corrosion problem is the external corrosion of pipes exposed to the corrosive soil [1].

Despite much research into investigating pitting corrosion of steels and buried cast iron pipes [2-7], only relatively limited attention has been given to the progression and the size of pits. The growth of pits in size can cause leaks through the pipe wall, as well as initiate cracking, which can then result in pipe bursts. As noted, these can lead to expensive repairs or require replacement. The extent and geometry of pits (or their plan areas) are also responsible for the failure and mechanical properties of other infrastructure [2, 8-12].

Reviewing the literature shows much research on the geometries of pits [8-13] is the result of experimental observation of short-term pitting corrosion (typically less than an hour). The literature review conducted for this study discovered a lack of information on the pit geometries of long-term corrosion, that is over many years, which is of most interest in infrastructure applications [4, 14]. A number of studies on the pit depths behaviour of variety of materials and environments [15-17] reported a difference between behaviour of short term and long term corrosion, such as is shown through a so-called bi modal model.

Kleiner et al. [2] investigated the geometry of pits through using the field data from exhumed ductile iron water mains in 4 different North American cities. Their results from a ring sampling scheme showed that the pit areas and pit volumes could be represented by a Weibull distribution, while a right-truncated Gumbel distribution was considered the best fit to the observed frequency distributions of the pit depths. Their analysis was mainly statistical, and there were no inferences based on a physicochemical understanding of the corrosion processes leading to pitting corrosion. In contrast, data for pitting of old cast iron pipes recovered after various periods of exposure in soils in Newcastle and in Sydney showed that the maximum pit depths of are inconsistent with a single Gumbel extreme value probability distribution [18, 19]. This behaviour is also consistent with the pattern noted earlier for the Extreme Value statistics of pitting in steels [20]. That work indicated that the pitting processes are not continuous in time, but rather that they occur in stages [21]. Therefore, if this is the case for pit depths in a ferrous metal such as steel, a generally similar pattern might be expected for the areas of pits formed in cast irons. The present study was designed to investigate the topography and pit areas of cast iron pipes that have been exposed for long periods of time.

2. EXPERIMENTAL DETAILS

A series of pieces of internally cement lined cast iron water mains were used to examine the pitting corrosion of cast iron pipes. As these pieces became available from different parts of the water supply networks, they can be considered as a set of random samples. The external surfaces of samples, approximately one metre in length, were grit blasted in-situ, with particular attention given to the pitted regions to remove all graphitizes material. Although some 37 pipes are available, herein only the results for 4 randomly elected examples are considered.

Each exposed clean cast iron pipe surface was then scanned using a commercial hand held laser scanner with a resolution for pit depth up to 40 microns and a spatial resolution of 50 microns. Before scanning, all pipes were marked with white targets both for identification and for orientation. The data stream generated by the scanner also was exported in Excel for further processing to determine numerical estimates for the depth of pits on the surface. For this a 2mm x 2mm grid over the surface was used.

In order to analyse the geometric properties of the corrosion pits, the pit depth data, based on the circumferential and longitudinal positions, were imported in MATLB programming to compute the contours of the pipe's external surfaces and then to measure the pit area data. Table 1 gives the information about the pipe details and the maximum pit depth and area for each of the 4 pipes considered herein.

Table 1. Pipe details and pit information

Pipe	Pipe type	Nominal diameter (mm)	Wall thickness (mm)	Exposure time (yr)	Maximum pit depth (mm)	Maximum pit area (cm2)
A	YS	300	11	46	4	1.1412
B	VC	600	22.2	83	10	123.75
C	VC	375	18	129	12	720.695
D	VC	375	18	129	10	1140.25

Pipe type: VC = Vertically Sand Cast, YS = Yennora Spun CI

3. RESULTS

In order to understanding of the growth of pits in the surface, the contour maps of the surfaces have been computed based on the pit depth data and their circumferential and longitudinal positions. The contour maps can show the topographies of the surfaces of pipes. Figures 1-4 give the contour maps of the surface of the pipes and illustrate the topography of the corroded surfaces.

As can be seen in Figures 1-4, the topography of the not so old pipe (pipe A) consists of low densities of corrosion pits along the pipe. The majority of the surface area of this pipe can be interpreted as being cathodic regions with no metal loss from corrosion. The pitted areas mostly also contain isolated single pits. An example of these single pits is shown in Figure 5. Figure 5 (a) is the enlarged view of a part of the contour map of pipe A. However, it should be noted that the horizontal and vertical scales are not the same. The shapes of the pits as shown, therefore, are different to the real pits. Figure5 (b) shows the cross-section view of the section A-A identified in Figure5 (a). The cross-section view of the pits displays that small single pits are formed close together and, as they are small, they have not interacted.

Comparisons between the topographies of the surfaces of the not so old pipe (Figure 1) with those of the older pipes (Figures 2-4) suggest that the cathodic regions have reduced in size with time. The topographies of the surfaces show some pits which appear to have grown in size, rather than in depth, resulting in the formation of relatively large shallow pits. Figure 6 illustrates an example of this category of pit. Figure 6 (a) shows an enlarged view of a part of the surface of pipe D and Figure 6 (b) is the cross-section view of the section A-A in Figure 6 (a). Similar to what was explained for Figure 5 the pit shown in Figure 6 (a) may not be exactly the same as the real pit because of the difference in the scales of the horizontal and vertical axes. These observations indicate that in the pitting corrosion of cast iron pipes, pits can grow either in depth or in size or they can grow simultaneously in both depth and size.

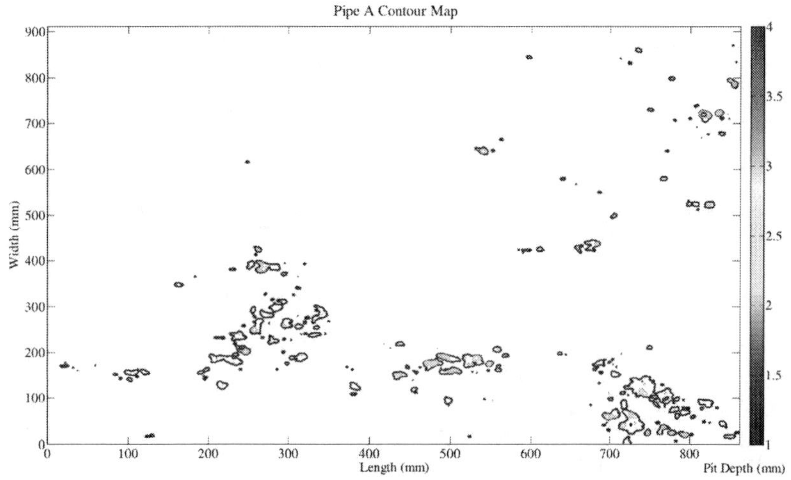

Figure 1: Pit contour map of pipe A (exposure time= 46 years).

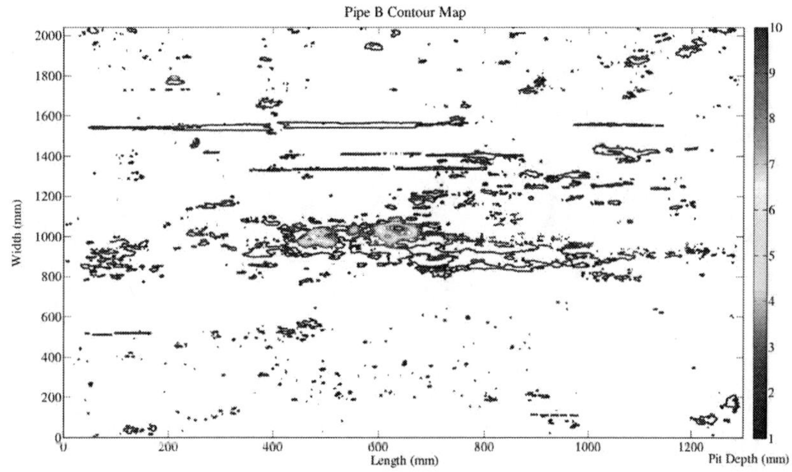

Figure 2: Pit contour map of pipe B (exposure time= 83 years).

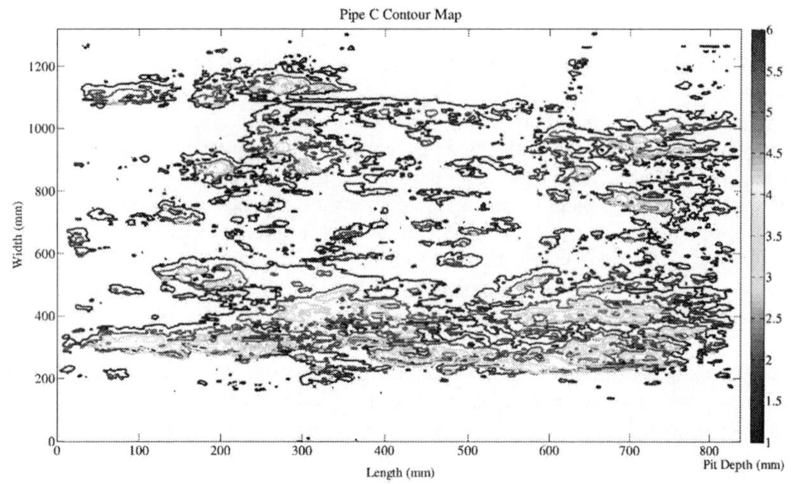

Figure 3: Pit contour map of pipe C (exposure time= 129 years).

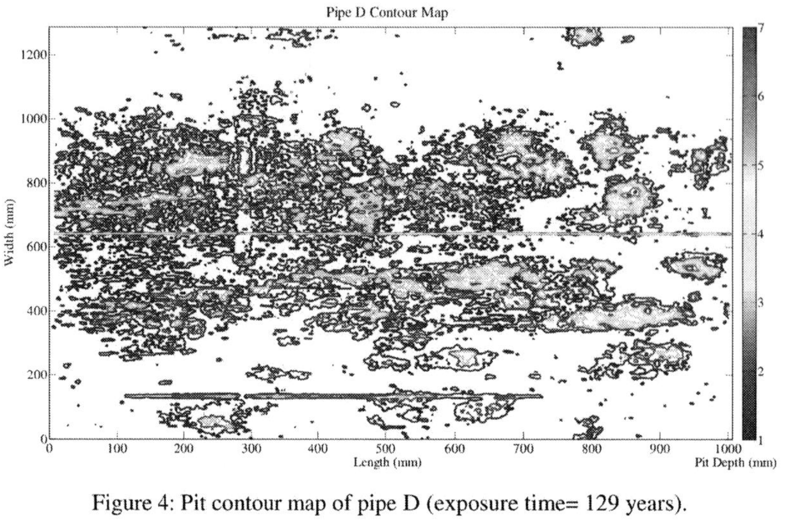

Figure 4: Pit contour map of pipe D (exposure time= 129 years).

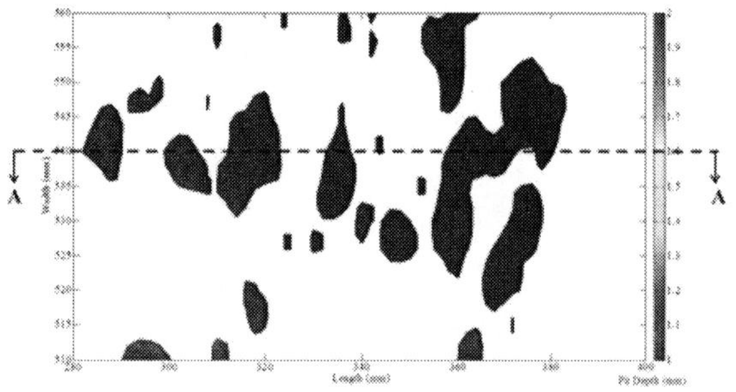

(a)Plan view

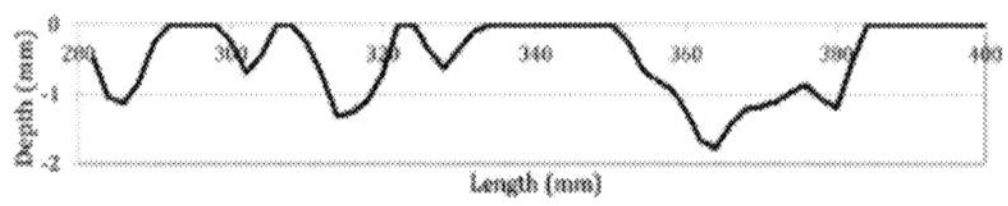

(b)Cross-section view of section A-A

Figure 5: Formation of single pits on the surface of pipe A.

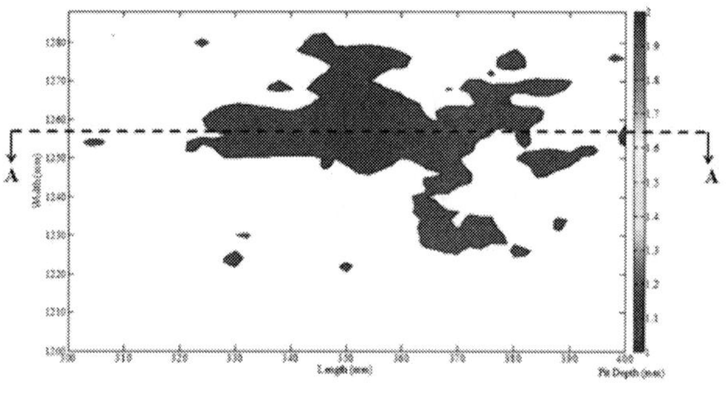

(a)Plan view

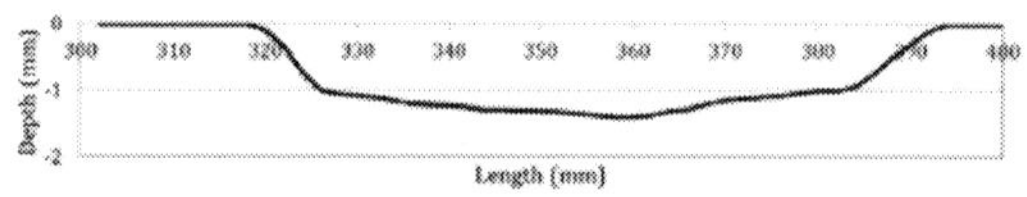

(b)Cross-section view of section A-A

Figure 6: Formation of single pits on the surface of pipe D.

The topographies of the surfaces of very old pipes (see Figures 2-4) show that after a very long time exposure to corrosive soil environments, in addition to the growth of the broad deeper pits, newer isolated pits begin to appear. A better view of the topographies of the pipe surfaces (see Figures 2-4), particularly the surfaces of very old pipe D, also showed that deep pits appeared inside the older, larger-area pits. The topographies of the surfaces in these sections are similar to benches [21]. As observed, some benches become quite deep and the sizes of the pits reduce for the deeper benches. It appears that the small pits form within the large shallow pits. In order to better illustrate this apparent grouping of pits, a sample of bench shape pits was selected from the surface of pipe D and is shown in Figure 7 (a) The cross-section view of a section of this pit (Figure 7 (b)) better demonstrates the bench shape of this category of pit.

The different populations of pits can be seen in the contour maps of very old pipes (see Figures 2-4). The topographies of the surfaces of these pipes show that even after a long period of exposure to corrosive soil (more than 80 years), there are still a significant number of single isolated pits on the surfaces of the pipes. This means that after a very long time exposure to a corrosive soil environment, in addition to growth of the deeper broad pits, newer isolated pits begin to appear. There is another possibility that these pits are related old pits which passivated over time.

There are also some broad large pits in the surfaces of these pipes. These pits may be formed through the coalescence of small pits, which are close to each other, into irregular larger broad pits, and eventually making clusters. Very large broad pits along the surfaces of the pipes, in particular old pipes, may be formed due to the interactions of smaller pits. For example, Figure 8 (a) depicts one sample of an irregular large broad pit, which may have formed by the interactions of multiple pits on the surface of pipe B. A cross-section view of this figure not only shows the interacting multiple pits, but also illustrates the inward stepped edges inward the original broad pits.

Based on the abovementioned results, the corroded surfaces of pipe commonly included different corrosion pit patterns, as shown in Figures 5-8 and as described above.

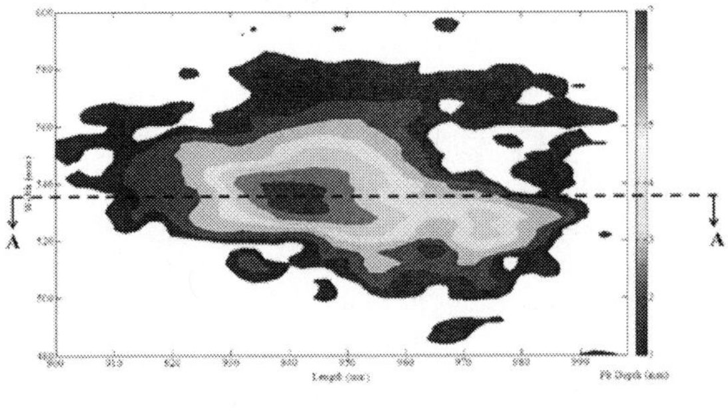

(a)Plan view

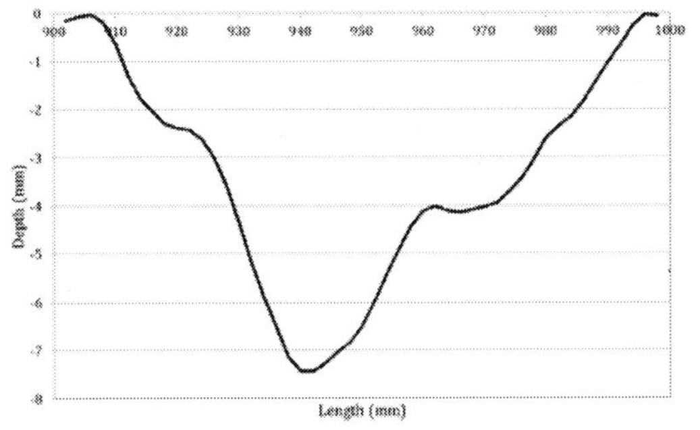

(b)Cross-section view of section A-A

Figure 7: Formation of a bench-shape pit on the surface of pipe D.

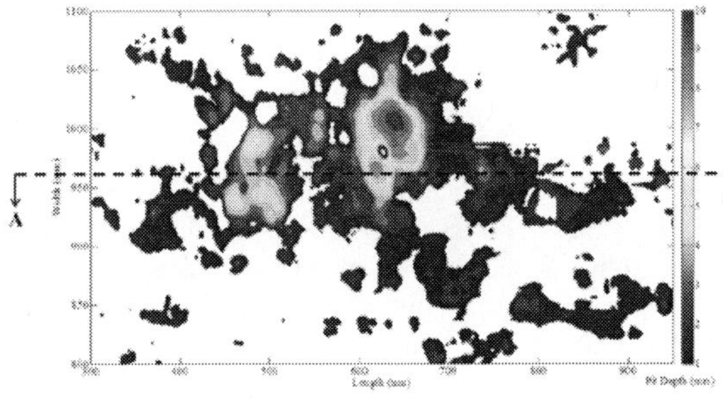

(a) Plan view

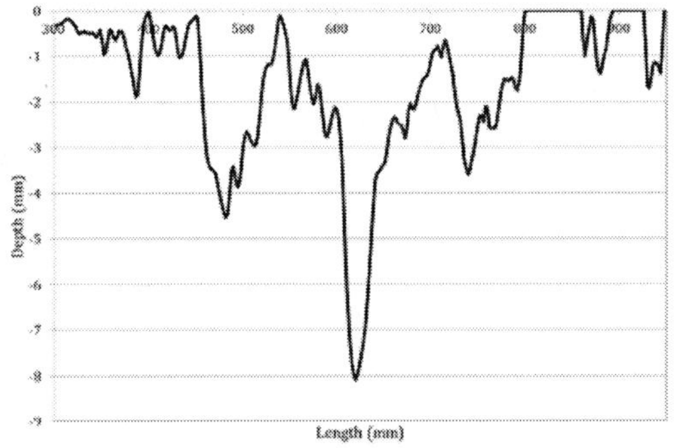

(b)Cross-section view of section A-A

Figure 8: The formation of an irregular large broad pit on the surface of pipe B.

4. DISCUSSION

The results of pit topography analysis (see Figures 1-4) can assist in better understanding the mechanisms of the long-term pitting corrosion of cast iron pipes. The topography results indicate that larger proportions of the surfaces of very old pipes (B, C and D) are corroded in comparison with the surfaces of not so old pipe (A). Furthermore, it has been shown that the areas of pits increase with increasing pit depths. It is possible that this is associated with the step-wise progression of pit depths also seen in steels [21] and is discussed further below.

The observations of the topographies of the scanned exterior pipe surfaces suggest the existence of different pit populations, as a result of long-term exposures. It would appear that this may be the result of the occurrence of a series of events during the development of the pitting corrosion on cast iron. Similar to other ferrous metals, when surfaces of cast iron pipes are exposed to a corrosive environment for the first time, a population of pits initiate and commence to grow [22, 23]. The corrosion loss in the early exposure period is relatively insignificant in comparison with later losses [24]. The cathodic regions can be inferred to decrease with longer time exposure, and that this causes conditions that are appropriate for the nucleation of new pits. The non-uniform topographies of the surfaces of 46 year old pipe (see Figure 1) indicate that the pitting processes occurs at different rates all over the surfaces. As can be seen from this figure, the number of isolated pits for these pipes is much greater than the number of broad pits. However, there are also a significant numbers of isolated pits on the surfaces of very old pipes, as seen in Figures 2-4. This can be explained under the assumption that these pits are initiated much later, and not immediately upon first exposure. It is also possible that these represent pits that have stopped growing, although this possibility is usually considered only in terms of meta-stable pits [25].

The continued corrosion of the surface also leads to the nucleation of new populations of pits, some inside the older, shallow pits. This phenomenon, which previously was observed on the corroded surfaces of mild steel in seawater [21], can be assumed to occur because pits tend to nucleate at surfaces with higher surface energy [26, 27]. This would allow the pits to continue to grow in depth, rather than in size. Therefore, after a time the three following interrelated scenarios are likely to develop. The first is that the cathodic regions gradually break up and more isolated pits initiate and grow. The second scenario is the development of broad pits, both in depth and in area, as a result of the coalescence of earlier pitting that is limited in depth by electrochemical potential limitations [17]. The third scenario is the generation of new pit populations inside the broad pits, which then result in bench shape pits. The different patterns of pits are shown in Figures 5-8. As noted, in addition to the pits growing horizontally and vertically in time, close pits may also join together and make large macro-pits and clusters.

5. CONCLUSIONS

The main scope of this study was attempted to develop a general understanding of how pits develop during pitting process after a long time. The results of the topography of the surfaces presented herein indicate that pits continue to grow to a certain depth, and then as the potential is exhausted grow in size. Furthermore, the non- uniform topography of the surface of the pipes suggests that pitting corrosion is not a continuous process as a function of time. Pitting development in depth and in area appears to occur in stages, which gradually changes the topography of the remaining surface.

6. ACKNOWLEDGMENTS

The financial support of a University of Newcastle Research scholarship and of the Australian Research Council is acknowledged with gratitude. The authors also are indebted to the Critical Pipes Project (CPP) for provision of data. The CPP is a research consortium of Australian Water utilities Sydney Water Corporation, Melbourne Water, Water Corporation (WA), South Australia Water Corporation, South East Water, Hunter Water Corporation and City West Water, together with UK Water Industry Research Ltd. and the Water Research Foundation of the USA. The research partners are Monash University, University of Technology, Sydney and The University of Newcastle.

7. REFERENCES

1. Petersen, R. and R. Melchers, Long-term corrosion of cast iron cement lined pipes. Centre for Infrastructure Performance and Reliability, 2012. 23: p. p1-10.

2. Kleiner, Y., B. Rajani, and D. Krys, Performance of ductile iron pipes. I: Characterization of external corrosion patterns. Journal of Infrastructure Systems, 2012. 19(1): p. 108-119.

3. Kreysa, G. and M. Schütze, DECHEMA Corrosion Handbook-Revised and Extended 2nd Edition. 2008: Dechema.

4. Rajani, B. and Y. Kleiner, Comprehensive review of structural deterioration of water mains: physically based models. Urban water, 2001. 3(3): p. 151-164.

5. Revie, R.W. and H.H. Uhlig, Uhlig's corrosion handbook. Vol. 51. 2011: John Wiley & Sons.

6. Rossum, J.R., Prediction of pitting rates in ferrous metals from soil parameters. Journal (American Water Works Association), 1969: p. 305-310.

7. Sheikh, A., J. Boah, and D. Hansen, Statistical modeling of pitting corrosion and pipeline reliability. Corrosion, 1990. 46(3): p. 190-197.

8. Ernst, P. and R. Newman, Pit growth studies in stainless steel foils. I. Introduction and pit growth kinetics. Corrosion Science, 2002. 44(5): p. 927-941.

9. Kleiner, Y. and B. Rajani, Performance of ductile iron pipes. II: Sampling scheme and inferring the pipe condition. Journal of Infrastructure Systems, 2012. 19(1): p. 120-128.

10. Kondo, Y., Prediction of fatigue crack initiation life based on pit growth. Corrosion, 1989. 45(1): p. 7-11.

11. Pidaparti, R.M. and R.K. Patel, Investigation of a single pit/defect evolution during the corrosion process. Corrosion Science, 2010. 52(9): p. 3150-3153.

12. Zhang, X.-Y., S.-X. Li, and R. Liang. Effect of corrosion pits on fatigue life and crack initiation. in ICF13. 2013.

13. Mankowski, J. and Z. Szklarska-Smialowska, The effect of specimen position on the shape of corrosion pits in an austenitic stainless steel. Corrosion Science, 1977. 17(9): p. 725-735.

14. Kleiner, Y. and B. Rajani, Comprehensive review of structural deterioration of water mains: statistical models. Urban water, 2001. 3(3): p. 131-150.

15. Melchers, R. and R. Jeffrey, The critical involvement of anaerobic bacterial activity in modelling the corrosion behaviour of mild steel in marine environments. Electrochimica Acta, 2008. 54(1): p. 80-85.

16. Melchers, R.E., Long-term corrosion of cast irons and steel in marine and atmospheric environments. Corrosion Science, 2013. 68: p. 186-194.

17. Melchers, R.E., Microbiological and abiotic processes in modelling longer-term marine corrosion of steel. Bioelectrochemistry, 2014. 97: p. 89-96.

18. Soltani Asadi, Z. and R.E. Melchers, Extreme value statistics for pitting corrosion of old underground cast iron pipes. Reliability Engineering & System Safety, 2017. 162: p. 64-71.

19. Soltani Asadi, Z. and R.E. Melchers, Pitting corrosion of older underground cast iron pipes. Corrosion Engineering, Science and Technology, 2017: p. 1-11.

20. Melchers, R.E., Extreme value statistics and long-term marine pitting corrosion of steel. Probabilistic Engineering Mechanics, 2008. 23(4): p. 482-488.

21. Jeffrey, R. and R.E. Melchers, The changing topography of corroding mild steel surfaces in seawater. Corrosion science, 2007. 49(5): p. 2270-2288.

1. Petersen, R. and R. Melchers, Long-term corrosion of cast iron cement lined pipes. Centre for Infrastructure Performance and Reliability, 2012. 23: p. p1-10.

2. Kleiner, Y., B. Rajani, and D. Krys, Performance of ductile iron pipes. I: Characterization of external corrosion patterns. Journal of Infrastructure Systems, 2012. 19(1): p. 108-119.

3. Kreysa, G. and M. Schütze, DECHEMA Corrosion Handbook-Revised and Extended 2nd Edition. 2008: Dechema.

4. Rajani, B. and Y. Kleiner, Comprehensive review of structural deterioration of water mains: physically based models. Urban water, 2001. 3(3): p. 151-164.

5. Revie, R.W. and H.H. Uhlig, Uhlig's corrosion handbook. Vol. 51. 2011: John Wiley & Sons.

6. Rossum, J.R., Prediction of pitting rates in ferrous metals from soil parameters. Journal (American Water Works Association), 1969: p. 305-310.

7. Sheikh, A., J. Boah, and D. Hansen, Statistical modeling of pitting corrosion and pipeline reliability. Corrosion, 1990. 46(3): p. 190-197.

8. Ernst, P. and R. Newman, Pit growth studies in stainless steel foils. I. Introduction and pit growth kinetics. Corrosion Science, 2002. 44(5): p. 927-941.

9. Kleiner, Y. and B. Rajani, Performance of ductile iron pipes. II: Sampling scheme and inferring the pipe condition. Journal of Infrastructure Systems, 2012. 19(1): p. 120-128.

10. Kondo, Y., Prediction of fatigue crack initiation life based on pit growth. Corrosion, 1989. 45(1): p. 7-11.

11. Pidaparti, R.M. and R.K. Patel, Investigation of a single pit/defect evolution during the corrosion process. Corrosion Science, 2010. 52(9): p. 3150-3153.

12. Zhang, X.-Y., S.-X. Li, and R. Liang. Effect of corrosion pits on fatig/ue life and crack initiation. in ICF13. 2013.

13. Mankowski, J. and Z. Szklarska-Smialowska, The effect of specimen position on the shape of corrosion pits in an austenitic stainless steel. Corrosion Science, 1977. 17(9): p. 725-735.

14. Kleiner, Y. and B. Rajani, Comprehensive review of structural deterioration of water mains: statistical models. Urban water, 2001. 3(3): p. 131-150.

15. Melchers, R. and R. Jeffrey, The critical involvement of anaerobic bacterial activity in modelling the corrosion behaviour of mild steel in marine environments. Electrochimica Acta, 2008. 54(1): p. 80-85.

16. Melchers, R.E., Long-term corrosion of cast irons and steel in marine and atmospheric environments. Corrosion Science, 2013. 68: p. 186-194.

17. Melchers, R.E., Microbiological and abiotic processes in modelling longer-term marine corrosion of steel. Bioelectrochemistry, 2014. 97: p. 89-96.

18. Soltani Asadi, Z. and R.E. Melchers, Extreme value statistics for pitting corrosion of old underground cast iron pipes. Reliability Engineering & System Safety, 2017. 162: p. 64-71.

19. Soltani Asadi, Z. and R.E. Melchers, Pitting corrosion of older underground cast iron pipes. Corrosion Engineering, Science and Technology, 2017: p. 1-11.

20. Melchers, R.E., Extreme value statistics and long-term marine pitting corrosion of steel. Probabilistic Engineering Mechanics, 2008. 23(4): p. 482-488.

21. Jeffrey, R. and R.E. Melchers, The changing topography of corroding mild steel surfaces in seawater. Corrosion science, 2007. 49(5): p. 2270-2288.

22. Burstein, G., et al., Origins of pitting corrosion. Corrosion Engineering, Science and Technology, 2004. 39(1): p. 25-30.

23. Butler, G., P. Stretton, and J. Beynon, Initiation and growth of pits on high-purity iron and its alloys with chromium and copper in neutral chloride solutions. British Corrosion Journal, 1972. 7(4): p. 168-173.

24. Melchers, R., Modeling of marine immersion corrosion for mild and low-alloy steels-part 1: phenomenological model. Corrosion, 2003. 59(4): p. 319-334.

25. Szklarska-Smialowska, Z., Pitting corrosion of metals. 1986: National Association of Corrosion Engineers.

26. Aziz, P., Application of the Statistical Theory of Extreme Values To the Analysis of Maximum Pit Depth Data for Aluminum. Corrosion, 1956. 12(10): p. 35-46.

27. Evans, U.R., The corrosion and oxidation of metals: scientific principles and practical application. 1960: Edward Arnold London.

BLUE GREEN WATER CORROSION OF COPPER PIPING – 15 YEARS ON

S. Furman
AECOM Australia Pty Ltd, Melbourne, Australia

SUMMARY: Blue green water (BGW) corrosion is a type of corrosion that predominantly affects new copper piping in cold potable water systems. It appears to be associated with specific a water quality that results in the formation of suspended insoluble copper corrosion products rather than protective copper oxides on the pipe walls. This type of corrosion is usually considered to be more of a water quality issue rather than a significant corrosion issue. While this corrosion process usually results in very slow uniform corrosion losses of the pipe wall, it can result in significant discolouration of the water particularly after periods of stagnation.

Over the years a number of management strategies have been development for copper pipe subject to BGW. These strategies range from regular flushing through to piping replacement. Regular flushing or flushing after period of stagnancy is a fairly common and inexpensive strategy. This paper reviews the current condition of a piping system that was first identified as being subject to BGW approximately 15 years ago. In the intervening period the BGW issue has been managed through a program of flushing. Recently an investigation was undertaken to assess the link between the historical BGW issues and the selected management practice, with the localized corrosion that has recently resulted in pinhole leaks in the piping.

Keywords: Copper, corrosion, pitting, blue-green water.

1. INTRODUCTION

In 2002 a commercial building in Melbourne was opened for use and within two months concerns were raised regarding the blue-green discolouration of the water. The diagnosis of the issue in the copper cold water pipe system was blue green water corrosion. At the time of this diagnosis, this type of corrosion was largely deemed to be a water quality issue associated with very low rates of general corrosion of the copper pipe.

Blue Green Water (BGW) corrosion is a type of corrosion that occurs predominantly in new potable water copper plumbing systems. It results in the formation of high levels of copper hydroxide or other largely insoluble copper corrosion products either on the walls of the pipe or suspended in the water. It is the suspended particulate matter that gives the water the characteristic partially opaque blue/green colouration. Following periods of stagnancy, the first daily use can contain the suspended particles as well parts of the deposited scale that have been loosened due to the water flow. The discolouration of the water is most apparent in the first daily use samples.

The business operations in this building can result in the intermittent use of a number of the facilities including the public toilets, function room toilets, function room tea-points and kitchen facilities. This results in periods of stagnancy in between short period of high use for many parts of the cold water piping network. To combat the BGW issues a decision was made to flush the pipes regularly to minimise the water quality issue. Chilled water units placed at each tea-point and kitchen provided drinking water. These units commonly contain micro-filtration which typically filters out particles to 0.2 μm in size which is considered adequate for the removal of the copper particles that discolour water [1, 2]. This effectively dealt with the water quality issue from both a consumption and a discolouration perspective.

This management strategy for the blue green water issue seemed to be working, though in 2013 the first perforation occurred in the copper network. The two problems were not immediately linked and the pitting corrosion events were dealt with in a reactive manner by the maintenance staff when they occurred. However, a review of the issue in early 2017 noted that the number of perforation events seemed to be increasing each year (Figure 1).

The continued use of a large number of rooms adjacent to low use toilet and tea-point facilities is considered critical to the function of the business in the building. Perforations and associated water leaks have the potential to interrupt the use of the facilities. The observed increase in the number of perforation events per annum is therefore seen as an increase in risk to the operations of the business. The first step in managing this risk was deemed to be an investigation into the cause of the perforations.

2. PIPING SYSTEM INFORMATION

The cold water piping network in the building is a distributed system with ten main droppers from the ring main and water storage tanks located at the top of the building. Each dropper feed a local section of piping on each floor and there is no interconnection between each dropper. Small bore horizontal pipework, predominantly 20mm diameter, forms the majority of distribution pipework which is located in the ceiling cavities. Short sections of vertical pipe form the final runs to each distortion point, e.g. tap or toilet, in the network.

Based on the markings on one of the failed pipe sections it has been assumed that the copper piping is Class B to Australian Standard AS 1432. According to this standard the material is phosphorous deoxidised copper, specifically grade UNS C12200. This alloy has a nominal composition of 99.9 % (min.) Copper and 0.015-0.040 % Phosphorous. A class B 20 mm nominal diameter pipe has a wall thickness of 0.88-1.17 mm.

The system has been constructed from pipe ranging in size from 15-100 mm diameter, though the 15 mm pipe has only been used in short vertical drops to a delivery point. All of the known pipe failures have been noted in the 20 mm piping, predominantly in the horizontal sections. Figure 1 shows the number of failures recorded each year since the piping was installed in 2002.

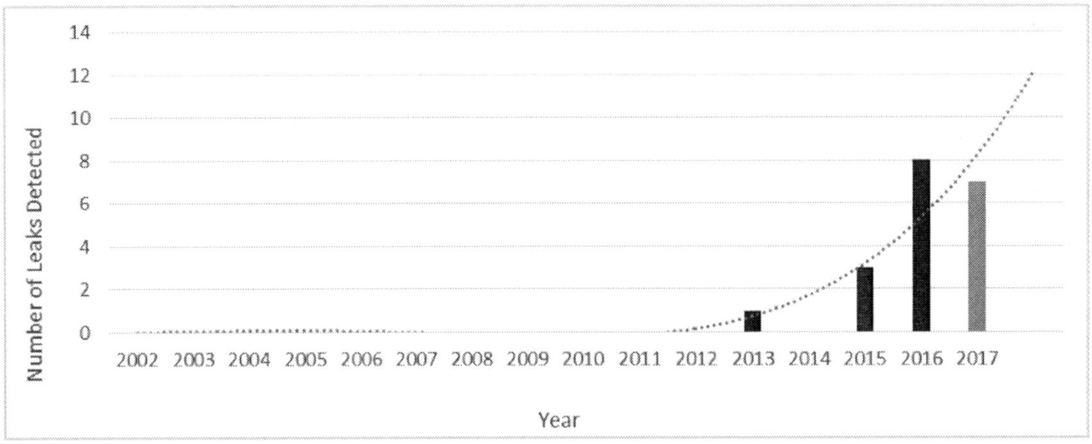

Figure 1. Recorded pinhole failures in the copper cold water piping system

3. FAILURE INVESTIGATION

Since approximately 2015 the facility management company had collected sections of the perforated pipe therefore a number of historic samples as well as several recent samples were available to be studied following the decision to investigate the cause of failure. Four of the more recent samples, all from horizontally oriented pipe, were selected to be analysed in the investigation. The four samples were from three floors in the building and from different parts of the network on each floor.

The investigation focused around the perforation site on each of the samples. Optical microscopy, scanning electron microscopy, energy dispersive x-ray analysis and Fourier transform infrared analysis techniques were employed to investigate the failures.

4. RESULTS

4.1 Visual examination

The perforation sites on each tube were clearly identified by the presence of a blue-green deposit on the external surface of the tube (Figure 2). Internal examination of the tube indicated the presence of extensive green, blue and black deposits that varied in thickness, texture and adhesion to the surface of the copper (Figure 3 to Figure 7). It is believed that the thicker deposits were located on the bottom of the horizontal tubes.

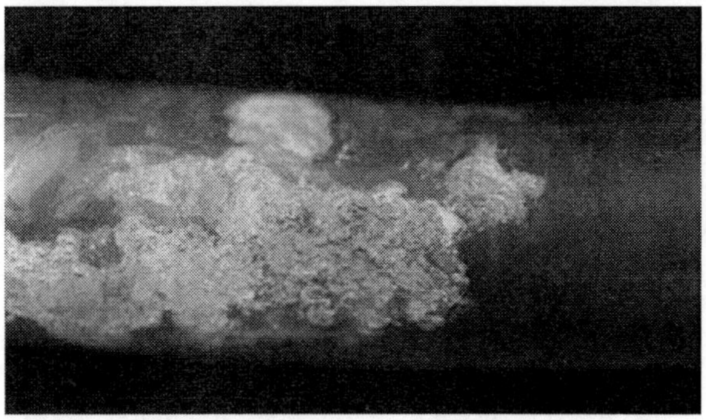

Figure 2. Suspected site of tube perforation

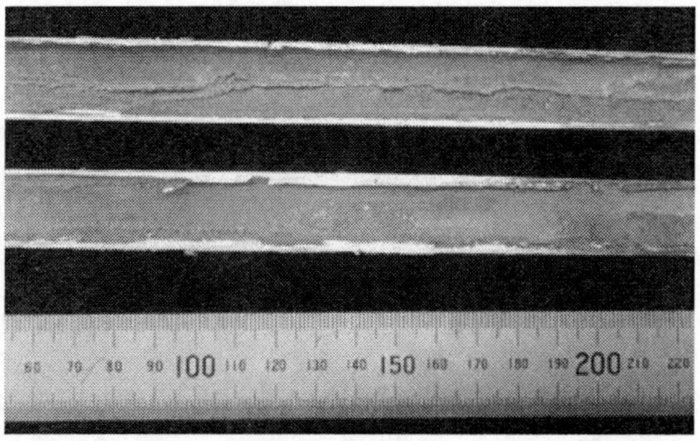

Figure 3. Variation in thickness of the deposits in the pipe

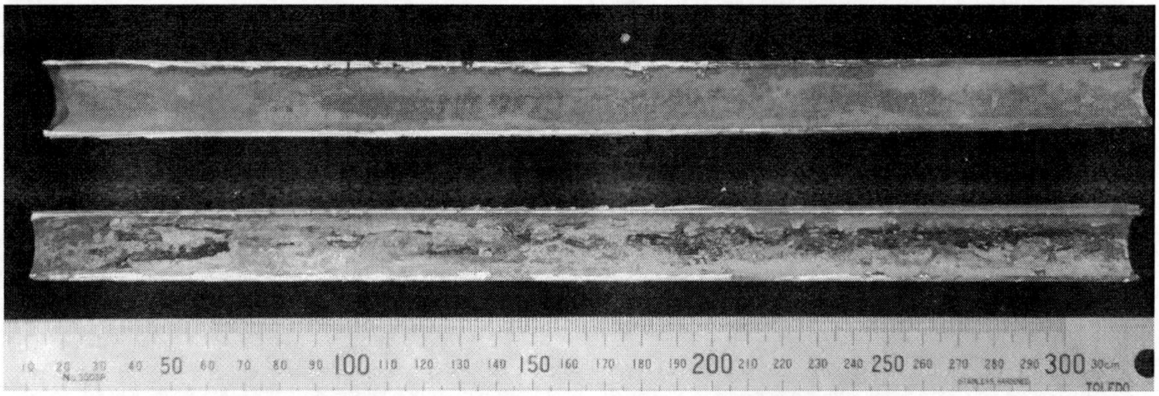

Figure 4. Variation in the colour of the internal deposits

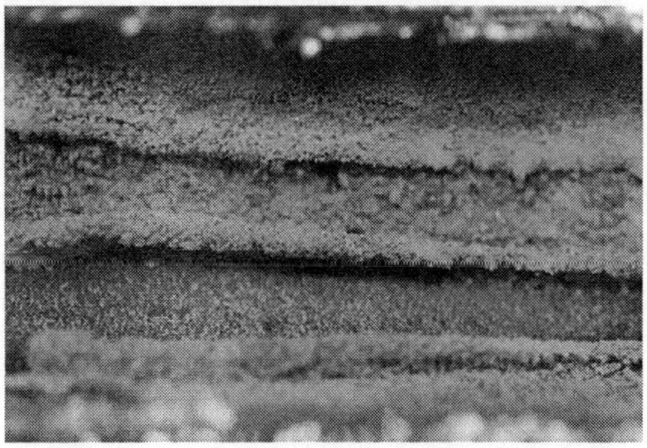

Figure 5. Close-up image of green surface deposits

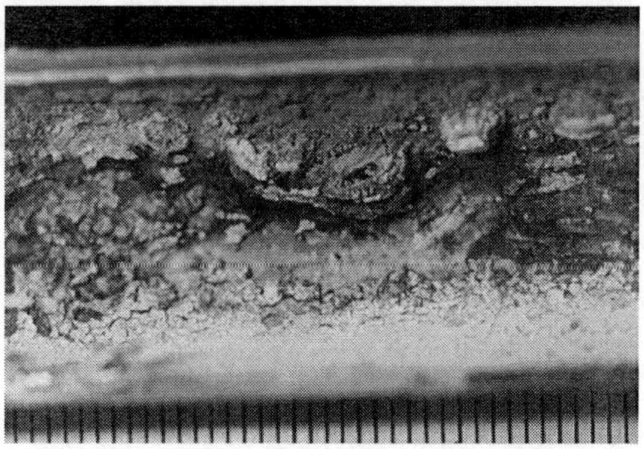

Figure 6. Close-up image of black surface deposits

At the internal locations coinciding with the sites of perforation removal of the locally thickened deposit exposed translucent red cubic crystals around and within the pit (Figure 9 and Figure 10). In the case of the perforated pits, the cubic crystals in the pits were obscured by acicular yellowish coloured deposits (Figure 8 and Figure 9). Removal of further green and blue deposits around the pits identified a network of shallow channels of red cubic crystals. The network was denser near the large pits and typically the channels were deeper under the locally thicker deposits.

A number of smaller pits were identified on the surface of the piping all at different stages of development. However only the largest pits, typically greater than 2mm in diameter were observed to contain acicular yellow deposits under the capping layer of the pit.

Investigation of the black deposit indicated although in many areas it was partly covered with a green surface deposit but that there did not appear to be any pits growing under the black deposit/products. This black deposit was identified as Cupric Oxide.

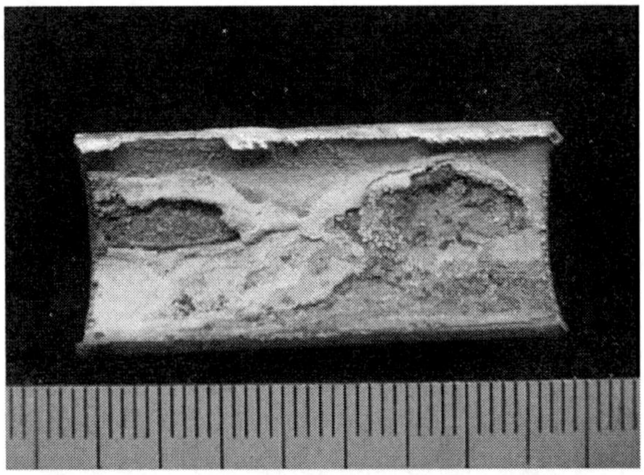

Figure 7. Removal of green surface deposits exposes blue product

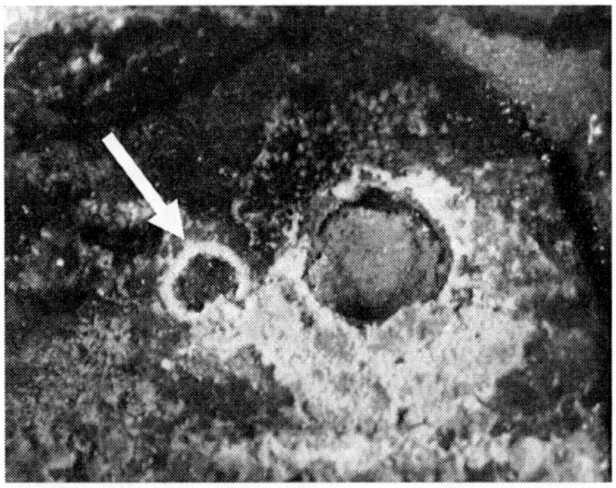

Figure 8. Yellow deposit exposed following removal of pit capping product from a perforated pit. (Secondary pit indicated by the white arrow)

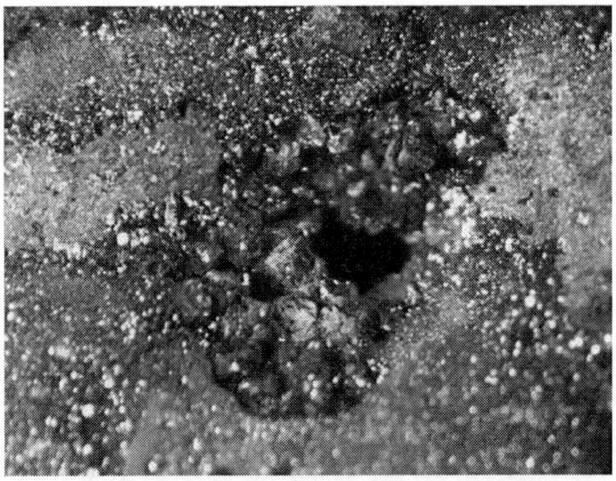

Figure 9. Red cubic crystal and remnants of acicular yellow deposit in a perforated pit

Figure 10. Shallow network of red crystals surrounding a perforated pit

4.2 Chemical analysis

Chemical analysis of the deposits on the internal surface of the pipe and within the corrosion pits was undertaken using the energy dispersive x-ray (EDX) analysis on a scanning electron microscope. The range of elements detected in the surface deposits and the corrosion deposits in the pits by EDX are noted in Table 1. No unusual elements were detected.

Fourier transform infrared (FTIR) analysis on loose deposits removed from the tube identified the presence of carbonates, hydroxides, silicates and possibly sulphates.

Table 1. EDAX SEM analysis of corrosion deposits

Element	Surface Deposits (wt %)	Deposits in cross sectioned pit (wt %)
Carbon	8.68-7.21	46.4-54.35
Oxygen	28.33-31.81	14.49-1.41
Magnesium	0-1.19	0
Aluminium	0.17-0.48	0
Silicon	1.00-5.35	0.17-0.46
Sulphur	0-2.14	0-0.15
Chloride	0-0.38	0-0.06
Calcium	0-0.22	3.41-7.04
Iron	0-0.47	0
Copper	49.40-55.46	22.28-34.25

The red crystalline deposit was identified a cuprite (cuprous oxide) and the black deposit as cupric oxide. The limited quantity of the acicular yellow product prevented a direct analysis but the description of the product, its colour and the trace amount of chloride detected near the yellow product in the pits, suggests that the product may be a crystalline form of copper chloride.

4.3 Micrographic analysis

The features of all the pits examined were consistent with wide shallow pits rather than narrow deep pits (Figure 11 and Figure 12). Detailed examination indicated that the large pits actually contained many sub pits. Smaller less mature pits were identified around the rim of the larger pits (Figure 8 and Figure 13). The extent of the deposit over the pit did not appear to correlate to the size of the pit beneath the deposit.

General attack and minor pits were identified around the circumference of the tube as would be expected to occur due to blue-green water corrosion (Figure 14).

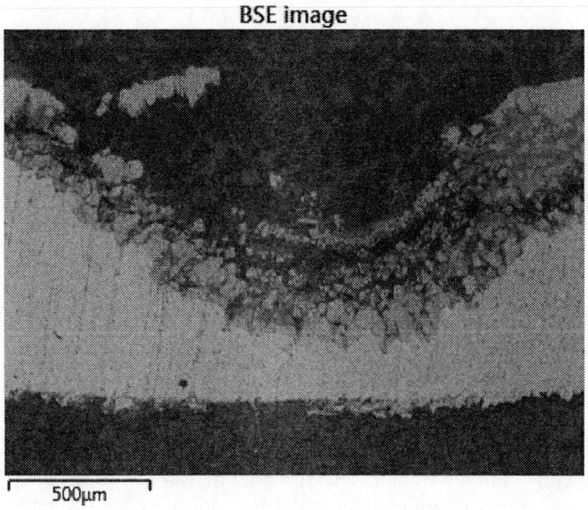

Figure 11. Back scattered image of a cross section through a corrosion pit

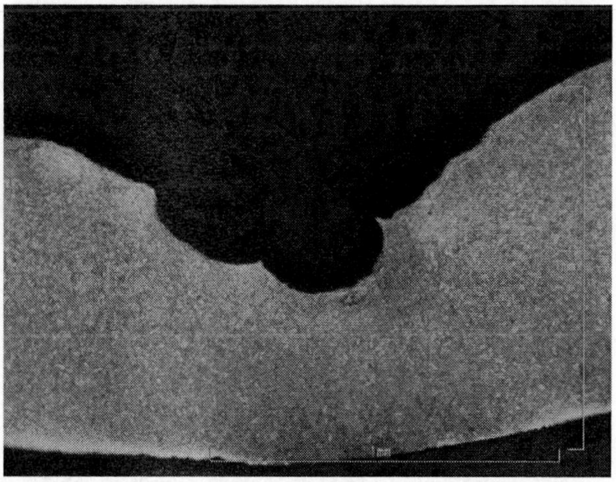

Figure 12. Optical image through a corrosion pit

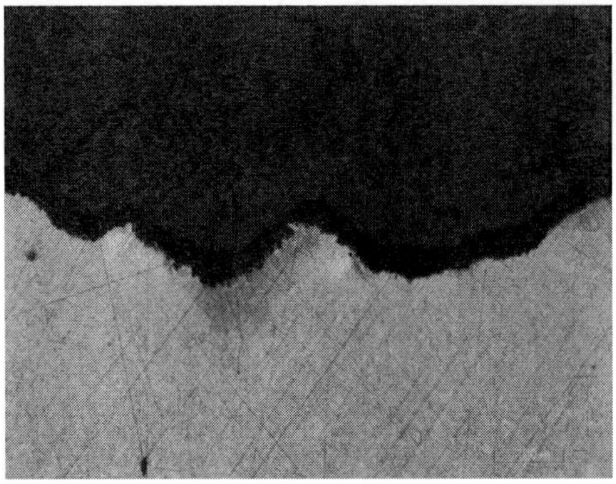

Figure 13. Smaller pits within a larger pit

Figure 14. General attack and minor pitting of the internal surface remote from the corrosion pits

4.4 Water Quality

Basic water quality testing undertaken in late 2016 and early 2017 indicated the incoming water to the building has a pH 7.4 – 7.9 and a total dissolved solids between 70-90 mg/L on the days of the measurement. Water samples collected on different floors of the building varied from pH 6.9 to pH 8.6 and TDS from 60-80 mg/L depending on the day and location of the measurements. The slight alkalinity of the water and the very low total dissolved solids are two of the key characteristics of water capable of supporting BGW corrosion [3].

Historic water quality measurements concentrated on copper levels which were shown to fluctuate up to 10 mg/L.

5. DISCUSSION

The investigation clearly indicated that the failure of the pipes examined was due to pitting corrosion yet BGW corrosion is still occurring in the cold water piping in the building. The majority of the literature on BGW corrosion describes this phenomenon as one of general corrosion of new piping that is considered to be more of a water quality issue than a significant corrosion issue [2-5]. This assertion is based on the discolouration of the water and the associated health issues with the release of high concentrations of insoluble copper products into the water particular during the first daily use of the system. However there are also a number of cases where water pipes suffering from BGW have over time developed pitting corrosion that has led to pipe perforation [1, 2, 6, 7]. One reported instance was of a 19 mm diameter pipe perforating in five years [7] and another of a pipe of unknown size perforating in six years [6].

BGW corrosion results in the formation of high levels of copper hydroxide or other largely insoluble copper corrosion products [5] either on the walls of the pipe or suspended in the water. It is the suspended particulate matter that gives the water the characteristic partially opaque blue/green colouration. Research has shown that it is most likely that a combination of microbiological, chemical and operational factors is involved in the initiation of BGW corrosion [1, 2, 7, 8].

One mechanism suggested in 1996 was that a locally adherent biofilm may stabilise corrosion processes on the surface of copper preventing the formation of stable protective oxides [8]. Other research presented in 1996 suggested that natural organic matter (NOM) can prevent the formation of stable surface films on copper and can increase copper release into the water and stabilise suspended particles [9]. But given that dissolved organic matter provides the energy source for bacterial growth the two concepts are possibly linked. Although there is strong evidence to support the microbiological contribution to BGW initiation, once established, it is appears that BGW corrosion can continue in the absence of microbiological factors [2, 7]. A variety of bacteria have been analysed from pipe with BGW including sulphate reducing bacteria [2], *Acidovarax spp* and *Sphingomonas sp.*[4] and *Pseudomonas paucimoblis*, *Rhodotorula spp.* and *Flavobacterium spp.* [8] though the exact role of these and other species that are likely to be present in potable water still needs to be determined.

Chemical factors that appear to promote BGW corrosion include low total dissolved solids, low buffering capacity, slightly alkaline pH, low residual free chlorine, high assimilable carbon content in the water. The term total dissolved solids (TDS) includes the inorganic salt predominantly of calcium, magnesium, potassium, sodium, bicarbonates, chlorides and sulphate as well as the organic matter that is found in water. Water with a TDS of less than 300 mg/L [3] is able to support BGW even with chloride and sulphates levels less than 7 mg/L [3] when free residual chlorine levels are less than 0.1 mg/L [6, 10]. Water containing 15-40 mg/L of calcium carbonate can be stated as having a low buffering capacity [10]. Water with a low buffering capacity will be subject to fluctuating pH due to biological or chemical processes [2] including corrosion processes or lime dissolution from cement lined pipe [6]. BGW corrosion has been observed in water with pH varying from

7.5 to 10 if the water has negligible free chlorine residual [1]. Evidence suggests it is possible that BGW corrosion will continue even if the pH of the potable water decreases [6]. This is supported by the water quality analysis of the cold water network in the building where local measurements indicated elevated copper levels were occurring at less than pH 7.5. One factor that may have contributed to the initiation of BGW corrosion in the buildings pipework may have been the installation of a new mild steel cement lined supply pipeline around the time the building was under construction. This may have resulted in high pH being used to commissioning the cold water pipe network in the building.

The presence of high levels of assimilable organic carbon (AOC) also appears to be a requirement for BGW [2]. This factor is related to the microbiological aspect of BGW. AOC refers to the fraction of the dissolved organic content in the water that is readily available to micro-organism to utilise as food/energy for growth. There is a complex relationship between AOC, chlorine, pH and bacterial growth. While disinfection with chlorine can reduce bacterial count in the water it can also react with the carbon compounds resulting in an increase in the fraction of AOC in the total organic content of the water [11, 12]. The pH of the water determines the balance between hypochlorous acid and hypochlorite ions, the two chemicals produced when chlorine reacts with water [13]. At pH greater than 7.5, hypochlorite ions are more stable than hypochlorous acid, but this form is significantly less effective as a disinfectant in water than hypochlorous acid. Importantly water that is biologically stable at pH less than 7.5 may be unstable above this pH resulting in the growth of bacteria. However not all bacteria will contribute to corrosion of the copper tubes. Only bacteria that form colonies on the surface of the pipe and affect either one of the half reactions of the corrosion process or alter the mechanism by altering the local environment can contribute to MIC of the copper pipe [14].

The main operational factor that is believed to have contributed to the initiation of BGW is stagnancy [2] of the water within a short period of the commissioning of the system [7]. Stagnancy in the piping in the early life of the piping can have a two-fold effect on the development of corrosion, specifically BGW on the pipe. Stagnancy can reduce dissolved oxygen available for oxide formation and it can promote formation of bacterial colonies on the surface of the pipe. The development of a stable oxide on the pipe requires exposure of the pipe to adequate dissolved oxygen in the water. As the water is circulated around the piping, the dissolved oxygen level reduces as it is consumed in the corrosion reactions to produce an oxide film on the copper. Water at the extremities of the network has the potential to have lower dissolved oxygen level.

To maintain pressure in a system, pipe diameters usually decrease at the extremities of the system, increase the ratio of surface area to volume in the pipe. So at the extremities of the network water with reduced dissolved oxygen content has a larger surface area which it has to effectively provide oxygen for film growth. In a stagnant system, the dissolved oxygen content can readily be consumed. If stagnation occurs early in the life of the system, there is inadequate oxygen for protective film to form. Stagnancy also benefits growth of bacteria, as bacteria can colonise pipe surface more readily under conditions of no flow. Additionally at the extremities of the system, chlorine residual levels are usually lower, so water with low residual chlorine levels when it enters the piping system may be negligible at the extremities of the system. These factors all contribute to the growth of bacterial colonies on the pipe at the extremities of the network [11], and once established the colonies have the ability to stabilise corrosion process [8, 10] and under the conditions defined above cause formation of insoluble copper corrosion products and BGW. Sudden flow conditions during irregular operations of the cold water pipe taps connected to the small diameter pipe at the extremities of the network can produce moderately high, though not necessarily uniform, shear forces on the pipe walls [10].

Build-up of the voluminous BGW insoluble corrosion products on the pipe surface due to stagnation may be sufficient to shield the surface from the bulk water allowing a decrease in pH due to ongoing corrosion processes under the deposits. A similar mechanism has been proposed for pitting corrosion in soft water [15, 16]. Under flowing conditions, shear stress on the pipe walls can remove some of the corrosion product, depending on flow patterns, resulting in the variation in film thickness observed on the tube walls (Figures 3-6). Under the thicker deposits it is hypothesised that the interfacial pH has reduced below pH 5 resulting in the formation of the observed red cubic cuprous oxide crystals. The development of stable conditions for propagation of pitting corrosion in soft water is dependent on the formation of these crystals.

While this mechanism seems feasible and it agrees with part of the overall pitting corrosion mechanism suggested for soft water, it is not the only proposed mechanism for reduction of interfacial pH. A number of studies have implicated microbiologically influenced corrosion in the pitting corrosion process [11, 17], as the biological activity can reduce the interfacial pH to low levels. Considering that the cold water piping in the building which is the subject of this paper is showing BGW corrosion, the water has the propensity to support biological activity. Therefore without adequate testing MIC cannot be discounted as being involved with the pitting process.

The appearance of the corrosion pits investigated agrees in principal with the mechanistic processes found in the literature [15, 16]. The different appearance of the pit as growth occurred is described including the formation of a yellow-green crystalline deposit, assumed to be cupric chloride, around the perimeter of the corrosion overlying the cuprous oxide crystals. In this investigation the yellow-green crystals were found to fully cover the cuprous oxide in the pit rather than just develop at the perimeter of the pit (Figure 8). This behaviour was only noted on the mature pits. This behaviour may be related to something specific in the chemistry of the potable water.

The investigation identified that the red crystal had formed in two locations. Dense large crystals were observed in the corrosion pits and arrow bands of small crystals were observed remote from the pit. The differences in the arrangement of the crystals may be related to the different patterns of accumulation of the cupric salts inside the tube. Both thick irregular longitudinal deposits and localised mounds were observed on inside these tubes. The differences in galvanic driving force that develop between the crystals and the copper surface in the mound and that in the longitudinal bands may be responsible for the higher rate of corrosion under the mounds compared with the minimal corrosion under the bands of crystals.

Although it has been suggested that the precipitation of red cuprous oxide crystals outside the pit can be related to the specific pitting mechanism [18], some of the crystals observed are a substantial distance from the pit. Therefore it is suggested that further investigation is undertaken to assess whether these crystal are associated with the corrosion pits.

During the drought which started in 2007 and the implementation of Stage 3 water restrictions for Melbourne, aerators were installed on each tap within the building to reduce the overall water usage. While this was an appropriate measure to deal with water conservation, it has possibly exacerbated the blue-green water corrosion issue by reducing flow rate and minimising the impact of flushing activities. During the removal of the aerators and subsequent flushing activities it became apparent that the aerators had effectively blocked a percentage of the copper particulates associated with blue green water. The initial flush activity resulted in the immediate discolouration of the water often to an almost dark blue colour (Figure 15 and Figure 16) with a gradual return to clear water. This intense discolouration indicated that the aerators and associated reduced flows were preventing a percentage of the particulate matter from leaving the network. As active flows are required to flush the particulate matter from the pipework, the temporary deposition of the product may well have contributed to the rate of pit formation.

Figure 15. Water discolouration after first flush following removal of aerators

Figure 16. Water discolouration after first flush following removal of aerators (Adjacent to sink in Figure 15)

The vivid discolouration noted above was a unique event associated with the first flush following removal of the aerators and the event lasted for only a less than one minute. This level of discolouration has not been observed again.

Although the installation of aerators may have contributed to the rate at which deposits accumulated in the pipe, it is unlikely that the aerators caused the initiation of pitting corrosion. The water quality in the pipe appears to support both BGW corrosion and pitting corrosion and the initiation of the BGW corrosion 15 years ago is the most likely cause of the pitting corrosion that resulted in perforation of the pipe. Although BGW corrosion is rare, pitting corrosion failures in pipe suffering from BGW corrosion have been reported in the literature [1, 2, 6, 7].

6. FUTURE APPROACH TO MANAGING THE COLD WATER NETWORK

Of the nineteen perforation and leak events recorded since 2013, seventeen of these were on offtakes to taps in the public toilets, function room toilets, function room tea-points or kitchen facilities. The other two were located in horizontal sections of piping prior to the offtake to a room with a single toilet and sink. Each of these situations corresponds to a potentially low use facility where the water is likely to remain stagnant for period of time in between periods of moderate use. The majority of these failures had the potential to impact business operations within the building.

A review of the locations of the failures noted they had occurred on five of the ten floors, with an additional failure on one of the basement levels. The failures had occurred in different parts of the building and the review noted that the water to the failures was not fed from a single dropper but rather from five of the ten dropper pipes. A number of other features were assessed but the only commonality of the failure locations was the pipe size, 20 mm diameter, and the locations near the end of the local distribution network.

The mechanisms driving both BGW corrosion and pitting corrosion are fairly autocatalytic in Melbourne's soft poorly buffered, low residual chlorine potable water once the processes have initiated. While research has indicated that BGW can be controlled through continual buffering and chlorination, effective methods to control pitting corrosion have not been developed. The acidic environment necessary to propagate pitting is maintained in the pit by the corrosion product cap that effectively shields the pit from the bulk water environment. Therefore treatments added to the water are unlikely to impact the processes occurring in the isolated pit.

The options for the management of the corrosion in the cold water network are limited to attempting to minimise the consequences of a perforation event, where possibly, or initiate a planned replacement program. However in the absence of any identifiable trends or links in the failure locations a management program may need to include gradual replacement of all of the 20 mm diameter pipe that has the potential to impact business operations. Due to the propensity for the potable water to cause similar corrosion issues with new copper pipe, alternative material should be considered for use in the replacement program.

Continuation of the regular flushing program will minimise the water quality issues associated with discolouration and may possibly slow corrosion product build up and development of new corrosion pits.

7. CONCLUSIONS

Around the time that the building was under construction the local sections of the water mains due to supply the building were replaced with a new mild steel cement lined pipe. The soft poorly buffered water typically found in Melbourne has the capacity to leach lime from newly installed cement lined water mains resulting in an increase in pH, and a potential increase in alkalinity and hardness. But possibly more importantly is the reduction in effectiveness of the chlorination as the pH increases. A decrease in the effectiveness of chlorination contributes to the increasing likelihood of microbial activity. The water quality combined with probable stagnancy following commissioning of the cold water system in the building is suspected to have caused the initiation of blue green water corrosion fifteen years ago.

The ongoing BGW corrosion issues over the last fifteen years ago is attributed as the most likely cause of the pitting corrosion of the cold water piping network. Although BGW corrosion is rare, pitting corrosion failures in pipe suffering from BGW corrosion have been reported in the literature.

Due to the increasing numbers of perforations that are occurring each year in the cold water network, there is an increasing risk to business operations in the building. A planned replacement program of the small bore copper pipe likely to impact on the critical function rooms could be undertaken to control the risk associated with water leaks due to pitting corrosion resulting in perforation of the pipe.

8. ACKNOWLEDGMENTS

Many thanks to Stephen Finger for undertaking the laboratory investigations for this work.

9. REFERENCES

1 O'Halloran, R; Smith, F; Taylor, R, Goodman, N: "Blue water corrosion of copper: Causes and implications for remedial treatment", Paper 030, Corrosion and Prevention 2001, Newcastle, Australia, 2001.

2 Edwards, M; Jacobs, S; Taylor, R: "The blue water phenomenon", American Water Works Association Journal, 92 (7), 72, 2000.

3 Nicholas, D; Carlyle, G: "The management of "blue water" issues-the Evans Head experience", Paper 39, Corrosion and Prevention 99, Newcastle, Australia, 1999.

4 Critchley, M.M; Pasetto, R; O'Halloran, R.J: "Microbiological influences in 'blue water' copper corrosion", Journal of Applied Microbiology, 2004, 97, 590-597.

5 Critchley, M; Taylor, R, O'Halloran, R: "Microbial contributions to blue water", Materials Performance Journal, 44, 6, 56, 2005

6 Taylor, R.J; O'Halloran, R.J; Cannington, P.H: "The causes of blue water in copper plumbing systems", Paper 144, 13th ICC, Melbourne, Australia, 1996.

7 Webster, B; Ryan, G; Dudley, K; Moody, S; Wilson, P: "A simple remediation treatment for blue water", Paper 090, Corrosion Control and NDT, Melbourne Australia, 2003

8 Webster, B.J; Wells, D.B; Bremer, P.J: "The influence of potable water biofilms on copper corrosion", Paper 294, Corrosion 96, NACE, 1996.

9 Korshin, G.V; Perry, S.A.L; Ferguson, J.F: "Influence of NOM on copper corrosion", American Water Works Association Journal, Jul 1996, Vol 88, Number 7.

10 Bremer, P.J; Webster, B.J; Wells, D.B: "Bio-corrosion of copper in potable water", American Water Works Association Journal, 93 (8), 82, 2001.

11 Cantor, A.F; Bushman, J.B; Glodoski, M.S; Kiefer, E; Bersch, R; Wallenkamp, H: "Copper pipe failure by microbiologically influenced corrosion", Materials Performance, Vol 45, Number 6, (June 2006),

12 Liu, X; Wang, J; Liu, T; Kong, W; Hw, X; Jin, Y; Zhang, B: "Effects of Assimilable carbon and free chlorine on bacterial growth in drinking water", PLoS ONE, June 2015.

13 Tuthill, A.H; Avery, R.E; Lamb, S; Kobrin, G: "Effect of chlorine on common materials in fresh water", Paper 708, Corrosion 98, NACE, 1998.

14 Little, B.J; Lee, J.S: "Microbiologically Influenced Corrosion", John Wiley and Sons, USA, 2007.

15 Harrison, D.B; Nicholas, D.M; Evans, G.M: "Mechanisms of pitting corrosion of copper tubes in soft drinking water", Paper 95, Corrosion and Prevention 2000, Auckland, New Zealand, 2000.

16 Harrison, D.B; Nicholas, D.M; Evans, G.M: "Pitting corrosion of copper tubes in soft drinking water: corrosion mechanisms", American Water Works Association Journal, Nov 2004, Vol 96, Number11.

17 Burleigh, T.D; Gierke, C.G; Fredj, N; Boston, P.J: "Copper tube pitting in Santa Fe municipal water caused by microbial induced corrosion", Materials, 2014, 7.

18 Claes, T: "Pitting corrosion of copper an equilibrium-mass transport study", Paper 141, 13th International Corrosion Congress, Melbourne, Australia, 1996

10. AUTHOR DETAILS

Sarah is an Associate Director in the Strategic Asset Management and Advanced Materials team at AECOM. She is a graduate of the Corrosion Science and Engineering Masters' program at UMIST in England and has approximately 25 years' experience in corrosion related consulting and research. She has specific experience in durability planning for new infrastructure, condition assessment and rehabilitation of aged infrastructure, performance assessments of materials and protective coatings, materials selection, specification of protective coatings, failure analysis and cathodic protection design.

CORROSION MITIGATION OF AGEING BURIEDSTEEL PIPELINES IN METROPOLITAN SYDNEY

J. Galanos
Corrosion Control Engineering (NSW) Pty. Ltd

SUMMARY: The on-going maintenance and corrosion protection of ageing buried steel pipelines in metropolitan Sydney is becoming more challenging due to a number of factors. These include an expansion of the DC powered rail and light rail networks, a significant increase in third party works around pipeline easements affecting pipeline access, ageing of pipeline coatings and the effectiveness of existing cathodic protection systems. These effects have been analysed on various pipelines and some specific examples are presented.

Keywords: cathodic protection, stray current mitigation, impressed current CP system, pipelines

1. INTRODUCTION

Cathodic Protection (CP) is a condition of licensing and operating buried steel pipelines that transport hazardous goods such as oil, gas and refined petroleum products [1] and is paramount for the corrosion mitigation of buried pipelines. Proving the operational effectiveness of CP systems is via routine testing. Other methods such as ILI (intelligent pigging), DCVG coating defects surveys, CIPS (Close Interval Potential Surveys) and other NDT tests are also used to verify the CP results and also to check for external corrosion due to cathodic shielding or other factors.

2. PIPELINE MAINTENANCE

Maintenance of buried oil and gas pipelines in metropolitan areas is critical to pipeline owners for reasons of safety, environmental risk, licensing and maintaining critical infrastructure for the supply of hydrocarbons into NSW. The following should form a significant part of pipeline maintenance programs. These maintenance procedures follow guidelines set out in the pipeline license and Australian Standards [1].

2.1 Routine Surveillance

Pipeline surveillance in the Sydney region is undertaken at daily, weekly and monthly intervals by the various pipeline operators. Surveillance programs are set up following detailed Safety Management Strategy (SMS) workshops that analyse the tasks associated with managing and licensing the pipeline. These workshops are mostly undertaken at yearly intervals and if required, surveillance program changes are made to reflect workshop outcomes. Routine surveillance is used to capture unauthorised third party activity, any significant changes to pipeline cover and vandalism.

2.2 Managing Third Party Activities

Third party activities are permitted in and around pipeline easements provided a strict set of guidelines are followed by the third party.

Adherence to these guidelines is critical in that:

- They firstly and most importantly, accurately locate the position and depth of the pipeline.
- They minimise, or preferably, eliminate any physical pipeline damage, including the external coating.
- They do not affect the operation of existing CP systems.
- They continue to provide pipeline access in the future.

2.3 Corrosion Mitigation

The effective operation of the CP system is mandatory in allowing the on-going use of the pipeline as per NSW State Government licensing requirements, including the Pipelines Act [2].

The ability of a CP system to fully protect a pipeline is influenced by:

- Coating type, age and condition
- DC stray currents

- Selection of appropriate CP system, the mode of operation and its location
- Foreign structures

The above factors are critical in maintaining the owner's asset, investment and business interests.

3. PIPELINE DETAILS AND CATHODIC PROTECTION

There are a number of pipelines built in the mid 1950's and early 1960's in Sydney, which are some of the earliest pipelines built in Australia. The pipelines in some cases follow electrified DC railway corridors. In all cases the pipelines are affected to some degree by DC stray currents. These pipelines were typically coated with coal tar enamel coating. Over the pipelines' life, the coal tar enamel coating has aged. Additionally, numerous coating repairs, using a variety of products have been used, including petrolatum tape, heat shrink sleeves and epoxy.

Ageing coal tar epoxy coating demonstrates the following characteristics:

- Becomes porous allowing both moisture and CP current to the steel surface.
- Disbonds over time especially if it is subject to high negative CP or stray current potentials.

The original CP systems consisted of sacrificial anodes, installed at numerous test points along the route. However, due to coating breakdown and stray current effects, Impressed Current Cathodic Protection (ICCP) systems were retrofitted.

The ICCP system for a particular pipeline in the Sydney metropolitan area consists of:

- 2-off ICCP systems and 1-off Railway Drainage Bond, upgraded to a TRAD (Transformer Rectified Assisted Drainage) system in the 1980's.
- An additional ICCP system was installed in 2015 at the upstream end of the pipeline.
- The installation of a further ICCP system at the downstream end of the pipeline is currently under review.

4. PIPELINE CATHODIC PROTECTION

4.1 CP Test Procedure

The CP systems are routinely tested as per AS2832.1 [3], AS2885.3[1], client specifications and licensing requirements. The test program may include the following:

- Daily check of the telemetry systems
- Monthly functional check of all CP units, including spot potential testing at all CP units.
- 6-monthly full data logger DC potential survey over approximately 24 hours at all test points.
- 6-monthly AC potential testing (spot survey) at all test points.

4.2 CP Test History

Our company became involved with the CP testing of the fore-mentioned pipeline after June 2014, with the first full data logger survey completed in October 2014.

A review of previous test reports and results showed the pipeline was at fully protected CP (Instant Off) potentials. However, at a number of locations, the CP (Instant Off) potentials were either approaching or at the minimum limit for full protection. Our review also flagged the following concerns:

- There had been a general positive shift towards unprotected potentials at a number of test locations over the previous 4-5 years, prior to June 2014.
- Previous DCVG testing showed numerous coating defects. Our concern was that some pipeline areas could be at below fully protected potentials at large defect locations and midway between test points. This is supported by ILI reports.
- As a number of coating repairs had previously been undertaken, localised overprotection was evident due to better quality coating and coating condition. Some potential imbalance exists.
- The existing coal tar coating was disbonding and possibly shielding CP current flow. That is, whilst potential testing using either the permanent or a portable reference electrode showed full protection, little or no CP current would flow behind the disbonded coating.
- Some CP units were operating at close to their registered maximum limits. An increase, under the existing registration, was not possible due to unacceptably high stray current interference effects to foreign-buried structures.

Prior to conducting the October 2014 data logger survey, the CP unit outputs were adjusted to optimise performance. Our first 6-monthly CP survey was undertaken in October 2014. The test results differed and in some cases significantly when compared to previous results as shown in Chart 1. The charts show the average recorded CP (On) potentials. The CP

(Instant Off) potentials, as expected, are slightly more positive. The CP unit changes made between June and October 2014, resulted in a more uniform spread of potentials along the pipeline route.

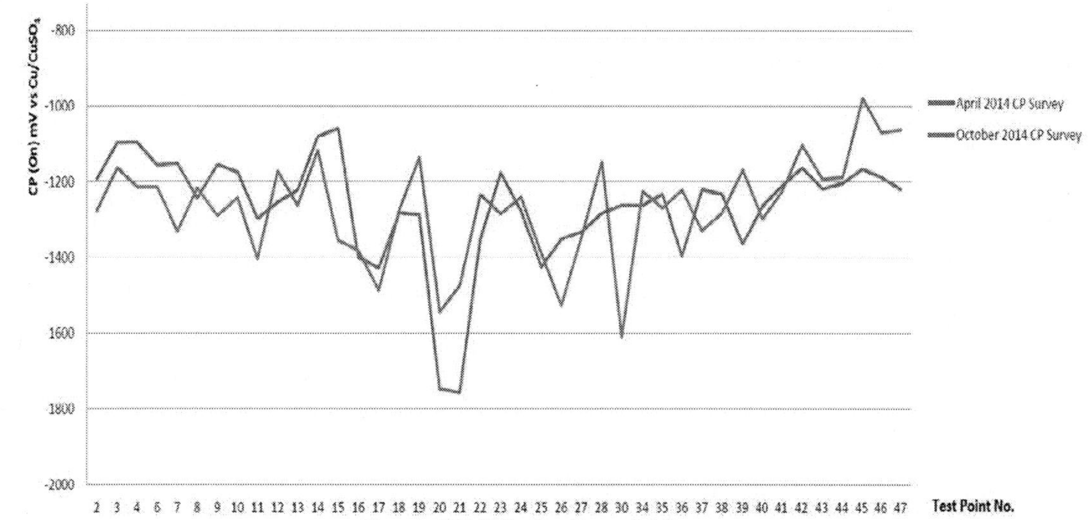

Chart 1: 2014 Pipeline Potential Summary

4.3 CP System Upgrades

Whilst there was an improvement in potential spread along the pipeline, further improvements were recommended. Due to continuing coating degradation and the associated positive shift in potentials, a number of smaller additional CP systems were recommended.

A new CP system was installed at the downstream end of the pipeline. The results showed an immediate improvement in protection levels at this end of the pipeline as shown in Chart 2. The second recommended ICCP system is yet to be installed. Once installed, significant improvements in protection levels and further tightening of potential spread along the pipeline is expected.

Due to the variation in potentials, coating type and condition, the installation of additional test points for additional potential monitoring is highly recommended. At present, the approach is to install new test points when the pipeline is exposed for other works, if the excavation is located more than 100m from an existing test point or if a metallic buried foreign structure is found.

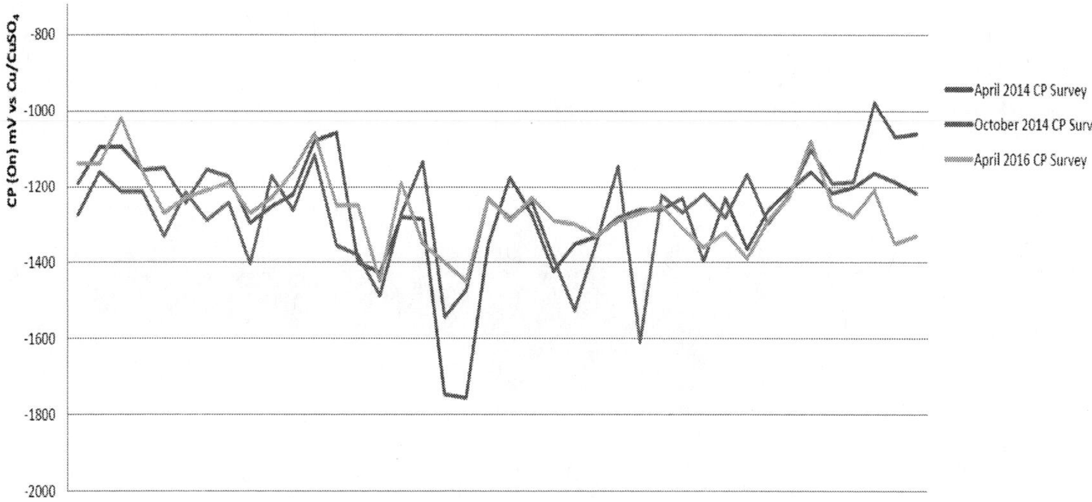

Chart 2: 2016 Pipeline Potential Summary

The existing CP unit registered limits were also re-examined. At one CP unit location (TP16 as shown in Charts 1 & 2 above), it was found that the installation of sacrificial anodes to the stray current (adverse) affected foreign structure would mitigate these effects and allow the TR unit to operate at a higher output. Once complete, the permitted output was more than doubled (from 6 Amps to 15 Amps).

In addition, the mitigation of the above stray currents also allowed the CP unit to operate in Auto (constant potential) mode rather than constant current. The aim of this mode change was to provide more uniform potentials by allowing the CP unit to regulate the current output and therefore provide more uniform potentials. Charts 3-5 show the pipeline potentials become significantly more uniform almost immediately after CP unit adjustment. Chart 6 shows the variation in CP unit current output to provide the above mentioned uniform potentials.

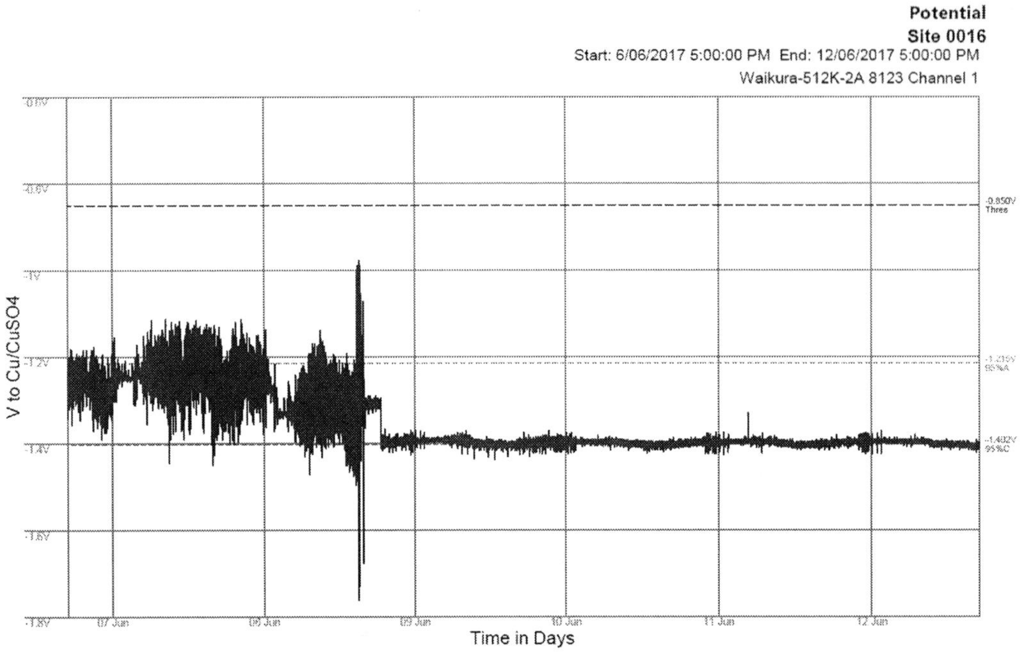

Standard	> Threshold	> Threshold	>Thres+50	>Thres+100	>Thres+850	Avg On	Min Neg (ON)	Max Pos (ON)
Actual	0.00%	0.00%	0.00%	0.00%	0.00%	-1.353V	-1.762V	-0.978V
Allowable	LP 10%	SP 5%	SP 2%	SP 1%	SP 0.2%			

Chart 3: Site 0016 Potential 2017

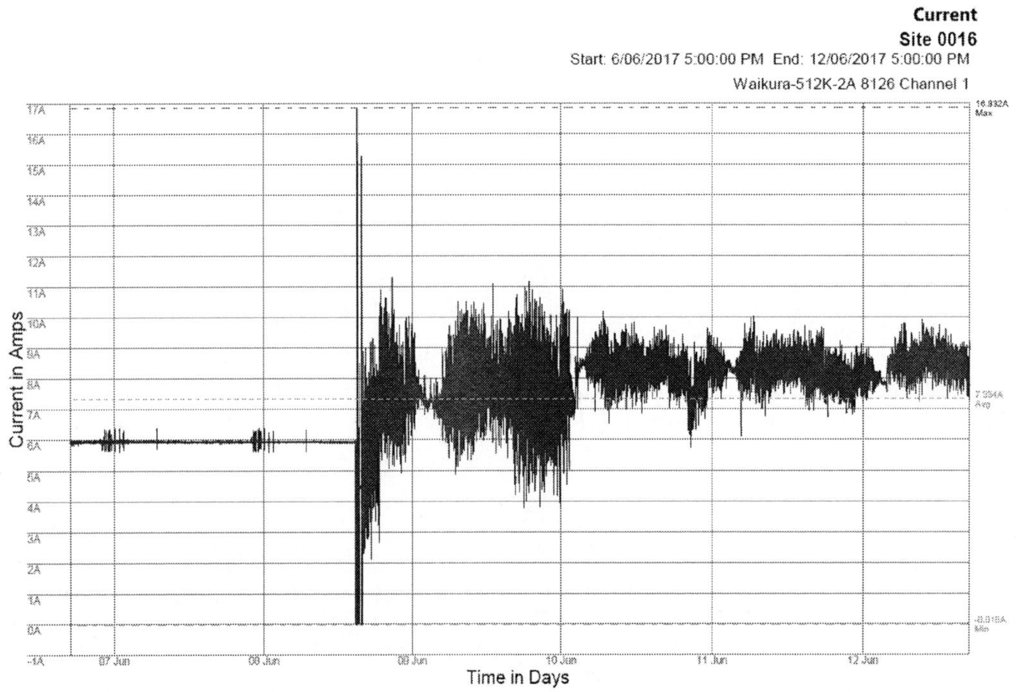

Readings above 0.1 Amps	Percentage above 0.1 Amps
51741	99.81%

Chart 4: Site 0016 Current 2017

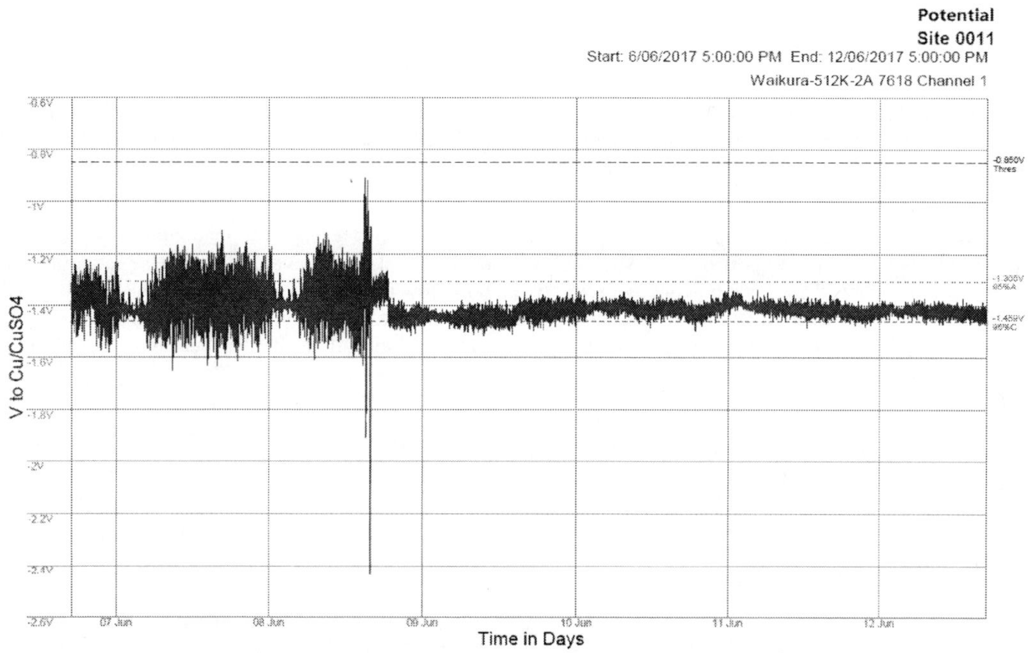

Standard	> Threshold	> Threshold	>Thres+50	>Thres+100	>Thres+850	Avg On	Min Neg (ON)	Max Pos (ON)
Actual	0.00%	0.00%	0.00%	0.00%	0.00%	-1.406V	-2.434V	-0.908V
Allowable	LP 10%	SP 5%	SP 2%	SP 1%	SP 0.2%			

Chart 5: Site 0011 Potential 2017

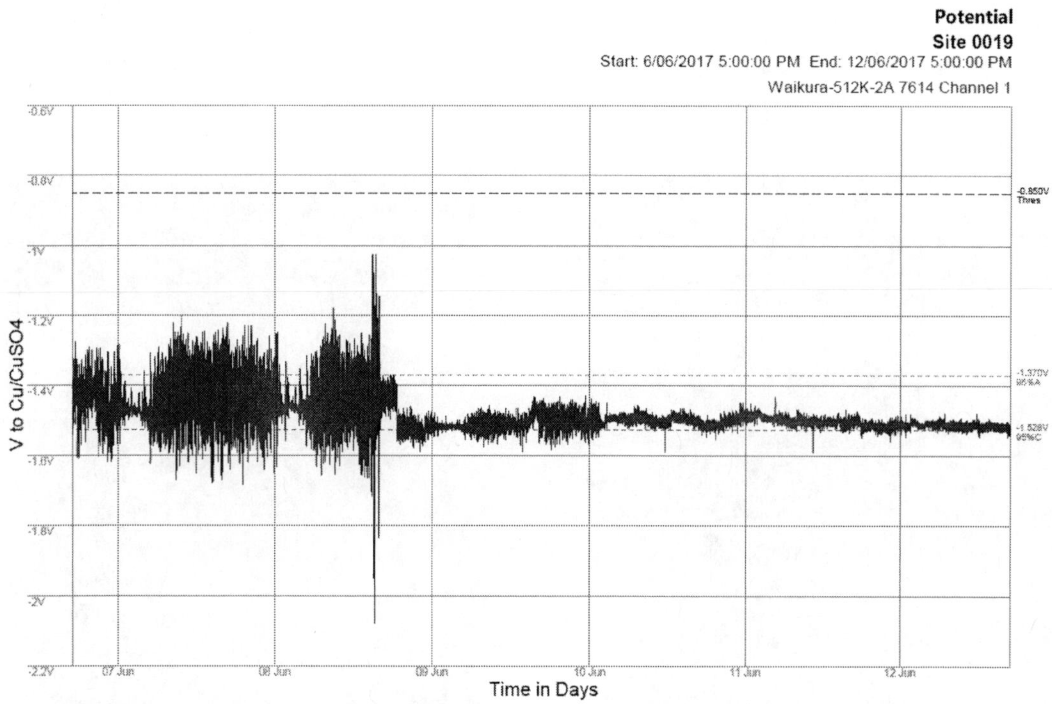

Standard	> Threshold	> Threshold	>Thres+50	>Thres+100	>Thres+850	Avg On	Min Neg (ON)	Max Pos (ON)
Actual	0.00%	0.00%	0.00%	0.00%	0.00%	-1.483V	-2.079V	-1.024V
Allowable	LP 10%	SP 5%	SP 2%	SP 1%	SP 0.2%			

Chart 6: Site 0019 Potential 2017

4.4 CP and Disbonded Coatings

In the past, it has been found that where CP current shielding, due to disbonded coating is occurring, CP (Instant Off) potentials in the higher (more negative) end of acceptable limits and can only provide limited protection.

This more negative shift in potentials is being implemented to the above pipeline, with the existing CP unit outputs, where permitted, increased to shift the pipeline potentials more negative. This has been undertaken continuously and is monitored monthly as part of CP unit functional tests as it is desirable to overprotect the pipeline. Chart 7 illustrates these changes as part of the 2017 survey.

In many cases, CP current shielding cannot be overcome by increasing output levels. The main aim is to protect as much of the pipe affected by shielding as possible and where full protection cannot be achieved, to reduce the rate of corrosion by as much as possible. Photo 1 shows a pipeline with no protection below the disbonded coating as the CP potentials were close to the minimum limit for full protection. Photo 2 shows some protection due to CP (in the form of calcareous deposits) below the disbonded coating. In this case, the pipeline CP (Instant Off) potentials were more negative than -1.0 Volts vs $Cu/CuSO_4$.

Chart 7: 2017 Pipeline Potential Summary

Photo 1: Corrosion Due to Disbonded Coating and Cathodic Shielding

Photo 2: Some Corrosion Protection due to CP at Disbonded Coating.

5. CONCLUSIONS

The on-going CP monitoring and maintenance of ageing pipelines, especially where coating breakdown and CP shielding occurs requires careful management and a dynamic approach to future CP system needs. CP system upgrades and a greater number of CP systems will almost certainly be required to provide adequate CP current and uniform potentials. That is, CP systems need to be able to "flat line" potentials along the pipeline route as much as possible.

The increase in stray current effects due to new electrified DC rail and light rail networks will also put additional strain on CP systems and their ability to maintain pipeline protection and uniform potentials. If DC stray currents can be managed, this will significantly reduce any accelerated corrosion due to DC current discharge off the pipeline. Additionally, DC stray current does not come into play at cathodically shielded locations, only at exposed defects.

The case study shows management of the above is not a simple or one-step task. The installation of a greater number of test points and CP systems, mostly operating well within their capacity limits are required to provide uniform potentials, particularly where stray currents are significant. It is important to note that it can be difficult to find suitable and acceptable locations for new ICCP systems in built up areas.

Some success has been achieved with horizontal linear anode beds installed parallel to and in close proximity to pipelines, especially where significant coating damage or cathodic shielding exists. This is of particular importance where cathodic shielding exists, as having good control of potentials, yet not overprotecting the pipeline, provides the best possible solution under these conditions. Therefore all tools available should be used to assess the condition of ageing pipelines. This includes constant monitoring of CP system potentials and the review of ILI data for corrosion losses and growth is critical in maintaining the pipeline's integrity. Additionally, this information should be verified with the assistance of DCVG and CIPS surveys.

When considering the use of pipeline integrity management software, it must be capable of both storing the relevant integrity data and providing summary reports that clearly show the status of the pipeline along its length.

6. ACKNOWLEDGEMENTS

Thank you to our Principal Engineer, John Kalis, and our colleagues for their input into this paper and for the photographs.

7. REFERENCES

[1] Standards Australia (2012) AS2885.3 Pipelines - Gas and Liquid Petroleum, Part 3: Operation and Maintenance
[2] NSW Government, Pipelines Act 1967
[3] Standards Australia (2015) AS2832.1 Cathodic Protection of Metals, Part 1: Pipes and Cables

8. AUTHOR DETAILS

Jim Galanos is the Engineering Manager at Corrosion Control Engineering's New South Wales office. He holds a Bachelor of Engineering in Industrial Engineering from the University of NSW as well as Australasian Corrosion Association Corrosion Technologist and NACE CP4 (Cathodic Protection Specialist) accreditation.

Jim has 20 years' experience in cathodic protection system design, supply, installation, commissioning and routine testing in both public and private sectors covering industrial, civil, marine, infrastructure and building structures. He has also prepared numerous cathodic protection technical specifications and undertaken many corrosion assessments.

COMPARISON OF CORROSION IN DIFFERENT PAINT DEFECTS, MATERIAL, AND ORIENTATION FOLLOWING PROHESION SALT FOG TESTING

S. Knight[1], A. Sudholz,[2] S. Jacob[1], J. Waldie[1]
[1]BAE Systems, Melbourne; [2]Defence Science and Technology Group, Melbourne

SUMMARY: Paint defects in aircraft can be present from manufacture or evolve as the paint coating degrades over time due to UV radiation, moisture, temperature, and stress. In addition, mechanical damage during maintenance can be a site for subsequent corrosion damage in an aircraft structure. In this study, paint defects generated by masking a 1 mm wide slot prior to paint application were compared with a 1 mm wide mechanical scribe. Paint coatings were made up of a praseodymium primer and a polyurethane topcoat. Pitting corrosion in defects was quantitatively analysed using optical profilometry to assess the differences in inhibitor protection and corrosion performance during prohesion salt fog testing. The results showed no significant difference in average corrosion volume between masked or scribed slots, but showed more variability in the scribed slots. Furthermore, the criteria to include pits greater than 30 µm, 50 µm, and 100 µm made no significant difference to the pit corrosion volume and pit density relationships over time. Of the two aluminium alloy substrates investigated, AA2024 showed a higher corrosion volume than AA7075. Furthermore, horizontal orientation was more severe than vertical orientation. Results of this study will be used to inform algorithms used in corrosion prediction models.

Keywords: AA2024, AA7075, Pitting, Paint defects, Sample orientation, Prohesion salt fog test.

1. INTRODUCTION

Corrosion-related issues are a major contributor to the cost of maintaining Australian Defence Force (ADF) aircraft. The ADF spends approximately $238 million annually on corrosion-related maintenance [1]. Costs can be incurred by unavailability due to unforeseen maintenance events or performing maintenance actions when they are not required. As such, the ADF are moving towards adopting condition-based maintenance practices to allow for more targeted maintenance activities. To facilitate this, BAE Systems and the Defence Science and Technology Group are developing corrosion prognostic health management (CPHM) systems to assist ADF aircraft maintainers in reducing costs, enhancing safety and increasing reliability. Key to achieving this goal is developing a quantitative understanding of the corrosion performance of aircraft materials when exposed to in-service conditions.

The aerospace industry use a variety of aluminium-based (Al) alloys for structural application owing to their high strength, low weight and good general corrosion resistance [2, 3]. Although resistance to general corrosion is good, Al alloys are highly susceptibility to localised forms of corrosion such as pitting [4]. A protective oxide forms on Al alloys in neutral solutions [5] but will suffer film breakdown by chloride absorption and this can lead to stable pit formation [6]. An effective mitigation of corrosion attack of aerospace Al alloys is the application of paint coatings, involving both barrier coatings and primers containing corrosion inhibitors. These coating systems have become an integral part of aircraft corrosion protection.

For decades, protection of aerospace Al alloys has been achieved by use of chromate-based corrosion inhibitors. Chromate inhibitors are highly effective in both preventing initiation of corrosion pits and limiting the growth of pits that have formed [7-9]. However, increasing awareness of the environmental impact and carcinogenic nature of chromium [10] has led to the implementation of a new generation of environmentally-friendly coating systems on modern aircraft [11, 12]. Such coatings include primers based on a praseodymium inhibitor, which has demonstrated appreciably different performance characteristics to its chromate-based predecessor [13]. Electrochemical investigation of $PrCl_3$ in solution has demonstrated the Pr inhibitor's ability to reduçe corrosion kinetics [14, 15] but little published information exists regarding the ability of Pr to limit pit initiation. An aim of the present study is to investigate the formation of pits on aerospace alloys while under the protection of a Pr-based inhibitor.

Although highly effective at preventing, coating systems are compromised inevitably during the service life of an aircraft. Factors causing coating breakdown include: ultra-violet (UV) light, moisture ingress, temperature fluctuations and mechanical stress, which can lead to coatings becoming brittle and cracking [16]. The breakdown of the coating system exposes the alloy substrate to the environment and subsequent corrosion attack. However, the introduction of moisture into sites of coating damage allows soluble inhibitors to leach from the primer coating layer and act to reduce corrosion of the substrate. To measure the effectiveness of this process, careful application of artificial damage to the coating system is required to simulate in-service damage in a reproducible way for laboratory testing. Testing of artificially-damaged chromate coatings has been reported [17] but no work has investigated a Pr-based coating systems in this way. The present work uses a prohesion environmental test chamber to provide a controlled accelerated test to compare both slots and scribes as a form of imparting artificial damage to the current coating system.

Testing of corrosion severity impact on coatings is usually made qualitatively by comparing corrosion staining of the substrate material against reference standards. However, this methodology is inherently subjective. More reliable and meaningful comparisons can be made by quantitative measurement of substrate corrosion in an artificially-damage coating system. The present work uses 3D optical profilometry to obtain quantitative information of corrosion damage to the substrate material in the presence of a damaged coating system.

The present work investigates two different aircraft Al alloys in AA2024 and AA7075, both being protected by a Pr-based primer coating system. The work investigates two different methods for introducing artificial damage into coatings systems and considers the effect of orientation during exposure. Damage to the substrate surface is measured using 3D optical profilometry and pitting corrosion is analysed by comparison of pit sizes and volumes. Practical considerations are also included regarding the use of a failure criterion for pitting corrosion relating to the diameter of the pits formed of the alloy at the site of the damaged coating. The present work aims to gain additional insight into in-service pitting corrosion to aircraft alloys protected by new coating systems for the purposes of developing CPHM tools for ADF aircraft.

2. EXPERIMENTAL DETAILS

2.1 Materials & Sample Preparation

Test plates with dimensions of 150 x 75 x 3 mm were manufactured from AA2024-T3 and AA7075-T6. The coating system consisted of a surface pre-treatment, a primer and a top coat. The surface treatment was an Alodine Pen 1132 chromate conversion coating. The non-chromate primer was Deft 44GN098 [18], which is an epoxy-polyamide primer containing a Pr-based corrosion inhibitor. The plates were finished with a white polyurethane topcoat.

To simulate damage to the coating system, the plates were prepared by the following methods:

- **Slots:** 4 slots made by masking, each 12.0 mm in length and 1.0 mm in width.
- **Scribes:** 3 scribes that were manually scored through the paint system into the base material, all of approximately 30 mm in length and ~1.0 mm in width.

Figure 1 shows corroded and cleaned slots and scribes used in these experiments. Each plate is individually numbered, with the slots/scribes individually identified. In total, there are 80 slots and 60 scribes. The scribes were carried out using conventional engineering tools (machinists scribe, centre punch or jewellers flat tip screwdriver). Surface of the manufactured slots were finished to ANSI 1200 SiC grit (Ra < 0.5 μm) whilst the scribes surface roughness following scribing is expected to be significantly higher.

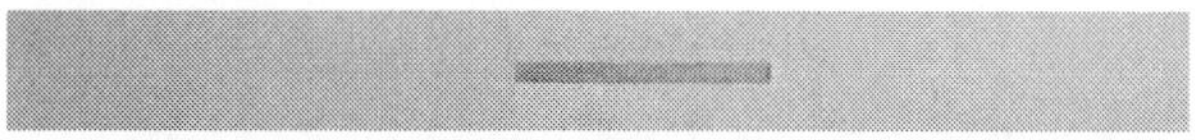

(a)

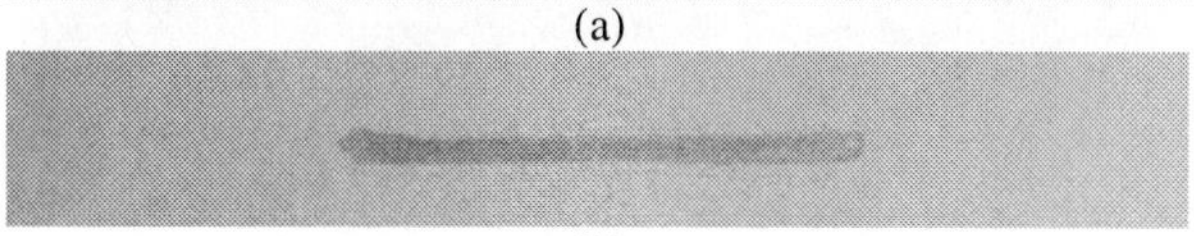

(b)

Figure 1. Example of prohesion test plate, including (a) a slot and (b) a scribe.

2.2 Prohesion Test Environment

The Prohesion test environment consists of two interchanging 1 hour cycles of a wet and dry environment. The wet environment is a continuous indirect salt spray of a salt solution (see below) at the ASTM B 117 fall-out rate [19]. Chamber conditions during the wet period are between 21°C to 27°C, at approximately 90% RH. Following the wet period, the chamber conditions change to a dry period of 1 hour at 35°C and ambient humidity conditions. The cycle repeats for the test duration. Prohesion test environment parameters are summarised in Table 1.

Table 1. Prohesion Test Environment – Experimental Parameters

Salt solution	0.05 % Sodium Chloride and 0.35 % Ammonium Sulfate sea spray
pH	5.0 – 5.4
Fallout rate	1.0 – 2.0 $ml/80cm^2/hour$
Temperature	27 ± 10°C or 35 ± 2°C
Humidity	Ambient conditions, up to 90 %RH

2.3 Prohesion Experimental Setup

Ten plates of each alloy were tested in a prohesion test environment (Vipac, Melbourne). During the test cycle, witness plate coupons will be supported in the vertical (0°) and horizontal (~90°) test orientations on the chamber rack. In the vertical orientation, the slot/scribe was oriented such that any condensation would roll along the slot, and not across the slot. The active region of the coupon faced towards the principle direction of the fog flow in the chamber. The fog collection was carried using two $80cm^2$ cones to ensure the rate of salt fog is in accordance with ASTM G 85 Annex 5 (i.e. 1 – 2 $ml/80cm^2/hour$) [20]. Horizontal oriented samples were removed between 45-955 h exposure, and vertical oriented samples were removed between 340-955 h exposure.

2.4 Sample Cleaning

The corrosion product from each coupon (in each slot and scribe) was removed in accordance with ASTM G1 Table A1.1, solution C1.2 (50% Nitric acid) [21]. Each coupon number was individually engraved on the back before cleaning. Figure 2 shows images of a witness plate coupon panel before and after cleaning in 50% Nitric Acid. Small regions of filiform corrosion were observed (top right in Figure 2A and 2B). This was evident by a blistering of the paint coating in Figure 2A which was removed after cleaning (Figure 2B).

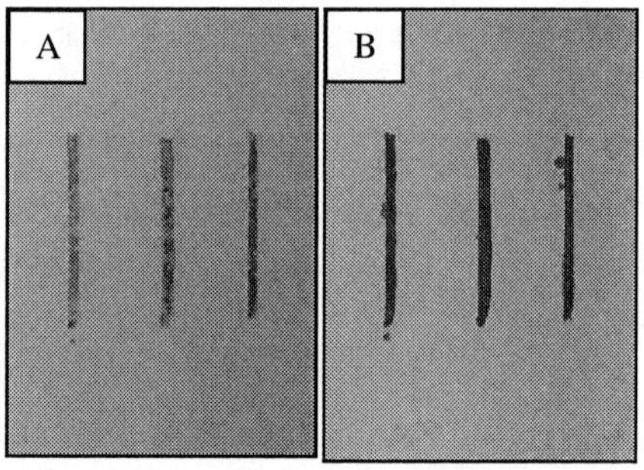

Figure 2. Photos of scribes in coupons (A) before cleaning and (B) after cleaning.

Cleaning of some of the test coupons (each slot/scribe) for more than 2 minutes seems to have removed some flakes of paint (top coat) as shown in Figure 3.

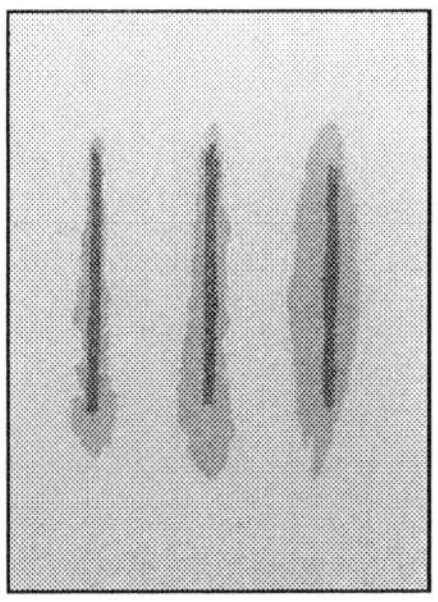

Figure 3. Example of scribes in coupon showing removal of flakes of paint due to cleaning process.

2.5 Measuring Pitting Corrosion

A Bruker CONTOUR GT-K1 3D optical profiler (Figure 4) was used to scan the surface of test plates and measure pitting corrosion damage to the alloy surface. A 3D optical profilometer is a non-contact imaging technique that uses an optical microscope with Mirau interference optics. It is based on the principle of vertical scanning interferometry. It is a non-destructive and highly sensitive technique with the capability to resolve large scanning areas with sub-nanometer vertical resolution and lateral resolution down to the sub-micrometer range.

A stitching mode was used to scan the slots and scribes as follows:

a) Slots with width 1.0 mm: They were all scanned with 5× objective and scanned length was 12 mm.
b) Scribes with width 1.0 mm: They were all scanned with 5× objective and scanned length was 30 mm.

Figure 4. Bruker CONTOUR GT-K1 3D optical profiler (Swinburne University, Melbourne).

The detailed analysis of stable pitting corrosion was carried out using Vision 64 software on all the images acquired in 80 slots and 60 scribes. For this work a pit was classified as being stable if both its depth is greater than 10 μm and any surface dimension (X-Y plane) was ≥ 30 μm, 50 μm, or 100 μm. Areas identified as filiform corrosion (earlier) were not included in the analysis of pitting corrosion. Although there was flaking of the paint following cleaning for both the masked and scribed slots, it was slightly more pronounced around the scribes possibly due to increased damage or delamination at the paint edge. Measurements were confined to the openly exposed aluminium alloy surfaces.

2.6 Metallographic sectioning

Sectioning of the masked slots was carried out (ALS Industrial Pty Ltd, West Melbourne) to determine whether primer would be exposed at the paint edge of the slots. The cutting process was carried out using a slow-speed water cooled saw. Sectioned slots were cold-mounted in epoxy, ground, and finally polished for examination with an optical microscope.

3. RESULTS

The following results were calculated normalised to scan area. The corrosion volume and the pit density per unit scan area were used as a measure of corrosion extent.

3.1 Aluminium Alloy 2024

Figure 5 shows the results for 2xxx alloy in the vertical position shows pit volume and pit density both increase with time. There was an incubation period before the onset of corrosion which occurred somewhere between 340-618 h.

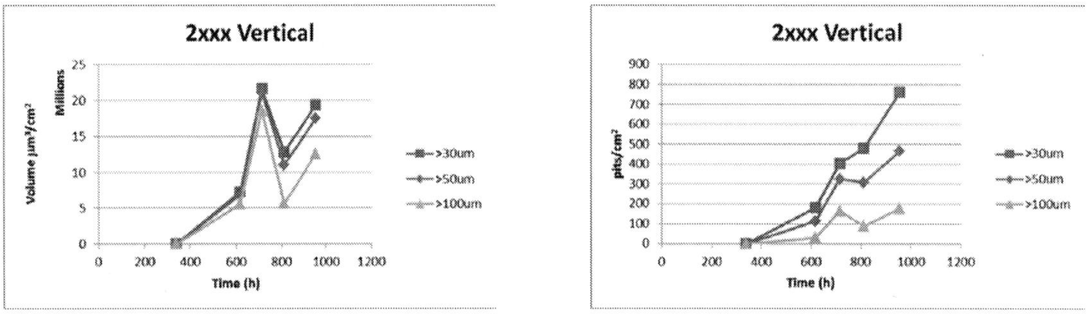

(a) (b)

Figure 5. (a) The corrosion volume and (b) pit density for the 2xxx alloy in the vertical position, counting pits greater than 30 μm, 50 μm, or 100 μm.

In the horizontal position, the 2xxx witness plate coupons showed an incubation period of ~618 h followed by relatively linear growth in corrosion volume (Figure 6). Pit density also showed an incubation period to between 340-618 h.

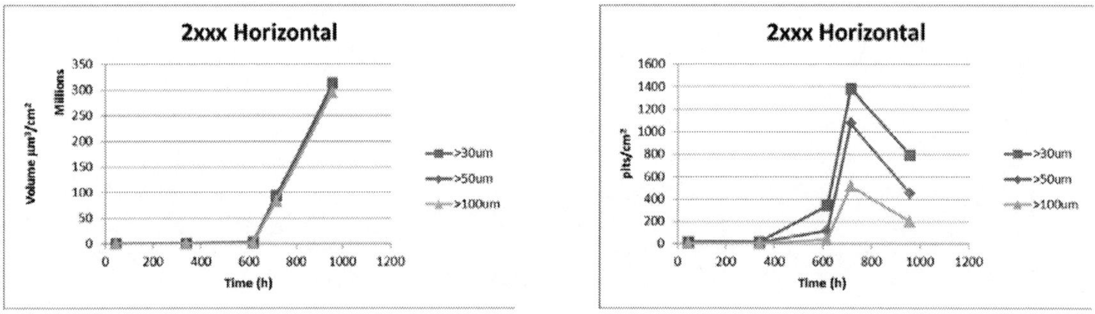

(a) (b)

Figure 6. (a) The corrosion volume and (b) pit density for the 2xxx alloy in the horizontal position, counting pits greater than 30 μm, 50 μm, or 100 μm.

3.2 Aluminium Alloy 7075

For the 7xxx alloy in the vertical position the corrosion volume was low for all exposure times and showed some variation with time and no clear trend (Figure 7). Pit density was low for the 7xxx alloy in the vertical position but typically showed an increase in density with time (most pronounced for the 30 μm criterion).

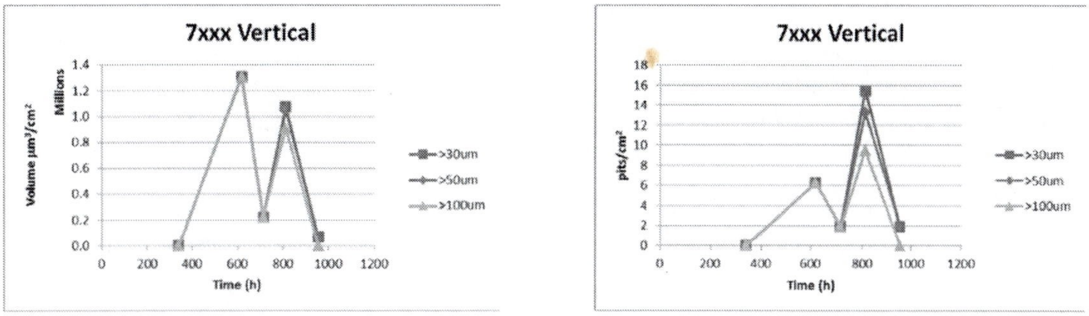

(a) (b)

Figure 7. (a) The corrosion volume and (b) pit density for the 7xxx alloy in the vertical position, counting pits greater than 30 µm, 50 µm, or 100 µm.

In the horizontal position, the 7xxx showed an incubation period before corrosion volume increased at times greater than 340 h (Figure 8). Pit density also increased at times greater than 340 h.

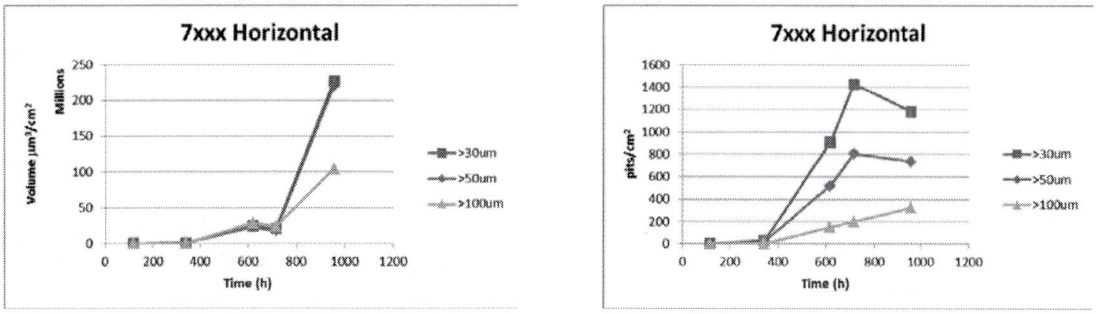

(a) (b)

Figure 8. (a) The corrosion volume and (b) pit density for the 7xxx alloy in the horizontal position, counting pits greater than 30 µm, 50 µm, or 100 µm.

3.3 Comparison of alloys 2024 and 7075

The horizontal position showed similar corrosion volume for both 2xxx and 7xxx series alloys in the 1 mm slot. The incubation period for corrosion was ~340 h for 7xxx and ~618 h for the 2xxx, but corrosion volumes increased more rapidly initially for 2xxx such that corrosion growth was similar for the two alloys over time (Figure 9). In the vertical position, the 2xxx corrosion volume increased at a rate approximately 104 times greater than the 7xxx series alloy after 340 h exposure.

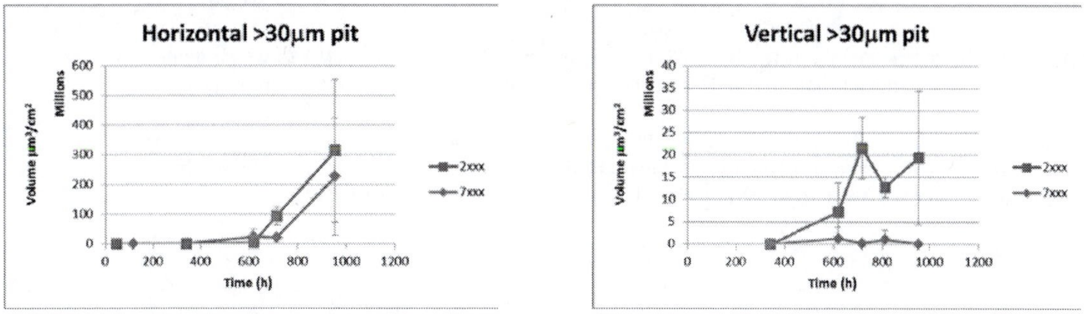

(a) (b)

Figure 9. The corrosion volume for the 2xxx and 7xxx alloy in (a) the horizontal position and (b) the vertical position, counting pits greater than 30 µm. Error bars represent one standard deviation from the mean.

3.4 Comparison of Orientation Horizontal and Vertical

For the horizontal slot of the 2xxx alloy, the corrosion volume showed an incubation time of ~618 h after which corrosion volume increased (Figure 10). In comparison, there was little corrosion volume growth in the vertical position over the same time. For the horizontal 7xxx alloy plate, there was an incubation period of ~340 h before the corrosion volume increased gradually, and showed a large increase between 714-955 h. In comparison, there was little or no corrosion growth in the vertical position over the same time.

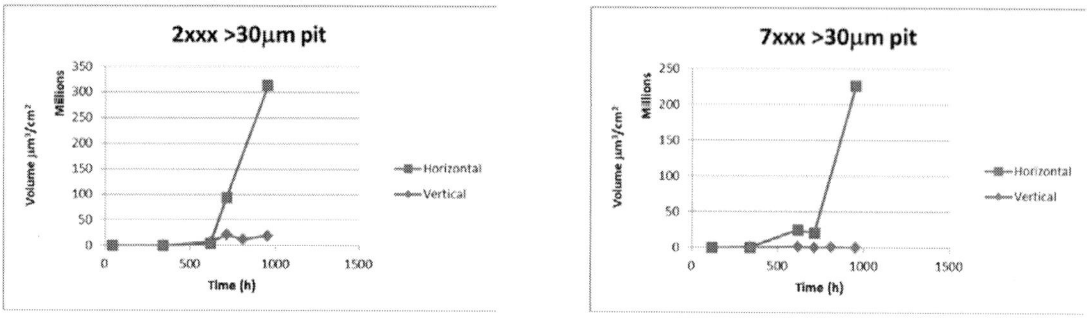

(a) (b)

Figure 10. The corrosion volume for the slots in (a) the 2xxx and (b) the 7xxx alloy in the horizontal position and the vertical position, counting pits greater than 30 μm.

3.5 Comparison of Slotted and Scribed Artificial Damage

The 2xxx series alloy in the horizontal position showed an incubation period before corrosion volume increased after ~618 h in the slot. The scribe showed similar increase in corrosion volume (Figure 11). In the vertical position for 2xxx, both the slotted and the scribed showed increasing trend of corrosion volume with time, there appeared to be higher variability associated with the scribed measurements.

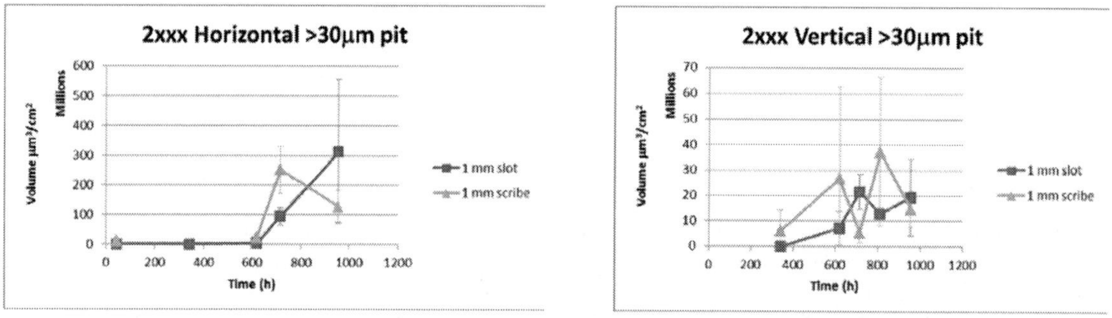

(a) (b)

Figure 11. The corrosion volume for 2xxx alloy in (a) the horizontal position and (b) the vertical position, counting pits greater than 30 μm. Error bars represent one standard deviation from the mean.

The paint coating edge has been metallographically sectioned and observed under an optical microscope, which showed that the primer is sufficiently exposed using the masking technique (Figure 12). The thickness of the primer at the edge of a slot was 120 ± 3 μm, whereas away from the edge the thickness of the primer was 50 ± 5 μm.

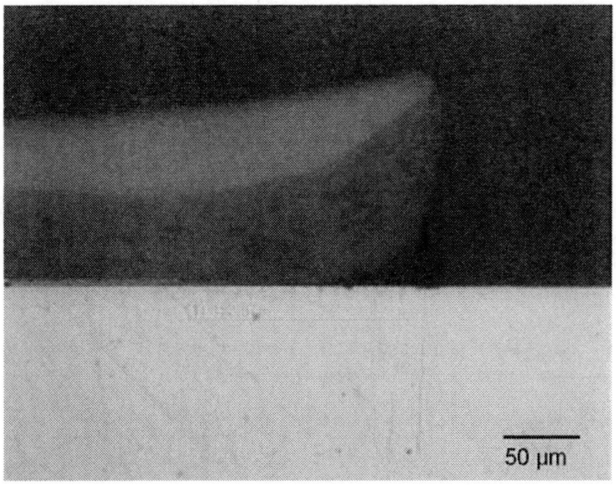

Figure 12. Optical micrograph showing the edge of a slot, featuring the primer and topcoat layers.

3.6 Horizontal/Vertical Volume ratio

For both the 2xxx and 7xxx series alloys the trend was for increasing horizontal/vertical ratio of corrosion volume with time (Figure 13). The horizontal/vertical ratio of corrosion volume was lower for the 2xxx series alloy compared with the 7xxx series alloy because corrosion volumes in the vertical orientation were higher for the 2xxx series alloy, whilst both alloys corroded similarly in the horizontal orientation.

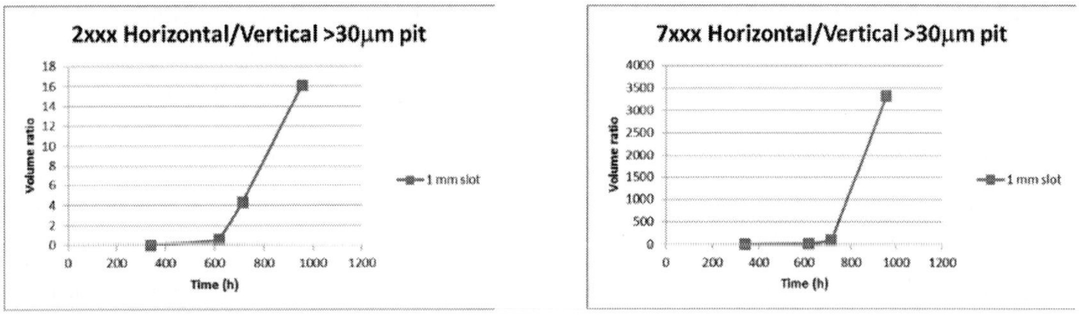

(a) (b)

Figure 13. The ratio of corrosion volume of the (a) 2xxx and (b) 7xxx alloys, counting pits greater than 30 µm measured in the horizontal and vertical orientations.

4. DISCUSSION

Accumulated corrosion volume and accumulated pit count were used to analyse the performance of coated 2024 and 7075 plates in a prohesion test environment. The presence of a 1mm defect in the coating system exposed the alloy substrate and revealed a coating system profile that would allow inhibitor from the primer to influence the corrosion. Various factors were considered including: the alloy substrate material, plate orientation within the test chamber, the method of artificially damaging the coating system (slot or scribe), and the size criterion for pitting. These factors are discussed below.

4.1 Comparison of different alloys: 2024 and 7075

The corrosion volume per unit area of the slot was measured for the 2xxx and 7xxx series alloys in both the horizontal and vertical orientations. In the horizontal orientation, the corrosion volume was similar for the two alloys over time despite a small difference in the apparent incubation. However, the different response of the two alloys is revealed by the 100µm criterion data for the 7xxx alloy recording a lower corrosion volume (Figure 8). At 960 h, the corrosion volume using the 100µm criterion is under half that of the other two criteria (Figure 8a). This result indicates corrosion volume in horizontal orientation is comprised of smaller pits for the 7xxx alloy when compared to the 2xxx. This is further supported by the pit counts for the respective alloys, with the 7xxx alloy recording few pits/cm^2 than the 2xxx alloy for a similar corrosion volume. Overall, the salt water pooling in the horizontal orientation renders both sets of plates under constantly wet (immersion) conditions and corrosion measured on both alloy was severe when compared to the vertical.

In the vertical orientation, there was an incubation period of >340 hr for the 2xxx series alloy. After ~340 hr, the 2xxx series alloy corrosion volume increased at a rate approximately 100 times greater than with the 7xxx series alloy. This result demonstrates a greater susceptibility of the 2xxx series alloy to corrosion than 7xxx series alloy, which is consistent with the data collected for plates in the horizontal orientation. However, it is worth noting, the corrosion volumes measured in the vertical orientation are still an order of magnitude less corrosion than was measured in the horizontal orientation. This observation highlights the dominant factor in the corrosion volume data is plate orientation, overshadowing alloy chemistry and microstructure.

In terms of alloy susceptibility alone, the higher susceptibility of 2xxx series alloy than 7xxx series alloys has been shown previously via electrochemical corrosion tests [22]. Previous tests showed that in NaCl environments, the limiting cathodic current density is higher for a 2024 alloy when compared with a 7075 alloy. which is an indication of corrosion current density in aluminium alloys [23], This finding is demonstrated by two different techniques, those being optical profilometry volume measurements and potentiodynamic electrochemical polarisation tests.

4.2 Comparison of Plate Orientation within Test Chamber

Plates of each alloy types were mounted in two different orientations within the prohesion test chamber. The positioning of these plates within the chamber allows for the comparison of surfaces horizontally and vertically aligned with the internal surfaces of an aircraft bay. Results in Figure 10 show that corrosion is more severe on the plates in the horizontal orientation, when compared to the vertical orientation. Increased corrosion on the horizontal plates occurs due to increases over time of accumulated water and salt, with little to no run-off. By contrast, the vertical orientation records little corrosion volume loss for the same period. This is likely caused by drying of the vertically orientated surfaces during the 'dry cycle'. The 'dry cycle' is unlikely to sufficiently evaporate of any pools of water accumulating on the horizontal plates, since temperatures only reach 35°C at ambient relative humidity for 1 hour. As such, any accumulated salt water on these witness plates can continue to cause corrosion.

Conversely, the vertical plates would experience significant run-off and only a thin film of electrolyte would remain on the substrate surface. The effect of increased run-off is demonstrated by a significantly lower accumulated corrosion volume and fewer pits counted over the same exposure time period for the vertical oriented samples compared with the horizontal oriented samples. This test did not attempt to employ fully the ASTM G-85 standard chamber test protocols [20], where the samples are usually oriented with 15-30 degrees from the vertical. The intention of the current work is to replicate the two extreme orientations found within a bay of a military aircraft. These two orientations demonstrate the influence of physical factors on corrosion severity likely to be experience by the internal aircraft structure. The results show clearly the susceptibility of horizontally orientated surfaces within the aircraft and indicate an increased level of inspection is potentially required.

4.3 Comparison of Artificial Damage: Slots and Scribes

The corrosion volume per unit scan area showed a reasonable correlation between average trends for slot and scribed damage. There was higher variability in the scribed data, which could be attributed to the surface roughness condition of the scribe and the slightly smaller sample size (three instead of four) per witness plate. The slots were finished to ANSI 1200 SiC grit (Ra < 0.5 μm), while the scribed surface roughness following scribing is expected to be significantly higher due increased shear of the surface and the uncontrolled nature of the scribing technique more generally. Thus, increased surface roughness could potentially create misleading data in the measurement of pit depths. Furthermore, the sampling size was slightly higher for the slots compared with the scribes, four versus three plates respectively.

Another key difference between a slot and a scribe is the condition of the exposed primer edge. The thickness of exposed primer edge for slots was measured to be 120 ± 3 μm, whereas away from the edge the primer thickness was 50 ± 5 μm and would be representative of the primer thickness at the edge of scribed damage. There would also be a more jagged edge along the length of a scribe compared with the masked slot. In terms of efficiency of inhibitor release when the slot is wet, it is believed that the masked slot would be more effective as there is a greater thickness of primer exposed compared with a scribed edge. However, a scribed edge is more jagged than the slot, which would increase the exposed area of primer in the scribe. The overall differences between these two techniques were not detectable from these results with this particular non-chromate inhibitor paint scheme.

There was a slightly higher incidence of paint flaking away from the surface during the cleaning process, which was attributed to the damage caused by scribing (e.g. delamination). However, any corrosion that occurred outside the boundaries of the slot (masked or scribed) was not considered in pitting corrosion statistic measurements. The filiform corrosion may be significant in terms of coating degradation mechanisms, however, it was sometimes covered with the coating (even after cleaning), and pitting corrosion was the dominant corrosion process that results in loss in section thickness or subsequently to other more damaging processes, such as fatigue cracking.

4.4 Comparison of Failure Criteria: 30, 50, and 100 μm

The impact on the data of changing the pit size criteria (30 μm, 50 μm, and 100 μm) did not significantly affect the relationships observed for the two alloys. The corrosion volume and pit density both decreased with increasing pit size threshold, as would be expected because the smaller (more frequent) corrosion sites were not included. Hence, corrosion volume was only marginally affected by increasing the pit size threshold criteria. In contrast, there were some clear differences in the pit density measurements with changing pit size criteria. The fact that corrosion volume did not decrease significantly with decreasing pit density, as a result of increasing pit size threshold, seems to suggest that it was the large pit size that contributed mostly to the corrosion volume measurements despite there being a significant number of small pits. A noticeable exception to this trend was the 7xxx alloy in the horizontal orientation (Figure 8). The corrosion volume was more than halved by changing the criterion to >100μm. In this condition, the corrosion volume comprises a greater number of smaller pits rather than was observed for other conditions. This is likely due to a delayed onset of pitting in the 7xxx alloy, where pits have formed in significant number but few have developed to the size of those measured on the 2xxx alloy.

For aluminium alloys in the prohesion environment, using the largest pit criterion is sufficient if corrosion volume is the measurement of interest. It appears from these tests that corrosion volume is the best indicator of corrosion extent under such conditions. Previous studies of a chromate-based inhibitor [24] used a threshold of >30 μm, which may not be applicable for non-chromate inhibitor systems. The present study demonstrated a different distribution of pit sizes can initiate and grow in the presence of a praseodymium-based paint scheme. The mechanism of protection by praseodymium implies that corrosion is required to induce its inhibiting effect [25], suggesting pits may grow to a greater size before the inhibitor arrests their development. As such, selection of a criterion that is too small may remove any apparent effect of the inhibitor in suppressing pitting corrosion.

4.5 Limitations of Measurement Technique

An important limitation of the optical profilometry technique is its difficulty in measuring intergranular corrosion. Optical profilometry is a line-of-sight technique and errors would be anticipated from undercutting and tortuous pathways of corrosion. Other work [23] has been carried out using X-ray tomography of the same two alloys, where corrosion morphology was identified. It was found that AA2024 was susceptible to pitting with the progression to intergranular corrosion, which could form fine networks of corrosion deep into the material. In comparison, AA7075 was also susceptible to pitting but the corrosion pathways were not as fine as for AA2024. These results indicate that the fine pathways formed by intergranular corrosion may have been missed in the present AA2024 data. However, the corrosion volume contribution of fine pathways is expected to be minor in comparison to the gross scale pitting.

In addition, the depth of corrosion underneath the paint coating (e.g. filiform corrosion) was not measured in this study. This corrosion could sometimes be covered by the coating making it impossible to measure by optical profilometry. However, pitting corrosion was the dominant corrosion mechanism observed in the slotted and masked coupons.

5. CONCLUSIONS

A prohesion testing environment was used for accelerated environmental testing of 2024 and 7075 plates. Each plate incorporated an artificially-damaged coating system, including a 1mm wide defect. Optical profilometry of the exposed substrate within these 1mm defects allowed for the collection of corrosion pitting data. The resulting pitting damage was analysed by considering the impact over time of plate orientation and alloy type on accumulated corrosion volume and accumulated number of pits of a certain threshold size. Results demonstrated that the horizontal orientation is more severe than the vertical orientation, which was attributed to salt water pooling. The 2xxx alloy showed a greater susceptibility to corrosion compared with the 7xxx alloy in the vertical orientation, but both alloys were highly susceptible to pitting in the horizontal orientation. The results also showed that the pit size criteria affected the pit density measurements to greater extent than the corrosion volume measurements, indicating that larger pits generally contribute the most to corrosion volume measurements. Comparison of techniques to artificially damage a coating system indicated that a paint masking technique can adequately represents damage caused by a scribe.

From a maintenance perspective, the results presented indicate that places where moisture can pool in an aircraft would be critical areas to inspect. Furthermore, areas within an aircraft including 2xxx series alloys may require more frequent inspections given that the rate of corrosion volume was higher than for 7xxx series alloys. Finally, a larger pit size criterion may be more applicable to the non-chromate based paint scheme, and this may be due to the different mechanism of protection of the praseodymium inhibitor.

6. ACKNOWLEDGMENTS

The authors would like to acknowledge the contribution of Dr Muhammad Awais Javed and Dr Scott Wade at Swinburne University of Technology for carrying out the optical profilometry scans and assisting with analysis of results. In addition, ALS Industrial Pty Ltd is acknowledged for carrying out metallographic sections of the slots, and Vipac Engineers and Scientists Ltd for prohesion testing in a chamber.

7. REFERENCES

1. Cassidy M, Palanisamy S, Waldie J. The Bottom Line: What Aircraft Corrosion Costs the Australian Defence Force. AIAC 16: Sixteenth Australian International Aerospace Congress; 24 February 2015; Melbourne, Australia, 2015.
2. Davis JR, Aluminun and Aluminum Alloys, ASM International, Materials Park, OH, 1994.
3. Polmear I, Light Alloys, Arnold, London, UK, 1995.
4. Davis JR, Corrosion of Aluminum and Aluminum Alloys, ASM International, Materials Park, 1999.
5. Pourbaix M, Atlas of Electrochemical Equilibria in Aqueous Solutions, Cebelcor, Brussels, Belgium, 1974.
6. Burstein GT, Liu C, Souto RM, Vines SP, Origins of pitting corrosion, Corrosion Engineering, Science and Technology 39(1) (2004) 25-30.
7. Kendig M, Jeanjaquet S, Addison R, Waldrop J, Role of hexavalent chromium in the inhibition of corrosion of aluminum alloys, Surface and Coatings Technology 140(1) (2001) 58-66.
8. Kendig MW, Buchheit RG, Corrosion inhibition of aluminum and aluminum alloys by soluble chromates, chromate coatings, and chromate-free coatings, Corrosion 59(5) (2003) 379-400.
9. Scholes FH, Furman SA, Hughes AE. Characterisation of self-healing behaviour in chromate-based aerospace paint systems, (In) Proc. Proceedings of the First International Conference on Self Healing Materials, 2007 18-20 April 2007, Noordwijk aan Zee, The Netherlands.
10. Beyersmann D, Hartwig A, Carcinogenic metal compounds: recent insight into molecular and cellular mechanisms, Archives of Toxicology 82 (2008) 493-512.
11. Frankel GS, Buchheit RG, Jaworowski M, Swain G. Scientific Understanding of Non-Chromated Corrosion Inhibitors Function: SERDP Project WP-1620, 2013.
12. Twite RL, Bierwagen GP, Review of alternatives to chromate for corrosion protection of aluminum aerospace alloys, Progress in Organic Coatings 33 (1998) 91-100.
13. Sudholz A. Leaching performance of a rare-earth-containing epoxy polyamide primer. 55th Annual Conference of the Australasian Corrosion Association 2015: Corrosion and Prevention 2015; Adelaide: ACA; 2015.
14. Sudholz A, Trueman A. Electrochemical characteristics of aluminium alloys in the presence of praseodymium(III) ions. 54th Annual Conference of the Australasian Corrosion Association 2014: Corrosion and Prevention 2014; Darwin2014.
15. Lopez-Garrity O, Frankel GS, Corrosion inhibition of aluminum alloy 2024-T3 by praseodymium chloride, Corrosion 70(9) (2014).
16. Bierwagen GP, The physical chemistry of organic coatings revisited – viewing coatings as a materials scientist, Journal of Coatings Technology and Research 5(2) (2008) 133.
17. Scholes FH, Furman SA, Hughes AE, Markley TA, Corrosion in artificial defects. I: Development of corrosion, Corrosion Science 48 (2006) 1812-1826.
18. PPG. Material Safety Data Sheet: 44GN098. 2014.
19. ASTM. ASTM B117 - 11 Standard Practice for Operating Salt Spray (Fog) Apparatus. American Society for Testing and Materials; 2011.
20. ASTM. ASTM G85-98 Standard Practice for Modified Salt Spray (Fog) Testing. American Society for Testing and Materials; 1998.
21. ASTM. ASTM G1-90 (1999) Standard Practice for Preparing, Cleaning, and Evaluating Corrosion Test Specimens. American Society for Testing and Materials; 1990.
22. Sudholz A, Knight S, Trueman A. Cathodically-limited corrosion kinetics on aircraft alloys in aerated chloride environments. 52nd Annual Conference of the Australasian Corrosion Association 2012: Corrosion and Prevention 2012; November 2012; Melbourne, Australia 2012.
23. Knight SP, Sudholz AD, Butler A, Palanisamy S, Dargusch MS, Trueman AR, The effect of droplet size on the localized corrosion of high-strength aluminum alloys, Materials and Corrosion 67(12) (2016) 1294-1307.
24. Sudholz A, Knight S, Trueman A. Comparative predictions of stable pit probability on aircraft Al alloys AA2024-T3 and AA7475-T7651 in aqueous environments. 53rd Annual Conference of the Australasian Corrosion Association 2013: Corrosion and Prevention 2013; 10-13 November 2013; Brisbane2013.

25. Treu BL, Pinc WR, Fahrenholtz WG, O'Keefe MJ, Morris E, Albers R, Characterization of transport processes in a praseodymium-containing coating, ECS Transactions 28(24) (2010) 229-237.

8. AUTHOR DETAILS

Dr Steven Knight is works as a Materials Engineer at BAE Systems Australia and a Post-doctoral Research Fellow at Swinburne University as part of the Defence Materials Technology Centre (DMTC). Steven has an interest in the metallurgy of high-strength aluminium alloys, and understanding the relationships between microstructure and material properties, in particular corrosion mechanisms and kinetics.

Dr Aaron Sudholz conducts research as a consultant materials engineer with the Defence Science and Technology Group. Aaron completed both his undergraduate and postgraduate studies at Monash University in the areas of chemistry, materials science and materials engineering. Aaron's research centres on the characterisation and modelling of materials properties in the area of corrosion science and engineering. Additionally, Aaron is studying a Master of Laws (Juris Doctor) and has a keen interest in intellectual property matters in the materials science space.

Mr Simon Jacob currently works at BAE Systems Australia as a Materials Engineer. During his post-graduate and defence industry experience, Simon has worked on a range of materials and processing topics relating to metal forming, chemical processing, finishing, NDI, painting, and manufacturing. Simon is currently working on developing a corrosion modelling capability that will be integrated in a corrosion prognostic health management system.

Dr James Waldie graduated with a Bachelor of Engineering (Aerospace) and Bachelor of Business from RMIT University. James continued on as a Research Scholar at the University of California in San Diego for his Masters, and earned his PhD from RMIT in 2005. James was also a Postdoctoral Fellow at the Department of Aeronautics and Astronautics at the Massachusetts Institute of Technology (MIT) for 2.5 years. James has worked for BAE Systems since 2002, most recently as the R&D Engineering Manager for the Corrosion Group.

AN OVERVIEW OF TECHNIQUES FOR MEASURING THE INTERACTION BETWEEN EROSION AND CORROSION

Yunze Xu[1], Jing-Li Luo[2] and Mike Yongjun Tan[1]
[1]School of Engineering and Institute for Frontier Materials
Deakin University, 75 Pigdons Road, Waurn Ponds, VIC 3216, Australia
[2]Department of Chemical and Materials Engineering
University of Alberta, Edmonton, Alberta, Canada, T6G 1H9

SUMMARY: In the multi-phase flowline environments such as in oil & gas pipelines, the mechanical impact of high-velocity fluids and particles has been found to act synergistically with electrochemical processes, resulting in sustained erosion corrosion damage to metal surface and localised penetration rates significantly higher than either erosion or corrosion acting alone. It is well known that erosion-corrosion occurs once the dense corrosion product film or passive film on the passive metal surface is removed by the multi-phase fluid or slurry, leading to a dramatic increase of the corrosion current. On the other hand, for active metals, the corrosion current may cause the metal surface to become rough and result in the reduction of surface hardness, which causes the acceleration of the erosion rate. Currently there are different views on the interactive effect of erosion and corrosion on metals. Since it is important for forecasting and preventing erosion-corrosion in engineering practice, over the past decades, the interaction between erosion and corrosion has been investigated using various electrochemical and non-electrochemical methods. Electrochemical test devices such as rotating disc electrode and slurry jet impingement facility are frequently used for the measurement of gravimetric mass loss, open circuit potential, potentiodynamic polarization, and electrochemical impedance spectroscopy. Non-electrochemical sensors based on the electrical resistance technique and acoustic sense technique are developed for the probing of erosion-corrosion in real life pipelines. Recently multi-electrode array sensor has also been used in the study of the relationship between erosion-corrosion and flow field distribution in combination with Computational fluid dynamics (CFD) simulations. Based on these test results, some theoretical and empirical models were established to explain both corrosion enhanced erosion and erosion enhanced corrosion. This paper presents an overview and critical discussion on major test methods used for the study of synergistic effect between erosion and corrosion in both laboratory testing and industrial application conditions. It has been found that 1) previous research heavily relied on the measurements of the overall and steady-state electrochemistry over metal electrode surfaces, rather than probing the dynamic electrochemical changes occurring at localised passive film breakdown sites, and therefore localised corrosion activities have not been sufficiently evaluated; 2) the performance of passive films was mostly studied under anodic polarisation conditions, rather than under natural passivation conditions, and therefore the erosion-corrosion behaviour of passive metals has not been completely understood; 3) the ex-situ analysis of corroded metal surfaces has not been directly linked to the localised corrosion processes. Limitations in erosion-corrosion measurement and monitoring methodologies are considered to be the main reasons for presently insufficient understanding of dynamic and localised passive film breakdown and restoration processes, which must be overcome before achieving effective control of the major factors influencing erosion-corrosion and its prevention.

Keywords: erosion-corrosion, interaction, electrochemical method, passive film, localized corrosion

1. INTRODUCTION

Multiphase fluids containing sand particles and gas bubbles are commonly found in the oil & gas pipelines where the oil and gas are transported from the reservoir formations[1]. When the flow rate is high, the abrasion between sand particles and pipeline surfaces may lead to the reduction of the wall thickness and this process is generally defined as erosion[2]. Moreover, once sand particles are mixed with the formation water, which may contain carbon dioxide, chloride ions and dissolved oxygen in the upstream, erosion-corrosion occurs and result in more serious damages to the pipeline system[3]. Corrosion of the pipeline inner surface is usually an electrochemical process, while erosion is considered as a wear process caused by mechanical actions. It is well known that mass losses due to the synergistic effect of erosion and corrosion can be dramatically higher than the summation of the mass loss caused by pure erosion and pure corrosion[4]. Understanding the synergism of erosion and corrosion is a key requirement for effective pipeline protection especially for pipelines where sand particles are observed.

Various laboratory experiments and field tests were conducted in past decades to identify the erosion enhanced corrosion and corrosion enhanced erosion. It is commonly suggested that the total materials loss of the metal can be divided into four parts during an erosion-corrosion process[5-7]:

$$W_t = W_c + W_e = (W_{c0} + W_c^e) + (W_{e0} + W_e^c) \qquad (1)$$

where W_t is the total material loss, W_c is the corrosion component with the presence of erosion, W_e is the erosion component with the presence of corrosion, W_{c0} is the pure corrosion rate without the appearance of erosion, W_c^e is the erosion enhanced corrosion, W_{e0} is the pure erosion rate without the appearance of corrosion, and W_e^c is the corrosion enhanced erosion.

Since the total material loss is caused by both electrochemical and mechanical processes, as suggested by equation (1), it is difficult to evaluate the total weight loss based only on Faraday current that is often measured in electrochemical experiments. Therefore, gravimetric mass loss measurements are always performed to calculate the overall weight loss, W_t, in the erosion-corrosion test[8]. As the corrosion process is an electrochemical process, the uniform corrosion rate W_c can be monitored based on traditional electrochemical methods such as liner polarization resistance (LPR) and electrochemical impedance spectroscopy (EIS)[9]. Then, the W_e also can be obtained from the difference between W_t and W_c. Furthermore, to figure out the synergistic effect W_c^e and W_e^c, the pure corrosion rate W_{c0} and pure erosion rate W_{e0} should be measured. In most cases, W_{c0} are decided in the working condition without sand particles to avoid the occurrence of erosion[10, 11], and W_{e0} are measured with cathodic protection or inhibitor to avoid the presence of corrosion[12]. Based on the methods mentioned above, the interaction between erosion and corrosion can be fundamentally evaluated.

However, this approach may not be suitable for evaluating dynamic erosion-corrosion processes occurring in real pipeline working conditions especially for pipelines made of passive metals such as stainless steels. Localized breakdown of the passive film caused by erosion may lead to a pitting corrosion rather than general electrochemical corrosion[13] and therefore conventional uniform corrosion based electrochemical methods may not be suitable for it. Meanwhile, passive films are mostly studied under anodic polarisation conditions during laboratory tests and therefore the results may not reveal passive film behaviours under natural conditions. Nevertheless, erosion-corrosion is usually happening under natural passivation conditions[14]. On the other hand, traditional electrochemical methods cannot be directly used in real pipelines for erosion-corrosion monitoring because it is unable to measure metal loss based on Faraday current and is unable to measure erosion induced metal loss. Currently practically applied sensors in pipeline industry are mostly designed based on electrical resistance (ER) and acoustic sensors[15]. These non-electrochemical sensors provide only overall weight losses and have limitations in providing detailed information of the erosion and corrosion processes and their synergistic effects.

This article provides an overview of the techniques used for the study of the interaction between erosion and corrosion in both industry applications and laboratory testing conditions with particular focus on internal pipeline conditions. The advantages and drawbacks of the non-electrochemical methods and traditional electrochemical methods are compared, and some theoretical and empirical models are introduced based on the previously reported results in the literature. Some new methods are suggested for the future study of erosion-corrosion through the probing of dynamic erosion induced localised corrosion processes by using the wire beam electrode (WBE) sensor system[16].

2. CURRENT TECHNIQUES FOR EROSION-CORROSION MONITORING IN THE PIPELINE INDUSTRY

In the oil and gas pipeline industry, the working condition keeps changing with variations in the pipeline pressure, flow velocity and the fluid temperature[17]. The erosion and corrosion rates could be very high and could vary significantly with time, and therefore they need to be on-line monitored in order to detect the potential risks associated with peak corrosion rate and localised corrosion[18]. Both electrochemical sensors and non-electrochemical probes are widely used in energy pipeline industry for the on-line monitoring of the corrosion processes. The instantaneous corrosion current can be easily obtained by using electrochemical methods such as LPR and EIS[19]. Though electrochemical sensors can gain the corrosion current easily and rapidly, they are not suitable for the erosion-corrosion monitoring in real pipelines due to the inability of measuring non-Faraday part of the metal loss. Therefore, ER probes become the most widely used method for the monitoring of total metal loss caused by erosion-corrosion in pipeline industry. However, compared with electrochemical methods, non-electrochemical method based probes such as ER probes need a period of response time to reflect the corrosion rate because the changing of the ER signal is based on the reduction of the wall thickness[20].

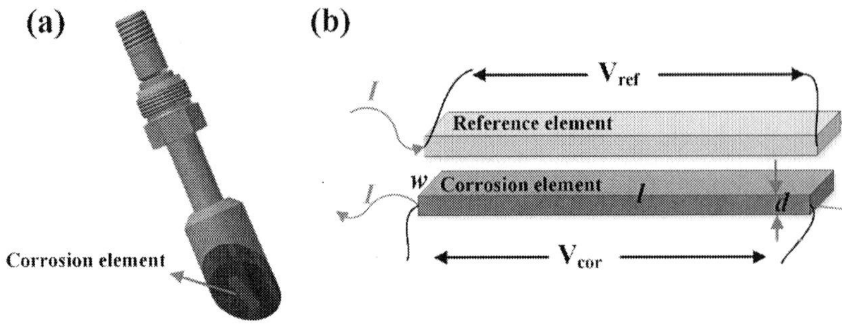

Fig. 1 The erosion-corrosion probe using ER technique (a), and the schematic diagram of the probe (b).

Fig. 1a shows a typical erosion-corrosion ER probe with its sensing part in a rectangular form. The measurement schematic diagram of the probe is shown in Fig. 1b. The sensing part of the probe is composed of a corrosion element which is directly exposed to the pipeline working condition and a temperature compensation element which is sealed inside of the probe. The resistance of each element can be obtained by four wires method[20] and the total corrosion depth x of the corrosion part can be expressed as[21]:

$$x = d\left(1 - \frac{R_{ref}/R_{cor}}{R_{ref0}/R_{cor0}}\right) \quad (2)$$

where d is the thickness of the corrosion element, R_{ref}/R_{cor} is the monitored resistance ratio between reference element and corrosion element, and R_{ref0}/R_{cor0} is the initial resistance ratio when no corrosion or erosion happens. It is seen from Eq. (2) that the total mass loss of the sensor only correlate to the geometry of the corrosion element which means the wear part of the mass loss can be easily measured based on ER technique. For different test locations, the ER probes can be placed at different directions to ensure the sufficient impingement of the sand particles to the corrosion element. Fig. 2 shows the arrangement of the ER probes at different locations of the pipeline in Fan's test[15]. It is seen that flush head probes were used at the elbow and the probe with a 45° angle was applied in the horizontal flow line.

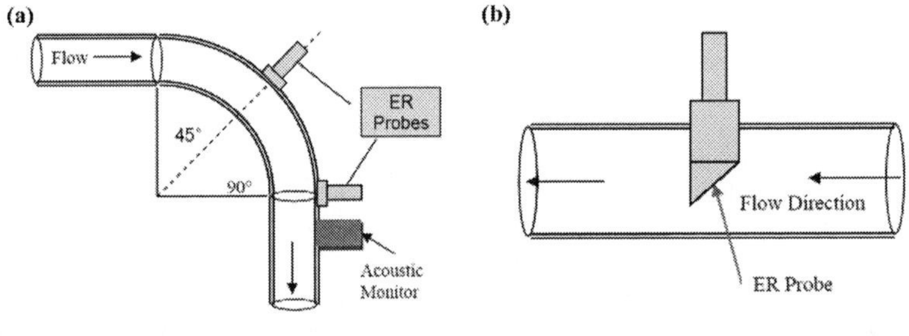

Fig. 2 The arrangement of ER probes for erosion-corrosion monitoring at different locations[15] (a) at elbow and (b) in the horizontal flow line.

It is reported in the previous studies that the flow distribution and sand concentration are not uniform in the pipeline[15, 22, 23]. Different monitoring data may be obtained when the erosion-corrosion probes are placed at different locations [15]. Therefore, some modified ER sensors are provided for the monitoring of non-uniform corrosion in pipeline systems[23-27]. Fig. 3 shows two kinds of modified ER sensors which are defined as field signature method (FSM) [26] and ring pair electrical resistance sensor (RPERS)[23], respectively. The basic principle of both sensors are dividing the pipeline into small sections and the resistance of each section can be on-line monitored. The localized erosion-corrosion rate of each segment can be reflected from the localized corrosion depth.

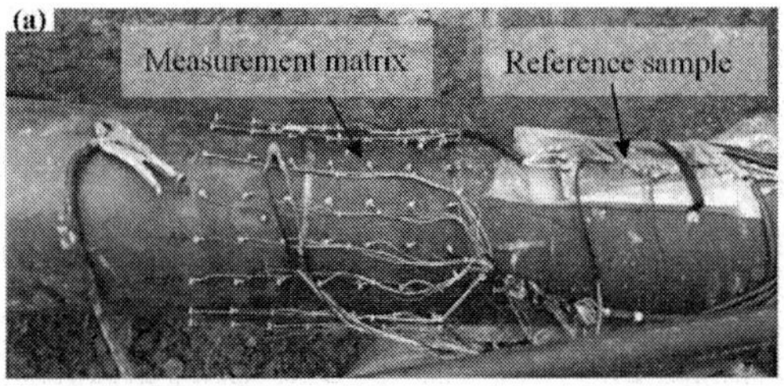

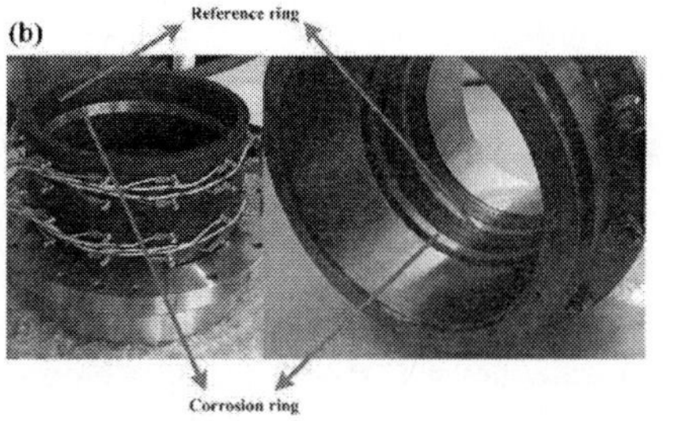

Fig. 3 The modified electrical resistance technique used for localized corrosion monitoring in pipeline working condition (a) FSM[26] and (b) RPERS[23].

The advantages and limitation of ER and FSM based metal loss monitoring probes has been discussed previously[28]. ER and FSM are often referred to as 'intelligent' weight-loss coupon tests. The ER probe monitoring metal loss by measuring the electrical resistance of a thin metal test wire (sensor element) since the resistance of the wire increases as the wire becomes thinner due to erosion or corrosion dissolution. An advantage of ER probe is that it provides cumulative metal loss values without the need to remove the samples from the service environment. Another advantage of ER probes is that the technique is applicable to both conductive and non-conductive corrosion environments. A major disadvantage of the ER technique is that it is unable to accurately detect localised corrosion since localised corrosion may neither lead to significant metal dissolution, nor noticeable change in electric resistance. ER measurements generally do not respond rapidly to a change in corrosive conditions, and it was reported to take four days to respond to 1 mpy corrosion rate[29]. The measurement sensitivity can be improved by decreasing the element thickness; however this would shorten the service life of the sensor element. It should also be noted that if conductive deposits such as iron sulphide are produced and attached on the sensor elements, the resistance of the elements could change and thus distorted readings could be obtained. This suggests that care has to be taken when applying ER in sour oil and gas systems where certain forms of microbial corrosion, such as SRB attack, can produce iron sulphide. FSM is also considered to be a cumulative corrosion monitoring technique developed from the commonly used principle of electrical resistance determination. FSM impress a current through several electrodes (sensing pins) on the outer surface of a structure and the potential drops between these electrodes are measured. The sensing pins are typically separated by a distance of 2-3 times wall thickness. Changes in the geometry in the form of cracks, general corrosion, erosion corrosion, or pitting will disturb the potential field in the metal. The change in the measured potential field pattern is related to a change in pipe wall thickness, and thus the development of corrosion or cracks can be detected. The FSM technique is non-intrusive and is used to measure corrosion damage over a relatively large section of a structure. Possible limitations of the technique include its relatively slow response to corrosion. It was reported to take one year to respond to 1 mpy corrosion rate[29, 30]. Although this measurement sensitivity could be improved with technical improvements, it is still difficult to provide real time erosion-corrosion rate date that are needed for building the relationship between the total metal loss and the impingements of the sand particles.

In order to overcome this limitation, acoustic sensor are used in combination with ER probes to monitor the sand rate[31]. It can be seen from Fig. 2a an acoustic sensor was applied in the downstream of the elbow in Fan's test[15] to provide the information of the sand rate. As the impact of the sand particles on the steel surface will cause a response signal in the metal around 100 kHz to 500 kHz, through the extraction of the signals in the certain range, the sand rate can be monitored when the sensor is arranged closed to the pipe wall. In recent studies, through analysing the generation of acoustic emission, the erosion rate also can be evaluated[32]. Nevertheless, the acoustic signal is easy to be disturbed by the background noise and the erosion rate calculated from the acoustic emission heavily relied on the property of the material. Therefore, acoustic sensor are almost implemented as an additional method in combination with ER probes in the probing of erosion-corrosion in pipeline industry.

Based on the modified ER sensors, the localized corrosion caused by erosion-corrosion could be solved in a large extent, especially for the erosion-corrosion at elbows. However, in the case of passive metals, pitting corrosion could initiate in the small section of the sensor and therefore the monitoring results could have a large measurement error which may lead to the underestimate of the local penetrating rate[33]. Another weakness of ER sensors is that the measurement of the erosion-corrosion rate may be significantly affected by the intensive fluctuation of the resistance values when the temperature has a large variation[15, 23, 34].

3. ELECTROCHEMICAL DEVICES AND METHODS FOR STUDYING EROSION-CORROSION

The total metal loss caused by erosion-corrosion is often the most concern for pipeline industry, and therefore ER methods is considered to be an effective technique for erosion-corrosion monitoring. However, ER sensors are unable to provide in-situ data for revealing the interaction of the erosion and corrosion, and the synergism of erosion and corrosion in the study of flow accelerated corrosion[35]. Consequently, some electrochemical devices were designed for the study of erosion-corrosion in laboratory tests in order to further understand the erosion enhanced corrosion and corrosion enhanced erosion issues. This understanding is useful for developing suitable technologies for the monitoring and inhibition of erosion corrosion. Two types of test facilities i.e. the rotating disc electrode[36-39] and the slurry jet impingement rig are commonly used for the study of erosion-corrosion in lab tests[40, 41]. It can be seen from Fig. 4 that two forms of rotating disc systems were applied in previous studies. Both test sets were composed of a three electrodes system and a rotating disc to simulate a hydrodynamic condition[42, 43]. The difference is that in the first set as shown in Fig. 4a, the test sample itself worked as the rotating disc and the test specimen in the second set was arranged at the edge of the rotating disc. In the first set, the linear velocity on the sample surface can be simply controlled by the rotate speed. For the specimen in the second set, the flow rate of the slurry on its surface cannot be seemed as the linear velocity of the disc due to the edge effect and the influence of the inserted specimen. Therefore, the corrosion behaviour of the test sample in the first set will be more uniform compared with that in the second set. However, the impact angle of the sand particles in the first set cannot be adjusted which indicates that all the mass loss caused by erosion in the first set are coming from the tangent direction of the disc. In contrast, the impact angle of the specimen can be adjusted from 0 to 90° in the second set. Thus, the effect of the impingement angle of the sand particles can be studied by using the second set.

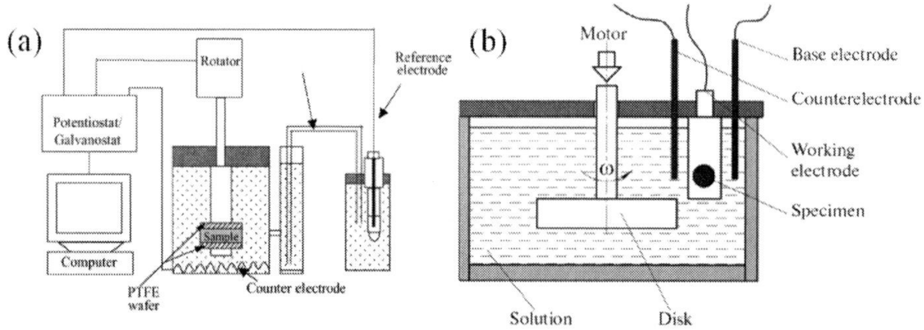

Fig. 4 Typical rotating disc electrode systems used for erosion-corrosion (a) with the electrode act as disc[42] (b) with electrode at the edge of the disk[43].

Guo[42, 44] and Lu[45] applied the rotating cylinder as shown in Fig. 4a in the study of the erosion-corrosion of carbon steel in both corrosive solution and slurry. In the test conducted in the corrosive solution without the sand particles, a constant current was carried out on the test sample[43]. The total anodic current was calculated with the compensation of the cathodic current through the extrapolation of the polarization curves. The corrosion caused by the anodic current follows the Faraday law and the weight loss caused by Faraday process can be calculated. It is surprised to find that the ratio of the total weight loss and the weight loss caused by corrosion current was above 1.6 after the rotation rate reached 9000 rpm. However, the erosion rate was close to zero when cathodic protection was applied to the test sample in the same rotation rate. It indicated that even in the corrosive solution without sand particles, corrosion still could significantly enhance the erosion when the flow rate is sufficiently high. When the carbon steel is under an active corrosion process, the corrosion rate is controlled by the diffusion of the dissolved oxygen in a high flow rate condition. The tests in both corrosive solution and slurry have verified that the erosion causes only a small increase in the corrosion rate in this situation[42, 44]. Conversely, corrosion may lead to an intensive rise of the erosion rate when no passive film forms on the carbon steel surface, especially when the corrosion current initiates the degradation of the surface hardness. Guo[46] studied the modification of the surface mechanical properties due to the electrochemical dissolution using in situ nanoindentation technique. A dramatic decreasing of the surface hardess was found with an 1 mA/cm^2 anodic current. However, without the anodic current, the surface hardness would recover to a large extent. She ascribed the deteriotation of the surface hardness during the anodic dissolution to the inhomogeneus distribution of the vacancies in the steel. The corrosion enhanced erosion can be expressed based on the experimental results and the theoretical derivation[42, 44, 47]:

$$\frac{W_e^c}{W_{e0}} = Z\log\left(\frac{i_a}{i_{th}}\right) \quad (3)$$

where Z is an experimental constant, i_a is the anodic current and i_{th} is the threshold anodic current density to induce the degradation of the surface hardness. In some recent studies[48, 49] by application of the cyclic erosion-corrosion which separate the erosion and corrosion processes to identify the interference of each process, it is found that the surface roughness of the test sample will have a 54% increase after the 2.5 h corrosion cycle. With a high impingement velocity of the sand particles which was over 31 m/s in erosion process, the rougness surface was more easily to be removed by the erosion process due to the vulnerable local peaks on the steel surface.

For metals in a passive state or covered by a protective layer, the corrosion rate may be relatively low in the corrosive solution. The impingement of the sand particles may damage the passive film and lead to the metal substrate directly exposed to the corrosive medium. In this case, erosion enhanced corrosion becomes the main factor for the metal loss. The impingement velocity and the impact angle of the sand particles have significant influences to the erosion-corrosion process. Some tests[43, 50] were conducted to study the effect of the impact angle on the passive metal by using the rotating disc electrode system as shown in Fig. 4b, however the flow rate of the slurry and the sand concentration cannot be accurately controlled in this system. Thereby, the slurry jet impingement rig was introduced into the study of erosion-corrosion process. Fig. 5 shows a typical slurry jet impingement rig[10, 51] which is composed of a three electrodes system and a slurry loop driven by an electro pump. Compared with the rotating disc electrode system, the wear of the pump will be an extra expense. However, the slurry jet impingement rig can provide a more benefit control of the impingement velocity and impact angle of the sand particles.

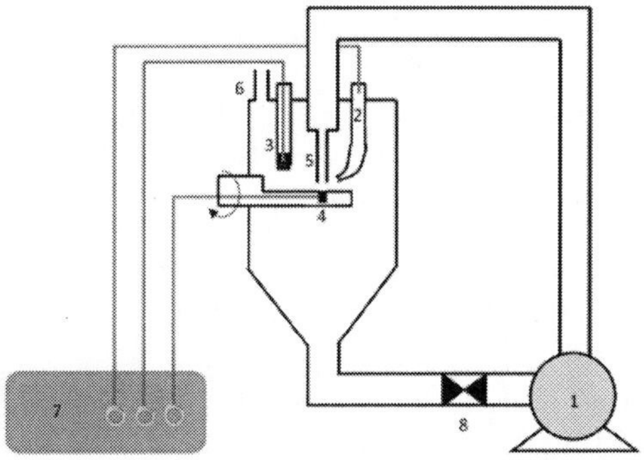

Fig. 5 Schematic of a typical slurry jet impingement rig[48], 1 electro pump, 2 reference electrode in capillary, 3 counter electrode, 4 sample, 5 nozzle, 6 slurry entrance to the container, 7 potentiostat, and 8 exit valve.

It is presented in previous studies that the pure erosion rate follows an empirical power law relationship with the flow velocity[8, 52]:

$$\text{Erosion rate} = kV^n \quad (4)$$

where V is the velocity of the sand particles, k is a constant, and n has values between 2 and 3.5 for metallic materials. In Lindsley's test[52] using a high speed slurry jet rig, it is found that n is related to the test conditions, and not target material properties for the alloys. Therefore, it is still hard to establish an effective model to forecast the pure erosion rate. In some other researches, it is found that Eq. (4) is not suitable for low flow rate conditions. It is pointed out in Tan's[12] studies that if the impact energy of the slurry is not sufficient enough to penetrate the passive film or the protective coating, the protective film on the steel surface can weaken the synergistic effect between erosion and corrosion. A critical velocity was proposed by Zheng[53] to identify the erosion-corrosion process. In his test, it is found that the open circuit potential (OCP) of the 304 stainless steel would show an obvious rise when the jet velocity of the slurry was 5 m/s and the metal loss remained at a low value until the jet velocity reached 12 m/s. The test results indicate that the passive process may be speeded up due to the acceleration of oxygen transfer with the increasing of jet velocity before it reaches the critical value. Once the jet velocity reaches the critical value, the passive film can be locally destroyed by the slurry and the substrate of the metal will be exposed to the slurry which can induce the erosion-corrosion. The erosion rate can be evaluated by using Eq. (4) when the flow rate is above the critical velocity. However, the erosion enhanced corrosion for passive metals is still hard to be assessed based on the above studies.

In the study of erosion-corrosion for stainless steel and carbon steel in a passive state, a restoration process of the passive film was investigated after the breakdown of the passive film in the tests conducted by Lu and Luo[7, 54, 55]. A theoretical model was proposed by Lu and Luo [56] to calculate the average anodic current density $\bar{i}$ caused by the breakdown of the passive film:

$$\bar{i} = i_s + i_{peak} \frac{\tau \left(\frac{C_p U \sin\theta}{m_p} \right) A_{crater}}{1 + \tau \left(\frac{C_p U \sin\theta}{m_p} \right) A_{crater}} \quad (5)$$

where i_s is the stable passive current density, i_{peak} is the peak current density induced by the local breakdown of the passive film, τ is the decay constant of the restoration process, C_p is the weight of solid particles in the slurry with the unit volume, U is the impingement velocity, and θ is the impact angle, m_p is the mass of a single particle, and A_{crater} is the impact area caused by once impingement. This theoretical model was established based on the assumptions that the passive film on the steel surface is uniform; the shape of the sand particles are in same sizes, the corrosion process on the steel surface has reached a stable status; and the energy of each sand particles are same. In the derivation of Eq. (5), the decay of the anodic current was assumed to follow the exponential form. While, for real passive metals, the decrease of the current may obey the bi-exponential decay law or power decay law [7]. The Eq. (5) was modified to suit for 304 stainless steel and the carbon steel in a passive state in Lu's further experimental studies[7, 55]. Through using the jet slurry rigs, this theoretical model was well fitted in various tests when the erosion-corrosion kept in a stable status.

In most of the studies on erosion-corrosion using electrochemical methods, it is investigated that the total weight loss are measured by gravimetric loss measurement, and the corrosion rate can be told by using LPR and EIS measurements. In the measurement of pure corrosion rate, the common method is to eliminate the sand particles from the slurry. Then, Faraday current caused by pure corrosion also can be measured by LPR and EIS methods. For the pure erosion rate measurement, the experimental studies are always conducted in the conditions such as corrosive medium free slurry, under cathodic protection or with inhibitor added. The average erosion rate was calculated from the weight loss. It seems that the contributions of both corrosion enhanced erosion and erosion enhanced corrosion can be evaluated based on these test methodologies, but serious measurement errors and data scatters were still found in the previous test[12]. The measurement accuracy of the four parts may be influenced by various factors during the electrochemical test and the main causes are concluded as follows:

1. The metal loss obtained from the gravimetric measurement is the accumulated material loss through the whole test period. Since it is found in the previous studies that the erosion-corrosion process keeps continuous changing with the modification of the surface status[13], the average erosion-corrosion rate may be not a stable value in the whole test, especially for the stainless steel when localized corrosion occurs on its surface. Therefore, using the erosion-corrosion rate calculated from the accumulated metal loss may lead to a wrong judgement of the erosion-corrosion process.

2. As the corrosion is a Faraday process, the corrosion rate in both corrosion and erosion-corrosion conditions can be on-line monitored by using electrochemical methods. It can guarantee the measurement accuracy of the corrosion rate in a large extent. However, there is still some problems may cause the measurement error occurring. In some cases, the impingements of the sand particles on the metal surface is not uniform[53, 57]. An internal circuit may initiate on the test sample during the polarization process due to the non-uniform potential distribution on the sample surface. This phenomenon may result in reduced current flow into the counter electrode and underestimate of the corrosion rate.

During the pure corrosion measurement, it is always thought that the erosion can be neglected without the appearance of sand particles in the corrosive solution. Nevertheless, it has mentioned in Guo's research[44] that the non-Faraday process is still exist in the solution without sand particles when the flow rate is sufficient high. It means the corrosion rate measured in the corrosive solution free of sand particles may be enhanced by the non-Faraday process as well, which may result in the overestimate of the pure corrosion rate.

3. Same with the measurement of the total metal loss, weight loss measurement is the most effective way to measure the pure erosion rate because of the non-Faraday process. The pure erosion rate can only be calculated from the accumulated metal loss as well. The evaluation of the pure erosion rate is a more complex problem compared with that of the total metal loss because the corrosion part should be removed in the pure erosion test. Firstly, the continuous introduction of noble gas in the solution can influence the flow distribution in the test cell. The bubble in the test solution may lead to cavitation and result in a high erosion rate. Secondly, when using cathodic protection to remove the corrosion part in sea water or tap water, the deposition of calcareous deposits[58] will affect the mass loss measurement and synergy calculations to a large extent. Finally, when using inhibitors, the thin organic protective film may be partially or totally eroded away during the tests in an uncontrollable fashion. Therefore, the pure erosion rate is the hardest part to be assessed in the study of erosion-corrosion.

4. In addition, the test duration of the erosion-corrosion in lab are always short to avoid the abrasion of the sand particles. As erosion-corrosion process is a long-term action in real energy pipelines, the test results in a short period may be different from the realistic situation. The empirical and theoretical models are almost established in considering that the erosion-corrosion process reaches a stable state. While in most practise cases, the erosion-corrosion is a dynamic process with a high penetration rate concentrating on a local area, especially for stainless steel. The local breakdown of the passive film may follow a restoration process at the initial stage which was well investigated from the potential and current transitions in the previous studies[7, 13]. Nevertheless, nearly no studies focused on the propagation process of the localized corrosion caused by the erosion-corrosion for stainless steel. With the forming of local anodic and cathodic areas on the steel surface, erosion-corrosion may concentrate on a certain area rather than the overall steel surface. In this case, the localized corrosion rate cannot be directly evaluated based on the calculation of the average anodic current and erosion rate. Currently, the breakdown of the passive films caused by erosion-corrosion are mostly studied under anodic polarisation conditions. Since most of the erosion-corrosion are occurring under natural passivation conditions, the anodic polarisation of the test sample may lead to different corrosion process of the stainless steel.

4. ADVANCES IN SIMULATING AND MONITORING EROSION-CORROSION

It is convenience for the experiments to be carried out on the erosion-corrosion tests by using jet impingement rigs and rotating disc electrode systems described above, however, the flow patterns and the hydrodynamics in both systems are quite different from real pipeline working conditions. It is difficult to transfer the test results or empirical model to large pipeline systems. Therefore, it is necessary to build some simulated pipeline flow loop to study the erosion-corrosion issue[59, 60]. Nevertheless, due to the high costs of the pipeline flow loop system, only limited tests were conducted in the previous studies[11, 61, 62].

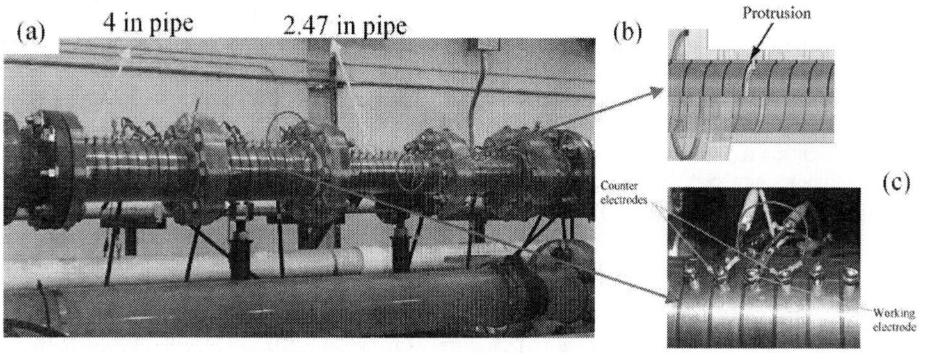

Fig. 6 An electrochemical test flow loop used for the study of erosion-corrosion[11] (a), the protrusion in the pipe (b), and the two electrode system in the loop.

Fig. 6 shows an electrochemical test system on erosion-corrosion in a simulated pipeline circle using special designed ring form electrode[11]. It is seen that different pipe sizes and a protrusion was arranged in the flow loop to establish changing conditions of the flow distribution. In this system, two electrodes system was applied for LPR test and the corrosion rate can be directly obtained from the LPR measurement results. The total weight loss of the electrodes in the erosion-corrosion test were measured by metal loss. The pure corrosion rate and pure erosion rate were measured in a sand particles free corrosive medium and in a non-corrosive slurry, respectively. It is found that for the carbon steel without forming the corrosion product layer on its surface, the corrosion enhanced erosion contributes to the most part of the metal loss caused by synergism. This test result is consistent with the model shown in Eq. (3). It is also investigated that the sudden change of the flow distribution will have a significant influence to the erosion-corrosion.

Though it is found that the protrusion in the pipeline loop can dramatically influence the interaction effect of erosion and corrosion, the accurate correlation between the erosion-corrosion and fluid dynamic is still hard to be assessed. To further study the relationship between the fluid dynamic and erosion-corrosion, Computational Fluid Dynamics (CFD) was applied to calculate the precise flow distribution in the test cell and pipeline circle in recent years[22, 63-65]. Zhang [65] and Zeng [22] studied the corrosion and erosion-corrosion behaviour of the carbon steel at different locations of the elbow in a simulated flow loop by using multi-electrode sensor[66] combined with CFD technique[67-69]. It is seen from Fig. 7a that multi-electrodes were adopted in Zeng's test [22] which were arranged at different locations of the elbow and the simulation of the sand concentration at the elbow can be precisely evaluated by CFD as shown in Fig. 7b. The total material loss and the corrosion rate were obtained by weight loss and EIS measurements, respectively. The test results indicate that the distribution of erosion-corrosion rate is in good agreement with the hydrodynamics and sand concentration distribution at the elbow. The combination of the multi-electrode sensor test and CFD provided a quantitative analysis of the relationship between erosion-corrosion and the flow distribution. In the further studies of the erosion-corrosion of stainless steel at the elbow by using coupled multi-electrode sensor system combined with the CFD simulation conducted by Zeng[13], it is found that the corrosion process of the stainless steel at the elbow was not a stable process due to the inhomogeneous distributions of the sand concentration and the flow rate. The galvanic currents among the electrodes at the different locations demonstrated that the location where confronted a high impingement frequency may become the anodic area and it can lead to a more serious erosion-corrosion process at this area. However, the erosion-corrosion test was only lasted for 4 h, the propagation of the localized corrosion cannot be obviously observed in this test. Meanwhile, the areas of the electrodes were nearly 1 cm^2 which means the anodic current may be not accurate due to the potential cathodic reactions on the anodic sample, and the gaps among the electrodes were over 1 cm which suggests the interaction of the multi-electrodes may be weaken because of the high electrical resistance among each electrodes.

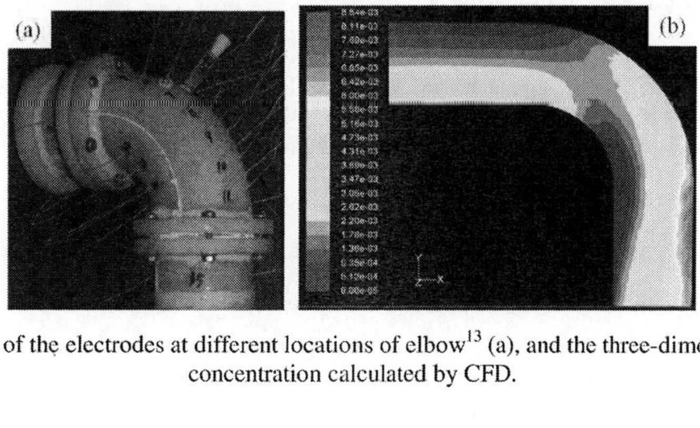

Fig. 7 The arrangement of the electrodes at different locations of elbow[13] (a), and the three-dimensional contour of sand concentration calculated by CFD.

It can be clearly found in the previous studies that the erosion-corrosion patterns on the metal surface were not uniform[53]. The continuous changing of the potential and galvanic current at the elbow area of the pipeline further demonstrated that the erosion-corrosion is a dynamic process rather than a stable static process with the breakdown and restoration of the passive film[13]. The localized corrosion is much easier to initiate on the stainless steel due to the localized breakdown of the passive film under the impingements of the sand particles. However, the propagation of the localized corrosion may be different from the conventional pitting or crevice corrosion[70] because local acidification is hard to be realized in a high flow rate condition. The inhomogeneous distribution of the flow field may become the driving force of the formation of localized corrosion in this situation. Thereby, it is necessary to capture the initiation and propagation process of the localized erosion-corrosion in a pipeline working condition. The traditional electrochemical methods such as EIS, LPR, and polarization curves measurement are also not suitable for the monitoring of the dynamic process of the localized corrosion. Though multi-electrode sensor has been introduced into the study of erosion-corrosion, the large size of the electrode cannot totally separate the anodic and cathodic areas. Meanwhile, the large distance between each electrode will also influence the restoration process of the passive film. Therefore, wire beam electrode (WBE) can be further applied as an effective method to understand the dynamic process of localized corrosion in the erosion-corrosion test. The WBE can simulate a one-piece electrode and the small area of the electrodes can separate the anodic area and cathodic area in a large extent[71]. The transition of the galvanic current among the electrodes can be used as an indicator to identify the damage and repair process of the passive film. Simultaneously, the formation of the stable localized corrosion area can be soon detected through the potential and current distribution maps[72]. As metal loss caused by erosion enhanced corrosion is the major contribution to the total material loss of stainless steel, the localized penetration rate also can be evaluated based on the anodic current of each wire.

5. CONCLUSIONS

The common probes used for the study of erosion-corrosion in pipeline industry are ER sensors and acoustic sensors. The total metal loss can be on-line monitored by ER sensors and the sand rate can be measured by acoustic sensors. However, the interaction effect of erosion and corrosion cannot be studied by using these methods and a measurement error will occur by using traditional ER method when localized corrosion happens. Some new kinds of sensors such as FSM and RPERS which are developed based on ER method can be used for the monitoring of localized erosion-corrosion. Currently, the rotating disc electrode system and slurry jet impingement rig are usually applied for the study of erosion-corrosion problems by using gravimetric mass loss and electrochemical measurements. However, the measurement error for both corrosion enhanced erosion and erosion enhanced corrosion cannot be avoided based on traditional methods. Most of the theoretical and empirical models are concern on the stable status of erosion-corrosion. The dynamic process of the erosion-corrosion cannot be evaluated based on these models and it is also hard to be monitored by using traditional electrochemical methods. Therefore, WBE can be considered as a potentially more effective method for the study of localized corrosion caused by erosion-corrosion in future studies.

6. REFERENCES

1. Wang X and Melchers RE, Corrosion of carbon steel in presence of mixed deposits under stagnant seawater conditions, Journal of Loss Prevention in the Process Industries 45 (2017) 29-42.
2. Zhang S, Wang S, Wu CL, Zhang H, Guan M and Tan JZ, Cavitation erosion and erosion-corrosion resistance of austenitic stainless steel by plasma transferred arc welding, Engineering Failure Analysis 76 (2017) 115-124.
3. Zheng Y, Yao Z, Wei X and Ke W, The synergistic effect between erosion and corrosion in acidic slurry medium, Wear 186 (1995) 555-561.
4. Bateni MR, Szpunar JA, Wang X, and Li DY, Wear and corrosion wear of medium carbon steel and 304 stainless steel, Wear 260 (2006) 116-122.
5. Tang X, Xu LY, and Cheng YF, Electrochemical corrosion behavior of X-65 steel in the simulated oil-sand slurry. II: Synergism of erosion and corrosion, Corrosion Science 50 (2008) 1469-1474.
6. Tian BR and Cheng YF, Electrochemical corrosion behavior of X-65 steel in the simulated oil sand slurry. I: Effects of hydrodynamic condition, Corrosion Science 50 (2008) 773-779.
7. Lu BT, Luo JL, Mohammadi F, Wang K and Wan XM, Correlation between repassivation kinetics and corrosion rate over a passive surface in flowing slurry, Electrochimica Acta 53 (2008) 7022-7031.
8. Tan KS, Wood RJK and Stokes KR, The slurry erosion behaviour of high velocity oxy-fuel (HVOF) sprayed aluminium bronze coatings, Wear 255 (2003) 195-205.
9. Zeng L, Zhang GA, Guo XP and Chai CW, Inhibition effect of thioureidoimidazoline inhibitor for the flow accelerated corrosion of an elbow, Corrosion Science 90 (2015) 202-215.
10. Pasha A, Ghasemi HM and Neshati J, Study of the pitting corrosion of superduplex stainless steel and X-65 carbon steel during erosion-corrosion by cyclic polarisation technique, Corrosion Engineering Science and Technology 51 (2016) 463-471.
11. Malka R, Nešić S and Gulino DA, Erosion–corrosion and synergistic effects in disturbed liquid-particle flow, Wear 262 (2007) 791-799.

12. Tan KS, Wharton JA and Wood RJK, Solid particle erosion–corrosion behaviour of a novel HVOF nickel aluminium bronze coating for marine applications—correlation between mass loss and electrochemical measurements, Wear 258 (2005) 629-640.

13. Zeng L, Shuang S, Guo XP and Zhang GA, Erosion-corrosion of stainless steel at different locations of a 90° elbow, Corrosion Science 111 (2016) 72-83.

14. Tan Y, Understanding the effects of electrode inhomogeneity and electrochemical heterogeneity on pitting corrosion initiation on bare electrode surfaces, Corrosion Science 53 (2011) 1845-1864.

15. Fan CL and Mclaury BS, Evaluation of Electrical Resistance Probes to Detect Pipeline Erosion and Sand Production in Low Liquid Loading Flow Conditions, (2011) SPE International.

16. Shi W, Wang T, Dong Z and Guo X, Application of wire beam electrode technique to investigate the migrating behavior of corrosion inhibitors in mortar, Construction and Building Materials 134 (2017) 167-175.

17. Kee KE, Richter S, Babic M and Nesic S, Experimental Study of Oil-Water Flow Patterns in a Large Diameter Flow Loop-The Effect on Water Wetting and Corrosion, Corrosion 72 (2016) 569-582.

18. Cho M, Huh C, Kang S and Baek J. Experimental Study on Corrosion Characteristics of CO_2 Pipeline for Offshore Transport and Geological Storage, In: Dixon T, Herzog H, Twinning S, (eds.) Energy Procedia, (2014) 2442-2447.

19. Wu JW, Bai D, Baker AP, Li ZH and Liu XB, Electrochemical techniques correlation study of on-line corrosion monitoring probes, Materials and Corrosion 66 (2015) 143-151.

20. Xu YZ, Zhu YS, Liu L, He LM, Wang XN and Huang Y, The study of the localized corrosion caused by mineral deposit using novel designed multi-electrode sensor system, Materials and Corrosion 68 (2017) 632-644.

21. Xu Y, Li K, Liu L, Yang L, Wang X and Huang Y, Experimental Study on Rebar Corrosion Using the Galvanic Sensor Combined with the Electronic Resistance Technique, Sensors 16 (2016) 1451.

22. Zeng L, Zhang GA and Guo XP, Erosion–corrosion at different locations of X65 carbon steel elbow, Corrosion Science 85 (2014) 318-330.

23. Xu Y, Huang Y, Wang X and Lin X, Experimental study on pipeline internal corrosion based on a new kind of electrical resistance sensor, Sensors and Actuators B: Chemical 224 (2016) 37-47.

24. Huang Y, Xu Y, Li B, Ying L, Yang F and Wang X, Novel electrical resistance method to measure underdeposit corrosion and its inhibition in pipeline steels, Corrosion Engineering, Science and Technology 51 (2016) 211-222.

25. Xu YZ, Yang LJ, He LM, Huang Y and Wang XN, The monitoring of galvanic corrosion behaviour caused by mineral deposit in pipeline working conditions using ring form electronic resistance sensor system, Corrosion Engineering, Science and Technology 51 (2016) 606-620.

26. Gan F, Wan Z, Li Y, Liao J and Li W, Improved formula for localized corrosion using field signature method, Measurement 63 (2015) 137-142.

27. Gan F, Tian G, Wan Z, Liao J and Li W, Investigation of pitting corrosion monitoring using field signature method, Measurement 82 (2016) 46-54.

28. Tan Y, An Overview of Principles and Issues in Structural Health Monitoring Using Corrosion Sensors, NACE 2017, New Orleans

29. Joosten MW, Fischer KP, Strommen R, and Lunden KC, Internal corrosion monitoring of subsea oil and gas production equipment, Materials Performance, April 1995, pp.44-48

30. Tan YJ, HuoY and Shu L, Monitoring and understanding the effects of high salt concentration and residue on copper and aluminium corrosion, in Solutions to environmental challenges through innovations in research, Asiatech Publishers Inc., New Delhi, India, (2013) 393-405.

31. Zhang P, Zheng S, Jing J, Zhou Y and Li Q, Wang K, Surface erosion behavior of an intrusive probe in pipe flow, Journal of Natural Gas Science and Engineering 26 (2015) 480-493.

32. Ukpai JI, Barker R, Hu X and Neville A. Exploring the erosive wear of X65 carbon steel by acoustic emission method, Wear 301 (2013) 370-382.

33. Li S, Kim Y, Jung S, Song H and Lee S, Application of steel thin film electrical resistance sensor for in situ corrosion monitoring, Sensors and Actuators B: Chemical 120 (2007) 368-377.

34. Kouril M, Prosek T, Scheffel B and Dubois F, High sensitivity electrical resistance sensors for indoor corrosion monitoring, Corrosion Engineering, Science and Technology 48 (2013) 282-287.

35. Ahmed WH, Bello MM, El Nakla M, Al Sarkhi A and Badr HM, Experimental investigation of flow accelerated corrosion under two-phase flow conditions, 267 (2014) 34-43.

36. Paolinelli LD, Brown B, Simison SN and Nesic S. Inhibition of CO_2 corrosion of carbon steel with 1% Cr, Materials Chemistry and Physics 136 (2012) 1092-1102.

37. Jiang X, Zheng YG and Ke W. Effect of flow velocity and entrained sand on inhibition performances of two inhibitors for CO_2 corrosion of N80 steel in 3% NaCl solution, Corrosion Science 47 (2005) 2636-2658.

38. Choi Y, Nesic S and Ling S, Effect of H_2S on the CO_2 corrosion of carbon steel in acidic solutions, Electrochimica Acta 56 (2011) 1752-1760.

39. Kondo M, Muroga T, Sagara A, Valentyn T, Suzuki A and Terai T, Flow accelerated corrosion and erosion–corrosion of RAFM steel in liquid breeders, Fusion Engineering and Design 86 (2011) 2500-2503.

40. Zhang GA and Cheng YF. Electrochemical characterization and computational fluid dynamics simulation of flow-accelerated corrosion of X65 steel in a CO_2-saturated oilfield formation water, Corrosion Science 52 (2010) 2716-2724.

41. Zhang GA, Xu LY and Cheng YF, Investigation of erosion–corrosion of 3003 aluminum alloy in ethylene glycol–water solution by impingement jet system, Corrosion Science 51 (2009) 283-290.
42. Guo HX, Lu BT and Luo JL. Interaction of mechanical and electrochemical factors in erosion–corrosion of carbon steel, Electrochimica Acta 51 (2005) 315-323.
43. Alegria-Ortega JA, Ocampo-Carmona LM, Suarez-Bustamante FA and Olaya-Florez JJ. Erosion-corrosion wear of Cr/CrN multi-layer coating deposited on AISI-304 stainless steel using the unbalanced magnetron (UBM) sputtering system, Wear 290 (2012) 149-153.
44. Guo HX, Lu BT and Luo JL. Non-Faraday material loss in flowing corrosive solution, Electrochimica Acta 51 (2006) 5341-5348.
45. Lu BT, Luo JL and Lu JF. An investigation on corrosion-enhanced erosion behaviour, In: Patnaik PC, (ed.), (2004) 449-464.
46. Guo HX, Lu BT, and Luo JL. Response of surface mechanical properties to electrochemical dissolution determined by in situ nanoindentation technique, Electrochemistry Communications 8 (2006) 1092-1098.
47. Lu BT and Luo JL. Synergism of Electrochemical and Mechanical Factors in Erosion−Corrosion, The Journal of Physical Chemistry B 110 (2006) 4217-4231.
48. Islam MA and Farhat Z. Erosion-corrosion mechanism and comparison of erosion-corrosion performance of API steels, Wear 376-377 (2017) 533-541.
49. Islam MA and Farhat Z. Mechanical and electrochemical synergism of API X42 pipeline steel during erosion–corrosion, Journal of Bio- and Tribo-Corrosion (2015) 1-26.
50. Dong CF, Xiao K, Li XG and Cheng YF. Erosion accelerated corrosion of a carbon steel–stainless steel galvanic couple in a chloride solution, Wear 270 (2010) 39-45.
51. Azarian NS, Ghasemi HM and Monshi MR, Synergistic Erosion and Corrosion Behavior of AA5052 Aluminum Alloy in 3.5 wt% NaCl Solution Under Various Impingement Angles, Journal of Bio- and Tribo-Corrosion (2015) 1-10.
52. Lindsley BA and Marder AR, The effect of velocity on the solid particle erosion rate of alloys, Wear 225 (1999) 510-516.
53. Zheng ZB, Zheng YG, Zhou X, He SY, Sun WH and Wang JQ, Determination of the critical flow velocities for erosion–corrosion of passive materials under impingement by NaCl solution containing sand, Corrosion Science 88 (2014) 187-196.
54. Lu BT, Mao LC and Luo JL, Hydrodynamic effects on erosion-enhanced corrosion of stainless steel in aqueous slurries, Electrochimica Acta 56 (2010) 85-92.
55. Lu BT, Luo JL, Guo HX and Mao LC, Erosion-enhanced corrosion of carbon steel at passive state, Corrosion Science 53 (2011) 432-440.
56. Lu BT, Luo JL and Ma HY, A Theoretical Model on Electrochemical Response of Passivated Metals to Solid Particle Impingement, Journal of the Electrochemical Society 154 (2007) C159.
57. Neville A, Reyes M and Xu H, Examining corrosion effects and corrosion/erosion interactions on metallic materials in aqueous slurries, Tribology International 35 (2002) 643-650.
58. Hoseinieh SM, Homborg AM, Shahrabi T, Mol JMC and Ramezanzadeh B, A Novel Approach for the Evaluation of Under Deposit Corrosion in Marine Environments Using Combined Analysis by Electrochemical Impedance Spectroscopy and Electrochemical Noise, Electrochimica Acta 217 (2016) 226-241.
59. Zheng D, Che D and Liu Y. Experimental investigation on gas–liquid two-phase slug flow enhanced carbon dioxide corrosion in vertical upward pipeline, Corrosion Science 50 (2008) 3005-3020.
60. Tang P, Yang J, Zheng J, Wong I, He S and Ye J, Failure analysis and prediction of pipes due to the interaction between multiphase flow and structure, Engineering Failure Analysis 16 (2009) 1749-1756.
61. Lotz u and Postlethwaite J, Erosion corrosion in disturbed 2 phase liquid particle flow, Corrosion Science 30 (1990) 95-106.
62. Zeng L, Zhang GA and Guo XP, Effect of Hydrodynamics on the Inhibition Effect of Thioureido Imidazoline Inhibitor for the Flow Accelerated Corrosion of X65 Pipeline Steel, Corrosion 72 (2016) 598-614.
63. Zhang GA, Zeng Y, Guo XP, Jiang F, Shi DY and Chen ZY, Electrochemical corrosion behavior of carbon steel under dynamic high pressure H_2S/CO_2 environment, Corrosion Science 65 (2012) 37-47.
64. Ahmed WH, Bello MM, El Nakla M, Al Sarkhi A and Badr HM, Experimental investigation of flow accelerated corrosion under two-phase flow conditions, Nuclear Engineering and Design 267 (2014) 34-43.
65. Zhang GA, Zeng L, Huang HL and Guo XP, A study of flow accelerated corrosion at elbow of carbon steel pipeline by array electrode and computational fluid dynamics simulation, Corrosion Science 77 (2013) 334-341.
66. Tan Y, Sensing localised corrosion by means of electrochemical noise detection and analysis, Sensors and Actuators B: Chemical 139 (2009) 688-698.
67. Shah A, Khan AH, Chughtai IR and Inayat MH, Numerical and experimental study of steam-water two-phase flow through steam jet pump, Asia-Pacific Journal of Chemical Engineering 8 (2013) 895-905.
68. Davis C and Frawley P, Modelling of erosion–corrosion in practical geometries, Corrosion Science 51 (2009) 769-775.
69. Stack MM, Abdelrahman SM and Jana BD, Some perspectives on modelling the effect of temperature on the erosion–corrosion of Fe in aqueous conditions, Tribology International 43 (2010) 2279-2297.
70. Frankel GS and Sridhar N, Understanding localized corrosion, Materialstoday 11 (2008) 38-44.
71. Tan Y, Fwu Y and Bhardwaj K, Electrochemical evaluation of under-deposit corrosion and its inhibition using the wire beam electrode method, Corrosion Science 53 (2011) 1254-1261.

72. Varela F, Tan MYJ and Forsyth M, An Electrochemical Method for Measuring Localized Corrosion under Cathodic Protection, ECS Electrochemistry Letters 4 (2014) C1-C4.

6. AUTHOR DETAILS

Yunze Xu, is an exchange PhD Student in the Institute for Frontier Materials at Deakin University. He graduated from Huazhong University of Science and Technology in 2012 and then, he becomes a PhD student in Dalian University of Technology. Now, he is working on the erosion-corrosion issue of the energy pipelines under the guidance of Prof. Mike Yongjun Tan. His project attempts to develop a new kind of erosion-corrosion senor which can probe the detailed information of the localized corrosion caused by erosion-corrosion.

Dr. Jing-Li Luo is a professor in the Department of Chemical and Materials Engineering at the University of Alberta, Canada. Dr. Luo received her B. Eng. Degree from the Department of Physics and Chemistry of Metals, University of Science and Technology Beijing in 1982 and her Ph.D in Materials Science and Engineering at McMaster University, Canada in 1992. Dr. Luo has conducted extensive research on corrosion control and the development of green energy systems and has published over 280 papers in refereed journals and holds five US patents. She has served as a member of International Corrosion Council since 2005.

Mike Yongjun Tan, is a Professor in Applied Electrochemistry and Corrosion Technologies at Deakin University in Australia. He is also a Research Program Leader of the Energy Pipelines Cooperative Research Centre. Dr Tan's principal teaching and research interests are in corrosion science and engineering and their applications for enhancing the reliability and durability of civil and industrial infrastructures. He contributed to electrochemical methods for corrosion testing, monitoring and prediction and corrosion inhibitor and anti-corrosion coating research. He is the author of some 150 publications and a book entitled 'Heterogeneous Electrode Processes and Localised Corrosion' (2012 John Wiley & Sons).

QUALIFICATION OF HIGH-TEMPERATURE PROTECTIVE COATINGS ON OFFSHORE PIPELINES

Lim J [1], Lam C [2]
[1]Shawcor, Singapore, [2]Shawcor, Toronto Ontario, Canada

SUMMARY: The recent completion of several large high-temperature insulated pipelines in Australia and the Deepwater projects in the Gulf of Mexico point to increasing demand for higher temperature pipelines in the next 3 – 5 years. Feasibility studies on pipelines operating at temperatures up to 150°C has taken an interest. The qualification of high temperature coatings has not been well established in the industry. This paper discusses the gaps in the current industry standards for qualifying materials and coating systems, the qualification process and test data of FBE and polyolefin coatings (3LPP/MLPP) withstanding design temperatures up to 150°C. A qualification protocol for offshore pipelines is proposed for industry review.

Keywords: Offshore, Pipeline, Coating, High-temperature, FBE, Qualification.

1. INTRODUCTION

High temperature polyolefin coatings such as 3-Layer Polypropylene (3LPP) or Multi-Layer Polypropylene (MLPP) are suitable as anti-corrosion coatings as they provide the moisture barrier under hydrostatic pressure, and mechanical protection for offshore applications and capable of withstanding much higher temperatures, as high as 140 - 150°C. The major component for a 3LPP/MLPP coating system for offshore pipelines is the high-temperature Fusion-bonded Epoxy (HT FBE), which is used as the base layer/primer in 3LPP/MLPP coatings. As the operating temperatures for pipelines move higher, the need for a qualification protocol to evaluate and qualify high-temperature FBE coatings which are fit for purpose becomes imperative.

The qualification process and requirements of such coatings have not been well established in the industry, except for some references provided on standard coating tests presented by the Canadian CSA Z245.20 Standard and rarely, by project specifications. To qualify such coating systems, in particular, the long-term stability of the materials and the stability of the bond between coatings and steel under the service environment, are major challenges due to the lack of industry standards. In previous work, M. Batallas, P. Singh. [1] evaluated methodologies to qualify FBE coating as part of an insulation coating system, intended for long term high temperature service up to 150°C and 30 years of service life.

This paper discusses the gaps in the current industry standards to qualify the high temperature polyolefin coating system, the qualification process and evaluation of high-temperature FBE and 3LPP/MLPP to withstand design temperatures up to 150°C. A qualification protocol for offshore pipelines is proposed for industry review.

2. INDUSTRY STANDARDS

2.1 Lack of Standard and Specification for High Temperature (>110°C) External Coatings for Offshore

There is a lack of national/international standards on external high temperature pipeline coatings that applies to service temperature up to 150°C for offshore applications.

Coating standards for external polyolefin coatings, such as ISO 21809-1:2011 standard, External coating for buried or submerged pipelines used in transportation systems – Part 1: Polyolefin coatings (3-Layer PE and 3-Layer PP), covers PP-based coatings in the design temperature range of -20 to +110°C. There is no provision for temperatures >110°C.

The Canadian Standards CSA Z245.20 Series-14, Plant-applied external coatings for steel pipe, apply to single-layer FBE with a Tg up to, and greater than 115°C (System 1A and System 1B respectively). The qualification requirements for FBE are the same for both systems, except for cathodic disbondment (CD) and adhesion tests, which are performed at the maximum temperature of 95°C, and a reduced acceptance criteria on flexibility. Test durations are specified for 24 hours and 28 days with test temperatures limited by the boiling point of water used for CD or hot water immersion (HWI) tests. The development for this standard was intended for coatings used in onshore applications.

The CSA Z245.20 standard does not apply to FBE as a primer in multi-layer polyolefin coatings. With a polyolefin topcoat such as PP, it provides the protective barrier to the coating system, allowing the FBE to perform at elevated temperatures without direct exposure to wet environment as it would be seen in subsea conditions. High temperature testing, such as CD or HWI tests, on standalone FBE coating does not represent the exposed conditions on the coating system, unduly subjecting the sample to severe conditions which do not exist in the field. In previous work, S.Guan and A.Kehr [2] reviewed the factors that affect high temperature CD tests of FBE. Tests for standalone FBE coating are performed on FBE coating as a quality control to ensure that the base layer is applied to known and established criteria, but does not demonstrate the performance of the coating system in higher operating temperature (i.e. > 95°C) and subsea conditions with hydrostatic pressure. Standards need to be developed or revised to address high temperature coatings, allow qualification requirements and test methods on high temperature coatings to be established.

2.2 Selection of High temperature FBE coatings

The ISO 21809-1:2011 standard, specifies the minimum glass transition temperature (Tg) of the FBE to be at least 5°C above the maximum pipeline design temperature. That is, with a pipeline design temperature of 145°C, the FBE requires a Tg of ≥ 150°C. Manufacturers would typically advise their customers to use a FBE coating that has a Tg 5 – 10 °C higher than the service or operating temperature.

It is known that the Tg of a FBE coating would decrease after moisture absorption. In a test of two HT FBE materials after immersed in hot water at 120°C for 7 days, the change in Tg was found to decrease by about 15°C. The absorption rate and permeability rate increase with increasing temperature. The risks of having both the coating material (or coating performance) and the bond between the coating and steel to deteriorate are relatively higher. On offshore applications, these risks would compound with the hydrostatic pressure depending on the water depth.

With a polyolefin topcoat acting as a protective barrier, it allows the FBE base coat to extend its lifetime and to perform at an elevated temperature. However, the effect of hydrostatic pressure on the coating properties is not known, further studies on the impact are required. The selection of a HT FBE considering the effect of service environment has not been quantified.

There are limited high temperature FBE coating materials with Tg ≥ 150°C commercially available and this is reflected by a lack of established test data for high-temperature FBE, in particular, both the long-term stability of the coating materials and the stability of coating adhesion under the service environment.

2.3 Qualification Requirements

As coating materials would deteriorate with time due to hydrothermal, thermal and thermal oxidative degradations, testing should evaluate the effect of service environment that associates to the different degradation types on coating performance. Therefore, for qualifying whether a coating system is fit for purpose, adhesion tests, cathodic disbondment tests, electrochemical spectroscopy (to assess barrier properties of a coating) and coating properties as discussed in Section 3.3, should be carried out after exposed to the maximum rated temperature, and/or hydrostatic pressure depending on the service environment.

2.4 Test Methods

High Tg FBE material has a different chemistry and service temperature capability, when compared to ambient temperature materials. Test procedures need to be revised to be applicable for high-temperature FBE and polyolefin materials.

2.4.1 Glass Transition Temperature

For a given service temperature, the stated Tg of the FBE may depend on the method of evaluation of the glass transition temperature. Manufacturers have traditionally used inflection point to determine the Tg of the material. However, in the recent CSA Z245.20-14 standard, the determination of Tg using pre-scan of the materials and the evaluation of the Tg at half-height has been specified. There is a need to ensure a more consistent interpretation method when selecting a FBE for a high temperature service line.

2.4.2 High temperature CD Test

High temperature CD test is performed on 3LPP at maximum design temperature with bulk electrolyte temperature controlled at 30 ± 5 °C, simulating the condition in subsea environment where the coating is exposed to low seawater temperature, and high internal service temperature, as suggested by NACE TM0115-2015 'Cathodic Disbondment Test for Coated Steel Structures under Cathodic Protection'.

2.4.3 High temperature Hot Water Immersion Test

High temperature hot water immersion test is performed on 3LPP at 120°C or max design temperature. Tests above boiling point of water is made by immersing sample in a pressurised vessel capable of maintaining temperature >100°C and the corresponding pressure. A 6 mm artificial holiday is drilled in the middle of the coating sample, and all exposed metal surfaces coated with a suitable material. Modifications to the test may be required to protect edges of the sample.

2.4.4 Thermal Ageing Test

Thermal ageing test by air oven ageing has been specified in polyolefin coating standards such as the ISO 21809-1:2011 standard, and DIN 30678:2013-09, Polypropylene coatings on steel pipes and fittings – Requirements and testing. The test simulates a hot, dry, and highly oxygenated environment, as occurs in air oven ageing is not applicable to the exposure conditions expected of the 3LPP/MLPP topcoat material in subsea operation. Generally, PP is prone to thermal oxidative degradation at elevated temperatures. Therefore, 3LPP/MLPP coating is limited to be used up to 110°C under oxygenated environment (i.e. for onshore application). For offshore applications, due to the minute oxygen content in subsea, MLPP is known to be used with a design temperature of 150°C. Bjorn Melve, Dana Ali [3] shares the qualification of thermal insulation coating for a project in the Norwegian Sea with a design temperature of 155°C, where MLPP was the only system that was found to satisfy all qualification requirements.

For subsea conditions, thermal ageing should be carried out under low oxygen content environment to simulate offshore conditions. CD, adhesion, EIS and coating property tests should be carried out after being subjected to the maximum service temperature to characterize the performance of coating under the thermal effect. For evaluating the change in coating properties after ageing, tensile test is recommended for the PP topcoat. Melt Flow Index (MFI) is not an appropriate metric as there is no clear failure point to indicate that the coating is unable to perform in service. It can only be appropriately used under controlled and repeatable conditions. For the FBE primer, Tg is a good measurement to characterize the change in FBE property.

2.4.5 Hydrostatic Thermal Ageing Test

Long term performance of the coating is difficult to establish without validating under representative conditions, and there are currently no industry-wide accepted criteria. To properly evaluate the coating system under service environment in offshore, hydrostatic thermal ageing tests should be carried out. The ISO 12736 standard outlines 1 year ageing test in a heated and pressurised in simulated seawater environment and requires evaluation of the test results but does not specify any acceptance criteria.

The purpose of the testing exposures in ISO 12736 is to evaluate the failure mode of the material in a worst-case scenario: at maximum operating temperature and pressure, in contact with water. The required metric for evaluation is measurement of water uptake via weight change and monitoring change in mechanical properties, i.e., tensile or flexural tests. To properly characterise the performance of the anti-corrosion coating after subjecting to thermal ageing conditions, CD, adhesion, EIS and coating property tests should be carried out.

3. QUALIFICATION OF HIGH TEMPERATURE FBE FOR 3LPP/MLPP COATINGS

The base layer/primer of the 3LPP/MLPP coating system, FBE, is a critical component that determines whether the coating system is fit for purpose. Prior to the use of high temperature FBE materials to be applied on production pipes, there is a need to select and qualify high-temperature FBE materials which are suitable for high temperature service. A qualification process was designed and developed which includes an initial material screening process and a plant qualification plan of high-temperature FBE for 3LPP/MLPP coatings.

3.1 Initial Material Screening

The qualification process starts by identifying FBE materials with the Tg within the range of the target design temperature. This is followed by a detailed desktop review of manufacturers' supplied laboratory reports, which shall demonstrate that the high-temperature FBE materials can meet all the requirements in the Material Specification as given in Tables 1 and 2 in

Appendix A. All candidates should pass the requirements as specified in these Tables. For HT FBE candidates, more focus would be put on the high temperature tests results in the initial screening.

Table 1 provides an example of test results for two different high-temperature FBE powders with Tg >150°C obtained from two FBE manufacturers.

Table 1. High Temperature Testing on Laboratory Coated Samples on two High Tg FBE Materials

Test	Requirement	Test Method	Material A	Material B
Hot Water Immersion Test at 95°C for 28 days	Rating 1 - 3	CSA Z245.20 -clause 12.14 ISO 21809-2 clause A.15	Rating 1	Rating 1
Hot Water Immersion Test at 120°C	Rating 1 - 3	CSA Z245.20- clause 12.14; ISO 21809-2 clause A.15 Sample immersed in vessel capable of maintaining temperature and withstanding the pressure.	Rating 1; No blistering, no holidays, and no disbondment of the coating was found after 20 weeks at 120°C	Rating 1 (CSA Z245.20) after 28 days at 120°C
Cathodic Disbondment Test at 28 days, -1.5V, 65°C	Max. 15 mm	CSA Z245.20 - clause 12.8 (Electrolyte replaced every 7 days with fresh solution preheated to test temperature.) ISO 21809-2 clause A.9	9.00 mm	11.00 mm
Cathodic Disbondment Test at 28 days, -1.5V, 95°C	Max. 15 mm	CSA Z245.20 - clause 12.8 (Electrolyte replaced every 7 days with fresh solution preheated to test temperature.) ISO 21809-2 clause A.9	8.80 mm	8.13 mm
Cathodic Disbondment Test at 28 days, -1.5V, 150°C	Max. 10 mm	ISO 21809-2 clause A.9 * Test specimen on hot plate or sand/steel grit bath with test temperature measured at coating-steel interface. Bulk electrolyte temperature controlled at 30 ± 5 °C	5.76 mm	8.38 mm

The outcome of the initial material screening allowed both candidates to be approved for the plant qualification stage.

3.2 Plant Qualification

Plant trials are necessary to ensure that the selected FBE can be successfully and consistently applied in the coating production facility. Due to the differences in the setup in each coating plant, coating parameters, blasting and application equipment and storage conditions, etc., which could affect the quality of the applied coating, it is an industry practice to conduct product qualification for each coating plant.

The high-temperature FBE and 3LPP system shall meet minimum test requirements to known and established acceptance criteria. Additional high temperature CD and hot water immersion tests, ageing tests are carried out to evaluate the stability of the coating at elevated temperatures.

3.2.1 HT FBE Qualification

Plant trials for Material B were carried out at two coating plants.

At Plant 1, a standalone high-temperature FBE was required for an onshore project and Material B was selected for plant trial. The project specification required the standalone high-temperature FBE to pass the requirements in CSA Z245.20.

Therefore, all qualification tests were carried out as per in CSA Z245.20 and the maximum duration of the hot water adhesion tested was only 28 days at 95°C. Long term ageing tests was planned. The FBE was applied at a minimum thickness of 355 microns (14 mils).

At Plant 2, a 3LPP coating system was required for potential offshore projects and Material B was applied as both standalone coating and as a base layer/primer for 3LPP for qualifications. For the standalone coating, tests were conducted based on procedures specified in ISO 21809-1 and ISO 21809-2. The FBE was applied at an average thickness of 393 microns (target 350 - 450 microns).

The test results obtained from the trials in Plant 1 and Plant 2 is provided in Table 2.

Both plants had specified different test plans and methods to satisfy specific location/projects requirements.

Results show that standalone tests performed on the Material B met known and established criteria for application and qualification requirements in Plant 1, but not in Plant 2 due to the failure in CD tests. CD test results obtained in Plant 2 failed the requirements in 28-day tests at 23°C and 95°C, obtaining and average results of 5.1 mm (maximum 5 mm) and 18.9 mm (maximum 15 mm), respectively.

Table 2. Plant Qualification results of High Temperature FBE (Material B)

Qualification Test	Conditions	Criteria	Plant 1	Plant 2
Thermal characteristics	-	ΔTg: -2 to +3 °C / ≥ 95%	≥ 98 % (Cure)	0.66 °C, -0.34 °C (ΔTg)
Porosity – Cross Section	-	Rating 1 - 2	Rating 2	Rating 2
Porosity – Interface	-	Rating 1 - 2	Rating 1 – 2	Rating 1
Adhesion ('V'-cut)	-	Rating ≤ 2 and breakaway ≤ 2 mm	Not applicable	1.11mm, 0.96 mm
Flexibility	Minimum 2.0 °/PD at -30 ± 3°C	No cracking	No cracking	Not applicable
	Minimum 2.0 °/PD at 0 ± 3°C	No cracking	No cracking	Not applicable
	Minimum 2.5 °/PD at 0 ± 3°C	No cracking	Not applicable	No cracking
	Minimum 2.5 °/PD at 20 ± 3°C	No cracking	Not applicable	No cracking
Impact resistance	Minimum 1.5 J/mm at -30 ± 3 °C	No holiday	No holiday	No holiday
	Minimum 1.8 J/mm at 23 ± 3 °C	No holiday	Not applicable	No holiday
	Minimum 1.5 J/mm at 50 ± 3 °C	No holiday	No holiday	Not applicable
Cathodic Disbondment Test	24 hrs, -3.5V, 65 ± 3 °C	≤ 5 mm	2.23 mm	Not applicable
	48 hrs, -1.5V, 65 ± 3 °C	≤ 5 mm	2.90 mm	Not applicable
	48 hrs, -3.5V, 65 ± 3 °C	≤ 5 mm	Not applicable	2.17 mm
	28 days, -1.5V, 23 ± 3 °C	≤ 5 mm	4.46 mm	5.13 mm
	28 days, -1.5V, 65 ± 3 °C	≤ 15 mm	12.99 mm	9.33 mm
	28 days, -1.5V, 95 ± 3 °C	≤ 15 mm	11.31 mm	18.87 mm
	28 days, -1.5V, 150 ± 3 °C	≤ 15 mm	9.63 mm	Not applicable
Hot water immersion test	28 days, 95 ± 3 °C	Rating 1 - 3	Rating 1 - 2	Rating 1
	28 days, 120 ± 3 °C	Rating 1 - 3	Rating 1	Rating 1 (Blisters)

3.2.2 3LPP (HT) Qualification

Plant trials for 3LPP using Material B as the base layer/primer were carried out at two coating plants.

At Plant 2, Material B was applied as the base layer of 3LPP (HT) on two pipes. The thickness of the FBE applied was an average of 412 microns (target 350 to 450 microns) and average 3.87 mm (minimum 3 mm) for 3LPP. Tests were conducted based on procedures specified in ISO 21809-1 and ISO 21809-2.

In another plant, Plant 3, Material B was as applied the base layer of 3LPP (HT) on three pipes. Nominal thickness of FBE is 210 – 220 microns and 3.2 – 3.3 mm for 3LPP. Tests were conducted based on procedures specified in CSA Z245.20, DIN 30678 or NF A 49-711.

Both plants had specified different test plans and methods to satisfy specific location/projects requirements.

In conducting the plant trial of 3LPP (HT), the coating was considered successfully qualified in Plant 2 and Plant 3. Summary of test results of the plant qualification is shown in the Table 3.

Additional long term thermal ageing tests were carried out on 3LPP coated samples from Plant 2. Results are discussed in Section 3.2.3.

Table 3. Plant Qualification results of 3LPP (HT) with High Temperature FBE (Material B) as base layer

Qualification Test	Conditions	Criteria	Plant 3	Plant 2
Peel Test	23 ± 2 °C	≥ 250 N/cm	304 N/cm*	461 N/cm
	95 ± 2 °C	≥ 50 N/cm	157 N/cm*	286 N/cm
	150 ± 2 °C	To be established	35 N/cm*	77 N/cm
Flexibility	Minimum 2.0 °/PD at 0 ± 3 °C	No cracking	Not applicable	No cracking
	4-point Bend at 23 °C	To be assessed when the first crack arise	10.5 °/PD	Not applicable
	4-point Bend at 0 °C	To be assessed when the first crack arise	10.5 °/PD	Not applicable
	4-point Bend at -30 °C	To be assessed when the first crack arise	10.5 °/PD	Not applicable
Impact Resistance Test	0 ± 2 °C	Min. 10 J/mm	19.9 J/mm, no holiday	10 J/mm, no holiday
	23 ± 2 °C	Min. 10 J/mm	21.6 J/mm, no holiday	10 J/mm, no holiday
Indentation Resistance Test	23 ± 3 °C	≤ 0.1 mm	Not applicable	0.03 mm
	110 ± 3 °C	≤ 0.4 mm	Not applicable	0.26 mm
Cathodic Disbondment Test	24 hrs, -1.5V, 65 ± 3 °C	≤ 5 mm	2.33 mm	2.15 mm
	48 hrs, -1.5V, 65 ± 3 °C	≤ 5 mm	Not applicable	1.92 mm
	28 days, -1.5V, 23 ± 3 °C	≤ 5 mm	Not applicable	3.94 mm
	28 days, -1.5V, 65 ± 3 °C	≤ 10 mm	5.69 mm	Not applicable
	28 days, -1.5V, 95 ± 3 °C	≤ 10 mm	3.68 mm	Not applicable

Qualification Test	Conditions	Criteria	Plant 3	Plant 2
High Temperature Cathodic Disbondment Tests	28 days, -1.5V, 150 ± 3 °C with bulk electrolyte temperature controlled at 30 ± 5°C	≤ 15 mm	Not applicable	5.50 mm
	28 days, -1.5V, 150 ± 3 °C with bulk electrolyte temperature controlled at 90 ± 5 °C	≤ 15 mm	Not applicable	5.37 mm
Hot Water Immersion Test	48 hrs, 95 ± 3 °C	Rating 1 - 3	Rating 1	No disbondment
	28 days, 95 ± 3 °C	Rating 1 - 3	Rating 2	Rating 1-2
High Temperature Hot Water Immersion Test	28 days, 120 ± 3 °C	Rating 1 - 3	Not applicable	Rating 1
	12 weeks, 120 ± 3 °C	Loss of adhesion ≤ 3 mm	Not applicable	1.20 mm

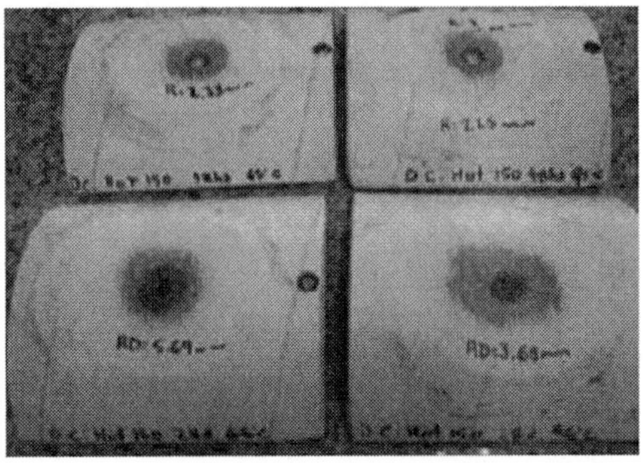

Figure 1 CD Test at Plant 3

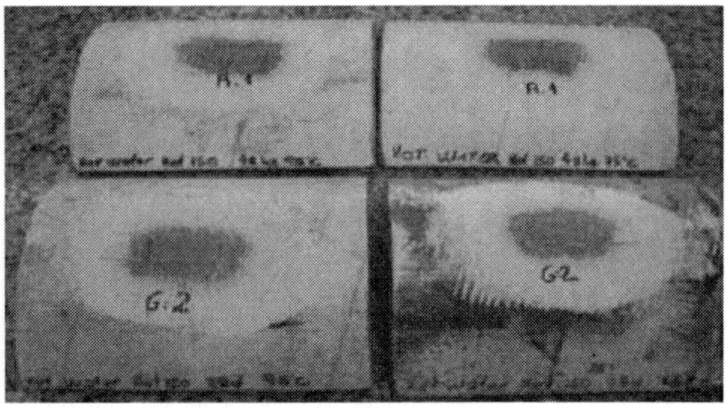

Figure 2 Hot water immersion test at Plant 3

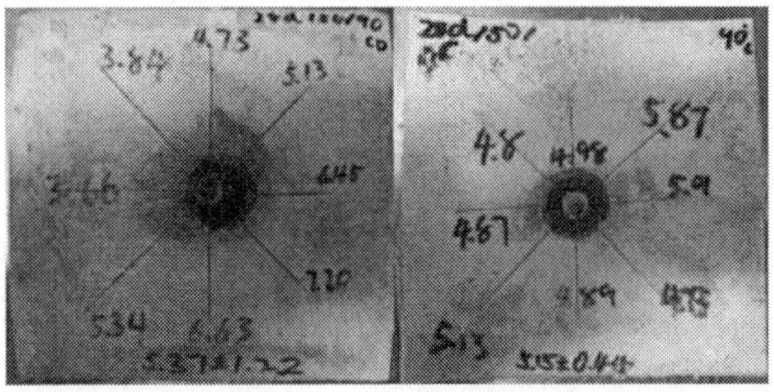

Figure 3 HT CD at 150°C, electrolyte controlled at 90°C

Figure 4 Hot Water Immersion Test on 3LPP at 120°C for 28 days

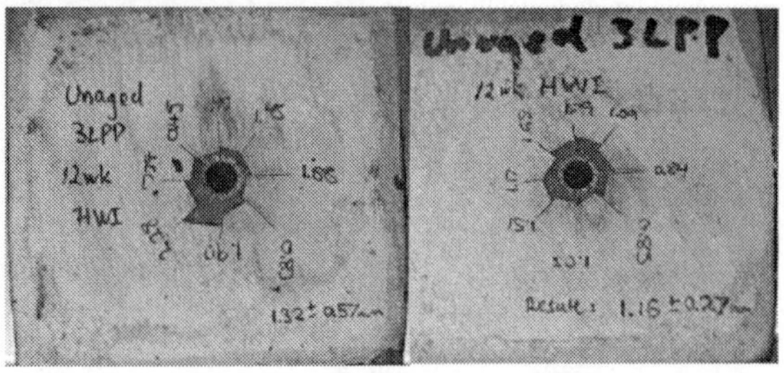

Figure 5 Hot Water Immersion Test on 3LPP at 120°C for 12 weeks

3.2.3 Long Term Thermal Ageing Tests

Long term stability tests are performed on selected powders approved from screening process.

A 2-year thermal ageing test at 150°C is carrying out in a Nitrogen purged oven ageing program with both standalone FBE (Material B) and 3LPP (HT). Aged samples were taken out from the test environment for evaluation every 6 months for evaluation as shown in Figures 6 - 7. Samples were recently obtained from the 1-year duration.

A summary of the observations and findings:

- For 3LPP (HT) with coating thickness approximately 4 mm, adhesion rating after 28-day HWI at 120°C was excellent (rating 1) before thermal ageing at 150°C. After thermal ageing under nitrogen at 150°C for 6 months, the same HWI tests were performed and the adhesion rating remained the same as those obtained from unaged samples. Similar results were found after 1-year aged in oven at 150°C. Results demonstrated that the FBE primer still function as a primer under the PP outer layer.
- For standalone FBE coating with coating thickness approximately 350 microns, adhesion rating after 28-day HWI at 120°C was excellent with unaged samples. After thermal ageing under nitrogen in 150°C for 6 months, the same HWI tests were performed and the adhesion for one sample was excellent (rating of 1), but blisters occurred on the other sample with adhesion rating of 2. After ageing for 1 year in 150°C oven, both samples exhibited blistering and adhesion rating decreased to 2 and 3.

EIS was conducted at 95°C for up to 120 days on 3LPP (HT) and standalone FBE samples before and after thermal ageing for up to 12 months. There was no significant change in impedance values, which indicates that the barrier properties for both 3LPP (HT) and standalone FBE had not changed after thermal ageing. However, standalone FBE coating exhibited lower impedance values (~E+08) compared to those obtained from 3LPP (HT) (~E+10). It is apparent that the 3LPP (HT) coating have better barrier property than the standalone FBE coating.

It is recommended that long term tests and material property tests ageing, EIS, etc. should be allowed to be qualified through historical test data.

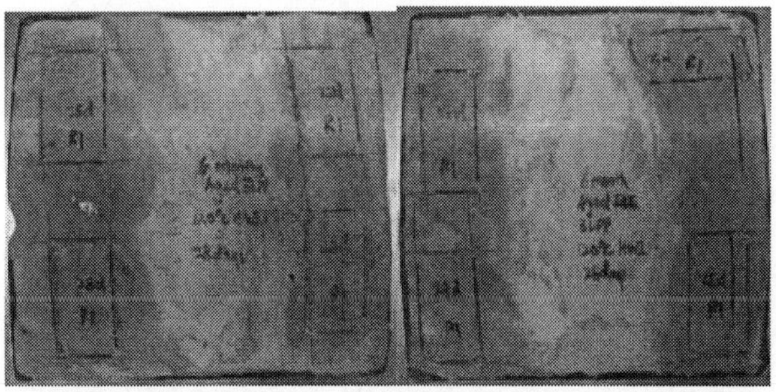

Figure 6 Aged HT CD at 150°C, electrolyte controlled at 90°C

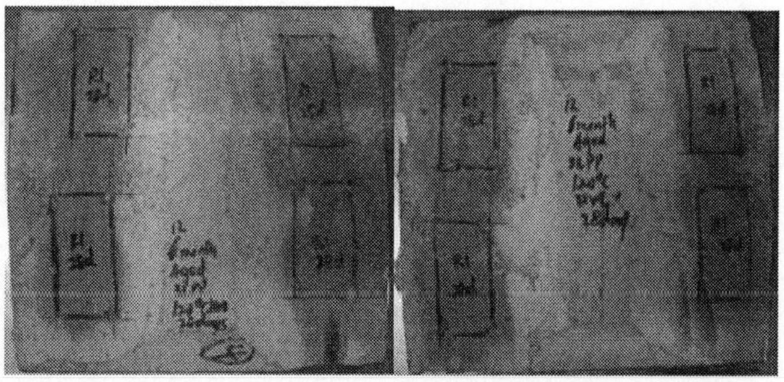

Figure 7 Aged Hot Water Immersion Test on FBE at 120°C for 28 days

4. PROPOSED COATING SYSTEM REQUIREMENT AND QUALIFICATION

A qualification protocol for offshore pipelines is proposed for industry review.

Table 4. Proposed Surface Preparation Requirements

Qualification Requirements	Acceptance Criteria
FBE thickness	Minimum 350 microns
Blast cleanliness	Minimum Sa 2.5 or better
Dust level	Rating 1 – 2, Dust class 2
Steel Surface roughness	Target Rz >70 microns, Rpc 20 – 40/cm
Thermal characteristics	FBE powder with Tg_0 +/- 3°C (by half-height) Coated film Tg -2 to +3°C (by half-height)

Table 5. Proposed qualification program of HT FBE to be used as base layer of 3LPP (HT)

Qualification Requirements	Acceptance Criteria
Cathodic Disbondment after 24 hrs, -3.5V, 65 ± 3 °C	≤ 5 mm
Cathodic Disbondment after 28 days, -1.5V, 65 ± 3 °C	≤ 15 mm
Cathodic Disbondment after 28 days, -1.5V, max design temperature, with electrolyte cooled to 30 °C ± 5 °C	≤ 15 mm
Hot water soak after 48 hrs, 80 ± 3 °C	Rating 1 - 3
Hot water soak after 28 days, 120 ± 3 °C or max design temperature	Rating 1 - 3
Flexibility at minimum 2.0 °ppd at -30 ± 3 °C	No cracking
Flexibility at minimum 2.5 °ppd at 0 ± 3 °C	No cracking
Impact resistance at minimum 1.5 J/mm at -30 ± 3 °C	No holiday

Table 6. Proposed qualification program of 3LPP (HT)

Qualification Requirements	Acceptance Criteria
Peel Test at 23 ± 2 °C	≥ 250 N/cm
Peel Test at 95 ± 2 °C	≥ 50 N/cm
Peel Test at 150 ± 2 °C	To be established
Flexibility at minimum 2.0 °ppd at 0 ± 3 °C	No cracking
Impact resistance, minimum 10 J/mm at 0 ± 2 °C	No holiday
Impact resistance, minimum 10 J/mm at 23 ± 2 °C	No holiday
Indentation resistance at 23 ± 3 °C	≤ 0.1 mm
Indentation resistance at 110 ± 3 °C	≤ 0.4 mm
Cathodic Disbondment after 48 hrs, -1.5V, 65 ± 3 °C	≤ 5 mm
Cathodic Disbondment after 28 days, -1.5V, 65 ± 3 °C	≤ 15 mm
Cathodic Disbondment after 28 days, -1.5V, max. design temperature, with electrolyte cooled to 30 °C ± 5 °C	≤ 15 mm
Hot water soak after 28 days, 80 ± 3 °C	Rating 1 – 3, loss of adhesion ≤ 3mm
Hot water soak after 28 days, 120 ± 3 °C or max design temperature	Rating 1 – 3

The following tests were proposed as additional tests on 3LPP (HT) after thermal ageing. Acceptance criteria needs to be established. Tests can be qualified as historical data.

- CD test at max design temperature, 28 days with electrolyte cooled to 30 °C ± 5 °C
- Hot water immersion test at max design temperature, 28 days
- EIS at 95°C
- Adhesion test - V-cut per ISO after shaving the coating system down to approximately 0.5 mm
- Tg of FBE primer
- Elongation of the PP topcoat

5. DISCUSSION

Due to the differences in FBE chemistries, high-temperature FBE with Tg > 130°C tend to be more sensitive to storage conditions and application process parameters. One of the reasons is the use of the low molecular weight resins in high-temperature FBE formulations. These resins are used to achieve higher crosslinking densities (i.e. higher Tg) in the final coatings. Theoretically, the low molecular weight resins are more soluble, and hence, they are more prone to dissolve in the curing agents and start to react. Depending on storage conditions, if material had aged before coating process, it can result in wetting issue during application and heterogeneity of the final crosslinked density of the FBE coating. Both would have significant impacts on the final performance of the coating in terms of bond stability and material stability. In this regard, proper storage, handling of the materials and the logistic of the material manufactured and coating application must be managed.

Qualification should be performed on full 3LPP/MLPP coating system and does not require passing as a standalone FBE. It is known that FBE manufacturers have developed specific grades of FBE that functions as a base layer/primer under a polyolefin topcoat, or a grade that would perform both as a standalone coating and as a base layer/primer. Testing the FBE as a standalone coating can provide a better understanding of the limitations of the material but the performance or stability of the coating is improved as a primer under a protective topcoat, allowing it to perform at elevated temperatures, without the direct exposure to the subsea environment.

6. CONCLUSIONS

To ensure a high temperature protective coating for offshore pipelines is fit for purpose, a relevant qualification protocol is required to be established in the industry. Generally, current test methods for evaluating and qualifying high temperature external FBE coatings for pipelines were intended for onshore applications. Some of test methods as discussed are not applicable to offshore environment. As the demands for high-temperature coating systems for offshore pipelines has increase, there is an imperative need to revised current standards and qualification procedures to allow qualification requirements to be established for offshore pipe coatings.

7. ACKNOWLEDGMENTS

The authors would like to thank Shawcor for providing data and permissions for this paper and its publication as well as the Research & Development team for their valuable inputs and supports. Special thanks to Mr. Marcos Mockel, Mr. Peter Mayes and Mr. Dave Conroy for their advice and technical insights.

8. REFERENCES

1. M. Batallas and P. Singh. Evaluation of Anticorrosion Coatings For High Temperature Service, NACE Corrosion Conference & Expo 2008, Paper No. 08039.
2. S.Guan and A.Kehr. High Temperature Cathodic Disbondment Testing for Pipeline Coatings, NACE Corrosion 2014, Paper 3860
3. Bjorn Melve, Dana Ali. Qualification of Thermal insulation for Subsea reeled Flowlines with a design temperature of 155°C, NACE Corrosion 2005, Paper No. 05004

9. AUTHOR DETAILS

J Lim is the Technical Solutions Engineer at Shawcor. She is responsible to provide technical support on coatings and various technical services in the Asia Pacific region since 2013. She is NACE CIP Level 2 - certified (No. 53355) and received the Bachelor of Technology Degree in Chemical Engineering from the National University of Singapore in 2008.

C Lam is the Technology Group Leader – Pipeline Coatings at Shawcor. She has worked in the pipe coatings industry for over 16 years with a focus on both mainline pipe coatings and field joint coatings for Shawcor. She is proficient in pipeline coating development, performance evaluation for anticorrosion and insulation systems, and surface treatment and modifications. She received a Master in Applied Science in Mechanical and Industrial Engineering at University of Toronto.

Appendix 1: Qualification of High-temperature FBE powders for Offshore Applications

Qualification of HT FBE powders for off-shore applications

1. Laboratory Qualification of FBE

Epoxy Powder Properties

Property	Acceptance Criterion	Test Method
Cure Time	<4 minutes @ application temperature	CSA Z245.20 - clause 12.1 ISO 21809-2 clause A.2
Gel Time	Within 20% of Manufacturer's specified nominal	CSA Z245.20 - clause 12.2 ISO 21809-2 clause A.3
Moisture Content	0.5% max 0.6% max	CSA Z245.20 - clause 12.3 CSA Z245.20 - clause 12.4 ISO 21809-2 clause A.5
Particle Size	3% max retain on 150μm mesh and 0.2% on 250μm mesh	CSA Z245.20 - clause 12.5 ISO 21809-2 clause A.6
Density	Within 50g/L of manufacturer's specification	CSA Z245.20 - clause 12.6 ISO 21809-2 clause A.7
Thermal Characteristics	Manufacturer's Specification	CSA Z245.20 - clause 12.7 ISO 21809-2 clause A.8

Coating Qualification requirements (Lab prepared specimens)

Property	Test parameters	Acceptance Criterion (3 specimens)	Test Method CSA or ISO
Thermal Characteristics		Manufacturer's Specification	CSA Z245.20 - clause 12.7 ISO 21809-2 clause A.8
Cathodic Disbondment (disbonded radius)	23°C, 28 days	8.0 mm max	CSA Z245.20 - clause 12.8 ISO 21809-2 clause A.9
	65°C, 2 days 14 days 28 days 56 days	 5 mm max 10 mm max 15 mm max Report values for Information	CSA Z245.20 - clause 12.8 (Electrolyte replaced every 7 days with fresh solution preheated to test temperature.) ISO 21809-2 clause A.9
	95°C, 14 days 28 days 56 days	 15 mm max 15 mm max Report values for Information	CSA Z245.20 - clause 12.8 (Electrolyte replaced every 7 days with fresh solution preheated to test temperature.) ISO 21809-2 clause A.9
	Maximum rated temperature*, 14 days 28 days 56 days	 10 mm max 10 mm max Report values for Information	ISO 21809-2 clause A.9 * Test specimen on hot plate or sand/steel grit bath with test temperature measured at coating-steel interface. Bulk electrolyte temperature controlled at 30 ± 5°C
Adhesion - Hot water Immersion @ 95°C	2 days 7 days 14 days 28 days 56 days	Rating 1-3 Rating 1-3 Rating 1-3 Rating 1-3 Report values for	CSA Z245.20 - clause 12.14 ISO 21809-2 clause A.15

		Information	
Adhesion Hot water Immersion @ 120 °C	14 days 28 days	Rating 1-3 Rating 1-3	CSA Z245.20-clause 12.14; ISO 21809-2 clause A.15 Sample immersed in vessel capable of maintaining temperature and withstanding the pressure.
12 weeks Adhesion (hot water immersion) @ max. rated temperature or 120 ± 3°C	mm disbondment from edge of holiday	3 max	Coat all exposed metal surface with a suitable material. Drill a 6 mm holiday in the middle of coating sample. Immerse in water in pressurized vessel capable of maintaining 120 ± 3°C
1.5° Strain Polarization	28 days, 23°C cathodic disbondment	No cracking	CSA Z245.20 - clause 12.13 ISO 21809-2 clause A.14
2° Flexibility	-30 °C or 0°C	No cracking	CSA Z245.20 - clause 12.11 ISO 21809-2 clause A.12
Impact	-30°C, 1.5J	No holidays	CSA Z245.20 - clause 12.12 ISO 21809-2 clause A.113
Porosity	Cross-section Interface	Rating 1-3 Rating 1-3	CSA Z245.20 - clause 12.10 ISO 21809-2 clause A.11
Electrochemical Impedance Spectroscopy	23°C 7 days 14 days 28 days 56 days	 $>10^{10}$ ohm-cm^2 $>10^{10}$ ohm-cm^2 $>10^{10}$ ohm-cm^2 Report values for Information	Measurement taken after immersion in 3% NaCl solution at specified temperature
	95°C 7 days 14 days 28 days 56days	 $>10^{7}$ ohm-cm^2 $>10^{7}$ ohm-cm^2 $>10^{7}$ ohm-cm^2 Report values for Information	

2. Plant trial - ring samples

Property	Test parameters	Acceptance Criterion (3 specimens)	Test Method CSA or ISO
Thermal Characteristics		Manufacturer's Specification	CSA Z245.20 - clause 12.7 ISO 21809-2 clause A.8
Cathodic Disbondment (disbonded radius)	23°C, 28 days	8.0 mm max	CSA Z245.20 - clause 12.8 ISO 21809-2 clause A.9
	65°C, 2 days 14 days 28 days	 5 mm max 10 mm max 15 mm max	CSA Z245.20 - clause 12.8 (Electrolyte replaced every 7 days with fresh solution preheated to test temperature.) ISO 21809-2 clause A.9
	95°C, 14 days 28 days	 10 mm max 15 mm max	CSA Z245.20 - clause 12.8 (Electrolyte replaced every 7 days with fresh solution preheated to test temperature.) ISO 21809-2 clause A.9
	Maximum rated temperature*, 14 days 28 days	 10 mm max 15 mm max	ISO 21809-2 clause A.9 * Test specimen on hot plate or sand/steel grit bath with test temperature measured at coating-steel interface. Bulk electrolyte temperature controlled at 30 ± 5°C
Adhesion - Hot water Immersion @ 95°C	2 days 28 days	Rating 1-3 Rating 1-3	CSA Z245.20 - clause 12.14 ISO 21809-2 clause A.15
Adhesion Hot water Immersion @ 120 °C	14 days 28 days	Rating 1-3 Rating 1-3	CSA Z245.20-clause 12.14; ISO 21809-2 clause A.15 Sample immersed in vessel capable of maintaining temperature and withstanding the pressure.
12 weeks Adhesion (hot water immersion) @ max. rated temperature or 120 ± 3°C	mm disbondment from edge of holiday	3 max	Coat all exposed metal surface with a suitable material. Drill a 6 mm holiday in the middle of coating sample. Immerse in water in pressurized vessel capable of maintaining 120 ± 3°C
1.5° Strain Polarization	28 days, 23°C cathodic disbondment	No cracking	CSA Z245.20 - clause 12.13 ISO 21809-2 clause A.14
2° Flexibility	-30 °C or 0°C	No cracking	CSA Z245.20 - clause 12.11 ISO 21809-2 clause A.12
Impact	-30°C, 1.5J	No holidays	CSA Z245.20 - clause 12.12 ISO 21809-2 clause A.113
Porosity	Cross-section Interface	Rating 1-3 Rating 1-3	CSA Z245.20 - clause 12.10 ISO 21809-2 clause A.11

PITTING IN WIND TURBINE GEARBOXES: CORROSION OR FATIGUE?

A Spark[1]
AECOM Australia Pty Ltd[1], Melbourne, Australia

SUMMARY: Wind energy is one of the most prevalent and accessible forms of renewable energy. This is of particular importance as the world works to reduce the effects of climate change and minimise global temperature increases. Of the three energy producing components of a wind turbine (blades, gearbox and generator), the gearbox is the most prone to failure. A dominant mode of failure of wind turbine gearboxes is through pitting of steel components. Components that fail by this mode include the rolling bearings of both the high speed shaft and planetary gears. In this review, the causes of pitting of rolling bearings are discussed with consideration to the diagnostic process for the root cause of the failure. There is much contention over whether the cause of pitting in wind gearboxes is due to corrosion or fatigue. Without fully understanding the cause of the failure, it is difficult for asset owners to both prevent further failures of a given asset and to assess the likelihood of other assets of a similar type failing, and to allow them to budget accordingly. Another key consideration is determining the point at which an asset must be taken out of service and the component replaced. Understanding the mechanism of failure is vital for this decision making process. As corrosion specialists it is important to understand all of the potential causes of pitting on steel components of wind turbine gearboxes.

Keywords: Steel, Wind energy, Gearbox, Fatigue, Pitting.

1. INTRODUCTION

Wind energy is one of the most prevalent forms of renewable energy production used worldwide. In 2009 it was the fastest growing component of the power generation sector (Barszcz & Randall, 2009). By 2030, the US is expected to generate 20% of its power needs from wind and in Europe that number is 25% (Lu, et al., 2009). In China, where especially in recent years there has been a strong push in favour of renewable energy as they phase out coal power production in efforts to combat air pollution problems, the target is 15% renewable energy by 2020. Of this a large proportion is wind energy as the government has introduced regulations in support of the industry at all levels (Liu, et al., 2010). In Australia, while there has been variation in the levels of support for wind energy and various public debates over the years, there has been a strong wind industry for the last 10 years with many more wind farms currently in the planning and development phase.

Modern wind turbines are essentially comprised of three major components: a set of blades, a gearbox, and a generator, collectively known as the drive train. Of the three major components, the generators are the most prone to failure (Dabrowski & Natarajan, 2017). However, generator failures are readily, and cheaply, repaired up-tower and do not pose much risk to owners and operators. The costliest component to replace or repair with the longest downtimes are gearboxes, with a new gearbox costing roughly 13% of the overall cost of a turbine (Feng, et al., 2013). While originally designed to last the life of a turbine (approximately 20 years) many operators are finding themselves with failed gearboxes before ten years is reached and in some cases less than five years (Barszcz & Randall, 2009). This often requires a gearbox to be replaced entirely, an expensive operation involving large cranes on site to lift the gearbox down from the top of the wind tower. The predominant reasons for the increased failures seen in wind gearboxes, both in comparison to other wind turbine components and other types of gearboxes, is an underestimation of the operating loads experienced and the large variability in wind turbine operational conditions, including sudden changes in wind speed, as well as overloading as the turbine is disconnected from the grid and during start-up of the turbine (Kandukuri, et al., 2016; Keller & Guo, 2016).

In recent years, a shift has begun in the wind industry from predominately reactive maintenance or time interval based maintenance to condition based maintenance, with the aim of managing early gearbox failures more proactively. This shift has been supported by the development of various condition monitoring techniques and systems which are able to detect potential problems and failures across the breadth of the turbine drive train (Link, et al., 2011). The conditions experienced by wind turbines that exacerbate gearbox failures also place challenges on monitoring systems that have had to be overcome to provide meaningful data to wind farm operators (Luo, et al., 2014). This is especially true of the components of the gearboxes which operate at much lower speeds than any component in traditional power generation from which most of the condition monitoring systems have been adapted. Unique algorithms for the monitoring system have had to be devised for wind turbines failure detection.

Rolling element bearings are the most commonly failed components of wind turbine gearboxes, including those in the low speed, high speed and planetary gear sections (Keller & Guo, 2016). Bearing failures account for some 70% of gearbox failures in the wind industry (Dabrowski & Natarajan, 2017) (Kandukuri, et al., 2016). Pitting and spalling is one mode of failure seen in bearings. As corrosion specialists, our instinct is to see pitting and assume corrosion, especially when pitting is seen early in the life of a bearing designed for twenty years of operation. However, there is also the potential for material, design, and construction defects to contribute to early and unexpected fatigue issues. In this paper we discuss the possible causes of pitting/ spalling failure seen in multiple high speed shaft roller bearings.

2. FAILURE EXAMPLE/ DETAILS

One hundred and ninety-nine 2 MW wind turbine gearboxes were surveyed. Each of these gearboxes was designed to last the full lifetime of the wind turbines, that is, approximately twenty years. They are three stage units comprised of one planetary low speed stage and two helical gear stages designated the intermediate shaft and high speed shaft stage. Each stage increased the speed of the wind turbine blades to allow a compact generator to be used, starting with the planetary stage next to the blades and ending with the high speed shaft which connects the gearbox to the generator. Two different gearbox models were used, with model A installed in 79 turbines and model B installed in the remaining 120 turbines. Failures were seen in both gearbox models, however the model A turbines showed a significant number of failures due to pitting and spalling, types of failures which were not seen in the model B gearboxes.

The pitting and spalling in model A occurred in predominately the inner race of the non-rotor end high speed shaft (HSS) bearings. These are the bearings closest to the generator. The pitting is detected via one of three ways. During routine maintenance, the oil filter of a gearbox is checked for metal particles. If metal particles are found either via a magnetic stick (Figure 1) or in an oil filter sediment cup (Figure 2), this triggers an endoscope inspection of the gearbox. Increases in particulate matter found during oil sampling can also trigger an endoscope inspection. Alternatively, an endoscope inspection can also be triggered by excessive vibrations or noise coming from the gearbox. If during the endoscope, pitting or spalling is seen on the HSS bearing, the bearing is replaced.

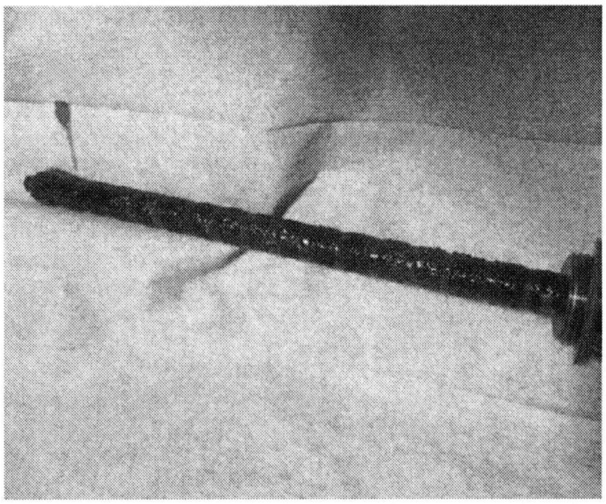

Figure 1 Metal particles on a magnetic stick from the oil filter indicating wear within the gearbox

Figure 2 Metal particles in an oil filter sediment cup indicating wear within the gearbox

Originally designed with a case carburised (CC) bearing of SAE-AISI 4320 steel, 31 of the 79 NRS bearings pitted within the initial warranty period of the turbines, within two years of commissioning. An example of the pitting is shown in Figure 3. When the pitting was initially detected, the turbine manufacturer worked with the gearbox manufacturer to replace and upgrade the pitted bearings with case carburised hot assembly (CCHA) bearings. The CCHA bearings had further surface treatment applied to the inner race. As these soon also began to pit, a case carburised hot assembly complete bearing (CCHACB) type, with the surface treatment applied to the entire bearing, was then introduced. Unfortunately, identical failures have been seen in this bearing type as well. To date, 61 of the 79 original bearings have been deemed failed and 12 gearboxes have experienced pitted HSS bearings twice. Of the repeat failures, eight were CCHA bearings and four were CCHACB bearings that failed. In some cases, these bearings have been seen to fail every two to three years of operation.

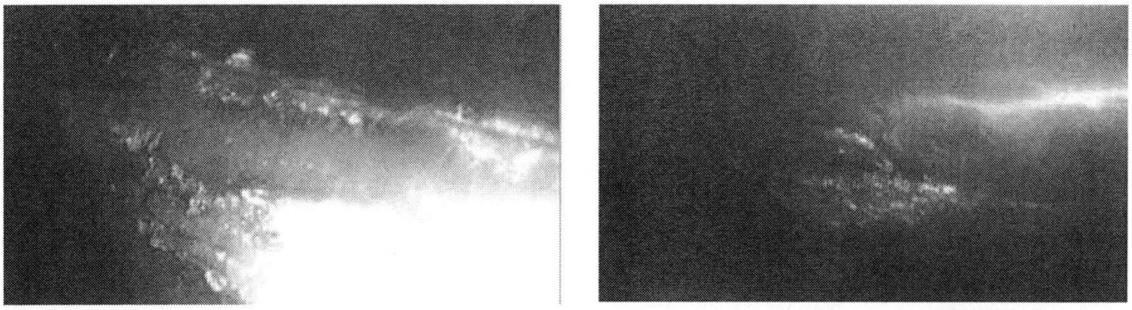

Figure 3 Endoscope images of pitting and 'cracking' seen along an original case carburised NRS HSS bearing

Initially the pitting was deemed to be due to an incorrectly specified material type/ surface treatment for the bearings, which led to the change in bearing type. While there is a strategy in place for approaching the problem, the continuing failures even after changing the bearing type, either to CCHA (Figures 4 and 5) or CCHACB (Figure 6), indicates that the root cause of the pitting has not been identified. Currently, as soon as pitting is seen the bearings are replaced. If left running for too long, there is a risk of the spalling metal causing damage to other components in the gear box including the intermediate shaft and planetary bearings. Without identifying the root cause of the pitting, it is difficult to monitor the bearings and manage the replacement of the bearings in such a way to maximise their use. This is not ideal for the owner of the wind turbines who would like to maximise the availability of the turbines for power production while minimising their maintenance costs.

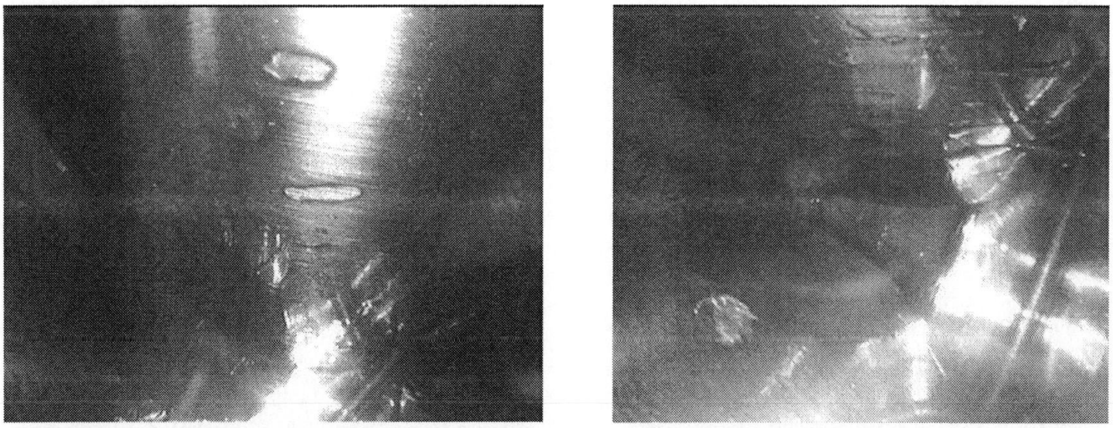

Figure 4 Endoscope images of pitting seen along a case carburised hot assembly NRS HSS bearing.

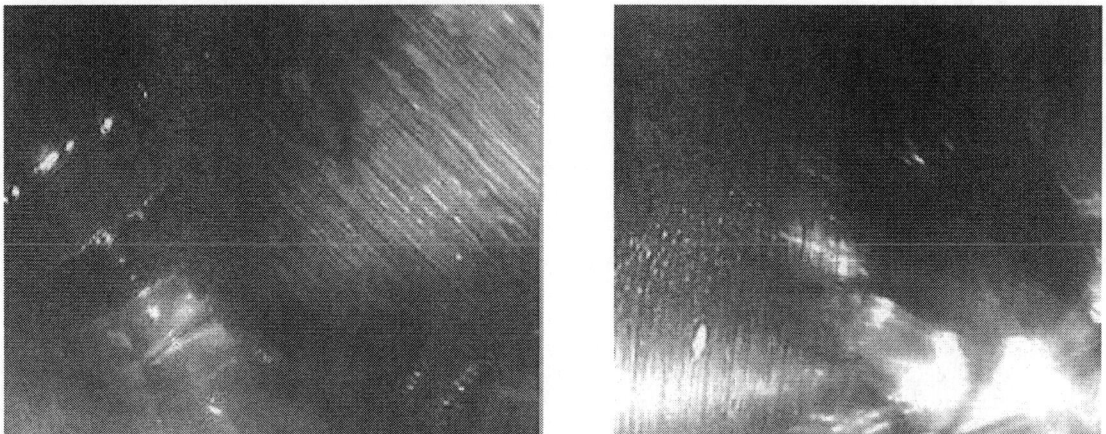

Figure 5 Endoscope images of pitting seen along a case carburised hot assembly NRS HSS bearing. The pitting seen was quite light, but due to the current understanding of the failure and monitoring processes in place, the bearing was replaced.

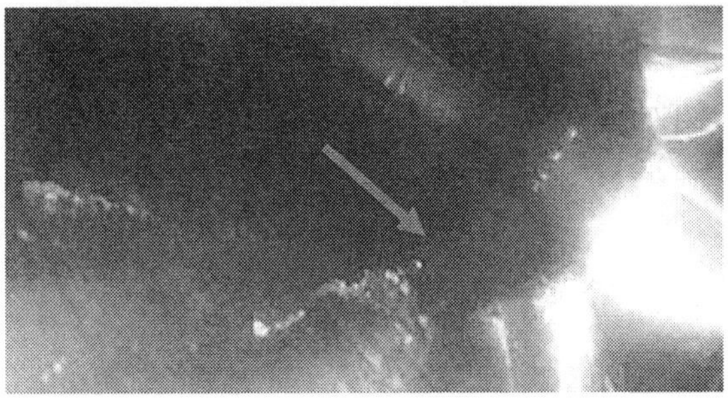 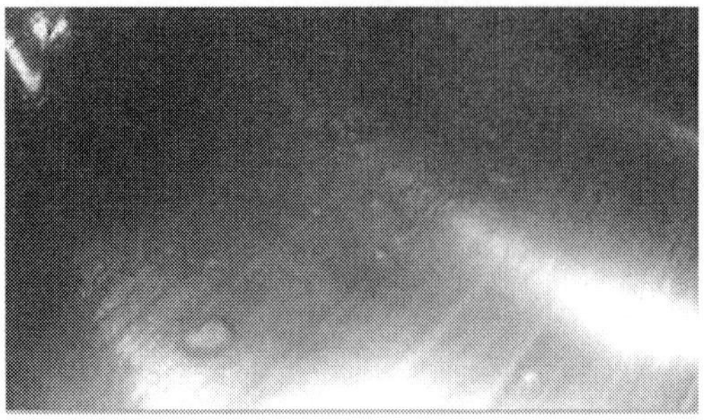

Figure 6 Endoscope images of pitting and 'cracking' seen along a case carburised hot assembly complete bearing. Similar cracking (marked with arrow) as with the other bearing types. The image on the right depicts an earlier pitting stage, pre cracking/ spalling.

One key difference in HSS bearing design has been established between the two models of gearbox. Model A uses surface hardened steel, as discussed in detail above. Model B's bearings are made from a through hardened, low alloy chrome steel. Case carburised bearings are designed to be more resistant than through hardened steels to fatigue, slippage, mechanical shock and particle over-rolling, all of which are important factors in wind turbine gearboxes (Schaeffler, 2015). Through hardened steel's main advantage over case carburised bearings is a greater resistance to corrosion. The dominant cause of pitting and spalling in roller bearings is rolling contact fatigue (Diederichs, et al., 2016), raising the question as to why these symptoms were only seen in the bearings made from the steel designed to be more resistant to fatigue. A literature review was undertaken to investigate if similar failures have been seen elsewhere and what factors could be contributing to the pitting.

3. POTENTIAL MECHANISM OF FAILURE

Much research has been conducted into the causes of rapid pitting and spalling failures in roller bearings, with a focus on wind turbine gearbox bearings. Roller bearings are well known to be affected by rolling contact fatigue (RCF) and this is a well understood phenomenon that bearings are readily designed to resist for their expected design life (Lai & Stadler, 2016). In some instances, roller bearings have been shown to fail prematurely at 5-10% of their calculated design lives, and due to the seemingly unpredictable nature of these failures research has focused on identifying the causes of these premature failures. Common to all the premature failures studied is a microstructure around pitting, spalling and cracking that appears white under optical microscopy when the steel surface is etched in nitric acid (Evans, 2013). The phenomenon is known as white etching cracks (WECs), white structure flaking and brittle flaking, depending on the publication. The term white etching cracks (WECs) will be used in this paper.

Rolling contact fatigue is an inevitable cause of spalling in roller bearings. The spalling is due to slow subsurface material changes as a result of changes in load in the bearings during operation causing sliding and impact between rollers (Kotzalas & Doll, 2010). These failures are influenced by pressure, material type and how the bearing is constructed and result in dark etching areas in the subsurface microstructure that transform to highly ordered low angle white bands and high angle white bands with time (Diederichs, et al., 2016). These white bands have been shown to be carbon free in their centres and be carbon rich at the edges with secondary carbides precipitating. WECs differ greatly in that the changes in microstructure occur in response to the WEC process, with plastic deformations accumulating and causing the migration of carbon and chromium from the area and stress placed upon the carbides in the matrix (Holweger, et al., 2015). This results in a rapid change toward WEC when this stressed area reaches saturation and the carbides deform and dissolve. WEC regions are subsurface areas composed of nanocrystalline body centred cubic (BCC) ferrite with inhomogeneous distributions of carbon and chromium around a network of random cracks (Diederichs, et al., 2016; Bhadeshia & Solano-Alvarez, 2015). One of the key identifying features is an increased hardness of the WEC regions compared to that of the surrounding matrix.

It is the rapid nature of the formation of WECs that is of particular concern. There are many hypotheses of what is contributing to these accelerated failures in certain applications, including in wind turbine gearbox bearings. All of the proposed mechanisms are focussed on explaining the microstructural changes seen with WECs and how sufficient energy is collected to cause the dissolution of carbides (Gould & Greco, 2016). Hypothesised mechanisms include hydrogen absorption, electrical current and electro thermal influences, non-metallic inclusions, bulk tensile stress and friction energy accumulation. Non-metallic inclusions including oxides have different coefficients of thermal expansion to the bulk metal leading to the introduction of tensile and compressive stresses (Evans, et al., 2013). Many WECs do not reach the steel surface, so there is some process occurring in the subsurface and in many studies inclusions are found within WECs. Non-metallics can be introduced through surface treatment processes, through deposition of components in the lubricant or through micro pitting on the surface due to corrosion where the lubricant layer has become insufficiently thick and small areas are exposed to air. One of the dominant theories is that the presence of hydrogen absorption leads to the development of WECs (Evans, 2013; Bhadeshia & Solano-Alvarez, 2015; Evans, et al., 2013; Holweger, et al., 2015; Tamada & Tanaka, 1996). Hydrogen is often found in association with WECs from field samples and has hydrogen charging of a steel surface has shown to increase the rates of WECs under laboratory conditions. It is proposed that the hydrogen increases fixing of grain boundaries and dislocations within the matrix, increasing the levels of localised plastic deformation in a manner similar to classical hydrogen embrittlement. Two sources of the hydrogen are proposed: water contamination of the lubricant, mostly believed to be unlikely, and through breakdown of compounds in the gearbox lubricant (Evans, et al., 2013). An alternative explanation for the presence of hydrogen at WEC sites is that the increased plastic deformation and cracks act as sinks for hydrogen once WECs are initiated (Diederichs, et al., 2016). More research is required to determine the extent of hydrogen's influence on WEC formation.

Electrical influences on WECs are often cited. There are many ways in which electrical current can be introduced to bearings, including a stray current from the generator, electrostatic potential build up and tribocharging due to friction between components (Ruellan, et al., 2017). Depending on the lubricant conditions, the presence of an electrical potential has been hypothesised to lead to either charge accumulation and arcing or electrochemical deposition of the lubricant. Both of these could contribute to WEC formation. Another theory is that the electrical currents act as a means of releasing hydrogen into steel (Evans, et al., 2013). Current has also been proposed as a driver for carbide decay where an imposed stress (such as a non-metallic inclusion) is present (Diederichs, et al., 2016; Ščepanskis, 2015). Differences in electrical and thermal properties of the various components of the steel microstructure lead to thermal induced stress which can push the carbides to deform. One study on electrostatic charges in rolling bearings found that WEC formation was amplified where there was a break in lubrication (Loos, et al., 2016). They also found that WEC formation only occurred when the bearing was acting as a cathode and that electrochemical processes were occurring similar to cathodic stress corrosion cracking. This finding is at odds with the understanding that pitting initiated by corrosion occurs due to loss of electrons from the anodic metal surface. It is likely that while the whole bearing was acting as a cathode, areas of the surface were acting as anodes. Most likely, this would be where there was a break in lubrication. The amplification in WEC formation due to breaks in lubrication seen by Loos et al was thought to be due to changes in hydrogen deposited directly on the steel surface.

An ongoing theme to all the contributing factors to WEC formation, is the introduction of localised stress. Sliding of the roller elements due to the dis-engagement/engagement of wind turbines generators to the grid is one means of introducing high magnitude shear stresses (Kotzalas & Doll, 2010). Wind turbines are particularly prone to large torques and other forces associated with their frequent turning on and off with changing wind conditions. The variation in operating conditions experienced is strongly linked to why WEC formation is so common in rolling bearing elements in wind turbines. However, the highly superficial area of bearings affected by sliding is not sufficient to drive the rate of fatigue crack growth seen in WEC formation (Lai & Stadler, 2016), which is why, while common, it does not occur in all wind turbines. A bulk cyclic tensile stress, not just a localised cyclic tensile stress, is required for fatigue crack growth and subsurface WEC formation. Fretting corrosion bands, as an indication of tensile stress on the shaft side of a bearing inner ring, have been found to align with WEC associated cracking on the other side of the inner ring (Lai & Stadler, 2016). In this case, the tensile stress was found to have been introduced by the process of the bearing being pressed onto the shaft.

Lubrication is stated to be a key contributor to WEC formation for many reasons, it's a source of hydrogen, it can increase or decrease friction, and depending on the components it can increase and decrease corrosion rates. Rolling bearings fail at their weakest point. If you have a poor lubricant and rough surface then the surface will fail, as it does in RCF (Lai & Stadler, 2016). RCF is well enough understood that good design prevents its formation and it is very rarely a concern. If a good lubricant is in use and the surface has been well prepared as per design standards to prevent RCF, then failure will occur due to WEC formation below the surface initiated by the presence of a material defect or high shear stress. WEC formation leads to failure rapidly and well within the design life of bearings.

4. DISCUSSION

There are several key characteristics of the HSS bearing failures seen that suggest they are likely to be due to WEC formation. Firstly, the pitting and spalling were first identified within 2 years of the turbines' commissioning. They have also, to date, affected roughly 80% of the original bearing population and failures of these bearings continue. Currently, the turbines are well within the approximate ten-year design life of the HSS bearings. Clearly, these failures are premature and unlikely to be due to rolling contact fatigue, which accumulates very slowly over the lifetime of a bearing.

Secondly, the pitting and spalling is only seen in the non-rotor side HSS bearings, which are the bearings closest to the generator. This suggests that the generator and the connection between the generator and high speed shaft is involved in the pitting process. One potential contribution from the generator is stray current flow affecting the bearings and contributing to the pitting and spalling formation. Stray current flow could affect the lubrication consistency across the bearing, increasing exposure of the surface to air and thus pitting corrosion; and increase the deposition of hydrogen from the lubricant, thus increasing non-metallic inclusions on the surface. These in turn could act as WEC initiation sites. An alternative reason for pitting to only be seen on the generator side bearings of the HSS is greater temperatures experienced near the generator. This would also affect the performance of the lubricant. It also indicates that the contribution of the generator is required for the pitting to occur, further strengthening the idea of WEC formation as a cause.

A third consideration is that cracking, when the pitting progresses to that point, is seen along the inner race in line with the circumference of the bearing. This is perpendicular to the shaft. Considering that tensile stresses have previously been seen to be introduced via the bearing fitting process and that when this occurred, consistent orientation of the pitting and cracking were seen (Lai & Stadler, 2016), the patterns of cracking seen here suggest something similar occurring in the studied bearings. There is also the potential for poor tolerance between the size of the shaft and the size of the bearings to introduce high tensile stresses.The original pitting in the case carburised bearings may have been contributed to by high tensile stress in the bearings associated with the way in which the bearing was fitted to the high speed shaft during construction. However, given pitting has also been seen in the updated bearings which were hot assembled and thus should not have been pressed onto the shaft, there are other factors at play leading to the pitting which include poor tolerances in the design.

Interestingly, these failures are only being seen in the gearboxes with case hardened bearings and not in the gearboxes with through hardened bearings. Both material types are known to be prone to WEC formation as roller bearings under certain conditions including the presence of stray currents, hydrogen absorption, cyclic bulk tensile stresses and breaks in lubrication. Based on their material properties, the case hardened bearings should be more resistant to fatigue, slippage, and particle over-rolling than the through hardened bearings (Schaeffler, 2015). This suggests that through hardened bearings are more likely to be affected by WECs than case hardened bearings, which is the opposite of what has been seen in this case study. Assuming the lubrication conditions and connection between HSS and generators are consistent between gearbox types, the one material property that may be contributing to the pitting seen is the decreased resistance to corrosion of case hardened steel used in these bearings.

During the operation of a wind turbine, the thickness of lubrication across a bearing changes due to the variation in forces applied to and temperatures experienced by the turbine. Where there are breaks in lubrication, WEC formation has been shown to increase (Loos, et al., 2016). While this has previously been discussed in relation to hydrogen absorption, it also increases the availability of the surface to oxygen and thus corrosion. The rapid changes in direction of the turbine have been shown to cause protective oxides formed on the surface by the lubricants to be removed (Grujicic, et al., 2016), providing the opportunity for corrosion to occur, especially if the material in question is more susceptible to corrosion, as the case hardened bearings are believed to be (Schaeffler, 2015). Corrosion of the surface forms micro pits. This can lead to both an increase in hydrogen deposition as the pits act as a hydrogen sink and an increase in localised stresses around the pit, both of which are amplified by the presence of an electrical current. Once localised stress build up is initiated, WEC formation is possible. Additionally, the process of case hardening works through the creation of residual compressive stresses in the surface layers of steel. The presence of the increased stress at the surface reduces the amount of external force required for cracking to occur once initiated, thus leading to increased rates of WEC propagation to spalling in case hardened bearings.

5. CONCLUSIONS

Large rates of pitting and spalling failures are being seen in case carburised HSS bearings of wind turbine gearboxes. This is likely to be due to WEC formation initiated by localised corrosion of the bearings under the influence of a hypothesised stray current and breaks in lubrication.

Having now reached a greater understanding of the potential mechanism of the failures seen, appropriate metallurgical testing can be performed to confirm this hypothesis. Condition monitoring systems can also be implemented to detect the failures and monitor their progression to maximise the outputs of the bearings. This will reduce costs and loss of availability of the asset for the owner.

6. ACKNOWLEDGMENTS

The author gratefully acknowledges the asset owner for their support of this research. She would also like to acknowledge Sarah Furman for her review and acting as sound board for the idea.

7. REFERENCES

Barszcz, T. & Randall, R., 2009. Application of spectral kurtosis for detection of a tooth crack in the planetary gear of a wind turbine. *Mechanical systems and signal processing,* Volume 23, pp. 1352-1365.

Bhadeshia, H. & Solano-Alvarez, W., 2015. Critical assessment 13: elimination of white etching matter in bearing steels. *Materials Science and Technology,* 31(9), pp. 1011-1015.

Dabrowski, D. & Natarajan, A., 2017. Identification of loading conditions resulting in roller slippage in gearbox bearings of large wind turbines. *Wind energy.*

Diederichs, A. et al., 2016. Study of subsurface initiation mechanism for white etching crack formation. *Materials Science and Technology,* 32(11), pp. 1170-1178.

Evans, M.-H., 2013. White structure flaking (WSF) in wind turbine gearbox bearings: effects of butterflies and white etching cracks (WEC). *Materials Science and Technology,* 28(1), pp. 3-22.

Evans, M.-H., Richardson, A., Wang, L. & Wood, R., 2013. Effect of hydrogen on butterfly and white etching crack (WEC) formation under rolling contact fatigue. *Wear,* Volume 306, pp. 226-241.

Evans, M.-H., Richardson, A., Wang, L. & Wood, R., 2013. Serial sectioning investigation of butterfly and white etching crack (WEC) formation in wind turbine gearbox bearings. *Wear,* Volume 302, pp. 1573-1582.

Evans, M.-H., Wang, L., Jones, H. & Wood, R., 2013. White etching crack (WEC) investigation by serial sectioning, focused ion beam and 3-D crack modelling. *Tribology international,* Volume 65, pp. 146-160.

Feng, Y. et al., 2013. Monitoring wind turbine gearboxes. *Wind energy,* Volume 16, pp. 728-740.

Gould, B. & Greco, A., 2016. Investigating the process of white etching crack initiation in bearing steel. *Tribology letters,* 62(26).

Grujicic, M. et al., 2016. Multiphysics computational analysis of white-etch cracking failure mode in wind turbine gearbox bearings. *Proc IMechE Part L: J Materials: Design and Applications,* 230(1), pp. 43-63.

Holweger, W. et al., 2015. White etching crack root cause investigations. *Tribology transactions,* 58(1), pp. 56-69.

Kandukuri, S., Klausen, A., Karimi, H. & Robbersmyr, K., 2016. A review of diagnostics and prognostics of low speed machinery towards wind turbine farm-level health management. *Renewable and sustainable energy reviews,* Volume 53, pp. 697-708.

Keller, J. & Guo, Y., 2016. *Gearbox reliability collaborative investigation of high-speed-shaft bearing loads,* Golden, CO: National Renewable Energy Laboratory.

Kotzalas, M. & Doll, G., 2010. Tribological advancements for reliable wind turbine performance. *Philosophical transactions of the royal society A,* Volume 368, pp. 4829-4850.

Lai, J. & Stadler, K., 2016. Investigation on the mechanisms of white etching crack (WEC) formation in rolling contact fatigue and identification of root cause for bearing premature failure. *Wear,* Volume 364-365, pp. 244-256.

Link, H. et al., 2011. *Gearbox reliability collaborative project report: findings from phase 1 and phase 2 testing.,* Golden, CO: National renewable energy laboratory.

Liu, W., Tang, B. & Jiang, Y., 2010. Status and problems of wind turbine structural health monitoring techniques in China. 2010. *Renewable Energy,* Volume 35, pp. 1414-1418.

Loos, J., Bergmann, I. & Goss, M., 2016. Influence of currents from electrostatic charges on WEC formation in rolling bearings. *Tribology transactions,* 59(5).

Lu, B., Li, Y., Wu, X. & Yang, Z., 2009. *A Review of Recent Advances in Wind Turbine Condition Monitoring and Fault Diagnosis.,.* IEEE Conference 24-26 June 2009, s.n.

Luo, H. et al., 2014. Effective and accurate approaches for wind turbine gearbox condition monitoring. *Wind energy,* Volume 17, pp. 715-728.

Ruellan, A. et al., 2017. Understanding white etching cracks in rolling element bearings: state of art and multiple driver transposition on a twin-disc machine. *Proc IMechE Part J: J Engineering Tribology,* 231(2), pp. 203-220.

Ščepanskis, M. J. A. K. I. e. a., 2015. The Numerical Model of Electro thermal Deformations of Carbides in Bearing Steel as the Possible Cause of White Etching Cracks Initiation. *Tribology letters,* 59(37).

Schaeffler, 2015. *Materials for Rolling Bearing Technology.* [Online]
Available at: http://www.schaeffler.com/remotemedien/media/_shared_media/08_media_library/01_publications/schaeffler_2/tpi/downloads_8/tpi_226_de_en.pdf[Accessed 31 05 2017].

Tamada, K. & Tanaka, H., 1996. Occurrence of brittle flaking on bearings used for automotive electrical instruments and auxiliary devices. *Wear,* 199(2), pp. 245-252.

8. AUTHOR DETAILS

Amy Spark is a Senior Engineer within the Strategic Asset Management and Advanced Materials team at AECOM in Melbourne, Australia.

Amy has eight years of experience as an engineer in a variety of environments and industries and holds a PhD in corrosion engineering. She has worked on numerous projects in durability/ materials engineering and asset management. Amy was awarded her PhD in 2016 having studied Microbiologically Influenced Corrosion (MIC) on the buried surface of steel water pipelines. She enjoys using her engineering expertise to improve public health and infrastructure outcomes in line with government regulations and international standards..

STRATEGIC INSPECTION AND REPAIR PROCEDURE FORMALISED FOR REPAIRS TO A MAJOR MARINE COMPLEX

I Khan, J Dyson, M Marosszeky & F Papworth
Building and Construction, Research and Consulting (BCRC), Australia

SUMMARY: Durability Plans for projects are seldom developed for repair of existing structures. However, fib Model Code 2020, which is in development, will include existing structures alongside new structures. This major step forward requires many durability developments, involving 30 working parties each with a group of leading experts, to bring service life design for repair up to the standard of that in MC2010 for new structures. This may incorporate durability planning procedures. CIA Recommended Practice Z7/01 is the first international document detailing durability design of new structures. The approaches, based on experience from 10 years of formal durability planning on major government infrastructure projects in Australia, can be applied to repair design but will need to be modified for the different status of repairs compared to new structures.

Durability plans are typically based on the project technical criteria (which are normally based on national codes) and the owner's standard specification as minimum requirements. For new structures, the durability plan may not typically make major changes to the structural design, although the durability consultant may help to identify significantly different options to the standard design requirements.

However, for the assessment and remediation of existing structures, the durability consultant's role is much more critical to the development of the primary design approach. As such the durability consultant needs to be involved prior to other experts to determine the nature and extent of deterioration and how that might influence the most economic design approach. Given the more uncertain nature of repairs they may also require time to undertake trials, particularly where cathodic protection systems are involved. Even where a design and construct project is envisaged, the durability consultant needs to identify parameters that will ensure that the repair solutions offered will closely meet the owner's needs and live up to design assumptions. There is a need for a much wider appreciation of materials properties than exists for new construction, where the range of materials used is usually more limited. Also, a much greater understanding of processes of assessment prior to repair design is required. This paper discusses a strategic inspection and repair procedure for a major marine complex.

Keywords: Durability planning, Existing structures, Inspection, Repair.

1 INTRODUCTION

On most infrastructure projects, preliminary durability design is sometimes left up to the design team and civil, structural and marine engineers rely on their own staff's knowledge to design for durability based on in-house standards, proprietary software, technical society publications and Australian Standards. Typically, the durability plan, developed by specialist durability consultants, is only used till the 85% design stage to finesse the drawings and specifications.

In some cases, particularly in the private sector, a specialist durability consultant is not employed. This practice can leave owners exposed to the risk their structures will not achieve the design life specified. Technical challenges include the fact that durability knowledge is rapidly developing, becoming more and more sophisticated and specialized. At the same time durability modelling is in a stage of rapid development. All this is against a backdrop of significant debate among international researchers and practitioners about specific theories such as, for example on the mechanism of chloride diffusion in concrete and its modelling. This means it is extremely difficu lt for traditional structural engineers to stay abreast of the field of durability design. Furthermore, standards have a long gestation period and therefore often lag best practice in rapidly developing fields.

For new structures, the durability plan may not typically make major changes to the structural design, although the durability designer may help to identify more effective options to the standard design requirements. However, for the assessment and remediation of existing structures, the durability consultant's role is much more critical to the development of the primary design approach. As such the durability consultant needs to be involved prior to other experts to determine the nature and extent of deterioration and how that might influence the most economic design approach. Given the more uncertain nature of repairs they may also require time to undertake trials, particularly where cathodic protection systems ar e involved. Even where a design and construct project is envisaged, the durability consultant needs to identify parameters that will ensure that the repair solutions offered will closely meet the owner's needs. There is a need for a much wider appreciation of materials properties than exists for new construction where the range of materials used is more limited. A much greater understanding of processes of assessment prior to repair design is required. Most structural engineers are not familiar with state of the art non-destructive testing (NDT) techniques and therefore might specify unnecessary destructive testing to correctly identify the type of testing needed.

2 FINANCIAL AND PERFORMANCE REQUIREMENTS

2.1 Reliability

Reliability is also a crucial issue and ensuring each zone of a structure is designed to an appropriate reliability (not over or under what is required) is key to unlocking best value for money for owners. For example, serious consideration is currently being given by code writers globally (fib and ISO) to the prospect that designing the underside (non-visible components) of a major wharf for a durability reliability of β=0.5 would be acceptable. This would translate to quite extensive deterioration in terms of concrete cracking and spalling within the design life, but no risk of structural failure. This would not be suitable for visible front faces of a wharf, where the required durability reliability may be much higher, say β=2.0 or higher. Designing to a durability reliability above what is required will mean the cost of construction is more than it needs to be, whereas designing to a durability reliability that is under what it should be may lead to significantly increased maintenance spend over the life of a structure. Table 1 gives the reliability requirements according to ISO 2394[2].

Table 1. Reliability (β) Requirements to ISO 2394 [2]

Relative cost of safety measures	Consequence of failure (e.g. Current Cost of Future Repairs)			
	Small	Some	Moderate	Great
High	0	1.5	2.3	3.1
Moderate	1.3	2.3	3.1	3.8
Low	2.3	3.1	3.8	4.3

2.2 Risk management

Risk is conventionally assessed by ISO 31000 [1] in two dimensions. A detailed version of this, giving examples of consequence and a description of likelihood is shown in Figure 1. The process of identifying risk, then evaluating and minimising the likelihood and/or consequences of potential element failure ensures financial and performance requirements of built assets are controlled to within the client's accepted risk profile. The allowable risk level might be defined by the client's specific requirements but in general it can be assumed that when assessing allowable risk of a civil engineering project, the acceptable risk is likely to be "Moderate". This is not necessarily the case with repairs where steeply rising costs to improve reliability may mean the acceptable risk profile changes, at least for serviceability limit states. In making the risk assessment it must also be recognised that the likelihood of failure may be more difficult to define as even the bas ic aspects, such as chloride ingress subsequent to repair, has no model. Other matters such as long-term bond failure of patches has only anecdotal information. Hence it is important for the durability consultant to understand the client's requirements and the client to understand the cost and reliability situation with repair as oppo sed to new construction.

Likelihood (During 50 year design life)		Consequence: 1 - Negligible	2 - Low	3 - Moderate	4 – Very High	5 - Extreme
		COST: Within normal maintenance expectations and budget. (eg <$10k OPERATIONS: Nil SHE: Nil	COST: Unexpected but will not effect cost of production (eg $10K-1m) OPERATIONS: May effect activities but not production SHE: Nil	COST: May have a minor effect on overall cost of production (e.g. cost $1m to 10m) OPERATIONS: May have a minor effect on production (e.g. <5days lost production) SHE: Nil	COST: May have a significant effect on overall cost of production (e.g. $ value 10m to 100m) OPERATIONS: May have an effect on production (e.g. 5-30days days lost production). SHE: May cause minor SHE outcome	COST: May effect ongoing viability (e.g. cost >100m) OPERATIONS: Problem may have a major effect on production (e.g. >30days lost production SHE: May cause significant SHE outcome
A - Almost Certain Very likely. The event is expected to occur in most circumstances. Possibly a history of occurrence.	>95%	Moderate	High	Extreme	Extreme	Extreme
B - Likely There is a strong possibility the event will occur. Possibly a history of occurrence.	50-95%	Low	Moderate	High	Extreme	Extreme
C - Possible The event might occur at some time as there is reported cases of it occurring.	5-50%	Very Low	Low	Moderate	High	Extreme
D - Unlikely Not expected but there is a slight possibility it may occur at some time.	1-5%	Negligible	Very Low	Low	Moderate	High
E - Rare Highly unlikely but it may occur in exceptional circumstances. It could happen, but probably never will.	<1%	Negligible	Negligible	Very Low	Low	Moderate

Figure 1 Risk Assessment Chart ISO 31000[1]

The optimum reliability approach shown in Figure 2 can be used to aid the risk assessment process by giving a better gauge of the optimum reliability and hence likelihood of failure sought. The analysis may show a large increase in repair cost for not much improvement in reliability or reduction in maintenance cost. The client may then prefer to accept a higher risk and lower reliability (higher likelihood of failure) consistent with scenario A in Figure 2. However, if the optimum situation is scenario B the higher construction cost is likely to be worthwhile.

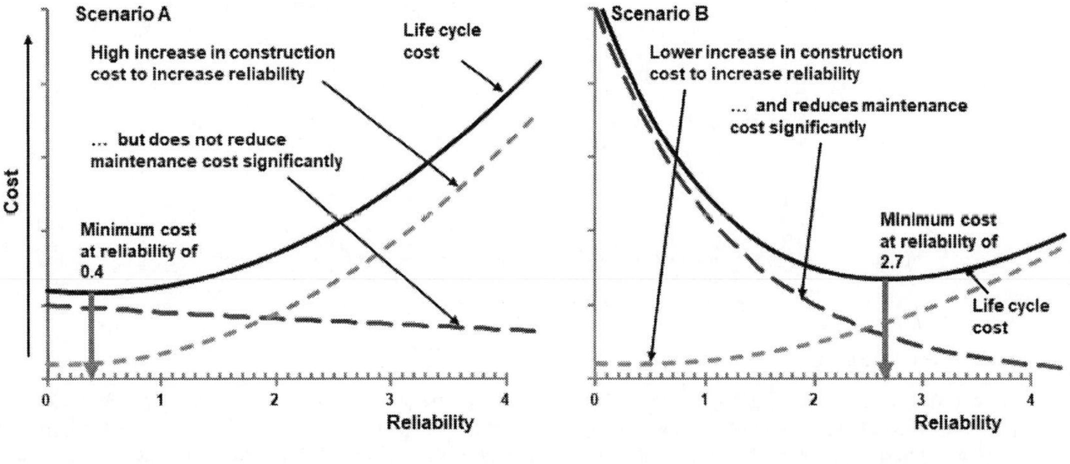

Figure 2 Optimum reliability from combination of construction cost and maintenance cost.

2.3 Service life

Very often concrete durability design of new structures is assumed to mean a service life period up until reinforcement corrosion initiation. Figure 3 shows the concrete condition deterioration phases. In Figure 3, "1" represents the corrosion initiation time when steel inside concrete gets depassivated either by chloride ions or by carbonations. "2" represents the cracking stage of propagation phase when corrosion products exert pressure on concrete resulting in cracking. "3" represents the stage where excessive cracking occurs which leads to spalling of concrete. "4" represents the stage when excessive spalling occurs eventually causing collapse of the structure. However, a real understanding of current durability code provisions shows that the corrosion propagation period (the time between corrosion initiation and either concrete cracking, spalling or collapse) is often included in design, as it should be (Figure 3). For example, relatively low concrete covers are often designated for seawater immersed structures, and yet the design life is achieved in spite of a relatively short corrosion initiation (t_0) period. This is because of the subsequent slow corrosion rate in the propagation phase due to oxygen starvation.

When dealing with existing structures the initiation period may already have been reached and the design life for repairs may be based on assuming a slowed penetration of the corrosion activation front and/or the slowed propagation period. There are no agreed methods for modelling these and no established criteria for use in modelling. It is more common to consider the anecdotal experience of how long repairs have been found to last historically, although there are many reasons why this will vary from case to case.

Since the 1980's there has been an increase in the occurrence of durability failure that has led to a growing business of concrete repair. This increase comes from a double issue of a growing stock of structures and a growing rate of failure. The latter may be due to inadequate understanding of the changing nature of materials and a reduction in quality management. This should provide an object lesson for repairs as there is a growing stock of structures that require repair, i.e.

a. Our understanding of repair materials, and the associated deterioration mechanisms, is lower than our understanding of the use of concrete in conventional construction. An example is the use of ribbon anodes in cementitious slots in splash zones. These were originally said to be suitable for a 50 year design life. After as little as 10 years some of these systems are failing due to an unforeseen mechanism, i.e. acidification at locations of low resistance. Understanding of repairs is less well understood than conventional construction and this leads to higher risks even when left to experts.
b. High quality concrete repair is more complex and difficult to achieve than high quality conventional concrete yet our systems of quality assurance are frequently at a lower level and often inexperienced but cheap labour is used. The later inevitably leads to premature failure. On one project patch repairs expected to last 20 years failed at 6 months.

This suggests that with repairs a high level of conservatism in design and great attention to quality assurance is required to achieve even a short design life, say 20 years, at a lower than normal SLS reliability.

Another aspect that is not well developed is how much corrosion and concrete damage is acceptable before structural failure occurs although if the design life is kept to 'occurrence of physical damage' this may not be critical.

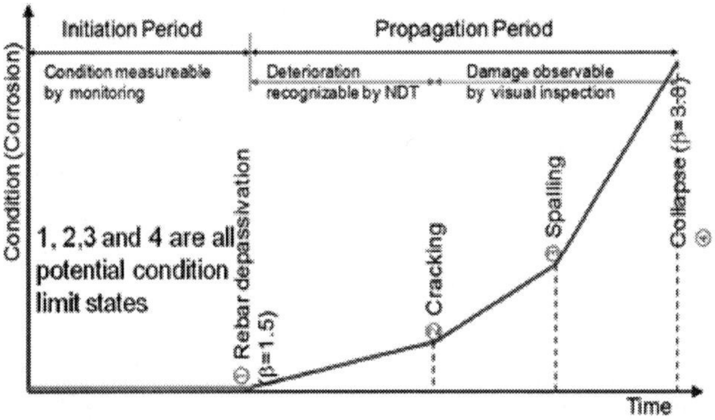

Figure 3: Concrete condition deterioration phases

3 EXPOSURE CLASSIFICATION

A complete exposure classification guide for all the elements of the project should be prepared. The Australian Codes only provide requirements for a limited number of exposure zones. Also, the exposure zones given in the codes are at times ambiguous. Therefore, durability consultants have developed several additional exposure classes and some of these are defined in the draft of CIA's Z7/02 [3]. Although not recognized within Australian Standards, these exposure zones are often referenced in durability plans. Examples include:

a. Concrete subject to long periods of seawater immersion followed by long periods of drying is recognized as a much more severe exposure than the splash zone, in some cases the exposure may be mitigated where the concrete has a high resistivity and the time open to the air is short. Australian code designations as a splash zone for seawater tank might over estimate requirements on some occasions and under estimate them on others.
b. Concrete in contact with ground is recognized as an area where capillary rise can occur, water can then evaporate off depositing salts. This is called 'evaporative concentration' and even water low in salt concentration can be aggressive.
c. Car parks are sometimes identified as internal structures with benign exposures. The low concrete performance and cover requirements designed in this context can lead to very early failure if the exposure is much higher. A common problem in Europe has been salt water from deicing salts being brought in on wet cars leading to very severe exposure and requirements for cathodic protection within parking garages. In a

recent case in Australia, a coastal basement car park suffered early corrosion damage due to marine salts from salt water seepage and in another case, another structure suffered salt contamination due to windblown salts from the ocean only a couple of hundred meters away.

d. Tunnels and carparks may also be exposed to high carbon dioxide levels. This is being recognized as a high risk as it causes accelerated carbonation of the concrete in the structure, and in the presence of moisture unacceptable corrosion risk. This is leading to detailed assessment of carbon dioxide levels, particularly in tunnels.

A contractor's durability consultant may identify these issues at tender stage but if the Contractor adds money to deal with these issues he may become uncompetitive. The risks to infrastructure arising from a deficient or limited client brief need further consideration, as when significant issues are not identified in the client brief, the contractual circumstances may preclude the consideration of critical durability requirements.

4 STRATEGIC INSPECTIONS OF MARINE STRUCTURES

For marine and coastal elements, where concrete is visually undamaged then an understanding of its current state is required. Figure 4 presents the general model for residual life and protection requirements. As shown in Figure 4, undamaged concrete could mean anything from no repair required to too late to coat and an electrochemical repair is required.

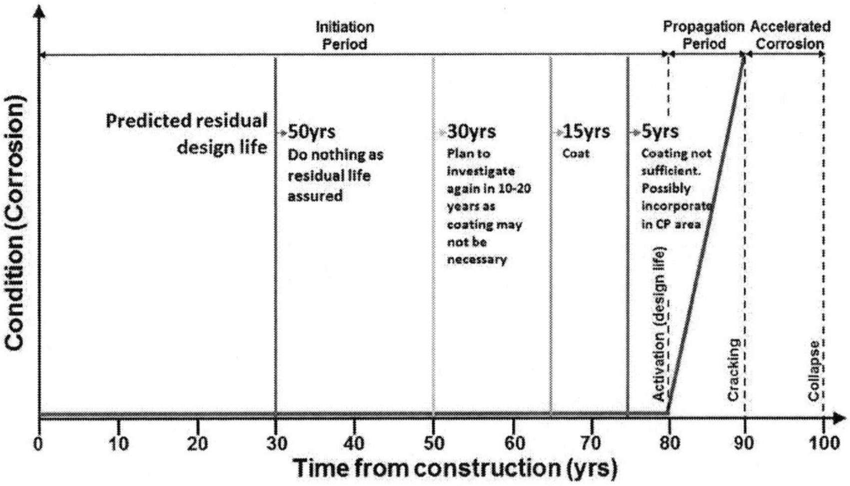

Figure 4: Residual life and protection requirements

This is to be determined ultimately for the entire concrete surface but for design purposes a statistically representative sample of each exposure zone needs to be assessed to determine:

a. The likely percentage of concrete that is actively corroding, i.e. past activation but not yet damaged - this will be assessed using electrical potentials with criteria developed specific to the project. These areas are likely targets for cathodic protection.

b. The residual life of concrete that is not yet corroding - the chloride diffusion coefficient will be estimated from Dapp, the age of the structure and cover. Dapp is estimated from the best fit diffusion curve through the chloride profiles. The residual life will then be assessed based on achieving a reliability suited to repair, possibly as low as
0.5. This low reliability is based on marine structure design being robust and aesthetics not being an issue. This might be re-assessed if the structural engineers determine areas where an undamaged cover is critical. From this and the cover depth the residual life will be calculated in accordance with fib Model Code 2010 [4].

Where there is no apparent damage, the modelling is undertaken using the methods described above to determine the residual life with no coating. Ideally the residual life after coating would also be modelled but there is no currently agreed model for how the chlorides in the cover zone redistribute within the concrete element once the source of chloride is removed. This is therefore reviewed based on crude estimates of whether chloride ingress will reach a relatively uncertain activation depth in the residual life of the structure. The assessment requires a deal of conservatism and engineering judgement.

There is no way of knowing what percentage of an undamaged areas will fall into which category shown in Figure 4 without testing as described in (a) and (b).

5 STANDARD PROTOCOL FOR TESTING

A standard routine for testing of an existing structure is shown in Figure 5.

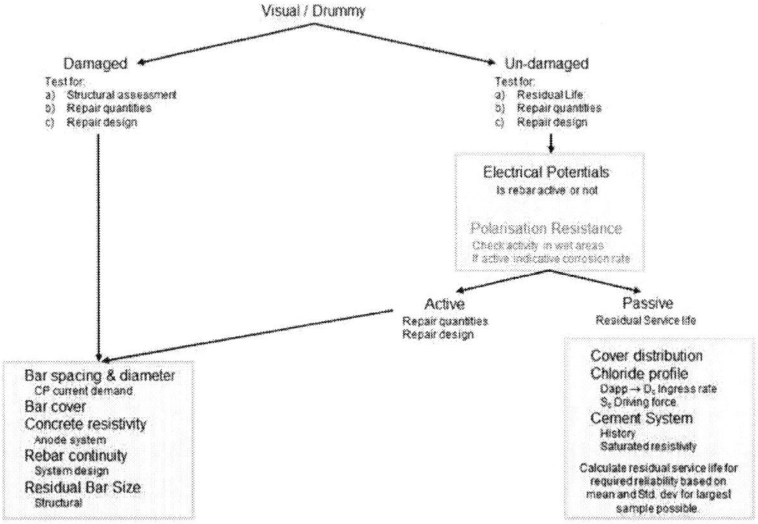

Figure 5: Standard protocol of testing

5.1 Assessment of reinforcement corrosion in undamaged areas

In areas where corrosion damage is already apparent and the cause is known (e.g. chlorides), there is little point in undertaking most types of testing because the prognosis is already apparent.

In areas where there is no damage apparent visual inspection does not give any guide to residual life. If the reinforcement is actively corroding, i.e. chloride has penetrated to the reinforcement in sufficient quantity to cause corrosion initiation, corrosion rate tests (e.g. resistivity and polarisation resistance) are undertaken to estimate how long it might be until cracking and spalling will occur. Tests are also undertaken to provide data for the design of a cathodic protection repair. If the reinforcement is not yet active, tests required to identify how long it will be before it is active include a cover survey and chloride profiling.

5.2 Extent of testing

The size of area to be tested varies depending on many factors but in general it should be no less than $1m^2$ and is unlikely to be larger than $10m^2$. Where an element type has different exposures in the same location, then testing in each exposure type is necessary. For example, a beam may have an exposed front face, a sheltered rear face and splashed soffit, in those cases a 1m wide length of beam can be selected and testing carried out on each exposure zone in the same operation. This may enable a reduction of points of electrical contact to the reinforcement, for example.

It is unlikely to be practical to test a truly statistically representative sample so decisions will be made based on testing of limited areas. Although this leads to inaccuracy the final assessment of repair extent is made based on additional testing at the time of repair. The preliminary inspection is to determine the type of repair and rough repair quantities.

5.3 Visual inspection

Detailed visual inspection should be undertaken using a standard set of terminology and defined meanings. Otherwise a statement like 'wide crack' could have almost any width and cause. ACI Code 201[5] gives a set of defect descriptions. It is also recommended that each inspector is trained in using photos in order that results are consistent survey to survey and inspector to inspector. Visual inspection before any testing is useful as it may provide information that will help identify causes for certain results that are obtained or even direct the inspector to modify, extend or abort the planned testing. The visual inspection should also provide some guidance on how representative it is of the general area being assessed and the degree of exposure in the test area and generally.

5.3.1 Drummy testing of concrete

Ideally, all surfaces of a structure with corrosion risk should be tested to determine if the concrete is delaminated. A soffit that looks pristine may well have a very large area of delaminated concrete. However limited testing of representative areas may be all that is viable on the preliminary survey. Testing can be undertaken using a delamination detector where a tapping tool on a long pole is listened to via a remote mike adjacent to the tapping tool. Where concrete is drummy, it is valuable to remove some of the drummy concrete, where safe to do so, in order to examine the reinforcement and establish the extent of loss of section.

5.3.2 Cover testing

Cover measurement is carried out using a covermeter or ground penetrating radar (GPR). A cover survey help s to explain why the structure is corroding and show which areas are most susceptible to corrosion due to low cover. Measurements are used in conjunction with corrosion activation threshold depths from chloride profiles to assess residual life. Cover measurements are taken at different locations to get the cover value using a probabilistic distribution.

5.3.3 Rebar condition

At a point where bars perpendicular to each other cross a core can be taken to the reinforcement depth. The core is retained for chloride analysis and carbonation depth tests. The condition of the reinforcement is examined and recorded. The cover to the bar and bar diameters are physically measured.

5.3.4 Electrical potential of reinforcement

Electrical potential defines the likelihood of corrosion activity. It is a quick and simple test once the connection is made to the reinforcement. Hence the extent of measurement should be extended as far as practical from the point of connection. Measurements should be taken on a 500mm x 500mm grid, and in more negative potential areas the grid spacing should be reduced to 250 x 250mm by taking intermediate readings. Half-cell potential readings are indicative of the probability of corrosion activity of the reinforcing steel located beneath the concrete surface. By mapping equipotential contours on the surface, those portions of the structure where there is a high likelihood of corrosion activity are identified by finding areas of more negative potential and steep potential gradients. The recommendations of RILEM TC 154-EMC-Half-cell potential measurements [6] should be followed for assessment of electrical potential measurements.

5.3.5 Resistivity

Resistivity of the concrete is highly affected by the concrete moisture condition. However, the base resistivity measurement also indicates the likely corrosion rate that can be sustained. In each test area four resistivity measurements should be taken. Resistivity measurements should be taken using the 4 probe Wenner method and the instrument must be calibrated before the commencement of testing. Tests must be undertaken so as to avoid testing over the reinforcement. The recommendations of RILEM TC 154-EMC-Test methods for onsite measurement of resistivity of concrete [7] should be followed.

5.3.6 Chloride analysis

Cores taken to establish chloride ingress should be cut into at least four depth increments and each depth should have the chloride content measured in accordance with AS 1012.20. Chlorides are measured by dissolving powder samples in acid. The samples can be taken from drillings or from ground cores. The chloride profile must be related to cover so that the extent to which rebars are exposed to high chlorides can be determined. Chloride profiles can also be used to determine the apparent diffusion coefficient at a specific location and thus predict the ongoing rate of chloride ingress at those location, subject to there be no change in the environment of to which the concrete is subjected. To do this a minimum of four depth increments are required for sensible curve fitting.

6 REPAIR PROCEDURES

6.1 General principles of repair design

ISO Technical Committee 71 (TC71) Subcommittee 7 (SC7) is responsible for standards on maintenance and repair of concrete structures. SC 7 has prepared a number of standards related to concrete repair. ISO 16475[8] provides guidelines for the repair of water-leakage cracks while ISO 16311[9] in four parts describes the overall approach to repair. Figure 6 sets out the structure of ISO 16311[9].

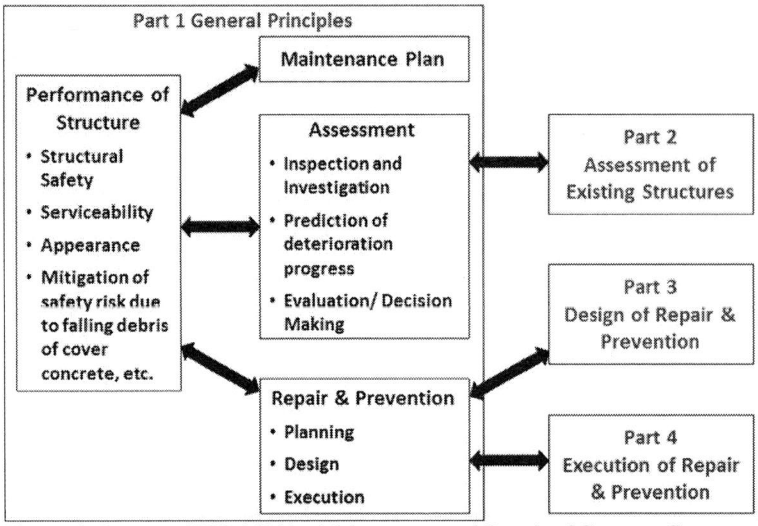

Figure 6 : Layout of ISO 16311 – Maintenance and Repair of Concrete Structures

The approach to concrete repair design should generally follow the principles of ISO 16311[9]. ISO 16311[9] follows the same general durability design principles as fib Model Code 2010[4] and fib Model Code 2020 is being developed to include many of the principles set out in ISO 16311[9] for existing structures which was not included in Model Code 2010[4].

6.2 Electrochemical protection

Cathodic Protection in its broadest definition is used in most professional repairs to control the impact of chloride induced corrosion damage, but the technical meaning of cathodic protection needs qualification because of the nuances applied. The following terms are used to define systems where anodes are used to polarize the reinforcement to a lesser or greater extent.

Cathodic Prevention	A low level of polarisation such that reinforcement that has not corroded will be prevented from corrosion. Typically used for incipient anode protection in full patch repairs.
Corrosion Control	A moderate polarisation of reinforcement that has corroded to reduce the corrosion rate often to such an extent that corrosion damage will not occur in a limited residual life. Typically used in full patch repairs.
Cathodic Protection	Polarisation of the reinforcement to achieve a level of polarisation given in the Australian Code to designate a state where all reinforcement in the protected area will have an insignificant corrosion rate. Typically used where only damaged concrete is to be removed.

The extent of polarisation has a profound effect on the repair method and cost of repair. The general layout of a sacrificial anode systems for cathodic prevention and cathodic protection are shown in Figure 7.

6.3 Incipient anodes

Cathodic prevention is used to overcome the development of incipient anodes at the edge of new patch repairs. Where concrete quality is locally a little inferior to the surrounding area or where the cover is a little bit less than the surrounding area, chlorides or a carbonation front will penetrate more quickly to the reinforcement and an anodic area will be formed. The surrounding area is still passive and it becomes the cathode. Chlorides will eventually reach this cathodic area but the adjacent anode provides cathodic prevention. So adjacent to an anode in concrete there is likely to be an area with slightly fewer chlorides that an adjacent area which is under cathodic prevention. The situation is accentuated because where there is a corrosion cell, the chlorides are drawn in the potential field towards the anode. When a spall occurs and the concrete is replaced with a new cementitious patch the reinforcement is re-passivated locally. This former anodic area now becomes cathodic to the surrounding area and the surrounding area becomes the anode and starts to corrode, hence the term incipient anode.

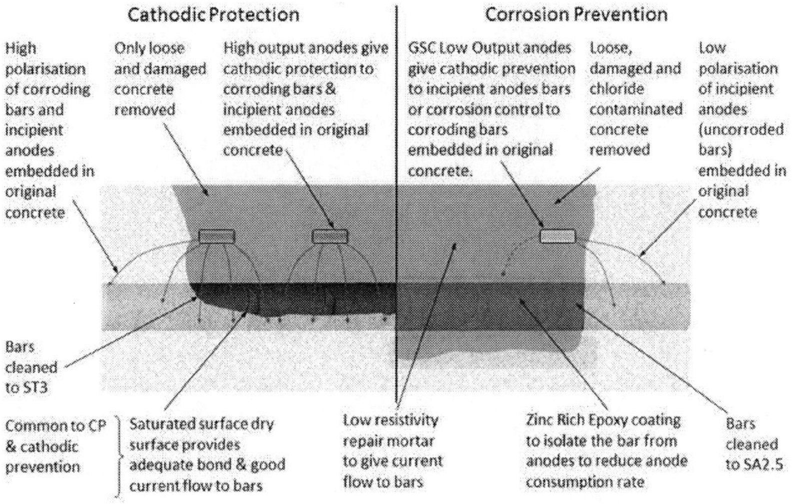

Figure 7: Cathodic protection and cathodic prevention type repairs

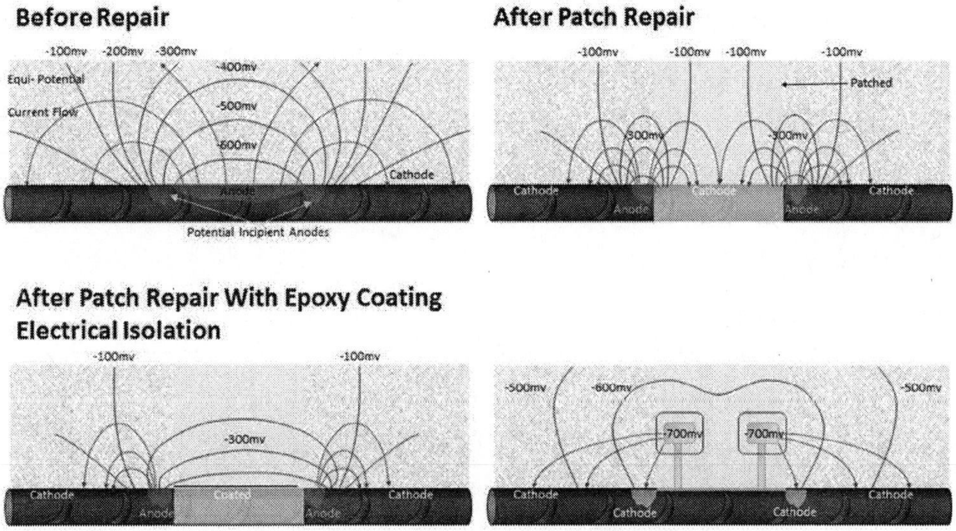

Figure 8: Development and control of incipient anode

There has been some debate about whether incipient anodes are real or are just a theoretical concept. Figure 9 shows a schematic of samples made by Barkey [10]. He was able to measure quite precisely the corrosion because his bar was made up of a number of separate sections so that he could measure where the current flowed. He also developed a theoretical model that matched the experimental data. The results showed that:

a. The corrosion rate is controlled by the resistance between the anode and cathode and hence potential difference is not a key control mechanism. The incipient anode area was quite small, just 5-10cm long, and it was quite close to the patch, 1-2cm away. This is significant because it confirms that sacrificial anodes within the patch are able to control the whole incipient anode.

b. The majority of current was towards the repair. Again, this was significant as it shows that epoxy coatings on the bar in the patch area will largely restrict the cathodic area and hence corrosion rate of the incipient anode.

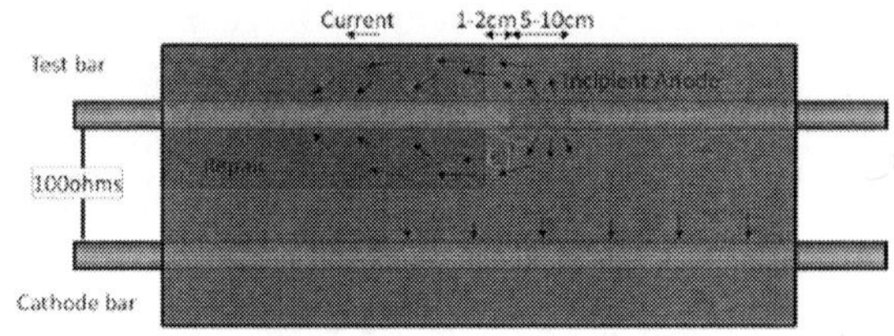

Figure 9: Incipient anode current flow [8]

6.4 Impressed current systems

ISO 12696[11], AS 2382.5[12] and CS TR 73[13] all provide information on cathodic protection installation testing and performance requirements, and provide information on each of the cathodic protection anode systems. However, none provide a process for selecting the most appropriate anode system for a particular repair and a particular client. In the absence of an accepted selection process or rationale, consultants and contractors influential in the selection process may select a system more suited to their interests than those of the client. Hence, a process for anode selection is needed, it should allow all suitable anode systems to be considered, with the final selection being based on the best fit with the limitations of the specific application, and considering the needs of the client. This will enhance competitive tendering.

Clients often do not appreciate that cathodic protection trials are required to optimize a system design so that the CP cost can be minimized. Where no trials are undertaken the designer will:

a. Design to the highest rebar current density possible - this may be several times the current density required.
b. Use maximum long term anode current densities for the short term - short term anode current densities can be two times the long term permitted current density without causing any long term issues.

The above leads to CP designs that have significantly more capacity than required. Not only is this expensive in terms of anodes as it will lead to exclusion of certain anodes specifically designed for the lower power requirements needed. It is also expensive in terms of control systems, resulting in more zones and more controllers than would otherwise be the case. By undertaking trials, the generic design can be more specific in terms of required current densities to be used.

6.5 Sacrificial anode systems

In small patch repairs, low output anodes can be used for protection of incipient anodes. The theory behind the use of low output anodes for cathodic prevention is shown in Figure 10.

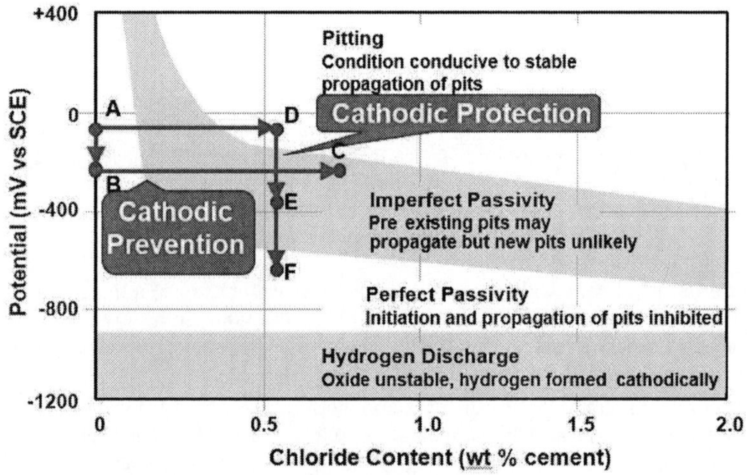

Figure 10: Application of high and low output anodes

The figure shows the zones of passivity based on chloride content of the concrete and the electrical potential of the reinforcement. After an element is constructed it starts at point A. No chlorides and passive potentials. As chlorides at the steel increase, the reinforcement moves to the area where pitting occurs (D). The electrical potential will decrease. If the passive reinforcement is polarized just a little (i.e. moves to point B) a much higher chloride content is needed to push the reinforcement into the pitting zone. This small shift provides cathodic prevention. If the pitted reinforcement were polarized just a little corrosion would continue, albeit at a slower rate, as pits continue to corrode in the zone of imperfect passivity. If the reinforcement were polarized sufficient to take it to the perfect passivity zone corrosion would stop.

For concrete repair for incipient anode protection, i.e. where prior to the repair the halo around the anode was protected by the anode itself, and hence had no pitting, the use of a low powered sacrificial anode can replace the active reinforcement that acted as the anode prior to repair. However, the small potential shift cannot provide protection to reinforcement that has already started to corrode. The large potential shift of corroded reinforcement requires cathodic protection. CP can be used to prevent the need for breakout of concrete around corroding bars and give a major reduction in breakout and patching cost. This difference in polarisation is important as different sacrificial anodes work in different ways.

6.6 Minor damage- patch repair with cathodic prevention

As this system is not intended to provide cathodic protection to corroding reinforcement in undamaged chloride contaminated concrete, the chloride contamination must be removed. This raises the question of how much concrete to remove behind the bar as breakout of undamaged concrete is expensive, and because the greater the breakout the greater the structural concern. This depends on the overall repair system as shown in Figure 11.

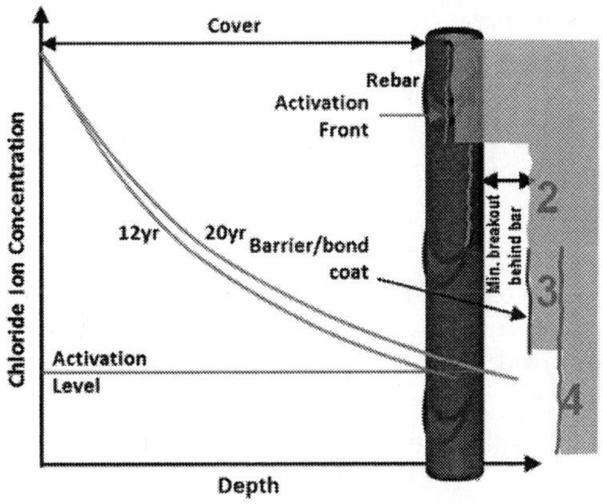

Figure 11: Amount of breakout required

Typical cathodic prevention repairs comprise:

a. Zinc rich epoxy coatings to protect the bar from chlorides and provide active protection
b. Concrete bond coatings to help prevent back diffusion of chlorides but not impede the polarisation from anodes c. Use of a repair mortar with high chloride diffusion resistance

The depth of breakout behind the bar is typically the greater of 20mm or 1.5 times the maximum repair mortar aggregate size. This enables the effective packing of mortar behind the bar. It is unlikely that more than a 10mm aggregate will be used for repairs but the 1.5 times limit is used to ensure no arching of aggregate behind the reinforcement.

6.7 Dispersed areas of corrosion risk

In many cases, it may be that corrosion damage and activation is quite extensive but is not sufficient, or is too dispersed, to use ICCP. It may also be the case that ICCP systems are just not convenient to install and maintain for a given project. Sacrificial anodes with high output are available and can be used to give cathodic protection, i.e. meet the depolarisation requirements given in AS 2382.5. For these high output anodes, the high polarisation means a high current output and much higher rates of zinc loss. This means the anodes must have a higher mass of zinc and be placed at close spacings, often making the anode cost higher overall. The offset is that only damaged concrete has to be removed and generally that means the reinforcement bond to the concrete is left intact and no structural issues arise.

Where sacrificial anode cathodic protection is used, the anode system may be trialled to demonstrate that the arrangement proposed will provide cathodic protection. However, history of previous successful performance is often taken as sufficient evidence that the system will work. Unlike ICCP systems where regular monitoring of the whole system is essential, it is not uncommon to manually monitor local areas of a large SACP system.

7 CONCLUSIONS

This paper provides information on strategic investigation and repair procedures required for the repair of a major marine complex and highlights several key points for inspection and repair contracts:

1) The new concept of required reliability introduced for new structures in CIA's Z7/01 is also a valuable part of assessing repairs of existing structures. However, in repairs it is likely that establishment of the optimum reliability may be useful in reviewing the risk of failure.
2) A close examination of the exposure class is required as in many repair situations it may be found that the exposure is not really adequately catered for in the codes. This can affect the decisions on repair materials.
3) Many experts contend that testing is always a vital part of repair assessment. This is not always strictly true there are occasions where the expense of testing is not warranted and the visual evidence is enough to define a reliable repair. However, in other cases the extent of testing often used is woefully inadequate in terms of extent and type to develop an appropriate repair method. It is vital to determine a suitable inspection strategy for the initial assessment of repair type and likely extent, and the testing during repair to define the extent of repair.
4) There is a woeful lack of understanding of the use of sacrificial anodes in concrete repair. Specifiers need to understand the differences between different types of sacrificial anodes and their performance in concrete repair for cathodic prevention, corrosion protection and cathodic protection or they are unlikely to specify the optimum repair approach.
5) One of the major costs in repair can be concrete breakout, particularly when it leads to the need for structural support and high operational costs. The reasons for specifying different levels of breakout and the significance to cost needs to be clearly understood.

The detailed testing and repair procedures described in this paper provide ample evidence of the benefits of sound durability advice for owners and contractors alike. They emphasise the need for careful durability planning. They also describe some of the many sources of risk that arise when inexperienced designers assume they have greater knowledge of durability and durability modelling than is the case.

Finally, it is economically negligent for an asset owner to commit capital expenditure without following through with the routine maintenance work that will guarantee the design life is achieved.

8 REFERENCES

[1]. ISO 31000:2009, Risk management – Principles and guidelines [2]. ISO 2394:2015, General principles on reliability for structures [3]. Z7/02 Concrete Institute of Australia— Exposure Classes
[4]. fib Model Code for Concrete Structures 2010, International Federation for Structural Concrete, 2010 [5]. ACI 201.1R-08 Guide for Conducting a Visual Inspection of Concrete in Service
[6]. RILEM TC 154-EMC Recommendations: Half-cell potential measurements - Potential mapping on reinforced concrete structures2003Materials and Structures461-471
[7]. RILEM TC 154-EMC Recommendations: Test methods for on site measurement of resistivity of concrete 2000 Materials and Structures 603-611
[8]. ISO/TR 16475: Preview Guidelines for the repair of water-leakage cracks in concrete structures, 2011 [9]. ISO 16311 Maintenance and repair of concrete structures, 2014
[10]. Dale P. Barkey, "Corrosion of Steel Reinforcement in Concrete Adjacent to Surface Repairs", ACI Material Journal, 2004, 101(4), 266-272
[11]. ISO 12696:2016 Cathodic protection of steel in concrete
[12]. AS 2832.5-2002 Cathodic protection of metals - Steel in concrete structures
[13]. CS TR 73 Cathodic protection of steel in concrete - including model specification, 2011

9 AUTHOR DETAILS

Inam Khan is durability Engineer at BCRC. Inam has over 12 years of research and professional engineering experience. His research has focused on corrosion of reinforcement, corrosion evaluation techniques, concrete durability in severe environments, early-age properties of concrete and time dependent deformations in concrete. Mr. Khan's PhD thesis is Experimental study of corrosion in reinforced concrete structures. Inam is experienced in assessing the durability of reinforced concrete in severe chloride environment. Inam is well versed with the state of the art techniques to monitor corrosion in reinforced concrete structures and material testing including non-destructive techniques.

Jonathon Dyson is NSW manager at BCRC. Jonathon has over 10 years professional engineering experience covering both contracting and consulting in the building, transport, water, mining and marine infrastructure sectors. Jonathon has strong technical consulting and project management skills and in depth knowledge of structure condition and durability assessments and reinforced concrete/structural steel design

Marton has been a Director of BCRC since 2006 and was an Executive Consultant at Evans & Peck between 2007 and 2012. His extensive experience as a qualified engineer spans road construction, structural engineering design, construction contractor to expert technical and litigation advice in relation to quality issues, and management consulting in construction and design process improvement and strategic planning.

His work in the durability and repair of concrete structures is nationally recognised. Also, his work in construction process improvement in relation to productivity, safety, quality and environmental management on building and construction projects is widely used on construction projects in NSW. He consults within Australia and overseas.

Since graduating as a civil engineer Mr Papworth has gained over 35 years' experience in concrete technology, durability research, and concrete inspection and repair consultancy. Mr Papworth's main interest is as a Durability Consultant for BCRC, a role he has filled on some of Australia's largest projects including tunnels, bridges, reservoirs, wharves and desalination plants. Mr Papworth is also Chairman of PCTE, Australia's leading supplier of concrete testing equipment and SRCP, Australia's most diverse supplier of anode systems for concrete. He was Chairman of the Concrete Institute of Australia (CIA) Durability Committee from

2010-2015, is Chairman of the International Structural Concrete Associate's (fib) commission on Concrete Durability and is Joint-Chair of the MC2020 Durability Action Group.

SURFACE PREPARATION APPLIED TO SHIPYARDS USING DESICCANT DRYERS. DESIGN AND ENERGY CONSIDERATIONS

G. Lacey BEng (Mech) MBA[1], C. Hipsley BEng (Mech)[1]
[1]DST Dryer Pty Ltd, Sydney

SUMMARY: Corrosion has long been a concern when building and maintaining ships as they work in a highly damaging environment. Dry air has been prominent in shipping applications for corrosion prevention in operation, preservation, and surface preparation and coating. In this paper, we look at how Korean shipyards use Desiccant dryers to improve automation of their construction, and the dehumidification design as the extreme climate in Korea is beneficial for the range of climate conditions in Australia. Using modelling techniques with various climate data we can look at selections and energy saving implications from this Korean Shipyard design and apply it to key Australian cities. We also look at the relationship between humidity and corrosion, the application of humidity control, and sizing methods for the use of humidity control in various defence and civilian applications.

Keywords: Desiccant, Dehumidifier, Relative Humidity, Dew Point, Psychometric, Absolute humidity

LIST OF ABBREVIATIONS AND TERMS

Absolute Humidity – Measure of moisture level regardless of temperature
ADF – Australian Defence Force
ARNG – Army National Guard
ASHRAE – American Society of Heating Refrigeration and Airconditioning Engineers
Binmaker – weather data program organised by city / country
Closed system – Dehumidification system where air is recirculated
Desiccant – A hygroscopic substance used to remove moisture
Dehumidifier – A piece of equipment that removes moisture in the air
DoD – Department of Defence
DX – Direct Expansion – uses refrigeration vapour expansion/ compression cycle to directly cool supply air
FMV – Försvarets Materielverk (Swedish Defence Material Administration; Procurement agency for the Swedish armed forces)
HR – Humidity Ratio – grams of water per kilogram of air
LTP – Long Term Preservation
MCDB – Mean coincidental Dry Bulb Temperature
Open system – Dehumidification system where fresh air is brought in from outside
OP – Operational Preservation
Relative humidity – the moisture in the air expressed as a percentage of the saturation level at that given temperature
RNZAF – Royal New Zealand Air Force
Sorption – The process of absorbing moisture like a sponge or filter
Thermal Lag/ Thermal Inertia – The process of temperature change in a metal taking longer than that of its surroundings
WB – Wet Bulb Temperature

1. INTRODUCTION

Awareness by the corrosion industry of the recent desiccant dehumidification trends is required for consultants and contractors to make informed decisions on surface preparation system selection. Korea is a leader in ship production, and their desiccant application during painting is analyzed here regarding climatic design. Since the mothballing of the US fleet after the second World war, the shipping industry has adopted dehumidification for preservation. Similarly, since the 1960s, sorption dehumidifiers have been used all over the world to protect various materials, components and critical systems against corrosion. If relative humidity in the air is not controlled, mould and corrosion can develop resulting in failure of sensitive electronic equipment. A dehumidifier ensures the humidity is efficiently controlled in surface preparation, with refrigeration heating and desiccants used for surface coating we look at the selection and energy costs of these, and the climate they operate in.

In the ship painting and surface preparation industry, desiccant technology is effective in creating a year-round humidity controlled environment to avoid corrosion. Corrosion rates have been shown to accelerate exponentially when the relative humidity (RH) is over 50%, whereas coating compliance for many paint specifications is 80%RH on metal surfaces. The effect of thermal inertia usually requires the surrounding air to be 20-30%RH less than the required surface humidity to compensate for the difference in temperature (a 4-6°C temperature difference is common).

Other methods of humidity control require full surface heating which carries very high energy costs. Refrigeration is only effective in climates which are humid all year round, and it is still unable to achieve the dryness required due to the surface humidity requirement.

Some of the benefits with dehumidification include avoiding corrosion, compliance with coating specification, minimal downtime due to unforeseen weather conditions and the ability to schedule works year-round with consistent ambient conditions.

This is mirrored in the marine industry with benefits such as; being able to conduct works in or out of water, the ability to store engine and electronics in place, effective mothball of temporary layup and it is also useful in composite and coat compliance.

The benefits from surface preparation and coating the internals of vessels and storage tanks are; year-round humidity control, being able to consistently hold the blast, helps to create a dry air coating across the entire surface at one time, and faster job duration with a better chance of a quality lining outcome.

1.1 Research

In this paper, we look at research that relates humidity to corrosion first identified by Vernon in 1930. We review the design of a typical Korean shipyard dehumidifier and the energy aspects when operating in their varied climates. Using weather data from Korea, we compare energy aspects of the dehumidification technologies in climates around Australia.

1.1.1 Technology

Desiccants and refrigeration dehumidification are conventional technologies in dealing with humidity issues, and these are reviewed to identify suitability in the climates identified.

1.1.2 Application to platform types.

Dehumidification is applied in open systems (for high leakage due to exhaust). Other systems such as ship and oil rig mothballing apply a closed system as they are sealed systems, and dehumidification is not effective in exposed systems. These are demonstrated with local and international examples.

Ship construction and shipping operations can benefit from dehumidification at all stages of their lifetime, from construction and painting to maintenance, storage and docking. Many of the critical areas such as sub-systems, platform-systems, electronic and storage rooms, and cargo holds all require some form of protection from corrosion in a salt laden environment.

2. RESEARCH AND EXPERIMENTAL DETAILS

Corrosion typically requires four components to implement the corrosive circuit. Oxidisation occurs at the anode, which is energized in the circuit. A reduction in energy occurs at the cathode at an equal rate to which the anode corrodes. The electrolyte (typically water or water vapour) and air for complete the circuit. The elimination of any of these contributions will halt the development of corrosion. This paper is focused on using the reduction of moisture in the form of water vapour as a technique to obtain an affordable quality of air to reduce corrosion rates to a negligible level.

2.1 Research conducted for the preservation of metals and design considerations

Research into the preservation of metals has gone back to the time of the early Egyptians, that dry atmospheres will create extended preservation of metals and fabrics. TH Vernon experiments indicated the relationship that relative humidity has with the corrosion rates of metals. This is displayed in the figure below, as the rate of corrosion begins to rise exponentially as the relative humidity starts to exceed 40%.

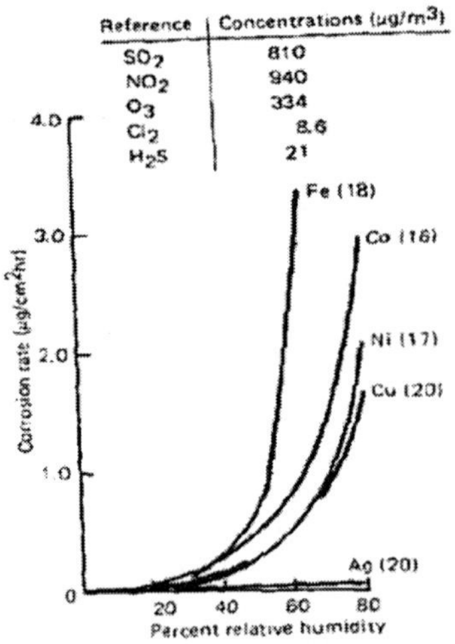

Figure 1: Corrosion rate of different metals at percent relative humidity (Rice, D. (1982) [10])

Ships that are still active can benefit with dehumidification with sub-systems, platform-systems, electronic and storage rooms, and cargo holds all being protected in a salt environment. Using dehumidification during maintenance periods, platform, weapon, and sensor systems can also keep equipment protected temporarily.

Similar to the Korean shipyard method, the US Navy has developed a shipboard dehumidifier for the reserve fleet. These were placed in logical locations, and then fed by a network of temporary ducts run throughout the ship. One dehumidifier was sufficient for a minesweeper or frigate, while six were needed for a large aircraft carrier. For dehumidified ships, maintaining an "external moisture boundary" was extremely important where a vertical hole the size of a half-dollar coin could introduce enough rainwater to overwhelm the dehumidifier.

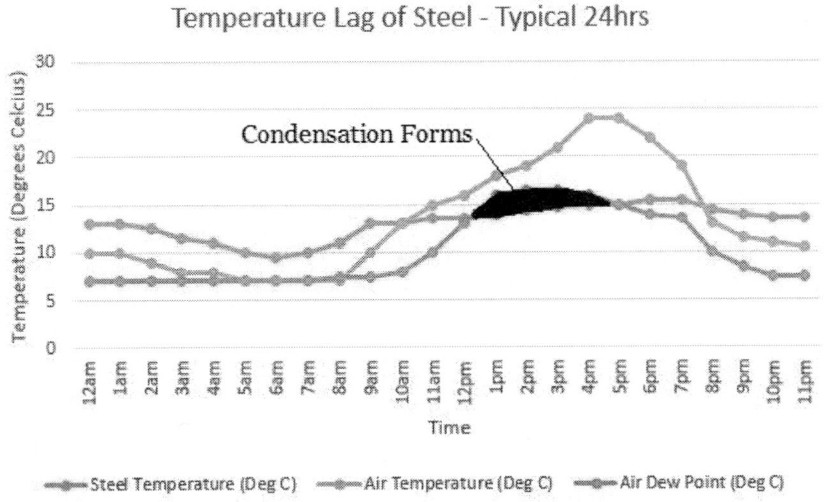

Figure 2: Thermal lag of a steel specimen over a 24-hour period

The thermal lag logs in figure 2 depicts the effect of thermal inertia in a heavy steel object where surface humidity differs from surrounding air humidity at differing temperatures. For surface preparation planning, the daily thermal swing should be considered when establishing a set point of both temperature and humidity.

The Journal of the Electrochemical Society contains a very succinct study of pure cobalt that demonstrates the effect of humidity as an isolated variable, as well as the effect of increasing humidity in conjunction with increasing levels of Sulphur Dioxide (Rice, D. 1982 [10]). After 100 hours, the weight gain due to corrosion was demonstrated to be proportional to the increase in humidity. At 30% RH, the weight gain was less than 10 nanograms per square centimetre, at 60% RH it was 40 NG and at 80% the volume was 120 NG. The interesting observation about air contaminants is that the corrosive effect of the contaminant is not related to temperature but rather relative humidity. This is demonstrated in the figure below.

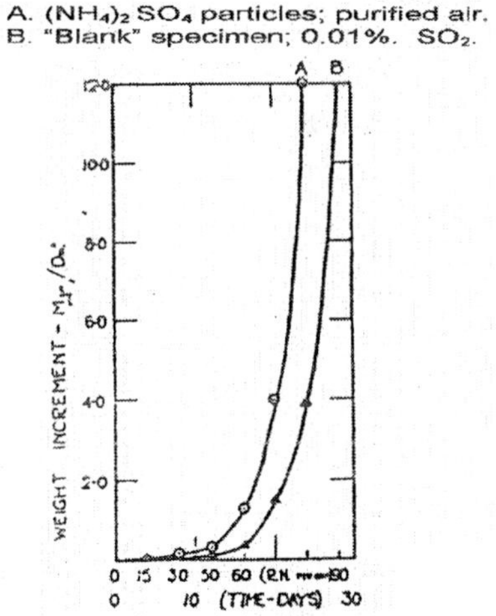

Figure 3: Weight increments of corrosion for a carbon sample over time (Rice, D. (1982) [10])

2.1.1 Technology – Desiccant wheel dehumidifiers (Dryers)

Desiccant dehumidifiers have been used to dry air for humidity control since the invention of the first desiccant dehumidifier (Pennington, N. 1951 [9]). This was later improved by Munters where they patented the idea of the desiccant rotor (Munters, C. 1968 [8]). More recently a silica gel technology has been developed for the use in rotors by Toshimi Kuma in 1984 as shown below.

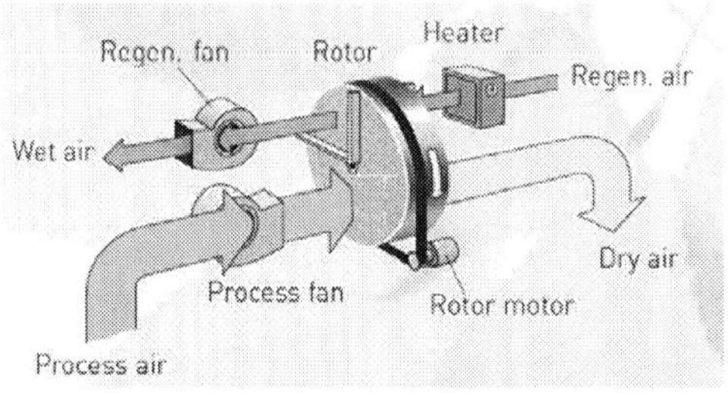

Figure 4: Desiccant wheel process (DST Seibu Giken (2017) [4])

As shown in figure 4, the desiccant wheel removes moisture from the air using a hygroscopic material, which attracts and holds water vapour, and is uniquely suited to removing moisture from air at low temperatures and humidity. Air passes through the desiccant impregnated flutes in the process air stream, and is dried as the wheel rotates slowly (6-9 revolutions per hour) between two air streams. The now humidity-laden wheel rotates slowly into a second, smaller heated air stream traditionally to 100°C above ambient. This exhaust reactivation air warms the desiccant wheel, allowing it to give off the moisture to the air which is discharged to atmosphere. Recent developments allow for low grade heat or solar hot water for regeneration with temperatures down to 40°C. Other developments use refrigeration condenser heat, but these are unsuitable for ambient temperatures below 18°C in winter.

2.1.2 Refrigeration

The first electric refrigerator was invented in 1902, where the refrigeration circuit provides an evaporator to cool, but more importantly for OA systems it is used to condense water by forcing the air to travel down the saturation curve on the psychrometric chart to an appropriate dew point. The limiting factor in this process is the fluid used in the evaporator (either chilled water or refrigerant). In an outside air system, a chilled water coil is more stable than refrigerant, where refrigerant can cause cycling due to the limited range of commercial compressors.

2.1.3 Heating

This is not considered in this review as heating is to be considered equivalent in all application comparisons, although refrigeration may have reverse cycle capabilities, and desiccants inherently heat the air. This is sometimes beneficial, but without pre-coolers it may provide excess heat.

2.2 Applications of platform types

2.2.1 Korean shipyard (open) systems

The methods of surface preparation and coating used in Korean shipyards have become the bench mark for large commercial ship construction, where sections are prefabricated and joined for final assembly. In the following section, we look at the desiccant dehumidification used in the design for Korean Ship building.

Figure 5: 3D model of the air handling unit (Sang, S. (2010) [11])

Typical unit supply airflow 15000m3/h (Ulsan climatic data)

Ambient 79.2°CWB 26.6 °CWb maximum design 91.6F (33.1°C)

HR 77.1°Fdp (25.0°5Cdp) 141.5grains/LB (20.2g/kg) 83.4°F(28.5°C) MCDB

Heating 20°F(-6.7°C)

Desiccant rotor SSCR 1520 (Seibu Giken 2017)

Nominal Design supply condition ambient less 10°C (16.6°C) 35%RH Used for calculation purposes.

Winter -6°C 35%RH

Reactivation heater 150kW

Reheat electrical 220kW

Typical pre-cooler of 200kWR with screw compressors and 150kW regeneration.

Figure 6: Weather controlled indoor blasting & painting facility, Hyundai, Ulsan, Korea (Sang, S. (2010) [11])

2.2.2 Modelling

In this section, we explore the three typical operating conditions, which are maximum moisture, maximum desiccant duty occurring at 13°C 100%RH, and winter minimum (100%RH).

These designs are used to simulate the times of peak load humidity in the space, as well as maximum power output needed during this time. The simulation allows us to customise the design to the specified conditions and improve the efficiency of the unit.

Summer design

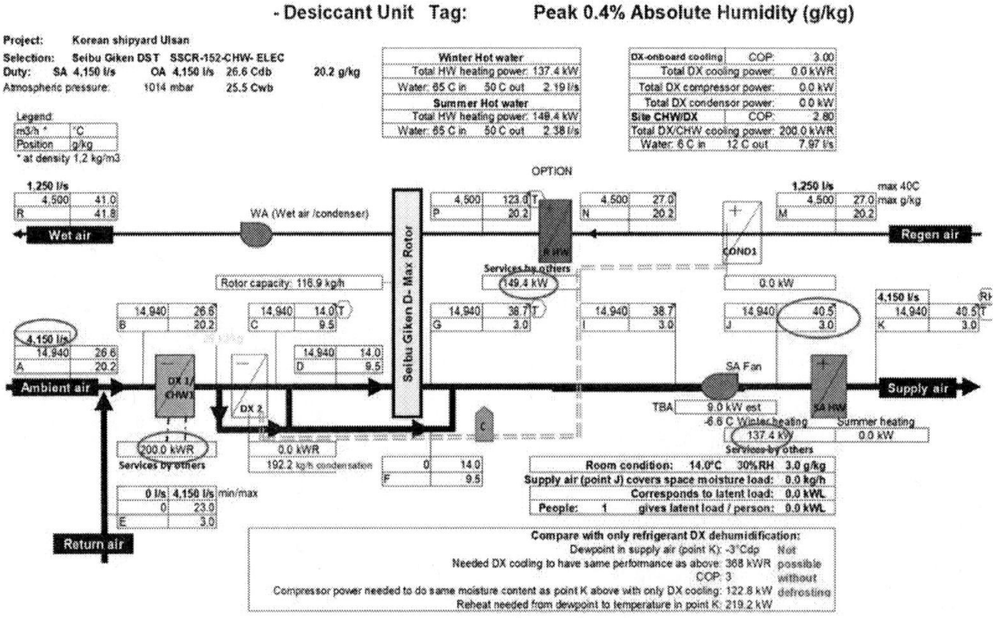

Figure 7: Summer design schematic of Desiccant Dehumidifier

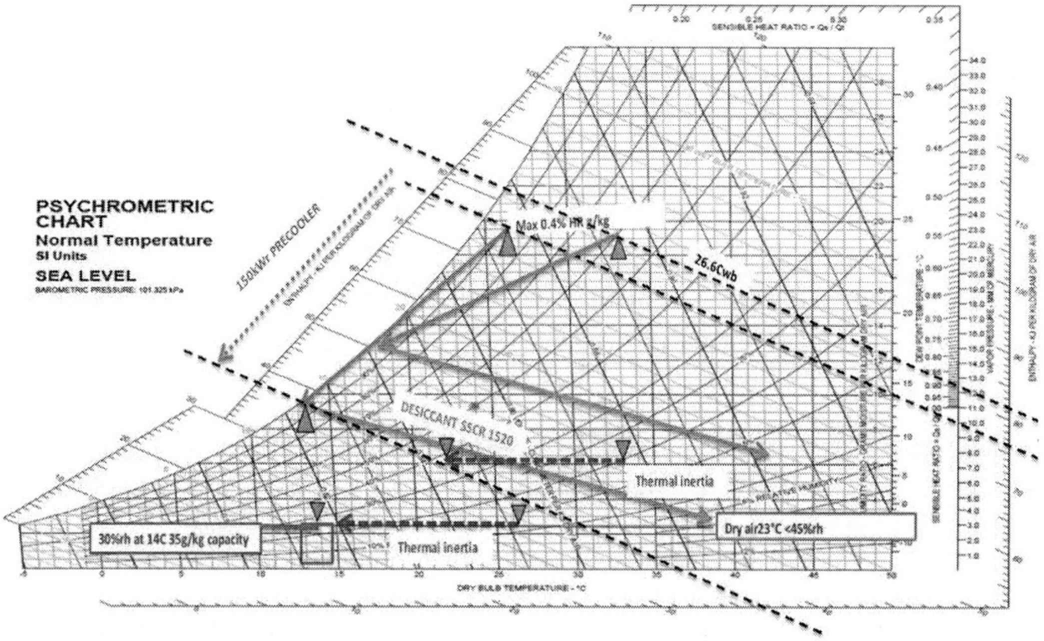

Figure 8: Summer design and energy savings of Desiccant Dehumidifier

Refrigeration systems for dehumidification (Figure 8 blue arrow) are limited to fluid temperatures, where 12°C is a common limitation before ice occurs and when compressors cycle. Desiccants (green line) are not limited to this temperature and are therefore effective in dehumidification below 12°C dew point (ie 24°C 50%RH). The dehumidification of an Ulsan (Korea) combined refrigeration desiccant unit is calculated in maximum 0.4% HR (humidity ratio) which considers humidiy and outdoor maximum temperature (wet bulb).

Maximum performance is demonstrated in the following system schematics, demonstrating peak and part load performance requirements of refrigeration and desiccant dehumidification. System selections rarely consider part load where refrigeration is not effective, and this is important , as shown in the schematics the heat load is still 137.4kW when the winter room desired conditions are higher than summer conditions. The latent load can go up to 7kWL, which needs to be dealt with before supplying supply air to the space.

Partload Max Desiccant duty -13°C 100%rh

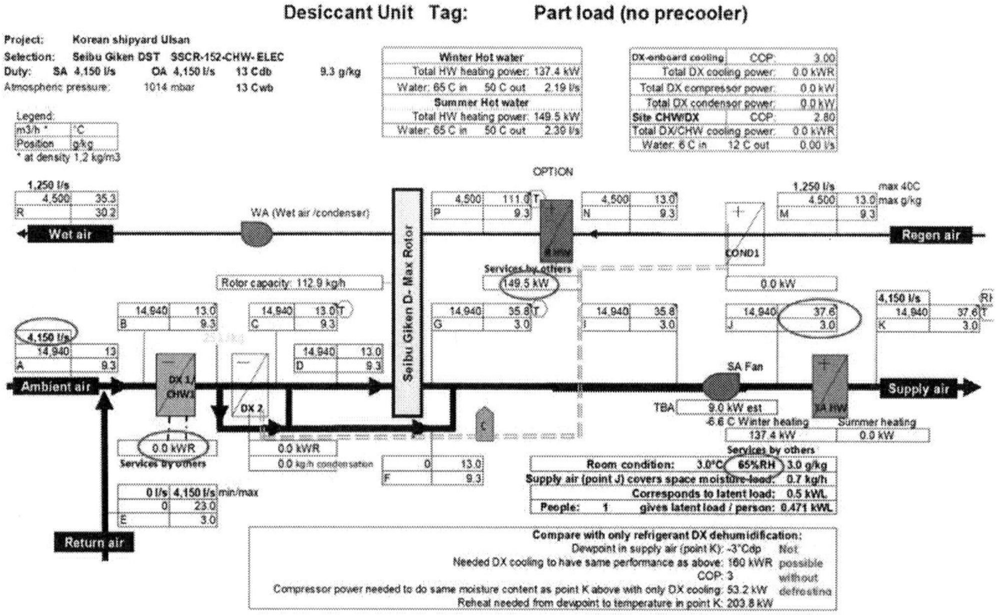

Figure 9: Part load design schematic of Desiccant Dehumidifier

Winter minimum -7°C 100%rh

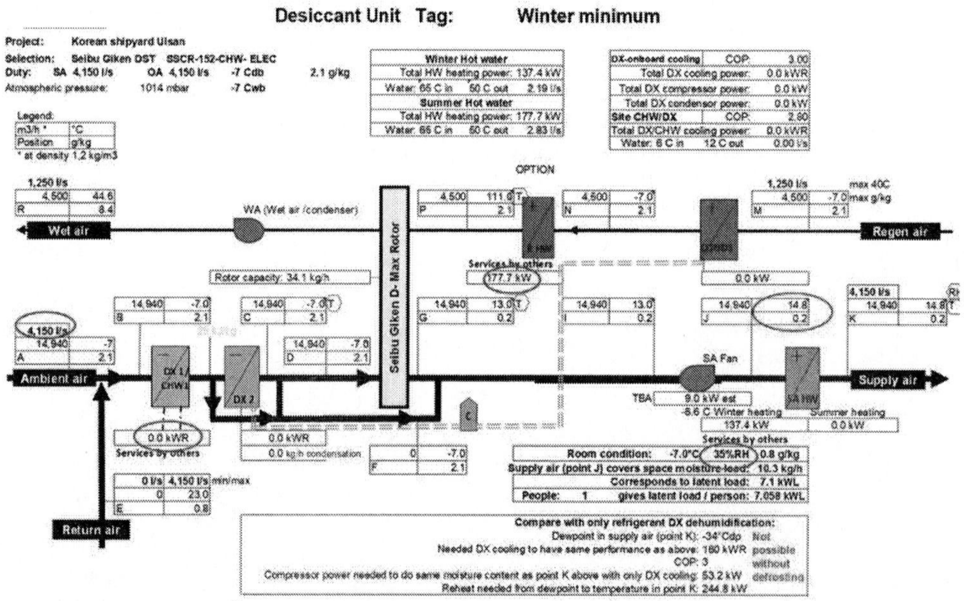

Figure 10: Winter design schematic of Desiccant Dehumidifier

2.2.3 Energy use modelling

Using binmaker global weather data for comparison of the Australian humidity ratio (g/kg) to the Ulsan shipyard climate in Korea, we can look at the yearly hours that different dehumidification technologies are the most effective. This can also model the annual refrigeraiton work and potential for Desiccant DX to save refrigeration energy using free condenser heat.

		Incheon	Darwin	Brisbane	Sydney	Melbourne	Adelaide	Technology
Calculations:-			Data insert, Binmaker					
	Totals >	8757	8760	8760	8760	8760	8760	
HR(g/kg)	HR(g/kg)	Hrs	Hrs	Hrs	Hrs	Hrs	Hrs	
Peak MC	27 to 28							Refrigeration
26.5	26 to 27							Refrigeration
25.5	25 to 26	2						Refrigeration
24.5	24 to 25	2	6					Refrigeration
23.5	23 to 24	6	31					Refrigeration
22.5	22 to 23	20	56					Refrigeration
21.5	21 to 22	37	378					Refrigeration
20.5	20 to 21	175	945	16				Refrigeration
19.5	19 to 20	254	1193	25	6			Refrigeration
18.5	18 to 19	223	944	27	45			Refrigeration
17.5	17 to 18	153	778	105	50	2		Refrigeration
16.5	16 to 17	214	733	202	105	2	4	Ref+ free Desicant
15.5	15 to 16	298	680	514	245	13	14	Ref+ free Desicant
14.5	14 to 15	401	608	695	491	35	30	Ref+ free Desicant
13.5	13 to 14	250	734	814	469	69	39	Ref+ free Desicant
12.5	12 to 13	292	393	1049	538	108	61	Ref+ free Desicant
11.5	11 to 12	419	228	687	752	180	132	Ref+ free Desicant
10.5	10 to 11	389	254	821	880	404	200	Ref+ free Desicant
9.5	9 to 10	444	130	947	745	368	547	Ref+ free Desicant
8.5	8 to 9	508	180	759	828	859	1018	Desiccant ONLY
7.5	7 to 8	513	130	592	786	1410	1846	Desiccant ONLY
6.5	6 to 7	612	133	545	720	1815	2344	Desiccant ONLY
5.5	5 to 6	676	138	441	1014	2169	1764	Desiccant ONLY
4.5	4 to 5	786	78	261	942	1184	652	Desiccant ONLY
3.5	3 to 4	705	8	180	141	127	98	Desiccant ONLY
2.5	2 to 3	904	2	75	3	5	10	Desiccant ONLY
1.5	1 to 3	474		5		10	1	Desiccant ONLY
0.5	0 to 1							Desiccant ONLY
Refrigeration		1384	5744	889	451	17	18	
Ref+ free Desicant		2195	2347	5013	3875	1161	1009	
Desiccant ONLY		5178	669	2858	4434	7579	7733	
Refrigeration		16%	66%	10%	5%	0%	0%	
Ref+ free Desicant		25%	27%	58%	45%	13%	12%	
Desiccant ONLY		60%	8%	33%	51%	88%	90%	

Figure 11: Moisture load hours by city (Binmaker, 2016 [2])

As we can see in the table above, Ulsan has high moisture content summers and low temperature winters, thus requiring refrigeration for peak moisture times and desiccant technology during cooler months. Figure 11 shows Darwin needing no cooling in winter but still has substancial hours (8% of year in dark green) where desiccant technology is required and there is scope for savings (light green hours 27% of year). Brisbane has a substancial period requiring desiccants (33%) and there is energy saving potential for 58% of year. Adelaide (88%) and Melbourne (90%) predominantly need desiccants with scope for the combined cooling desiccant to also provide cooling for hot days without saturated air. (Binmaker, 2016 [2])

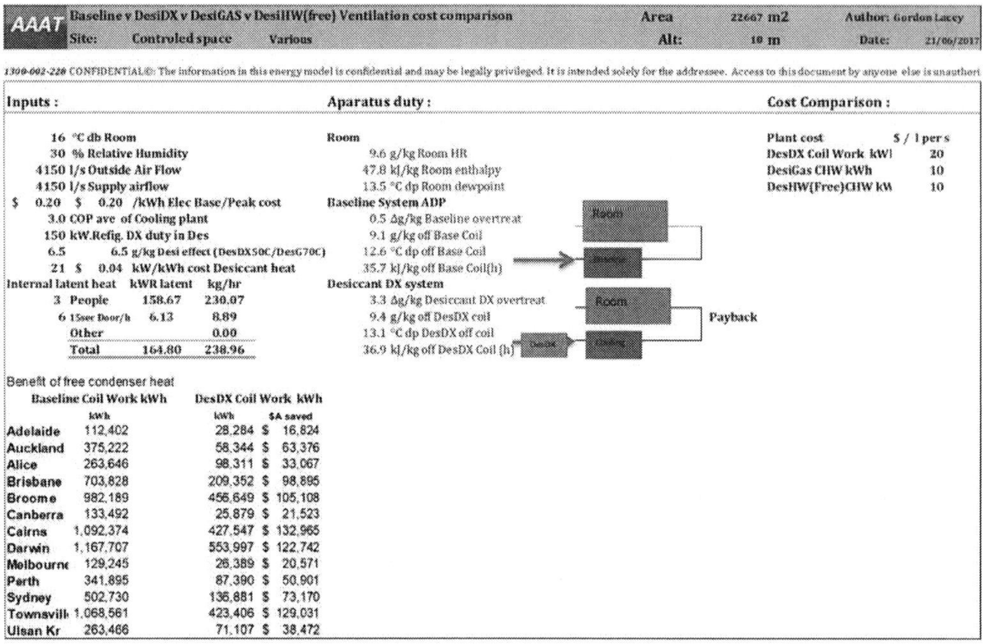
AAAT Baseline v DesiDX v DesiGAS v DesiHW(free) Ventilation cost comparison — Area 22667 m2 — Author: Gordon Lacey
Site: Controled space Various — Alt: 10 m — Date: 21/06/2017

1300-002-220 CONFIDENTIAL©: The information in this energy model is confidential and may be legally privileged. It is intended solely for the addressee. Access to this document by anyone else is unauthori

Inputs :

16 °C db Room
30 % Relative Humidity
4150 l/s Outside Air Flow
4150 l/s Supply airflow
$ 0.20 $ 0.20 /kWh Elec Base/Peak cost
3.0 COP ave of Cooling plant
150 kW.Refig. DX duty in Des
6.5 6.5 g/kg Desi effect (DesDX50C/DesG70C)
21 $ 0.04 kW/kWh cost Desiccant heat

Internal latent heat	kWR latent	kg/hr
3 People	158.67	230.07
6 15sec Door/h	6.13	8.89
Other		0.00
Total	164.80	238.96

Aparatus duty :

Room
9.6 g/kg Room HR
47.8 kJ/kg Room enthalpy
13.5 °C dp Room dewpoint
Baseline System ADP
0.5 Δg/kg Baseline overtreat
9.1 g/kg off Base Coil
12.6 °C dp off Base Coil
35.7 kJ/kg off Base Coil(h)
Desiccant DX system
3.3 Δg/kg Desiccant DX overtreat
9.4 g/kg off DesDX coil
13.1 °C dp DesDX off coil
36.9 kJ/kg off DesDX Coil (h)

Cost Comparison :

Plant cost	$ / l per s
DesDX Coil Work kWl	20
DesiGas CHW kWh	10
DesHW(Free)CHW kW	10

Benefit of free condenser heat

	Baseline Coil Work kWh	DesDX Coil Work kWh	
	kWh	kWh	$A saved
Adelaide	112,402	28,284	$ 16,824
Auckland	375,222	58,344	$ 63,376
Alice	263,646	98,311	$ 33,067
Brisbane	703,828	209,352	$ 98,895
Broome	982,189	456,649	$ 105,108
Canberra	133,492	25,879	$ 21,523
Cairns	1,092,374	427,547	$ 132,965
Darwin	1,167,707	553,997	$ 122,742
Melbourne	129,245	26,389	$ 20,571
Perth	341,895	87,390	$ 50,901
Sydney	502,730	136,881	$ 73,170
Townsvill	1,068,561	423,406	$ 129,031
Ulsan Kr	263,466	71,107	$ 38,472

Figure 12: Benefits of condenser heat by city, adapted from Binmaker software

By modelling the binmaker softwaredata from figure 11, it is possible to analyse the combined technology applied to various climate to predict fuel savings when technology and control methods are applied. Figure 12 shows seasonal bindata as a prediction of the fuel consumption in various months and the hours of operation. This analysis provides far more accuracy in the estimation of fuel consumption and viability of control techniques., but this is a trade off between reliability and energy cost of continous operation of the dehumidifier.

3. RESULTS – APPLICATION TO PLATFORM TYPES

From this research, we can look at some field installations, and review them to demonstrate practical applications for each platform type.

3.1 Open system (full fresh air) installations.

Open systems installations are applied where system readiness, ease of access or excessive leakage can be expected. For a system of full fresh air, using accurate maximum design weather data from sources such as the ASHRAE Handbook is important, mainly using 0.4% Dehumidification DP/MCDB and HR (Sydney 21°Cdp 16.1g/kg 25.2°C MCDB exceeded 55hrs in a year) maximum design humidity ratio. If refrigeration pre-coolers are used, the 0.4% Evaporation WB/MCDB, and Sydney 23°CWB 36.3°CDB is appropriate so is still under designed for critical installations, as Sydney airport occasionally exceeds 40°C in summer. Consideration should also be taken for the thermal inertia of metals, which is usually an estimate based on the daily temperature range to ensure the surface of the preserved item is under the nominated relative humidity (nominal 35% for electronics and 50% for metal items). Some paint specifications nominate conditions such as 80%RH or a dew point differential to avoid condensation in the paint, which must be considered when designing open systems that will allow fresh air into the space.

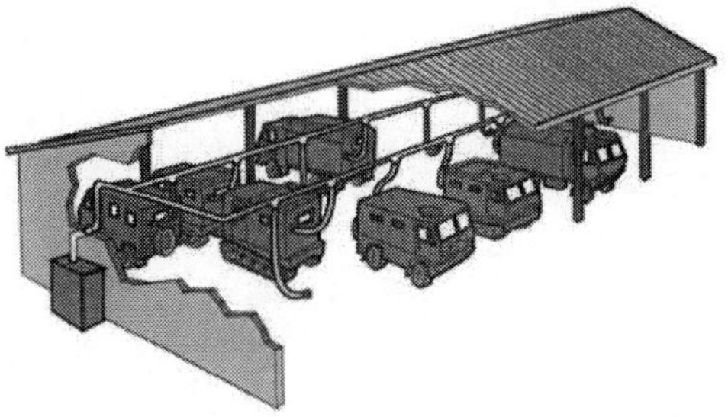

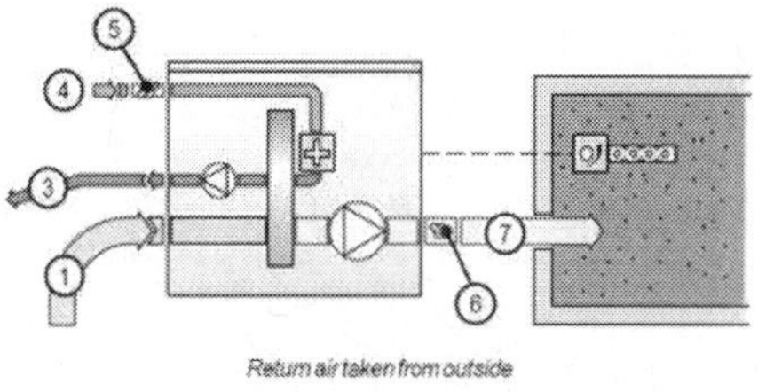

Figure 13: FMV an example of an open system (FMV 2007 [3]) Figure 14: General duct layout of open systems (DST 2017 [4])

3.2 Closed systems

Closed systems employ recirculation of air in a tight enclosure which is a capital and energy efficient method of storage, when items are stored or able to have a supply (dry air) and return (process air) connection.

Calculation of infiltration of the structure and door openings as well as occupant ventilation (10L/s per person AS1668 code) needs to be considered. Calculation methods such as Moisture Load calculation (DST, 2017 [4]) or desiccant system vendor will assist this design process.

Figure 15 shows an example of a closed system application, and figure 16 shows the general duct layout of a closed system, where the process airstream (1) and optional outside air (8) are being dehumidified, which is then supplying dry air (7) to space, Reactivation (3) of the desiccant wheel removes moisture via wet air (4).

Advanced control using surface temperature probes and measurement of surrounding dewpoint allow for compliance confirmation to humidity specification, and provides the facility to control the desiccant and refrigeration to save energy consumption. The capability of advanced control techniques is demonstrated on the psychrometric chart in figure 17.

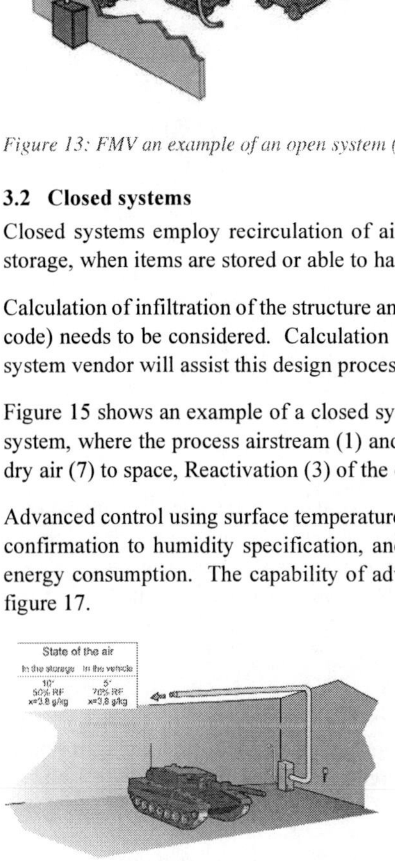

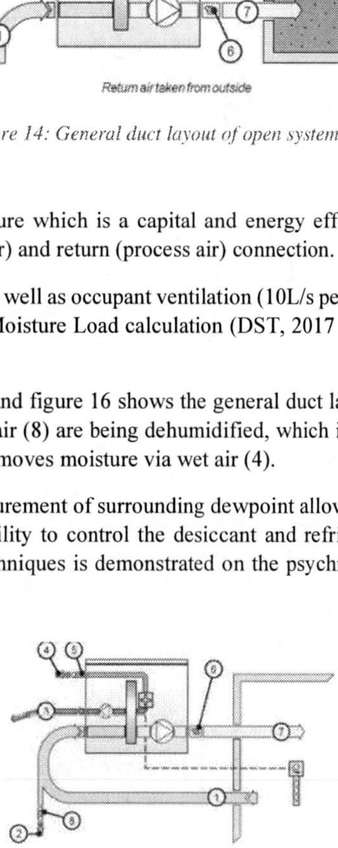

Figure 15: Example of Closed System (FMV 2007 [3]) *Figure 16: General duct layout closed system (DST 2017 [4])*

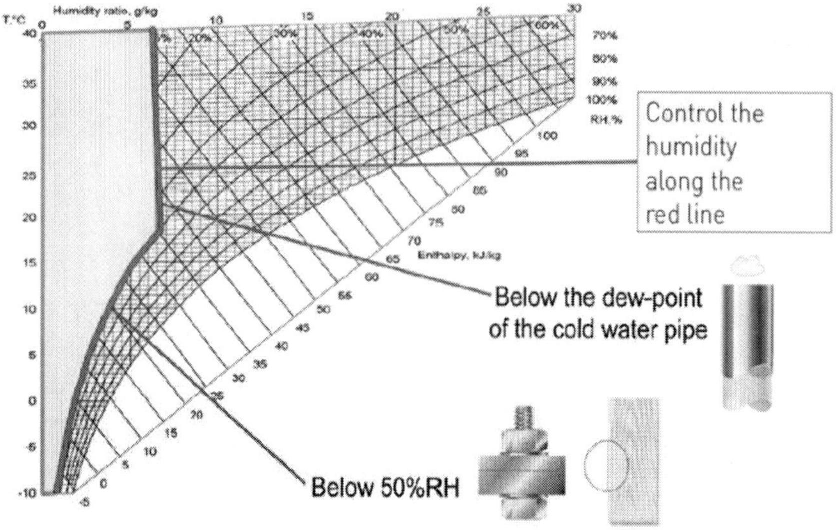

Figure 17: IX (European) diagram, psychometric control for PLC with surface probe (DST 2017 [4])

4. DISCUSSION

By comparing the climate data of different locations around the world, we demonstrate selection suitability of dehumidification systems to save a substantial amount in energy consumption, the cost of rental equipment and prevent failed coating application. The use of desiccants is more effective in cooler climates, and are more energy efficient at moderate climates. The use of desiccant cooling systems is now available to cover variable climates like the data demonstrated in Korea and Brisbane, and this is very useful in the design of desiccant dehumidifiers in the shipyard, maintenance and construction process.

5. ACKNOWLEDGMENTS

Author would like to thank DST and FMV for support and the modelling research conducted into metal corrosion at varying humidity levels. DST has provided many scenarios of desiccant drying, and the benefits to using dry air as a preventative measure for reducing corrosion.

6. REFERENCES

[1] ASHRAE (2005). *ASHRAE Handbook.* 2nd edn, Adelaide: ASHRAE. Pages 2.33-2.53

[2] Binmaker. (2016). New Hampshire: Linric corporation

[3] FMV (2007). *Handbook for the production of equipment maintenance schedules*, Sweden: Vårdsystem FM Swedish Defence.

[4] DST Seibu Giken (2017). *Desiccant Rotors, Sorption Principles and Energy Saving*. Journal [online] Available at: http://www.dst-sg.com/ [Accessed: 24th May 2017].

[5] JWH (2016). *Mothballing the US Navy after WWII*, Journal [Online] Available from: http://www.tscforum.org/TSCF/bfiles/2010/Paper%205.1%20PSPC-%20Preparation%20and%20Application%20in%20Korean%20Shipyard(0).pdf. [Accessed: 1st June 2017].

[6] LtCol Schweitz, H. (1986). *Instruction for maintenance during long term storage*. Sweden: FMV.

[7] LtCol Swietz, K., Lacey, G. & Mojet, G. (2008). *Logistics technology – The Dry Air Method use for storage, forward positioning and deployment*, ADF Land Warfare conference. Volume 1

[8] Munters, C. (1968). *Inorganic, Fibrous, Gas Conditioning Packing for Heat and Moisture Transfer*. U.S. Patent. 3377225

[9] Pennington, N.A. (1951). *Heat-Exchanger for AirConditioning Unit*, U.S. Patent 2,563,415, USA.

[10] Rice, D., Cappell, R.J., Phipps, P.B.P. & Peterson, P. (eds) (1982). *Indoor Atmospheric Corrosion of Copper Silver, Nickle, Cobalt and Iron.* Atmospheric Corrosion, Volume 1. Chapter 46.

[11] Sang, S. (2010). *Preparation and Application in Korean shipyard, TSCF 2010 Shipbuilders Meeting,* [Online], volume 1. Available at: http://www.tscforum.org/TSCF/bfiles/2010/Paper%205.1%20PSPC-%20Preparation%20and%20Application%20in%20Korean%20Shipyard(0).pdf [Accessed: 1st June 2017]

[12] Vernon, W.H.J. (1927). *First and second report on atmospheric corrosion research committee*, Transactions of the Faraday Society. Volume 23.

7. AUTHOR DETAILS

G. Lacey is the Chief Engineer at DST Dryer Pty Ltd Australia, and has had vast exposure to desiccant dehumidifiers with over 30 years in the industry. He is responsible for the design, selection and implementation of Desiccant Dehumidifiers, and holistic drying solutions for the company.

C. Hipsley is a Mechanical Engineer at Dry Dryer Pty Ltd Australia, and has had 2 years' experience in the design and implementation of Desiccant Dehumidifiers in industries such as Shipyards, Power stations, Military and Surface Preparation and Coating.

HIGH-CURRENT PIPELINE CATHODIC PROTECTION DESIGN - ACHIEVING 0.35OHM LOOP RESISTANCE ONSHORE

N. Doblo
APA Group, Brisbane

SUMMARY: This paper is a case study, from an asset owners' perspective, into achieving cathodic protection on some of Australia's older and high current demand onshore pipelines. Over the many decades of service these pipelines have seen large increases in the current density required to provide cathodic protection.

To meet these challenges in providing cathodic protection, special consideration must be given to a range of design factors in particular, loop resistance, the quantity of carbon backfill used in anode beds, and anode to pipe separation distances. It has been shown by practical experience that the use of 1135kg (2500lb) of carbon backfill per anode provides good resistivity for longer term output and minimizes current requirements of the power source. Loop resistance of 0.35 ohm and lower can be regularly achieved with such a design.

Similarly, separation between the anode beds and the structure equivalent to a calculated potential rise of 0.2V on the primary structure (as per AS2832 requirements for a foreign structure) provides better current distribution along the pipeline, allowing fewer CP units at higher output levels.

Keywords: Cathodic Protection, Pipeline, Low resistance, Carbon Backfill, Anode bed

1. INTRODUCTION

As pipelines age, depending on their coating type and various other factors their demand for CP current can increase significantly. This paper reviews some of the design methodologies on a number of Australia's pipelines which have shown significant current demand since commissioning.

Maintaining good levels of cathodic protection for such a system requires some non-traditional approaches and approaches which are contrary to traditional design principles. This paper will examine some of these design variations and provide alternatives currently in use.

2. HISTORY & SYSTEM OVERVIEW

2.1 Pipeline

The pipeline landscape in Australia really kicked off in the late 1960s, with many pipelines from this era still being in service today. A review of their CP systems will show some significant differences in their current requirements. These differences appear to be primarily related to the soil type and weather conditions experienced by the pipelines as well as the coating type. In the early days of the Australian pipeline industry, it was common for line pipe to be transported by sea, without coating, as deck cargo.

The pipe was than surface prepared, often by basic mechanical means without abrasive blasting, and coated over the trench prior to lowering and burial. Common coating types at the time were polyethylene based tape wrap and coal tar enamel coatings. Time and in particular wet/dry cycling of clay soil types has led to the degradation of these pipeline coatings and the commensurate increase in required current density.

Many of these older lines around the country have been subsequently duplicated or expanded for additional capacity, along with having common maintenance practises such as coating refurbishment and relocations. As such CP system design for these pipelines requires considering not just one type of pipeline coating but most of the list below.

- Numerous tape wrap systems from the various suppliers including primer, inner and outer wrap solutions.
- Coal Tar Enamel (CTE)
- Extruded PE (yellow jacket)
- UHB liquid-applied epoxy – Factory and field applied
- Fusion Bonded Epoxy
- Heat shrink sleeves
- 3 Layer PE / Trilaminate

2.2 Existing System overview

The available electronic data for three pipelines' CP current demand has been collated and mapped to illustrate the significant growth in current required and with a 20-year projection of likely future demand as displayed in Figure 1

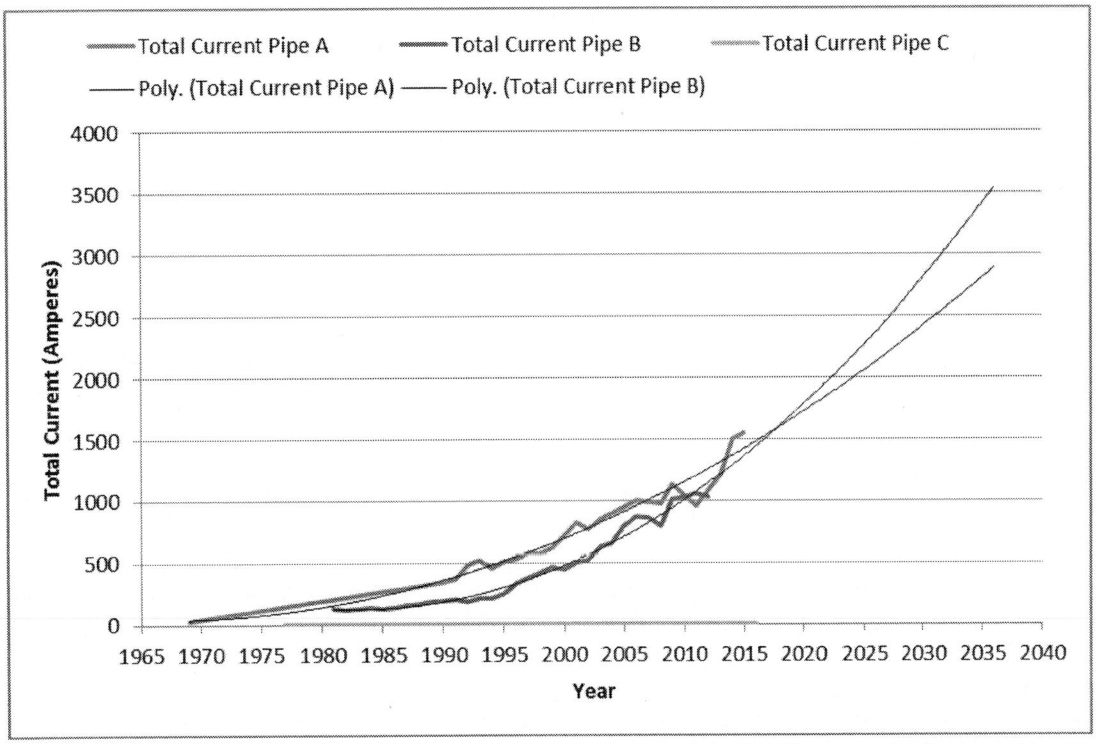

Figure 1: System Current Demand

Figure 1 graphically illustrates the total current demand of three of Australia's older pipelines. Of notable interest is the two pipelines with high current demand having similar increasing demand curves both modelled with exponential fit curves, while the 3rd is almost invisible on the bottom axis. Pipe C (shown for comparative purposes) has two critical differences, the first is that it was factory coated rather the field coated. The second is it doesn't see the same wet/dry cycling in clay soils as Pipe A and B.

Where demand has increased, and continue to increase significantly, changes to the design methodologies of the CP systems are required to ensure efficient distribution of current and adequate protection levels are achieved.

3. DESIGN CONSIDERATION FOR HIGH CURRENT DESIGN

3.1 Standard anode bed design philosophy breaks down around 25 Amps.

The high current density requirement pipelines noted above have both featured fairly typical anode bed designs over their life and had good performance from them. One such design, used commonly to achieve 25Amp in farmed black soil with low resistivity consisted of;

- 8 off impressed current anodes made from a high silicone content iron alloy.

- 8 off 44 Gallon of drum of metallurgical coke breeze, Metallurgical breeze is a by-product from steel coke mills and is much coarser than the calcined petroleum coke which is produced primarily for use as anodes in the aluminium smelting process.

- a watering / vent system and of around 200m from pipe to bed.

A bed of this design recently failed after 15 years in service. The current output at the time of failure was 50 Amps and it had been running over 25 Amps for at least 10 years of service with the help of regular watering.

Traditional carbon utilisation convention on the amount of carbon backfill required per anode appear not to hold at significantly higher anode bed output currents. The traditional designs of 6 to 8 x 22.7kg (50lb) bags per anode, or even 1x 44 Gallon drum of carbon backfill per anode proves to be insufficient. Significantly increasing the amount of carbon backfill to around 1100kg or '1 pallet' per anode has shown to help reduce loop resistance and provide sufficient carbon to retain a sustainably high output current. This is believed to be due to a significantly larger volume of carbon holding more water to allow better ion diffusion

3.2 Balance the cost of low resistance beds against power consumption

At high current demands, the electrical power involved in operating each system can become significant. When power is needed to be supplied by batteries and solar panels in particular, design optimisation should consider whether lowering the loop resistance below the minimum design constraints will provide a net benefit. If anode bed resistance can be halved, so is the total site power consumption halved. The savings of halving the power consumption can be significant on solar sites exposed to harsh desert environments where battery storage capacity is heavily affected by the storage temperature.

For grid connected sites, the cost savings are generally less significant. Savings in these situations come from operating the equipment at lower voltages and heat outputs along with the power supply cost savings. However, as CP units approach 6.5kw loadings on single phase (approx. 30 Amps) even grid connect power costs can become significant.

3.3 Auto Potential CP units aren't just for well coated / stray current affected pipelines.

The original CP units, as available at the time were all constant voltage. Mostly this control was via variac control which has proved a reliable and useful control type. While very adjustable they still suffered from the same constraint as all voltage controlled CP units. That is, they are unable to adjust for changes in loop resistance caused by wet / dry cycles.

The next step up is constant current output CP units. These are a significant improvement over a constant voltage unit, being able to account for anode beds drying out, flooding, etc. It can be argued that these units should be sufficient for pipelines without stray current or telluric interference.

However, the pipeline potentials at a location are often dependant on 2 to 3 CP units in either direction. With increased number of CP units providing influence on potentials, the likelihood of one of those units not working appropriately increases significantly. The use of auto potential / constant potential CP units allows nearby units to attempt to correct for the failure of a nearby unit.

Constant Potential CP units once used to cost a premium for the functionality and resulted in additional installation costs for the permanent reference cells. Most manufacturers now supply constant potential at no additional cost and most remote pipeline in the industry have fulltime monitoring making use of the reference cell. As such there is no longer any additional cost of implementing constant potential units.

3.4 Isolated parallel pipelines don't work at very high current densities.

Running a well-coated pipeline in parallel to such a high current demand pipeline and trying to keep it electrically isolated has not worked well in practice.

The traditional thought is, understandably, to avoid cross bonding the nice new, well coated pipeline to an older, higher current density pipeline. By cross bonding you expose the new pipeline to higher voltages and increased risk of under protection if something happens to an old-line CP unit. However, in these high-current pipelines, that design philosophy was not appropriate for a number of reasons.

- Parallel / Looped pipelines often cross at a number of points providing current discharge points where interference can occur
- The new pipeline can frequently be located between the anode bed and the older pipeline again leading to current pickup and discharge, i.e. CP interference on the new pipeline

- The new, well coated pipeline can be an unrealised resource, by acting as a well coated current distribution conductor. Frequent cross bonding can allow the CP units to operate at higher current outputs by distributing the CP current around the attenuation causing coating defects on the old line.

3.5 Anode bed access can be hard to get

Private negotiated access for anode beds, or the 'few cartons of beer method' worked well in the past where development was low and public perception of the gas industry was better. Changes in legislation around easements, titles, cultural heritage and environmental protection along with public perception changes of the gas industry and pipelines in general has resulted in land access for anode beds, even like-for-like replacement of existing beds, to be challenging or impossible. This factor becomes critical on ageing pipelines, as the distance between anode beds reduces while the physical size of the bed increases. Often, in broad acre cropping and grazing, moving the bed away from a difficult land holder is not an option and a win/win solution must be found.

3.6 Low midpoints can be due insufficient anode bed separation rather than attenuation

Figure 2 below shows the CP 'Off' potentials from three different annual surveys of a pipeline section. A common first assumption on reviewing this data would be that the spacing between CP units is too wide for the current density required, due to attenuation, and that additional infill CP will need around the 68/89, 72, 76, etc. While this would be an effective solution, the root cause is actually that the anode beds are too close. This could result from using a standard offset distance unsuitable for the local soil conditions, or the bed being designed for a lower operating output than designed.

This can be seen when viewing the survey results around locations 81 and 83, particularly comparing the 2015 and 2016 results. In 2016, prior to the survey both anode beds were replaced. Both beds were 40 Amp designs, with identical components, achieving 0.33 and 0.34 ohms loop resistance respectively. In both cases output was increased from 12-13 Amps in 2015 to 24 Amps (CP Unit maximum) for the 2016 survey.

The bed at 81 was replaced in an existing easement with 230m separation. The bed at MP 83 was installed in a new easement, with 600m separation to the pipe. This was the required to achieve a 200mV shift at the primary structure at 40 amp output.

The potential curve from 2016 around location 83 is much broader with a lower relative peak voltage. This allowed a higher current injection without causing potentials at the CP unit to be excessively negative, and also improved mid-point potentials without the need for additional costly infill sites.

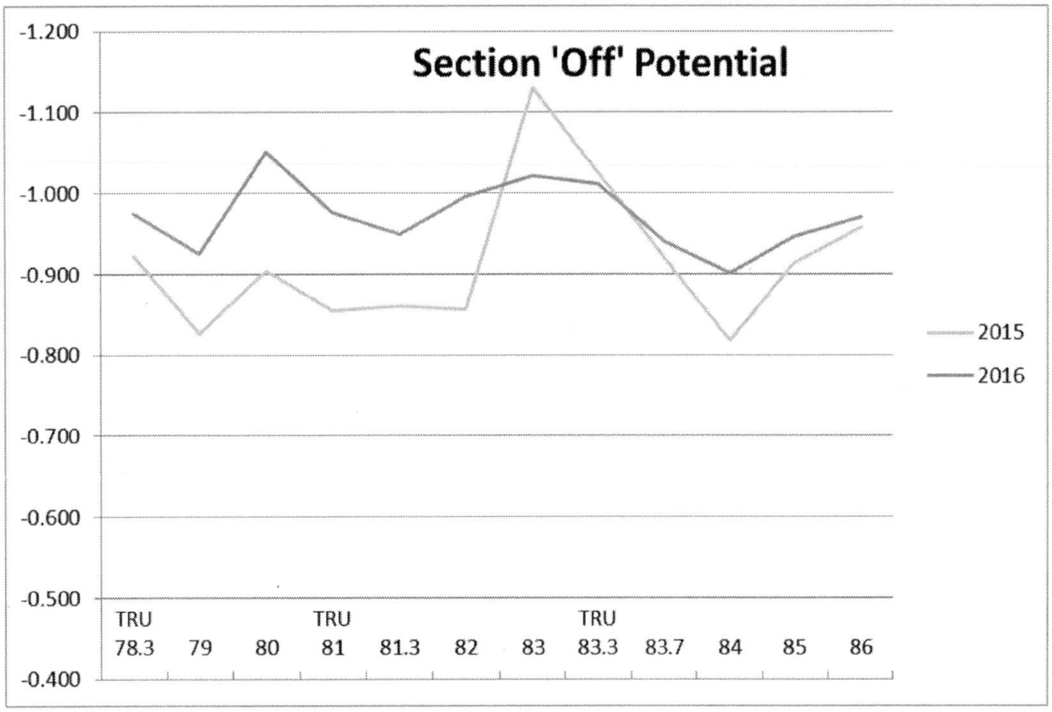

Figure 2: Off Potentials on a pipeline section 2015 & 2016

4. DESIGNING FOR THE FUTURE

Following extremely heavy rainfall in 2011 and subsequent current demand increases on pipeline A, the 40-amp CP unit and anode beds suddenly quickly changed from being the nominated upgrade path to operating near full capacity. A higher capacity design was required, but to what was unclear, 40 Amps was already considered large for a pipeline CP unit. This lead to two important questions, (1) What is possible?; and (2) What does the pipeline need?

The review from that questioned produced both Figure 1: System Current Demand and a current density calculation on the highest current density part of the pipeline. The average current density in the selected region was $23mA/m^2$ of pipeline surface area, notably above the widely accepted value of 20 mA/m^2 for bare steel in soil, despite there still being coating on the pipeline. This showed an answer to maximum projected current requirement wasn't available from theoretical current density requirements.

Reviewing the total current demand, a fit curve was applied with a forward projection of 20 years, the anticipated design life of a CP system on this pipeline. This curve can be seen in Figure 1 which predicts a total current demand approaching 3000 Amps, approximately double the existing output. This estimation is purely a fit curve based on historical data and may require frequent reanalysis. It is expected that at some point a peak current density should be reached resulting in a flattening of the curve.

Doubling of the total output current then indicates that a doubling of the individual CP unit and anode bed design currents would be required, if feasible from a current distribution perspective. At the time of this review a number of 80 Amp TR units were already on trial at locations where 40 Amp units had been running at capacity. This showed 80 Amps per site, where high current demand existed, to be feasible pending the correct anode bed. Due to legislative requirements, a driving voltage of 50VDC could not be exceeded necessitating a design loop resistance of 0.625 Ohm or better.

Field commissioning results from lower output beds indicated that loop resistance below this were comfortably achievable and hence an 80Amp design and even 120Amps was feasible from a current output perspective.

5. DESIGNING FOR 0.35-OHM LOOP RESISTANCE.

The following designs, as described in Table 1, evolved through the mix of theoretical design calculation and field trials on pipelines A and B.

Pipeline A arrived at the design by way of an trial of using higher carbon backfill volumes on a recently replaced anode bed. This bed had been installed with much more coke per anode, specifically 1 shipping pallet (2500lb or 1134kg) of calcined petroleum coke per anode. The bed performed very well and required significantly less maintenance watering than the previous designs.

Using this information, some scalable designs were developed and tested at 25 and 40 Amp anodes beds due for upgrade. In each case the bed was designed and installed for one level higher than the system was currently operating. That is a 40Amp bed was installed where a 25 Amp system was currently running and an 80 Amp bed where a 40 Amp unit was in place. The specifics are noted below in Table 1. The implementations have proved successful against design in all instances tried thus far, though no 120 Amp bed has yet been installed or tested.

Of the two pipelines. Pipeline A has the higher current density requirements, significant lengths of the line in lower resistivity, high soil stress black soil coupled with a lower mechanical strength coating.

Table 1: Anode Bed installation detail for Pipeline A

Current Rating	Pipeline A - 40A	Pipeline B – 40A	Pipeline A - 80A	Theoretical 120A*
Nominal Anode Type (pending design calc)	5 Amp x 20 years MMO, 25mm x 1500mm, end connect. 6mm2 tail	Silicon Iron Anodes – 50mm x 1500mm, end connect with 16mm2 tail.	5 Amp x 20 years MMO, 25mm x 1500mm, end connect. 6 & 16mm2 tail	5 Amp x 20 years MMO, 25mm x 1500mm, end connect. 6 & 16mm2 tail
Junction Boxes	1	1	1	3
Number of Anodes	10	24	20	30
Carbon Backfill	10 x Pallets (11.3t)	19 x Pallets 21T	20 x Pallets (22.6t)	30 x Pallets (33.9t)
Trench Length	35m Nominal, 30m Min	110m Nominal	70m Min	120m Min
Trench Depth	3 Meters – horizontally laid			
Trench width	500mm Nominal			
Cable size	35mm2 PE/N/PE unless a long run	35mm2 PE/N/PE unless a long run	70mm PE/N/PE or higher pending Voltage drop	70mm PE/N/PE or twin 70m2 pending V drop
In use	Yes	Yes	Yes	No
Pipeline separation	As calculated by soil resistivity with 0.2V allowable gradient at the primary structure. 80 and 120A locations are often 700-1000m separation.			
Venting & watering	Made to order system using 100mm Dia slotted and sheathed AGI pipe with a 150mm macro slotted poly water pipe. 150mm Dia risers protected at ground level by gal steel riser.			
*120 Amp anode bed design is theoretical only at stage and may have regulatory compliance issues. QLD currently limits output to 100Amps on shore in a 1000m2 surface area, measured radially, from the centre of the bed.				

6. FUTURE OPPORTUNITIES

Traditionally shallow depth, horizontally installed anode beds have been very cost-effective solutions, however at these current densities there are some potential limitations. These are

- Significant separation required to the primary structure
- Significant physical space required for the trench
- Not suitable in congested areas

In some cases this design of anode bed is unlikely to be applicable.

Alternative solutions are being considered for these cases, primarily replaceable deep well anode beds and linear anodes. It is intended to run trials of both of these options in the future.

Preliminary indications suggest a linear anode system will be relatively expensive for the length of pipeline protected, however in heavily congested areas requiring relatively high current densities, there aren't many technical solutions available and this may be the best option.

A deep well design will potentially make land access easier by facilitating a much larger output anode bed in an existing easement while keeping voltage gradient reasonable. Access to a lower soil resistivity and higher moisture may allow a reduction in the volume of carbon and length of large diameter copper cable being used. If so, these savings may offset some of the higher installation costs typically incurred with this method.

7. CONCLUSION

This paper has reviewed the CP systems on some of the oldest operating gas pipelines in Australia which have most coating types in place. Over their operational life two pipelines have seen huge increases in the current density required to provide cathodic protection.

To meet these challenges in providing cathodic protection special consideration must be given to a range of design factors in particular loop resistance, the quantity of carbon backfill and anode to pipe separation distances. It has been shown that horizontal installation with 1100kg of carbon backfill per anode, and a design of 2 to 4 amps per anode, provides good resistivity for longer term output and minimizes current requirements of the power source. Loop resistance of 0.33 ohm and lower can be regularly achieved with such a design.

Similarly, separation between the anode beds and the structure equivalent to calculated potential rise of 0.2V (as per AS2832 for a foreign structure) has been shown to provide better current distribution along the pipeline allowing fewer CP units at higher output levels.

The above designs have been shown to work well in broad acre areas but deep well and linear anodes methodologies are being investigated as technical and economic solutions in more difficult areas.

8. AUTHOR DETAILS

Nick Doblo is a Corrosion Engineer based in Brisbane for APAs East Coast Grid Integrity team, a position he has held for four years. He is responsible for the engineering management of a number of pipeline CP systems in the East Coast Grid including survey, maintenance and capital upgrade work.

CATHODIC PROTECTION AT AUSTRALIA'S LARGEST LPG STORAGE FACILITY

D. Sunjaya
Corrosion Control Engineering, Sydney

SUMMARY: Cathodic protection of buried and immersed steel structures in the oil and gas industry is essential to prevent corrosion that can lead to catastrophic failures and potentially affect human life, the environment, corporate reputation, expensive repair costs and lost production time.

AS NZS 3778: Pressure Equipment – In service inspection and AS NZS 1596: The storage and handling of LP Gas stipulate mandatory cathodic protection be installed and maintained to prevent corrosion on the external buried and/or immersed surfaces of any vessel containing LPG. This paper covers the numerous cathodic protection systems installed on various structures at Australia's largest LPG storage facility, The Elgas Cavern in Sydney.

Keywords: Impressed Current Cathodic Protection, Sacrificial Anodes, Oil and Gas, Coupons.

1. INTRODUCTION:

The Elgas Cavern in Sydney, Port Botany is Australia's largest Liquid Petroleum Gas (LPG) (Propane) storage facility in Australia with a capacity of 65,000 tonne. The Cavern took 4 years to construct and was commissioned in the year 2000. Upon opening, it increased Australia's LPG reserve from 3 days to over 60 days. It is the first cavern in Australia and was built using 40 year old technology that was already used in 100 locations worldwide. It has a design life of 50 plus years.

The Cavern consists of four main galleries, 230m long, 11m high and 14m wide (See Figure 1). The cavern is cut out from solid Hawkesbury sandstone and is built 130 metres below Port Botany (See Figure 2). The LPG is contained under the pressure of the groundwater which is sufficient to keep LPG in liquid state at ambient rock temperature of 15°C. The construction was an engineering challenge as it was built in rock below sea level. The maintenance of the cavern is another challenge as there has been no access to the bottom of the shaft since its commissioning.

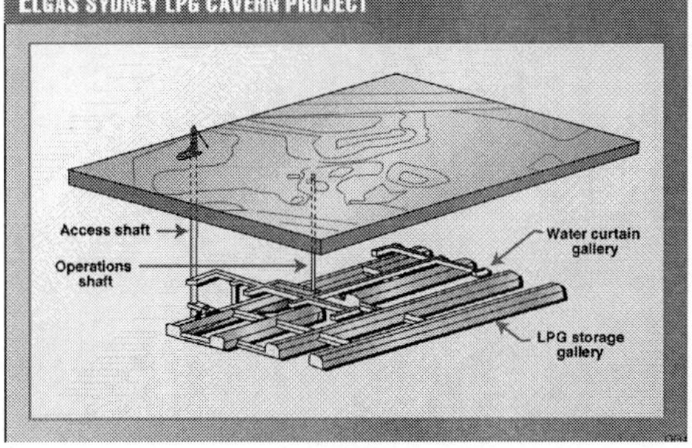

Figure 1: Diagram showing cavern gallery layout

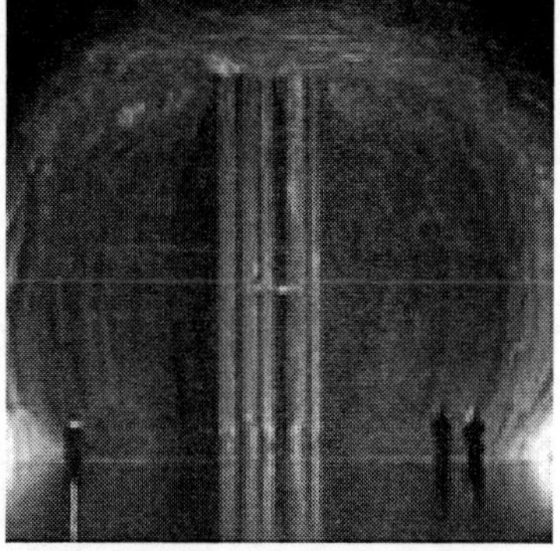

Figure 2: Photograph inside the cavern during construction

At the Elgas Port Botany Terminal, LPG is delivered via ship to the adjacent Bulk Liquids Berth. LPG is transferred from the ship to the facility via a loading arm and buried pipework. Within the Terminal, LPG is transported to the buried cavern via a shaft containing vertical pipework. A concrete plug combined with the pressure of approximately 130 metres of water gives the head pressure which keeps the LPG stored within the below ground sandstone cavern. From the cavern, LPG is delivered to ships via the bulk liquids berth, or trucks via a loading gantry and a pipeline to a neighbouring facility. Whilst filling the LPG into road tankers, the vapour returns via a pipeline to a mounded 60 tonne steel vapour recovery tank.

Cathodic Protection (CP) is used to protect a variety of metallic structures within the facility including the buried transfer pipework, the shaft pipework and the 60 tonne mounded vapour recovery vessel. This paper discusses the CP history of the protected structures at this critical facility including the design, installation and routine monitoring over its 17 year life.

2. CATHODICALLY PROTECTED STRUCTURES

There are three CP systems installed at the Elgas Cavern Terminal. The external surfaces of the buried transfer pipework from the Bulk Liquids Berth to the Elgas Terminal has a sacrificial system installed. The cavern shaft pipework and the concrete reinforcement in the cavern plug are protected by an Impressed Current CP (ICCP) system. The 60 tonne mounded vapour recovery vessel also has a sacrificial anode CP system.

3. PROTECTION CRITERIA

The protection criteria outlined in the Australian Standards (1), (2) and (3) have been adopted for the structures at this facility.

4. CP SYSTEM DESCRIPTION

A sacrificial anode CP system is installed on the approximately 60 metres length of buried pipelines from the Bulk Liquids Berth, crossing Fishburn Road, to the terminal. Pipelines include the 450mm NB LPG line, 100mm NB vapour recovery line and the 25mm NB air line. The system uses 2-off buried magnesium anodes that connect to the Pipelines via an above ground test point. The pipelines are coated and isolated from foreign structures using isolating flanges at the air to soil interface at both ends.

Gas is transferred in and out of the cavern via pipework passing down the vertical shaft. The steel pipework is epoxy coated and immersed in water. Due to the overall high current demand required for protection, an ICCP system has been installed. This CP system also protects the concrete reinforcement in the cavern plug and 3-off corrosion coupons.

Drawings of the installation are shown in Figures 3 and 4. Originally a 30Amp Transformer Rectifier Unit (TRU) with a SCADA system linked to the control room was installed. This was upgraded to a 50A TRU at a later stage due to an increase in current requirement. 20-off Mixed Metal Oxide (MMO) anodes are suspended evenly around the perimeter of the pipework at varying depths, distributed evenly along the length of the shaft. This provides an even current distribution across the entire length of all the pipework. The anodes are 6mm diameter rod with 1.5m effective length. Anode connections are encapsulated in epoxy and suspended by their steel wire armoured cable tails. This design allows the anodes to be replaced easily as they reach the end of their service life which is approximately 25 years. The four (4) upper most anodes are suspended above the water line and have been disconnected from the system. 3-off submersible zinc references are permanently installed at varying depths for monitoring purposes. The upper most zinc reference is suspended above the water line. The CP design was based on the predicted water level within the shaft. Following installation, the water level was lower than expected, and hence some CP components are suspended above the water level.

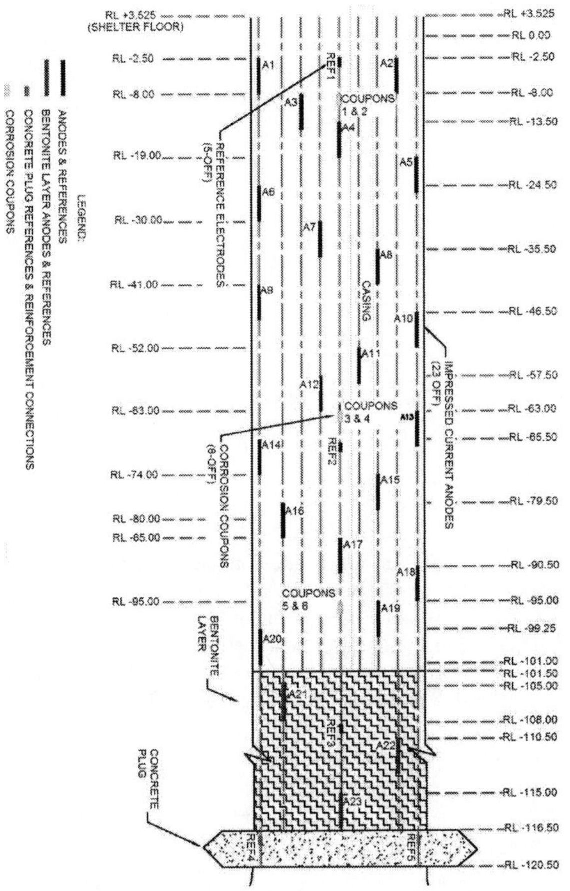

Figure 3: Section view of shaft CP system

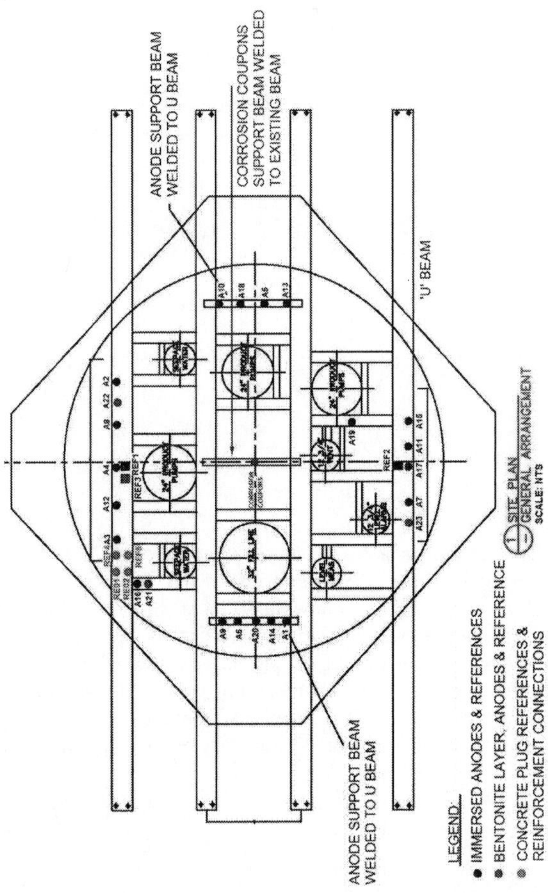

Figure 4: Plan view of shaft CP system

Additionally, 8-off corrosion coupons are installed at varying depths to assess the effectiveness of the CP system. The coupons are made from ASTM A106B steel which is the same material as the shaft pipework. The coupons are 200mm long x 100mm wide x 3mm thick with a 50x50mm tab at the end for a connection. Two (2) of the coupons are deliberately left in the atmosphere above the water line as a control. Of the six (6) immersed coupons, three (3) are bonded to the CP system at varying depths. One (1) metre above each of the CP coupons, an unprotected coupon is suspended.

The 60 tonne mounded vapour recovery vessel is coated with a high build epoxy coating and backfilled in clean sand to minimise damage to the coating during backfilling and to provide a uniform backfill around the tank. The resistivity of the sand backfill was measured to be 14,500 Ωcm at the time of installation. A sacrificial anode system is installed using high potential magnesium rod anodes distributed evenly around the perimeter of the vessel. Cables are terminated within an EXD flame proof junction box as it is located in a hazardous area within the terminal. Six (6) permanently installed zinc monitoring electrodes are installed around the tank for monitoring purposes. The mound that houses the vessel does not have a roof and is covered in gravel. This allows rainwater to enter the mound minimising drying of the mound.

5. CP SYSTEM PERFORMANCE

The CP on the Fishburn Road pipelines was commissioned on the 5th March 1999. The pipelines had natural potentials of -0.52 Volts Vs $Cu/CuSO_4$ for the 450mm NB LPG line, -0.51 Volts Vs $Cu/CuSO_4$ for the 100mm NB Vapour Recovery Line and -0.22 NB Volts Vs $Cu/CuSO_4$ for the 25mm air line. Since commissioning, on the 5th March 1999, the CP on the Fishburn Road pipelines have been inspected every six (6) months with test results showing that the pipelines have always remained fully protected from corrosion with instant off potentials consistently more negative than -0.85 Volts Vs $Cu/CuSO_4$. With the anode current output consistently less than 1mA, the coating is in excellent condition. (refer to Figure 5).

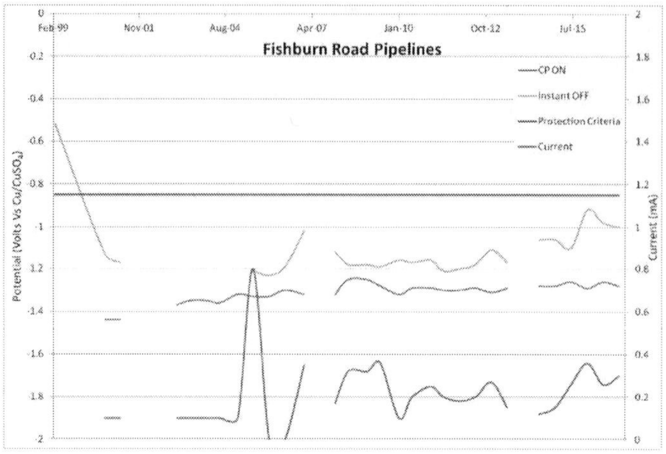

Figure 5: Fishburn Road pipelines historical CP performance

The shaft pipework had a natural potential of -0.71 Volts Vs Ag/AgCl and was commissioned on the 18th December 2000. At commissioning, the pipework shifted to fully protected potentials of -0.94 Volts Vs Ag/AgCl with the TRU operating at 13.9 Amps. Over the following 30 months the isolation between the pipework and other elements of the shaft broke down increasing the total surface area to be protected. The system after this period settled to a relatively constant current output of 47 Amps. Subsequent testing on the 8th July 2003 measuring protected shaft potentials of -0.84 Volts Vs Ag/AgCl. All testing since this date has measured protected potentials ranging from -0.82 to -1.07 Volts Vs Ag/AgCl with little TRU adjustment required. The pipework potentials down the depth of the shaft have remained within fully protected potentials over this period. A historical summary of the shaft pipework test results is shown in Figure 6.

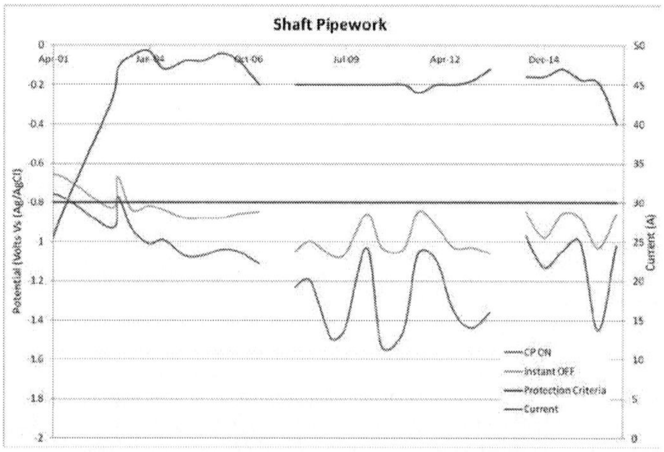

Figure 6: Shaft Pipework historical CP performance

The TRU operates in constant current mode as the water resistivity within the shaft varies depending on rainfall. The current distribution along the shaft has remained fairly uniform with individual anodes outputting similar current. All anodes have remained operating satisfactorily at fairly uniform outputs since installation with no replacements necessary.

Pairs of coupons are installed within the shaft to validate the effectiveness of the CP system. Two are suspended above the water line to act as a control. The three (3) other pairs of coupons are installed at the top, middle and bottom of the shaft. At each location, one coupon is connected to the CP system and one is not. Over the life of the Sydney Cavern, coupons have been replaced with the latest set of eight (8) identical coupons installed on the 8th September 2009.

The cleaning and weighing procedure stipulated by the client's consultant involves retrieving the coupons annually and weighing each coupon individually after cleaning with a wire brush and an acid solution. The atmospheric coupons showed consistent minor weight loss every year. The submerged coupons with CP applied have never shown a weight lower than the original installation weight. The CP applied coupons are all heavier than the original weight due to the calcareous deposit and generally increase in weight every year. The acid solution stipulated by the client's consultant has not been effective in completely removing the calcareous deposit. This cleaning process needs to be examined to see if the calcareous deposit can be removed. The submerged unprotected coupons consistently showed major weight loss and as of August 2016, some were at half their original weight, with pitting corrosion evident. The latest retrieval in August 2016 showed pits penetrating the coupon's wall. Refer to Table 1 for a historical summary of coupon weights.

Table 1. Historical summary of coupon weights

Coupon No.	1	2	3	4	5	6	7	8
Shaft Location	Bottom	Bottom	Middle	Middle	Top	Top	Atmosphere	Atmosphere
Cathodic Protection	Protected	Unprotected	Protected	Unprotected	Protected	Unprotected	Unprotected	Unprotected
Original Weight (g) 8/9/09	483.938	483.920	481.958	487.234	481.451	428.105	481.411	481.379
Recovered Weight (g) 29/7/10	526.270	422.005	491.081	421.643	530.590	413.344	478.458	481.090
Recovered Weight (g) 6/7/11	543.472	387.030	514.679	403.246	557.779	381.741	454.612	479.174
Recovered Weight (g) 6/7/12	550.104	383.177	491.441	402.95	585.232	380.605	451.165	478.525
Recovered Weight (g) 18/7/13	597.27	371.38	532.45	381.14	619.56	372.97	445.86	475.82
Recovered Weight (g) 31/7/14	631.66	360.40	572.32	363.54	633.54	359.93	439.87	474.94
Recovered Weight (g) 14/8/15	647.02	306.07	593.79	293.17	656.34	302.55	436.85	468.28
Recovered Weight (g) 16/8/16	N/A	N/A	567.65	227.43	663.63	269.07	429.33	464.01
Recovered Weight (g) 26/8/16	700.37	231.09	N/A	N/A	N/A	N/A	N/A	N/A
Weight change Since Installation (g)	+216.432	-252.83	+85.692	-259.804	+182.179	-159.035	-52.081	-17.369

Due to the extreme corrosion on the submerged unprotected coupons, July 2017 will be the final coupon retrieval for the September 2009 installed coupons, with new coupons installed to reset the process. Refer to Figure 7 and 8 for representative photos of the coupons.

Figure 7: Top protected coupon retrieved in August 2016 after cleaning

Figure 8: Top unprotected coupon retrieved in August 2016 after cleaning

The natural potential of the 60 tonne mounded vapour recovery vessel was measured to be -0.38 Volts Vs $Cu/CuSO_4$ prior to commissioning on the 5th March 1999. Since commissioning, the vessel has consistently been at fully protected potentials of approximately -1.26 Volts Vs $Cu/CuSO_4$ with an anode current output of approximately 3mA. There have been slight variations in test results over the 18 year life of the vessel and the suspected cause is the water content/resistivity of the mound/electrolyte surrounding the tank. Refer to Figure 9 for a summary of the test results. In November 2010, the vessel was exposed for an inspection. The inspection revealed no corrosion on the buried sections of the vessel with the coating remaining in excellent condition. Although the CP potentials indicate slight over protection as per AS2832.2, the coating remained intact with no evidence of coating disbondment. Minor corrosion was found in exposed areas of the manway covers and pipe flanges. These locations were above ground (out of the electrolyte) and hence did not receive any CP current.

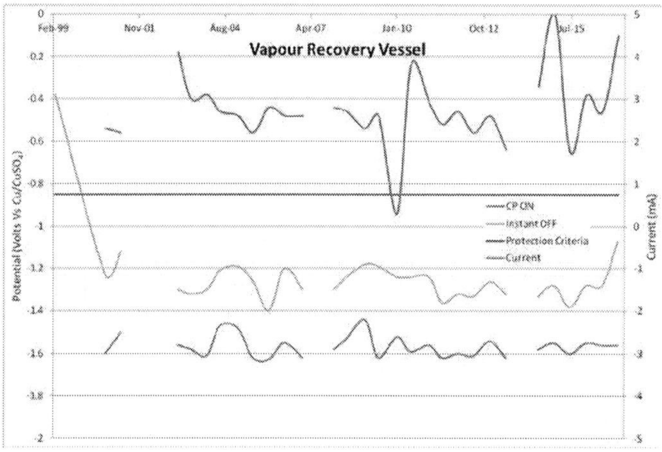

Figure 9: Vapour Recovery Vessel historical CP performance

6. FUTURE EXPECTATIONS

The CP system installed on the Fishburn Road pipelines is operating satisfactorily with the pipelines remaining fully protected since commissioning. Although the pipelines are relatively short, the pipelines may be inspected using inline inspection techniques and Direct Current Voltage Gradient to identify any coating defects.

The ICCP system on the shaft pipework remains operating satisfactorily with all areas of the shaft pipework at fully protected potentials. Although none of the anodes have failed to date, the anodes may eventually fail as they reach the end of their service life which would be observed in the routine monitoring where individual anode currents are measured. When this occurs, the failed anodes can be readily replaced to maintain uniform protection.

In July 2017, the September 2009 coupons will be retrieved and weighed for the final time. It is expected similar results will be observed with the protected coupons gaining weight due to calcareous deposit and the unprotected coupons losing weight. The new coupons will be installed in July 2017 and it is expected that they will perform in a similar manner to the September 2009 coupons.

The CP on the 60 tonne mounded vapour recovery vessel has been constant since its commissioning with little fluctuation in the current and tank potentials. The CP on the vessel is expected to last for many more years with routine monitoring in place to identify any variances.

7. CONCLUSIONS

All CP systems installed at the Elgas Caverns have been operating satisfactorily with all structures remaining fully protected over the majority of their life.

The Fishburn Road pipelines have been fully protected since commissioning with the existing system expected to last for several more years.

The shaft pipework originally showed some protection concerns, possibly due to breakdown of isolation from other shaft components. As a result, the TRU was upgraded and current output increased. Once the correct current output was selected, the pipework shifted to fully protected potentials and has remained protected since.

The corrosion coupons have validated the effectiveness of the CP system with unprotected coupons losing weight and protected coupons not experiencing any weight loss.

The effectiveness of the CP system installed on the 60 tonne mounded vapour recovery vessel has been validated when the tank was exposed approximately 10 years after its original installation. The inspection revealed no corrosion in areas where CP had been applied.

The operating design life of the Elgas Sydney Cavern is 50 plus years. Over the first one third of the asset's life, fully protected CP potentials have been maintained and coupons have proven no metal loss. Given the proven track record, Elgas could consider options of extending the operating life of the asset.

8. ACKNOWLEDGEMENTS

We thank Neil Tarin from Elgas Ltd for permission to publish this paper.

9. REFERENCES

1. Standards Australia 2015, *Australian Standard AS2832.1 Cathodic Protection of Metals, Part 1: Pipes and Cables,* Standard Australia, Sydney

2. Standards Australia 2003, *Australian Standard AS2832.2 Cathodic Protection of Metals, Part 2: Compact Buried Structures,* Standard Australia, Sydney

3. Standards Australia 2005, *Australian Standard AS2832.3 Cathodic Protection of Metals, Part 3: Fixed Immersed Structures,* Standard Australia, Sydney

10. AUTHOR DETAILS

David Sunjaya is a corrosion engineer at Corrosion Control Engineering (NSW) Pty Ltd. He holds a Bachelor of Science (Honours) in Nanotechnology from the University of Technology Sydney and is a current member of the Australasian Corrosion Association (ACA) NSW branch committee. He has also been an active member of the ACA Young Corrosion Group since 2012. David is also NACE Coating Inspector Level 1 – Certified.

David has over 9 years cathodic protection experience and has been involved in all aspects of cathodic protection including design, project management, installation, commissioning, routine testing, troubleshooting, repairs and reporting. Areas of CP expertise include pipelines, wharves, bridges, shipping, buried tanks, reservoirs, industrial plants and concrete steel reinforcement.

EXPERIENCE OF STAINLESS STEEL AS A STRUCTURAL MATERIAL IN CIVIL INFRASTRUCTURE

J McCray[1], S H Mameng[2], G Gedge[1] & C Tigerstrand[2]
[1]Arup, Sydney, Australia, [2]Outokumpu Stainless AB, Avesta, Sweden

SUMMARY: Australian infrastructure asset owners are demanding very durable structures with minimum maintenance. A specified design life of 100 years is common. Stainless steel is an attractive choice as the ultimate low-maintenance structural material for bridges and other infrastructure. Only a small number of major stainless steel structures exist in Australia but they are more common overseas. From a corrosion perspective, stainless steel grades are chosen to avoid deterioration and it is important to get this grade selection correct. Under-specification can lead to corrosion; over-specification is needlessly costly. Economical specification is key.

Twelve major stainless steel structures in Sweden, Spain and the UK have been visited and inspected. These structures represent a range of environments from the seafront at Menorca (Spain) and Blackpool (United Kingdom), to inland, rural Flen (Sweden). The selected structures also use a range of stainless steel grades including austenitic, duplex and lean duplex. They also range in age. This work presents the results of the inspections. Results are analyzed and discussed in term of the risk of corrosion that would damage structural integrity as well as superficial, aesthetic corrosion. The structures represent successful examples of stainless steel used in structural applications. Although the results are from European environments, observations of stainless steel performance and other lessons to be learnt are relevant to the structural use of stainless steel in Australia.

Keywords: Stainless steel, grade selection, field performance.

1. INTRODUCTION

Australian infrastructure asset owners are demanding very durable structures with minimum maintenance. Stainless steel is a potential candidate structural material where a minimal maintenance approach is preferred. Structural design codes for stainless steel exist, particularly EN 1993-1-4 (European Committee for Standardisation, 2006) and the AISC Steel Design Guide 27 (American Insitute of Steel Construction, 2013). Another guide has also been published by the (UK) Steel Consutrction Institute. In Australia, a stainless steel structure was opened in April 2016 at 'The Gap' in Torndirrup National Park near Albany in Western Australia (Outokumpu, 2017). This structure is a viewing platform at the edge of a granite cliff overlooking the Southern Ocean. The Newcastle Memorial Walk in New South Wales also makes structural use of stainless steel (Northrop, 2016). Other stainless structures, including road and pedestrian bridges, have been built in Europe and South East Asia. Most of these structures utilise duplex stainless steel for structural components.

Many of these structures make use of new lean duplex alloys which offer high strength and reasonable atmospheric corrosion resistance at lower cost than duplex alloy 1.4462 (also known as 2205).

Laboratory test data and exposure trial data is available for these newer alloys – see (Burket, Lehmann, Burkert, Mietz, & Gumpel, 2014) and (Mameng, Pettersson, Leygraf, & Wegrelius, 2016) among other sources. However there is less published research available about the in-service performance of actual, built structures made of these stainless steel grades.

This research project, undertaken collaboratively by Arup and Outokumpu, aimed to collect field experience of stainless steel in the built environment. We visited a selection of major projects utilising stainless steel structurally which have been completed in Europe in the last fifteen years. We observed and recorded the condition and performance of the various stainless steel elements at the sites.

It is intended that the results, published here, will inform the design of future stainless steel structures by providing a reference point for past successful applications.

2. METHODOLOGY

The project was conceived as a series of field inspections of completed, structures across Europe. All use stainless steel as a primary structural element. The purpose of the project was to document the condition of these structures through a standardised inspection methodology and relate the environmental exposure conditions to those given in EN1993-1-4 'Design of steel structures. General rules. Supplementary rules for stainless steels' Annex A.

2.1 Selection of inspection sites

A list of stainless steel structures (many of them bridges) was prepared. Selection criteria for sites would be visited as part of this research project were:

- Exposure environment
- Location (ease and efficiency of access)
- Alloys used
- Age of the structure

Each potential structure was assessed qualitatively against these criteria and twelve sites were chosen for the field inspection stage. They are detailed in Table 1. For the convenience of Australian readers, an approximate equivalence guide between the European grade numbers used in this work and the American (ASTM/SAE) grade designations is given in Table 2.

Selected sites have stainless steel elements ranging from 2 – 15 years old at the time of inspection. All of these structures are intended to have a service life much longer than that. For stainless steel durability design, an alloy is chosen that will not meaningfully deteriorate in the service environment throughout the service life. Specifically, stainless steel relies on the formation of a 'passive film' of chromium oxide for its corrosion protection. Durability problems for stainless steel occur when this passive film is disrupted and cannot self-repair. For stainless steel components, good performance in the first few years of service indicates likely good performance throughout the service life, assuming there is no change to the service environment. Therefore we consider the age of the structures inspected to be adequate for making assessment of the suitability of the stainless steel alloys selected for the service environments in which they have been placed in the long term.

In order to comment on appropriateness of grade selection in different environments those environments must be classified. The obvious choice of classification system might be the 'C' class corrosivity classifications defined in standards like ISO 9223 and AS 4312. However we decided to use the 'CRC' system defined in EN 1993-1-4 Annex A. The annex was published in 2015. This standard is the Eurocode governing the structural use of stainless steel. Annex A defines a methodology for selecting stainless steel for use in structural applications in the built environment. This methodology assigns a required 'corrosion resistance class' (CRC, I to V) for a particular stainless steel element. The CRC is a combined assessment of exposure to chloride ions (either from road de-icing salts or the sea), exposure to sulphur dioxide pollution, and any washing by rain or a cleaning program for the element in question. The standard then suggests grades of stainless steel appropriate for those corrosion resistance classes.

We believe this system offers three key advantages over the 'C' classes:

- It was developed specifically for stainless steel
- It takes account of whether or not the steel is washed, which is an important factor in the performance of stainless steel
- In our experience, engineers without specialist knowledge of corrosion tend to misinterpret the qualitative descriptions of 'C' classes given in standards like those mentioned above. We believe the CRC system is less open to misinterpretation.

The CRC system does not take account of parameters such as surface roughness, finish and treatment which do need to be considered in applications with aesthetic performance requirements.

Table 1: List of sites visited as part of this project

Structure number	Structure name	Location	Co-ordinates	Year of construction	Stainless steel grade for the main construction	CRCs observed
01	Sölvesborg Pedestrian and Cycle Bridge	Solvesborg, Sweden	56°03'04.7"N 14°35'34.7"E	2013	1.4162	II, III
02	Orrhammarvägen Bridge	Flen, Sweden	59°05'01.7"N 16°34'23.9"E	2009	1.4162	II
03	Sicklauddsbron (bridge)	Stockholm, Sweden	59°18'17.4"N 18°06'09.0"E	2002	1.4462	II, III
04	Nynashamn road bridge	Nynashamn, Sweden	58°55'53.6"N 17°58'02.2"E	2011	1.4162	IV
05	Millennium Bridge	York, United Kingdom	53°56'41.6"N 1°04'55.6"W	2000	1.4462	II
06	Blackpool seafront (various elements)	Blackpool, United Kingdom	53°47'32.1"N 3°03'29.9"W	Circa 2000-2010	1.4462, 1.4401, 1.4301	IV, V
07	Celtic Gateway Bridge	Holyhead, United Kingdom	53°18'35.9"N 4°37'54.8"W	2006	1.4362	III
08	Sant Fruitós Bridge	Manresa, Spain	41°44'46.1"N 1°51'40.2"E	2009	1.4162	II
09	Zumaia footbridge	Zumaia, Spain	43°17'31.5"N 2°15'29.6"W	2008	1.4462	III
10	Anorga bridge	San Sebastian, Spain	43°17'18.8"N 1°59'52.5"W	2011	1.4162	II, III
11	Puerto Arrupe (bridge)	Bilbao, Spain	43°16'11.2"N 2°56'12.8"W	2003	1.4362	II
12	Cala Galdana bridge	Menorca, Spain	39°56'22.6"N 3°57'34.8"E	2005	1.4462	IV, V

Table 2: Equivalent† ASTM/SAE grades to European grades

European grade	Equivalent† ASTM/SAE grade
1.4161	2101
1.4301	304
1.4362	2304
1.4401	316
1.4462	2205

† It must be emphasised that there is no strict direct equivalence between the European and ASTM/SAE stainless steel grading system with different composition limits and sub-grades specified by both systems. However the 'equivalent' grade listed here are the 'closest match' and expected to have practically equivalent performance in service.

2.2 Procedure at inspection sites

At each inspection site, the following procedure was followed:

- An initial visual inspection to identify the different elements of the structure being viewed (e.g. for bridges: structural beams, balustrades, cover plates, etc.)
- The condition of typical examples of each element was visually assessed from available and safe vantage points. Digital photographs were taken to record the condition
- Where access was possible, measurements were taken with a hand-held X-ray spectroscopy analyser to confirm the stainless steel alloy used for that element
- Where access was possible, surface replicas were taken with a Struers kit for subsequent analysis for surface roughness in the laboratory
- For sites adjacent to a body of water, a sample of that water was taken where possible for subsequent analysis for chloride content in the laboratory.
- Where access was possible and potentially interesting contamination was observed, an adhesive scanning electron microscope stub was placed onto the contamination and pulled off. Some contamination tended to stick to the stub. These traces were analysed in a scanning electron microscope (SEM) to try to ascertain the nature of the contamination

Field work was undertaken over the summer and autumn of 2015.

2.3 Procedure on return to laboratory

2.3.1 Surface roughness measurements

Two dimensional roughness measurements and three dimensional surface texture characterization were performed on surface replicas with a NanoFocus µsurf optical confocal microscope. The roughness evaluation was performed as close as possible in accordance to the recommendations given in ISO 4288 [6]. The evaluation lengths ranged from 1.25 up to 12.5 mm while the cut-off lengths ranged from 0.25 up to 2.5 mm depending on roughness. A needle filter was used. The evaluated two-dimensional parameters have been calculated as an average of 512 profiles. The 10XS and 20XS objectives were used during the roughness evaluation. Stitching was done when the required evaluation length was longer than one objective field length.

2.3.2 Water sample analysis for chloride ion concentration

For sites where water samples were taken, the chloride ion concentration in water samples was analysed by DIONEX DX-120 ion chromatography for anions.

2.3.3 Analysis of the contamination on stainless steel surface

Stub with adhered contamination were examined by SEM and analysed with energy-dispersive X-ray spectroscopy (EDS) to identify the nature of the contamination.

3. RESULTS

The results from the project are arranged by environment classification (CRC). Overall images of each structure visited are shown in Figure 1.

Figure 1a Sölvesborg Pedestrian and Cycle Bridge

Figure 1b Orrhammarvägen Bridge

Figure 1c Sicklauddsbron pedestrian bridge

Figure 1d: Nynashamn road bridge

Figure 1e Millennium Bridge

Figure 1f Blackpool seafront

Figure 1g Celtic Gateway Bridge

Figure 1h Sant Fruitós Bridge

Figure 1i Zumaia footbridge

Figure 1j Anorga bridge

Figure 1k Puerto Arrupe

Figure 1l Cala Galdana bridge

3.1 CRC II

This class covered the greatest proportion of stainless steel elements inspected during this work. It includes rain washed elements located greater than 1km from the coast in areas with low sulphur dioxide pollution, and elements that are not rain washed located greater than 10km from the coast in areas with low sulphur dioxide pollution. The class covers the majority of the land area of Europe and therefore it is unsurprising that this was the most commonly observed exposure class.

Alloys listed in EN 1993-1-4 Annex A as being suitable for this service environment include alloy 1.4301 ('304').

This class covers environments with relatively low corrosivity. Unsurprisingly, during this research all the austenitic and duplex stainless steels used in these environments were showing good performance. Alloys most commonly present in this class were 1.4162 ('lean duplex'), 1.4301 ('304') and 1.4401 ('316').

In this environment, relatively rough surface finishes have been applied successfully. Measured surface roughness in the range of 2.5 – 5 µm Ra were common. These rough surfaces were performing well with none of the discolouration ('tea-staining') that might be expected on stainless steel elements in marine climates with rough finishes like these.

Some discolouration was observed where steel detailing caused a crevice to be formed. This typically appeared as a minor orange stain adjacent to the location of the crevice. The phenomenon could even be seen occasionally on some of the higher performance alloys used in this environment, for example the grade 1.4462 balustrading at the Sicklauddsbron, Stockholm (Figure 14). The importance of avoiding crevices in the design and detailing of structures is discussed further below.

Figure 2: Grade 1.4462 balustrading at the Sicklauddsbron, Stockholm.
Note the staining at the tube-plate interface which is a crevice.

3.2 CRC III

This class includes, in low sulphur dioxide environments, elements greater than 250m from the coast but within 1000m of the coast. Both rain washed and non-rain washed elements at this distance are included. It also includes elements within 250m of Baltic Sea coasts and the North Sea coast of Germany which are rain-washed or regularly cleaned – these regions are known to be less corrosive than other coastal parts of Europe (exposed to open oceans).

The alloys listed in EN 1993-1-4 Annex A as being suitable for this service environment include alloys 1.4401 ('316'), 1.4162 and 1.4362 (both forms of 'lean duplex').

The key observations falling into the environment category are those made at the Nynashamn road bridge in Sweden, the Celtic Gateway bridge in Holyhead and the Zumaia footbridge in Spain.

The alloy used at Nynashamn is 1.4162. It appears to be performing well with no staining observed despite a rough surface finish of approximately 6-8 μm Ra. The only aesthetic problems observed were mild orange staining near the southern abutment on horizontal surfaces (Figure 3a) and a small number of very isolated shallow pits with associated orange stain on the vertical surfaces (Figure 3b).

Figure 3a A mild tea-staining like appearance near the southern abutment on horizontal surfaces

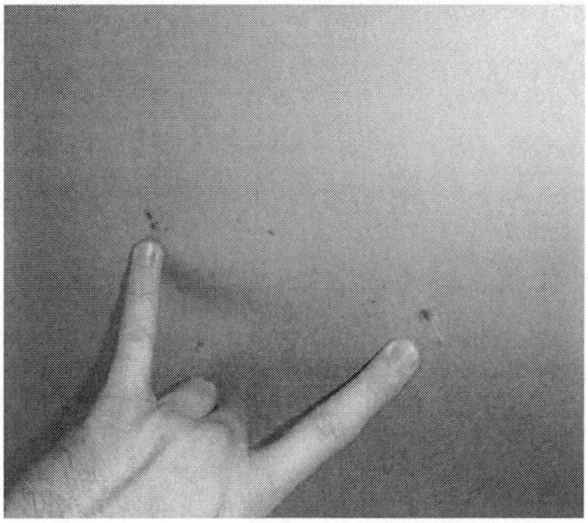

Figure 3b A small number of very isolated shallow pits with associated orange stain on the vertical surfaces

The Celtic Gateway bridge is made predominately of alloy 1.4362 with some 1.4401 used for balustrading, etc. This steel is performing well. The Celtic Gateway Bridge uses cementitious materials in its deck build-up which has led to efflorescence. This has not had a detrimental effect on the stainless steel.

The bridge at Zumaia is constructed predominantly from alloy 1.4462 ('2205') with 1.4401 used only for nuts and bolts. Both these alloys were observed to be performing well in this environment. The 1.4462 material represents a potential over-specification for corrosion resistance, and it is not known why this grade was specified.

3.3 CRC IV

This class includes, in low sulphur dioxide environments, rain washed elements less than 250m from European coasts. It also includes non-rain washed elements within 250m of Baltic and German North Sea coasts.

The alloys listed in EN 1993-1-4 Annex A as being suitable for this service environment include alloy 1.4462 ('2205').

In this project stainless steel elements with this classification were located at Blackpool and Cala Galdana.

The bridge at Cala Galdana uses two alloys for major components. 1.4462 duplex steel is used for the bridge arches and balustrades. 1.4401 is used for a tubular barrier between vehicle and pedestrian areas of the deck. Both alloys are finished in a similar surface roughness (approximately 3.5–4.0 µm Ra) however the performance is noticeably different. The 1.4401 alloy is covered with orange staining whereas the 1.4462 alloy areas are not. The contrast can be seen in figure 4a. Adjacent galvanised bolts (shown in figure 4b) are not performing well. The poor performance of the galvanised bolts demonstrates the high corrosivity of the environment at Cala Galdana.

Figure 4a 1.4401 alloy tubing is covered with orange staining whereas the 1.4462 alloy arch is not, Cala Galdana

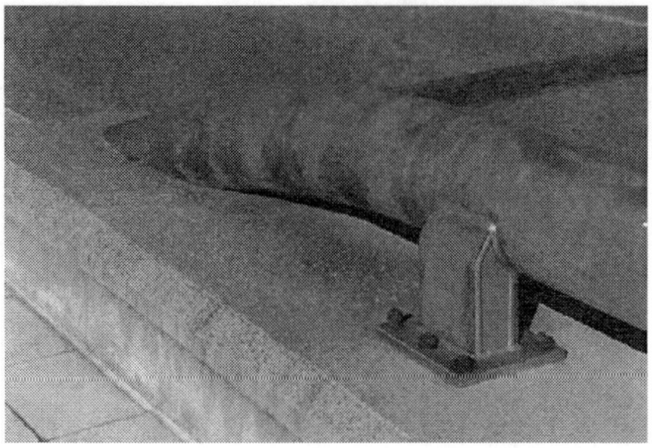

Figure 4b Galvanised bolts and 1.4401 alloy tubing, Cala Galdana

At Blackpool the same two alloys (1.4401 and 1.4462) were commonly in use. The structures inspected here were a mix of various stainless steel items located along the seafront promenade. Installation dates varied. Performance was similar to that observed at Cala Galdana.

Alloy 1.4462 was performing well at Blackpool with the only staining on the surfaces observed was a golden-coloured sheen. the removable 'sheen' mentioned above. The sheen could be wiped off with alcohol, as shown in figure 5a. A very small number of individual pits were observed across the totality of the items inspected. On some items where detailing of the metal created a crevice, a small associated orange stain was seen (figure 5b). This is despite a relatively rough surface finish to come components, e.g. Ra approx. 3μm in the case of 1.4462 alloy curved lighting poles.

Figure 5a The lower part of this light-pole has been cleaned with alcohol, to remove the 'golden sheen'. The upper part has not. Blackpool.

Figure 5b Crevice with associated small rust stain. Blackpool.

Alloy 1.4401 was not performing as well. Orange 'tea staining' was commonly seen. Even on the way-finding signage, which had a measured surface roughness of 0.450 ± 0.058 μm Ra, staining was observed. Conventional wisdom is that this alloy should be specified with Ra < 0.5 μm to avoid tea staining (Australian Stainless Steel Development Association, 2017). This has not worked in this case. Rougher examples of alloy 1.4401 were performing worse. In particular bad staining was observed at created crevices in bench seating (figure 6a). Supporting struts for the seating behind the 'dolphin tail' wind shelters were also performing poorly (figure 6b).

Figure 6a Bench seating, alloy 1.4401, staining. Blackpool.

Figure 6b Alloy 1.4462 wind shelter performing well. Alloy 1.4401 struts with staining. Blackpool.

4. DISCUSSION

'Lessons learnt' for detailing of stainless steel on future projects are discussed here.

4.1 Creation of corrosion crevices

It is well known that crevice situations, where two surfaces are brought in close contact and a differential aeration cell is created between those surfaces and adjacent exposed surfaces, create a locally more severely corrosive environment than the general environment in which the crevice is located. Crevices can be created by detailing of connections. For example, crevices can form under bolt heads in snug-tight joints or at surface features in welds, for example at weld toe undercut. Consistently, crevices observed on these structures were associated with stains. It is good detailing practice to avoid crevices. Details such as seal welds are useful here.

4.2 Contamination with mild steel

Specialist stainless steel fabricators are well aware of the problems that can be caused by contamination of stainless steel with mild steel. Contamination from parts being in contact, grinding swarf falling onto adjacent materials, and use of cutting and grinding tools on stainless steel that have previously been used for mild steel are all well known phenomena.

We observed contamination that had led to aesthetic damage of stainless steel components in service.

One more unusual example of contamination was observed at the Flen bridge. Regular spaced contamination on the bottom flange of one beam, illustrated in figure 7, leads us to suspect that the beam had been rested on steel trestles at some point during its fabrication, transportation or installation.

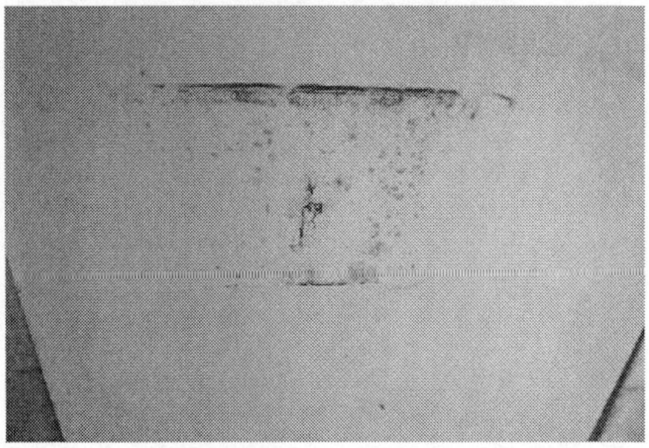

Figure 7a Stain observed at the Flen bridge likely due to contamination with mild steel

Figure 7b Regularly spaced contamination stains on beam soffit

The risk of contamination is not a new finding, but the field observations confirm that the care taken by specialist contractors to avoid contamination is justified and necessary. Specifications should reflect this.

4.3 Selection of surface finishes for alloys

Tea staining is a phenomenon that has effected sea- and water-side architectural applications for stainless steel. Tea staining refers to the discoloration of the stainless steel surface by corrosion processes. Metal loss is negligible, so structural performance is not impaired. EN1993-1-4 Annex A explicitly does not deal with this issue.

As mentioned above, conventional wisdom suggests the use of alloy 1.4401 (also known as 316) with a surface finish of Ra 0.5μm or less. At Blackpool, this was observed to not be sufficient to prevent tea staining. Elements made of alloy 1.4462 (duplex 2205) were not tea stained even with a measured surface roughness an order of magnitude greater than the rule-of-thumb. Blackpool is exposed directly to rough sea and experiences persistent on-shore winds. As such it has a very salty atmospheric environment. This experience may be transferable to Australian cities that are also extremely exposed and salty, like the Gold Coast or Fremantle.

4.4 Vandalism and general service damage

Some of the structures inspected had been subject to vandalism and damage in service. Vandalism included painted tags. Damage included dents and scratches from, for example, bicycle crashes. We observed one such crash take place during the inspection program. All objects placed in the public realm are potentially subject to this sort of damage and stainless steel is not immune. On stainless steel, such damage is typically cosmetic and has no real effect on the durability of the material. For other material types this sort of damage can have a significant adverse effect on the durability of the item. The obvious example of this is painted mild steel. If the paint coating is damaged then corrosion of the underlying steel, which was being protected by the coating, will occur.

We did observe at Bilbao that areas with tags painted onto the stainless steel had been painted over, presumably as a way of attempting to hide the tags. This is not the best method for removing unwanted painted tags from stainless steel. It would have been more preferable to chemically clean off the tags, revealing the stainless steel surfaces that remain underneath. Chemical cleaning treatments can also be used to remove staining caused by contamination.

4.5 Channelling of water run-off

On two of the bridges, the Millennium Bridge at York and the bridge at Cala Galdana, the shape of the bridges was such that water falling onto the bridge tended to concentrate and run-off from a single point. At these points, a stain was observed (figure 8). This stain may be caused by the washing of concentrations of contaminants over these points. The stains are really very minor in their nature, without apparent detrimental effects on the structure. Thoughtful detailing can avoid such concentrations of run-off altogether.

It is important to note that this phenomenon was observed in the relatively benign CRC II environment of York and the more severe CRC IV environment of Cala Galdana. The effect can be seen regardless of the corrosivity of the environment. In both cases the alloy 1.4462 was used. However there is no reason to assume that other (more lean) alloys would not also be affected. Unfortunately we were not able to get close enough to any of these stains to see if they could be wiped off.

Figure 8a Stain occurring at a location of concentrated rainwater run-off, Millennium bridge, York

Figure 8b Stain occurring at a location of concentrated rainwater run-off, Cala Galdana bridge, Minorca

5. CONCLUSIONS

We have found that the vast majority of the stainless steel structures inspected are performing very well. Each of the structures inspected could be used as a case study for the successful application of stainless steel structures in civil engineering. There were some specific problems observed that were causing staining. In particular crevices were leading to staining across the environment visited and alloys used. There were also localised areas of observed contamination with mild steel.

It is very tempting to focus on small details that are not performing as well as they might and can detract from the overall success of a project. Attention to these details in design, fabrication and maintenance is important. This is not a new message for stainless steel or, arguably, other materials. Guidance on these points exists. These minor examples should not detract from the overall success of the structures inspected.

6. ACKNOWLEDGMENTS

This work was performed within the Stainless Steel in the built environment project at Avesta Research Center, Outokumpu Stainless AB. Special thanks to Jan Björk, Andy Backhouse and Alberto Aliende for support during inspections in Sweden, UK and Spain.

7. REFERENCES

American Insitute of Steel Construction. (2013). *Design Guide 27: Structural Stainless Steel.* Chicago, IL, USA.

Burket, A., Lehmann, J., Burkert, A., Mietz, J., & Gumpel, P. (2014). *Mater. Corros., 65*, 1080-1095.

European Committee for Standardisation. (2006). *Eurocode 3—Design of steel structures – Part 1-4: Genaral Rules – Supplementary rules for stainless steels.*

European Committee for Standardization. (2006+A1:2015). Eurocode 3 — Design of steel structures — Part 1-4: General rules — Supplementary rules for stainless steels. *EN1993-1-4.*

Mameng, S. H., Pettersson, R., Leygraf, C., & Wegrelius, L. (2016). *Berg-und Huttenmännischefte, 161*, 33-43.

Northrop (2016, December 6). *Newcastle Memorial Walk wins Australian Engineering Excellence Award.* Retrieved from http://northrop.com.au/201612/newcastle-memorial-walk-wins-australian-engineering

Outokumpu. (2017, June 10). *The Gap in Western Australia gets stainless walkways and viewing platforms.* Retrieved from http://www.outokumpu.com/en/stainless-news/cases/the-gap-in-western-australia-gets-stainless-walkways-and-viewing-platforms/Pages/default.aspx

8. AUTHOR DETAILS

Jonathan McCray is a materials engineer for Arup in Sydney and has worked for Arup in Sydney and London since 2012

Jonathan specialises in durability, maintainability and maintenance of built assets. His work covers specification of new build projects as well as inspections and characterisation of the condition of existing buildings and infrastructure. He also designs remedial measures and supervises their implementation.

HOT DIP GALVANIZING OF QUENCH AND TEMPERED STRUCTURAL GRADE SPECIALTY STEELS

W. McLean[1], A. Sheehan[1], P. Golding[1]
[1]Galvanizers Association of Australia

SUMMARY: An increasing amount of high strength structural steel is being specified in Australia each year, allowing engineers and architects the freedom to lower steel member size and mass while achieving equivalent strength. High strength quench and tempered structural steels were previously known to readily achieve a sound and continuous hot dip galvanized coating for durable corrosion protection. Technical development in the steel making industry has seen the silicon content of quench and tempered steels rise over the past ten years, providing increased hardenability without sacrificing performance.

This study explores the effect this new steel chemistry has on the potential for high strength, quench and tempered, structural grade specialty steels to be hot dip galvanized as a means of corrosion protection. Steels that are higher in silicon content are known to produce a thicker galvanized coating, however it can be more brittle in nature due to the influence on the coating's microstructure, leading to higher susceptibility of peeling and flaking which affects its corrosion protection. Various other factors known to influence the hot dip galvanizing coating formation were also investigated, including the steel thickness, different types of cutting processes, the sharpness of cut edges and surface pre-treatment prior to galvanizing. This paper reports observed average coating thickness of the hot dip galvanized samples, as well as the impact of previously mentioned influencing factors on corrosion performance. From these results, recommended precautions and procedures are provided to aid in the successful hot dip galvanizing of high strength quench and tempered structural grade specialty steels.

Keywords: Protective coating, High strength steel, Hot dip galvanizing, Quench and tempered.

1. INTRODUCTION

An increasing amount of high strength structural steel is being specified in Australia each year, allowing engineers and architects the freedom to lower steel member size and mass while achieving equivalent strength. Typical applications for the steel include high rise structures, mining equipment, bridge construction and lifting and transport equipment. The increased yield strength of the steel is usually obtained through a heat treatment process consisting of quenching and tempering. Over the past ten years the silicon content of these high strength, structural grade, quench and tempered steels has risen, allowing for increased hardenability without sacrificing weldability.

Hot dip galvanizing of steel involves a cleaning process before the steel is submerged in a molten zinc bath. The reaction between the steel and zinc creates the metallurgical bonded galvanized coating which provides corrosion protection to the steel. High strength steels with a silicon content around 0.20% by mass were known to readily achieve a continuous and thick hot dip galvanized coating for corrosion protection, however the newer chemical composition of speciality high strength steel containing a maximum of 0.60% silicon by mass had not been tested. Steels with a higher silicon content are known to produce a thicker galvanized coating, however it can be more brittle in nature due to the coating's microstructure and subsequently more prone to peeling and flaking during handling which affects its corrosion protection.

This paper explores the effect this new steel chemistry has on the ability to hot dip galvanize high strength quench and tempered structural grade specialty steels as a means of corrosion protection. Various factors known to influence the galvanizing coating formation were investigated, including the steel thickness, different types of cutting processes, the sharpness of cut edges and surface pre-treatment prior to galvanizing.

2. QUENCH AND TEMPERED STRUCTURAL GRADE SPECIALTY STEELS

2.1 Manufacture of Feed Plate

The manufacture of high strength quench and tempered steel requires high quality steel plate to put through the heat treatment process, as unwanted impurities in the steel's crystalline lattice structure can lead to defects and undesirable mechanical and chemical properties (1). Refining processes used to manufacture high quality feed plate include hot metal deoxidation and desulphurisation, making alloy additions under vacuum, argon gas bubble stirring, calcium injection to remove inclusions and vacuum degassing (2). The chemical and mechanical properties of the feed plate are carefully engineered to ensure the feed plate is optimised for use with the heat treatment process (3).

2.2 The Quench and Tempering Process

The quench and tempering process employed during the manufacture of high strength structural steels uses different heats for carefully controlled timing to achieve the desired microstructure (1). For high strength steels, quenching to a martensitic lattice microstructure and tempering to a bainitic microstructure is a commonly used method to achieve a combination of strength and ductility. Small additions of molybdenum, chromium and boron added during the manufacture of the feed plate promote a martensitic structure when the steel is quenched. The time and temperatures used for austenitising, quenching, and tempering depend on the chemical composition of the steel, with different time temperature transformation graphs applicable for different steels.

In a typical heat treatment process the steel is heated in the austenitising furnace to around 900°C to convert the microstructure from a body centred cubic (BCC) to a face centred cubic (FCC) austenitic structure and dissolve any iron carbide. The steel is then quenched at a carefully controlled temperature for a controlled amount of time to achieve a martensitic microstructure, which has a body centred tetragonal (BCT) atomic arrangement. The lattice strain created by carbon atoms trapped in the martensitic BCT microstructure gives the steel its high strength, but reduces its ductility and impact properties.

During tempering the steel is heated to roughly 600°C for around an hour and allowed to cool slowly. This process results in precipitation of very fine carbide particles and gradually transforms the crystal structure from BCT to BCC, with the microstructure called tempered martensite (1). Tempering improves the ductility and toughness of the martensite, accompanied by a slight reduction in strength and hardness. The high strength steel can be tempered using an open flame that may result in modified carbon and silicon content at the steel's surface. The steel is cooled at a controlled rate after tempering and the mechanical properties remain unchanged unless reheated above the temperature used in the tempering process.

2.3 Chemical Composition of Final Product

High strength steels are typically low alloy steels, with only a small percentage by mass of elements other than iron present. The open flame tempering process or other high heat processes such as thermal cutting or welding can cause chemical composition and microstructure changes at the steel's surface (4) (5). Common chemical changes which may influence the success of the hot dip galvanizing process are the formation of a decarburized layer and a lowering of silicon content at the steel's surface. The typical weight percentages of elements in quench and tempered structural steel manufactured to Australian Standards is given in Table 1 (3).

Table 1. Typical Weight Percentages of Elements in Structural Quench and Tempered Steel to AS 3597

	Maximum weight %							
Thickness (mm)	C	P	Mn	Si	S	Cr	Mo	B
5 - < 8	0.18	0.025	1.2	0.25	0.008	0.25	0.25	0.002
≥ 8 - < 30	0.18	0.025	1	0.6	0.008	1	0.25	0.002
≥ 30 - ≤ 80	0.18	0.025	1.3	0.6	0.008	1	0.25	0.002
> 80 - 100	0.18	0.025	1.5	0.6	0.008	1.2	0.25	0.002

2.4 Mechanical Properties of Final Product

The mechanical properties of high strength quench and tempered structural steels depend on the quality of the feed plate, as well as the temperature and time the feed plate is exposed to during austenitising, quenching and tempering. After the heat treatment process the yield strength can be over three times higher than the yield strength of typical mild steel. The required mechanical properties for Grade 700 manufactured to the Australian Standard for quench and tempered structural steel (6) are listed in Table 2.

Table 2. Mechanical Properties of Grade 700 Steel to AS 3597

Property	Specification
0.2% Proof Stress	≥ 690 MPa
Tensile Strength	790 - 930 MPa
Elongation in 50 mm G.L.	≥ 18%
Charpy Impact (Longitudinal)	≥ 40 J

2.5 Fabrication using High Strength Steels

When fabricating with high strength steel, the same fabrication methods for carbon steel can be used if slight modifications accounting for the higher strength are made. General information regarding the influence of fabrication and heat application on the mechanical properties of high strength structural grade steels can be found in Australian Standards (6), however the steel manufacturer may provide fabrication advice specific to their products.

Oxy-LPG and oxy-acetylene cutting processes can usually be used for all thicknesses of high strength steel when the travel speed is reduced to around 70% of speed used for the same thickness of carbon steel (3). The flame cutting process will usually produce a heat affected zone (HAZ) which results in higher hardness less than 2mm from the cut edge, a lower hardness from approximately 2 to 3 mm from the cut edge and distances beyond 3mm typically outside the HAZ. Air cooling after flame cutting is recommended to prevent a loss of ductility in the welded area. Plasma cutting, laser cutting and power sawing are also possible when slight modifications are made to the processes, with the high heat applied during laser cutting preventing the reduction of hardness in the HAZ. The use of water jet cutting does not create a HAZ and avoids adding internal stresses into the steel.

High strength quench and tempered structural steels can usually be cold formed using brake press bending or plate rolling techniques if sufficient machine power is available to accommodate the higher yield strength (3). The minimum recommended bending radius varies depending on the plate's grade, thickness, and the bending direction with respect to the axis the steel was rolled on, with a minimum recommended bending radius above 4 times the plate thickness in some configurations. For grades up to and including AS/NZS 3597 Grade 700, cold shearing up to a 25mm thickness is possible. To achieve this, guillotine blades must be very sharp and 2 to 3 times the shearing force may be required compared to that for plain carbon steel. Drilling, countersinking and tapping are all possible when using suitable high powered and rigid drilling equipment. Cobalt type, high speed steel drills, or drills using replaceable carbide inserts are recommended for drilling high strength structural grades of steel, with an adequate water and oil emulsion mixture required as the coolant.

Australian Standard AS/NZS 1554.4 (7) gives comprehensive information on the welding of structural high strength quench and tempered steels, with most high strength quench and tempered steels manufactured with a controlled carbon equivalent for optimal welding (3). They can usually be successfully welded to themselves and a range of other steels, provided low hydrogen consumables are used to prevent hydrogen embrittlement. The heating and cooling of the weld requires special attention to avoid modifying the microstructure of the steel in the HAZ, with successful welding requiring preheating of the steel, control of pass temperature and careful heat input.

3. HOT DIP GALVANIZING

3.1 Hot Dip Galvanizing Process for Typical Carbon Steel

The hot dip galvanizing process includes pre-treatment stages to produce the required cleaned and deoxidized steel surface which allows it to react with the molten zinc. For typical carbon steel, chemical pre-treatment stages are used to clean the steel, as seen in Figure 1. When surface contamination exists that cannot be removed by the conventional alkaline degreaser or acid pickle, or if a high risk of hydrogen embrittlement exists from exposure to the acid pickle, then a mechanical pre-treatment is undertaken, which generally involves abrasive blast-cleaning to a cleanliness of AS/NZS 1627.4 (8) Class Sa 3. The final stage of cleaning sees the steel dipped in a flux solution of zinc ammonium chloride to help prevent oxidation of the steel before exposure to the molten zinc and provide optimal conditions for the metallurgical reaction. Immersion of the steel into the zinc lowers the temperature of the zinc bath due to the temperature of the steel prior to dipping being much lower than that of the molten zinc at approximately 450°C. The steel temperature rises to the same temperature as the zinc, during which the metallurgical reaction takes place and then the steel is withdrawn. The steel thickness is a major influence on the time the steel is immersed in the zinc, as thicker steels take longer to reach the temperature of the zinc. For this reason, steels of similar thicknesses are typically dipped together. The steel is usually quenched to cease the metallurgical reaction and assist with cooling for handling. In Australia, the quenching stage also facilitates a chromium based passivation forming on the coating's surface to help prevent any initial rapid corrosion of the fresh zinc.

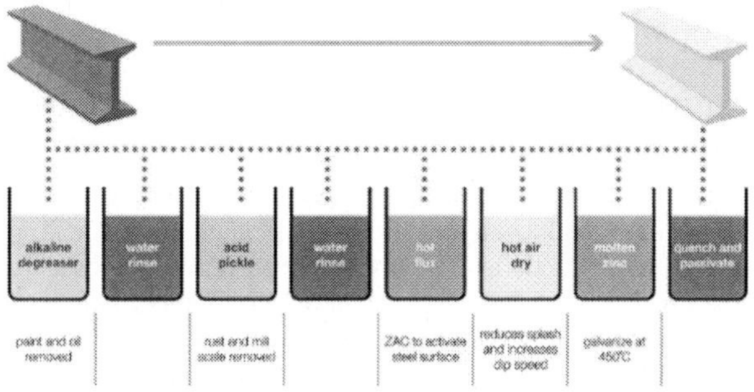

Figure 1. Diagram of the typical galvanizing process for mild steel

3.2 Galvanized Coating Properties of Typical Carbon Steel

The metallurgical reaction between the steel and the zinc results in zinc-iron alloy layers forming at the steel surface that are harder than typical carbon steel, as seen in Figure 2 (9). The molten zinc covers the alloys when the steel article is withdrawn from the bath, generally solidifying in place to form the pure zinc top layer of a typical galvanized coating with an initial shiny appearance. However, the composition of the steel, its initial surface condition and cooling rate can influence the coating structure. The silicon content of the steel is a driving factor of the reactivity, as the silicon triggers layer growth in the reaction (10). A graph of the relationship between the coating thickness and silicon content is shown in Figure 3 (10). The silicon contents that result in higher than average coating thicknesses are more likely to have alloy layers continuing to form using the pure zinc layer, resulting in zinc-iron alloys at the surface of the coating and a dull grey initial appearance. When abrasive blasting is used as the surface cleaning method, the thickness of the alloy layers is similar to those observed in a typical coating however a thicker outer layer of zinc is obtained. The time spent in the bath also contributes to the structure of the coating, with longer dip times typically resulting in larger alloy layer thickness and thicker overall coating.

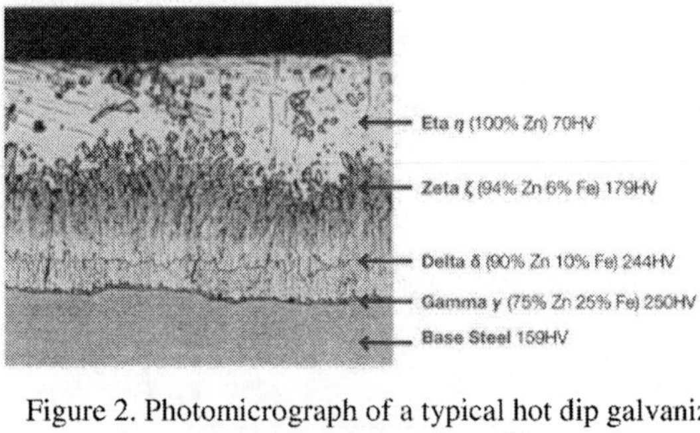

Figure 2. Photomicrograph of a typical hot dip galvanized coating, adapted from (9).

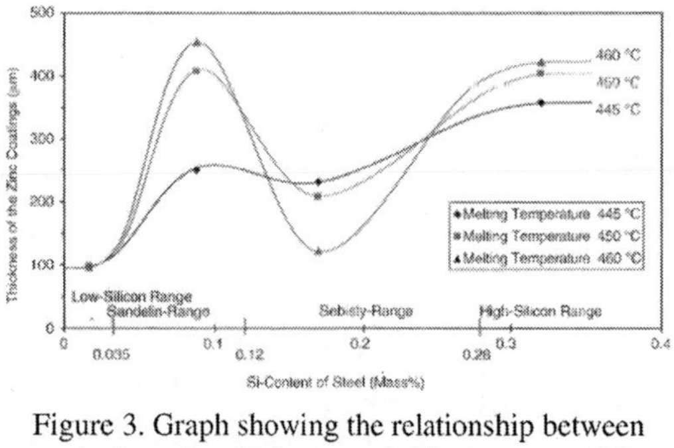

Figure 3. Graph showing the relationship between typical coating thickness and silicon content (10)

3.3 Considerations for High Strength Steels

When treating high strength steel, hydrogen embrittlement is a potential issue if hydrogen can be absorbed and trapped in the steel such as the use of chemical pre-treatment involving picking in acid (11). This may cause the steel to lose its ductility and suffer from brittle fracture. The Australian Standard for the design and durability of hot dip galvanized steels, AS/NZS 2312.2 (12) warns steels harder than 325 HB are susceptible to hydrogen embrittlement. Although most grades of high strength structural steel have a hardness value lower than this, in temperature affected areas resulting from thermal cutting or in the HAZ of welds the hardness is typically above this value. For this reason, the recommended surface preparation method for high strength steel is abrasive blasting to a cleanliness of Class Sa 3, as per AS/NZS 1627.4 (8), before fluxing and galvanizing.

As noted in section 2.2 of the paper, the mechanical properties of quench and tempered steels would only be altered by the galvanizing process if the bath temperature exceeded the tempering temperature of the steel (11). This does not occur for structural grades of high strength steels as they are typically tempered at 600°C and the temperature of a galvanizing bath is generally around 450°C. Wear resistant grades of high strength steel have much higher hardness to resist abrasion and are generally tempered under 400°C, making them unsuitable for the galvanizing process.

Quench and tempered structural grade steels with a typical silicon content of 0.2% by mass have been galvanized previously and achieved a sound and continuous coating (3) (11). Figure 3 shows this silicon content is not in one of the higher reactivity regions and although the coating thickness is expected to be comfortably above AS/NZS 4680 requirements it is unlikely to have any peeling or flaking issues related to excessively thick coatings, however this is a potential issue for higher silicon high strength quench and tempered steels.

3.4 Hot Dip Galvanizing to Australian Standards

In Australia, AS/NZS 4680 (13) is most commonly specified when performing batch hot dip galvanizing. For the product to conform to this Standard, it must first be cleaned and fluxed to a condition that will allow the galvanizing reaction to take place. The coating must also be continuous and as smooth and evenly distributed as possible and must be free of defects that may affect the stated end use of the article. It is noted that steels of certain compositions or articles cooled slowly after galvanizing may cause the finish to become partly or wholly grey in colour. Provided the coating has adequate adhesion, the grey finish is not detrimental and is not a cause for rejection. There is some allowance for damaged or uncoated areas but their total must not exceed 0.5% of the total surface area or 250cm^2, whichever is the lesser, and no individual damaged or uncoated areas shall exceed 40cm^2 unless they were caused by unavoidable airlocks or prior contamination of the steel. The coating must be sufficiently adherent to withstand normal handling without peeling or flaking and the average coating thickness required determined by the thickness of the steel, with minimum values listed in Table 3.

Table 3. Coating Thickness Requirements for Non-Centrifuged Articles (13)

Article Thickness (mm)	Average Coating Thickness Minimum (μm)	Average Coating Mass Minimum (μm)
≤ 1.5	45	320
> 1.5 ≤ 3	55	390
> 3 ≤ 6	70	500
> 6	85	600

AS/NZS 2312.2 (12) is a guide Standard commonly used for design advice and to estimate the durability of various zinc coatings in different atmospheric exposures. The Standard uses the same atmospheric corrosivity categories and first year corrosion rates as ISO 9223 (14), with the first-year corrosion rates of zinc listed in Table 4. Using the coating thickness of an article and by assessing the corrosivity of the environment into which it will be placed, the life to first maintenance can be estimated.

Table 4. First Year Corrosion Rates of Zinc in Different Atmospheric Environments (12)

Corrosivity Category	First Year Zinc Corrosion Rate (μm/year)
C1 (Very Low)	≤ 0.1
C2 (Low)	> 0.1 ≤ 0.7
C3 (Medium)	> 0.7 ≤ 2.1
C4 (High)	> 2.1 ≤ 4.2
C5 (Very High)	> 4.2 ≤ 8.4
CX (Extreme)	> 8.4 ≤ 25

4. EXPERIMENTAL DETAILS

4.1 Fabrication of the Samples

Grade 700 high strength steel to AS 3597 (6) was used for testing the suitability of hot dip galvanizing as a corrosion protection method. Four different steel thicknesses were chosen, being 6mm, 25mm, 40mm and 60mm, and four 150x150mm samples of each thickness were tested. For the 6mm, 25mm and 40mm, two samples of each thickness were cut using an oxy-acetylene flame and the other two cut using a water jetting method. The 60mm thick samples had two oxy-acetylene flame cut edges and two water jet cut edges. A combination of sharp edges and edges rounded to roughly a 2mm radius were present on the samples.

A 10mm diameter hole was placed in one corner of each sample, as shown in Figure 4, to act as a hanging point during the galvanizing process. The chemical composition and hardness values of the high strength steel samples tested are shown in Table 5 and meet the chemical composition and mechanical property requirements of the Australian Standard for Grade 700 quench and tempered steel, as set out in Table 1 and Table 2.

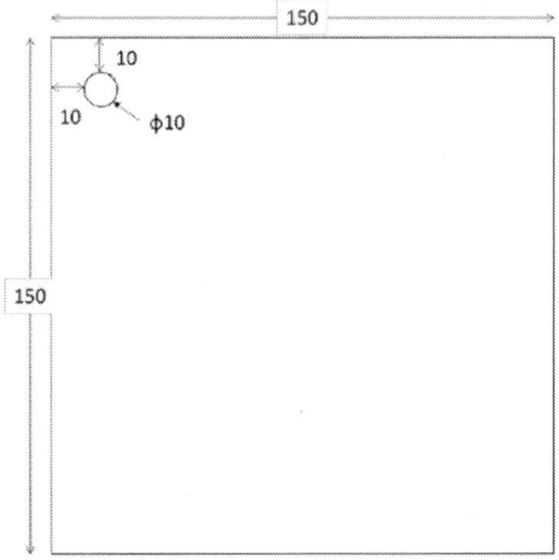

Figure 4. Diagram of the sample's geometry

Table 5. Chemical Composition and Hardness Values of High Strength Steel Samples for Testing

	Weight %								Hardness
Thickness (mm)	C	P	Mn	Si	S	Cr	Mo	B	HBW 10/3000
6	0.166	0.02	1.10	0.21	0.001	0.021	0.21	0.0016	255
25	0.150	0.008	0.39	0.45	0.002	0.730	0.19	0.0015	250
40	0.161	0.014	0.79	0.5	0.001	0.820	0.20	0.0014	259
60	0.158	0.018	0.99	0.52	0.003	0.800	0.210	0.0011	249

4.2 Hot Dip Galvanizing Process

4.2.1 Identification of samples

The cutting method used was marked on each sample using a water based marking pen and the 6, 25 and 40mm samples can be seen on a mesh tray prior to the blasting procedure in Figure 5. To ensure the cutting method could be identified after galvanizing, an extra piece of hanging wire was added through the hole of the water jet cut samples, which is shown in Figure 6.

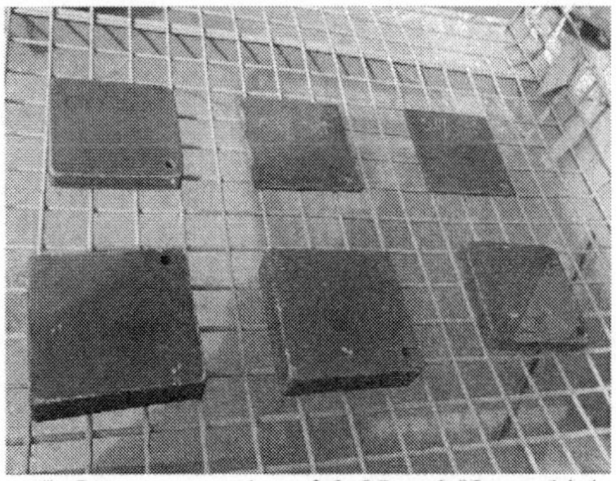

Figure 5. Oxy cut samples of 6, 25 and 50mm thickness.

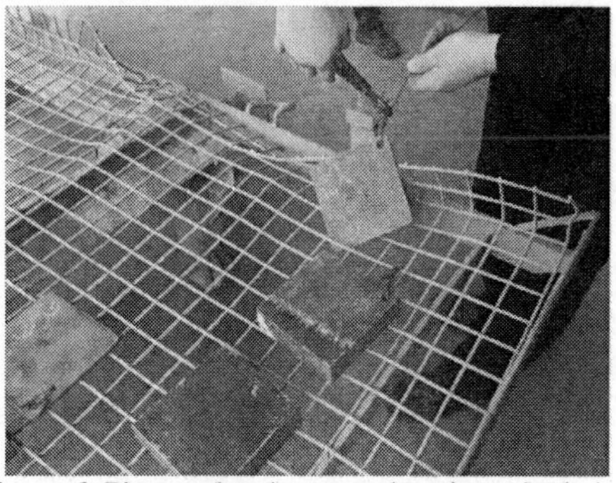

Figure 6. Picture showing extra hanging wire being placed in water jet cut samples

4.2.2 Pre-treatment/Cleaning Method

The samples were cleaned via abrasive blasting to Class Sa 3 with garnet used as the blast media. A blasting pressure of 100 psi was used at a 45-degree angle, then the garnet turned off and any remaining blast media blown off the surface using high pressure air cleaning.

On some samples, small regions were identified where the blasting cleanliness was not at a Class Sa 3 standard, with examples shown in Figure 7 and Figure 8. These areas on cut edges were generally due to the area being missed by the blast operator, and other areas on the main 150x150 faces may have been handled with dirty gloves after blasting, contaminating the surface. While these areas would generally be expected to be re-blasted following the post blasting inspection, it was decided to leave them in this condition to test the effect of different cleanliness levels on the formation of the hot dip galvanized coating.

Figure 7. Small area of missed blasting on 25mm sample

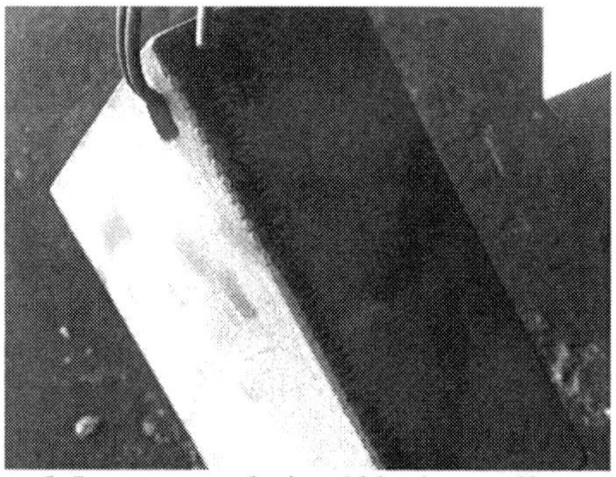

Figure 8. Larger area of missed blasting on 60mm thick hung sample

4.2.3 Dipping Procedure

The samples were galvanized in four separate dips, with the first dip commencing approximately 15 minutes after the conclusion of blasting. For each dip, the four samples of the same steel thickness were hung on a galvanizing jig and processed together to provide a consistent treatment time. The samples were first dipped in the zinc ammonium chloride flux solution before being immersed in the molten zinc and then quenched, with the time spent submerged during each stage given in Table 6.

Table 6. Time Samples were submerged in the Flux solution, Zinc and Quench

Dip Order	Sample Thickness (mm)	Time in Flux solution(s)	Time in Zinc (m:s)	Time in Quench (s)
1	40	20	2:43	16
2	25	14	6:20	19
3	6	14	6:27	16
4	60	>60	3:53	35

5. RESULTS AND DISCUSSION

5.1 Abrasive Blasting

After blasting the surface profile was measured using replica tape and a spring micrometer, as seen in Figure 9 and Figure 10. The profile measurements obtained are listed in Table 7 and show a smaller profile was achieved on the cut edges tested compared to the 150x150mm faces. The surface profile of the 60mm samples were not measured due to time constraints on the day.

Figure 9. Replica of the blasted profile being created using a burnishing tool

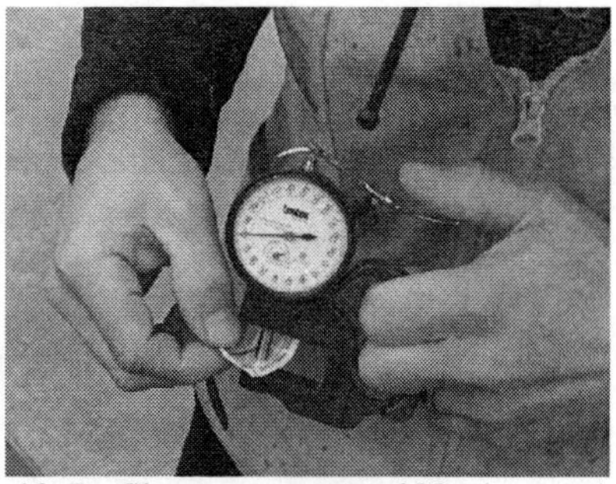

Figure 10. Profile measurement of 90 microns obtained on the square surface

Table 7. Blast Profile Achieved on Different Samples

Sample Thickness (mm)	Location of Measurement	Profile Achieved (μm)
6	150x150 face	80
25	150x150 face	95
40	150x150 face	90
40	Water jet cut edge	75
40	Oxy-acetylene cut edge	75

5.2 Coating Appearance

A smooth and continuous hot dip galvanized coating was observed on all samples where sufficient surface preparation from the blasting was achieved. The 150x150 faces of each sample had continuous hot dip galvanized coatings, although zinc runs and some yellow staining from the chromate was visible on samples of each thickness except 6mm (Figure 11). These were more pronounced as the sample thickness increased, with the effects seen on a 60mm sample in Figure 12. The runs and staining occurred because the hole was orientated horizontally during the galvanizing process so the zinc and quench water drained out slower from this location as a result. Zinc runs of this nature can usually be avoided by withdrawing the article from the bath slowly, which helps to prolong the time until the zinc cools and solidifies therefore allowing more time for drainage. The zinc runs are mostly aesthetic in nature, with the measured coating thickness in those area showing insignificant difference in the coating thickness of the surrounding areas. The chromate staining is aesthetic in nature and will disappear over time with exposure to the atmosphere. The coating had a shiny appearance for all thicknesses except the 6mm samples, indicating the coating structures included a pure zinc layer on top of the zinc-iron alloy layers. The mottled appearance, seen in Figure 11, of the 6mm samples represents areas where the alloy layers have grown to the surface. This is likely due to the samples being immersed and at the temperature of the zinc for longer than normal practice for this thickness of steel.

Figure 11. Picture of a typical 6mm sample without visible runs or chromate staining

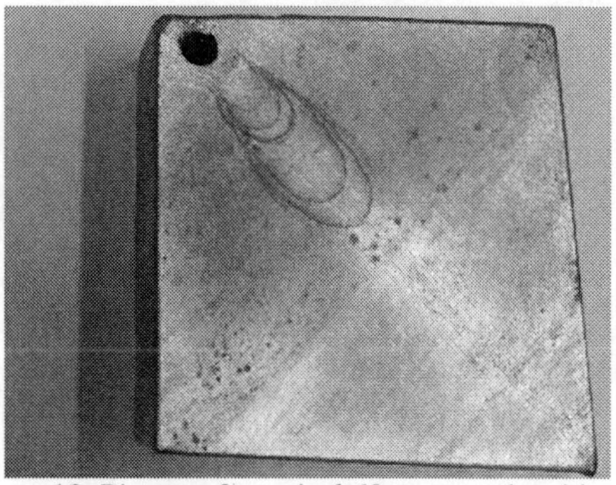

Figure 12. Picture of a typical 60mm sample with zinc runs along the diagonal and quench staining on top face

The appearance of the edges differed depending on the cutting method used, with the typical appearance of an oxy-acetylene edge shown before and after galvanizing in Figure 13 and Figure 14 respectively and the typical appearance of a water jet cut edge before and after galvanizing in Figure 15 and Figure 16 respectively. The oxy acetylene cut edge had a rougher texture, with the galvanized coating conforming to the texture of the steel substrate. The characteristic lines spanning the edge between faces were present after galvanizing and can be seen in Figure 14. Towards the top and bottom of the edge in Figure 14 a difference in appearance can be seen. These areas felt thicker when touched and were most likely due to a variation in the chemical composition at the steel's surface, with the flame cut edge having a thinner coating from depleted silicon levels and higher silicon levels with a thicker coating obtained where edges have been ground. The variation in coating thickness and texture of the flame cut edge can be prevented using a suitable grinding method to remove roughly 2mm of steel before the blasting procedure takes place. In contrast, the water jet cut edges were quite smooth with no variation in texture compared to the square face of the sample.

Figure 13. 40mm thick sample with flame Cut Edge

Figure 14. Typical appearance of an oxy-acetylene cut edge, 25mm sample.

Figure 15. 40mm thick sample with water jet cut edge

Figure 16. Typical appearance of a water jet cut edge, 25mm sample

In areas where the steel surface's cleanliness was lower than Sa 3, there were bare spots or the coating formed was rough and inconsistent, as seen in Figure 17 and Figure 18. The uncoated area in Figure 18 occurred in the same location of inadequate blasting seen in Figure 8, indicating the blasting was the most likely reason for the surface condition. It was noted that the blasting imperfections only occurred on the oxy-acetylene cut edges, where the rougher texture of the flame cut edges potentially needed more attention during blasting to ensure the surface was properly cleaned. Since blasting due to surface contamination is usually used in conjunction with the chemical pre-treatment process in the galvanizing plant, the importance of complete and satisfactory surface preparation when abrasive blasting is used as the sole method of surface preparation should be stressed in the specification.

Figure 17. Picture showing uncoated areas on a 6mm flame cut sample

Figure 18. Picture showing uncoated areas on the edge of a 60mm thick flame cut sample

5.3 Coating Thickness

An Elcometer 456 with a separate ferrous straight probe was calibrated and used to measure the thickness of the galvanized coating. 20 measurements were taken in random locations on both 150x150mm faces of each sample with averages reported in Table 8 and 10 measurements were taken on each cut edge of the 25, 40 and 60mm samples with averages reported in Table 9. Measurements on the edges of the 6mm samples were not performed due to the exceedingly rough texture of the oxy-cut on these samples and width of the measurement are being too close to the gauge's minimum sample diameter.

The cutting method used had no significant difference on the coating thickness of the 150x150mm faces, with all measured averages being well above the requirements of AS/NZS 4680 (13) (Table 3). Based on literature and practical experience, the thicker steel samples were expected to have a thicker zinc coating than thinner samples, which was not the observed result in this case. The absence of these effects was most likely due to the extended time spent immersed in the molten zinc, allowing the zinc and the steel to react for longer and form a thicker zinc coating, especially on the 6mm samples. The average standard deviation of measurements was much larger for the thicker steels, representing the effect of more zinc solidifying while draining over a larger portion of the face.

Table 8. Average Coating Thickness Measurements of 150x150mm Faces

Edge Cutting Method	Sample Thickness (mm)	Average Coating Thickness	Average Standard Deviation
Oxy-Acetylene	6	243	13
Water Jet	6	242	11
Oxy-Acetylene	25	232	14
Water Jet	25	240	17
Oxy-Acetylene	40	232	45
Water Jet	40	252	37
Both	60	153	41

The cutting process used on the edges significantly influenced the coating thickness measurements obtained. The coating thickness on oxy-acetylene cut edges varied depending whether the edge had been ground or not, with well ground edges having coating thicknesses similar to that of the water jet cut edges and 150x150 faces. Given the coating thickness obtained on the 150x150mm faces was typically greater than the cut edges, the open flame used during the tempering process did not seem to have significantly influenced the galvanizing process and the silicon content at the tempered surface would be thought to match that of the rest of the steel.

The coating thickness on the water jet cut edges was larger than the oxy-acetylene cut edges for all thicknesses, however the 40mm samples had an unexpected coating thickness 130μm lower than their adjacent 150x150mm faces. This was most likely a result of the shorter dipping time used for these samples, with a longer immersion times in the zinc potentially required to form a coating thickness on water jet cut edges equivalent to the other faces. The average coating thickness on the oxy-acetylene cut edges was below the 85μm requirement of AS/NZS 4680 (Table 3). Given AS/NZS 4680 states the coating thickness measurements must be carried out over 10mm from edges, corners and flame cut surfaces (13), the sample would comply. These coating thickness values highlight the importance of the recommendation in AS/NZS 2312.2 (12) for the fabricator to grind off flame cut surfaces prior to the galvanizing pre-treatment process to provide sufficient coating thickness.

Table 9. Average Coating Thickness Measurements of Cut Edges

Edge Cutting Method	Sample Thickness (mm)	Average Coating Thickness	Average Standard Deviation
Oxy-Acetylene	25	75	11
Oxy-Acetylene with Grinding	25	233	19
Water Jet	25	256	8
Oxy-Acetylene	40	73	10
Water Jet	40	122	11
Oxy-Acetylene	60	113	13
Water Jet	60	138	19

5.4 Optical Imaging

Optical imaging was performed on two samples of each steel thickness, one of each cutting method. For each sample, two pieces were extracted[1] and prepared for imaging, which provided over 30 imaging locations. Micrographs were taken of the 150x150mm faces, water jet cut edges, oxy-acetylene cut edges as well as both sharp and rounded corners. An image taken of the top surface of a 25mm thick sample can be seen in Figure 19 and Figure 20 with different magnification. Both Figures show different regions of alloy layers which formed to create the coating. Comparing the structure to the typical coating for carbon steel shown in Figure 2, this sample had a similarly thin gamma layer formed at the interface of the base steel, followed by a relatively thin delta layer and a significantly thicker zeta layer with a much coarser structure. The sample's coating was topped with a relatively thin eta layer, which provided it with a shiny appearance. The coating thickness and structure of the 25mm sample's water jet cut edge (Figure 21) is very similar to the top surface. The oxy-acetylene cut edge (Figure 22) is much thinner with a finer microstructure with a more typical ratio of alloy layers and eta layer thickness.

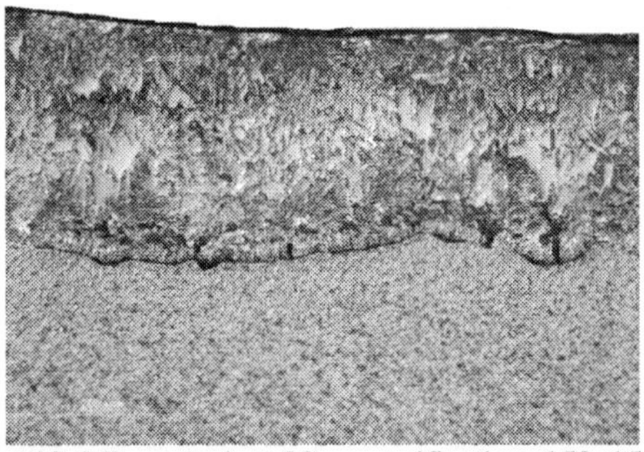

Figure 19. Micrograph at 20x magnification, 150x150mm face of 25mm thick water jet cut sample.

Figure 20. Micrograph at 20x magnification, 150x150mm face of 25mm thick water jet cut sample.

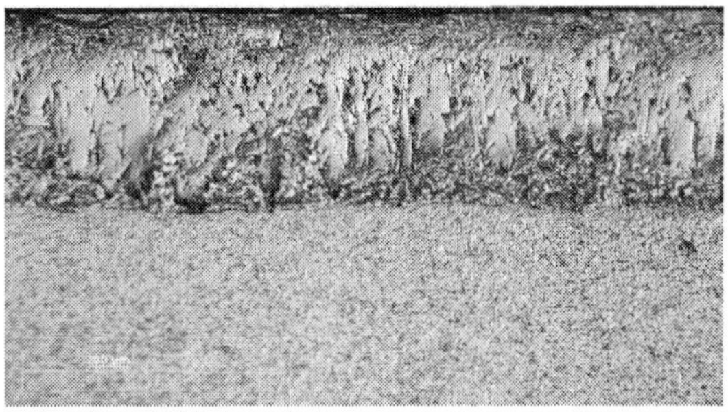

Figure 21. Micrograph of a water jet cut edge of a 25mm sample at 20x magnification

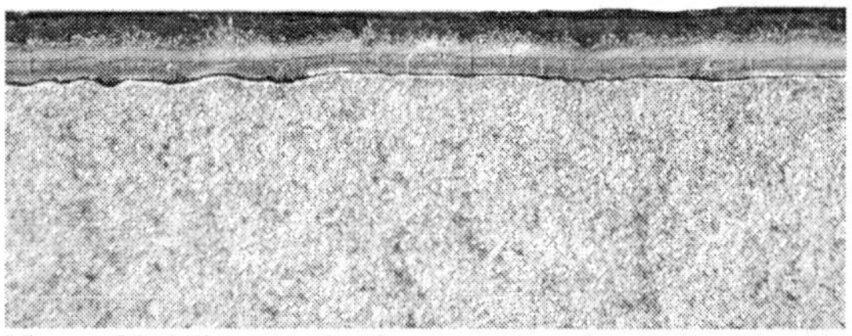

Figure 22. Micrograph of an oxy-acetylene cut edge of a 25mm sample at 20x magnification

Images were taken of round and sharp corners from the 6mm water jet cut and oxy-acetylene cut samples respectively to ensure the thick coating completely bonded with the base steel. Figure 23 shows the rounded corner of a water jet cut sample and Figure 24 shows the sharp edge of an oxy-acetylene cut sample. A continuous coating was found to have formed on both the sharp and round edges, with no voids between the coating and the steel. From the micrographs, the coating thickness on the round edge was estimated to be 200μm and approximately 160μm on the sharp edge. Importantly, the coating thickness at the corners are similar to coating thickness of the flat faces surrounding them and will have comparable corrosion protection to the rest of the sample. Figure 24 shows a different coating structure on each flat edge adjacent the corner, the horizontal surface being the oxy-acetylene cut edge and the vertical surface being the 150x150 face. This highlights how the cutting process used can influence the galvanizing of the steel.

[1] Optical imaging was performed on one 60mm sample only and four pieces were extracted, two on a water jet cut edge and two on an oxy-acetylene cut edge.

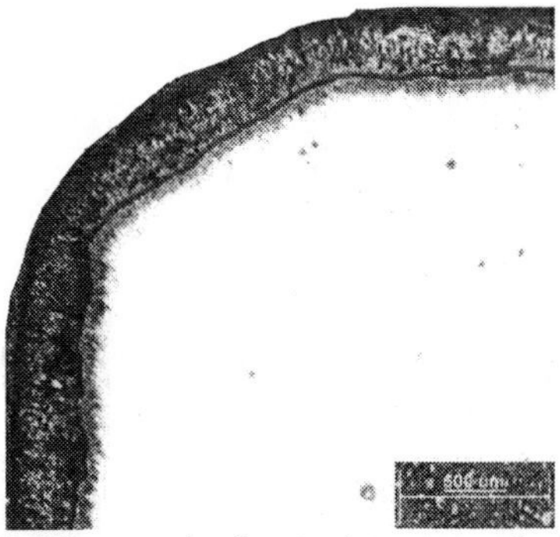

Figure 23. Micrograph of water jet cut round edge of 6mm sample, 5x magnification.

Figure 24. Micrograph of oxy-acetylene cut sharp edge of 6mm sample, 5x magnification.

5.5 Expected Durability

Using the first year corrosion rates of zinc from Table 4 and sample's measured averaging coating thicknesses (Table 8 and Table 9), the expected life to first maintenance was estimated and seen in Table 10.

Table 10. Estimated Life to First Maintenance (Years) Based on First Year Corrosion Rates

		Atmospheric Corrosivity Category				
Average Measurement Location on Sample	Average Coating Thickness (μm)	C2	C3	C4	C5	CX
150x150mm Face of 6, 25, 40mm	240	>100	>100	57 - >100	28 - 57	9 - 28
150x150mm Face of 60mm	153	>100	72 - >100	36 - 72	18 - 36	6 - 18
Oxy-Acetylene Cut Edge of 25 and 40mm	75	>100	35 - >100	17 - 35	8 - 17	3 - 8
Water Jet Cut Edge of 40 and 60mm	130	>100	61 - >100	30 - 61	15 - 30	5 - 15
Oxy-Acetylene Cut Edge of 60mm	138	>100	65 - >100	32 - 65	16 - 32	5 - 16

6. RECOMMENDED FABRICATION AND GALVANIZING PROCEDURE

Based on the results obtained from the testing undertaken, some basic recommendations can be made to achieve the successful hot dip galvanizing of high strength quench and tempered speciality steels. If using flame cutting during the fabrication process the edges should be ground to ensure the coating formed conforms to AS/NZS 4680 (13). Water jet cutting should be used where possible to avoid introducing a HAZ or internal stresses and a slightly higher dip time than usual should be used, especially for thick steels, to ensure a thick coating can form on edges.

An abrasive blast to Class Sa 3, as per AS/NZS 1627.4 (8), or equivalent must be used to prepare the steel for galvanizing. Since blasting is the only method of cleaning the steel, care must be taken to ensure it is evenly prepared over the surface and handled carefully after blasting to prevent surface contamination. The steel should be galvanized within 4 hours of being blasted and transported in a covered or enclosed area when possible to prevent excessive oxidation from occurring on the surface.

7. CONCLUSIONS

High strength quench and tempered structural grade steel with high silicon content could be successfully hot dip galvanized for corrosion protection with good adhesion, coating structure and thickness. Due to the higher sensitivity of high strength steels to hydrogen embrittlement, abrasive blasting needs to be used to clean and prepare the steel's surface for a reaction with the molten zinc, with examples of poor coating obtained when there was inadequate cleanliness from blasting. The average coating thickness measured on the large faces of samples was over 230μm on most samples. This corresponded to a life to first maintenance well over 50 years for atmospheric environments with a C3 corrosivity. Edges cut using a water jet prevented a HAZ from affecting the coating thickness and similar thicknesses to uncut surfaces were obtained when longer immersion time in the molten zinc was given. Edges cut using an oxygen-acetylene flame had a much lower coating thickness and reduced durability as a result, highlighting the importance of grinding flame cut edges before the hot dip galvanizing pre-treatment process.

8. FURTHER RESEARCH

Knowing that high strength structural grade quench and tempered steels can be successfully galvanized, it is suggested that further research be done to test situations likely to be encountered when galvanizing fabrications of high silicon, high strength structural steel grades. Testing welded samples using commonly used high strength welding electrodes would be useful to establish whether the weld material forms a continuous galvanized coating and confirm that the heat of the bath won't cause undesired expansion of the weld material leading to distortion. The high silicon, high strength steel should be tested using air cooling instead of quenching to investigate the potential for continued alloy formation, which can result in excessively brittleness, peeling or flaking. Although research into heat treated steels has shown the mechanical properties of the steel won't be affected unless raised to the tempering temperature, the hot dip galvanized high silicon steel samples could have their mechanical properties tested to confirm this. Similarly, the assumption that hydrogen embrittlement will occur in the HAZ of high strength structural grade steels in the acid pickling could be tested using acids with different inhibitors and without inhibitors to determine whether using blasting as the steel cleaning method is required. Dipping samples of the same thickness for different amounts of time and investigating the effect on coating properties and structure is also suggested, as different thicknesses of steel can be present in single fabrications and therefore need to be dipped together.

9. ACKNOWLEDGMENTS

This research was supported by Bisalloy®, who manufactured and fabricated the samples and Geelong Galvanizing, who performed the blasting and hot dip galvanizing of the samples. Thank you to Deakin University's Institute for Frontier Materials in Waurn Ponds, who performed the optical imaging.

10. REFERENCES

1. **Groover, Mikell P.** *Fundamentals of Modern Manufacturing: Materials, Processes and Systems.* Hoboken, NJ : John Wiley & Sons, 2010. ISBN 978-0470-467002.

2. **Wang, Lawrence K, Shammas, Nazih K and Hung, Yung-Tse.** *Waste Treatment in the Metal Manufacturing, Forming, Coating and Finishing Industries.* New York : CRC Press, 2009. ISBN 978-1-4200-7223-5.

3. **Bisalloy.** *BISPLATE Techincal Manual.* Wollongong : Bisalloy, 2006.

4. **Kuklik, Vlastimil.** *Hot Dip Zinc Coatings on Burnt Areas.* Ostrava : 17th Hot Dip Galvanizing Conference, 2011. Volume 13.

5. **Kennon, N R and Kohler, G R.** *Microstructures and Cracking in the Heat Affected Zone of Oxygen-Acetylene Cut Steels.* Woolongong : Australian Welding Research, 1979.

6. **Standards Australia.** *Structural and pressure vessel steel - Quenched and tempered plate.* Sydney : Sai Global, 2008. AS 3597.

7. **Standards Australia/Standards New Zealand**. *Structural Steel Welding, Part 4: Welding of high strength quench and tempered steels.* Sydney : Standards Australia, 2014. AS/NZS 1554.4:2014.

8. **Standards Australia**. *Metal Finishing - Preparation and Pretreatment of surfaces, Part 4: Abrasive Blast Cleaning of Steel.* Sydney : SAI Global, 2005. AS 1627.4.

9. **American Galvanizers Association.** The HDG Coating. *American Galvanizers Association.* [Online] [Cited: 30 06 2017.] https://www.galvanizeit.org/hot-dip-galvanizing/what-is-hot-dip-galvanizing-hdg/the-hdg-coating.

10. **Maass, Peter and Peissker, Peter.** *Handbook of Hot-Dip Galvanization.* Leipzig : WILEY-VCH, 2011. ISBN 978-3-527-32324-1.

11. **Sepper, S, Peetsalu, P and Saarna, M.** *High Strength Steel Behaviour During Hot Dip Galvanizing.* Senec : 18th Hot Dip Galvanizing Conference, 2012.

12. **Standards Australia/Standards New Zealand.** *Guide to the protection of structural steel against atmospheric corrosion by the use of protective coatings - Part 2: Hot dip galvanizing.* Sydney/Auckland : SAI Global, 2014. AS/NZS 2312.2.

13. **Standards Australia/Standards New Zealand**. *Hot-dip galvanized (zinc) coatings on fabricated ferrous articles.* Sydney/Auckland : SAI Global, 2006. AS/NZS 4680.

14. **International Organization for Standardization.** *Corrosion of Metals and Alloys - Corrosivity of Atmospheres - Classification, Determination and Estimation.* Geneva : ISO, 2012. ISO 9223.

11. AUTHOR DETAILS

Will is the Market Development Engineer for the Galvanizers Association of Australia. His responsibilities include responding to engineering enquiries, preparing technical documentation for members, staying abreast of new Australian Standards, and presenting to engineers, architects and students Australia wide. Will keeps his knowledge up to date by attending relevant conferences and is a member of various Victorian young professionals committees, including Engineers Australia, Australian Steel Institute and Australasian Corrosion Association.

Ann completed a Bachelor of Applied Chemistry in 2010 and went on to obtain First Class Honours in Applied Science from RMIT University in 2012. She started working part-time for the Galvanizers Association of Australia (GAA) in 2009 and became full-time at the start of 2012. Ann is currently the Corrosion and Sustainability Officer for the GAA with one of her main roles being to provide technical guidance to users of hot dip galvanizing. Ann has become both a NACE Certified Level 2 Coating Inspector and a certified ACA HDG Inspector.

Peter Golding is the Chief Executive Officer of the Galvanizers Association of Australia. Peter joined the GAA in April 2011. Prior to being named CEO, Peter worked for nearly 20 years in the Australian steel industry in various product engineering, manufacturing and management roles. He has overall responsibility for the marketing and technical development of hot dip galvanized steel and represents the Association on various industry and Standards bodies. Peter earned his MBA and completed his undergraduate studies at the University of Adelaide, where he received an Honours degree in Mechanical Engineering.

ON THE HYDROGEN-INDUCED CORROSION OF AISI 410

S. Thomas[1,*], K. Wang[1,2], R. L. Liu[1], N. Birbilis[1]
[1]Department of Materials Science and Engineering, Monash University, Victoria, Australia
[2]School of Materials Science and Engineering, Central South University, Changsha 410083, Hunan, China.

SUMMARY: The martensitic stainless steel (SS) AISI 410 is prone to sulphide stress cracking (SSC) in "sour" service, such as H_2S environments, due to hydrogen (H) embrittlement. The presence of absorbed H within SS makes them more susceptible to pitting, corrosion and environmentally assisted cracking, however the mechanistic aspects of such H-induced corrosion has not been fully elucidated in research to date. Herein, a zero resistance ammeter (ZRA) was used to investigate the impact of absorbed H on the corrosion of AISI 410, by galvanically coupling H-charged and uncharged SS specimens. A mechanism of H-induced corrosion was proposed based on the evidence of hydrogen oxidation, metal dissolution and passive film reduction from the H-charged specimens during such ZRA tests.

Keywords: Hydrogen embrittlement, AISI410, sulphide stress cracking, Hydrogen charging

1. INTRODUCTION

The martensitic stainless steel (SS) AISI 410 is typically used in turbo-machinery that operate in H_2S-containing environments, also termed as "sour" environments [1-2]. This (high strength) steel is susceptible to sulphide stress cracking (SSC) caused by hydrogen embrittlement (HE) [1]. Hydrogen (H) enters such steels during corrosion, in addition to the generation of hydrogen from the cathodic hydrogen evolution reaction (HER) [3]. The HER consists of multiple steps according to the Volmer-Tafel mechanism (1-3). Firstly, protons are reduced to absorbed H on the metal surface (M) forming MH_{ads} (1), whilst in (2) the adsorbed H atoms typically recombine to form molecular H_2 thus completing the HER. However, sulphide species (or H_2S) act as a poison for (2) and thus in "sour" environments a small proportion of the adsorbed H atoms are absorbed into the metal (3) forming MH_{abs} [4].

$$M + H^+ + e^- \longrightarrow MH_{ads} \quad (1)$$

$$MH_{ads} + MH_{ads} \longrightarrow 2M + H_2 \quad (2)$$

$$MH_{ads} \longleftrightarrow MH_{abs} \quad (3)$$

Absorbed H could remain in the vicinity of the surface (as surface H) or propagate as diffusible H (H_{Diff}) within steels [5]. The H_{Diff} internally diffuses towards regions with high tri-axial stresses and could concurrently embrittle steel resulting in HE/SSC [5]. The precursor to SSC is usually pit initiation/propagation in the SS and since AISI 410 contains about 12-13% chromium (Cr), it has been difficult to fathom the mechanism by which pit initiation occurs upon this SS in H_2S-containing saline solutions. It is not clear whether H induces pitting (local passivity breakdown) in SS, which then transitions into cracking or whether SSC is totally independent of the electrochemical effects of H. The presence of H decreases the corrosion potential (E_{corr})/pitting potential (E_{pit}) of stainless steels [6] and also increases the corrosion rate of carbon steels [6, 7]. The most popular mechanism ratifying the H-induced corrosion of steel is that, absorbed H could reduce the chromium oxide (Cr_2O_3) or iron oxide (Fe_2O_3) film formed upon such steels; thus enhancing their susceptibility to localised corrosion [8-12]. Recently, Thomas et al. performed on-line inductively coupled plasma-atomic emission spectroscopy (ICP-OES) on H-charged and uncharged 1018 carbon steel, when electrochemically polarised in acidic solutions [7]. It was observed that the metal dissolution rate as reconciled by ICP-OES analysis, from H charged steels was much higher than that of uncharged steels. Such observations led the authors to rationalise the findings by proposing that interstitial H weakens metallic bonding and therefore, could intrinsically enhance metal corrosion rather than via destabilisation of the passive film [7]. In spite of several such developments, a clear understanding on the mechanism of H-induced corrosion eludes the research community. It is not clear as to why absorbed H tends to lower the E_{corr} of steel nor on what is the driving force causing the enhanced metal dissolution in H-charged steels. The enhanced corrosion of H-charged steels is evident during anodic polarisation tests [6] however, rarely has the effect of H on SS corrosion been assessed at or near open circuit conditions.

The diffusible hydrogen concentration ($C_{H,\ diff}$) in H-containing steels is typically estimated using the so-called Barnacle cell method, wherein the steel specimen is coupled to a Ni/NiO electrode in an alkaline solution thus polarising the steel sample to 0.35 V_{SCE} [13]. At this potential, absorbed H is oxidised and the $C_{H,\ diff}$ could be estimated from the recorded steady-state H-oxidation current densities [14, 15]. However, the Barnacle cell method is a destructive method and the anodic polarisation experienced by H-charged steel specimens during the test itself could aggravate electrochemical reactions upon their surfaces. The $C_{H,\ diff}$ estimated on the basis of anodic polarisation of H-charged steel specimens might be not be as accurate as that estimated using specimens maintained at or near OCP. Herein, an electrochemical method was devised to clarify electrochemical effects of H on the corrosion of steel, near its OCP. Both un-charged and H-charged AISI 410 specimens were coupled using a zero resistance ammeter (ZRA) to measure the galvanic current flowing between the two specimens. The resultant currents were then used to estimate the "true" $C_{H,\ diff}$ and also to realise a mechanism for H-induced corrosion for SS.

2. EXPERIMENTAL

The AISI 410 specimens were machined from AISI 410 peeled bars (as received from Interalloy, Australia). The stocked peeled bars are typically hardened in the temperature range 950 ^{o}C-1010 ^{o}C and then tempered between 600-700 ^{o}C, followed by cooling in air. The AISI 410 specimens were firstly prepared to a 2000 grit finish using silicon carbide (SiC) paper. They were then subject to cathodic charging (for 24 hours) in 0.1 M NaCl (adjusted to pH 10 using NaOH), at different current densities, namely 1, 10 and 100 mA/cm^2. The H-charged specimens were retrieved and galvanically coupled to an uncharged specimen (using the ZRA) in 0.1 M NaCl adjusted to pH 3.5 using citric acid.

The galvanic current flowing between the H-charged and uncharged specimens were used to determine the $C_{H,\ diff}$. After the ZRA test, aliquots from the electrolyte were retrieved and subject to ICP-OES analysis using an Optima 8000 (Perkin-Elmer). The concentrations of iron (Fe) and chromium (Cr) were correspondingly determined using ICP-OES analysis. In separate experiments, the H-charged specimens were polarised at 0.33 V_{SCE} in 0.1M NaOH under de-aerated conditions (using the Barnacle cell method [15]). The H-oxidation current densities realised during these measurements were also used to independently determine the $C_{H,\ diff}$ values.

3. RESULTS AND DISCUSSION

The open circuit potential (OCP) values of H-charged and uncharged AISI 410 specimens are shown in Figure 1. It is evident that the H-charged specimens are significantly less noble than that of the uncharged specimen. The OCP values were also found to decrease with an increase in the cathodic charging current densities (i_c). The equilibrium electrode potential (E^o) of the H/H$^+$ redox reaction in a solution at pH 3.5 is around -2 V_{SCE} [16]. Therefore, as H enters the SS its OCP tends towards the E^o of the H/H$^+$ reaction resulting in such negative values of H-charged SS specimens.

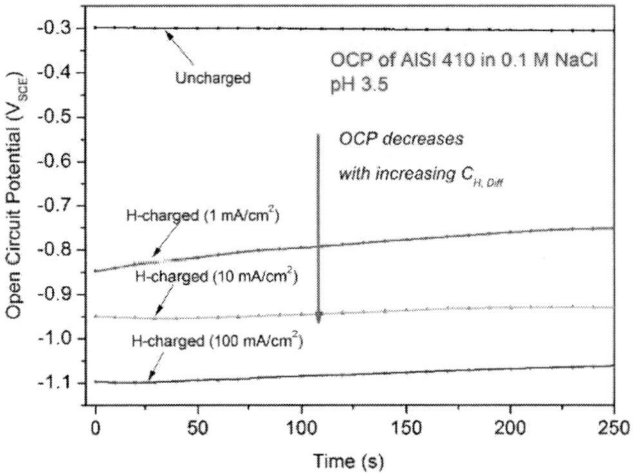

Figure 1: The OCP of uncharged and H-charged AISI 410 specimens during immersion in 0.1 M NaCl (with pH 3.5).

The H-oxidation currents measured from each of the three H-charged specimens (following the Barnacle cell method) are presented in Figure 2a.

Similarly, the galvanic currents recorded during ZRA measurements between the different H-charged and uncharged specimens, in 0.1 M NaCl are shown in Figure 2b. The OCP values of the H-charged specimens are well below that of the uncharged specimen and therefore, during the ZRA, the H-charged specimen tends to become the anode and the uncharged specimen the cathode (Figure 2b). A positive current correspondingly flows from the H-charged specimens to the uncharged specimen and this current in fact mainly corresponds to the oxidation of absorbed H. These H-oxidation current densities were also found to increase with an increase in the i_c values in both the Barnacle cell and ZRA tests. The steady state H-oxidation current densities (at a time of 400 s), were used to compute the $C_{H,\,diff}$ using (4). Here, $C_{H,\,diff}$ is in mol/cm^3, i_t is the current at a certain time, F is the Faraday constant, z is 1, and $D_{H,\,eff}$ (the diffusion coefficient of H in AISI 410) is taken as 10^{-7} cm^2/s [17].

$$C_{H,diff} = \frac{i_t}{zF}\left(\frac{D_{H,eff}}{\pi t}\right)^{\frac{1}{2}} \qquad (4)$$

The $C_{H,\,diff}$ values as estimated using the Barnacle cell and ZRA tests are presented in Figure 3a-b. The values of $C_{H,\,diff}$ as estimated using the ZRA tests are higher than that estimated by the Barnacle cell method. This difference could be due to the fact that the solution used in the ZRA test is acidic, whilst the solution used in the standard Barnacle cell test method is alkaline. However, the ZRA test serves to identify the $C_{H,\,diff}$ in the environment of interest (such as pH 3.5, 0.1 M NaCl solution) without aggressively polarising or corroding the specimens, whilst the Barnacle cell method is typically only used in the conventional deaerated 0.1 M NaOH solution. The ZRA test could therefore be a more suitable method to estimate $C_{H,\,diff}$ when compared to the Barnacle cell method. The metal ion (Fe and Cr) concentrations in the electrolyte after the ZRA tests, as estimated using ICP-OES analysis are presented in Figure 4. It is worthwhile to note that although the OCP of the H-charged specimen is less noble or cathodic to the OCP of an uncharged AISI 410 specimen, Fe dissolution still takes place. This suggests that absorbed H intrinsically promotes Fe dissolution, perhaps even by weakening the Fe-Fe bonding in steels [6]. The reduction of Cr_2O_3 film is evidenced by detection of Cr^{3+} ions in the electrolyte, using ICP-OES (Figure 4). This reveals that H-oxidation is usually accompanied by reduction of the Cr_2O_3 film, which could destabilise passivity and initiate localised corrosion. These results indicate that absorbed H enhances the corrosion rate of this steel by destabilising or reducing the passive film and also intrinsically promotes Fe dissolution.

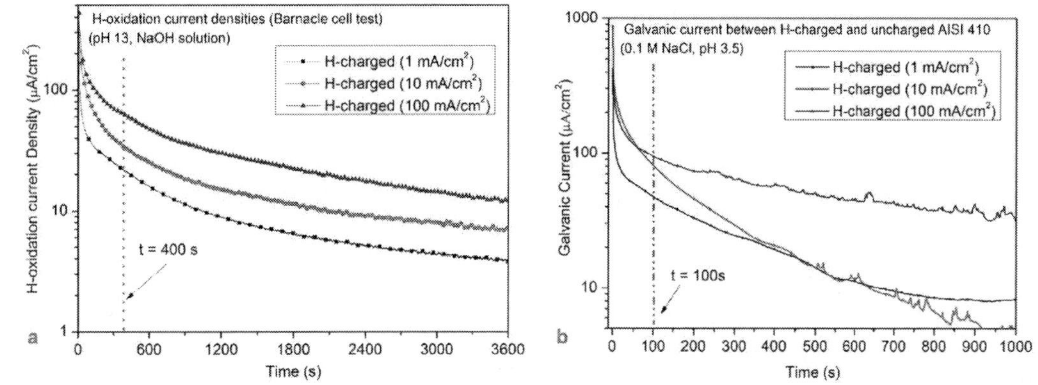

Figure 2: a) The H-oxidation current densities emanating from H-charged specimens during the barnacle cell tests in 0.1 M NaOH solution and b) during the ZRA tests in 0.1 M NaCl (pH 3.5)

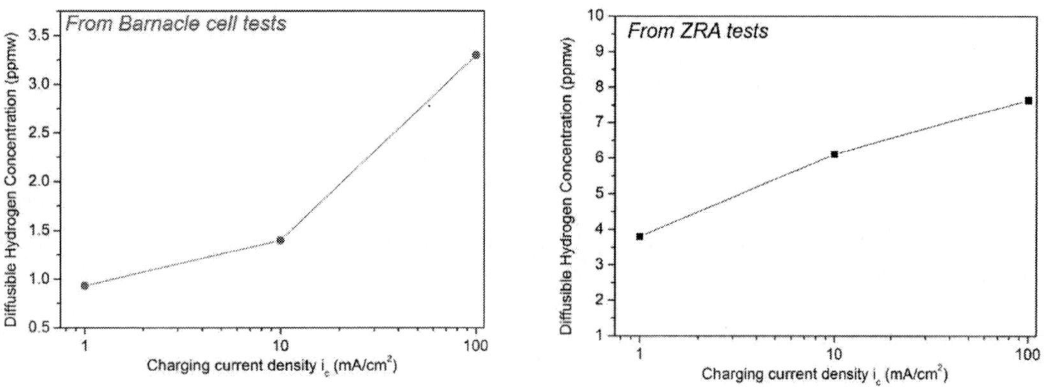

Figure 3: a) The $C_{H, diff}$ values as estimated using the Barnacle cell tests and b) ZRA tests for the three H-charged specimens

The results from the ZRA tests (as summarised in Figs. 2b and 3b) reveal that H-oxidation is the primary electrochemical reaction that takes place upon the surface of an H-charged SS specimen. Correspondingly, the H-oxidation current densities and $C_{H, diff}$ values are both found to increase with the charging current densities (i_c) (Figure 2a-b, 3a-b). Interestingly, metal dissolution during the ZRA test is found to be the highest for the sample cathodically charged at 1 mA/cm^2 and lowest for that charged at 100 mA/cm^2 (Figure 4). The equilibrium electrode potential for the Fe/Fe^{2+} redox reaction at pH 3.5 is around -0.75 V_{SCE} [16]. The OCP values of the H-charged specimens (with i_c at 10-100 mA/cm^2) are below the redox potential of Fe/Fe^{2+} when compared to the sample charged at 1 mA/cm^2. It is therefore plausible that the Fe dissolution rates from H-charged samples may not be substantial if the OCP of the samples are below the redox potential of the Fe/Fe^{2+} reaction.

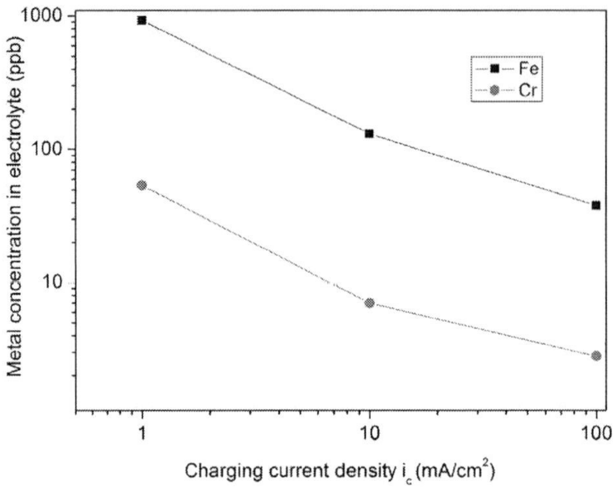

Figure 4: The Fe/Cr concentrations as detected in 0.1 M NaCl (pH 3.5) using ICP-OCP analysis after the ZRA tests.

The entry of H at certain regions in the steel makes those regions anodic with respect to the pristine (or uncharged) steel surface. The resulting galvanic coupling between the H-charged and uncharged areas in the AISI 410 specimens. This causes the oxidation and egress of absorbed H, with the exiting H atoms undermining passivity by destabilising/reducing the passive film. The dissolution of Fe takes place through a de-passivated surface, with absorbed H activating Fe dissolution when the E_{corr} of the H-containing SS is above the equilibrium potential of the Fe/Fe^{2+} redox reaction. The rate of metal dissolution/film reduction is lower if the E_{corr} of the H-charged steel is below the redox potentials of Fe dissolution/Cr_2O_3 reduction respectively.

4. CONCLUSIONS

The impact of H on the corrosion of AISI 410 was studied using ZRA testing, wherein H-charged specimens were galvanically coupled to uncharged specimens to interpret the resulting galvanic currents. It was observed that the H-charged specimens are less noble than that of the uncharged specimen and take the role of the anode in such ZRA tests. The predominant electrochemical reaction taking place upon H-charged or H-containing steels is H-oxidation. The dissolution of Fe and the Cr_2O_3 film also concurrently takes place in H-containing steels, albeit at lower rates when compared to H-oxidation. The reduction of the Cr_2O_3 film (or depassivation) mainly takes place when H exits the steels (as the counter reaction), however, Fe dissolution is more pronounced when the OCP of the H-containing steel is anodic to the equilibrium potential of the Fe/Fe^{2+} redox reaction.

5. ACKNOWLEDGMENTS

We are grateful for support by Woodside Energy. The technical assistance of Cory Wong and Andrew Speer is also gratefully acknowledged.

6. REFERENCES

1) A. Turnbull, A. Griffiths, Corrosion and cracking of weldable 13 wt-% Cr martensitic stainless steels for application in oil and gas industry, British Corrosion Journal, 1996, 31, 39.
2) A. Turnbull, M. Saenz De Santa Maria, N. D. Thomas, The effect of H_2S concentration and pH on hydrogen permeation in AISI 410 stainless steels in 5% NaCl, Corrosion Science, 1989, 29, 89.
3) R. N. Iyer, H. W. Pickering, M. Zamanzadeh, Analysis of hydrogen evolution and entry into metals for the discharge-recombination process Journal of Electrochemical Society, 1989, 136, 2463.
4) E.G. Dafft, K. Bohnenkamp, H.J. Engell, Investigations of the hydrogen evolution kinetics and hydrogen absorption by iron electrodes during cathodic polarisation Corros. Sci., 1979, 19, 591.
5) R. L. S. Thomas, D. M. Li, R. R. Gangloff, J. R. Scully, Trap-governed hydrogen diffusivity and uptake capacity in ultrahigh-strength AERMET 100 steel, Met. Trans. A, Phys. Metall. Mater. Sci., 2002, 33, 1991.

6) S. Thomas, G. Sundararajan, P. D. White, N. Birbilis, The effect of absorbed hydrogen on the corrosion of steels: Review, Discussion and Implications, Corrosion, 2017, 73, 426.
7) S. Thomas, N. Ott, R. F. Schaller, J. A. Yuwono, P. Volovitch, G. Sundararajan, N. V. Medhekar, K. Ogle, J. R. Scully, N. Birbilis, The effect of absorbed hydrogen on the dissolution of steel, Heliyon, 2016, 2, e00209.
8) S.-L. Pyun, C. Lim, R.A. Oriani, Corrosion Science, The Role of Hydrogen in the Pitting of Passivating films on Pure Iron, Corrosion Science, 1992, 33, 437.
9) Y.M. Zeng, J.L. Luo, P.R. Norton, Initiation and propagation of pitting and crevice corrosion of hydrogen-containing passive films on X70 micro-alloyed steel, Electrochimica Acta, 2004, 49, 703.
10) S. X. Mao, B. Gu, N. Q. Wu, L. Qiao, The mechanism of hydrogen facilitated anodic dissolution type stress corrosion cracking: Theories and experiments, Philosophical Magazine, 2001, 81, 1813.
11) H. Yashiro, B. Pound, N. Kumagai, K. Tanno, The effect of permeated hydrogen on the pitting of type 304 stainless steel, Corrosion Science 1998, 40, 781.
12) L. J. Qiao, J. L. Luo, Hydrogen-Facilitated Anodic Dissolution of Austenitic Stainless Steels, Corrosion, 1998, 54, 281.
13) R. F. Schaller, S. Thomas, N. Birbilis, J. R. Scully, Spatially resolved mapping of the relative concentration of dissolved hydrogen using the scanning electrochemical microscope, Electrochemistry Communications, 2015, 51, 54.
14) Schaller, R. F.; Scully, J. R. Measurement of effective hydrogen diffusivity using the Scanning Kelvin Probe, Electrochemistry Communications, 2014, 40, 42.
15) ASTM Standard F1113-87, Standard Method for Electrochemical Measurement of Diffusible Hydrogen in Steels, ASTM International, West Conshohocken, PA, 2005.
16) M. Pourbaix, Atlas of Electrochemical Equilibria in Aqueous Solutions, 2nd edition, Houston, TX, NACE International, 1974, 307.
17) G. Hinds, J. Zhao, A.J. Griffiths, A. Turnbull, Hydrogen Diffusion in Super 13% Chromium Martensitic Stainless Steel Corrosion, 2005, 61, 348.

6. AUTHOR DETAILS

Sebastian Thomas is a postdoctoral research fellow in the Department of Materials Science and Engineering, Monash University. He presently works on the corrosion of additively manufactured alloys, coatings, corrosion under insulation and hydrogen induced corrosion/cracking.

Kangyan (Ken) Wang is a visiting undergraduate student to Monash University from the School of Materials Science an Engineering, Central South University, China.

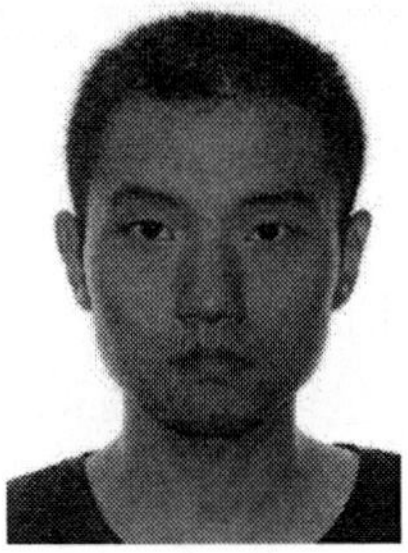

Ruiliang (Tony) Liu is a PhD student in the Department of Materials Science and Engineering, Monash University. He holds a B.Eng (Hons.) from Monash University in Materials Engineering. His research is in the development of corrosion resistant Mg-alloys.

Nick Birbilis is Professor and Head of Department, Material Science and Engineering, at Monash University. His research interests include the broad areas of corrosion and corrosion control.

USING INTUMESCENT COATINGS TO ACHIEVE BUILDING CODE COMPLIANCE ON STEEL STRUCTURES

Allan Jowsey MEng CEng MIFireE MSPFE
Akzo Nobel, Stoneygate Lane Felling NE10 OJY

SUMMARY: Intumescent Coatings are the first and natural choice for structural steel fire protection on commercial projects in Europe due to them being a cost effective and set & forget solution when the correct system is selected. Their use in the Australasian region is dramatically increasing as alternative building code compliance, fire engineering and structural steel solutions are proposed for projects. However Intumescent coatings are often incorrectly understood and specified, leading to the project not knowing exactly what is required and less than optimum costing of projects.

Keywords: Steel, Fire, Cellulosic, Intumescent, Specification,

1. INTRODUCTION

The ever-increasing use of steel as a construction material has led to enhanced flexibility in design, as well as significant time savings in the construction industry. However, its use in social infrastructure and buildings has brought an additional challenge - that of fire safety. In the event of a fire, steel can lose its strength and collapse, resulting in damage to property and in the worst case, loss of life.

Almost all buildings incorporate some fire safety measures. In the event of a fire, structures are required to maintain their stability for a reasonable period of time to enable occupants to evacuate and to provide safety to fire-fighters.

There are numerous products available to designers to satisfy the fire resistance requirements of their projects. Understanding passive fire protection technology and the impact of legislation is important to ensure compliance with standards for fire and life safety.

Intumescent coatings provide a cost effective and environmentally method to protect structural steelwork, compared to other traditional methods such as vermiculite spray.

As with any fire protection product, it is important to understand its basis to ensure correct specification and one that is fit for purpose. This paper outlines some of the key issues that architects, engineers, fabricators, applicators and Approving Authorities should be aware of when dealing with intumescent coatings with the intention of raising awareness and providing an increased level of understanding with Intumescent coating technology.

Understanding Intumescent technology, structural response, fire types, fire resistance ratings, anti-corrosion performance, specification, and certification is key to ensuring the correct system is documented and applied.

The importance of fireproofing materials in the event of a fire in a building is critical. However, for many designers little thought is generally given to specific details relating to fireproofing structural steelwork. This article provides a high-level overview of approaches being adopted in the UK and Europe that combine structural engineering approaches, testing of fireproofing materials and collaborative efforts between designers and manufacturers to arrive at optimized, efficient, robust and safe designs. Such approaches take advantage of steel design and over the last two decades have played an important role in making steel the preferred choice for the construction industry.

2. OBSERVATIONS

2.1 Choice of Fire Protection Material

Thin-film intumescent coatings now dominate the UK structural fireproofing market in new buildings as shown in Figure 1. This can be attributed to the commitment on the part of the manufacturers to research and design. This has been helped and encouraged by the growth of in-shop application and structural fire engineering by consulting engineers.

In-shop application is more expensive than most other forms of fire protection in terms of initial cost. However, for projects where speed is of the essence, health and safety creates difficulties, access is difficult, the weather may cause problems, disruption of other trades on site, etc., this premium can pay big dividends later. It is mainly used where medium to large buildings are constructed quickly. Estimates in the UK suggest that in-shop application accounts for about 15% market share of all fireproofing materials, with around 25% of all thin-film intumescent coating applied this way.

Twenty years ago, intumescent in the UK had a relative low usage with an approximate overall 20-25% market share. Today that figure is in excess of 70% and manufacturers are now starting to look at structural fire engineering approaches which assist clients to reduce cost, increase efficiency and promote the use of structural steelwork.

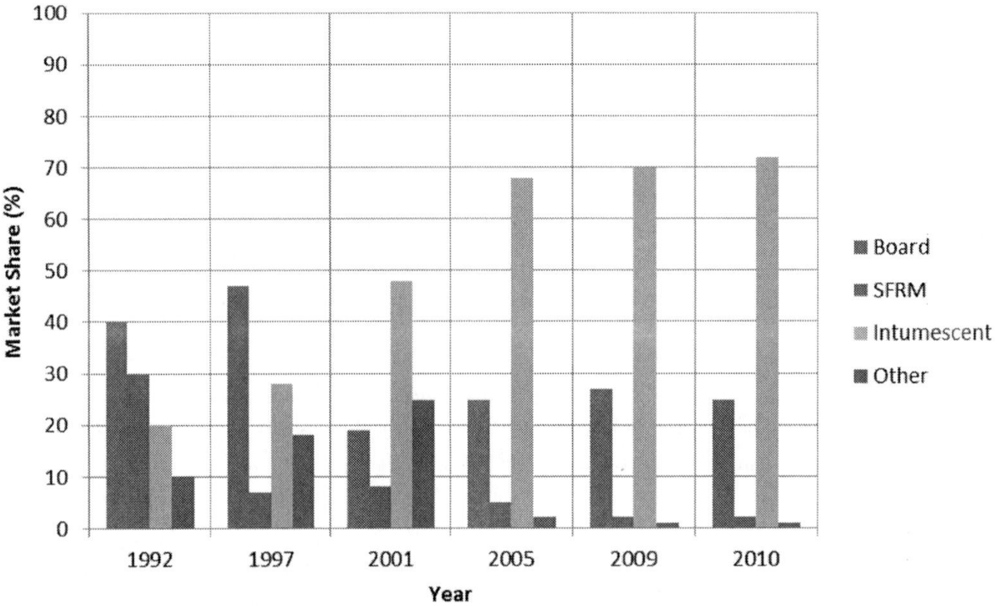

Figure 1: Fireproofing material usage by date in the UK for new building construction. *Courtesy of the British Constructional Steelwork Association (BCSA) and Tata Steel.*

SFRM – Sprayed fire resistant material
Board – Fire rated board
Other – Blankets etc

2.2 Key Considerations for Specifiers

The specifier should consider the environment in which the intumescent coating system is going to be applied and also the environment it must withstand with respect to the life of the asset in question. Usually the life of the coating system is measured in terms of the years before first major maintenance as described in ISO 12944.

A range of intumescent coating technologies are available to choose from depending on the environment type and the expected design life of the system. Thin film intumescent coatings such as solvent or water-borne acrylics are generally suited for interior conditioned spaces where as comparatively thicker epoxy coatings are more suited for aggressive exposed conditions.

An intumescent coating system has three main components – a primer, a basecoat (intumescent) and a sealer coat, which should all be supplied from one manufacturer to avoid any compatibility or performance issues. In some cases the sealer coat may not be necessary.

The primer assists with the adhesion of the basecoat to the steel, while the sealer protects against degradation of the basecoat. The basecoat is the key component that reacts to and provides protection from fire.

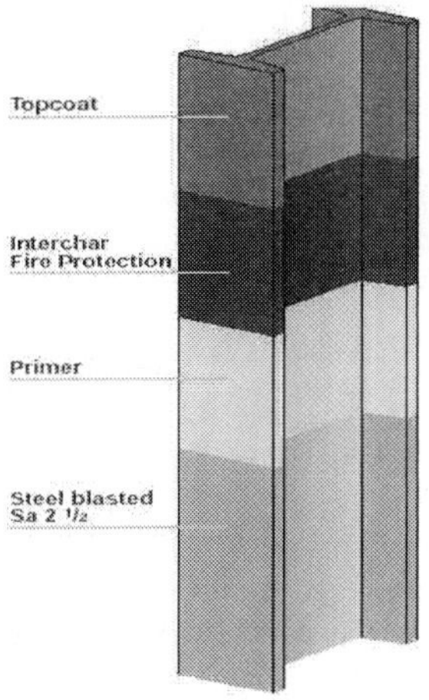

Figure 2: Intumescent basecoats usually consist of the following:

- A catalyst, which decomposes to produce a mineral acid e.g. Ammonium polyphosphates are common catalysts.
- A carbonific, such as pentaerythritol, which combines with the mineral acid to form a carbonaceous char.
- A binder, or resin, simultaneously softens and helps char adhere to steel.
- A spumific agent that decomposes, to liberate large volumes of non-flammable gases such as carbon dioxide, ammonia and water vapour. The gases cause the binder to foam and expand to provide an insulating char many times the original coating thickness.

A typical specification for intumescent coating system will include: -

- The steel substrate and its preparation standard e.g. Swedish Standard SA 2.5
- The primer, its generic type and DFT (dry film thickness)
- The intumescent coat, its generic type and DFT
- The sealer coat, where applicable to include colour and finish

The fire resistance rating required for the structural steel should be arrived after understanding the requirements prescribed in the building codes (Australia AS1530.4 / AS4100 and New Zealand BS476) and the recommendation from the approving authority. In some instances structural robustness in fire may be assessed by a competent structural fire engineer or specialist consultancy that has in turn justified and advised an alternate fire resistance rating required for all or part of the building in question.

In order to establish the correct intumescent basecoat thicknesses required to meet the specified fire resistance periods, the following information must be made available to a fire protection manufacturer:

- Required Fire Resistance Level (FRL) of the structural members
- Steelwork drawings to show the elements to be protected
- Steelwork section sizes and BoQ (Bill of quantity)
- Steelwork exposure e.g. 3-sided beam or 4-sided column
- Environment exposure based on ISO 12944 classifications
- Selection of the preferred application system (onsite or offsite)

In many situations, fast construction schedules can lead to a limited time for applying fire protection. On-site application is likely to cause disruption due to sequencing and construction schedules and could have a direct implication on cost.

Evidence suggests that the most significant cost element in steel framed buildings is not the frame itself but the items built around and attached to it. Consequently any delays to the frame installation will delay the critical path and affect the project completion schedule.

The application dry film thickness (DFT) and drying properties of intumescent coatings is a function of time required for application. The DFT required is a function of temperature at which the steel is assumed to lose its ability to maintain its load-bearing capacity (also known as limiting or critical core temperature). In many parts of the world fire design methodologies allow designers to calculate the limiting temperatures based on structural calculations under specific loading combinations. This permits optimised DFTs for a given fire resistance period. This also means, in some instances especially at high fire ratings, intumescent coatings will become more acceptable to customers as the cost per installed unit area is lowered.

2.3 Structural Fire Engineering

There are three common methods of structural fire engineering: -

1. **Reducing fire resistance periods**

Prescriptive codes such as the IBC or NFPA 101 define periods of fire resistance for elements of structure. These are typically periods of up to 180 minutes in 30 minute intervals. A fire engineer however, can look at the anticipated fuel within a building, the compartment geometry, the potential ventilation and the use of suppression systems to arrive at a realistic transient fire and therefore define a performance-based fire resistance period. It is not uncommon in the UK for high-rise buildings to reduce from 120 minutes (the maximum fire rating in the UK) to 90 minutes fire resistance. This allows for greater choice for fireproofing materials and can reduce cost significantly.

2. **Limiting steel temperature**

Fire resistance periods should ideally be complemented by a limiting steel temperature (the temperature that the steel will reach whilst still maintaining enough strength to carry an amount of load and thus prevent collapse) for every single structural member. This should be determined by a competent structural engineer and can be used by the fireproofing manufacturer to assess the appropriate thickness of insulation. In the UK and Europe, fireproofing materials are tested to a range of temperatures 662°F to 1382°F (350°C to 750°C). The limiting steel temperature is a function of the member capacity and the degree to which it is loaded at the time of a fire. Structural codes such as the Eurocodes provide a methodology to calculate the limiting steel temperature. This approach frequently results in higher failure temperatures than the defaults assumed by fireproofing manufacturers in the absence of it being calculated. Therefore, a reduced thickness of material is required and often substantial cost savings are achieved. At present however, the limiting steel temperature for the North American market is implicit within the UL 263 fire test and is set at 1000°F (538°C) for columns and 1100°F (593°C) for beams. These temperatures may appear to be conservative, but there are a number of scenarios in which they may not be safe.

3. **Optimization by weight**

The greater mass or weight of a steel section, the more slowly it will heat in comparison to a lighter steel section. Large sections will therefore require a comparatively thin thickness of fireproofing material to achieve a specific fire resistance period when compared to smaller sections. Structural engineers often design for the lightest, most efficient steel section; however this can result in high fireproofing costs. Designers in the UK are now looking at optimization approaches which account for the combined cost of steel, fireproofing material and application rates. In almost all cases, this approach demonstrates that the lightest steel design is not the most economic. This process is illustrated in Figure 4

Figure 3 shows the cost breakdown for elements forming the structural frame of a typical multi-storey building. It is shown that the cost of fire protection to the steelwork can be in the order of 20% of the frame costs. This is not an insignificant value and as such, designers are now looking for innovative ways to reduce cost on projects. Structural fire engineering can provide a solution in this respect.

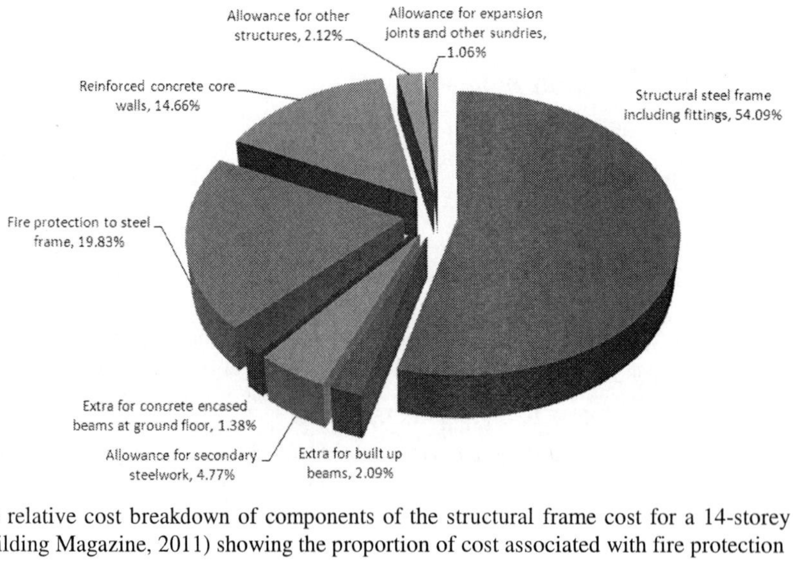

Figure 3: Example relative cost breakdown of components of the structural frame cost for a 14-storey office building in central London (Building Magazine, 2011) showing the proportion of cost associated with fire protection

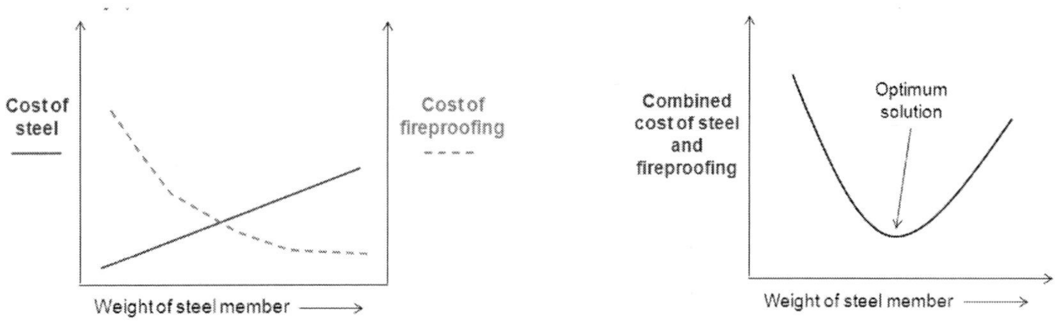

Figure 4: (Left) Illustration of the typical steel and fireproofing costs associated with an increase in the weight of a steel member, (right), combined cost of steel and fire proofing, showing optimum solution with respect to steel weight.

2.4 Building codes

Building code effectiveness, their implementation and a strong regulatory environment are very important while designing and constructing a building. Not all countries in the Asia Pacific region have comprehensive building codes, and the ones which do often struggle with compliance and implementation within a weak regulatory framework. The recent fire tragedies in Bangladesh, India, and Pakistan have exposed the level of inefficiencies and lack of fire safety in public buildings in the region. One of the key issues the region faces is the lack of awareness amongst the general public. In certain instances, the authorities are not sufficiently trained or equipped to fight these death traps when they occur.
Historically, many countries in the Asia Pacific region have developed their national building codes to comply with or align closely with British Standards. Several countries within South-East Asia are now starting to adopt US-based design standards for fire safety such as National Fire Protection Association (NFPA) codes, the International Building Code (IBC) or derivatives of these. Structural design codes have largely been based on British Standards although a number of countries are starting to recognise and adopt Eurocodes. With respect to structural fire design, the Eurocodes are globally recognised as the most advanced set of codes and precedents have been set for their use on projects considering a performance-based, structural fire engineered solution. Various countries in the region including India have held conferences as a part of global promotion of Eurocodes as they recognise no further updating of their equivalent British codes will happen.

Australia and New Zealand have their own well established building codes; testing of products is similar across both countries but the actual assessment methods used in each country are very different which usually results in more higher film thicknesses in the Australian market due to the regression analysis required in AS4100.

The code or standard used to design the fire protection strategy for a building will state that materials used to achieve the required fire resistance must be tested in accordance with a specific fire test standard. Typically, this standard defines the definition and classification of fire-resistant coatings for steel structures including technical requirements, test methods, assessment criteria and specifications relating to packing, marking, labelling, storage, transportation and the product description. The following standards are typically referenced and requested for intumescent coatings in the Asia Pacific region:

- AS 1530.4 / AS 4100 (Australian Building Code Compliance)
- BS 476 Parts 20-22 (Historically common due to the influence of British Standards used in New Zealand)
- EN 13381-8 (Relatively new and comprehensive standard used across Europe)
- ASTM E-119 / UL 263 / NFPA 251 (US-based tests referenced to due increased use of US design codes)
- GB 14907 (Chinese fire test standard)
- GOST (Russia and CIS region

2.5 Importance of Testing and Certification

Intumescent coatings must undergo a wide range of fitness for purpose testing which includes fire performance to a stipulated fire curve, environment and durability testing, mechanical testing, etc. to assess their performance to a wide range of conditions and parameters.

In the majority of global fire test standards, the size and construction of a fire test specimen would ideally represent the element in its intended position in a building. In such tests, typically a loaded beam is tested horizontally, with protection applied to three sides and with the top flange directly in contact with a floor slab. Columns are tested vertically, with the protection applied to all sides. This leads to the terms "3-sided" and "4-sided" exposure when dealing with fire protection to steelwork. Such fire test standards define a test programme for unloaded sections to explore the relationship between fire resistance, coating film thickness and structural section size. A typical programme may include 10-40 sections, including short beams, short columns, and tall columns. To address the issue of adhesion to the substrate under deflection (stickability), additional loaded member tests are required to complement the unloaded member tests.

Methods of assessing the performance of fire protection materials have been developed that enable the thickness of protection for a wide range of situations to be predicted, based on a limited number of specimens as defined in a fire test. This approach can then be applied to the test results, which enables predictions of required coating thickness to be made for a range of structural sections. (www.asfp.org.uk Yellow Book)
Leading manufacturers also commit themselves to independent 3rd party verification and assessment of the test data via various schemes by reputed organisations to give confidence to the customers that the tested material is identical to that supplied for the project.

2.6 Influencing the Steel Market

The feature of the fire protection market which has enabled the UK steel construction industry to solve the problem of fire is illustrated in Figure 5 and is often referred to as a virtuous circle. In the UK, the key event was the appearance of lightweight fire protection systems in the 1980s. These reduced prices considerably which improved the economies of steel construction which in turn increased the amount of steel in use. This then encouraged more fireproofing companies into the market which then encouraged research and innovation (in the form of structural fire engineering and product development) and decreased prices. Ultimately this approach improved the economics of steel construction and the circle began again.

Figure 6 shows the change in usage of the main framing options in multi-storey, non-domestic construction, over the last thirty years. It is shown that steel has increased its market share hugely since the early 1980's. One of the key reasons for this is attributed the reduction in cost of fire protection for fabricated steelwork.

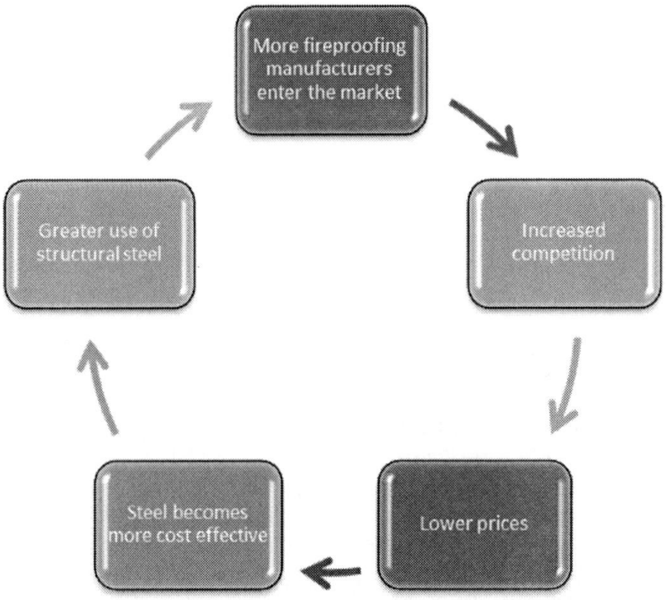

Figure 5: Virtuous circle showing interaction between fireproofing manufacturers and increased use of structural steelwork. *Courtesy of the British Constructional Steelwork Association (BCSA) and Tata Steel.*

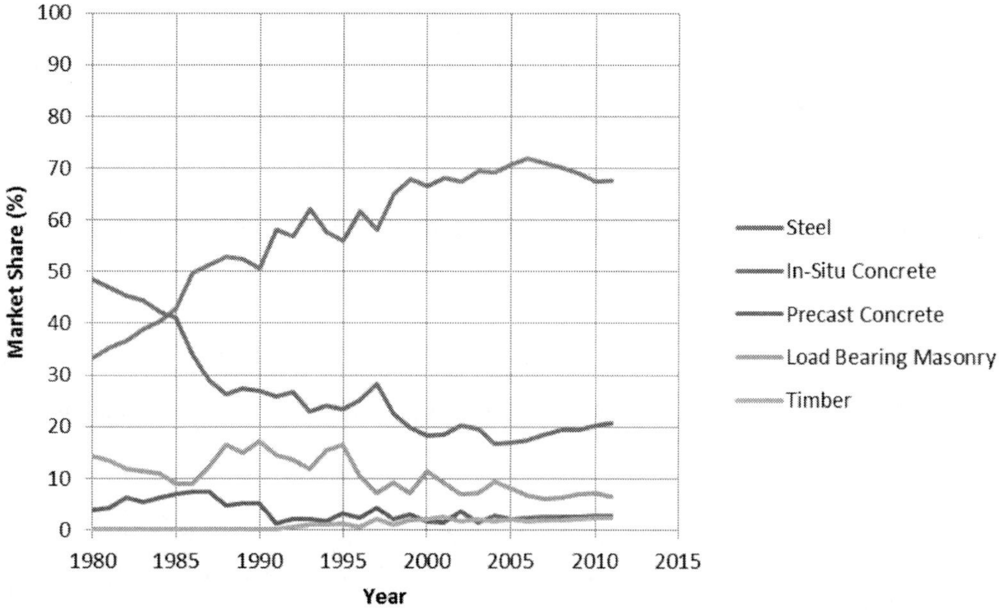

Figure 6: Structural material usage for multi-storey buildings by date in the UK. *Courtesy of Construction Markets Annual Survey, the British Constructional Steelwork Association (BCSA) and Tata Steel.*

3. SUMMARY

Fire protection is an important life safety issue and warrants careful attention to detail at the planning and design stage of a project. Intumescent coating technology has evolved over the last decade to provide more efficient, cost effective solutions to customers around the world compared to other forms of passive fire protection. However, in certain parts of the Asia Pacific region there are still large knowledge gaps in this area which has a detrimental effect on the industry. This knowledge gap needs to be addressed via sustained, effective and enhanced communication amongst key stakeholders.

Ensuring that a specification for passive fire protection to structural steel has been written correctly is very important. Specifiers are encouraged to open dialogue with intumescent coating manufacturers for advice on how best to select the most appropriate solution for their project.

There are a number of methods of undertaking structural fire engineering, each with the potential to bring substantial cost savings to a project. Importantly though, they can be used to quantify structural performance in the event of a fire rather than assuming performance, implicit from a fire test alone.

Fireproofing manufacturers in the UK and Europe now employ qualified structural and fire engineers and are beginning to align with consulting engineers, steelwork fabricators and steel construction institutes to demonstrate added value by incorporating fireproofing characteristics into steelwork designs. This approach is welcomed by the steel industry as it helps to promote steel as a construction material and ultimately leads to robust and safe designs in the event of a fire.

4. AUTHOR DETAILS

Dr. Allan Jowsey MEng CEng MIFireE MSPFE

Fire Engineering Manager – Akzo Nobel.

allan.jowsey@akzonobel.com

ICCP ADVANCES FOR MAJOR WHARF REPAIR

I Godson
Infracorr Consulting Pty Ltd, Melbourne, Australia

SUMMARY: The Webb Dock East Berth 4&5 repair project is Australia's largest and most technically advanced Impressed Current Cathodic Protection (ICCP) System, completed for the Port of Melbourne in late 2016 for an approximate cost of $35m including the extensive concrete and steel repairs required.

The paper describes the investigation, design and technical aspects of the Webb Dock East 4&5 concrete ICCP and water anode ICCP systems by comparison to the CP based repair of a very similar wharf structure, the Swanson Dock West ICCP system, completed in 2009.[1] The comparison of the ICCP designs by the same designer, illustrates the technical advances in this relatively short period.

The concrete ICCP systems both utilize similar design current densities and discrete anode arrangements, with the main differences being the extent of zoning and control utilized with the more recent project via use of a distributed rectifier remote control system.

The Water anode ICCP systems are both designed to pass approximately 3500Amp via titanium tubular anodes to protect steel circular and sheet pile that run the length of the wharves. The method of current distribution however is vastly different, with the Webb Dock East system utilizing AC power distribution to a distributed rectifier system and the Swanson Dock system based on high voltage DC power distribution. Both systems achieved a high degree of electrical efficiency but with very different technology.

In summary, this paper illustrates the recent developments and benefits of ICCP design for concrete and water systems for the efficient management of large scale CP systems.

Keywords: Cathodic Protection, ICCP, Concrete Repair, Distributed Rectifier, Remote Control

1. PROJECT BACKGROUND

The Port Capacity Project (PCP) in the Port of Melbourne involved significant infrastructure repair to the existing berths at Webb Dock East, and the construction of a new 1km long wharf at Webb Dock West.

As Webb Dock East berths 4 and 5 have been in service since the 1970's, they had suffered significant corrosion related deterioration. As one distinct part of the PCP, the structural capacity of these two structures needed to be returned to the original capacity.

1.1 Structure Background

The two Berths of Webb Dock East were nominated for rehabilitation under the contract –Berth 4 which was constructed in 1975, and Berth 5, which was completed in 1982. The overall combined length of the two Berths totals approximately 560m, with width 30m. Both berths are half slab on-ground and half suspended over-water, and are supported on concrete filled steel piles.

The steel piles and interlocking sheet piles support precast concrete transverse beams which in turn are spanned by precast deck planks used as in-situ formwork for a cast in-situ concrete deck. The Swanson Dock West Berth 3&4 are of almost identical construction but are somewhat younger being completed in 1988. A typical section of the Berths is provided in figure 1.

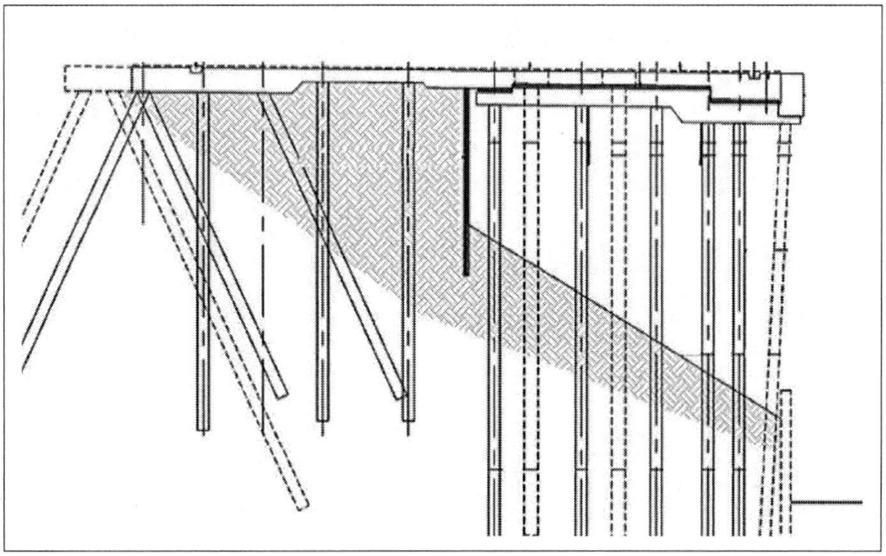

Figure 1. The typical section of Webb Dock East Berths 4&5 is virtually identical to the Swanson Dock W Berths 3&4.

2. CONDITION ASSESSMENT

A detailed investigation of Webb Dock East 4&5 was carried out from mid 2004 to confirm the condition of the berths. This process was designed to establish the state of the berths to a high degree of accuracy. This information was then used in conjunction with element-specific deterioration modelling and in depth structural analysis to ascertain the projected operational life expectancy of each Berth. Where remaining operational life was found to fall below the required 30 year design life, a rehabilitation design was developed to ensure the specified design life would be achieved.

The condition assessment of elements of the structure are addressed under Concrete Elements and Steel Elements below.

2.1 Concrete Elements

The concrete elements requiring condition assessment and repair included:

1. Deck slab (surface) – In situ reinforced concrete slab varying in thickness from 740mm to 1320mm.
2. Precast deck soffit panels – Used as a sacrificial formwork system for construction of the insitu deck areas.
3. Precast transverse beams – Approximately 14.5m in length, 1.0m wide at an average spacing of 4.5m.
4. Precast pile caps – Heavily reinforced element.
5. Pile jackets – Petrolatum pile jackets had been retrofitted mainly to Berth 4 piles in the early 2000's.

The Webb Dock East precast transverse beams were a major focus of the investigation due to their poor condition and their structural significance. The majority of the 135 beams are tidally immersed twice per day for the front 5m of their span.

A 100% delamination survey (figure 2) was used to determine the extent of the beams that had spalled and delaminated. On average this initial estimate found delamination of 45% along the beam soffits and 22% along the beams northern and southern faces. Combined, these figures culminate in an initial concrete repair estimate of approximately 2000m^2.

The average concrete cover to steel reinforcement was found to exceed 55mm in the vast majority of locations, indicating that low cover was not deemed to have been a factor in such a large area becoming corrosion affected.

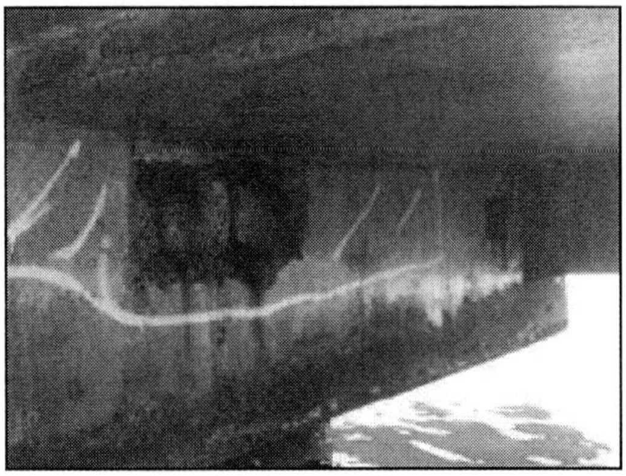

Figure 2 – Typical spalling and delamination markup on transverse beams of Webb Dock East. The front section of the beams are tidally affected in both Swanson and Webb Dock East.

Extensive testing of the precast beams and other concrete elements was undertaken to establish the extent of chloride contamination, and the locations of active corrosion. Equipotential mapping (half cell potential) surveys (refer figure 3) formed an important aspect of this testing, indicating likely corroding areas. The results of the equipotential testing were then used to target locations for chloride content sampling, resistivity testing, and corrosion rate testing. All the test results were combined to determine the type and extent of remediation required to protect the elements from ongoing corrosion.

The electrical continuity of the reinforcement was generally excellent throughout the concrete precast beams but was generally not continuous between the beam reinforcement and the concrete filled steel piles.

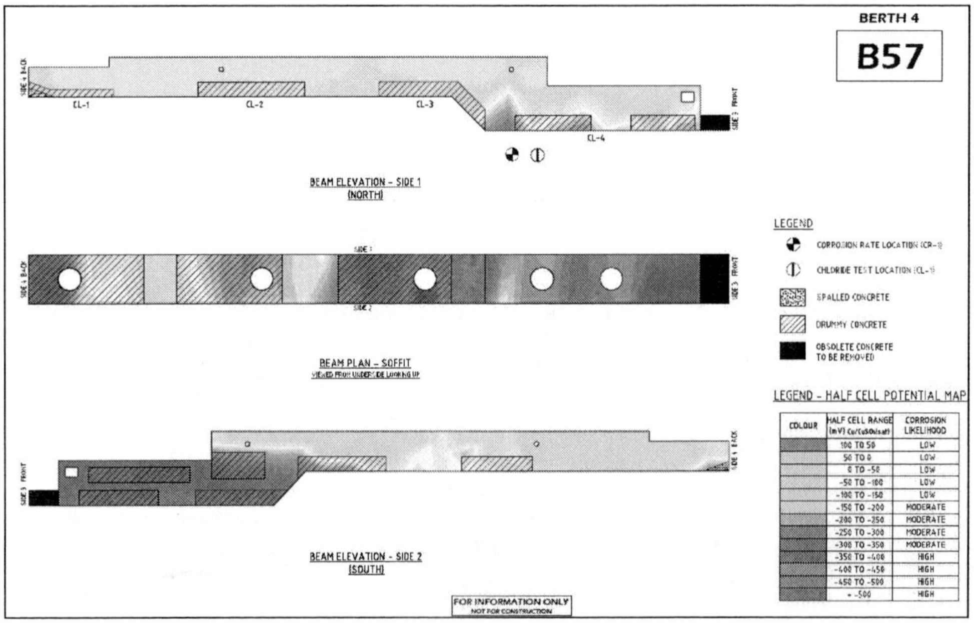

Figure 3 – Typical half cell potential plot of Webb Dock East transverse Beam B57 (red and pink indicating potential <-250mV and likely corroding areas) and hatching showing locations of delamination.

2.2 Steel Elements

The steel elements for both Webb and Swanson Dock included concrete filled, tubular steel piles of 600mm diameter as well as sheet-piling running the length of the wharves.

The condition assessment for Webb Dock East involved cleaning 100% of the marine growth and heavy rust from the surface of the piles in order to allow a 100% visual assessment above and below water. This assessment identified defects, significant corrosion loss and holes. Ultrasonic thickness testing (UTT) was then conducted at over 20,000 individual locations at set spacings on both the circular and sheet-piles above and below water. Where the visual assessment identified holes or defects, additional UTT measurements were taken to confirm the steel area below the specified minimum steel thickness.

Extensive areas of thin steel were identified on both the circular and sheet-piles, predominately above the mid-tide level. However holes were also commonly found at significant depths all the way down to bed level. Steel plate welding and reinforced concrete pile jackets for significantly corroded circular piles were selected for the rectification works.

3. ICCP BASED CONCRETE REPAIR

Impressed Current Cathodic Protection was selected as the preferred system for corrosion control in the transverse concrete beams in both the Swanson Dock and Webb Dock East projects. The design and installation methods adopted are described below.

3.1 Concrete Repair for CP

The concrete repair for the Webb Dock East transverse beams was completed prior to ICCP installation. Generally, the design required hydro-demolition of all delaminated and spalled concrete to expose the affected reinforcement. Main bar reinforcement with section loss greater than 10%, or ligatures with any corrosion loss at all, required additional steel to be appended by welding to restore the original structural capacity. (Refer figure 4) The exposure of the reinforcement provided an ideal opportunity to test for electrical continuity and to install a 316 stainless steel continuity bar from the reinforcement to each steel pile.

Dry sprayed gunite was the preferred method of concrete replacement with the approved repair material having a resistivity less than 25,000 Ohm-cm and being applied without the use of reinforcement treatments or concrete bonding agents.(Refer figure 5)

Figure 4 – Steel reinforcement repair (mainly ligatures) are welded into position and electrical continuity lugs welded to each steel pile.

Figure 5 – Guniting using the dry process was utilized for the concrete repair

3.2 Impressed Current Cathodic Protection (ICCP)- Transverse Beams

The Webb Dock East transverse beams all required ICCP to their entire length. A conservative ICCP design was completed to the requirement of the Australian Standard (AS2832.5)[2]. A design current of 20mA/m^2 of steel reinforcement surface area was selected. A typical 14.5m long transverse beam was calculated to require ~1.2 Amp of current. The distribution of the current to the various parts of the beams was zoned to avoid "current dumping", especially in areas regularly wetted such as the tidally affected front of the beams and the rear splashed sections. Concrete resistivity test results from the initial condition assessment were utilised to assist in defining each of the zone arrangements. The design was finalised with four zones per beam as set out in Figure 6. Each zone was designed to be monitored by a reference electrode pair, consisting of a Manganese dioxide reference electrode and a mixed metal oxide coated titanium pseudo electrode.

By comparison, the Swanson Dock design incorporated only 2 zones per beam, with each of these zones shared between five (5) beams (see figure 7)

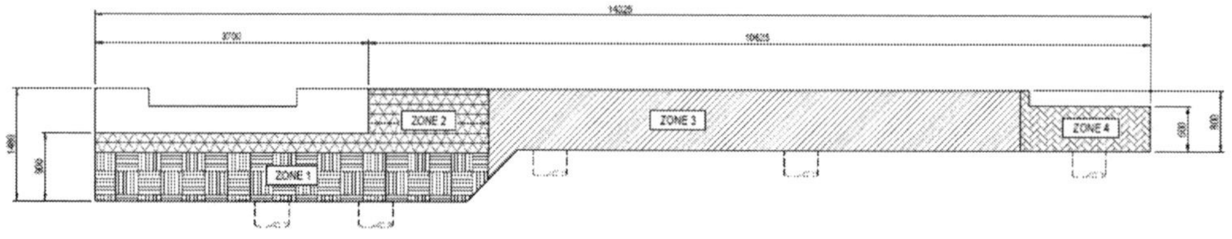

Figure 6 – A typical Webb Dock East Beam zoning arrangement has four (4) independently controlled zones per beam.

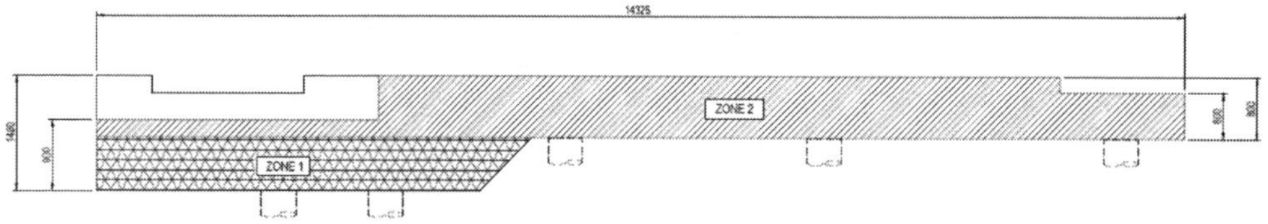

Figure 7 – A typical Swanson Dock West Beam zoning arrangement has only two (2) zones per beam with these outputs shared between five (5) beams.

The anodes selected for both the Webb and Swanson Dock projects were discrete, titanium Mixed Metal Oxide (MMO) anodes. They were installed in ~30mm diameter holes drilled into the concrete and backfilled with a low resistivity mortar. The anode was constructed with MMO titanium ribbon, spot welded into a 400mm long active length and with a titanium tail that was in turn spot welded to a titanium conductor bar linking all anodes within a zone. Each 400mm long anode was rated to output a current of 5.7mA. (at 110mA/m2 anode surface area). A typical Webb Dock East beam had approximately 230 anodes, installed from both sides of the beam to provide the design current of ~1.2Amp. (refer figure 8)

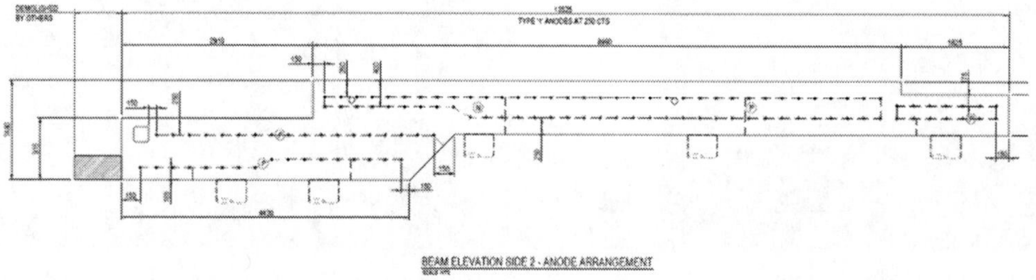

Figure 8 – A typical Webb Dock East Transverse beam anode arrangement, with 400mm long discrete anodes installed into drilled holes on both sides of the beam. Similar discrete anodes were utilized at Swanson Dock

3.3 Cathodic Protection Installation

Installation of the discrete MMO coated titanium mesh anodes was performed post-concrete repair, and began with an initial surface cover-meter survey to establish the location of embedded steel elements within the beams and allow the anode positions to be marked in appropriate locations on the concrete surface. Holes were then drilled using pneumatic drills followed by down-hole cover-meter testing to ensure that appropriate separation between the anode and embedded steel reinforcement was maintained. (figure 9) The final step in this process involved the installation and grout backfilling of the anodes using a specially developed acidification resistant grout, formulated for maximum life expectancy. A "current injection" test was specified and completed on each anode, to ensure the elimination of short circuits. The conductor bar was then welded to the anode tails in each zone. (Refer figure 10)

Figure 9 – Webb Dock East Transverse beam with anode holes drilled and conductor bar slots cut

Figure 10 – Webb Dock East Transverse beam with anodes installed and welded to conductor bar in "zones", prior to filling slots with mortar

3.4 Water Based Cathodic Protection

Prior to the ICCP installation, both projects utilized welded steel plates in order to re-establish the structural capacity of circular piles and sheet piles. For the Webb Dock East project, the original ultrasonic thickness testing investigation was utilized to size and locate the welded steel plates over the low thickness steel areas. Over 350m^2 of steel plate welding was completed to submerged and tidal splash areas. In the instances where the circular piles were severely corroded, reinforced concrete pile jackets were installed. The steel piles in the atmospheric zone were protected from corrosion by the use of petrolatum tape jacket systems from 1m below low-tide level to the top of pile.

A water based Cathodic Protection system via impressed current water anodes was selected to protect all steel below the mid-tide level on both the Webb Dock East and Swanson Dock projects. The design was completed to the Australian Standard (AS2832.3) [3] with the design current required to protect all submerged steel piles and sheet-piles calculated at 120mA/m^2 of exposed steel surface area. The protection requirements for the 2007 designed Swanson Dock project was -800mV to -1150mV (AgAgCl). The Webb Dock East project required an increased level of polarisation between -900mV to -1150mV (AgAgCl) in accord with the PIANC guideline [4], to overcome the likely occurrence of Accelerated Low Water Corrosion.

Similar MMO coated tubular titanium anodes (25mm diameter x 1.2 long) were used for both Swanson and Webb Dock East projects to provide the ~3500 Amps of current required for both systems. However the design methodology for the anodes was vastly different with a substantially higher maximum design current of ~45Amp with shields used for the Swanson design versus ~22Amp and no shields utilized for the Webb Dock East design. As a result the Webb Dock East system utilized substantially more anodes (150#) compared to the Swanson Dock (77#).

The method of fixing the anodes was also substantially different with the Swanson Dock system utilizing the weighted, shielded anodes being suspended by their cables from brackets mid way between the piles. (Figure 11). The Webb Dock East project aimed for a more durable design with the anodes mounted on PVC supports attached directly to piles via smart bands. The shielding to the steel pile was provided by the extension of the petrolatum tape and jacket system to RL -3m. (Figure 12)

Figure 11 – Swanson Dock utilized 25mm by 1.2m long Ti tubular anodes suspended between piles with weighted PVC "shields" to provide ~45amp output per anode

Figure 12 – Webb Dock East utilized the same tubular titanium anodes fixed via PVC supports to the pile with "smart bands". The petrolatum pile wrap was extended to RL-3.5m to "shield" the steel pile.

The electrical continuity of the steel piles was also an issue on both projects, generally with no continuity between the steel piles and the reinforcement in the precast beams. Swanson Dock utilized several methods for establishing continuity to piles, eventually settling on welded steel reinforcing bars installed underwater across each pile bent to the sheet pile wall. This was replaced in the Webb Dock East project by the welding of a 316 stainless steel bar from the beam reinforcement to each pile as part of the concrete repair process.

4. THE ICCP CONTROL SYSTEMS

The ICCP control systems represent the largest technology change between the 2009 Swanson Dock and 2016 Webb Dock East systems as described below.

4.1 The Swanson Dock Control System.

The Swanson Dock concrete control system was a manual T/R based system mounted in an air-conditioned control room on the wharf deck, with the numerous cables from the beams fed along a cable tray mounted below the deck. The arrangement of two (2) zones per beam, shared between five (5) beams, required a junction box (refer Figure 13) mounted on the central beam of the five (5) beams below the wharf deck. The positive, negative and reference electrode cables from each of the five (5) beams were connected to the junction box. Main cables from the junction box, including main positive, negative and a multicore twisted pair instrumentation cable for the references was passed on the cable tray back to the control room T/R's. Accordingly, each of the five (5) beams was in itself a ICCP system, controlled by two (2) T/R outputs. Nine of these systems were installed to protect the 45 beams in the Swanson Dock West system.

Being effectively a manual system, monitoring of the CP required site visits to record the currents, "Instant Off" and depolarized potential from each zone from the above deck control room. However, individual sub-zone currents to each of the two (2) zones in each beam could only be obtained by accessing each junction box below the tidally affected wharf, typically requiring approximately 3 days of technician access from a small boat. Accordingly, monitoring was a relatively slow and costly exercise.

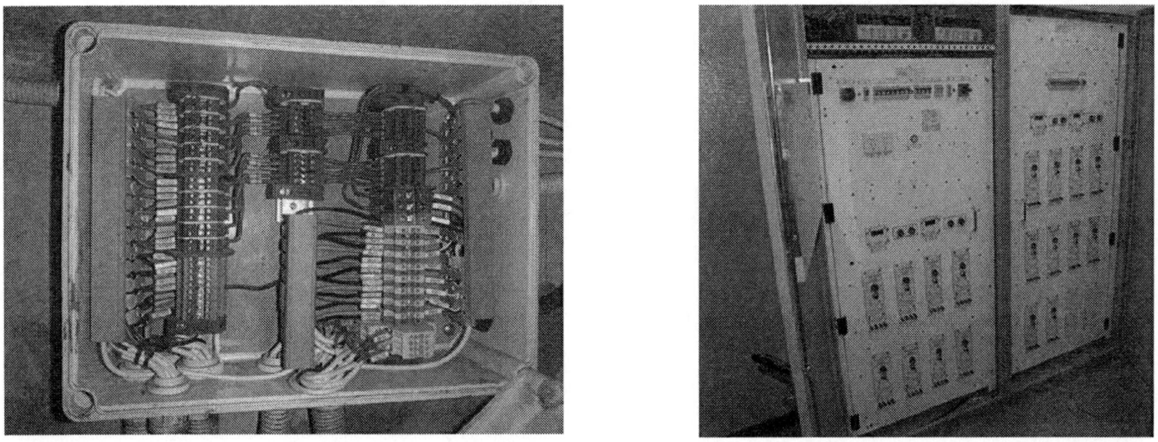

Figure 13 – The Swanson Dock Concrete control system utilized manual T/R's in the above deck control room and long main cables to junction boxes (JB's) placed under the wharf. The JB's shared 2 T/R outputs between 5 transverse beams

The Swanson Dock water anode control system consisted of rack mounted T/R's in the above deck control room. Due to the long cable runs and corresponding high extent of voltage drop, a system was designed to distribute power at high voltage DC to minimize the voltage drop. The T/R's passed up to 200V DC via up to 500m of cables to the "multiplier" (Refer Figure 14) located at the anode location. The multiplier converted the high DC voltage at a set conversion of 20:1 to ~ 9V at 45 Amp to the hung water anodes.

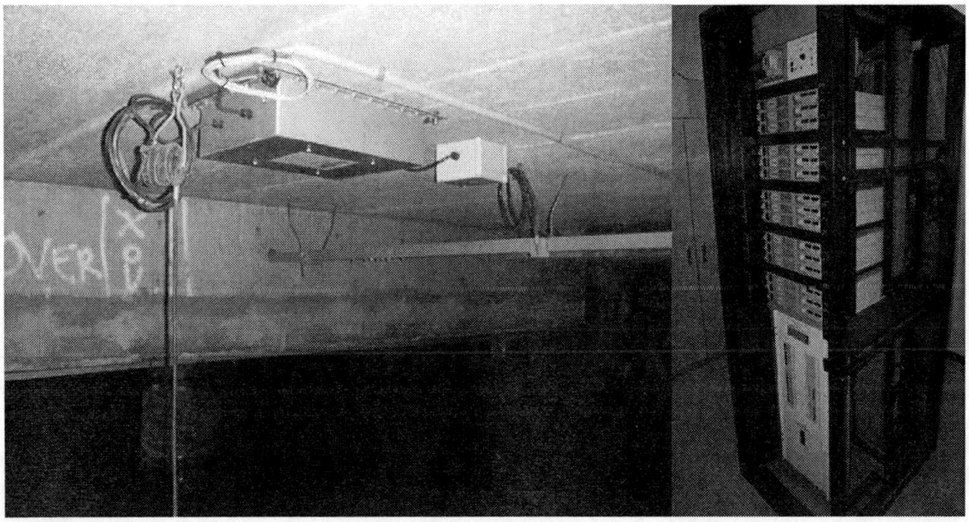

Figure 14 – The Swanson Dock water anode control system consisted of rack mounted T/R's in the above deck control room, passing up to 200V DC via up to 500m of cables to the "multiplier" located at the anode location. The multiplier converted the high DC voltage to ~ 9V at 45amp to the hung water anode.

4.2 The Webb Dock East Control System

The cathodic protection control system for the Webb Dock East project was an under wharf, 3-phase AC distributed system. This was developed to minimize the total amount of copper electrical cabling required, and avoid the need to install large gauge, long run DC cables. The concrete ICCP control system was required to provide the required current to each zone and allow monitoring of the reference electrodes and remote adjustment.

Each transverse beam is provided cathodic protection in four independent, individually controlled zones, making for a total of 552 zones of concrete CP. The water ICCP system operates in 128 individual zones giving a high level of adjustment to the 147 water anodes in the system.

With the overall system required to control ~680 independent ICCP zones and accurately report on the ~1150 reference electrode readings, a sophisticated remote monitoring system was developed which allows local and remote set-up, adjustment and monitoring, with the majority of control and monitoring completed remotely via web access to the system.

The ICCP control systems required the installation of the 316 stainless steel enclosures for the transformer rectifiers (TRUs) to be mounted at selected locations to the beam below the wharf deck. (Refer figure 15) These included:

- Central Control Unit – 1 unit to control both concrete & water ICCP systems,
- Water Transformer Rectifiers – 64 No (2 channel per unit),
- Concrete Transformer Rectifiers – 46 No (12 channel per unit),
- Concrete Junction boxes – 46 No,
- Supply Junction Boxes – 46 No, 3-phase to single phase,
- Communications Junction Boxes – 46 No.

Only the Central Control Unit (refer figure 16) was located on the top side, as it could be installed away from the trafficked areas of the Berth.

The under-wharf location of the electrical equipment cabinet's necessitated the careful sealing of all conduits and penetrations in the boxes. To control this aspect, a bespoke pressure test was developed to ensure all electrical cabinets maintained the specified IP67 rating. All penetrations were glanded and filled with sealant to ensure the integrity of the box would not be compromised in the event of conduit damage.

To adhere to the requirements of the project, the components within the TRU's are modular and are able to be easily removed and replaced without the need to dismount the housing and disconnect the cabling.

The CP systems were commissioned remotely via web access, allowing the combined total of 680 zones of concrete and water CP and ~ 1150 monitoring references to be managed efficiently.

Figure 15 – Completed Webb Dock East CP to transverse beams with distributed CP control system installed below deck. An instrumentation cable linked the distributed system to the Central Control Unit.

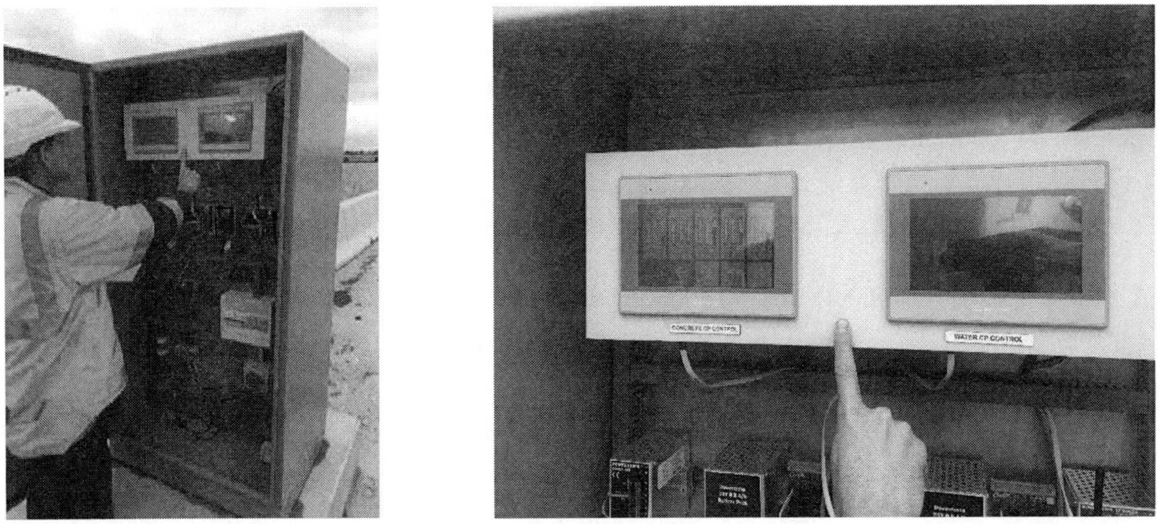

Figure 16 – The Webb Dock East Central Control Unit (CCU) allows top of wharf based adjustment or monitoring of the Concrete and Water Anode CP systems. Remote access via the web mirrors the local CCU ability with enhanced data-logging and alarm facilities via text messaging.

5. CONCLUSIONS

The Webb Dock East Berth 4&5 ICCP system is an advanced design incorporating over 30,000 concrete discrete anodes, 147 titanium water anodes, ~680 CP zones and ~1150 reference electrodes to control the concrete and water anode systems with a total output over 3700 Amp. While the large number of zones and references provide a high degree of control and adjustment of the system, the commissioning and monitoring of such a system would be unmanageable without the incorporation of a similarly advanced remote monitoring system. The web based control system allows efficient monitoring, diagnosis and control of the system without site access being required and the text based alarms notify the operators immediately of any out of tolerance results allowing rapid correction.

The distributed nature of the control system also provides a highly efficient system, vastly reducing the power consumption compared to traditional ICCP systems using large copper cables to distribute and balance the current.

6. ACKNOWLEDGEMENTS

The authors wish to acknowledge the management and staff of the following organizations for their assistance with the project: Port of Melbourne Corporation, McConnell Dowell Constructors (Aust.) Pty Ltd, Jacobs Engineering Group, Freyssinet Australia Pty Ltd, Infracorr Consulting Pty Ltd.

7. REFERENCES

1. IB Godson. Cathodic Protection to Swanson Dock West in Melbourne, Australia. Eurocorr 2009 Paper No 8447.
2. AS2832.5 2008 Cathodic Protection of Metals. Steel in Concrete Structures
3. AS2832.3 2005 Cathodic Protection of Metals. Fixed Immersed Structures
4. PIANC Guideline 2005. Accelerated Low Water Corrosion, International Navigation Association

MODELLING A RETROFIT CP SYSTEM FOR AN ABOVE GROUND TANK, A METHODOLOGY AND IMPLEMENTATION

S Mostafavi[1], R Salazar[1], G Mckinnell[2]
STORK, Perth, Australia[1], Fareham, UK[2]

SUMMARY: This paper presents a systematic approach to troubleshooting and enhancement of an existing cathodic protection system, for an above ground storage tank. The problem of localized under-protection on a tank floor is investigated by field tests. Insulation gasket and soil current mapping tests were conducted to ensure that current leakage or discharge was not occurring on the tank floor. However, soil resistivity tests have revealed that high resistance due to underground rock embedment around the under-protected area inhibits the tank floor from receiving sufficient CP current. A Finite Element simulation was carried out using Comsol Multiphysics software. The model demonstrated that soil inhomogeneity plays a critical role in current distribution on the tank floor. The current distribution and potential difference between the center of the tank and its periphery is modelled and a retrofit CP System was designed and commissioned based on the results. The modelling provided a clear picture of the potential profile and current distribution on the tank floor, and helped to achieve compliant cathodic protection potentials over the entire tank floor.

Keywords: Finite Element Modelling, Cathodic Protection, Above Ground Storage Tank, Current Distribution.

1. INTRODUCTION

One of the challenges with maintaining effective cathodic protection for existing above ground storage tanks is being able to provide a uniform cathodic protection current to the tank floor [1]. The delivery of the current to the centre of tank floors is influenced by the terrain inhomogeneity, concentration profiles of oxygen underneath the tank bottom plate, void formation and bulging of tank floor plate due to cyclic filling and un-filling [2], installation of secondary spill containment, etc. The non-uniformity of the current distribution increases as the tank diameter increases.

Installation of corrosion probes and reference electrodes in perforated tubes, and a network of ribbon anodes underneath the tank floor during construction, or even remote deep well anodes, could help to ensure more uniform current distribution. However, for existing tanks in congested plants, there will be limited options available. Monitoring old CP systems of existing tanks without centre reference electrodes, usually consists of measuring tank to soil potentials at the perimeter. In addition to that, measuring the current output of anode strings, as well as the rectifier inspection will be other tasks of the inspection program. However, monitoring the perimeter potentials is not sufficient to ensure that tank floor is receiving compliant cathodic protection.

This paper presents a systematic approach for troubleshooting and retrofitting of a common cathodic protection system of two adjacent above ground tanks using finite element modelling. Figure 1, shows the schematic of the CP system including the anode arrangement, current output and potential values measured on the perimeter of the tanks. As shown an under-protected section was found on the perimeter, (southern side) of one of the tanks, named TK B in this paper.

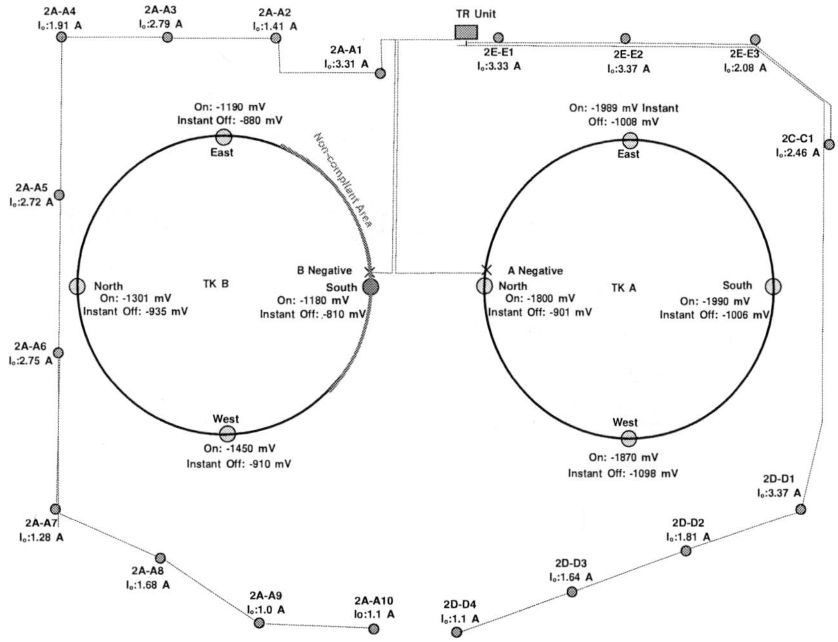

Figure 1. Schematic of the CP system of the tanks

Finite element modelling, and field investigations were used to design a CP system to mitigate the under-protection problem and achieve compliant potentials over the entire tank floor.

2. EXPERIMENTAL DETAILS

2.1 Field Tests

A series of field tests were conducted to investigate all the possible scenarios such as electric short, current leaking to other structures and current discharge or shielding. The earthing system and also the possible CP current leakage to the adjacent structures were tested using a Swain Meter. The results show that the decouplers and isolation gaskets at the process piping flanges provide sufficient electrical isolation to the tank. Soil voltage gradient mapping was also conducted around the tank to detect any possible current discharge areas. The result show that the direction of CP current in the soil is toward the tank at all points, including the under-protected area, hence there was no current discharge point. 4-pin Wenner soil resistivity tests were performed around the tank and it was found that the soil resistivity at the southern side of TK B is significantly higher than other areas. The excavation later revealed that there is a rock embedment under the surface layer at the southern side of the tank. Table 1 presents the soil resistivity results.

Table 1. The Soil Resistivity values around Tank B

Location / Layer (m)	North (ohm-cm)	West (ohm-cm)	South (ohm-cm)	East (ohm-cm)
0-1	990	1014	**3016**	1001
1-2	902	891	**2793**	861
2-3	890	861	**2710**	819
3-4	803	821	**2050**	801

2.2 Finite Element Model

The CAD geometry of the tanks and anodes in electrolyte were created in COMSOL Multiphysics. The electrolyte around TK B was discretized into four zones in order to assign different resistivity values. An appropriate meshing is essential especially at the interface of the tank floor, and anodes with soil. Figure 2 shows the geometry of the model.

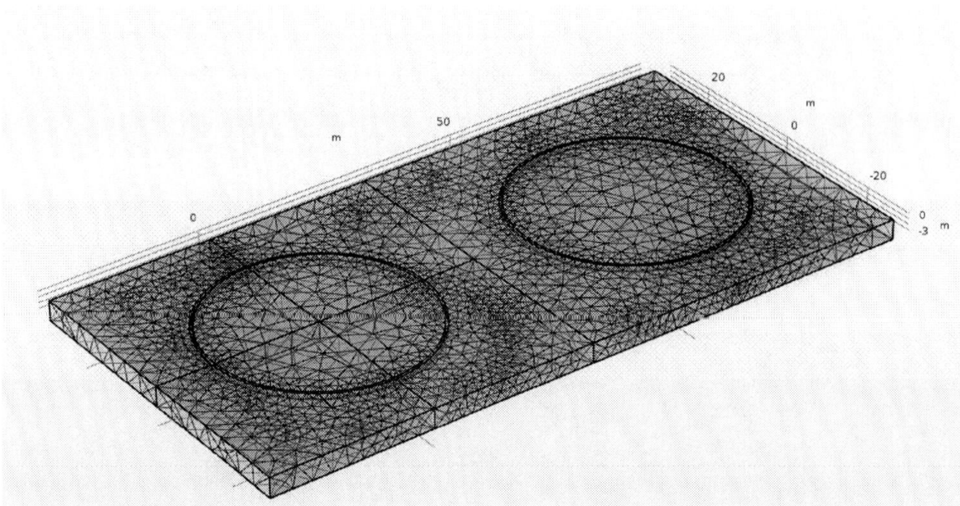

Figure 2. Geometrical discretization of the tank floors, electrolyte and anodes

Three different corrosion reactions were considered to be occurring on the tank floor including iron oxidation, oxygen reduction and hydrogen evolution. The kinetics of these reactions were described by Tafel expressions as shown below [3]:

$$i_{Fe} = i_{0,Fe} . 10^{\frac{\eta_{Fe}}{A_{Fe}}} \quad (1)$$

$$i_{O_2} = -\frac{c_{O_2}}{c_{O_2,ref}} . i_{0,O_2} . 10^{\frac{\eta_{O_2}}{A_{O_2}}} \quad (2)$$

$$i_{H_2} = -i_{0,H_2} . 10^{\frac{\eta_{H_2}}{A_{H_2}}} \quad (3)$$

The parameters in Table 2 were used for the above equations:

Table 2. Reaction Parameters

Parameter	Unit	Fe	O_2	H_2
Equilibrium Potential, E_{eq}	V	-0.76	0.189	-1.03
Exchange current density, i_0	A/m^2	7.1E-5	7.7E-7	1.1E-2
Tafel Slope, A	V/decade	0.41	-0.18	-0.15

The oxygen reduction reaction underneath the tank floor depletes the oxygen concentration in the soil, however, the bellows action of the tank floor as it is emptied replenishes the oxygen again [1] . The anode current output values from figure 1, were used in the boundary condition of electrolyte.

3. RESULTS

Figure 3 shows the tank to soil potential profile from the existing CP system. The potential contours indicate that tank to soil potentials drop toward the centre of the tanks. As can be seen, the contours of tank B (left), are notably non-symmetric with the potential levels on the south west quarter significantly lower (electro-negatively) than other areas. Figure 3 demonstrates that most of the external surface of the tank floors do not receive sufficient protection.

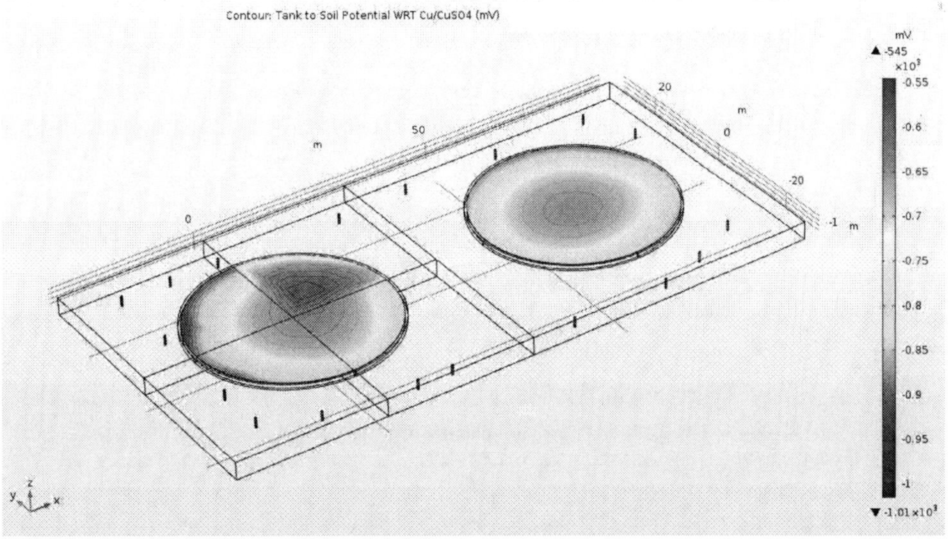

Figure 3. Tank to soil potential contours on the external surface of the tank floors

Based on this potential profile, the decision was made to install a retrofit CP system to improve the CP potentials of the tanks. A new model was created by adding the retrofit cathodic protection system. The new CP system consisted of four horizontal anodes with current output of 4x1.25A located on the southern side of TK B, connected to a new power supply (30VDC / 30ADC) in order to overcome the high resistance of the soil. The computation was repeated for the new configuration of the CP systems. Figure 4 presents the location of the new anodes and tank to soil potential contours after installing the new anodes. As can be seen the lowest (electro-negatively) potential on the tank floor is brought to -850 mv thus AS 2832 criterion is satisfied on the entire external surface.

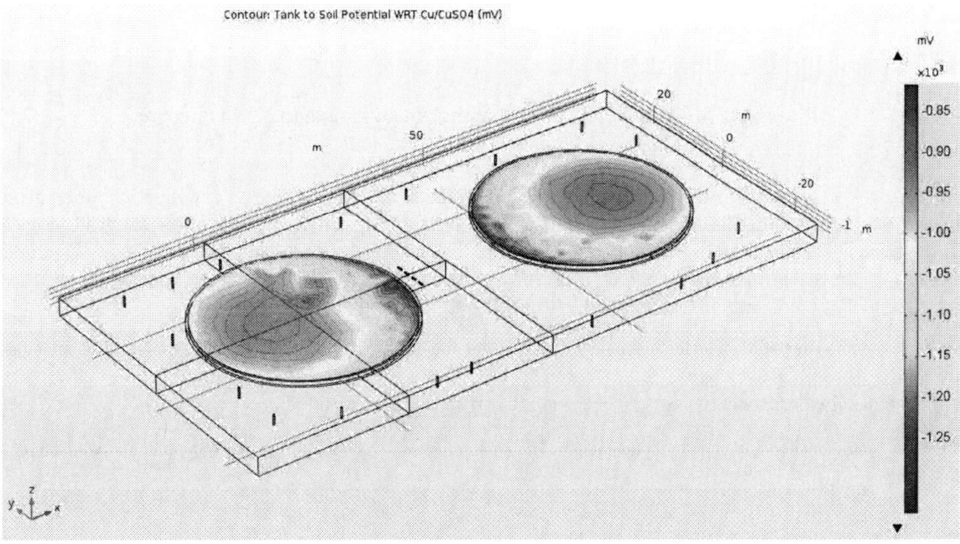

Figure 4. Tank to soil potential contours on the external surface of the tank floors after installing new anodes

Figure 5 shows TK B's current distribution along the x axis (south to north direction) before and after installation of retrofit CP system. It also compares the modelling results with a closed form relationship (equation 4) for a primary current distribution on a disk electrode of radius r_0, embedded in an infinite insulating plane [4]:

$$\frac{i_{local}}{i_{avg}} = \frac{0.5}{\sqrt{\left(1-\left(\frac{r}{r_0}\right)\right)}} \qquad (4)$$

Where i_{local} is the current density at a distance r from the centre of the disk in A/m^2 and i_{avg} is the average current density which for this case is the total current output of all anodes including retrofit anodes divided by the surface area of both tanks.

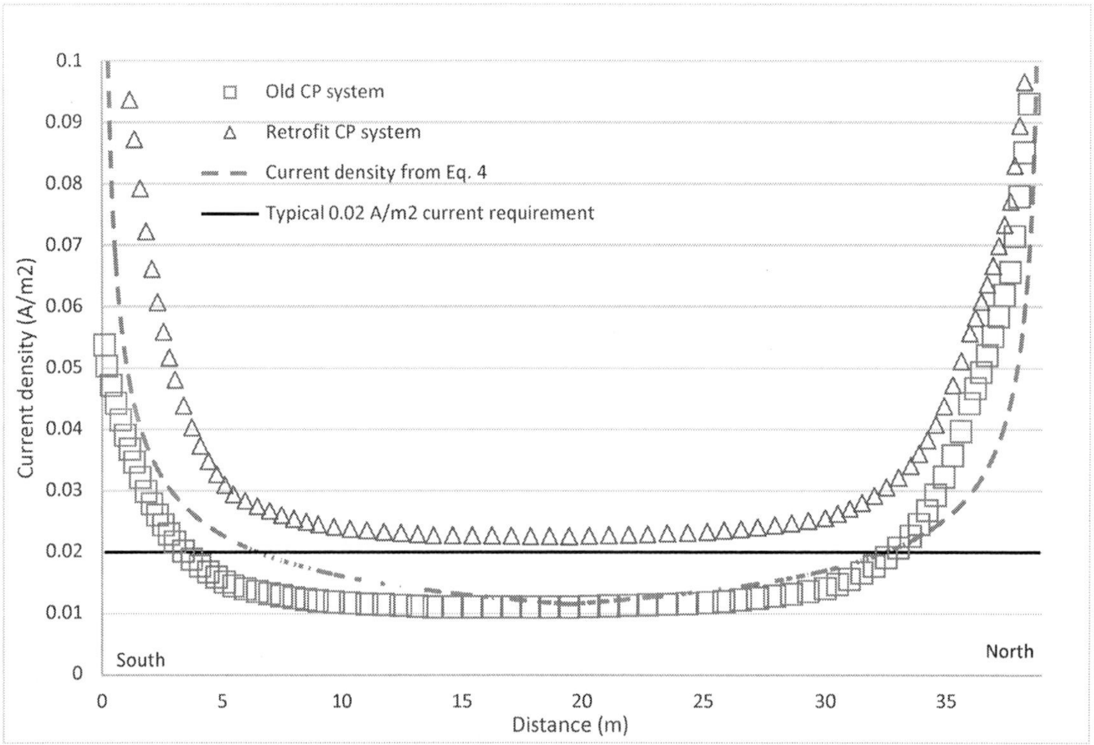

Figure 5. TK B current density along X-axis (South to North direction)

The tank floor has a non-uniform primary current distribution with the current density exponentially increasing toward the edge. In fact, according to the disk electrode model (eq. 4) the current density is infinite at the electrode edge. Figure 5 presents the current density profiles in comparison with the disk electrode model and also a typical value of current requirement for bare steel in a neutral soil, i.e. 0.02 A/m^2. As can be seen in Figure 5, the entire surface of the tank bottom with the retrofit CP system satisfies the min 0.02 A/m2 current requirement.

Figure 6 presents the potential profile of the tank floor along the x axis for TK B. The under-protected areas are improved after adding the retrofit CP system and the potential levels comply with AS 2832.2 criterion [5], i.e. minimum -850 mV instant off potential with respect to copper/copper sulphate reference electrode.

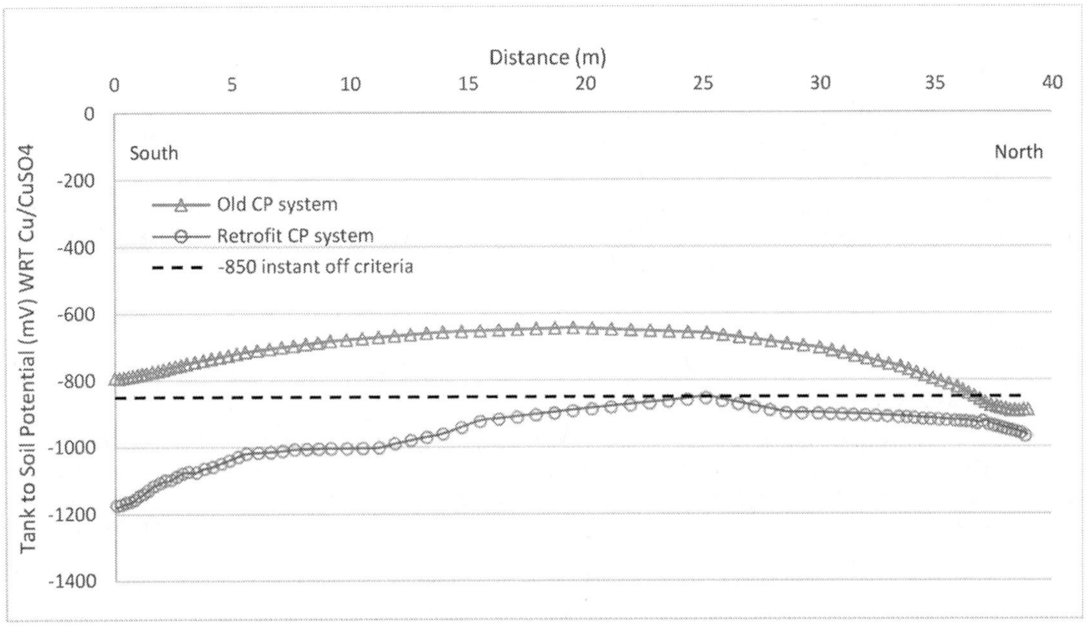

Figure 6. Tank (TK B) to soil potential profile along X-axis (South to North direction)

After installation of the retrofit CP system, the potentials were measured around the tank perimeter and compared with the simulation result (Figure 7). GPS synchronized interrupters were used to interrupt CP Systems. The tank to soil potentials measured on field before and after the installation of retrofit CP system appear to be more negative than the modelling results. The deviation in readings could be the result of voltage drops other than across tank/soil interface [6] e.g. the instrument, test leads or reference electrode internal resistance, or soil gradient due to other CP systems in the plant incorporated in the field readings.

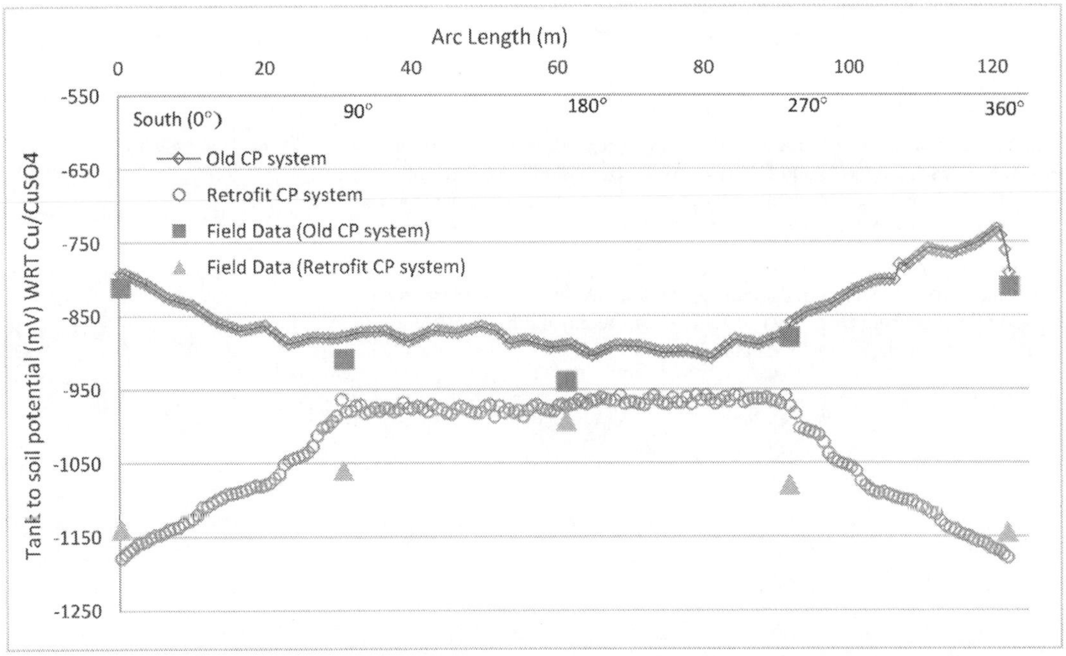

Figure 7. Tank (TK B) to soil potential profile along perimeter

4. DISCUSSION

The troubleshooting methodology consisted of:

- Verification of the integrity of insulation gaskets and DC decouplers
- Soil current mapping to ensure that there was no local current discharge area on the tank floor
- Soil resistivity profile around the tank.

The soil resistivity values revealed that terrain inhomogeneity escalated the non-uniformity in potential profile and resulted in under-protection of a large section of the tank floor. Since no reference electrodes were installed underneath the tank floor, it was not possible to know potential levels and the extent of the under-protection on the tank bottom. Hence, a FEM model was set-up from the geometry, current output and the soil resistivity values. The current density and potential profile graphs from the model, were used to design an additional CP system so that entire tank floors receive compliant cathodic protection.

5. CONCLUSIONS

Finite element modelling demonstrates a clear picture of the current density and potential profile for the complex boundary conditions and complex geometries where monitoring options or using empirical relations have limitations. For this case, installation of reference electrodes underneath the tanks was tedious and costly. Current distribution and potential contours from the FEM model helped to design a CP system that enhanced the protection level of the entire surface of the tank floors.

6. REFERENCES

1. Riemer, D.P. and M.E. Orazem, A mathematical model for the cathodic protection of tank bottoms. Corrosion Science, 2005. 47(3): p. 849-868.
2. Arash Zamani, Y.W., An Elastic Plastic Finite Element Case Study of Void Presence Underneath the Tank Bottom Plate, in The 8TH Australian Congress on Applied Mechanics (ACAM 8). 2014: Melbourne, Australia.
3. COMSOL MULTIPHYSICS Reference Manual. 2013, COMSOL MULTI-PHYSICS.
4. Kelly, R.G., Electrochemical Techniques in Corrosion Science and Engineering. Corrosion Technology, ed. J.R. Scully, D. Shoesmith, and R.G. Buchheit. 2002, Hoboken: Hoboken : Taylor and Francis.
5. Standard, A., Cathodic Protection of Metals, Part 2: Compact Buried Structures. 2003.
6. International, N., Measurement Techniques Related to Criteria for Cathodic Protection on Underground or Submerged Metallic Piping Systems. 2012.

7. AUTHOR DETAILS

Seyed Mostafavi is a CP specialist (NACE CP4) with +10 years' experience in corrosion, fitness for service and failure analysis. He has a PhD degree in Mechanical Engineering from University of New South Wales and has conducted several research projects in materials modelling using Finite Element Method.

Ramon Salazar is a Senior Corrosion Engineer with +20 years' experience in corrosion control, especially in cathodic protection. He has a Master Degree in Corrosion Science from University of Zulia (Venezuela) and has experience working in Venezuela, Middle East and Australia.

ADDRESSING SHEET PILE MIC FRESH WATER CORROSION ON OHAU DIVERSION WALL

R El Sarraf[1], W L Mandeno[2] & M Townsend[3]
Opus International Consultant Consultants Ltd, Auckland[1], New Zealand, Opus International Consultants Ltd, Wellington[2], New Zealand & Bay of Plenty Regional Council[3], Tauranga, New Zealand

SUMMARY: The Ohau Diversion Wall is a 1275m long submerged steel wall consisting of interlocking sheet piles, and king piles up to 50m long driven into the bed of Lake Rotoiti; that is severely affected by MIC corrosion after only 10 years of service. Concerns were raised about the continued performance of the wall, and whether it would last the remaining 40 years of its design life. Opus International Consultants, were commissioned by the Bay of Plenty Regional Council to investigate this and develop alternative remediation solutions.

This paper presents the findings of this investigation project to date, and a comparison between the alternative remediation solutions. These included the use of cathodic protection (sacrificial or impressed current), the use of protective coatings, patch repair, and full replacement using vinyl sheet piles or impermeable polymeric screens. An environmental review of the effects of each option, and development of mitigation measures will also be presented. The paper concludes with a net present value cost analysis between the different options, which assisted in determining a preferred solution.

Keywords: Steel, Sheet piles, Microbial Influenced Corrosion, Cathodic Protection, Fresh Water Corrosion.

1. INTRODUCTION

1.1 Background

The Ohau Diversion Wall, which was built in 2008, is located in Lake Rotoiti, near Rotorua, New Zealand. The purpose of the wall is the diversion of nutrient rich water flowing out from Lake Rotorua, through the Ohau Channel into the outlet of Lake Rotoiti that is the head of the Kaituna River as illustrated in Figure 1. This has improved the water quality and reduction of algal blooms since the wall's installation.

Figure 1 Aerial view of the Ohau Diversion Wall.

The Ohau Diversion Wall was constructed under the Rotorua Lakes Protection and Restoration Programme which is a partnership of the Rotorua Lakes Council, the Bay of Plenty Regional Council (BoPRC) and Te Arawa Lakes Trust. The asset is maintained and managed by BoPRC for both condition and regulatory requirements.

1.2 Diversion Wall Design

The Ohau Diversion Wall was designed in 2005, and was constructed over 2007/2008. It consists of a 1275m long sheet pile wall running from the outlet of the Ohau Channel to Te Akau Point in Lake Rotoiti. The wall is located ~75 metres offshore and runs parallel to State Highway 33. The wall is made of interlocking steel sheet piles, at water depths between 0.5 and 7m, and steel king piles that are driven up to 50m below the lake bed in places. It has been designed with a visual low profile which extends 500mm above the water surface as shown in Figure 2.

Figure 2 Example of the Diversion Wall visual low profile on the lake.

The wall was designed for a 50 year design life, based on a chosen corrosion allowance (Beca 2014) of 0.04mm/year for the uncoated steelwork immersed in fresh water, and 0.015mm/year for embedded steelwork in soil, in accordance with the piling design standard AS 2159:1995. In total, the uncoated steelwork surface area was calculated as being ~50,000 m^2; roughly split between the steel piles immersed in lake water (~22,000 m^2) and the piles embedded in the lake bed (~28,000 m^2).

As mentioned above, the diversion wall comprises of interlocking sheet piles and king piles, the steel thicknesses are:

- ESC S8 Sheet Pile– 8mm wall thickness;
- ESC H60/30A-1 King Pile – 12mm web wall thickness, 17mm flange wall thickness;
- ESC H70/30A-1 King Pile – 13mm web wall thickness, 20mm flange wall thickness.

1.3 Investigating the Cause of Corrosion

In 2014, during an inspection of the diversion wall by divers, it was found that the steel piles were corroding at a faster rate than originally allowed for (Greenfields 2014). Based on limited spot measurements of the underwater steelwork, using an underwater ultrasonic thickness gauge, the Greenfields report concluded that:

- Steel thickness of all submerged components estimated to be degrading at an average rate of approximately 0.5mm/year
- Steel sheet piles exhibiting an average thickness loss of 2.4mm (~29.1%) based on an original assumed pile wall thickness of 8.5mm (sic).
- Steel king (H) piles exhibiting an average thickness loss of 2.17mm (~17.35%) based on an original assumed web wall thickness of 12.5mm (sic).

Further investigation into the cause of corrosion was undertaken by the University of Waikato (Hartland et al 2015), who concluded that the corrosion was *"consistent with a form of accelerated microbially-induced corrosion (MIC) of the steel under aerobic conditions"*. They also noted that the presence of manganese and sulphur rich inclusions in the steel *"were capable of providing sites for preferential corrosion and are likely to be a contributing factor in the accelerated corrosion"*. Additional in situ and laboratory thickness measurements at one section of the Diversion Wall, indicated that the measured corrosion rates were between 0.24 to 0.28mm/annum for the sheet piles and 0.36 to 0.37 mm/annum for the king piles. When compared with the upper bound adopted corrosion rate of 0.04mm/annum as given in AS 2159 for moderately corrosive fresh water, it appeared that the corrosion rate was 7.5 times (sheet piles) to 9.25 times (king piles) greater.

The (Hartland et al 2015) report also advised that the use of cathodic protection should be considered as a way of "reducing the corrosion rate up to full inhibition". The findings of the University report were agreed by the peer reviewer (LBA 2015), who also advised that *"it is necessary to retro-fit a CP system (likely to be ICCP)"*; i.e. impressed current cathodic protection.

In addition to the above, the report (Hartland et al 2015) also indicated that the *"remaining life of the steel diversion wall (without intervention) is estimated to be about 7 years for shallow steel components, to about 12 years for mid-water components".*

2. REGISTRATION OF INTEREST REVIEW

Based on the guidance given to the BoPRC, discussed above, the Council decided to approach contractors and interested parties with their views and cost estimates to reinstate the Ohau Diversion Wall under a Registration of Interest (ROI) process.

Opus International Consultants were initially commissioned by the BoPRC to undertake an independent review of ROI tenders, which was extended to include considering and developing environmental mitigation measures, as well as comparing the different solutions based on a whole of life net present value (NPV) cost modelling, taking into account expected future maintenance costs (Opus 2015). The NPV modelling was undertaken in accordance with ASTM A1068-10.

The options considered were:

2.1 Do Nothing

This option was given by the designer (Beca 2014), where the structure is allowed to continue deteriorating until it is no longer fit for purpose due to excessive perforation or structural failure. This was estimated to occur in 7 years' time (Hartland et al 2015). At that time, assuming that leaving the collapsed diversion wall in situ is not acceptable, the wall is fully removed. However BoPRC indicated that the wall function of separating the two bodies of water was still required for the original 50 year duration, so this option was considered to be unacceptable.

2.2 Full Replacement

Another scenario is to do nothing until the wall starts to collapse (as above), and then it would be progressively removed and reinstated with a similar steel sheet pile wall but with a suitable corrosion protection system. This system was assumed to need to be inspected on a 10 yearly cycle, and maintained where necessary if damaged.

2.3 Monitor, Maintain and Replace (MMR)

Maintain and replace corroded members over the remaining 40 year life, as required. This is a reactive strategy that is already being undertaken by the replacement of corroded bolts. A number of maintenance and repair options are available, from welding on steel plates to close perforations, welding on plates to strengthen weakened areas, to welding plates between the sheet pile outpans and filling them with concrete (see the next option below for further details).

These options and a recommended Risk-Based Maintenance Strategy are discussed in further detail in (INA 2005).

The cost of this option will vary depending on the required maintenance operation for the given year, in addition to a biennial inspection and monitoring cost. This will also be affected on whether an underwater welder is required, or the use of a coffer dam to provide a dry and safe working environment. Also, mitigation measures to limit the periodic expected environmental effects of any such maintenance operation will need to be considered and allowed for, which may be limited by the current or any future Resource Consent. Therefore to provide a reliable cost estimate at this time for this option is very difficult, however, the model assumed the use of a geofabric silt curtain to provide the water separation barrier.

However, it should be noted that this option carries a large level of uncertainty, as it is not known when the maintenance and replace operations will be required. This will greatly affect the NPV model, with a large risk of significantly increasing the nett cost if work is required earlier than assumed. Another factor to consider is the periodic impact of such operations over the life of the structure on the local environment, which may not be acceptable to the Bay of Plenty Regional Council, the lake users and the local community.

2.4 Concrete Encasement

This option was given in the designer's report (Beca 2014), as well as discussed in *Accelerated Low Water Corrosion* (INA 2005); for which we assessed 2 alternatives in this option, either concrete encasing the top third of the wall where the most severe corrosion currently is (and could also include the above water portion), or full concrete encasement down to lake bottom.

This option appeared to provide the most practical engineering solution to reinstate the wall, with minimal ongoing environmental risk, other than during the initial installation, where strict measures will be required to prevent spillage or leakage of concrete or grout. However this option was discounted due to the potential risk of cement leaking into the lake environment during construction, and issues with the removal of the wall at the end of its design life.

2.5 Protective Coatings

As above, there are also 2 alternatives in this option, the partial coating of the top third portion of the wall (which should include above water portion), or fully coat the entire wall down to lake bed level. For either case, a coffer dam will be required for in situ application of the coating system. This is required to achieve a dry environment to prepare and protect the surface with an immersion grade epoxy system, and may be impractical to keep dewatered at a depths of up to 8m. It should be noted any coating will degrade over time, and so will also require cathodic protection with increasing current over the life of the wall.

This option has the highest NPV cost of the options considered, with a high risk of an ongoing disturbance to the environment, due to the long coating period expected. At this cost, it is more cost effective to rebuild the wall.

2.6 Cathodic Protection

Cathodic protection (CP) was promoted as the most cost effective method of protecting poorly coated buried or immersed structures (INA 2005), (Beca 2014), (Hartland et al 2015) and (LBA 2015). There are 2 options that could theoretically be used, sacrificial anode (SACP) or impressed current cathodic protection (ICCP).

As the large (>50,000 m^2) uncoated steel surface (i.e. the steel sheet piles and king piles) forming the Ohau Diversion Wall is operating in a fresh water lake, with high resistivity (taken as 5,000 Ωcm) water and ignimbrite bedrock, the use of sacrificial anode cathodic protection (SACP) is not considered to be practical to provide the high current required (~80mA/m^2 assumed) and is not recommended (the estimated annual anode consumption is 10 tonnes of magnesium with an estimated cost of $200,000/annum). Also the current density required to protect against MIC in fresh water and resistivity of the biological slime coating has not been well documented so a trial to confirm design parameters will be required. Early indications for the installation of an impressed current cathodic protection (ICCP) is up to $1.5M (as estimated in the ROI submissions). However, the ongoing electricity cost is expected to be of the order of $80,000 /year, and as this is a fresh water lake, the power cost over the life of the system is significantly higher than a comparable system in marine environments where a protective calcareous scale forms reducing current requirements. Costs assume that the power is supplied by the National Grid, instead of from an alternative power source such as a dedicated wind turbine and/or solar panels; which due to the size of the wall and power required may also not be feasible:

- 24 volts (DC) x 2000 Amps = 48,000 Watts
- 48kW x (365 days x 24 hours) 8760 hrs = 420,480 kWh/a
- 420,480 kWh x 19 cents/kWh = $79,891/annum

Therefore, when adding the annual cost of providing power to the ICCP system to the initial installation cost of that system, the overall cost of this option is significantly higher than may have been assumed. Also, not considered is the cost of protecting fish from the effects of the high voltage gradients (>0.1V/cm) (Beaumont et al, 2002) that would be created near the wall.

The benefit of this option is that most of the work is undertaken in the first year, which includes the bulk of any environmental impact. However, the ongoing power cost and the risk of stray current corrosion from or of adjacent structures (that still needs to be resolved), have increased the whole of life cost of this option.

2.7 Cost Comparison

The expected NPV cost between the different options are outlined in Table 1.

Table 1: Expected NPV cost between the different remediation options.

Option	NPV Cost	Comments
Do Nothing	$1.3M	While the lowest cost, not being able to separate the 2 bodies of water is unacceptable.
Concrete Encasement	$1.6M	Has a high risk of potential environmental damage. Option disqualified.
Monitor, Maintain and Replace	**$1.8M**	Further investigation required.
ICCP	**$2.8M**	Further investigation required.
Full Replacement	$8.1M	Too expensive.
Protective Coatings	$18.1M	Prohibitively expensive.

The outcome of the initial review indicates that additional research is required into an impressed current cathodic protection and the Monitor, Maintain and Replace option.

3. DETAILED OPTIONS ANALYSIS

Based on the findings of the ROI Review a detailed option analysis was undertaken on the two potential remediation options, ICCP and the Monitor, Maintain and Replace. The findings of this report were given in (Opus 2016) and are summarised herein.

3.1 Monitor, Maintain and Replace (MMR)

Several different maintain methodologies were considered, which included solutions outlined in Section 6 of *Accelerated Low Water Corrosion* (INA 2005); of which the following solutions were proposed:

- Maintain: HPDE Perforation Patch

- Maintain: Steel Doubler Plates – Welded
- Maintain: Steel Doubler Plates – Bolted
- Replace: Vinyl Sheet Piles and King Pile Channels
- Replace: Impermeable Polymeric Screen

This investigation included an in-depth assessment of each solution, its pros and cons, and rough order cost. As well as developing environmental mitigation measures, such as the removal of fish from coffer dams, and limiting the amount of lake bed disturbance by the use of turbidity curtains.

This option was developed based on (Hartland et al 2015) corrosion rates, where the structural capacity of the sheet piles and associated king piles, can withstand the applied loading between 3 to 9 years for the sheet piles and 25 to 30 years for the king piles (depending on the cantilever depth of the component) as determined in (Opus Sept 2010). This also included calculating the minimum critical steel thickness of the sheet pile as being 5.2mm, which is the minimum thickness required to ensure the continued structural performance of the wall in resisting the applied loading acting on it. Hence it was assumed that certain sections of the Diversion Wall is to undergo local maintenance/strengthening over those periods, with the introduction of an impermeable polymeric screen in year 23. This includes strengthening of the king piles at that point in time.

To assist in developing the net present value (NPV) cost model for this option, different scenarios were developed, which was based on the assumed percentage corrosion on the wall that requires maintenance over the 23 year period. An example of such a scenario is given below in Table 2, where it is assumed that 10% of the surface area of the wall requires local maintenance/strengthening. It was assumed that 1100m^2 of the wall requires this strengthening, which equates to 192 doubler plates. From this the whole of life net present value cost comparison was calculated. This was also done for a 25% and 50% surface area of the wall requiring strengthening. Then at year 23 a polymeric impermeable screen in introduced to provide the main water separation barrier, with the king piles supporting the screen and the sheet piles are allowed to corrode.

Table 2: Example of a possible scenario when assess the NPV cost analysis.

Year	Option	Estimated Repair Area	Cost	NPV Cost
3	Maintain: Steel Doubler Plate - Bolted	Install the first 100m length of the wall	$235,000	$197,000
8		Install the remaining 1150m length of the wall	$703,000	$441,000
13		Install the first 100m length of the wall	$235,000	$110,000
18		Install the remaining 1150m length of the wall	$703,000	$246,000
23	Replace: Polymeric Impermeable Screen	Install a new screen along the full 1250m length of the wall; this includes strengthening the King Piles.	$4,765,000	$1,250,000
33		Remove and install a new screen along the full 1250m length of the wall.	$1,600,000	$233,000
		Total	$8.24M	$2.4M

The main limitation when developing the scenarios is the lack of detailed information on the actual condition of the wall, and the extent of corrosion. Hence, the need to assume the extent of corrosion breakdown on the Diversion Wall, based on the only available information was what was provided by (Greenfields 2014) and (Hartland et al 2015). Thus, one of the main recommendations was the request for a detailed condition assessment of the diversion wall to be undertaken.

Another version of this option was the introduction of a polymeric impermeable screen at year 0 and its replacement every 10 years. While simple in theory the long-term performance of the screen is not known at this stage, and there is a medium to high risk to damage from the lake users (such as entangled anchors, and the like); that needs to be accounted for.

3.2 Cathodic Protection

As discussed earlier in Section 2.6, the use of SACP was considered to be impractical and costly. This is mainly due to the large surface area of unprotected steelwork (~50,000m^2), and the high resistivity of the fresh water and soil. While in seawater the resistivity is typically between 20 to 25 Ωcm, the fresh water in Lake Rotorua and Rotoiti ranged between 4500 to 6000 Ωcm. Hence a larger driving voltage is needed to provide the required current density, which can only be supplied by an ICCP.

As part of our review, there are two important aspects must be considered, these are:

3.2.1 Environmental Impact of ICCP on the Lake Environment

As with SACP, there is little published literature regarding the effects of ICCP on the fresh water aquatic environment, which makes it difficult to envisage how installing ICCP on the Ohau Diversion Wall will affect the natural and physical environment of Lake Rotoiti. An electrical potential is applied to the metal structure that creates a voltage field in the water column of the lake, which may have some effect on the freshwater biota in the lake. There are examples of high voltage fields that have been used to repel passage of pest fish, which are able to detect and avoid the electric field when approaching from beyond the affected area (Beaumont et al, 2002). If the ICCP method is adopted, it was recommended that a field trial be carried out to determine how it will affect freshwater biota. Due to the lack of published information, without some applied field research, it will not be possible to accurately determine effects on the environment.

Depending on where the anodes are located, the current will create an electric field within the water. This would create a situation similar to that produced by an electric fishing machine, which utilises water to complete the circuit between the anode and the cathode in order to temporarily stun fish within the electrified area. It is possible that fish within the area between the cathode and anode, in contact with the anode or cathode, or in close proximity to the electric field, may be affected by the ICCP system when electrical current is applied. However, it is noted that it is considered safe for divers to be in the water while ICCP systems are operating as regulations limit the driving voltage to less than 50V DC. Therefore, the effect of the ICCP system on fish will depend on the individual fish species electricity sensitivity. There is limited research as to the sensitivity of New Zealand freshwater aquatic life to electrical fields.

The highest electrical gradient will occur immediately adjacent to the anodes and fish will only be affected for a very short time once the ICCP system has been turned on. Once on, the electric field may discourage fish from coming into close proximity meaning the effect on fish is likely to be low. Due to the lack of published information on ICCP in fresh water fisheries, it was not possible to make any conclusions regarding the effects of CP on the environment at this time. Trials should be conducted before its use is committed to.

3.2.2 Efficiency of ICCP Arresting Corrosion on the Diversion Wall

The other main concern was the efficiency of the ICCP system arresting the corrosion acting on the Diversion Wall. While the resistivity of water was known (taken as 4500 to 6000 Ωcm), the resistivity of the silt and ignimbrite bedrock was unknown and additional testing is required to confirm that value.

Thus upon discussing the current limitations of protecting the Diversion Wall in such difficult conditions (when compared against a marine environment), the ROI registrants whom proposed ICCP all agreed that a large scale trial required. Their main concern was that ICCP may not be able to completely arrest the corrosion occurring on the wall, at best it may be able to slow the corrosion rate to that originally considered by the designer, and at worst it may be insufficient to provide complete protection.

3.3 Cost Comparison

The following table outlines the expected NPV cost between the different options.

Table 3: Expected NPV cost between the different remediation options.

Option	NPV Cost	Comments
MMR: Steel Doubler Plate - 10%	$2.4M	Assume 10% of the wall requires strengthening.
Polymeric Impermeable Screen	$3.3M	
ICCP	$4.7M	Cost estimate increased when risk cost taken into account
MMR: Steel Doubler Plate - 25%	$4.8M	Assume 25% of the wall requires strengthening.
MMR: Steel Doubler Plate - 25%	$7M	Assume 50% of the wall requires strengthening.

3.4 Discussion

Up until this stage, several solutions were promoted by different parties, based on a limited set of available information. Most of the available guidance, such as (INA 2005) and (CIRIA 2005), promotes the use of cathodic protection, especially ICCP; as one of the optimum methods of addressing accelerated microbially-induced corrosion (MIC). However, all of this guidance is based on accelerated low water corrosion in marine environments, with lower water resistivity, where ICCP has been successful with retarding or even fully arresting the MIC.

The Ohau Diversion Wall is different due to fact that it is in a fresh water lake, with a near constant level and over 225x the resistivity of seawater, with a very large portion (~28,000m^2) of the unprotected steel in lakebed material of volcanic origin with an unknown resistivity value. Traditionally, steel structures in such fresh water environments are simply designed with a corrosion allowance, as given in AS 2159:1995 and NZS 3404.1:2009. Unfortunately, in this case, this was not sufficient.

Thus the main outcome of this review was the need for additional information to be gathered, especially in regards to assessing the different MMR scenarios and their expected cost. When comparing between MMR and Impermeable Polymeric Screens for example, where the MMR scenario at 16% breakdown will provide a comparable cost as the screens. As such, the findings of the review indicate that there was an urgent need to confirm the extent of ALWC throughout the Diversion Wall, and if needed, confirming the resistivity of the lake bed soil. Once this information is known, then the promoted and developed solutions can be re-evaluated.

4. DETAILED INSPECTION

A detailed inspection of the Diversion Wall was undertaken in June 2016. The underwater inspection was undertaken by scuba divers, under the supervision of the author and a BoPRC representative. The wall thicknesses were measured using a Cygnus M1-DIVE ultrasonic thickness gauge, with the pit depths measured using an anvil micrometer. Measurements were taken at 50m intervals along the full length of the wall, at alternating surfaces. Examples of the retrieved MIC corrosion products are given below.

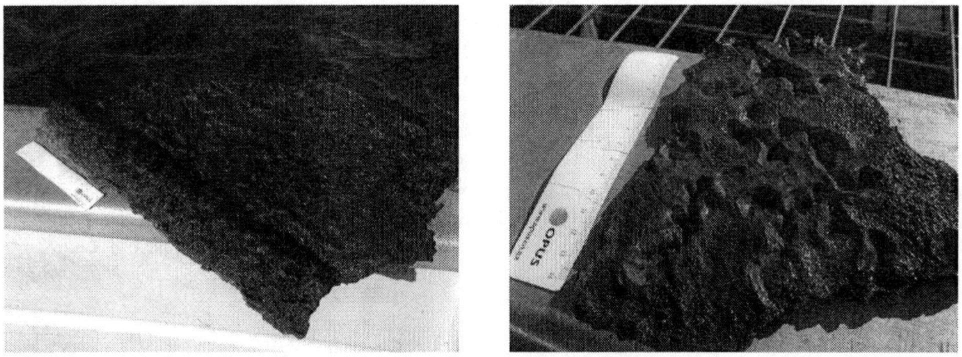

Figure 3: Example of MIC corrosion products.

4.1 Inspection Findings

A summary of the average, minimum and maximum thickness readings from all 24 sites is given in the following tables:

Table 4: Lake Rotorua Side Steel Thickness Measurements.

Lake Rotorua Side	King Piles		Sheet Piles
	Flange	Web	
Top (0.5m from waterline)	$17.64mm_{21.50}^{16.50}$	$9.83mm_{10.90}^{8.40}$	$5.07mm_{6.50}^{3.10}$
Mid Depth	$17.96mm_{21.70}^{16.50}$	$10.66mm_{11.50}^{9.30}$	$6.01mm_{6.50}^{5.40}$
Lake Bed	$17.59mm_{22.00}^{10.50}$	$10.27mm_{11.10}^{9.70}$	$5.85mm_{6.80}^{4.40}$

Table 5: Lake Rotoiti Side Steel Thickness Measurements.

Lake Rotoiti Side	King Piles	Sheet Piles
	Flange	
Top (0.5m from waterline)	$17.22mm_{18.90}^{16.10}$	$5.43mm_{6.00}^{4.00}$
Mid Depth	$17.68mm_{18.70}^{16.30}$	$6.15mm_{6.60}^{5.40}$
Lake Bed	$17.38mm_{18.60}^{16.40}$	$6.04mm_{6.50}^{5.30}$

Based on the specified and current steel thickness measurements, the estimated corrosion rates were determined for the different locations along the diversion wall, over the 9 year immersion period.

Table 6: King Pile Steel Thickness Loss.

King Pile Location			Average Section Loss (mm)	Average Loss/year (mm/y)	Opus 23/9/15 Report Assumed Average Loss/year (mm/y) *
Lake Rotorua Side	Flange	Top (0.5m from waterline)	2.36	0.262	0.32
		Mid Depth	2.04	0.227	
		Lake Bed	2.41	0.268	
	Web	Top (0.5m from waterline)	3.17	0.352	
		Mid Depth	2.34	0.260	
		Lake Bed	2.73	0.303	
Lake Rotoiti Side	Flange	Top (0.5m from waterline)	2.78	0.309	
		Mid Depth	2.32	0.258	
		Lake Bed	2.63	0.292	

*Based on (Beca 2016).

Table 7: Sheet Pile Steel Thickness Loss.

Sheet Pile Location	Average Section	Average Loss/year	Opus 23/9/15 Report Assumed

	Loss (mm)	(mm/y)	Average Loss/year (mm/y)*
Top (0.5m from waterline)	2.75	0.305	0.24
Mid Depth	1.92	0.213	
Lake Bed	2.06	0.228	

*Based on (Beca 2016).

4.2 Site Observations

The following site observations were made:

- The sheet pile wall thickness was found to be uniform along the full length of the wall, on average within ± 0.5mm, with the top third of the wall exhibiting the highest section loss due to corrosion rate.
- The corrosion product on the Lake Rotorua side, was found to be harder, and well adhering, with a rougher pitted surface underneath it. Whereas on the Lake Rotoiti side, the corrosion product was softer and easy to remove, with a smoother less pitted surface. However, it appears that the well adhering corrosion product may have resulted in the corrosion rate plateauing over time on the Lake Rotorua side. Whereas the weakly adhering corrosion product on the Lake Rotoiti side is continuously exposing a fresh steel surface, so the corrosion process is more active on that side. This is discussed further below.

4.3 Corrosion Product and Water Composition

In addition to assessing the extent of corrosion and the condition of the Diversion Wall, it was also found that the corrosion product was either hard and brittle (Lake Rotorua side) or soft and crumbly (Lake Rotoiti side). The hypothesis was that the level of nutrients (phosphates and nitrates) on the Lake Rotorua side are higher, effectively providing a continuous source of food for the microbes, that are also expected to be more active. To confirm this, corrosion product (see Figure 3 below) and water samples from both sides of the Diversion Wall were collected by Bay of Plenty Regional Council staff on 23 June 2016. These were transported in water tight containers in a cool box, as instructed by the testing laboratory.

In summary, it was found that:

- Chemical Composition:
  - Under X-Ray Fluorescence (XRF) sampling method, in both samples ferric oxide (Fe_2O_3) was found. This was 72.72% of the sample on the Rotorua side, and 86.11% on the Rotoiti side.
  - The ferric oxide was further analysed using X-Ray Diffraction (XRD) where it was found that on the Rotorua side goethite (FeOOH) was measured at 62%, magnetite (Fe_3O_4) was measured at 30% and lepidocrocite (FeOOH) was measured at 8%. While on the Rotoiti side magnetite (Fe_3O_4) was measured at 57%, goethite (FeOOH) was measured at 31% and lepidocrocite (FeOOH) was measured at 12%.
- Water Results:
  - The Rotorua water exhibited higher overall microbial levels than the Rotoiti water (e.g. 340 vs 8 cfu/ml of Pseudomonas), suggesting higher nutrient levels in the Rotorua water.
  - Light levels of filamentous iron-encrusted bacteria were detected in the water sample from the Rotorua side while only sporadic levels of other non-encrusted filamentous bacteria were observed from the Rotoiti side.
- Corrosion Product (described as Sludge in the Biodet Report) Results:
  - Sulphate-reducing bacteria and Clostridium were at higher levels in the Rotorua sludge sample than in the Rotoiti sludge sample, suggesting biofilms were more established on the Rotorua side of the wall. These microorganisms have the ability to produce corrosive organic acids.
  - The Rotorua sludge exhibited much higher levels of iron-precipitating bacteria than the Rotoiti sludge, again suggesting higher nutrient levels in the Rotorua water.
  - Filamentous bacteria were not obvious in either of the sludge samples, and as these are aerobic bacteria, it suggests that anaerobic conditions within the sludge may have prevented the proliferation of these microorganisms on the surfaces of the wall.

The above results confirm the hypothesis that the corrosion products are more established on the Rotorua side, with an active level of corrosion (more pits and a rougher surface) observed. However, the Rotoiti side is just as active, as the steel substrate is continuously being re-exposed through the regular removal of the corrosion product due to stronger wave action; thus it seems to be experiencing a more uniform general corrosion with a smoother steel surface and less pitting.

As such, the corrosion rate of the sheet pile is taken as being similar on either side of the wall, even though the water chemistry and activity of the microbial influence corrosion is different.

4.4 Discussion

The main outcome of the detailed site inspection, was that the variation of corrosion with depth can be taken as being similar along the length of the Diversion Wall. The corrosion rate can also be taken as being similar on both sides of the wall. Furthermore, the critical minimum thickness of 5.2mm (Opus Sept 2015), appear to have been reached in some cases, thereby indicating that remediation action on the Diversion Wall is required sooner than the originally assumed 3 year grace period.

5. REMEDIATION OPTIONS REVIEW

Based on the site inspections findings, the previously preferred Monitor, Maintain and Replace option is now deemed to be uneconomical, as that option would have been cost effective if the extent of corrosion was less than 16 % of the wall surface area, with the focus of protecting the top third of the wall. However, the findings of the detailed condition assessment indicate that 100% of the top third of the Diversion Wall (or 33% of the total sheet pile surface area) is now of concern, making this option uneconomical and impractical in the short to medium term (5-10 year) period.

Furthermore, since ICCP (in our view) would first need an expensive trial installation to confirm it to be a practical and effective solution, alternative options were investigated that could be implemented more quickly.. The findings of (Opus 2017) are summarised below.

5.1 Stiffening Walers

With better understanding of the actual condition of the Diversion Wall, an alternative structural solution was developed. It was proposed that this would assist in stiffening the sheet pile by bracing it at the top, using a steel waler. This would take advantage of its trapezoidal configuration, and the fact that such sheet piles are up to 9000 times stiffer in their vertical "strong axis"; than what it is currently erected and supported along its "weak axis".

The results of the structural modelling indicate that supporting the top of the sheet pile with a 150 PFC waler (located just above the water line), will reduce the stresses acting on the sheet pile wall. This will also result in reducing the minimum critical thickness for piles at water depths >5m, from the current 5.2mm (Opus Sept 2015) to 2 mm (i.e. 3.2mm improvement).

Localised failures (tearing) at the clutches are still possible; but this risk is considered to be significantly less than if the diversion wall was not supported at the top. Based on the above corrosion rates given in Section 2.2, the expected increase in the time before major refurbishment is estimated as 3.2mm/0.305mm/year ≈ 10 years.

A sketch of this Stiffening Waler concept is given below.

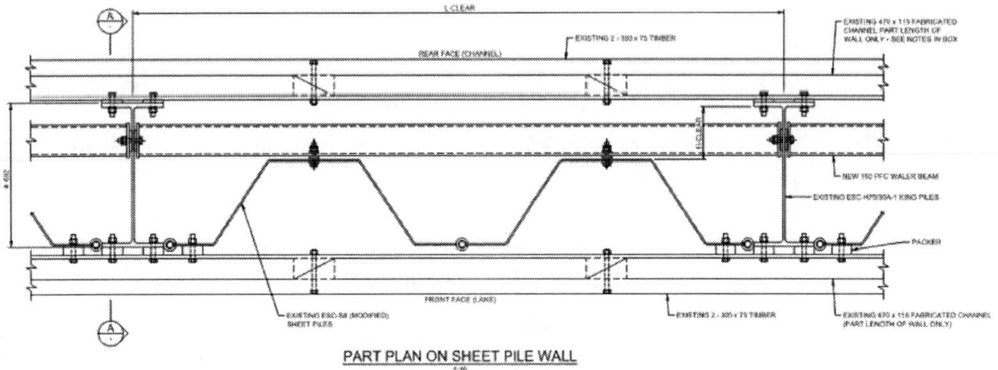

Figure 4: Stiffening waler concept.

Thus by installing a Stiffening Waler in 2018, it is expected that the structural capacity of the Diversion Wall can be extended until 2028, by which time the installation of Impermeable polymeric Screens can be considered. Schematics of how a screen can be attached to the Diversion Wall is given in Figure 5 below. Note the second row of new piles that may be needed.

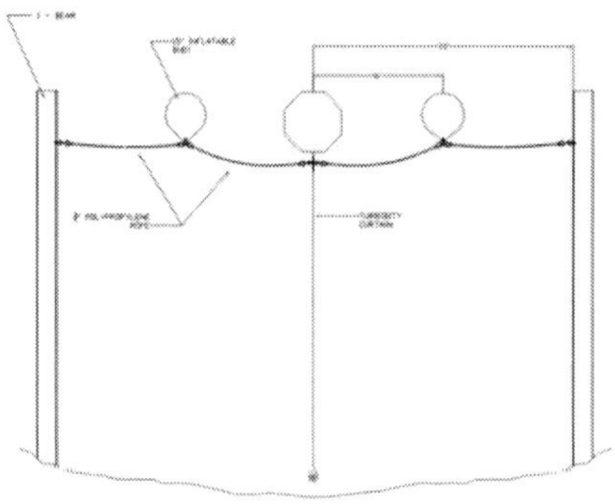

Figure 5: Example of how the screen may be attached to the Diversion Wall.

The structural review also considered the structural performance of the king piles, based on their corrosion rate, and the applied loadings. As part of installing the Stiffening Walers, it was also recommended that the last king pile (designated End Pile) at chainage 1275m is to be strengthened at the same time.

As for the remainder of the king piles, the applied loadings are expected to change once the screens are installed, and whether the sheet piles are to remain or be removed. For example, if it was after installation the screens the applied loading on the king piles are reduced (expected to be by 50%), then any remediation of the king piles are expected to be postponed by year 20, which may include the removal of the sheet piles at the same time. While if it was decided to remove the sheet piles in year 10, then it is practical to strengthen the king pile at the same time.

As seen before on this project, to decide on which solution to proceed with, further information is required. In this case, a hydrodynamic analysis is required to confirm the loadings acting on the wall, and their direction, after the installation of the Impermeable Polymeric Screens. If it was found that the wave load reduction is more than 50%, this will assist in extending the date to pull out the sheet piles, possibly providing further cost savings by postponing this activity further into the future. Also, if it was found that the loads on the king piles are acting on the strong axis, instead of its weak or on both axis, then strengthening of the remaining king piles may also not be required.

This analysis is also required to assist in the design of the Impermeable Polymeric Screens, to confirm the number of piles required to anchor the screen and loading acting on the screen to ensure that the correct screen weight, cables and fixings are allowed for.

5.2 A New Set of Remediation Options

Based on the latest findings, it is believed that the Stiffening Walers will provide up to 10 years continued structural performance of the Diversion Wall. The next set of questions are mainly to do with what to do from year 10 onward?

- Shall we remove the sheet piles before installing the Impermeable Polymeric Screen or keep them?
- When shall we strengthen the king piles?
- What are the risks that we need consider?

Hence, different scenarios were considered, which are summarised below including their NPV cost.

Table 8: Cost Comparison Table.

Ranking	Option	NPV Cost
1	Scenario 4: Independent Impermeable Polymeric (IP) Screen	\$2.9M
2	Scenario 2A: IP Screen – Remove Sheet Piles in Year 20	\$3.4M
3	Scenario 2B: IP Screen – Remove Sheet Piles in Year 20 + Strengthening King Piles	\$3.6M
4	Scenario 2C: IP Screen – Remove Sheet Piles in Year 10	\$3.7M
5	Scenario 2D: IP Screen – Remove Sheet Piles in Year 10+ Strengthening King Piles	\$4.1M
6	Scenario 3: Impermeable Polymeric Screen – Full Diversion Wall Removal	\$4.9M
7	Scenario 1: Impermeable Polymeric Screen + Strengthening King Pile	\$6.2M

Based on the above cost comparison, it appears that Scenario 4, with a fully Independent Impermeable Polymeric Screen, leaving the existing Diversion Wall to remain provides the lowest whole of life NPV cost at \$2.9M. However, it is unlikely that this option will be acceptable to the public, the current Resource Consent and the BoPRC. Therefore, this option will not be considered further, but was included here to demonstrate the range of costs that may be expected for the continued separation of the waters of Lake Rotoiti and the Ohau Channel.

The next lowest cost option is Scenario 2A, with an Independent Impermeable Screen, and removal of the sheet piles in Year 20, at \$3.4M. However, as discussed above, the future performance of the king piles is dependent on the wave loading and its direction, confirming whether this option is possible. Next is Scenario 2B which includes the strengthening of king piles at $3.6M. This is followed by the removal of the sheet piles (Scenario 2C) and strengthening of the king piles at Year 10 (Scenario 2D), at $3.7M and $4.1M respectively.

The full removal of the diversion wall, covered by Scenario 3 is another costly option, with its own challenges. While removal of the sheet piles is considered to be relatively simple, it is the removal of the king piles that is a challenge. This could either be done by the full removal of the 50m long king piles, or by cutting them off at just below the lake bed level. The latter is understood to be a cost effective option in comparison to their full removal, but in both cases this option costs more than what is listed in Scenario 2.

What the above findings portray is that it is difficult at this stage to provide a single optimum solution. The only common denominator for all cases, is that the Stiffening Waler and End Pile Strengthening is required in 2017/2018, which will allow for the installation of the Impermeable Polymeric Screen in 2028. During this period additional investigation is required which will assist in confirm which remediation option is required for the king piles.

In addition to the above, 3 yearly monitoring of the Diversion Wall should be undertaken to monitor the corrosion rate of the steel at the designated sites, and the overall condition of the wall.

6. DISCUSSION

The remediation of the Ohau Diversion Wall is an ongoing process, that keeps on changing as new information becomes available. It is a good example of continuous improvement using corrosion management and the principles of Asset Integrity Management.

Typically, when attempting to reinstate a deteriorating structures, the main focus is on rehabilitating that structure, whether it is a bridge, pipeline or retaining wall. However, in this case the main function of the Diversion Wall is the separation of two bodies of water, preventing their mingling to retain the quality of one of them.

Hence, while every effort was made at the beginning of this project to rehabilitate the existing unprotected steel sheet piles and king piles, it became very clear that task is impossible. Our belief that MIC in fresh water cannot be economically addressed by impressed current cathodic protection, due to the much higher water resistivity (when compared with seawater), resulted in pursing alternative solutions.

The main learning that should be taken from this case study, is that knowledge is key to a good design and developing optimum solutions. For example, the original designer of the Diversion Wall, using AS 2159:1995, adopted the recommended corrosion rate for steelwork in fresh water. MIC was not considered as most of the international guidance on this corrosion mechanism were just being published at the time of the design (circa 2005). Furthermore, all of the MIC on sheet piles guidance is written for marine seawater environment. Thus there was no known reason to believe that the adopted corrosion rate would not allow the Diversion to meet its 50 year design life.

In hindsight, some form of corrosion protection would have been beneficial. However, the "standard of the day" did not highlight the risk of accelerated MIC in fresh water. If this risk was known and allowed for, then the outcome would have been different.

It should be noted that the latest version of AS 2159:2009 gives in Section 6.5.3 a range of uniform corrosion rates for different exposure classifications, and notes that "localized pitting corrosion" may occur on piles subject to accelerated low water corrosion; no additional guidance on possible corrosion rates for MIC is given. Also the actual parameters that encourage the establishment of microbial influenced corrosion in fresh water are still not clearly understood. It appears that additional research into the causes of such growth is required.

Another example of why knowledge is power, is the way the remediation solutions evolved as additional information became available. The original assumption was that ICCP would economically provide the protection needed. Upon closer consideration of the electrolyte resistivity, this was demonstrated to be not the case. Later on, a more detailed condition assessment of the wall demonstrated the actual extent and rate of corrosion throughout the wall, resulting in another change in direction.

While we were originally engaged to review the remediation options, a combination of our environmental and structural engineers, in addition to hydrologist and asset managers are also now involved. This has resulted in the development of a practical long term solution that may need to be revised again, depending on the results of future monitoring of the Ohau Diversion Wall.

7. CONCLUSIONS

Our conclusions are:

- Accelerated MIC of uncoated steel sheet piles in fresh water is unlikely to be mitigated economically by an impressed current cathodic protection, and definitely not with sacrificial anodes.

- Ensure the actual condition of the affected structure is known, prior to approaching contractors and interested parties, or embarking upon the detailed development of remediation options.
- Continuous improvement, via reviewing the available data, is good Asset Integrity Management Practice. Self-criticise, and question yourself - are you developing the optimum solution?
- Do not hesitate to approach other disciplines, that may provide alternative practical solutions.
- Finally, Corrosion Engineers should become familiar with the principles of Corrosion Management and Asset Integrity Management, especially the use of Whole of Life Net Present Value cost modelling, and Risk Assessment when developing remediation solutions.

8. ACKNOWLEDGMENTS

The authors would like to acknowledge Beca and the Bay of Plenty Regional Council for the permission to write this paper, ensuring the dissemination of the lessons learned. Also, to thank Opus International Consultants for their support to present the results of this investigation to this conference.

9. REFERENCES

- AS 2159-1995/2009. *Piling - Design and installation,* Standards Australia
- ASTM A1068-10 (2015). *Standard Practice for Life-Cycle Cost Analysis of Corrosion Protection Systems on Iron and Steel Products,* ASTM International, West Conshohocken, PA, USA.
- Beaumont, W.R.C., Taylor, A.A.L., Lee, M.J., and Welton, J.S. 2002, *Guidelines for Electric Fishing Best Practice*, R&D Tech Report W2-054/TR, Environment Agency, WRc, Bristol, UK.
- Beca, November 2006, Ohau Channel Diversion Wall Tender Drawings, Auckland, NZ.
- Beca, January 2007. Ohau Channel Diversion Wall: Design Criteria Summary, Auckland, NZ.
- Beca, 16 July 2014, Ohau Diversion Wall- Corrosion Investigation- Rev 2. Auckland, NZ
- CIRIA, 2005, *Update to C634 Accelerated Low Water Corrosion*, Report No P2951. London UK.
- Greenfield Diving Services, 28 May 2014, Report on the inspection of the underwater components of the sheet pile diversion wall in Lake Rotoiti, Tauranga, NZ.
- Greenfield Diving Services, 19 October 2015, Report on the inspection of the underwater lake bed components of the sheet pile diversion wall in Lake Rotoiti, Tauranga, NZ.
- Hartland, A., Lee, K.C., Pickering, K., Zhang, Y., Barker, S. and Cary, C, 2015, Corrosion of the Ohau Channel Diversion Wall. Environmental Research Institute Report No. 50, Environmental Research Institute, Faculty of Science and Engineering, The University of Waikato, Hamilton, NZ.
- INA, 2005, *Accelerated Low Water Corrosion*, International Navigation Association, Brussels, Belgium.
- LBA and Associates, 15 June 2015, *Review of UoW report on Corrosion of the Ohau Channel Diversion Wall,* Auckland, NZ.
- NZS 3404.1:2009. *Steel structures Standard. Part 1: Materials, fabrication and construction*, Standards New Zealand, Wellington, NZ.
- Opus, 23 September 2015, *Ohau Diversion Wall Corrosion Mitigation – Task 2: Review and Investigation – Rev 2*, Auckland, NZ.
- Opus, 28 September 2015, *Ohau Diversion Wall Corrosion Mitigation – Above Water Condition Assessment*, Auckland, NZ.
- Opus, 16 March 2016, *Ohau Diversion Wall Corrosion Mitigation – Detailed Analysis Report*, Auckland, NZ.
- Opus, 3 April 2017, *Ohau Diversion Wall – Development of a Structures Management Plan*, Auckland, NZ.

10. AUTHOR DETAILS

Raed El Sarraf is a Corrosion and Asset Integrity Consultant at Opus International Consultants Ltd in their Auckland office in New Zealand. He is a Chartered Professional Engineer, has been an ACA member since 2003 and is the current ACA NZ Branch President.

He has co-authored a number of articles, papers and guidance documents in both New Zealand and Australia, such as the *Australian Steelwork Corrosion and Coatings Guide*, *Weathering Steel Design Guide for Bridges in Australia* and the NZ Transport Agency document entitled *Protective coatings for steel bridges*.

Willie Mandeno is the former Technical Principal: Corrosion Engineering and Protective Coatings of Opus International Consultants Ltd, based in Wellington, New Zealand. He is a Chartered Professional Engineer, an IPENZ Fellow, an ACA Corrosion Medallist and ACA Life Member, and a NACE Corrosion Specialist and CIP Instructor.

Photo not available

Mark Townsend is the Engineering Manager at the Bay of Plenty Regional Council. He manages the maintenance of different structures operated by the Regional Council.

IRATA UNLEASHED - GROWTH IN THE USE OF ROPE ACCESS IN CORROSION MITIGATION AND REPAIR

A. Caddy
Absafe Pty Ltd, Melbourne, Australia

SUMMARY: The use of rope access to aid corrosion repair efforts has grown remarkably in the last 20 years, but to what extent are rope access technicians capable of performing qualitative corrosion remediation? This paper discusses the evolution of the industry and identifies the challenges associated with ensuring that repairs are completed to a consistently high standard. It advocates how industry professional entrusted with difficult access works must aspire to the principle that 'the harder it is to get to, the better the repair has to be', and presents a grounded framework to guide this endeavor. By combining two indispensable quality control systems; Industrial Rope Access Trade Association (IRATA) and Painting Contractor Certification Program (PCCP), rope access technicians can secure successful outcomes that don't compromise on safety or quality during any project delivery.

Keywords: Rope Access, Corrosion Repair, Corrosion Mitigation, Qualitative, IRATA, PCCP

1. INTRODUCTION

The common problems associated with carrying out works at height stem from securing the safety of workers, public and plant. Variables such as time, weather delays, slow production rates and the cost of using expensive access systems combined with slow processes, also contribute significantly to the total cost of works. Enter rope access, a form of work positioning that provides workers trained in its use, the advantage of accessing difficult or high areas more easily and efficiently. Of course even if a labourer is trained, it begs the question as to what standard corrosion mitigation and repair can be performed and what mechanisms exist to ensure the safety of the system. The viability of rope access methods is demonstrated by the Industrial Rope Access Trade Association (IRATA) which refers to the peak governing body that oversees rope access in Australia and abroad. To address IRATA's role in enhancing corrosion mitigation and repair efforts, an overview of its methodology and presence both in Australia and worldwide, will be presented. IRATA has grown significantly in Australia, and signals how the rope access industry as a whole is maturing and fast becoming the preferred access system. But is it capable of providing qualitative outcomes? The answer is a resounding 'yes', provided that qualitative systems and controls are maintained, along with adequate training, plus an ethos of innovation. Rope access industry professionals entrusted with difficult access works, must aspire to the principle that 'the harder it is to get to, the better the repair has to be', and present a grounded framework to guide this endeavor. Two real-world examples of innovative use of rope access to solve corrosion and repair issues are given, which cover crucial quality control systems including IRATA and also the Painting Contractor Certification Program (PCCP).

2. INDUSTRIAL ROPE ACCESS TRADE ASSOCIATION (IRATA)

Rope Access is a form of work positioning initially developed from techniques used in caving and climbing. It applies practical rope-work for the purpose of granting access into hard-to-reach locations without the use of traditional scaffolding, swing stages or elevated work platforms. IRATA has been established in Australia for over 20 years. From its initial beginnings in the UK as a Trade Association for rope access companies working predominately on North Sea Oil rigs, IRATA has grown to become a worldwide organization that is the premium body controlling rope access. Marked by voluntary membership, IRATA member companies are active across a range of industries including property, engineering, construction, telecommunications, mining and gas, plus energy sectors. IRATA is the sole global rope access association in the 'working at heights' sector, and IRATA Australia is the peak industry association in Australia. IRATA International supplies a formal operator training certification scheme and grading structures that all IRATA members are obliged to use. The IRATA ICOP (International Code of Practice) dictates all operating procedures and methods, with IRATA companies subject to external audit and accreditation processes similar to the ISO provisions 9001, 18001 and 22846.

Currently, IRATA represents the interests of more than 300 member companies. It employs more than 50,000 rope access technicians, trains more than 2,000 rope access technicians annually and services more than 5,000 clients annually.

Table 1: Breakdown of IRATA Membership

IRATA membership worldwide	
Technical Grade	*No. of Technicians*
IRATA Level 1	38,029
IRATA Level 2	6,898
IRATA Level 3	7,410
Total Active	**52,337**
Total Registered (non-active)	**100,102**

IRATA membership in Australia	
Technical Grade	*No. of Technicians*
IRATA Level 1	2551
IRATA Level 2	492
IRATA Level 3	662
Total Active	**3,705**

Source: (Marage, 2016)

IRATA has undergone major changes in recent years with its focus shifting to a more international platform, and an awareness that 25 years since its inception there is now a group of highly skilled workers who no longer desire to maintain 'active' certification. To this end they are creating a training based competency program to enable experienced rope access workers to transition to management roles within organisations. The two paths currently being discussed are the roles of Rope access manager and Rope Access Safety Supervisor.

Figure 1: IRATA representation by continent (Marage, 2016)

3. OTHER QUALITATIVE PROCESSES

Despite rope access being a relatively new approach, there has been a rather substantial shift in its perception and use by industry. Today, the predominant workforce engaged in rope access is divided between trades-qualified or semi-skilled labourers. The growth in trades-based rope access workers has been particularly pronounced, which is witnessed by the volume of traditional service based companies incorporating rope access divisions in such fields as protective coatings, boiler making/steel trades as well as construction trades and mining and plant maintenance. Rope access has always held a large contingent of trained testing personnel including NACE inspectors, but is now expanding to include qualified engineers, with a number of traditional engineering firms now offering rope access trained personnel. The use of qualitative systems such as PCCP and ISO Certification, and the added training that can be accessed by such organisations as the Australian Corrosion Association (ACA) or the Australasian Concrete Repair Association (ACRA), translates to a consolidation of knowledge plus a series of checks and balances in the form of ITP's, checklists and other qualitative documentation controls.

3.1 Accessing of Research and Development (R&D) grants

Access to development of R&D is of primary importance in assisting rope access providers to utilise and provide practical applications of new technologies. Current expenditure on R&D by Absafe is 4% of total income. Both of the project case studies outlined in the proceeding sections of this paper, incorporated R&D grants to facilitate innovation and design.

3.2 Emergence of new technologies

Rope access is ideally positioned to take advantage of new technologies as they develop. Advances in plant and equipment with an emphasis on suppling easier operator use and increased output mean that a rope access tech working at 120 metres can be supplied with the same tools and equipment and achieve similar production rates as platform or ground workers.

Two examples of advancing technology being complemented by rope access are the advances in pump technology that now allows us to carry out shotcreting works at up to 150 metres in height, and the use of light weight yet extremely strong fabrics that can be used to provide full encapsulation with minimal requirement or without the need for scaffold. These situations are outlined in the following two projects implemented by Absafe.

4. CASE STUDIES

4.1 Case study one: Narracan Spillway Gates

Southern Rural Water required a series of maintenance and corrosion mitigation tasks to be carried out on the dam spillway gates located on the Narracan Reservoir. This contract involved maintenance painting of the spillway gates and overhauling the gate rollers and side seals at Narracan Dam. The works involved abrasive blasting back to class 2.5, removal and repair of corroded sections, pressure cleaning and application of a 2 pack epoxy coating system. The dilemma lay in the extremely environmentally sensitive area in which the works were to occur. The spillway gates were to be raised above the ogee crests and would allow a constant flow of water into the Morwell River, making scaffolding a very difficult and ultimately untenable option. Our chosen method after some experimentation was to rig two screens of Monarflex, a very lightweight and extremely strong fabric, as side curtains and use a custom made heavy duty PVC base as a catchment area. Full-length containment sheeting could then be hung either side of each gateway as panels attached to tensioned cables, with an impervious catchment pond sheet of heavy duty vinyl sheeting suspended underneath the works area to capture all debris. The base had a filtered drain to allow collection of waste materials. The panels suspended on tensioned 8mm cable provided a total encapsulation system and the side panels could also be pulled back to allow greater accessibility if and when required. Monarflex was also used as the roof of the system.

Figure 2: Encapsulation System in place Narracan reservoir (Absafe, 2016)

With these mechanisms in place, Absafe was then able to carry out all the works using IRATA rope access methods. The spillway gates were raised into position and held in place by bolting a resting plate to each side of the gateway which lined up with the existing support plates. The resting plate also prevented lateral movement after removal of the rollers. Following encapsulation, both upstream and downstream sections of the gateway could be access via rope access. The system also allowed for full access to all areas including undersides. Blasting hoses, air lines and water pressure hoses could then be independently rigged and fed through at the required heights into the works area.

Figure 3: Installation of side sheeting to PVC floor (Absafe, 2016)

4.2 Case study two: Hazelwood Power Station

The Hazelwood Power Station is a brown coal-fueled base-load thermal power station located in the Latrobe Valley of Victoria, Australia. Built between 1964 and 1971, the 1,600 megawatt capacity power station supplied up to 25% of Victoria's base load electricity and more than 5% of Australia's total energy demand. When confronting the scope of works required after the initial inspection, Absafe and Hazelwood Power Station looked for an innovative solution to an increasingly common issue on these ageing chimney stacks; increasing amounts of external shell repairs required to maintain the integrity of the structure. Hazelwood has 8 chimneys each being 137 meters high. Each chimney has a diameter of 11 meters at the base, 7.4 meters at the top and is made from self-supporting reinforced concrete lined with fire brick. Delamination of the chimneys had been increasing in part due to their age and had been carefully monitored via regular inspections. Rather than conventional 'form and pour' or handpacking methods, Absafe with the help of BASF and Antec, successfully carried out shotcreting as the repair method to the external shell of the chimney. We had begun extensively researching the potential for shotcreting at height after discounting the feasibility of conventional methods which proved relatively slow and too laborious an undertaking for the given surface area. After consultation with our affiliate, a repair spec was drawn up for the use of shotcreting as the repair mortar placement system. Expert advice in equipment selection was sought and the specification of a German manufactured shotcrete machine was found to be the best solution leaving this means as the most feasible methodology given the circumstances. The process relies on a dry-spray method whereby the dry concrete mix is carried via a large air compressor system to the work face. Water is then injected into the mix at the nozzle and impacted onto the surface at pressure to allow the bond to form.

It should be noted that to our knowledge, this method has never before been applied at this height off either ropes or swing stages. The potential production benefits of shotcrete outweighed the uncertainty created by the extreme height, and of working close to the specified limits of the shotcrete machine.

Figure 4: Shotcrete application at Hazelwood Power Station (Absafe, 2016)

All works were carried out by harnessing rope access as the primary access method. To help with the volume of concrete to be removed, the decision was made to deploy swing stages suspended from the upper gantry stanchions. These swing stages were a custom modular design of three two meter sections combining to make a 6m swing stage that that can be angled to fit around the curve of the chimney. They were also able to be altered continuously to adjust for the pronounced taper of the stacks. The swing stage was positioned directly underneath the patch being broken out and was used to help capture removed concrete. During the shotcreting process they were utilised as a working platform for rope access technicians rather than an access method in itself. By having the swing stage in the middle of a work area, it allowed works to be carried out on either side of it by providing stability and support to rope access techs at either end. This allowed for a much broader sweep across the face of the chimney at each drop and allowed for continuous placement of shotcrete to even the largest patches.

Figure 5: Custom swing stage with rubbish chute system Hazelwood (Absafe, 2016)

Another innovation on this project was the use of a garbage chute system well above the height usually advocated. Following consultation with the manufacturer and analysis by our engineers, a method of installation was devised that spread the load adequately across the whole length of the chimney. Reinforcing plates were also added to any required curves to add protection. The rubble, though dropping at an extremely fast rate, was captured effectively in covered steel skips that had been prepared with an old rubber conveyer belt and a preloaded rubble slope for protection. Absafe was also responsible for designing a system by which the shotcreting hose could easily be manoeuvred into position and handled by an abseiling 'Nozzleman'. This was achieved by the use of an independently suspended custom frame supporting the hose system and manoeuvred into position via powered winch. The pump used was a German machine by Clever and Co., and offered the following advantages:

•Low-rebound shotcrete processing by means of even delivery flow
•Dust-free machine operation due to fully contained system
•Conveying distances up to 1,500m, conveying heights up to 150m
•Stepless conveying capacity setting from 0.3m3/h to 15m3/h
•A machine for small applications which can also be used for large quantities of shotcrete
•Low wear costs due to cooling and lubrication of the wear parts

5. CONCLUSION

As the rope access industry matures, so too does the ability to provide corrosion repair efficiently and effectively. Rope access will continue to capitalize on the increased growth in operator numbers and attraction of qualified professionals to adopt its inventive methods. The advent of new technologies and processes for corrosion control combined with increased training, accreditation and other quality control measures, will drive investment in rope access as stakeholders recognise that large scale qualitative works can be completed to exceedingly high standards. As established throughout this paper, IRATA-based rope access methods offer a proven benefit to the practice of corrosion mitigation, enabling greater mobilization of resources and access to structures that avoid many of the financial and time-based encumbrances of conventional methods. By forging new frontiers in the accessibility of difficult worksites and the ability to withstand heightened scrutiny, rope access can be credited with making corrosion mitigation and repair a more cost-effective, and thus proactive endeavour.

6. REFRENCES

Marage, R. (2016), IRATA Australia AGM 2016 Address, April 2016.

7. AUTHOR DETAILS

Andrew Caddy is the Director at Absafe Pty Ltd, a leading specialist contractor that provides industrial rope access inspection, remedial engineering, coatings and maintenance services to all forms of façade and structures, plus height safety services. Actively involved in the industry since 1996, he is a registered building practitioner (Victoria) and advanced rigger. His qualifications include a Diploma of building construction, Cert IV in building Construction, and Cert IV Occupational Health and Safety. Also testament to his expertise, Andy holds positions on the Australian Regional Advisory Committee of IRATA and the ACA's newly formed Applicators Technical Group.

DC TRACTION ELECTROLYSIS MODELLING OF A REINFORCED CONCRETE RETAINING WALL

A. Vinnell, & U. Kreher
Aurecon Australasia Pty. Ltd., Melbourne, Australia

SUMMARY: This paper investigates electrolysis corrosion issues impacting reinforced concrete piled retaining walls constructed as part of level crossing removal projects. A series of electrical models were developed to allow an estimation of the magnitude of stray current discharged in different scenarios including a direct short to earth caused by a failed spark gap. In the failed spark gap scenario, the model determined rail to earth resistance and soil resistivity surrounding the structure were critical features controlling the magnitude of current discharged. Based on the magnitude of current discharged from the retaining wall corrosion rates of the reinforcing were estimated.

Keywords: Electrolysis, DC Traction, corrosion design, Level Crossing, Stray Current, Concrete reinforcing corrosion

1. INTRODUCTION

A number of level crossing removal projects are currently underway across metropolitan Melbourne. These major construction projects are being undertaken to remove road level crossings from the suburban electrified train network to improve vehicular traffic flow, reduce road congestion, reduce hazards to drivers and pedestrians and increase the capacity of the train network. These projects will make significant modifications to local areas of the train network and as the network operates as a 1500V DC traction system, electrolysis corrosion due to stray currents must be considered in the design.

This paper discusses some elements of the electrolysis mitigation strategy developed as part of the design for one such project which involved lowering the existing track into excavated cuttings. The cuttings are supported by piled reinforced concrete retaining walls with infill mesh and shotcrete between piles running along the cutting.

This paper will focus on the assessment and management of electrolysis risk for these walls and cover some of the modelling that was conducted to assess possible electrolysis risks to the walls as a result of direct injection of rail voltage. Discussion of the complete strategy is outside the scope of this paper

2. GENERAL PROJECT BACKGROUND

The population of the Greater Melbourne region has experienced a long period of sustained growth; 800,000 additional people moved into the region over the last ten years, with the 2016 population reaching 4.64 million [3]. Growth is continuing with the region recording growth of 2.4% in 2015-2016, the largest of any capital city region in Australia. In addition 2015-2016 has seen significant growth in outer suburban areas such as Cranbourne East (26.2%), Officer (15.4%) and Craigieburn (9.3%) [3].

This growth has resulted in a large increase in commuter volumes on the Melbourne Metropolitan train network which covers of 869 km of track across 15 lines [1]. Additional train services are required however one of the barriers to increasing the number of train services is the 178 road level crossings [2] that exist within the network.

These crossings range from small suburban streets to major arterial roads and create a conflict between rail traffic, road traffic, and pedestrians. During the 2 hour peak period, boom gates at some intersections close more 20 times and are down for more than 60% that time. This creates significant road congestion and delays [2]. There have also been serious collisions resulting from pedestrians and vehicles driving around closed boom gates or being trapped on the crossing due to traffic congestion.

3. LEVEL CROSSING REMOVAL DESIGN FEATURES

A variety of different construction methodologies have been utilised in the removal projects including:

- Lowering the rail below the road and constructing a road bridge over the tracks. This generally involves one or more of the following:
    - Soil nail walls or terraced soil
    - Retaining wall consisting of reinforced concrete piles
    - Sheet pile walls
- Constructing a long via-duct style rail bridge along the rail alignment over the road crossing.

This paper will focus on a project based on the rail-under-road methodology utilising reinforced concrete piled retaining walls.

Figure 1 – Level crossing removal structure based on piled wall concept

3.1 Construction Methodology - Piled Retaining Wall Design Principles

The piled wall construction methodology consists of bored steel reinforced concrete piles and steel mesh concrete infill panels between the piles. Shotcrete is then used to encapsulate the infill and seal the connection between the infill and the pile. An electrically continuous capping beam is cast on top of the piles and electrically bonds all piles in one section. Structural movement joints are provided at typically 30m intervals which provide electrical insulation between each section. Welding is used to make all reinforcement in each section electrically continuous. The general concept is illustrated in Figure 2 and the finished site pictured in Figure 1.

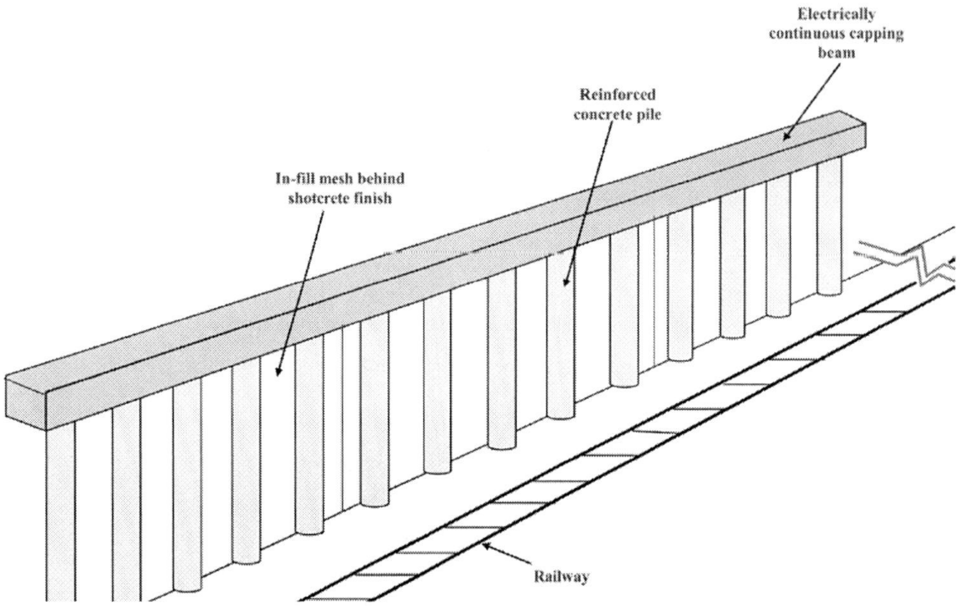

Figure 2 - General arrangement of pile walls

3.2 Overhead Support Structures and

Within the rail corridor overhead wire support structures (OHWS) support the high voltage catenary, contact wire and other connections such as 22kV AC, electrolysis feeders and side feeders (auxiliary 1500V lines) above the track. These structures are typically installed on independent piled footings alongside the track. However in this project they were required to be mounted on top of the capping beams due to rail corridor width and timeframe constraints. To structurally support these elements the capping beams were increased in size. This resulted in each overhead wiring structure becoming electrically continuous with the corresponding 30m section of retaining wall.

3.3 Spark Gaps

DC Traction systems operate on an unearthed (floating) rail return system which does not reference the rail voltage or high voltage overhead systems to earth. While this arrangement provides the best achievable stray current reduction strategy, it introduces electrical safety risks as a fault may not be detected and removed. For example in this system if an overhead conductor operating at 1500V failed or was entangled with a train and dropped onto one of the overhead wire support structures, minimal current would flow and the 1500V conductor would stay live along with the structure. In an AC system referenced to earth a dropped conductor would result in high currents tripping protective devices such as circuit breakers.

To manage this risk spark gap devices are installed which bond between the rail (return conductor) and any metallic element or structure within the Current Collector Zone or Overhead Contact Line Zone as described by EN50122.1 [8] and local rail authority standards shown in Figure 3. This zone is considered to encapsulate the zones of high risk of impact by a dropped overhead 1500V line or entanglement by a damaged pantograph from a train. In Victoria every overhead wire support structure is typically bonded to rail via a spark gap.

Spark gaps (shown in Figure 4) operate as surge arrestor devices and electrically close (short circuit) when a high voltage is applied across them. They are then able to conduct current from a dropped or entangled overhead conductor back to the rail which will trip the protection in the DC substation(s).

However these systems typically are not re-closable i.e. they will only operate one time under large fault currents, and will remain in the electrically closed position. While being electrically closed mitigates electrical hazards it also introduces stray current corrosion problems due to creating a direct short between rail and the attached structure. They can also operate as a result of a lightning strike and have been known to close occasionally with no confirmed cause.

The rail authority conducts an annual routine maintenance inspection of these devices, to test and replace as required.

Alternative systems such as Voltage Limiting Devices are becoming more readily available and may replace some spark gaps in the future. These devices can provide better reliability and the capability to re-open once the fault is removed preventing excessive stray current corrosion. However there are no 'type approved' devices (permitted by the rail authorities) currently available for general use in Victoria and available devices are still cost prohibitive for use on the wider train network.

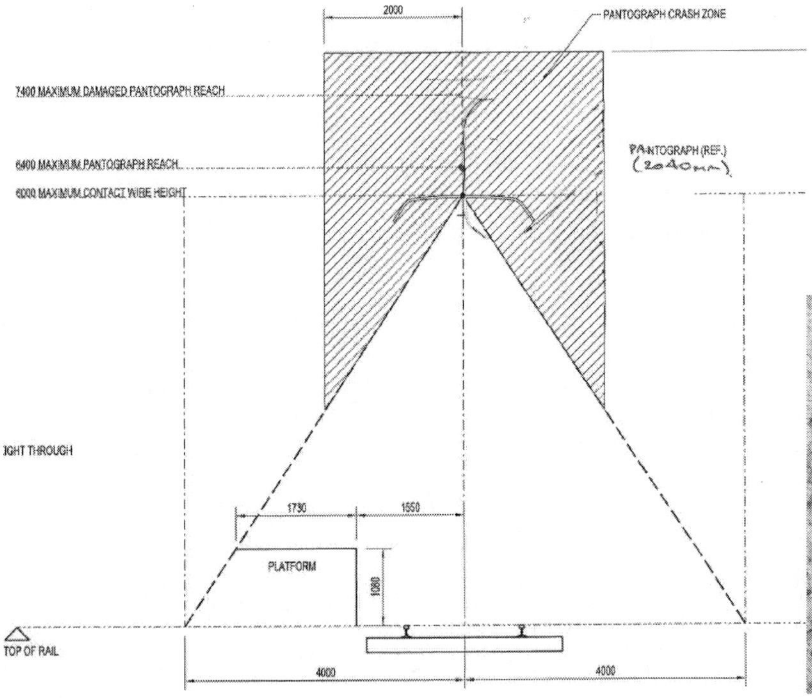

Figure 3 -Overhead contact wire line zone (below dotted lines) and current collector zone (hatched) defined by Victorian Rail Authorities

Figure 4 - Spark gap

4. ELECTROLYSIS CORROSION MECHANISMS

4.1 Traction System Operation

The Melbourne Metropolitan train network utilises overhead 1500V DC conductors to energise the trains. The DC current is provided by a number of substations along the rail easement which convert high voltage AC supply from the electricity grid into the 1500V DC used for the trains. The return current back from the train to the substation is via the rails and these should provide the lowest resistance path to the substation.

However the rails are resistive at about 0.0330 ohms/km [5] for the 60kg/m rail used in Melbourne. While this may seem a low resistance, return currents are can be greater than 2000A for each accelerating train. The result of the high current and resistive rail is development of a volt drop between the substation and location of the current draw (train).

The target maximum rail voltage for the network is 60V, however higher voltages can occur at peak network loadings and during fault situations and the maximum permissible track voltage is 120V DC as per EN50122.1 [8]. Track voltages can be lowered by:

- Providing additional substations
- Installing bonding conductors in parallel with the rail
- Bonding multiple parallel tracks together to reduce track resistance

However the rail voltage is highly variable and dependent on network loads.

4.2 Common Electrolysis Mechanisms

In a perfect world rails would be perfectly insulated from the earth (electrically floating), and rail voltage does not cause any stray current corrosion problems. However there is always some finite rail resistance to earth even in modern tracks with insulated fixings, although the insulation does provides significant improvements. Rail to earth resistance is lowered further as the track becomes dirty, insulations break down and ballast, dirt or other debris such as brake dust build up around the rail and create resistive connections between the earth and the rail.

As a result of this resistive connection to earth and the elevated rail voltage, current jumps off the rails using the parallel low resistance path into the earth and is again picked up by rails nearer to the substation as illustrated in Figure 5. At any location where current is discharged electrolysis corrosion will occur.

Other metallic assets between the pickup and discharge points such as a pipeline (or a long continuous retaining wall) can become part of this lower resistance return path and also pickup and discharge current. At the current discharge point the metal becomes anodic and experiences metal loss at a rate proportional to the current flow. According to Faradays Law, 9.2kg of steel can be lost if a current of one (1) ampere is discharged for one (1) year.

For a reinforced concrete structure this metal loss can cause:

1. Reduction of structural capacity and
2. Cracking or spalling of the cover concrete surrounding the reinforcement resulting from tensile pressure build-up of steel corrosion product around the reinforcement which occupies a larger volume than the metal.
3. Reduced structural durability and service life due to cracking or spalling of the cover concrete introducing external aggressive agents to the reinforcing, i.e. Chlorides, water or a direct carbonation path etc.

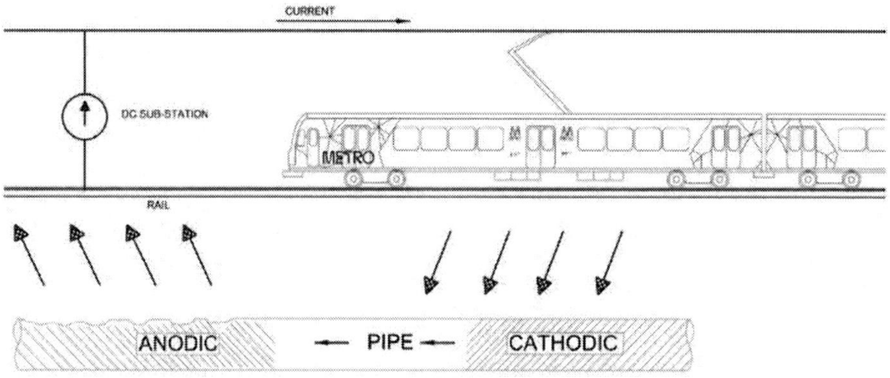

Figure 5 - Typical electrolysis corrosion mechanism [9]

One of the features of the electrolysis mitigation strategy used in this project was electrical segregation the long piled walls into typically 30m lengths. This approach reduced the voltage gradient experienced by each wall and hence the magnitude of current discharge that would have otherwise occurred at the extreme ends of the structure.

Electrolysis impacts to foreign assets and other elements of the level crossing structure itself considered in the overall project design strategy and are outside the scope of this paper.

4.3 Piled Wall Corrosion Mechanism - Spark Gap Failure

A potentially more serious electrolysis corrosion mechanism was created by mounting the overhead wire masts onto the piled retaining walls coupled with the requirement for the masts to be bonded to rail via a spark gap to mitigate high voltage fault scenarios.

If a spark gap activates (electrically closes or shorts) it directly bonds the overhead structure to rail. While this in itself may not be a significant concern, rail authority standard [4] requires the hold down bolts of overhead masts to be electrically continuous with the reinforcement within the supporting retaining wall.

In this situation an activated spark gap can result in rail voltage (60V or more) being applied directly to the wall and pile reinforcing. The piled retaining walls provide good contact with soil and hence a low electrical resistance path to earth. The low resistance path and direct rail voltage could result in significantly more electrolysis current being discharged from the wall and subsequent corrosion of the wall reinforcement (and foreign assets) than expected under the typical electrolysis model. The system layout in this scenario has been illustrated in Figure 6 below.

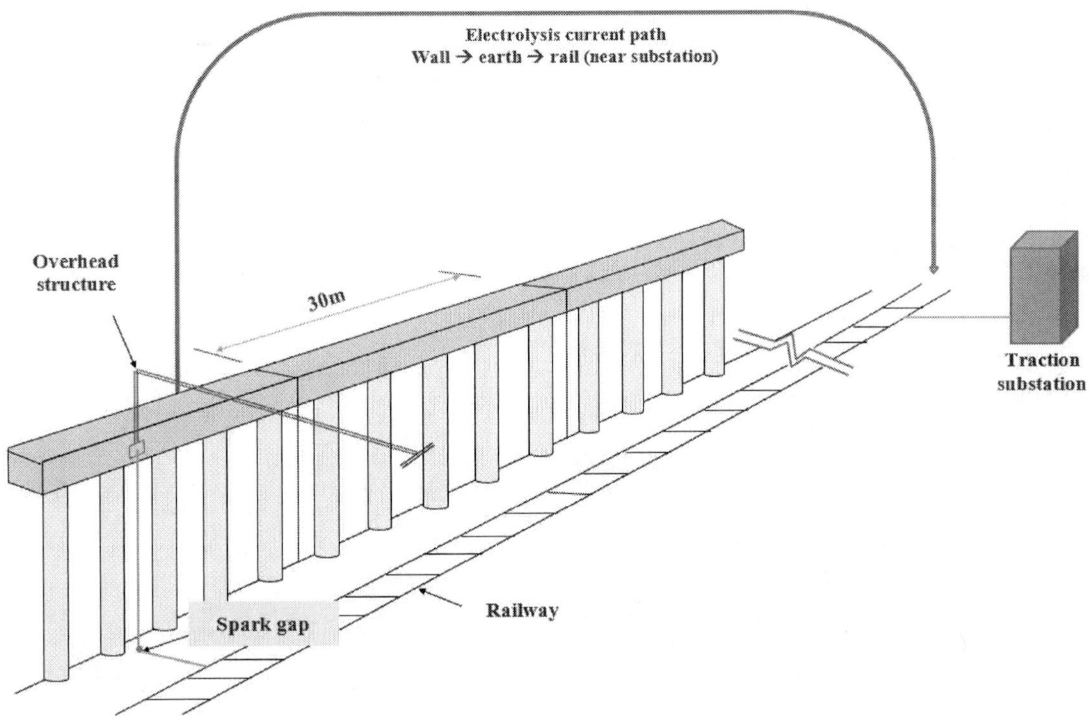

Figure 6 - Electrolysis corrosion mechanism due to shorted spark gap

5. PILED WALL ELECTROLYSIS ASSESSMENT

The potential electrolysis corrosion significance of a failed spark gap and project durability requirements for service life of 100 years resulted in further investigation being undertaken to assess the corrosion impact of a failed spark gap.

5.1 Track and Electrolysis Network Modelling

Several simplified electrical models of the DC traction system and associated connections to earth and retaining walls were developed to determine the controlling elements in the electrolysis corrosion of the piled reinforced concrete wall. These included the best case scenario of an infinite rail-earth resistance, a common electrolysis scenario of rail discharge and pickup and the case of a failed spark gap applying rail voltage to the retaining wall.

Many factors that control the electrolysis current cannot be determined until a system is operational. The models presented here utilise a number of assumptions to assess the magnitude of the DC current being discharged and subsequent corrosion. A list of the assumptions included in the model are shown in Table 1. These values were selected based on predicted network conditions, typical values from standards and other literature, and expected environmental parameters.

Table 1 - Electrolysis Model Base Parameters

Parameter	Value	Source/Reference
Train distance from nearest substation	1km	Model assumption. No other substations or trains in the area
Overhead wire resistance	0.0991 ohm	Assumed 1km of $185mm^2$ cable. AS/NZS 1125
Rail resistance	0.0165 ohm	2x1km long 60kg/m rails in parallel (0.0330 ohms each) [5]
Rail resistance to earth	50ohm/km	Assumed standard clean ballast [6].
Substation voltage	1500V	Operation voltage in Victoria
Return current	2000A	Assumed maximum current demand of single train
Steel corrosion loss rate	9.2 Kg/A/year	Faraday's law for iron
Steel density	$7850\ kg/m^3$	

5.2 No Current Loss from Rails

The electrical schematic shown in Figure 7 illustrates the DC traction system where all current is returned to the substation via the rail. This is the 'ideal' condition where no current is lost through a stray current path and would only occur where there is infinite resistance between the earth and the rails, or the rails are infinitely conductive.

In this model the traction substation outputs current through the overhead wires, through the train and the current returns through the low resistance rails. The assumed resistance of the train is adjusted to provide a 2000A circuit.

This model arrives at a track voltage of 33V over a distance of 1km with two rails used in the return circuit.

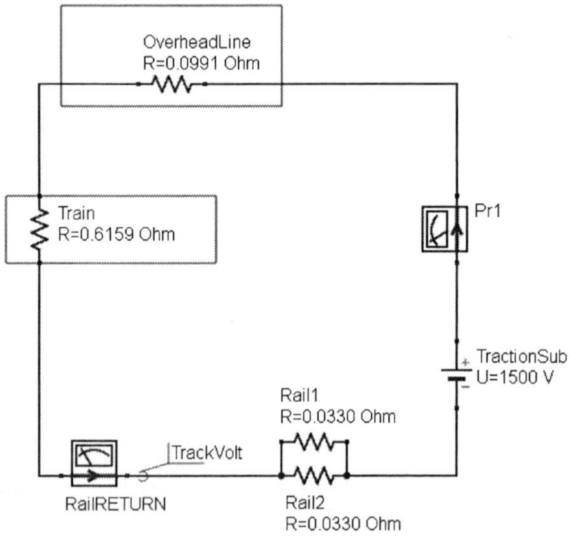

Figure 7 - Ideal traction system with no losses

5.3 Common Electrolysis Mechanism

Adding a track resistance to earth as shown in Figure 8 provides a stray current path. Based on a uniform resistance between rail and earth of 50ohm/km this model assumes 500m of track discharges (100ohms) current and 500m picks up current (100ohms).

Corrosion would occur over any areas discharging current. Any assets buried in the earth between the discharge and pickup areas could also be subject to stray current corrosion.

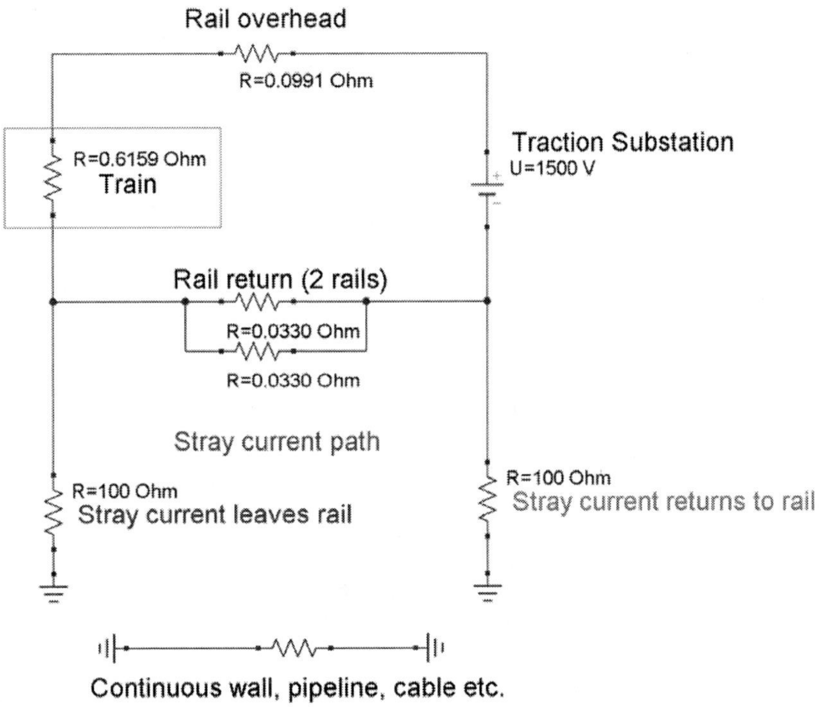

Figure 8 – Common electrolysis mechanism

5.4 Failed Spark Gap Scenario

In the scenario of a failed spark gap shorting the rail to the piled retaining wall and hence the earth, further assessment is required to determine the expected amount of current that will flow. This model has been developed assuming one 30m section of the electrically segregated retaining wall has been shorted and this short has occurred at the midpoint between substations where the track voltage is highest (worst case scenario).

5.4.1 Calculation of resistance between piled wall and remote earth

The first step was to calculate the resistance of the wall to remote earth. This was completed by modelling the wall in an electrical finite element simulation software package CDEGS.

For simplification the wall was modelled as a vertical flat plate electrode 30m x 6m placed 0.5m below grade with the flat plate being made up of a series of 450mm diameter steel conductors as shown in Figure 9. One face of this plate was in contact with the earth.

The simulation was repeated for a range of soil resistivity values to obtain a profile of resistance to remote earth vs soil resistivity.

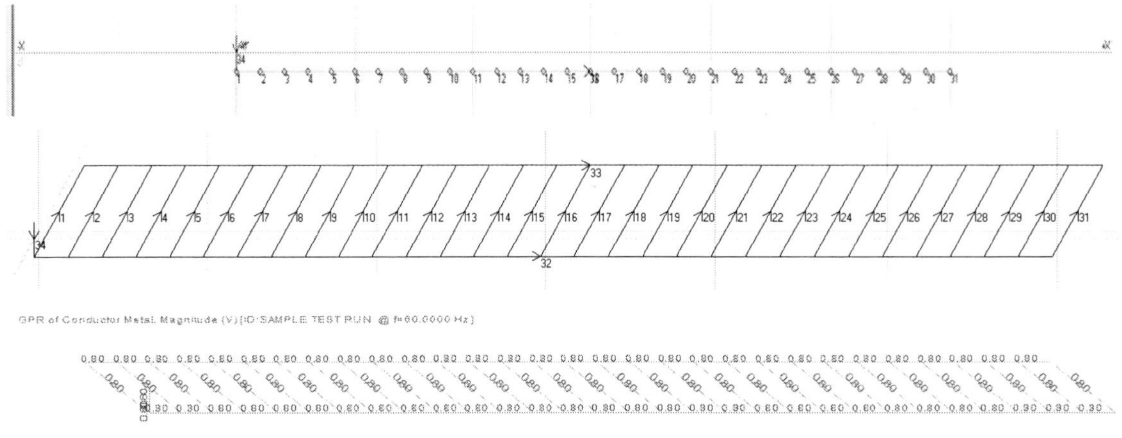

Figure 9 - CDEGS Model of structure resistance

The use of cylindrical steel conductors in the model may seem like an overly conservative approach due to the additional resistance provided by the concrete. However IEEE80 [7] reports that when buried in soil, concrete behaves as a semiconducting medium of 30-90 ohm.m. This value was similar to the local soil resistivity at the site, and therefore the concrete to soil resistance was not considered significant in the model.

5.4.2 Calculation of track and earth circuit stray currents

Using the resistance values obtained from the CDEGS modelling the electrical circuit model was updated as shown in Figure 10. This model is similar to the common electrolysis model (Figure 8) however the discharging resistance to earth is much lower due to the good contact between the piled retaining wall and earth.

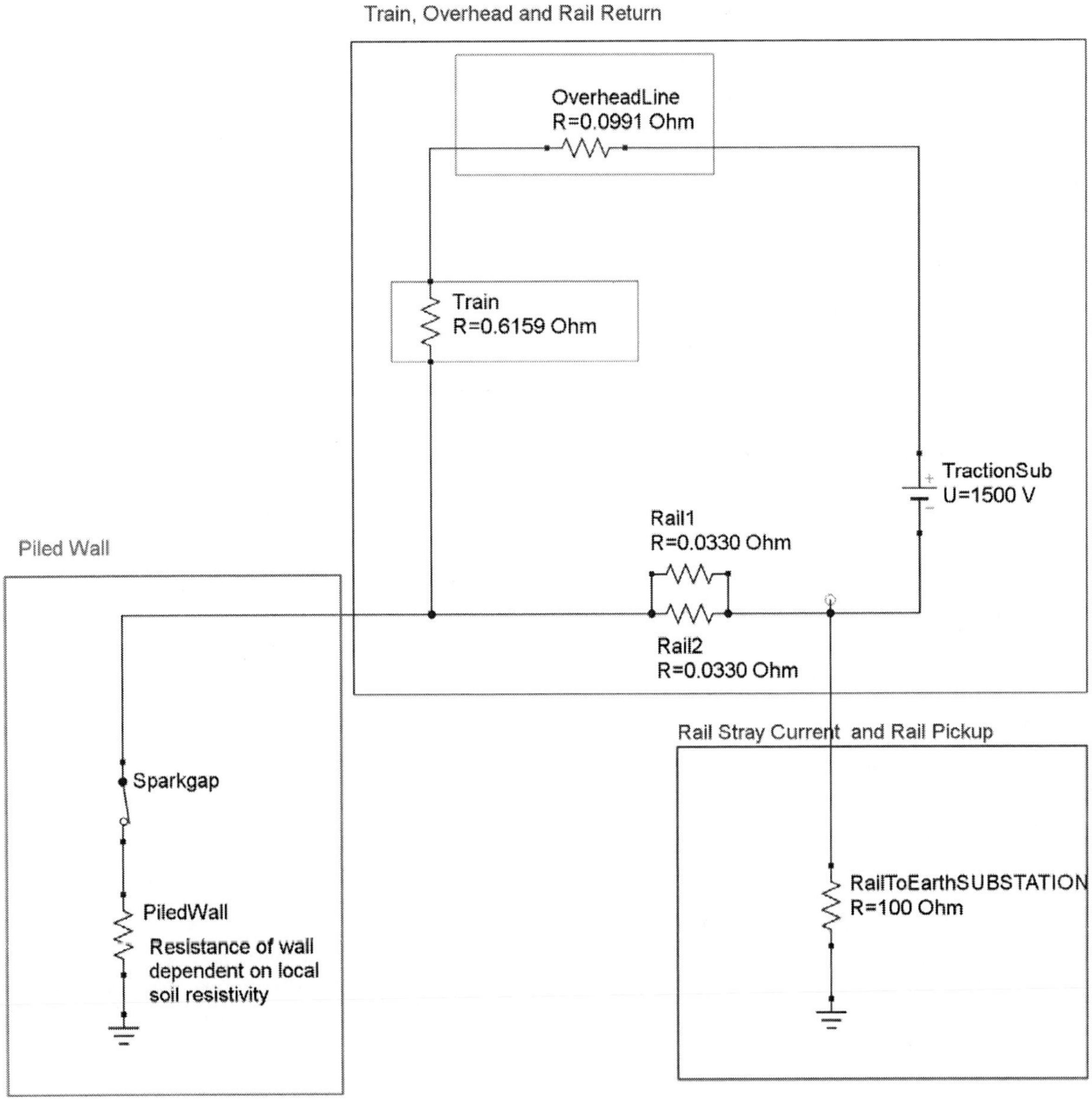

Figure 10 - Traction power model with path to earth via spark gap and piled wall

5.4.3 Variation of soil resistivity and rail to earth resistance

The amount of stray current discharged calculated by the model is strongly dependent on two main factors:

- Rail to earth resistance
- Piled wall to earth resistance which is largely dependent on local soil resistivity

These two factors were used in the model to generate a 3D surface plot which indicates the magnitude of stray current flow leaving the wall expected in different environments.

5.5 Calculation of steel corrosion rate

Based on estimates of current flow in the electrolysis circuit the mass and therefore volume of steel lost due to corrosion can be calculated based on Faraday's Law.

$$m = \frac{Q}{F} \times \frac{M}{z}$$

Equation 1 - Faraday's Law

m	=	Mass of the substance dissolved at an electrode in grams
Q	=	Total electric charge passed through the substance
F	=	Faraday constant (= 96,485 C/mol)
M	=	Molar mass of the substance
z	=	Valency number of ions of the substance

The actual corrosion rate (section loss) of steel is determined by distributing the mass/volume of steel lost evenly across the steel surface area impacted by the stray current electrolysis corrosion. This assumes discharge is evenly distributed over the full 30m length across a 6m average pile height.

However the ratio of *surface area of steelwork within concrete* to the *surface area of overall structure in contact with earth* is not 1:1. This is further complicated as the infill panels are lightly reinforced, while the piles are heavily reinforced.

To make a more accurate assessment of corrosion loss the reinforcement density of the walls was calculated and an 'average' density of *reinforcement surface area* to *concrete surface area* was calculated and used as a multiplier to distribute the known amount of current discharged from the wall. An average ratio of 0.91:1 (steel to concrete) was calculated for the retaining wall constructed as part of this project.

6. RESULTS

6.1 Common Electrolysis Mechanism

Based on the model parameters a track voltage of 33V was developed which resulted in 169mA of stray current flow between the pickup and discharge locations over the 1km run.

6.2 Failed Spark Gap Scenario

The stray current flow resulting from the failed spark gap and consequent short between the piled wall and rail was then calculated based on a range of soil resistivity and rail-earth resistance. The resulting current flow is graphically presented in Figure 11.

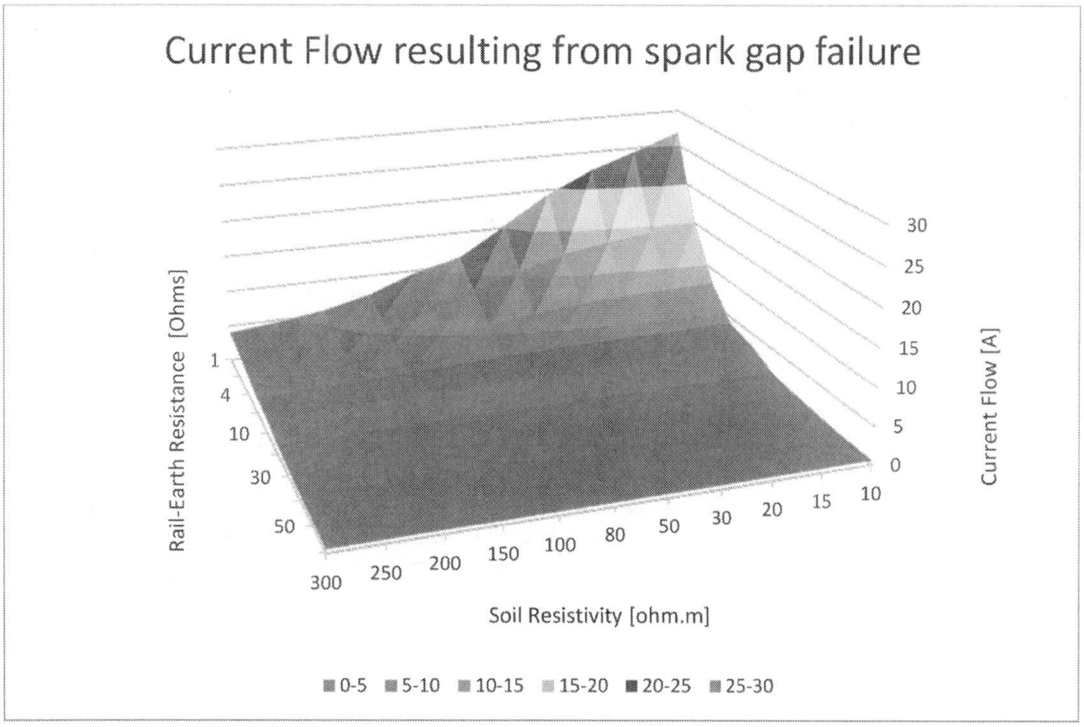

Figure 11 –Stray current flow resulting from a spark gap failure

6.3 Corrosion Rate (Metal Loss)

The stray current flow combined with Faraday's Law and the steel to concrete surface area ratio was then used to estimate the steel corrosion rates expected from the wall reinforcement.

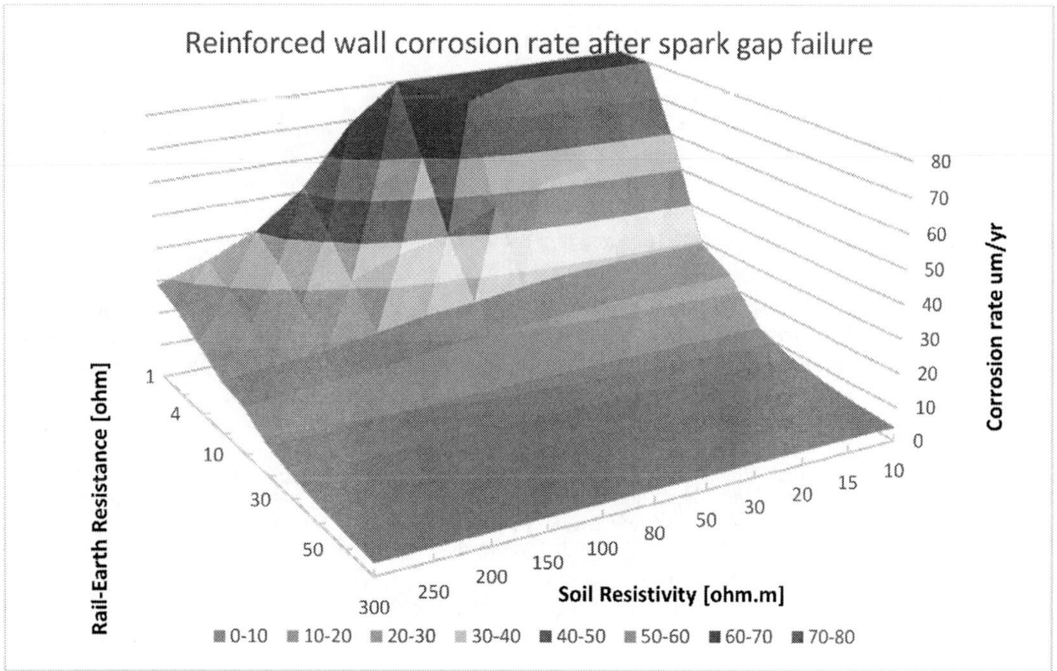

Figure 12 -Metal loss due to electrolysis corrosion

7. DISCUSSION

7.1 Common Electrolysis Mechanism

The common electrolysis mechanism with a spark gap open and operating normally, resulted in 169mA of stray current equating to a corrosion rate of 1.21µm distributed over the reinforcing of the wall. While this could have an impact on external assets it was not considered a risk to the retaining wall and has not been discussed further in this paper.

7.2 Reinforcement Corrosion Durability Impacts

This study was undertaken primarily to assess the significance and impact of stray current corrosion on the piled retaining wall structure resulting from a failed spark gap. The corrosion was assessed considering the structure may be shorted to rail for up to a year at a time due to the rail authority maintenance periods and the potential for this to occur several times during the 100 year structure design life. Two main concerns were considered on this basis:

a) Section loss of reinforcement and resulting impact on structural integrity
b) Secondary durability impacts due to local cracking of the concrete caused by the expansion of corrosion products

If typical values of soil resistivity of 30 ohm.m and track earth resistance of 25ohm are used, the model estimated a corrosion rate of approximately 9µm per year. The corrosion allowances made in the structural design were many times this value and therefore there was no impact to structural integrity even if the spark gap was shorted for many years.

The durability concern occurs as the stray current discharge can result in the formation of corrosion products around the reinforcement that are larger than the natural bar which may result in cracking and spalling of the cover concrete from the pile. These cracks provide a direct path to the reinforcement for aggressive agents, and hence a reduction in the design life of the pile due to early onset of corrosion.

The expansion rate of the steel and resulting tensile forces applied to the concrete are difficult to quantify and are known to vary considerably due to concrete strength, cover and type of oxide product formed [10]. However, corrosion losses of 100 to 150 micron of steel are typically accepted industry values at which corrosion can initiate cracking of a reinforced concrete element. Based on these assumptions the spark gap can be shorted to rail for 11 years before cracking would be a concern.

However it should be noted that corrosion due to stray currents is rapid and can form little expansive corrosion products due to the lack of oxygen in concrete. This corrosion mechanism therefore results in less risk of cracking or visible structural damage compared to mechanisms where the corroding reinforcing expands in volume.

7.3 Applicability of Model Relationship to Other Structures

This model followed the case study of a corrosion rate assessment on an electrically continuous reinforced piled retaining wall as a result of an applied rail voltage due to a failed spark gap. While this model was focussed on this particular wall, critical factors were determined to be rail-earth resistance and soil resistivity surrounding the structure. The corrosion *rates* determined by the model are independent of the mass of reinforcement, therefore (with exclusion of mutual interference effects from a large structure) the corrosion *rate* for any area of a continuous structure connected to the rail can be assessed in terms of the soil resistivity, local track to earth resistance and track voltage at the local area.

On this basis other structures on the network connected via spark gaps i.e. piled overhead wire support structures could be assessed using a similar method and would be expected to have a similar corrosion *rate* for equivalent external conditions. However each structure would have a different corrosion current due to different reinforcement density, mass and layout.

7.4 Future Investigation - Current Measurement

Although beyond the scope of this paper, further work could be performed to measure the actual rail voltage and more accurately calculate the current injected into the retaining wall as a result of traction currents under failed spark gap conditions. Based on a known reinforcement layout, a corrosion rate could be calculated from this and an assessment of the severity of corrosion made. This would require data logging to capture the different network conditions and current flows over the course of each day and week.

7.5 Model Assumptions

This model is focussed on rail voltage rise resulting from return current from operating trains and the consequent current discharge due to a finite rail to earth resistance. While this concept is suitable for an initial assessment there is a large number of external factors that could significantly affect the actual corrosion rates experienced by the structure. Unfortunately the majority of these are dependent on installed conditions and cannot be accurately predicted prior to construction. For example the model assumes the following:

a) No track crossbonding is considered (between adjacent tracks) however this is likely to be provided in most cases where there is more than one set of rails. Bonding to multiple tracks can significantly reduce track voltage by lowering the rail resistance.

b) No regenerative train braking has been considered, this could reduce the average rail voltage although the effects would be variable and difficult to model due to the 'moving substation' affect.

c) The load applied by the train in this model would be near the maximum normal operational value for a single train. The model has assumed this to be the steady state condition which would significantly overestimate the corrosion loss as the time weighted average voltage rise would be significantly less than this when taking into consideration the quiescent period (overnight) and that the trains are only accelerating heavily for a short time over the course of the day. Typically time weighted average rail voltage is measured following track commissioning although modelling of the overall track network can be completed.

d) There is uniform soil resistivity across each site resulting in uniform corrosion rates along the wall. In reality there will be varying strata that will cause differential corrosion *rates* across the retaining wall structure. The actual corrosion rate will be dependent on the soil resistivity in contact with the local area of the structure even if the structure is continuous.

e) The retaining wall has been made electrically continuous, therefore all reinforced elements are assumed to be equally subject to corrosion and contribute to the low resistance circuit.

f) Corrosion rates across all reinforcement are assumed to be uniform. In addition to the soil resistivity and continuity factors described above, mutual interference effects will be caused due to the wall acting as a large flat plate anode. This will cause lower resistances between soil and the reinforcement near the edges of the wall and accelerated corrosion at these locations.

8. CONCLUSIONS

This paper investigated electrolysis corrosion issues impacting reinforced concrete piled retaining walls constructed as part of level crossing removal projects. Simplified electrical models were developed to consider stray current corrosion impacts of the following scenarios:

- Ideal DC traction system with infinite rail to earth resistance
- Typical electrolysis mechanism with current discharged and picked up by the traction return rails
- Electrolysis of a reinforced concrete piled retaining wall directly shorted to rail via a closed spark gap

The piled wall with a shorted spark gap scenario was found to cause the most significant stray current concerns due to the direct bond to rail. Modelling of this scenario identified that rail to earth resistance and soil resistivity surrounding the structure were critical features controlling the magnitude of current discharged. Providing the highest rail to earth resistance possible was the best way to reduce stray current flow and resulting corrosion.

Based on the predicted stray current discharge from the wall shorted to rail, corrosion rates of the reinforcing within the wall were calculated to be 9μm per year. This rate could be tolerated by the structure for several years before causing cracking and was well below the limits which could cause an impact to structural integrity.

Future work could involve undertaking site testing of completed installations to validate the assumptions and predictions of these models.

9. ACKNOWLEDGMENTS

The author would like to thank Aurecon Australasia for the support and the ACA for the opportunity to present this paper.

10. REFERENCES

1 Metro Trains Melbourne, Who We Are [Internet]. Melbourne, Victoria [cited 10 June 2017]. Available from: http://www.metrotrains.com.au/who-we-are/

2 Level Crossing Removal Authority. Program Business Case; May 2017; 287 pages

3 Australian Bureau of Statistics. 3218.0 - Regional Population Growth, Australia, 2015-16. Canberra, 30/03/2017

4 L1-CHE-STD-010 Electrical Networks Standard Railway Bridges Electrical Protection and Bonding, Metro Trains Melbourne, Version 1, 18 June 2013

5 ESD-07-02, Track Bonding & Signalling Infrastructure in Electrified Areas, Australian Rail Track Corporation, Version 1.2 13 August 2010.

6 Hook N, Bradshaw G. T HR EL 12002 GU, Guide Electrolysis from Stray DC Current, Transport for New South Wales, Version 1.0 1 May 2014

7 IEEE-SA Standards Board, IEEE Std 80-2000 IEEE Guide for Safety in AC Substation Grounding, 30 January 2000

8 European Standards Board, EN50122.1-2011 Railway applications – Fixed installations - Electrical safety, earthing and the return circuit -- Part 1: Protective provisions against electric shock

9 Energy Safe Victoria, VEC Resource Manual, TRIM REF: DOC/14/11519, July 2014, 110 pages

10 Hwan B, Kim K, Jang B. Critical Corrosion Amount to Cause Cracking of Reinforced Concrete Structures, ACI Materials Journal. 2009 July-August; 106-M38

11. AUTHOR DETAILS

Adrian is a Materials Engineer with ten years experience within the corrosion management field. He has a diverse skill set including, condition assessment, cathodic protection, electrolysis management, protective coatings, materials selection and durability design and assessment. With a particular focus on cathodic protection he has worked through all stages of the project lifecycle from feasibility studies, to design, construction and remedial condition assessments.

Adrian's experience also encompasses a wide range of industries from ports, mining, oil and gas, water and waste water treatment and civil structures with a variety of scope such as new infrastructure design, desktop studies and remedial system design through to site inspection, condition assessment, project management and auditing.

Ulf Kreher is an Associate of Aurecon and the leader of the Materials Technology Group. He holds a PhD in Chemistry, is an ACA accredited Corrosion Technologist and certified as NACE Cathodic Protection Specialist (NACE CP4). His experience spans from the design and troubleshooting of cathodic protection systems for buried and immersed steel structures, concrete structures, pipelines and tanks, to concrete condition assessment and remediation including development of concrete corrosion monitoring devices.

He is heavily involved in stray current electrolysis assessments and mitigation design on major DC rail infrastructure projects and in the preparation of an Earthing and Bonding Standard for local authorities.

CASE STUDY: REMOTE IN-FIELD ENVIRONMENT AND CORROSION MONITORING

A. Sudholz[1], H. Thanh[2], J. Waldie[2], J. Talevski[1], C. Loader[1], A. Butler[1]
[1]Defence Science and Technology Group, Melbourne, [2]BAE Systems, Melbourne

SUMMARY: Corrosion modelling of aircraft materials is intended to enable condition-based maintenance practices, which can lead to increased availability of aircraft and a reduction in the cost burden of their maintenance. This system is known as Corrosion Prognostic Health Monitoring. To validate predictions of atmospheric corrosion, a significant amount of data is required that represents the in-field corrosion damage experienced by various aircraft alloys across a broad range of environments. The present programme involves the installation of monitoring stations in the field to record environmental and corrosion data necessary to validate corrosion prediction modelling to be used in shaping Defence aircraft maintenance practices.

Keywords: Corrosion monitoring, Environment monitoring, Sensors, Modelling, Aircraft alloys, Protective coating.

1. INTRODUCTION

To assist maintainers of military aircraft the Defence Science and Technology Group (DST) and BAE Systems Australia (BAE) are developing corrosion prediction models that will allow for targeted inspections, limiting unnecessary and time consuming maintenance. Key to building the maintainer's confidence is validation of modelling outputs with in-field data. Collection of in-field environmental data can be very expensive, due to the certification required for aircraft-specific on-board monitoring. Additionally, the corrosion on aircraft typically occurs whilst it is on the ground. Internal aircraft bays are inherently difficult to inspect, because access is restricted, making them ideal locations for condition-based monitoring.

Many aircraft bays are closed but will experience some external environment ingress via changes in atmospheric pressure and temperature through drainage holes and internal air circulation during their time on the ground. DST Group and BAE have developed a series of ground-based stations (GBS) to record local environmental data, monitor corrosion of typical aircraft aluminium alloys and collect data for predicting the translation of external environments to internal conditions. GBS comprise environmental sensors and witness plates of a selection of typical aircraft aluminium alloys located both externally and internally in a semi-sealed compartment. The semi-sealed nature of the GBS will allow for the correlation of the external environments to internal conditions expected within aircraft bays. GBS are installed at five Australian sites representing a broad range of environments relevant to the Royal Australian Air Force (RAAF) and a sixth site embarked on a ship for helicopter deployments.

The sensors are used to record the environmental parameters determined to be critical to prediction of corrosion in metallic components: air and surface temperatures, relative humidity, and time-of-wetness (ToW) [1]. The environmental data collected is transferred periodically using an automated process, while witness plates are returned every three months for corrosion damage assessment.

If a correlation is established the external environmental data can be used to predict the internal condition data which in turn acts as inputs to predict and model the corrosion. Corrosion modelling predictions using environmental inputs can therefore be compared to the corrosion experienced by each aluminium alloy. This comparison to in-field data will assist in validating the predictive capability of current corrosion models and build confidence for use in condition-based maintenance.

2. INSTALLATION SITES

Six GBS were installed across five land sites and on board a ship. The five land sites are: Edinburgh Parks (SA), Nowra (NSW), Williamtown (NSW), Cowley Beach (QLD), and Tindal (NT). The sixth site is on board a Hydrographic Ship, which will spend most of its time out at sea and represents conditions experienced by embarked aircraft. An investigation was conducted into base corrosion severity (Figure 1) to determine the best selection of sites to install GBS. This investigation used the Geographic Corrosion Index (GCI) developed by DST Group [2] to predict the average Classify Industrial and Marine Atmospheres (CLIMAT) response at potential sites and to classify these sites in terms of wet/dry (based on mean relative humidity) and hot/cold (based on mean temperature). The sites selected represent the broad range of corrosion severity experienced across Australian Defence sites (Figure 2).

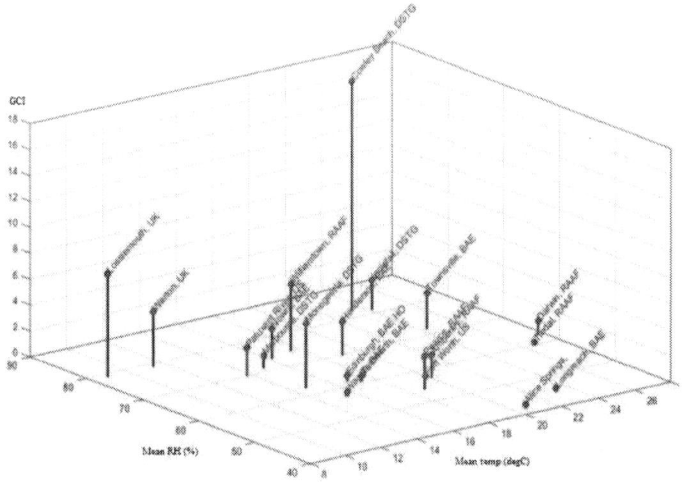

Figure 1. Corrosion severity of various candidate sites.

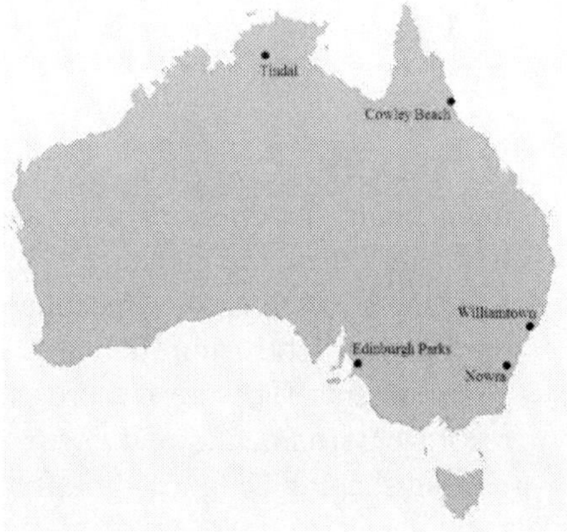

Figure 2. Installation sites around Australia.

3. SHELTER DESIGN

An important part of each GBS installation is the shelter. The shelters are designed to support the mounted cabinets and provide a covering representative of a full-size aircraft shelter. The shelter is designed to enable the GBS to replicate the daily humidity and temperature cycling of an aircraft positioned under a shelter.

The shelter itself (Figure 3) comprises two vertical posts, with four beams mounted horizontally between the two posts. Both the Sensor Cabinet (SC) and the Battery and Data Cabinet (BDC) are mounted on these horizontal beams, facing back-to-back. Atop of the vertical posts is an awning that provides control of direct rain or sunlight experienced by the cabinets. The angle of the awning matches that used in full-size shelters. The GBS shelter structure supports power hardware to deliver to the BDC. Some sites can accommodate a 240 V mains cable to the BDC, whereas other sites require a solar panel to be mounted on the awning to power the BDC. Interruptions to the primary power supply are addressed by the backup battery, which can maintain data collection for approximately two weeks without the primary power supply.

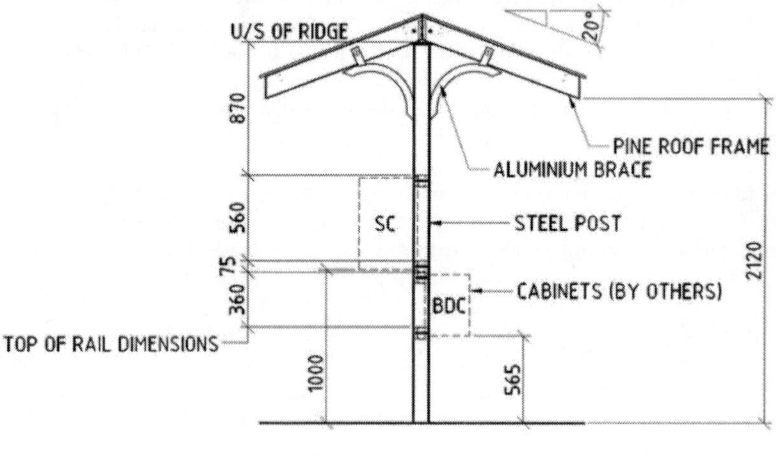

Figure 3. Design of shelters for GBS.

The Hydrographic Ship installation is located on the lower deck of the ship (Figure 4d). The GBS cabinets are mounted side-by-side on a large steel plate that is fixed to the ship. A 240 V AC mains supply is located adjacent to the sensor cabinet for convenient access. The upper deck provides some shelter for this installation, although the position was not intended to replicate aircraft shelters, as with the other installations.

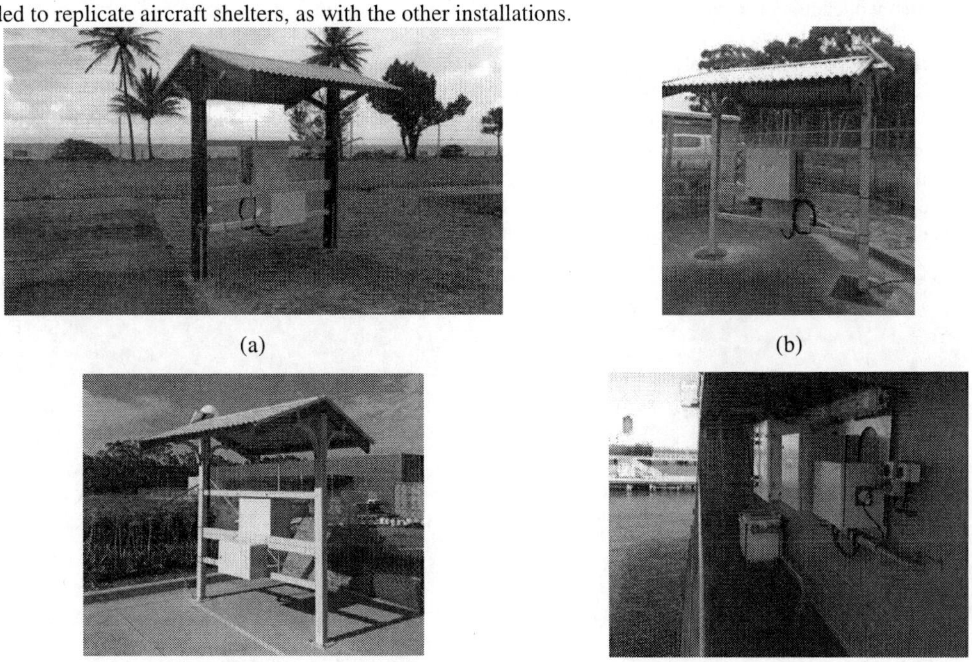

(a) (b) (c) (d)

Figure 4. GBS installations at different sites around Australia:
(a) Cowley Beach, (b) Williamtown, (c) Nowra, and (d) a Hydrographic Ship.

4. CABINET DESIGN

4.1 Battery and Data Cabinet

The BDC contains the electrical components necessary to operate the GBS. These components include: a data logger, a solar power regulator, a backup battery and central power switches (Figure 5). The BDC receives a large cable from the SC containing connections to sensors fitted to the SC. Separating the BDC and the SC is intended to prevent heat output from the battery, solar power regulator and data logger in the BDC from influencing the internal conditions measured in the SC. The data logger continually monitors the sensor outputs.

Figure 5. BDC includes data logger, primary power connection and back-up battery.

4.2 Sensor Cabinet

The SC houses the internal sensor panel and a selection of witness plates. There are two configurations of the SC, which are defined by the particular selection of witness plates contained within. A sensor panel is located on the rear wall of the SC and is identical for both SC configurations. Three monthly services will be undertaken on witness plates from both configurations, which will capture a three month period of corrosion on the relevant witness plates.

Configuration A (Figure 6a) contains a polypropylene panel mounted on the rear of the SC with eight multi-alloy witness plates (MAWP) (see Section 6.1) attached with a 10 mm nylon bolt. Each service event will involve the removal of one MAWP from the panel. Configuration B (Figure 6b) comprises a polypropylene rack for holding 18 painted artificially-damaged witness plates (ADWP) (see Section 6.2), positioned upright on their elongated edge. The 18 ADWPs includes nine 2xxx series aluminium material (top row) and nine 7xxx series aluminium material (bottom row). This configuration also includes nine MAWPs adhered to the reverse of side of the 2xxx series aluminium ADWPs. Each service event will involve one of each ADWP being removed, which also includes a MAWP.

Figure 6a. Inside SC with Configuration A.

Figure 6b. Inside SC with Configuration B.

The external environment can enter internal bays of aircraft via small drainage holes. Therefore, an important aspect of the SC design is their semi-sealed nature, where a 22 mm diameter hole in the base of the cabinet allows the external environment to drive the internal conditions in much the same way as a semi-sealed aircraft bay.

5. SENSORS

Each GBS has installed a variety of sensors for monitoring environmental data relevant to corrosion. Environmental sensors are installed both internally and externally to the semi-sealed cabinet. A BAE Systems Advanced Technology Centre (ATC) developed ToW sensor is only located internally. The sensor plate installed in each SC allows for additional sensors to be included for corrosion and environmental monitoring. Wiring for all sensors is routed through a single cable connecting the SC to the BDC.

Modelling inputs require environmental data such as air and surface temperatures and relative humidity. Additionally, modelling uses a ToW to confirm wetting of the relevant surface. Each of these inputs is recorded both internally and externally to the SC. Figure 7 shows the sensors and data logger used for monitoring the environment.

(a) (b) (c) (d)

Figure 7. Environmental monitoring via (a) temperature/relative humidity probe, (b) surface temperature, (c) time-of-wetness sensor, and (d) a multi-channel data logger.

A humidity and temperature probe is used to record air temperature and relative humidity (Figure 7a). Surface temperature is recorded via a surface-mount temperature sensor (Figure 7b) attached to an aluminium plate to replicate witness plate coupon material. ToW is monitored by two ToW sensors: an ATC sensor (internal only) and a NAVAIR sensor (both internal and external). The ATC ToW sensor measures voltage and is highly sensitive to initial wetness. The NAVAIR sensor (Figure 7c) measures current and its response scales with surface wetness. Data is collected from all GBS sensors by a data logger (Figure 7d) located in the BDC.

The data logger has sufficient memory to store collected data for the full length of the programme. However, during the programme the data logger automatically transfers sensor data via an integrated cellular modem in the form of an email each month. This feature allows use of the data for model refinement during the collection period and also allows for remote monitoring of the current status of each of the GBS in the field.

Data is being collected via environmental data from sensors and corrosion damage captured on witness plates, both internal and external to the cabinets. Table 1 summaries the environmental sensors used on all GBS.

Table 1. Environmental sensors used on GBS

Location	Sensor	Units
Internal	ATC ToW	mV
	NAVAIR ToW	mA
	Surface temperature	°C
	Air temperature	°C
	Relative Humidity	%
External	NAVAIR ToW	mA
	Surface temperature	°C
	Air temperature	°C
	Relative Humidity	%

6. WITNESS PLATES & CLIMAT UNIT

6.1 Multi-alloy Witness Plates

The MAWP captures corrosion data relating to the different material performances within the same environment at each GBS site. The exposure of bare aluminium alloy coupons will allow for collection of corrosion information during the relevant period (Figure 8). Environmental data collected during the relevant exposure period can be used by corrosion models to predict corrosion, which can be compared to corrosion damage observed on each witness plate for verification and validation.

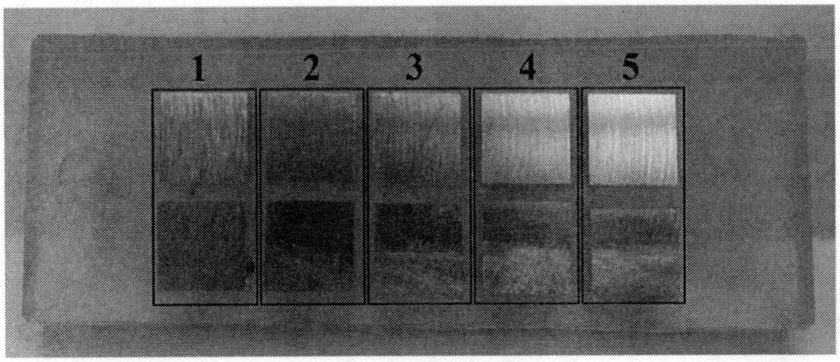

Figure 8. An example of a multi-alloy witness plate (MAWP) including five aircraft aluminium alloys: (1) AA2024-T351, (2) AA2124-T851, (3) AA7050-T7451, (4) AA7075-T651, and (5) AA7085-T7651.

Aluminium alloy coupons are encapsulated in epoxy resin to expose a defined area of 10 mm x 10 mm. The witness plate contains five alloys in duplicate, including: AA2024-T351, AA2124-T851, AA7050-T7451, AA7075-T651, and AA7085-T7651. The exposed alloy surfaces are prepared to a 1200 grit finish and cleaned with isopropyl alcohol. MAWPs are to be returned for analysis at three-monthly intervals. Additional MAWPs will be returned following exposures of 3, 6, 9, 12 and 24 months. These exposure intervals will allow for comparison of cumulative damage with corrosion from specific time periods. This feature will allow for validation of modelling predictions over different time periods.

6.2 Artificially-Damaged Witness Plates

ADWP (Figure 9) are installed in SC of Configuration B at three GBS locations: Williamtown, Cowley Beach and Tindal. The plates are slotted into a rack contained within the SC. The plates are positioned horizontally, resting on their thin edge along the longest side. The ADWPs are made of two different aluminium alloys: AA2124-T851 and AA7050-T7451. There are nine witness plates of each alloy type in each of the three GBS. The 2xxx series aluminium plates also include a MAWP attached to the reverse side of the plate (as shown).

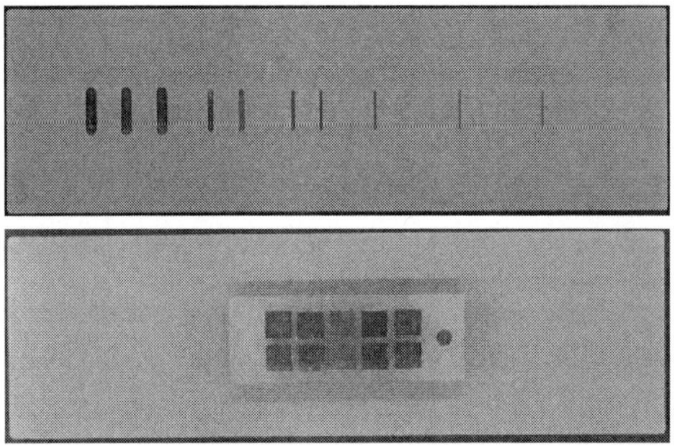

Figure 9. An example of an artificially-damaged witness plate (ADWP) showing the milled slots of different sizes

The ADWPs are coated with a typical aircraft coating system: surface treatment, primer and topcoat. The surface pre-treatment consists of a Thin Film Sulfuric Acid anodised film (Type 1C) with a chromic acid seal. The primer used in this work contains a new environmentally-friendly inhibitor [3] containing a praseodymium oxide and is Deft 44GN098. The outer layer is a polyurethane topcoat (MIL-PRF85285). This coating configuration represents a modern corrosion protection coating system.

The ADWPs include machined slots in the coating system to represent a damaged coating. These slots are machine milled: 1 x 0.4 mm, 2 x 0.6 mm, 2 x 1.0 mm, 2 x 2.0 mm, 3 x 4.0 mm. Slots in the coating system expose the metal substrate with a thin profile of the primer layer along its edges as a source of corrosion inhibitor. Artificial damage allows atmospheric corrosion of the substrate material in the presence of the inhibitor layer and analysis of ADWPs will allow validation of corrosion modelling that incorporates inhibitor performance.

6.3 Classify Industrial and Marine Atmospheres

A method for comparing base corrosion severity uses ASTM G116-99(2015) 'Standard Practice for Conducting Wire-on-Bolt Test for Atmospheric Galvanic Corrosion' [4]. This method is more commonly known as CLIMAT, which is useful for measuring the relative atmospheric corrosion severity of different locations [5, 6]. One CLIMAT unit (Figure 10) will be installed at each GBS location.

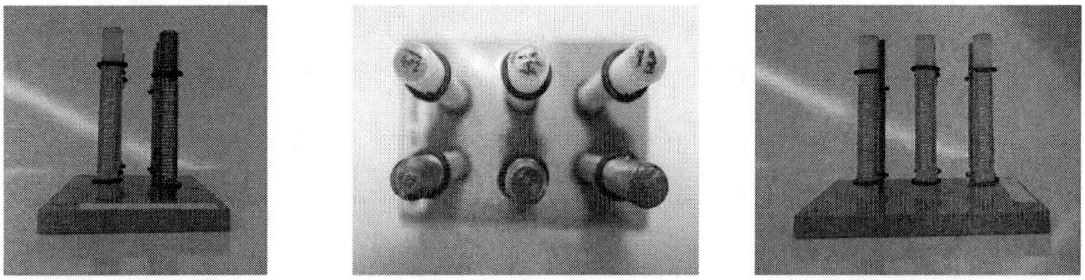

Figure 10. CLIMAT unit showing bolts and wire setup for application of ASTM G116-99(2015).

A CLIMAT unit comprise a baseplate with three copper bolts and three nylon bolts mounted vertically. An aluminium wire is wrapped around each of the six bolts, which is weighed before and after exposure to determine its mass-loss (after removal of corrosion product). Corrosion causing mass-loss is driven by the galvanic coupling of aluminium and copper. The nylon bolts, providing no galvanic couple, serve as a reference. The exposure period is normalised to 90 days and expressed as a percentage mass-loss per meter of wire with respect to the reference. Exposed CLIMAT units will be removed every three months and replaced with a pristine CLIMAT unit. Analysis of data collected from CLIMAT units will used to rank the severity of each location at different times of the year and provide a reference to historic data through the Geographic Corrosion Index [2].

6.4 External Holder

The External Holder is an assembly made of polypropylene components and nylon bolts. The purpose of the External Holder is to provide a corrosion-resistant mounting device for MAWPs, a CLIMAT unit and external environmental sensors (Table 1). The External Holder is to be serviced every three months, when two MAWPs are removed and replaced with pristine MAWPs.

The External Holder has two sections: one upper and one lower. The lower section allows a MAWP to be removed and replaced, representing a three-month sample for the preceding period. Adjacent to the removable MAWP is a non-removable MAWP, which will remain in place for the full two-year duration of the programme. The upper section allows MAWPs to be rotated by removing the bottommost MAWP and inserting the pristine replacement into the top of the upper section so as to move the remaining MAWPs down one position. The effect of this rotation procedure is to collect MAWPs with accumulated damage from 3, 6, 9 and 12-month periods.

Figure 11. Externally mounted MAWPs, CLIMAT and environmental sensors.

External sensors monitor air and surface temperatures, relative humidity and ToW. The surface temperature and ToW sensors are mounted on aluminium coupons encased in epoxy, which occupy the bottom two witness plate positions of the External Holder. The air temperature/relative humidity sensor is partially covered by an open-ended PVC tube located under the CLIMAT bracket. This allows the sensor to monitor the atmosphere without being inundated from direct contact with water. The external sensors allow for analysis of the role external environment will have on the internal conditions of a semi-sealed aircraft bay and allow for the development of a translating algorithm to predict internal conditions from the external environment.

7. DATA COLLECTION & ANALYSIS

With installation of GBS completed in the early part of 2017, some environmental and corrosion data has been collected from the installation sites. This data includes information provided by sensors and captured on witness plates. Some early data is presented here by way of example of the type of information that is being collected. Ultimately, the complete data set collected by the programme will provide greater insight into site-specific and material-specific features of atmospheric corrosion over time.

7.1 Sensor Data

The equilibrium concentration of sodium chloride in a droplet varies with relative humidity fluctuations above its deliquescence point. Additionally, the rate of pit germination in aluminium alloys varies substantially with sodium chloride concentration [7]. As such, the pitting corrosion caused by the deliquescence of sodium chloride on aluminium alloys can be driven by relative humidity fluctuations. It is therefore useful to corrosion modelling to predict the relative humidity of the internal aircraft bay from external environmental data.

Environmental sensor data, both internal and external to the SC, will be used to inform algorithms that translate external environmental data to internal aircraft bay conditions. An example of data collected to develop such a translation algorithm is shown in Figure 12, which depicts a comparison of relative humidity fluctuations recorded both internal and external to the GBS located at RAAF Williamtown. It clearly shows that the rise in the relative humidity internally lags the external relative humidity.

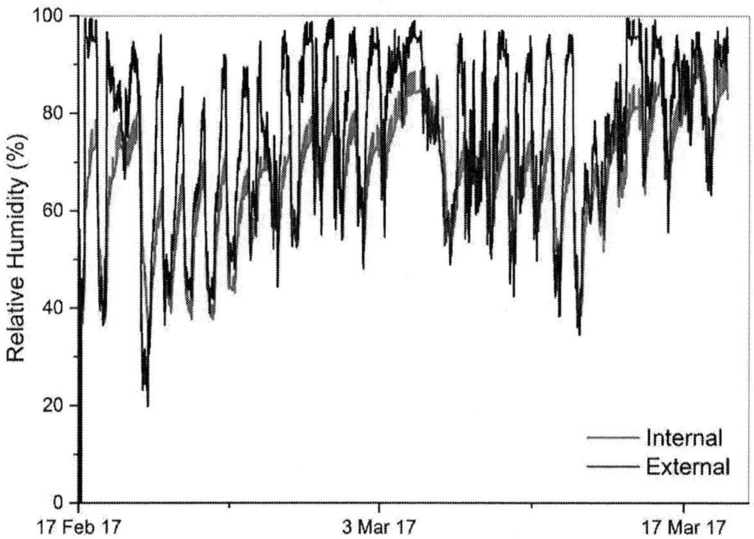

Figure 12. Relative humidity data collected from GBS located at RAAF Williamtown.

Environmental data collected allows for comparison of Bureau of Meteorology (BOM) data with external sensor environmental data to verify BOM data can be used as reliable inputs to corrosion modelling. A translation algorithm will allow BOM data for non-monitored sites to be used as the environmental data inputs to a corrosion predictive modelling capability.

7.2 Witness Plate Data

The corrosion damage from all witness plates, both internal and external, will be characterised by use of a profilometer. A profilometer uses light and optics to scan a sample's surface and accurately image the topography of the surface. This feature makes it particularly useful for measuring corrosion damage on exposed alloy surfaces. Pitting can be described by a distribution of discrete pit volumes, which can be used to approximate pit depths/diameters. Corrosion modelling outputs can also predict pit size distributions for a given alloy and environmental history, thus allowing for direct comparison with data collected from profilometry conducted on MAWPs.

Figure 13. MAWP (externally mounted) after three months exposure at Williamtown.

The surface area of the MAWP will be swabbed to collect data regarding salt loading of the location. This can be compared with previously collected salt candle information from the same location, as well as aircraft swab data. The amount of salt loading can also be inferred from the data collected from ToW sensors on the GBS. A correlation between salt loading on samples and ToW output can assist in validating input data to corrosion modelling.

The exposed alloy surface of the ADWP will be scanned to quantify the corrosion damage in terms of size distributions of volume of material removed from the surface. This data can be compared with corrosion damage predicted by modelling. Corrosion damage measured on the ADWPs will be influenced by artificially-damage coating system, which will include the influence of the praseodymium inhibitor from the primer. Current corrosion modelling capability incorporates the electrochemical impact of inhibitor concentration [8] and the primer's leaching performance [9]. Corrosion modelling outputs can predict the occurrence of pitting, the size to which pits grow and the likely size distribution of these initiated pitting events under such conditions. The surface scans provided by the profiler will allow for validation of these predicted outputs based on the environmental data.

8. SUMMARY

The installation of GBS at various sites around Australia aims to capture environmental data across a broad range of climactic locations for the purposes of improving atmospheric corrosion predictive modelling tools. To achieve this, GBS have incorporated a suite of environmental sensors and witness plates to record both corrosion-relevant information and actual corrosion damage of aircraft aluminium alloys. GBS will be serviced every three months and the current programme will run for two years. The data collected over the course of this programme will be used to develop further corrosion predictive modelling and assist in validating existing corrosion modelling capabilities.

9. ACKNOWLEDGMENTS

The authors gratefully acknowledge JSF Division and Lockheed Martin Aeronautics for research funding, John Nasser (BAE Systems) for assisting with GBS installations, and All Life Pty Ltd for fabrication work of cabinets in consultation with BAE Systems and DST Group.

10. REFERENCES

1. Trueman T, Trathen P, Begbie K, Davidson L, Hinton B, The development of a corrosion prognostic health management system for Australian Defence Force aircraft, Advanced Materials Research 38 (2008) 182-200.
2. Bitcon JC. Atmospheric Corrosivity at Australian and Overseas Airbases and Airports: DSTO-TN-1320, 2014 Contract No.: DSTO-TN-1320.
3. PPG. Material Safety Data Sheet: 44GN098. 2014.
4. Compton KG, Mendizza A, Galvanic couple corrosion studies by means of the threaded bolt and wire test, Symposium on Atmospheric Corrosion of Non-Ferrous Metals STP 175 (1955) 116-125.
5. Doyle DP, Wright TE. Rapid methods for determining atmospheric corrosivity and corrosion resistance. In: Ailor WH, editor. Atmospheric Corrosion: John Wiley and Sons; 1982.
6. Doyle DP, Godard HG, A rapid method for determining the corrosivity of the atmosphere at any location, Nature 200(4912) (1963) 1167-1168.

7. Sudholz A, Knight S, Trueman A. Comparative predictions of stable pit probability on aircraft Al alloys AA2024-T3 and AA7475-T7651 in aqueous environments. 53rd Annual Conference of the Australasian Corrosion Association 2013: Corrosion and Prevention 2013; 10-13 November 2013; Brisbane2013.
8. Sudholz A, Trueman A. Electrochemical characteristics of aluminium alloys in the presence of praseodymium(III) ions. 54th Annual Conference of the Australasian Corrosion Association 2014: Corrosion and Prevention 2014; Darwin2014.
9. Sudholz A. Leaching performance of a rare-earth-containing epoxy polyamide primer. 55th Annual Conference of the Australasian Corrosion Association 2015: Corrosion and Prevention 2015; Adelaide: ACA; 2015.

11. AUTHOR DETAILS

Dr Aaron Sudholz conducts research as a consultant materials engineer with DST Group. Aaron completed both his undergraduate and postgraduate studies at Monash University in the areas of chemistry, materials science and materials engineering. Aaron's research centres on the characterisation and modelling of materials properties in the area of corrosion science and engineering. Additionally, Aaron is studying a Master of Laws (Juris Doctor) and has a keen interest in intellectual property matters in the materials science space.

Hue Thanh is a Systems Engineer at BAE Systems. Hue graduated from the University of Melbourne with bachelor's degree in electrical/electronic engineering, and holds a Master of Information Technology from Swinburne University. Hue's fields of interest include systems and network security, human factors in system design.

Dr James Waldie graduated with a Bachelor of Engineering (Aerospace) and Bachelor of Business from RMIT University. James continued on as a Research Scholar at the University of California in San Diego for his Masters, and earned his PhD from RMIT in 2005. James was also a Postdoctoral Fellow at the Department of Aeronautics and Astronautics at the Massachusetts Institute of Technology (MIT) for 2.5 years. Dr Waldie has worked for BAE Systems since 2002, most recently as the R&D Engineering Manager for the Corrosion Group.

Jovan Talevski is a Technical Specialist undertaking research in the Corrosion Sciences team at DST Group. Jovan has vast experience in corrosion science research, having contributed to a large number of corrosion protection projects for a variety of ADF aircraft platforms.

Chris Loader is a Research Scientist at DST Group.

Dr Andrew Butler is a research scientist contracted to DSTO since 2001. Andrew has worked on many corrosion prevention and remediation projects associated with Australian Defence Force platforms, as well as contributing to the development and validation of sensor systems for prognostic health monitoring. Andrew has been involved in the development of Supersonic Particle Deposition for use on the JSF and ADF platforms. Prior to DSTO Andrew completed a PhD in high resolution laser spectroscopy, held a post-doctoral research position in laser physics and worked at TRL (UK) on corrosion of reinforcing steel in concrete bridges.

THE USE OF EPOXY AND POLYMER MODIFIED CEMENTITIOUS COATINGS TO RE-LINE OLD CONCRETE AND BRICK MANHOLES AND SEWER STRUCTURES

David Johnstone
Akzo Nobel Pty Limited

Summary: Concrete manholes are an integral part of all sewerage systems. They are however exposed to corrosion from Hydrogen Sulfide (H2S) gas which, if not addressed leads to sectional loss of concrete and can ultimately lead to a catastrophic failure of the system.

Corrosion to concrete sewage infrastructure is estimated to cost around $91 million per year in Australia (*Report commissioned by the Australasian Corrosion Association (ACA), the report, Corrosion Challenges – Urban Water Industry*).

Protective linings such as Polyurea, Vinyl Ester, and Novolac epoxies provide excellent barrier protection when applied correctly, however these systems are difficult to apply as often the application environment has very high humidity, with the concrete structure being damp and/or subject to water pressure from the outside face of the structure.

Extending the long term durability of the structures and providing chemical resistance to Hydrogen Sulfide (H2S) attack is very challenging, particularly when the expected design life of the structure can be >25 years. Issues such as reinstatement of the effective concrete cover, fast return to service capabilities, the presence of surface moisture and high atmospheric humidity during coating application all contribute to the complexity, cost and overall result of the coating application. This paper explores the issues of applying protective coatings to degraded manholes and sewer pipes and reviews the use of advanced, two-component cementitious coatings as an environmentally friendly, cost saving solution providing long term protection from Hydrogen Sulfide (H2S) attack.

Keywords: Concrete, Sectional loss of Concrete, Long Term Durability, Design Life, Chloride Diffusion, Cementitious Coatings

THE CORROSION PROCESS

When dissolved oxygen in a system falls below 0.1mg/L. Reference – Jason Kane, McBerns Paper from 39th Annual Queensland Water Industry Operations Conference & Exhibition "Hydrogen Sulphide Gas in Sewers – The Challenges of Odour and Corrosion" water becomes anaerobic allowing bacteria to reduce dissolved sulphates in the system to Hydrogen Sulfide Gas (rotten egg smell).

Hydrogen Sulfide is then converted to Sulphuric Acid by Thiobacillus Bacteria.

This Thiobacillus bacteria thrives in warmer temperatures and the higher the temperature the faster the rate of H2S production. The acid produced from the Thiobacillus bacteria attacks OPC (cement paste) resulting in major corrosion of concrete structures, and ultimately corrosion of the reinforcing steel if present. The acid attack also corrodes mass placed concrete without reinforcement.

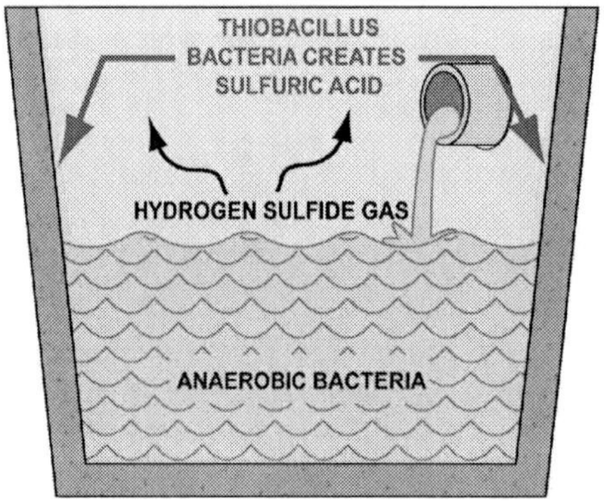

Figure 1: The Corrosion Process

FACTORS TO CONSIDER WHEN SELECTING A PROTECTIVE COATING SYSTEM

When specifying a lining, asset owners and engineers need to be mindful of the many factors which will have an effect on the application and long term performance of the lining system. These considerations include;

- Manholes have very high humidity and are generally damp – *Is the lining system tolerant of surface moisture and high humidity?*
- Concrete repairs are generally required to reinstate the concrete cover – *Can the lining system be applied to freshly applied repair mortars?*
- Reinstatement of concrete can delay the application of the protective coating – *Can the asset be off line for a length of time while the concrete repairs cure and the moisture levels reduce to a level where coatings can be applied?*
- Generally structures need to be back on line quickly so costs associated with sewer diversion and traffic management is minimised – *Is the lining system designed for fast return to service?*
- Environmentally friendly systems (VOC Free) – *Is the lining system compliant?*
- Fit for purpose system – *Is the specified coating resistant to acid attack and will the specified coating provide a long term corrosion barrier?*

LIMITATIONS OF OTHER LININGS TECHNOLOGIES

When specifying a lining, asset owners and engineers need to be mindful of the limitations some types of technologies have, which may affect the application tolerances and long term performance. These considerations include;

Novolac Epoxy

- Can't be applied to damp substrates or over freshly placed (green) repair mortars (return to service time is compromised)
- Can't be applied in environments subject to water pressure from the outside face of the structure (restricts use on manhole applications)

Polyurea

- Fast gel time (4-5 seconds) limits adhesion performance to concrete and delamination is a common
- Can't be applied to damp substrates or in humid environments, or over freshly placed repair mortars (limits use on manhole applications)
- Can't be applied in environments subject to water pressure from the outside face of the structure (restricts use on manhole applications)

Calcium Aluminate Cement

- Generally need to be applied at a minimum thickness of 25mm (thickness of applied system is not cost efficient)
- Sacrificial systems, which generally deteriorate over time (asset will require relining after 10 years' service)
- Limited abrasion resistance (limits the long term performance of the product)

ADVANCED CEMENTITIOUS COATINGS TECHNOLOGY

Advanced cementitious coatings are a two-component, cementitious/epoxy modified systems which have very high chemical resistance as well as high abrasion resistant qualities.

A formulation of cement and polymer materials designed to minimise porosity (total volume of voids within the matrix) and water permeability (rate of flow of liquid through the matrix) thus maximising the linings physical characteristics.

This coating formulation provides manhole and sewer infrastructure with the following technical benefits;

- High chemical resistance to Hydrogen Sulfide Gas ($H2S$) up to 200ppm, Hydrochloric Acid (20%) and Sulphuric Acid (20%)
- Application to green and damp concrete in humid environments
- Application over repair mortars as soon as the mortar has achieved final set (generally 6 hours)
- Application to Concrete, Masonry / Brick and Steel substrates
- Fast Return to Service with the asset able to be put back on line after 1 hour cure
- Service design of >25 years
- Resistant to 10 bar (100m) positive/negative head of water pressure
- VOC / Isocyanate free and water based technology
- High abrasion resistance of <0.05mm (tested using aaccelerated abrasion method in accordance with BS 8204: Part 2: 2002 Annex B)
- Application by brush, trowel and airless spray equipment
- High resistance to freeze thaw cycles (-36°C to 180°C).

When applied at 2mm thick, advanced cementitious coatings have been tested to provide the equivalent of an additional 100mm effective concrete cover for resistance from Chloride induced corrosion (Taywood Engineering / Vinci Testing)

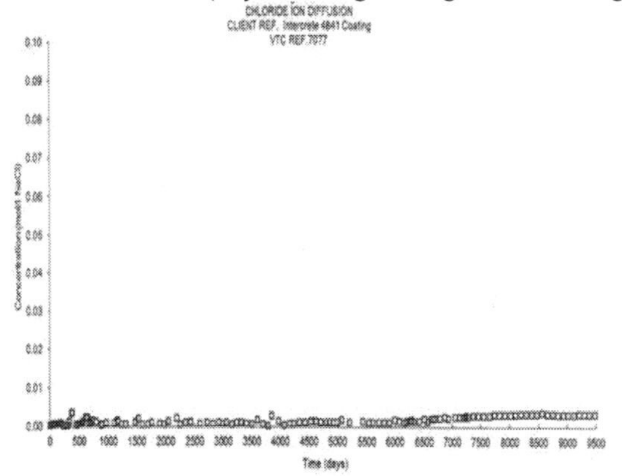

Figure 2. Chloride Ion Diffusion

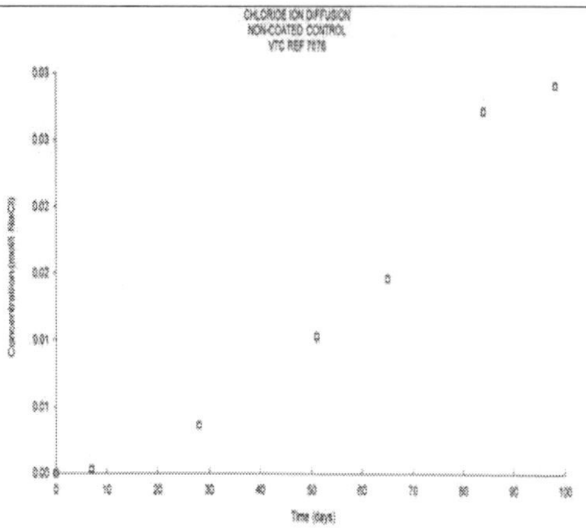

Figure 3. Chloride Ion Diffusion non coated control

EXAMPLE PROJECTS

5.1 Melbourne Water, Western Treatment Plant (25 WEST – Sewage Pond No.1 Ring Beam) – 2015

Figures 3 & 4: Depict exposed aggregate surface from H2S attack

Epoxy and Polymer modified Cementitious coatings were used to re-line a 1.2km long (960m2) concrete ring beam surrounding a sewage sediment pond. A floating liner is installed over the sediment pond to reduce odour, however Hydrogen Sulfide gas build up under the liner has attacked the concrete ring beam. Epoxy and Polymer modified Cementitious coatings were specified due to the build-up of acidic gas between the floating cover and the concrete.

5.2 Banana Shire Council, Queensland – Manhole Re-Lining – 2014

Figure 5: Manhole with significant deterioration from H2S attack

2014 – 35 manholes suffering from Hydrogen Sulfide (H2S) attack, with up to 25mm (nominal) sectional loss of concrete were originally specified with Calcium Aluminate Cement (CAC). Epoxy and Polymer modified cementitious coatings were chosen due to the lower thickness build-up of 2mm compared to a minimum requirement of 25mm for CAC, thus significant cost savings were achieved. The manholes have been inspected one year after application with no degradation from acid attack reported.

5.3 Melbourne Water, Eastern Treatment Plant, Victoria – Digester No:5 – 2015

Figures 6 & 7: Overview of the digester

Figure 8: Old membrane being removed

Figure 9: Prepared substrate with primer applied

Figure 10: Applicators inlaying geotextile fabric

Figure 11: Mixing the waterproofing membrane

Figure 12: Geotextile installed, awaiting top coat

Figure 13: Digester back in service

Cementitious coatings were used to waterproof and protect a 1200m2 floating digester roof. The existing rubber liner has severe degridation from UV exposure and was leaking. Epoxy and Polymer modified Cementitious coatings were specified due to the high moisture content within the concrete and resistance to Hydrogen Sulfide (H2S). This technology has very good adhesive bond strength to damp / SSD concrete. The concentration of H2S was minimal due to the structure having an open top.

5.4 Tweed Heads Shire Council, NSW – Manhole Re-Lining – 2013

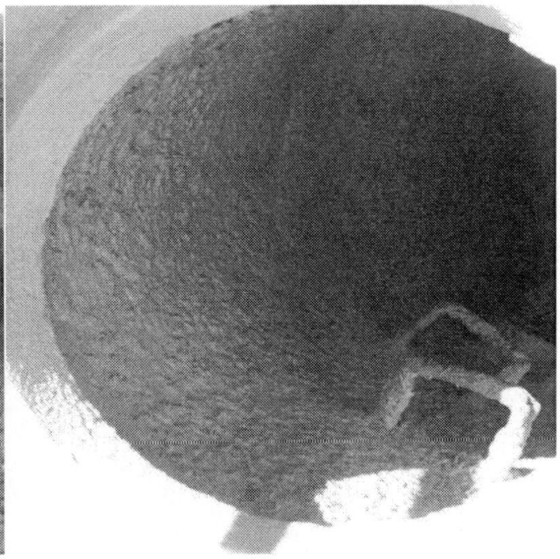

Figure 14: Manhole with exposed aggregate from H2S attack

Figure 15: Manhole relined

40 manholes suffering from Hydrogen Sulfide (H2S) attack, with up to 30mm (nominal) sectional loss of concrete were originally specified with Polyurea. The majority of the manholes had issues with water pressure from the outside face of the structure producing a very damp surface, thus epoxy and polymer modified cementitious coatings were chosen due to their ability to bond to damp surfaces and resistance to 10 bar (+/-) head of water pressure.

5.5 Merri Creek, Victoria – Manhole Re-Lining – 2016

Figure 16: Brick manhole with lime mortar degraded from H2S attack

48 manholes (combination of concrete and brick) suffering from Hydrogen Sulfide (H2S) attack. The brick manholes had no mortar remaining and the concrete manholes has up to 20mm (nominal) sectional loss. The manholes were originally specified with Calcium Aluminate Cement (CAC), however Melbourne Water wanted to extend the design life of the structures by an additional 25 years. Calcium Aluminate Cement (CAC) only had an anticipated design life of 10 years, thus epoxy and polymer modified cementitious coatings were chosen as they could meet Melbourne Water's design life expectations.

5.6 *North Warrandyte, Victoria – Wet Well Re-Lining – 2016*

Figure 17: Manhole relined with Advanced Cementitious Coating

Structure suffering from Hydrogen Sulfide (H2S) attack, with up to 15mm (nominal) sectional loss of concrete was originally specified with Polyurea. The project was scheduled for repair in July 2016 and being winter in Melbourne the structure had a very damp surface, thus epoxy and polymer modified cementitious coatings were chosen due to the ability to bond to damp surfaces and resistance to 10 bar (+/-) head of water pressure.

SUMMARY

Correctly formulated and tested resin modified cementitious coatings have more than 29 years of proven performance track records in the sewage and waste water industry.

This performance proves that this technology is the ideal solution to protect reinforced concrete and masonry structures from chloride induced corrosion and Hydrogen Sulfide (H2S) gas attack.

Cementitious/epoxy modified lining systems provide reinforced concrete structures with effective concrete cover increase for resistance against carbonation and chlorides induced corrosion. These cementitious systems ensure the intended design life of the structure is both achieved and extended.

ACKNOWLEDGEMENTS

Christopher Lloyd and Graham James from Flexcrete Technologies, Leyland, UK.

CASE STUDY: CONDITION ASSESSMENT OF AGED UTILITY PIPES IN UNDERGROUND MINES

V. Afshari
GHD Pty Ltd, Australia

SUMMARY: This paper will discuss the method and findings of the inspection and condition assessment of pipe services for a WA underground mining operation with respect to potential corrosion integrity threats. There is currently increasing pressure in the mining industry to minimize costs while improving the reliability and availability of plant including the piped services in underground. The piped services for the proposed mine have been in service for over 25 years. During this study a number of direct examination techniques (i.e. visual inspection and non-destructive test) and in-direct inspection methods (i.e. sample collection and laboratory tests) were utilised for the purpose of condition assessment to verify pipe integrity, and corrosion risk issues. The potential corrosion risk will be discussed through a developed risk matrix, which provides a tool to focus attention on the most critical areas and pipe services.

Keywords: Condition Assessment, Mine waters, Corrosion Inspection.

1. INTRODUCTION

Process grade water is a basic commodity for the mining industry. Bore water is usually the main source of process water in the mining industry, which is recovered from ground water resources. The prime purpose for water in underground mining operations is for rock drilling and dust suppression after blasting. Bore waters generally contains high total dissolved solids (TDS) and may be associated with high alkalinity, bicarbonates and bacteria, especially in Australian Mining regions.

Pumping contaminated underground water (mine water) to the surface through dewatering pipework have also a detrimental effect on the service piping that form a main artery in the infrastructure of a mine. Mine water is typically hard with high levels of salinity. In general, increasing total salinity of mine waters tends to increase the corrosiveness of this water [1]. The source of mine water may be from surface accumulations, ground water i.e. open or confined aquifers, drainage, etc.

The great majority of piping used in mines for transporting process water and drainage/dewatering water is made of carbon steel. The majority of utility pipes of underground mining operations in WA are well into their anticipated design life. The reliability and availability of the water pipes in underground operating is a major concern for asset owners, particularly for olds mining operations. Corrosion is the major problem with most mine piping and occurs internally, externally or both. Preempting corrosion failures would eliminate post-failure corrosion costs that have significant adverse economic impacts on the mining operation. Corrosion in carbon steel piping systems increases costs due to the reasons listed below:

- Repair and labour costs due to pipe failures i.e. leakage;
- Build-up of internal corrosion products then decrease system efficiency due to increasing frictional losses; and
- Deferred production cost.

All this results in increasing plant operation downtime, pressure losses and decreased system performance. Eliminating or reducing potential corrosion failures can be achieved to a large extent through understanding the key parameters of corrosion causes and proper application of corrosion management including implementation of corrosion protection measures.

This paper presents the results of a study carried out in the condition assessment of utility pipe services for a WA underground mining operation with respect to potential corrosion integrity threats. An evaluation of the corrosivity of the process water pipes and dewatering pipes is assessed to ascertain the corrosion risk issues and to verify integrity of the piped services in underground.

2. METHODOLOGY

For the purpose of this study, the following types of direct and in-direct inspection methods were considered during the course of the investigation.

2.1 Visual Inspection

A visual inspection was carried out on all accessible areas of the underground services steel pipework. Any defect finding was recorded including location and type of imperfection. An on-site internal visual inspection was completed on selected locations to determine the internal condition of the piping services. The inspection was conducted by opening the flange joint connections, after isolation/drainage of the internal water at proposed locations.

2.2 Ultrasonic Thickness Measurements

Spot ultrasonic thickness measurements were accomplished on the representative locations of the service pipes to evaluate the general metal loss and degree of corrosion. The test either was performed on locations to be a representation of the overall condition of the system, or were deemed to be at higher risk of corrosion.

2.3 Chemical Composition Analysis

From earlier visual examination, certain areas were identified for further investigation, which may provide valuable information as to the cause of the corrosion.

2.3.1 Water Samples

The corrosivity of water is significantly influenced by the concentration of dissolved species, pH, temperature, suspended matter and bacteria [2]. The water samples were collected at representative locations to determine the water chemistry effect on internal corrosion.

2.3.2 Corrosion Products

Energy Dispersive Spectrometer Analysis (EDS) was used to determine the chemical nature of internal pipe deposits and corrosion products.

3. SYSTEM DESCRIPTION

3.1 Process water

The process water is supplied via gravity discharge from the Process Water Tank into both the main shaft and the decline. The source of process water is from bore water sourced from nearby the site mine.

3.2 Drainage/dewatering water pipe

Drainage / dewatering water is collected in the sumps located around the underground mine and gravity drained or pumped to the dams located on different levels. Each dam has a dedicated pump station that transfers the water via steel pipework in the mine shaft, then pumps to the evaporation pond on the surface.

3.3 Material specifications

The materials of the piped services within the shaft are summarised below:

- Process water : API 5L Grade B, DN150 (Sch. 80)
- Drainage/dewatering water pipe: API 5L Grade B, DN150 (Sch. 80)

From information provided by client, it is understood that the pipework is nominally 25 years old at the time of inspection. Some small sections of the pipework have been replaced as part of ongoing maintenance but most of the pipework, including the shaft is original.

4. RESULTS

4.1 Visual Inspection

The results of the visual inspection revealed that the external surfaces of the service pipes were covered by a uniform solid scale in most areas within the shaft. Significant rust was observed on the external surface of the service pipes in several locations where the ground water splashes onto the surface. Shallow pitting was detected under the deposit formations in the wet area. A typical photo of the external surface of the service pipes are shown in Figures 1 and 2.

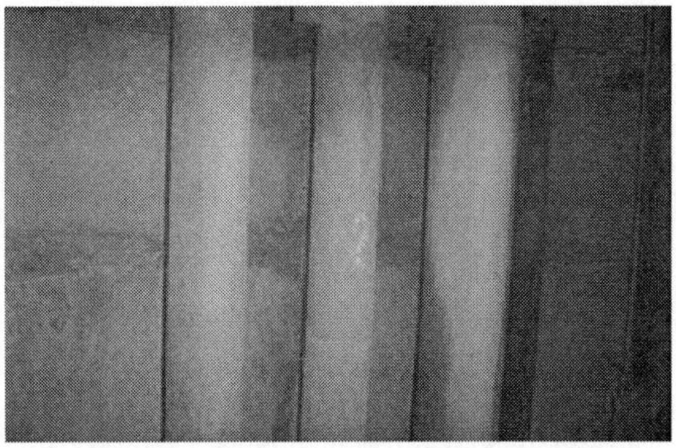

Figure 1 Scale formation on the external surface of service pipes in dry area

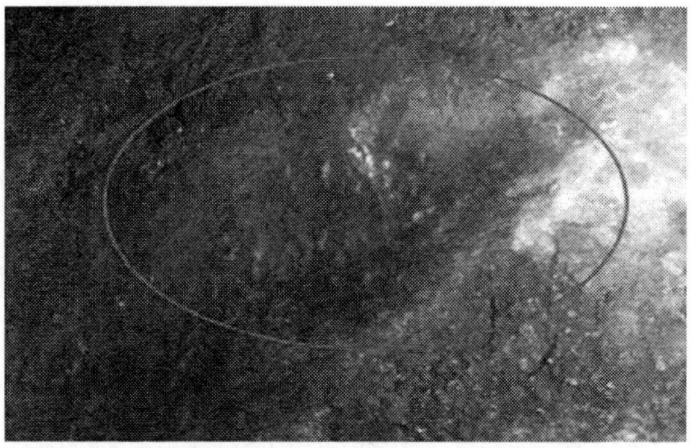

Figure 2 Close view of corrosion indicates an irregular steel surface (shallow pitting) under surface scale of service pipe in wet area

4.2 Pipe-wall thickness measurements

An ultrasonic thickness examination of the service pipe wall was completed to evaluate the general metal loss. The results indicate that there are some wall thinning occurring at the areas which experience a wet and dry cycle due to ground water splashes onto the surface. The average loss of the wall thickness in the proposed wet area is calculated as 16.1%.

The external surface of the service pipes had considerable pitting visible to the naked eye in some locations, which was a sure sign of severe soluble salt contamination. The localised pit depth was measured and the results are presented in Table 1.

Table 1. External localised corrosion (pitting) depth measurements

Pit Depth (mm)	Maximum depth of pitting and pitting rate
1.44	
2.72	
1.22	Maximum pit depth: 3.67 mm
1.78	****************************
1.84	Maximum external pitting rate = Maximum pit depth /Life time
3.67	= 3.67/25 = 0.15 mm/year
1.97	
2.06	
1.51	
2.00	
1.99	

4.3 Water Chemistry

The results of water quality analysis of the collected water samples are presented in Table 2.

Table 2. Chemical Analysis of Collected Water Samples

Sample ID	Max. TDS (ppm)	Total Hardness as $CaCO_3$ (ppm)	Total Alkalinity as $CaCO_3$ (ppm)	Max. Sulphate (ppm)	Max. Chloride (ppm)	pH
Process Water	1400	270	250	170	450	7.5 ~ 8.1
Dewatering	47000	14000	40	1400	26000	7.3 ~ 7.6

4.4 Microbiological Analysis

The biological analysis of water samples are summarised in Table 3.

Table 3. Microbiological Analysis of Collected Water Samples

Sample ID	Sulphate Reducing Bacteria (cfu/mL)	Other Bacteria (cfu/mL)
Process Water	<1	41500
Dewatering	14	138

4.5 EDAX Analysis

A sample of external deposit on the process water pipe service within the shaft and corrosion product formed on the external surface of the dewatering service pipe were removed for laboratory testing to determine the chemical nature of deposits and corrosion products. EDS was used to determine the chemical nature of external deposits and corrosion products.

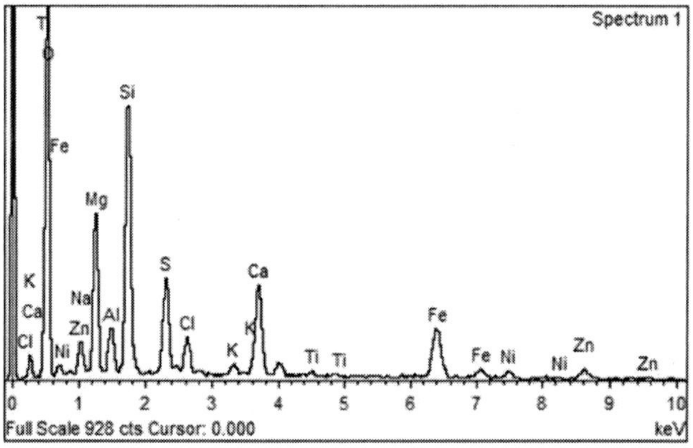

Element	Weight%	Atomic%
O K	49.21	67.18
Na K	1.43	1.36
Mg K	7.97	7.16
Al K	1.96	1.59
Si K	12.03	9.35
S K	4.75	3.23
Cl K	2.04	1.26
K K	0.51	0.28
Ca K	6.08	3.31
Ti K	0.25	0.11
Fe K	8.94	3.50
Ni K	1.26	0.47
Zn K	3.58	1.20
Totals	100.00	

Figure 3 EDS Analysis of external pipe deposit from mine water

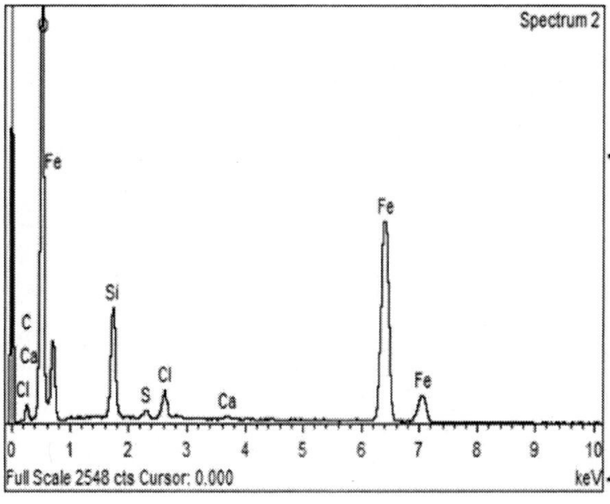

Element	Weight%	Atomic%
C K	7.00	13.52
O K	43.37	62.93
Si K	5.85	4.83
S K	0.41	0.30
Cl K	1.55	1.01
Ca K	0.16	0.09
Fe K	41.67	17.32
Totals	100.00	

Figure 4 EDS Analysis of external corrosion products

5. DISCUSSION

5.1 External Corrosion

Atmospheric corrosion is the natural degradation of material exposed to air and its pollutants. The rate of atmospheric corrosion is influenced by the relative humidity (the ratio of the quantity of water vapour present in the atmosphere to the saturation quantity at a given temperature) [3]. Atmospheric corrosion is further accentuated by the presence of pollutants such as mineral dust and exposure to mineral water. All of these conditions are present in the underground pipe services inside the shaft. Solid mineral deposits on the service pipes in dry areas create a barrier and result in lower corrosion rates over time. However, the service pipes suffered from external corrosion in several locations where the ground water splashes onto the surface. Shallow pitting was observed under the deposit formations in the wet area. Shallow pitting usually occurs by the action of macro-galvanic cells. In a macro-galvanic cell, the anodic and cathodic areas are macroscopic, and their locations are fixed, leading to a shallow pitting form of corrosion. The macro-galvanic cells are formed by differential aeration and differential pH that leads to the formation of passive-active cells.

5.2 Corrosion Products Analysis

The results of EDS analysis of the external deposit on the pipe surface shows one of the main components of scale is crystals of calcium sulfate mineral and iron sulphide which could be a good indication of bacterial activity in the mine water.

The results of EDS analysis of external corrosion products revealed that it was composed primarily of oxygen, carbon and iron. Significant concentrations of chlorine was observed on the corrosion products, which is most likely due to mine water salt contamination.

5.3 Water Chemistry Analysis

The two ways in which the parameters of the transferred fluid will affect the integrity of the pipe are corrosivity and scale tendency. Several corrosive and scale indices have been developed to evaluate the corrosivity and scale tendency of water toward steel surfaces as described in detail elsewhere [3]. The results of water corrosion and calculation of scale tendency indices of representative water samples are summarised in Table 4.

Table 4. Water Quality Analysis of Collected Water Samples

Sample Location	LSI	RSI	PSI	L-S	Scale Tendency of Water	Corrosivity of Water
Dewatering	0.8	5.6	5.97	1032	Heavy Scale Forming	Very Aggressive
Process Water	0.3	7.0	6.54	3	Slightly Scale Forming	Aggressive

LSI: Langelier Saturation Index, RSI: Ryznar Stability Index, PSI: Practical Saturation Index, L-S: Larson-Skold Index

The results of analyses of dissolved matters in the dewatering system shows the water is very hard. In hard waters that contain high concentrations of calcium and bicarbonate, the natural deposition of calcium carbonate on the steel surface provides an effective diffusion barrier to oxygen diffusion, greatly decreasing corrosion. Low to moderate amount of bacterial activity was detected in the dewatering line.

Issues arising in the dewatering lines primarily relate to scale deposition rather than corrosion, which can result in a reduction of flow through pipes due to reduced internal cross section. The pump type (positive displacement) limits the effect of the reduced cross section until it reaches a critical level. The scale formation will be accelerated in areas of the dewatering system that have higher temperatures and lower velocity.

The results of the saturation and stability index show that the process water is slightly scale forming. Chloride and Sulfate ions have essentially no effect on the internal corrosion of process water as long as a protective uniform scale presence on the internal pipe surface is formed.

5.4 Microbiological Analysis

The biological analysis of collected dewatering water samples revealed the presence of microbial communities (aerobic and anaerobic), with slime formed and metal reducing bacteria. However, the population of microbial colonies are measured to be low to moderate and the risk of MIC is considered to be low.

The results of the bacterial analysis revealed a significant amount of bacteria presence within the process water lines. The bacterial greatly accelerates the formation of tubercles by accumulating a large amount of ferric hydrate and significantly increases the risk of localised corrosion (e.g. pitting) in areas with low water velocity. Bacteria tend to settle non-uniformly on the metal surface. They may form discrete colonies, which later completely cover the metal surface. Under ideal conditions, the bacteria are reported to grow best in a temperature range 15 to 45 °C and pH of 6 to 8 [4]. The majority of the bacteria can form extracellular polymeric material known as slime. The slime is involved in attaching the bacteria to the metal surface, trapping and concentrating nutrients.

5.5 Internal Corrosion Risk Assessment

The water scale tendency and the corrosivity of water were combined to rank each location and determine the areas of concern on a 4×4 matrix as shown in Figure 5. The assessment is based on average values. This ranking provides a tool to focus attention on the most critical areas and pipe services.

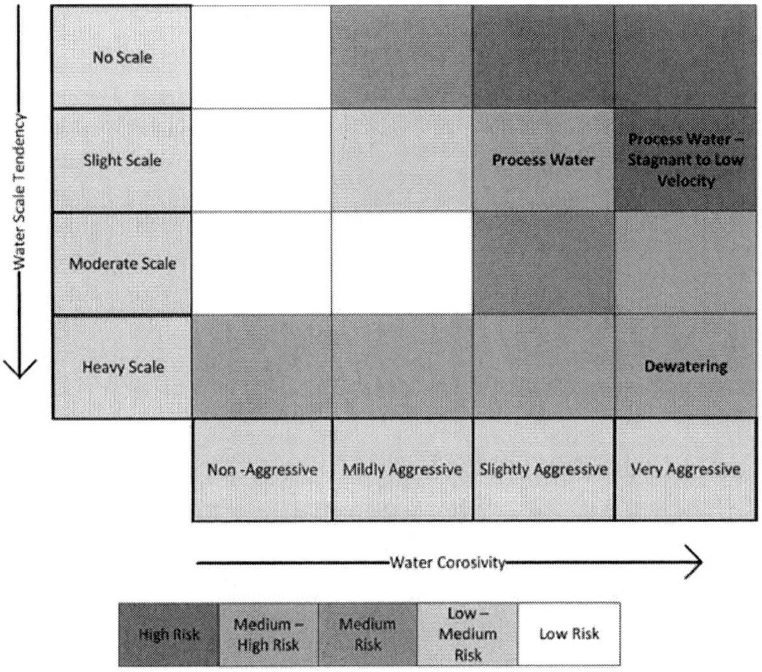

Figure 5 Water Quality Analysis Matrix

The area of most concern is the water qualities that fall in red zone (primarily corrosive waters) with no protective scale. It should be noted that corrosion issues in the red zone could be addressed by either materials selection (changing the steel pipework to HDPE) or treatment of the water supply.

Some process water lines whose characteristics fall into the orange zone squares in Figure 5 are also considered aggressively corrosive, but not as highly corrosive as the red zone. Some form of planned corrosion mitigation is also appropriate in these zones.

The dewatering pipes are considered as "hard" waters, having high concentrations of calcium and often also magnesium, and high buffering capacity – hence the tendency to deposit calcium based scales such as calcium carbonate, especially when heated.

Issues arising in dewatering line waters primarily relate to scale deposition, so generally will require resolution by treatment of the water supply.

Although deposition of scale may provide some corrosion protection, it also reduces the cross sectional area of the pipe and increases turbulence, potentially resulting in a restricted flow. For this reason, the water with this characteristic was categorised as medium to high risk in the matrix.

6. CONCLUSIONS

An investigation was performed to assess the condition of the pipework for a WA underground mining operation with respect to potential corrosion integrity threats and to establish the level of deterioration. The piped services for the mine have been in service for 25 years. For the purpose of this study, a number of direct examination techniques (i.e. visual inspection and non-destructive tests) and in-direct inspection methods (i.e. sample collection and laboratory tests) were utilised to verify pipe integrity and corrosion risk issues.

The external surface of the piped services in the underground mine has experienced general and pitting corrosion resulting from salt contamination of surfaces in area exposed to wet/dry cycles due to mine water splash and inadequate, or failure of , corrosion protection measures. The external corrosion of the service pipes were particularly intense in areas where it is regularly splashed by mine water or areas, which experience condensation or high humidity. High levels of chlorides in mine water splash, and accumulation of dust increases severity and corrosion potential.

The results of the saturation and stability index show that the process water is slightly scale forming. Chloride and Sulfate ions have essentially no effect on the internal corrosion of process water as long as a protective uniform scale presence on the internal pipe surface occurs. The results of the bacterial analysis of process water revealed a significant amount of bacteria presence within the process water lines. The bacterial greatly accelerates the formation of tubercles by accumulating large amount of ferric hydrate and significantly increases the risk of localised corrosion (e.g. pitting) in areas with low water velocity.

The results of the investigation show that the primary problem in the dewatering line is related to scale deposition. The heavy scale formation would probably stifle the internal corrosion of the dewatering line, but also reduces the cross sectional area of the pipe and increases turbulence, potentially resulting in a restricted flow. The pump technology of the dewatering pump (positive displacement) and the low demand variation means that the scale will not have an impact on the system performance until it reaches a critical level of build-up.

The biological analysis of collected water samples revealed the presence of microbial communities (aerobic and anaerobic), with slime formed and metal reducing bacteria. However, the population of microbial colonies are measured to be low to moderate and the risk of MIC is considered to be low.

The most likely locations of microbiological induced corrosion attack in dewatering line will be associated with the weld joint connection near the bends. The change to flow direction provides disruption to the water flow pattern, with the local turbulence trapping the solids present in the water, including bacteria food, around it.

The potential corrosion risk is discussed through a developed risk matrix, which provides a tool to focus attention on the most critical areas and pipe services.

7. ACKNOWLEDGMENTS

The author would like to thank GHD Pty Ltd for the support and permission in publishing this paper.

8. REFERENCES

[1] J.N. de la Vergne, "Hard Rock Miner's Handbook",Stantec Consulting, 2008.

[2] R. Winston Revie, "Uhlig's Corrosion Handbook", John Wiley & Sons Inc., 2008.

[3] V. Afshari," ACA Conference, Corrosion & Prevention 2012, Paper 155, Melbourne, Australia, November 2012

[4] M. Davies & P. J. B. Scott, “Oilfield Water Technology”, NACE, 2006.

9. AUTHOR DETAILS

Vahid is a Senior Materials and Corrosion Engineer in GHD Perth office. He is PhD in Metallurgy and Materials Science from University of Tehran. He has over 18 years of experiences in materials selection and corrosion engineering consideration at asset design stage and corrosion management throughout asset life cycle. He is experienced in materials selection, corrosion engineering, chemical treatment, specification of painting system, coating and lining, cathodic protection, failure analysis, welding metallurgy and NDT.

A NEW CLASS OF POLYMER LININGS – COMMONLY USED SYSTEMS BECOME OUTDATED IN FAVOUR OF INSTANT GRATIFICATION AND RELIABLE RESTORATION

Nick Subotsch
Peerless Industrial Systems, Melbourne

SUMMARY: Concrete and steel structures are an integral part of society. Today all water systems and the treatment facilities are constructed from concrete and steel, both susceptible to corrosion under many conditions.

Considerable effort goes into the design and construction of assets whether they are equipment of infrastructure, the concrete pipes, sewerage pumping stations and storage facilities. These represent considerable financial investment and protection of these assets and preserving the protection will provide maximum operational life.

There are various ways to protect concrete and steel from corrosion. This is a simple overview of the common systems in use and outlines the various products strengths and weaknesses, and introduces a new class characterizing all that the modern world seeks in relation to instant results, instant outcomes, performance and gratification.

While steel is generally the primary material that comes to mind when corrosion is mentioned, this paper places a significant emphasis on concrete in the water industry.

Keywords: Corrosion, Steel, Concrete, Rehabilitation, High-Build, Ultra.

1. INTRODUCTION

Sewer mains, man holes, wet wells and other associated assets are all referred to here without making any differentiation. Concrete has been the predominant construction material of choice. It is relatively inexpensive, easy to cast or shape allowing it to be precast or laid in-situ, and can be connected using a variety of techniques.
As a construction material, it is ideal but concrete can be degraded by corrosion from waste and effluent which often consists of a variety of chemicals or complex compounds. To protect the concrete from deterioration, admixtures, linings, coatings, and other fixes are applied as a membrane or barrier. Many protection systems have a brief moment in the spotlight and many others are attempts to create new trends or initiatives, questionable whether they have really amounted to very much.
This paper discusses a basic overview of the existing assets and rehabilitation, the fixes over the past 50 years with focus on the materials of the past 20 years, the reasons for limited success and risks to long term service. It will conclude with the innovative Ultra, a means of increasing reliability in providing the protection through simply revisiting history.

2. DEGRADATION OF CONCRETE

Specifically the area of focus here is corrosion from internal flows, and does not include the corrosion that can arise from non-service related characteristics like surrounding sulphate soils, or internally from AAR (Alkali Aggregate Reaction). The subject is very complex and is a science that will not be delved into beyond the basics to avoid misunderstanding or errors. Corrosion resulting from internal flows can be expected to arise from chlorides and nitrates in trade or industrial waste, however sulphur is the most common or predominant cause encountered. Sulphur from various organic and general waste materials carried in our sewerage is derived from the sulphates present, interacting with microbial materials. The compound is often loosely referred as Sulphurous Acid but many different events and compounds can be present.

$S_2^{=} + 2H^+ \rightarrow H_2S$ because under anaerobic (bacteria slime) conditions $SO_4^{=}$ can form $S^{=}$

The H_2S gas rises to the vacant space and in the presence of oxidizing bacteria, have the H_2S react with airborne O_2 to form H_2SO_4

$H_2S + 2O_2 \rightarrow 2\ H_2SO_4$

H_2SO_4 can then react with the Calcium Hydroxide (Hydrated lime) resulting in Calcium Sulphate, being the loss of concrete structure.

$H_2SO_4 + Ca(OH)_2 \rightarrow CaSO_4 + 2H_20$

H_2SO_4 is not in solution, but produced on the open damp areas of concrete and amount of acid produced (bacteria dependant) is in low concentrations and localized. The reactions take time; it is not a significantly large scale effect.

The following simplified diagram [2] represents the activity in a pipe but is equally relevant in other assets.

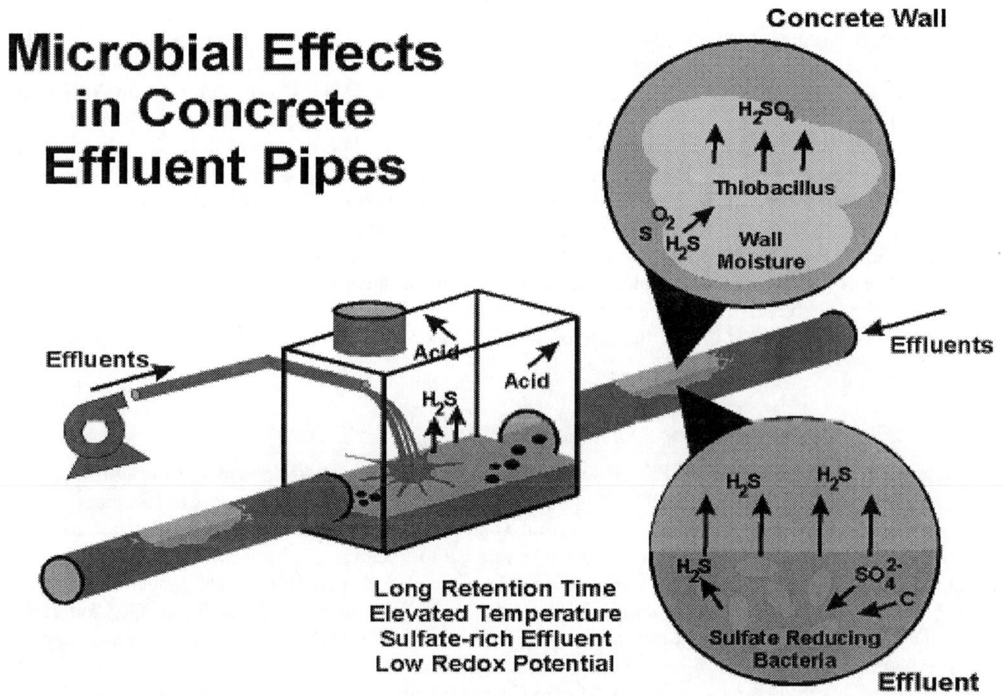

3. PROCTECTING CONCRETE FROM DEGRADATION BY CORROSION

Although various options exist in protecting concrete from corrosion, those relating to dosing the flows depend on consistent monitoring and accurate dosing with respect to the levels of corrosive compounds present. It is impractical unless batch treatment is possible and given the potential volume of sewerage involved, it becomes an expensive effort. Preference over time has always come back to coatings or linings to provide a barrier that separates the effluent from the substrate.
Even treating the sewerage at a designated WWTP relies on coatings and linings especially when enclosed facilities or odour control is practiced. The concrete is best protected by a barrier. Odour control is a relatively new endeavour and has hastened the introduction of coatings and linings into WWTP.

There are many materials capable of providing a barrier including thermoset and thermoplastic materials, rubbers, cement compounds, and a whole host of hybrids. Steels and their alloys are rarely used due to the inability to cover the range of chemicals in the exposure stream, installation can be difficult, and those that are suitable can be very expensive.

Typically, the materials more commonly used are the following:
1/ Epoxy – 2 pack coatings and linings, predating the 70s
2/ PVC – Polyvinylchloride preformed plastic, showing increased attention during the 80s and 90s
3/ PE & HDPE – Polyethylene and high density polyethylene, increased use after the 90s
4/ Cement, Polymer Cement or Aluminate – Lime and related reactive cement bases, intermittent use post 90s with increased attention during the past decade
5/ Polyurethane and Polyurea – 2 pack single pump and subsequently plural spray, the urethane gaining attention through the 80s and the urea attracting more attention as plural equipment becomes available after the 90s

It is accepted there are many more varieties used in different parts of the world, and many hybrids of those mentioned but the greatest proportion in very simple terms are the 5 types mentioned here and could comprise over 90% of the market today. Different regions favour different systems and to source definitive volume or usage is not possible.

4. FAILURES OF THE COMMON MATERIALS

There is reference to failures as the significant aspect of this paper because although each material has been found to be acceptable in providing protection in service, in sewerage rehabilitation it is the practical installation and ability for that material to sustain integrity over a broad range of conditions that limit life. Conditions that can cause premature failure is essentially the weak link and from which the comments focused.

4.1 Epoxy

These are by far the most common materials both historically - predating the 1970s - and in modern use of all the non-sacrificial types that demonstrate both versatility and resistance to the exposures. Originally simple 2 pack paints, the difficulty in covering severely degraded surfaces resulted in failures through pinholes, loose substrate and outgassing when overcoating voids and bony concrete. These failures originally called for fairing by epoxy mortar and later for economics to smoothing the substrate with sacrificial materials like cement render that altered the mode of failure to coating delamination from the render due to either contamination of the cement from the environment vapour during cure, the render itself letting go of the primary concrete due to poor adhesion, or the retained water in the render blowing off the lining. Subsequently, higher build epoxy was introduced which still has difficulty providing a proper barrier over extremely damaged concrete. Further developments in the 90s led to fast cure epoxy types using plural spray that allowed heavy builds over the semi cured epoxy allowing the possibility of very high thickness but these systems are invariably exothermic in nature and the heat would draw moisture from within the concrete to the surface compromising adhesion. Alternatively the system itself on cooling would be contracting and the stress relief results in significant tension with a risk of failure as mud cracking, or simple delamination from the existing substrate.

4.2 PVC

In many ways, these materials appeared ideal and it was probably the first real move to thermoplastics during the 80s. Shaped or preformed and able to be welded, they could be slipped into place and back filled if necessary to bed into the surrounding asset. Spiral winding and placement was also an innovation. Over time, degradation of PVC through plasticizer migration led to a more brittle material and the lack of adhesion to the substrate led to fractures that hide corrosive Hydrogen Sulphide ingress and degradation of base substrate. Thus inspection is never able to be properly carried out. The longer term damage can be extreme with collateral impact on the asset or adjacent structures. Another issue with PVC is the inability to terminate properly the liner so the incidence of Hydrogen Sulphide vapour able to find a way behind the liner occurs as condensation of the more corrosive acid results in a similar risk of asset or collateral damage.[1]
Some question arises of other potential mechanisms like PVC breaking down over time liberating Hydrochloric Acid and therefore itself being responsible for degradation of the asset, similar to that seen on steel roofing. Because that type of failure is centred on moist conditions and sunlight, there is very little documented technical proof pertaining to this type of failure but it remains for the purposes of this paper technically feasible.

4.3 PE & HDPE

Polyethylene followed on from the PVC gathering momentum post 1990s, as a material that overcame greatly the longer term degradation and embrittlement issues of PVC. Over a brief period of time some failures caused by the PE creeping with thermal changes in the environment resulted in fractures of the welds, or large areas letting loose till fatigue led to ingress of Hydrogen Sulphide or sewerage inflow. These failures were quickly identified and the introduction of HDPE followed. Many of these liners included anchoring points or mushroom shaped clips that were welded to the back of the HDPE and encapsulated by the back fill mortar after installation, providing greater stability across large areas. Instances of weld failure were still a potential risk, and ingress at the termination of the HDPE however the advent of fabric backed HDPE which was welded to the existing liner allowed the fabric backed section of the liner to be terminated using epoxy adhesives and fibreglass.[5]

There are now many types and variations of this type of liner being installed. In terms of resistance to the corrosive conditions in sewerage environments, they are marvellous. History is yet to decide the longer term suitability of the HDPE liners with many maintenance teams expressing the same misgivings as for any preformed liner, how do you check over time the ingress of inflow behind the liner and be able to address any seepage before substrate degradation and collateral damage on adjacent structures - roadways and buildings - occurs. If it cannot be monitored or assessed, the old proverb "what you cannot measure you cannot manage" applies.
In discussion with many asset owners today, concern over instances of failed HDPE "Systems" are becoming more common with ground movement or asset upgrade and the risks or inability to manage and maintain following several years' service is proving to be a concern.

4.4 Cements, Polymer Cements or Aluminates

Although these materials do not possess the chemical resistance to provide long term protection, they are finding renewed favour in rehabilitation work for the reasons that are not altogether clear technically. It is possibly the trends of change that come full circle. Modern materials are giving way to the traditional cements and their hybrids referred to as polymer cements and aluminate. The primary aim of these materials is to rehabilitate the assets prior to loss of structural integrity irrespective of their sacrificial to chemical nature.
The essential characteristic is the alkalinity as a characteristic of the binder. Therefore at some point over time in acidic service these materials are compromised and with the reduction of alkalinity in the binder decay. This is not necessarily unacceptable, the technical treatise of steel pipework in refineries where the corrosion rate is measured and time to replacement planned. In a modern refinery these practices are as much a process requirement as a safety issue however it is questionable if these materials are ever applied to assets with the same monitoring and management protocols. In the past decade these materials have come to include hybrid epoxy that is water dispersible. Cost and the suggestion of epoxy base does attract attention.[3]
The economics and operational parameters limit viability in the utilities domain. Continual change in the environment and maintenance of live assets around a bustling community has limitations.

4.5 Polyurethane and Polyurea

It could be said that polyurethane came hot on the heels of the epoxy, to overcome substrate movement, cracking and provide some flexibility. The 80s saw a strong drive from the industry to incorporate them across the water industry. The low tolerance to moisture and need for primers led to many installation problems and whilst polyurethane is still in use, the industry has demonstrated some reluctance to use them again. It wasn't long before the polyol cured polyurethane gave way to the amine cured polyurea – Isocyanate reinvented. This was greatly supported by plural metering and spray systems that accommodated the short pot life characteristic of amine cure isocyanate.
The requirement for primer still exist depending on who you talk to, the tolerance to moisture and chemical resistance improves with the significant advantage being the time to cure before return to service.
The use of very expensive plural equipment to mix and apply the polyurea is still a hurdle today and one that excludes a lot of contractors and increases the complexity of the work program. Together with the high exotherm experienced with cure, and the bubbling or blistering associated with thermal expansion or outgassing of air in blow holes and bony concrete, the secondary issue of cure shrinkage leads to many instances of delamination.
Today there are further complications with the hybrids that exist notwithstanding the confusion in nomenclature. The market continues to introduce new concepts.[9]

It is clear that each rehabilitation material has its own advantages and disadvantages. The period of installation history for many is short in years so monitoring and reporting on the success of each is in the hands of individual choice and the technical debates influenced by marketing and financial concerns run wild.

5. INNOVATIVE NEW MATERIALS – BEST PRACTICE

The trends of change on occasion come full circle and expect trends will continue to be re-introduced time and time again cyclic yes, but at each cycle something more comes from the older material being reintroduced.
The times when epoxy was seen as the ideal material are always present with a select section of the market and often those that witnessed firsthand the success. But over time something new creeps in with more value and advantage than ever before but how do we recognize when it happens? Long life and the ability to monitor the integrity of any asset is a basic necessity, just like a medical check-up.
The advent of renewed developments including plural heated systems focused on a better strategy. It provided a springboard to using technology that was fundamentally sound but pushed the edge of physical installation.
Adapt fundamental installation practices to install linings with enhanced polymer, take traditionally sound or proven materials that are known to perform, and focus on how to place them easily and quickly.

Take known performance, make it easy to apply without expensive equipment, allow it to be applied at 0.5mm or 20mm without sagging or curtaining, and be able to touch-up with brush, or trowel around edges and corners within the hour and the advantage may not immediately be clear but it is novel. Innovative thinking applied to epoxy, certainly if using conventional spray equipment. It is far more versatile than any of the conventional epoxy systems and here referred to as Ultra.

The main considerations have always been:

1/ Application to sound concrete substrate
2/ Application fast enough to eliminate the effects of contaminants within the liner system
3/ Application thick enough to provide adequate coverage that ensures a barrier free of imperfections
4/ A material that has low exotherm so no tension or stresses are present
5/ A material that can be quickly patched if ever required without total removal and replacement

Deteriorated Concrete

One application of Ultra, spray and trowel touch-up

Addressing each of these points more specifically.

1/ Modern techniques of concrete preparation allow significant improvements in speed and thoroughness. The historical research and reporting on concrete preparation allows many available options providing benefit moving forward. Expansive gypsum or ettringite or other anomalies in the concrete can be addressed to support coatings, and aggressive media can be added to high pressure water preparing the substrate, and to quickly remove sulphurous build-up or simply leave the concrete clean. Fans and dehumidification equipment can be employed to rapid dry surfaces or captive blasting to keep surfaces dry.

2/ Cement type liners can be placed in a very short space of time using modern equipment, and so too can coatings limiting intercoat exposure to the environment. The use of more conventional spray equipment for coatings makes it more economically viable over heated lines and plural spray where access or location is challenged. Added complexity can compromise reliability, increase cost, and reduce or limit the choice of contractor. Reliance on specialist equipment to undertake work is considered more a matter of economics until the issue of equipment failure or availability is introduced. Installing preformed plastic liners can be time consuming but the continuity of the liner as it is installed limits environmental influences and so they are suitable especially as piping.

3/ All materials are able to be applied at thickness to eliminate risk in continuity of the lining - i.e.: PVC, PE, HDPE, Epoxy and their mortars or those loosely based on cement type mortars. Epoxy and Polyurethanes often require recoating and it is the delay between recoats that introduced adhesion problems, contamination and potential failure. More recent epoxy types can quite easily be applied at high thickness overcoming some of the shortcomings of cements appreciating that the cements are sacrificial to the corrosion.

4/ The latest and newest epoxy, referred to as Ultra uses the same background polymer systems as in the past, with the added feature or ability to be able to be applied at extremely high thickness to provide coverage over bad surfaces or bony concrete.
The proven performance experienced as a barrier to resist the corrosion over the past 40+ years remains so the record over this same 40 years is well placed to predicting asset reliability and life. Avoiding the use of plural heated systems alleviates exotherm, a critical factor in high thickness application and the resulting stress from contraction and tension.

5/ In time, any lining must undergo some form of damage or incident leading to a compromise of its continuity. Be it a retrofitted ladder, screen, or other event. The cement type liners and epoxy based have demonstrated that simple patching or repair can be carried out without a great deal of effort, skill, or equipment that makes such an exercise within the scope of the general contractor. Welding of plastics in repair is possible but the skills of the workforce require significant design and training. The level of preparation is not always simple to get good adhesion or reduce weld fatigue, notwithstanding the question of termination of the plastic.

	Ease of Application	Effects of Corrosive Conditions	High Thickness	Low Internal Stress	Ability to Patch
Traditional Epoxy	X	X		X	X
PVC		X	X	X	
PE & HDPE		X	X	X	
Cement based Mortars	X		X	X	X
Polyurethane/Polyurea[9]		X	X		
Epoxy Type Ultra	X	X	X	X	X

Reviewing the table, the new Ultra type demonstrates the versatility across the important criteria listed. It is a subjective review but each point has been considered by reference to industry (reference section 7, Acknowledgements) and papers. The cement based mortars deserve special mentions because of ease of inspection and maintenance but the new type of Ultra has a significant advantage. If the long term resistance to sewerage environment is accepted as equal to traditional epoxy, the Ultra is certainly a material that deserves attention.

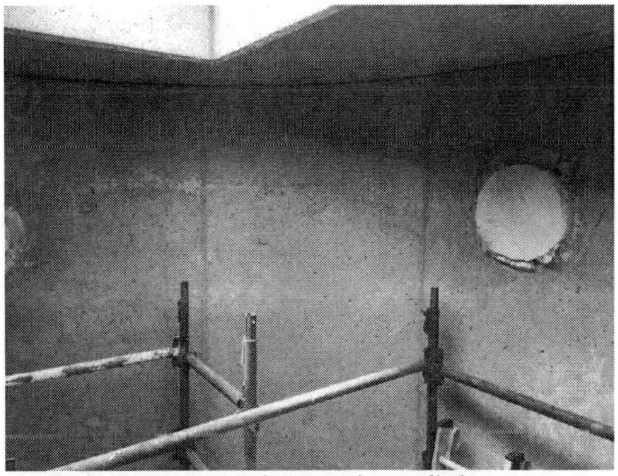

Typical concrete state to have coating applied

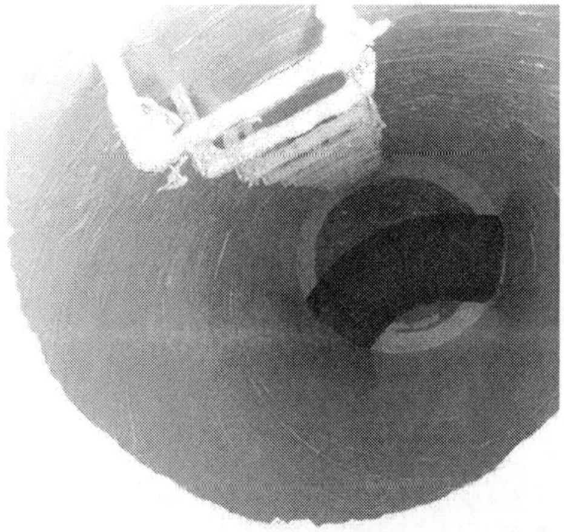

Adjacent structure coated and checked for holidays

6. CONCLUSIONS

There are many materials as either hybrid types of the aforementioned, or consisting of entirely different technology that have been used in the rehabilitation of sewers. These have, from time to time included methylmethacrylates, polyesters and vinylesters, and fibreglass laminates. Unfortunately they have not taken to the industry well and suffer from either confined space issues pertaining to flammable solvents or require primers or are water sensitive. Failures and difficulty with installation make them unsuitable for use across a broad range of practical considerations.

We still see today the majority of discussion and proposals based around the same few materials that have been in use for decades, and will continue to be used for decades to come. None have the history and respect that epoxy polymers have earned.

The motivation to introduce new materials has always been based on the limitations of existing systems but like solvent based epoxy gave way to solvent free, and the plural made the application of high build repairs practical, the Ultra is a class that has never been considered possible.

The newer preformed plastics will continue to be offered to the market and appear to have a place during new construction however ongoing issues pertaining to termination and how to arrest any seepage behind the liner is still a concern, and the ability to identify this before major damage occurs is not reported.

The Ultra type technology has none of the shortcomings of earlier systems and has the demonstrated chemical resistance to be a firm advantage, not relying on the promise of 50 years maintenance free life – yet quite possibly achievable. That is assuming the structural integrity of the asset itself is not compromised. Given the fact it lends itself to be able to be repaired in the field with relative ease leaves it the most modern practical consideration available.

7. ACKNOWLEDGMENTS

Daniel Leeming, PIS, WA Aust
Graham Thomson, Barwon Water Vic, Aust,
Daniel McElligott, Mac Coatings Geelong Victoria
Roy Orr, Qld Aust
Michael Arnott, McElligott & Partners, Vic Aust

8. REFERENCES

[1], G Thomson (Barwon Water). "Corrosion & Rehabilitation of Concrete Access Chambers (WIOA 2000)
[2], M Tullman Corrosion Club (www.corrosion-club.com/ 2010)
[3], Lafarge Fondu International. "Lafarge Studies On Corrosion Principles" (2004).
[4], Baker. C.A (1974) "Chemical Corrosion of Concrete". Australian Corrosion Association Conference, Hobart. 1974.
[5], R Orr, AGRU & STEULER, Technical Officer & Advisor, (2000 – 2013)
[6], Shiwei William Guan, Ph.D., Madison Chemical Industries Inc.(AUCSA May 2001), "Corrosion Protection by Coatings for Water and Wastewater Pipelines
[7], Handbook of Composites, S.T.Peters, 1998
[8], Handbook of Epoxy Resins, Henry Lee; Kris Neville, 1967
[9], Kyle Flanagan, Freedom Chemical Corporation, "The History and Industry Adoption of Polyurea Coating Systems", www.corrosionpedia.com December 2015

9. AUTHOR DETAILS

Nick Subotsch began career at Peerless in 1978, in the laboratory and undertook studies at University at night. He progressed his work in research and development on plastics, hot melts, detergent and emulsion chemistry. In 1987 he assumed responsibility for the Peerless Polymer division, focused on epoxy resin technology. With various responsibilities over the years, and still involved in the laboratory today, he spends a lot of time in the field bringing together test tube and trowel, or beaker and spray for realistic solutions. Recent accomplishments include the development of a new internationally patented polymer fabric called KIS, and the introduction of unique new Ultra high-build coating system for simple installation on rough or badly corroded surfaces in time critical applications. Nick holds membership of Australasian Corrosion Association and was Chair for the production of the associations 50 years Jubilee Corrosion Video. He also holds membership in the SSPC, NACE and the SSPC Concrete Coatings Inspector L2, and NACE 3.

PRACTICAL EXAMPLES FOR THE MANAGEMENT OF ASSETS WITH CORROSION

T. Pape, F. Blin, S. Furman
AECOM Australia

SUMMARY: There is increasing pressure on infrastructure owners and manager to ensure their asset base is safe and reliable, and maintains acceptable levels of service for the duration of the expected life of the asset. This requires balance with often limited budgets, requiring optimized approaches to be developed and implemented. This balance becomes critical when a requirement to extend the life of an asset must be actioned. With the release of ISO 55000 'Asset management - Overview, principles and terminology' [1], and other supporting guidelines, there is improved guidance on the strategic whole-of-life management of an asset base, and there are a number of techniques that can be adopted to achieve this goal. When an asset is affected by corrosion, this can threaten its longevity and serviceability, with consequences for the functionality of the individual asset and the larger network which it supports. This paper provides practical examples where such issues were identified and discusses the various asset management and corrosion-control /management strategies that were developed and implemented to mitigate the corrosion risks. Examples have been provided for an ageing asset and for a young asset. The requirements for developing and applying appropriate asset management strategies that include corrosion management are also discussed.

Keywords: asset management, corrosion, deterioration, optimisation, case studies.

1. INTRODUCTION

A diverse range of infrastructure currently exists in Australia, covering sectors such as transport, energy, water and communications. Figure 1 highlights the increasing level of infrastructure spending over the last few decades [2]. This increase ultimately adds to the pool of existing aging assets that require maintenance and management to ensure its long-term functionality and safety over its serviceable life. Subsequently, increasing funds are required. In the 2015-16 budget, the Australian Federal government allocated over $6.8 billion to the States and Territories for infrastructure expenditure, which included both capital and operational costs [3]. Of this amount, a significant portion is directed towards asset maintenance. As an example, annual expenditure on road maintenance, renewal and minor improvements in NSW is approximately $1.46b, with $280m spent in the Sydney region alone [3]. Another example is the observance of a $1.8b backlog for water, sewerage and drainage assets across 152 Councils in NSW, indicative of asset maintenance underspend in regional areas [4]. Increasing pressure on asset owners is likely to occur in this scenario, as the number of new assets increases, combined with existing aging infrastructure being required to exceed original design lives [4].

A variety of issues initiate such maintenance requirements, such as fatigue, redundancy of the asset, unplanned damage, normal wear-and-tear from use and condition deterioration. For the latter issue, corrosion is a significant contributor. In the United States it is estimated to cost in excess of $2.5 trillion per annum [5]. In Australia, it is estimated to be approximately $32 billion per annum, with $8 billion attributed to avoidable corrosion damage [6]. It is possible, however, that if appropriate asset management strategies are implemented, it is possible to restore an asset to its original condition and maintain its functionality for its remaining service life and potentially beyond.

This paper first presents an overview of asset management principles and strategies and corrosion management approaches within it. It also discusses practical examples on the management of infrastructure that has been affected by corrosion or designed and constructed to minimise corrosion-related risks. Drawing on methodologies and strategic principles outlined in the recently updated in the international standard ISO 55000 series on 'Asset management' [1], the asset management issues for each case study is explored. Resulting action plans and strategies for the restoration and management of the asset in the long term are discussed, and key issues are noted to provide guidance and overarching principles for those seeking to manage assets that have undergone corrosion.

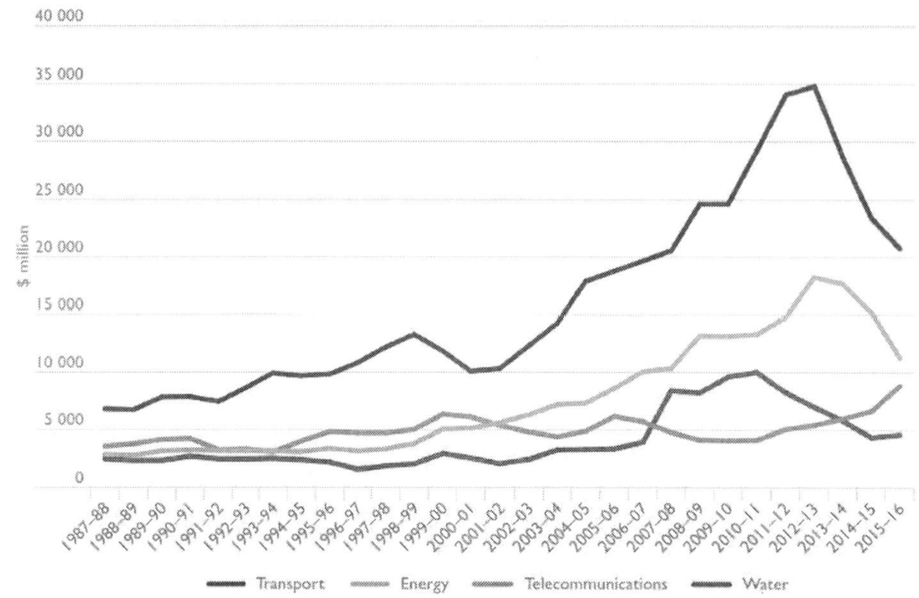

Figure 1. Infrastructures construction activity in Australia since 1987 [2]

2. OVERVIEW OF ASSET MANAGEMENT PROCESSES

The concept of asset management is not new and has been in operation either formally or informally for several decades. It has, however, evolved from a focus on asset maintenance to one that embraces broader business objectives that seeks to proactively optimise the use its assets to meet demand requirement and customer expectations [1, 7].

By definition [1], asset management is the coordinated activity of an organisation to realise value from their assets, including:

- cost savings (to capital and operational expenditure)
- reduced risk of asset failure on service provided
- improved performance and tailored levels of service

With a focus on long term objectives and purpose, investment decisions are made to build in cost effective maintenance into the initial design of the asset.

A key outcome for asset management is the cost effective & efficient delivery of the desired levels of service to customers & community.

It is important to understand what function (or Level of Service) an asset needs to achieve over its lifecycle, with those considerations to be taken into account during its planning, design, construction, operation and decommissioning. This includes consideration towards the impact of an asset failure on the asset itself and on the broader network it supports. Planning for an asset failure enables the development and implementation of appropriate risk management strategies to safeguard the asset and the network it supports (such as a road network or pipeline infrastructure) to avoid interruptions and consequences.

Failure modes can include:

- Asset deterioration (expected and predictable asset / material degradation over time)
- Growth in demand to or that exceeds original asset design capability (including unexpected catastrophic events, changed regulatory requirements, etc.)
- Change in use that exceeds original design capability (changing market conditions that require a shift in focus for a business and the use of its' assets)
- Economic failure (through obsolescence, more cost effective solutions for delivery of the same service, new disruptive technologies, etc.) and/or
- Human error

ISO 13823 [8] and ISO 15686 [9] relate the performance of an asset to an engineering evaluation of the asset, whereas a business asset management approach has been to link those technical parameters to levels of service and investment decisions. As the interaction between an asset and its environment throughout its life can have a significant impact on its performance and therefore the levels of service it can provide; the concept of durability management has received a growing focus across the world [10].

ISO 13823 defines durability as "the capability of a structure or any component to satisfy, with planned maintenance, the design performance requirements over a specified period of time under the influence of the environmental actions, or as a result of a self-ageing process". Therefore, durability is linked to the value or return on asset that will be provided from a whole-of-life perspective, i.e. the lifecycle cost for an asset can often be closely related to its durability characteristics [9].

In that context, adopting a 'do more for less' approach for pro-active, planned and reactive maintenance activities may improve budget optimisation by allocating prioritised funding towards critical assets, as well as addressing the remaining non-critical assets. This optimisation will lead to an increased focus on selectively monitoring the condition of assets, improved understanding of their modes of failure and developing tools and processes to better assess current and predicted future performance respectively. In return, this will enable better understanding on how to improve the management of asset information, how to identify and use data that is really needed to enable decision making processes.

The challenge and opportunity of the current business climate is therefore to involve organisational governance, knowledge, processes, tools and resources synergistically to find more creative ways to best balance the performance of their services with their costs and risks. Succeeding in achieving this and being able to demonstrate it would set organisations and wider societies on a more sustainable course for the future.

3. DEVELOPMENT OF ASSET AND CORROSION MANAGEMENT STRATEGIES

At the earliest stage (typically planning / business case) of the asset creation process the definition of the services that an asset project is intended to provide is key in determining the need for a particular portfolio of assets. This is because, in alignment with ISO 55000 and the International Infrastructure Management Manual (IIMM) [11] assets only exist to deliver a required service for an acceptable level of risk and lifecycle costs. This is a very important phase as several factors can impact the subsequent phases of the assets' life such as for instance:

- The ground conditions that the asset will be exposed to and the durability risks that will have to be managed accordingly,
- Potential changes to the exposure conditions resulting from operation of the asset. For example production of brine in a desalination plant, or elevation of acidic gases due to exhaust emissions in road traffic tunnels.
- The type of contract (design and construct, public private partnership etc.) may lead to more or less reflections and/or tensions between the short term (construction) and long term (operations) considerations.
- The available funding / budget may constrain the ability to build all the tunnels' components with long design life resulting in likely higher operational budgets (OPEX) to deliver on the targeted CAPEX.

While at this stage the assets are only conceptually designed their base form and functionality are defined to support the preparation of cost estimates for the business case that ultimately drives the decision to proceed to asset creation. At this early stage of an assets life, the main modes of potential asset failures (including corrosion) are typically considered as part of the risk assessment process. This allows such risks to be taken into account, typically by the engineering team, and reflected into early cost estimates for both asset creation and operation.

In the next stage of an assets' lifecycle the assets is designed and constructed to achieve specified criteria.

- Functionality as defined with the asset owner and users including key requirements, serviceability and availability,
- Technical quality comprising safety, security, social, environmental, durability, sustainability, aesthetics, constructability, inspectability, maintainability, etc.

There is a strong technical and engineering focus at this stage to deliver assets that comply with the project scope requirements and this presents two major risks as far as the asset management strategy is concerned:

- Underspending / reducing the CAPEX to the point that it can significantly impact the OPEX over the asset's life. An example would be the underestimation of the corrosion of galvanised steel coating caused by accumulation of sulphur dioxide levels from vehicle exhaust emissions. This would lead to the implementation of a coating / repair program earlier than the original design life. This is itself may or may not have a great impact on the services delivered by the tunnels depending on the criticality of the components and the associated risk of failure on operations (e.g. the access requirements for the safe and effective coating of the components could lead to unplanned shutdowns).

- Overspending ('gold plating') on the CAPEX to minimise risks of asset failure below what could be an acceptable manageable level. This can be achieved through over-engineering solutions where the worse consequences of asset failure drives the material selection process for instance as opposed to a risk based approach (that takes into account both the likelihood and consequence of failure and wherever possible in a probabilistic rather than deterministic approach). An example would be the use of hot dip galvanised coated reinforcements, together with high quality concrete mixes that contain corrosion inhibitors and/or permeability reducing admixture together a coating of the concrete surface.

There could also be a perverse situation whereby in an attempt to mitigate durability risks a design & construction team could increase the materials requirements and yet achieve an undesired outcome. An example would be the increase in cementitious content in reinforced concrete mixes beyond an acceptable range on the incorrect assumption that more cement equates greater durability (through greater strength and lower permeability presumably). However, high quantity of cements can render the mix very difficult to handle as well as cause cracks from the heat of hydration and temperature differentials between the core of the concrete and the its outside perimeter in contact with the atmosphere. The cracking itself could weaken the asset from a structural and durability perspective.

The risks listed above can be mitigated by setting up an asset management framework during the design and construction phase that brings asset owner, designer, constructor and the Operations & Maintenance (O&M) teams together in the decision making process. By collectively identifying the short and long term risks and, where possible, estimating the costs of asset failure and required intervention the acceptable balance of CAPEX vs. OPEX can be achieved. An example of such risk assessment approach would include estimating the consequence cost of potential failures and multiplying it by the probability of failure over a given period of time to determine whether this can be afforded or whether the risk is too great. This risk cost can include direct costs such as that of access including traffic management or shutdown as necessary and the repairs themselves as well as indirect costs (e.g. loss of revenue due to tunnel shutdown, reputational impact, and customer dissatisfaction). If the risk is deemed to high (the risk cost too expensive) then mitigation strategies can involve for instance increasing the durability at the onset or designing an asset monitoring that would for instance involve testing to assess the rate of corrosion over time.

As many corrosion professionals have experienced it can be challenging to estimate corrosion rates when the original steel thickness is not accurately known. Moreover, it can also be difficult at times to comment on whether the pace of materials degradation is progressing as designed or whether it is slower (the service life being therefore likely to exceed the design life) or faster (in which case more maintenance may be required than originally allowed to achieve the required design life or the timing for renewal may need to be brought forward). This emphasises the importance of data management throughout the lifecycle of an asset. More and more data is considered by organisations as an asset and managed accordingly. This includes being strategic about data collection (including asset condition assessments), storage, analysis and communication as it underpins decision making and the ability to best balance service levels, risk and lifecycle costs. In that context asset creation projects that invest early on in the development of an as-designed asset register, that is subsequently updated to become an as-built register, provide future asset managers with an invaluable record and baseline to make informed decisions.

4. PROJECT EXAMPLES

To illustrate the application of asset management principles for assets affected by or predicted to be affected by corrosion, two practical examples are presented. The first case study concerns an aging asset where issues of corrosion were identified and found to be diminishing the functionality of the individual asset and the larger network that it supports. The second case study concerns an asset that is early in its design life, though it is expected that various sub-assets are likely to corrode during the overall asset life. The case studies deal with a variety of infrastructure types, and various asset management and corrosion-control/management strategies that were developed and implemented to mitigate the corrosion the assumed or actual risks.

4.1 Ageing Asset - Rooney's Bridge

4.1.1 Background

Rooney's Bridge is located in Townsville and is a vital link within the city's transport network. Constructed in 1950, the bridge facilitates local access from the CBD to southern residential and commercial areas, and leads onto the Bruce Highway. It is a steel-concrete composite bridge which has 16 spans of 12.2 metres length, and a total length of 198 metres. The deck has an overall width of 8 metres, with a width of 7.4 metres between kerbs, and comprises to two traffic lanes with an asphalt wearing surface. The superstructure comprises five simply supported steel I-girders with steel cross-girders at quarter points, and a continuous composite reinforced concrete (RC) deck. The piers and abutments comprise steel headstocks supported by four square precast RC driven piles. It underwent major refurbishment in the early 2000's, carrying out concrete repairs and pile encasements, repainting of girders, replacement of corroded bearings, and alteration of some of the existing steel sliding bearings to a fixed state to facilitate improved load distribution across the bridge for braking loads.

Figure 23. Overview of Rooney's Bridge

Recent inspections have highlighted that the structure is in a deteriorated state (Condition Rating 3 / Poor, 2016), with severe cracking, spalling and reinforcement corrosion observed in the piles, and protective coating failure and significant corrosion in most components, including the recently replaced bearings. Taking into consideration the current condition of the bridge and potential future traffic requirements, the implications that condition will have on its long and short term functionality and load capacity needs to be understood to inform future management strategies. TMR sought to understand the implications of its condition and explore the risks and strategies to manage this structure in the short and long term, with the intention to ensure no reduction in structural capacity over its remaining required life. Two separate preservation strategies were developed to facilitate this understanding to appropriately manage Rooney's Bridge over its intended remaining life. These were:

- **short term strategy**: five year plan to maintain the bridge at a safe level of service until it is replaced as part of proposed upgrade works
- **long term strategy**: twenty year plan to restore and maintain the bridge to a safe level of service and integrate it into the proposed upgrade works, with the intent that it be replaced after 20 years.

4.1.2 Development of Strategies

A desk top review of existing inspection data was completed, from which an approach to current maintenance requirements, along with deterioration predictions, was developed. Deterioration predictions were based on historic inspection findings and condition ratings.

A streamlined structural assessment of the bridge was undertaken to determine the critical elements and inform the strategic management of the structure. The assessment concluded that the girders had sufficient capacity in bending and in shear for a T44-76 design vehicle, which included some cross-sectional area losses due to corrosion in critical locations (i.e. mid-span). Piles were found to be sufficient for vertical loads, however pile bending results were more sensitive to corrosion losses depending on assumptions (such as pile lengths, load distribution and corrosion location).

For the short-term 5-year plan, considerations were more on minimising capital expenditure and seeking to maintain its serviceability until the end of the five years. Based on the rate of deterioration on key structural components (such as the piles, headstocks, and girders), it was noted that a minimal approach of monitoring and access management would likely be sufficient. The monitoring regime included an increased frequency of inspection, with particular focus on critical components and measurement of defects to identify areas of structural distress. The development of a structure management plan was also recommended, to capture key issues and risks identified such as structure critical components, key defects and permit/heavy vehicle access management. Recommended intervention levels were provided in relation to corrosion section losses to enable TMR to assess where a failure in the level of service was imminent.

For the longer 20-year strategy, different considerations were required, such as projected deterioration, future use of the bridge and what maintenance/intervention activities would be required. Based on the estimated projected deterioration, it was evident that significant maintenance would be required at some stage. It was assumed that the first five years would follow the same intensive monitoring program recommended for the short-term strategy, which would enable any permits and approvals to be developed prior to designing repair or maintenance works to ensure its prolonged service. Several maintenance interventions were considered. For the piles, this involved cathodic protection, full encasements, targeted concrete patch repairs and crack injection. For the steelwork, it involved complete blast/repaint all components, a component-specific program of blast/repaint or targeting deteriorated areas only. A multi-criteria analysis (MCA) identified the optimal maintenance solution of concrete repairs (with some crack injection of the encasements) and targeted repaint of deteriorated steelwork on the girders, bearings and headstocks. Estimated life-cycle costs were determined for each strategy, and took into account operational costs, such as the frequency of inspections and routine replacement of items such as joints.

Relatively low costs of approximately $126,000 were estimated for the short-term strategy, based minimal maintenance and operational costs (such as increased frequency of inspections) and with a greater proportion of costs earlier in the strategy (**Error! Reference source not found.**~~Figure 4~~). Estimated costs for the longer-term strategy were developed based on the preferred maintenance solutions identified from the MCA, with overall costs in the order of $2.5 m (**Error! Reference source not found.**~~Figure 5~~). The minimalistic maintenance approach of the short-term strategy was proposed for the first 5 years of this program, followed by the first of two significant interventions to reinstate critical condition-poor components to a serviceable state. Based on deterioration modelling, a second intervention was recommended to allow for repairs on components not previously repaired.

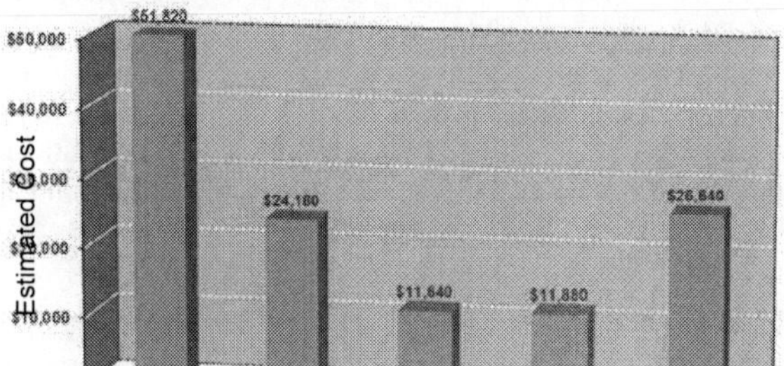

Figure 34 Estimated costs for short-term preservation strategy

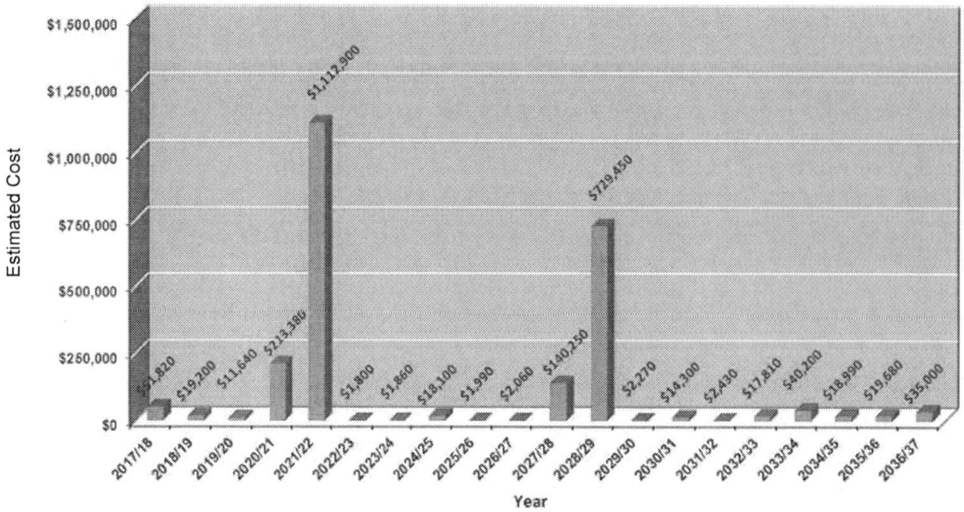

Figure 45 Estimated costs for long-term preservation strategy (preferred repair options)

4.1.3 Other Observations

During the course of the review, a number of key pieces of information were identified and influenced the provision of the advice given. These are discussed below.

Where is the corrosion?

Several areas across the new and refurbished bearings, the reinforced concrete piles and the headstocks were identified with significant corrosion section losses. Notably, the girders were in relatively good condition with minimal losses despite the onset of corrosion being evident. The corrosion in the sliding bearings was significant and had appeared to cause some of the bearings to seize.After reviewing the corrosion impact on the overall performance of the bridge, the fixity of the bearings was noted to be more beneficial in cateringfor braking loads. For longer term management of the bridge, a monitoring regime was recommended for the bearings to review the impacts of the corrosion on the bridge's ab ility to accommodate thermal movements.

For the headstocks, the location of the corrosion had occurred towards the ends of the member and were not part of the direct load path from the supertructure. And whilst additional section losses were noted in the headstock stiffeners, these are elements that are only critical for jacking up the bridge during bearing replacement activities – something that is not likely to occur now the majority of the bearings have been fixed retrospectively.

In contrast, significant corrosion of the reinforcing bars at the pile at the welded interface connecting to the headstock support required review due to the amount and location of the corrosion. This connection point represents a key mechanism in the transfer of loads from the deck to the piles, particularly for bending, which the piles were assessed to be sensitive. However, it was observed that not all piles in the same pier were in this condition. Due to the degree of redundancy at this location (i.e. the alternative load paths offered by the number of piles), the structural significance of the observed defects was more accurately considered in the assessment of the bridge and in the recommendation for future inspection and maintenance requirements.

(a) existing piles

(b) bearings and adjacent steelwork

Figure 56. Examples of deterioration (2016)

Finally, most severe cases of corrosion were found to coincide with water leakages either from the water main service attached to the bridge or via failed bridge joints, allowing water to pass through from the deck to the substructure below. Repairing such components can reduce the risk of ongoing corrosion in these locations due to the elimination of water.Such remedial measures are an effective and relatively inexpensive strategy to implement in a preservation plan to ensuring ongoing service of the bridge. As such, these options were recommended for the long-term strategy provided.

How much corrosion?

To explore the impact of corrosion on the bridge, a structural review was conducted on the piles and the girders. A number of section loss scenarios were trialled in structure-critical areas, with each trial showing progressively more severe corrosion losses. It was found that for the girders almost 50% section loss would be required in the bottom flange to trigger an unsatisfactory assessment ratio for both shear and bending. For the piles, concrete section losses at each corner was considered, however in compression it made minimal difference to the overall compression capacity of the pile, which is not unexpected [12]. Corrosion of the reinforcement also was found to be non-critical, however as previously discussed the location of corrosion at the welded headstock interface was found to compromise the integrity of the pile.

The impact of corrosion was also considered from a failure mechanisms perspective. The amount of corrosion can alter the ductility of steel, therefore promoting a brittle, sudden failure [13]. However due to the number of piles and steel girders in the structural system, the presence of cross-girders (facilitating the even distribution of load), the distribution of corrosion across the various components, and the composite concrete-steel superstructure, a reasonable degree of redundancy is noted.

Condition of the Girders:

The most recent inspection report noted a 10 mm sag in the girders of all spans, which were subsequently marked to be in a severe condition. It was originally thought that these girders had been damaged by the previous heavy vehicle regime permitted to access the bridge. However upon closer inspection of the drawings, it was noted that this sag in the girders was identified in refurbishment drawings, flagging an intentional detail.

Additional discrepancies were observed when documenting the definition and severity of defects recorded in the previous inspection reports, with some defects recorded as severe when they were by definition not, and vice-versa. Observations from AECOM's site inspection confirm these discrepancies. The variability in the historic condition ratings reinforces the sometimes subjective nature of inspections.

This observation highlights the requirement for care to be taken when reporting on component condition and the importance of critical review of data before it is used.

4.1.4 Key Messages

In summary, various options and strategies were developed to provide appropriate asset management processes were implemented in accordance with the required level of service for the required duration. This study highlights the considerations required and options available for the management of an asset with significant corrosion.

Key learnings from the assessment included:

- Suitable strategies can be developed to enable ongoing serviceability and life-extension of the asset. Key to this is understanding the location and severity of the corrosion in relation to the structural and functional operation of individual components and the bridge as a whole, which can therefore influence the development of appropriate life-cycle maintenance requirements.
- Understanding the requirements of your level of service and service-life will enable the development of tailor-made asset management strategies.
- Simple maintenance activities such as repairing leaking water fixtures and faulty bridge joints can lead to asset life extension and optimised life cycle costs.

4.2 Young Asset - Seawater Reverse Osmosis Desalination Plants

4.2.1 Background

Over the past decade a number of large seawater reverse osmosis (SWRO) desalination plants have been constructed and are now in operation near major Australian cities., At the time it was built in 2010 the Victorian desalination plant (VDP), south-east of Melbourne, was the largest SWRO desalination plant in the world [14]. Under full operation this plant has the capacity to provide 150 GL of treated water per annum. SWRO desalination plants have a variety of corrosive environments including atmospheric, immersed and buried environments. The assets that were constructed for this project included: seawater inlet structure, filtration system,

Figure 67. Victorian Desalination Plant [Ref: The Age, 9/10/2012]

dual pass reverse osmosis trains, remineralisation system, a water storage facility, a transfer pipeline, power supply system and a seawater outlet structure [14].

4.2.2 Development of Strategies

During the concept design phase of a desalination plant, it is commonly recognised that corrosion may impact the performance and functionality of the assets unless appropriate measures are embedded into the design and an appropriate asset management strategy developed well in advance of completion of the detailed design phase of the project. In general, the approach to managing assets in desalination plants depends on whether the asset is physically involved with the production of desalinated water or not. For production related assets, for example pipelines and process vessels, inspection and maintenance needs to co-ordinate with the production cycles and the operational availability requirements of the individual assets. Then consideration is given to accessibility and deterioration due to environmental exposure or due to physical deterioration associated with hours of operation. For non-production related assets, for example buildings, inspection and maintenance activities are more dependent on accessibility and predicted deterioration in the durability of the asset. The construction materials used in desalination plants include steel, stainless steel, aluminium, concrete and glass reinforced polymers (GRP). These all react differently to the various exposure conditions, that is, they all have different inherent durability.

The approach taken at the VDP was to optimise the durability in consideration of; the maintenance schedule that had been developed based on operational availability, the accessibility; the risk to the operation of the plant should the asset fail; as well as the physical requirements of the construction material and any specific construction constraints [14]. Durability planning typically commences early in the design phase and evolves as the design progressed through to completion of the detailed design. For VDP the process was undertaken in co-ordination with the Operations and Maintenance team as well as the related design disciplines [14]. This enables durability to be embedded into all aspect of the project. The advantage of involving all teams in the process enabled a clear strategy for life cycle asset management to be developed.

A major consideration in the design process is the "durability risk", that is the risk that the asset or asset component has inadequate durability for the nominated design life and proposed maintenance regime. The durability risk is often greatest in the more aggressive exposure conditions like brine, seawater and chemicals, the inaccessible permanently immersed or buried environments or for the assets that are critical to the operations of the plant and the production of desalinated water. During the detailed phase assets with a high durability risk are subject to an intense optimisation of the life cycle durability requirements and different strategies for mitigating durability risk may be considered during this process. The durability risk can be mitigated with a number of different strategies including [14]:

- substituting the materials initially selected with one with a greater inherent durability and a reduced maintenance demand
- minimising the consequences of a failure but adding redundancy in the system
- increasing the inspection frequency for maintainable items

Due to the range of materials, environmental exposures, asset design life, accessibility for maintenance, availability for maintenance, durability risk, predicted rate of deterioration under operational or standby conditions, a range of asset management strategies can be successfully utilised for assets at desalination plants (Table 1~~Table 1~~).

Table 1: Examples of asset management strategies used at desalination plants

Asset Strategy	Description of strategy
Do nothing	Provide the structure with an appropriate corrosion allowance or an appropriate concrete mix design to provide adequate durability for the design life.
Combination of protective coatings and corrosion allowance	Apply a protective coating system at the time of construction. Following deterioration of the protective coating rely on a corrosion allowance to product adequate durability for the remaining design life
Combination of protective coatings plus cathodic protection (CP) system	Apply a protective coating system at the time of construction and install a cathodic protection to protect the structure at defects in the coating. Maintain, and replace as required, the CP system to ensure the durability of the asset is maintained for the design life.
Maintenance of protective coatings	Apply a protective coating system at the time of construction. As the coating deteriorates, maintain or replace the coating system as required over the life of the asset.
Materials substitution	In the design, phase if a material is considered to have inadequate durability the material can be substituted for one that is more durable in the specific application.

A couple of specific examples of where these can be used include:

- Assets like the buildings are considered to be accessible and should be available as required for maintenance as these are not process asset. So a cost effective asset management strategy for the steel components is "Maintenance of Protective Coatings". However, it is important for the recoating to be undertaken at a specific time. If a maintenance cycle is missed, this can lead to significant cost increase in the life cycle cost as significantly more surface preparation may be required compared with the planned maintenance.
- Assets like the seawater inlet and outlet structures are fairly inaccessible and therefore extremely difficult to maintain. Therefore a "Do Nothing" strategy was the selected option for the asset management. The concrete mix of these assets was optimised to ensure that corrosion of the steel reinforcing did not initiate until the remaining life was less than 10 years [15].

4.2.3 Key Messages

The specific assets including process assets, buildings, power assets, water storage and pipeline systems all have different maintainability based on accessibility and operational availability constraints. Therefore in consideration of the durability and maintainability of the assets a variety of strategies were required for the management of these various assets over the design life of the plant.

5. CONCLUSIONS

A wide variety of asset management strategies can be utilised for the design and operation of assets affected by corrosion. In developing strategies for new assets the balance of CAPEX and OPEX cost needs to considered in relation to the availability of the asset for maintenance, and the safety, consequence of corrosion, and reliability, likelihood of corrosion, and the required design life. In support of an overarching asset management strategy a durability plan is a critical tool that clearly outlines likely corrosion-related risks and agreed mitigation approaches as early as possible in the asset's lifecycle (e.g. planning/design). Coupled with an as-built asset registers it can allow for a useful baseline to monitor the condition and performance of the assets against and therefore tailor asset management plans accordingly.

In developing asset management strategies for existing ageing assets the maintenance needs to be optimised between short and long term strategies to minimise disruption to operations while providing a safe and reliable asset appropriate for future requirements. Understanding the relevance of the impact of the corrosion on structural performance is critical for the development of appropriate asset management strategy for existing assets. Being able to assess what corrosion issues can actually lead to a decrease in the levels of service provided by the assets allows optimising the extent and timing of future intervention and the associated costs.

6. ACKNOWLEDGMENTS

We would like to acknowledge our clients including QTMR and Watersure, specifically in co-authoring the OzWater 2016 paper which has been used in the development of this paper.

7. REFERENCES

1 International Organization of Standards, 2014, "Asset Management", ISO 55000, 55001 and 55002, Geneva.

2 Bureau of Infrastructure, Transport and Regional Economics (BITRE), 2016, "Australian infrastructure statistics yearbook 2016", Commonwealth of Australia, Canberra, ACT.
Available at: https://bitre.gov.au/publications/2016/yearbook_2016.aspx

3 Infrastructure Australia, "Australian Infrastructure Plan: Priorities and reforms for our nation's future", 2016. Report Available at:
http://infrastructureaustralia.gov.au/policypublications/publications/files/Australian_Infrastructure_Plan.pdf

4 GHD: "Infrastructure maintenance-A report for Infrastructure Australia", Infrastructure Australia, 2015
Available at:
http://infrastructureaustralia.gov.au/policy-publications/publications/files/GHD-Infrastructure-Maintenance.pdf

5 Koch, G; Varney, J; Thompson, N; Moghissi, O; Gould, M; Payer, J: "International Measure of Prevention, Application, and Economics of Corrosion Technologies Study", NACE International, 2016
Available at: http://impact.nace.org/documents/Nace-International-Report.pdf

6 Ratcliff, S: "Research shows corrosion costs the local economy" - News and Events | Curtin University, Perth, Western Australia [Internet]. News and Events. 2009 [cited 2017 Jul 4]. Available from: http://news.curtin.edu.au/media-releases/research-shows-corrosion-costs-the-local-economy/ "Research shows corrosion costs the local economy"

7 International Infrastructure Management Manual, Institute of Public Works Engineering Australasia, 2011

8 ISO 13823:2008 - General principles on the design of structures for durability.

9 ISO 15686, 2001-2011, Buildings and constructed assets - Service life planning, parts 1 – 9.

10 Blin, F: Furman, S; and Mendes, A: "Durability Design of Infrastructure Assets - Towards a Uniformed Approach", Proceedings of the Corrosion & Prevention Conference, Australasian Corrosion Association, Paper 212, Perth, Australia, 2011

11 National Asset Management Steering Group, 2015, "International infrastructure management manual", Johnsonville, Wellington

12 Snyder, D: "Ultimate Capacity of Damaged Prestressed Piles", PhD Thesis, University of South Florida, 2000

13 Du, Y.G; Clark, L.A; Chan, A.H: "The effect of corrosion on ductility of reinforcing bars", Magazine of Concrete Research, Volume 57, Issue 7, September 2005, pp407-419, Institution of Civil Engineers.

14 Gabbedy, J; Turvey, J; Blin, F; Furman, S; Penson, P: "Lifecycle Asset Management Optimisation of Desalination Plant", Paper 375, Proceedings of the OzWater '16 Conference, May 2016, Melbourne, Australia.

15 Blin, F; Furman, S: "Durability Guidelines for Materials in Aggressive Brine Exposures", Paper 032, Proceedings of the Corrosion and Protection Conference, Nov. 2010, Adelaide, Australia.

8. AUTHOR DETAILS

Dr Torill Pape has over 17 years' experience as a structural engineer across a number of sectors, including consulting, construction, public service and academia. She holds a PhD on the validity of inspection techniques when assessing corrosion-affected concrete structures, providing a sound understanding the relationship between condition and structural performance. Throughout her career she has built up extensive experience in asset management of road and rail structures, with technical and managerial experience in structural and condition inspections, durability assessments, the development of life-cycle management strategies, and linking component condition with the overall performance of a structure. In her previous role in ARRB Group providing specialist technical advice and applied research solutions for various road authorities, she has acquired a unique perspective of the challenges facing infrastructure owners.

Dr Frédéric Blin has 17 years' experience with a PhD in corrosion engineering and leads AECOM's Strategic Asset Management and Advanced Materials team. Frédéric has worked on and led numerous projects in the field of asset management and durability/materials engineering. Frédéric can draw upon his experience on the management of infrastructure assets throughout their lifecycle to support the project in achieving whole-of-life outcomes that satisfy stakeholders' requirements. Frédéric is passionate about partnering with clients to help optimise business outcomes through sustainable lifecycle asset management. He enjoys drawing from his knowledge of the ageing of physical assets to inform tactical and strategic plans and business decisions.

Sarah is an Associate Director in the Strategic Asset Management and Advanced Materials team at AECOM. She is a graduate of the Corrosion Science and Engineering Masters' program at UMIST in England and has approximately 25 years' experience in corrosion related consulting and research. She has specific experience in durability planning for new infrastructure, condition assessment and rehabilitation of aged infrastructure, performance assessments of materials and protective coatings, materials selection, specification of protective coatings, failure analysis and cathodic protection design.

DEPLOYMENT OF SMART ULTRASONIC SENSORS FOR INTERNAL CORROSION MONITORING USING INTERNET OF THINGS

Nestor Sequera
SN Integrity Pty. Ltd., Adelaide, Australia.

SUMMARY: Equipment inspection and monitoring for any signs of internal corrosion are vital activities for safe operation in the Oil and Gas industry. The aim is to determine corrosion rate, location and likely causes to ensure Fit for Service (FFS) or if early intervention is needed. A well-managed plan is set to avoid an uncontrollable release of flammable products with the potential to cause a fire, explosion, environmental disaster, and loss of personnel. These incidents can have a huge effect on company reputation, budget and its ability to stay in business.

Since the dawn of the industry there has been a need for procedures / devices able to ensure that internal corrosion was not an issue. At the time, procedures to create a "mechanical fuse" were developed. Tell-Tale holes were common, aiming to fail before a major release occurred.

Later, new intrusive devices were invented: Corrosion Coupons, Corrosion Probes. Their main objective was measuring a Corrosion Rate (CR) from inside the equipment and to some extent (by coupons) types of corrosion (i.e. pitting or uniform). These were combined with non-intrusive apparatus like Ultrasonic Thickness Measurement (UTM) Meters able to measure equipment's remaining thickness. Therefore, FFS calculation could have been performed.

New Deployment of Smart Ultrasonic Sensors are now available taking advantages of what is called Internet of Things (IoT). This basically improves a proven method (UTM), by removing the operator (major source of data errors) and adding data logging, allowing analysis and trending of data, online on a cloud environment from anywhere in the world using a Laptop, Tablet or Smartphone. Reports can be customise including flagging alarms of integrity status without dispatching technicians to dangerous, remote or inaccessible sites improving asset integrity and productivity while increasing safety.

1 INTRODUCTION

Since the foundation of the Oil and Gas Industry, inspection and monitoring for any sign of internal corrosion have been considered vital activities to ensure equipment integrity. The aim has been to confirm if internal corrosion is occurring, and if it is, to calculate the rate of corrosion, locations where it is expected to arise, and likely root cause (s). This led to the importance of inspections and collection of quality monitoring data. Using this data, Fitness for Service (FFS) calculation for vessels, pipelines or other plant equipment are made. It is therefore possible to determine if it is safe to operate and when maintenance/replacement is necessary.

In the early days, engineers confronted with internal corrosion recognized the need for procedures / devices capable of indicating the "health" of pressure containment equipment to effectively manage process safety and business risks. Our peers at the time came out with the idea of a "mechanical fuse" in the form of a Tell-Tale Hole (TTH) drilling. This is a pre-drilled hole of certain depth into the pressure envelope of vessels from the outside diameter surface. Once internal wastage reaches this depth due to heavy corrosion, erosion or mechanical abrasion, a small leak will occur. The ultimate goal of a Tell-Tale hole was to avoid an uncontrollable release of flammable products which potentially could cause a fire, explosion, environmental disaster and loss of personnel. A leakage from any of these purposed "pilot" or "weeper" hole provides a "controlled release warning" that the pressure containment has been compromised to the predetermined allowable minimum wall thickness. This process is still used to this day in some refineries [1].

During the forties and fifties, a lot of research occurred with the focus of developing reliable methods for monitoring internal corrosion. Intrusive systems like Corrosion Coupons (CCs) started to become popular after 1950, when the 2" access fitting was introduced together with the first high-pressure retriever tool making on line extraction of them possible [2]. The main objective of a coupon is to estimate corrosion rate of the system. A coupon is a piece of material of known shape and weight (with composition similar to the equipment being monitored), expose to equipment operating conditions for a period of time. After the set time elapsed, the coupon is retrieved and re-weighted with any loss of mass associated to internal corrosion. Over the years, standards procedures were developed to normalize approach of this industrially accepted method, for manufacturing, installing, and examining results. This included the calculation of corrosion rates from corrosion coupons [3, 4].

Another intrusive method developed at a later stage (taking advantages of the 2" access fitting and retrieval tool mentioned) was the Electrical Resistance Corrosion Probes, normally referred to as "ER probes". They are based on the principle that a wire will increase its resistance to current flow upon reduction of its cross section. As with corrosion coupons, the probe is equipped with a sensing element which is a material having similar characteristics to that of the equipment being monitored. A baseline resistance reading of the sensing element is taken upon installation as a reference reading. Any increase of sensing element resistance going forward is associated to metal loss and a corrosion rate can be estimated.

Over the last 50 years different Non-Destructive Testing (NDT) methods for internal corrosion monitoring have been applied like Ultrasonic Thickness Measurement (UTM), Radiographic Testing (RT) or In Line Inspection (ILI) Tool for piping (Pigging). Also, process stream sampling for chemical analysis to measure key parameters like Iron counts, pH, chlorine, conductivity, Total Dissolved Solids (TDS), O2 content, bacteria enumeration and other elements have been applied as monitoring tool to gather data about process corrosiveness. Today, any process plant will have a combination of the above (intrusive and non-intrusive) techniques for its assets monitoring and inspections to ensure internal corrosion is not a threat for pressure containment.

This paper will focus on the application of non-intrusive Ultrasonic Thickness Measurement, for internal corrosion monitoring, emphasizing the advantages of the Deployment of Smart Ultrasonic Sensors using the Internet of Things (IoT) to manage the data collected. This new monitoring approach, basically, improves a proven method (UTM) that has played a key role in the Oil and Gas industry since its first patent was granted back in 1940 [5]. This new system removes the NDT technician (major source of data errors). variability of equipment used and thickness monitoring location (another known source of errors) whilst also adding data logging. This innovation allows for analysis and trending of data, online in a cloud environment from anywhere in the world by connecting to the internet. Reports can be customised to include flagging alarms of integrity status without dispatching technicians to dangerous, remote or inaccessible sites improving asset integrity and productivity whilst increasing safety.

2 NON-INTRUSIVE INTERNAL CORROSION MONITORING ULTRASONIC SMART SENSORS USING INTERNET OF THINGS (IoT)

The use of ultrasonic thickness measurement for internal corrosion monitoring is a well-known and industry accepted approach with over 70 years of use. It is a solid pillar to the foundation of any plant monitoring system. It is simple and provides a direct non-intrusive measurement of thickness remainder, therefore, UTM is key for corrosion rates calculation as well as Fitness for Service assessment.

The new system described in this paper pertain to the application of "conventional" ultrasound for thickness gauging coupled with the use of Internet of Things (IoT) in a Cloud environment.

Before describing the system, a definition of IoT is necessary. Recommendation ITU-T Y.2060 [6] from ITU[1] defined IoT as:

> 'A global infrastructure for the information society, enabling advanced services by interconnecting (physical and virtual) things based on existing and evolving interoperable information and communication technologies'.

The Internet of Things is rapidly becoming popular, as, through the installation of low-cost sensors and actuators into equipment (for data collection, monitoring, decision making, including process optimization) and bringing them into the Internet, businesses are able to gain a competitive edge [7]. However, it is not a recent discovery. Gathering equipment data relating to their operation and condition using sensors has been on the market for more than 20 years [8]. Internet of Things is in the spotlight today due to the massive growth in mobile devices and applications, in parallel with widespread wireless connectivity. These facts are responsible of a massive expansion of worldwide coverage, bringing prices down rapidly, and changing the way to communicate over Internet Protocol (IP) networks.

Another significant factor triggering this popularity has been the advent of the Cloud as a lane to easily store, share and process large volumes of data. The Cloud provides an avenue to use internet-based analytics technologies, allowing businesses to customize how they manage and extract relevant information from huge volume of data rapidly, with minimal or no local software or processing power needed, thus making the entire process very cost-effective [7, 8].

Internet of Things and the Cloud are considered "Disruptive Technologies"[2] with an estimate forecast impact to the world economy in 2025 of US $ 2.3 to 6.2 trillion and 1.7 to 6.2 trillion respectively, with a combine effect of US$ 4 to 12.4 trillion (Figure 1) [7]. A business survey conducted in September 2014 on early adopters of these platforms asked about potential areas being considered for the deployment of Internet-of- Things-based solutions during the next 18 months [8]. The five top areas on the list were as follow:

- Remote Asset Management and/or Asset tracking,
- Security,
- Energy Data Management,
- Condition-Based Monitoring and
- Fleet Management.

Figure 2 demonstrates the results. It can be seen that 36% of total surveyed are looking at Internet of Things initiatives for Remote Asset Management and/or Asset tracking, and 21% are doing the same for Condition- Based Monitoring proposals [8]. This reinforces the relevance and timing of the "Deployment of Smart Ultrasonic Sensors for Internal Corrosion Monitoring using Internet of Things" presented in this paper.

[1] International Telecommunication Union
[2] Technologies that have "the potential to disrupt the status quo, alter the way people live and work, rearrange value pools, and lead to entirely new products and services"[7]

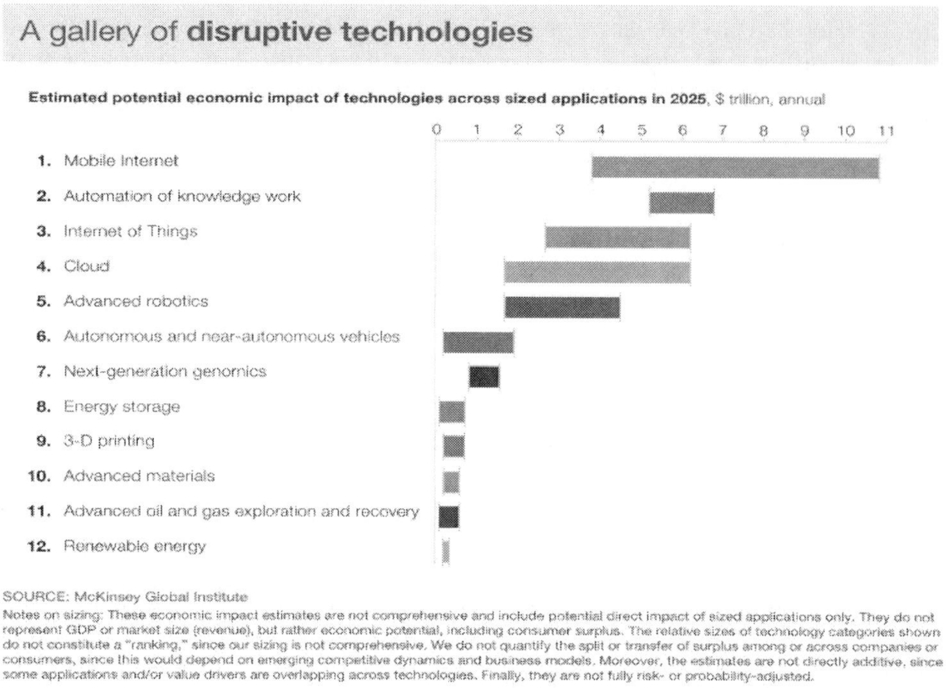

Figure 1.- Disruptive Technologies Estimated Potential Impact to World Economic in 2025. Reprinted from *Disruptive Technologies: Advances that will Transform Life, Business and The Global Economy. Slideshow* (exhibit 1) by McKinsey Global Institute Report, May 2013, McKinsey Global Institute. Copyright 2013.

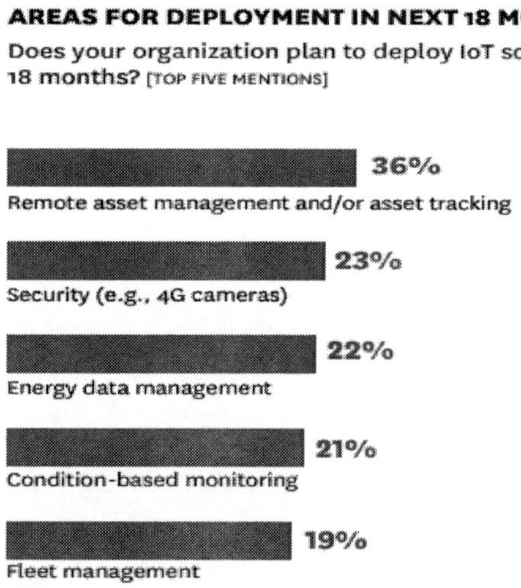

Figure 2.- Top Five Business areas for Internet of Things (IoT) based initiatives Deployment from a September 2014 Survey on early IoT adopters Companies. Reprinted from *Internet of Things: Science Fiction or Business Fact* (p.4) by hbr.org/hbr-analytic-services, September 2014, Harvard Business Review. Copyright 2014.

2.1 ULTRASONIC THICKNESS MEASUREMENT (UTM) USING SMART ULTRASONIC SENSORS

As stated, UTM using smart ultrasonic sensors is based on the same theory that has been used over the years for ultrasonic thickness gauging and corrosion rate monitoring. It will propagate ultrasonic pulse- waves through the material been tested by the use of a sensor or transducer attached to the surface either permanently or temporarily. Figure 3 below shows the ultrasonic principle on which the system is based. A transducer -made of a crystal with the capacity to convert electrical energy to high-frequency acoustic or ultrasonic energy and vice- versa- is used in intimate contact to the surface of the object being tested. The transducer is excited by an electrical pulse to generate a stress wave through the object. This wave will travel until it reaches the back wall of the object or a defect, returning echoes back which are converted to voltage. This reading is measured and recorded by the ultrasonic instrumentation that is attached to the transducer. The transit time between the initial pulse and return echoes (or between successive echoes) is used to calculate wall thickness, when the velocity of sound on the object being tested is known. Physical indications such as the actual object wall thickness, or the remainder thickness on a defect or corroded area (e.g. pit, crack, blister) can be very accurately measured with this technique.

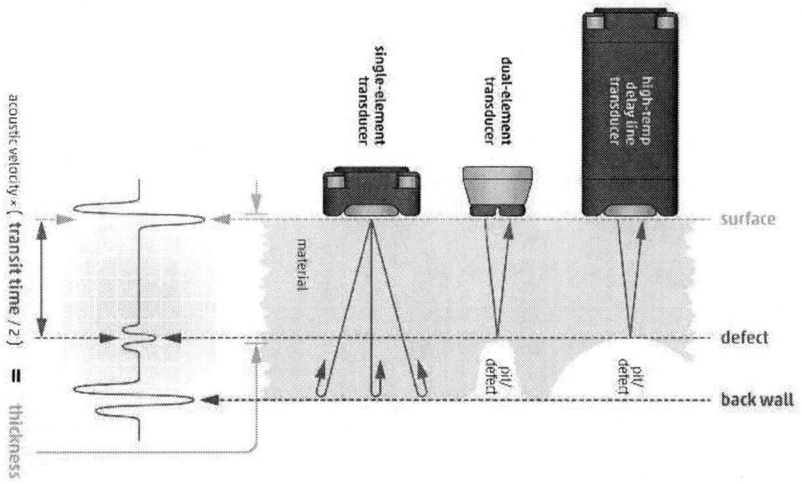

Figure 3.- Ultrasonic Thickness Measurement principle measures transit time between the initial pulse and return echoes to calculate wall thickness. Physical indications such as actual wall thickness, or remainder wall thickness on a corroded area (e.g. pit, crack, blister) can be very accurately measured [9].

2.1.1 System Description

The concept was to develop a flexible and cost-effective system capable of handling deployment of hundreds to thousands of ultrasonic transducers (sensors) for standard ultrasonic thickness measurement data collection. This data is used to assess internal corrosion by continuous monitoring. The equipment that is being monitored will be fitted with one or more transducers (sensors). These sensors will connect to each other and to systems that can understand or present information from the sensor's data feeds. Connection will be by means of IoT solutions, in both wired and wireless versions, using industry-standard communication protocols. This allows for integration with existing plant equipment and Data Control Systems (DCS) or a backhaul to an Internet Provider network, like a Cloud-based database using cellular or wireless connectivity. A schematic view including wired and mobile (cellular) wireless transmitters, transducer (Sensor) types and tablet PC for either manual or fully-automated data collection is shown in Figure 4. The explanation of each system component shown in this figure is presented on points 2.1.1.1 to 2.1.1.7.

2.1.1.1 System controller

A tablet PC, industrial computer, or remote terminal unit (RTU) depending on whether the data collection will be manual, automated or integrated with a plant control system can be acting as the system controller. The system controller communicates with the network of Digital Sensor Interfaces or "DSIs" via Modbus over a RS-485 cable, which is a well-known industrial instrumentation practice. The system controller then transmits the data to an IP server where it can be viewed, analyzed and/or stored for trending. For wireless systems, the system controller is not needed, as data is sent to the IP server by a cellular (mobile) transmitter.

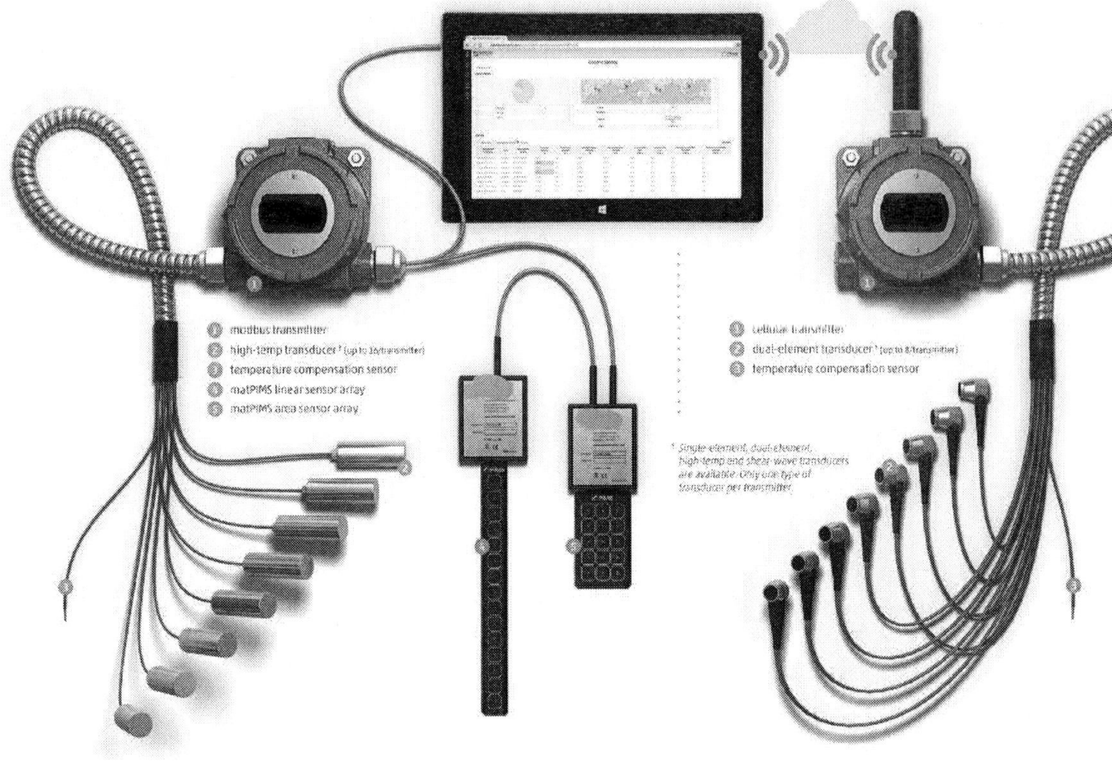

Figure 4.- Schematic view including wired and cellular (mobile) wireless transmitters, transducer types and tablet PC for either manual or fully-automated data collection [10].

2.1.1.2 Digital Sensor Interface (DSI)

It is a dedicated instrument box for ultrasonic data collection equipped with a CPU[3] that can handle up to 16 transducers. The box is programmed for pulsing and receiving data from the transducers, digitizing the collected data and converting the digital ultrasonic data into thickness values then transmitting it to the system controller. The DSIs can be set up in a multi-drop network of up to 32 DSIs daisy-chained, giving room for a maximum of 512 individual transducers. When a cellular (mobile) is selected, as mentioned, there is no need for a system controller as the DSI will have room for a SIM[4] Card and data will be uploaded to the Internet Provider network by wireless connection through the protocol of the mobile service provider.

2.1.1.3 Sensors (Transducers)

The ultrasonic sensors (transducers) are just piezo-electric[5] ceramic materials that upon an excitation pulse will transmit and receive ultrasound. They can be manufactured in several models and types, such as according to equipment operational conditions, mechanical properties and likely corrosion mechanisms expected. For example, dual-element sensors are the optimal selection for thin measurements (< 2.5 mm or 0.100") and highly-pitted back-wall surfaces. Sensors can be set randomly, or in square or linear array, allowing the corrosion engineer to install them based on known trouble locations or expected damaged mechanisms (e.g. elbow extrados, bottom of lines / vessels, injection nozzles, among others). They can be magnetically attached or clamped to the monitored piping or equipment. Table 1 below presents the different smart sensors models available including their technical specifications and recommended application.

[3] Computer Processing Unit
[4] System Identity Module
[5] A material that can convert electric pulse into mechanical waves of sound and vise-versa

Table 1: Technical Specifications of Different Smart Sensors Available

	single-element contact	dual-element contact	delay-line contact	matPIMS
model	XD-101	XD-301	XD-201	XD-401
application	general purpose	severe pitting	ultra-high temp	general wall loss
frequency	5 MHz	5 MHz	7 MHz	7.5 MHz
active area (dia.)	0.25"/6.35mm	0.375"/10mm	0.375"/10mm	0.25"/6.35mm
overall (dia. x h)	1.0 × 1.0" 25.4 × 25.4 mm	0.75 × 0.75" 19 × 19 mm	0.8 × 2.25" 20.3 × 57.2 mm	1.0 × 9.12" 25.4 × 231.6 mm
# of transducers	1-16	1-8	1-16	16 (1 reference, 15 active)
resolution	0.001"/0.025mm	0.001"/0.025mm	0.001"/0.025mm	0.001"/0.025mm
thickness range†	0.200-6.0" 5.1-150.0mm	0.040-6.0" 1.0-150.0mm	0.125-1.0" 3.0-25.0mm	0.125-6.0" 3.0-150.0mm
temp range	-5 to +150 °F -20 to +80 °C	-5 to +300 °F -20 to +150 °C	-5 to +932 °F -20 to +500 °C	-5 to +150 °F -20 to +80 °C
attachment	magnet/adhesive	magnet/adhesive	mechanical clamp	adhesive

†minimum resolutions stated as typical values, but will vary with pipe condition

2.1.1.4 Couplants

It is known that for an ultrasonic thickness measurement system to work, the sensor must have an intimate contact with the surface of the material being tested without any air gap. As the sensors are intended to stay in place for long periods of time (regardless of if they are considered permanent or temporary installed), a liquid couplant normally applied during manual UTM cannot be used. The smart sensor system will stop taking any reading as soon as the liquid dried out. To overcome this issue, different solid couplants have been assessed. Three options have been chosen to cope with the range of temperatures that the sensors can handle, as follow: dry coupled to the pipe using a proprietary elastomeric material (for standard temperatures i.e. -20 to 65 °C), adhesively bonded (for temperature up to 150 °C), or clamped with gold metal foil (for temperatures up to 500 °C).

2.1.1.5 Temperature measurement device

To achieve the most accurate and precise readings using ultrasound, it is necessary to correct the ultrasonic velocity for changes that occur due to fluctuations in asset temperature (Refer to Table 2). Changes in temperature have an effect in the velocity of sound, which therefore varies the value from the true thickness. The system incorporates a Resistance Temperature Detector (RTD) that can be placed on the asset adjacent to the ultrasonic transducers to correct the readings.

2.1.1.6 Data reporting

The system includes a cloud-based software for data management allowing access to thickness and corrosion-rate data from any connected device with an internet browser. Each sensor can be configured with up to six hierarchy levels of tagging. As a result, accurate and near real-time data reporting from the sensor locations is achievable. Furthermore, the system is able to issue alerts via email or text message, identifying the precise sensor (s) where the thickness or corrosion rate is triggering the alarm level. Hosting may be carried out by any internet cloud provider or on a company's intranet.

Data collection can be done manually using a portable tablet PC solution, allowing data and report files to be emailed from the device or sent to the cloud-based software. This can be completed when back at the office. In the case of Sensors Cellular (Mobile) Wireless System, data is sent to the cloud-based software by "hands-shaking" between the Digital Sensor Interface unit and the cellular (mobile) tower at the frequency set (e.g. 2X day, 1X day, 1X Week). The regularity of data transmission is based on a compromise between minimum frequency needed and battery life usage.

2.1.1.7 Data Accuracy, and Precision

Ultrasound thickness gauging equipment uses different methods to do the measurement calculation. This is normally programmed as a routine into the circuitry of the gauge.

For the deployment of smart sensors a tablet loaded with a software called DataPIMS is used for commissioning and data logging of Modbus Wired System (refer to figure 5). For Cellular (Mobile) wireless system, the DataPIMS loaded tablet is only needed for commissioning, as data will be going straight to the internet provider network (refer to figure 6).

It is known that Ultrasonic Thickness Measurement can provide very accurate and precise thickness values of equipment tested when all factors that impact on quality of these parameters are controlled to reduce variability. These devices typically have a display resolution of 0.025mm (0.001") with precision gauges usually showing a resolution as low as 0.0025mm (0.0001") [10]. However, the UTM field accuracy and precision is another story, as it has been shown to be orders-of-magnitude worse, primarily due to operator and equipment variability [10]. In Table 2 a list of the most relevant factors affecting precision and accuracy of UTM field measurements are presented. It is noted that majority of factors affecting accuracy and precision can be eliminated by the installation of Smart Ultrasonic sensors. Other factors (like echo quality and surface roughness) can be controlled, as a case by case, through optimization of the application during commissioning and selection of sensors based on corrosion type [10].

Table 2.- Factors Affecting Gauge Accuracy and Precision in the Field* [10]

Accuracy	Precision
Operator variability	Operator variability
Sound velocity and acoustic zero calibration	Velocity and acoustic zero cal (msmt to msmt)
Echo quality	Echo quality
Sound velocity uniformity	Electronic or ultrasonic noise
Surface roughness	Transducer placement variability
Transducer coupling variability	Transducer coupling variability
Temperature variation	Temperature Variation

* Parameters in green are eliminated or reduced with smart Ultrasonic sensors

Figure 5 shows a schematic view of the typical arrangement used in the Deployment of a Smart Sensors Modbus wired Manual System. As illustrated, the smart sensors system allows for data collection outside of hazardous or difficult to reach areas, which would be unavoidable for normal manual ultrasonic thickness measurements. Data collection from sensors can be done by running a wired from the Digital Sensor Interface (DSI) to a Nema box installed in a "friendly zone", providing easy and safe access to the data.

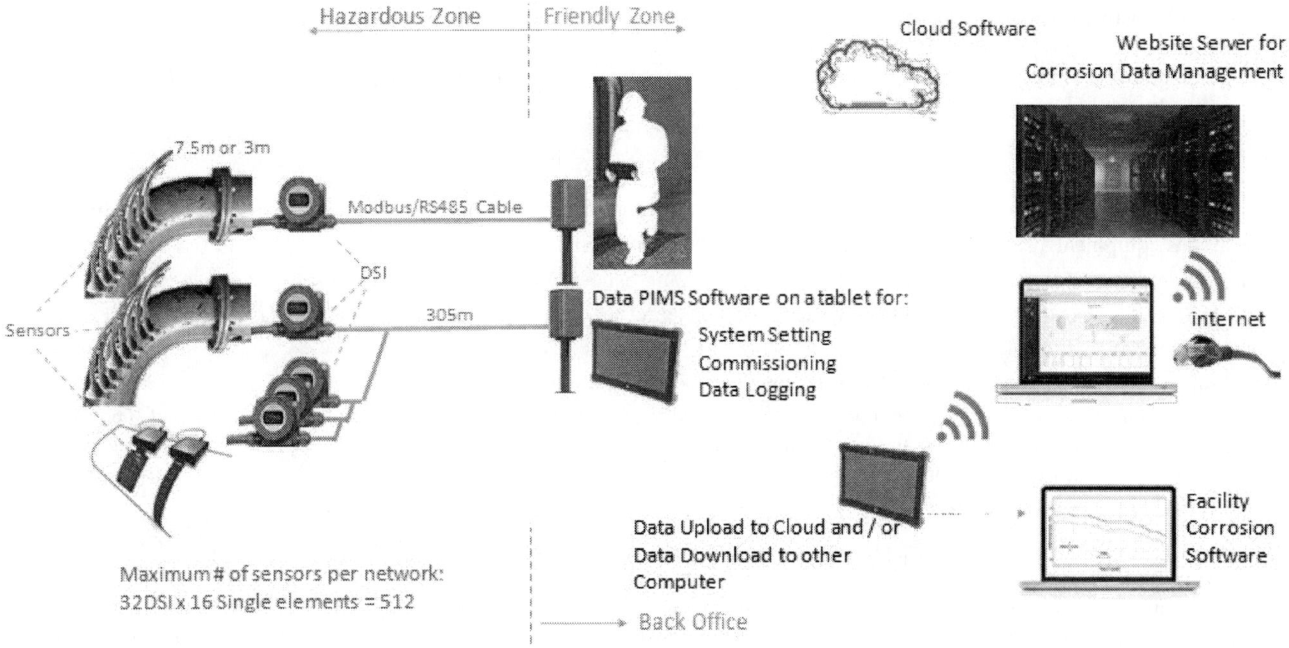

Figure 5.- Schematic view showing typical components integration for deployment of a Smart Sensors Wired Modbus System

While Figure 6, shows a schematic of typical arrangement for Deployment of a Smart Sensors Cellular (Mobile) wireless Automatic System. It is noticed, that on-site installation is faster than the Modbus wired system as there is no need to run cables and install a Nema box for data collection. Data starts to flow to the internet provider network, customer data server, or intranet, immediately at the frequency deemed necessary by the asset owner.

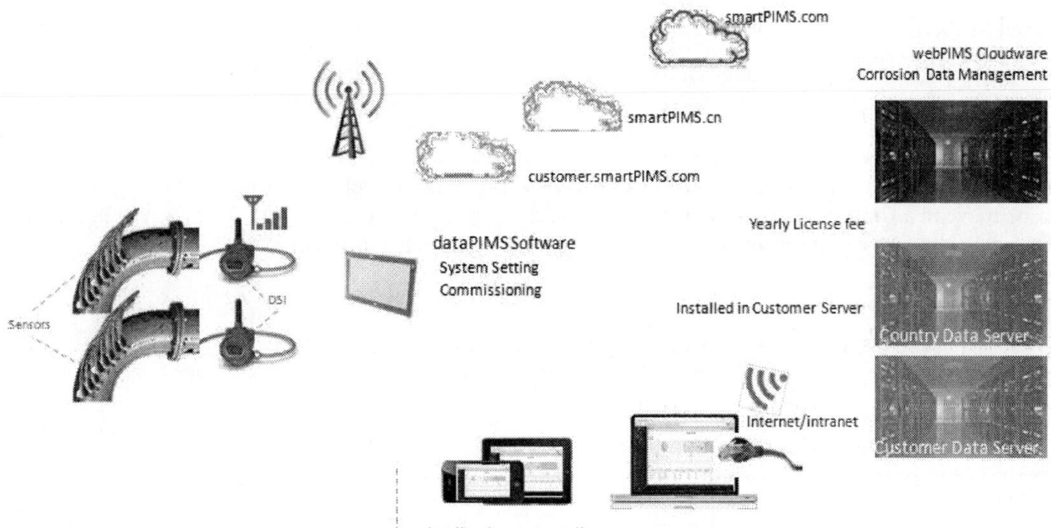

Figure 6.- Schematic view showing typical components integration for deployment of a Smart Sensors Cellular (Mobile) Wireless System.

3 ADVANTAGES OF SMART ULTRASONIC SENSORS SYSTEMS USING INTERNET OF THINGS

Smart ultrasonic sensors are emerging as a new technology to compete with manual Ultrasonic Thickness Measurement inspections and existing or traditional corrosion-rate intrusive monitoring solutions like corrosion coupons and ER probes. The main benefits are the use of the proven UTM technique for quality high frequency data collection, whilst combining a non-invasive installation with the ability to operate remotely without operator interaction, subsequently cutting personnel access costs over time by using Internet of Things.

3.1.1 High-Frequency & High-Quality Data

A major advantage of smart ultrasonic sensors is the ability to collect a larger quantity of high-quality thickness data. This is possible as the equipment used and location targeted after the sensors are installed is set. Also is related to the fact that an ultrasonic gauge meter measures thickness by very precisely timing echoes. In order to convert these time measurements into thickness values, the instrument must be programmed with the sound velocity in the test material as well as any necessary zero offset that is required by the instrument, transducer type, or echo shape. This procedure is commonly referred to as velocity/zero calibration [9]. During commissioning, each sensor is individually programmed / calibrated to ensure the quality of the data. This is backed up by the 3000-value storage capacity built into the Digital Sensor Interface units. Having 1X per day or even 2X per day reading is not an issue for these smart ultrasonic sensor systems. This would be impractical or too costly from manually-collected measurements.

It is known that corrosion metal losses are rarely constant and can vary between periods of virtually zero corrosion to episodic events causing high corrosion rates [10]. The smart sensor systems (either when permanently or temporarily installed), can produce large amount of data that allows visibility to the dynamics of the wall thickness reduction process. Smart sensors allow for setting targets for minimum wall thickness or acceptable corrosion rate. After a target is hit, an alarm is triggered, and an e-mail is sent to asset owners for action almost in real-time. More quality data also means that numerical tools such as data filtering and linear regression are easily deployed in the cloud-solution software to do corrosion rate calculations.

3.1.2 Internet of Things

Cloud data allows for alarm settings, e-mail warnings, flagging potential issues for maintenance planning practically in real-time. Data can be accessed from anywhere in the world, using any device that can connect to an internet provider network. Alternately, data can be integrated with company corrosion software as it is readily available either on the network for wireless cellular (mobile) systems, or from a table for wired manually systems (refer to Figure 5 & 6). Integrity status can be known without dispatching technicians to dangerous, remote or inaccessible sites.

3.1.3 Installation

Instrumentation and sensors can be deployed on the equipment in a permanent or temporarily mode and access to locations inside hazardous or difficult to reach areas is only needed during initial installation (including wired Modbus systems), as the collection point can be move to a friendly zone away from these challenging areas (refer to Figure 5). The system can also be rapidly installed without welding, ensuring sensors can be easily moved for short term installation.

3.1.4 Save in Opex budget

The system allows for asset managers to use resources only when needed, eliminating the repetitive cost of access – scaffolding, lagging /delagging, offshore access (e.g. Flights, accommodation, training), personnel costs (crew daily rates).

3.1.5 Improve Ultrasonic Thickness Measurement Accuracy and Precision

Removes significant sources of error due to the fixed sensor position and instrumentation. The variability on results caused by operator-to-operator, probe-to-probe, and instrument-to-instrument is eliminated
(Refer to 3.1.1 & Table 2).

4 FIELD APPLICATION SHOWING BENEFITS OF DEPLOYMENT OF ULTRASONIC SMART SENSORS USING INTERNET OF THINGS

4.1 Atmospheric Gas-oil Line

4.1.1 Overview

During a shutdown this line was found to have an area with low wall thickness. It was not part of the scope of the turn-around. Asset owner wanted to extend life to next turn-around and make sure that line is no longer corroding.

4.1.2 Details

Atmospheric gas-oil, operation temp. 270 °C, 80 mm Schedule 40 (5.5 mm Nominal Wall Thickness), subject to severe but uniform corrosion, most piping is < ½ nominal thickness (2.7mm).

4.1.3 Product Used

- A Digital Sensor Interface Cellular (Mobile) wireless system with 4 High Temperature Sensors (SmartPIMS) probes temporarily attached, managed by on-site service provider.
- Monitoring interval set: 4 hours.
- Data is monitored & trended daily using cloud based software (webPIMS) available through Internet Provider network.
- Installation time: 6 hours.

4.1.4 Outcome

By installing the smart sensors system the refinery achieved the following benefits:

- Personnel can safely monitor process piping internal condition that otherwise would have required to perform inspection at high temperature.
- Replacement of line can be planned, avoiding the need to extend turn-around schedule and losses due to production deferrals.
- More data available to monitor trend and plan repairs that would have been risky, impractical or too costly by manual inspection.

Figure 7.- Gas-Oil Line Smart sensors Installation. Left: general view of location of sensors installed. Top right: details of Digital Sensor Interface cellular (mobile) installation. Bottom right: clamp used to install one of the High Temperature sensors.

4.2 Crude Overhead Line

4.2.1 Overview

Customer installed new overhead lines connecting units. Lines ware in un-accessible area. Many of these lines have no platform access. Customer wanted data on corrosion rates and inspection needs.

4.2.2 Details

Crude Overhead Line 100°C-38°C

- 300 mm Sch. 40 … all nominal wall thickness 10.1 mm +/- 12%

4.2.3 Product Used

- Customer installed permanently sensors (SmartPims) to monitor pipe intrados, extrados, top and bottom locations.
- A Digital Sensor Interface Cellular (Mobile) wireless system with 8 dual element sensors (probes) permanently attached.
- Monitoring interval: 1 reading every 2 days, transmission every 6 days. Estimated battery life ~4 yrs.
- Digital Sensor Interface unit bolted to hand rail and Unistrut, cables run to thickness monitoring locations

4.2.4 Outcome

- Cut Inspection costs – lift, scaffolding, or rope access required to reach locations at 12 m off the ground.
- Process control – access to higher volume of more accurate and quality data to trend corrosion rates, including impact of process excursion within days of occurring.

Figure 8. - -Overhead Lines Smart Sensors Installation. Top: general view of line including installation of Digital Sensor Interface (the blue box). Bottom Left: Showing sensors installed on an elbow extrados with one sensor at 6 o'clock. Bottom Right: showing sensors installed on an elbow intrados with one sensor at 12 o'clock.

4.3 Gas Pipeline

4.3.1 Overview:

An In Line Inspection (ILI) report showed a number of low spots at three separate locations along a 30 m stretch of gas pipeline. When the previous ILI was performed 7 years ago, these low spots did not exist. The operator not only wanted data to tell if the corrosion was episodic in nature or active but also did not want to fix/repair the pipeline and did not want to perform another ILI on a shorter interval. After mapping the internal corrosion on the pipeline, installed sensors were deployed to monitor the low spots along the excavated line.

4.3.2 Details:

Monitoring pits low spots instead of fix/repair

- 750 mm natural gas transmission line nominal wall thickness ~ 7.6 mm
- Low spots ranging from 3.1 to 6.1 mm– buried

4.3.3 Product Used:

Sensors (SmartPIMS) Modbus wired configuration with/ 8 dual element sensors (probes) (7.6 m length cable) permanently attached and buried to monitor "low spots" as identified by manually ultrasonic thickness measurement screening.

- Sensors were attached via epoxy & viscoelastic putty used to seal sensors before wrapped and buried
- A Nema enclosure used to house Digital Sensor Interface unit & act as collection point for techs
- Operator will vary frequency of manual readings.

4.3.4 Outcome:

- Condition monitoring of "low spots" without the need of dig out the area to access the points or run an in-line inspection tool
- Process control – access to higher volume of more accurate and quality data to trend corrosion rates and confirm if corrosion was episodic or is still active.

Figure 9. – Gas Pipeline Smart sensors Installation. Top Left: Showing sensors installed on the line. Top Right: Showing one sensor now totally covered by the putty in preparation for wrapping the area. Centre: General view after sensors have been installed before the final wrap is applied. Bottom Left: showing the section finally wrapped in preparation to buried the line back. Bottom Right: Nema Box for Digital Sensor Interface unit installation and data downloading going forward.

5 CONCLUSIONS

Equipment inspection and monitoring for any signs of internal corrosion are vital activities for safe operation in the Oil and Gas industry. New Deployment of Smart Ultrasonic Sensors are now available taking advantages of what is called Internet of Things (IoT). This basically, improves a proven method (UTM), by removing the operator (major source of data errors) and adding data logging, allowing analysis and trending of data, on line on a cloud-based software environment from anywhere in the world using a laptop, tablet or smartphone. Reports can be customise including flagging alarms of integrity status without dispatching technicians to dangerous, remote or inaccessible sites improving asset integrity and productivity while increasing safety.

Deployment of Smart Ultrasonic Sensors can offer:

- Higher-fidelity and higher-frequency data collected.
- Internet of Things: Data on Cloud available, alarm settings, e-mail warnings, potential issues flagging for maintenance planning.
- Integrity status without dispatching technicians to dangerous, remote or inaccessible sites.
- Lower cost-per-point for thickness monitoring location (TML) with a wide temperature range of applications
- Upon installation system offer data that can be monitored locally or remotely.
- Can be temporarily installed without welding - sensors can be easily moved.
- Short Term Installation/Data: Re-useable smartPIMS Sensors
    - Quick Deploy, Experimental research, Evaluation, among others applications.
- More reliable and safer for personnel than manual Ultrasonic Thickness Measurement inspection.
- Save in Opex budget by using resources only when needed
- Integrate with current corrosion intrusive monitoring devices like coupons and ER probes, as well as, customer corrosion monitoring and / or Risk Based Inspection software, enhancing Plant integrity.

6 REFERENCES

1. Badeen, Mike 1998, 'Tell Tale Holes (TTH). A Complement to Existing Piping & Fitting Inspection Practices', *Inspectioneering Journal*, Journal Archives, viewed 14 June 2017, <https://inspectioneering.com/journal/1998-05-01/331/tell-tale-holes-tth>.
2. Cosasco, *Home*, About Cosasco, viewed 13 June 2017, http://www.cosasco.com/>.
3. NACE International, *Standard Practice SP-0775-2013*, Preparation, Installation, Analysis, and Interpretation of Corrosion Coupons in Oilfield Operations
4. ASTM, *Standard G4-95,* Standard Guide for Conducting Corrosion Coupon Tests in Field Application
5. Wikipedia, *Ultrasonic Testing*, History, viewed on 23rd of June 2017, https://en.wikipedia.org/wiki/Ultrasonic_testing#History
6. International Telecommunication Union, *Recommendation ITU-T Y.2060*, Overview of Internet of Things, viewed on 18th of June 2017, http://www.itu.int/rec/T-REC-Y.2060-201206-I
7. Manyika, J., Chui, M., Bughin, J., Dobbs, R., Bisson, P., Marrs, A., 2013, 'Disruptive Technologies: Advances that will Transform Life, Business and The Global Economy' McKinsey Global Institute Report, May 2013, viewed 24 June 2017, http://www.mckinsey.com/business-functions/digital- mckinsey/our-insights/disruptive-technologies
8. Harvard Business Review, *Harvard Business Review Analytic service report*, Internet of Things: Science Fiction or Business Fact, viewed on 23rd June 2017, http://www.verizonenterprise.com/resources/reports/rp_hbr-internet-of-things-science-fiction-or-business-fact_en_xg.pdf
9. Olympus, *Ultrasonic Flaw Detection Tutorial*, 3.6 Calibration Concepts, viewed 31 July 2017, http://www.olympus-ims.com/vi/ndt-tutorials/flaw-detection/calibration-concepts/.
10. Strachan, S., Feydo, M., Pellegrino, B., Nugent, M., NACE 2017, Non-Intrusive Ultrasonic Corrosion- Rate measurement in lieu of manual and intrusive methods, March 2017. New Orleans, Louisiana USA.

CORROSION BEHAVIOUR OF PIPES IN SOIL AND IN SIMULATED SOIL SOLUTION

Muhammad Wasim, Chun-Qing Li, Dilan Robert, Mojtaba Mahmoodian
RMIT University, Melbourne, Australia

SUMMARY: Corrosion is a serious concern for asset managers and owners of the buried pipelines. Millions of dollars are spent every year on the repair and maintenance of the existing buried water pipes which were laid underground many decades ago. Recent investigations revealed that the external corrosion can be the root cause of failure for buried cast iron pipelines. Hence, various research works have been reported in the literature to discover the corrosion effect on pipes buried either in real soils or simulated soil solutions. However, the corrosion models developed on the basis of pipes in these two environments are not largely comparable. In the current research, the carefully controlled test program was executed to investigate the corrosion behavior between the pipes buried in soil and simulated soil solution. Firstly, a series of experiments was conducted using cast iron pipe coupons buried in homogenous real soil environments having a constant void ratio, the degree of saturations and pHs. Secondly, cast iron pipe coupons having similar material composition were submerged in simulated soil solution which was created from the corrosive components of the same soil. Results revealed a substantial difference in the metal corrosion rates between the different interaction mediums. The research findings can be helpful to develop a correlation model to compare the corrosion rates of cast iron metals between soil and simulated soil solution mediums. Such models will support more reliable and convenient service life prediction of cast iron pipes when the corrosion experiments are based on simulated soil solutions.

Keywords: Cast iron, clay soil, simulated-soil solution, pH, moisture content and corrosion rates.

1. INTRODUCTION

It has been reported that the annual estimated cost of corrosion in the US has exceeded to over 1 trillion dollars in 2016 in all the various sectors of the country (G2MT Labs, 2016). Alarmingly out of this whooping cost; the highest percentage is incurred in the repair and maintenance of existing buried water and sewer pipes as reported earlier in the NACE reports (NACE 2002). In Australia, 80% of the water utilities are buried and more than 70% of these utilities are ferrous metals pipes (Worthington, 2011). The main failures of these buried metal pipes are leaks and rupture which cause significant financial losses every year. The reason of such leaks in buried pipes is reported as the external corrosion by several researchers (Kleiner and Rajani 2001, Melchers 2013 and Wasim et al., 2016).

The external corrosion of buried pipes is not preventable despite the use of advanced corrosion protective techniques such as sacrificial anodes, coatings and impressed current (Kim and Kim 2001, Sparks 2003, Yan et al., 2008). Several field studies have been conducted by researchers in the past to investigate the external corrosion of buried metal pipes (Petersen et al., 2013, Li and Mahmoodian 2013 and Mahmoodian and Li 2016). Besides the field investigation, several laboratory investigations have also been reported in literature. Most of the laboratory studies are limited to investigate the effect of soil moisture content on the external corrosion of metal pipes (Wan et al., 2013, Noor and Al-Moubaraki, 2014, Murray and Moran 1989, Gupta and Gupta 1979).

On the other hand, considerable research work has been conducted in the past on corrosion behavior of buried pipes using simulated soil solutions (Yang and Frank 2016, Wu et al., 2010, Liu et al., 2009a, Liu et al., 2009b) etc. The simulated soil solutions used in those studies were developed by extracting key chemical and minerals present in the available soils to develop simulated soil solutions of near neutral pH. But theoretically, these developed solutions do not simulate the real physical and chemical properties of the soils such as soil's particles, density, porosity, resistivity etc. Moreover, there is very limited research in the literature explaining and comparing the long term corrosion behavior of pipes in soil and its simulated soil solution.

Considering the limited work undertaken in labs with reference to pH, depth of burial, and the knowledge about corrosion of buried pipes in general, a comprehensive experimental program was developed to investigate the correlation between corrosion of pipes in soil and its simulated soil solution. The purpose of this study was to observe the corrosion behavior and then develop correlation models which can be used to assess the corrosion of buried metals when corrosion of pipes is examined using simulated soil solution.

2. EXPERIMENTAL DETAILS

The experimental program of the current research comprised of two series of tests on corrosion of cast iron coupons. The first set of test setups were the laboratory-controlled simulations of underground external corrosion environment by burying the coupons in the highly corrosive soil mediums. The second sets of test setups were also external corrosion simulation, but a simulated soil solution was used as a corrosive medium to investigate and compare the corrosion behavior of cast iron between the two mediums. For both the test setups in soils and simulated soil solutions mediums; cast iron specimens of same material composition, size, and their total numbers with similar coatings procedure and same surface orientation in exposures were used. Moreover, pHs of both the test setups were also kept similar i.e., 2.5, 3.5 and 5 respectively for deriving a correlation between the two mediums. Figure 1 illustrates the experimental program of the current research. The overall test plan is shown in Table 1, while more description about the test setups preparations is discussed in the later sections of this paper.

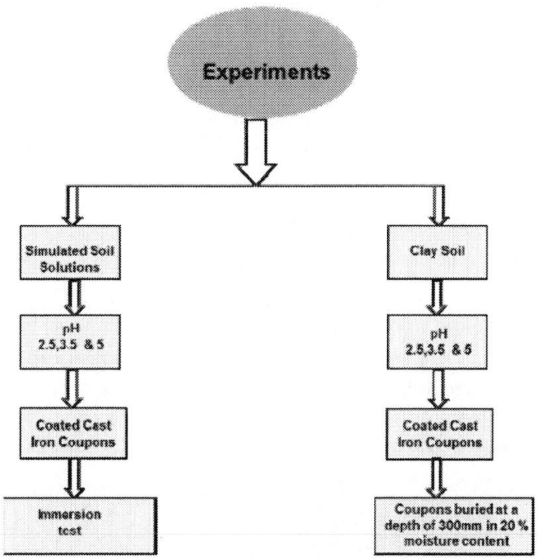

Figure 1 Experimental program for the current research

Table 1. The overall test plan for experiments

Material	Medium	pH	Moisture content %	Measurements of Mass loss Time interval (months) 6	12
Cast iron	Soil	2.5	20	3	3
		3.5	20	3	3
		5	20	3	3
	Simulated soil solution	2.5	-	3	3
		3.5	-	3	3
		5	-	3	3

2.1 Design of test specimen

It has been reported that the majority of the existing aged buried water pipes are made up of cast iron in various parts of the world (Mahmoodian and Li 2016) with a high frequency of failures. The test specimens used in this experimental study were coupons cut from cast iron pipes replicating the composition and wall thickness as of old exhumed cast iron pipes obtained from several water utilities in Melbourne, Australia. The material composition of the cast iron coupons are showed in Table 2.

Table 2. Chemical composition of cast iron specimens

C %	Mn %	Si %	S %	P %	Ni %	Cr %	Mo %	Cu %	Ti %	Al %	Mg %
3.58	0.74	2.48	0.06	0.067	0.01	0.01	<0.01	0.02	0.18	<0.01	<0.01

The dimension of the polished test specimens were selected as 100 x 20 x 10 mm with the intension of carrying out fracture toughness testing as per ASTM 1820-13 at certain exposure of times (i.e. to capture the time dependent corrosion effect on fracture resistance). The schematic representation of the specimen is shown in Figure 2 with the notch at the center which was made in all the specimens after the completion of test duration. The selected thickness of specimen in Figure 2 was also replicating with the old exhumed cast iron pipes of Melbourne. In the current paper, only corrosion results are presented as the tests are in progress.

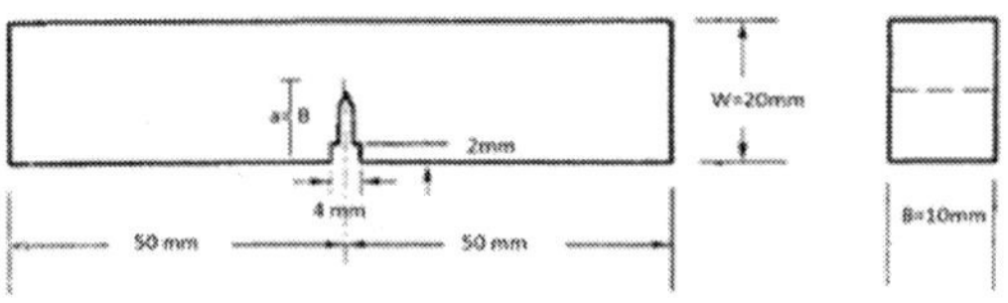

Figure 2 The schematic representation of the test specimens with dimensions

The specimens used in the current research for corrosion in soil and in simulated soil solutions were coated with corrosion protection epoxy and then wrapped with acid resistant plastic tape (Figure 3(a)) except one surface exposed to simulate external corrosion environment. Their initial masses (in grams) were measured before coating and wrapping for mass loss measurements which were carried out after corrosion at specified intervals of time. The coated and plastic wrapped specimen used in the current research is shown schematically in Figure 3 (b). Some traces of coatings were left on the exposed surface of the specimens which were removed by acetone prior to corrosion testing.

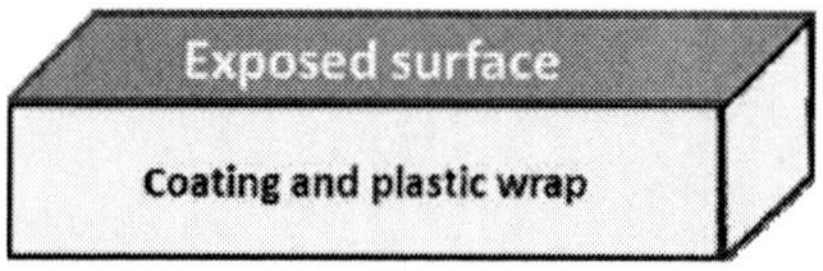

a) White acid resistant coatings and plastic wrapping of the specimens b) Schematic illustration of the coated and plastic wrapped of a specimen with an exposed surface

Figure 3 Coating procedures of the specimens

2.2 Corrosion of cast iron in clay soil

From previous research studies, low pH and clay soil is reported to be the most corrosive medium for buried cast iron pipes (Petersen et al., 2013, Romanoff 1957, Schwerdtfeger, 1953). Therefore, clay soil was used in the current study as a source of accelerating corrosion along with low pH.

The clay soil used in the experiments was obtained from a land fill site in Melbourne. The resistivity and pH of the soil was

23.46 Ωm and 8.17 respectively. The other chemical and mineral compositions are shown in Table 3.The soil received from the field was crushed to uniform size by a crusher and further sieved using BS 410 having an aperture size of 2.36 mm to avoid any variation in corrosion results due to different size soil particles.

Table 3. Test soil composition and its properties

Cl (mg/kg)	(SO_4) (mg/kg)	(NO_3) (mg/kg)	Organic matter (mg/kg)	K (mg/kg)	Ca (mg/kg)	Mg (mg/kg)	Na (mg/kg)
2.48	50	23.05	0.26	39.1	333	197	119

The soil pH was adjusted for corrosion acceleration using hydrochloric acid (HCl) to achieve target pHs of 3.5, 2.5 and 5, selected on the basis of their corrosivity. To achieve the target pHs; several trials have been conducted using small quantity of soil and a varying quantity of acid. Based on these trails, 187g of diluted 4% HCl is added with water to 10 kg of soil to achieve target pH of 3. Similarly, 150g and 300g of same concentration of acid are mixed with water to achieve pH of 2.5 and 5 respectively. The moisture content was maintained at 20% which included both acid and water for 3.5, 2.5 and 5 pH set ups. Moreover, to avoid moisture loss and uniform saturation of acid into the pores, the soil of known pH is kept in plastic bags for 2 days. The soil preparation stages for the experiments are shown in Figure 4. For observation of long-term corrosion behavior over time, three-time intervals were selected i.e., 180, 365 and 545days (or 6, 12 and 18 months) respectively in addition to initial time. Thus, there were four points in time in total. These times were selected on the basis of the literature reviewed and research experience to ensure the measurable corrosion with time (e.g.,Mohebbi and Li 2011). For statistical studies, three specimens for each time interval with the designated test variables were used (Table 1). In the current publication, 12 months of observations are presented as the current research is in progress.

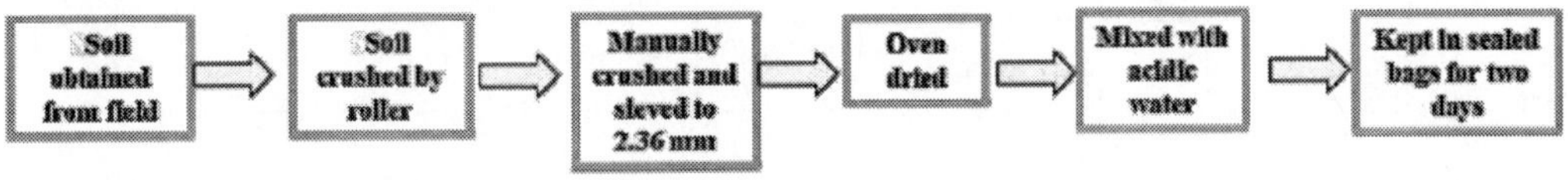

Figure 4 Acidic soil preparations for experimental set up

2.2.1 Test containers preparations

A plastic container, non-reactive to acidic chemicals, is used in the test setup. Soil was prepared in the container following careful control measurements to maintain a uniform density throughout the container. Firstly, a bedding layer is prepared in the container with 80mm height at the target dry density (1600kg/m^3) representing most of the field soil conditions reported in literature (Petersen et al., 2013). Then the coupons, coated on all the sides except the top surface, were placed on the bedding layer simulating the real corrosive environment for the external pipe wall of the buried pipes. Having placed the coupons on soil, the container was filled with soil in layers. Each layer was filled with weighed quantity of soil with uniform compaction to maintain the homogeneity in the soil. The purpose of uniform compaction was to achieve uniform density in each layer to avoid variation in aeration and permeability. The test containers were divided into three compartments to minimize the disturbance to the soil and coupons while removing the coupons for testing. It was planned to test the three coupons of each test set up at every 6 month intervals. The schematic representation of test set up with embedded moisture/temperature sensors along with data collection system is shown in Figure 5. The initial conditions of the test are given in Table 4.

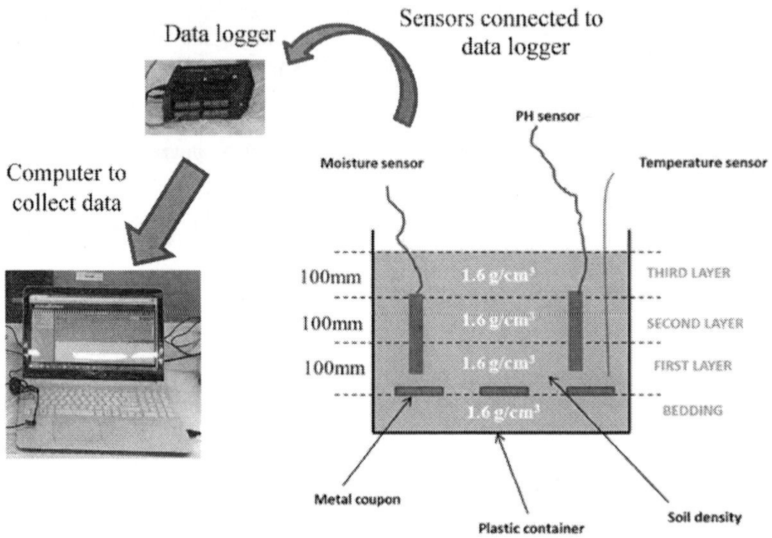

Figure 5 Schematic representation of the test set up with sensors and over all test assembly

Table 4. Initial conditions of the test

Ambient humidity	Ambient temperature	Initial dry density of the soil	Moisture content of soil (including acid)	Degree of saturation	Specific gravity	Void ratio
43%	23 Celsius	1600kg/m^3	20%	80%	2.66	0.66

2.3 Corrosion of cast iron in simulated soil solution

In order to develop a correlation between the corrosion in simulated soil solution and in real soil, three test setups of simulated soil solution with pH of 2.5, 3.5 and 5 were prepared. Simulated soil solution developed using key corrosive elements by following previous research as discussed in the introduction section of this paper. The composition of the developed simulated soil solution used in this experiment is shown in Table 5. The pH of the simulated soil solution was found to be 7.5-8. It was adjusted to 2.5, 3.5 and 5 using diluted HCl.

Table 5. Chemical composition of simulated soil solution (in g/l)

$NaHCO_3$	$MgSO_4\,7H_2O$	$CaCl_2\,2H_2O$	KCl	$NaNO_3$
0.695	0.109	1.476	0.389	0.505

Similar to corrosion in soil experiments, the polished cast iron specimens with the dimension of 100 x 20 x 10 mm were used for simulated soil experiments. Figure 6 shows an experimental set up of specimens in simulated soil solution. As it can be seen from Figure 6, the top of the test containers was covered with plastic to avoid loss of moisture due to environment, but for the oxygen diffusion, some equally spaced holes were made in the plastic cover for corrosion to take place. The pH of the solution was timely adjusted by adding diluted HCl.

Figure 6 Set up for cast iron specimens in simulated soil solution

3. RESULTS AND ANALYSIS

3.1 Corrosion of cast iron in soil

3.1.1 Monitoring of moisture, pH and temperature

The monitoring of various key parameters such as moisture, pH and temperature was conducted for corrosion of cast iron experiments in soil medium. As found from the literature, it was noted that these parameters can highly influence the corrosion behavior of metal pipes. Based on one year of observations, it can be noted that the moisture and soil pH were maintained at the target values with the uniform temperature throughout the test duration. The monitoring of pH indicated that it was held fairly uniform for the whole test duration (Figure 7). Water was added on weekly basis to keep the moisture at 300mm depth remained at the desired level indicated by moisture sensors (Figure 7). The embedded thermometers show fairly consistent temperature for the whole duration of testing indicating no significant change in the temperature of the soil at the burial depth of coupons (Figure 8).

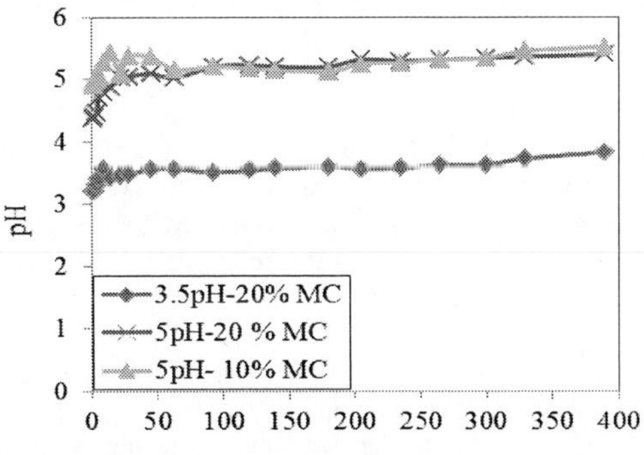

Figure 7 pH monitoring of various test set ups

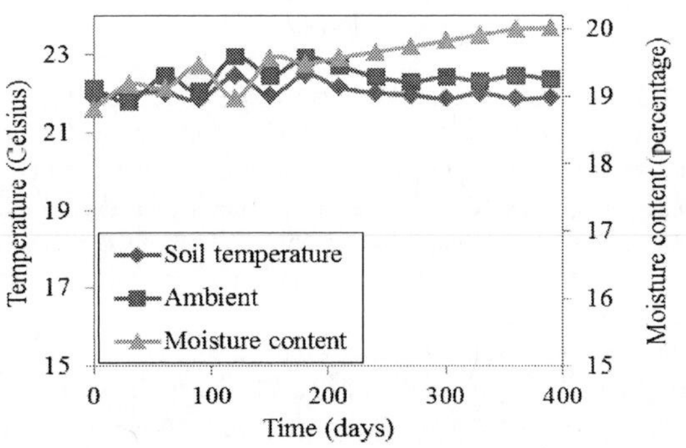

Figure 8 Moisture and temperature monitoring of test set ups

3.1.2 Mass loss measurements of specimens and corresponding corrosion rates

Physical mass loss measurements were carried out of the specimens dug out from the test setup for investigating the actual corrosion damage and its severity. Some of the corrosion products from all the different specimens were kept in a sealed bag for XRD analysis, results of which is not presented in this paper. The corrosion products were removed using Clarks' solution (Wasim et al., 2016). The metal coupons were immersed in Clarks' solution for 1 min before removing the corrosion products by using a brush. Soon after the corrosion products removal, the specimens were weighed to observe the mass loss due to corrosion.

The three mass loss measurements of specimens from each of the soil environment (i.e., 2.5, 3.5 and 5 pH with 20 % moisture content) were carried out and their corresponding corrosion rates were calculated. The corrosion rates of three specimens in each soil environment were found fairly similar authenticating the homogeneity of soil, consistency and the accuracy of the measurements. The average of the three readings of corrosion rates from each soil environment is shown in Figure 9.

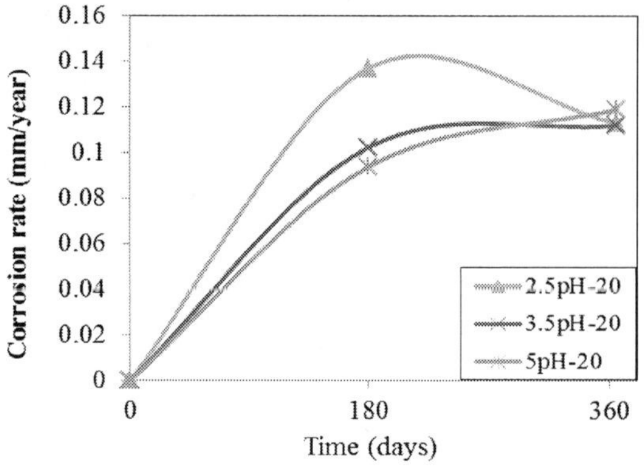

Figure 9 Corrosion rates of the cast iron coupons buried in various soil environments

The highest mass losses were measured from specimens buried in 2.5 pH and 20 % moisture content after 180 days of observations as evident from Figure 10. The average corresponding corrosion rate of 0.137 mm/year was calculated using ASTM G-01 based on mass loss. This result indicates very high corrosion in short span of 180 days of corrosion. The reason for such significant mass loss can be explained as a contribution from various factors causing accelerated corrosion of buried coupons. Firstly, the moisture content has a dominant effect in corrosion as previously explained in the manuscript. The 20 % moisture content of soil (i.e. 80 % degree of saturation) used in the current research allowed the diffusion of oxygen through the pores of the soil to the metal coupons, and hence the corrosion process continued accelerating. Due to this acceleration process, high corrosion mass loss and corresponding corrosion rates were obtained. Secondly, the low pH prevents the formation of rust layer on the metals due to its high acidic environment. Hence due to the coupled effect of low pH and high moisture, corrosion process aggravates for short term as observed in the current study (Handbook, 1987, Romanoff, 1957a). However, immediately after reaching peak values, the corrosion rates were declined and reached to 0.112 mm/year after one year of corrosion. The reason of the decline in corrosion rates could be due to the formation of corrosion products which hindered further diffusion of oxygen and moisture to the metal surface.

Mass losses and corresponding corrosion rates were also calculated for the specimens buried in pH 3.5 and 5 with 20 % moisture content. The average corrosion rates of 0.094 and 0.089 mm/year were obtained from specimens of 3.5 and 5 pH respectively with higher rate in 3.5 pH than 5 pH due to its more corrosivity for the duration of 180 days (Fig. 10). These rates were found lower than specimens buried in 2.5 pH specimens as of low pH compared to the pH 3.5 and 5 cases though the moisture content was the same in all of them. However, after one year of corrosion, the corrosion rates (0.119 mm/year) of specimens buried in soil pH 5 with 20% moisture content were found similar to the corrosion rates of soil pH
2.5 and 3.5 with same moisture content. The possible reason for this observation can be explained as follows. As of less corrrosivity in the pH 5 than pH 2.5 and 3.5, the corrosion rates were initially low as compared to other low pH cases. However, as the time progressed, the rates started to decline in low pH cases due to the development of heavy rust. In contrast, the specimens buried in 5 pH with 20 % moisture content, the corrosion process continued and eventually the rates attained higher after one year.

3.2 Corrosion of cast iron in simulated soil solutions

3.2.1 Mass loss measurements of specimens and corresponding corrosion rates

Similar to the corrosion experiments in soils, the mass loss measurements were recorded for the specimens immersed in three different solutions of pH 2.5, 3.5 and 5 using Clark's method after 6 and 12 months. Using the mass loss of each of the specimen from the test set ups, the corrosion rates of the specimens were calculated using ASTM G-01. The calculated corrosion rates are shown in Figure 10. It can be seen from Figure 10 that the specimens immersed in 2.5 pH solutions were heavily corroded with maximum corrosion rates of 0.117 mm/year after 6 months of corrosion. Further the corrosion rates of the specimens immersed in 3.5 and 5 pH solutions were assessed as 0.087 and 0.086mm/year respectively at this age. These corrosion rates were found to be pH dependent as the lowest pH caused more corrosion rates. The corrosion rates of the specimens in high pH i.e., 3.5 and 5 pH solution were comparatively lower than 2.5 pH case. Thus, it can be seen that the corrosion rates followed the trend of 2.5 pH> 3.5 pH> 5 pH after 6 months of the exposure period. These results are in compliance with the previous studies of metal corrosion in simulate solutions, which showed the highest corrosion rates of ferrous metals are caused by the lowest pH solutions (Hou et al., 2016, Wu et al., 2010).

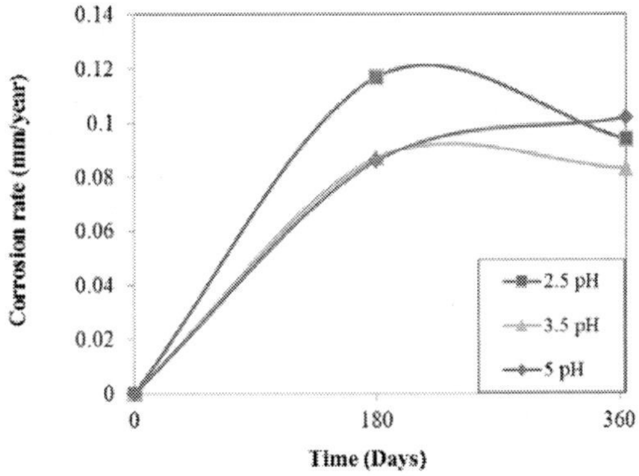

Figure 10 Corrosion rates of cast iron specimens immersed in simulated soil solutions

However, after 12 months, a different corrosion trend was observed. The corrosion rates of the specimens in a solution of 2.5 pH declined from 0.117 mm/year to 0.096 mm/year as shown in Figure 10. This was due to the development of corrosion products on the specimens which hindered the access of oxygen and water to the exposed surface of the specimens. In contrast, the corrosion rates of the specimens in 3.5 pH solution remained stable at 0.087 mm/year after one year. Interestingly, the corrosion rates of the specimens immersed in 5 pH solution were increased from 0.086 to 0.109 mm/year on reaching 12 months similar to what was found in the same pH experiment in soil with 20 % moisture content. The reason for this rise in corrosion rates was the high pH of the solution due to which the rust layers were not fully developed and rates continued to rise till 365 days of corrosion.

3.3 Comparison between corrosion of cast iron in soil and simulated soil solution

The corrosion rates of the specimens immersed in simulated soil solutions are compared in Figure 11 with the specimens buried in similar pHs soils with 20 % moisture content. The results indicated that the corrosion in soil was higher as compared to the simulated simulation at 6 and 12 months intervals. The corrosion rates of specimens in soils of 2.5, 3.5 and 5 pHs were found 15%, 15% and 8 % higher than the specimens in the simulated soil solutions of same pH respectively after 180 days of corrosion. Upon completion of one year of corrosion, these percentages increase to 17%, 26% and 15% in soil pH 2.5, 3.5 & 5 respectively compared to that immersed in solution. To clarify this further, the percentage change in the corrosion rates of the specimens buried in soil from their respective soil solutions is highlighted in Table 6.

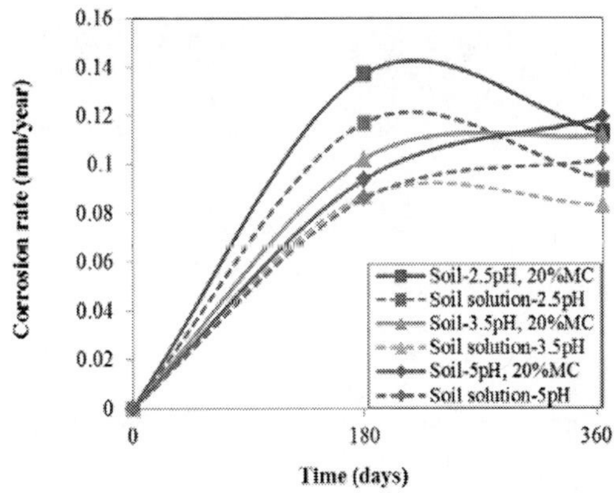

Figure 11 Comparison of corrosion behavior between in soils and simulated soil solutions

Table 6. Comparison of corrosion rates of specimens in soil and soil solution mediums

pH	Days	Corrosion rates in soil	Corrosion rates in soil solution	Percentage difference in corrosion rates
2.5	180	0.137	0.117	15%
	360	0.113	0.094	17%
3.5	180	0.102	0.087	15%
	360	0.112	0.083	26%
5	180	0.094	0.086	8%
	360	0.119	0.102	14%

Two possibilities can be identified for the high corrosion rates in soil as compared to the simulated soil solutions;

- The porous structure of the soil allows more diffusion of oxygen into the pores which can result in the high corrosion rates as compared to the solution with less dissolved oxygen. Also, it is well known that the presence of oxygen is a key factor in corrosion of metals (oxidation process) to take place along with moisture (reduction process).
- Alternate drying and wetting of soil setups (water addition to keep moisture constant) at constant temperature and humidity could be another reason of high corrosion rates in soil medium. The alternate drying and wetting is also a common procedure for corrosion acceleration in laboratory practice. Alternate wet and dry cycles may result in the formation of partially dry corrosion products that can further cause the accumulation of corrosive agents by absorption of moisture and the creation of under deposit corrosion.

4. DISCUSSION

From the tests results and analysis, further observation and discussion can be made for the comparison and correlation of corrosion in soil and simulated soil solution. The visual observations of the corroded specimens can provide some insight into the behavior of corrosion in different environments of corrosive mediums. From Figures 12, it can be seen that a mild change can be observed in terms of change of color (rust) of specimens taken out from soil environments. After one year of corrosion, the specimens taken out from the soil of 2.5 pH with 20 % moisture content can be seen as heavily corroded, followed by the corrosion of the specimen under 5 pH with 20 % moisture content soil. The least corroded seems to be a specimen from 3.5 pH soil. The color of the corrosion is different in all the specimens, which is due to the reaction of soil constituents with metal specimens under varying pH resulting in the formation of different corrosion products.

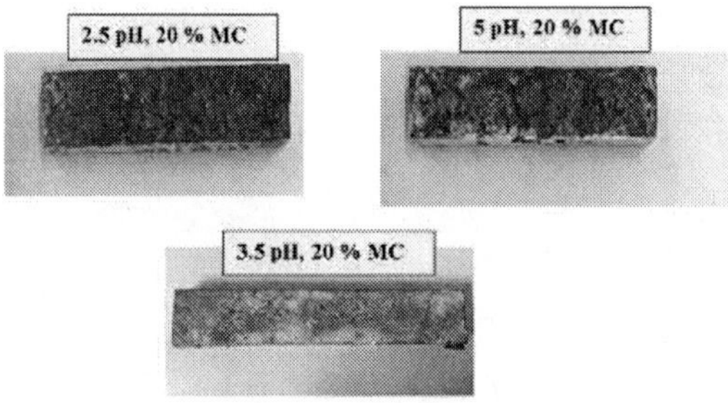

Figure 12 Visual observations of corroded specimens taken out from various soil environments

Corrosion pattern (such as pitting and its distribution) is clearly visible on the samples buried in soil (Fig. 12). The specimen taken out from the soil of the lowest pH (2.5) has a high density of pits with varying sizes. While, the specimen in a soil of higher pH i.e 5, has small evenly distributed pits around its exposed surface. The specimens dug out from the soil of 3.5 pH has random pits with uniform corrosion as shown in Figure 13. In contrast, the rust color is almost the same in all the specimens taken out from various solutions and the corrosion seems to be uniform in all the specimens (Fig. 13). From this observation, it can be noted that the solution is providing a uniform environment for corrosion resulting in uniform corrosion.

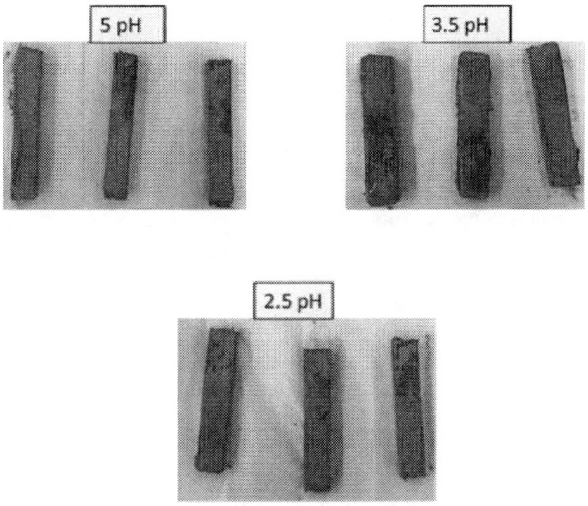

Figure 13 Visual observations of the specimens removed from various simulated solutions

Interestingly upon a comparison of the monitoring observations of soil and simulated soil set up, the pH of the soil remained fairly uniform for longer times as compared to the pH of the solutions during testing. The reason may be the free movement of various ions in solutions that reduced the hydrogen ions concentration as compared to the soil where its resistivity prevented the change in pH for longer times.

In overall, results revealed that the influence of the key elements, such as pH and soil moisture for corrosion of buried metal pipes are well captured by the soil medium in contrast to the simulated solution, resulting 15-26% difference in corrosion rates. While the contribution of soil particles and depth of burial are found as less significant for corrosion rates of buried pipes, however, it is important to highlight that variation of burial depth can generate different corrosion rates (Chesnaux, R., 2009).

5. CONCLUSIONS

The current research investigates the long-term evaluation of corrosion of cast iron in soil and simulated soil medium using an extensive laboratory experimental program. The results indicated that the corrosion rates of the specimens buried in soil were considerably higher than the corrosion rates of specimens submerged in simulated soil solution having the same pH. i.e., the corrosion rates of specimens buried in soils were 15%, 15% and 8 % higher than that observed in specimens immersed in solution at the identical pH's of 2.5, 3.5 and 5 respectively. Upon completion of one year of corrosion, these percentages increase to 17%, 26% and 15% for 2.5 pH, 3.5 pH and 5 pH respectively. Due to such considerable difference of corrosion rates in solutions, careful attention should be devoted when assessing the condition of buried metal pipes based on corrosion idealizations in solutions to avoid any inaccuracy in failure prediction of pipes. The solution based corrosion rates should be co-related with the realistic corrosion rates that are anticipated in buried soil environment using the results such as derived from the current study. The research is being extended to identify a correlation model to compare the corrosion rates of cast iron metals between soil and simulated soil solution mediums. The outcomes of this research will result in reliable, easier, quicker, cost effective and less labor extensive simulations of external corrosion of buried ferrous metal pipes in laboratories. The development of such correlation models will facilitate more convenient and reliable service life prediction of cast iron pipes using corrosion assessments in simulated soil solutions.

6. ACKNOWLEDGMENTS

The financial support from Australian Research Council under DP140101547, LP150100413 and DP17010224 is gratefully acknowledged.

7. REFERENCES

ASTM G1 - 03(2011). Standard practice for preparing, cleaning, and evaluating corrosion test specimens. *ASTM International.*

ASTM E1820 – 2013. Standard test method for measurement of fracture toughness. *ASTM International.*

CHESNAUX, R., 2009. How groundwater seepage and transport modelling software can be useful for studying gaseous transport in an unsaturated soil. Water and Environment Journal, 23(1), 32-40.

G2MT LABS. http://www.g2mtlabs.com/corrosion/cost-of-corrosion/ HANDBOOK, A. M. 1987. Volume 13. *Corrosion, ASM International.*

KIM, J.-G. & KIM, Y.-W. 2001. Cathodic protection criteria of thermally insulated pipeline buried in soil. *Corrosion Science,* 43, 2011-2021.

KLEINER, Y. & RAJANI, B. 2001. Comprehensive review of structural deterioration of water mains: statistical models. *Urban Water,* 3, 131-150.

LI, C.Q. AND MAHMOODIAN, M., 2013. Risk based service life prediction of underground cast iron pipes subjected to corrosion. *Reliability Engineering & System Safety,* 119, pp.102-108.

LIU, Z. Y., LI, X. G., DU, C. W. & CHENG, Y. F. 2009a. Local additional potential model for effect of strain rate on SCC of pipeline steel in an acidic soil solution. *Corrosion Science,* 51, 2863-2871.

LIU, Z. Y., LI, X. G., DU, C. W., LU, L., ZHANG, Y. R. & CHENG, Y. F. 2009b. Effect of inclusions on initiation of stress corrosion cracks in X70 pipeline steel in an acidic soil environment. *Corrosion Science,* 51, 895-900.

MAHMOODIAN, M., & LI, C. Q. (2016). Serviceability assessment and sensitivity analysis of cast iron water pipes under time-dependent deterioration using stochastic approaches. *Journal of Water Supply: Research and Technology- Aqua,* 65(7), 530-540.

MOHEBBI, H. & LI, C. 2011. Experimental investigation on corrosion of cast iron pipes. *International Journal of Corrosion,* 2011.

MURRAY, J. N. & MORAN, P. J. 1989. Influence of Moisture on Corrosion of Pipeline Steel in Soils Using In Situ Impedance Spectroscopy. *Corrosion,* 45, 34-43.

NACE, 2002. Pipeline external corrosion direct assessment methodology. Houston, TX: *NACE* .

NOOR, E. A. & AL-MOUBARAKI, A. H. 2014. Influence of Soil Moisture Content on the Corrosion Behavior of X60 Steel in Different Soils. *Arabian Journal for Science and Engineering,* 1-15.

PETERSEN, R., DAFTER, M. & MELCHERS, R. 2013. Long-term corrosion of buried cast iron water mains: field data collection and model calibration. *Water Asset Management International,* 9, 13-17.

ROMANOFF, M. 1957. Underground corrosion, National Bureau of Standards Circular 579. *US Government Printing Office, Washington,* 25.

SCHWERDTFEGER, W. 1953. Laboratory Measurement of the Corrosion of Ferrous Metals in Soils. *National Bureau of Standards.*

SPARKS, D. L. 2003. *Environmental soil chemistry,* Academic press.

WAN, Y., DING, L., WANG, X., LI, Y., SUN, H. & WANG, Q. 2013. Corrosion Behaviors of Q235 Steel in Indoor Soil. *Int. J. Electrochem. Sci,* 8, 12531-12542.

WORTHINGTON, A. C. 2011. Productivity, efficiency and technological progress in Australia's urban water utilities. *Waterlines Report Series, National Water Commission, Canberra (forthcoming).*

WU, Y. H., LIU, T. M., LUO, S. X. & SUN, C. 2010. Corrosion characteristics of Q235 steel in simulated Yingtan soil solutions. *Materialwissenschaft und Werkstofftechnik,* 41, 142-146.

YAN, M., WANG, J., HAN, E. & KE, W. 2008. Local environment under simulated disbonded coating on steel pipelines in soil solution. *Corrosion Science,* 50, 1331-1339.

Yang, Yao, and Y. Frank Cheng. 2016. Effect of stress on corrosion at crack tip on pipeline steel in a near-neutral pH solution. *Journal of Materials Engineering and Performance,* 25, 4988-4995.

8. AUTHOR DETAILS

Muhammad Wasim is a PhD student in the School of Engineering of RMIT University, Melbourne Australia. His research interest is related to corrosion and its effect analysis on the mechanical properties of buried pipes. Prior to his current position he worked as a structural engineer for two year in top consulting firm in Pakistan and almost 5 year as a research engineer/associate in the Centre of Excellence for Concrete Research and Testing (CoECRT) of King Saud University, Saudi Arabia.

Chun-Qing Li is a professor of civil engineering in the School of Engineering of RMIT University, Melbourne Australia. His research areas include risk and reliability analysis of engineering works, service life prediction of civil infrastructure, steel corrosion in concrete and its effects on concrete structures, deterioration of materials and structures, methodology of whole life design and assessment of civil infrastructure, fracture mechanics of buried pipes and construction management. Prof. Li has published over 200 peer reviewed technical papers in above said research areas and currently have over $2.5 m grants including three ARC grants.

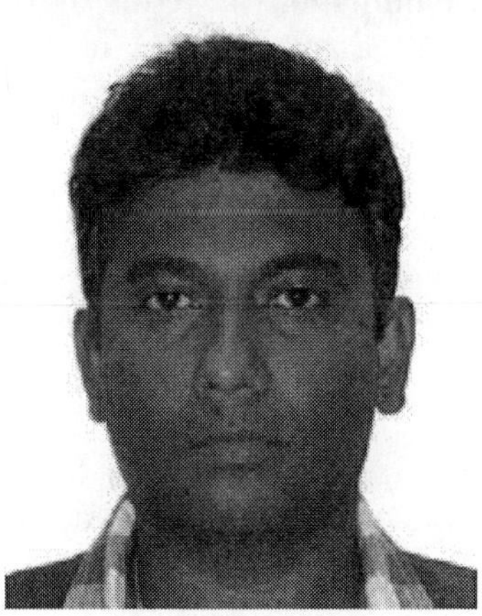

Dr Dilan Robert is a lecturer in Civil Engineering at RMIT University. He obtained his first degree (2005) from Moratuwa University, Sri Lanka and the PhD degree (2011) from University of Cambridge, UK in 2011. Prior to his current position at RMIT, he has worked 2 years as a geotechnical engineer in the UK and 2 years at Monash University, Australia as a research fellow. Dr Robert has published over 50 peer reviewed technical papers in soil-structure analysis since 2011.

EPOXY BASED MATERIALS FOR ELEVATED TEMPERATURE END USES: WHAT DO WE REALLY KNOW?

Surojit B Mukherjee*, Adam Ovington[+]
***International Paint Ltd., Protective Coatings, 6001 Antoine Drive, Houston, Texas, 77091, USA**
[+]International Paint Ltd., Protective Coatings, Stoney gate Lane, Felling, Tyne & Wear, NE10 0JY, UK

SUMMARY: Epoxy phenolic based coatings have become the industry standard for CUI protection, widely respected for their temperature and corrosion resistance, while alternative epoxy technologies have received little attention. As a result epoxy phenolic technology takes a key place in the NACE SP0198_2016 CUI guideline offering protection across the critical CUI range of temperatures. This paper will take a fresh look at epoxy phenolic technology, reviewing chemistry, terminology and performance when subjected to a wide range of operating conditions, both insulated and un-insulated with a view to characterizing performance and understanding why it has become the first choice for many in terms of CUI protection. The paper will subsequently benchmark epoxy phenolic performance against that of alternative epoxy technologies in order to better understand their limitations in terms of application, temperature resistance and corrosion protection in both insulated and un-insulated service conditions. The paper will conclude with a review of the guidance offered in NACE SP0198 regarding conventional epoxy and epoxy phenolic technologies and the terminology used to describe them.

Key Words: Epoxy, Phenolic, Novolac, CUI, SP0198, benchmark, corrosion.

INTRODUCTION

Epoxy based chemistry has been used in the coating industry over the past seven decades. The creation of specification positions (using chemistry terms interchangeably) has created a confusing number of generic chemistries such as epoxy, epoxy phenolic, epoxy novolac, epoxy phenolic novolac, etc. There are however significant structural and functional differences between these materials, as established in the published literature. Table 1 & Figure 1 below list some of these intricacies including differing epoxy functionality of the generic types, associated differences in viscosity and the different chemical structure for bisphenol-A epoxy resin, bisphenol-F epoxy resin and epoxy phenol novolac (the latter 2 often referred to as epoxy phenolic in product data sheets and specifications). From Table 1 and Figure 1 it can is clear that bisphenol-F epoxy resin and epoxy phenol novolac epoxy resin, while often referred to as being generically similar differ in the key areas of epoxy functionality (higher for epoxy phenol novolac which speaks to high temperature resistance), viscosity (directly linked to application benefits) and chemical structure. In pursuit of these terminology driven specification positions, benefits from simpler (application benefits) and cost effective solutions are often overlooked. The primary focus of this paper is to review the performance of epoxy chemistry up to 232°C (450°F).

The NACE SP0198 [1] guideline clearly defines the background to many of the topics touched on in this paper and itself provides evidence to support the discussion. Field experience has shown many instances of CUI on equipment operating at temperatures up to 175°C (350°F), and it has been recognized that carbon steel operating with a skin temperature in the temperature range of –4°C (25°F) to 175°C (350°F) has the greatest likelihood of CUI. Equipment that operates continuously below –4°C (25°F) usually remains free of corrosion. Corrosion of equipment operating above 175°C (350°F) is reduced because the carbon steel surface is warm enough to remain dry. However, corrosion tends to occur at those points of water entry into the insulation system where the temperature is below 175°C (350°F) and when the equipment is idle.

Table 1 – typical epoxy resin types and key property

Epoxy type	Typical functionality	Relative viscosity
Bisphenol A (DGEBA)	1.9	Medium
Bisphenol F (DGEBF)	2.1	Low
Epoxy (phenol) novolac	2.6 – 3.5	High
Epoxy cresol novolac	2.7 – 5.4	Very high
Resorcinol DGE	~ 2	Very Low

Figure 1 – molecular structure differences for key epoxy resin types

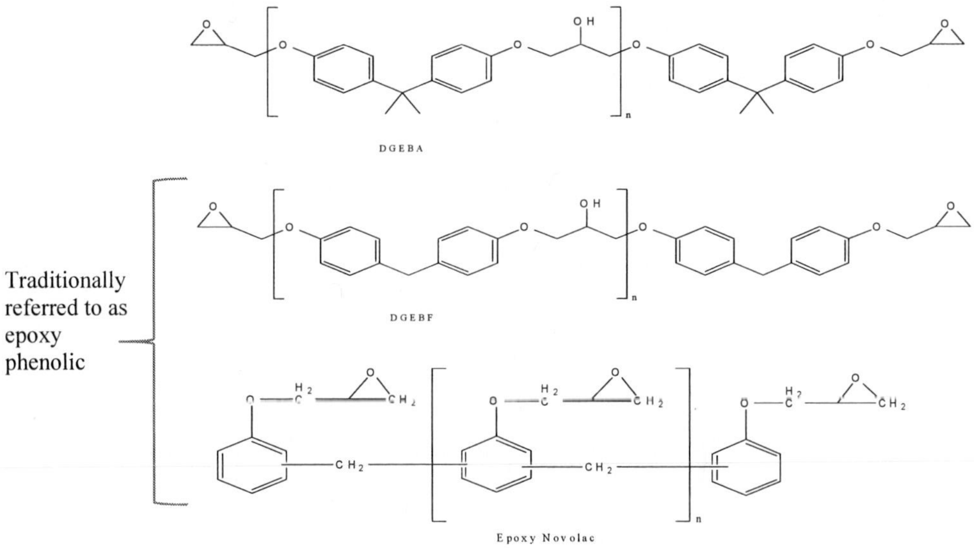

Variability in functionally affects CUI performance significantly, however there are numerous other intricacies to consider. The confusion is compounded by industry standards and established commercial products available off-the-shelf. See Table 2 for some examples of the confusion in the market.

Table 2 – variation in industry standards for high temperature epoxy products

Standard	Epoxy Product Specified	Specified Temperature
NACE SP0198	Novolac	205°C
NACE SP0198	Phenolic	150°C
CUSTOMER A	Phenolic	200°C
NORSOK M501[2]	Phenolic	150°C
CUSTOMER B	Phenolic	200°C
CUSTOMER B	Novolac	200°C

The key factor to note is within the terminology of 'Epoxy Phenolic' where there is variability in the temperature range. This sends mixed signals to the owner / specifier and raises the question 'what material is being supplied?'. While it is true that chemistry type can give an indication of performance there are many other formulation factors that will impact performance, e.g. filler package, additives used to react/fuse into the polymer network and type of amine curing agent.

Experimental

In order to maintain consistency in testing and evaluation surface preparation and application was kept identical. Both pipe sections and flat panels were utilized for testing. Performance evaluation was carried out using three test protocols as listed below –

1. Dry heat testing
2. Corrosion under insulation (CUI)
3. High heat tolerance testing

All tests were performed on carbon steel substrate and surface preparation was done by grit blasting to Sa 2.5 with a profile of 50 – 65 microns (2 – 2.6 mils). Coating material was airless spray applied to a thickness as specified on published product data sheets.

DRY HEAT TESTING

The objective of this test is to identify the maximum dry temperature resistance that a coating can withstand. The coating to be tested was applied on carbon steel pipe sections of 60 cm length and 4mm thickness. The pipe was insulated to minimize heat loss. The hot plate was set at a desired temperature to create a temperature gradient along the pipe. The test was run for 30 days continuously with no temperature cycling. After 30 days the pipe was removed and the coating evaluated for discolouration, cracking, flaking and disbondment. Figure 2 shows a typical set up while Figures 3 and 4 show various specimens after 30 days of dry heat testing.

Figure 2 - Typical pipe test set up for dry heat testing

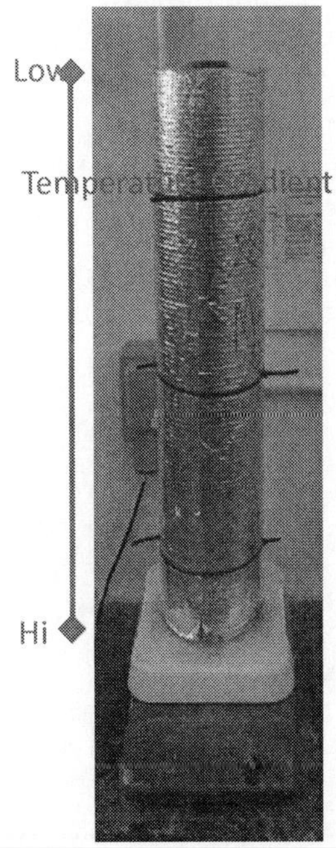

Figure 3 – Comparison of epoxy Novolacs after dry heat testing

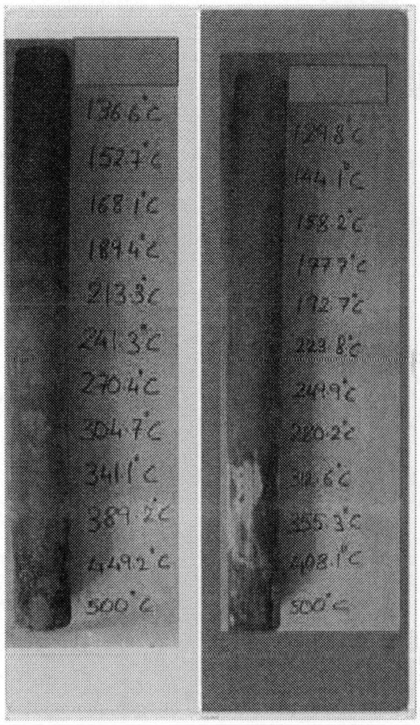

Figure 4 – Comparison of Bis A Epoxies after dry heat testing

Figure 3 shows two different products, both based upon epoxy novolac technology. From the calibration temperature noted up the right hand side of the pipe it can be seen that the left hand pipe fails before the 213°C mark, while the right hand pipe maintains its integrity up to 223°C. This simple test demonstrates that relying upon generic chemistry type alone will not give an accurate picture of maximum temperature resistance. The same picture is presented in Figure 4 where two different Bisphenol A based epoxy coatings are tested to the same method and here you can see that the left hand pipe demonstrates much higher temperature resistance than the right one.

Corrosion Under Insulation Testing

Using a similar test set-up shown previously, pipe sections were prepared and subjected to cycling testing. The pipes were insulated to minimize heat loss. The insulated pipe was then subjected to the following cycle:

- Cycle description:
  - Add 1 litre of salt water (1% NaCl Solution) to the thermal insulation
  - Apply heat for 8 hours to produce a thermal gradient
  - Add one more litre of salt water
  - Allow to cool to ambient for 16 hours

Figures 5 and 6 show the set up and cyclic process carried out. The test was continued for 30 days, i.e. 30 cycles, and then as previously the pipes were evaluated for discolouration, cracking, flaking and disbondment.

Figure 5 - Typical pipe test set up for CUI testing

Figure 6 – Test cycle for CUI testing

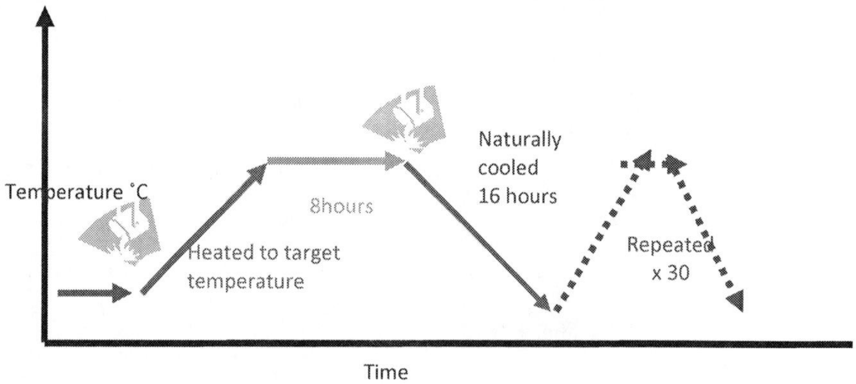

Figures 7 and 8 show the performance of two different epoxy chemistries tested under identical conditions. In Figure 7 both systems are epoxy novolac based but candidate 2 outperforms candidate 1 in this corrosion under insulation test. Both formulations differ in curing agent (amine) chemistry and subsequently cure characteristics. In Figure 8 both formulations are based on Bis A epoxy but candidate 2 significantly outperforms candidate 1.

Figure 7 - Comparison of epoxy Novolacs after CUI testing

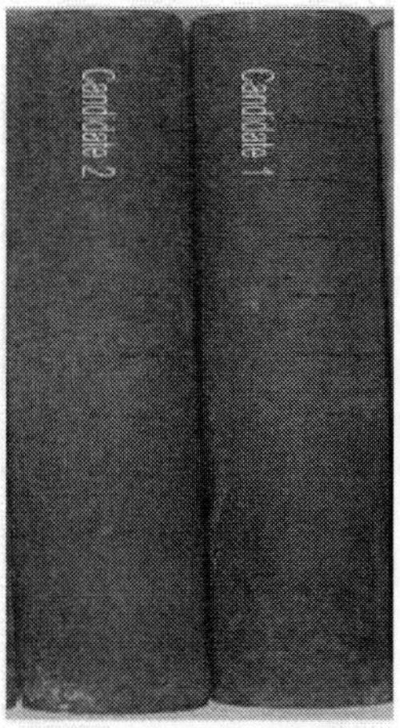

Epoxy Novolac Systems

Figure 8 - Comparison of Bis A Epoxies after CUI testing

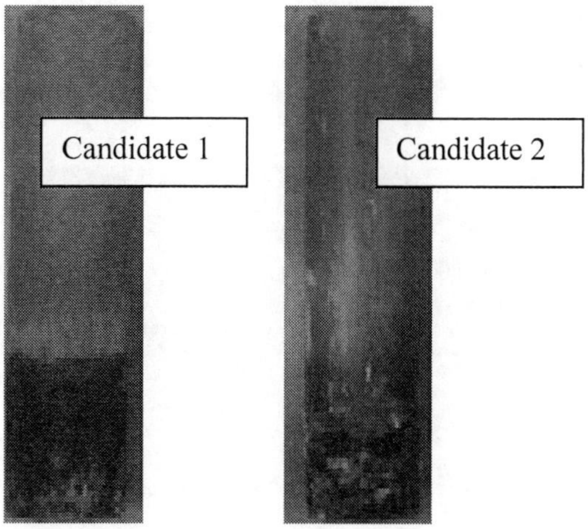

Bis A based Systems

The above work shows that careful consideration of coating chemistry is necessary prior to selection of a suitable coating.

HIGH HEAT TOLERANCE TESTING

This test program included high temperature cycling with 5 cycles up to 200°C (392°F). For panels that appeared intact following the high temperature cycling further anti-corrosive testing was conducted to assess the ability of the temperature cycled coating to continue to offer corrosion protection. Figure 9 details the test procedure and durations. Flat panels were utilized. Sample preparation and application was done as per the previously described method.

Figure 9 – test cycle for high heat tolerance testing

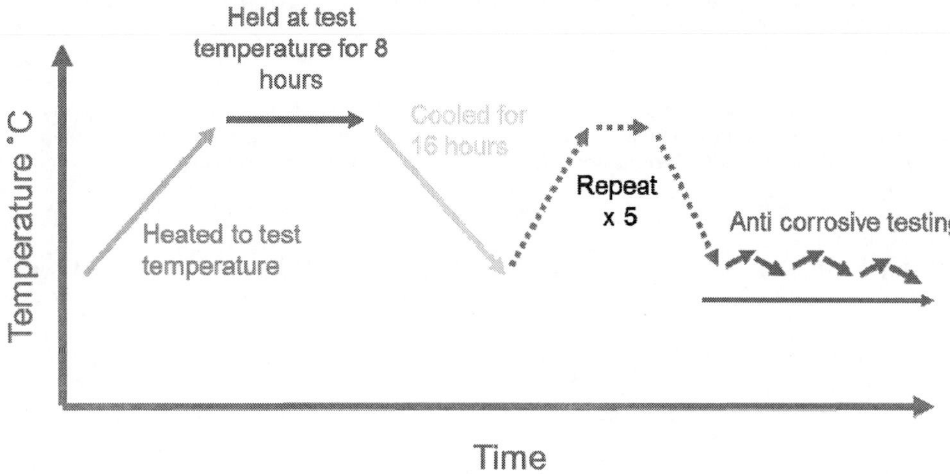

4 different products were put through this test. As before two different commercially available materials described as epoxy novolac based and two commercially available materials described as Bis A epoxy resin based were tested (see Figures 10 and 11, respectively). Both the epoxy novolac and the Bis A materials were applied at the same dfts.

Figure 10 - Comparison of epoxy Novolacs tolerance testing

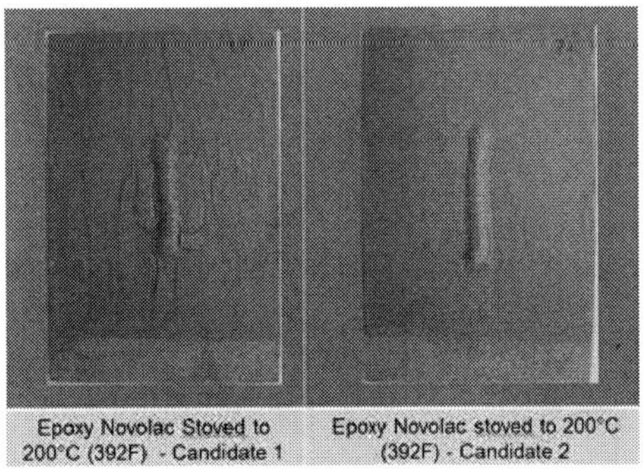

Figure 11 - Comparison of Bis A Epoxies after high heat high heat tolerance testing

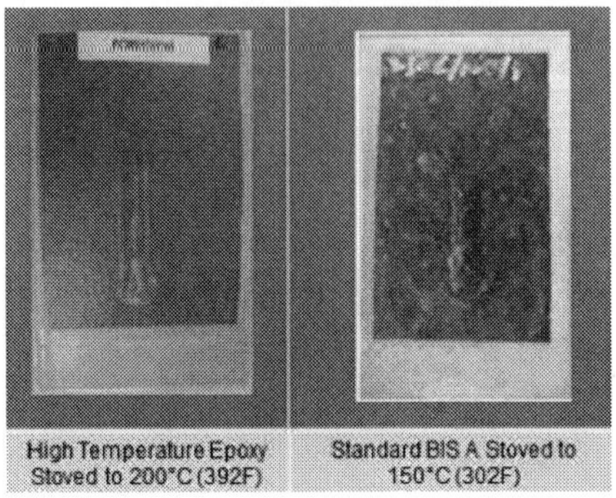

Figure 10 clearly demonstrates that although generically the same, under a simple cyclic temperature test Candidate 2 offers superior elevated temperature resistance compared to candidate 1. Figure 11 demonstrates that for two products based upon Bis A epoxy resin technology, the left hand panel which has been specifically formulated to withstand elevated temperature vastly outperformans a standard Bis A epoxy based technology. What is more, the high temperature epoxy in Figure 11 outperforms the candidate 1 Epoxy Novolac in Figure 10!

CONCLUSIONS

From the extensive work carried out in the laboratory, it has been demonstrated that expectations of elevated temperature performance cannot be based upon resin chemistry alone. There are numerous other ways by which a material's temperature resistance can be improved outside of just generic resin type choice. As such this once again raises the issue of generic resin based coating specifications such as those offered in SP0198, where following correct generic resin type guidance can lead to the use of materials which are not fit for purpose.

To answer this issue thorough lab and field evaluation of a given product is essential to provide confidence in performance for a given elevated temperature scenario. Furthermore consideration should be given to new ways of thinking and approaching coating formulation development that challenges the conventions based upon generic resin type and pushes the boundaries of other materials in order to deliver the next generation of elevated temperature coatings.

REFERENCES

1. NACE SP0198-2016 - Control of Corrosion Under Thermal Insulation and Fireproofing Materials—A Systems Approach

2. NORSOK M501 – Surface preparation and protective coating (Edition 6, February 2012)

FAST CURING POLYCYCLAMINE CURED TANK LININGS

M. Winter
International Paint LLC., Protective Coatings, USA

SUMMARY: Traditional epoxy tank linings are limited in their applications due to minimum curing temperatures and the need for long cure times at low temperatures before being returned to service. Polycyclamine cured epoxies allow for the production of very high solid content, fast curing single coat tank linings systems that can be both applied at low substrate temperatures and cure rapidly at low temperatures to allow for a fast return to service which minimizes downtime for the asset owner.

Furthermore, the use of this technology does not compromise chemical resistance or high service temperature resistance, unlike other low temperature curing systems.

This paper discusses the chemistry and performance characteristics of these systems and demonstrates how the use of such systems can save an owner time and money.

Keywords: Tank Linings, Chemical resistance, Epoxy, Novolac, Polycyclamine, fast return to service

1. INTRODUCTION

Epoxy resin based coatings have been used for many years to line the inside of chemical storage tanks and process vessels. They are particularly suited for services that require resistance to hydrocarbons, water immersion, caustics and certain mineral acids. Generically the coatings are often identified by the terms "epoxy", "epoxy phenolic (or phenolic epoxy)" and novolac epoxy phenolic", although is there is no accepted definition of what this means[1]. The term "phenolic epoxy" is especially abused and has been used for coatings containing Bisphenol A, Bisphenol F and novolac epoxy resins, or combinations thereof.

Apart from the inaccurate use of generic terms for epoxy resins, the use of this terminology misses out on a key part of an epoxy lining formulation – the curing agent. There is a relative handful of epoxy resins to choose from when formulating an epoxy coating, but there is a huge choice of curing agents. The experienced tank lining formulator knows that he or she can change the coating properties more through the choice of curing agent than by choice of epoxy resin. Curing agent selection plays a critical role in the chemical resistance, speed of curing, temperature of curing, ease of application and mechanical performance of the finished product.

2. EPOXY RESIN CHEMISTRY

There are three types of epoxy resin typically used in tank lining formulations – those based on Bisphenol A, Bisphenol F and epoxy phenol novolac resins. All of these resin types are available in a range of molecular weights (& therefore viscosities) and functionalities. Bisphenol F resins have a slightly higher functionality (typically 2.1) and a lower viscosity than equivalent Bisphenol A epoxy resins. Epoxy Phenol Novolac resins have a higher functionality – and therefore can give more highly crosslinked cured films – but also have higher viscosities which makes formulating high solids or solvent free paints more difficult.

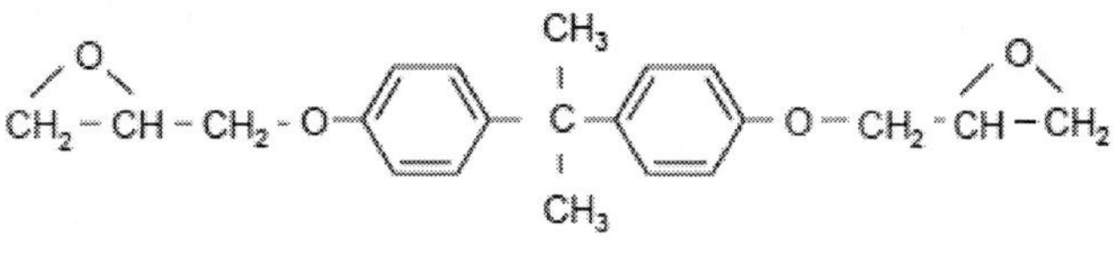

Figure 1: Bisphenol A based epoxy resin

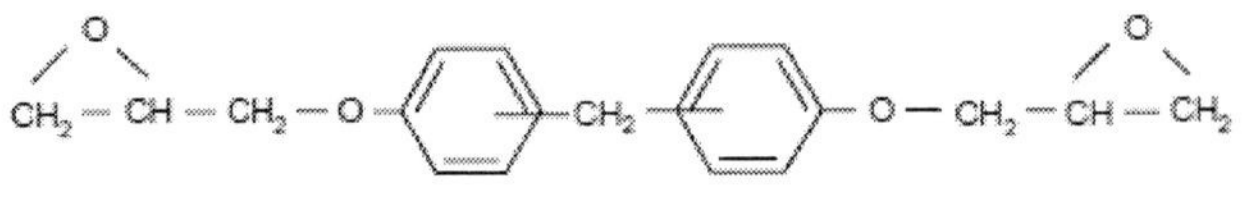

Figure 2: Bisphenol F based epoxy resin

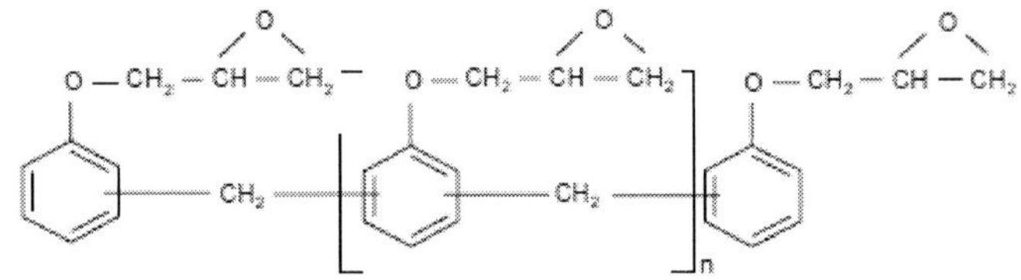

Figure 3: Epoxy phenol novolac resin

In principal, increasing the crosslink density will increase the chemical resistance (while reducing flexibility), at least to solvents, of the cured film, so that typically the chemical resistance imparted by these resins is as follows: epoxy phenol novolac resins > Bisphenol F resins > Bisphenol A. However, the reality of chemical resistance is much more complicated than this and cannot be simply defined according to generic epoxy resin types.

There are certain chemical exposures where the epoxy resin type plays a key role – one example is in concentrated sulfuric acid resistance. The possibility of Bisphenol A epoxy resins to form a stabilized tertiary carbocation in strong acidic conditions reduces their resistance compared to Bisphenol F or epoxy phenol novolac resins – see Figure 4.

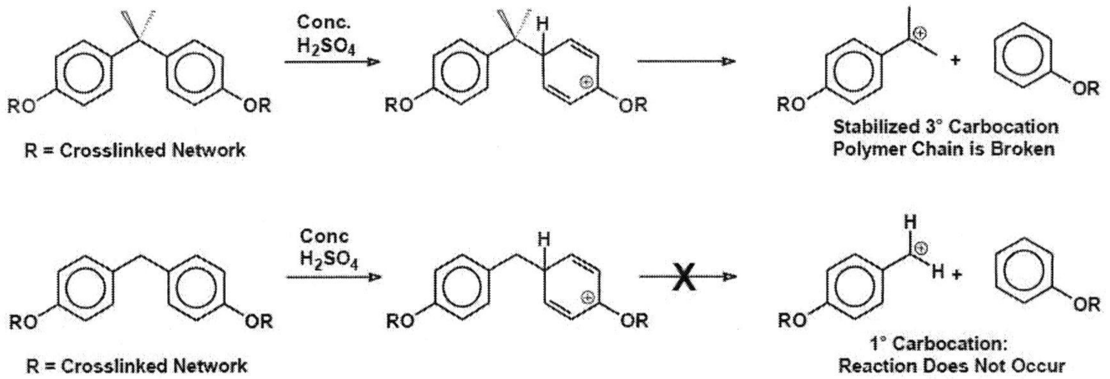

FIGURE 4: Concentrated acid resistance of Bis-A (top) vs Bis-F and Novolac resins (bottom).

3. CURING AGENTS FOR EPOXY RESINS

While there is only three main classes of epoxy resin used in coatings, the formulator can choose from a huge array of curing agents[2]. The original epoxy curing agents were unmodified aliphatic amines, such as diethylene triamine (DETA – see Figure 5), triethylene tetramine (TETA – see Figure 6) and triethylene pentamine (TEPA – see Figure 7). These curing agents can provide high levels of chemical resistance, but when used unmodified give films that are rather brittle and the curing agents have poor compatibility with epoxy resins, potentially causing problems with amine blush. Pre-reacting these amines with epoxy resins to create adducts improves the compatibility, but has little effect on film properties.

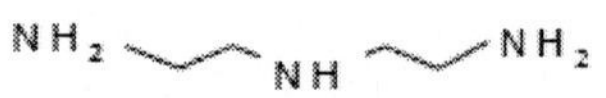

Figure 5: Diethylene triamine

Figure 6: Triethylene tetramine

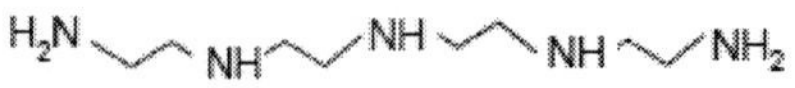

Figure 7 – tetraethylene pentamine

The brittleness of films produced from aliphatic amine curing agents resulted in the development of polyamides curing agents. These are produced by condensing an aliphatic amine with a dimer acid. Although the resulting resins are polyamides, they retain amine functionality which is reactive towards epoxy resins (see Figure 8).

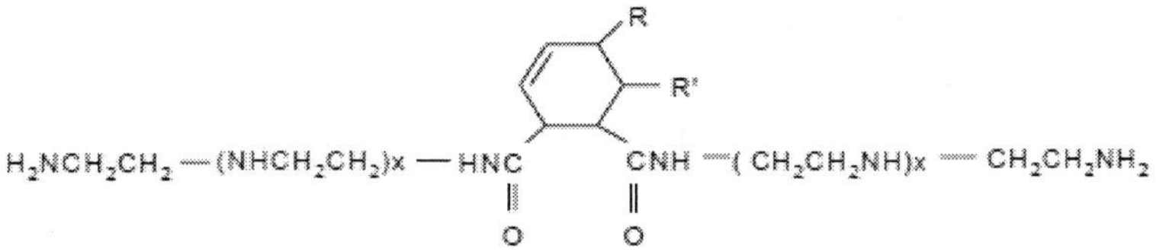

Figure 8: Example of polyamide curing agent

While polyamide curing agents give an increase in flexibility and water resistance compared to aliphatic amine films, they have a lesser resistance to strong solvents and acids and their high viscosity makes it hard to formulate low VOC/high solids coatings.

A later development of polyamide chemistry was to use a monofunctional acid in place of a dimer acid to produce an amidoamine condensation product (see Figure 9). The principal advantage of amidoamines over polyamides is their somewhat lower viscosity that allows for formulation for higher solids systems.

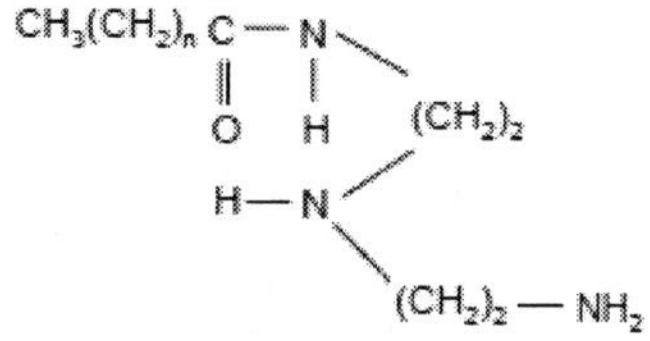

Figure 9: Example of amidoamine curing agent

Aromatic amine curing agents, such as methylene dianiline (MDA – see Figure10) were used for many years as curing agents for epoxy tank linings due to the very high chemical resistant properties that could be obtained. The aromatic backbone of MDA (and other aromatic amines) contributes to the high Tg (glass transition temperature) as well as the excellent acid and solvent resistance that can be obtained using these materials. Aromatic usually require a higher curing temperature to achieve full crosslinking.

Unfortunately, MDA is a suspect carcinogen and the toxicological concerns over the use of this and other aromatic amines has led to them being phased out and formulators have had to search for alternative curing agents to provide the best chemical resistance.

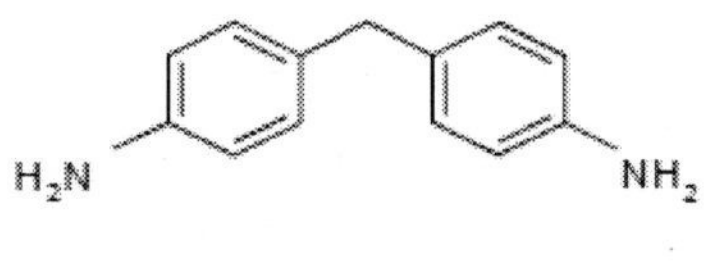

Figure 10: Methylene dianiline

One curing agent that has proved useful in formulating chemical resistant coatings without using aromatic amines is meta xylene diamine (MXDA – see Figure 11). This combines some features of aromatic and aliphatic amines as it has an aromatic ring, but the amine groups are separated from this by methylene groups. Good performance and handling properties are usually achieved by adducting with epoxy resin.

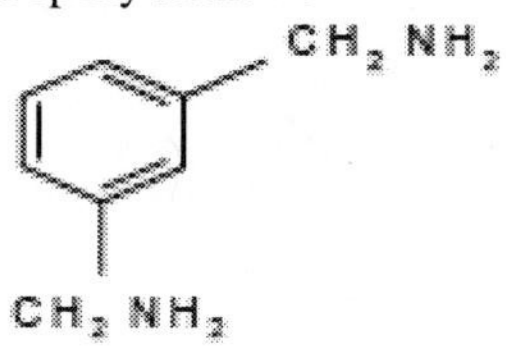

Figure 11: Metaxylyene diamine

Today the most commonly used class of curing agents for formulating high solids and solvent free chemical resistant coatings are cycloaliphatic amines. Examples are isophorone diamine (IPD – see Figure 12), cyclohexane diamine (see Figure 13) and 4,4 – bis (para amino cyclohexyl) methane (see Figure 14). Cycloaliphatic amines are very useful for formulating very low VOC and high solids epoxy coatings with high chemical resistance. It is possible to make coatings with chemical resistance properties approaching those of aromatic amine cured systems, but with improved cure speed and lower temperature curing capability. The downside of these materials can be tendency for amine blushing.

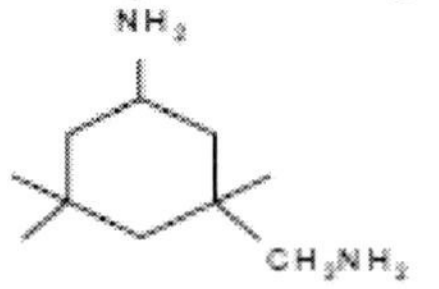

Figure 12: Isophorone diamine

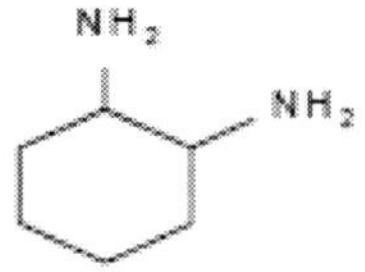

Figure 13: Cyclohexane diamine

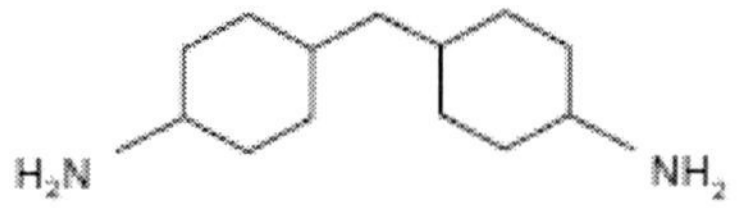

Figure 14: 4,4 bis (para amino cyclohexane)

A more recent development has been the introduction of polycycloaliphatic polyamines (polycyclamines for short), see for example Figure 15. These amines have multiple reaction sites spaced along a cycloaliphatic or partially aromatic backbone and when adducted or otherwise modified can provide even greater performance and cure response than the conventional cycloaliphatic amine curatives. Coatings based on these materials have been reported to show excellent performance in Canadian oil field environments[3]. These curing agents can also be used, either as adducts or blended with other curing agents such as cycloaliphatics, to produce rapid curing, fast return to service tank linings with high level of chemical and temperature resistance.

Figure 15: Examples of polycycloaliphatic polyamine (polycyclamine) curing agents

4. CROSSLINK DENSITY

In the section on epoxy resins an example was given on how the chemistry of the polymer backbone affects chemical resistance. However, the resistance to certain chemicals, especially strong solvents, is largely dependent on the crosslink density in the cured film. In principal, the higher the crosslink density of a cured film, then the greater the resistance to permeation and swelling from solvents or water (permeation can also be modified based on the choice of pigmentation used in the coating). Figure 16 shows how the resistance to methylene chloride for a particular system increases as the degree of cure and therefore crosslink density increases. In this case, the same system was cured at three different temperatures – higher curing temperatures give rise to a higher degree of cure and a higher glass transition temperature for the cured film. The lesser cured films showed a rapid absorption of methylene chloride, followed by a weight loss as some of the unreacted components leach out of the coating film. The more highly crosslinked sample gives a much slower weight gain, indicating a greater permeation resistance.

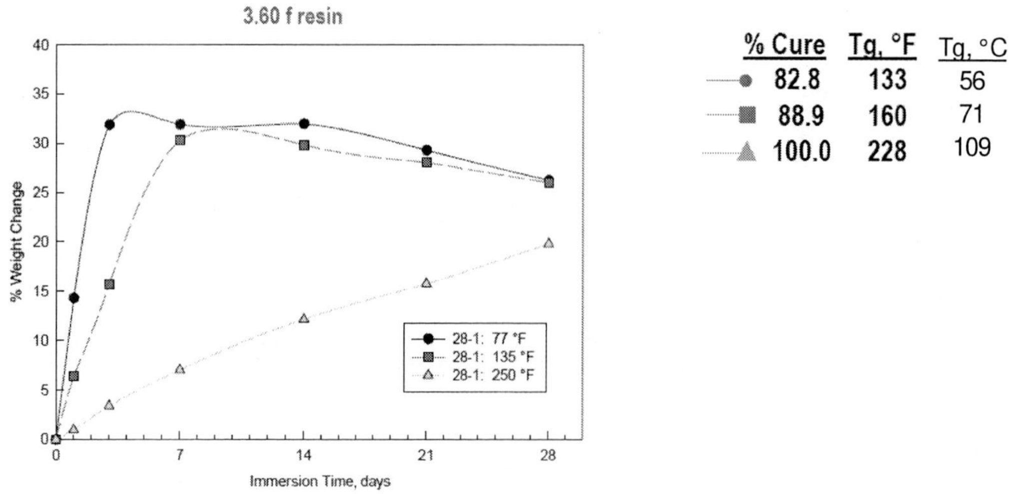

Figure 16: Methylene chloride resistance as a function of degree of cure

It would therefore stand to reason that choosing epoxy resins and curing agents with the highest functionality and the shortest distance between crosslinking sites would result in a cured film with the highest crosslink density and therefore the highest degree of chemical resistance. If only it was that simple……

With high functional resins curing at normal ambient temperatures, there is a tendency for the coating film to gel before complete reaction has occurred. When this happens, the film is left in a partially cured state, with a significant amount of unreacted epoxide and amine functionality. In addition to not achieving a fully crosslinked state, the unreacted groups may further compromise chemical resistance by being available to chemical attack from certain chemical species. Figure 17 gives graphical depiction of a coating film becoming "frozen" before full crosslinking has occurred.

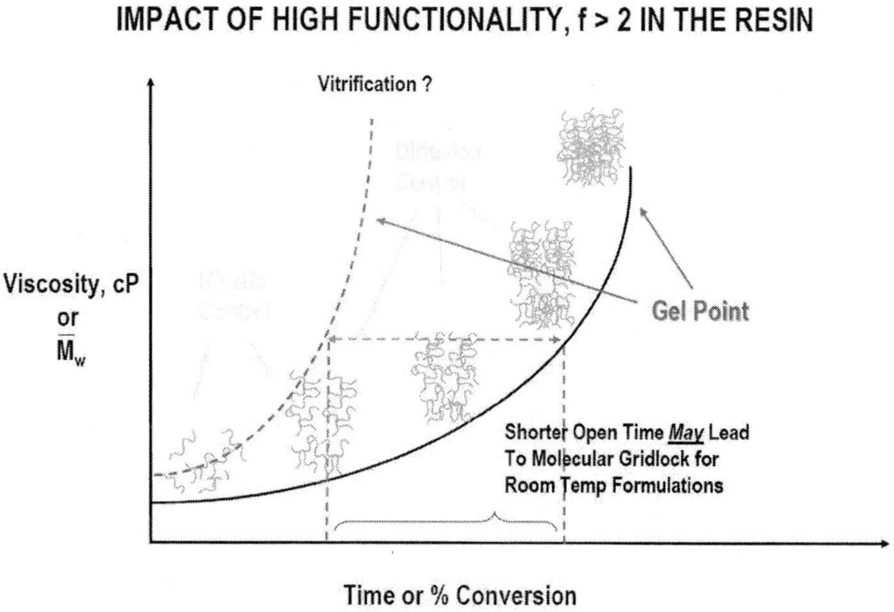

Figure 17: Gelation of a coating before full reaction.

To get around this problem, experienced coating formulators know how to modify formulations using various reactive and non-reactive diluents (these may be benzyl alcohol, nonylphenol and various low molecular weight glycidyl ethers) to increase molecular mobility during curing of the coating film (see Figure 18). There are those in the industry that maintain that using diluents decreases chemical resistance or otherwise impacts long term performance – indeed, inappropriate use of diluent, use of the wrong diluent or adding too much may have a deleterious effect – but the judicious use of the right material may significantly aid in the development of a crosslinked network and therefore help to obtain the maximum degree of chemical resistance. At the end of the day, the properties of the total formulation as applied and cured are most important, rather than the individual components and the coating formulator must understand how to maximize the benefits that may be obtained from a certain material, such as a polycyclamine curing agent, by using the correct type and amount of other formulation components.

IMPACT OF DILUENTS / PLASTICIZERS

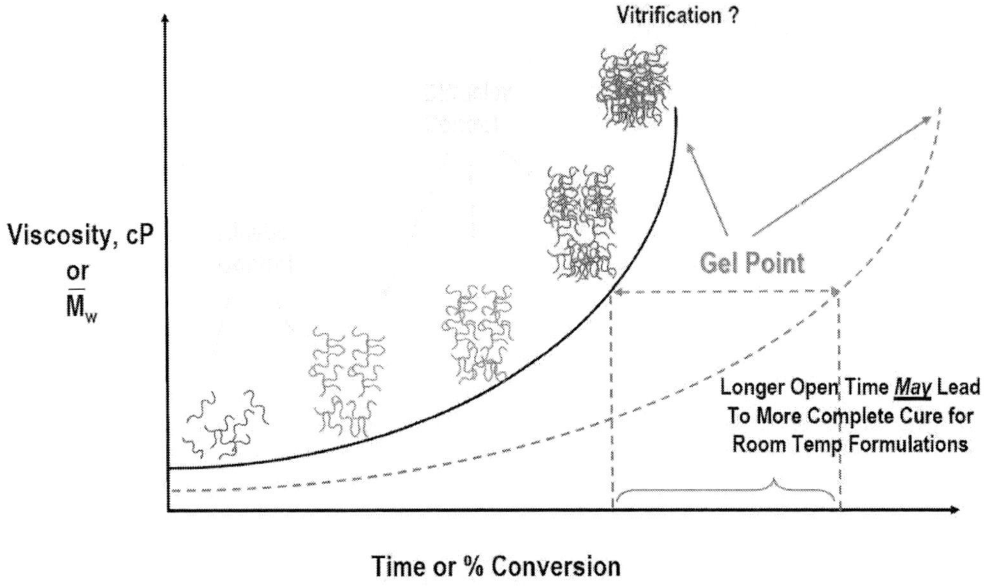

Figure 18: Effect of diluents on cross linking

5. TANK LININGS – RETURN TO SERVICE

To work properly, a tank lining must be resistant to the chemicals that will be stored or processed in the vessel. However, an owner/operator and a painting contractor may also have other requirements that have to be taken into consideration.

Outside of coating selection, the most important factor affecting performance is surface preparation. For chemical tank linings it is critical that the surface is prepared to the required degree of cleanliness, with the correct surface profile for the coating being installed. For aggressive environments, such as hot water services, it may be necessary to wash the surface several times in order to achieve the right degree of cleanliness.

The coating system must be practical in install in the minimum time period possible to allow the equipment to be returned back into service, thus minimizing costly downtime. Historically tank linings were left to cure for periods of 7-10 days at temperatures of 25°C and above before being returned to service. Over time, formulators learnt how to reduce this time period to 3-4 days at 25°C. Now, using well formulated coatings based on polycyclamine curing agents, it is possible to return linings into service after 24 hours or less at 25°C. Extreme chemical services may still require a longer cure time, or a force cure schedule, but a large amount of chemical services can now be handled with epoxy lining systems that are applied in a single coat one day and are placed in service the following day. With the correct selection of epoxy resin type and/or blending with other curing agents, polycyclamine curing agents can be used to produce coatings with resistance to strong solvents, such as methanol or concentrated mineral acids.

The maximum benefit of fast return to service systems is obtained when the coatings are formulated to be applied in a single coat. This requires the coating to be very high solids or solvent free to minimize solvent entrapment when applying a thick film, this may be from 16 to 60 mils thick, depending on product and service environment.

6. FAST CURE POLYCYLAMINE SYSTEMS – LABORATORY PERFORMANCE

Polycyclamine cured systems have been tested in the laboratory against a wide range of chemicals. The chemical resistance profile of an individual coating depends upon all the formulation components, but polycyclamines have been used to create coatings with some of the following interesting chemical resistance properties.

A polyclamine based coating was placed in hot (95^{0}C) deionized water immersion for one year. After one year the coating showed no blisters or other defects (see Figure 19) and no sub film corrosion (see Figure 20).

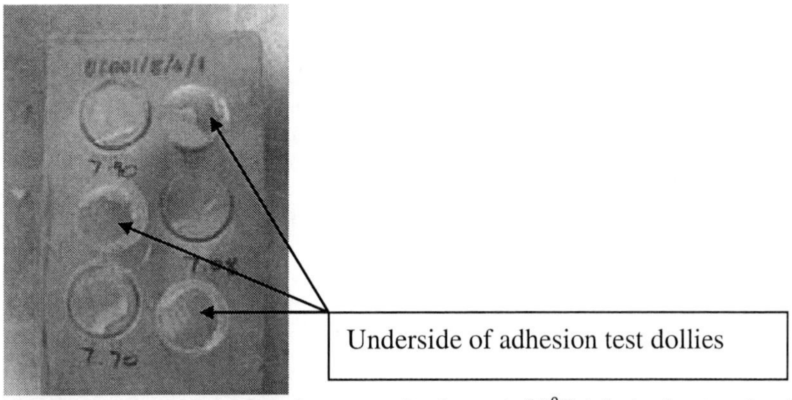

Figure 19: Polycyclamine novolac epoxy after 1 year in 95^{0}C deionized water, showing results of adhesion pull tests indicating good post test adhesion.

Figure 20: No sub-film corrosion after 1 year in 95 C deionized water

The same polycyclamine cured novolac epoxy coating was tested in an autoclave with deionized water at 150^{0}C/5 bar pressure for 30 days (see Figure 21). The coating showed no blistering, flaking or other defects and excellent post-test adhesion.

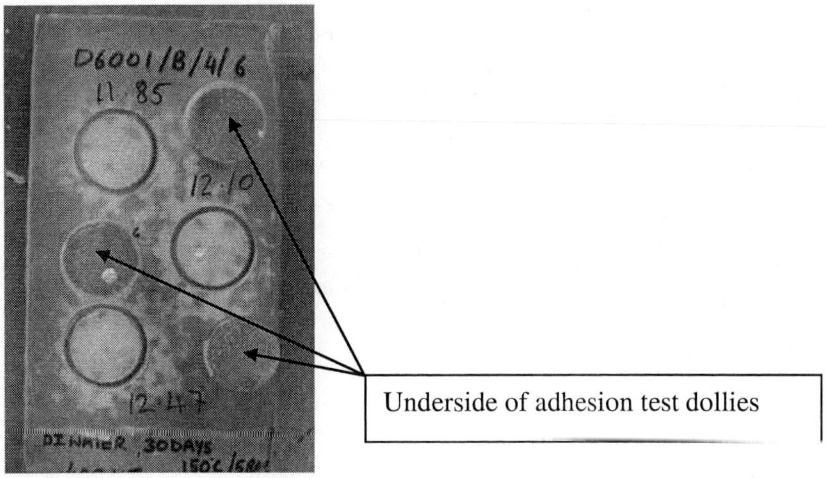

Figure 21: Polycyclamine novolac epoxy after 30 days in deionized water at 150^{0}C/5 bar

A different polyclamine cured novolac epoxy was immersed for 12 months in concentrated (98%) sulfuric acid at 40^{0}C for one year. This coating was post cured at 50^{0}C to achieve maximum chemical resistance. After one year immersion, the coating had discolored, but was otherwise in excellent shape. A similar coating was immersed in 10% sulfuric acid at 60^{0}C for one year without post curing and showed excellent performance, with only a slight discoloration of the coating noticed.

Another different formulation of a polycyclamine cured novolac epoxy was immersed in methanol at 40^{0}C for 2 years without any adverse effect. Again, this particular coating was post cured to achieve maximum chemical resistance.

7. FAST CURE POLYCYLAMINE SYSTEMS – FIELD HISTORY

As mentioned above, polycyclamine curing agents can be formulated together with Bisphenol A, Bisphenol F or novolac epoxy resins (or combinations thereof) depending on the desired properties of the product. Some real life case histories described below illustrate how the fast curing characteristics, extreme chemical resistance and high temperature immersion resistance of polycyclamines can be effectively utilized in practice: A Bisphenol A epoxy/polycyclamine system that is formulated for fast return to service and good resistance to hydrocarbon solvents, gasolines and ethanol was used to coat an ethanol storage tank. During the application the ambient temperature fluctuated from a high of 55°F (13°C) during the daytime down to 35°F (2°C) at night. Despite the cool conditions, the lining was cured sufficiently to be placed back in service after only 3 days.

A novolac epoxy/polycyclamine system formulated for good resistance to sulfuric acid was used to line the internals of a tank used to store 93% sulfuric acid. After 4 years, the lining was inspected and found to be in perfect condition.

A novolac epoxy/polycyclamine coating formulated for high temperature immersion service was used to line produced water vessels that see service temperatures of up to 190-200 °F (88 - 93°C). These vessels were placed back in service 24 hours after being coated. The lining was inspected after 18 months service and found to be in perfect condition (see Figure 22).

An epoxy/polycyclamine coating formulated for acid resistance and high temperature immersion resistance was applied to a deionizer vessel, operating at 215–225 °F (101-107°C) and periodically dosed with 7% HCl and 20% NaOH. After 6 years the lining was inspected and found to be in excellent condition (see Figure 23).

Figure 22: Polycyclamine cured novolac epoxy after 18 months service in hot produced water

Figure 23: Polycyclamine cured epoxy after 6 years in deionizer vessel.
Black streaks are tar balls on the surface of the coating.

8. CONCLUSIONS

Raw material developments have provided coatings formulators with a large tool box of ingredients from which to work. The use of traditional generic descriptions of epoxy resins to categorize performance of chemical resistant linings is a gross oversimplification. Curing agent chemistry, together with other formula ingredients, plays a large role in determining the performance of a particular coating. The use of the most recent advances in curing agent chemistry, such as polycyclamines, can offer significant advantages to an owner in both chemical resistance and rapid return to service when properly formulated into a coating and supported by the correct test data and field experience.

9. REFERENCES

1 See, for example, *"Ask the Coatings Experts – Understanding Epoxies and Where to Use Them"* Edited by Lou Vincent, Materials Performance Vol 49, No 5, May 2010

2 See, for example, *"Protective Coatings, Fundamentals of Chemistry and Composition"* Clive Hare, Chapter 15 Epoxy systems

3 *"Field Performance versus Laboratory Testing: A Study of Epoxy Tank and Vessel Linings used in the Canadian Oil Patch",* Mike O'Donoghue, Ron Garrett, Ron Graham and V.J. Datta

INVESTIGATION OF CORROSION EFFECT ON UNDERGROUND METAL PIPES

C.Q. Li, W.G. Wang, D.J. Robert and A. Zhou
School of Engineering- Civil and Infrastructure Discipline, RMIT University, Melbourne, Australia

SUMMARY: Underground metal pipes are essential infrastructure for water, oil and gas transport. The presence of localized pitting corrosion has been identified as one of the main deterioration mechanisms for this pipe material. These corroded pipes, when exposed to external loadings, can easily fail due to intensified stresses at the corrosion pits. Disruptions to pipelines not only greatly affect the life of citizens, but also cause severe economic loss and safety risk, and hence, accurate prediction of safe design life of buried pipes becomes significant. The main objective of this paper is to investigate the effect of corrosion on bearing capacity of corroded pipes. Two types of exhumed pipes made of cast iron and steel were used for parallel-plate loading tests. The corrosion parameters of the corroded pipes were thoroughly examined by various techniques, such as 3D scanning measurements and secondary electron microscopy (SEM). The results showed a significant reduction of rupture modulus of the cast iron pipes with the increase of exposure time and corrosion depth. This research work will enhance the current knowledge on corrosion to better understand the deterioration of metal pipes for facilitating safe life prediction of buried pipelines.

Keywords: Corrosion, old pipes, cast iron, steel, underground, pit depth, deterioration.

1. INTRODUCTION

Underground pipelines are essential infrastructure that have been widely used for water, gas and sewage transmission. In Australia, there are about 210,000 km of buried pipes and most of pipes are made of metals, e.g., cast iron and steel (Cole and Marney 2012). Due to these pipe have been exposed to aggressive environment such as corrosive soils for long term, material and structural deterioration of metal pipes have caused an unexpected high rate of failures (Hou et al. 2016). For example, over 20 pipe breaks per 100 km were reported in Australia per year, and Australian urban water industry need pay over $ 90 million to replace the pipes (National Water Commission Australia 2009). This figure in Canada is as high as 39 breaks per 100 km per year as reported by Makar (2000). The increasing number of failures in present times not only can greatly disrupt the daily life of citizens, but also cause great economic loss, environmental pollution and social impact. Therefore, it is imperative to thoroughly examine the effect of corrosion on the integrity of pipe structure.

Experience and investigation of pipe failures suggest that corrosion is a main cause of deterioration of buried pipes, resulting in either decrease of effective wall thickness or increase in stress concentration (Seica et al. 2004 and Makar 2000). Localised corrosion, as a result of the formation of aeration cells, is especially regarded as the most hazardous corrosion type for the pipes (Romoff 1957). Depending on material, corrosion type and loading condition, failure of pipelines can occur in different modes (Makar et al. 2001), e.g. leakage, rupture and collapse etc. One of main failure mechanisms is that corrosion induced pits can amplify the stresses in pipe, and they can act as crack initiators in brittle material, e.g. cast iron (Seica et al. 2004). Also, the effect of corrosion induced pits on the residual strength of pipes is important because geometrically interacting pits can further reduce the wall thickness and increase their risk of failure.

In the past decades, several researchers have investigated the effects of corrosion on the integrity of metal pipes (e.g. Rajani 2000, Conlin et al. 1999, Kiefner and Vieth 1989 etc.). More recently, a test program was designed by Hou et al. (2016) to investigate the effect of corrosion on tensile strength and fracture toughness of metals in three different pH environments. It was found that the corrosion reduces both the tensile strength and fracture toughness of metals. Mohebbi and Li (2011) investigated the effect of microstructure morphologies on corrosion behaviour of gray cast iron. They concluded that localised corrosion is the primary form of corrosion of cast iron water pipes and corrosion rate for ferritic cast iron pipes slightly lower than that of the pearlitic ones. Also, a series of full-scale tests were conducted by Dewanbabee and Das (2012) to investigate the effect of corrosion on the axial load-carrying capacity of steel pipes. Artificially machined defects with rectangular and square geometries were used to represent corrosion, and the results revealed that the 'corroded' steel pipe has a softer load-deformation response than that of the non-corroded pipes. Oh et al. (2009) proposed a series of expressions of plastic limit loads for steel pipes with internal local wall thinning under various loading conditions, i.e. internal pressure, bending load and the combination conditions.

Although these earlier studies in general are scientifically significant and provide useful knowledge for investigation of corrosion effect on pipes, the corrosion assessments on corroded pipes dug from field are relatively limited with the exception of few published works (e.g. by Harrison et al. 2004, Seica et al. 2004, Rajani 2000 and Conlin et al. 1999). A thorough review of published literature also reveals that almost no research has been conducted to correlate the reduction of pipe structural strength with the statistics of corrosion depths. Thus, the intention of this paper is to present the results obtained from parallel-plate loading tests on exhumed metal pipes to investigate the effect of corrosion on the reduction of pipe strength. Both cast iron and steel pipe sections obtained from in-service water pipelines in separate sites were used for tests, and a 3D scanning measurement technique was employed to visualize corrosion distribution and compute the depths of corrosion induced pits. The relationship between corrosion depth and the reduction of strength has been correlated. The results presented have the potential to contribute in the service life prediction of underground pipelines.

2. EXPERIMENTAL PROGRAM

2.1 Test Parameters and Material

The common performance limit of most rigid pipes, e.g. cast iron pipes, is pipe wall rupture due to internal or external loads, or the combination effects. It is well known that stresses due to external loads on cast iron pipes are usually not directly considered (Moser and Folkman 2001). Instead, rupture modulus, i.e. crush strength, is often determined and acts as a basic parameter which characterises the loading capacity of pipe against the prism load by including the vertical gravitational action of the soil above the pipe and traffic loads (Seica et al. 2004). From the experimental investigations on coupons or arbitration bars, it can be seen that a ring cut out of pipe was found more closely represent the material (Talbot 1908). The rupture modulus is commonly determined in the laboratory by either a parallel-plate loading test or a three-edge test, and can be expressed as follows (Seica et al. 2004)

$$R = 954\frac{P_{max}\left(I+\bar{t}\right)}{L\cdot\bar{t}^{2}} \qquad (1)$$

where R is modulus of rupture. $P_{\max}$ is the load applied at fracture. L is the mean length of pipe section. I is the mean internal diameter of the ring, mm. $\bar{t}$ is the mean wall thickness of ring measured at fracture position, mm.

For flexible pipes buried in soil, since the deflection capability to carry the imposed soil load must be considered, the performance limits for these pipes, e.g. steel pipes, are usually deflection-related. Thus, pipe stiffness rather than rupture modulus is the critical pipe property, which is commonly represented as (Moser and Folkman 2001)

$$S = \frac{PD}{\Delta L} \qquad (2)$$

where S is defined as pipe stiffness, P is also readily determined by a parallel-plate load at ring deflection $\Delta/D = 5\%$, Δ is decrease in vertical diameter of ellipse from a circle of equal circumference (Watkins and Anderson 1999) and D is diameter corresponding to the neutral surfaces of the cross section of the wall. For most pipe analyses, it is sufficiently accurate to use the mean of outer and inner diameter. L is the length of pipe. It can be seen that pipe stiffness has same unit as the modulus of elasticity. The use of parallel-plate load to determine pipe stiffness can also be justified by the fact that the long term load on flexible pipes, e.g. steel pipe, approaches the prism load (Moser and Folkman 2001).

In the present research, a total of eight exhumed pipes were used , comprising of five cast iron pipes and three steel pipes. All pipe samples were removed from full-length pipelines in service from separate job sites in Melbourne (see Table 1). These exhumed pipes exhibit varying degrees of corrosion with a wide range of service time. The pipe samples chosen for current tests have an inner diameter of 100 mm. The nominal wall thickness varies from 4.8 mm to 11mm. The element compositions of all pipes were determined and presented in Table 2. It can be seen that cast iron is uniquely characterized by an alloy of carbon content over 2% and the steel have carbon under 2%. All of the steel pipes are low carbon steel (under 0.2% carbon) (Keil and Devletian 2011). Also, microstructure analysis has been conducted through a Quanta 200 scanning electron microscope (SEM) to confirm the material type and discriminate the microstructure morphologies. Small samples, approximately 10 mm×10mm× pipe wall thickness were cut from all pipe samples. These samples were mounted in conducting Bakelite and then subjected to metallographic preparation resulting in a 0.1 μm surface finish. The Figure 1 shows typical morphologies of microstructure of pipes used in the current study. As can be seen, Figure 1a shows a type A "scythe/sickle" graphite morphology obtained from cast iron pipe 1 and Figure 1b shows a type B "rosette" graphite morphology obtained from cast iron pipe 2 based on the classification of ASTM A 247 (2004). The results also show an obvious variation in the size and distribution of graphite flake morphology (carbon-rich phases) between different cast iron pipes. The Figures 1c - 1d show the morphologies of two steel pipes with larger grain size observed in pipe 6 in contrast to smaller grain size observed in pipe 7.

Table 1. Summary of pipe information

Pipe No.	Year of burial	Age (yr)	Location	Nominal Thickness	Material	Internal diameter
1	1957	59	Pascoe Vale, Melbourne	11 mm		
2	1963	53	Pascoe Vale, Melbourne	11 mm		
3	1960	56	Footscray, Melbourne	9 mm	Cast iron	
4	1956	60	Eltham, Melbourne	8.6 mm		100 mm
5	1964	52	Dallas, Melbourne	10.7 mm		
6	1964	52	Craigieburn, Melbourne	5 mm		
7	1961	55	Reservoir, Melbourne	4.8 mm	Steel	
8	1953	63	Braybrook, Melbourne	5 mm		

Table 2. Element composition of cast iron and steel pipes

Pipe No.	C	Si	Mn	P	S	Mg	Al
1	3.58	2.48	0.74	0.67	0.06	<0.01	<0.01
2	3.47	2.29	0.41	0.84	0.08	<0.01	<0.01
3	3.59	2.43	0.34	0.53	0.05	<0.01	0.01
4	3.25	2.31	1.30	1.23	0.07	<0.01	<0.01
5	3.60	1.37	0.49	0.12	0.07	<0.01	0.01
6	0.12	0.03	0.35	0.02	0.02	---	<0.01
7	0.07	0.03	0.42	0.02	0.02	---	<0.01
8	0.10	0.09	0.41	0.04	0.04	---	0.02

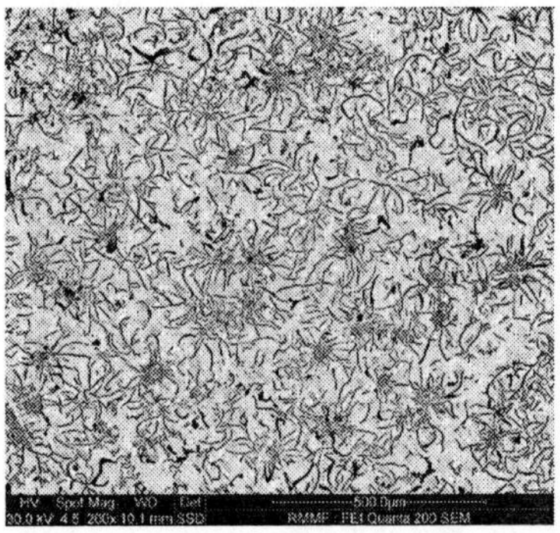

(a)

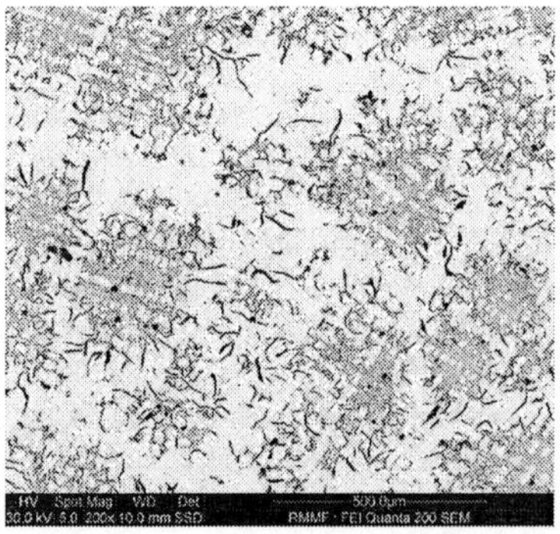

(b)

(c) (d)

Figure 1 Microstructure morphology for (a)-(b) cast iron and (c)-(d) steel pipes

2.2 Specimens for parallel-plate loading test

The specimens used for determining rupture modulus of pipes are based on sections (in 120 mm long) cut from the pipe samples. The selection of these sections within the pipe was made randomly based on the received pipe samples. All pipe sections have a length of 120 mm, which is consistent with published literature (Seica et al. 2004, Rajani 2000 and ASA/AWWA 1962) in which at least 50 % of the pipe diameter was used. Two to three pipe sections were prepared from each of pipe samples based on the availability of material. In total, ten cast iron pipe sections and six steel pipe sections were used for the parallel plate loading test. Same pipe sections were used for 3D scanning measurement.

2.3 Test Setup and Procedure

Before mechanical tests, all pipe sections were cleaned by sandblasting, and then the external surfaces of pipe sections were scanned by Artec Eva spider 3D scanner, with point accuracy up to 0.03 -0.05 mm. To ensure a sufficient accuracy of measurement, a non-corroded pipe section with 11mm nominal wall thickness was scanned by the 3D scanner and the wall thickness information at different locations of ends was extracted (see Figure 2). The 3-D scanner measured wall thicknesses were then compared with the results obtained by a digital calliper ruler. The comparison between 3d scanner data and digital caliper readings is presented in Table 3, showing that that the measurement by the 3D scanner is in good agreement with the calliper ruler measured values with a maximum difference less than 4.5%. This scanner has also been proved to have high resolution and ability to see sharp edges, e.g. pit front. Having verified the accuracy of 3D scanning data, corrosion measurements were performed on pipe sections, and the pit depths were measured by projecting all surface points to a corresponding reference surface that was generated based on the coordinates of non-corroded areas.

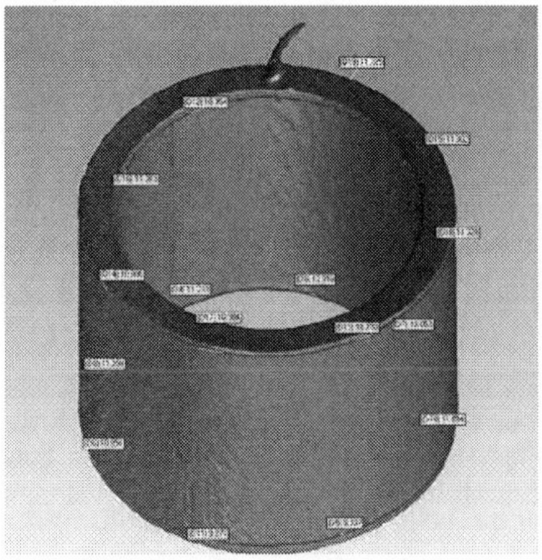

Figure 2. 3D scanned image of the non-corroded pipe section

Table 3. Comparison of 3D scanner and calliper ruler on non-corroded pipe

Digital Vernier Caliper	11.03	11.03	11.17	10.98	10.86	10.52	10.91	11.21
3D scanner	10.95	11.23	11.30	10.77	10.38	10.88	11.38	10.95
Difference (%)	0.69	-1.77	-1.18	1.91	4.38	-3.42	-4.3	2.28

All the tests were conducted in a 1000KN MTS machine and the pipe sections were loaded at a displacement of 0.3 mm/min. A photograph of the parallel-plate test setup used in this research program are shown in Figure 3. The orientation of most severe corrosion areas was made either on top or bottom during the test in order to reduce variations in the rupture loads. To ensure a good contact between loading plate and ring specimens, a resin cushion of about 0.5 mm thick was laid on the bed and bottom of loading plates of the machine. The specimen was put in place so that the bottom element rested in cushion. The load was applied by running down the head of the machine. At each appearance of crack, its location and size were recorded. The loading was continued until the pipe had reached complete failure, i.e. rupture for cast iron pipes and sufficient large deflection for steel pipes. After completion the tests, wall thickness of pipe sections at fracture was measured to the nearest 0.01 mm. The compressive load F (for steel pipe sections) is readily determined until the vertical deflection reach 5 percent of the diameter.

(a) (b) (c)

Figure 3. Parallel-plate loading test (a) before flattening, (b) after flattening for steel pipe and (c) after flattening for cast iron pipe

3. RESULTS AND ANALYSIS

3.1 3D corrosion measurement

The results of 3D corrosion measurement were reported in the form of color-coded contour maps as shown in Figure4-5. For example, the front and back view of pipe No 2 can be seen in the left and right sides of Figure 4a, respectively. A number of on-screen callouts that describe the deviation information at the specific points of interest can be seen in each graph. The reported deviation in these figures are average of the deviations within a user-defined radius (i.e. 0.2 mm) around the inspection point. From the results, it can be seen that the exhumed pipe sections exhibit a range of corrosion depths. This is not unexpected considering the exhumed pipe sections were cut from pipelines at different locations and corrosion of buried iron has been found as correlated to soil types, i.e. it varies from an almost unchanged rate in some soils to nearly stopped in another soils (Schwerdtfeger 1953). Further, distinct differences between cast iron and steel pipes were observed with regards to pit distribution. For example, a widely distributed and deeper corrosion pits were more often resulted in cast iron pipes than in steel pipes (see Table 4). The corrosion induced pits in all steel pipes are shallow and blunt. This implies that although pitting corrosion and uniform corrosion commonly co-exist at the surface of pipe, localised corrosion are prone to become the primary form of cast iron material deterioration.

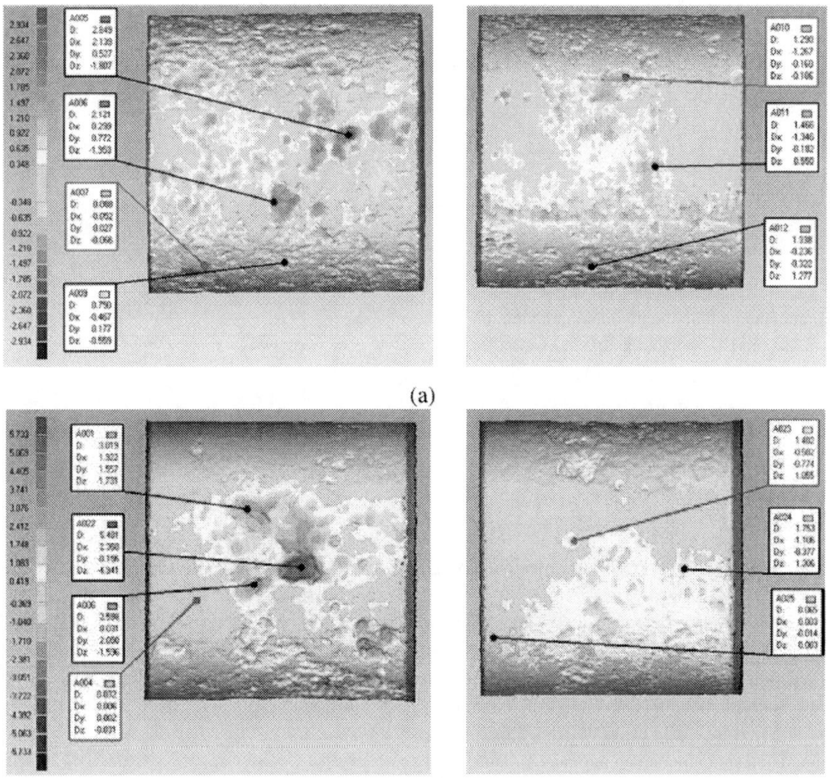

(a)

(b)

Figure 4 3D measurement of cast iron pipe (a) pipe 2 and (b) pipe 3, unit:mm

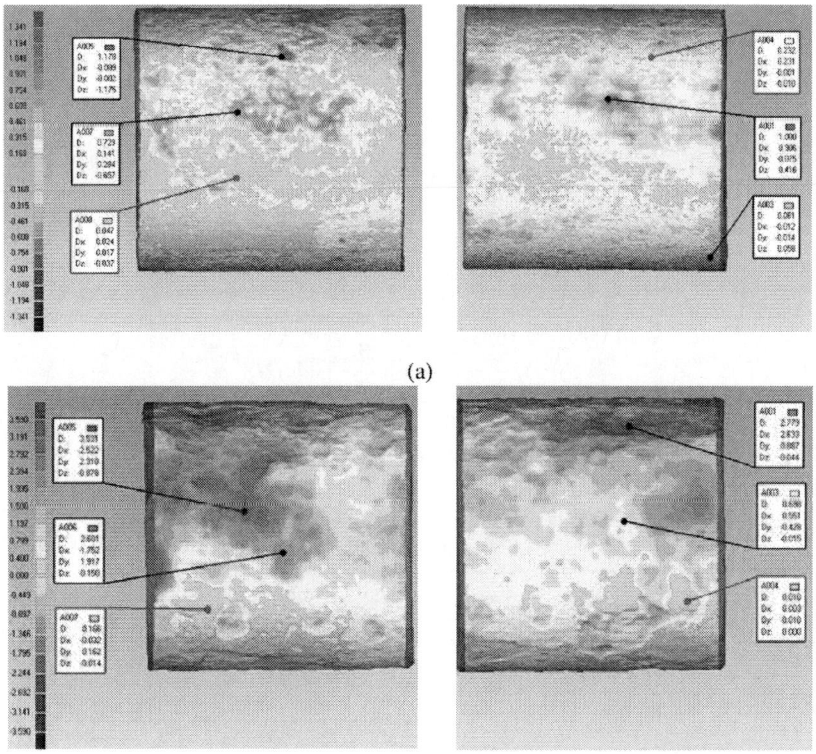

(a)

Figure 5 3D measurement of steel pipe (a) pipe 7 and (b) pipe 8, unit:mm

Table 4. Statistical data on pitting depths for each pipe

Pipe NO.	Maximum(M.)	Average(A.)	A./M.
1	5.785	2.259	0.39
2	2.414	0.3115	0.12
3	6.036	0.5925	0.1
4	3.072	0.926	0.3
6	3.072	0.926	0.3
7	3.667	0.9815	0.27
8	3.625	0.626	0.17

3.2 Load-Deflection Behaviour of Pipes

Due to the difference in intrinsic material properties, different load deflection behaviours are expected for cast iron and steel pipes. From the observation of tests in current research, it has been observed that at a load over 75% of the maximum, a crack initiates either at crown or invert locations for cast iron pipes. An increase of load has caused this crack immediately extended throughout the wall of pipe and rupture was resulted subsequently in every case (See Figure 3c). When the machined was continuously run down after the occurrence of rupture, the pipe section gave further deflection but the load sustained became much less than the load at first breaking and pipe section finally broke again at springline locations. On the other hand, steel specimens demonstrated a continuous deformation under compressive plate loading. Because of the flexibility of the steel pipes, the pipe sections tend to deflect into a primarily ellipse with a decrease in vertical diameter. Examples of force–deflection relationships from pipe No 1 (cast iron) and pipe No 6 (steel) are presented in Figure 5.

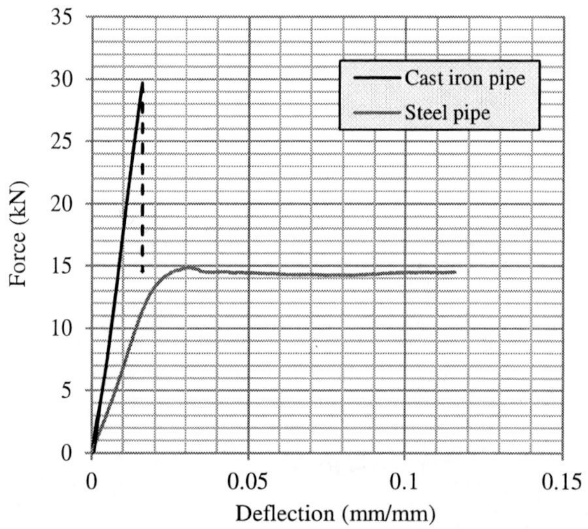

Figure 5 Typical force-deflection response of cast iron and steel pipe section under parallel plate load

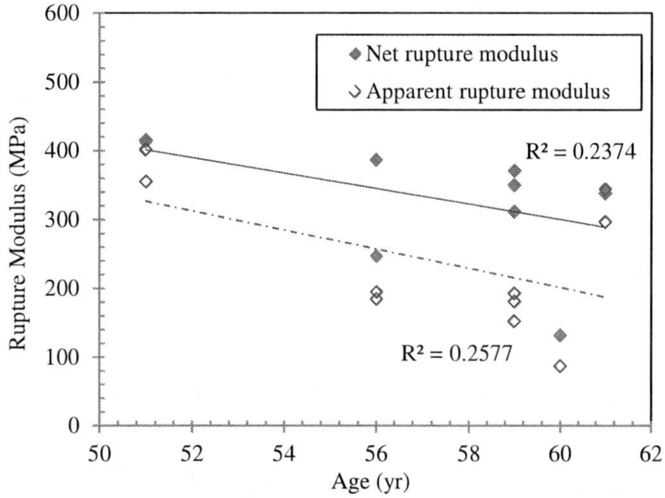

Figure 6 Comparison of the pipe net and apparent rupture modulus versus the service age

3.3 Effect of corrosion depth on the reduction of rupture modulus

Results of the current tests were further analysed to investigate the effect of corrosion on pipe rupture modulus. Two rupture modulus were calculated: the apparent rupture modulus based on the gross wall thickness, and the net rupture modulus calculated by the actual wall thickness. In current study, the gross thickness, i.e. the original non-corroded wall thickness, was approximated by the nominal pipe thickness. As expected, the cast iron pipe sections give a range of rupture modulus due to various chemical composition and microstructure of the material (Conlin et al. 1999). However, most of the pipes exhibited values of net rupture modulus that were close to those given in the specifications and published literature (e.g. by Seica et al. 2004 and Rajani 2000). Also, a lower rupture modulus (apparent) was obtained for pipe sections based on the gross thickness. A strength reduction was then determined as the percentage difference between the apparent and net modulus of rupture. Quantitatively, the reduction of rupture modulus due to corrosion was approximately between 3% and 49 %. It can be seen from Figure 6 that although the data pertain to just five pairs of pipe samples, there is a clear reduction trend of rupture modulus over the service ages and between the two rupture modulus. In order to explore the relationship between corrosion and the reduction of rupture modulus, both the maximum and averaged pit depths obtained by the 3D scanner were plotted against the values of rupture modulus reduction (Fig. 7). It can be seen from the results that both the maximum and averaged pit depths are positively related with the rupture modulus loss. Comparison of the slopes of two trend lines indicates that the reduction of rupture modulus is more sensitive to the increase of averaged pit depth than that of maximum pit depth.

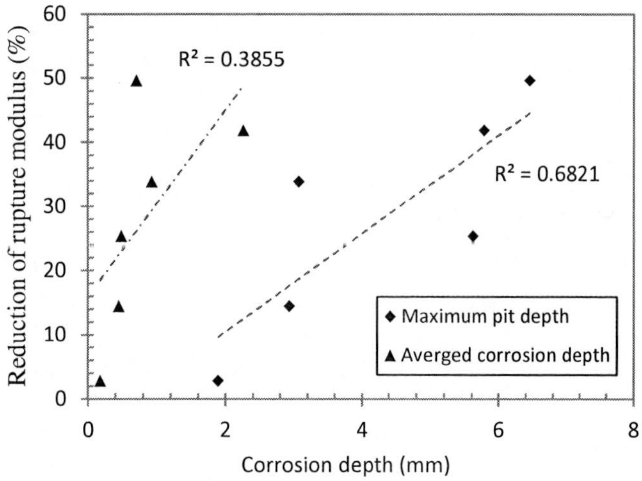

Figure 7 Relationship between the reduction of rupture modulus and corrosion depth (maximum and averaged) for cast iron pipes

4. DISCUSSION

Further discussion and observation can be made on the basis of the results from the tests in current study. Results revealed that most of pipe sections exhibited net rupture modulus that are close to those given in the reported literature (Seica et al. 2004), ranging from 250 MPa to 450 MPa. However, an exception was given to cast iron pipe No. 4 with a lower value of rupture modulus (See Figure 6), although corrosion examination shows a minor level of corrosion occurred in this pipe. An inspection of material composition information for this pipe shows a considerably higher phosphorus level (1.23%) than any other pipes. It is known that adding suitable phosphorus is a customary way in old manufacturing method to promote the fluidity of the molten metal and the soundness of the castings. However, it has been found that a percentage of phosphorus over 1% in material can result its intrinsic embrittlement increasing, and strength decreasing rapidly (MacKenzie 1928). This coincides well with the mechanical property obtained from pipe No. 4 in the current test.

The 3D scanning technique has been successfully used in current tests to locate and quantify the corrosion induced pits on pipe surface. As one of powerful non-destructive testing methods, the 3D scanning technique is uniquely characterized by the capability of collecting shape data and constructing digital three-dimensional models. This provides an opportunity of investigating the corrosion effect on pipe structural response by numerical simulation, e.g. finite element analysis.

It was observed that with a small deflection, cast iron pipes experienced rapid failure while steel pipes exhibited a high ductility. This is a direct result of the fact that cast iron is characterized by the presence of graphite flakes in the matrix of iron (Conlin and Baker 1991). These graphite flakes cannot transform stress in tension but acts as stress concentrators and crack initiators (Seica et al. 2004 and Conlin and Baker 1991). Also, it was noted that crack in cast iron pipes starts at either top or bottom in every case. This phenomenon coincides with the theoretical knowledge that the bending moment at 12 o'clock and 6 o'clock is around 16/9 times of that at 9 o'clock and 3 o'clock (Watkins and Anderson 1999). Overall, the developed test program produced effective outcomes for investigating the corrosion induced mechanical property deterioration. The results from the current study are not only useful for the research community on experimenting buried metal corrosion but also can facilitate the asset managers for better service management through qualitative corrosion assessments.

5. CONCLUSIONS

The strength degradation as a result of corrosion, has been investigated in this paper. Parallel plate loading tests were conducted on cast iron and steel pipe sections cut from ex-service water pipelines. A 3D scanning measure technique was employed to measure the depth of corrosion induced pits. From the analysis of the test results, a relationship between corrosion depth and the reduction of rupture modulus for cast iron pipe has been developed. It has been found that the corrosion patterns for cast iron and steel pipes are generally different, with cast iron pipes often characterized by relatively wide distributed and deeper corrosion pits and shallow and blunt pits in steel pipes. It has also been found that the failure mode of cast iron pipes under compressive parallel-plate load is fragmentation due to brittle material nature, while steel pipes tend to continuously deflect until reach its performance limit. It has also been found that although with the presence of significant composition and microstructure variation in sample pipes, a reduction trend of rupture modulus for cast iron pipes was observed with increase of service age. This work is expected to enrich current knowledge on deterioration of pipe materials and facilitate the evaluation of remaining strength of corroded pipe.

6. ACKNOWLEDGMENTS

Financial support from Australian Research Council under DP140101547, LP150100413 and DP170102211 is gratefully acknowledged.

7. REFERENCES

American Water Works Association. (1962). American Standard for Cast Iron Pipe Centrifugally Cast in Metal Molds for Water or Other Liquids. ASA/AWWA A21. 6/C106, 62.

ASTM, A. (2004). 247-67,"Standard Test Method for Evaluating the Microstructure of Graphite in Iron Castings". 13.

Cole, I. S., and Marney, D. (2012). The science of pipe corrosion: A review of the literature on the corrosion of ferrous metals in soils. Corrosion Science, 56, 5-16.

Dewanbabee, H., & Das, S. (2012). Structural behavior of corroded steel pipes subject to axial compression and internal pressure: Experimental study. Journal of Structural Engineering, 139(1), 57-65.

Harrison, D. B., Nicholas, D. M., & Evans, G. M. (2004). Pitting corrosion of copper tubes in soft drinking waters: corrosion mechanism. Journal (American Water Works Association), 96(11), 67-76.

Hou, Y., Lei, D., Li, S., Yang, W., & Li, C. Q. (2016). Experimental investigation on corrosion effect on mechanical properties of buried metal pipes. International Journal of Corrosion, 2016.

Kiefner, J. F., and Vieth, P. H. (1989). A modified criterion for evaluating the remaining strength of corroded pipe (No. PR-3-805). Battelle Columbus Div., OH (USA).

Keil, B., & Devletian, J. (2011). Comparison of the Mechanical Properties of Steel and Ductile Iron Pipe Materials. In Pipelines 2011: A Sound Conduit for Sharing Solutions (pp. 1301-1312).

Makar, J. M. (2000), "Preliminary Analysis of Failures in Grey Cast Iron Water Pipes." Engineering Failure Analysis 7(1):43–53.

Makar, J. M., Desnoyers, R., & McDonald, S. E. (2001). Failure modes and mechanisms in gray cast iron pipe. Underground Infrastructure Research, 1-10.

MacKenzie, J. T. (1928). INFLUENCE OF PHOSPHORUS ON STRENGTH PROPERTIES OF CAST IRON PIPE. Journal (American Water Works Association), 19(6), 747-759.

Moser, A. P., & Folkman, S. L. (2001). Buried pipe design (p. 57). New York: McGraw-Hill.

National Water Commission Australia. National performance report 2009-10. http:// archive.nwc.gov.au/__data/assets /pdf_file/0015/11265/NPR_urban.pdf

Oh, C. K., Kim, Y. J., & Park, C. Y. (2009). Effects of local wall thinning on net-section limit loads for pipes under combined pressure and bending. Nuclear Engineering and Design, 239(2), 261-273.

Seica, M. V., and Packer, J. A. (2004). Mechanical properties and strength of aged cast iron water pipes. Journal of materials in civil engineering, 16(1), 69-77

Schwerdtfeger, W. J. (1953). Laboratory measurement of the corrosion of ferrous metals in soils. *National Bureau of Standards.*

Rajani, B. (2000). Investigation of grey cast iron water mains to develop a methodology for estimating service life. American Water Works Association.

Romanoff, M. (1957). Underground corrosion. Washington, DC: US Government Printing Office.

Watkins, R. K., & Anderson, L. R. (1999). Structural mechanics of buried pipes. CRC press.

Talbot, A. N. (1908). Tests of cast-iron and reinforced concrete culvert pipe. University of Illinois at Urbana Champaign, College of Engineering. Engineering Experiment Station.

8. AUTHOR DETAILS

Professor Chunqing Li is a Professor of Civil Engineering in the School of Engineering at RMIT University. Professor Li's research areas include risk and reliability analysis of engineering works, service life prediction of civil infrastructure, steel corrosion in concrete and its effects on concrete structures, deterioration of materials and structures, methodology of whole life design and assessment of civil infrastructure, stochastic modeling of loading and structural resistance, and construction management. Professor Li's research in risk-based service life prediction of civil infrastructure was rated "international leading" (i.e., 4* - the highest rating) in UK's 2008 Research Assessment Exercise (RAE). Professor Li has authored or co-authored around 180 international peer reviewed publications to date including journal articles and conference papers.

Weigang Wang is a PhD candidate in School of Engineering-Civil, Environmental &Chemical Engineering Disciplines, RMIT University. He holds a BE (Safety Engineering, 2013) from the China University of Petroleum. He started his PhD study at RMIT University, Australia in July, 2014. His PhD research was in the area of failure analysis of buried pipeline subjected to corrosion.

Dr Dilan Robert is a lecturer in Civil Engineering at RMIT University. He obtained his first degree (2005) from Moratuwa University, Sri Lanka and the PhD degree (2011) from University of Cambridge, UK in 2011. Prior to his current position at RMIT, he has worked 2 years as a geotechnical engineer in the UK and 2 years at Monash University, Australia as a research fellow. Dr Robert has published over 50 peer reviewed technical papers in soil-structure analysis since 2011.

Dr Annan Zhou is a senior lecturer in geotechnical engineering at RMIT University. He was awarded the PhD from the University of Newcastle in 2011and then the ARC DECRA Fellow from 2013. He was awarded the Hugh Trollope Medal from AGS in 2014. He has published over 40 journal papers and his h-index is 14.

ELECTROPOLYMERISATION OF ANILINE ON ALKALINE PRE-TREATED MAGNESIUM ALLOY AND CORROSION EVALUATION

A. Baloch, M. Bobby Kannan
Discipline of Chemical Engineering, College of Science and Engineering, James Cook University, Townsville, Queensland 4811, Australia

SUMMARY: In this work, electropolymerisation of aniline was carried out on alkaline pre-treated AZ91 magnesium alloy. The pre-treatment of the alloy was done in 3M sodium hydroxide solution at a constant potential. Electropolymerisation was performed using a cyclic voltammetry technique in an electrolyte containing aniline and sodium salicylate. The corrosion protectiveness of the polyaniline coating on the alloy was evaluated using potentiodynamic polarization technique in 0.5 g/L sodium chloride solution. The electrochemical results showed that the polyaniline coating reduced the corrosion rate of the alloy by an order of magnitude. However, the post-corrosion analysis revealed delamination of the coating. It is suggested that the polyaniline did not adhere well to the hydroxide/oxide layer formed on the alloy during the pre-treatment.

Keywords: Magnesium, Alkaline Pre-treatment, Protective coating, Corrosion.

1. INTRODUCTION

Magnesium alloys such as AZ91, AM50 and ZE41 are attractive for engineering applications due to their light weight and good mechanical properties [1,2]. They are lighter than aluminium-based alloys and hence preferred where light weight is critical. However, the corrosion rate of magnesium alloys is higher as compared to most engineering metals in moist or chloride-containing environments. [3,4].

Generally, surface coating is a popular method for corrosion protection of metals. In recent years, conducting polymers are receiving high interest for their electrical conductivity, which is useful for inhibiting corrosion. A significant amount of work has been done on coating conducting polymers such as polyaniline (PANI) and polypyrrole on low carbon steel and aluminium alloy [5-12]. Generally, galvanostatic, potentiostatic or cyclic voltammetry technique is used for electropolymerisation [5,6,7,11]. The synthesis of conducting polymers on metallic surface is done in an electrolyte containing oxidising agents. However, the polymerisation is often preceded by dissolution of the metal, which is an issue for magnesium-based alloys since they dissolve rapidly in oxidising environments.

Surface treatment of the magnesium alloys could potentially reduce the dissolution of the alloy during the electropolymerisation and facilitate better polymer coating. It is well known that at alkaline pH magnesium hydroxide forms on magnesium which reduces the dissolution of magnesium in corrosive environments [3,4, 13]. Hence, in this study, a magnesium alloy (AZ91) was alkaline pre-treatment and then electropolymerisation of aniline was done. The coating performance was tested using potentiodynamic polarisation technique in a chloride-containing solution.

2. EXPERIMENTAL DETAILS

The chemical composition of the magnesium alloy (AZ91) used in this study is given in Table 1. The test samples (dimensions: 20 mm × 20 mm × 2mm) were ground with SiC papers up to 2500 grit followed by ultrasonic cleaning in ethanol. The pre-treatment of the samples was carried out using a potentiostat (Model: VersaSTAT3 PAR) and a three-electrode cell, i.e., sample as working electrode, silver/silver-chloride (Ag/AgCl) electrode as reference electrode and graphite as a counter electrode. A constant potential of -195 mV was applied to the sample for 3600 s in 3 M sodium hydroxide. Cyclic voltammetry technique was used to electropolymerise aniline on the AZ91 magnesium alloy in an electrolyte bath containing 0.3M aniline + 0.1M sodium salicylate ($C_7H_5NaO_3$). The coated samples were baked in an oven at 80° C for 24 h to remove any electrolyte residue. The coating performance was investigated in a chloride-containing solution (0.5 g/L NaCl) using potentiodynamic polarisation technique. Prior to the test the samples were kept immersed in the electrolyte bath for 1 h and then polarised at a scan rate of 0.5 mV/s. The tests were done in triplicate and compared to test carried out with uncoated samples.

The coatings were examined using an optical microscopy (Model: Nikon Eclipse 50i POL) to analyse the morphology of the coating before and after potentiodynamic polarisation. The performance of the coating in corrosive environment was evaluated and the degree of protection was calculated using the below equation:

$$DP = \frac{i_{corr}\text{ (uncoated sample)} - i_{corr}\text{ (coated sample)}}{i_{corr}\text{ (uncoated sample)}} \times 100 \quad (1)$$

Table 1. Chemical composition of AZ91 magnesium alloy

Al	Zn	Mn	Fe	Cu	Si	Mg
8.81	0.79	0.21	0.003	0.003	<0.01	Bal

3. RESULTS AND DISCUSSION

The macrographs of the bare AZ91 magnesium alloy and the pre-treated alloy in 3 M NaOH solution are shown in Fig. 1. Pre-treatment at -195 mV exhibited a whitish surface confirming the formation of magnesium hydroxide layer.

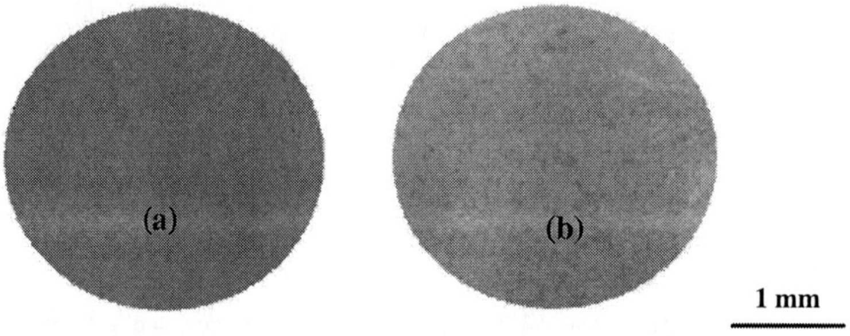

Figure 1. Macrographs of: (a) bare alloy, and (b) alkaline pre-treated alloy.

The cyclic voltammetry curve of alkaline pre-treated AZ91 magnesium alloy in 0.3 M aniline and 0.1 M sodium salicylate solution is shown in Figure 2. The voltammetry curves revealed that the first cycle of the alloy exhibited a low peak current density of 0.002 A/cm^2, which can be attributed to passive film formation on the alloy. However, the current density gradually increased by each cycle as the polymer coating formed.

The potentiodynamic polarisation curves of the bare metal and polyaniline coated AZ91 magnesium alloy in 0.5 g/L NaCl solution are shown in Fig. 3. The electrochemical corrosion data for the samples are given in Table 2. The corrosion current density (i_{corr}) of the coated alloy was 0.177 μA/cm^2 and the bare metal was 1.03 μA/cm^2. The polyaniline coating reduced the corrosion current by an order of magnitude. The degree of protection provided by the polymer was ~83%.

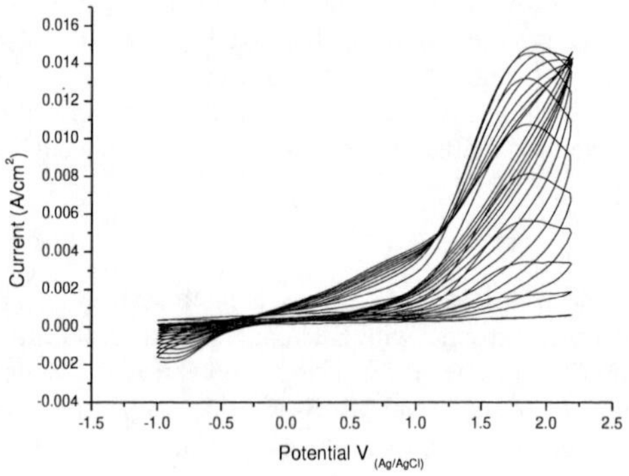

Figure 2. Cyclic voltammetry curves of alkaline pre-treated AZ91 magnesium alloy in 0.3 M aniline and 0.1 M sodium salicylate solution.

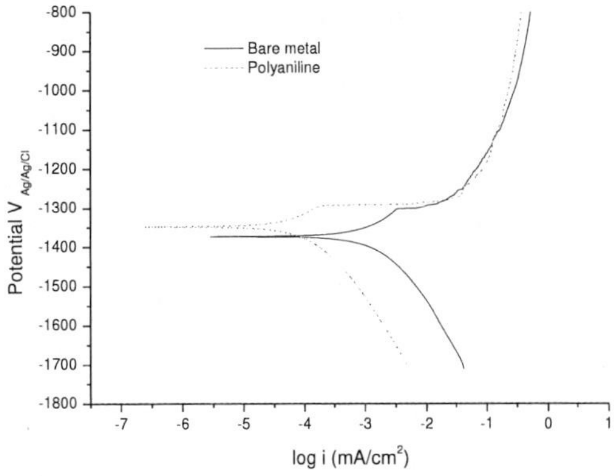

Figure 3. Potentiodynamic polarisation curves of the bare metal and polyaniline coated AZ91 magnesium alloy in 0.5 g/L NaCl solution.

Table 2. Electrochemical corrosion data from the potentiodynamic polarisation curves of the bare alloy and polyaniline coated alloy in 0.5 g/L NaCl solution.

.

Parameter	Bare Alloy	Polyaniline Coated
E_{corr} (mV)	-1296 ± 63.23	-1349 ± 79.37
i_{corr} (μA/cm^2)	1.03± 0.28	0.177 ± 0.08
Corrosion rate (mpy)	2.02	0.32
Degree of protection (DP)	—	82.81 %

Figure 4 shows the macrographs of the coated alloy before and after potentiodynamic polarisation in 0.5 g/L NaCl solution. The macrograph of the sample before polarisation reveals no visible pores in the coating. However, after potentiodynamic polarisation, the coating showed evidence of delamination. White corrosion products were also visible. This suggests that alkaline pre-treatment is not a suitable step for electropolymerisation of aniline on magnesium alloy. The bonding between the hydroxide/oxide layer and polymer was weak and hence failed after polarisation.

Figure 4. Macrographs of coated alloy: (a) before and (b) after potentiodynamic polarisation in 0.5 g/L NaCl solution

4. CONCLUSIONS

Alkaline pre-treatment was done on AZ91 magnesium alloy to reduce the dissolution of the alloy during electropolymerisation and improve the overall performance of the polyaniline coating. Potentiodynamic polarisation showed that the coating significantly reduced the corrosion rate of the alloy. However, post-corrosion analysis suggests that bonding between the hydroxide/oxide layer and polymer was weak and hence showed delamination features.

5. REFERENCES

[1] Mordike B.L and Ebert T, Magnesium properties, applications and potential. Materials Science and Engineering, 2001, A302, 37–45.

[2] Easton M, Beer A, Barnett M, Davies C, Dunlop G, Durandet Y, Blacket S, Hilditch T and Beggs P. Magnesium alloy applications in automotive structures. The Journal of the Minerals, 2008, JOM. 60 No. 11, 57-62.

[3] Song G and Atrens A. Understanding magnesium corrosion, a frame work for improved alloy performance. Advanced engineering materials 5, (2003) 12.

[4] Ghali E, Dietzel W, and Kainer. K. General and localized corrosion of magnesium alloys: A critical review. Journal of Material Engineering and Performance 13, (2004) 7-23

[5] Kazum O and Kannan M B. Galvanostatic polymerisation of aniline on steel: Improving the coating performance in chloride-containing environment. Synthetic Metals, 2013, 180, 54-58.

[6] Raotole P, Patil PP and Gaikwad Ab. Polypyrrole coatings on low carbon steel from aqueous oxalate solution. Int. Journal of Emg. Tech and Adv. Tech 3, issue 11 (2013)

[7] Cascalheira A C, Aeiyach S, Lacaze P C , Abrantes L M. Electrochemical synthesis and redox behaviour of polypyrrole coatings on copper in salicylate aqueous solution. Electrochimica Acta 48 (2003) 2523-2529

[8] Guo X W, Jiang Y F, Zhai, C Q, Lu C and Ding W J. Preparation of even polyaniline film on magnesium alloy by pulse potentiostat method. Synthetic Metals 135-136 (2003) 169-170

[9] Martins N C T, Moura e Silva T, Montemor M F, Fernandesa J C S, Ferreira M G S. Electrodeposition and characterization of polypyrrole films on aluminium alloy 6061-T6. Electrochimica Acta 52, (2008) 4754-4663

[10] Jude O. Iroh, and Wencheng Su. Corrosion performance of polypyrrole coating applied to low carbon steel by an electrochemical process. Electrochimica Acta 46 (2000) 15–24

[11] Arenas M A, Gónzalez L, Bajos, J.J. Damborenea de, Ocón P. Synthesis and electrochemical evaluation of polypyrrole coatings electrodeposited onto AA-2024 alloy. Progress in organic coatings 62 (2008) 79-86

[12] Liu X, Zhang L, Li Y.A, Bian S.U, Zhang L. Electropolymerization of aniline in aqueous solutions at pH 2 to 12. Journal of Materials Science (2005).

[13] Banerjee P.C and Singh Raman R.K. Electrochemical impedance spectroscopic investigation of the role of alkaline pre-treatment in corrosion resistance of a silane coating on magnesium alloy, ZE41. Electrochimica Acta 56 (2011) 3790-3798.

6. AUTHOR DETAILS

Asif Baloch is a post-graduate student in Chemical Engineering at James Cook University, Australia.

Bobby Mathan is an Associate Professor (Chemical Engineering) at James Cook University, Australia.

GAS ARRESTORS TRUS & THE VALIDITY OF ASSUMPTIONS ABOUT LIGHTING DAMAGE

Graeme Gummow
G Squared Solutions, Melbourne

SUMMARY: Working though the management of most risks and hazards is reasonably intuitive as most can be readily conceptualised and easily understood; however some things are counterintuitive and not readily understood and cannot be rationally managed without proper analysis and the application of good science. At the counterintuitive end of the spectrum are more esoteric issues like lightning and electrical hazards on pipelines where conventional thinking or fixed perceptions can lead to invalid conclusions. A significant case in point is a seemingly popular view based on common misinterpretations of observations that leads to an invalid belief that a lighting strike can travel for an appreciable distance along a pipeline to reach and damage equipment. This paper explores electric signal transmission line science in the context of lightning and pipelines in an attempt to modify or extend conceptual understandings within the pipeline industry about electric hazards on pipelines, especially in relation to lighting. The objective of the paper is to facilitate better informed understandings in order that safer and more responsible policies and procedures are adopted for management of lightning and electrical hazards on pipelines.

Keywords: Lightning, LFI, Pipeline Safety, Energy, Surge Diverters, Transmission Lines.

1. INTRODUCTION

The two main concerns about lightning in relation to pipelines are 'what damage can arise from a direct lightning strike to a pipeline', and 'what electrical safety hazards can be propagated along a pipeline as a consequence of lightning strikes'.

Why lightning may be attracted to above ground pipework and what happens when struck by lightning is reasonably obvious and intuitive by comparison to buried pipelines. Conceptual understandings necessary to explain the apparent prevalence of lightning strikes damaging buried pipelines and the consequences of that are not so intuitive or straight forward, and this has recently become an issue of growing concern and interest within the pipeline industry [2].

Of greater concern is the correlation between lightning related electrical surges on pipelines that lead to equipment damage, especially where power line lightning rods have been installed in very close proximity to pipelines. The intuitive explanation about lightning strikes and associated propagation of electrical hazards down a pipeline is contentious.

It is commonly observed that electrical surges associated with lightning often travel reasonable distances along pipelines, but the science says that only a relatively small amount of lightning energy can travel down pipelines and then only for short distances. Where large amounts of energy travel down a pipeline for any reasonable distance (more than several tens of metres at most) then it is almost certain that the electrical surge traveling along the pipeline is lightning triggered Low Frequency Induction (LFI) and not the lightning energy itself.

A conceptual understanding that interprets the observations as lightning triggered LFI has profound implications on what the real hazards are, how the risks should be managed, and the common usage of gas arrester surge diverters and observed failure rates. The application of electrical engineering transmission-line science shows that lightning triggered electrical hazards (LFI) observed to propagate a kilometre or so along a pipeline to damage equipment could just as easily travel tens of kilometres along pipelines depending upon the pipeline's electrical characteristics.

1.1 The challenge of communicating complex issues as simple ideas

Two quotes from author Sir Terence David John "Terry" Pratchett (OBE). Born 28th April 1948. Died 12th March 2015:

- *It's very hard to talk quantum using a language originally designed to tell other monkeys where the ripe fruit is.*
- *Trouble is, just because things are obvious doesn't mean they're true.*

Both these quotes allude to some of the difficulties in developing and communicating practical and reliable conceptual models for complex and sometimes esoteric concepts. Arguably the complexity and counter intuitive behaviour of lighting and Low Frequency Induction similarly evokes challenges for communicating the details as a simple idea.

2. LIGHTNING STRIKES TO EARTH

The location at which lightning strikes the earth (soil or structures) is largely determined by upward streamers [1]. Having a better understanding of where lightning strikes the earth logically must rely on having a better understanding of the characteristics of streamers and what factors influences their behaviour.

2.1 Lightning and Upward Streamers

Lightning strikes are not quite as simple as just a stroke of lightning coming down from the sky and hitting the soil or some structure on the ground. Australian / New Zealand Standard AS/NZS 1768 Lightning, Section B2.2 *The lightning attachment process* provides a concise and suitable description of the strike process.

> The first stroke of a ground flash is normally preceded by a downward-progressing low-current leader discharge that commences in the negatively charged region of the cloud and progresses towards the earth, depositing negative charges in the air surrounding the leader discharge channel. When the lower end of the leader is roughly 100 m from the earth, electrical discharges (streamers) are likely to be initiated at protruding earthed objects, and to propagate upwards towards the leader discharge channel. Several streamers may start, but usually only one is successful in reaching the downcoming leader. The high current phase (return stroke) commences at the moment the upward moving streamer meets the downcoming leader. The position in space of the lower portion of the lightning discharge channel is therefore determined by the path of the successful streamer, i.e. the one that succeeded in reaching the downcoming leader. [1]

As can be seen from AS/NZS 1768 B2.2 (above) the point at which lightning hits the earth is largely determined by streamers (intense ionising ribbon-like electric fields extending up to meet the earth-bound discharge channel leader.

2.2 Electric Field of Streamers and Attraction to Pipelines

As upward streamers form to initiate a discharge path for downward leaders, the upward streamer current is relatively low, typically between about 30 Amps to 300 Amps with an average in the order of around 100 Amps [3]. By comparison to the full discharge current surge (typically about three orders of magnitude greater), these streamer currents are relatively small.

For negatively charged lightning (the more common form), in conjunction with the relatively low streamer initiating current is the formation of a positive field on the ground surface that is almost equal and opposite in electric field strength to the negative field imposed by the downward leader [7].

In deep homogeneous flat surfaced soil it is reasonable to assume that the downward leader will have predominant influence over the location of upward streamers. Where the soil is naturally elevated at some location, or artificially raised by a structure, (e.g. tower or tree) then these will have a secondary influence on modifying where streamers are more disposed to form and therefore where the lightning will strike.

It is also reasonable to suppose that buried structures may influence streamers. As an hypotheses it is proposed that if a streamer begins to form within a few tens of metres of a pipeline, then the large rate of change in electric field adjacent to the pipeline $\delta V/\delta t$ will make the coating appear somewhat transparent (capacitive effect) thereby inhibiting the rate of change in electric field in close proximity to the pipeline providing a relatively less positive local field around the pipeline.

In effect the top of the positive upward streamer is attracted to the downward negative leader while the bottom of the positively charged leader is attracted toward the comparatively less positively charged area adjacent to the pipeline, i.e. the area in which the pipeline limits rate of change of electric field in the soil at the pipeline surface is attractive to the base of the developing positive streamer. It is proposed that this mechanism provides a credible explanation as to why lightning in flat open areas with otherwise relatively even soil morphology appears to be attracted to buried pipelines.

A 1993 test program monitoring the environment 20 m from the point of 'triggered lightning' by R.J. Fisher and G.H. Schnetzer [7], indicates a substantial development of the electric field (many thousands of Volt per metre) within soil during streamer formation. Their data derived from lightning triggered by small rocket devices; if (as would seem reasonable) the same observations apply to natural formation of lightning and has a depth component sufficient to reach the pipeline, then it is also plausible that initial breakdown of pipeline coating might not occur when the negative discharge hits the soil but instead might be initiated as the surround soil becomes intensely positive during streamer formation.

The implications of these possibilities is that not only is there a mechanism to attract upward streamers and therefore lightning to a pipeline, but that this attraction is disposed to producing coating dielectric breakdown which would in effect imply that upward streamer formation extends through the soil to the pipeline surface. If this is correct then it could have a significant impact on ionisation pathways and how the final high current discharge might engage with the pipeline.

While this explains why lighting appears to be attracted to pipelines in flat rural or remote areas, it does not apply to pipelines in metropolitan areas or pipelines that share corridors with other features such as the electrical power network

2.3 Soil Ionisation where Lightning meets the Ground

Popular diagrams of electric fields and current paths for lightning hitting the soil are typically consistent with a model that incorporates electric grid earthing systems (e.g. earth stakes); however for 80% of cases where current exceeds 5 kA and all cases where the current approaches or exceed 15 kA then this model is no longer valid as the associated intense electric field creates ionisation of the soil at the surface [8].

An elegant explanation and photos are provided by Moore et al in an examination of the physics of what happens when lightning strikes the ground [8] and ionises the soil surface. A more detailed explanation is given on the same ionising phenomena as it applies to lightning earthing rods by Daniel et al based on a transmission line modelling method to represent soil ionisation in grounding systems [5].

Once the electric field strength at the point of a lightning strike exceeds the dielectric strength of the soil then ionisation occurs. Where there is no earth stake involvement the ionisation forms a relatively flat disk. Anecdotal evidence suggests that a disk with a radius of about 8 m should not be surprising. The ionised soil should be assumed to be as conductive as metal [6][3], as such the lightning behaves as if it has struck a large steel plate sitting on the surface.

There appears to be some debate between technical papers on the degree of soil ionisation that should be expected. Given the broad disparity between results of investigations into soil 'loss factor' i.e. the imaginary part of electrical permittivity of soil, then perhaps it is not surprising that different researches find similar disparity in soil ionisation characteristics [13][14][15][16].

Where a buried earth stake is involved with a lightning strike then the ionisation is formed as lateral projections (side flashes) from the earth stake which diminish with depth. The shape of the ionised field approximates a cone with a length profile looking a bit like an exponential decay. This can be modelled as a rod with stepped diameters [5].

The effects of high current ionisation of soil is evident in Figure 1 by earth stake resistance values provided in AS/NZS 1768 *Lightning* Table C1: *Examples of reduction of resistance of earth connection under impulse conditions.*

TABLE C1

EXAMPLES OF REDUCTION OF RESISTANCE OF EARTH CONNECTION UNDER IMPULSE CONDITIONS

Soil resistance characteristic	No. of rods and arrangement			
	One isolated rod		Four rods at corners of square, 3.05 m sides	
Soil resistivity, Ω.m	10^2	10^3	10^2	10^3
Resistance at low currents, Ω	30	300	10.5	105
Resistance at current peak, Ω	11.3	54	6.8	37
Ratio of resistance at current peak/resistance at low current	0.38	0.18	0.65	0.35

NOTE: The table depends on the following earthing dimensions and conditions:
Diameter of rods = 10 mm   Depth in earth = 3.05 m
Peak current injected = 80 kA   Time to current crest = 4 μs

Figure 1 AS/NZS 1768 Indication of resistance reduction due to soil ionisation with peak currents

As can be seen above, the soil ionisation effect of a 3 m long (Ø10 mm) earthing rod electrode reduces the effective resistance by nearly a factor of three (1/3rd) in 100 Ωm soil and by a factor of just over five (1/5th) in 1000 Ωm soil which is pretty impressive given that the resistance of a rod electrode is much more dependent on its length than its diameter, as such ionisation predominantly and substantially increases the earth stakes effective diameter rather than its effective length.

The implications of earth stake rods being effectively increased in size to behave like a very wide cone through soil ionisation might even have far reaching implications that extend to legislation such as the Victorian Road Management Act 2004 Management of infrastructure in road reserves (Code of Practice) in managing asset separation requirements. If so the same implications would extend to other jurisdictions.

3. A PIPELINE AS AN ELECTRICAL TRANSMISSION LINE

Crucial to progressing any conceptual understanding of what might occur if lightning directly hits an above ground pipeline or a buried pipeline via topsoil cover; or else by way of an earth stake in close proximity to a pipeline, requires an understanding about lightning's propensity to travel down a buried pipeline. For our purposes any strike to above ground pipework shall be viewed as a short section of a continuous long pipeline that is otherwise buried (e.g. a creek crossing).

Electrical engineering analysis was performed for a pipeline as a transmission line. Similar results (not recorded herein) were calculated for a buried cable consistent with a cable connection from a pipeline to a cathodic protection unit or TRU (transformer rectifier unit) i.e. results are based on a case study.

3.1 Choosing a model pipeline

Although a wide range of pipelines could have been assessed and compared the generality of results deems it appropriate to select a typical pipeline (if such a thing can be said to exist) and then change the most dominant variable of coating quality while look at the attenuation of voltages across a frequency spectrum with variations of pipeline length.

A welded construction steel pipeline with external diameter of 200 mm (DN200) is assumed with an 8.2 mm wall thickness, a 1.5 mm coating, and buried to a centre line depth of 1.8 m in a homogenous equivalent soil of 200 Ωm resistivity. The relative electrical permittivity is assumed to be 11 for the soil and 3.2 for the coating and the pipeline steel resistivity is assumed to be 180 nΩm while a relative magnetic permeability of 300.

These parameters arguably could be representative of a gas, water, or petrochemical pipeline. Two 1.5 mm thick coating qualities are considered, 10^8 Ωm for a good quality and 10^5 Ωm for a poor quality coating.

3.2 Analysis method adopted and limitations

Electrical transmission line circuit theory mathematics was used to make judgements about the propensity for different spectral components of lightning to travel down a pipeline.

3.2.1 Fourier approach

While there are more appropriate means for calculating how a specific and clearly defined impulse changes as it propagates along a pipeline (e.g. FDTD: Finite Difference Time Domain analysis), a circuit theory model that makes course assumptions about the Fourier components of an impulse provides a better general understanding about the propensity of lightning to propagate along a pipeline and this is the approach adopted. CIGRE 543 technical bulletin which provides guidance on numerical analysis methods for electromagnetic surge phenomena [9].

For simplicity and as general indicators, frequency components analysed were restricted to 2 MHz, 100 kHz and 20 kHz.

3.2.2 Soil Permittivity: Absolute or Complex

At frequencies much above 1 MHz, calculations (especially for shunt admittance of a pipeline) start to fall apart. Once radio frequencies get into the region of High Frequency (HF i.e. > 3 MHz), it becomes necessary to use the complex values of soil permittivity whereas for most pipeline transmission calculations (Low Frequency), use of the absolute value is acceptable.

At HF frequencies absolute permittivity values produce invalid calculations of negative soil resistance. The appropriate correction is to use complex values for soil permittivity rather than absolute values; however complex values vary widely and unpredictably with subtle variations of soil chemistry and moisture. In any event the calculated soil resistance at these frequencies is particularly low and favours scattering of currents to earth rather than conduction along conductors.

The obvious patterns and conclusions drawn are largely unchanged by use of complex values, and frequencies much beyond 1 MHz have very low relevance and do not change the final conclusions. As the absolute value of electrical permittivity of soil does not significantly vary and justifying any specific value for the complex components of soil permittivity could be contentious, calculations are limited to an upper level of 2 MHz and the absolute value of typical soil permittivity is used.

3.2.3 Shunt Admittance: Coating Only or Coating and Soil

As a cautionary note, software typically used for transmission line calculations are designed for electrical cables where the resistive (conductive) path between a buried cable and remote earth is assumed to be determined by the resistance of the coating to the exclusion of any resistance through the soil.

This is not so much of a problem if the pipeline has an excellent coating, but leads to very substantial errors for calculations involving pipelines with degraded coatings. This is a common trap and Low Frequency Induction (LFI) software that overlooks this issue underestimates the real level of hazardous voltage for any pipelines designated as having poor coatings.

Calculations used did included soil resistance (conductance) as recommended by CIGRE 95 technical bulletin on high AC voltages and their influence on pipelines Appendix G3 [11].

3.2.4 Recursive Nature of Impedance Calculations

Although approximations of the essential inputs to voltage attenuation (i.e. characteristic impedance, shunt admittance, series impedance, and propagation coefficient) are readily calculated, accurate calculations are more challenging because series impedance and the propagation coefficient have recursive interdependencies which also affect characteristic impedance. To resolve this, recursive iterations were employed until calculations converged to produce reliable values.

3.3 Characteristic length

The characteristic length for a given electrical signal frequency in a transmission line (in this case a pipeline) is the distance over which signal halves in magnitude. The value is derived as the reciprocal of the real part of the propagation coefficient which varies with frequency.

If the characteristic length is x metres then over a pipeline length of y metres the voltage would decrease by a factor of $2^{(y/x)}$ (e.g. over 30 m a signal with a characteristic length of 10 m would reduce by a factor of 8, i.e. if it started at 24 V it would be reduced to 3 V after 30 m).

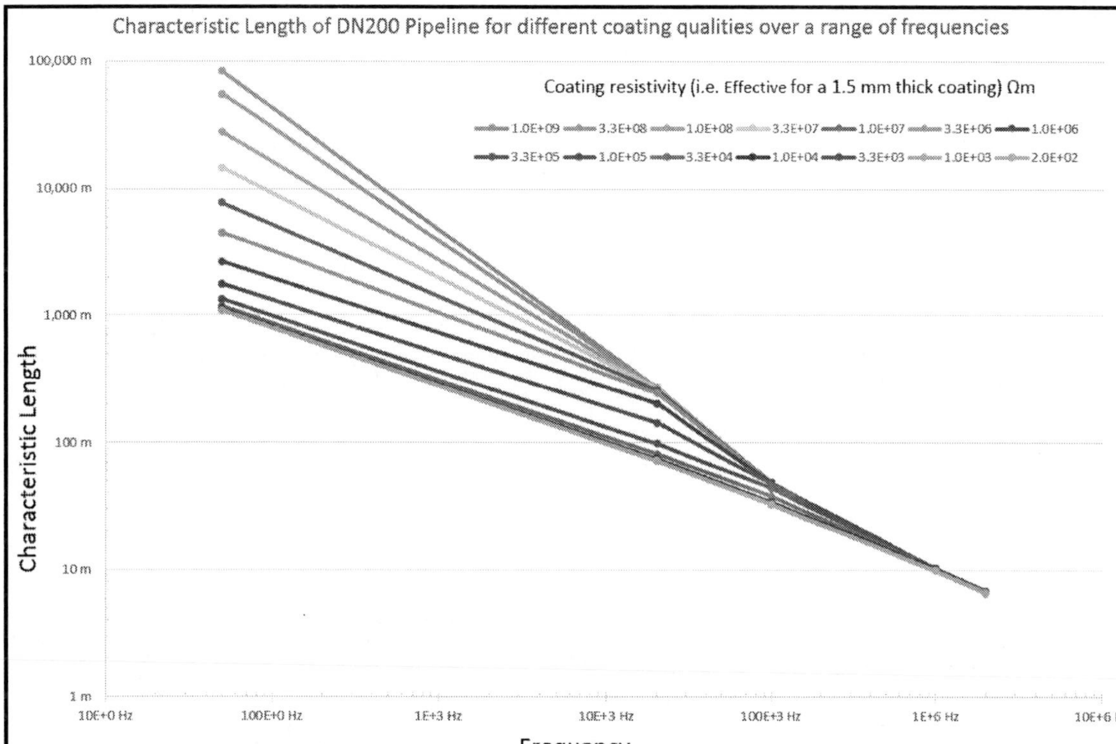

Figure 2 Changes in characteristic length of pipeline depending upon frequency and coating quality

4. ENERGY PICKUP ANALYSIS

As can be seen above, once frequency components get much above 10 kHz the characteristic length of the pipeline becomes quite short and converge on the behaviour of a bare steel pipeline (i.e. where coating resistivity is 200 Ωm = soil resistivity).

At 100 kHz and above all characteristic lengths are below 50 m.

4.1 Maximum Voltage on Pipeline

It is proposed that once the voltage of a pipeline exceeds the dielectric strength of its coating then insulation properties are effectively lost (i.e. it starts to behave as though it has no coating) and under those circumstance a consistent spattered pattern of pinhole perforations should be expected in the coating where breakdown occurs. Reported incidents of lightning striking buried sheathed cables are consistent with this phenomenon; however pipeline observations do not seem to report this phenomenon suggesting that pipelines are less likely to develop voltages much in excess of their coating's dielectric strength.

If a pipeline is at a voltage much beyond the coating's dielectric strength then the ratio of current staying on the pipeline compared with the amount leaving the pipeline is determined by the ratio of the pipeline's resistance (or impedance) to earth compared with its characteristic impedance. It is proposed that a few tens of metres of pipeline (behaving as if it has little to no coating i.e. above dielectric strength rating) would find a lower resistance direct pathway to earth through the soil rather than the impedance for current to continue down the pipeline. As such as lighting energy goes up the immediate loss to soil increases disproportionately to induce a degree of non-linear plateauing of voltages on pipelines

It is proposed that if a pipeline had a voltage much above its coating dielectric strength then it is very unlikely that the voltage would remain beyond the coating dielectric strength for more than a few tens of metres. This of course assumes that a very large voltage is on the pipeline. Accordingly propagation analysis assumes a maximum of about 20 kV on a pipeline once you get much beyond the intense electric field at the strike location.

It is proposed that if the pathway for current to get onto a pipeline was for a direct strike through the soil to a buried pipeline, then then the vast majority of the current emerging from the disk of ionised soil would find an easier path directly to remote earth by comparison to getting onto and then traveling along a pipeline.

If an earth stake is virtually touching a pipeline and an ionised pathway is formed, then it would initially seem that the preferred pathway is through the pipeline, but if side flashing of the earth rod occurs to produce a very large diameter cone of ionised soil with metal-like conductivity, then suddenly the pipeline looks like a much less attractive pathway to remote.

The preceding assertions are dependent on the characteristic impedance of a pipeline and whether or not it looks like an almost perfect pathway to earth by comparison to a direct pathway through the soil to remote.

4.2 Characteristic Impedance of a Pipeline

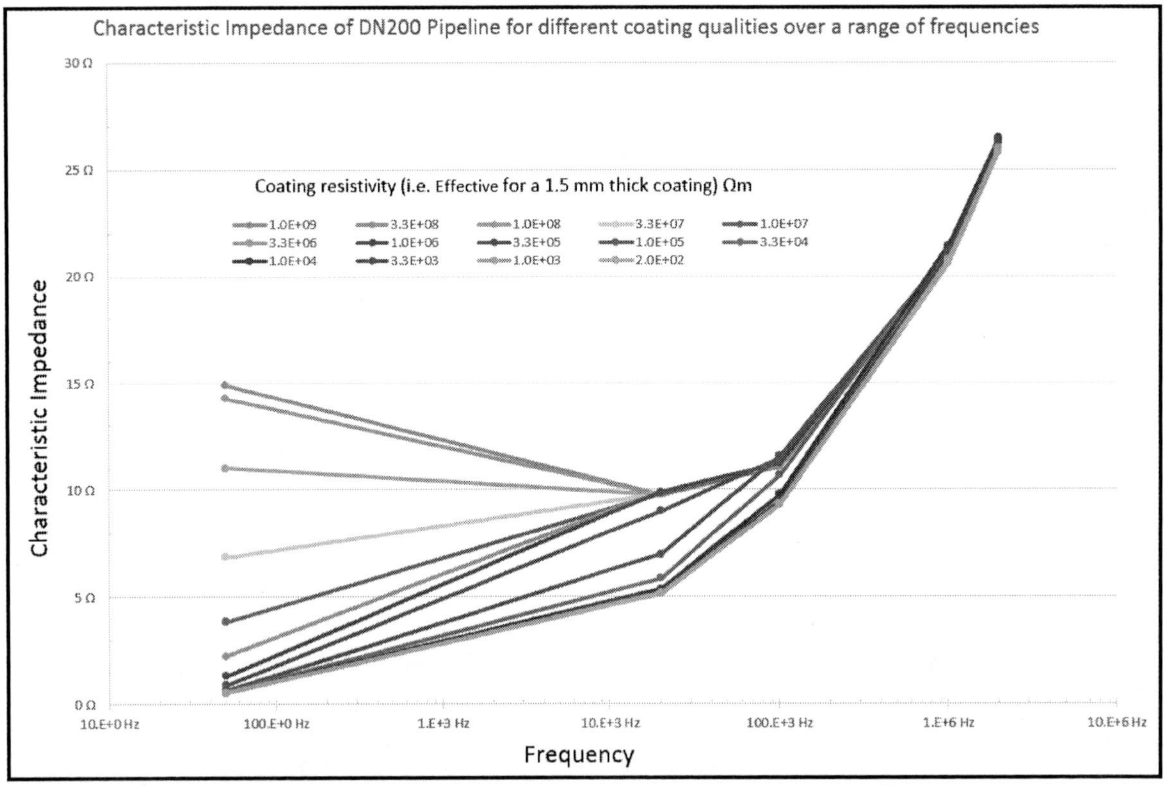

Figure 3 The absolute value of complex characteristic impedance of a long pipeline for various coatings and frequencies

At 10 kHz, even a bare steel pipeline has a characteristic impedance that looks like about 5 Ohms (Note: Fig 2 is for 200 Ωm soil) and a well coated pipeline looks like about 10 Ohms with impedance increasing as frequencies increase.

As frequencies increase, currents only travel in the outer surface of a conductor, this is called skin depth. Although skin depth cannot be reliably applied to all components of lightning in a pipeline, calculations of skin depth still give instructive insight into the fact that high frequency signal components are constrained to a very thin section of the pipelines outer surface.

Table 1 Conductive Skin Depth of Pipe Steel at Frequencies

Current Skin Depth of Frequencies in Pipeline Steel (i.e. 180 nΩm)	
Frequency (Hz)	Skin Depth
50 Hz	1.74 mm
20 kHz	87.2 µm
100 kHz	39.0 µm
1 MHz	12.3 µm

4.3 Summary Conclusion of Pipeline/Lightning Transmission Line Characteristics

When lightning hits soil then either a large (up to about 20 m diameter) disk of ionised soil is produced with an effective conductivity of metal is generated, or else if the strike is via an earth stake then a similar ionised structure is formed in the soil with a reduced diameter but increased depth as a cone of ionised soil with metal-like conductivity.

Given the characteristic impedance of a pipeline, entry of current via ionised pathway (arcing to a point on the pipeline) will be sufficient to create an initial capacitive surge onto the pipeline, but beyond that the pipelines characteristic impedance and skin effects do not make the pipeline a particularly attractive pathway for current to flow to remote earth.

Taking these factors into account and also considering the breakdown in insulation that occurs once a pipelines voltage exceeds its dielectric limits, beyond a few tens of metres from a lightning strike a pipeline potential wouldn't exceed about 20 kV. These factors substantially limit the amount of energy that might get onto a pipeline.

The characteristic length data clearly indicates that high levels of energy won't travel for several kilometres without being substantially attenuated. The remaining question is 'how much energy might travel over a distance of say 200 m or 1 km.

Energy calculations have to start with some assumptions about the duration of lightning getting onto or entering a pipeline.

4.4 Duration of Lightning and Sum Energy Input

On the basis of AS/NZS 1768 Lightning, Appendix Table B1 we shall assume a worst case of 15 strokes in a lightning flash. Note that the Footnote 1 of Table B1 states '*Values at the 1 percent and 99 percent levels are very uncertain, and are given only to indicate an order of magnitude*'.

SUMMARY OF THE FREQUENCY DISTRIBUTIONS OF THE MAIN CHARACTERISTICS OF THE LIGHTNING FLASH TO GROUND

Item No	Lightning characteristic	Percentage of events having value of characteristic greater than value shown below (see Note 1)							Unit
		99	90	75	50	25	10	1	
1	Number of common strokes	1	1	2	3	5	7	12	-

Figure 4 Extract from AS/NZS 1768 Table B1

In an attempt to conservatively overstate the level of energy of lightning getting onto a pipeline we shall assume that each stroke has a duration of 600 µs, for energy collation we shall ignore the intervals between strokes and assume a net sum duration of 15x 600 µs = 9.0 ms, a value considered to significantly overstate the magnitude of duration.

In addition we shall assume that all the listed frequency components exist at 20 kV for the net duration of 9.0 ms.

5. ENERGY TRANSFER ANALYSIS

Energy transfer is calculated on the basis of the following configuration with a Transformer Rectifier Unit (TRU) connected almost immediately to the pipeline (i.e. additional energy losses from cabling to a pipeline are ignored:

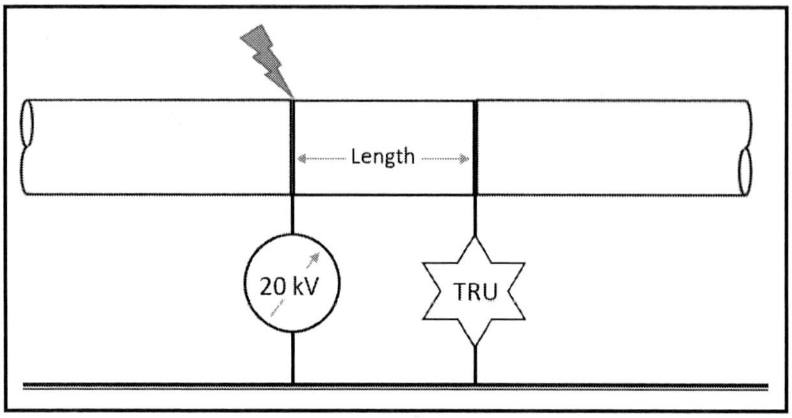

Figure 5 Equivalent Circuit Analysis Diagram for Pipeline Transmission from a Point Source

The Voltage at the TRU is given as:

$$V_{TRU} = 20kV \times \left(\frac{2}{e^{\gamma L}+e^{-\gamma L}}\right) \times \left(\frac{(Z_{TRU}//Z_0)}{Z_{Th}+(Z_{TRU}//Z_0)}\right) \qquad (1)$$

$$Where: (Z_{TRU}//Z_0) = \frac{Z_{TRU}\times Z_0}{Z_{TRU}+Z_0} \qquad (2)$$

$$and\ Z_{Th} = Z_0 \times \frac{e^{\gamma L}-e^{-\gamma L}}{e^{\gamma L}+e^{-\gamma L}} \qquad (3)$$

An arbitrary value of the TRU impedance Z_{TRU} of 5 Ω was chosen as a fixed resistance that over the range of calculations uniformly provided a credible value and higher overall level of energy transfer (i.e. good impedance matching).

The proposed maximum likely voltage to get onto a pipeline is assumed to be limited to 20 kV and hyperbolic attenuation equations are based on the pipeline's Propagation Coefficient γ and the length of pipeline L between the lightning energy attachment and the TRU.

Calculations are performed for *lengths* of 200 m and 1 km, for *frequencies* of 2 MHz, 100 kHz, 20 kHz, and then for 50 Hz for comparison with attenuation of Low Frequency Induction. A better coating quality is assumed of 10^8 Ωm resistivity and a lesser coating quality of 10^5 Ωm resistivity for a 1.5 mm coating thickness.

Table 2 Analysis of frequency dependant energy transmission along steel pipelines

Soil Rho = 200 Ωm, coating = 1.5 mm thick, DN200 pipeline 8.2 mm wall, TRU as 5 Ω load (Ref: Fig 5)									
	10^8 Ω.m (Good coating)				10^5 Ω.m (Poor Coating)				
Voltage	**20 k**			**5 k**	**20 k**			**5 k**	**V**
Duration	**9 m (i.e. 15x 600 µ)**			**800 m**	**9 m (i.e. 15x 600 µ)**			**800 m**	**s**
Freq	**2 M**	**100 k**	**20 k**	**50**	**2 M**	**100 k**	**20 k**	**50**	**Hz**
Length = 200 m									
V_{TRU}	545e-12	149.2	4.33 k	4.91 k	559e-12	95.6	1.56 k	4.26 k	V
I_{TRU}	109e-12	29.8	866	982	112e-12	19.1	311	851	A
$Power_{TRU}$	59.5e-21	4.5 k	3.75 M	4.82 M	62.5e-21	1.83 k	484 k	3.62 M	W
$Energy_{TRU}$	535e-24	40.1	33.7 k	3.86 M	562e-24	16.5	4.36 k	2.90 M	J
Calories	2.24e-21	168	141 k	16.1 M	2.35e-21	65.8	18.2 k	12.1 M	Cal
δ H_20 dm³	0.000	0.170	141	16140	0.000	0.069	18.2	12120	°C / l
δ H_20 m³	0.000	0.000	0.141	16.14	0.000	0.000	0.018	12.120	°C / kl
Length = 1 km									
V_{TRU}	51.9e-63	9.10 µ	242	4.55 k	58.9e-63	1.05 µ	443 m	2.26 k	V
I_{TRU}	10.4e-63	1.82 µ	48.3	910	11.8e-63	209e-9	88.6 m	451	A
$Power_{TRU}$	538e-126	16.5e-12	11.7 k	4.14 M	693e-126	219e-15	39.3 m	1.02 M	W
$Energy_{TRU}$	4.84e-126	149e-15	105	3.31 M	6.23e-126	1.97e-15	353 µ	814 k	J
Calories	20.3e-126	623e-15	439	13.8 M	26.1e-126	8.26e-15	1.48 m	3.41 M	Cal
δ°H_20 dm³	0.000	0.000	0.439	13845	0.000	0.000	0.000	3408	°C / l
δ°H_20 m³	0.000	0.000	0.000	13.845	0.000	0.000	0.000	3.408	°C / kl

To assist making the energy units more tangible to a conceptual undeerstanding, expression of units have been included in terms of how much the calculated energy would heat a litre of water (δ°H_20 dm³) as degrees per litre (°C/l), and also how much it would heat a cubic metre of water (δ°H_20 m³) in units of (°C/kl).

To draw a comparison between propagation of lightning frequencies and Low Frequency Induction (electric mains frequency), calculations are also performed for 50 Hz.

5.1 Interpretation of Energy Transfer

As can be seen from the Table 2 above, if 20 kV is on a long well coated pipeline, then on the basis of the assumptions provided after only 200 m transmission along the pipeline, the amount of energy at any of the lightning related frequencies would be enough to heat a litre of water beyond boiling point at the lowest end to the nominated frequency spectrum (20 kHz); and for a poorly coated pipeline only 18°C. By comparison only 5 kV of LFI would heat a cubic metre of water (a thousand litres) more than 16°C and 12°C respectively.

To put this into context the suggestion of lightning having a sustained 20 kV low frequency (20 kHz) component for a net duration of 9 ms is an intentional overstatement, but still worthy of consideration for comparison purposes.

Giving the same considerations after 1 km transmission along a well coated pipeline, for lightning related frequencies the highest energy transfer is also at the lowest nominated end of the frequency spectrum (20 kHz) and is not enough to heat a litre of water 0.5°C, while after 1 km a poorly coated pipeline has not transferred enough energy to heat a litre of water a one thousandth of 1°C. By comparison the LFI for the well coated pipeline at 1 km has enough energy to heat a cubic metre of water (one thousand litres) nearly 14°C and nearly 3.5°C for the poorly coated pipeline.

5.2 Criticisms of method employed

There are numerous technical criticisms of the methods employed above. Some attempts have been made to overstate the effective duration of surges and full amplitudes have been assigned at specific frequencies where a more correct representation would be to distribute the energy with significantly reduced amplitudes over a broad aggregated spectrum.

Lightning has a large DC component and analysis limited to frequency domain ignores the level of current that might continue between strokes within a lightning strike, nonetheless the frequency analysis provides reasonable insight into a pipelines resistance to the rapid rates of voltage rise and fall and the associated current flow.

Even a moderately small amount of energy over a very small time can lead to very concentrated and dramatic energy release. Nonetheless devices such as gas discharge diverters (or arresters) are designed to absorb and handle the kind of energy shock that might normally be expected to travel from lightning along a pipeline, they are not however designed to handle the amount of energy that (as calculated above) might be expected from a lightning triggered LFI event.

The calculations above assume an infinitely long pipeline extending past the TRU. This assumption does affect calculations and tends to reduce the amount of energy that would be expected at the TRU for frequencies having a long value for *characteristic length*. As such the principal criticism of overstating the trailing length is that predominantly calculations for mains frequency (50 Hz: which has a relatively long propagation coefficient) will tend to be understated. The affect to nominated lightning frequency calculations is much less due to their comparatively short *characteristic length*.

6. A RE-THINK ON CONCEPTUAL UNDERSTANDINGS FOR LIGHTNING AND PIPELINES

The preceding data and analysis is not intended to produce accurate predictive models, instead it is designed to reflect the order of magnitude of energy transfer with trends and patterns to be considered so that prevailing conceptual understandings can be challenged and reviewed.

While the approach provided has limitations, it is hoped that the perspective helps free people from what they might be expected to assume and to facilitates a degree of freedom that comes with having an additional mental model or renewed way of how to think about and approach the problems associated with lightning and hazardous voltages on pipelines.

6.1 Lightning Hitting a Buried Pipeline

The discussion above extends the view that downward streamers are attracted to lightning to suggest that corresponding upward streamers are also be attracted to pipelines due the way in which pipelines limit the electric field in soil adjacent to the pipeline's surface.

It is credible that ionisation of soil might extend to the pipeline surface during streamer formation and this could account for substantial burns (and damage) to buried pipelines from direct lighting strike to soil events. While soil ionisation might help current reach a pipeline, the extent of ground surface ionisation also assists in providing substantial low resistance current pathway to remote earth away from rather than through (or along) the pipeline.

If the pipeline potential is substantially raised beyond dielectric strength, then the coating becomes a good conductor for discharge of pickup, and the discussion above supports the view that only limited energy is transported along a pipeline and then only for a limited distance.

Where substantial amounts of lightning energy appear to be transmitted along a pipeline, especially over any distance, then the discussion above suggests that lightning triggered LFI is the most credible culprit.

6.2 Lightning Hitting a Buried Pipeline via an Adjacent Power Pole Earth Stake

When lightning goes to earth via a power line earth stake in very close proximity to a pipeline, then substantial ionisation occurs around the earth stake. This is a similar scenario to the previous subsection but is restricted to those situations where earth stake positioning and the formation of ionised fields from the lightning strike is close enough to contact the pipeline.

Where current density is sufficient to trigger soil ionisation (with or without an earth stake) a low resistance pathway to remote earth other than through the pipeline is formed, and in both cases it would seem unlikely that lightning energy would be disposed to traveling far along a pipeline or preferring the pipeline as a competing pathway to remote earth for current.

6.2.1 The Addition of Power Line Flashover

Where things do substantially change is for those circumstances where the lightning is associated with flashover.

Flashover is where a lightning strike creates an electrical connection between one (or more) of the power line phases and earth by means of an ionised pathway in the air between the phase cable and earth. In addition to the lightning current to earth stake, there is also substantial power line fault condition current that will be maintained until a power grid circuit break clears the fault, this is typically around one second or less.

As can be seen from the Table, while lightning frequencies from the earth stake are not disposed to traveling along the pipeline, mains frequencies (50 Hz) is conducted quite well. Any residual soil ionisation would tend to aid 50 Hz current flow from the earth stake to the pipeline. The ionised soil would effectively enhance the earth stake performance for 50 Hz.

In many cases it is reasonable to consider that the pipeline would simultaneously be subject to LFI as a consequence of the power line fault condition. Depending upon the length of common route it is also conceivable for the energy transfer of LFI to exceed the energy transfer arising from the mains frequency point pickup of fault current at the earth stake.

It is proposed that in a worst case situation, the effect of LFI voltage on a pipeline in conjunction with a low resistance path to the high voltage (flashover effected) earth stake would combine to produce severe arcing to the pipeline from the earth stake for the fault condition duration which could cause substantially more damage to the pipeline wall than might be expected from lightning alone.

7. SUMMARY CONCLUSION

Just as physicists might look at light as being either a particle or a wave depending upon the problem being approached, the conceptual insights provided above have individual benefits and limitations. Nonetheless the discussion is provided to help convey a conceptual model for understanding aspects of hazardous voltages on pipelines and the underlying mechanisms.

It is hoped that the concepts provided help break down over reliance on simplistic 'lightning only' models that ignore the overwhelming likelihood of LFI being involved where equipment (TRUs or Surge Diverters) are damaged, and that this facilitates a perspective which allows a greater range of hazards and risks to be recognised and addressed more readily.

It would seem that in the field, too often lightning alone is blamed for damage to equipment our analysis above does not support that conjecture. Lightning may be guilty, but not as charged. Too often it is much more likely that lightning has triggered an LFI event and the damage incurred is not from the lightning itself but from lightning triggered LFI instead.

Where lightning is incorrectly held reasonable for equipment damage (i.e. were assumptions preclude LFI involvement) then the likelihood of the same event producing hazardous voltages on a pipeline for many tens of kilometres or more and at a large number of appurtenances is inadvertently overlooked and ignored. The 'lightning only' model underestimates the number of appurtenances that might be hazardous to personnel during an incident event.

The often prevailing conceptual model of equipment damage being associated with lightning in an area overlooks and ignores the possibility (perhaps even likelihood) that LFI might be the real culprit for equipment damage might be triggering by lightning up to 100 km away or more, as advised by the Australian Pipelines and Gas Association (APGA) [10].

Where it is assuming on an anecdotal basis that equipment damage must have been caused by lightning rather than having being caused by LFI, this leads to an assumption that damaging and hazardous events are totally dependent on the presence of lightning (i.e. safe in the absence of lightning), yet only about 70% of LFI events are associated with lightning 'flashover' triggers. Other common causes of LFI events include flashover from bird excrement over insulators and from animals on cross arms or equipment failures. As such seemingly lightning dependant hazards are not dependent upon lightning.

Although the discussions above consider that there may be significant effects arising from the impact of soil ionisation pathways directly to a pipeline from a lightning strike to a buried pipeline (if indeed such an ionisation pathway exists) and ionisation surrounding an earth stake on a power line adjacent to a pipeline, the extent to which these effects actually exist and any sense of predicting what those effects are is elusive. We propose that a better mental model of these ionisation effects would greatly contribute to the pipeline industry's understanding and management of lightning related pipeline hazards.

8. REFERENCES

[1] AS/NZS 1768:2007 Lightning

[2] Mahbub Rabbani, Amanullah MT Oo and Alex Stojcevski (Deakin Uni): Analysis of Lightning Current Characteristics to Investigate Lightning Strike Damages to Energy Pipeline, 2014 International Conference on Lightning Protection (ICLP), Shanghai, China

[3] J. Schoene, M. A. Uman, V. A. Rakov, V. Kodali, K. J. Rambo, and G. H. Schnetzer: "Statistical characteristics of the electric and magnetic fields and their time derivatives 15 m and 30 m from triggered lightning", Journal of geophysical research, vol. 108, no. D6, 4192, doi: 10.1029/2002JD002698, 2003

[4] X. Qie, R. Jiang, P. Laroche: Triggering Lightning Experiments: an Effective Approach to the Research of Lightning Physics, AerospaceLab Journal Issue 5 - December 2012

[5] Daniel S. Gazzana, Arturo S. Bretas, Guilherme A. D. Dias, Marcos Telló, Dave W. P. Thomas, and Christos Christopoulos: "The Transmission Line Modeling Method to Represent the Soil Ionization Phenomenon in Grounding Systems" *IEEE Transactions on Magnetics*, Vol. 50, No. 2, February 2014

[6] M.I. Lorentzu, N.D. Hartziargyriou, B.C. Papadias, S. Sekioka: Modeling of Soil Ionisation for Calculation of the Transient Response of Grounding Electrodes. IPST Int'l conference on power systems transients June 24 – 28, 2001

[7] Richard J. Fisher, George H. Schnetzer: 1993 Triggered Lightning Test Program: Environments within 20 metres of lightning channel and small area temporary protection concepts. Sandia National Laboratories SAND94-0311 UC-706

[8] C. B. Moore, G. D. Aulich and William Rison: An Examination of Lightning-Strike-Grounding Physics. Langmuir Laboratory for Atmospheric Research, New Mexico Tech, Socorro, NM 87801

[9] CIGRE Working Group C4.501: CIGRE Technical Bulletin 543, Guideline for numerical electromagnetic analysis method and its application to surge phenomena, 2013

[10] Guidelines for Electrical Hazards (Version 2: 15/1/14), the Australian Pipeline Industry Association Ltd (APGA)

[11] CIGRE Working Group 36.02: CIGRE Technical Bulletin 95, Guide on the Influence of High Voltage AC Power Systems on Metallic Pipelines, (Electromagnetic Compatibility with telecommunication circuits, low voltage networks and metallic structures), 1995

[12] V. Komarov, S. Wang, J. Tang: Permittivity and Measurements, Encyclopedia of RF and Microwave Engineering, Edited by Kai Chang, ISBN 0-471-27053-9 © 2005 John Wiley & Sons, Inc.

[13] V.V. Navar khele, A.A. Shaikh and R.S. Ramshetti: Dielectric properties of black soil and inorganic matters at microwave frequencies, Indian Journal of Radio & Space Physics, col 38, April 2009 pp. 112-115

[14] G. Ayalew, N.M. Holden and S.M. Ward: Microwave dielectric properties of horticultural peat products Mires and Peat, Volume 2 (2007), Article 03, http://www.mires-and-peat.net/, ISSN 1819-754X © 2007 International Mire Conservation Group and International Peat Society

[15] W.G. Fano and V. Trainotti: Dielectric properties of soils, IEEE 0-7803-7053-8/1/ Annual report conference on electrical insulation and dielectric phenomena, 2001

[16] Ben K. Steinberg and Tsylya M. Levitskaya: Electrical parameters of soils in the frequency range from 1 kHz to 1 GHz, using lumped-circuit methods, Radio Science Vol 36 No 4 , Pages7 09-719, July/August 2001, © the American Geophysical Union, 2001.

9. AUTHOR DETAILS

Graeme Gummow is the Managing Director and Principal Consultant at G Squared Solutions predominantly working in the area of compliance consultancy for cathodic protection and hazardous voltages on pipelines where he provides specialised services for analysis and mitigation of electrolysis corrosion and low frequency induction (LFI) on pipelines.

Calculations for pipeline parameters are calculated by routines within Geographical Information System (GIS) software developed in conjunction with G Squared Solutions for Low Frequency Induction (LFI) analysis and are built upon previous Research and Development employment and experience in electronics.

QUALIFYING TANK COATINGS FOR BROAD SERVICE CAPABILITIES

Stephen Bowditch
A. W. Chesterton, Groveland, MA, USA

SUMMARY: Tank farms are required to store and transfer various feedstocks, intermediates as well as end products. Often these tanks are lined with a protective coating suitable for service under one set of conditions but, depending on tank capacity demands, it may be exposed to a different, and more corrosive conditions. If this occurs the tank lining may fail prematurely and expose the tank to corrosion and possible extensive mechanical repairs. Since tank farm capacity is measured in available barrels the unplanned loss of availability or decommissioning of a tank can have significant repercussions on overall terminal/plant capacity. If a tank can be lined with a broad-spectrum coating, capable of handling an increased range of corrosive conditions, then the tank farm has maintained and, depending on the lining system chosen, quite possibly increased its tank capacity for a broader range of products while still maintaining overall tank integrity. This paper describes techniques used to qualify coatings for broad immersion service in tank farms associated with gas to liquid plants utilizing the Fischer-Tropsch process.

Keywords: coatings, tank, corrosion, alkylate, demineralized, wastewater.

1. INTRODUCTION

Tank farm operators face many challenges; high among them is the compatibility of the vessel with an ever-widening range of compounds. For example, a tank lining selected to resist light oils is not necessarily compatible with alkylates. When PetroSA's Mossel Bay facility, one of the largest gas-to-liquid (GTL) processing plants in the world, was faced with a demand for increased tank capacity, they determined that a major concern impacting capacity and tank availability was the suitability of the installed corrosion-resistant linings for the intended process stream. With a total capacity of 36,000 bbl./day, the plant uses the Fischer-Tropsch process, to convert natural, methane-rich, gas into ultra-clean, low-sulphur, low-aromatic synthetic fuels and high-value products such as unleaded petrol, kerosene (paraffin), diesel, propane, liquid oxygen and nitrogen, distillates, eco-fuels, process oils and alcohols. Historically, many of the tanks had been lined with systems optimized for a specific chemical but not generally suited for the wider range of intermediate and finished products being processed. To maintain tank integrity, many tanks were also being relined every three to five years due to premature lining failure. The cost to isolate, drain, clean and ventilate gas free, , inspect, mechanically repair and then prepare the surfaces and apply/cure a lining system to a mid-size tank is typically $150–200K USD. Factoring in a five-year life, the invested asset would depreciate at a rate of $40,000 USD/year. Due to these high costs, the plant decided to see what other lining options were available to increase the mean time between failures (MTBF) for lined surfaces of tanks. At the same time, they wanted to increase the availability of their lined tanks to handle a broader range of raw stock, intermediate and finished good.

2. EXPERIMENTAL DETAILS

To assess suitability of recommended linings an expanded list of chemicals being processed, transferred and stored was established as well as what the periodic inspection and cleaning/maintenance practices were being used. Testing was required to validate performance in service as well as during periodic tank cleaning and inspection. The primary qualification test was NACE TM 0174 Laboratory Methods for the Evaluation of Protective Coatings and Lining Materials on Metallic Substrates in Immersion Service; Method A (for applications where a thermal gradient was expected) was tested at 65°C, based on the estimated maximum possible service conditions of 60°C for the tanks. This test method was selected based on the ability to simulate actual tank conditions with representative immersion and vapor phase exposures and thermal gradients. Test panels, having been coated on one side simulate an internal tank liner in actual service. Testing involved a minimum of 16 weeks in service with total test runs extending out to 24 weeks. A set of 6 test panels for each product was considered as statistically viable for each test solution.

Each panel was then coated with pre-screened products to manufacturers recommended dry film thickness which was verified per SSPC PA2 standard Procedure for Determining Conformance to Dry Coating Thickness Requirements, then tested per NACE SP 0188 - Testing of New Protective Coatings on Conductive Substrates using a Tinker & Rasor High Voltage Spark Tester Model #AP-W to verify films were correctly and completely applied within test area.

Once properly cured each test panel was bolted to test cylinder and test solutions were added and then heated for Method A to 65°C (Figure 1). The test solutions were a weak alkylate solution containing 2-3% hydrofluoric acid (HF) provided by the client and demineralized water. For this project the presence and degree of blistering (>8D size and frequency, increasing to 6 or lower indicates failure), loss of hardness (>5% change from control indicating softening or plasticization breakdown of polymer structure), and reduction of adhesion (>10% change indicating softening or plasticization breakdown of polymer).

Evaluation of test areas took place immediately following "failure" indicated by red highlighting, or at successful conclusion of 24-week test cycle or failure, whichever occurred first and included both immersion and vapor phases. Yellow highlighting indicates the point when the film exhibited the first sign of transitioning from a rating of 10 per ASTM D714. Evaluation methods compared to control involved;

- ASTM D714 Standard Test Method for Evaluating Degree of Blistering of Paints
- ASTM D2240 Standard Method for Rubber Property – Durometer Hardness – Scale D
- ASTM D4541 Standard Test Method for Pull Off Strength of Coatings Using Portable Adhesion Testers

Figure 1 Test Cell

3. RESULTS

The following tables illustrate the test results gleaned over 24 weeks following NACE TM 0174 Method A, exposed to a test solution of alkylate rundown water with a 2% concentration of hydrofluoric acid at 65°C with a ΔT of 20-25°C for a 24-week duration. The number, followed by the letters IP (immersion phase) and VP (vapor phase) indicate the blister size and density at the week noted. For example, Wk 1; 10IP/VP no blisters noted. A ranking of 6D in immersion and 8F in vapor phase at Wk. 24 indicates the immersion phase showed blisters equal to a 6 in size with D for dense coverage in immersion phase test area and blisters equating to 8 in size and F for few in frequency. If the panel exhibited no blisters in immersion or vapor phase, then a ranking of 10IP/VP was recorded. A box highlighted in yellow indicates the first shift down from a 10 ranking. If the box then changes to red this indicates progressive deterioration.

Table 1a - Alkylate
NACE TM 0174 Method A
ASTM D714
IP=Immersion Phase / VP=Vapor Phase

System 1	Wk 1	Wk 2	Wk 3	Wk 4	Wk 5	Wk 6	Wk 7	Wk 8	Wk 9	Wk 10	Wk 11	Wk 12	Wk 13	Wk 14	Wk 15	Wk 16	Wk 20	Wk 24
Panel 1	10 IP VP	10 IP VP	10 IP VP	10 IP VP	10 IP VP	10 IP VP	10 IP VP	10 IP VP	10 IP VP	10 IP VP	10 IP VP	10 IP VP	10 IP VP	10 IP VP	10 IP VP	10 IP VP	10 IP VP	10 IP VP
Panel 2	10 IP VP	10 IP VP	10 IP VP	10 IP VP	10 IP VP	10 IP VP	10 IP VP	10 IP VP	10 IP VP	10 IP VP	10 IP VP	10 IP VP	10 IP VP	10 IP VP	10 IP VP	10 IP VP	10 IP VP	10 IP VP
Panel 3	10 IP VP	10 IP VP	10 IP VP	10 IP VP	10 IP VP	10 IP VP	10 IP VP	10 IP VP	10 IP VP	10 IP VP	10 IP VP	10 IP VP	10 IP VP	10 IP VP	10 IP VP	10 IP VP	10 IP VP	10 IP VP
Panel 4	10 IP VP	10 IP VP	10 IP VP	10 IP VP	10 IP VP	10 IP VP	10 IP VP	10 IP VP	10 IP VP	10 IP VP	10 IP VP	10 IP VP	10 IP VP	10 IP VP	10 IP VP	10 IP VP	10 IP VP	10 IP VP
Panel 5	10 IP VP	10 IP VP	10 IP VP	10 IP VP	10 IP VP	10 IP VP	10 IP VP	10 IP VP	10 IP VP	10 IP VP	10 IP VP	10 IP VP	10 IP VP	10 IP VP	10 IP VP	10 IP VP	10 IP VP	10 IP VP
Panel 6	10 IP VP	10 IP VP	10 IP VP	10 IP VP	10 IP VP	10 IP VP	10 IP VP	10 IP VP	10 IP VP	10 IP VP	10 IP VP	10 IP VP	10 IP VP	10 IP VP	8F IP 10 VP	8M IP 10 VP	8D IP 10 VP	6D IP 8F VP

Table 1b – Alkylate
NACE TM 0174 Method A
ASTM D714
IP=Immersion Phase / VP=Vapor Phase

System 2	Wk 1	Wk 2	Wk 3	Wk 4	Wk 5	Wk 6	Wk 7	Wk 8	Wk 9	Wk 10	Wk 11	Wk 12	Wk 13	Wk 14	Wk 15	Wk 16	Wk 20	Wk 24
Panel 1	10 IP VP	10 IP VP	10 IP VP	10 IP VP	10 IP VP	10 IP VP	10 IP VP	8F IP 10 VP	8F IP 10 VP	8D IP 10 VP	6D IP 8F VP	4D IP 8D VP						
Panel 2	10 IP VP	10 IP VP	10 IP VP	10 IP VP	10 IP VP	10 IP VP	10 IP VP	10 IP VP	10 IP VP	8D IP 10 VP	6D IP 8F VP	4D IP 8D VP						
Panel 3	10 IP VP	10 IP VP	10 IP VP	10 IP VP	10 IP VP	10 IP VP	10 IP VP	10 IP VP	10 IP VP	10 IP VP	10 IP VP	8F IP 10 VP	8D IP 10 VP	6D IP 8F VP				
Panel 4	10 IP VP	10 IP VP	10 IP VP	10 IP VP	10 IP VP	10 IP VP	10 IP VP	10 IP VP	10 IP VP	10 IP VP	8F IP 10 VP	8F IP 10 VP	8D IP 10 VP	6D IP 8F VP				
Panel 5	10 IP VP	10 IP VP	10 IP VP	10 IP VP	10 IP VP	10 IP VP	10 IP VP	10 IP VP	10 IP VP	10 IP VP	10 IP VP	10 IP VP	10 IP VP	8F IP 10 VP	8D IP 10 VP	6D IP 8M VP		
Panel 6	10 IP VP	10 IP VP	10 IP VP	10 IP VP	10 IP VP	10 IP VP	10 IP VP	10 IP VP	10 IP VP	10 IP VP	10 IP VP	10 IP VP	10 IP VP	10 IP VP	8F IP 10 VP	8M IP 10 VP	8D IP 10 VP	6D IP 8F VP

Table 2a - Demineralized Water
NACE TM 0174 Method A
ASTM D714
IP=Immersion Phase / VP=Vapor Phase

System 1	Wk 1	Wk 2	Wk 3	Wk 4	Wk 5	Wk 6	Wk 7	Wk 8	Wk 9	Wk 10	Wk 11	Wk 12	Wk 13	Wk 14	Wk 15	Wk 16	Wk 20	Wk 24
Panel 1	10 IP VP	10 IP VP	10 IP VP	10 IP VP	10 IP VP	10 IP VP	10 IP VP	10 IP VP	10 IP VP	10 IP VP	10 IP VP	10 IP VP	10 IP VP	10 IP VP	10 IP VP	10 IP VP	10 IP VP	10 IP VP
Panel 2	10 IP VP	10 IP VP	10 IP VP	10 IP VP	10 IP VP	10 IP VP	10 IP VP	10 IP VP	10 IP VP	10 IP VP	10 IP VP	10 IP VP	10 IP VP	10 IP VP	10 IP VP	10 IP VP	10 IP VP	10 IP VP
Panel 3	10 IP VP	10 IP VP	10 IP VP	10 IP VP	10 IP VP	10 IP VP	10 IP VP	10 IP VP	10 IP VP	10 IP VP	10 IP VP	10 IP VP	10 IP VP	10 IP VP	10 IP VP	10 IP VP	10 IP VP	10 IP VP
Panel 4	10 IP VP	10 IP VP	10 IP VP	10 IP VP	10 IP VP	10 IP VP	10 IP VP	10 IP VP	10 IP VP	10 IP VP	10 IP VP	10 IP VP	10 IP VP	10 IP VP	10 IP VP	10 IP VP	10 IP VP	10 IP VP
Panel 5	10 IP VP	10 IP VP	10 IP VP	10 IP VP	10 IP VP	10 IP VP	10 IP VP	10 IP VP	10 IP VP	10 IP VP	10 IP VP	10 IP VP	10 IP VP	10 IP VP	10 IP VP	10 IP VP	10 IP VP	10 IP VP
Panel 6	10 IP VP	10 IP VP	10 IP VP	10 IP VP	10 IP VP	10 IP VP	10 IP VP	10 IP VP	10 IP VP	10 IP VP	10 IP VP	10 IP VP	10 IP VP	10 IP VP	10 IP VP	10 IP VP	10 IP VP	10 IP VP

Table 2b – Demineralized Water
NACE TM 0174 Method A
ASTM D714
IP=Immersion Phase
VP=Vapor Phase

System 2	Wk 1	Wk 2	Wk 3	Wk 4	Wk 5	Wk 6	Wk 7	Wk 8	Wk 9	Wk 10	Wk 11	Wk 12	Wk 13	Wk 14	Wk 15	Wk 16	Wk 20	Wk 24
Panel 1	10 IP VP	10 IP VP	10 IP VP	10 IP VP	10 IP VP	10 IP VP	10 IP VP	10 IP VP	10 IP VP	10 IP VP	10 IP VP	8F IP 10 VP	8F IP 10 VP	6D IP 8F VP				
Panel 2	10 IP VP	10 IP VP	10 IP VP	10 IP VP	10 IP VP	10 IP VP	10 IP VP	10 IP VP	10 IP VP	10 IP VP	8F IP 10 VP	6D IP 8F VP						
Panel 3	10 IP VP	10 IP VP	10 IP VP	10 IP VP	10 IP VP	10 IP VP	10 IP VP	10 IP VP	10 IP VP	10 IP VP	10 IP VP	8F IP 10 VP	8D IP 10 VP	6D IP 8F VP				
Panel 4	10 IP VP	10 IP VP	10 IP VP	10 IP VP	10 IP VP	10 IP VP	10 IP VP	10 IP VP	10 IP VP	10 IP VP	10 IP VP	8F IP 10 VP	8D IP 10 VP	6D IP 8F VP				
Panel 5	10 IP VP	10 IP VP	10 IP VP	10 IP VP	10 IP VP	10 IP VP	10 IP VP	10 IP VP	10 IP VP	10 IP VP	10 IP VP	10 IP VP	10 IP VP	8F IP 10 VP	8F IP 10 VP	8M IP 10 VP	6D IP 8F VP	
Panel 6	10 IP VP	10 IP VP	10 IP VP	10 IP VP	10 IP VP	10 IP VP	10 IP VP	10 IP VP	10 IP VP	10 IP VP	10 IP VP	10 IP VP	10 IP VP	8F IP 10 VP	8F IP 10 VP	8M IP 10 VP	8D IP 10 VP	6D IP 8F VP

Table 3a -Alkylate
NACE TM 0174 Method A
(ASTM D2240 Shore D Hardness)
IP=Immersion Phase / VP=Vapor Phase

System 1	Week Reading	System 2	Week Reading
Panel 1	24 84 IP/VP	Panel 1	10 78 IP/82 VP
Panel 2	24 84 IP/VP	Panel 2	10 78 IP/82 VP
Panel 3	24 83 IP/84 VP	Panel 3	13 78 IP/82 VP
Panel 4	24 84 IP/84 VP	Panel 4	13 76 IP/80 VP
Panel 5	24 84 IP/84 VP	Panel 5	15 78 IP/82 VP
Panel 6	24 82 IP/84 VP	Panel 6	20 76 IP/80 VP
Control	84	Control	82

Table 3b
Demineralized Water
NACE TM 0174 Method A
(ASTM D2240 Shore D Hardness)
IP=Immersion Phase / VP=Vapor Phase

System 1	Week Reading	System 2	Week Reading
Panel 1	24 84 IP/84 VP	Panel 1	14 81 IP/82 VP
Panel 2	24 84 IP/84 VP	Panel 2	12 80 IP/82 VP
Panel 3	24 83 IP/84 VP	Panel 3	13 80 IP/82 VP
Panel 4	24 84 IP/84 VP	Panel 4	13 82 IP/82 VP
Panel 5	24 84 IP/84 VP	Panel 5	20 79 IP//82 VP
Panel 6	24 83 IP/84 VP	Panel 6	20 78 IP//82 VP
Control	84	Control	82

Table 4a
Alkylate
NACE TM 0174 Method A
(ASTM 4541 Tensile Adhesion)
IP=Immersion Phase

System 1	Wk/Reading	System 2	Wk/Reading
Panel 1	24/4000 psi	Panel 1	12/3000 psi
Panel 2	24/3900 psi	Panel 2	12/3050 psi
Panel 3	24/3980 psi	Panel 3	14/2800 psi
Panel 4	24/4100 psi	Panel 4	14/2500 psi
Panel 5	24/4250 psi	Panel 5	16/2550 psi
Panel 6	24/3750 psi	Panel 6	24/2650 psi
Control	4200 psi	Control	3200 psi

Table 4b
Demineralized Water
NACE TM 0174 Method A
(ASTM 4541 Tensile Adhesion)
IP=Immersion Phase

System 1	Wk/Reading	System 2	Wk/Reading
Panel 1	24/4000 psi	Panel 1	14/2940 psi
Panel 2	24/3950 psi	Panel 2	12/3050 psi
Panel 3	24/3900 psi	Panel 3	13/2800 psi
Panel 4	24/4100 psi	Panel 4	13/2750 psi
Panel 5	24/4200 psi	Panel 5	20/2100 psi
Panel 6	24/4150 psi	Panel 6	20/1950 psi
Control	4200 psi	Control	3200 psi

4. DISCUSSION

Blister formation is an excellent indicator of permeation and chemical resistance and can be useful in establishing relative suitability for continuous immersion or vapor exposure. When a hot solution permeates a lining, and meets a cool surface (steel test panel) condensation can occur, leading to localized corrosion sites and eventual blistering and loss of adhesion. Under extreme situations blisters can grow, in size and density, to show complete loss of adhesion. Once test panels exhibited progressive failure exceeding a blister ranking of 8D with progressive growth to 6 the test panel was taken out of service for evaluation. A blister with a size ranking of 10 indicates no signs of blistering with the naked eye while blisters rated at 8 is the size nominally considered to be visible to the naked eye. For this project the presence and degree of blistering (≥8 in size and D in frequency, increasing to 6 or lower) indicates failure. As ranking number drops from 10 to 8 and then to 6 a failure was declared as the blisters increase in size and density until the entire evaluation areas is covered blisters ranging from 8-2. We determined that any panel exhibiting blisters greater than 8 in size and "D" in density of appearance would be recorded as unsuitable as the test was short term in relation to the 10-year service life being requested.

Chemical permeation of a test solutions into the film may cause plasticization or softening of the film so hardness, as measured by ASTM D2240 "D" Scale, was considered indicative of attack as well. This was a secondary ranking method, after blister size and frequency, was selected to correlate hardness losses with plasticization and blistering. A loss of hardness of >5% from the control indicated sufficient softening, or plasticization of polymer structure, to indicate deterioration beyond acceptable limits. Yellow indicates losses in hardness compared to control while red indicates losses exceeding 5%.

The last evaluative method selected to run on test panels was tensile adhesion. Reduction in tensile adhesion indicates losses in a critical physical property which would translate into reduced long term performance expectations. Since the immersion phase of the test panels experienced the highest degree of blister formation and growth we elected to test in that area only and established that a reduction of adhesion >10% change would be indicative of breakdown of polymer.

In System 1 five of the six test panels exhibited no blistering or significant loss of hardness or tensile adhesion values after 24 weeks run time. System 2 exhibited progressive blister failure within a span of 8-15 weeks with a commensurate loss in hardness and tensile adhesion indicating significant reduction of critical properties. The end user was requiring a 10-year expected service life for recommended lining system and based on our test results System 1 performed the best overall as seen in figure 2A, 2B and 2C.

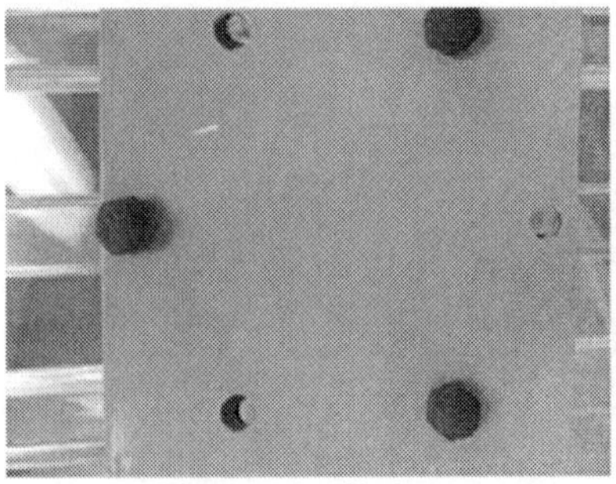

Figure 2A Alkylate
System 1 Test Panel 1 – 24 Weeks

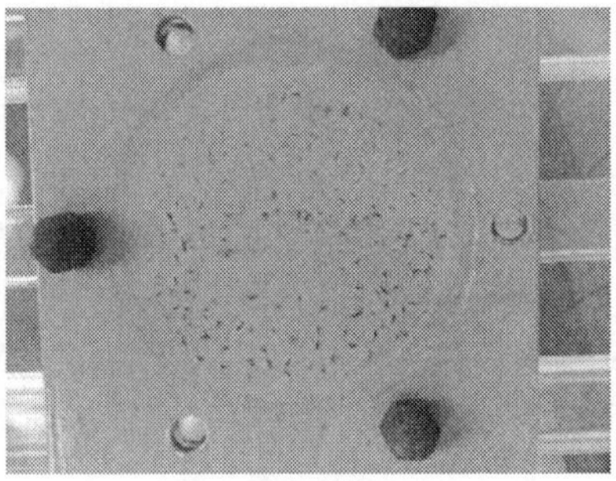

Figure 2B Alkylate
System 1 Test Panel 6 – 24 Weeks

Figure 2C Demineralized Water
System 2 Test Panel 5 – 20 Weeks

5. CONCLUSION

Based on test results System 1 was recommended to end user. It was installed in the summer of 2006 into three separate tanks. Over ten years' time each of these tanks saw intermittent service as alkylate run down storage, demineralized water storage and weak wastewater storage. The first tank was opened for evaluation at 1 year, the second was opened at 5 years and the third was opened at 9 years and 10 months. The witness group included plant personal and their inspection services group, as well as representatives from the manufacturer of System 1 and one other vendor. The inspection revealed no issues in any of the tanks and they were put back into service with an extended inspection cycle of 5 years. By adopting a lining system qualified for multiple service conditions the tank owner has been able to develop a protective coatings program that meets the expectations of service while significantly extending inspection and maintenance cycles. Additionally, the benefit of maintaining of multiple tanks capable of handling three service conditions extends total site tank capacity without added costs.

6. ACKNOWLEDGMENTS

The author would like to acknowledge the plant personnel at PetroSA's, Mossel Bay plant; G.B. Dube – Corrosion Engineer, G.F. Filmalter – Manager of Inspection and Metallurgy, Johan Gerber – Chief Mechanical Engineer. In addition the author would like to acknowledge the services of Chesterton Test Technician Thavalong Prum.

7. AUTHOR DETAILS

Stephen Bowditch is the Global Product Manager for AW Chesterton's ARC Protective Coatings business unit where he is responsible for supporting market initiatives across all industries related to protective coatings. He is a NACE CIP Level III Coatings Inspector and member of numerous NACE and ASTM technical committees related to protective coatings technology and their uses in industry. He holds a Bachelor of Science from the University of Massachusetts in Chemical Engineering with a specialization in polymer sciences.

AUTHOR INDEX